Flora of Pathanamthitta

(Western Ghats, Kerala, India)

FLORA OF PATHANAMTHITTA

(Western Ghats, Kerala, India)

N. Anil Kumar
M. Sivadasan
N. Ravi

2005
DAYA PUBLISHING HOUSE
DELHI-110 035

ISBN 81-7035-589-3

Published by : **Daya Publishing House**
1123/74, Deva Ram Park
Tri Nagar, Delhi - 110 035
Phone: 27383999
Fax: (011) 23244987
E-mail: dayabooks@vsnl.com
Website: www.dayabooks.com

Showroom : 4762-63/23, Ansari Road,
New Delhi - 110 002
Ph.: 23245578, 23244987

Laser Composed by : **Vaishnav Graphics & Systems**
New Delhi - 110 055

Printed at : **Chawla Offset Printers**
Delhi - 110 052

PRINTED IN INDIA

Foreword

This book contains an excellent description of the flora of the Pathanamthitta district of Kerala. The authors Dr. N. Anil Kumar, Dr. M. Sivadasan and Prof. N. Ravi have found nearly 1250 species of plants in the district. Among them, 7 species are described for the first time in literature. An analysis of the flora of the district showed that 260 species are endemic. There is a very high concentration of endemic species (85/100 sq. km. as against 1.7/1000m sq. km. in Peninsular India as a whole) in this district. Also the studies have shown that about 200 species are rare, while 175 are severely threatened with extinction.

Soon our country will have a Biodiversity Act designed to promote concurrent attention to conservation, sustainable use and equitable sharing of benefits. Making a resource inventory of the areas is an important step in enabling the proposed Panchayat level Biodiversity Management Committees and the State Biodiversity Boards.

Thus, this publication is a timely one and meets a felt need. Our gratitude goes to Dr. Anil Kumar, Dr. Sivadasan and Prof. N. Ravi for the thoroughness with which they have undertaken the study and prepared this book. Our gratitude also goes to Daya Publishing House for bringing out such a valuable publication.

Prof. M. S. Swaminathan
UNESCO Cousteau Chair in Ecotechnology &
Chairman, M.S. Swaminathan Research Foundation
3rd Cross Street, Taramani Institutional Area,
Chennai (Madras) - 600 113, INDIA

Acknowledgements

The authors wish to express their sincere gratitude to :

- Dr. M.S. Swaminathan, Chairman of the M.S. Swaminathan Research Foundation (MSSRF), Chennai for whole-hearted encouragements.
- Dr. K. Unnikrishnan, former Head of the Department of Botany, University of Calicut for rendering necessary help for the work.
- Dr. M.V.N. Panikkar, retired Professor of Botany, S.N. College, Kollam, and Dr. P. Pushpangadan, the then Director of the Tropical Botanical Garden and Research Institute, Thiruvananthapuram and presently the Director, National Botanical Research Institute, Lucknow for helpful suggestions and encouragements.
- The late Dr. V.V. Sivarajan, Professor of Botany, Department of Botany, University of Calicut for fruitful discussions and suggestions.
- Dr. A.N. Henry and Dr. E. Vajravelu, Retd. Scientists of the Botanical Survey of India, Southern Circle, Coimbatore for their valuable suggestions and encouragements.
- Dr. A.K. Pradeep, Curator of the Calicut University Herbarium, Dr. N. Sasidharan, Scientist, Kerala Forest Research Institute, Peechi, Thrissur and Dr. D. Narasimhan, Lecturer in Botany, Madras Christian College, Chennai for various help.
- Mr. M.S. Kiran Raj, Senior Research Fellow (CSIR), Department of Botany, University of Calicut for various help including proof reading and corrections.

The second author expresses his sincere gratitude to the Ministry of Environment & Forests, Government of India for conferring on him the **Chair of Taxonomy** (Plant Science), the first and the only one of its kind instituted by it, and for the Grant provided for promotion of research in Plant Taxonomy.

Contents

Foreword (v)

Acknowledgements (vii)

Introduction 1-2

Area of Study 3-7

Vegetation 8-11

Review of Earlier Work 12

Present Work 13-15

Floristic Analysis 16-22

Threats of the Flora 23

Proposed Area for Conservation 24

Systematic Treatment 25-591

References 593-596

Index 597

INTRODUCTION

It is an accepted fact that without a basic knowledge of plant distribution, it is impossible to plan for the conservation of plant diversity. Inventories of plant resources are the corner stones of plant conservation. For this, floristic studies can provide efficient and convenient information about individual plants. But, often this branch of botany in which taxonomy is an integral part, is considered as old-fashioned and out-dated. It is not strange to hear the opinion that taxonomy is not a science but simply a mechanistic hobby of 19th century botanists, and is a subject limited to the study of 'bits of hay' stuck on herbarium sheets and stored in pigeon holes of cupboards (Subramanyam & Nayar, 1971). But till the turn of this century, Botany was synonymous with this 'mechanistic hobby' (Henry & Daniel, 1988). Now, there is a great revival for this 'hobby' in view of the alarming erosion of species and ecosystems. But, unfortunately the number of the systematic botanists is shrinking, with the result that the extent of ignorance on the number of species of organisms on earth is becoming vast (Swaminathan, 1991). Our Knowledge about the diversity and distribution of plant species is indeed very poor and inadequate that we still do not know exactly how many species exist in our earth. There are estimations rainging from 10 million to over 50 million species as distributed on our earth. Wilson (1988) calculated that the total number of named species is about 1.4 million. Because of the lack of adequate knowledge, we do not even know as to how many species we are losing. Current extinction rate is 100-1000 times of what they would be in unperturbed nature (Reid *et al.,* 1989). By the destruction of tropical forests alone we may lose about 25% of the total species by the year 2015 (Raven, 1988).

The tropical forests forming a small green band around the equator, harbours more than half of the world's species. Myers (1990) has estimated that the 18 places identified on earth as 'hot spots' support nearly 50,000 endemic plant species which form about 20% of the world's total of 2,50,000 flowering plants. Most of these 'hot spots' are in tropical countries. The Western Ghats and Eastern Himalayas are the 'hot spots' of our country. It is also estimated that Western Ghats alone harbour about 4000 species out of country's total of 15000-17000 species of flowering plants (Nayar, 1996).

Since the publication of the *Flora of British India* (Hooker, 1872-1897), over 3400 species have been added to the flora of the present Indian territory (Henry & Daniel, 1988). Over 150 new species and 15 new genera have been discovered during last 3 decades from the Western Ghats alone (Nair & Daniel, 1986). Still more and more new species and new genera are being described, which include even tree species (Ravi & Anil Kumar, 1990; Sasidharan & Vink, 1991). From Agasthyamala region alone more than 25 new taxa have been described (Henry *et al.,* 1984; Sivadasan, 1989; Sivadasan & Mohanan, 1991; Sivadasan *et al.,* 1989, Mohanan & Sivadasan, 2002). Sreekumar and Nair (1991) have listed 350 species of grasses, including 30 new species and 2 new genera, from Kerala state. Thus, more and more new species are being reported, and simultaneously already known ones are becoming more and more rare or even becoming extinct. Arora and Nayar (1986) recognised more than 300 wild relatives of crop plants from the country. They are of the opinion that sufficient detailed investigations are to be carried out to know the precise nature of interrelationship with cultivated taxa. But disgracingly this hasn't yet been done, possibly due to negligence and lack of taxonomic, morphologic and biosystematic knowledge of our plants. Unfortunately still many of our forest areas, especially of southern part of Western Ghats, are left underexplored or even unexplored.

All these show the immediate need of an intensive botanization of the area to know our plant resources, their correct identity, distribution, present status and extent of threat if any, endemism, the wild relatives of known economic plants and their interrelationships, the dynamism of flora, the economic importance of species, and so on. An up-to-date comprehensive Flora is an essential prerequisite for achieving this goal. In this context, Botanical Survey of India, the apex institution for plant exploration, encouraged district-wise floristic studies. Often the "Florulas" contain more data than large Floras. It is with this intention that the present work on Seed plants of Pathanamthitta district, Kerala State, was undertaken and hope that this Florula would enable us to achieve the above said goal.

AREA OF STUDY

Location

The name 'Pathanamthitta' was derived from the location of the town Pathanamthitta near river-bank or stream-bank. In Malayalam, the language of Kerala State, 'Pattanam' (which became modified into 'Pathanam') means town, and 'thitta' refers to the river-bank; and the town located on the bank of river Pamba obtained the name Pathanamthitta. The district by the name 'Pathanamthitta' with its head quarters at Pathanamthitta came into existence on November 1, 1982. This district was carved out from the portions of other three districts viz., Kollam, Alappuzha and Idukki. It lies between 9° 5' and 9° 29' N latitude and 76° 38' and 77° 16' E longitude with many parts in the Western Ghats, and hedged in by hills. This district is endowed with flagrant natural beauty with its vast, unending stretches of forests, rivers and rural landscapes. It is bound on the north by Idukki and Kottayam districts, on the south-east by Tirunelveli district of Tamil Nadu state, on the south by Kollam district, and on the west by Alappuzha district (Map 1). In Kerala, Pathanamthitta is the 7th largest district in land-area size, having an area of 2697 sq. km. The official records of 15 years old show that the forest cover is about 1466.85 sq. km, which forms about more than half of the total area. The total population, according to the 1986 census, is 11.87 lakhs. The district consists of 5 taluks viz., Ranni, Kozhencherry, Adoor, Tiruvalla and Mallappally.

Topography and Climate

Pathanamthitta consists of three natural divisions as in the case of most of the other districts of Kerala, viz., the lowlands, midlands and highlands. But the lowland area is small in size and is restricted to Western borders of the district. The highland (the Ranni taluk and some parts of Kozhencherry taluk) stretches through the Western Ghats and descends to the midland (Adoor & Mallapppally taluks) in the centre. The topography of the district is highly undulating with hills and valleys. It starts from the higher hillslopes with their lush greenery on the east along mountains down to the valley and small hills, to the flat-land of coconut trees in the west. The altitude ranges from (5) 10 to 1200 m; the lowest represented by Tiruvalla taluk (5-10 m) and highest by Kakki hills and Devarmala (1000-1200 m). The other prominent hills in the area are Chelikkalkarmala (c. 997 m), Thunethumala (721 m), Kottampara (520 m), and Sabarimala (500 m).

The district has a tropical humid climate as present elsewhere in the other hilly districts of the State with dry season from December to February and hot season from March to May.

Temperature

Temperature varied from 24° C to 36° C in the plains and 15° to 32° C in the hills. March to May are the hottest months, wheras December and January are the coldest.

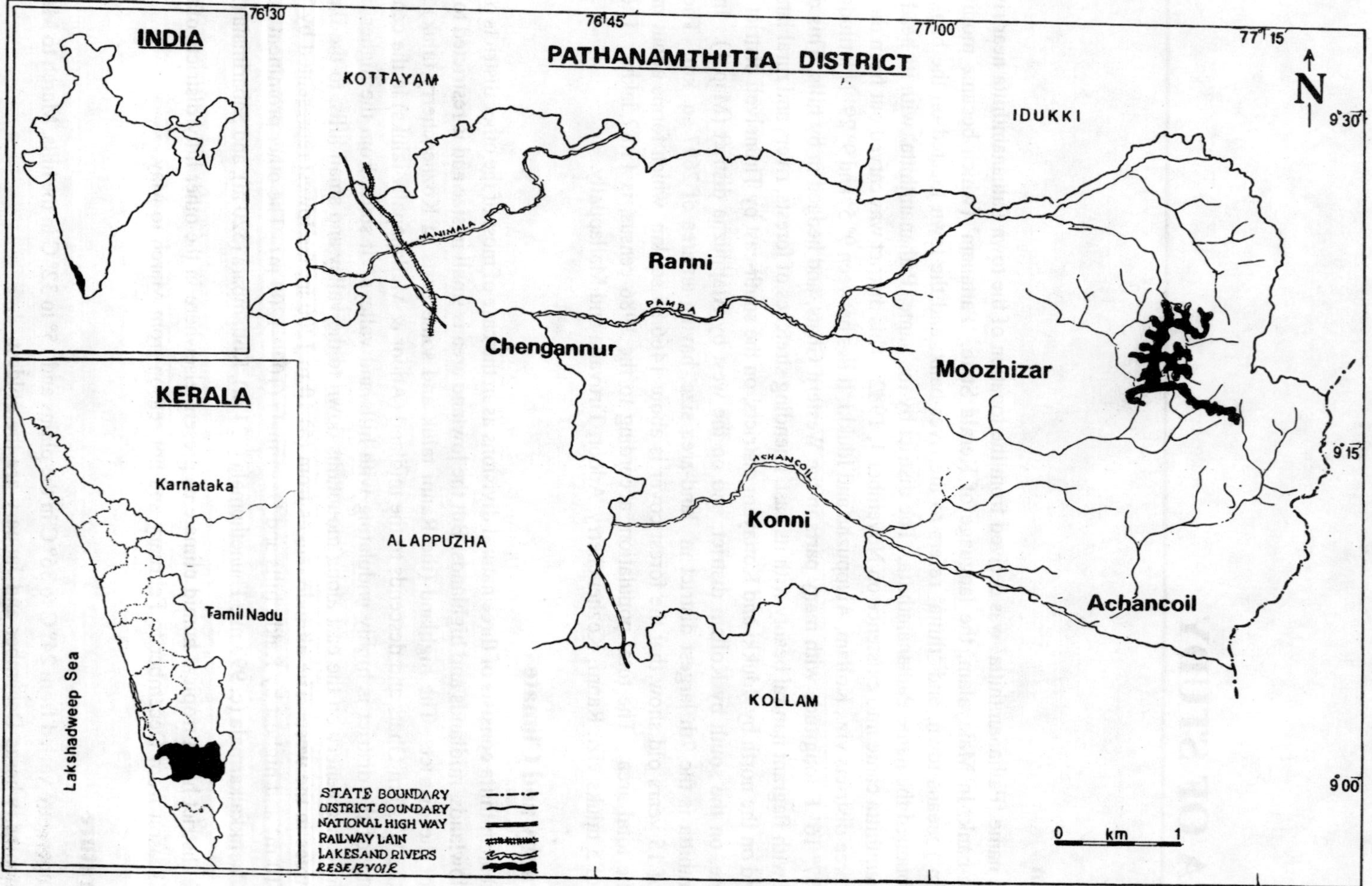

Map 1. : Pathanamthitta District

Rainfall

The South-West monsoon from June to September and North-East monsoon from October to November provide fairly good rain. The South-West monsoon is usually very heavy and about 75% of the annual rain is received from this. Average annual rainfall was recorded to be 240mm. Average annual rainfall was recorded to be 240 mm. Average maximum rainfall of about 590 mm was recorded during the month of June, and the minimum of about 20 mm during the month of December.

Geology and Soil

This district along with other parts of Peninsular India is also geologically very old. The parent rock is of igneous origin. The rock formation consists of Charnockite, Pyroxen, Granulite, Garnet-biotite gneiss, Cordierite gneiss, acidic intrusives like granites, quartz and pegmatic veins. The disintegration of the rocks give rise to various types of soil which are mainly lateritic and red loamy in character turning clayey in certain areas. On the high ridges coarse soil is mixed with large amount of quartz grains, but in the valleys there is a fine textured loam. In the hills the crystalline rock consists mainly of the minerals, silica, feldspar, muscovite and biotite with small amount of accessory ferromagnesium minerals. Fine sandy loam, rich in organic matter, are seen on the river banks. Depth of soil varies from place to place.

Rivers

There are 3 important rivers flowing through this district. They are the Achankovil river, Manimala river and the Pamba river. The Pamba river with a length of about 110 miles is formed by the confluence of Pambayar, Arudiayar, Kakkadayar, Kakkiar and Kallar. It descends from Sabarimala, flows mainly through various parts of Ranni taluk, enters Alappuzha district after joining with river Manimala and river Achankovil (80 miles), and empties into the Vembanadu lake. The river Achankovil is formed by the confluence of several small streams originating from Rishimala, Pasukkidiamettu and Ramakkalteri and joins the Pamba river at Veeyapuram in Alappuzha district.

One-third of the electricity produced by the State comes from this district. The power is generated from the Sabarigiri Hydro-electric Project situated at the Pamba basin in the district.

Developmental Activities in the District

Agriculture

Agriculture is the main occupation of the people. About 80% of the population depend on it directly or indirectly. The main crops raised in the district are rubber, paddy, tapioca, sugarcane, arecanut, pepper and banana. The total cultivated area in the district is estimated to be 1086. 53 sq. km.

Pathanamthitta district accounts for the maximum quantity of rubber produced in the State. Rubber cultivation is concentrated mainly in the hilly eastern zone; and one cannot see even a single house in the district without a rubber tree in its compound. Rubber occupies a prominent position among the commercial crops of the district. Tapioca is a major food crop next to paddy. The important crops and the extent of area of cultivation are given in the table 1.

Table 1 : Crops and Area of Cultivation

Name of Crops	Area of Cultivation
Rubber	46773 ha
Coconut	25636 "
Paddy	21318 "
Pepper	8215 "
Banana	6061 "
Vegetables	4683 "
Tapioca	2088 "
Arecanut	2074 "
Cocoa	2000 "
Ginger	1482 "
Sugarcane	1450 "
Cashew	1358 "
Turmeric	290 "
Pineapple	113 "

Eventhough tapioca is cultivated in all parts of the district, it is mainly concentrated in Mezhuveli, Konnni, Konnithazhaom in Kozhencherry taluk, Thonnalore and Enath of Adoor taluk. Coconut also enjoys an important position in the agricultural economy of the State. Paddy is cultivated in two seasons viz., Virippu (first crop, from April-May to August-September) and immediately followed by Mundakan (second crop, from September-October to December-January). Both are rain-fed and hence are extensively cultivated.

Plant-based Industries

Pathanamthitta district is not industrially much advanced. At present there are five large-scale and one medium-scale plant-based industries in the district. Their details are given in table 2.

Table 2 : Industries in the district

Sl. No.	Name of the Industry
1.	United Vineers (P) Ltd., Kaviyoor.
2.	Vineers and Laminations, West Othara.
3.	Travancore Sugars and Chemicals, Tiruvalla.
4.	Mannam Sugar Mills, Pandalam.
5.	Wood System India (P) Ltd., Mallappally.
6.	Prime Wood Industries, Nellimala, Tiruvalla.

Irrigation

Pathanamthitta has two major irrigation projects. They are the Pamba Irrigation Project and the Kallada Irrigation Project. They are partly commissioned and further work is in progress. As there are three major rivers flowing through this district, it has further scope for implementation of more irrigation projects.

The Kallada Irrigation Project is the biggest multipurpose project being undertaken by the State Government. The estimated cost of the project is Rs. 83.3 crores and is expected to benefit an area of 1.05 lakh hectares (gross). The project comprises of a masonry dam of 335 m long with a maximum height of 81 m at Parappar in Thenmala across Kallada river to form a reservoir, a pick up 'weir' of 118 m long and 9 m high and sluices at Ottakal, 4.6 km down stream from the dam, a right bank canal of 69 km long, and a left bank canal of 57-75 km long, both taking off from the pick up 'weir', and having network of minor canals.

The estimate cost of the Pamba project is Rs. 33.51 crores and the expected benefit is for 41683 hectares (gross). It comprises of a barrage at Maniyar across Kakkad river having a length of 115.22 m and height of 16.76 m, and a network of canals taking off from barrage.

There are some medium and minor irrigation projects comprising wells, small tanks and reservoirs to cover areas which do not come under the purview of the major irrigation projects mentioned above. Most of these schemes are now under various stages of execution.

Forest Produces

Timber is the most important forest produce of the district as elsewhere. Quality timber is used for constructing buildings and making furniture. In fact, forest is the main source of raw materials for wood-based industrial units in the district. *Tectona grandis* (teak), *Dalbergia latifolia* (rosewood), *Artocarpus hirsuta* (anjili), *A. integrifolia* (jacktree), *Haldenia cordifolia* (manjakadambu), *Alstonia scholaris* (pala), etc. are some of the important timber yielding species available in this district. Apart from providing raw materials for rayon, newsprint, plywood, etc., the forests are a source for a number of useful minor products like bamboo, reeds, honey, and medicinal plants and herbs.

VEGETATION

The vegetation can be broadly divided into 3 categories such as lowland vegetation, midland vegetation and upland or highland vegetation.

Lowland vegetation (5)-(10-20 m)

The vegetation is mainly of herbaceous and bushy type; occasionally sacred groves are seen in certain areas. This land is mainly cultivated with coconut. The entire Tiruvalla taluk including the places like Kadapra, Nedumpuram, Peringara, Kizhakkambhagom, Tiruvalla, Kavumbhagom, Eraviperoor, Thottapuzhassery and Koippuram, is having this type of vegetation. The common herbaceous species are : *Abutilon indicum, Ammania baccifera, Cleome viscosa, Desmodium heterophyllum, D. triflorum, Hibiscus surattensis, Mitracarpus verticllatus, Scoparia dulcis, Sida acuta, S. rhombifolia, Tridax procumbens,* and aquatic species like *Blyxa echinosperma, Eichhornia crassipes, Hydrilla verticillata, Monochoria vaginalis,* etc. The common sedges and grasses are *Cyperus haspan, C. iria, Fimbristylis dichotoma, F. difformis, F. haspan, Pycreus polystachyos, Dactyloctenium aegyptium, Digitaria adscendens, Eleusine indica,* etc. The common sacred-grove species are *Holigarna arnottiana, Hopea ponga, Hydnocarpus pentandra, Ichnocarpus frutescens, Pothos scandens, Syzygium zeylanicum,* etc. The other common tree species in this area are *Garcinia gummi-gutta, Hibiscus tiliaceus* and *Lannea coromandelica.*

Midland vegetation (20-40 m)

Most of the portions of Adoor and Mallappally taluk harbour this type of vegetation. The soil is coarse and often reddish. Most of the areas are under rubber cultivation. The main attractive feature of this area is the presence of quite a large number of small sacred -groves, particularly in Adoor taluk. Presence of these sacred-groves manifest the once existed dense lowland evergreen forests of this area. Some of the common species of this area are: *Amischophacelus axillaris, Anisomeles indica, Blumea mollis, Cardiospermum halicacabum, Chrysopogon aciculatus, Cleome viscosa, Cyclea peltata, Cyperus haspan, C. iria, Ficus tsjakela, F. hispida, Fimbristylis dichotoma, F. haspan, F. tenuispica, Fioria vitifolia, Garcinia gummi-gutta, Justicia japonica, Hedyotis auricularia, H. corymbosa, Heteropogon contortus, Hybanthus enneaspermus, Ichnocarpus frutescens, Limnophila aromatica, L. indica, Microcos paniculata, Mimosa pudica, Ziziphus oenoplia,* etc.

Upland or highland vegetation (40-1200 m)

According to Champion and Seth (1968) this type of vegetation is primarily of 3 categories: Tropical evergreen including small sholas of upper hills, Tropical semi-evergreen, and Tropical moist-deciduous forests.

Tropical evergreen forests and sholas

This type of forests are seen at altitudes from 200 to 1200 m. Most of the forest areas under Ranni and some of the areas of Konni forest division come under this category. In Kakki hills, Chelikkalkar, Chembalakar, Thunathumala, Kodamala, Moozhiar, Sabarimala, Arampa, Kottampara, etc. this type of forests with good canopy are seen.

The typical attributes of this forest type are its wonderful plant diversity. There are many different kinds of trees, shrubs, herbs, climbers, stragglers, epiphytes, etc. It is very difficult to see a dominant species in any of these forms and also the number of individuals per species is very limited. Associations of two or more species can be readily observed. *Cullenia-Mesua-Palaquium* association, as Meher-Homji (1982) reported, is observable in Kakki hill-Pampa dam area. The top canopy of this forest is extremely dense with lofty trees, reaching to 40-50 m height. Often these tall trees are with slender boles and buttresses. The characteristic buttressed trees are *Dimocarpus longan, Dipterocarpus bourdilloni, Elaeocarpus tuberculatus* and *Mesua nagassarium*. The top storey is often represented by species like *Aglaia perviridis, Bombax ceiba, Cullenia exarillata, Diospyros bourdilloni, Dysoxylum malabaricum, Hopea parviflora, Mesua nagassarium, Palaquium ellipticum, Pterygota alata, Semecarpus travancorica, Syzygium malabaricum*, etc. The second storey is composed of medium-sized trees. There are overlapping of some of the species of top canopy. The characteristic species of this layer are *Artocarpus hirsuta, Chionanthus mala-elengi, Cinnamomum malabatrum, Diospyros candolleana, D. paniculata, Garcinia spicata, Harpullia arborea, Holigarna arnottiana, Litsea coriacea, L. floribunda, Lophopetalum wightianum, Persea macrantha, Semecarpus anacardium, Syzygium laetum, S. mundagam, Turpinia malabarica, Vitex altissima*, etc. The third storey usually consists of small trees and shrubs such as *Acronychia pedunculata, Antidesma menasu, Atalantia racemosa, Baccaurea courtallensis, Calamus hookerianus, C. thwaitesi, C. vattayila, Desmodium ferrugineum, Dichapetalum gelonioides, Gomphandra coriacea, Ixora leucantha, I. nigricans, Leea indica, Mallotus rhamnifolius, Meiogyne pannosa, Pavetta indica, P. zeylanica*, etc.

The ground flora mainly occur where the top canopy is not dense. In many of the forest areas top canopy becomes less dense due to forest fire, fall of trees and such other happenings. Following are some of the common undershrubs and herbaceous species : *Abelmoschus manihot, Atylosia lineata, Barleria courtallica, Calamus thwaitesii, C. vattayila, Costus speciosus, Desmodium microphyllum, D. heterocarpon, D. velutinum, Flemingia macrophylla, F. congesta, Habenaria longicorniculata, Hibiscus lunarifolius, Murdannia dimorpha, M. zeylanica, Murraya paniculata, Nilgirianthus ciliatus, N. heyneanus, Phyllanthus rheedii, P. urinaria, P. virgatus, Thottea siliquosa, Tournefortia heyneana, Triumfetta pilosa, Zingiber macrostachyum*, etc. The common herbaceous and woody climbers include *Artabotrys zeylanicus, Chilocarpus malabaricus, Erythropalum populifolium, Jasminum azoricum, Naravelia zeylanica, Myxopyrum smilacifolium, Pothos scandens*, etc.

The epiphytic flora includes many orchidaceous members, species of Loranthaceae, ferns, etc. The common epiphytic species are *Pholidota imbricata, Oberonia brunoniana, O. platycaulon, O. proudlockii, Rhynchostylis retusa, Bulbophyllum neilgherrense, Helixanthera obtusata, H. wallichiana, Scurrula parasitica, Viscum orientale, Remusatia vivipara*, etc.

Forest border and clearings have been encroached by some noxious weeds . The indigenous common peripheral species are : *Stephania japonica, Abelmoschus moschatus, Cissus javana, Crotalaria walkeri, C. clarkei, Desmodium microphyllum, D. heterophyllum, Indigofera spicata, Torrenia bicolor, Justicia japonica, Leucas biflora, L. hirta, Murdannia dimorpha, M. japonica, Scleria levis, S. lithosperma, Carex baccans, C. filicina, Rhynchelytrum repens, Setaria palmifolia,* etc.

Typical sholas can be seen in Kakki hill-top region (c. 1200 m). The trees are small with closed canopy and short boles. The dominant species include *Debregeasia ceylanica, Gordonia obtusa, Scolopia crenata, Vaccinium neilgherrense, Symplocos macrophylla, Ardisia pauciflora, Syzygium malabaricum, Premna coriacea, Euonymus crenulatus, Eria dalzellii,* etc.

Tropical Semi-evergreen forests

Most of the parts of Konni forest division are coming under this category or the next category. Because of the indiscriminate felling of evergreen tree species, the evergreen forests of the past are now changing to this category or becoming deciduous type. Here the dominant species are deciduous followed by the evergreen species. The top canopy is not dense and trees attain a size not more than a height of 40 m. The species diversity is less when compared with that of evergreen forests, but the number of individuals per species is dominating. Species diversity discussed in the following category is broadly applicable to this forest type also.

Tropical Moist-deciduous forests

This type of forest occurs from low hills to an elevation of 250 m. Most of the forest areas of Konni forest division are either moist-deciduous or semi-evergreen or a mixed-up of both of these. The general canopy is open or less dense. Trees are (20) 25-30 m tall. The common tree species are *Dillenia pentagyna, Sterculia balanghas, S. guttata, Schleichera oleosa, Pterocarpus marsupium, Albizia chinesis, Xylia xylocarpa, Terminalia crenulata, T. paniculata, Caryea arborea, Holarrhena pubescens, Tabernaemontana alternifolia, Stereospermum colais, Ficus hispida, Grewia tiliifolia, Artocarpus hirsutus, Dimocarpus longan,* etc. The shrubby species include *Glycosmis mauritiana, G. pentaphylla, Cipadessa baccifera, Helecteres isora, Ziziphus oenoplia, Cassia alata, Melastoma malabathricum, Leea indica,* etc. The common climbers are *Abrus precatorius, Calycopteris floribunda, Ziziphus rugosa, Jasminum rottlerianum,* etc. The ground cover species diversity is very rich and includes *Sida fryxellii, S. rhombifolia, Triumfetta annua, Urena lobata, Naregamia alata, Crotalaria pallida, Desmodium heterophyllum, D. triflorum, Tephrosia purpurea, Cassia absus, Passiflora foetida, Hedyotis auricularia, H. nitida, Spermacoce articularis, S. hispida, Blumea membranacea, B. mollis, Spilanthes calva, S. ciliata, S. paniculata, Hemidesmus indicus, Solanum torvum, Justicia japonica, J. procumbens, J. prostrata, Asystasia dalzelliana, Anisomeles indica, Hyptis suaveolens, Tragia hispida, Cyperus haspan, Fimbristylis dichotoma, Scleria lithosperma* and several grass species.

In addition to these major categories of forests, vegetation of special habitats include the following types :

Grasslands

Extensive grasslands occur in Kakki hills, Chelikkalkar and Devarmala regions. The common grass species are : *Arundinella purpurea, Chrysopogon hackleii, C. aciculatus, Dimeria ornithopoda, Eragrostis cilianensis, E. uniloides, Eulalia trispicata, Heteropogon contortus, Themeda triandra,* etc. The other associated herbaceous species are *Polygala bolbothrix, P. chinensis, P. elongata, Linum mysorense,*

Impatiens pusilla, Alysicarpus buplerifolius, Crotalaria calycina, C. clarkei, C. evolvuloides, Tephrosia pulcherrima, Drosera burmanni, Exacum sessile, E. tetragonum, Fimbristylis cinnamometorum, Eriocaulon dianae, E. quinquangulare, E. thwaitesii, Habenaria heyneana, H. longicorniculata, Striga asiatica, etc.

The riparian or riverine vegetation

This vegetation along streams and rivulets is quite interesting. It includes trees, shrubs, herbs, climbers, etc. *Hydnocarpus pentandra, Calophyllum calaba, Hopea glabra, Pongamia pinnata, Humboldtia vahliana, Mastixia arborea, Madhuca longifolia, Ochrenauclea missionis, Trevia polycarpa, Vitex leucoxylon,* etc. are some of the tree species of this vegetation. The climbing species include *Grewia umbellifera, Hiptage benghalensis, Dalbergia pseudo-sisso, Derris canarensis, Combretum latifolium, Entada rheedei,* etc. The shrubby species are *Homonoia riparia, Ficus heterophylla, Rotula aquatica, Ochlandra travancorica,* etc. The common herbaceous species are *Lagenandra toxicaria, Cryptocoryne retrospiralis, Floscopa scandens, Eleocharis retroflexa* and members of Podostemaceae.

Aquatic and Semi-aquatic Vegetation

In ponds, pools, puddles, paddy fields, marshy places, etc. this type of vegetation can be met with. The common aquatic species are : *Eichhornia crassipes, Hydrilla verticillata, Lemna perpusilla, Monochoria vaginalis, Najas graminea, Nelumbo nucifera, Nymphaea nouchali, Ottelia alismoides, Utricularia flexuosa* and *Blyxa echinosperma.* On rocks the podostemacean members like *Polypleurum stylosum, Zeylanidium olivaceum* are common. Along the paddy fileds the species like *Cyperus difformis, C. iria, Mollugo pentaphylla, Coldenia procumbens, Heliotropium indicum, Cleome viscosa, Ludwigia perennis, Eclipta alba, Polygonum barbatum, P. glabrum, Hygrophila angustifolia, Fimbristylis dichotoma, F. littoralis, F. miliacea, Panicum repens* are common.

REVIEW OF EARLIER WORK

It is very difficult to analyse the previous work done in the district alone. But the Southern- Western Ghats to which the present-study area belongs to has attracted many veteran botanists because of its enormous floristic richness and diversity. The history of botanization starts from Garcia de Orta (1565) who enumerated the medicinal plants of India. After a century of this work van Rheede, the then Dutch Governor of Malabar surpassed all the previous work by compiling his monumental and historic work in 12 volumes, the *Hortus Indicus Malabaricus* (1678-1703). Many of the plates contained in this work were the base material for Linnaeus to describe many new Indian species. This show the taxonomic and historic importance of the work. Among the 19th century plant collectors of this region, Robert Wight (1830-1853) made a systematic approach and brought out a series of books of botanical illustrations. This was followed by Richard Beddome who also published data on plants in the form of illustrations (1868-1874). From 1872-1897, Joseph Dalton Hooker for the first time, brought out floristic data of Indian plants in the form of a flora, the *Flora of British India*, in 7 volumes. This distinguished work inspired many botanists in accelerating further explorations. This resulted in publication of many regional floras. Among this, mention must be made about the *Flora of the Presidency of Madras* by Gamble and Fischer (1915-1936). This work still remains as the basic identification manual of plant taxonomic studies in Peninsular India. Records show that Gamble made several collections in Konni forest division of the district. The notable plant collectors who concentrated their work in this part were T.F. Bourdillon and Rama Rao who brought out the publication, the *Forest Trees of Travancore* (1908) and the *Flowering Plants of Travancore* (1914), respectively. Their study areas included Konni and Ranni forest divisions of the district. Their collections are now housed in TBGT, and the University College Hebarium (UCT), Thiruvananthapuram. Another notable plant collector of this period was K.C. Jacob who brought out many publications on new plant species from this region.

Among the present-day plant collectors of the district mention must be made of C.N. Mohanan (1984) who studied the flora of Quilon district. A considerable extent of area of the present study was part of erstwhile Quilon district. But Mohanan had concentrated mainly in Kulathoopuzha, Achankovil and Konni areas. His collections from the species-rich Moozhiar hills and Arampa area are comparatively sparse. The other plant collectors of this regions were the scientists like E. Vajravelu, N.C. Nair, K. Vivekananthan, A. G. Pandurangan and M. Chandrabose of the Botanical Survey of India (Southern Circle), Coimbatore.

PRESENT WORK

Several field trips of duration ranging from 2 to 10 days were made at regular intervals to various parts of the district extending over a period of more than 4 years, and made intensive and extensive collection of plants. A total of about 3100 plants have been collected. Collection trips were repeated till full data on each species have been gathered. Usually 5 specimens of each species were collected from all the available localities in view to have a knowledge on range of variation, the local distribution and commonness/rarity of the taxa.

For collection and preservation, the procedures given by Jain and Rao (1977) and van Balgooy (1987) were generally followed. Difficulties were often faced for collection of specimens from tall trees and large lianas. In such cases tree climbers were employed, or used long aluminium, adjustable pole with hook. Normally each collection contained either flowers, fruits or both. In the case of dioecious species both male and female specimens were collected separately. The parasitic plants were most commonly collected along with respective hosts, except in certain cases of root parasites as *Balanophora*. In such cases detailed notes on host species were made. Special techniques as given by Fosberg and Sachet (1965) were employed for collecting the plants of families like Araceae, Arecaceae, Pandanaceae and the sub-family Bambusoideae. In some cases as in orchids, live specimens were also collected and have grown in green house of the Sree Naryana College, Kollam, for future observations and studies. Besides the routine details like locality, habit, habitat, exuadate if any, flower-colour, the field notes included special features viz., ecological and biological features, species frequency and dominance. Pollination of flowers, dispersal mechanism of fruits, etc. were also noted for certain selected groups. Local uses if any, of plants along with method of preparation, administration, if available were also noted.

For preservation of specimens, the wet method (Fosberg & Sachet, 1965) using 70% methylated alcohol was employed. It was experienced that this method was time saving, space saving and effective. For drying and poisoning the standard method was followed. Descriptions and drawings were most often made from herbarium specimens. For this purpose, one or two flowers were boiled to soften the material and used, except in the case of Cyperaceae and Poaceae members. For these families, the spikelets were soaked overnight in detergent solution prior to dissection. Analytical descriptions were made for the taxa which seemed to be rare. Illustrations of some selected rare/endemic or otherwise interesting taxa have been made. Often compound light microscope was used for observation of microscopic characters like indumentum, placentation details, etc. Almost all collections cited were self-made except a very few which were mainly of BSI scientists. The herbarium specimens collected by the author were deposited in CNH, MH, and SNCH (Kollam). Initially identifications were provisionally done by using the *Flora of the Presidency of Madras* (Gamble & Fischer, 1915-1936), *Flora of Hassan District* (Saldanha & Nicolson, 1976), *Flora of the Tamil Nadu Carnatic* (Matthew, 1983), *A Revised Handbook to the Flora of Ceylon* (Dassanayake & Fosberg, 1980-1991), *Flora of Silent Valley* (Manilal, 1988) and available monographs and other relevant literature.

For confirmation of identification, the specimens were matched with that of authenticated ones available in MH. The critical specimens were identified by experts of respective families, notably from Kew Gardens, Rijksherbarium, Texas A & M University, and Southern and Central circle offices of the Botanical Survey of India.

Presentation of Data

For general format of Flora, the format given by Radford *et al.*, (1974) was followed. The families were classified according to Bentham and Hooker's system (1862-1883) with necessary alterations consequent on recognition of split-up treatment of various families as proposed by Hutchinson (1926, 1934, 1959 & 1973). Artificial keys to groups and families of seed plants have been provided. Keys are strictly dichotomous and indented. As far as possible the vegetative, floral and fruit characters which were readily observable were employed in the key. Full citation of generic names has been given referring *Index Nominum Genericorum* (Farr *et al.,* 1979). Genera with more than one species were provided with a short generic description covering only the characters of the included species. In cases of genera with only one species, the genus description was omitted and a detailed description of the species was given. Key for species has been provided under each genus with more than one species. Treatment of genera within a family and species within a genus followed alphabetical order. Each species was provided with detailed nomenclatural citations including reference to the *Flora of British India* (Hooker, 1872-1897), *Flora of the Presidency of Madras* (Gamble & Fischer, 1915-1936), *Flora of Hassan District* (Saldanha & Nicolson, 1976), *Flora of the Tamil Nadu Carnatic* (Matthew, 1983), *Flora of Silent Valley,* (Manilal, 1988), *A Revised Handbook to the Flora of Ceylon* (Dassanayake & Fosberg, 1980-1991 Dassanayake, Fosberg & Clayton, 1995; Dassanayake & Clayton, 1996-2000) and *Flora Malesiana series* (Van Steenis, 1948-58; 1972-76). The synonyms relevant to Peninsular Indian flora were given in italics. *Icones* of Wight (1839-1853) and Beddome (1868-1874) were cited wherever applicable. Local names, if available, were given in italics at the end of nomenclatural citation. Abbreviations of the periodicals are mainly according to those given in *Botanico Periodicum Huntianum* (B-P-H) (Lawrence *et al.*, 1968). The nomenclatural citations often included the references to recent monographic work, if any. Nomenclatural citation was followed by a precise and diagnostic species description in the following sequence : Habit; Leaves; Inflorescence; Flowers; calyx; corolla; stamens; ovary; Fruit and Seed. The characters like glabrous and entire have not been specified unless they were key characters. Regarding the terminology, the term 'lanceate' has been consistently used in cases where basal 1/3 portion is wider and the rest gradually tapers towards apex; lanceolate is also used in the sense of narrowly elliptic. Species description was followed by data on phenology; the world-wide distribution, information on frequency, habitat, rarity/endemism if applicable, and specimen citation. Notes on medicinal or other economic importance, taxonomic or nomenclatural problems if any, have been provided at the end in separate paragraph. Ornamental or other cultivated plants seen in the district were named under the respective genera or at the end of respective families.

The following abbreviations and symbols are used in the book :

AGP	-	A. G. Pandurangan
BDS	-	B. D. Sharma
BINU	-	S. Binu
BKN	-	B. K. Nayar
BSI	-	Botanical Survey of India

cm	-	Centimeter
CNH	-	Central National Herbarium
CNM	-	C. N. Mohanan
DEB	-	D.B. Deb
EV	-	E. Vajravelu
ha	-	hectare
KFRI	-	Kerala Forest Research Institute
KV	-	K. Vivekananthan
m	-	Meter
mm	-	Millimeter
MH	-	Madras Herbarium
MCB	-	M. Chandrabose
NCN	-	N. C. Nair
PR	-	Pandurangan
SNCH	-	Sree Narayana College Herbarium, Kollam
TBGT	-	Tropical Botanic Garden & Research Institute Herbarium, Thiruvananthapuram
UCT	-	University College Herbarium, Thiruvananthapuram.
ϕ	-	Across (broad)
α	-	Many
0	-	Absent
±	-	More or less
c.	-	About

FLORISTIC ANALYSIS

It is an established view that Indian land mass had past connection with South America, Africa, Malesian Islands, Madagascar, Australia and Antartica of the Gondwanaland and has floristic similarities between these regions. Nayar (1982) had brought out several examples to sustantiate this view and the origin of Indian flora. He was of the view that characteristic flora of India is that of Peninsular India which actually is relicts of a bygone age as demonstrated by the presence of more paleoendemics and the affinity of the flora of the Western Ghats in Peninsular India with that of Malesian region and Sri Lanka. The present study also substantiate these views.

Floristic Diversity

There are 1249 species distributed in 658 genera belonging to 148 families in a total area of 2697 sq. km. A better estimation shows that the entire Kerala region with 38,863 sq. km. may harbour 4679 species of flowering plants (Sasidharan, 2004). Analysis of Flora of Tamil Nadu showed that the state with 1,30,069 sq. km. harbours c. 5640 species, whereas the whole country is estimted to have a total of 17000 species distributed in an area of 32,87,263 sq. km. Pathanamthitta district contains roughly 460 species/1000 sq. km. The Kerala region contains about 51 species/1000 sq. km, whereas Tamil Nadu contains only 43 species/ 1000 sq. km. The country's total species concentration is only 5.1/1000 sq. km. From this analysis the floristic richness of the district relative to the state and the whole country is very clear. The Table III shows the ten top dominant families of the area of the present study. Each family was further analysed with regard to the number of genera, and number and rarity of species . Three dominant families were critically analysed with respect to their species diversity, endemism, rarity and special features, if any.

Table 3 : Ten dominant families with their diversity and rarity

Name of Family	No. of genera	No. of species	No. of Rare species	% of Rarity
Fabaceae	47	117	21	24.0
Poaceae	50	81	17	18.0
Euphorbiaceae	29	65	9	13.0
Rubiaceae	25	58	13	22.1
Cyperceae	14	54	9	15.1
Orchidaceae	30	50	8	15.6
Asteraceae	32	48	10	22.2
Acanthaceae	22	44	10	23.0
Scrophulariaceae	16	31	6	19.6
Convolvulaceae	10	35	7	14.7

Fabaceae

In the order of dominance the family Fabaceae ranks first with 117 species belonging to 47 genera (Table III). Of this, the sub-family Faboideae alone contain 89 species under 36 genera.

This family is with a high degree of diversity in habit and habitat. It includes trees, herbs and climbers. The tree species are : *Dalbergia latifolia, Erythrina stricta, Pongamia pinnata* and *Pterocarpus marsupium*. There are 21 climbing or twining species of which the notable are : *Butea parviflora, B. purpurea, Canavalia virosa, Centrosema pubescens, Calopogonium mucunoides, Dalbergia pseudosisso, Derris canarensis, D. scandens, D. brevipes, D. thothathrii, Dolichos trilobus, Mucuna gigantea, M. hirsuta, M. monosperma, Pueraria phaseoloides, Rhynchosia acutissima, Shuteria vestita* and *Teramnus mollis.* The largest genus is *Crotalaria*, having 16 species, followed by *Desmodium* with 14 species. In *Faboideae* there are about 21 species which are very rare with restricted population. These are *Butea purpurea, Crotalaria clarkei, C. evolvuloides, C. multiflora, C. salicifolia, C. subperfoliata, Dalbergia pseudo-sisso, Derris canarensis, D. brevipes, D. thothathrii, Desmodium ferrugineum* ssp. *wynandense, D. microphyllum, Indigofera constricta, I. galegoides, I. longiracemosa, Kunstleria keralensis, Mucuna monosperma, Rhynchosia acutissima, Smithia blanda, Tephrosia tinctoria* and *Vigna pilosa*. About 24% of the total species of this subfamily is rare . Among this, *Crotalaria clarkei* and *Rhynchosia acutissima* are not represented in MH. *Derris thothathrii* and *Indigofera constricta*, are extremely rare or vulnerable. The latter species has already been listed in Red Data Book (Nayar & Sastry, 1987). *Dalbergia pseudo-sisso, Derris canarensis* and *D. thothathrii* are being reported for the first time from Kerala. Out of the total 89 species, 17 are endemic to Peninsular India, particularly to Western Ghats. These species are *Atylosia lineata, Butea purpurea, Crotalaria clarkei, C. salicifolia, C. walkeri, C. subperfoliata, Derris brevipes, D. thothathrii, Desmodium ferrugienum* ssp. *waynadense, Flemingia semialata, Indigofera constricta, Kunstleria keralensis, Mucuna hirsuta, Smithia bigemina, S. blanda, Vigna pilosa,* and *Zornia quilonensis*. The endemic species common to Peninsular India and Sri Lanka are about 15, viz., *Butea parviflora, Crotalaria barbata, C. evolvuloides, C. multiflora, C. umbellata, C. nana, Derris canarensis, Desmodium ferrugineaum* ssp. *ferrugineum, Dolichos trilobus, Geissaspis cristata, Mucuna monosperma, Psuedarthria viscida, Rhynchosia acutissima, Tephrosia pulcherrima* and *Tephrosia tinctoria*. A total of 32 species out of the 89 are endemic to Indian subcontinent, indicating high degree of endemism (about 35%). The Indo-Malesian species of this subfamily are *Alysicarpus monilifer, Crotalaria sericea, Desmodium motorium, Flemingia strobilifera, Indigofera galegoides*, and *Mucuna gigantea*. Almost all other species have distribution in Malesia, Trop. Africa, Java, S. China, Sri Lanka, S.America and North Australia.

The subfamily Caesalpinioideae is represented by 16 species including 2 endemic species viz., *Cassia intermedia* and *Humboldtia vahliana. Saraca asoca*, another species of this subfamily is almost endangered in the wild. Only one tree could be located during the present study. The subfamily *Mimusoideae* is represented by 11 species of which almost all species are pantropical in distribution.

Poaceae

This famiy is the second dominant one represented by 49 genera and 81 species including the new species described recently by Ravi and Anil Kumar (1992), Mohanan and Ravi (1996), Ravi and Mohanan (1997), and Ravi *et* Shaju (1998). The 16 endemic species viz., *Arundinella mesophylla, A. nervosa, A purpurea, Bambusa arundinacea* var. *gigantea, Chrysopogon hacklei, Dimeria idukkiensis, D.*

kurumthotticalana, D. lawsoni, D. mooneyi, D. namboodiriana, D. ornithopoda, D. sivarajanii, Garnotia courtallensis, Ochlandra travancorica, Ischaemum quilonensis, and *Urochloa panicoides* of Peninsular India are present in the district. The newly described species from the district viz., *Themeda sabarimalayana* (Sreekumar & Nair, 1989), *Dimeria idukkiensis* (Ravi & Anil, 1992) *D. sivarajanii* (Mohanan & Ravi, 1996), *D. namboodiriana* (Ravi & Mohanan, 1997), *Ischaemum quilonensis* (Ravi *et al.*, 1998) are hitherto known only from Southern Western Ghats. This family showed about 18% of endemism. Of the above species *Dimeria kurumthotticalana* has been colleted for the first time since its type collection in 1946. The protologue of this species has been amended after studying a good number of plants (Ravi & Anil Kumar, 1992). The collection of *Dimeria mooneyi,* another rare grass from the district, forms a new record for Kerala which has so far been reported only from Orissa State. All the endemic species are with very restricted distribution and most of them are rare. *Dimeria kurumthotticalana, D. mooneyi* and *Garnotia courtallensis* are very rare or vulnearble. *Themeda sabarimalayana,* a new species reported by Sreekumar and Nair (1989) from Sabarimala couldn't be located during the present study. The most common grass species include *Alloteropsis cimicina, Oplismenus compositus, Chrysopogon aciculatus, Dactyloctenium aegyptium, Brachiaria ramosa, Eragrostis uniloides,* etc.

Euphorbiaceae

The third and diverse family represented by 30 genera and 69 taxa, is Euphorbiaceae. The representative species exhibit all froms of habit and occur in varied habitats. Among these, *Antidesma zeylanicum, Baliospermum solanifolium, Croton malabaricus, Glochidion malabaricum, G. tomentosum, Phyllanthus bailloniana, Sauropus saksenianus, Tragia muelleriana, T. involucrata* var. *angustifolia* are extremely rare. *Sauropus saksenianus* is a newly described species from Silent Valley (Manilal, 1988). The endemic species of the family are *Baccaurea courtallensis, Croton malabaricus, Drypetes oblongifolia, Glochidion malabaricum, G. ellipticum, G. tomentosum, Mallotus resinosus* var. *resinosus, M. resinosus* var. *muricatus, Phyllanthus bailloniana, P. gageana, Sauropus saksenianus, Tragia involucrata* var. *angustifolia,* and *Trevia polycarpa.* In Peninsular India, the family has 18.8% endemism.

Endemism

Ahemedullah and Nayar (1986) listed 1932 taxa out of the estimated 6500 species as endemic to Peninsular Indian region, which form about 29.7%. Present analysis of the flora of Pathamanthitta shows that 260 species which fo·m 22% of the total 1214 species (excluding alien species) are endemic. An analysis made to assess the area wise concentration of endemic species demonstrated that Pathanamthitta district harboured 85 species per 1000 sq. km. as against 1.7 species per 1000 sq. km. of Peninsular India (1932 taxa in an area of 110000 sq. km). This data manifest the high degree of concentration of endemic species in the area.

Some families were analysed for their endemism (Table IV). Among those, Dipterocarpaceae showed maximum representation of endemic taxa with 100% endemism, followed by Annonaceae with 69.2%, and Balsaminaceae with 50%. Out of the 260 endemic taxa, 60 are tree species. All the tree species are paleo-endemics. Of the paleo-endemic species, some are recent findings as exemplified by *Kunstleria keralensis* (Mohanan & Nair, 1982), the only representative species of the genus in whole of the country, and the phytogeographicaly interesting species, *Julostylis polyandra* (Ravi & Anil Kumar, 1990).

Until the discovery of the species, the genus *Julostylis* was treated as monotypic, and as endemic to Sri Lanka. The discovery of the speies from Southern Western Ghats formed a record of the extended distribution of the genus, and further substantiate the view of Gondwanalad origin of Peninsular indian flora.

Table 4 : Top ten families with high percentage of endemism

Sl. No.	Family	Total species in Pathanamthitta district	No. of Endemic species	% of Endemism
1.	Dipterocarpaceae	5	5	100
2.	Annonaceae	13	9	69.2
3.	Balsaminaceae	12	6	50
4.	Acanthaceae	45	18	40
5.	Apocynaceae	15	4	26.6
6.	Rubiaceae	58	14	24
8.	Euphorbiaceae	65	13	20
7.	Poaceae	81	16	19.8
9.	Orchidaceae	50	9	18
10.	Fabaceae	117	17	14.6

The majority of the endemic species are with very critical number of populations having very narrow distribution. The present area is an isolated fragment of the Western Ghats showing a high rate of endemism and rarity; and any sort of disturbance would totally wipe out majority of the species.

Rarity/Vulnerability of species

The species of the flora of the area of present study were also subjeted to critical analysis of their rarity/vulnerability on the basis of their range of distribution in the past and present, present population density, and trends of threat to extant populations. The analysis showed that about 175 species are severely threatened and are rare plants. Most of these species are local endemics also. The percentage of total rarity of species in the present flora is 19.2. This is comparatively a high percentage and demand immediate conservation measures to save the valuable species. During the course of study, repeated searches were made for *Sageraea grandiflora* Dunn, an endangered species listed in the Red Data Book (Nayar & Sastry, 1987). This plant was first collected by Bourdillon on 2nd Dec., 1884 from Konni forest area which in the past was a dense evergreen forest. Now in Konni area, there are only patches of some semi-evergreen forests. The result of the present searches has forced to conclude that this species has met with extinction in the district. Further analysis showed that 30 species are endangered (Table 5). Most of those species are known only from single collection and no recent report of their occurrence in any of the neighbouring regions are available.

Table 5 : Endangered plants in Pathanamthitta District

Sl. No.	Name of Species	Sl. No.	Name of species
1.	*Acampe ochracea*	16.	*Goniothalamus rhyncantherus*
2.	*Aglaia perviridis*	17.	*G. wightii*
3.	*Bentinckia condapanna*	18.	*Hedyotis membranacea*
4.	*Bombax scopulorum*	19.	*Indigofera constricta*
5.	*Ceropegia metziana*	20.	*Litsea travancorica*
6.	*Ceropegia fimbriifera*	21.	*Nageia wallichiana*
7.	*Colubrina travancorica*	22.	*Nothopegia travancorica*
8.	*Cordia octandra*	23.	*Phaeanthus malabaricus*
9.	*Cucumella silentvalleyii*	24.	*Phrynium capitatum*
10.	*Derris thothathrii*	25.	*Pogostemon travancoricus*
11.	*Dimeria kurumthotticalana*	26.	*Salacia malabarica*
12.	*Dimeria mooneyi*	27.	*Sphenodesme involucrata*
13.	*Dipterocarpus bourdilloni*	28.	*Syzygium travancoricum*
14.	*Eria dalzellii*	29.	*Thottea dinghoui*
15.	*Ficus travancorica*	30.	*Willisia selaginoides*

Maximum rarity of species is met within the family Annonaceae accounting for 61.5%. In this family *Goniothalamus rhyncantherus, G. wightii, Phaeanthus malabaricus* are critically endangered. The first and the last of the species were already listed in the Red Data Book. During the present study, *Salacia malabarica* (Gamble, 1916), a species believed to have become extinct could be relocated after 75 years of its type collection. Incidentally, the species was located at a proposed dam-site in the district. *Rotala ritchei,* another threatened species known from only two collections (Clarke, 1879; Janardhanan, 1979), also could be collected during the present study. The table VI shows the top 6 families with high rate of rarity among their species.

Table 6 : Top Six families with high percentage or rarity

Sl. No.	Family	Total species in the district	No. of Rare/ threatened species	% of Rarity
1.	Annonaceae	13	8	61.5
2.	Balsaminaceae	12	4	33.3
3.	Acanthaceae	45	10	22.2
4.	Rubiaceae	59	13	22.1
5.	Dipterocarpaceae	5	1	20.0
6.	Orchidaceae	51	8	15.6

Alien species

Many exotic species have become naturalized in lowlands and forest areas. There are 43 species in this category. Most of these plants are native of America. The family Asteraceae is the largest in this regard, with 14 alien species. Among these, *Mikania cordata* has become a serious threat to native species. It has established in almost all disturbed forest areas. *Ageratum haustonianum* and *A. conyzoides* are other 2 members of this family which are well established on hills. Another troublesome weed is a newly introduced species, *Clidemia hirta.* This species is well established in hilly areas like Pampa-Thriveny area. On the waysides to Pampa, this species form dense populations just like a monoculture. This Malesian element might have reached here via Sri Lanka through the pilgrims to the holy shrine of Lord Ayyappa situated in the Sabari hills. *Lantana camara* var. *aculeata,* another well established alien species has occupied almost all waysides towards Kakki hill area. *Hyptis rhomboidea,* a new alien, is slowly becoming established in uphills. The other well estabished weedy species are *Chromolaena odorata, Croton bonplandianus, Eichhornia crassipes, Hyptis suaveolens, Monochoria vaginalis, Scoparia dulcis, Synedrella nodiflora* and *Tridax procumbens.*

Wild relatives of Crop Plants

About 90 species which are closely realted to our cultivated crop plants have been collected from the district. Among these, 3 are wild relatives of cereals and millets, 9 species are of leguminous crops of which 3 species are endemic and rare. They are *Atylosia lineata* (a close wild relative of pigeon pea), *Mucuna hirsuta* (a close relative of sword bean), and *Vigna pilosa* (a wild blak gram). Rich diversity is noticed in fruit crops. There are about 23 fruit crops excluding the wild jack and mango varieties. In fruit crops *Madhuca neriifolia* is very rare and endemic to this tract. Many species belonging to Cucurbiataceae, Amaranthaceae, Solanaceae and Dioscoreaceae are wild relatives of existing cultivars. *Momordica dioica* is a close relative of bittergourd. There are two species of *Trichosanthes* which are closely related to the cultivated snakegourd. *Luffa aegyptiaca* is another wild relative of smoothgourd. Among oil seed crops, *Sesamum mulayanum* run wild and is very common in plains. In this species the grains are much larger than that of gingelly, and the plant seems to be very resistant to drought and pests. Among spices, a relative of black pepper viz., *Piper argyrophyllum* is with very long spikes (c. 20 cm) and the wild *P nigrum* is with considerably large fruits. A close relative of nutmeg (*Myristica fragrans*), is *Myristica malabarica* with large fruits and brightly coloured aril, and is very rare and endemic.

Medicinal Plants

A large number of medicinal palnts also occur in the district. The medicinal application of these species ranges from their use in simple ailment like headache to even cases of cancer. Many of the medicinal plants are becoming more and more rare because of their over exploitation by drug industries. For example, the wild populations of the plants like *Baliospermum solanifolium, Rauvolfia serpentina, Sida rhombifolia* ssp. *retusa, Mucuna hirsuta, Drosera burmannii, Saraca asoca, Myxopyrum smilacifolium, Phyllanthus amarus, Gloriosa superba, Cymbopogon citratus,* etc. have become critically reduced.

Floristic Novelties

Present study indicates that the flora of the district is rich, diverse and with a good percentage of endemism; and at the same time facing serious threat from different sources. This study has revealed that this area was under-explored or even unexplored as evidenced by the following findings of new species, and reports on new distributional records.

New Species

The present study yielded 7 species new to science (Table 7). Two of the new species are tree members. These are *Julostylis polyandra* (Ravi & Anil Kumar, 1990) and *Dysoxylum swaminathanianum* (Anil Kumar *et al.*, 2001). The former species is of great significance in many ways. This discovery forms a record of extended distribution of the genus *Julostylis,* which was till then considered to be monotypic and occurring only in Sri Lanka. Two of the other new species viz., *Sida ravii* and *Hibiscus sreenarayanianus* are undershrubs. The latter was first collected from the heart of Kollam town indicating the lack of intensive plant exploration in the past. The other species are herbaceous and neo-endemics.

Table 7 : New species discovered from the district

Sl. No.	Name of species	Family
1.	*Dimeria idukkiensis*	Poaceae
2.	*Dysoxylum swaminathanianum*	Meliaceae
3.	*Fimbristylis angamoozhiensis*	Cyperaceae
4.	*Fimbristylis pseudonarayanii*	Cyperaceae
5.	*Hibiscus sreenarayanianus*	Malvaceae
6.	*Julostylis polyandra*	Malvaceae
7.	*Sida ravii*	Malvaceae

New Distributional Records and Reports

The findings of the species *Aeschynomene americana* var. *glandulosa* (Fabaceae), *Passiflora foetida* var. *foetida* (Passifloraceae) during the present study from the district formed new reports of the taxa from the country; and the findings of *Dalbergia pseud-sisso, Derris canarensis* (Fabaceae), *Balanophora abbreviata* (Orobanchaceae), *Dimeria mooneyi* (Poaceae) and *Rotala ritchei* (Lythraceae) formed new reports for Kerala State.

Some newly described species like *Sauropus saksenianus, Hydnocarpus pendulus, Cucumella silentvalleyii, Smithsonia straminea, Pothos crassipedunculatus, Sida fryxellii,* and *Aristolochia krisagathra* from other districts of the State were collected from Pathanamthitta district during the present study. The first three species were so far known only from Silent Valley, but the present collections form record of their extended distribution.

THREATS TO THE FLORA

As anywhere else, the most serious threat to the flora is from man. The forests have been drastically disturbed and destroyed by his direct and indirect activities. There are five dams in this small district, of which four are hydro-electrical and one is irrigation. Of the four hydro-electrical dams, 3 are in the species–rich Kakki hill area. A vast area of forest has been removed for the construction of the dam. Two- third of State's electricity is generated from this district alone. The power generation unit is in Moozhiar and it is about 36 km away from the dam site. The whole stretch of forest between the dam and the power generation site has been removed. The cleared areas are now occupied by the alien species which pose threat to the remaining native species. The majority of Kakki hill-top area has been encroached by people, especially cardamom plantation labourers. Recently, some Sri Lankan refugees have also been rehabilitated in this region. This area is witnessing all sorts of primary developments and this upland forest area has become easily accessible from all directions and is facing destruction at a very fast rate.

Clearing the forests for plantations also pose threat to the forests. It seems that Forest Department is more interested in planting economically important trees than conservation of our forest wealth. Over 50% of the forest cover has been cleared for the monoculture plantations of Teak, Eucalyptus and Acacia. *Acacia auriculiformis,* an exotic species has been introduced even in the primary grasslands of the district. The grasslands of the Ponnambalamedu region has been very severely destructed by planting *Acacia* and *Eucalyptus*. The entire wayside from Lahai estate to Plappally is now with dense populations of Acacia. Every year the area of teak plantation is being increased. The sides of almost all rivers in hilly areas have been converted into teak plantations. Pathanamthitta and the neighbouring Kottayam district are the top-most rubber producers of the state; the entire forests of the midland and a large portion of highlands have been cleared for rubber plantation.

Tourism is another very dreadful threat to the flora. The entire Sabarimala area with dense luxuriant vegetation having many rare wild life-forms in the past has drastically changed now. As the holy temple of Lord Ayyappa is situated there, lakhs of pilgrims from all over India and abroad come here every year for the annual ritual festivals 'Mandala pooja' and 'Makara vilakku'. During these periods trees are cut for using as firewood. The Sabarimala, Karimala and Neelimala hill tracts are now drastically cleared and these areas harbour quite a lot of alien species. The proposed developmental activities including the construction of roads and a helipad in the Sabarimala area would result in further destruction of the remaining forests.

Human settlement in hills is one of the common threats to the flora of the district. The Government has legalized all encroachments of forest lands by giving title deeds to many of the illegal settlers owing to political affiliation and pressures. The indiscriminate clearing of forests for construction of roads which are criss-crossing the hilly areas as a part of developmental work of the district has removed a vast extent of forests.

PROPOSED AREAS FOR CONSERVATION

It is well known that endemic species are more threat-prone. The analysis of the flora has manifested that about 20.9% of endemism prevails among species of this area; and a vast majority of those endemic species are rare.

Almost all these endemic species were collected from Moozhiar, Angamoozhy, Kakki hills of Goodrical forest range of Ranni forest division, and Arampa of Mannarappara forest range of Konni forest division. These areas are with typical evergreen forest and are now under various forms of threats. Angamoozhy, a small township is now surrounded by degraded, but comparatively species-rich forests. In Moozhiar and Kakki hills the power house and dams respectively are situated. Arampa is the main collection centre of reeds. All the above places are provided with jeepable roads which aid in extensive pilferage of timber, and poaching.

Incidentally, of the listed 175 rare plants, about 90% were collected from the above mentioned localities. Among those, Kakki hill area and Moozhiar hill area harbour maximum number of rare and endemic species and calls for immediate protection of these areas.

SYSTEMATIC TREATMENT

KEY TO THE SEED PLANTS OF PATHANAMTHITTA DISTRICT

Key to the Major Categories of Seed Plants

1. Ovules/seeds enclosed within megasporophyll/fruit wall. .... ***Angiospermae***
1. Ovules/seeds not enclosed within megasporophyll/fruit wall . ... ***Gymnospermae***

Key to the Categories of *Angiospermae*

1. Most commonly leaf-venation reticulate; flowers usually 5-merous or 4-merous; embryo almost always with cotyledons. .... ***Dicotyledones***
1. Most commonly leaf-venation parallel; flowers usually 3-merous; embryo almost always with 1 cotyledon. .... ***Monocotyledones***

Key to the Categories of *Dicotyledones*

1. Perianth whorls distinguishably 2-seriate : outer sepaloid, inner petaloid (except *Magnoliaceae* and *Annonaceae*). .... ***Dichlamydeae***
1. Perianth whorls not distinguishable and may be 1-seriate, 2-seriate or 0. .... ***Monochlamydeae***

Key to the Categories *Dichlamydeae*

1. Petals distinct, free from stamens. .... ***Polypetalae***
1. Petals connate, fused with stamens. .... ***Gamopetalae***

Key to the families of *Gymnospermae*

1. Scandent shrubs. .... **148. Gnetaceae**
1. Trees.
    2. Palm-like trees; leaves pinnate. .... **146. Cycadaceae**
    2. Profusely branched trees; leaves simple. .... **147. Podocarpaceae**

Key to the families of *Polypetalae*

1. Stamens hypogynous (except *Nymphaeaceae*).
    2. Sepals usually imbricate in bud.
        3. Sepals free to base.
            4. Stamens α (more than 10).
                5. Sepals 2 or indistinguishable.
                    6. Trees; sepals indistinguishable. .... **3. Magnoliaceae**
                    6. Herbs; sepals 2, distinguishable. .... **16. Portulacaceae**
                5. Sepals 4 to α.

7. Aquatic plants.
 8. Carpels free, superior. ....7. **Nelumbonaceae**
 8. Carpels connate, 1/2 inferior or 1/2 superior. ....6. **Nymphaeaceae**
7. Terrestrial plants.
 9. Sepals persistent.
  10. Leaves often sheathing at base. ....**2. Dilleniaceae**
  10. Leaves not sheathing at base.
   11. Leaves alternate.
    12. Stamens free from petals; ovary embedded on large lobed disc. ......**34. Ochnaceae**
    12. Stamens adnate with petal base; ovary not embedded on disc. ....**19. Theaceae**
   11. Leaves opposite (rarely verticillate).
    13. Trees; flowers usually polygamous. ....**18. Clusiaceae**
    13. Herbs; flowers bisexual. ....**17. Hypericaceae**
 9. Sepals deciduous.
  14. Climbers. ....**1. Ranunculaceae**
  14. Non-climbers.
   15. Ovary usually raised on a stalk (gynophore). ....**9. Capparaceae** (p.p.)
   15. Ovary not raised on a stalk.
    16. Herbs; flowers bisexual. ....**10. Cleomaceae**
    16. Trees; flowers polygamo-dioecious. ....**12. Flacourtiaceae**

4. Stamens few (less than 10).
17. Stamens 6.
 18. Climbers; flowers mostly 3-merous, dioecious. ....**5. Menispermaceae**
 18. Non-climbers; flowers 4-5-merous, bisexual.
 19. Ovary 1-locular; disc 0 or annular. ....**10. Cleomaceae** *(Cleome monophylla)*
 19. Ovary usually 2-locular; disc of 4 glands. ....**8. Brassicaceae**
17. Stamens either less than or more than 6.
 20. Ovary 1-locular.
  21. Trees; anterior petal keeled. ....**14. Xanthophyllaceae**
  21. Herbs; anterior petal not keeled.
   22. Corolla irregular; anthers spurred. ....**11. Violaceae**

22. Corolla regular; anthers not spurred. .... **15. Caryophyllaceae**

20. Ovary 2-5-locular.

23. Staminal filaments free.

24. Trees; leaves odd-pinnate. .... **46. Staphyleaceae**

24. Herbs; leaves simple or 3-foliolate.

25. Flowers regular; leaves 3-foliolate. .... **30. Oxalidaceae**

25. Flowers irregular; leaves simple. .... **31. Balsaminaceae**

23. Staminal filaments connate.

26. Seeds caruncled; flowers irregular. .... **13. Polygalaceae**

26. Seeds non-caruncled; flowers regular.

27. Erect annuals; stamens 5. .... **27. Linaceae**

27. Trees; stamens 10-12. .... **28. Erythroxylaceae**

3. Sepals connate at base.

28. Leaves gland-dotted. .... **32. Rutaceae**

28. Leaves not gland-dotted.

29. Leaves opposite. .... **29. Malpighiaceae**

29. Leaves alternate.

30. Trees with resinous or acrid juice.

31. Stamens free; leaf-base equal.

32. Calyx accrescent or winged; flowers bisexual. .... **20. Dipterocarpaceae**

32. Calyx not accrescent or winged; flowers polygamous. .... **48. Anacardiaceae**

31. Stamens connate; leaf-base usually oblique. .... **36. Meliaceae**

30. Trees or shrubs or rarely herbs without resinous or acrid juice.

33. Stamens alternate with petals.

34. Leaves compound.

35. Filaments free; leaflets alternate. .... **33. Simaroubaceae**

35. Filaments connate; leaflets opposite. .... **36. Meliaceae**

34. Leaves simple.

36. Ovary 1-2-locular; ovules pendulous.

37. Petals 2-lobed. .... **37. Dichapetalaceae**

37. Petals entire. .... **40. Icacinaceae**

36. Ovary 2-3 (5)-locular; ovules erect.

38. Stamen 4-5; seeds arillate. .... **41. Celastraceae**

38. Stamens 3; seeds non-arillate. .... **42. Hippocrateaceae**

33. Stamens opposite the petals.

39. Plants with hooked branches. .... **21. Ancistrocladaceae**

39. Plants without hooked branches.

40. Leaves estipulate.
41. Tendrillar climbers; staminodes present. ....**39. Erythropalaceae**
41. Trees; staminodes absent. ....**38. Olacaceae**
40. Leaves stipulate.
42. Petals valvate; stamens 4-5, all fertile.
43. Tendrillar climbers; ovary 2-locular. ....**44. Vitaceae**
43. Non-tendrillar erect shrubs; ovary 3-6- locular. **45. Leeaceae**
42. Petals imbricate; stamens 5, 2 fertile. ....**46. Sabiaceae**
2. Sepals usually valvate in bud.
44. Sepals free; perianth usually 3-merous (sepaloid). ....**4. Annonaceae**
44. Sepals connate; perianth usually 5-merous.
45. Staminal filament variously connate.
46. Stamens poly-adelphous; leaves digitately compound. ....**23. Bombacaceae**
46. Stamens monadelphous; leaves simple or lobed.
47. Anther 1-celled; staminal column basally coherent with petals. ....**22. Malvaceae**
47. Anther 2-celled; staminal column free from petals. ....**24. Sterculiaceae**
45. Stamens free.
48. Leaves simple.
49. Stamens enclosed by petals; ovary immersed in the disc. ....**43. Rhamnaceae**
49. Stamens not enclosed by petals; ovary free from disc.
50. Petals ciliate or laciniate. ....**26. Elaeocarpaceae**
50. Petals not ciliate or laciniate. ....**25. Tiliaceae**
48. Leaves compound. ....**35. Burseraceae**
1. Stamens epi-or perigynous.
51. Leaves highly reduced or absent; stem flat, fleshy and with spines. .... **68. Cactaceae**
51. Leaves normal; stem herbaceous or woody without spines.
52. Ovules axile or basal.
53. Flowers unisexual.
54. Plants tendrillar-climbers.
55. Petals connate; corona 0; ovary inferior. ....**65. Cucurbitaceae**
55. Petals free; corona present; ovary superior. ....**64. Passifloraceae**
54. Plants non-tendrillar herbs of trees.
56. Trees; flowers regular; ovary wingless. ....**67. Datiscaceae**
56. Herbs; flowers irregular; ovary 3-winged. ....**66. Begoniaceae**
53. Flowers bisexual.
57. Insectivorous plants. .... **54. Droseraceae**
57. Non-insectivorous plants.
58. Leaves usually stipulate; pistil monocarpellary. ....**51. Fabaceae**

58. Leaves usually estipulate (except *Rosaceae*); pistill multicarpellary.
59. Carpels free.
60. Stamens a; leaves stipulate. .... **52. Rosaceae**
60. Stamens 5-10; stipules 0.
61. Climbers; leaves alternate. .... **50. Connaraceae**
61. Erect succulent herbs; leaves opposite. .... **53. Crassulaceae**
59. Carpels connate.
62. Petals 0.
63. Trees; ovary 1-locular. **56. Combretaceae**
63. Herbs; ovary 2-5-locular. .... **69. Molluginaceae**
62. Petals present.
64. Calyx-lobes valvate; fruit a capsule.
65. Ovary free from calyx-tube. .... **61. Lythraceae**
65. Ovary adnate to the calyx tube. .... **62. Onagraceae**
64. Calyx-lobes imbricate; fruit a berry.
66. Stamens more than 12.
67. Leaves usually gland-dotted. .... **57. Myrtaceae**
67. Leaves not gland-dotted. .... **58. Lecythidaceae**
66. Stamens less than 12.
68. Ovary 4-5-locular; leaves 3-5-nerved from base. .... **59. Melastomataceae**
68. Ovary 1-locular; leaves penni-nerved. .... **60. Memecylaceae**

52. Ovules pendulous.
69. Ovules 2 or more per locule.
70. Leaves usually 2- glandular near base. .... **64. Turneraceae**
70. Leaves eglandular near base.
71. Ovary 2-6-locular. .... **55. Rhizophoraceae**
71. Ovary 1 locular.
72. Ovary inferior; ovules 2-3. .... **56. Combretaceae**
72. Ovary superior; ovules α. .... **63. Samydaceae**
69. Ovule 1-per locule.
73. Inflorescence usually an umbel.
74. Herbs; ovary 2-locular; fruit a mericarp. .... **70. Apiaceae**
74. Shrubs/trees; ovary 5-6-locular; fruit non mericarp. .... **71. Araliaceae**
73. Inflorescence other than umbel.
75. Scandent shrubs; petals 10. .... **72. Alangiaceae**
75. Trees; petals 5. .... **73. Cornaceae**

Key to the families of *Gamopetalae*

1. Ovary completely or partly inferior.
   2. Inflorescence an involucrate head (*Capitulum*); calyx reduced to pappus. .... **76. Asteraceae**
   2. Inflorescence other than an involucrate head; calyx normal with sepals.
      3. Stamens α; ovary 3-locular. .... **82. Symplocaceae**
      3. Stamens few; ovary 1-10-locular.
         4. Leaves opposite.
            5. Stipule 0; ovary 1-3-loccular. .... **74. Caprifoliaceae**
            5. Stipule present (Mostly interpetiolar); ovary 2-10-locular. .... **75. Rubiaceae**
         4. Leaves alternate.
            6. Stamens opposite to petals; ovary 1-locular. .... **79. Myrsinaceae**
            6. Stamens alternate to petals; ovary 2 or 3-locular. .... **77. Lobeliaceae**

1. Ovary superior.
   7. Ovary 1-locular.
      8. Corolla irregular; stamens 2; leaves modified into insectivorous bladders. .... **95. Lentibulariaceae**
      8. Corolla regular; stamens 4-6; leaves not modified. .... **79. Myrsinaceae**
   7. Ovary 2-α-locular.
      9. Stamens opposite to petals.
         10. Corolla urceolate or tubular-campanulate; stamens free.
            11. Corolla urceolate; flowers bisexual; ovules a per locule. .... **78. Vacciniaceae**
            11. Corolla tubular-campanulate; flowers unisexual; ovule 1 per locule. .... **81. Ebenaceae**
         10. Corolla salver-form; stamens epipetalous. .... **80. Sapotaceae**
      9. Stamens alternate with petals.
         12. Corolla regular.
            13. Leafless parasitic climbers. .... **91. Convolvulaceae**(*Cuscuta*)
            13. Leafy, non-paratisic non-climbers.
               14. Leaves opposite, rarely whorled.
                  15. Stamens 2. .... **83. Oleaceae**
                  15. Stamens 4 or more.
                     16. Carpels free.
                        17. Pollen grains aggregated into masses. .... **85. Asclepiadaceae**
                        17. Pollen grains not aggregated into masses. .... **84. Apocynaceae** (p.p.)

16. Carpels connate.
    18. Terrestrial plants.
        19. Stipule present; corolla-lobes imbricate or valvate. .... **86. Loganiaceae**
        19. Stipule 0; corolla-lobes usually twisted.
            20. Corolla hypocrateriform, lobes twisted to right. .... **84. Apocynaceae** (*Chilocarpus, Carissa*)
            20. Corolla campanulate or funnel-form, lobes twisted to left. .... **87. Gentianaceae**
    18. Aquatic plants. .... **88. Menyanthaceae**

14. Leaves alternate.
    21. Most commonly climbers (except *Evolvulus alsinoides* var. *decumbens*). .... **91. Convolvulaceae**
    21. Other than climbers.
        22. Ovules α-per locule; fruit usually a berry, or capsule. .... **92. Solanaceae**
        22. Ovules 2 or 1-per locule; fruit of 4 dry nutlets or drupaceous.
            23. Trees or erect shrubs. .... **90. Cordiaceae**
            23. Herbs or subscandent shrubs. .... **89. Boraginaceae**

12. Corolla irregular.
    24. Ovule(s) 1-2-per carpel.
        25. Ovary usually entire; style terminal. .... **100. Verbenaceae**
        25. Ovary usually 4-lobed; style gynobasic. .... **101. Lamiaceae**
    24. Ovules α -per carpel.
        26. Fruits elastically dehiscent. .... **99. Acanthaceae**
        26. Fruits not elastically dehiscent.
            27. Ovary 1-locular.
                28. Leafless parasitic non-chlorophyllous herbs. .... **94. Orobanchaceae**
                28. Leafy non-parasitic chlorphyllous herbs. .... **96. Gesneriaceae**
            27. Ovary 2-locular.
                29. Trees; fruits to 50 cm long; seeds winged. .... **97. Bignoniaceae**
                29. Herbs; fruits to 5 cm or less long; seeds not winged.
                    30. Leaves 2-glandular at base; seeds exalbuminous. .... **98. Pedaliaceae**
                    30. Leaves eglandular at base; seeds albuminous. .... **93. Scrophulariaceae**

Key to the families of *Monochlamydeae*

1. Ovary superior.
    2. Flowers bisexual.
        3. Perianth present.
            4. Anther - dehiscence valvular. ....**111. Lauraceae**
            4. Anther- dehiscence otherwise.
                5. Indumentum scally. ....**112. Elaeagnaceae**
                5. Indumentum not scally.
                    6. Plants thalloid, submerged. ....**106. Podostemaceae**
                    6. Plants not thalloid, terrestrial.
                        7. Stipule ocreaceous. ....**105. Polygonaceae**
                        7. Stipule not ocreaceous, or absent.
                            8. Trees; seeds arillate. ....**110. Myristicaceae**
                            8. Herbs or undershrubs; seeds not arillate.
                                9. Perianth petaloid. ....**102. Nyctaginaceae**
                                9. Perianth not petaloid.
                                    10. Perianth and bracts scarious; stamens connate below. ....**103. Amaranthaceae**
                                    10. Perianth and bracts not scarious; stamens free. ....**104. Chenopodiaceae**
        3. Perianth absent.
            11. Scandent shrubs/succulent herbs. ....**108. Piperaceae**
            11. Erect shrubs. ....**109. Chloranthaceae**
    2. Flowers unisexual or polygamous.
        12. Parasitic achlorophyllous plants. ....**116. Balanophoraceae**
        12. Non-parasitic chlorophyllous plants.
            13. Perianth absent; flowers in cyathia. ....**117. Euphorbiaceae**(*Euphorbia*)
            13. Perianth present. Flowers not in cyathia.
                14. Ovary most commonly 3-locular, rarely 2-locular. ....**117. Euphorbiaceae**
                14. Ovary 1-locular.
                    15. Style undivided.
                        16. Ovule apical. ....**119. Moraceae**
                        16. Ovule basal. ....**120. Urticaceae**
                    15. Style divided; ovule 1, apical. ....**118. Ulmaceae**
1. Ovary inferior.
    17. Parasitic plants.
        18. Branchlets flattened; anther - dehiscence porous. ....**114. Viscaceae**
        18. Branchlets terete; anther - dehiscece longitudinal. ....**113. Loranthaceae**

17. Non-parasitic plants.
    19. Ovary 4-6 locular; ovules a; stamens 6 or more. ....107. **Aristolochiaceae**
    19. Ovary 1-locular; ovules 1-3; stamens 3-5. ....115. **Santalaceae**

Key to the families of *Monocotyledones*

1. Perianth ± petaloid.
   2. Flowers unisexual (staminate), dioecious or rarely monoecious.
      3. Climbers.
         4. Tendrillar climbers; inflorescence an umbel. ....132. **Smilacaceae**
         4. Non-tendrillar climbers or twiners; inflorescence otherwise. ....130. **Dioscoreaceae**
      3. Non-climbers.
         5. Terrestrial plants with pseudostem (Flowers monoecious). ....126. **Musaceae**
         5. Aquatic plants without pseudostem. ....121. **Hydrocharitaceae**
   2. Flowers bisexual or pistillate.
      6. Pistill 3 or more (Aquatic plants). ....141. **Alismataceae**
      6. Pistil 1.
         7. Ovary superior.
            8. Perianth-lobes 3.
               9. Inflorescence a bracteate head. ....134. **Xyridaceae**
               9. Inflorescence other than a bracteate head. ....135. **Commelinaceae**
            8. Perianth-lobes 6.
               10. Aquatic plants. ....133. **Pontederiaceae**
               10. Terrestrial plants.
                   11. Plants herbaceous; leaves succulent or membranous. ....131. **Liliaceae**
                   11. Plants woody; leaves coriaceous.
                       12. Climbers.
                           13. Leaf-tip tendrillar; stem smooth. ....136. **Flagellariaceae**
                           13. Leaf-petiole often tendrillar; stem prickly. ....132. **Smilacaceae**
                       12. Non-climbers (Leaves often crowded towards apex). ....28. **Agavaceae** (*Dracaena terniflora*)
         7. Ovary inferior.
            14. Aquatic plants. ....121. **Hydrocharitaceae**
            14. Terrestrial plants.
                15. Climbers or twiners. ....130. **Dioscoreaceae**

15. Neither climbers nor twiners.
16. Leaf-margin spiny. .... **128. Agavaceae** *(Furcraea foetida)*
16. Leaf-margin smooth.
17. Perianth-tubular, split on one side; leaves ± 2 m long. .... **126. Musaceae**
17. Perianth otherwise, not split on one sie; leaves at the most 1 m long.
18. Flowers regular; fertile stamens 3 or 6.
19. Fertile stamens 6.
20. Perianth-lobes connate. .... **127. Amaryllidaceae**
20. Perianth-lobes free. .... **129. Hypoxidaceae**
19. Fertile stamens 3 (Scapose leafless herbs). .... **122. Burmanniaceae**
18. Flowers irregular; fertile stamen(s) 1(2).
21. Staminodia petaloid; stamens and style distinct.
22. Petiole pulvinate above; ovule(s) 1-per locule. .... **126. Marantaceae**
22. Petiole not pulvinate above;ovules α-per locule.
23. Leaves not spirally arranged. .... **124. Zingiberaceae**
23. Leaves spirally arranged. .... **125. Costaceae**
21. Staminodia not petaloid; stamens and styles connate to form a column. .... **123. Orchidaceae**
1. Perianth scaly, bristly or 0.
24. Plants woody.
25. Leaves pinnately or palmately compound. .... **137. Arecaceae**
25. Leaves simple.
26. Leaves alternate, spiral, margin spiny. .... **138. Pandanaceae**
26. Leaves 2-ranked, margin smooth. .... **145. Poaceae** (*Bambusa bambos*)
24. Plants herbaceous.
27. Plants thallus-like, minute. .... **140. Lemnaceae**
27. Plants not thallus-like or minute.
28. Flowers subtended by glumaceous bracts (glumaceous spikelets).
29. Inflroescence a solitary head. .... **143. Eriocaulaceae**
29. Inflorescence otherwise.
30. Style 1; seeds (2-3)-angled. .... **144. Cyperaceae**
30. Styles 2; seeds terete. .... **145. Poaceae**
28. Flowers not subtended by glumaceous bracts.
31. Flower(s) solitary.
32. Leaves denticulate, base auriculate. .... **142. Najadaceae**
32. Leaves entire, base not auriculate. .... **121. Hydrocharitaceae**
31. Flowers densely packed in spikes. .... **139. Araceae**

1. RANUNCULACEAE

1. Tendrillar, stout climbers; petals 4-12; achenes linear. .... **Naravelia**
2. Non-tendrillar, slender climbers; petals 0; achenes ovoid. .... **Clematis**

CLEMATIS Linnaeus
Sp. Pl. 543.1753.

Clematis gouriana Roxb. ex DC., Syst. 1:139. 1817; Wight, Ic. t. 933 & 934, 1845; Hook. f. & Thoms. in Hook.f., Fl. Brit. India 1:4.1872; Dunn in Gamble, Fl. Pres. Madas 3.1915; Gandhi in Sald. & Nicols., Fl. Hassan Dist. 62.1976; Matthew, Ill. Fl. Tam. Carnatic t. 2.1982; Matthew & Britto in Matthew, Fl. Tam. Carnatic 3(1):1.1983; Manilal, Fl. Silent Valley 1.1988. *Nikida kodi.*

Slender climbers. Leaves opposite, 2-3-pinnate; leaflets variable in shape, usually ovate-lanceate, 2-6 by 1-2 cm, base rounded to subcordate, margin coarsely dentate to entire, apex gradually acuminate, both surfaces sparsely pubscent, basally 3-5- nerved. Inflorescence a terminal panicle. Flowers dense, bisexual, regular, c.1 cm ϕ; sepal 4, cream coloured or pale greenish; stamens α, α-seriate, free; carpels α, distinct, 1-locular; ovule pendulous. Achenes ovoid, sessile.

Fl. & Fr.: Dec.-Mar. *Distr.*: India, Myanmar, W. & C. China. Less common. In moist evergreen forests, 750-1200 m. *NAK 2543* (Kakki hills, c. 1100 m).

In young plants rarely petiole base with small stipuliform leaves present. Leaves, fresh roots and stems bruised and used against cold and skin diseases.

NARAVELIA A.P. de Candolle
Syst. Nat. 1:129, 167.1817.

Naravelia zeylanica (L.) DC., Syst. 1:167. 1817; Hook.f. & Thoms. in Hook. f., Fl. Brit. India 1:7. 1872; Dunn in Gamble, F1. Pres. Madras 3.1915; Gandhi in Sald. & Nicols., Fl. Hassan Dist. 62.1976; Matthew, Ill. Fl. Tam. Carnatic t. 2. 1982; Matthew & Britto in Matthew, Fl. Tam. Carnatic 3(1) : 2.1983; Manilal, Fl. Silent Valley 1.1988; Mohanan & Sivad., Fl. Agasthyamala 48.2002. *Atragene zeylanica* L., Sp. Pl. 542.1753; Roxb., Pl. Corom. t. 188. 1805 & Fl. Ind. 2:670.1832. *Karuppa kodi, Vatham kolli.*

Stout climbers. Leaves opposite, 2-3- foliolate, often terminal leaflet modified into a tendril; leaflets ovate to elliptic-ovate, 4-11 by 2-7 cm, base rounded or acute, apex acute-apiculate, lower surface tomentose, basally 3-5- ribbed. Panicle terminal or axillary, large. Flowers lax, regular, 1-2 cm ϕ; sepals 4-5; petals 4-12, white; stamens α, free; carpels α, distinct, 1-locular; ovule 1, pendulous. Achenes linear, stalked, spirally twisted.

Fl. & Fr.: Nov.-Mar. *Distr.*: Tropical Himalaya, E. Nepal, N. Myanmar, East India to Peninsular India & Sri Lanka *NAK 288* (Upper Moozhiar, 350 m), *2273* (Mannarappara, 250 m). Spreading on thickets in deciduous forests and moist semi-evergreen forests, 250-950 m.

Smelling crushed roots relieves head-ache.

2. DILLENIACEAE

1. Herbs; stipules fugacious; flowers yellow. .... **Acrotrema**

1. Trees or climbing shrubs; stipules 0; flowers yellow or white.
   2. Trees; leaves to 50 by 20 cm; flowers yellow. .... **Dillenia**
   2. Climbing shrubs; leaves to 10 by 5 cm; flowers white or cream coloured. .... **Tetracera**

ACROTREMA W. Jack
Malayan Misc. 1(5) : 36. 1820.

Acrotrema arnottianum Wight, Ill. Ind. Bot. 1:t. 3. 1840; Hook. f. & Thoms. in Hook.f., Fl. Brit. India 1:32. 1872; Dunn in Gamble, Fl. Pres. Madras 6. 1915; Mohanan & Sivad., Fl. Agasthyamala 49.2002. **(Fig. 1)**

Rhizomatous herbs. Leaves simple, closely alternate; stipule foliaceous, fugacios; lamina obovate to panduriform, 10-20 by 5-13 cm, base cordate, margin distantly denticulate, apex rounded or retuse, lateral veins, c.20 pairs, prominent. Racemes to 10 cm long, axilary. Flowers lax, bisexual, regular, c. 15 mm ϕ; pedicel 4-8 cm long; sepals 5; petals 5, bright yellow, narrowly obovate, caducous; stamens α, in 3-bundles; carples 3, coherent; style slender; stigma capitate; ovules 2-per locule. Follicle ovoid. Seeds arillate.

Fl. & Fr.: Sep.-Dec. *Distr.*: Endemic to Western Ghats. Rare. Along moist hill slopes in open shady evergren forests to 500 m. *NAK 1110* (Lower Moozhiar, c.250 m), *1832* (Angamoozhy, c.200 m).

DILLENIA Linnaeus
Sp. P1. 535.1735.

Dillenia pentagyna Roxb., Pl. Corom. 20.1795; Hook.f. & Thoms. in Hook. f., Fl Brit. India 1:38. 1872; Dunn in Gamble, Fl. Pres. Madras 7.1915; Hoogland, Blumea 7:117. 1952; Ramamoorthy in Sald. & Nicols., Fl. Hassan Dist. 114. 1976; Manilal, Fl. Silent Valley 2.1988; Mohanan & Sivad., Fl. Agasthyamala 49.2002. *Malampunna, Vazhappunna, Pattipunna.*

Trees. Leaves often clustered towards apex of branchlets, elliptic-oblanceate, 30-50 by 10-20 cm, base cuneate, margin serrulate, apex acute, lateral veins 40-45 pairs, prominent, thin-coriacous. Flowers few, in axillary fascicles, c. 4 cm ϕ; sepls 5, fleshy, persistent, petals 5,yellow, caducous; stamens α-seriate, inner whorl longer; carpels 5, ovoid, coherent. Fruts berry-like, globose. Seeds non-arillate.

Fl. & Fr.. Oct.-Dec. *Distr.*: China to Indo-Malesia. Common in moist deciduous forests. *NAK 1613* (Chathanthara, Ranni R.F., c.100 m).

Wood is soft and suitable for match boxes and splints; leaves used as manure.

TETRACERA Linnaeus
Sp. Pl. 533. 1753.

Tetracera akara (Burm. f.) Merr., Philip. J. Sci. 19:366. 1921; Hoogland, Reinwardtia 2:208. 1953; Nicols., Suresh & Manilal, An Interpr. Hort. Malab. 101.1988; Mohanan & Sivad., Fl. Agasthyamala 50.2002 *Calophyllum akara* Burm. f., Fl. Ind. 121. 1768. *Tetracera laevis* auct., non Vahl, 1794: Hook.f. & Thoms. in Hook. f., Fl. Brit. India 1:31. 1872; Dunn in Gamble, Fl. Pres. Madras 6.1915. *T. rheedei* DC., Syst. 1:402. 1818, ("*rheedii*"); Wight, Ic.t. 70. 1838. *Acara-patsjotti* Rheede, Hort. Malab. 5: 15, t. 8. 1685. *Nennal valli.* **(Fig.2)**

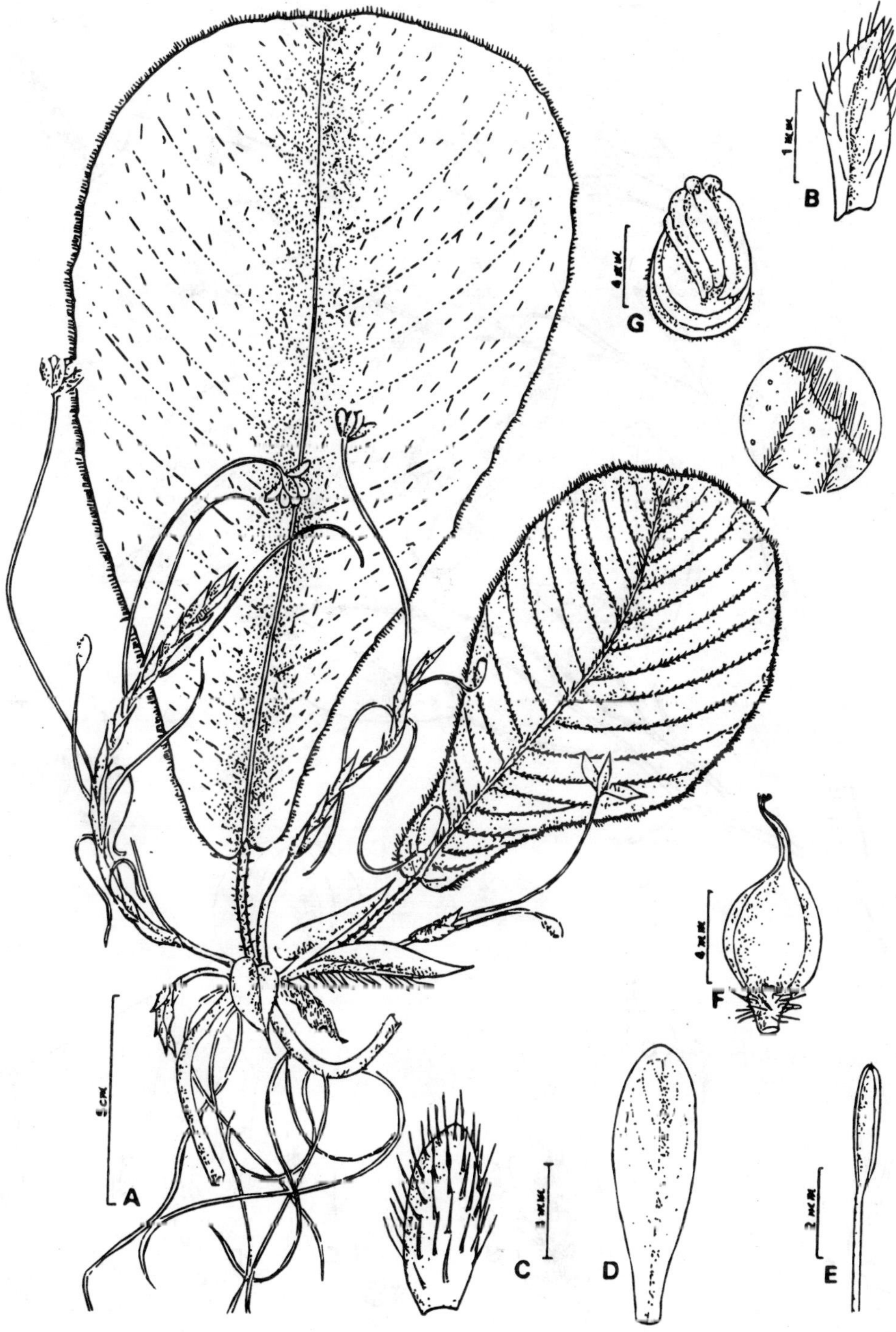

Fig. 1. ***Acrotrema arnottianum*** **Wight : A. Plant; B. Bract; C. Sepal; D. Petal; E. Stamen; F. Single pistil; G. Seed.**

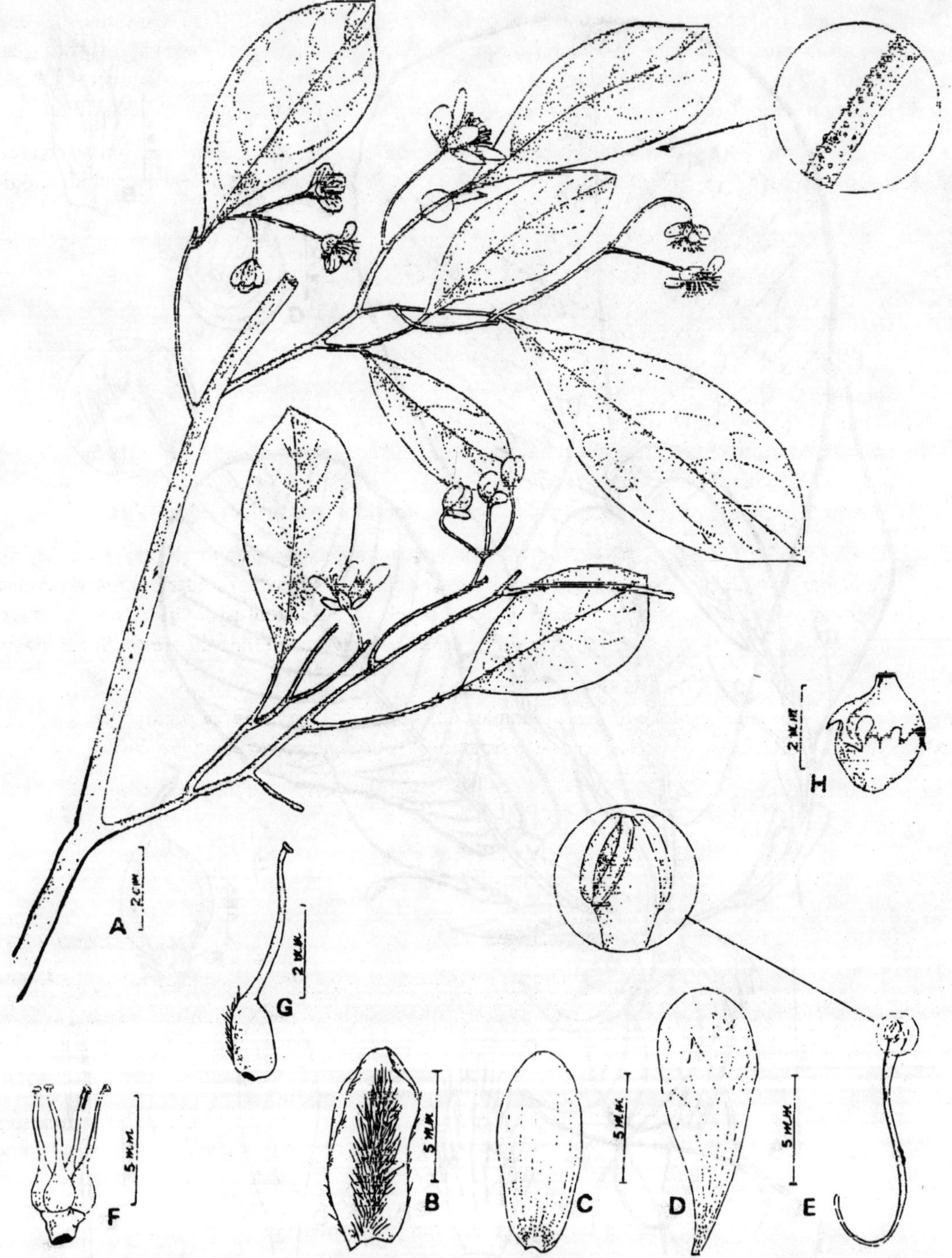

Fig. 2. ***Tetracera akara*** **(Burm.f.) Merr.: A. Twig; B. Sepal - adaxial view; C. Sepal- abaxial view; D. Petal; E. Stamen; F. Pistils; G. Single pistil; H. Seed.**

Climbig shrubs, branchlets scabrid. Leaves obovate to elliptic, 4-10 by 2-5 cm, base cuneate, apex acute to shortly acuminate, lateral veins 6-8 pairs, prominent, distant. Panicle usually ends in leafy branches. Flowers few to α, c.2 m ϕ; sepals 4-6, silky inside; petals 4-6, white; stamens α, filaments dilated towards apex, anthers distant; carpels 3-5, coherent at base, ovules α. Follicle globose, shiny. Seeds arillate.

Fl. & Fr.: Mar-May. Distr.: Peninsular India, Sri Lanka, Java to Borneo. Common in semi-evergreen and deciduous forests. *NAK 608,* (Thekkuthode, c. 70 m), *1585* (Chathanthara, 150 m). Stem used for making baskets.

3. MAGNOLIACEAE

MICHELIA Linnaeus
Sp. Pl. 536. 1753.

Michelia champaca L., Pl. 536. 1753; Roxb., Fl. Ind. 2:656. 1832; Wight, Ill. Ind. Bot. 13. 1840; Hook. f. & Thoms. in Hook. f., Fl. Brit. India 1:42. 1872; King, Ann. Roy. Bot. Gard. (Calcutta) 3 : t. 64. 1891; Dunn in Gamble, Fl. Pres. Madras 9.1915; Gandhi in Sald. & Nicols., Fl. Hassan Dist. 34. 1976; Matthew, Ill. Fl. Tam. Carnatic t. 3. 1982; Matthew & Britto in Matthew, Fl. Tam. Carnatic 3 (1): 3.1983; Nicols., Suresh & Manilal, An Interpr. Hort. Malab.169 1988. *Champacam* Rheede, Hort. Malab. 1:31-32, t. 19.1678.

Trees. Leaves simple, spiral; stipule prominent, deciduous. Lamina lanceate or elliptic, 10-25 by 5-8 cm, base cuneate, margin undulate, apex acute to acuminate, chartaceous. Flower(s) solitary, axillary; perianth lobes 11-15 or more, yellow; stamens α, free, α-seriate; carpels α, on a stout gynophore, sessile, spiral; ovary 1- locular; ovules α on ventral suture; style short. Follicles aggregate, ellipsoid, warty. Seeds scarlet coloured.

Fl. & Fr.: May-Jun. *Distr.*: Nepal, India, Myanmar, Thailand, Indo-China. Occasionally planted. *NAK 1227* (Moozhiar, c. 250 m). An evergreen tree with fragrant flowers; sacred for Hindus.

Decoction of root is used in cases of unconsiousness and the bark has febrifugal properties.

4. ANNONACEAE

1. Stragglers.
    2. Peduncle hooked; petals lanceate; ovules 2. .... **Artabotrys**
    2. Peduncle straight; petals orbicular; ovules α. .... **Uvaria**
1. Trees or erect shrubs.
    3. Inner petals arching over the torus.
        4. Inner petals arching over the torus by the base only; calyx cupular. .... **Cyathocalyx**
        4. Inner petals arching over the torus by the margin or tip only; calyx not cupular.
            5. Petals cohering throughout the margin; stamens a.
                6. Petals clawed; leaf-veins faint; flowers greenish-yellow. .... **Goniothalamus**
                6. Petals clawless; leaf-veins prominent; flowers cream coloured. .... **Phaeanthus**

5. Petals cohering by the tip only; stamens six. .... **Orophea**

3. Inner petals not arching over the torus.

7. Ovules 2-8 per carpel on marginal placenta; monocarp 2-a-seeded. ....**Meiogyne**

7. Ovules 1-2 per carpel on basal placenta; monocarpseeded. ....**Polyalthia**

ARTABOTRYS R. Brown
Bot. Reg. 5: 423. 1820.

Artabotrys zeylanicus Hook. f. & Thoms., Fl. Ind. 128.1855 & in Hook. f., Fl. Brit. India 1:54. 1872; King, Ann. Roy. Bot. Gard. (Calcutta) 4:43, t.53. 1893; Dunn in Gamble, Fl. Pres. Madras 14. 1915; Gandhi in Sald. & Nicols., Fl. Hassan Dist. 35.1976; Manilal, Fl. Silent Valley 2.1988; Mohanan & Sivad., Fl. Agasthyamala 51.2002. *Manoranjini valli.* **(Fig. 3)**

Stragglers. Leaves simple, alternate, elliptic-oblong, 6-18 by 3-8 cm, base rounded, apex shortly acuminate, shiny, thin-coriaceous. Fascicles leaf-opposed; peduncle hooked. Flowers bisexual, regular, fragrant, c. 4 cm ϕ; pedicel, c. 1 cm long; sepals 3, tomentose without; petals 6, yellow, 3+3, outer longer spreading, inner shorter, bent at the base over torus; stamens α, connective truncate, concealing the anthers; carpels c. 12, obovoid or clavoid; ovules 2. Fruit a cluster of globose berries, Seed solitary.

Fl. & Fr.: May (Jul.) - Sep. *Distr.*: Peninsular India to Sri Lanka. Common in the moist semi-evergreen forests. Extensive large climbers. *NAK 793* (Kolakutty, Ranni R.F., 150 m), *1783* (Angamoozhy, 350 m).

A. hexapetalus (L.f.) Bhandari is seen cultivated as an ornamental plant for its highly fragrant flowers (*NAK 2298*).

CYATHOCALYX Champion ex J. D. Hooker et T. Thomson
Fl. Ind. 126. 1855.

Cyathocalyx zeylanicus Champ. ex Hook. f. & Thoms., Fl. Ind. 127. 1855 & in Hook. f., Fl. Brit. India 1.53. 1872; Bedd., Ic. t. 47. 1868-1874; Dunn in Gamble, Fl. Pres. Madras 13. 1915; Mohanan & Sivad., Fl. Agasthyamala 52.2002. **(Fig. 4)**

Trees. Leaves oblong-lanceate, 15-25 by 5-8 cm, base cuneate, apex acuminate. Flowers large; sepals 3, connate, cupular, 3-toothed, bent over the torus; stamens α, connective truncate; carples few; stigma clavate; ovules α. Berry subglobose. Seeds rugose.

Fl. & Fr.: Apr.-Aug. (Nov.) *Distr.:* Peninsular India. Rare. In moist-evergreen forests. *NAK 2400* (Moozhiar, c. 300 m). Bark yield strong fibre.

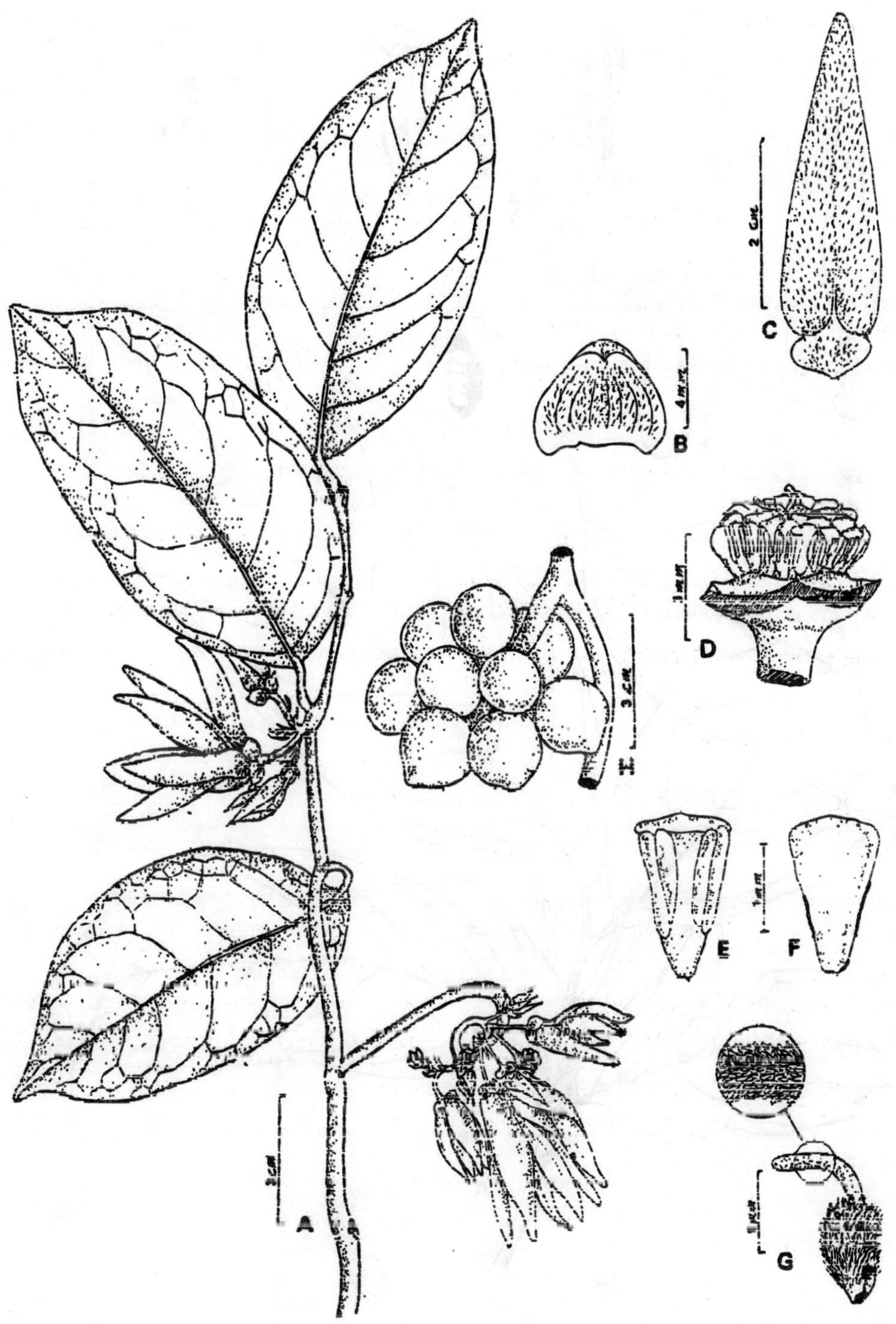

Fig. 3 . ***Artabotrys zeylanicus*** **Hook. f. & Thoms.: A. Twig; B. Sepal; C. Petal; D. Torus with stamens and pistils; E. Stamen - abaxial view; F. Stamen - adaxial view; G. Single pistil; H. Fruits.**

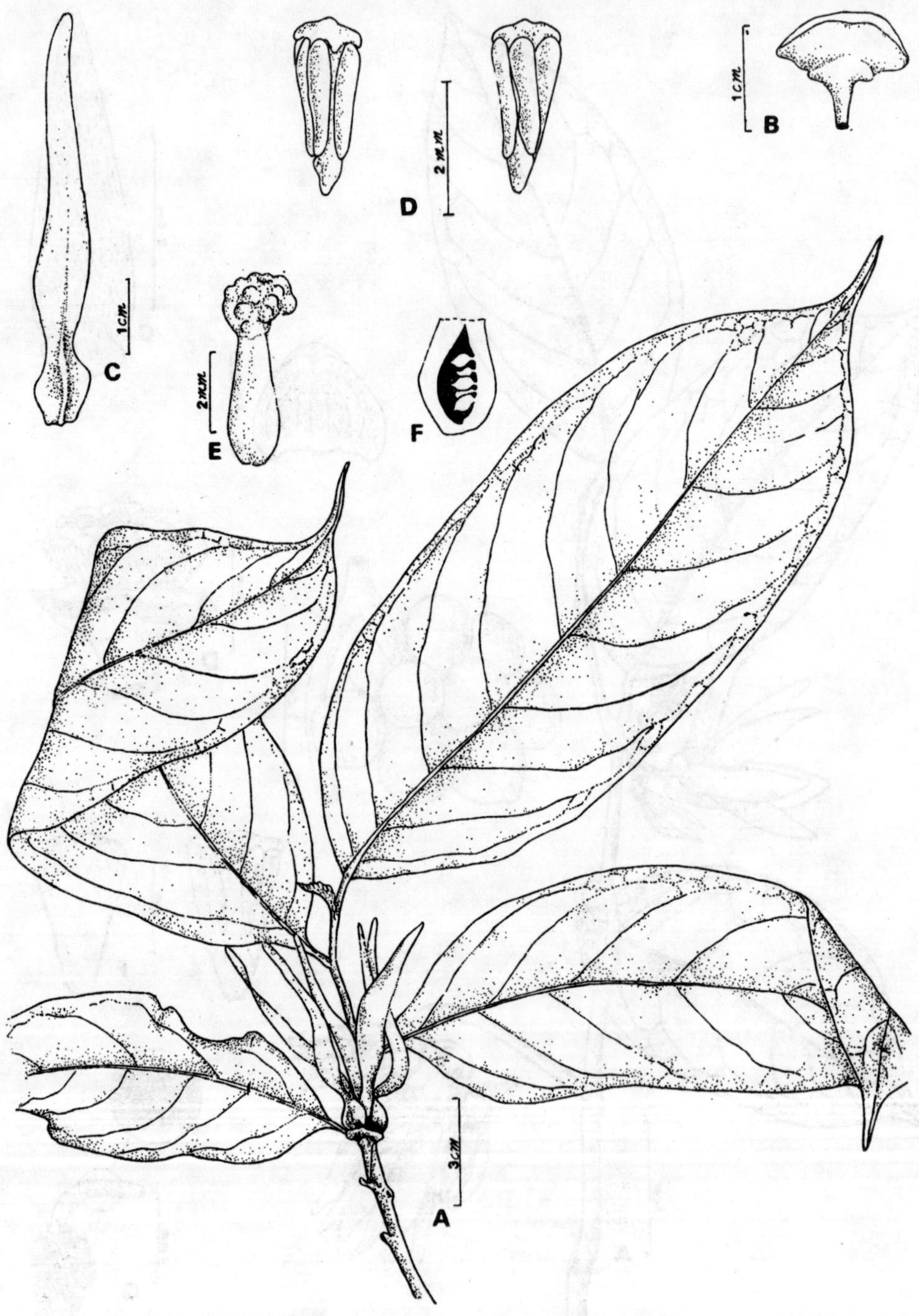

Fig. 4 . *Cyathocalyx zeyianicus* Champ. ex Hook. f. & Thoms.: A. Twig; B. Calyx; C. Petal; D. Stamen - different views; E. Single pistil; F. Ovary - L.S.

GONIOTHALAMUS (Blume) J.D.Hooker & T. Thomson
Fl. Ind. 105. 1855.

Trees. Leaf-veins slender, faint, forming marginal loops. Flower(s) solitary or few in fascicles, extra-axillary or sub-terminal; sepals 3, valvate; petals 6, 3+3, clawed, inner ones cohering, arching over torus, cone-like; stamens α, anthers linear, connective beaked or truncate; carpels α; style 1-or 2-fid; ovule(s) 1-2. Berries globose. Seed 1.

1. Leaves oblong-elliptic; outer petals glabrescent; style 1. .... **1. G. rhyncantherus**
1. Leaves linear-lanceate; outer petals tomentose; style 2-fid. .... **2. G. wightii**

1. **Goniothalamus rhyncantherus** Dunn, Kew Bull. 1914:182.1914 & in Gamble, Fl. Pres. Madras 19.1915; Mohanan & Sivad., Fl. Agasthyamala 54. 2002. **(Fig. 5)**

Leaves oblong-elliptic, 8-20 by 3-6.5 cm, base cuneate, apex acuminate, lateral veins 12-14 pairs, chartaceous. Flower(s) solitary or in pairs, c. 2.5 cm ϕ, extra-axillary or sub-terminal; sepals broadly ovate, basally connate; corolla greenish, outer petals ovate-lanceate, inner broadly ovate, long- clawed, connective beaked; carpels linear, villous; style 1, slender. Fruits not seen.

Fl. & Fr.: Mar.-Jun. *Distr.*: Endemic to Western Ghats. Rare. In moist evergreen forests. *NAK 2388* (Moozhiar, 300 m).

2. **Goniothalamus wightii** Hook. f. & Thoms., Fl. Ind. 106. 1855 & in Hook. f., Fl. Brit. India 1:76. 1872; Bedd., Ic.t. 63. 1868-1874; King, Ann. Roy. Bot. Gard. (Calcutta) 4:92, t. 122. 1893; Dunn in Gamble, Fl. Pres. Madras 18.1915; Mohanan & Sivad., Fl. Agasthyamala 56.2002.

Leaves linear-lanceate, 7-13 by 2-3.5 cm, base and apex acute, lower surface brownish, pellucid-dotted, chartaceous. Flowers, c. 15 mm ϕ; sepals broadly ovate; corolla greenish, outer petals ovate, to 2.5 cm long, tomentose, inner ones smaller; carpels oblong, stalked; style 2-fid. Berries globose.

Fl. & Fr.: Mar.-Jun. *Distr.*: Western Ghats. Rare. In moist evergreen forests. *BDS 43943* (MH), (Pachakkanam, 1000 m).

MEIOGYNE Miquel
Ann. Mus. Bot. Lugduno-Batavi 2: 12. 1865.

Shrubs or small trees. Leaves prominently veined beneath. Flower(s) solitary, sessile, leaf-opposed, cream coloured; sepals 3, ovate, valvate; petals 6, 3+3, rarely inner 0; torus concave; stamens α, connective truncate; carpels α; ovules 2-8. Berries α, globose or pyriform.

1. Fruitlets globose; carpels soft velvetty. .... 1. **M. pannosa**
1. Fruitlets pyriform or ovoid; carpels puberulous. .... 2. **M. ramarowii**

1. **Meiogyne pannosa** (Dalz.) Sinclair, Sarawak Mus. J. 5:604. 1951; Mohanan & Sivad., Fl. Agasthyamala 56.2002. *Unona pannosa* Dalz. in Hooker's J. Bot. Kew Gard. Misc. 3:207. 1851; Hook. f. & Thoms. in Hook.f., Fl. Brit. India 1 : 58, 1872; Bedd., Ic. t. 52. 1868-1874; Dunn in Gamble, Fl. Pres. Madras 14.1915; Das, Bull. Bot. Surv. India 5:41. 1963. *Panthal maram.*

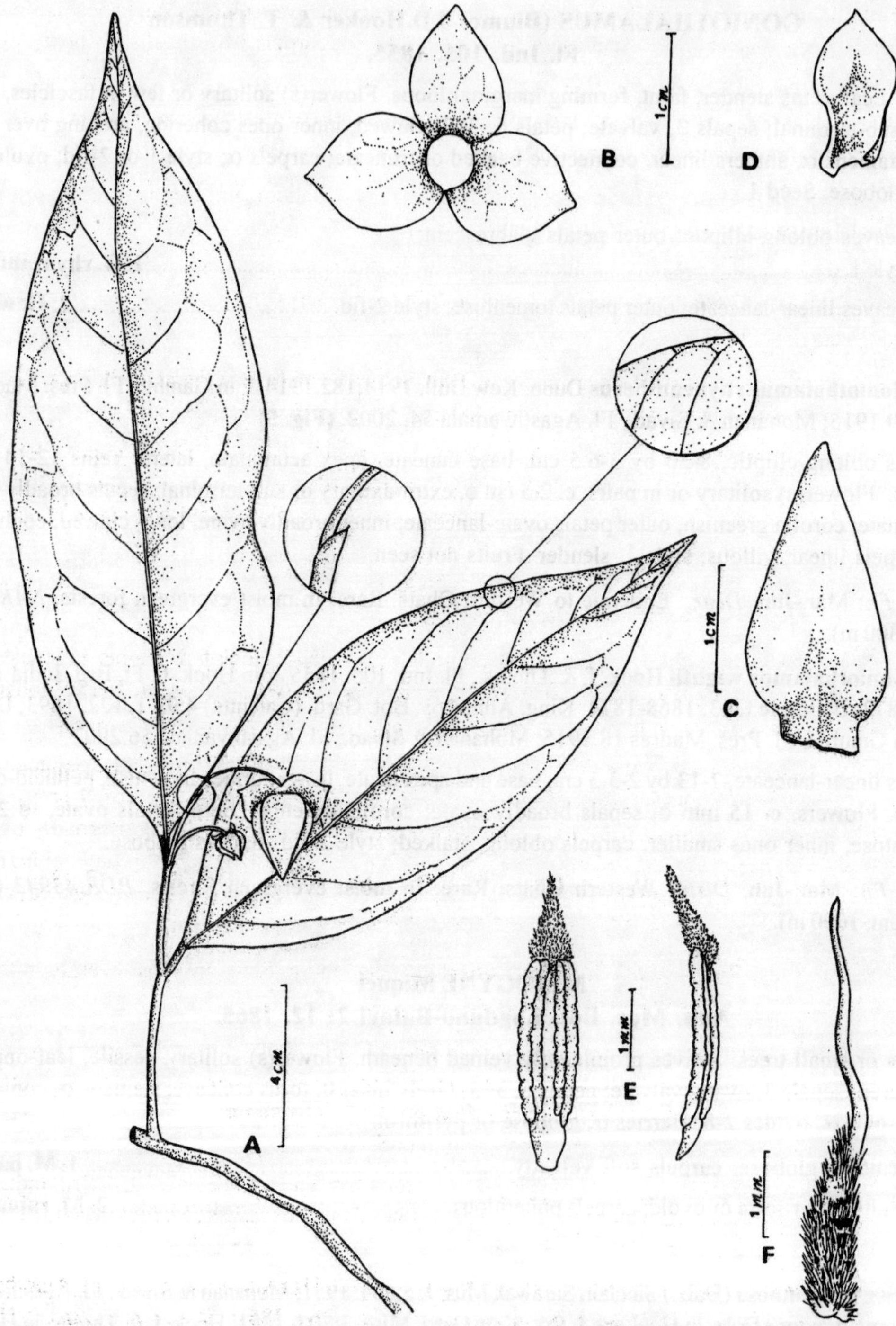

Fig. 5. ***Goniothalamus rhyncantherus*** **Dunn: A. Twig; B. Calyx; C. Inner petal; D. Outer Petal; E. Stamen - different views; F. Single pistil.**

Leaves elliptic-oblong, 6-12 by 2.5 cm, base acute, apex acuminate, veins, pubescent beneath. Flower(s), c. 4 cm ϕ; petals linear-lanceate, to 4 cm long, golden tomentose; carpels ovoid, softly velvetty. Berries globose, glabrescent.

Fl. & Fr.: Feb.-May. *Distr.*: Peninsular India. Rare. In riverine forests of lower ghats. *NAK 2279* (Mannarapara, c. 250 m). Bark yields strong Fibre.

Note : **M. pannosa** is closely related to **M. ramarowii** (Dunn) Gandhi, and detailed studies of the two may render the latter to be a subspecies of former.

2. **Meiogyne ramarowii** (Dunn) Gandhi in Sald. & Nicols., Fl. Hassan Dist. 38. 1976. *Unona ramarowii* Dunn, Kew Bull. 1914:183. 1914 & in Gamble, Fl. Pres. Madras 15.1915. *Desmos ramarowii* (Dunn) Das, Bull. Bot. Surv. India 5:42. 1963.

Leaves elliptic-oblong to elliptic-ovate, 6-16 by 3-6 cm, base rounded or broadly cuneate, apex long-acuminate, nerves puberulous beneath, chartaceous. Flower(s) 1-3 cm ϕ; petals lanceate, c. 3 cm long; anthers linear; carpels 8-12, nearly cylindric or pyriform, puberulous; stigma yellowish. Berries ovoid or pyriform.

Fl. & Fr.: Jan.-Apr. (May). *Distr.*: Peninsular India. Common in wet semi-evergreen forests. *NAK 706* (Pandalam, 50 m), *1478* (Kodumudi-Chittar, 200 m), *1611*(Chathanthara, Ranni, R.F., 150 m).

OROPHEA Blume
Bijdr. 18. 1825.

Orophea erythrocarpa Bedd., Trans. Linn. Soc. London 20:5. 1846; Hook.f. & Thoms. in Hook. f., Fl. Brit. India 1:91. 1872; Bedd., Ic.t. 67. 1868-1874; Dunn in Gamble, Fl. Pres. Madras 24. 1915; Mohanan & Sivad., Fl. Agasthyamala 59.2002. **(Fig. 6)**

Trees. Leaves elliptic-oblong to elliptic-ovate, 4-10 by 2-4.5 cm, base acute to rounded, apex acuminate, pubescent on nervature below, chartaceous. Flowers in clusters of 2-3, axillary or terminal, 1 cm ϕ; sepals 3, broadly ovate, tomentose without; petals 6, 3+3, dark purple or cream coloured, outer broadly ovate to oblong ovate, 4 by 3 mm, tomentose without, inner clawed, claw, c.4 mm long, limb trullate, cohering by their margins at tip forming a cap; stamens 6, ovoid, anthers contiguous, connective shortly beaked; carpels 6, pyriform, sessile, hispid; ovules 4. Monocarp cylindric, c.1.5 cm long, pubescent.

Fl. & Fr.: Jan.-May. Rare. *Distr.*: Sri Lanka and Peninsular India. Rare. In moist evergreen forests. *NAK* 1438 (Arampa, 550 m), *2394* (Moozhiar, 50 m).

Trees with black coloured bark. New foliage yellowish-green, striking.

PHAEANTHUS J.D. Hooker &. T. Thomson
Fl. Ind. 146. 1855.

Phaeanthus malabaricus Bedd., Ic. t. 76. 1868-1874; Hook.f. & Thoms. in Hook. f., Fl. Brit. India 1:72. 1872; King, Ann. Roy. Bot. Gard. (Calcutta) 4: 154, t. 201 B. 1893; Dunn in Gamble, Fl. Pres. Madras 17.1915; Mohanan & Sivad., Fl. Agasthyamala 59. 2002. **(Fig. 7)**

Shrubs or small trees. Leaves elliptic-oblong to oblong, 6-20 by 2-6 cm, base rounded or subcordate, apex acuminate, puberulous on nerves beneath, chartaceous. Flower(s) solitary or very few in extra-axillary fascicles, c. 15 cm ϕ, cream coloured; sepals 3, subdeltate, pubescent; petals 6, 3+3, coriaceous, outer ovate-lanceate, pubescent without, cohering by their margins, arching over the torus; torus convex; stamens α, anthers linear, connective truncate; carpels α, botuliform, villous. Fruits not seen.

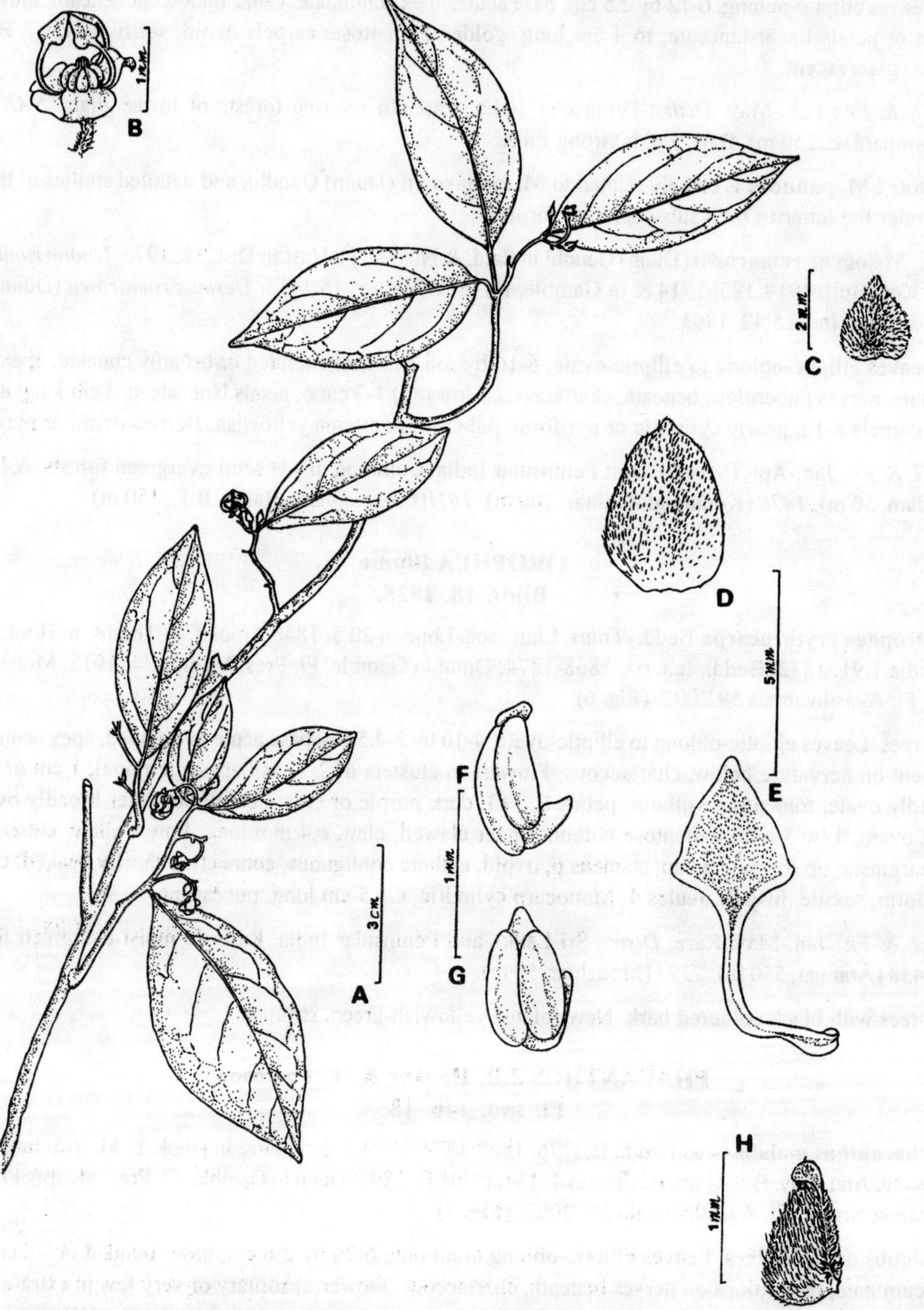

Fig. 6 . *Orophea erythrocarpa* Bedd.: A. Twig; B. Flower; C. Sepal; D. Outer petal; E. Inner petal; F. Stamen-side view; G. Stamen - abaxial view; H. Single pistil.

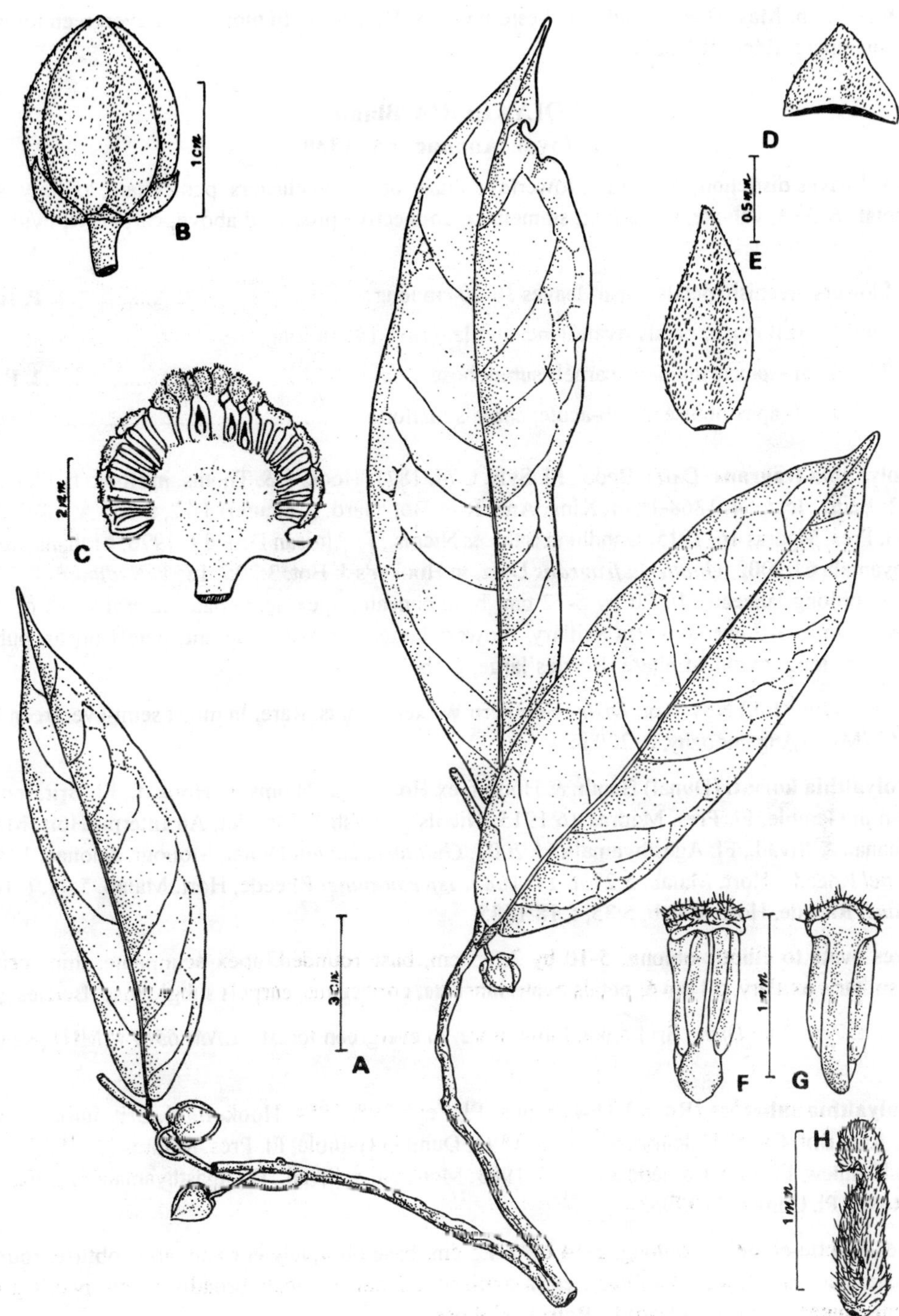

Fig. 7 . *Phaeanthus malabaricus* Bedd.: A. Twig; B. Flower; C. L.S. through torus with carpels and stamens; D. Sepal; E. Petal; F. Stamen - abaxial view; G. Stamen - adaxial view; H. Single pistil.

Fl. & Fr.: Feb.-May. *Distr.*: Southern-Western Ghats. Very Rare. In moist semi-evergreen forests. *NAK 1598* (Chathanthara, Ranni R.F., 200 m).

POLYALTHIA Blume
Fl. Javae Anonac. 68. 1830.

Trees. Leaves distichous or spiral. Flower(s) solitary or few in clusters, pedicellate, axillary; sepals 3, valvate, petals 6, 3+3, valvate, spreading; stamens α, connective produced above; carpels α; ovule 1. Fruit a ring of 1-seeded berries.

1. Flowers greenish; petals linear; leaves 10-20 cm long. .... **1. P. fragrans**
1. Flowers yellowish; petals ovate-lanceate; leaves 3-14 cm long.
    2. Leaf -apex acuminate; carpels subglobose. ....**2. P. korinti**
    2. Leaf- apex obtuse to sub-acute; carpels fusiform. ....**3. P. suberosa**

1. **Polyalthia fragrans** (Dalz). Bedd., Fl. Sylv. t. 74. 1871; Hook. f. & Thoms. in Hook. f., Fl. Brit. India 1:63. 1872; Bedd., Ic. t. 54. 1868-1874; King, Ann. Roy. Bot. Gard. (Calcutta) 4:72, t. 100 A. 1893; Dunn in Gamble, Fl. Pres. Madras 16. 1915; Gandhi in Sald. & Nicols., Fl. Hassan Dist. 39. 1976; Mohanan & Sivad., Fl. Agasthyamala 61.2002. *Guatteria fragrans* Dalz. in Hooker's J. Bot. 3:206. 1851. *Nedunar.*

Leaves oblong-lanceate, 10-20 by 5-12 cm, base cuneate, apex acuminate, lateral veins α, oblique, chartaceous. Cymes a-flowered, extra axillary. Flowers large, greenish, fragrant; petals linear, pubescent; carpels, to 3 cm long, ovoid, stipitate. Berries large, globose.

Fl. & Fr.: (Jun.-Jul.) Nov.-Apr. *Distr.*: Southern Western Ghats. Rare. In moist semi-evergreen Forests. *MCB 49163* (MH) (Angamoozhy, c. 200 m).

2. **Polyalthia korinti** (Dunal) Benth. & Hook.f. ex Hook. f. & Thoms. in Hook. f., Fl. Brit. India 1:64. 1872; Dunn in Gamble, Fl. Pres. Madras 16.1915; Nicols., Suresh & Manilal, An Interpr. Hort. Malab.50. 1988; Mohanan & Sivad., Fl. Agasthyamala 61.2002. *Guatteria korinti* Dunal, Monogr. Anonac. 134. 1817. *Corinti-panel* Rheede, Hort. Malab. 5:27, t. 14. 1685. *Tsjerou-panel* Rheede, Hort. Malab. 5:31, t. 16.1685. *Katsjan-panel* Rheede, Hort. Malab. 5:35, t. 18.1685.

Leaves ovate to elliptic-oblong, 5-10 by 2.5-5 cm, base rounded, apex acuminate, thin-coriaceous. Flower(s) solitary, axillary, c.1 cm ϕ; petals ovate-lanceate, coriaceous; carpels subglobose. Berries globose.

Fl. & Fr.: Apr.-Jun. *Distr.*: Sri Lanka, India. Rare. In evergreen forests. *CNM 65074* (MH) (Ranni R.F., 300 m).

3. **Polyalthia suberosa** (Roxb.) Thw., Enum. Pl. Zeyl. 398. 1864; Hook. f., Fl. Brit. India 1:66. 1872; King, Ann. Roy. Bot. Gard. (Calcutta) 4: t. 77B. 1893; Dunn in Gamble, Fl. Pres. Madras 16. 1915; Matthew & Britto in Matthew, Fl. Tam. Carnatic 3 (1): 13. 1983; Mohanan & Sivad., Fl. Agasthyamala 62.2002. *Uvaria suberosa* Roxb., Pl. Corom. 34.1795.

Leaves elliptic-oblong to oblong, 5-14 by 2-4.5 cm, base obliquely cordate, apex obtuse, rounded or subacute, chartaceous. Flower(s) solitary, extra-axillary, c.1 cm ϕ; sepals broadly ovate; petals greenish-yellow, ovate-lanceate; carpels clavoid. Berries globose.

Fl. & Fr.: Sep.-Mar. *Distr.*: Sri Lanka, India, Myanmar, Indo-China. Rare. An evergreen species. *NAK 1888* (Adoor, c.50 m). Wood durable.

Polyalthia longifolia (Sonn.) Thw. is commonly planted in gardens and as avenue trees.(*NAK 2296*).

UVARIA Linnaeus
Sp. Pl. 503. 1753.

Scandent shrubs. Leaves thin-coriaceous. Flower(s)solitary, terminal or leaf- opposed, brown or purplish; sepals 3, connate at base; petals 6, 3+3, suborbicular, often connate at base, imbricate; stamens α, connective truncate, concealing the anther; carpels α, linear-oblong; ovules α , Berries globose.

1. Leaves to 15 cm long; flowers brownish, c.3 cm ϕ or more. .... **1. U. narum**
1. Leaves to 10 cm long; flowers purplish, c.2 cm ϕor less. .... **2. U. zeylanicum**

1. **Uvaria narum** (Dunal) Wall. ex Wight & Arn., Prodr. 9. 1834; Wight, Ill. Ind. Bot. t. 6. 1840; Hook. f. & Thoms. in Hook.f., Fl. Brit. India 1:50. 1872; King, Ann. Roy. Bot. Gard. (Calcutta) 4:27, t. 21.1893; Dunn in Gamble, Fl. Pres. Madras 13. 1915; Gandhi in Sald. & Nicols., Fl. Hassan Dist. 39. 1976; Matthew, Ill. Fl. Tam. Carnatic t. 10. 1982; Matthew & Britto in Matthew, Fl. Tam, Carnatic 3(1) : 14. 1983; Nicols., Suresh & Manilal, An Interpr. Hort. Malab. 51. 1988. *Unona narum* Dunal, Monogr. Anonac. 99. 1817. *Uvaria hookeri* King, Ann. Roy. Bot. Gard. (Calcutta) 4:26, t. 19. 1893. *Narum-panel* Rheede, Hort. Malab. 2:11-12, t. 10. 1679.

Leaves oblong-lanceate, 8-15 by 2-5 cm, base cuneate or acute, apex acuminate. Flowers brownish, c. 3 cm ϕ , terminal; sepals orbicular, stellate-tomentose; petals ovate to suborbicular, coriaceous, apex incurved; carpels narrowly obconic; stigma diffuse; ovules 2-10. Berries globose.

Fl. & Fr.: Nov.-Apr. *Distr.*: Sri Lanka, Peninsular India. Common in sacred groves. *NAK 122* (Adoor-Koottungal Kavu, 50 m), *1061* (Kodumon, 70 m).

Juice of leaves boiled with oil and is used for eczema; root decoction is best medicine for jaundice and fever.

2. **Uvaria zeylanica** L., Sp. Pl. 536. 1753; Hook. f., & Thoms. in Hook. f., Fl. Brit. India 1:51. 1872; Bedd., Ic. t. 78. 1868-1874; King, Ann. Roy. Bot. Gard. (Calcutta) 4:26, t. 19. 1893; Dunn in Gamble, Fl. Pres. Madras 13. 1915; Nicols., Suresh & Manilal, An Interpr. Hort. Malab.51. 1988. *Gautteria malabarica* Dunal, Monogr. Anonac. 134. 1817. *Kaltsjerou - panel* Rheede, Hort. Malab. 5:33, t. 17. 1685.

Leaves lanceate to oblong-lanceate, (3) 6 9 by 2-3 cm, base acute, apex acuminate, Flowers purplish, c.2 cm ϕ, usually leaf-opposed or terminal; sepals orbicular-ovate; petals oblong or narrowly elliptic, apex acute or obtuse,straight. Carples ovoid or globose, ovules α. Berries globose.

Fl. & Fr.: Sep.-Mar. *Distr.*: Southern Western Ghats. Peninsular India, Sri Lanka. Apparently scarce, occasionally in certain sacred groves. *NAK 2769* (Kodumon, c. 70 m).

Cananga odorata (Lam.) Hook. f. & Thoms. is cultivated for its fragrant flowers (*NAK 1070*). **Annona reticulata** L. and **A. squamosa** L. are cultivated in most of the places for their edible fruits.

5. MENISPERMACEAE

1. Woody climbers; inflorescene cauline; seeds globose.
    2. Leaf nerve-axils tufted hairy below; stamens connate; drupe pedicelled. .... **Anamirta**

2. Leaf neve-axils glabrous below; stamens free; Drupe sessile. .... **Diploclisia**

1. Herbaceous climbers; inflorescence not cauline (Except in *Tinospora*); seeds not globose.

3. Leaf nerve-axils glandular papillate below; inflorescence axillary and cauline; seeds curved. .... **Tinospora**

3. Leaf nerve-axils glabrous below; inflorescence axillary; seeds horse-shoe shaped.

4. Leaves ovate; flowers bright yellow; carpels 3-9. .... **Tiliacora**

4. Leaves otherwise; flowers pale green; carpel 1.

5. Sepals free; petals 4.

6. Leaves reniform; sepals 4 in male; petals connate in female; stamens 4. .... **Cissampelos**

6. Leaves cordiform; sepals 4+4 in male; petals free in female; stamens 6-8. .... **Stephania**

5. Sepals connate; petals 4-8. .... **cyclea**

ANAMIRTA Colebrooke
Trans. Linn. Soc. London 13:52,66. 1821.

Anamirta cocculus (L.) Wight & Arn., Prodr. 446. 1834; Hook. f. & Thoms. in Hook. f., Fl. Brit. India 1:98. 1872; Dunn in Gamble, Fl. Pres. Madras 26. 1915; Gandhi in Sald. & Nicols., Fl. Hassan Dist. 66. 1976; Forman, Kew Bull. 32:329. 1978; Matthew, Ill. Fl. Tam. Carnatic t. 11.1982; Matthew & Britto in Matthew, Fl. Tam. Carnatic 3(1) : 17. 1983. *Menispermum cocculus* L., Sp. Pl.340. 1753; Roxb., Fl. Ind. 3 : 807. 1832. *Polla.*

Woody climbers. Leaves ovate-cordate, 12-18 by 10-13 cm, base truncate, apex acute, 5-nerved from base; nerve-axils densely tufted hairy below, thin-coriaceous. Panicles large, from old wood. Flowers dioecious, pale yellow or white, c.6 mm ϕ; sepals 6, deciduous; petals 0; stamens connate; staminodes c.9, clavate; carpels 3. Drupes stalked.

Fl. & Fr.: Sep.-Mar. *Distr.*: Sri Lanka, India, Thailand. Common. In moist riverine forests and in sacred groves. *NAK 2073* (Angamoozhy, 250 m).

Dried fruits are narcotic and yield an oil used for poisoning fish. Seed oil reported to be effective for skin diseases.

CISSAMPELOS Linnaeus
Sp. Pl. 1031. 1753.

Cissampelos pariera L., Sp.Pl. 1031. 1753; Hook.f. & Thoms. in Hook. f., Fl. Brit. India 1:103. 1872; Gamble, Fl. Pres. Madras 30.1915; Matthew, Ill. Fl. Tam. Carnatic t. 12. 1982. var. **hirsuta** (Buch. -Ham. ex DC.) Forman, Kew Bull. 22: 356. 1968; Gandhi in Sald. & Nicols., Fl. Hassan Dist. 66. 1976; Matthew & Britto in Matthew, Fl. Tam. Carnatic 3 (1) : 18. 1983; Manilal, Fl. Silent Valley 4.1988; Mohanan & Sivad., Fl. Agasthyamala 64. 2002. *C. hirsuta* Buch.-Ham. ex DC., Syst. Nat. 1:535. 1817 & Prodr. 1:101. 1824. *Malathangi*

Herbaceous climbers. Leaves orbicular-reniform, 2-5 by 2-4 cm, base cordate to truncate, apex obtuse or retuse, sparsely tomentose, 5-7- nerved, chartaceous. Male flowers in pendulous subcorymbose cymes; sepals 4, free; petals 4, connate; stamens 4, connate. Female flowers in thyrsoid clusters; bracts foliaceous; petals connate. Drupe subglobose. Seeds curved.

Fl. & Fr.: Jun.-Aug. *Distr.*: Tropics, Thailand. Commonly spreading on wayside thickets and forests borders. *NAK 653* (Thekkuthode, 50 m).

Whole plant decoction is used for cough. Leaves bruised in coconut-milk and taken internaly for veneral diseases. A paste of roots with cotyledons of *Entada rheedei* is given twice daily after mensturation as a contraceptive.

CYCLEA Arnott ex R. Wight
Ill. Ind. Bot. 1:22. 1840.

Cyclea peltata (Poir.) Hook. f. & Thoms., Fl. Ind. 201. 1885 & in Hook.f., Fl. Brit. India 1:104. 1872, p.p.; Dunn in Gamble, Fl. Pres. Madras 31.1915; Santapau & Janardhanan, Bull. Bot. Surv. India 10:368. 1969; Gandhi in Sald. & Nicols., Fl. Hassan Dist. 67. 1976; Matthew, Ill. Fl. Tam. Carnatic t. 14. 1982; Matthew & Britto in Matthew, Fl. Tam. Carnatic 3(1): 19. 1983; Manilal, Fl. Silent Valley 5. 1988; Nicols., Suresh & Manilal, An Interpr. Hort. 179. 1988; Mohanan & Sivad., Fl. Agasthyamala 66.2002. *Menispermum peltatum* Poir. in Lam., Encycl. 4:96. 1797. *Pada-kelengu, Pada-valli* Rheede, Hort. Malab. 7:93, t. 49. 1688.

Climbers. Leaves deltoid, peltate, 5-13 by 4-8 cm, base truncate, apex acute, upper surface glabrescent, lower surface tomentose, thin-coriaceous. Panicle to 20 cm long (in male), shorter in female; sepals 4 in male flowers, 1 in female; petals greenish, 4-8 in male, one in female; carpel 1. Drupe subglobose, pendulous. Seeds curved.

Fl. & Fr.: Jan.-Mar. *Distr.*: Sri Lanka, India, Malesia. Common on thickets of plains and hills. (10 to 1000 m). *NAK 127* (Adoor, c. 50 m), *541* (Angamoozhy, 350 m), *1497* (Kodumudi-Chittar, 250 m).

Root powdered and mixed with honey or in decoction taken internally for jaundice, stomach-pain, fever, vomiting, leprosy, sores, liver complaints,worms and as a good nerve tonic.

DIPLOCLISIA Miers
Ann. Mag. Nat. Hist. (Ser. 2) 7:37, 42. 1851.

Diploclisia glaucescens (Bl.) Diels in Engl., Pflanzenr. IV. 46:225. 1901; Dunn in Gamble, Fl. Pres. Madras 28.1915; Gandhi in Sald. & Nicols., Fl. Hassan Dist. 68.1976; Matthew, Ill. Fl. Tam. Carnatic t. 15. 1982; Matthew & Britto in Matthew, Fl. Tam. Carnatic 3(1): 20. 1983; Manilal, Fl. Silent Valley 6.1988; Nicols., Suresh & Manilal, An Interpr. Hort. Malab. 179. 1988; Mohanan & Sivad., Fl. Agasthayamala 66.2002. *Cocculus glaucescens* Bl., Bijdr. 25.1825. *C.macrocarpus* Wight & Arn., Prodr. 13. 1834; Hook. f. & Thoms. in Hook. f., Fl. Brit. India 1 : 101. 1872. *Nsisjutum, Batta-valli* Rheede, Hort. Malab. 7:1-2, t.1, 1688. *Cattu-valli, Batta-valli* Rheede, Hort. Malab. 11:127, t 62. 1692.

Climbers. Leaves suborbicular, 5-10 by 5-11 cm, base truncate to cordate, apex obtuse to retuse, lower surface glaucous, basally 5-nerved, thin-coriaceous. Panicle from old wood. Flowers yellowish, c.4 mm ϕ; sepals 6, (3+3); petals 6; stamens 6, free; carpels 3; staminodes 6. Drupe obovoid, c.2 cm ϕ. Seed curved.

Fl. & Fr.: May-Jul. *Distr.*: Sri Lanka, India, Myanmar, Thailand, S. China, Inod-China and Malesia. Common in hills of evergreen forests and in sacred goves. *NAK 2689* (Kakki hills, 1050 m).

Leaves pounded and taken with milk is reported to cure syphilis and gonorrhoea.

STEPHANIA Louriero
Fl.Cochinch. 598, 608. 1790.

Stephania japonica (Thunb.) Miers, Ann. Mag. Nat. Hist. (ser.3) 18:14. 1866; Dunn in Gamble, Fl. Pres. Madras 29.1915; Forman, Kew Bull. 11:49.1956; Gandhi in Sald. & Nicols., Fl. Hassan Dist. 68.1976; Matthew, Ill. Fl. Tam. Carnatic t.17. 1982; Matthew & Britto in Mathew, Fl. Tam. Carnatic 3(1) : 22. 1983; Manilal, Fl. Silent Valley 6. 1988; Mohanan & Sivad., Fl. Agasthyamala 67.2002. *Menispermum japonicum* Thunb., Fl. Jap. 193. 1784. *Clypea hernandiifolia* Wight & Arn., Prodr. 14. 1834. *Stephania hernandiifolia* auct., non Walp, 1842: Hook. f. & Thoms. in Hook.f., Fl. Brit. India 1:103. 1872.

Slender twiners. Leaves widely ovate, 4-10 by 3-7 cm, peltate, base truncate to rounded, apex acute to shortly acuminate. Umbels axillary. Flowers small, greenish; sepals and petals 4+4, free; petals 4; stamens 6-8, coherent; sepals and petals 3-5 in female; staminodes 0. Drupes obovoid. Seeds horse-shoe shaped.

Fl. & Fr.: Dec.-Mar. *Distr.*: Sri Lanka, Peninsular India Malesia, Korea and Japan. Rare. On thickets, evergreen forest borders, etc. *NAK 1265* (Lower Moozhiar, 250 m).

TILIACORA Colebrooke
Trans. Linn. Soc. London 13:53, 67.1821.

Tiliacora acuminata (Lam.) Miers ex Hook. f. & Thoms., Fl.Ind. 187.1855; Dunn in Gamble, Fl. Pres. Madras 28.1915; Forman, Kew Bull. 30:90.1975; Matthew, Ill. Fl. Tam. Carnatic t. 18. 1982; Matthew & Britto in Matthew, Fl. Tam. Carnatic 3(1) : 23. 1983; Nicols., Suresh & Manilal, An Interpr. Hort. Malab. 180. 1988. *Menispermum acuminatum* Lam., Encyl. 4: 101. 1797. *M. radiatum* Lam., Encyl. 4: 100. 1797. *Tiliacora racemosa* Colebr., Trans. Linn. Soc. London 13:67. 1821: Hook.f. & Thoms. in Hook. f., Fl. Brit. India 1:99. 1872. *Valli-caniram* Rheede, Hort. Malab. 7:5-6, t.3. 1688. **(Fig. 8)**

Climbers. Leaves ovate-lanceate, 6-13 by 3-6 cm, base truncate to rounded, apex acuminate-apiculate, basally 3-5-nerved, coriaceous. Flowes in axillary racemes; sepals 3+3, free; petals 6, bright yellow; stamens 6, free; carpels 6; staminodes 0. Drupes obovoid. Seeds horse-shoe shaped.

Fl. & Fr.: Sep.-Mar. *Distr.*: Sri Lanka, India, Myanmar, Nepal. Less common. Extensive climbers, on wayside thickets in plains. *NAK 1645* (Tiruvalla, c.10 m).

Crushed root with water given as an antidote in snake poison. Stem used for making baskets by locals.

TINOSPORA Miers
Ann. Mag. Nat. Hist. (Ser. 2) 7:35. 1851.

Tinospora cordifolia (Willd.) Miers ex Hook. f. & Thoms., Fl. Ind. 1:184. 1855 & in Hook.f., Fl. Brit. India 1:97.1872; Dunn in Gamble. Fl. Pres. Madras 26.1915; Forman, Kew Bull. 36:403. 1981; Matthew, Ill. Fl. Tam. Carnatic t. 19. 1982; Matthew & Britto in Matthew, Fl. Tam. Carnatic 3(1) : 23. 1983; Nicols., Suresh & Manilal, An Interpr. Hort. Malab. 180. 1988. *Menispermum cordifolium* Willd., Sp. Pl. 4:826. 1806; Roxb., Fl. Ind.3: 811. 1832. *Cit-amerdu* Rheede, Hort. Malab. 7:39, t.21.1688.

Stout climbers. Leaves broadly ovate to orbicular, 4-8 by 4-7 cm, base deeply cordate, apex acuminate, lower surface with glandular papillose patches in nerve axils, basely 5-7 nerved, chartaceous. Racemes axillary, terminal or cauline. Flowers greenish; sepals 3+3, free; petals 6; stamens 6, free; carpels 3; staminodes 6, subulate. Drupe globose. Seeds curved.

Fl. & Fr.: Feb.-May. *Dsitr.*: Sri Lanka, India, Bangladesh. Common. Spreading on wayside thickets, fences in plains and in hills. *NAK 2253* (Mannarappara, 250 m).

Decoction of powdered stem taken along with ghee strengthens brains and improves memory, and promotes digestion. Dried fruits powdered and mixed with honey in used against jaundice and rheumatism.

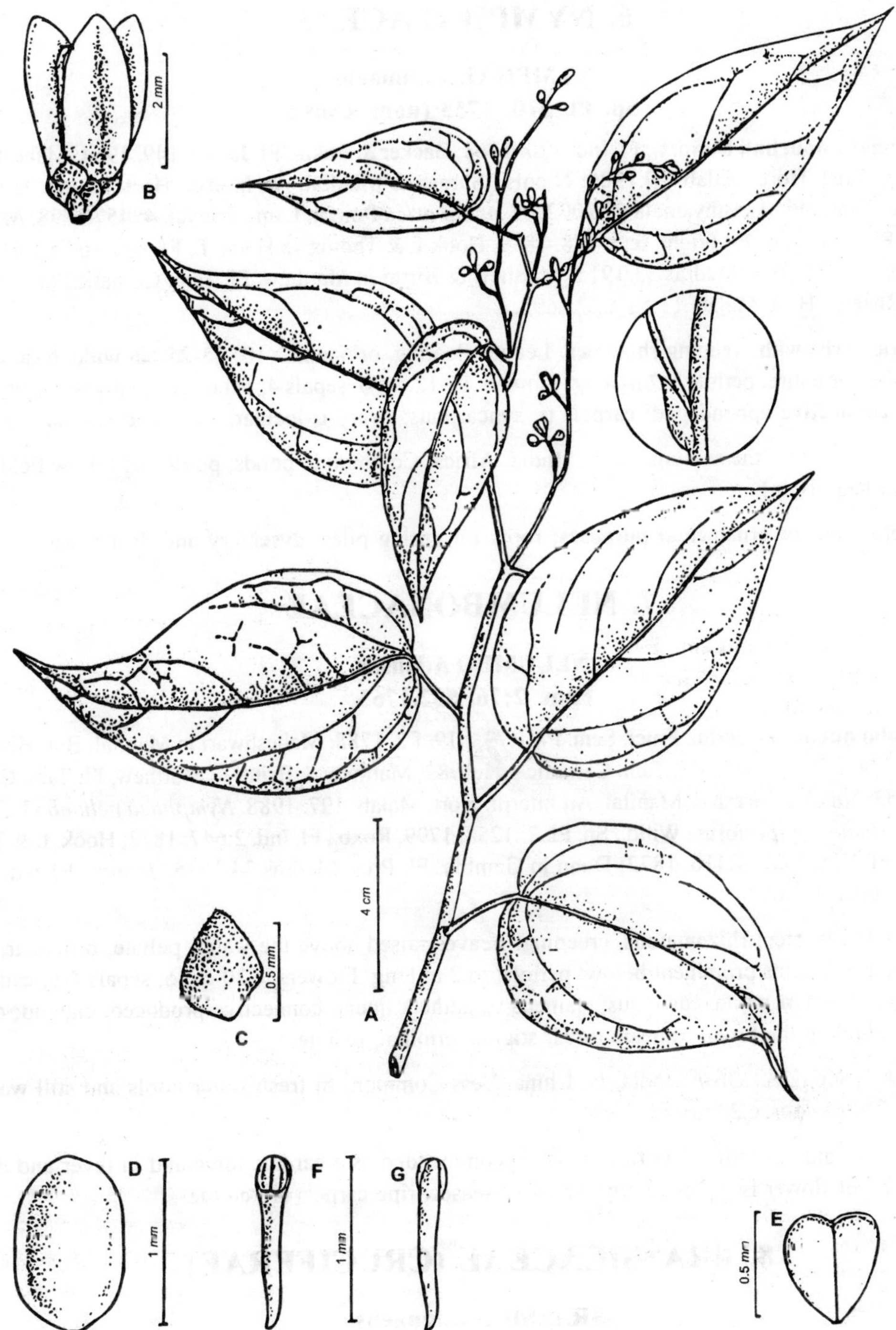

Fig. 8. *Tiliacora acuminata* (Lam.) Hook. f.: A. Twig; B. Flower; C. Outer sepal; D. Inner sepal; E. Petal; F. Stamen - abaxial view; G. Stamen - adaxial view.

6. NYMPHAEACEAE

NYMPHAEA Linnaeus
Sp. Pl. 510. 1753 (nom. cons.)

Nymphaea nouchali Burm. f., Fl. Ind. 120. 1768; Backer & Bakh., Fl. Java 1:149. 1964; Maheshwari in Manilal, Bot. Hist. Hort. Malab. 29.1980; Nicols., Suresh & Manilal, An Interpr. Hort. Malab. 198. 1988; Mohanan & Sivad., Fl. Agasthyamala 69.2002. *N. malabarica* Poir. in Lam., Encycl. 4: 457.1798. *N. stellata* Willd., Sp. Pl. 2:1153. 1799; Wight, Ic. t. 178. 1839; Hook.f. & Thoms. in Hook. f., Fl. Brit. India 1:114.1872; Dunn in Gamble, Fl. Pres. Madras 33.1915; Matthew & Birtto in Matthew, Fl. Tam. Carnatic 3(1): 28.1983. *Cit-ambel* Rheede, Hort. Malab. 11:53, t.27.1692.

Aquatic herbs with creeping rhizomes. Leaves floating, orbicular, (10) 15-25 cm wide, base cordate, margin toothed or entire, petiole 1-2 m long. Flowers 10-12 cm ϕ; sepals 4; petals a, usually white or yellow; stamens α, connective appendaged; carpels α, syncarpous; ovary α-locular, ovules α; stigma radiating.

Fl. & Fr.: Most of the seasons. *Distr.*: India, Africa. Common in ponds, pools and fallow fields. *NAK 1047* (Tiruvalla, c. 10 m).

Flowers used for ornamental purposes; roots for curing piles, dysentery and dyspepsia.

7. NELUMBONACEAE

NELUMBO Adanson
Fam. 2:76, 582.1763.

Nelumbo nucifera Gaertn., Fruct. Sem. Pl. 1:73, t.19, f.2. 1788; Maheshwari in Manilal, Bot. Hist. Hort. Malab. 123.1980; Matthew, Ill. Fl. Tam. Carnatic t.21.1982; Matthew & Britto in Matthew, Fl. Tam. Carnatic 3(1): 26. 1983; Nicols., Suresh & Manilal, An Interpr. Hort. Malab. 197. 1988. *Nymphaea nelumbo* L., Sp. Pl. 511. 1753; *Nelumbium speciosum* Willd., Sp. Pl. 2: 1258. 1799; Roxb., Fl. Ind. 2:647: 1872; Hook. f. & Thoms. in Hook. f., Fl. Birt. India 1:116. 1872; Dunn in Gamble, Fl. Pres. Madras 34.1915. *Tamara* Rheede, Hort. Malab. 11:59-60, t. 30.1692.

Aquatic herbs; stem rhizomatous, creeping. Leaves raised above the water, peltate, orbicular, 30-90 cm wide, upturned; veins prominent below; petiole, to 2 m long. Flowers, c.20 cm ϕ; sepals 4-5, caducous; petals α, pinkish-red or white, caducous; stamens α, anthers linear, connective produced, clavoid; carpels α, free, embedded in the broad truncate torus; stigma terminal, peltate.

Fl. & Fr.: Sep.-Dec. *Distr.*: India, N. China. Less Common. In fresh water pools and still waters in plains. *NAK 250* (Adoor, c.20 m).

The sacred lotus of India. The flowers are recommended as a cardiac tonic and in fever and disease of liver. Honey of flower is good remedy for eye diseases; ripe carpels are edible.

8. BRASSICACEAE (CRUCIFERAE)

BRASSICA Linnaeus
Sp. Pl. 666. 1753.

Brassica juncea (L.) Cosson, Bull. Bot. Soc. France 6:609. 1859; Hook.f. & Thoms., J. Linn. Soc. Bot. 5:170.1861 & in Hook.f., Fl. Brit. India 1:157. 1872; Dunn in Gamble, Fl. Pres. Madras 38. 1915; Gandhi in

Sald. & Nicols., Fl. Hassan Dist. 188. 1976; Matthew, Ill. Fl. Tam. Carnatic t. 25. 1982; Matthew & Britto in Matthew, Fl. Tam. Carnatic 3(1): 32. 1983. *Sinapis juncea* L., Sp. Pl. 668. 1753.

Herbs. Leaves spiral, lyrate, 7-14 by 3-7 cm, terminal lobe ovate, margin irregularly dentate or serrate; upper leaves lanceate, membranous. Racemes terminal, to 25 cm long. Flowers 4-merous, bisexual; sepal 4, subequal; petals 4, yellow, narrowly obovate; stamens 2+4; ovary 2-locular; stigma capitate; ovules α. Siliqua beaked. Seeds in a row, ovoid, finely alveolate.

Fl. & Fr.: May-Jul. *Distr.*: Central Asian in origin. Now cultivated throughout the world. Run wild along waste places in plains and hills. *NAK 683* (Pandalam, c. 10 m), *1808* (Angamoozhy, c. 250 m).

9. CLEOMACEAE

CLEOME Linnaeus
Sp. Pl. 671. 1753.

Herbs. Leaves alternate, 1-3 (-7)-foliolate. Flower(s) solitary or few, in racemes, regular, hypogynous; sepals 4; petals 4; subequal, clawed; stamens 6 to α, free or on androphore; ovary sessile or on a gynophore, ovules α; stigma capitate. Capsule cylindric, beaked. Seeds reniform or subglobose, ridged, crested.

1. Leaves 3-5-foliolate; capsule stout.
    2. Glabrous herbs; flowers blue; fruiting pedicel to 2.5 cm long. .... **1. C. burmanni**
    2. Viscous herbs; flowers yellow; fruiting pedicel to 2.5 cm long. ....**3. C. viscosa**
1. Leaves 1-foliolate; capsule slender. ....**2. C. monophylla**

1. **Cleome burmanni** Wight & Arn., Prodr. 22. 1834; Hook.f. & Thoms. in Hook. f., Fl. Brit. India 1 : 170. 1872; Dunn in Gamble, Fl. Pres. Madras 41. 1915; Babu & Majumdar, J. Bombay Nat. Hist. Soc. 71 : 630. 1976.

Leaves 3-foliolate; leaflets ovate-lanceate, 2-3 by 0.8-1.5 cm, base cuneate, apex acute, membranous. Flower(s) solitary, axillary; pedicel slender; corolla bluish-purple; stamens 6; ovary on a short gynophore. Capsule narrowly fusiform, striate. Seeds ridged.

Fl. & Fr.: Feb.-Aug. *Distr.*: India, Sri Lanka. Less Common. On exposed sandy areas and on waste place. *NAK 692* (Pandalam, c. 10 m).

2. **Cleome monophylla** L., Sp. Pl. 672. 1753; Roxb., Fl. Ind. 3:129. 1832; Hook.f. & Thoms. in Hook. f., Fl. Brit. India 1 : 168. 1872; Dunn in Gamble, Fl. Pres. Madras 41. 1915; Ramamoorthy in Sald. & Nicols., Fl. Hassan Dist. 187. 1876; Britto & Matthew in Matthew. Fl. Tam. Carnatic 3(1) : 48. 1983; Nicols., Suresh & Manilal, An Interpr. Hort. Malab. 78. 198. *Tsjeru-vela* Rheede, Hort. Malab. 9 : 63, t. 34. 1689.

Leaves oblong-lanceate, 2-5 by 1-2.5 cm, base rounded, truncate or subcordate, apex acute, pubescent, chartaceous. Flower(s) solitary, axillary; bracts foliaceous; petals pale purple, long-clawed; stamens 6. Capsule slender, 8-10 cm long. Seeds subglobose.

Fl. & Fr.: Mar.-Aug. *Distr.*: Tropical Africa & India. Less common. On exposed dry hill. *NAK 638* (Puthukkulam, c. 100 m).

3. **Cleome viscosa** L., Sp., Pl. 672. 1753; Roxb., Fl. Ind. 3:128. 1832; Hook.f. & Thoms. in Hook. f., Fl. Brit. India 1 : 170. 1872; Dunn in Gamble, Fl. Pres. Madras 41. 1915; Ramamoorthy in Sald. & Nicols., Fl.

Hassan Dist. 187. 1976; Britto & Matthew in Matthew, Fl. Tam. Carnatic 3(1) : 49. 1983; Nicols., Suresh & Manilal, An Interpr. Hort. Malab. 78. 1988; Mohanan & Sivad., Fl. Agasthyamala 69.2002. *Aria-vela* Rheede, Hort. Malab. 9:41, t. 23. 1689.

Leaves 3-5-foliolate; leaflets obovate to elliptic, 2-4 by 1-1.5 cm, base cuneate, apex acute, chartaceous. Flower(s), c. 15 mm ϕ; sepals oblong; petals yellow, obovate; stamens 12-18. Capsule slender-fusiform. Seeds reniform, closely ridged.

Fl. & Fr.: Most of the seasons. *Distr.*: India, Tropical Africa, Tropical Australia & Malesia. Common along waysides, waste places, fallow fields, etc. *NAK 691* (Pandalam, c. 20 m).

Juice of the plant boiled with oil smeared over body for rheumatism, ear-diseases, diseases of the brain and stomach-ache. Seeds carminative and stimulent.

10. CAPPARACEAE

CRATEVA Linnaeus
Sp. Pl. 444. 1753.

Crateva nurvala Ham., Trans. Linn. Soc. London 15 : 121. 1827; Wight & Arn., Prodr. 23. 1834; Blatter *et al.,* Beautif. Ind. Trees (ed. 2) 47. 1954; Jacobs, Blumea 12 : 194. 1964; Ramamoorthy in Sald. & Nicols., Fl. Hassan Dist. 187. 1976; Nicols., Suresh & Manilal, An Interpr. Hort. Malab. 78. 1988. *C. magna* auct., non (Lour.) D.C., 1824: Jacobs, Blumea 12:206. 1964; Matthew, Ill. Fl. Tam. Carnatic t. 35. 1982; Britto & Matthew in Matthew, Fl. Tam. Carnatic 3(1) : 51. 1983. *C. religiosa* auct., non Foster f., 1786: Bedd., Fl. Sylv. t.116. 1871; Dunn in Gamble, Fl. Pres. Madras 47. 1915. *Niirvala* Rheede, Hort. Malab. 3 : 49-50, t. 42. 1682.

Trees. Leaves 3-foliolate; leaflets ovate to lanceate, 10-20 by 5-7 cm, base cuneate, apex gradually caudate, lateral veins c. 20 pairs, prominent, chartaceous. Racemes terminal. Flowers c. 4 cm ϕ; petals white, turning yellow, long-clawed; androphore short, fused with gynophore. Berry globose, c. 4 cm ϕ. Seeds α.

Fl. & Fr.: Apr.-Dec. *Distr.*: India, Myanmar, China, Indo-China, Malesia. Common. Along streamsides. *NAK 1621* (Pampa Valley, 250 m).

Bark of root and stem dried and powdered and given with honey as a remedy for hydrocoele, congestion in abdomen, worms, heart disease and in ear-ache. Juice of leaves boiled with oil; good for curing brain diseases.

11. VIOLACEAE

HYBANTHUS N.J. Jacquin
Enum. Pl. Carib.2.17. 1760 (nom. cons.).

Hybanthus enneaspermus (L.) F.V. Muell., Fragm. Phyt. Austr. 10 : 81. 1876; Tennant, Kew Bull. 16 : 431. 1963; Jacobs & Moore in Steenis, Fl. Males. I, 7 : 197. 1971; Ramamoorthy in Sald. & Nicols., Fl. Hassan Dist. 165. 1976; Grey-Wilson, Kew Bull. 36 : 103. 1981; Matthew, Ill. Fl. Tam. Carnatic t. 37. 1982; Matthew & Britto in Matthew, Fl. Tam. Canatic 3 (1) : 5. 1983; Nicols., Suresh & Manilal, An Interpr. Hort. Malab. 267. 1988; Mohanan & Sivad., Fl. Agasthyamala 70.2002. *Viola enneasperma* L., Sp. Pl. 937. 1753. *Viola suffruticosa* L., Sp. Pl. 937. 1753; *Ionidium suffruticosum* (L.) Roem. & Schult., Syst. Veg. 5 : 394. 1891;

Wight, Ic. t. 308. 1840; Hook.f., Fl. Brit. India 1:185. 1872; Dunn in Gamble, Fl. Pres. Madras 4. 1915. *Nelamparenda* Rheede, Hort. Malab. 9 : 117, t. 60.1689.

Diffuse to erect herbs. Leaves alternate, simple, linear-lanceate or oblanceate, 1-4 by 0.3-2 cm, base subattenuate, margin faintly serrate, apex acute or obtuse, chartaceous. Flower(s) axillary, solitary; pedicel articulated; sepals 5, basally connate; petals 5, rose coloured or pink, unequal, lower one orbicular, larger, clawed, saccate at base; stamens 5, connate, anterior filaments appendaged, anthers pubescent; ovary superior, 1-locular, ovules α, on 3 parietal placentae; stigma oblique. Capsule subglobose, obtusely 3-angled Seeds striate.

Fl. & Fr.: Most of the seasons. *Distr.*: Sri Lanka, India, S.E. China, Africa, Madagscar & Tropical Australia. Common along waysides, waste lands etc. in plains. *NAK 169* (Nellimukal-Adoor, c. 40 m).

Roots used as a remedy in gonorrhoea and infections of the urinary organs. Whole plant pounded and taken along with milk to improve memory and vitality and as a remedy in asthma, fever, leprosy and to reduce blood sugar.

12. FLACOURTIACEAE

1. Flowers unisexual; leaves alternate.
   2. Armed trees. .... **Scolopia**
   2. Unarmed trees.
      3. Sepals 4-6; petals 0; ovary faintly 3-5-locular. .... **Flacourtia**
      3. Sepals 5; petals 5; ovary 1-locular. .... **Hydnocarpus**
1. Flowers bisexual; leaves distichous.......... **Casearia**

CASEARIA N.J. Jacquin
Enum. Pl. Carib. 4. 1760.

Casearia ovata (Lamk.) Willd., Sp. Pl. 2(2) : 629. 1799; Mukh., Bull. Bot. Surv. India 14:184. 1972; Nicols., Suresh & Manilal, An Interpr. Hort. Malab. 146. 1988. *Anavinga ovata* Lamk., Encycl. 1:148. 1783. *Vareca zeylanica* Gaertn., Fruct. 1:290, t.60, f.6.1788. *Casearia anavingu* Pers., Syn. Pl. 1:485 1805. *C. esculenta* Roxb., (Hort. Beng. 90.1814, nom. nud.) Fl. Ind. 2:422. 1832; Clarke in Hook.f., Fl. Brit. India 2:592. 1872; Gamble, Fl. Pres. Madras 521.1919; Ramamoorthy in Sald. & Nicols., Fl. Hassan Dist. 162. 1976; Mohanan & Sivad., Fl. Agasthyamala 73.2002. *C. zeylanica* (Gaertn.) Thw., Enum. 19.1858. *Ana-vinga* Rheede, Hort. Malab. 4:101-102, t.49. 1683. *Malampavetta.*

Shrubs or very small trees. Leaves distichous, elliptic-lanceate, 8-15 by 3-6 cm, base cuneate or rounded, apex acuminate, coriaceous. Flowers in axillary clusters, bisexual, regular; pedicels articulate near base; sepals 4-5, greenish; petals 0; stamens 8-10, basally connate with alternating staminodes; ovary superior, 1-locular; ovules few on 3 parietal placentae; stigma capitate. Capsule succulent. Seeds arillate.

Fl. & Fr.: Jan.-May. *Distr.*: Sri Lanka, India. Rare. In moist deciduous to semi- evergreen forests. *NAK 2378* (Padam, c. 150 m).

Root is used as a remedy for piles and considered as very effective in diabetes.

FLACOURTIA Commerson ex L' Heritier de Brutelle Stirp. Nov. 59. 1786.

Trees, usually armed. Leaves alternate, often clustered towards apex, margin crenate-serrate, basally 3-5- nerved. Flowers 1-sexual; sepals 4-5, petals 0; stamens α; disc divided into glands; ovary faintly 3-5-locular; ovules in pairs on 3-5-placentae. Drupe globose.

1. Evergreen trees; leaves clustered towards apex; petiole to 5 mm long. .... 1. **F. indica**
1. Deciduous trees; leaves alternate-spiral; petiole to 10 mm long. .... 2. **F. ramontchi**

1. **Flacourtia indica** (Burm. f.) Merr., Interpr. Rumph. Herb. Amboin. 377. 1917; Sleumer in Steenis, Fl. Males. I, 5 : 76. 1954; Matthew & Britto in Matthew, Fl. Tam. Carnatic 3(1) : 59. 1983; Nicols., Suresh & Manilal, An Interpr. Hort. Malab. 147. 1988. *Gmelina indica* Burm. f., Fl. Ind. 132. 1768. *Flacourtia sepiaria* Roxb., Pl. Corom. t. 68. 1796 & Fl. Ind. 3:835. 1832; Hook.f. & Thoms. in Hook.f., Fl. Brit. India 1: 194. 1872; Dunn in Gamble, Fl. Pres. Madras 54.1915. *Couro-moelli* Rheede, Hort. Malab. 5:77, t.39.1685.

Leaves clustered towards apex, alternate towards base, elliptic-ovate, 2-4 by 1-2 cm, base attenuate, apex acute to obtuse; petiole c.5 mm long. Racemes axillary. Flowers cream coloured, c.3 mm ϕ; stamens unequal; ovary 3-locular, ovules 2 per locule. Drupe globose. Seeds obovoid.

Fl. & Fr.: Nov.-Mar. *Distr.*: Africa, India, S.E. Asia & Malesia. Rare. On open areas. *CNM 55528* (MH) (Pathanamthitta, c.50 m).

Berries edible. As the plant is thorny it can be used for fences.

2. **Flacourtia ramontchi** L' Herit., Stirp. Nov. 3:59. 1785; Wight & Arn., Prodr. 29.1834; Wight, Ic. t. 85. 1838; Hook.f. & Thoms. in Hook.f., Fl. Brit. India 1:193. 1872; Dunn in Gamble, Fl. Pres. Madras 54.1915; Matthew, Ill. Fl. Tam. Carnatic t. 41.1982; Matthew & Britto in Matthew, Fl. Tam. Carnatic 3(1) : 60.1983; Nicols., Suresh & Manilal, An Interpr. Hort. Malab. 147. 147.1688. *Talir kkara* Rheede, Hort. Malab. 5.75, t. 38. 1685.

Leaves oblong or obovate, 10-15 by 3.5-5 cm, base cuneate, margin crenate-serrate, apex gradually acuminate, thin-coriaceous; petiole c. 10 mm long. Male flowers in short panicles; sepals cream coloured; petals 0; stamens α, anthers versatile; ovary globose, faintly 5-locular. Drupe globose.

Fl. & Fr.: Sep.-Mar. (Apr.) *Distr.*: India, Madagascar & Malayan Archipelago. Rare. In deciduous to semi-evergreen forests. *NAK 2653* (Appupanthode, c. 400 m).

Fruits edible. *Flacourtia montana* Grah. is seen cultivated occasionally for its edible fruits.

HYDNOCARPUS J. Gaertner Fruct. 1:288, t. 60. 1788.

Trees. Leaves simple, alternate, coriaceous. Flowers 1-sexual, in cymose or racemose clusters, axillary; sepals 5, imbricate; petals 5, appendaged with a scale inside; stamens 5-15; ovary 1-locular, ovules α, on parietal placentae. Berry globose, large. Seeds α, embedded in the pulp.

1. Leaves lanceate or oblong-lanceate, margin entire; flowers (male) long-pedicelled, pendulous. .... 1. **H. pendulus**

1. Leaves elliptic or elliptic-oblong, margin crenate-serrate; flowers (male) short-pedicelled, erect. .... 2. **H. pentandra**

1. **Hydnocarpus pendulus** Manilal, Sabu & Sivar., Trop. Plant Sci. Res. 1(4) : 35, f.1-8. 1983; Manilal, Fl. Silent Valley 11. 1988.

Leaves lanceate or oblong-lanceate, 10-20 by 3-4.5 cm, base acute, apex gradually acute to slightly acuminate. Flowers polygamous, in axillary racemes, pendulous. Male flowers : pedicel slender, 2-3 cm long; sepals oblong or elliptic, rusty pubescent without; petals white, linear-lanceate, longitudinally strongly inflexed, enclosing a linear oblong scale; stamens 5, erect, filaments short. Bisexual flowers : pedicels stouter and shorter; ovary ovoid, tomentose; ovules α, on 5 parietal placentae. Fruit globose, c. 8 cm ϕ. Seeds α, embedded in the pulp.

Fl. & Fr.: *Jan- Apr. Distr.*: So far reported from Silent Valley only (Palghat Dist.). Rare. In moist evergreen forests. *NAK 2278* (Arampa, c.400 m).

2. **Hydnocarpus pentandra** (Buch.-Ham.) Oken, Allg. Naturgesch. 3(2) : 1381. 1841; Mukh., Bull. Bot. Surv. India 14:183. 1972; Nicols., Suresh & Manılal, An Interpr. Hort. Malab. 147. 1988. *Chilmoria pentandra* Buch.-Ham., Trans. Linn. Soc. London 13 (2) : 501. 1822. *Hydnocarpus wightiana* Bl., Rumph. 4 22. 1849; Hook.f. & Thoms. in Hook.f., Fl. Brit. India 1: 196. 1872; Dunn in Gamble, Fl. Pres. Madras 52.1915. *Hydnocarpus laurifolia* (Dennst.) Sleum., Bot. Jahrb. 69:33.1938; Ramamoorthy in Sald. & Nicols., Fl.Hassan Dist. 164. 1976. *Munnicksia laurifolia* Dennst., Schluss. Hort.Ind. Malab. 13, t. 27. 1818, nom. nud. (cf. Taxon 17:496.1968). *Marotti* Rheede, Hort. Malab. 1:65-66, t. 36.1678.

Leaves elliptic-oblong, (5) 10-20 by (3.5) cm, base rounded to obliquely cuneate, margin crenate-serrate, apex acuminate, lateral nerves prominent below, pubescent on nerves below. Flowers in axillary cymes. Male flowers : sepals 5, broadly ovate, tomentose without; petals 5, yellow; scales brown, broadly ovate, margin fimbriate; stamens 5, free; pistillode 0. Female flowers : ovary globose, wooly. Berry globose, 5-6 cm ϕ.

Fl.: Most of the season (male flowers) *Fr.*: Oct.-Dec. *Distr.*: Western Ghats. Very common in riverine forests. *NAK 448, 508, 574* and *1601* (Chathanthara, c. 150 m).

Seeds yield an oil, used for lamps, and used internally and externally for leprosy and worms. Seeds pound with sandal and applied to sores and ulcers.

SCOLOPIA Schreber
Gen. 335. 1789 (nom. cons.).

Scolopia crenata (Wight & Arn.) Clos, Ann. Sci. Nat. Bot. (Ser. 4) 8:250. 1857, exc. specim. Philippin.; Bedd., Fl. Sylv. t. 78. 1871; Hook.f. & Thoms. in Hook.f., Fl.Brit. India 1 : 191. 1872; Dunn in Gamble, Fl. Pres. Madras 52. 1915; Sleumer, Blumea 20:39. 1972; Ramamoorthy in Sald. & Nicols., Fl. Hassan Dist.164. 1976; Matthew, Ill. Fl. Tam. Carnatic t.42. 1982; Matthew & Britto in Matthew, Fl. Tam. Carnatic 3 (1) . 61.1983; Mohanan & Sivad., Fl. Agasthyamala 75.2002. *Phoberos crenatus* Wight & Arn., Prodr. Fl. Ind. Or. 29.1834.

Armed trees; spines branched, to 5 cm long. Leaves simple, alternatc. Lamina broadly elliptie, 5-8 by 3-4 cm, base rounded or broadly cuneate, margin very faintly glandular-crenate, apex obtusely short-acuminate, coriaceous. Racemes, c. 2 cm long, axillary, pubescent. Flowers bisexual; bracts lanceate; petals 5, white, ciliate; stamens α, anthers versatile, connective appendaged; ovary globose, sessile, 1-locular; style stout; stigma 4-lobed. Berry globose, c. 2 cm ϕ. Seeds ovoid.

Fl. & Fr.: Feb.-Apr. *Distr.*: Peninsular India. Rare.In evergreen forests. Fruits edile. *NAK 2331* (Aikadu sacred grove, c. 100 m). Leaves variable in size and shape.

13. POLYGALACEAE

1. Flowers small; keel petal crested; stamens 8. .... **Polygala**
1. Flowers minute; keel petal hooded; stamens 4-5. .... **Salomonia**

POLYGALA Linnaeus
Sp. Pl. 701. 1753.

Diffues to erect herbs. Leaves spiral, simple, margin often revolute. Flower(s) in short or long racemes or solitary, axillary or extra-axillary, 5-merous, zygomorphic, bisexual; bracts and bracteoles present; sepals 5, unequal; outer 3 smaller, inner 2 (wing sepals) petaloid, persistent; petals dissimilar, 3 basally connate and adnate with staminal sheath, lateral 2 variously shaped, posterior keel-shaped with a dorsal crest; stamens 8, monadelphous, filaments variously connate, anthers ellipsoid, poricidal; ovary laterally compressed, 2-locular, ovule 1 per locule; style variously curved; stigma lateral, oblique. Capsule laterally compressed, margined, apex emarginate. Seeds pendulous, densly villous, carunculate; caruncle 3-toothed or appendaged.

1. Erect herbs, to 80 cm high; wing sepals brodly ovate; caruncular appendages,broad, membranous, more or less equal to seeds. .... 5. **P. javana**
1. Diffuse or erect herbs, to 50 cm high; wing sepal elliptic; caruncular appendages linear, about half as long as seed.
   2. Staminal sheath bearing middle 2 free filaments and lateral two bundles of 3 connate filaments; style hooded and toothed at apex. .... 7. **P. telephioides**
   2. Staminal sheath bearing 8 free filaments; style horse-shoe shaped at apex.
      3. Flowers yellow. Capsule oblong, obliquely oblong or rhomboid.
         4. Keel crested with forked appendages; caruncle with 2 apendages and a tooth. .... 6. **P. rosmarinifolia**
         4. Keel crested with fimbriate appendages; curucle 3-toothed.
            5. Racemes to 2 cm long; style apex hooked; capsule symmetric. .... 1. **P. arvensis**
            5. Racemes to 10 cm or more long; style apex horse-shoe shaped; capsule asymmetric. .... 4. **P. elongata**
      3. Flowers other than yellow; capsule suborbicular.
         6. Wing sepals glabrous except at the margin; caruncle with 2 appendages and a tooth. .... 3. **P. chinensis**
         6. Wing sepals sparsely pubescent; caruncle with 3 appendages. .... 2. **P. bolbothrix**

1. **Polygala arvensis** Willd., Sp. Pl. 3(2) : 876. 1802; Burtt, Notes Roy. Bot. Gard. Edinburgh 32:404. 1973; Ramamoorthy in Sald. & Nicols., Fl. Hassan Dist. 410. 1976; Chandrabose & Nair, Proc. Ind. Acad. Sci. (Plant Sci.) 90 : 123. 1981; Britto & Matthew in Matthew, Fl. Tam.Carnatic 3(1) : 65. 1983; Manilal, Fl. Silent Valley 14.1988; Nicols., Suresh & Manilal, An Interpr. Hort. Malab. 210. 1988. *P. chinensis* auct., non L. 1753: Benn. in Hook.f., Fl. Brit. India 1: 204. 1872, p.p.; Dunn in Gamble, Fl. Pres. Madras 58. 1915; Mukh., Bull. Bot. Soc. Bengal 12: 38. 1958, p.p. *Katu-vistna-clandi* Rheede, Hort. Malab. 9:119, t.61. 1689. *Kodatsjeri* Rheede, Hort. Malab. 9:131, t. 67. 1689.

Leaves oblanceate to obovate or elliptic, 4-20 by 2-6 mm, base cuneate, margin thickened, apex subacute-mucronulate, lower surface pubescent. Flower(s), c. 4 mm long, solitary or in short racemes; outer sepals elliptic, lanceate; wing sepals ovate-elliptic, glabrous or puberulous without, apex mucronate, 5-7-nerved; corolla yellow, keel petal crested, subapical; filaments connected at the same level; style curved, apex hooked. Capsule ovate-oblong, apex obliquely emarginate. Seeds cylindric; caruncle equally 3- toothed.

Fl. & Fr.: Most of the seasons. *Distr.*: Sri Lanka, India, S. China & Malesia. Common. In open places and occasionally as a weed in cultivated fields. *NAK 1852* (Kodumon, c. 100 m), *1936* (Vallicode, c.150 m). Roots given in fever and dizziness.

2. **Polygala bolbothrix** Dunn, Kew Bull. 1916:63. 1916 & in Gamble, Fl. Pres. Madras 58. 1915; Mukh., Bull. Bot. Soc. Bengal 12:37.1958; Chandrabose & Nair, Proc. Ind. Acad. Sci. (Plant Sci.) 90:120. 1981; Britto & Matthew in Matthew, Fl. Tam. Carnatic 3 (1): 66.1983; Manilal, Fl. Silent Valley 15. 1988. *P. ciliata* Wight & Arn., Prodr. 38.1834, non L., 1753. *P. ramaswamiana* Mukh., J. Bombay Nat. Hist. Soc. 53:55. 1955 (incl. var. *devicolamensis* & var. *paliensis*).

Vegetative parts, sepals outside softly bulbous hairy. Leaves obovate or eliptic oblong, 10-20 by 5-7 mm, base cuneate, apex subacute, acute or rarely emarginate, mucronulate. Racemes c. 6 cm long, extra-axillary. Flowers c. 6 mm long, outer sepals ovate-lanceate; wing sepals obliquely elliptic-ovate, apex aristate, c.7-nerved; corola pink, keel crested with filiform appendages; filaments free to middle; ovary suborbicular; stylar apex widened and horse-shoe shaped. Capsule suborbicular. Seeds ovoid; caruncle equally 3-toothed.

Fl.& Fr.: Jan.-May. *Distr.*: Peninsular India. Rare. In moist grass lands of hills. *NAK 193* (Konni, 100 m), *2758* (Kakki hills, 1200 m).

3. **Polygala chinensis** L., Sp. Pl. 704. 1753; Mukh., Bull. Bot. Soc. Bengal 12:38.1958, p.p.; Chandrabose & Nair, Proc. Ind. Acad. Sci. (Plant Sci.) 90:118.1981. *P. glomerata* Lour., Fl. Cochinch. 426. 1790; Benn. in Hook.f., Fl. Brit. India 1: 206. 1872; Sundar., J. Bombay Nat.Hist. Soc. 53:523. 1956; Adema, Blumea 14:270. 1966; Burtt, Notes Roy. Bot. Gard. Edinburgh 32:404. 1973.

Leaves elliptic to oblanceate or obovate, 1.5-6 by 0.8-2 cm, base cuneate, apex acute, subacute or retuse, mucronulate, puberulous or glabrous, chartaceous. Raceme 2-4-flowered, extra-axillary. Flowers c. 8 mm long; outer sepals subequal, triangular-ovate; wing sepals oblong-ovate, margin cliate, .7-nerved; corolla pale pink turning white, keel crested with filiform appendages; filaments free to middle; ovary obliquely orbicular. Capsule suborbicular, margin faintly winged. Seeds ovoid; caruncle with 2 appendages (often exceeding the middle of seed) and a tooth.

Fl. & Fr.: Apr.-Dec. *Distr.*: India, Eastern Himalaya, Myanmar & China. Less common. In moist shady lower hills. Abundant in forests clearings. *NAK 585* (Maniyaar, c. 70 m), *802* (Ranni R.F., c.100 m).

4. **Polygala elongata** Klein ex Willd., Sp. Pl. 3:879.1802; Benn. in Hook.f., Fl. Brit. India 1:203. 1872; Dunn in Gamble, Fl. Pres. Madras 58.1915; Mukh., Bull. Bot. Soc. Bengal 12:36. 1958; Adema, Blumea 14:273. 1966; Ramamoorthy in Sald. & Nicols., Fl. Hassan Dist. 412.1976; Chandrabose & Nair, Proc. Ind. Acad. Sci. (Plant Sci.) 90:125. 1981; Birtto & Matthew in Matthew, Fl. Tam. Carnatic 3(1) : 66. 1983. **(Fig. 9)**

Leaves narrowly obovate, narrowly elliptic to linear, 7-15 by 3-3 mm, base obtuse or subacute, apex obtuse, subacute or acûte, mucronulate, glabrous or sparsely pubescent. Racemes 13-10 cm or more long, extra-axillary or axillary. Flowers yellow, 6 mm long; outer sepals ovate wing sepals obliquely elliptic-ovate, glabrescent, c.7-nerved, apex mucronate, keel crested with filiform appendages; filaments free to middle. Capsule obcordate, margin with hooked hairs. Seeds ellipsoid; caruncle equally 3-toothed.

Fl. & Fr.: Aug.-Dec. Distr.: Sri Lanka, Peninsular India. Rare. In moist rocky grass lands. *NAK 2020* (Kokkathode, c. 400 m).

5. **Polygala javana** DC., Prodr. 1:327. 1824; Benn. in Hook.f., Fl. Brit. India 1:201. 1872; Dunn in Gamble, Fl. Pres. Madras 58.1915; Mukh., Bull. Bot. Soc. Bengal 12:44. 1958 (incl. var. *angustifolia* Thw.); Adema, Blumea 14:261.1966; Chandrabose & Nair, Proc. Ind. Acad. Sci. (Plant Sci.) 90:112. 1981; Britto & Matthew in Matthew, Fl. Tam. Carantic 3 (1) : 70. 1983;Mohanan & Sivad., Fl. Agasthyamala 77.2002.

Leaves obovate, oblanceate or elliptic-oblong, 0.5-5 by 0.2-2 cm,base cuneate, apex obtuse to emarginate, pubescent, glabrescent. Racemes to 9 cm long, lateral. Flowers yellow with pink crest, c.7 mm long; outer sepals unequal, ovate-lanceate, pubescent, wing sepals broadly ovate, 4-10 by 6-9 mm, puberulous, keel with filiform appendages; filaments free to middle. Capsule oblong; caruncle 2-appendaged, 1-toothed.

Fl. & Fr.: Sep.-Mar. *Distr.*: Sri Lanka, S.E. Continental Asia & Malesia. Rare. In moist hill slopes among grasses. *NAK 1095* (Vallicode, 150 m).

6. **Polygala rosmarinifolia** Wight & Arn., Prodr. 37. 1834; Benn. in Hook.f., Fl. Brit. India 1:204. 1872; Dunn in Gamble, Fl. Pres. Madras 58.1915; Mukh., Bull. Bot. Soc. Bengal 12:38.1958; Chandrabose & Nair; Proc. Ind. Acad.Sci. (Plant Sci.) 90: 117. 1981; Britto & Matthew in Matthew, Fl.Tam. Carnatic 3 (1) : 74.1983;Mohanan & Sivad., Fl. Agasthyamala 78.2002.

Leaves linear-oblong or elliptic-oblong to oblanceate, 0.5-5 by 0.1-0.6 cm, base cuneate, apex acute or obtuse, apiculate, chartaceous. Racemes lateral or extra-axillary. Flowers yellow, to 3.5 mm long; outer sepals lanceate; wing sepals obliquely ovate-elliptic, apex acute or acuminate, keel crested with shortly forked filiform appendages, hooked. Capsule rhomboid or obliquely oblong. Seeds cylindric; caruncle 2-appendaged, 1-toothed.

Fl. & Fr.: Most of the seasons. *Distr.*: Sri Lanka, Western Peninsular India. Common in grasslands of upper hills. *NAK 2743* (Kakki hills, c. 1100 m).

7. **Polygala telephioides** Willd., Sp.Pl.3:876.1802; Roxb., Fl. Ind. 3:218.1832. Wight & Arn., Prodr. 36.1834; Benn. in Hook.f., Fl. Brit. India 1:205.1872; Dunn in Gamble, Fl. Pres. Madras 59.1915; Chandrabose & Nair, Ind. Acad. Sci. (Plant Sci.) 90: 117. 1981; Britto & Mathew in Matthew, Fl. Tam. Carnatic 3(1) : 74.1983;Mohanan & Sivad., Fl. Agasthyamala 79.2002. **(Fig. 10)**

Leaves narrowly obovate to elliptic-oblong, 4-15 by 2-6 mm, base cuneate, margin revolute, apex subacute, mucronulate, midvein prominent below, thick-chartaceous. Racemes, c.1 mm long, leaf-opposed, lateral or axillary. Flowers c.3 mm long, violet or blue; outer sepals subequal, wing sepals elliptic, glabrescent,

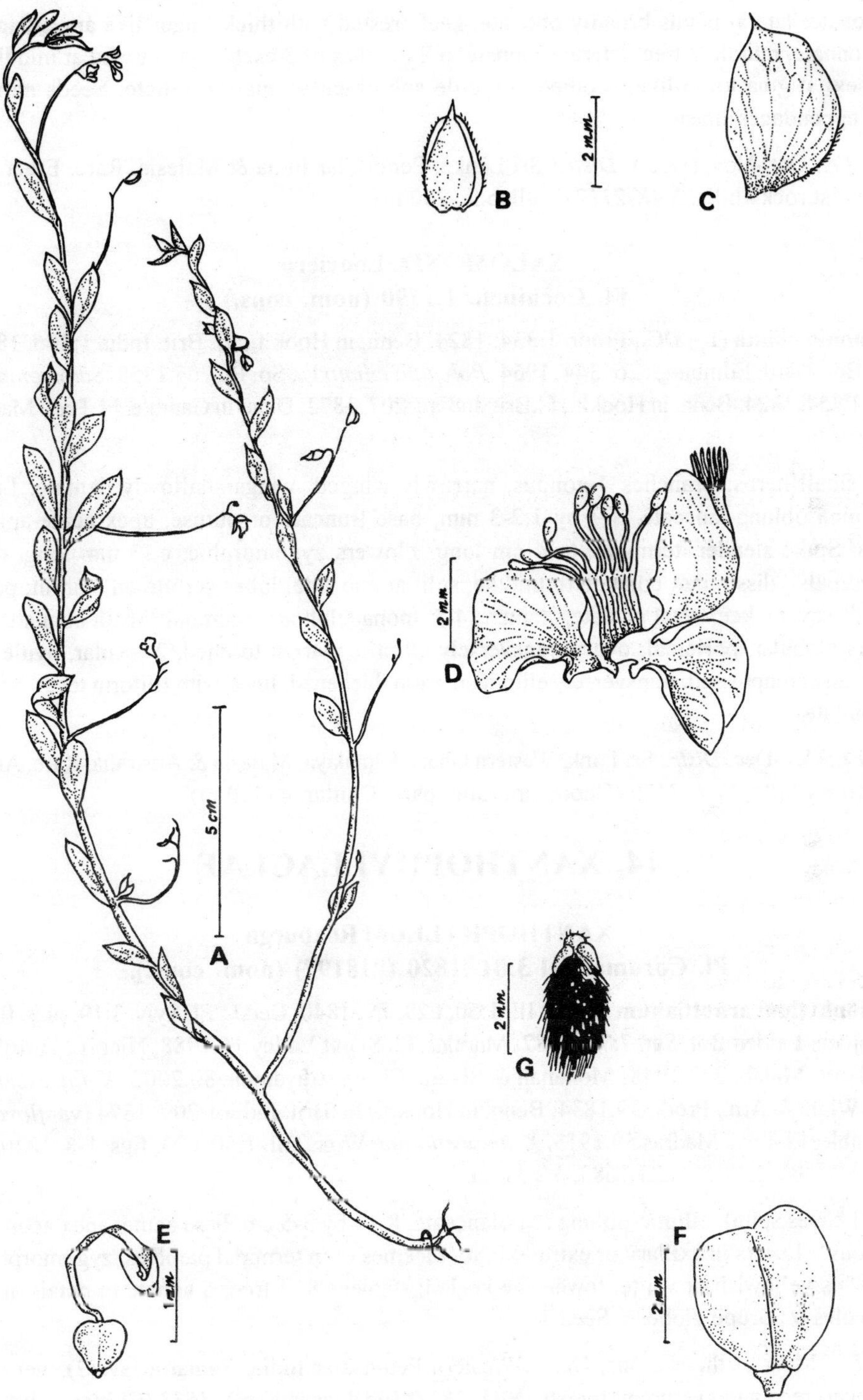

Fig. 9. ***Poygala elongata*** **Klein ex Willd.: A. Plant; B. Outer sepal; C. Wing sepal; D. Petals with staminal column; E. Pistil; F. Fruit; G. Seed.**

apex mucronate; lateral petals broadly obovate, keel crested with thick finger like appendages; filaments variously connate, middle 2 free, laterals connate in 2 bundles of 3 each; style curved at middle, broadened towards apex; stigma knob-like, toothed. Capsule suborbicular, margin ciliate. Seeds narrowly ovoid, caruncular appendages linear.

Fl. & Fr.: Oct.-Nov. (Dec.). *Distr.*: Sri Lanka, Peninsular India & Malesia. Rare. Erect herbs among grasses in moist rocky hills. *NAK 2117* (Vallicode, 200 m).

SALOMONIA Louriero
Fl. Cochinch. 1.1790 (nom. cons.).

Salomonia ciliata (L.) DC., Prodr. 1:334. 1824; Benn. in Hook.f., Fl. Brit. India 1:206. 1872; Lauener, Notes Roy. Bot. Gard. Edinburgh 26: 344. 1964. *Polygala ciliata* L., Sp. Pl. 705.1753. *Salomonia oblongifolia* DC., Prodr. 1:354. 1824; Benn. in Hook.f., Fl.Brit. India 1:207. 1872; Dunn in Gamble, Fl. Pres. Madras 56.1915. **(Fig. 11)**

Erect small herbs; branches 4-gonous, narrowly winged, wings shallowly dentate. Leaves spiral, sessile. Lamina oblong-lanceate, 3-7 by 1.2-3 mm, base truncate or obtuse, apex acute-apiculate, thick chartaceous. Spike slender, terminal, to 4 cm long. Flowers zygomorphic, c. 3 mm long, dense, bracts lanceolate; sepals , dissimilar, connate to middle, split at one side, lobes serrate on margin; petals 3, pink, lateral broadly ovate, keel petal galeate; stamens 4-5, monadelphous, staminal sheath connate to laterals at base, anthers globular, poricidal; ovary transversely elliptic, margin toothed, 2-locular, ovule 1per locule. Capsule laterally compressed, transversely elliptic, margin thickened, lined with filiform teeth. Seeds obovoid, black, strophilate.

Fl. & Fr.: Oct.-Dec. *Distr.*: Sri Lank, Western Ghats, Himalaya, Malesia & Australia. Rare. Among grasses in exposed rocky hills. *NAK 2170* (Choothuporuthu para, Chittar, c. 400 m).

14. XANTHOPHYLLACEAE

XANTHOPHYLLUM Roxburgh
Pl. Coromandel 3:81. 1820 ("1819") (nom. cons.).

Xanthophyllum arnottianum Wight, Ill. 1:50, t.23, f.9. 1840; Bedd., Fl. Sylv. 3:19, pl.3, fig. 2,2. 1869; Van der Meijden, Leiden Bot. Ser. 7:79. 1982; Manilal, Fl. Silent Valley 16.1988; Nicols., Suresh & Manilal, An Interpr. Hort. Malab. 210.1988; Mohanan & Sivad., Fl. Agasthyamala 80.2002. *X. flavescens* auct., non Roxb.1820: Wight & Arn., Prodr. 39.1834; Benn. in Hook.f., Fl. Brit. India 1:209. 1874 (var.*flavescens,* p.p.); Dunn in Gamble, Fl. Pres. Madras 59.1915. *X. angustifolium* Wight, Ill. 1:50, t.23, figs. 1-8. 1840. *Karin-kara* Rheede, Hort. Malab.4:49-50, t.23.1683. *Madukka.*

Trees. Leaves spiral, elliptic-oblong to oblanceate, 8-14 by 3-5 cm, base acute, apex acuminate, shiny, thin-coriaceous. Flowers in axillary or extra-axillary racemes or in terminal panicles, zygomorphic; sepals 5, subequal; petals yellowish or white, lower one keeled; stamens 8, 2 free, 6 adnate to petals at base; ovary 1-locular; ovules α. Drupe globose. Seed 1.

Fl. & Fr.: Most of the seasons. *Distr.*: Western Peninsular India, Sumatra,Java(?). Very common in moist semi-evergreen and evergreen forests. *NAK 282* (Moozhiar, 400 m), *1533* (Chittar, c.400 m).

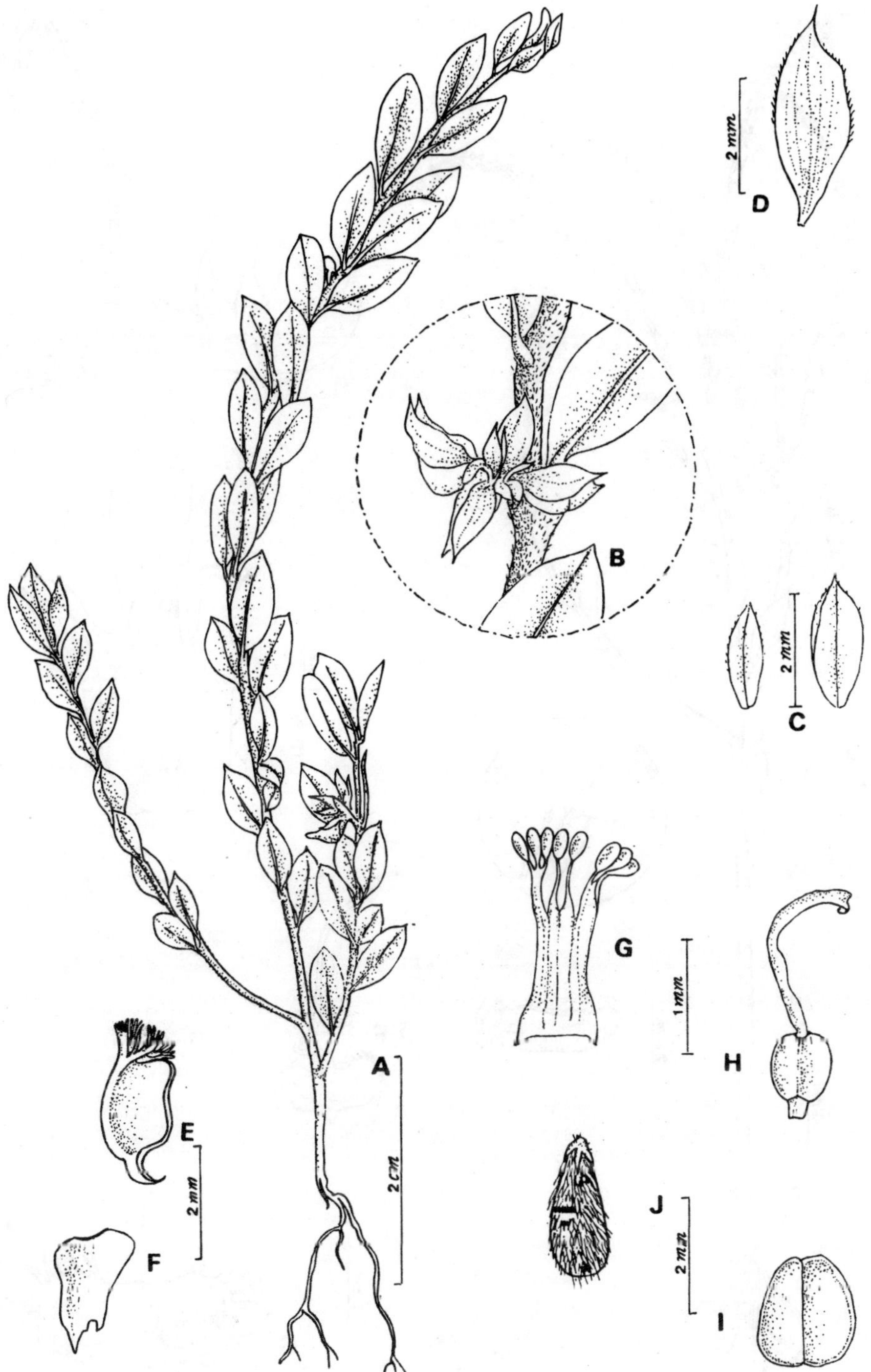

Fig. 10. *Polygala telephioides* Willd.: A. Plant; B. A portion of flowering shoot; C. Outer sepals; D. Wing sepal; E. Keel petal; F. Lateral petal; G. Staminal column; H. Pistil; I. Fruit; J. Seed.

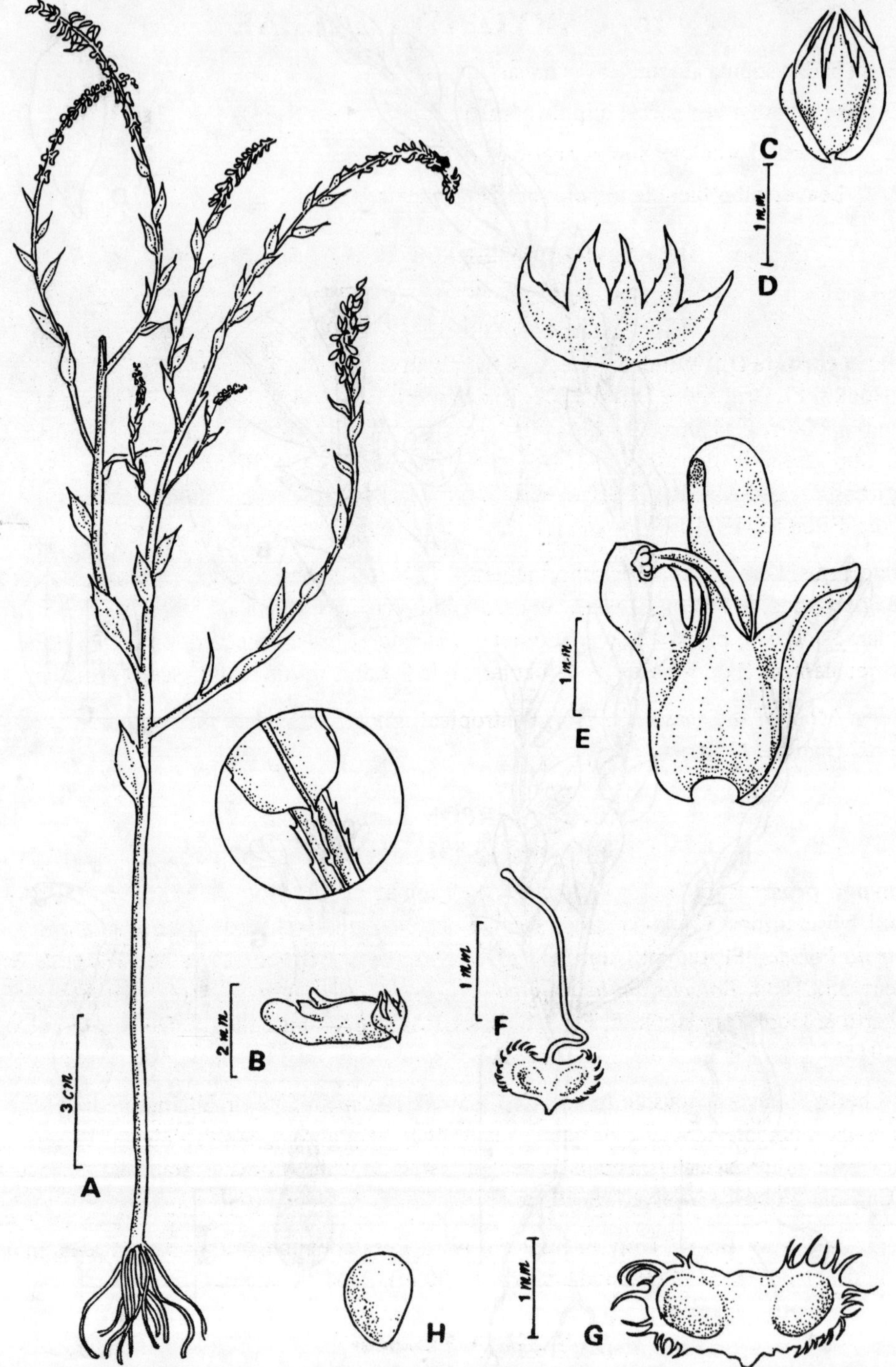

Fig. 11. *Salomonia ciliata* (L.) DC.: A. Plant; B. Flower; C. Calyx; D. Calyx -split open; E. Corolla with androecium; F. Pistil; G. Fruit; H. Seed.

15. CARYOPHYLLACEAE

1. Erect herbs; stipule absent; leaves linear. .... **Sagina**
1. Diffuse or prostrate herbs; stipule present; leaves otherwise.
    2. Leaves spathulate; stipule scarious; petals entire. .... **Polycarpon**
    2. Leaves suborbicular; stipule setaceous; petals 2-partite. .... **Drymaria**

DRYMARIA Willdenow ex J.A. Schultes in J.J. Roemer & J.A. Schultes, Syst. Veg. 5: 406. 1819-1820.

Drymaria cordata (L.) Willd. ex Roem. & Schult. in L., Syst. Veg. (ed. 16) 5:406. 1819/20; Edgew. & Hook.f., in Hook.f. Fl. Brit. India 1:244. 1872; Ramamoorthy in Sald. & Nicols., Fl. Hassan Dist. 97. 1972; Dunn in Gamble, Fl. Pres. Madras 63.1915; Matthew, Ill. Fl. Tam. Carnatic t. 45. 1982; Matthew & Britto in Mathew, Fl. Tam. Carnatic 3(1): 76. 1983; *Holosteum cordatum* L., Sp. Pl. 88. 1753. ssp. *diandra* (Blume) Duke, Ann. Missouri Bot. Gard. 48 : 253. 1961; Majumdar, Bull. Bot. Surv. India 10:294. 1969. *Drymaria diandra* Blume, Bijdr. 62. 1825.

Prostrate herbs. Leaves opposite, suborbicular, 4-12 mm wide, base obtuse, subtruncate or subcordate, apex obtuse or rounded, 3-5-nerved, chartaceous; stipules bristly, setaceous. Flowers in axillary or terminal cymes, regular, 5-merous, bisexual; sepals, oblong, connate at base, 3-nerved; corolla white or greenish, petals 2-partite; stamens 2 or 3(5); ovary 1-locular; style 3-fid. Capsule ovoid. Seeds orbicular.

Fl. & Fr.: Most of the seasons. *Distr.*: Pantropical. Locally abundant in shady moist areas of upper hills. *NAK 1402* (Pampa, 1100 m).

POLYCARPON Linnaeus Syst. ed. 10.881,1360.1759

Polycarpon prostratum (Forssk.) Asch. & Schweinf., Oester. Bot. Zeitschr. 39:128,1889; Milnoe-Redhead, Bull. Misc. Inform.1948:451. 1948; Matthew & Britto in Matthew, Fl. Tam. Carnatic 3(1):78. 1983. *Alsina prostrata* Forssk., Fl.Aegypt.-Arab.207.1775. *Hapalosia loeiflingiae* Wall.ex Wight & Arn., Prodr. Fl. Ind. Orient. 358.1834. *Polycarpon loeflingiae* (Wight & Arn.) Benth. & Hook. Gen. Pl. 1:153.1862, in note; Edgeworth & Hook.f. in Hook. f., Fl. Brit. India 1:245.1874, nom. illegit.; Dunn in Gamble, Fl. Pres. Madras 64. 1915.

Diffuse herbs. Leave deccusate to whorled, subsessile, spathulate or oblanceate, 6-15 by 2-3 mm, base attenuate,apex obtuse, subacute or acute, puberulous or glabrous, subsucculent. Flowers in axillary clusters, subsessile; bracts scarious; sepals 5, keeled; petals 5, white, obovate; stamens 3; ovary 1-locular; style 3-fid. Capsule globose, 3-valved. Seeds α.

Fl. & Fr.: Jan.-Mar. *Distr.*: Trop. & Subtrop. Asia. Less common. Along streamsides, moist places, near paddy fields, etc. *NAK 1596* (Perunthenaruvi, c.100 m), *2304* (Kodumon, 70 m).

SAGINA Linnaeus Sp. Pl. 128.1753.

Sagina saginoides (L.) Karsten, Deutsch. Fl. 539. 1882; Mizushima, J.Japn. Bot. 35 (7) : 194.1960. *Spergula saginoides* L., Sp. Pl. 1:441. 1753. *Sagina procumbens* auct., non L. 1753: Edgew.& Hook.f. in Hook. f., Fl. Brit. India 1:242.1874, p.p.; Dunn in Gamble, Fl. Pres. Madras 63. 1915.

Small herbs. Leaves opposite, basally connate, often falcate, to 1 cm long, base obtuse, margin revolute, apex acute, veins faint, Flower(s) solitary, c.3 mm ϕ, globose, axillary and terminal; pedicel slender, to 7 mm long; sepals 5, equal; petals 5 yellow; stamens 5-10; ovary 1-locular; style 5-fid; ovules α. Capsule globose, 5-valved. Seeds α., reniform.

Fl. & Fr.: Nov.-Jan. *Distr.*: North & South temperate zones. Occasional. In moist shades of upper hills. *NAK 1399* (Pampa, c.1100 m).

Note : The Species *S. procumbens* is reported to occur in India only in Sikkim.

16. PORTULACACEAE

1. Prostrate herbs; flowers to 1 cm ϕ; capsule-dehiscence circumscissile. .... **Portulaca**
1. Erect herbs; flowers to 2 cm ϕ; capsule-dehiscence valvular. .... **Talinum**

PORTULACA Linnaeus
Sp. Pl. 445. 1753.

Portulaca oleracea L., Sp. Pl. 445.1753; Roxb., Fl. Ind. 2: 463.1832., Wight & Arn., Prodr. 356.1834., Dyer in Hook.f., Fl. Brit. India 1: 246. 1874; Dunn in Gamble, Fl. Pres. Madras 66.1915; Gees., Blumea 17:292. 1969 & in Steenis, Fl. Males. I, 7: 129. 1971; Ramamoorthy in Sald. & Nicols., Fl. Hassan Dist. 100. 1976; Matthew, Fl. Tam. Carnatic 3(1) : 81.1983; Manilal, Fl. Silent Valley 17.1988; Nicols., Suresh & Manilal, An Interpr. Hort. Malab. 212. 1988. *Kara-tsjira* Rheede, Hort. Malab. 10:71, t.36.1690.

Prostrate subsucculent herbs. Leaves alternate or subopposite with scarious nodal appendages, obovate to spathulate, 1-3 by 0.5-1 cm, base cuneate or attenuate, apex obtuse or truncate. Flower(s) sessile, 3-6, clustered, terminal, encircled by a whorl of leaves, 4-or 5-merous, bisexual; sepals 2, connate below; petals 5, yellow; stamens 12, unequal; ovary ½ inferior, 1-locular; ovules α, on free-central placenta. Capsule ovoid, dehiscence circumscissile. Seeds tubercled.

Fl. & Fr.: Most of the seasons. *Distr.*: Pantropical. Common in clayey waste soil along waysides in plains. *NAK 1024* (Tiruvalla, c.10 m).

The leaves and seeds are proposed to be useful in diseases of the kidney, urinary bladder and lungs. Seeds used in dysentery.

TALINUM Adanson
Fam. 2:245.1763.

Talinum portulacifolium (Forssk.) Asch. ex Scheinf., Bull. Herb. Boissier 4, App. 2:172. 1896; Chitt., Dict. Gard. (ed. 2) 4:2077. 1956; Matthew, Ill. Fl. Tam. Canatic t.50. 1982; Matthew & Britto in Matthew, Fl. Tam. Carnatic 3(1) : 84. 1983. *Orygia portulacifolia* Forssk., Fl.Aegypt.-Arab. 103.1775. *Portulaca cuneifolia* Vahl, Symb. Bot. 1:33. 1790. *Talinum cuneifolium* Willd., Sp. Pl.2: 864. 1799; Roxb., Fl. Ind. 2 : 465. 1832; Thistelton in Hook. f., Fl. Brit. India 1 : 247. 1874; Dunn in Gamble, Fl. Pres. Madras 66. 1915.

Succulent herbs. Leaves alternate, obovate to oblanceate, 2-5 (8) by 1-2 (3) cm, base attenuate, apex obtuse, succulent. Flowers large, in terminal panicles; bracts liner; sepals 2; petals 4-5, pink, obovate; stamens a, basally connate; ovary superior, 1-locular, ovules α, on free-central placenta. Capsule globose, dehiscence valvular or irregular. Seeds striate.

Fl. & Fr.: Feb.-Aug. *Distr.*: India, Arabia & Africa. Occasional. In moist shady places of plains. *NAK* 2738 (Seethathode, c. 200 m).

Leaves are edible.

17. HYPERICACEAE

HYPERICUM Linnaeus
Sp. Pl. 783. 1753.

Hypericum japonicum Thunb. ex Murray in L., Syst. Veg.(ed. 14) 702. July 1784; Thunb., Fl. Jap. 295, t.31. Aug. 1784; Wight & Arn., Prodr. 99. 1834; Thiselton in Hook. f., Fl. Brit. India 1:256. 1874; Dunn in Gamble, Fl. Pres. Madras 70. 1915; N. Robson, Blumea 20: 267. 1972 & in Steenis, Fl. Males. I, 8:27. 1974; Gandhi in Sald. & Nicols., Fl.Hassan Dist.127. 1976; Matthew, Ill. Fl. Tam. Carnatic t. 54.1982; Matthew, & Britto in Matthew, Fl. Tam. Carnatic 3 (1) : 92. 1983.

Small herbs. Leaves opposite below, verticillate towards apex, sessile, lanceate or elliptic, base auricled, apex acute, 3-nerved, glandular-punctate. Flowers bisexual, 5-merous, in terminal panicles; sepals lanceate; petals yellow, oblanceate; stamens α, coherent below; ovary superior, 3-lobed, 1-locular; ovules α; styles 3. Capsule ovoid, septicidal. Seeds α, minute.

Fl. & Fr.: Apr.-Aug. *Distr.*: Sri Lanka, India, Malesia, Japan, S. Korea, S.E. China, Australia, New Zealand. Rare. On exposed grassy lands in marshes. *NAK 2734* (way to Ponnambalamedu, c. 1100 m).

18. CLUSIACEAE
(GUTTIFERAE)

1. Leaf-nerves α, parallel, close; flower(s) solitary or in racemes; ovary 1-2-locular.
   2. Flowers in racemes; ovary 1-locular, ovule. .... **Calophyllum**
   2. Flower(s) solitary, or in pairs; ovary 2-locular, ovules 4. .... **Mesua**
1. Leaf-nerves not so; flowers in umbellate clusters; ovary 2-12-locular. .... **Garcinia**

CALOPHYLLUM Linnaeus
Sp. Pl. 513. 1753.

Calophyllum calaba L., Sp. Pl. 514. 1753; Lam., Encycl. 1:553.1785; Nicols., Suresh & Manilal, An Interpr. Hort. Malab. 81. 1988 ; Mohanan & Sivad., Fl. Agasthyamala 81.2002. *C. apetalum* Willd., Ges. Naturf. Freunde Berlin Mag. Neuesten 5:79. 1811; Maheshwari, Bull. Bot. Surv. India 2:143. 1960; Gandhi in Sald. & Nicols., Fl. Hassan Dist. 124. 1976; Steevens, J. Arnold Arb. 61: 253.1980. *C. decipiens* Wight, Ic. t. 106.1839 & Ill. 1: 128. 1840; Dunn in Gamble, Fl. Pres. Madras 76.1915. *C. wightianum* Wall. (List 4847-1831, nom.) ex Planchon & Triana, Ann. Sci. Nat. Bot. Ser. 4, 15:256. 1861; Anderson in Hook. f., Fl. Brit. India 1:274. *Tsjerou-ponna* Rheede, Hort. Malab. 4:81, t. 39. 1683. **(Fig.12)**

Trees. Leaves decussate, simple, obovate, 5-7 by 3-4.5 cm, base obtuse, cartilaginous, apex obtuse, rounded to retuse, veins α, close, parallel, coriaceous. Racemes axillary, usually on axils of fallen leaves, to 5 cm long; bracts lanceate. Flowers polygamous, mostly bisexual; perianth 4, white, ovate to elliptic-oblong, dissimilar; stamens α, filaments slender, anthers yellow, oblong; ovary superior, ellipsoid, 1-locular; style slender; stigma peltate; ovule 1. Drupe globose, c.12 mm ϕ.

Fl. & Fr.: Jun.-Apr. *Distr.*: Western Ghats. Common along streamsides and riverein forests. *NAK 549* (Thekkuthode, c.70 m), *620* (Thannithode, c.75 m).

The sun-dried seed kernels yield an oil having a characteristic odor and a bitter taste and in used in rheumatism and leprosy. The stem yield timber.

GARCINIA Linnaeus
Sp. Pl. 443. 1753.

Trees; branches often drooping. Leaves simple, opposite-decussate, coriaceous. Flowers polygamo-dioecious (females few) in axillary or terminal umbellate or paniculate clusters, hypogynous; sepals 4-5, imbricate; petals 4-(5), imbricate; stamens α, free and variously connate below; staminodes minate; ovary superior, 2-12-locular; style short or 0; stigma broadly peltate; ovule 1-per locule. Berry globose, pulp juicy. Seeds covered with pulp.

1. Flowers 4-merous.
    2. Leaves lanceate or elliptic, apex acute or subacute; stamens 10-20; anthers bilocular. .... 1. **G. gummi-gutta**
    2. Leaves linear-lanceate, apex caudate; stamens 1-6;anthers mono-locular. ....3. **G. wightii**
1. Flowers 5-merous. .... 2. **G. spicata**

1. **Garcinia gummi-gutta** (L.) Robson, Brittonia 20: 103. 1968; Kosterm., Ceylon J. Sci. 12:55. 1976; Manilal, Fl. Silent Valley 19. 1988; Nicols., Suresh & Manilal, An Interpr. Hort. Malab. 82.1988; Mohanan & Sivad., Fl. Agasthyamala 84.2002. *Cambogia gummi-gutta* L., Gen. Pl.ed.5, (unnumb. final page 522). 1754 "*Cambogia G. gutta*"; *C. gutta* L., Syst. ed. 10, 1073. 1759, nom. illegit. (incl. type of *C. gummi-gutta* L., 1754). *Mangostana cambogia* Gaertn., Fruct. 2:106. t. 105, 1790, nom. illegit. (incl. type of *C. gummi-gutta* L., 1754). *Garcinia cambogia* Desr. in Lam., Encycl. 3:701. 1792, nom. illeg. (incl. type of *C. gummi-gutta* L. 1754); Anderson in Hook.f., Fl. Brit. India 1:261.1874; Dunn in Gamble, Fl. Pres. Madras 73. 195. 1915; Maheshw., Bull. Bot. Surv. India 6: 128. 1965; Gandhi in Sald. & Nicols., Fl. Hassan Dist. 126.1976; Matthew & Britto in Matthew, Fl. Tam. Carnatic 3(1): 94. 1983; *Coddam-pulli* Rheede, Hort. Malab. 1:41-42, t. 24. 1678.

Leaves deccusate, lanceate or elliptic, 7-15 by 2-6 cm. base cuneate, apex acute, subacute or very shortly acuminate, shining. Male flowers : in short axillary umbellate clusters; sepals 4, coriaceous except the margin, petals 4, yellowish; stamens 12-20 or more, anthers bilocular. Bisexual flowers : in axillary or terminal clusters; stamens 10-20, often few sterile; ovary subglobose, 2 or 3-locular; stigma 2-3-lobed. Berry yellow, c.5 cm ϕ.

Fl. & Fr.: Jan.-Aug. *Distr.*: Sri Lanka, Peninsular India. In evergreen forests and in cultivation. *NAK 2310* (Kodumon, c.50 m).

The 'gummi-gutt' or gamboge is used medicinally as a purgative and emetic agent. Fruit is very acidic and eaten raw or pickled. Dried rind used as a condiment for flavouring curries. Seed oil used as medicine.

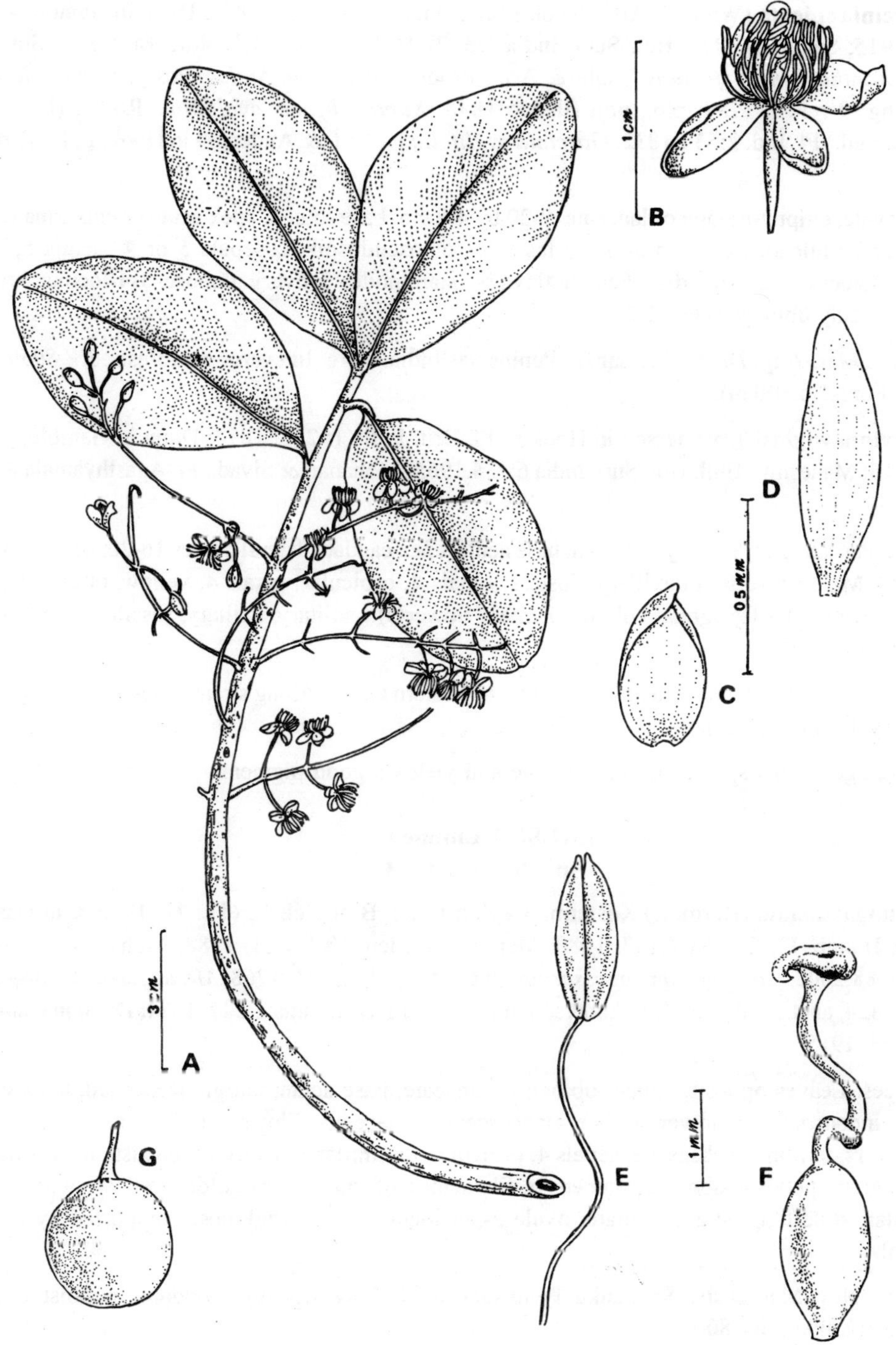

Fig. 12. ***Calophyllum calaba*** **L.: A. Twig; B. Flower; C. Outer tepal; D. Inner tepal; E. Single stamen; F. Pistil; G. Fruit.**

2. **Garcinia spicata** (Wight & Arn.) Hook.f., J. Linn. Soc. 14 : 486. 1875; Dunn in Gamble, Fl. Pres. Madras 73. 1915; Santapau, Rec. Bot. Surv. India (ed. 2) 16:14. 1960; Maheshw., Bull. Bot. Surv. India 6:112. 1965. *Xanthochymus spicatus* Wight & Arn., Prodr.1: 102.1834. *Stalagmitis cambogioides* Murr., Comm. Gotting. 9 : 173. 1789, p.p., non Blanco 1846. *Xanthochymus ovalifolius* Roxb., (Hort. Beng. 42.1814, nom. nud.) Fl. Ind. 2:632.1832. *Garcinia ovalifolia* Hook.f. & Anderson in Hook. f., Fl. Birt. India 1:269.1874.

Leaves ovate, elliptic-oblong or lanceate, 9-20 by 4-8 cm, base rotund, apex acute or emarginate, lateral veins prominent. Male flowers : in axillary fascicles or pseudospikes; sepals 5 or 4; petals 5, yellow, membranous; stamens α, in 5 bundles. Female flowers : staminodes 5; ovary globose, 3-4-locular; stigma 5-lobed. Berry broadly oblong. Seeds 1-3.

Fl. & Fr. : Mar.-Aug. *Distr.* : Sri Lanka, Peninsular India. Rare. In evergreen forests. *EV 80595* (MH) (Aappancovil Forest, c.300 m).

3. **Garcinia wightii** T. Anderson in Hook.f., Fl. Brit. India 1: 265. 1874; Dunn in Gamble, Fl. Pres. Madras 74.1915; Maheshw., Bull. Bot. Surv.India 6:134.1965; Mohanan & Sivad., Fl. Agasthyamala 87.2002. *Attukaruka.*

Leaves linear-lanceate, 9-14 by 2-2.5 cm, base acute, apex caudate, lateral nerves 16-20, parallel, forming marginal loops. Male flowers : in axillary clusters; sepals 4, orbicular; petals 4, yellow, obovate; stamens c.20, basally connate in a tetragonal column. Female flower(s) : solitary, axillary, sessile; stigma 4-lobed. Berry subglobose.

Fl. & Fr.: Nov.-Mar. *Distr.*: Endemic to Southern Western Ghats. Along streamsides of hills. *BDS 42497* (MH) (Thriveny, Pampa, 250 m).

The gamboge of this species is very soluble and yields a good pigment.

MESUA Linnaeus
Sp. Pl. 515. 1753.

Mesua nagassarium (Burm. f.) Kosterm., Ceylon J. Sci. Biol. Sci. 12 (1) : 71. 1976 & in Dassan. & Fosberg, Rev. Handb. Fl. Ceylon 1:107. 1980; Manilal, F. Silent Valley 21. 1988; Mohanan & Sivad., Fl. Agasthyamala, 88.2002. *Calophyllum nagassarium* Burm. f., Fl. Ind. 121. 1768. *Mesua ferrea* Choisy in DC., Prodr. 1:562. 1824, auct., non. L., 1753; Anderson in Hook.f., Fl. Brit. India 1:277. 1874; Dunn in Gamble, Fl. Pres. Madras 77. 1915.

Lofty trees. Leaves opposite, linear-oblong or lanceate, base acute, margin thickened, apex caudate, veins α, fine, inconspicuous, lower surface glaucescent, coriaceous. Flower(s) bisexual, regular, usually solitary or in pairs, terminal, subsessile; sepals 4, orbicular, dissimilar, coriaceous, persistent; corolla white, 8-10 cm ϕ,, fragrant; petals 4, obovate; stamens α, filaments filiform, anthers golden-yellow; ovary superior, ovoid, 2-locular; style long; stigma peltate; ovule 2-per locule. Fruit subglobose, supported by persistent enlarged sepals.

Fl. & Fr. : Nov.-Mar. *Distr.*: Sri Lanka, Peninsular India. Rare. Locally abundant. In moist evergreen forests. *NAK 59* (Kakki hills, 800 m).

19. THEACEAE

1. Flowers unisexual, 4-5 mm ϕ, in fascicles; fruit a berry. ....**Eurya**
1. Flowers bisexual, 4-5 cm ϕ, solitary; fruit a capsule. ....**Gordonia**

EURYA Thunberg
Nova. Gen. Pl. 67.1783.

Eurya nitida Korthals in Temminck, Nat. Gesh. Bot. 3:115, t.17, f. 13. 1940; Manilal, Fl. Silent Valley 22.1988; Mohanan & Sivad., Fl. Agasthyamala 91.2002. *Eurya japonica* auct., non Thunb., 1784: Bedd., Fl. Sylv. t.92. 1871; Dyer in Hook.f., Fl. Brit. India 1: 284; Dunn in Gamble, Fl. Pres. Madras 79.1915; Ramamoorthy in Sald. & Nicols., Fl. Hassan Dist. 121.1976; Matthew & Britto in Matthew, Fl. Tam. Carnatic 3(1) : 97.1983. *Kattu-theyila.*

Shrubs or small trees. Leaves simple, alternate, elliptic to elliptic-oblong, 3-8 by 1.5-3 cm, base cuneate, margin serrulate, apex acuminate, coriaceous. Flowers dioecious, in axillary fascicles, c.5 mm ϕ; bracts and bracteoles prominent; sepals 5, suborbicular, free, persistent in female; petals 5, white, orbicular, free; stamens 12, distinct; ovary 3-4- locular; styles 3-4, free. Berry globose, apiculate. Seeds α, tuberculate.

Fl. & Fr.: Apr.-Oct. *Distr.*: Sri Lanka, Peninsular India. Rare. In moist-evergreen forests. *NAK 1298* & *1347* (Kakki hills, c.1000 m).

GORDONIA J. Ellis
Philos. Trasns. 60 : 520. 1771 (nom. cons.).

Gordonia obtusa Wall. ex Wight & Arn., Prodr. 87. 1834; Dyer in Hook. f., Fl. Brit. India 1:291. 1874; Dunn in Gamble, Fl. Pres. Madras 79. 1915; Ramamoorthy in Sald. Nicols., Fl. Hassan Dist. 121. 1976; Mohanan & Sivad., Fl. Agasthyamala 91.2002. *Kattukarana.*

Trees. Leaves shortly petioled, narrowly elliptic to oblong-lanceate, 5-12 by 2-4 cm, base cuneate, margin crenate, apex obtuse or obtusely acuminate, shiny, coriaceous. Flower(s) axillary, solitary, 4-5 cm ϕ, usually towards branch-apices; sepals 4-5, dissimilar, orbicular; petals white, broadly orbicular, apex notched; stamens α, usually variously adelphous, basally connate with petals; ovary 3-locular; styles 3; ovules 2-8 per locule. Capsule 3-valvular. Seeds winged.

Fl. & Fr.: Nov.-Mar. *Distr.*: Western Ghats. Less common. In Shola and in moist evergreen forests. *NAK 2682* (way to Ponnambalamedu, c. 1200 m).

Leaves often used as substitute for tea. Wood durable.

Camellia sinensis (L.) Kuntze, was cultivated in Perumalai of Konni R.F. Now it is replaced by Rubber Plantations.

20. DIPTEROCARPACEAE

1. Calyx accrescent; connective of stamens produced above; capsule, to 2 cm ϕ.
   2. Infloresense spreading, accrescent; calyx-lobes lanceate; stipular scars encircling the stem. ....**Dipterocarpus**

2. Inflorescence secund, accrescent; calyx-lobes oblong; stipular scars not encircling the stem. .... **Hopea**

1. Calyx not or scarcely accrescent; connective seldom produced above; capsule c.4 cm ϕ. .... **Vateria**

DIPTEROCARPUS C.F. Gaertner
Fruct. 3:50. 1805.

Dipterocarpus bourdilloni Brandis in Hook., Ic. Pl. t. 2403.1896; Dunn in Gamble, Fl. Pres. Madras 81. 1915. *Karanjili.*

Large trees. Leaves simple, alternate, stipulate; stipules amplexicaul, caducous, leaving a circular scar. Lamina elliptic-oblong, base obtuse, apex acute, lateral nerves marginally looped, tomentose below, coriaceous. Panicle large; calyx-lobes 5, tube inflated, 5-winged, 2 lobes accrescent; petals white, pubescent; stamens α; ovary superior, 3-locular. Nut enclosed by accrescent calyx-tube, wings 2, lanceate, to 10 cm long.

Fl. & Fr.: Jan.-Jun. *Distr.*: Endemic to Western Ghats. Rare. In evergreen forests. One plant of this species has been specially protected by Forest officials near Pampa (Sabarimala). Flowering or fruiting twigs could not be collected.

HOPEA Roxburgh
Pl. Coromandel 3:1811 ("1819") (nom. cons.).

Trees; stem with resinous ducts. Leaves alternate, coriaceous. Racemes paniculate, usually secund. Flowers regular, bisexual; sepals 5, imbricate; petals 5, free; stamens 10-15, connective apically produced into an awn; ovary globose, lobed above, 3-locular, ovules 2-per locule; style short. Nut enclosed in calyx-tube; accrescent calyx-lobes 2, oblong, to 9 cm long. Seed 1.

1. Panicles lax, glabrous; leaves 8-20 cm long.
   2. Leaves narrowly elliptic-lanceate, bas acute, apex acuminate or acute. .... 1. **H. glabra**
   2. Leaves oblong, base truncate to subcordate, apex obtuse. .... 3. **H. ponga**

1. Panicles dense, tomentose; leaves 4-6 m long. .... 2. **H. parviflora**

1. **Hopea glabra** Wight & Arn., Prodr. 1:85. 1834; Dyer in Hook.f., Fl. Brit. India 1 : 309. 1874; Fischer in Gamble, Fl. Pres. Madras 1867. 1936; Manilal, Fl. Silent Valley 23.1988. *Ilapongu.* **(Fig. 13)**

Leaves narrowly elliptic-lanceate, 4-17 by 2-5 cm, base acute, apex acute to mostly acuminate, lateral nerves 5-8, prominent. Panicle to 12 cm long, glabrous; sepals reddish, lanceate; petals 5, white; stamens 15; ovary globose, constricted above; style subulate. Nut globose, wing oblong, to 5 cm long.

Fl. & Fr.: Jan.-Jul. *Distr.*: Western Ghats. Rare, but locally abundant. Along streamsides. *NAK 1729* (Thannithode, c. 70 m). Wood strong and durable.

2. **Hopea parviflora** Bedd., Fl. Sylv. t. 7. 1869; Dyer in Hook.f., Fl. Brit. Inida 1308. 1874; Dunn in Gamble, Fl. Pres. Madras 82. 1915; Mohanan & Sivad., Fl. Agasthyamala 93.2002. *Kampakam.*

Leaves lanceate, 4-6 by 2-2.5 cm, base rounded to acute, apex acute, nerve-axils often glandular. Panicle to 8 cm long, tomentose, mostly terminal. Flowers small; bracts and bracteoles caducous; sepals lanceate, tomentose; stamens c.15, anthers globose; ovary globose; style subulate; calyx-wing oblong, to 4 cm long

Fl. & Fr.: Jan.-Jun. *Distr.*: Western Ghats. Less Common. In evergreen forests. *EV 80547* (MH) (Kallar Valley). Wood is used for railway sleepers. Soaked wood bark is useful to ward off forest leeches while trekking.

3. **Hopea ponga** (Dennst.) Mabb., Taxon 28: 587. 1979; Nicols., Suresh & Manilal, An Interpr. Hort. Malab. 101. 1988. *Artocarpus ponga* Dennst., Schlussel 15, 18, 30. 1818. *Hopea wightiana* Wall. ex Wight & Arn., Prodr. 1:85. 1834; Dyer in Hook.f., Fl.Brit.India 1:309. 1874; Dunn in Gamble, Fl. Pres. Madras 82. 1915; Ramamoorthy in Sald. & Nicols., Fl. Hassan Dist. 118.1976. *Ponga, Pongu* Rheede, Hort. Malab. 4:73-74, t.35. 1983.

Leaves oblog, 10-18 by 4.5-6.5 cm, base subcordate to truncate, apex obtuse or abrubptly very short-acuminte. Panicle to 25 cm long, axillary or terminal. Flowers shortly pedicelled; bracts ovate; bracteoles lanceate; sepals reddish, ovate and oblong; petals white, lanceate; stamens 15, anthers globose; ovary ovoid; style subulate; calyx wing oblong or spathulate, to 6 cm long.

Fl. & Fr.: Mar.-Jun. *Distr.* : Western Ghats. In evergreen forests and common in sacred groves. *NAK 1729* (Aikadu, c. 40 m).

VATERIA Linnaeus
Sp. Pl. 515. 1753.

Vateria indica L., Sp. Pl. 515. 1753; Roxb., Pl. Corom. t. 288. 1820; Dyer in Hook.f., Fl. Brit. India 1 : 313. 1874; Dunn in Gamble, Fl. Pres. Madras 85. 1915; Ramamoorthy in Sald. & Nicols., Fl. Hassan Dist. 119. 1976; Manilal, Fl. Silent Valley 24. 1988; Nicols., Suresh & Manilal, An Interpr. Hort. Malab. 102. 1988; Mohanan & Sivad., Fl. Agasthyamala 94. 2002. *Penone* Rheede, Hort. Malab. 4: 33-34, t. 15.1683.

Trees. Leaves elliptic-oblong or oblong to narrowly obovate, 7-18 by 3-8.5 cm, base rounded to subcordate, apex obtuse, rarely obtusely acuminate,lateral veins c.20 pairs, prominent, coriaceous. Panicle terminal and/or axillary, c.15 cm long, stellate pubescent. Flowers c. 2 cm ϕ, bisexual, regular; sepals 5, ovate-lanceate, white-stellate tomentose without, persistent; petals 5, white; stamens α , anthers linear; ovary sessile, conical, 3-locular, ovule 2-per locule. Capsule large, c. 4 cm ϕ, fleshy.

Fl. & Fr.: Mar.-Jul.*Distr.*: Western Ghats. Common in sacred groves. *NAK 1075* (Kodumon, c. 50 m).

The resin from the trunk makes good varnish like copal, and emit good smell on burning.

21. ANCISTROCLADACEAE

ANCISTROCLADUS Wallich
Num. List. 1052. 1929 (nom. cons.).

Ancistrocladus heyneunus Wall. ex Graham, Cat. Pl. Bombay 28. 1839; Wight, Ic.tt. 1987, 1988. 1853; Dyer in Hook. f., Fl. Brit. India 1: 313. 1874; Dunn in Gamble, Fl. Pres. Madras 85. 1915; Ramamoorthy in Sald. & Nicols, Fl.Hassan Dist. 172. 1976; Manilal, Fl. Silent Valley 24. 1988; Nicols., Suresh & Manilal, An

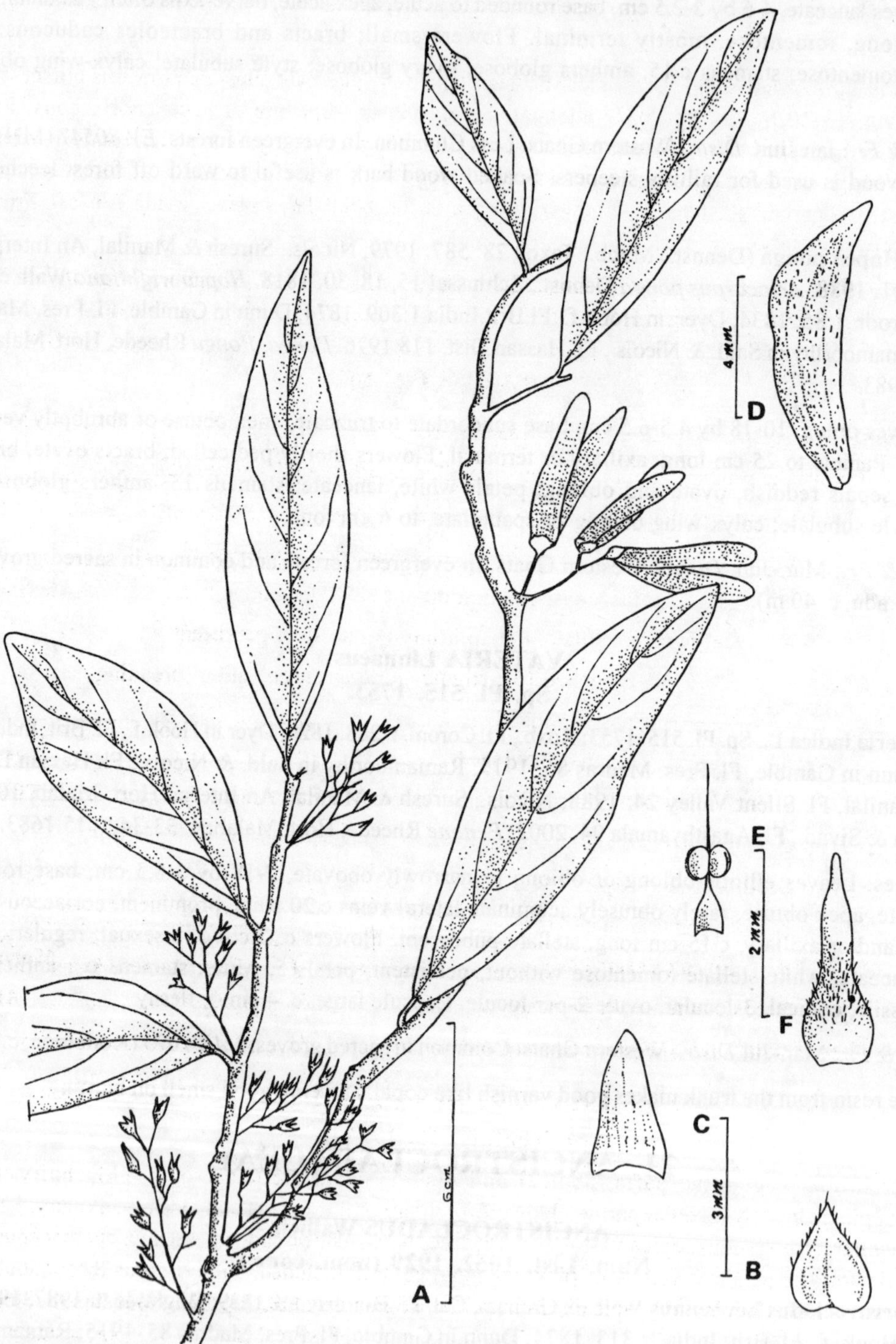

Fig. 13. ***Hopea glabra*** **Wight & Arn.: A. Twig; B. Bract; C. Sepal; D. Petal; E. Stamen; F. Pistil.**

Interpr. Hort. Malab. 49.1988; Mohanan & Sivad., Fl. Agasthyamala 95. 2002. *Modira-valli* Rheede, Hort. Malab. 7:87, t.46. 1688.

Climbing shrubs with hooks. Leaves sessile, simple, closely alternate, clustered towards branch-apices, elliptic-oblong, 13-20 by 4-7 cm, base attenuate, apex obtuse, subacute or acute, coriaceous. Panicle dichotomous, terminal or lateral. Flowers caducous, regular, bisexual, epigynous; pedicel articulated above; calyx yellowish-green, lobes 5, connate at base and adnate with ovary, accrescent; petals 5, orange-red, unequal, basally coherent; stamens 10, 5 longer; ovary inferior, 1-locular; styles 3, ovule 1, basal. Fruit with 5-winged accrescent calyx.

Fl. & Fr.: Feb.-Apr. *Distr.*: Western Ghats. Less Common. *NAK 2638* (Manniera,Thannithode R.F., c. 100 m). Large vines in semi-evergreen forests.

22. MALVACEAE

1. Number of stylar branches equal or less than the number of carpels.
   2. Fruit a capsule; staminal column 5-toothed at apex, antheriferous throughout.
      3. Flowers in dense large panicles; epicalyx-lobes large, persistent. .... **Julostylis**
      3. Flower(s) solitary or very few in short panicles; epicalyx-lobes not so.
         4. Calyx spathaceous, splitting on one side, deciduous. .... **Abelmoschus**
         4. Calyx campanulate, not splitting on one side, persistent.
            5. Style branched (5); stigma splitting on one side, spreading.
               6. Capsule winged. .... **Fioria**
               6. Capsule not winged. .... **Hibiscus**
            5. Style unbranched; stigma coherent. .... **Thespesia**
   2. Fruit a schizocarp; staminal column truncate, antheriferous at the apex only.
      7. Seed 1per mericarp.
         8. Epicalyx segments 0. .... **Sida**
         8. Epicalyx segments 3. .... **Malvastrum**
      7. Seeds 2-or more per mericarp. .... **Abutilon**
1. Number of stylar branches twice as many as the carpels. .... **Urena**

ABELMOSCHUS Medikus
Malvenfam. 46. 1787.

Undershrubs; vegetative parts, pedicel and outside calyx stellately or prickly hairy. Leaves alternate, simple, palmately lobed or partite. Flower(s) solitary or few in terminal racemes, axillary, 5-merous, bisexual, regular, hypogynous; epicalyx-lobes 4-16, free, persistent or caducous; calyx spathaceous, fairly 5-toothed, deciduous; petals 5, yellow or pink; staminal column included, antheriferous throughout; ovary 5-locular; ovules a per locule; style apically 5-branched; stigma discoid. Capsule ovoid, loculicidal. Seeds a, reniform.

1. Epicalyx-lobes 4-6, ovate; branches prickly hispid. ....1. **A. manihot** ssp. **tetraphyllus**

1. Epicalyx-lobes 6-10, linear; branches softly tomentose. ....2. **A. mochatus**

1. **Abelmoschus manihot** (L.) Medik., Malv. 46.1787; Hochr., Candollea 2:87. 1924; Sivar. & Pradeep, Malv. South. Penins. India 67.1996. ssp. **tetraphyllus** (Roxb. ex Hornem.) Borss., Blumea 14:97. 1966; Paul & Nayar, Fl. India Fasc. 19:75.1988. *Hibiscus manihot* L., Sp. Pl. 696. 1753. *Hibiscus tetraphyllus* Roxb. ex Hornem., Hort. Hafn. 661. 1815. **(Fig. 14)**

Branches prickly hispid. Leaves broadly ovate to orbicular, 4-15 by 3.5-13 cm, 3-5(7)-lobed or palmately 3-5- fid, base cordate, margin dentate to serrate, apex acuminate, basally 5-(7-) nerved, chartaceous. Flower(s) solitary or in terminal racemes, c.6 cm ϕ; epicalyx-lobes ovate-lanceate, c.12 mm wide; corolla yellow or pale pink, with a purple centre; ovary ovoid; style slender; stigma diffuse. Capsule beaked, villous; calyx-lobes shorter than capsule. Seeds black, stellate-hairy.

Fl. & Fr.: Nov.-Mar. *Distr.*: India, Pakistan, S. China, Malesia &North Australia. Less Common. *NAK 1117, 1120, 1364, 1472* and *1538* (Kakki hills, c. 1200 m). In forest borders among ndergrowth in upper hills.

2. **Abelmoschus moschatus** Medic., Malv. 46. 1787; Hu, Fl. China (Fam. 153) 38. 1955; Borss., Blumea 14:90. 1966; Paul & Nayar, Fl. India Fasc. 19:76. 1988; Nicols., Suresh & Manilal, An Interpr. Hort. Malab. 170. 1988; Mohanan & Sivad., Fl. Agasthyamala 96. 2002. *Hibiscus abelmoschus* L., Sp. Pl. 696. 1753; Mast. in Hook.f., Fl. Brit. India 1:342. 1874; Dunn in Gamble, Fl. Pres. Madras 97. 1915. *Cattu-gasturi* Rheede, Hort. Malab. 2:71-72, t.38. 1679.

Vegetative parts softly tomentose. Leaves broadly ovate to orbicular, palmately 3-7-lobed, 4-18 by 3-20 cm, lobes ovate or obovate-oblong; stipule linear-lanceate. Flower(s) axillary, c.8 cm ϕ; pedicel, c.6 cm long; epicalyx-lobes 6-10, linear, to 15 mm long; corolla yellow with a dark purple centre, petals broadly obovate. Capsule ovoid, beaked, hirsute. Seeds concentrically ribbed, minutely stellate-hairy.

Fl. & Fr.: Aug.-Dec. *Distr.*: India, China, Thailand, Malesia & Fiji Islands. Rare. In exposed rocky hills. *NAK* 2026 (Kokkathode, c. 450 m).

Seeds yield an oil used in perfumery and as a flavouring agent for coffee.

ABUTILON P. Miller
Gard. Dict. Abr. (ed. 4.) 1.1754.

Erect shrubs; branchlets stellate-tomentose. Leaves palmately 7-9-nerved, base cordate, long petioled. Flower(s) solitary or in terminal racemes; pedicel articulate above the middle; epicalyx 0; calyx campanulate, 5-lobed; petals 5, yellow; staminal column shorter than petals; ovary 5-a-locular; styles as many as locules. Schizocarp subglobose; mericarps laterally compressed. Seeds 2-3-per locule.

1. Flower(s) solitary; carpels and mericarps 15-25. ....1. **A. indicum** ssp. **indicum**

1. Flowers in terminal panicles or racemes, rarely solitary; carpels and mericarps 5. ....2. **A. persicum**

1. **Abutilon indicum** (L.) Sweet, Hort. Brit. 1:54. 1826; Don, Gen. Hist. 1:504. 1831; Wight, Ic. t. 12. 1838; Mast. in Hook.f., Fl.Brit. India 1:326. 1874; Dunn in Gamble, Fl. Pres. Madras 91.1915; Ramamoorthy

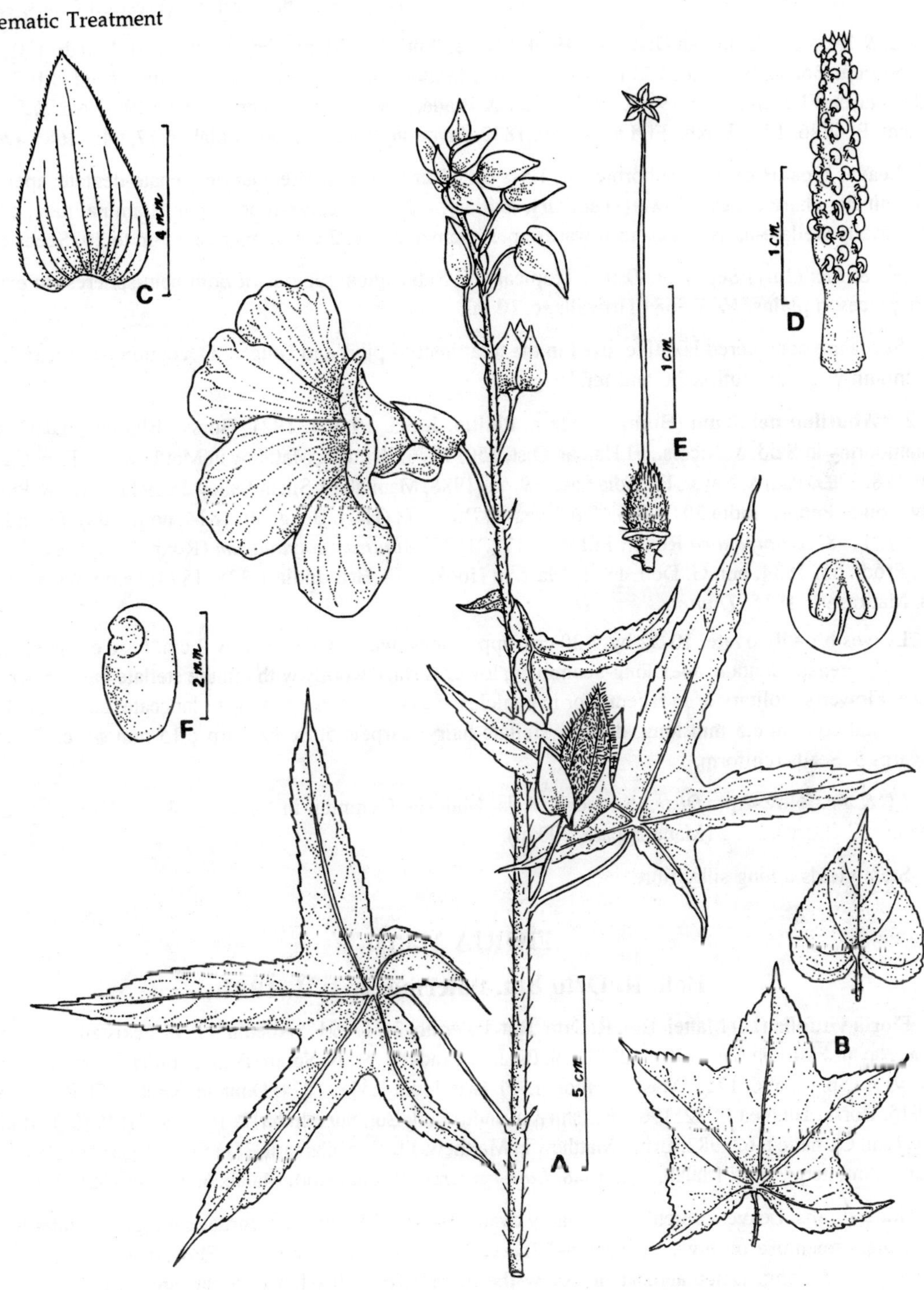

Fig. 14. *Abelmoschus manihot* (L.) Medik. ssp. *tetraphyllus* (Roxb. ex Hornem.) Borss.: A. Twig; B. Leaves - showing variation; C. Sepal; D. Androecium; E. Pistil; F. Seed.

in Sald. & Nicols., Fl. Hassan Dist. 149. 1974; Nicols., Suresh & Mani., An Interpr. Hort. Malab. 170. 1988. ssp. **indicum** Borss., Blumea 14:171. 1966; Britto & Matthew in Matthew, Fl. Tam. Carnatic 3(1) : 107. 1983; Paul & Nayar, Fl. India, Fasc. 19:88. 1988; Sivan. & Pradeep, Malv. South. Penins. India 193.1996. *Sida indica* L.; Cent. Pl. 2:26. 1756 Roxb., Fl. Ind. 3: 179. 1832. *Beloeren* Rheede, Hort. Malab. 6:77, t.45. 1686. *Ooram.*

Leaves broadly ovate to suborbicular, 1.5-8 by 2-7 cm, base; ordate, margin crenate-dentate, apex acute to acuminate, chartaceous. Flower(s) solitary, axillary, c.2 cm ϕ; calyx-lobes ovate; staminal tube, c.7 mm long, basally stellate-hairy, glabrous towards apex. Schizocarp c. 2 cm ϕ; mericarps up to 20. Seeds ovoid.

Fl. & Fr.: (July) Sep.-Apr. *Distr.*: Tropical and Subtropical regions of both hemispheres. In exposed waste places in plains. *NAK* 868 (Tiruvalla, c. 10 m).

Seeds are considered laxative; used in the treatment of piles and coughs. Decoction from bark is used as a mouth-wash in tooth-ache and tender gums.

2. **Abutilon persicum** (Burm. f.) Merr., Phillip. J. Sci. 19:364. 1921; Borss., Blumea 14:163. 1966; Ramamoorthy in Sald. & Nicols., Fl.Hassan Dist. 150. 1976; Britto & Matthew in Matthew, Fl. Tam. Carnatic 3(1) : 108. 1983; Paul & Nayar, Fl. India Fasc. 19: 91. 1988; Manilal, Fl. Silent Valley 25.1988; Sivar. & Pradeep, Malv. South. Penins. India 201.1996. *Sida persica* Burm. f., (Hort. Beng. 50. 1814, nom. nud). Fl. Ind. 148, t.47, f.1. 1768. *S. polyandra* Roxb., Fl.Ind. 3:173. 1832. *Abutilon polyandrum* (Roxb.) Wight ex Wight & Arn., Prodr. 55. 1834, non G. Don, 1831; Mast. in Hook.f., Fl. Brit. India 1:325. 1874; Dunn in Gamble, Fl. Pres. Madras 91. 1915. *Thutthi.*

Leaves broadly ovate, 10-22 by 9-20 cm, upper ones ovate-lanceate, 4 by 2 cm, base cordate, margin irregularly crenate-dentate, apex long-acuminate, lower surface woolly with minute stellate hairs, 5-9 nerved at base. Flower(s) solitary or in racemes or in panicles; calyx cupular, lobes ovate-lanceate; corolla, c.4.5 cm ϕ; staminal column c.5 mm long, densely stellate-hairy; carpels 5. Schizocarp c.15 mm ϕ; c.15 mm ϕ; mericarps 5. Seeds reniform.

Fl. & Fr.: Nov.-Apr. *Distr.*: S. and E. Asia, Malesia. Common in forest outskirts. *NAK 284* (Upper Moozhiar, C.500 m), *2252* (Manarappara, 300 m).

Stem yields a long silky fibre.

FIORIA Mattei

Boll. R. Orto Bot. Palermo. 2, 2:71. 1917.

Fioria vitifolia (L.) Mattei, Bol. R. Orto. Bot. Palermo. 2:71. 1917; Abedin, Fl. W. Pakistan 130:5. 1979; Paul & Nayar, Fl. India Fasc. (19): 109. 1988; Sivar & Pradeep, Malv. South. Penins. India 75. 1996;*Hibiscus vitifolius* L., Sp. Pl. 696. 1753; Mast. in Hook. f., Fl. Brit. India 1:338. 1874; Dunn in Gamble, Fl. Pres. Madras 98. 1915; Borss., Blumea 14:82. 1966; Rakshit & Kundu, Bull. Bot. Surv. India12: 166. 1972 ("1970"); Matthew, Ill. Fl. Tam. Carnatic t. 62. 982; Britto Matthew in Matthew, Fl. Tam. Carnatic 3(1) : 119. 1983; Nicols., Suresh & Mani., An Interpr. Hort. Malab. 172. 1988. *Katu-beloeren* Rheede, Hort. Malab. 6:79, t.46.1686. *Velli-oeren.*

Undershrubs. Leaves orbicular to broadly ovate, 3-10 by 3-9 cm, base cordate, margin crenate-dentate, apex acute, tomentose on both surfaces, 5-7- nerved at base, thin-coriaceous. Flower(s) solitary, axillary; pedicel c. 5 cm long, articulated at or below the middle; epicalyx-lobes linear, spreading; calyx-lobes ovate,persistent; corolla yellow, c.5 cm ϕ, purplish inside at base; staminal tube antheriferous throughout; ovary orbicular, 5-angled, 5-locular; ovules 4-per locule. Capsule suborbicular, 5-winged, beaked, prominently reticulate. Seeds reniform.

Fl. & Fr.: Apr.-Dec. *Distr.*: India, Sri Lanka, Myanmar & Australia. Locally abundant in exposed waste places and along roadsides in plains. *NAK 997* (Tiruvalla, c.10 m). Stem yields a good fibre.

HIBISCUS Linnaeus
Sp. Pl. 693. 1753 (nom. cons.)

Herbs, shrubs or small trees; vegetative parts usually stellate-pubescent. Leaves simple, alternate-spiral, lobed or entire. Flower(s) solitary, axillary or in terminal racemes, hypogynous, regular, bisexual; epicalyx-lobes 4-12, free or connate at base, persistent; corolla showy, petals 5, connate at base with staminal tube; staminal tube dentate at apex, antheriferous throughout or in upper half only, anthers reniform; ovary 5-locular; ovules 3 or more per locule; style distally 5-fid. Capsule 5-valved, loculicidal. Seeds reniform.

1. Herbs or undershrubs; leaves palmately 3-7- lobed.
   2. Stem stellate-scarbrous; epicalyx-lobes simple.
      3. Calyx-lobes c. 3-nerved; pedicel 5-10 mm long. ....2. **H. lunariifolius**
      3. Calyx-lobes c. 7-nerved; pedicel 5-8 cm long. ....3. **H. sreenarayanianus**
   2. Stem prickly; epicalyx-lobes appendaged.
      4. Stipules lanceate, linear. ....1. **H . hispidissimus**
      4. Stipules ovate, auriculate, foliaceous. ....4. **H. surattensis**
1. Trees; leaves entire, orbicular. ....5. **H. tiliaceus**

1. **Hibiscus hispidissimus** Griffith, Not.Pl. Asiat. 4:521. 1854; Pradeep & Sivar., Taxon 40:637. 1991; Sivar. & Pradeep, Malv. South. Penins. India 106.1996. *Hibiscus aculeatus* Roxb., Fl. Ind. (ed. Carey) 3:206. 1832; Paul & Nayar, Bull. Bot. Surv. India 22(1-4) : 194. 1982 & in Fl. India Fasc. 19:123.1988; Manilal, Fl. Silent Valley 26. 1988. *Hibiscus furcatus* Roxb. (Hort. Beng. 51.1814, nom. nud.) ex DC., Prodr. 1:449. 1824, non Willd., 1809; Roxb., Fl. Ind. (ed. Carey) 3:204. 1832; Mast. in Hook. f., Fl. Brit. India 1:335. 1874; Dunn in Gamble; Fl. Pres. Madras 97.1915; Rakshit & Kundu, Bull. Bot. Surv.India 12:161. 1972("1970"); Ramamoorthy in Sald. & Nicols., Fl. Hassan Dist. 151. 1976; Britto & Matthew in Matthew, Fl. Tam. Carnatic 3(1): 113.1983.

Rambling undershrubs, stem, petiole, leaf-veins on below and pedicel clothed with small prickles; stipules linear-lanceate. Leaves usually 3-5- lobed or entire towards apex of branch, 5-8 by 4-8 cm, base cuneate, margin crenate-serrate, apex acute. Flower(s) axillary; pedicel c.5 cm long; epicalyx-lobes 8-12, apex divided into 2 leafy portions; calyx-lobes ovate-lanceate; corolla greenish-yellow with a purplish centre, 6-8 cm f. Capsule ovoid, enclosed in enlarged calyx. Seeds reniform.

Fl. & Fr.: Sep.-Feb. *Distr.*: India, Sri Lanka, Myanmar & South Africa. Very Common in forest clearings, teak plantations, etc. *NAK 1489* and *2665* (Punnamada, Manniora R.F., 450 m)

Juice of leaves mixed with honey and applied in eye diseases. Decoction of root-bark is said to cure internal poison and swellings in the kidneys. Bark yield good strong fibre.

2. **Hibiscus lunariifolius** Willd., Sp. Pl. 3:811. 1800; Mast. in Hook. f., Fl. Brit. India 1 : 338. 1874; Dunn in Gamble, Fl. Pres. Madras 98. 1915; Rakshit & Kundu, Bull. Bot. Surv. India 12:173. 1972 ("1970"); Britto & Matthew in Matthew, Fl. Tam. Carnatic 3(1): 114. 1983; Paul & Nayar, Fl. India Fasc. 19:135. 1988; Manilal, Fl. Silent Valley 27.1988; Sivar & Pradeep, Malv. South. Penins. India 138. 1996; *Malankoova-paruthi.*

Undershrubs; whole parts covered with yellow stellate-indumentum mixed with simple hairs. Leaves 3-5- lobed to entire, 4-12 by 2.5-10 cm, base cordate to rounded, margin crenate-serrate, apex acuminate, chartaceous. Flower(s) solitary, axillary or few in terminal racemes; epicalyx-lobes 5, linear-lanceate, calyx-lobes prominently 3-nerved; staminal column c.2 cm long, antheriferous throughout; ovary obovoid. Capsule globose. Seeds α, reniform.

Fl. & Fr.: Oct.-Jan. *Distr.*: India, Trop. Africa, Malesia. Less common. In forest outskirts of upper hills. *NAK 285, 1208* (Kakki hills, c. 1000 m).

3. **Hibiscus sreenarayanianus** Anil & Ravi, Rheedea 4:129. 1994; Sivar. & Pradeep, Malv. South. Penins. India 144.1996. **(Fig. 15)**

Annual herbs, up to 3 m tall; stem terete, tomentose with (a) thin, long, simple hairs (up to 5 mm long), (b) straight, stiff, simple, bulbous based and stellate bristly hairs shorter than the former, (c) short, simple, thin hairs and (d) short or long-stalked capitate glandular hairs, all intermixed. Leaves simple, alternate, stipulate; stipules linear, c. 5 mm long; deciduous; petiole slender, terete, hairy as the stem, c. 12 cm long; lamina broadly ovate, 8-12 by 7-15 cm, base deeply cordate, lobes overlapping, margin dentate to dentate-serrate, apex acute to acuminate, palmately 5-lobed towards the base, progressively unlobed and narrower towards branch-apices, glandular-punctate, puberulous on both surfaces, prominently so on the lower surface, sparsely with large, simple and stellate hairs, all over on the lower surface, and on the major veins on the upper surface, broader than long, chartaceous. Flower(s) solitary, axillary, often in terminal racemes through decrescens of the upper leaves; pedicels slender, terete, hairy as the stem, c. 5 cm long, accrescent to 8 cm long, articulate; epicalyx-lobes 8-10, linear-subulate, c.8 mm long, hairy, more or less as the stem, persistent; calyx gamosepalous, puberulous on the outside; densely intermixed with longer and thinner simple and stiffer stellate hairs, the former more abundant towards the margins of the lobes, sericeous inside, c.1.5 cm long, accrescent to 2 cm, 1.8 cm f, lobes triangular, acute, 7-nerved, 8 mm long, accrescent to 1.5 cm by 7 mm at the base; corolla yellow with a reddish-brown centre; petals broadly obtuse at apex, obliquely cuneate towards base, 5 by 3.5 cm, very sparsely shortly hairy on both surfaces; staminal column antheriferous throughout, 2 cm long, triangular, obtuse; staminal column and filaments glandular with stipitate, elliptic yellow glands; ovary orbicular-ovoid, adpressed-pubescent with longer hairs towards the apex; chambers 5; ovules α, in two vertical rows; style 1-1.5 cm long; stigmatic branches 5, short, 2 mm long, purplish; stigmas discoid, purplish, hairy. Capsules enclosed by a shortly exceeding calyx, orbicular-ovoid, ± abruptly acute, adpressed pubescent with 5 thick longitudinal lines of hirsute hairs along the back of the loculi, glabrous inside. Seeds α, brownish black, compressed, obliquely triangular-reniform,verruculose with reticulately pitted surface.

Fl. & Fr. : Sept.-Dec. *Distr.*: Plains of Kerala (Kollam and Pathanamthitta). Heliophilous. Along wayside bushes and waste places. NAK 3120 (Ezhammile, Adoor, c.20m)

Note: Paul and Nayar (1988) have recognised 9 sections under the genus *Hibiscus* to include the Indian species. But *H. sreenarayanianus* could not satisfactorily be included under any of those sections owing to its some unique characters. One of the sections namely, *H.* sec. *Ketmia* is characterised by having 5-loculed capsules, entire persistent epicalyx segments, 3-nerved calyx lobes, shorter staminal column, and tubercled seeds. The present species agrees in having all the characters except in the character of the calyx lobes which is 7-nerved. There are 3 species under *Hibiscus* sect. *Ketmia viz., H. caesius, H. obtusifolius* and *H. lunariifolius. H. sreenarayanianus* very much resembles with *H. lunariifolius,* but differs in having strikingly different indumentum, articulated long pedicels, 7-nerved calyx lobes, and long and simple haired capsules.

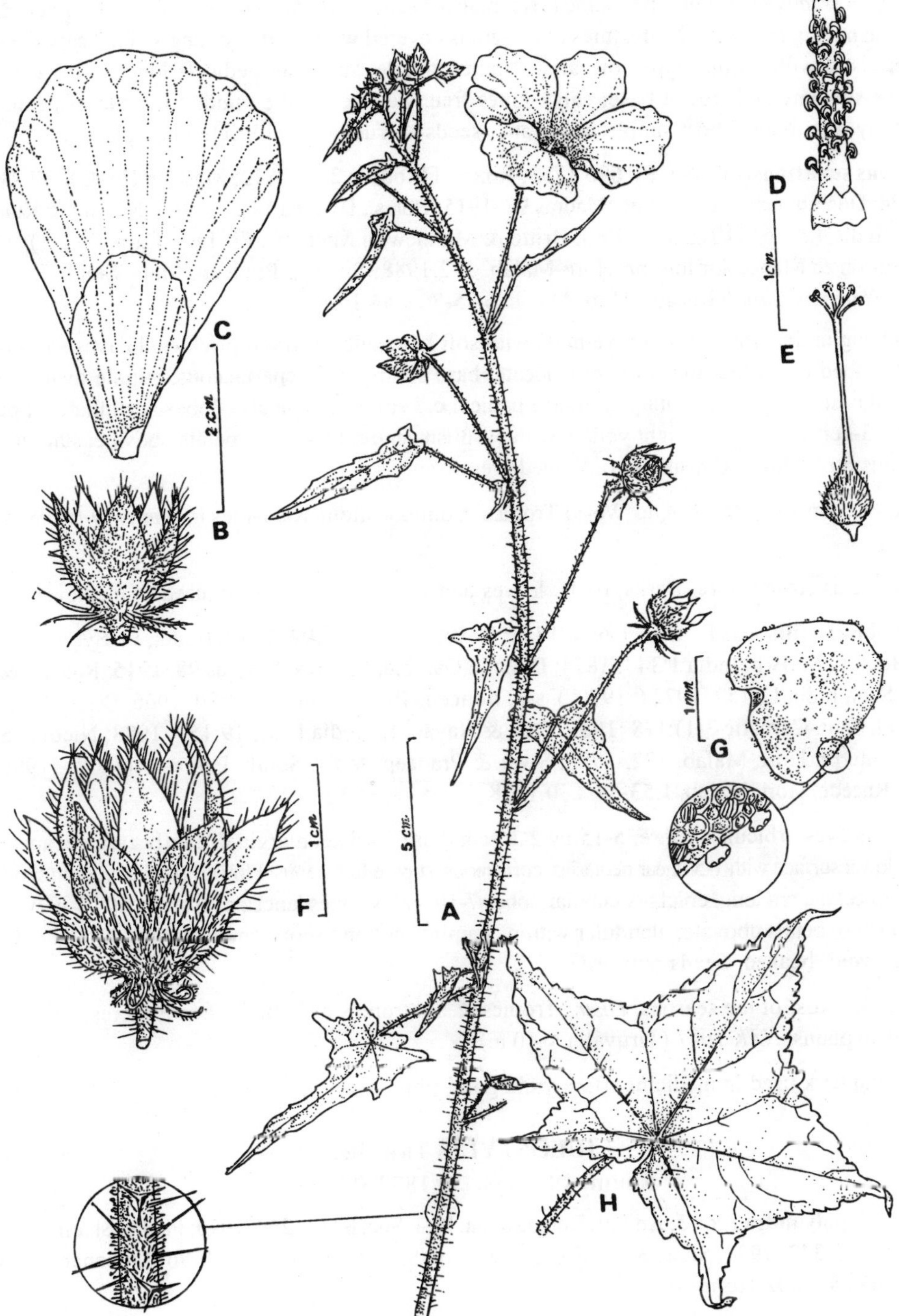

Fig. 15. ***Hibiscus sreenarayanianus*** **Anil & Ravi: A. Twig; B. Calyx and Expicalyx; C. Petal; D. Staminal tube; E. Pistil; F. Capsule; G. Seed; H. Lower leaf.**

Hibiscus sreenarayanianus has some resemblance with *H. panduraeformis* of *H.* sect. *Trichospermum*, but differs in having the following features : the stem is covered with short and long stalked capitate glandular hairs intermixed with 3 other types of hairs; stipules simple, subulate; pedicels long, to c.5 cm; epicalyx lobes linear-subulate and free at base; calyx lobes free, well below the middle; outside of petals covered with sparsely distributed short simple hairs; and, seeds verruculose with reticulate pittings.

4. **Hibiscus surattensis** L., Sp. Pl. 696.1753; Roxb., Fl.Ind. 3: 205. 1832; Mast. in Hook.f., Fl. Brit. India 1:334.1874; Dunn in Gamble, Fl. Pres. Madras 97. 1915; Borss., Blumea 14:57. 1966; Rakshit & Kundu, Bull. Bot. Surv. India 12: 161. 1972 ("1970"); Britto & Matthew in Matthew, Fl. Tam. Carnatic 3(1): 118.1983; Nicols., Suresh & Mani., An Interpr. Hort. Malab. 172.1988; Sivar & Pradeep, Malv. South. Penins. India 119. 1996; *Narinam- poulli* Rheede, Hort. Malab. 6:75-76, t.44.1686.

Rambling undershrubs; vegetative parts with soft hairs and recurved prickles. Leaves suborbicular to ovate, 3-5- lobed or partite, lobes linear-lanceate, base subtruncate, chartaceous; stipule ovate, foiaceous, auriculate at base. Flower(s) solitary, axillary; pedicel c.7 cm long; epicalyx-lobes spathulate, appendaged; calyx-lobes 3-nerved; corolla bright yellow with purplish centre, to 6 cm ϕ; petals obovate; staminal column antheriferous throughout. Capsule ovoid. Seeds reniform.

Fl.& Fr.: Jun.-Oct. *Distr.*: Old World Tropics. Common along roadsides in plains. *NAK 2825* (Adoor, c. 20 m).

Stem yields strong fibre. Barks, roots, leaves and flowers are variously used in medicine.

5. **Hibiscus tiliaceus** L., Sp. Pl. 694. 1753; Roxb., Fl. Ind. 3:192. 1832; Bedd., Fl. Sylv. t. 4, f.1.1869; Mast. in Hook.f., Fl. Brit. India 1:343. 1874; Dunn in Gamble, Fl. Pres. Madras 98. 1915; Rakshit & Kundu, Bull. Bot. Surv. India 12:157. 1972 ("1970"). ssp. **tiliaceus** Borss., Blumea 14:30. 1966; Britto & Matthew in Matthew, Fl. Tam. Carnatic 3(1):118. 1983; Paul & Nayar., Fl. India Fasc. 19:157. 1988; Nicols., Suresh & Mani., An Interpr. Hort. Malab. 172. 1988; Sivar & Pradeep, Malv. South. Penins. India 91. 1996; *Pariti, Tali-pariti* Rheede, Hort. Malab. 1:53-54, t.30.1678.

Trees. Leaves orbicular, entire, 5-15 by 2-15 cm, base cordate, margin distantly crenate, apex acute to acumiante, lower surface with 1-5 linear nectaries, coriaceous; stipule foliaceous. Flower(s) solitary, or in short terminal racemes; pedicel c.2 cm long, epiclayx cupular, lobes 7-10; calyx-lobes lanceate, prominently 3-nerved; corolla yellow, c.5 cm ϕ, petals obovate, glandular within; staminal column short, antheriferous throughout. Capsule globose or ovoid, beaked. Seeds reniform.

Fl. & Fr.: Most of the seasons. *Distr.*: Tropics & Subtropics of both the hemispheres. Common along streamsides in plains. *NAK 1041* (Tiruvalla, c.10 m).

Root and bark used in medicine. Bark yield good fibre.

JULOSTYLIS Thwaites
Enum. Pl. Zeyl. 30.1858 ('1854').

Julostylis polyandra Ravi & Anil., J. Bombay Nat. Hist. Soc. 87.(2):260. 1990; Paul in Sharma & Sanjappa (eds.), Fl. India 3:342. 1993; Sivar. & Pradeep, Malv. South. Penin. India 152. 1996; Mohanan & Sivad., Fl. Agasthyamala 98. 2002.*Vazhukkal.*

Trees. Leaves alternate-spiral; suborbicular, broadly ovate or obovate, 4-21 by 2-18 cm, palmately 5-angled or shallowly 5-lobed towards the upper half, base rounded or cuneate, margin subentire, apex acuminate or acute, both surfaces sparsely stellate pubescent, glabrescent, primary nerves 5, palmately from the base with nectaries on inner two lateral ribs at lower surface, chartaceous. Flowers in large terminal and axillary panicles, c. 2 cm f; epicalyx-lobes 4-6, connate at base, accrescent; calyx-lobes 5, connate to the middle, persistent; corolla yellow with a purple centre, petals 5, narrowly obovate, alternate with small clavate outgrowths at base; stamens 17-20, staminal column short, 5- partite at apex, antheriferous around the middle; ovary 2-locular, ovules 2-per locule; style divided above the middle; stigma thick-peltate. Capsule globose. Seeds reniform.

Fl. & Fr.: Nov-Mar. *Distr.*: Hitherto known from South-West Travancore hills (Kerala). In moist evergreen forest from 200 to 1100 m. *NAK 70, 364* (Kakki hills, c. 1000 m), *1520* (Gurunadhan mannu 400 m); *AGP* 79289 (MH), (Kakki dam area, c. 1050 m), *BKN et.al.*, K 10527 (CALI) (Kallippara, Sabarigiri, c. 1000 m). Bark yields strong fibre.

Note: The genus *Julostylis* described by Thwaites was based on a single species, *J. angustifolia* from Sri Lanka, characterised by lanceate, entire leaves with reticulate venation and having only 10 stamens. The present species will broaden the range of variability of the genus.

MALVASTRUM A. Gray
Mem. Amer. Acad. n.s. 4:21.1849 (nom. cons.).

Malvastrum coromandelianum (L.) Garcke, Bonplandia 5:295. 1857; Dunn in Gamble, Fl. Pres. Madras 88.1915; Bross., Blumea 14:152. 1966; Ramamoorthy in Sald. & Nicols., Fl. Hassan Dist. 153. 1976; Matthew, Ill. Fl. Tam. Carnatic t.65. 1982; Britto & Matthew in Matthew, Fl. Tam. Carnatic 3(1):121. 1983; Paul & Nayar, Fl. India Fasc. 19: 180. 1988; Sivar. & Pradeep, Malv. South. Penins. India 219. 1996. *Malva coromandaliana* L., Sp. Pl. 687. 1753. *M. tricuspidata* R. Br. in Aiton f., Hort. Kew. 4:210. 1812. *Malvastrum tricuspidatum* (R.Br.) A. Gray, Pl. Wright 1:16. 1852; Mast. in Hook.f., Fl. Brit. India 1:321. 1874.

Woody herbs to undershrubs; branchlets appressed stellate-tomentose. Leaves lanceate to oblong, 4-8 by 2-3.5 cm, base acute, margin dentate-serrate, apex acute, 5-nerved at base, both surfaces pubescent, chartaceous. Flower(s) solitary or few in clusters, axillary; epicalyx segment 3, linear; calyx-lobes 5, ovate, connate below; corolla c.15 mm ϕ, yellow, petals 5, obliquely obovate; staminal column antheriferous at apex; ovary 13-15 locular, ovule 1-per locule. Schizocarp globular; mericarps c. 15, awned, awn apically hooked. Seeds reniform.

Fl. & Fr.: Mar.-Sep. *Distr.*: Pantropical. In exposed sunny waste areas. Not Common. *NAK 1963* (Tiruvalla, c.10 m).

SIDA Linnaeus
Sp. Pl. 683. 1753.

Prostrate, decumbent herbs to erect shrubs or subshrubs. Leaves simple, spiral, short-petiolate. Lamina very variable; ovate-cordate, lanceate to ovate, rhombic or elliptic, margin basally entire to crenate-serrate throughout. Flower(s) mostly solitary or in diffuse panicles, axillary, sometimes aggregated apically; calyx 10-ribbed; corolla yellow. Schizocarp exceeding the calyx or not; mericarps 5-12, trigonous, muticous or bi-awned at apex, laterally reticulate or not. Seeds turgid, glabrous or hairy at apex.

It is the most difficult genus of family with highly variable constituent species.

1. Erect woody herbs or subshrubs.
    2. Mericarps 5; flowers initially solitary, finally in condensed racemes or in panicles. .... 6. **S. mysorensis**
    2. Mericarps 6-12; flowers usually solitary, rarely in terminal clusters.
        3. Leaves ovate, oblong, elliptic-oblong or obovate-orbicular, margin dentate throughout.
            4. Leaves ovate to elliptic-oblong, rarely orbicular, base cordate, apex mostly acute.
                5. Mericarp-awn c.4 mm long, retrorsely hairy. .... 4. **S. cordifolia**
                5. Mericarp-awn c. 1 mm long, glabrous or sparsely hairy (never retrose). .... 7. **Sida ravii**
            4. Leaves obovate-orbicular to ovate or elliptic, base rounded or acute, apex mostly obtuse or retuse. .... 5. **S. fryxellii**
        3. Leaves elliptic, rhombic to linear, (obovate in *S. rhombifolia* ssp. *retusa*), margin distally dentate and entire towards base.
            6. Stipules of each pair dissimilar, usually falcate or linear, a-veined; leaves and branching pattern distichous. .... 1. **S. acuta**
            6. Stipules of each pair similar, linear, 1-(3)-veined; leaves spirally disposed. .... 8. **S. rhombifolia**
1. Prostrate or decumbent herbs.
    7. Prostrate slender herbs, rooting at lower nodes; accrescent pedicel to 6 cm long, articulate far above the middle; mericarps awnless. .... 2. **S. beddomei**
    7. Decumbent to sub-erect stout herbs, never or occasionally rooting at nodes; accrescent pedicel to 3 cm long, articulate at or near the middle; mericarp minutely awned or apiculate. .... 3. **S. cordata**

1. **Sida acuta** Burm. f., Fl. Ind. 147. 1768, emend. K. Schum. in C. Martius, Fl. Bras. 12(3):326. 1891; Dunn in Gamble, Fl. Pres. Madras 90. 1915; Borss., Blumea 14:186. 1966; Ramamoorthy in Sald. & Nicols., Fl.Hassan Dist. 154. 1976; Matthew, Ill. Fl. Tam. Carnatic t. 68.1982; Britto & Matthew in Matthew, Fl. Tam. Carnatic 3(1) : 127. 1983; Paul & Nayar, Fl. India Fasc. 19:202.1988; Mohanan & Sivad., Fl. Agasthyamala 99. 2002. Manilal, Fl. Silent Valley 28. 1988; Nicols., Suresh & Mani., An Interpr. Hort. Malab. 173. 1988;Sivar & Pradeep, Malv. South. Penins. India 152. 1996. *Tsjeru-parua* Rheede, Hort. Malab. 10:105, t.53. 1690.

Subshrubs. Leaves lanceate, lanceolate to linear, 4-9 by 0.6-3.5 cm, base obtuse to acute, margin dentate-serrate, apex acute, glabrescent, chartaceous; stipules of each pair dissimilar. Flower(s) solitary, axillary or in clusters, corolla, c.10 mm ϕ; staminal column short, antheriferous towards the apex. Mericarps 6-10, awns to 1.5 mm long. Seeds obtusely 3-gonous.

Fl. & Fr.: Most of the seasons. *Distr.*: Pantropical. A gregarious weed on waysides, wastelands from plains to hills. *NAK* (Kakki hills, c. 1100 m).

Juice of leaves boiled in oil and applied in elephantiasis and swellings of the testis.

2. **Sida beddomei** Jacob, J. Bombay Nat. Hist. Soc. 47:51. 1950; Karthik. *et al.,* Bull. Bot. Surv. India 22 : 235. 1982 ("1980"); Sivar & Pradeep, Malv. South. Penins. India 266. 1996. **(Fig. 16)**

Prostrate herbs, rooting at lower nodes. Leaves broadly ovate to suborbicular, 2-5 by 2-4.5 cm, base cordate, margin crenate-serrate, apex acute to shortly acuminate, basally 5-7- nerved, sparsely pubescent, chartaceous. Flower(s) solitary, axillary; pedicel 3 to 6 cm long, slender, articulate near apex; calyx-lobes short, tube 8-9 mm long; corolla c.10 mm ϕ; staminal column c.2 mm long; mericarps 5, each with a pubescent muticous process. Seeds ovoid, glabrous except the hilum.

Fl. &Fr.: Jan.-Mar. *Distr.*: Endemic to Southern-Western Ghats. Commonly found among moist forest undergrowth in cleared areas. *NAK 1481* (Kodumudi-Chittar, 200 m), *1572* (Velithode-Moozhiar, c.300 m).

Note : In the recent revision of the family Malvaceae of India by Paul and Nayar (1988) *Sida beddomei* has been treated as conspecific with *S. cordata.* But as noted by Sivarajan and Pradeep (1996) detailed studies clearly reveal that the former is quite distinct from the latter especially in the mericarp characters.

3. **Sida cordata** (Burm.f.) Borss., Blumea 14:182. 1966; Ramamoorthy in Sald. & Nicols., Fl. Hassan Dist. 155. 1976; Britto & Matthew in Matthew, Fl. Tam. Carnatic 3(1):128. 1983; Nicols., Suresh & Mani., An Interpr. Hort. Malab 173. 1988; Sivar. & Pradeep, Malv. South. Penin. India 266.1996. *Melochia cordata* Burm. f., Fl. Ind. 143. 1768. *Sida veronicifolia* Lam., Encycl. 15. 1783 ("veronicaefolia"); Dunn in Gamble, Fl. Pres. Madras 89. 1915. *Sida radicans* Cav., Diss. 1 : 9, t. 10, f.2.1785. *S. humilis* Cav., Diss. 5:277, t. 134, f.2. 1788, var. *veronicifolia* (Lamk.) Mast. in Hook.f., Fl. Brit. India 1: 322. 1874. *Nela-vaga* Rheede, Hort. Malab. 10: 137, t.69. 1690.

Diffuse or ascending stout herbs, very rarely rooting at lower nodes. Leaves orbicular or broadly ovate, base cordate, 3-6 by 2.5-5 cm, apex acuminate or acute, basally 5-7- nerved, both surfaces pubescent, chartaceous. Flower(s) solitary, axillary; pedicel to 3-4 cm long; corolla 8-9 mm ϕ; staminal column c.2 mm long. Mericarps 5, shortly unequally awned, rarely awnless. Seeds ovoid.

Fl. & Fr.: Jan.-Apr. *Distr.*: Pantropical. In shady areas along waysides in plains. *NAK 2245* (Adoor, c.20 m).

Bruised leaves applied to cure cuts and given for diarrhoea in pregnancy in women.

4. **Sida cordifolia** L., Sp. Pl. 684. 1753; Roxb., Fl. Ind. 3:177. 1832; Mast. in Hook.f., Fl. Brit. India 1:324. 1874; Dunn in Gamble, Fl. Pres. Madras 89. 1915; Borss., Blumea 14:199. 1966; Ramamoorthy in Sald. & Nicols., Fl. Hassan Dist. 155. 1976; Britto & Matthew in Matthew, Fl. Tam. Carnatic 128. 1983; Paul & Nayar, Fl. India Fas. 19: 207. 1988; Nicols., Suresh & Mani., An Interpr. Hort. Malab. 173. 1988; Sivar & Pradeep, Malv. South. Penins. India 256. 1996 *Katu-uren* Rheede, Hort. Malab. 10: 107, t. 54. 690. *Anakurunthoti.*

Woody herbs. Leaves ovate to elliptic-oblong, rarely orbicular, 3-6 by 2-3.5 cm, base cordate, margin crenate-serrate, apex acute or obtuse, basally 5-7- nerved, chartaceous. Flower(s) usually solitary, sometimes few in clusters towards apex; calyx lobes triangular; corolla c.1.5 m ϕ; staminal column c.3 mm long. Mericarp 8-10, awns to 4 mm long, retrorsely hairy. Seeds reniform, except the hilum.

Fl. & Fr.: Most of the seasons. *Distr.*: Pantropical. Less Common. On exposed lands and along wasides in plains. *NAK 2308* (Adoor, c. 20 m).

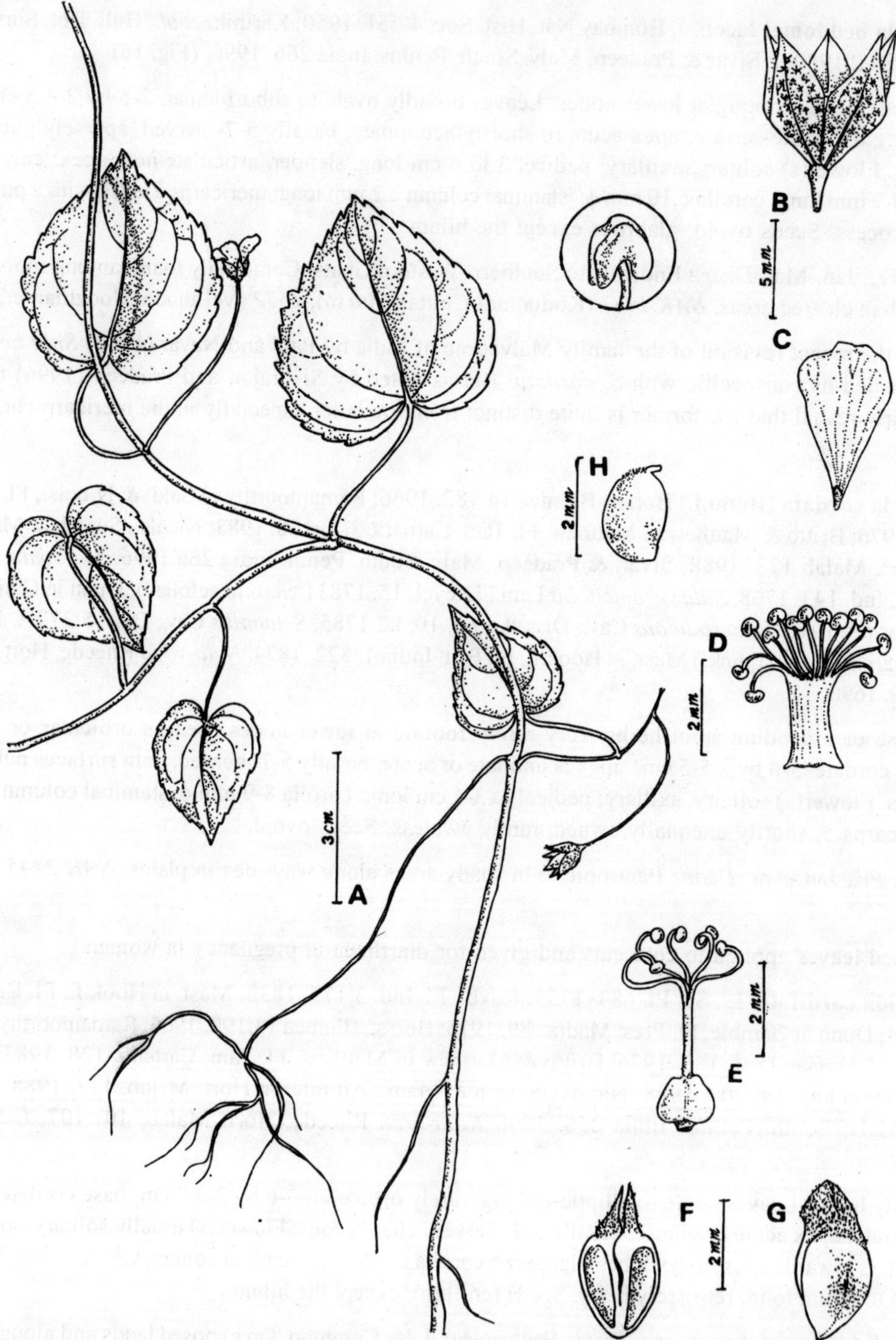

Fig. 16. ***Sida beddomei*** **Jacob: A, Twig; B. Calyx; C. Petal; D. Staminal column; E. Pistil; F & G. Mericarp- different views; H. Seed**

Infusion of roots is coolant, astringent and tonic, given in nervous and urinary diseases.

5. **Sida fryxellii** Sivar. & Pradeep, Kew Bull. 45(4): 725-727. 1990, Malv. South Penins. India 259. 1996.

Erect undershrubs, densely soft stellate-pubescent. Leaves dimorphic, early ones obovate-orbicular, later ones ovate or elliptic, 3-5 by 3-7 cm, base rounded or acute, margin crenate-serrate, apex mostly retuse or obtuse to acute, lateral veins, c.6 pairs, prominent below, thin-coriaceous. Flower(s) solitary, axillary, occasionally few in terminal racemes; calyx 5-8 mm ϕ, lobes triangular; corolla to 1.5 cm ϕ; staminal column to 3 mm long, stellate-hairy. Mericarps 8-10, trigonous, awns to 2 mm long, retrorsely hairy. Seeds sub-reniform, glabrous except the hilum.

Fl. & Fr.: Nov.-Mar. *Distr.*: South-West Peninsular India. A gregarious subshrub seen large populations on waysides both in plains and hills. *NAK 969* (Sabarimala hills, c. 250 m).

Note : Possibly this is a descendant form of *Sida cordifolia* L.

6. **Sida mysorensis** Wight & Arn., Prodr. 59. 1834; Mast. in Hook.f., Fl. Brit. India 1:322. 1874; Borss., Blumea 14:180. 1966; Paul & Nayar, Fl. India Fasc. 19:209. 1988. *S. glutinosa* Roxb., Hort. Beng. 97.1814 (nom.nud.). Roxb., Fl. Ind.(Carry *ed.)* 3:172.1832, non Cav., 1785; Dunn in Gamble, Fl. Pres. Madras 89.1915.

Erect viscid undershrubs; vegetative parts densely pubescent with minute stellate, simple and glandular hairs. Levaes broadly ovate, 3-8 by 2.5-6 cm, base cordate, margin crenate-serrate, apex acuminate or acute, basally 5-7- nerved, chartaceous. Flower(s) solitary, ultimately in condensed racemes or panicles, axillary; pedicel to 1 cm long, articulated near apex; calyx, c.5 mm ϕ, lobes triangular; corolla, c.1 mm ϕ. Mericarps without awns and reticulations. Seeds ovoid.

Fl. & Fr.: Feb.-Apr. *Distr.*: India, Sri Lanka, Pakistan, Myanmar, China & Malesia. Rare. In disturbed or deciduous forests. *NAK 2256* (Mannarappara R.F., c.250 m).

7. **Sida ravii** Sivad. & Anil., Willdenowia 25: 651.1996 **(Fig. 17)**

Erect, profusely branched undershrubs, c. 1.5 m tall; branchlets subterete; vegetative parts, pedicels and outer side of calyx with soft minute stellate indumentum; stipules 2, isomorphic, linear, 5-7 mm long; petiole cylindirc, 5-15 mm long, slightly thickened below the lamina; lamina ovate-lanceate, oblong-elliptic or lanceate, 2-7 cm by 1-4 cm, base more or less cordate or rarely rounded, margin crenate serrate up to base, apex acute, sparsely stellate pubescent a long the midvein on the upper surface and all over the lower surface of the lamina, glabrescent, lateral veins 6-8 pairs, basally 5-nerved, chartaceous. Flower(s) solitary or very few (3-4) on aging in racemes, axillary, 15-20 mm ϕ; pedicels erect, cylindirc, bent upwards, 1-3 cm long, articulated near calyx-base; calyx widely campanulate, 8-10 mm ϕ, 6-8 mm long, lobes 5, deltoid, 5 by 5 mm, apex acuminate, faintly 10-veined, glabrous inside; corolla yellow, 15-20 mm f, petals 5, obliquely obovate, 10-14 by 7-12 mm, apex shallowly emarginate, soft- simple hairy without except the half covered protion; staminal tube cylindric, 3-4 mm long, filaments c. 1.5 mm long, both simple glandular hairy; ovary broadly ovoid; styles 8-10, scarlet coloured, glabrous; stigma capitate. Schizocarp c. 7 mm ϕ; mericarp 8-10 (12), radially compressed, 3 by 2 mm, outer surface reticulately ribbed, glabrous, black, apex divergent, mucronate on a distinct shoulder, glabrous or rarely sparsely stellate-hairy (never retrorse). Seeds sub-reniform, 2 by 1.5 mm, glabrous except the hilum.

Fl. & Fr.: Jul. - Nov. *Distr.*: Hitherto known only from the type locality and Adoor, Pathanamthitta District. Heliophilous. Occuring along open waysides in small populations. *NAK* 2831 (Adoor, c. 20m).

Note: The mericarps of *S. rhombifolia* L. var. *scabrida* (Wt. & Arn.) Mast. of *Sida* sect. *Sidae* look more or less similar to those of *S. ravii,* but the former has mostly rhombic leaves with base obtuse or cuneate, margin entire in basal third (or more), different indumentum and also the acroscopic mucro of the mericarps which are without distinct shoulder.

8. **Sida rhombifolia** L., Sp. Pl. 684. 1753; Mast. in Hook.f., Fl.Brit. India 1:323.1874; Borss., Blumea 14:193.1966; Sivar & Pradeep, Malv. South. Penins. India 230. 1996 Mohanan & Sivad., Fl. Agasthyamala 100.2002.

Woody herbs or undershrubs; branchlets, leaves, petiole and pedicels minutely stellate-pubescent, glabrescent. Leaves more or less rhombic to lanceate or elliptic oblong to obovate, margin basally entire, distally crenate-serrate, apex acute, obtuse or retuse, chartaceous. Flower(s) solitary, axillary; corolla 10-15 mm ϕ; staminal column c. 3 mm long, glabrescent. Mericarps 6-13, laterally compressed, reticulate, apically 2-awned, awns never retrorsely hairy. Seeds reniform.

Key to the subspecies

1. Erect undershrubs or woody herbs; leaves ovate to oblong; rhombic or lanceate to oblong, pedicel to 4 cm long; corolla, c.17 mm ϕ. ....**Sida rhombifolia** ssp.**rhombifolia,** var. **scabrida**

1. Diffuse or ascending woody herbs. Leaves obovate to orbicular; pedicel to 3 cm long, corolla 20-25 mm ϕ. ....**Sida rhombifolia** ssp. **retusa**

1. **Sida rhombifolia** L. ssp. **rhombifolia var. scabrida** (Wight & Arn.) Mast. in Hook.f., Fl. Brit. India 1:324. 1874. *S. scabrida* Wight & Arn., Prodr. 57.1834.

Subshrubs, vegetative parts with rigid simple stellate hairs. Leaves rhombic-lanceate Mericarps 9-12, apex bicuspidate.

Fl. & Fr.: Most of the seasons. *Distr.*: Peninsular India. This is the very common variety found gregariously growing in forest clearings in upper hills. *NAK 1157, 1209, 1492, 1513* (Kakki hills, c.1100 m).

2. **Sida rhombifolia** L. ssp. **retusa** (L.) Borss., Blumea 14:198. 1966. *S. retusa* L., Sp. Pl. ed. 2,961.1763; Roxb., Fl. Ind. 3:175. 1832; Wight & Arn., Prodr. 58. 1834. *S.rhombifolia,* var. *retusa* (L.) Mast. in Hook.f., Fl. Brit. India 1:324. 1874; Nicols., Suresh & Mani., An Interpr. Hort. Malab.173.1988. *Kurundoti* Rheede, Hort. Malab. 10:35, t.18. 1690.

Diffuse to ascending subshrubs. Leaves obovate, apex obtuse or retuse. Mericarps 2-awned.

Fl. & Fr.: Oct.-Dec. *Distr.*: Tropical. Rare. In exposed sandy areas. This is one of the commonest plants used in Ayurvedic medicine. *NAK 305* (Adoor, c. 50 m).

Note : Now this species is becoming rare because of the over exploitation for the medicinal uses.

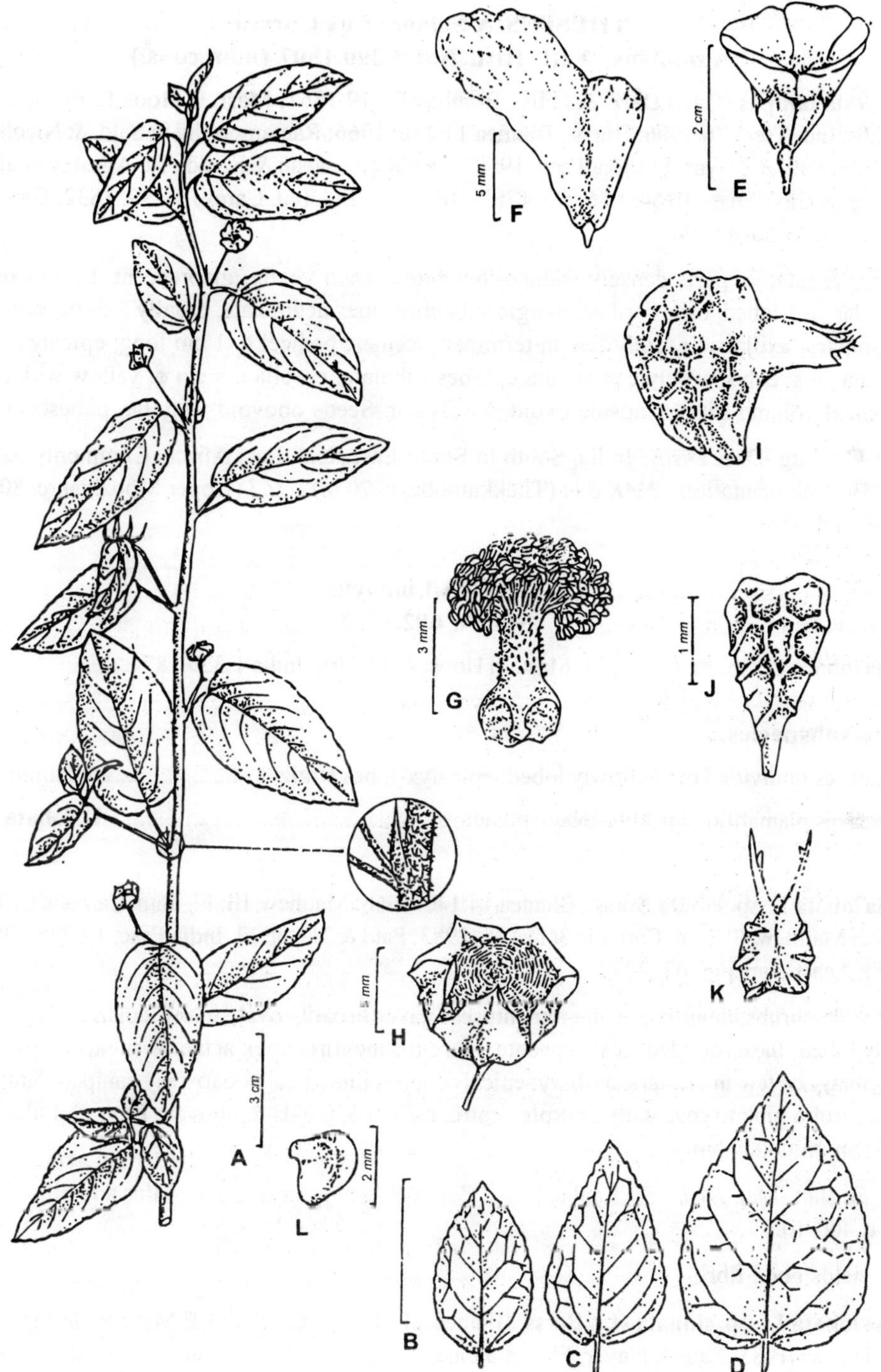

Fig. 17. *Sida ravii* **Sivad.& Anil. A. Twig; B, C. & D. Various forms of leaf; E. Flower; F. Petal; G. Staminal column; H. Schizocarp; I, J & K. Mericarp - side view, back view and top view respectively; L. Seed.**

THESPESIA Solander ex Correa
Ann. Mus. Natl. Hist. Nat.9:290.1807 (nom.cons.)

Thespesia lampas (Cav.) Dalz. & Gibs., Bombay Fl. 19.1861; Mast. in Hook.f., Fl. Brit. India 1:345. 1874; Brandis, Ind. Trees 76.1906; Borss., Blumea 14:116. 1966; Ramamoorthy in Sald. & Nicols., Fl. Hassan Dist. 156.1976; Paul & Nayar, Fl.India Fasc. 19: 222.1988; Mohanan & Sivad., Fl. Agasthyamala 101. 2002. *Hibiscus lampas* Cav., Diss. 3:154, t.56, f.2. 1787; Roxb., Fl. Ind. (ed. Carey) 3:197. 1832; Gamble, Fl. Pres. Madras 98.1915. *Kattu-paruthi.*

Shrubs; vegetative parts densely stellate-tomentose when young, glabrescent. Leaves ovate, deltoid or suborbicular, 3-5-lobed, base cordate, margin subentire, apex acuminate, basally 3-5- nerved, chartaceous. Flower(s) solitary, axillary or rarely few in terminal racemes; pedicel c. 1 cm long; epicalyx-lobes 5, free, subulate, caducous; calyx widely campanulate, lobes subulate; corolla c.5 cm ϕ, yellow with a dark purple centre; staminal column short. Capsule ovoid, 5-valvular. Seeds obovoid, angular, pubescent.

Fl. & Fr.: Aug.-Dec. *Distr.*: India, South to South-East Asia, East Africa. Commonly found in forest celaring and in teak plantations. *NAK 834* (Thekkuthode, c. 70 m), *1163* (Upper Moozhiar, c. 800 m). Young twigs yield good fibre.

URENA Linnaeus
Sp. Pl. 692.1753.

Urena lobata L., Sp. Pl. 692.1753; Mast. in Hook. f., Fl. Brit. India 1:329.1874.

Key to the subspecies.

1. Leaves undivided or shallowly lobed; epicalyx-lobes straight ....**U. lobata** ssp. **lobata**
1. Leaves plamatifid; epicalyx-lobes spreading. .... **U. lobata** ssp. **sinuata**

Urena lobata L.ssp. **lobata** Borss., Blumea 14:140. 1966; Matthew, Ill. Fl. Tam. Carnatic t.70.1982; Britto & Matthew in Matthew, Fl.Tam. Carnatic 3(1):135. 1983; Paul & Nayar, Fl. India Fasc. 19:228.1988; Mohanan & Sivad., Fl. Agasthyamala 102. 2002.

Erect undershrubs, densely stellate tomentose. Leaves broadly ovate to orbicular, 2-8 by 2-6 cm, entire to shallowly lobed, base rounded to subcordate, margin subentire, apex acute or subacute, thin-coriaceous. Flower(s) solitary or few in clusters, axillary; epicalyx-lobes linear; calyx narrowly campanulate, lobes 5, free to the base; corolla pink to rose with a purple centre, c.2 cm ϕ, petals 5, obovate Fruit globular, glochidiate. Seeds reniform, minutely hairy.

Fl. & Fr.: Jan.-Mar. *Distr.*: Sri Lanka, India. Common along waysides in plains and hills. *NAK 172, 347* (Plappally, c. 400 m).

Stem yields good fibre.

Urena lobata L. ssp. **sinuata** (L). Borss. Blumea 14 : 142. 1966; Britto & Matthew in Matthew, Fl. Tam. Carnatic 3(1) : 135.1983; Paul & Nayar, Fl. India Fasc. 19 : 230. 1988; Mohanan & Sivad., Fl. Agasthyamala 103. 2002. U. *sinuata* L., Sp. Pl. 692.1753; Dunn in Gamble, Fl. Pres. Madras 92.1915.

Undershrubs. Leaves palmatifid, lobes elliptic. Flowers pink; epicalyx-lobes spreading.

Fl. & Fr.: Dec.-Mar. *Distr.*: Sri Lanka, India. Common in Forest clearings. *NAK 2739* (Maniyaar, c. 70 m).

23. BOMBACACEAE

1. Leaves palmately compound; stem prickly; flower(s) large, solitary or in clusters towards the ends of branchlets. .... **Bombax**
1. Leaves simple; stem smooth; flowers small, in clusters on old wood. .... **Cullenia**

BOMBAX Linnaeus
Sp. Pl. 511. 1753.

Deciduous trees. Leaves palmately 5-7-foliolate, alternate. Flowers precocious, solitary, in pairs or in cluster towards the ends of branchlets; calyx cupular, lobes 3-5, irregular; corolla large, blood-red or rose coloured, 10-15 ϕ, stamens α, adelphous in 4-5 whorls, filaments basally connate, anthers 1-celled, hypocrateriform; ovary superior, 5-locular, ovules a per locule. Capsule woody, loculicidal.

1. Flowers blood-red coloured, to 8 cm long; stamens c.80; large trees. .... 1. **B. ceiba**
1. Flowers rose coloured, to 12 cm long; stamens, c.600; small trees. .... 2. **B. scopulorum**

1. **Bombax ceiba** L., Sp. Pl. 511. 1753, p.p.; Robyns, Taxon 10:160. 1961; Dassan. & Fosberg, Rev. Handb. Fl. Ceylon 1:64.1980; Sald.& Nicols., Fl. Hassan Dist. 2. 1976; Britto & Matthew in Matthew, Fl. Tam. Carnatic 3(1). 136. 1983; Nicols., Suresh & Mani., An Interpr. Hort. Malab. 57. 1988; Mohanan & Sivad., Fl. Agasthyamala 104. 2002. *B. malabaricum* DC., Prodr. 1 : 479. 1824; Mast. in Hook.f., Fl.Brit. India 1:349. 1874; Dunn in Gamble, Fl. Pres. Madras 99.1915. *Salmalia malabarica* (DC.) Schott & Endl., Melet.Bot. 35. 1832. *Moul-elavou* Rheede, Hort. Malab. 3: 61-62, t.52. 1682.

Large trees with buttress roots; trunk with conical prickles. Leaves 5-7-foliolate, to 15 cm wide, leaflets elliptic or oblong, 18 by 6 cm, apex acuminate. Flowers blood-red coloured, solitary or fascicled. Capsule ovoid, c.15 cm long.

Fl. & Fr.: Jan.-Apr. *Distr.*: Tropical Asia & New Guinea. Large deciduous trees with very showy flowers. Common in evergreen and deciduous forests. *NAK 146* (Moozhiar, c. 250 m).

Cotton used in stuffing pillows, root has stimulant and tonic properties.

2. **Bombax scopulorum** Dunn in Gamble, Fl. Pres. Madras 100.1915; *B. insigne* sensu Bourd., For. Trees Travancore 45. 1908, non Wall., 1830. *Kal-ilavu.*

Small trees; trunk with clustered prickles. Leaves 5-7- foliolate; leaflets elliptic, thin-coriaceous. Flowers showy, rose-red coloured, solitary or in pairs or in clusters; stamens α (c.600) grouped in 5 phalanges. Capsule oblong.

Fl. & Fr.: Nov.-Jan. *Distr.* : Western Ghats. Rare. On exposed rocky hills. *NAK 219* (Choothuporuthu para, Chittar, c. 200 m).

CULLENIA R. Wight
Ic. Pl. Ind. Or. 5(1) : 22, t. 1761-62. 1851.

Cullenia exarillata Robyns, Bull. Jard. Bot. Nat. Belg. 40: 249. 1970; Manilal, Fl. Silent Valley 30.1988; Mohanan & Sivad., Fl. Agasthyamala 105.2002. *C. excelsa* Wight, Ic. t.1761 & 1762. 1852, p.p.; Mast. in Hook.f., Fl. Brit. India 1:350. 1874; Dunn in Gamble, Fl. Pres. Madras 101. 1915. *Vediplavu, Karayani.*

Trees; young parts lepidote. Leaves simple, oblong-lanceate to elliptic, 10-20 by 3-5.5 cm, base rounded or subacute, apex acuminate, veins pinnate, α, coriaceous. Flowers yellow, 1 cm ϕ, in dense fascicles on old wood; calyx tubular; petals 0; stamens in 5 bundles, anthers clustered; ovary 5-locular, ovules 2-per locule; style woody, stigma capitate. Fruit a large globose capsule, 15 by 12 cm, spinescent, dehiscing 5-valvular. Seeds large, arillate.

Fl. & Fr.: Jan.-Nov. *Distr.*: Endemic to Southern Western Ghats. Rare. In moist evergreen forests. *NAK 1295* (Kakki hills, c.1100 m).

Fruits are favourite food of monkeys.

24. STERCULIACEAE

1. Trees or shrubs.
    2. Flowers polygamous; petals 0.
        3. Leaves penni-nerved, divided or undivided; follicles oblong; seeds not winged. .... **Sterculia**
        3. Leaves palmi-nerved at base, undivided; follicles globose; seeds winged. .... **Pterygota**
    2. Flowers bisexual; petals 5.
        4. Flowers crimson coloured, irregular; fruit a spirally twisted; follicle. .... **Helicteres**
        4. Flowers white or yellow, regular; fruit a smooth; capsule. .... **Pterospermum**
1. Herbs or subshrubs.
    5. Flowers pink; calyx obscurely lobed; petals clawless. .... **Melochia**
    5. Flowers yellow; calyx deeply lobed; petals clawed. .... **Waltheria**

HELICTERES Linnaeus
Sp. Pl. 963. 1753.

Helicteres isora L., Sp. Pl: 963. 1753; Roxb., Fl. Ind. 3: 143. 1832., Wight, Ic.t. 180. 1839; Mast. in Hook.f., Fl. Brit. India 1:365.1874; Dunn in Gamble, Fl. Pres. Madras 107. 1915; Ramamoorthy in Sald. & Nicols., Fl. Hassan Dist. 141. 1976; Matthew, Ill. Fl. Tam. Carnatic t.72. 1982; Britto & Matthew in Matthew, Fl. Tam. Carnatic 3(1) : 144. 1983; Manilal, Fl. Silent Valley 31. 1988; Nicols., Suresh & Mani., An Interpr.

Hort. Malab. 253. 1988; Mohanan & Sivad., Fl. Agasthyamala 107. 2002. *Isora-murri* Rheede, Hort. Malab. 6: 55-66, t.30. 1686. *Idampiri-valampiri.*

Shrubs. Leaves simple, alternate; stipule caducous. Lamina obovate or suborbicular, 5-15 (20) by 4-15cm, base rounded or shallowly subcordate, margin coarsely serrate, apex acuminate or rotund, basally 5-7-nerved, thin-coriaceous. Flowers bisexual, irregular, few in axillary cymes; calyx persistent; corolla crimson coloured; gynandrophore present stamens 10, basally connate; staminodes 5; ovary 5-locular; ovules a per locule. Follicles 5, spirally twisted to left or right sides. Seeds wrinkled.

Fl. & Fr.: Sep.-Mar. *Distr.*: Sri Lanka, Peninsular India, Malaya Peninsula, Australia. Very common in disturbed forest areas, in teak plantations, etc. *NAK 4* (Moozhiar, 250 m), *819* (Ranni, 100 m).

Stem yields a good fibre used for making sacks and bags. Fruit used in medicine for colic, and juice of root -bark as a remedy for disorders of stomach.

MELOCHIA Linnaeus
Sp. Pl. 674 ('774'). 1753.

Melochia corchorifolia L., Sp. Pl. 675. 1753; Roxb., Fl. Ind. 3:139. 1832; Mast. in Hook. f., Fl. Brit. India 1:374. 1874; Dunn in Gamble, Fl. Pres. Madras 110. 1915; Ramamoorthy in Sald. & Nicols., Fl. Hassan Dist. 143. 1976; Matthew, Ill. Fl. Tam. Carnatic t.80. 1982; Britto & Matthew in Matthew, Fl. Tam. Carnatic 3(1):147.1983; Nicols., Suresh & Mani., An Interpr. Hort. Malab. 254. 1988; Mohanan & Sivad., Fl. Agasthyamala 108. 2002. *Tsjeru-uren* Rheede, Hort. Malab. 9:143, t.73. 1689. *Niruren* Rheede, Hort. Malab. 10: 109, t.55. 1690.

Subshrubs. Leaves alternate, simple, ovate-lanceate, 1.5-5 by 1-2.5 cm, base truncate, margin serrate, apex acute, chartaceous. Flowers bisexual, in terminal capitate clusters; bracts and bracteoles linear; calyx-tube cupular, 5-toothed; corolla pink, c. 1 cm ϕ; petals 5, free, obovate; stamens 5, connate below; staminodes 0; ovary sessile, 5-locular; ovules 2-per locule; styles 5, connate below. Capsule 5-valvular. Seeds ovoid.

Fl. & Fr.: Aug.-Dec. *Distr.* : Pantropical. Common. Along banks of paddy fields and wet waysides. *NAK 174* (Adoor, c.20 m), *671* (Pandalam, c.40 m).

Stem yields a fibre.

PTEROSPERMUM Schreber
Gen. 2:461. 1791 (nom. cons.).

Pterospermum diversifolium Blume, Bijdr. 88. 1825; Mast. in Hook.f., Fl. Brit. India 1 : 367. 1874; Dunn in Gamble, Fl. Pres. Madras 108. 1915; Nicols., Suresh & Mani., An Interpr. Hort. Malab. 254. 1988. *P. glabrescens* Wight & Arn., Prodr. 9. 1834; Mast. in Hook.f., Fl. Brit. India 1 : 369. 1874. *Solda* Rheede, Hort. Malab. 6:103, t.58. 1686.

Trees, branchlets stellate-tomentose. Leaves simple, 3-5- angled, alternate, obovate oblong, 8 15 by 7-10 cm, base cuneate, subcordate or subpeltate, apex acuminate, basally 5-nerved, lower surface densely golden stellate-tomentose. Flower(s) bisexual, axillary, solitary or few in clusters, c.10 cm ϕ; calyx deeply 5-lobed, sepals linear-oblong, to 10 cm long; petals 5, white, oblong, ± equal to sepals, deciduous; receptacle elongated into a gynophore; stamens 15, connate at base and adnate with gynandrophore; ovary 5-locular, ovules α. Capsule ovoid, c. 12 cm long. Seeds flat, winged.

Fl. Fr.: Dec.-Apr. *Dsitr.*: Southern Peninsular India. Rare. In moist evergreen forest. *NAK 2630* (Moozhiar, c.300 m).

PTERYGOTA Schott & Endlicher
Melet. Bot. 32. 1832.

Pterygota alata (Roxb.) R. Br. in Bennet & Brown, Pl. Jav. Rar. 234. 1844; Dunn in Gamble, Fl. Pres. Madras 104. 1915. *Sterculia alata* Roxb., Pl. Corom. 3: 84, t. 287. 1820; Mast. in Hook.f., Fl. Brit. India 1:360. 1874. *S. paynii* Bedd., Fl. Sylv. t.230. 1872. *Ahnathondi.*

Trees. Leaves simple, broadly ovate-oblong, 10-15 by 7-12 cm, 5-7- nerved at base, thin-coriaceous. Panicle axillary to fallen leaves. Flowers polygamous, c.2 cm ϕ; calyx-lobes shortly connate below, woolly without; petals 0; staminal column cylindric, bearing 5 phalanges of 5 anthers; ovary sessile, 5-locular, distinct; ovules α; style short, recurved. Follicles 5, large, obovoid or globose, woody. Seeds winged.

Fl. & Fr.: Jan. (Feb.) - Nov. *Distr.*: Sri Lanka, India. Rare. in semi-evergreen forests. *NAK 1426* (Mannarpara, c.300 m).

STERCULIA Linnaeus
Sp. Pl. 1007. 1753.

Trees. Leaves crowded towards apices of stem, alternate, simple, digitate or lobed. Panicles terminal or axillary. Flowers polygamous or unisexual; calyx campanulate or cupular, lobes 5; petals 0, receptaecls elongate; stamens 10 to α; sessile, clustered on the staminal column; ovary 5-locular, ovules α. Follicles 5, woody, oblong. Seeds wingless.

1. Leaves elliptic-oblong or ellipti-ovate; flowers reddish-brown; calyx-lobes linear. .... 1. **S. balanghas**
1. Leaves oblong-ovate or ovate to obovate; flowers flesh-coloured; calyx-lobes broadly ovate. .... 2. **S. guttata**

1. **Sterculia balanghas** L., Sp. Pl. 1007. 1753; Wight, Ill. 1: t.30. 1840; Mast.in Hook.f., Fl. Brit. India 1:358. 1874; Dunn in Gamble, Fl. Pres. Madras 106. 1915; Nicols., Suresh & Mani., An Interpr. Hort. Malab. 254. 1988. *Cavalam* Rheede, Hort. Malab. 1:89-90, t.49. 1678.

Leaves elliptic-oblong or elliptic-ovate, 7-17 by 4-8 cm, base obtuse to rounded, apex obtusely short-acuminate, lateral nerves 10-12 pairs, thin-coriaceous. Panicles c.15 cm long; bracts small, caducous. Male flowers c. 1 cm ϕ, reflexed, reddish hairy; androphore slightly curved, Follicles oblong. Seeds black.

Fl. & Fr. : Jun.-Aug. *Distr.*: Sri Lanka, India. In groves and planted elsewhere. *NAK 2783* (Adoor, c. 20 m).

2. **Sterculia guttata** Roxb. ex DC., Prodr. 1:482. 1824; Roxb., Fl. Ind. 3:148. 1832; Wight, Ic. t.487. 1841; Bedd., Fl. Sylv. t. 105. 1871; Mast. in Hook.f., Fl. Brit. India 1:355. 1874; Dunn in Gamble, Fl. Pres. Madras 106. 1915; Ramamoorthy in Sald. & Nicols., Fl. Hassan Dist. 144. 1976; Britto & Matthew in Matthew, Fl. Tam. Carnatic 3(1). 152. 1983; Mohanan & Sivad., Fl. Agasthyamala 111. 2002. *Pee-naari, Potta-cavalam.*

Leaves oblong-ovate or ovate to obovate, 15-30 by 9.5-14 cm, base subcordate, apex acuminate, stellately pubescent on nerves beneath, lateral nerves c.12 pairs, thin-coriaceous. Racemes, c.20 cm long, terminal. Flowers flesh-red, c.15 mm ϕ; calyx stellate-tomentose without, lobes oblong-ovate or broadly ovate; stamens 10, gynandrophore curved. Follicle 2-3-lobed. Seeds ellipsoid.

Fl. & Fr.: Sep.-Mar. *Distr.*: Sri Lanka, India (Western Ghats). Common. In semi-evergreen forests. *NAK* 789 (Ranni, 100 m).

Bark yields a strong fibre for paper making.

WALTHERIA Linnaeus
Sp. Pl. 673. 1753.

Waltheria indica L., Sp. Pl. 673. 1753; Wight & Arn., Prodr. 67. 1834; Mast. in Hook.f., Fl. Brit. India 1:374. 1874; Dunn in Gamble, Fl. Pres. Madras 111. 1915; Matthew, Ill. Fl. Tam. Carnatic t.83. 1982; Britto & Matthew in Matthew, Fl. Tam. Carnatic 3(1): 154. 1983; Mohanan & Sivad., Fl. Agasthyamala 112. 2002, *W. americana* L., Sp. Pl. 673. 1753; Ramamoorthy in Sald. & Nicols., Fl. Hassan Dist. 144. 1976.

Woody herbs; branchlets, leaves and peduncle stellately pubescent. Leaves simple, ellipitc, 2-5 by 0.7-2.5 cm, base cordate, margin serrate, apex subacute, lower surface stellately woolly, basally 3-5- nerved, coriaceous. Flowers bisexual, aggregated, axillary or terminal; calyx-tube long, lobes 5; corolla yellow, c.5 mm ϕ, petals 5, clawed; stamens 5, filament connate at base, staminodes 0; ovary sessile, 1-locular, ovules 2, basal. Capsule obconic, 2-valvular. Seed wrinkled.

Fl. & Fr.: Most of the seasons. *Distr.*: Tropics. A weed of waysides and fields in plains. *NAK 173* (Nellimukal, c. 30 m).

25. TILIACEAE

1. Erect or scandent shrubs or trees; fruit a drupe.
   2. Inflorescence a large terminal panicle. .... **Microcos**
   2. Inflorescence an axillary short cyme or umbellate cluster. .... **Grewia**
1. Erect herbs or subshrubs; fruit a capsule.
   3. Petals glandular at base; leaf margin at base appendaged; capsule unarmed. .... **Corchorus**
   3. Petals glandular at base; leaf margin at base unappendaged.; capsule prickly or bristly. .... **Triumfetta**

CORCHORUS Linnaeus
Sp. Pl. 529. 1753.

Woody herbs or subshrubs. Leaves simple, alternate, basally 3- or 5-nerved, basal serrature of leaf margin appendaged. Flowers in axillary or leaf-opposed cymes, 5-merous, bisexual; sepals 5, oblong, mucronate; petals 5, yellow, glandular; stamens 10 to α, inserted on a short torus; ovary 2-5- locular, ovules a per locule. Capsule narrowly oblong, angled or terete, septate. Seeds α.

1. Woody herbs to subshrubs, 50-75 cm tall; capsule 6-angular, wings 3. .... 1. **C. aestuans**
1. Subshrubs, to 1 m tall; capsule 10- ribbed, wing 0. .... 2. **C. olitorius**

1. **Corchorus aestuans** L., Syst. Nat. (ed.10), 1079. 1759; Sant., Fl. Khandala (ed. 3), 27. 1967; Ramamoorthy in Sald. & Nicols., Fl. Hassan Dist. 132. 1976; Britto & Matthew in Matthew, Fl. Tam. Carnatic (1) : 157. 1983. *C. acutangulus* auct., non Forssk., 1775: Wight, Ic. t. 739. 1844; Mast. in Hook.f., Fl.Brit. India 1:398. 1874; Dunn in Gamble, Fl. Pres. Madras 121.1915.

Leaves ovate, 2-5 by 1-2.5 cm, base rounded, margin crenate, apex acute to shortly acuminate, basally 3-5-nerved, chartaceous. Flowers small in leaf-opposed clusters; sepals lanceate; petals narrowly obovate or spathulate; stamens c.20; ovary 3-locular. Capsule cylindric, 6-angled and 3-winged, beak 3-fid. Seeds truncate.

Fl. & Fr.: Most of the season. *Distr.*: Sri Lanka, India, Pakistan, Bangladesh, Myanmar, Malesia, Australia, Trop. Africa & C. America. Common in exposed wastelands in plains. *NAK 1078* (Kodumon, c.50 m).

2. **Corchorus olitorius** L., Sp. Pl. 529. 1753; Roxb., Fl. Ind. 2: 581. 1832; Wight & Arn., Prodr. 73. 1834; Mast. in Hook.f., Fl. Brit. India 1:397. 1874; Dunn in Gamble, Fl. Pres. Madras 122. 1915; Matthew, Ill. Fl. Tam. Carnatic t.85. 1982; Britto & Matthew in Matthew, Fl. Tam. Carnatic 3(1):160.1983.

Leaves ovate-lanceate, 5-8 by 2.5-4 cm, base obtuse, margin serrate, apex acuminate, basally 3-5- nerved. Cymes opposite to leaves. Flowers c.5 mm ϕ; sepals oblong; petals oblong-spathulate; stamens α; ovary 5-locular. Capsule 10-ribed, beak entire. Seeds trigonous.

Fl. & Fr.: Most of the seasons. *Distr.*: Tropics. A weed of wastelands. *NAK* 1846 (Adoor, c. 30 m).

GREWIA Linnaeus
Sp. Pl. 964. 1753.

Trees or weak shrubs; branchlets stellate-tomentose. Leaves simple, basally 3-5- nerved. Flowers in cymes or umbellate clusters, leaf-opposed or terminal, bisexual; sepals 5, linear-oblong, free; petals 5, yellow, very small, glandular at base, free; stamens α on a raised torus; ovary 2-4 -locular, ovules 2-α; style subulate. Drupe 2-4-lobed. Seed 1-2 per locule.

1. Leaves 5-nerved at base; stipule falcate, base sub-auricled. .... 4. **G. tiliifolia**
1. Leaves 3-nerved at base; stipule lanceate, base not so.
    2. Peduncle shorter than or equal to petioles.
        3. Leaves glabrous, base subcordate or rounded, margin irregularly serrate or lobed. .... 1. **G. abutilifolia**
        3. Leaves tomentose, glabrescent, base mostly acute, margin crenate or regularly serrate.
            4. Leaves elliptic or elliptic-oblong; cymes umbellate, a-flowered; pedicels slender; fruit usually 4-lobed. .... 5. **G. umbellifera**
            4. Leaves lanceate or oblong-lanceate; cymes not umbellate; pedicel stout; fruit unlobed. .... 3. **G. obtusa**
    2. Peduncle 2 or 3 times as long as petioles. .... 2. **G. disperma**

1. **Grewia abutilifolia** Vent. ex A.L. Juss., Ann. Mus. Nat. 4:92. 1804; Wight & Arn., Prodr. 79.1834, p.p.; Mast. in Hook.f., Fl. Brit. India 1:390. 1874, p.p.; Matthew, Ill. Fl. Tam. Carnatic t. 86. 1982; Britto & Matthew in Matthew, Fl. Tam. Carnatic 3(1): 163. 1983. *G. aspera* Roxb., Fl. Ind. 2:591. 1832; Dunn in Gamble, Fl. Pres. Madras 119. 1915.

Shrubs or small trees. Leaves broadly ovate to rotund, 5-18 by 3-13 cm, base rounded or subcordate, margin irregularly serrate or dentate, lobed near apex, apex acute-acuminate, upper surface scabrous, lower surfce stellate-pubescent, baslly 3-veined. Cymes umbellate; peduncle c.3 mm long. Flowers c. 1 cm ϕ; sepals oblong-lanceate; petals white, gland rounded; torus- 5-angled; ovary globose, villous, 2-locular, ovule 1-per locule. Drupe globose, faintly 4-lobed.

Fl. & Fr.: Feb.-May. *Distr.*: Peninsular India. Less common. In open forest areas and streamsides. *NAK 558* (Thekkuthode, c. 70 m), *2357* (Kalleli, c.250 m).

2. **Grewia disperma** Rottler ex Sprengel in L., Syst. Veg. (ed. 16) 2:579. Jan.-May, 1825; Dunn in Gamble, Fl. Pres. Madras 118. 1915; Ramamoorthy in Sald. & Nicols., Fl. Hassan Dist. 135. 1976; Hara, Enum. Flow. Pl. of Nepal 2:71. 1979; Britto & Matthew in Matthew, Fl. Tam. Carnatic 3(1) : 166. 1983. *G. glabra* Bl., Bijdr. 115. June-Dec. 1825; Ghafoor, Fl. W. Pakistan 75:11. 1974; Manilal, Fl. Silent Valley 33.1988. *Aana kotti-maram.*

Trees. Leaves lanceate or oblong lanceate, 8-14 by 2.5-4.5 cm, base subacute, margin serrate, lower serrature glandular, apex acuminate, basally 3-nerved. Cymes axillary; peduncle, c.2 cm long. Flower buds ovate, 2.5 cm ϕ; sepals oblong-lanceate; torus terete; ovary globose, 2-locular; ovule 1 per locule. Drupe deeply 2-lobed.

Fl. & Fr.: Jul.-Sep. *Distr.*: India, Myanmar, Tropical Africa. Less common. In moist semi-evergreen forests. *NAK 747* (Pampa Valley, c.200 m).

Bark yields a fibre for cordage.

3. **Grewia obtusa** Wall. ex Dunn in Gamble, Fl. Pres. Madras 117. 1915; Ramamoorthy in Sald. & Nicols., Fl. Hassan Dist. 136.1976. *G. bracteata* auct., non Heyne ex Roth, 1821: Wight & Arn., Prodr. 76.1834; Mast. in Hook.f., Fl. Brit. India 1:389. 1874, p.p.

Large shrubs; branchlets rusty stellate-tomentose. Leaves elliptic-ovate to oblong-ovate, 5-15 by 2-5 cm, base rounded or broadly cuneate, margin serrate, lower few serratures glandular, apex long-acuminate, upper surface glabrescent, lower surface densely tomentose on primary veins, basally 3-nerved. Cymes axillary or leaf-opposed, peduncles slightly to twice as long as petiole; pedicel stout. Flower buds ovoid - obtuse. Flowers yellow inside; ovary 2-lobed and 2-locular; ovules 2 per locule. Fruits could not be collected

Fl. & Fr.: April onwards. *Distr.*: Peninsular India. Rare. In moist evergreen forests. *NAK 2390* (Lower Moozhiar, c.250m).

4. **Grewia tiliifolia** Vahl, Symb. Bot. 1:35.1790; Roxb., Fl.Ind. 2:587.1832; Wight & Arn., Prodr. 80.1834; Mast. in Hook.f., Fl.Brit. India 1:386.1874, p.p.; Dunn in Gamble, Fl.Pres.Madras 118.1915; Britto & Matthew in Matthew, Fl.Tam. Carnatic 3(1):173.1983; Manilal, Fl.Silent Valley 34.1988; Nicols., Suresh & Mani., An Interpr. Hort. Malab. 257.1988; Mohanan & Sivad., Fl. Agasthyamala 115. 2002. *Pai-paroea, Couradi* Rheede, Hort. Malab. 5:91-92, t.46.1685. *Unnam, Chadachi*

Trees. Leaves ovate-oblong, 14-14 by 2-8 cm, base obliquely subcrodate, margin crenate-serrate, apex acuminate or acute, upper surface glabrescent, lower surface greyish-tomentose, basally auricled. Cymes axillary, peduncles c. 1 cm long. Flowers c. 1 cm ϕ, deep yellow; sepals linear-oblong; petals narrowly spathulate; gland obovoid; ovary globose, 2-locular. Drupe globose.

Fl. & Fr. : Feb.-Jun. *Distr.*: Sri Lanka, India, Trop. Africa, Myanmar. Very common in deciduous and semi-evergreen forests. *NAK 526* (Thekkuthode, c. 70 m), *1623* (Pampa Valley, c. 250 m).

Bark externally applied in cow-itch, and given along with water as a remedy in dysentery.

5. **Grewia umbellifera** Bedd., For. Man. Bot. 37.1871; Mast. in Hook.f., Fl.Brit. India 1:393.1874; Dunn in Gamble, Fl.Pres. Madras 117.1915; Ramamoorthy in Sald. & Nicols., Fl. Hassan Dist. 137.1976; Manilal, Fl.Silent Valley 34.1988; Mohanan & Sivad., Fl. Agasthyamala 116. 2002. *Bhasma-valli.*

Scandent shrubs. Leaves elliptic-oblong or oblong ovate, 3-16 by 2-8 cm, base rounded, margin crenate-serrate, apex acuminate, glabrescent, 3-nerved at base. Cymes umbellate, axillary, solitary or fascicled, up to 20-flowered; peduncle 3-5 cm long. Flowers white to yellow, c. 2 f; sepals linear-lanceate; petals oblong-lanceate; glands ovoid; torus 5-angled; ovary globose, 4-lobed and 4-locular. Drupe globose, 4-lobed.

Fl. & Fr.: Mar.-May. *Distr.*: Peninsular India. Most common species of the genus. Along streamsides. *NAK 551, 1101* (Thekkuthode, Pampa Valley etc.)

MICROCOS Linnaeus
Sp. Pl. 514.1753.

Microcos paniculata L., Sp. Pl. 514.1753; Burret, Notizbl.Bot. Gart. Berlin-Dahlem 9:773.1926; Merr., Trans. Amer. Phil.Soc. 24(2):257.1935; Nicols., Suresh & Mani., An Interpr. Hort.Malab. 257.1988 ; Mohanan & Sivad., Fl. Agasthyamala 116. 2002. *Grewia microcos* L., Syst. ed. 12, 602.1767, nom. illegit. (incl. *Microcos paniculata* L., 1753); Mast. in Hook.f, Fl.Brit.India 1:393. 1874; Ramamoorthy in Sald. & Nicols., Fl.Hassan Dist. 136.1976; Dunn in Gamble, Fl.Pres. Madras 116.1915. *Schageri-cottom* Rheede, Hort. Malab. 1:105-106, t.56.1678. *Cotta.*

Shrubs to very small trees. Leaves elliptic-lanceate or ovate-lanceate, 6-22 by 2-7 cm, base rounded, margin obscurely crenate, apex acuminate, glabrescent, basally 3-nerved, chartaceous. Flowers in terminal panicle or rarely in upper leaf-axils; bracts linear. Flowers white, c.1 cm ϕ; sepals 5, oblanceate; petals 5, ovate-lanceate, glandular at base; stamens α, on a raised torus; ovary globular, 2-locular; ovule 1 per locule. Drupe globose, c. 1 cm ϕ.

Fl. & Fr.: Mar.-Sep. *Distr.*: Sri Lanka, India, Java & China. Common in exposed hill cuttings and sacred groves. *NAK 662* (Pandalam, c. 20 m), *1646* (Tiruvalla, c. 10 m).

Stem yields a fibre. Paste of fruits applied to remove swellings on the genital organs.

Note : It is seen that **Microcos** has been recognised as a section of **Grewia** (Mast., l.c.) and in all Indian Floras *Grewia microcos* L. has been accepted as the name for the common species of the area, which actually is an illegitimate name.

TRIUMFETTA Linnaeus
Sp. Pl. 444.1753.

Subshrubs. Leaves simple or lobed. Flowers in axillary or leaf-opposed cymes or racemes; sepals 5, oblong, apiculate or beaked; petals 5, yellow, glandular at base; stamens 5-15, free, raised on a short lobed torus; ovary 5-locular; ovules 2 per locule. Capsules globose, usually prickly or bristly.

1. Leaves undivided; capsule 10-20 mm ϕ (including bristles).
    2. Glabrescent subshrubs; capsule, to 10 mm ϕ; bristles glabrous throughout .... 1. **T. annua**
    2. Tomentose or woolly subshrubs; capsules, to 20 mm ϕ; bristles apically glabrous, basally hispid. .... 2. **T. pilosa**
1. Leaves angled or deeply lobed; capsules 3-5 ϕ (including prickles). .... 3. **T. rhomboidea**

1. **Triumfetta annua** L., Mant. Pl. 1:73.1767; Mast. in Hook.f., Fl.Brit. India 1:396.1874; Dunn in Gamble, Fl. Pres. Madars 120.1915; Britto & Matthew in Matthew, Fl. Tam. Carnatic 3(1):175.1983; Manilal, Fl. Silent Valley 34.1988; Mohanan & Sivad., Fl. Agasthyamala 117. 2002.

Branchlets bulbous based hairy. Leaves ovate-lanceate, 4-9 by 2-4 cm, base rounded, margin serrate, apex acuminate, basally 3-5- nerved, chartaceous. Flowers in axillary clusters; sepals oblong; petals yellow, narrowly obovate; stamens 8; torus small; ovary papillose. Capsule globose, to 10 mm ϕ. Seeds globose.

Fl. & Fr.: Sep.-Mar. *Distr.*: India Pakistan, Tropical & S. West Africa. Rare.Along forest edges in upper hills. *NAK 219* and *1151* (Kakki hills, c. 1000 m).

2. **Triumfetta pilosa** Roth, Nov. Pl. Spec. 223.1821; Wight & Arn., Prodr. 74.1834; Mast. in Hook.f., Fl. Brit. India 1:394.1874; Dunn in Gamble, Fl.Pres. Madras 120.1915; Ramamoorthy in Sald. & Nicols., Fl. Hassan Dist. 137.1976; Matthew, Ill. Fl. Tam. Carnatic T.87.1982; Britto & Matthew in Matthew, Fl.Tam. Carnatic 3(1):176. 1983; Manilal, Fl.Silent Valley 35.1988.

Branchlets densely tomentose or wooly. Leaves ovate-lanceate, 3-10 by 1.5-5.5 cm, base rounded to subcordate, margin serrate, apex acute to acuminate, lower surface stellate-woolly, basally 3-5- nerved, thin-coriaceous. Flowers in umbellate cymes; sepals oblong; petals yellow, obovate; stamens 10; ovary globose. Capsule globose, bristles hooked, hispid towards base. Seeds ovoid.

Fl. & Fr.: Nov.-Feb. *Distr.*: Sri Lanka, India Nepal, Malaya, China, Trop. Africa. Along forest clearing in upper hills. *NAK 337, 1159, 1174* (Upper Mozhiar-Kallippara, 550 m).

3. **Triumfetta rhomboidea** Jacq., Enum.Syst. Pl.22.1760; Mast. in Hook.f., Fl.Brit. India 1:395.1874; Dunn in Gamble, Fl.Pres. Madras 120.1915; Ramamoorthy in Sald. & Nicols., Fl.Hassan Dist. 138.1976; Britto & Matthew in Matthew, Fl.Tam. Carnatic 3(1):176. 1983; Manilal, Fl.Silent Valley 35.1988; Mohanan & Sivad., Fl. Agasthyamala 118. 2002.

Leaves broadly suborbicular, apically 3-angled to deeply 3-5-lobed, lobes narrowly obovate or rhomboid or oblanceate, 1.5-4 by 0.7-2 cm, base rounded, margin serrate, apex shortly acuminate, pubescent. Cymes condensed, leaf-opposed. Flowers yellow; sepals oblong; petals obovate; stamens 10; ovary 4-locular. Capsule prickly, 3-5 mm ϕ. Seed 1- per locule.

Fl. & Fr.: Sep.-Feb. *Distr.*: Pantropical. Common in exposed wastelands in plains. *NAK 233* (Thekkuthode, c. 70 m) *1010* and *1011* (Tiruvalla, c. 10 m).

Stem yields a soft useful fibre.

26. ELAEOCARPACEAE

ELAEOCARPUS Linnaeus
Sp. Pl. 515. 1753.

Trees. Leaves simple, alternate; branchlets usually with persistent leaf scars. Flowers white, bisexual, in axillary or lateral racemes; sepals 4-5, free; petals 4-5, lanceate; stamens α, inserted on the disc, anthers oblong, often bearded or ciliate apically, poricidal; ovary superior, 3-5- locular; ovules 2 per locule. Drupe globose or ellipsoid. Seeds tuberculate or smooth.

1. Leaves to 15 cm long, apex caudate or obtusely acuminate or acute; flowers less than 2 cm ϕ.
    2. Leaves elliptic-ovate, base truncate, apex caudate;anthers aristate. ....1. **E. munronii**
    2. Leaves elliptic or obovate, base never truncate, apex obtusely acuminate to acute; anthers bearded. ....2. **E. serratus**
1. Leaves to 20 cm long, apex rounded or obtuse; flowers, c.2 cm ϕ. ....3. **E. tuberculatus**

1. **Elaeocarpus munronii** (Wight) Mast. in Hook.f., Fl. Brit. India 1:407.1874; Dunn in Gamble, Fl.Pres. Madras 124.1915, ("munroii"); Ramamoorthy in Sald. & Nicols., Fl.Hassan Dist. 129.1976; Manilal, Fl.Silent Valley 35.1988.; Mohanan & Sivad., Fl. Agasthyamala 120. 2002. *Monocera munronii* Wight, Ill.1:84. 1840, Ic.t.952.1845. *Kal-rudraksham.*

Leaves elliptic ovate, 4-10 by 2-5 cm, base rounded to truncate, margin crenate-serrate, apex acuminate, thin-coriaceous. Racemes few-flowered. Flowers c. 1 cm ϕ, drooping; sepals lanceate; petals obovate, tomentose; anthers c. 2 mm long, arista c. 1 mm long; ovary 2-locular; ovules 2 per locule. Drupe oblong. Seeds tubercled.

Fl. & Fr.: Sep.-Apr. *Distr.*: Western Ghats. Rare. In moist evergreen forests. *NAK 1349* (Kakki hills, c. 1100 m).

2. **Elaeocarpus serratus** L., Sp. Pl. 515.1753; Wight & Arn., Prodr. 82.1834; Mast. in Hook.f., Fl.Brit. India 1:401.1874; Dunn in Gamble, Fl.Pres. Madras 123.1915; Ramamoorthy in Sald.& Nicols., Fl.Hassan Dist. 131.1976; Matthew, Ill. Fl. Tam. Carnatic 88.1982; Britto & Matthew in Matthew, Fl.Tam. Carnatic 3(1):179.1983; Nicols., Suresh & Mani., An Interpr. Hort. Malab. 103.1988; Mohanan & Sivad., Fl. Agasthyamala 121. 2002. *Perin-kara* Rheede, Hort. Malab. 4:51-52, t.24.1683.

Leaves elliptic or obovate, 5-12 by 3-6.5 cm, base obliquely cuneate to obtuse, margin distantly crenate-serrate, apex acute to obtusely acuminate, glabrescent, thin-coriaceous. Recemes axillary, peduncle reddish, c.7 cm long. Flowers c.1 cm ϕ; sepals lanceate; petals obovate; stamens c. 20, anthers bearded or ciliate at apex; ovary 3-locular. Drupe oblong. Seeds tubercled.

Fl. & Fr.: Apr.-Sep. *Distr.*: India, Trop. Himalaya. Common in semi-evergreen forests. *NAK 941* (Plappally, c. 350m), *155* (Velithode, c. 400 m).

Fleshy outer portion of fruits edible. Paste of leaves applied to cure ulcers.

3. **Elaeocarpus tuberculatus** Roxb., (Hort. Beng. 93.1814, nom. nud.) Fl.Ind. 2:594.1832; Mast. in Hook.f., Fl.Brit.India 1:404.1874; Dunn in Gamble, Fl.Pres. Madras 124.1915; Ramamoorthy in Sald. & Nicols.,

Fl. Hassan Dist. 131.1976; Manilal, Fl.Silent Valley 37.1988; Mohanan & Sivad., Fl. Agasthyamala 121. 2002. *Adraksham, Rudraksham.*

Leaves brodly obovate, 8-20 by 5-9.5 cm, base cuneate, margin faintly crenate-serrate, apex rounded or obtuse to retuse, lower surface tomentose, lateral veins c. 14 pairs, prominent below, thin-coriaceous. Racemes usually clustered towards upper leaf-axils. Flowers white, c. 2.5 cm ϕ; sepals lanceate; petals obovate, tomentose, deeply laciniate; stamens c. 25, anthers aristate; ovary 3-locular. Drupe ovoid. Seeds ellipsoid, coarsly tubercled.

Fl. & Fr.: Dec.-Aug. *Distr.*: Indo-Malesia. Common in moist evergreen and semi-evergreen forests. *NAK 529* (Angamoozzhy, 250 m), *1571* (Velithode, c. 300 m).

Nuts powdered and taken as a remedy in rheumatism, typhoid, epilepsy and syphilis. Nuts often worn by Hindus as religious ornament.

27. LINACEAE

LINUM Linnaeus
Sp. Pl. 277.1753.

Linum mysorense Heyne, Edward's Bot.Reg.16:sub t.1326.1830; Wight & Arn., Prodr.134.1834; Wight, Ic.t.60.1840; Hook.f., Fl.Brit.India 1:41.1874; Dunn in Gamble, Fl. Pres. Madras 125.1915; Gandhi in Sald. & Nicols., Fl.Hassan Dist. 406.1976; Matthew & Britto in Matthew, Fl. Tam. Carnatic 3(1):181.1983.

Slender annual herbs. Leaves alternate, simple, sessile, narrow elliptic to oblanceate, 2 by 0.5 cm, 3-veined from the base, base cuneate, apex acute. Inflorescence a terminal corymbose panicle. Flowers 5-merous, bisexual; bracts leafy; sepals 5, free, elliptic; petals 5, yellow, free; stamens 5, connate at base with a gland adnate to staminal tube; staminodes setaceous; ovary superior, 5-locular; styles 5, connate below; ovule(s) 1or 2 per locule. Capsule globose, opening 10-valvular.

Fl. & Fr.: Nov.-Dec. *Distr.*: Sri Lanka, India, W.Himalayas. Rare. In exposed upper hills. Collected only once from Kakki hills. *NAK 1832* (Kakki hills, 1200 m).

28. ERYTHROXYLACEAE

ERYTHROXYLUM P. Browne
Civ. Nat. Hist. Jamaica 278.1756.

Erythroxylum monogynum Roxb., Pl. Corom. t.88.1798 & Fl.Ind. 2:449.1832; Hook.f., Fl.Brit. India 1:414.1874; Dunn in Gamble, Fl.Pres. Madras 127.1915; Ramamoorthy in Sald. & Nicols., Fl.Hassan Dist. 406.1976; Matthew, Ill. Fl. Tam. Carnatic t.91.1982; Matthew & Britto in Matthew, Fl.Tam. Carnatic 3(1):182.1983. *Vella devadaram.*

Trees. Leaves simple, alternate, some times fascicled on short shoots, narrowly obovate, 2-6 by 0.8-2 cm, base cuneate, apex obtuse-retuse, thin-coriaceous. Flower(s) solitary or 2-3 in axillary fascicles, 5-merous, bisexual; calyx cupular, lobes 5, broadly ovate, imbricate; corolla pale yellow, c. 5 mm ϕ; petals obovate with a corona like ligule attached on front side; stamens 10, alternating shorter and longer, adelphous below; ovary superior, 3-locular, only one fertile; ovule 1 per locule; styles 3. Drupe cylindric or clavoid, apiculate. Seed 1, smooth.

Fl. & Fr.: Apr.-Aug. *Distr.*: Western Ghats. Rare. On dry patches of forests in Arampa. Collected only once. *NAK 2452* (Arampa, c. 450 m).

29. MALPIGHIACEAE

HIPTAGE J. Gaertner
Fruct. 2:169.1790.

Hiptage benghalensis (L.) Kruz, J. Asiat. Soc. Bengal, pt.2, Nat. Hist. 43:136.1874; Jacobs in Steenis, Fl. Males. I, 5 : 132. 1955; Gandhi in Sald. & Nicols., Fl. Hassan Dist. 410.1976; Matthew, Ill. Fl.Tam. Carnatic t.92.1982; Matthew & Britto in Matthew, Fl. Tam. Carnatic 3(1):183.1983; Manilal, Fl. Silent Valley 38.1988. *Banisteria benghalensis* L., Sp. Pl. 427.1753. *Hiptage madablota* Gaertn., Fruct. 2:169, t.116.1790; Hook.f., Fl.Brit. India 1:418.1874; Dunn in Gamble, Fl. Pres. Madras 128.1915.

Woody climbers. Leaves decussate, often glandular at base, ovate or elliptic-ovate, 7-9 by 4-6 cm, base obtuse, apex acuminate. Flowers 5-merous, bisexual, in axillary racemes; sepals 5, oblong with a large gland at the base outside; petals 5, white, clawed, reflexed, margin crisped; stamens 10, unequal, connate at base; ovary 3-lobed, 3-locular; ovule 1 per locule. Fruit a 3-winged samara. Seed 1, globose.

Fl. & Fr.: Jun.-Nov. *Distr.*: Sri Lanka, India, China & Malesia. Rare. In evergreen forests. *Deb 30471* (MH) (Pamba to Vandiperiyar, c. 1025 m).

30. OXALIDACEAE

1. Leaves paripinnate; capsular valves spreading. .... **Biophytum**
1. Leaves 3-foliolate; capsular valves cohering with axis. .... **Oxalis**

BIOPHYTUM A.P. de Candolle
Prodr. 1:689.1824.

Annual or perennial herbs. Leaves crowded towards the apex of stem, even-pinnate, some what sensitive to touch; Leaflets 12-18 pairs, opposite, inequilateral, rachis bristly at apex. Flowers in terminal pesudo-umbels, 5-merous, bisexual; sepals 5, free, persistent; petals 5, yellow; stamens 10, connate at base; ovary 5-lobed, 5-locular; ovules a-per locule; styles 5. Capsule ovoid, loculicidal. Seeds ridged or tubercled.

1. Leaflets 8-20 pairs; pedicel 3-10 mm long; sepals shorter or equal to capsule. ..1. **B. reinwardtii**
1. Leaflets 6-12 pairs; pedicel 1-2 mm long; sepals much exceeding the capsule. ....2. **B. sensitivum** var. **candolleanum**

1. **Biophytum reinwardtii** (Zucc.) Klotzsch in Peters, Reise Mossamb. Bot. 1:85.1862; Edgew. & Hook.f. in Hook.f., Fl.Brit. India 1:437.1874, incl. var. *metziana;* Gamble, Fl.Pres. Madras 133.1915. *Oxalis reinwardtii* Zucc., Abh. Math. Phys. Cl. Koenigl. Bayer Akad. Wiss. 1:274.1829-1830. *Mukkutti.*

Leaves c.9 cm long; leaflets 12-20 pairs, ovate-oblong, 4 by 3 mm, terminal ones obovate to oblong-ovate, 15 by 6 mm, base truncate or cuneate, apex acute-apiculate, glabrous or hirsute, lateral nerves 12-16 pairs, faint; rachis pubescent. Umbels 3-10, each upto 8-flowered; peduncle to 12 cm long. Flowers, c. 1 cm ϕ; sepals lanceate, strongly nerved, pubescent; petals obovate; stamens 10. Capsule obovoid or ellipsoid. Seeds spirally tubercled.

Fl. & Fr.: Most of the seasons. *Distr.*: S.W. Sri Lanka, India, China, Malesia. Common in plains to hills. *NAK 460* (Vetchuchira, c. 100 m) *824* (Chellikuzhy-Ranni, c. 350 m).

2. **Biophytum sensitivum** (L.) DC., Prodr. 1:690.1824, var. **candolleanum** (Wight) Edgew. & Hook.f. in Hook.f., Fl.Brit. India 1:437.1874; Ramamoorthy in Sald. & Nicols., Fl.Hassan Dist. 399.1976; Manilal, Fl.Silent Valley 38.1988. *B. candolleanum* Wight, Ill. 1:161, t.62.1840, "*candolianum*"; Gamble, Fl. Pres. Madras 133.1915; Nicols., Suresh & Mani., An Interpr. Hort. Malab. 155.1988. Matthew & Britto in Matthew, Fl. Tam. Carnatic 3(1):188.1983. *Todda-Vaddi* Rheede, Hort. Malab. 9:33-34, t.19.1689.

Leaves c. 6 cm long; leaflets c. 13 pairs, oblong, terminal pair obovate, 3-10 by 3-4 mm, base obtuse, apex rounded, apiculate, upper surface hairy on nerves, glabrescent, lateral nerves 10-12 pairs, slender. Umbels 4-8, each up to 6- flowered; peduncle c. 7 cm long. Flowers, c.1 cm ϕ; sepals 6-(7)- nerved; stamens nearly free; styles barbellate. Capsule obovoid. Seeds concentrically 7-ridged, tubercled.

Fl. & Fr.: November onwards. *Distr.*: Sri Lanka, Peninsular India. Rare. In hill cuttings. *NAK 2229* (Thazhoor kadavu, Pathanamthitta, c. 50 m).

OXALIS Linnaeus
Sp. Pl. 433. 1753.

Oxalis corniculata L., Sp. Pl. 435. 1753; Roxb., Fl. Ind. 2:457.1832; Wight, Ic. t.18.1838; Edjew. & Hook.f. in Hook.f., Fl.Brit.India 1:436. 1874; Gamble, Fl. Pres.Madras 132. 1915; Ramamoorthy in Sald. & Nocols., Fl. Hassan Dist. 399. 1976; Matthew, Ill. Fl. Tam. Carnatic t. 97.1982; Matthew & Britto in Matthew, Fl. Tam. Carnatic 3(1):190.1983; Mohanan & Sivad., Fl. Agasthyamala 128.2002. *Puliyarila.*

Trailing herbs; branchlets rooting at lower nodes. Leaves alternate, deigitately 3-foliolate; leaflets obcordate, 10-15 by 10-20 mm, base cuneate margin ciliolate, apex emarginate, membranous. Flowers bisexual, 5-merous, in pseudo-umbels; pedicel c. 1 cm long; sepals 5, imbricate, persistent; petals 5, yellow, twisted; disc present; stamens 10, filaments unequal; ovary 5- locular; ovule(s) 1-α per locule. Capsule cylindric. Seeds ellipsoid, transversely ridged.

Fl. & Fr.: Most of the seasons. *Distr.*: Cosmopolitan. Common along moist shady places in plains to hills. *NAK 2217* (Chittar, c. 200 m).

The plant is coolant and has an acidic taste; paste of leaves given with milk as a specific medicine for jaundice and dysentery.

31. BALSAMINACEAE

1. Lateral petals free; capsule subglobose or ellipsoid. .... **Hydrocera**
1. Lateral petals united in pairs; capsule fusiform or linear. .... **Impatiens**

HYDROCERA Blume
Bijdr. 241. 1825.

Hydrocera triflora (L.) Wight & Arn. in Wight, Cat.28.1833 & Prodr. 140.1834; Hook.f., Fl.Brit.India 1:483.1875; Gamble, Fl.Pres.Madras 146.1915. *Impatiens triflora* L., Sp.Pl. 938.1753.

Aquatic branched erect herbs; stem 5-angular, fistular in the internodal region, rooting at lower nodes. Leaves simple, alternate, linear-lanceate, 8-12 by 1.5-3 cm, base cuneate, margin serrate, lower serratures glandular, apex acuminate, chartaceous. Inflorescence an axillary cyme. Flowers showy, zygomorphic, bisexual, bracteate; sepals 5, petaloid, resupinate, posterior one produced into a spur; petals 5, pink, free, disc 0; stamens 5, united towards the connective, anthers connate around the pistil; ovary 5-locular; ovule 2-3 per locule. Capsule purplish, subglobose, fleshy, dehiscing septicidally. Seeds curved.

Fl. & Fr.: Aug.-Dec. *Distr.*: Sri Lanka, India, Myanmar. Very rare. In pools and ponds. *NAK 1975* (Tiruvalla, c. 10 m).

IMPATIENS Linnaeus
Sp. Pl. 937.1753.

Succulent herbs. Leaves simple, opposite, alternate or verticillate, petiole often glanduliferous. Flowers in lateral racemes, umbellate clusters or rarely solitary, irregular, showy, resupinate; sepals 3-5, imbricate, posterior one produced into a spur; petals 5, lateral petals united in pairs, each petal entire or 2-lobed; stamens 5, filaments fused, anthers coherent and adnate to the ovary; ovary superior, 5 - locular; ovules a per locule. Capsule fusiform to linear, loculicidally 5-valvular. Seeds α, tubercled.

1. Acaulescent scapigerous herbs; leaves radical. ....8. **I. scapiflora**
1. Caulescent non-scapigerous herbs; leaves cauline.
    2. Leaves opposite.
        3. Spur 5 mm - 4 cm long.
            4. Leaves sessile; flowers c.3.5 cm ϕ. .... 2. **I. chinensis**
            4. Leaves petioled; flowers c. 2 cm ϕ.
                5. Leaves ovate, long-petioled.
                    6. Spur inflated, longer than lip. ....11. **I. viscosa**
                    6. Spur not inflated, shorter or longer than lip. ....4. **I. goughii**
                5. Leaves linear-lanceate, short petioled. ....7. **I. minor**
        3. Spur 0-2 mm long.
            7. Glabrous herbs; lip keel-like.
                8. Leaves ovate-oblong. ....6. **I. inconspicua**
                8. Leaves narrowly linear. ....5. **I. herbicola**
            7. Tomentose herbs; lip sac-like. ....9. **I. tomentosa**
    2. Leaves alternate or verticillate.
        9. Leaves verticillate; flowers scarlet coloured. ....10. **I. verticillata**
        9. Leaves alternate; flowers pink cloured.
            10. Capsule tomentose; seeds smooth. ....1. **I. balsamina**
            10. Capsule glabrous; seeds minutely papillose. ....3. **I. flaccida**

1. **Impatiens balsamina** L., Sp. Pl. 938.1753; Wight & Arn., Prodr. 135.1834; Hook.f., Fl.Brit. India 1:453. 1874; Gamble, Fl.Pres.Madras 142.1915; Ramamoorthy in Sald. & Nicols., Fl.Hassan Dist. 400.1976; Matthew & Britto in Matthew, Fl. Tam. Carnatic 3(1):195.1983.

Leaves alternate, linear-lanceate, 2-5 by 0.3 cm, båse alternate, margin serrate, apex acute; petiole glanduliferous. Flower(s) solitary or 2-3 in axillary fascicles; lateral sepals ovate; lip saccate, spur curved, c.2.5 cm long; petals pale pink or rose coloured, standard broadly ovate, lateral suborbicular. Capsule ellipsoid. Seeds tubercled.

Fl. & Fr.: Oct.-Mar. *Distr.*: Indo-Malesia, China. Locally abundant in exposed clayey rocks. *NAK 157* (Pandalam, c. 30 m).

2. **Impatiens chinensis** L., Sp. Pl. 937.1753; Hook.f., Fl.Brit. India 1:444.1874; Gamble, Fl.Pres.Madras 139.1915; Ramamoorthy in Sald. & Nicols., Fl.Hassan Dist. 402.1976; Matthew & Britto in Matthew, Fl.Tam. Carnatic 3(1):196.1983; Manilal, Fl.Silent Valley 39.1988; Nicols., Suresh & Mani., An Interpr. Hort. Malab. 70. 1988; Mohanan & Sivad., Fl. Agasthyamala 130.2002. *Onapu* Rheede, Hort. Malab. 9:89-90, t.47.1689.

Leaves opposite, lanceate to elliptic, 3-6 by 0.5-1 cm, base obtuse, margin serrate, apex obtuse or acute, whitish below, pinkish-nerved. Flowers 2-3, in clusters, rarely solitary; pedicel to 4 cm long; lateral sepals oblanceate; spur to 4.5 cm long; petals pink, standard broadly ovate. Capsule ellipsoid.

Fl. & Fr.: Oct.-May. *Distr.*: India, Myanmar, China. Common. On exposed marshes, streamsides, abandoned paddy fields, etc. *NAK 2317* (Kodumon, c. 50 m).

3. **Impatiens flaccida** Arn. in Hook., Comp. Bot.Mag. 1:322.1835; Hook.f., Fl.Brit. India 1:457.1874; Gamble, Fl.Pres. Madras 143.1915; Matthew & Britto in Matthew, Fl.Tam. Carnatic 3(1):196.1983.

Leaves alternate, ovate to ovate-lanceate, 2-9 by 1.5-4 cm, base cuneate, margin crenate, glanduliferous, apex acuminate; glandular. Flowers usually in pairs; pedicel to 3 cm long, lateral sepals lanceate; spur to 3 cm long; petals white to pale rose coloured. Capsule ellipsoid. Seeds papillose.

Fl. & Fr.: Jul.-Oct. *Distr.*: Sri Lanka, Peninsular India. Rare. In the outskirts of moist shady forests. *NAK 56* (Moozhiar, c. 250 m), *960* (Sabarimala, c. 200 m).

4. **Impatiens goughii** Wight, Ill. 1:160.1840 & Ic.t.1603.1850; Hook.f., Fl.Brit.India 1:452.1874; Gamble, Fl.Pres. Madras 144.1915. *I. anamallayensis* Bedd., Ic.t. 150.1868-1874.

Leaves opposite to rarely alternate, ovate-lanceate, 1.5-4 by 1-2 cm, base rounded to subacute, glandular, margin serrate, apex acuminate. Flowers in false umbles, or in short racemes, lateral sepals linear-lanceate; spur curved, c.5 mm long; petals pink, standard orbicular. Capsule ovoid. Seeds tubercled.

Fl. & Fr.: Sep.-Dec. *Distr.*: Western Ghats. Rare. In upper hills, usually on wet rocky areas. *NAK 83* and *105* (Upper Moozhiar, c. 550 m).

5. **Impatiens herbicola** Hook.f., Bull.Misc. Inform. 1911:354.1911; Gamble, Fl.Pres.Madras 141.1915; Ramamoorthy in Sald. & Nicols., Fl.Hassan Dist. 402.1976.

Leaves opposite, linear-lanceate, 2-6 by 0.2-0.3 cm, base and apex acute, margin distantly serrulate. Flowers 2-3, in clusters or solitary; pedicel to 1.5 cm long; sepals subulate; spur very short; petals white, lip cymbiform, saccate, standard broadly ovate. Capsule ellipsoid. Seeds ovoid.

Fl. & Fr.: Nov.-Jan. *Distr.*: Western Ghats (Anamalais & Travancore hills). Rare. In exposed fields of upper hills. *NAK 34* (upper Moozhiar, 550 m).

6. **Impatiens inconspicua** Benth. ex Wight & Arn, Prodr. 139. 1834; Wight, Ic. t. 970.1845. *Impatiens pusilla* Heyne ex Hook.f. & Thoms., J.Linn. Soc. Bot. 4:122. 1859. *I. inconspicua* Benth. in Wall., Cat. 4741.1845; Hook.f., Fl.Brit. India 1:447.1874; Gamble, Fl.Pres. Madras 140.1915.

Leaves opposite, elliptic, 1-4 by 0.5-1.5 cm, base and apex acute, margin bristly. Flower(s) solitary or in pairs; pedicel to 7 mm long, very slender; lateral sepals linear-lanceate; spur absent; petals pale pink, standard ovate. Capule linear. Seeds globose.

Fl. & Fr.: Sep.-Dec. *Distr.*: Western Ghats. Rare. *KV 48347* (MH) (Pamba-Anathode).

7. **Impatiens minor** (DC.). Suresh in Nicols., Suresh & Mani., An Interpr. Hort. Malab. 71.1988. *Balsamina minor* DC., Prodr. 1:686.1824. *Impatiens kleinii* Wight & Arn., Prodr. 1:140.1834; Hook.f., Fl. Brit. India 1:445.1874; Gamble, Fl.Pres.Madras 140.1915; Ramamoorthy in Sald. & Nicols., Fl.Hassan Dist. 403.1976; Manilal, Fl.Silent Valley 40.1988. *Man-onapu* Rheede, Hort.Malab. 9:99, t.51.1689. **(Fig. 18)**

Leaves opposite, ovate-oblong, 2-8 by 0.7-2 cm, base cuneate or attenuate, margin faintly serrate, apex acute. Flower(s) solitary; pedicel, to 2 cm long; sepals subulate; spur long, slightly curved; petals purple, pink or white, standard ovate. Capsule fusiform. Seeds tubercled.

Fl. & Fr.: Aug.-Dec. Distr.: Western Ghats. Rare. In exposed grassy areas of upper hills. *NAK 822* (Chelikuzhy-Ranni, c. 100 m).

8. **Impatiens scapiflora** Heyne ex Roxb., Fl.Ind. 2:464.1824; Hook.f., Fl.Brit. India 1:443.1874; Gamble, Fl.Pres.Madras 138.1915; Ramamoorthy in Sald. & Nicols., Fl.Hassan Dist. 403.1974; Manilal, Fl.Silent Valley 41.1988; Mohanan & Sivad., Fl. Agasthyamala 133.2002.

Leaves radical, orbicular-ovate, 5-10 by 4-8 cm, base deeply cordate, margin crenate, apex obtuse. Flowers in scapose racemes; scape to 25 cm long; pedicel c.2.5 cm long; petals white or rose coloured, wing petals deeply divided into 3 lobes; spur curved, to 4 cm long. Capsule ellipsoid.

Fl. & Fr.: Jul.-Dec. *Distr.*: Western Ghats. Most common species of the genus in hills. In upper hills, on dripping rocks. *NAK 87* (Moozhiar, c. 250 m).

9. **Impatiens tomentosa** Heyne ex Wight & Arn., Prodr. 139.1834; Wight, Ic.t. 749.1844; Hook.f., Fl.Brit. India 1:449.1874; Gamble, Fl.Pres.Madras 141.1915; Mohanan & Sivad., Fl. Agasthyamala 133.2002.

Branchlets rusty tomentose. Leaves opposite, sessile, oblong-lanceate, 2-4 by 0.4-0.5 cm, base amplexicaule, margin serrulate, revolute, apex acute-mucronate, upper surface green, scabrous, lower surface white-tomentose. Flower(s) solitary, rarely in pairs; pedicel c.3 cm long; lateral sepals linear-lanceate; spur curved upwards, c.1 cm long; petals rose or pink coloured, standard broadly obovate. Capsule narrowly elliposid. Seeds smooth, shiny.

Fl. & Fr.: Sep.-Dec. *Distr.*: Western Ghats. Less common. In swamps and along wet waysides in upper hills *NAK 407* and *1328* (Kakki hills, c. 1100 m).

10. **Impatiens verticillata** Wight, Madras J. Lit. Sci. Ser.1, 5:15. 1837; Hook.f., Fl. Brit. India 1:452.1874; Gamble, Fl.Pres. Madras 144.1915; Mohanan & Sivad., Fl. Agasthyamala 135.2002.

Leaves verticillate, narrowly oblanceate, 4-12 by 1-1.5 cm, base attenuate, margin distantly serrulate, apex acute to acuminate, lower surface glaucous, chartaceous. Flowers in pseudo-umbels, terminal or axillary; peduncle c. 3 cm long; pedicel c. 2 cm long, lateral sepals ovate-lanceate; petals scarlet coloured; standard broadly ovate or suborbicular. Capsule elliposid. Seeds smooth.

Fl. & Fr.: Sep.-Apr. *Distr.*: Western Ghats. Rare. Very beautiful woody herbs along streamside rocks in hills. *NAK 2415* (Moozhiar, c. 250 m).

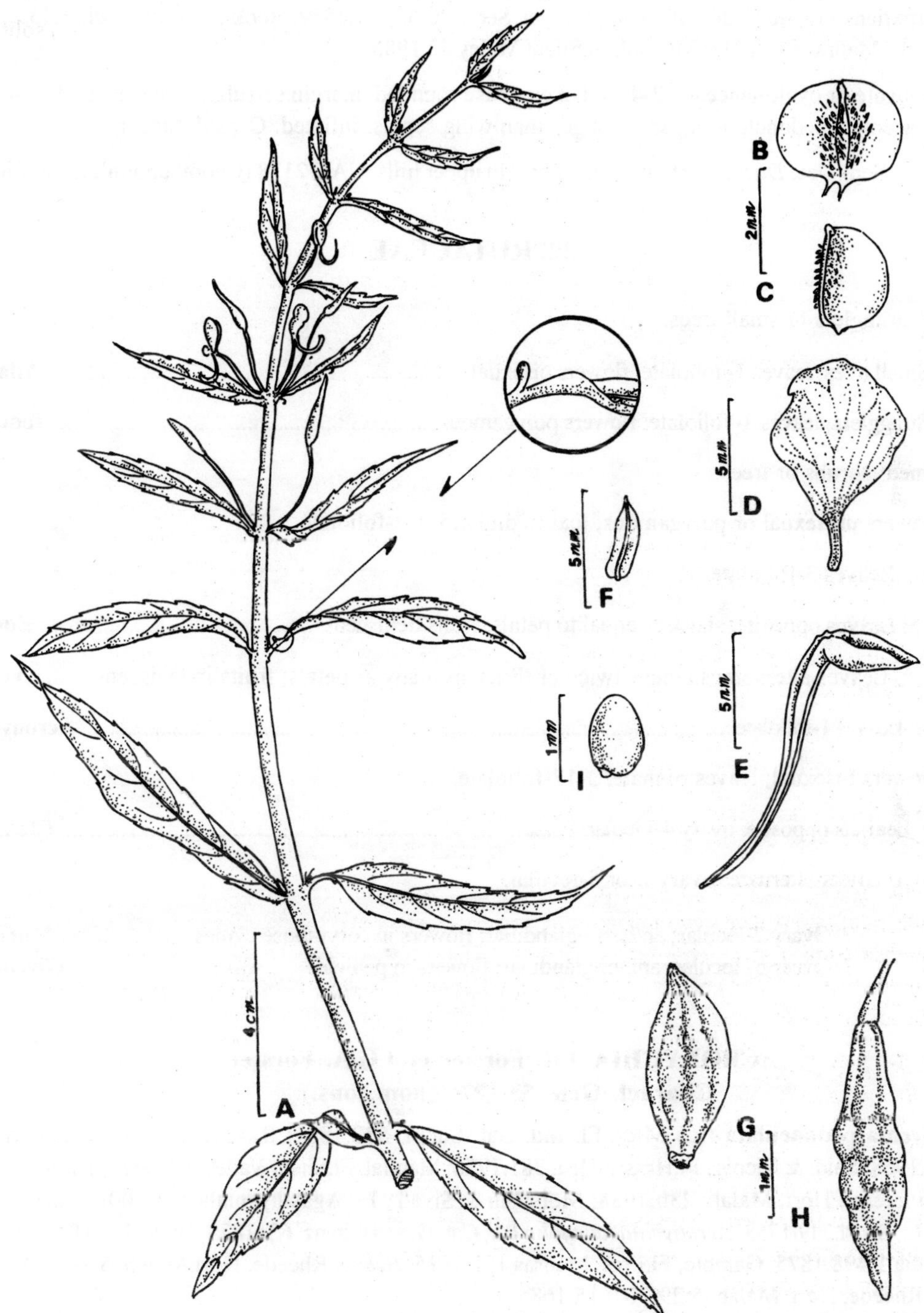

Fig. 18. ***Impatiens minor*** **(DC.) Suresh: A. Twig; B. Posterior sepal; C. Posterior sepal- side view; D. Petal; E. Dorsal sepal; F. Lateral sepal; G. Pistil; H. Fruit; I. Seed.**

11. **Impatiens viscosa** Bedd., Madras J. Lit. Sci. Ser. 2,10:66, t. 7.1859; Hook.f., Fl.Brit. India 1:453.1874; Gamble, Fl.Pres.Madras 144.1915; Manilal, Fl.Silent Valley 42.1988.

Leaves ovate or ovate-lanceate, 2-4 by 1-2 cm, base rounded, margin serrulate, apex acute. Flowers α, in terminal racemes; peduncle long; spur longer than wing sepals, inflated. Capsule linear.

Fl. & Fr.: Sep- Dec. *Distr.*: Western Ghats. Rare. In upper hills. NAK 2178 (Choothuporuthupara, Chittar, c. 400 m).

32. RUTACEAE

1. Armed stragglers or small trees.
    2. Small trees; leaves 1- foliolate; flowers bisexual. .... **Atlantia**
    2. Stragglers; leaves 3- foliolate; flowers polygamous. .... **Toddalia**
1. Unarmed shrubs or trees.
    3. Flowers unisexual or polygamous; leaves digitate, 1-3-foliolate.
        4. Leaves 3-foliolate.
            5. Leaves opposite; stamens equal to petals; fruits dehiscent. .... **Euodia**
            5. Leaves alternate; stamens twice or thrice as many as petals; fruits indehiscent. .... **Vepris**
        4. Leaves 1-foliolate. .... **Acronychia**
    3. Flowers bisexual; leaves pinnate, 3-17-foliolate.
        6. Leaflets opposite; ovary 4-locular. .... **Clausena**
        6. Leaflets alternate; ovary 2-or 5-locular.
            7. Ovary 2-locular; anthers eglandular; flowers in corymbose cymes. .... **Murraya**
            7. Ovary 5- locular; anther glandular; flowers in panicles. .... **Glycosmis**

ACRONYCHIA J.R. Forster et J.G.A. Forster Charact. Gen. 53.1776 (nom.cons.).

Acronychia pedunculata (L.) Miq., Fl. Ind. Bat. Suppl. 532.1861; Rau, Bull. Bot. Surv. India 10 (Suppl.2):18.1968; Sald. & Nicols., Fl.Hassan Dist. 381.1976; Manilal, Fl.Silent Valley 42.1988; Nicols., Suresh & Mani., An Interpr. Hort. Malab. 230.1988; Mohanan & Sivad., Fl. Agasthyamala 137.2002. *Jambolifera pedunculata* L., Sp.Pl.349.1753. *Acronychia laurifolia* Bl., Cat. Gew. Buitenz. 63. 1823, Bijdr. 245.1825; Hooker f., Fl.Brit. India 1:498.1875; Gamble, Fl.Pres. Madras 152.1915. *Beenel* Rheede, Hort.Malab. 5:7-8, t.4.1685. *Perin-panel* Rheede, Hort. Malab. 5:29-30, t.15.1685.

Trees. Leaves opposite to alternate, 1-foliolate, elliptic-obovate, 8-13 by 4-6 cm, base cuneate, apex obtusely short-acuminate, thin-coriaceous, pellucid-punctate. Inflorescence an axillary or terminal corymb; peduncle c.5 mm long. Flowers polygamous, c. 6 mm ϕ, 4-merous; sepals 4, connate below; petals 4, yellow, spreading, tomentose within; stamens 8, alternately shorter and longer, disc prominent, 8-10-angled; ovary superior, 4-locular; ovule(s) 1-or 2-per locule; stigma 4-groved. Capsule loculicidal, ellipsoid.

Fl. & Fr.: May-Oct. *Distr.*: Indo-Malesia. Occasional. In evergreen forests. *NAK 1183* (Upper Moozhiar, c. 550 m). Bark made into a paste and applied to sores and ulcers.

ATLANTIA Correa
Ann. Mus. Natl. Hist. Nat. 6:383,385,1805 (nom.cons.).

Atalantia racemosa Wight & Arn., Prodr. 91.1834; Oliv., J. Linn. Soc. 5. (Suppl.2):24. 1861; Hook. f., Fl. Brit. India 1:512. 1875; Gamble, Fl.Pres. Madras 159. 1915; Matthew & Britto in Matthew, Fl.Tam. Carnatic 3(1):201.1983; Manilal, Fl. Silent Valley 43. 1988.

Armed or umarmed trees. Leaves 1-foliolate, oblong-lanceate, 6-10 by 2-4 cm, base acute to cuneate, margin subentire, apex obtuse-emarginate, thin-coriaceous. Flowers in racemes, axillary, 4-merous, bisexual; calyx-lobes 4; petals 4, cream coloured; stames 8, basally connate; ovary 2-locular; ovules 2-per locule. Berry ellipsoid or globose.

Fl. & Fr.: Oct.-Jan. *Distr.*: Sri Lanka, Peninsular India. Less common. In semi-evergreen forests, towards streams. *NAK 1698* (Pampa- Thriveny, c. 200 m).

CLAUSENA N.L. Burman
Fl.Ind. 87("Claucena") t.29,f.2.1768.

Clausena austroindica Stone & Nair, Nordic J. Bot. 14 (5): 491. 1994; Mohanan & Sivad., Fl. Agasthyamala 138. 2002. *Clausena heptaphylla* sensu Hook.f., Fl.Brit. India 1: 504. 1875; Gamble, Fl. Pres. Madras 155. 1915; Manilal, Fl. Silent Valley 44.1988, non (Roxb.) Wight & Arn.,1834.

Shrubs. Leaves odd-pinnate, c. 30 cm long; leaflets alternate or subopposite, lanceate to oblanceate, 6-20 by 3.5-8 cm, base oblique, margin crenate, apex acuminate, lateral nerves 8-12 pairs, chartaceous, profusely glandular-punctate. Inflorescence a thyrsoid-panicle, axillary or subterminal, c. 20 cm long. Flowers bisexual, 4-5 merous, c. 7 mm ϕ; calyx-lobes 4-5, glandular without; petals 4-5, yellow; disc elongate; stamens 8-10, alternately unequal; ovary 4-locular; ovuale 1 per locule. Berry globose.

Fl.: & Fr.: Jan.-Apr. *Distr.*: Peninsular & Eastern India. Common in the outskirts of semi-everygreen forests. *NAK 602* (Thekkutode, c.70 m), *1652* (Pampa -Thriveny, c. 200 m).

EUODIA J.R. Forster et J.G.A. Forster
Charact. Gen. 13. 1776.

Euodia lunu-ankenda (Gaertn.) Merr., Philipp. J. Sci. Bot. 7:378. 1912, ('*Evodia lunur-ankenda*'); Gamble, Fl. Pres. Madras 148. 1915 ('*Euodia lunu-akenda*'); Gandhi in Sald. & Nicols., Fl. Hassan Dist. 384. 1976; Matthew & Britto in Matthew, Fl. Tam. Carnatic 3(1):204. 1983; Manilal, Fl. Silent Valley 44. 1988; Mohanan & Sivad., Fl. Agasthyamala 140. 2002. *Fagara lunu akenda* Gaertn., Fruct. 1:334, t.68, f.9.1788. *Evodia roxburghiana* Benth., Fl. Hong. 59. 1861; Hook.f., Fl. Brit. India 1:487. 1785. *Kambili, Kanala.*

Trees. Leaves digitately 3-foliolate, ovate-lanceate or elliptic, 10-14 by 4-6 cm, base obliquely cuneate, apex acuminate, lower surface pubescent, thin-coriaceous. Inflorescence a paniculate-cyme. Flower 4-merous; sepal 4, basally connate; petals 4, white; disc 4-5- lobed; stamens 4; inserted below the disc; pistillode short-lobed; ovary superior, 4-lobed, 4-locular; ovules 2 per locule. Drupe ellipsoid of 2 valved cocci.

Fl. & Fr.: Feb.-Oct. *Distr.*: Sri Lanka, India, Malaya Peninsula, Malesia. Occasional. In evergreen forests. *NAK 710* (Plappally, c. 350 m), *1552* (Moozhiar, c.250 m).

Juice of leaves boiled with oil is used in fever. Wood suitable for match industry.

GLYCOSMIS Correa
Ann. Mus. Natl. Hist. Nat. 6:384. 1805.

Unarmed shrubs. Leaves alternate, rarely opposite, odd-pinnate; leaflets 1-15, alternate, faintly glandular- punctate. Inflorescence usually a panicle, rarely cyme, axillary. Flowers 4-5 merous, bisexual; stamens 8-10, unequal; disc 4-lobed; anthers apically glandular; ovary superior, 2-5-locular. Fruits globose or ellipsoid.

1. Leaflets 3-5 (usually 3), margin entire, lateral nerves 5-8 pairs; inflorescence up to 3 cm long. .... 1. **G. mauritiana**
1. Leaflets usually 3-7, margin subentire, lateral nerves c.12 pairs; inflorescence upto 8 cm long. .... 2. **G. pentaphylla**

1. **Glycosmis mauritiana** (Lam.) Tanaka, Bot. Not. 28:159. 1928, Bull. Soc. Bot. France 75:708. 1928, J. Bot. 68:226. 1930 & J. Indian Bot. Soc. 16:229. 1937; Brizicky, J. Arnold Arbor. 43:90. 1962; Matthew, Ill. Fl. Tam. Carnatic t. 107. 1982; Stone in Matthew, Fl. Tam. Carnatic 3(1):205.1983; Manilal, Fl. Silent Valley 45. 1988; Mohanan & Sivad., Fl. Agasthyamala 143. 2002. *Limonia mauritiana* Lam., Encycl. 3:517. 1788(1792). *L. pentaphylla* auct., non Retz., 1789: Roxb., Fl. Corom. t.84. 1798 & Fl. Ind. 2:381. 1832.

Leaves 3-5-foliolate; usually 3; leaflets lanceate or oblanceate, 3-10 by 1.5-3.5 cm, base cuneate, apex acute to shortly acuminate, lateral nerves 5-8 pairs, chartaceous. Panicle axillary, to 3 cm long; sepals ovate, margin scarious; petals white, elliptic-ovate, glandular without; anthers lanceoid, apically glandular, dorsal side without gland. Fruit subglobose.

Fl. & Fr.: Sep.-Mar. *Distr.*: Sri Lanka, India, Coastal S.E. Asia. Common in exposed abandoned fields. *NAK 1941* (Tiruvalla, c. 10 m).

2. **Glycosmis pentaphylla** (Retz.) DC., Prodr. 1:538. 1824 (nomen solum) Tanaka, Bot. Not. 28:159. 1928; Mitra & Subram., J. Arnold Arbor. 50:155. 1969; Stone in Matthew, Fl. Tam. Carnatic 3 (1):207. 1983 & in Dassan. & Fosberg, Rev. Handb. Fl. Ceylon 5:449. 1985; Nicols., Suresh & Mani., An Interpr. Hort. Malab. 232. 1988 ; Mohanan & Sivad., Fl. Agasthyamala 144. 2002. *Limonia pentaphylla* Retz., Obs. Bot. 5:24. 1788. *L. arborea* (Roxb.) DC., Prodr. 1:538.1824; Brizicky, J. Arnold Arbor. 43:91. 1962; Gandhi in Sald. & Nicols., Fl. Hassan Dist. 384. 1976. *G. cochinchinensis* auct., non Pierre *ex* Engl., 1896: Gamble, Fl. Pres. Madras 153. 1915, p.p. *Panel* Rheede, Hort. Malab. 2:9-10, t. 1679.

Leaves 3-7- foliolate; leaflets elliptic, 10-20 by 4-8 cm, base oblique, margin crenate, apex obtuse or obtusely acuminate, lateral veins to 12 pairs, chartaceous. Panicle axillary, to 8 cm long; sepals ovate, margin scarious, ciliate; petals white; anthers ovate, apically and dorsally glandular. Fruit subglobose.

Fl. & Fr.: Most of the seasons. *Distr.*: Sri Lanka, India, S.E. Asia and W. Malesia. Very common in exposed waste lowlands and forest edges. *NAK 1942* (Tiruvalla, c. 10 m).

Decoction of root improve blood and given in diarrhoea, rheumatism, and twigs used as tooth brush.

MURRAYA J.G. Koenig ex Linnaeus
Mant. 2:554, 563. 1771 ("Murraea"); corr. Murraya Syst. Veg. ed. 13, 331. 1774 (nom. *et* orth. cons.).

Murraya paniculata (L.) Jack, Malayan Misc. 1. (5) : 31. 1820; Alston in Trimen, Fl. Ceylon 6 (Suppl.) : 37. 1931; Gandhi in Sald. & Nicols., Fl. Hassan Dist. 386. 1976; Matthew, Ill. Fl. Tam. Carnatic t. 109. 1982; Stone in Matthew, Fl. Tam. Carnatic 3(1) : 214. 1983; Mohanan & Sivad., Fl. Agasthyamala 145. 2002. *Chalcas paniculata* L., Mant. Pl. 68. 1767. *Murraya exotica* auct, non L., 1771: Roxb., Fl. Ind. 2:374. 1832; Wight, Ic.t. 96. 1838; Hook.f., Fl. Brit. India 1:502. 1875; Gamble, Fl. Pres. Madras 155. 1915.

Shrubs. Leaves 3-5- foliolate, rarely 1-foliolate; leaflets ovate or ovate-elliptic, 3-8 by 1.5-3.5 cm, base cuneate, apex acuminate, lateral veins c.8 pairs, chartaceous. Panicle few-flowered, axillary. Flowers 5-merous, bisexual, regular; sepals ovate-lanceate; petals white, elliptic or oblong-elliptic, disc annular; stamens 10, subequal; ovary 2-5- locular; ovules 2 per locule; stigma 2-lobed. Berry ovoid. Seeds pubescent.

Fl. & Fr.: Mar.-Sep. *Distr.*: Sri Lanka, India, Indo-China, Malesia to Australia. Less Common. In forest borders, usually in shades. *Binu 3058* (TBGT) (Attathode-Plappally, c. 350 m).

TODDALIA A.L. Jussieu
Gen. 371. 1789 (nom. cons.)

Toddalia asiatica (L.) Lam., Tabl. Encycl. 2:116. 1792, var. **floribunda** Gamble, Fl. Pres. Madras 151. 1915; Gandhi in Sald. & Nicols., Fl. Hassan Dist. 388. 1976; Matthew, Ill. Fl. Tam. Carnatic t. 113. 1982; Matthew & Britto in Matthew, Fl. Tam. Carnatic 3(1) : 214. 1983; Manilal, Fl. Silent Valley 46. 1988. *Paullinia asiatica* L., Sp. Pl. 365. 1753. *Toddalia aculeata* Pers., Syn. Pl. 1:249. 1805; Wight, Ic. t. 66. 1840; Hook.f, Fl. Brit. India 1:497. 1875. *Kakka-Thodali.*

Stragglers; stem profusely prickled. Leaves 3-foliolate; leaflets oblanceate, 4-10 by 1.5-4.5 cm, base cuneate, margin crenate-glandular, apex acuminate, glandular punctate, lateral nerves α, ascending, coriaceous. Panicle terminal or axillary, c.9 cm long. Flowers unisexual; pentamerous; sepals triangular; petals yellow, linearly elliptic; stamens 5, free; pistillodes prominent; ovary globose, 4-7-locular; ovules 2-per locule. Berry globose. Seeds angular, reniform.

Fl. & Fr.: Sep.-Mar. *Distr.*: Peninsular India. Not common. In evergreen forests. *NAK 1185* (Upper Moozhiar, 550 m), *2622* (Moozhiar, c.250 m).

Root is used as a remedy in diarrhoea which possess active stimulant and carminative properties.

VEPRIS Commerson ex A.H.L. Jussieu
Mem. Mus. Hist. Nat. 12 : 509. 1825.

Vepris bilocularis (Wight & Arn.) Engl. in Engl. & Prantl, Pflanzenfam. 3(4): 178. 1896; Gamble, Fl. Pres. Madras 151. 1915; Gandhi in Sald. & Nicols., Fl. Hassan Dist. 388. 1976. *Toddalia bilocularis* Wight & Arn., Prodr. 149.1834; Bedd., Ic. t. 167. 1874; Hook.f., Fl. Brit. India 1:497. 1875. *Kar-Agil.*

Dioecious trees. Leaves 1-3- foliolate, alternate, leaflets elliptic - oblong, 10-20 by 4-7 cm, base cuneate, apex acuminate, thin-coriaceous. Inflorescence a panicle, axillary or terminal. Flowers small, unisexual; calyx minute; petals 2-4, white; stamens 4-8; disc ridged; pistillode 2-lobed; ovary globose, 2-4 -locular; ovules 2-per locule. Berry oblong or globose. Seeds 2-grooved.

Fl. & Fr.: Feb.-May. *Distr.*: Western Ghats. Rare. In evergreen forests. *NAK 1791* (Angamoozhy, c. 275 m).

Decoction of wood along with oil is used in eye and ear diseases, and rheumatism.

Aegle marmelos (L.) Corr. *Citrus* spp. and *Murraya koenigii* (L.) Spreng. are cultivated in the study area.

33. SIMAROUBACEAE

BRUCEA J.F. Miller
Icon. Anim. Pl. t. 25. 1779-1780 (nom. cons.).

Brucea sumatrana Roxb., Fl. Ind. 1:449. 1832; Hook.f., Fl. Brit. India 1:521. 1875.

Shrubs. Leaves odd-pinnate, c. 50 cm long, alternate; leaflets 4-6 pairs, elliptic, 10-20 by 3-6 cm, base oblique, margin coarsely toothed, apex acuminate, lower surface pubescent. Flowers very small, bisexual, rarely polygamous, 4 - merous, grouped in small cymes which are in axillary panicles; calyx minute, lobes 4, petals 4, yellow, disc prominent, 4-lobed; stamens 4, outside the disc; ovary superior, 4-locular; ovule 1 per locule. Drupes 4, globose.

Fl. & Fr.: Jul.-Dec. *Distr.*: Eastern India, S. China, Sumatra, Java, Philippines. Rare. Collected as wild from a destructed easily accessible forest area. *NAK 655* (Puthukkulam, c. 40 m).

The whole plant is very bitter and it is proposed as a good antidysenterical medicine.

Ailanthes excelsa Roxb. is planted for soft wood.

34. OCHNACEAE

GOMPHIA Schreber
Gen. 291. 1789.

Gomphia serrata (Gaertn.) Kanis, Taxon 16:422. 1967 & Blumea 16:53. 1968; Nicols., Suresh & Mani., An Interpr. Hort. Malab. 199. 1988. *Meesia serrata* Gaertn., Fruct. 1:344, t. 70, f. 6. 1788. *Gomphia angustifolia* Vahl, Symb. Bot. 2:49. 1791; Bennett in Hook. f., Fl. Brit. India 1:525. 1875. *Ouratia angustifolia* (Vahl) Baill. ex Laness., Fl. Util. Col. Fr. 607. 1886; Gamble, Fl. Pres. Madras 167. 1915. *Ouratia serrata* (Gaertn.) Robson, Taxon 11:51. 1962. *Tsjocatti* Rheede, Hort. Malab. 5:95-96, t.48. 1685. *Poea-tsjetti* Rheede, Hort. Malab. 5:103-104, t.52. 1685.

Shrubs. Leaves simple, alternate, elliptic-oblong 8-18 by 2.5-5 cm, base acute, margin sharply serrulate, apex shortly acuminate, shiny, lateral nerves a, close, parallel, coriaceous. Inflorescence an axillary or terminal panicle; bracts triangular. Flowers c.15 mm ϕ, 5-merous, bisexual; sepal 5, free, elliptic-ovate; petals 5, bright yellow, narrowly obovate; disc lobed; stamens 10, subsessile, arising outside the disc; anthers poricidal; ovary deeply 4-5-lobed, reddish; ovule 1 per locule. Drupes 4-5 seated on disc.

Fl. & Fr.: Jan.-Aug. *Distr.*: Sri Lanka, Peninsular India & Philippines. Occasional. Seen in sacred groves. *NAK 148 & 2335* (Kodumon, c. 50 m).

Roots and leaves bitter, commonly used as decoction in milk or water as a tonic for stomach disorders.

35. BURSERACEAE

CANARIUM Linnaeus
Amoen. Acad. 4:121. 1759.

Canarium strictum Roxb., Fl.Ind. 3:138. 1832; Wight & Arn., Prodr. 175. 1834; Bedd., Fl. Sylv. t.128. 1871; Bennett in Hook.f., Fl. Brit. India 1:534. 1875; Gamble, Fl. Pres. Madras 172. 1915; Ramamoorthy in Sald. & Nicols., Fl. Hassan Dist. 371. 1976; Matthew, Ill. Fl. Tam. Carnatic 3(1):225. 1983; Mohanan & Sivad., Fl. Agasthyamala 147.2002.

Trees. Leaves odd-pinnate, alterante. Inflorescence an axillary panicle. Flowers polygamous, 3-merous; disc annular; stamens 6, connate into a tube; ovary 3-locular; ovule 1 per locule. Drupe oblong.

Fl. & Fr.: Mar.-Apr. *Distr.*: India, Upper Myanmar. Rare. In evergreen forests.

Plant yields a good resin which is dried and powdered and given internally as a remedy for rheumatism, cough, fever, epilepsy, asthma and blood impurities.

Note :Failed to collect the flowering/fruiting twigs of this plant, but observed a few in a deep forest near Aluvamkudi, Gurunadhan Mannu Forest range. The local people use these trees for resin.

36. MELIACEAE

1. Herbs; leaves digitately 3-foliolate; petiole winged. .... **Naregamia**
1. Trees (rarely shrubs); leaves pinnately a-foliolate; petiole not winged.
    2. Leaves even-pinnate.
        3. Leaves 2-3- pinnate; leaf margin dentate. .... **Melia**
        3. Leaves 1-pinnate; leaf margin entire.
            4. Stamens connate, inserted at the base of staminal tube; flowers yellowish-cream; ovary 2-locular. .... **Aphanamixis**
            4. Stamens tree, inserted on a fleshy torus; flowers white; ovary 5-locular. .... **Toona**
    2. Leaves odd-pinnate.
        5. Fruit a capsule; staminal tube apically 5-0- toothed or rarely entire.
            6. Staminal tube cylindric; disc cupular. .... **Dysoxylum**
            6. Staminal tube globose; disc annular or obscure.
                7. Lobes of staminal tube 2-fid; leaflets opposite; flowers bisexual. .... **Trichilia**
                7. Lobes of staminal tube entire or absent; leaflets subopposite; flowers polygamo-dioecious. .... **Aglaia**
        5. Fruit a drupe; staminal tube apically 2-toothed. .... **Cipadessa**

AGLAIA Loureiro
Fl. Cochinch. 173. 1790 (nom. cons.).

Trees. Leaves 3-13-foliolate, odd-pinnate, alternate; leaflets opposite or subopposite. Inflorescence an axillary panicle. Flowers polygamo-dioecious, 5-merous; calyx 5-lobed; petals 5, free, imbricate; stamens 5; staminal tube subglobose or urceolate; apically toothed or entire; disc 0; ovary superior, 1-2-locular; ovules 1 or 2 per locule. Berry globose, tomentose. Seeds ellipsoid.

1. Flowers in large panicles; leaflets 5-6 pairs, drying black. ....2. **A. perviridis**

1. Flowers in small panicles; leaflets 2 or 3 pairs, drying not so. ....1. **A. elaegnoidea**

1. **Aglaia elaeagnoidea** (Juss.) Benth., Fl. Australia 1:383. 1863; Backer & Bakh. f., Fl. Java 2: 128. 1965; Ramamoorthy in Sald. & Nicols., Fl. Hassan Dist. 392. 1976; Mabb., Taxon 26: 530. 1977; Nicols., Suresh & Mani., An Interpr. Hort. Malab. 177. 1988; Mohanan & Sivad., Fl. Agasthyamala 149. 2002. *Nemedra elaegnoidea* Juss., Mem. Mus. Hist. Nat. Paris 19:259, t. 14.1830. *Aglaia roxburghiana* Miq., Ann. Mus. Lugd.- Bat. 4:41. 1868; Gamble, Fl. Pres. Madras 180. 1915. *Nyalel, Nialel* Rheede, Hort. Malab. 4:37-38, t. 16. 1683.

Leaflets 7, oblong-lanceate, 6-12 by 2-4 cm, base obtuse, apex shortly acuminate, drying brownish. Inflorescence lepidote; calyx-lobes minute, scaly; petals pale yellow, oblong. Berry subglobose. Seeds ex-arillate.

Fl. & Fr.: Jan.-Apr. *Distr.*: Indo-Malesia. Less common. In evergreen and semi-evergreen forests. *NAK 2790* (Pampa-Anathode, c.1000 m).

2. **Aglaia perviridis** Hiern in Hook. f., Fl. Brit. India 1:556. 1875; Panell, Kew Bull. Add. Ser. 16:198. 1992. *Aglaia maiae* Bourd., J. Bombay Nat. Hist. Soc. 12:349. 1898; Gamble, Fl. Pres. Madras 180. 1915; Ramamoorthy in Sald. & Nicols., Fl. Hassan Dist. 392. 1976. *Cheru chokla.* **(Fig. 19)**

Leaves c. 30 cm long; leaflets subopposite, elliptic-oblong or ovate-oblong, 6-14 by 4-6 cm, base very oblique, apex obtuse or obtusely acuminate, drying black. Panicle c. 30 cm long; calyx-lobes ovate; petals yellow, narrowly obovate; staminal tube obscruely 5-toothed. Berry globose. Seeds ellipsoid.

Fl. & Fr.: Apr.-Sep. *Distr.*: Endemic to Western Ghats. Rare. Medium sized trees in evergreen forests. *NAK 1707* (Sabarimala, c. 200 m).

Wood scented like sandal wood.

APHANAMIXIS Blume
Bijdr. 165. 1825.

Aphanamixis polystachya (Wall.) Parker, Ind. For. 57:486. 1931; Mabb., Taxon 26:528. 1977; Ramamoorthy in Sald. & Nicols., Fl. Hassan Dist. 393. 1976; Manilal, Fl. Silent Valley 48. 1988 ; Mohanan & Sivad., Fl. Agasthyamala150. 2002. *Aglaia polystachya* Wall. in Roxb., Fl.Ind. Orient. 119. 1834; Bedd., Fl. Sylv. t. 132. 1871; Hiern in Hook.f., Fl. Brit. India 1:559. 1875; Gamble, Fl. Pres. Madras 181. 1915. *Chemmarom.*

Trees. Leaves odd-pinnate; leaflets up to 13, opposite, ovate-oblong or oblong-lanceate, 10-18 by 6-8 cm, base obique, margin subentire, apex obtuse or obtusely acuminate lateral veins prominent, coriaceous. Flowers polygamo-dioecious. Male flowers in panicles. Female flowers in long drooping spikes; calyx-tube

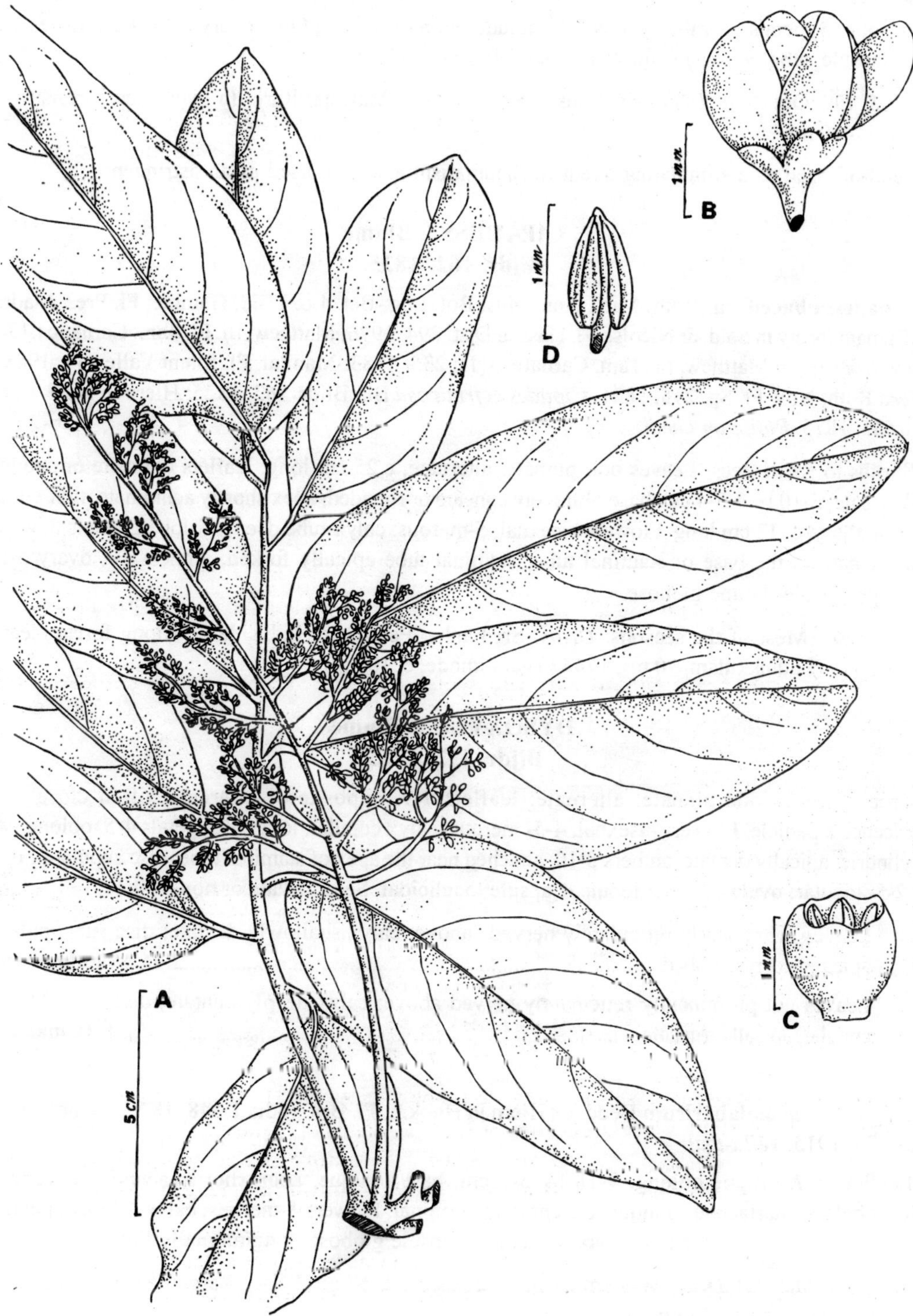

Fig. 19. ***Aglaia perviridis*** **Heirn : A. Twig; B. Flower; C. Stamial column; D. Stamen.**

cupular, lobes 5; petals 3, white; stamens 6, included in the staminal tube; ovary 3-locular; ovules 1or 2 per locule. Capsule subglobose, loculicidal. Seeds oblong.

Fl. & Fr.: Oct.-Mar. *Distr.*: Sri Lanka, N.E. India to Malesia. Rare. In evergreen forests. *NAK 499* (Nilakkal, c. 400 m).

Seed-oil used as a stimulating agent in rheumatism and bark used as an astringent.

CIPADESSA Blume
Bijdr. 162. 1825.

Cipadessa baccifera (Roth) Miq., Ann. Mus. Bot. Lug.-Bat. 4:6. 1168; Gamble, Fl. Pres. Madras 176. 1915; Ramamoorthy in Sald. & Nicols., Fl. Hassan Dist. 394. 1976; Matthew, Ill. Fl. Tam. Carnatic t.127.1982; Matthew & Britto in Matthew, Fl. Tam. Carnatic 3(1): 235. 1983; Manilal, Fl. Silent Valley 49. 1988. *Melia baccifera* Roth, Nov. Pl. Sp. 215. 1821. *Cipadessa fruticosa* Bl., Bijdr. 162. 1825; Hiern in Hook.f., Fl. Brit. India 1:545. 1875. *Pulippan-chedi.*

Shrubs to small trees. Leaves odd-pinnate, alternate, c.25 cm long; leaflets opposite or subopposite, ovate-lanceate, 3-10 by 1.5-4 cm, base obliquely cuneate or rounded, apex shortly acuminate, thin-coriaceous. Panicles axillary, c. 12 cm long. Flowers bisexual, 5-merous; calyx-tube deeply 5-lobed; petals 5, white; disc cupular, adnate to the base of staminal tube; staminal tube apically forked, anthers 10; ovary 5-locular; ovules 2 per locule. Drupe globose.

Fl. & Fr.: Most of the seasons. *Distr.*: Sri Lanka, Peninsular India, Java. Along forest clearings in hills. *NAK 646* (Puthukkulam, 70 m), 1744 (Thannithode, c. 100 m).

DYSOXYLUM Blume
Bijdr. 172. 1825.

Trees. Leaves odd-pinnate, alternate; leaflets 7-9, opposite or alterante, often closely veined. Inflorescence a panicle. Flowers bisexual, 4-5- merous; calyx cupular, lobes 4-5; petals 4-5, oblong; staminal tube cylindric, apically dentate, anthers 8-10, attached near the apex of staminal tube; disc cupular or cylindric; ovary 2-5- locular; ovules 1-2 per locule. Capsule loculicidal. Seed arillate or non-arillate.

1. Leaves prominently reticulately nerved above; calyx shallowly dentate; disc subcupular, apically coarsely lobed. .... 2. **D. swaminathanianum**
1. Leaves not prominently reticulately nerved above; calyx deeply dentate, disc cupular, apically truncate. .... 1. **D. malabaricum**

1. **Dysoxylum malabaricum** Bedd. ex Hiern in Hook.f., Fl. Brit. India 1:548. 1875; Gamble, Fl. Pres. Madras 178. 1915. *Vella-akil.*

Leaflets 5-7, elliptic-oblong, 9-18 by 5-7 cm, base oblique, apex shortly acuminate, reticulations prominent below, chartaceous. Panicle c.5 cm long, terminal. Flowers 4-merous; calyx 4-lobed; petals white; staminal tube cylindric, anthers 8; ovary 4-locular. Capsule globose, longitudinally lined.

Fl. & Fr.: Mar.-Jul. *Distr.*: Western Ghats (Travancore & Nilgiri Hills). Rare. In evergreen forests. *NAK 2354* (Appuppanthode, c. 450 m).

Oil prepared with wood cures ear and eye disease. Wood is a valuable timber.

2. **Dysoxylum swaminathanianum** Anilkumar *et* Sivad., Rheedea 11(2): 116. 2001. **(Fig. 20)**

Trees; branchlets obtusely angled, thinly pubescent. Leaves irregularly odd-pinnate or even-pinnate, c.30 cm long; petiole and rachis sharply angled; leaflets 6(7)-8(9), alternate, elliptic-oblong or elliptic-ovate, 10-18 by 4.5-7 cm, base oblique, apex shortly acuminate or obtuse, glabrescent, nervature finely reticulate, prominent, chartaceous; petiole c.5 mm long. Panicle axillary, c.15 cm long, finely pubescent. Flowers c. 1 cm ϕ; pedicel jointed at middle; calyx-tube cupular, ± 4 mm long, c. 5 mm ϕ, lobes irregular, 4 or 5; petals 5, c. 13 mm long, cream cloured, oblong, recurved, apex thick, 3-angled; disc cupular, exceeding the ovary, c. 3 mm long, coarsely (8-)-lobed; staminal-tube cylindric, c.1 cm long, 8-lobed, anthers 8, included, linear; ovary ovoid, c.2 mm long, 4-locular; ovules 2 per locule; style slender, c.8 mm long, puberulous near base, stigma capitate. Fruits could not be collected.

Fl.: Apr.- May. *Distr.*: Hitherto known only from Angamoozhy, the type locality in moist evergreen forests. *NAK 1818* (MH; SNCH). Angamoozhy, 200m.

MELIA Linnaeus
Sp. Pl. 384. 1753.

Melia azedarach L., Sp. Pl. 384. 1753; Roxb., Fl. Ind. 2.395. 1832; Wight & Arn., Prodr. 117. 1834; Wight, Ic.t. 160. 1839; Hiern in Hook.f., Fl. Brit. India 1:544. 1875; Gamble, Fl. Pres. Madras 176. 1915; Matthew, Ill. Fl. Tam. Carnatic t. 128. 1982; Matthew & Britto in Matthew, Fl. Tam. Carnatic 3(1):237. 1983; *Vempu, Kattu-veppu.*

Trees. Leave usually bipinnate or tripinnate, alternate; leaflets, c.5 pairs, opposite, ovate-lanceate, 2-7 by 1-2 cm, base obliquely cuneate, margin deeply serrate, apex acuminate. Panicles axillary, c.20 cm long. Flowers c. 1.5 cm ϕ, bisexual, regular; sepals 5, lanceate; petals 5, lilac, narrowly obovate or oblong; staminal tube cylindric, ribbed, dark blue or purple, apically 10-toothed, teeth linear, bifid; disc annular; ovary 5-locular; ovules 2 per locule. Drupe oblong or globose.

Fl. & Fr.: Oct.-Mar. *Distr.*: Native of W. Himalaya, now naturalised in India, Pakistan, Myanmar, China, Iran. Planted. Occasionally run wild. *NAK 1046* (Tiruvalla, c. 10 m).

Juice of leaves given internally as an anthelmintic, diuretic, astringent; injurious to insects.

NAREGAMIA R. Wight et Arnott
Prodr. 116. 1834 (nom. cons).

Naregamia alata Wight & Arn., Prodr. 117. 1834; Wight, Ic. t. 90. 1838; Hiern in Hook.f., Fl. Birt. India 1:542. 1875; Gamble, Fl. Pres. Madras 175. 1915; Nicols., Suresh & Mani., An Interpr. Hort. Malab. 178. 1988; Mohanan & Sivad., Fl. Agasthyamala 151.2002. *Nela-naregam* Rheede, Hort. Malab. 5:7-8, t.4.1685.

Woody herbs. Leaves digitately 3-foliolate, alternate; petiole winged; leaflets sessile, obovate or rhombic, 1-4 by 0.7-1.5 cm, base attenuate, margin entire or distantly dentate-serrate, apex obtuse, chartaceous. Flower(s) solitary or in pairs, axillary, 5-merous, bisexual, regular; calyx-lobes 5, triangular; petals 5, white, free, narrowly spathulate, c. 2 cm long; disc annular; staminal tube clawed, apex obscruely 10-toothed; anthers 10, appendaged above; ovary 3-locular; ovules 2 per locule; style long, filiform. Capsule globose, 3-winged. Seeds muricate.

Fl. & Fr.: Sep.-Dec. *Distr.*: Peninsular India. Common along moist hill cuttings and exposed hills. *NAK 1084* (Vallicode, c. 100 m).

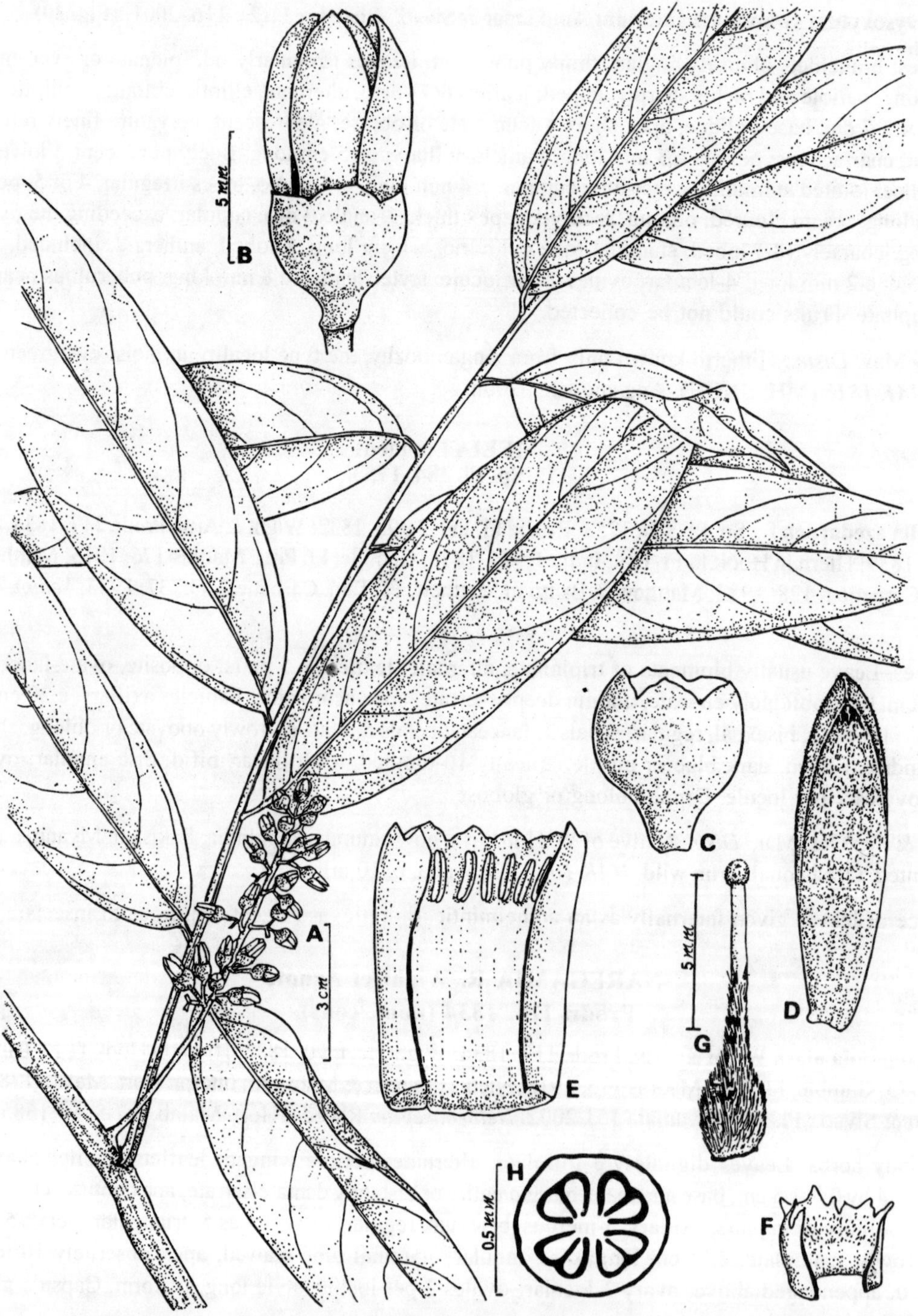

Fig. 20. ***Dysoxylum swaminathanianum*** **Anilkumar** ***et*** **Sivad. : A. Twig; B. Flower; C. Calyx; D. Petal; E. Staminal tube split open; F. Disc; G. Pistil; G, H. Ovary - C.S.**

Juice of roots boiled with oil and used in rheumatism, swellings and treatment of impure blood, and given internally for acute dysentery.

TOONA (Endlicher) M.J. Roemer
Fam. Nat. Syn. Monogr. 1:131, 139. 1846.

Toona ciliata Roemer, Syn. Hesper. 139. 1846; Ramamoorthy in Sald. & Nicols., Fl. Hassan Dist. 13951. 1976; Matthew, Ill. Fl. Tam. Carnatic t. 131. 1982; Matthew & Britto in Matthew, Fl. Tam. Carnatic 3(1) : 240. 1983; Manilal, Fl. Silent Valley 50. 1988. *Cedrela toona* Roxb. ex Rottl. & Willd., Ges. Naturf. Freunde Berlin Neue Sehriften 2: 198. 1803; Wight, Ic. t. 161. 1839; Hiern in Hook.f., Fl. Brit. India 1: 568. 1875, p.p.; Gamble, Fl. Pres. Madras 186. 1915. *Mathagiri-vembu.*

Trees. Leaves odd-pinnate, alterante; leaflets α, opposite or subopposite, ovate-lanceate, 6-9 by 2.5-4 cm, base oblique, apex acuminate, chartaceous. Inflorescence a terminal or subterminal panicle. Flowers bisexual, 5-merous; calyx-lobes 5; petals 5, white, ciliate on margins; stamens 4-6, free, inserted on the disc; ovary 5-locular; ovules a per locule. Capsule ellipsoid, septicidal. Seeds oblong, winged on both ends.

Fl. & Fr.: Dec.-Mar. *Distr.*: India, Pakistan, Myanmar through Malesia, Australia. Rare. In evergreen forests. *NAK 2629* (Upper Moozhiar, c. 400 m).

Bark astringent; gives a resinous gum. In fever and diarrhoea gum is used as a febrifuge.

TRICHILIA P. Browne
Civ. Nat. Hist. Jamaica 278. 1756 (nom. cons.).

Trichilia connaroides (Wight & Arn.) Bentvelzen, Acta Bot. Neerl. 11:13. 1962; Ramamoorthy in Sald. & Nicols., Fl. Hassan Dist. 396. 1976; Matthew, Ill. Fl. Tam. Carnatic t. 132. 1982; Matthew & Britto in Matthew, Fl. Tam. Carnatic 3(1) : 241. 1983; Manilal, Fl. Silent Valley 51. 1988; Mohanan & Sivad., Fl.Agasthyamala 152. 2002. *Zanthoxylum connaroides* Wight & Arn., Prodr. 148. 1834, ("*Zanthoxylon*"). *Heynea trijuga* Roxb. ex Sims, Bot. Mag. 41: t. 1738. 1815, non *Trichilia trijuga* Vell., 1825; Hiern in Hook.f., Fl. Brit. India 1 : 565. 1875; Gamble, Fl. Pres. Madras 183. 1915. *Kora kkadi.*

Trees. Leaves odd-pinnate, c. 50 cm long; leaflets alternate, ovate-lanceate or oblong-lanceate, 10-20 by 4-8 cm, base oblique, apex caudate, lower surface glaucous, thin-chartaceous. Panicle large, axillary or terminal. Flower bisexual, c.5 mm ϕ, 5-merous; calyx-lobes 5, triangular; petals 5, white, linear-oblong; stamens 10, connate near half-way, filaments broad, bufurcate at apex, disc annular; ovary 2 locular; ovules 2 per locule. Capsule globose.

Fl. & Fr.: Dec.-Apr. *Distr.*: Peninsular India, N.E. India through Malaya Peninsula to Malesia. Rare. In Moist evergreen and semi-evergreen forests. *NAK 2359* (Moozhiar, c. 300 m).

37. DICHAPETALACEAE

DICHAPETALUM L.M.A.A. du Petit-Thouars
Gen. Nova Madag. 23. 1806.

Dichapetalum gelonioides (Roxb.) Engler in Engler & Prantl, Pflanzenf. 3(4): 348. 1896; Gamble, Fl. Pres. Madras 188. 1915; Mohanan & Sivad., Fl.Agasthyamala 153. 2002. *Moacurra gelonioides* Roxb., Fl.

Ind. 2:69. 1832. *Chailletia gelonioides* (Roxb.) Bedd., For. Man. Bot. 59, Anal. Gen. t. 9/1.1871; Hook.f., Fl. Brit. India 1:570. 1875. *Cherumaram.*

Shrubs or trees. Leaves simple, alternate; oblanceate, narrowly obovate or elliptic, 7-14 by 2.5-3.5 cm, base cuneate, margin subentire or repand, apex caudate-acuminate, chartaceous. Flowers polygamo-monoecious, in axillary clusters, 5-merous; sepals 5, unequal, slightly connate at base; petals 5, yellow or white; apically 2-fid; stamens 5, adnate to petals at base; disc of 5 scales, opposite the petals; ovary 2-locular; ovule(s) 1 or 2 per locule. Drupe globose, densely tomentose.

Fl. & Fr.: Mar.-Jun. *Distr.*: Peninsular India, N.E. India, Sri Lanka. Common. In semi-evergreen forests. *NAK 1547* (Sabarimala forest, 250 m), *1599 & 1661*. (Ranni, c. 250 m).

38. OLACACEAE

STROMBOSIA Blume
Bijdr. 1154. 1826-1827.

Strombosia ceylanica Gard., Calcutta J. Nat. Hist. 6:350. 1846; Bedd., Fl. Sylv. t. 137. 1871; Mast. in Hook.f., Fl. Brit. India 1:579. 1875; Gamble, Fl. Pres. Madras 191.1915.

Trees. Leaves simple, alternate, or elliptic, 10-20 by 4.5-7.5 cm, base rounded or broadly cuneate, apex subacute to shortly acuminate, nerves basally obscurely palmately 3-veined, punctate, coriaceous. Flowers c.3 mm ϕ, 5-merous, aggregated in axillary fascicles, bisexual; calyx cupular, lobes 5, margin ciliate; petals 5, free, valvate, recurved, hairy within; stamens 5, adherent to the base of petals; staminodes 0; ovary inferior or superior, imperfectly 4-5- locular; ovule 1-per locule, pendulous. Drupe ellipsoid or pyriform.

Fl. & Fr.: Jan.-Apr. *Distr.*: Sri Lanka, India (Southern Western Ghats). Rare. In evergreen forests. *NAK 1528* (Gurunadhan mannu, c. 400 m), *2360* (Kokkathode, 450 m).

39. ERYTHROPALACEAE

ERYTHROPALUM Blume
Bijdr. 921. 1826.

Erythropalum populifolium (Arn.) Mast. in Hook.f., Fl. Brit. India 1:578. 1875; Gamble, Fl. Pres.Madras 191. 1915; Gandhi in Sald. & Nicols., Fl. Hassan Dist. 300. 1976; Manilal, Fl. Silent Valley 51: 1988; Mohanan & Sivad., Fl.Agasthyamala 153. 2002. *Mackaya populifolia* Arn., Mag. Zool. Bot. 2:531. 1838.

Tendrillar climbers. Leaves simple, alternate, ovate-lanceate, subpeltate, 5-15 by 2.5-8.5 cm, base truncate, or rounded, apex acuminate, glaucous below, basally 3-nerved, thin-coriaceous. Cymes axillary, lax, c.9 cm long. Flowers 5-merous; c.4 mm ϕ, bisexual; calyx cupular, 5-toothed; petals 5, yellowish, slightly connate at base; stamens 5, coherent with petals; staminodes alternating the stamens; ovary half-inferior, adherent to disc, 1-locular; ovule(s) 1 or 3, pendulous. Drupe oblong.

Fl. & Fr.: Dec.-Mar. *Distr.*: Western Ghats. Common. Among wayside thickets in evergreen forests. *NAK 1204 & 1442* (Moozhiar, c.300 m).

40. ICACINACEAE

1. Climbers.
   2. Slender climbers; leaves palmately-veined; inflorescence umbelliform. ....**Miquelia**
   2. Stout climbers; leaves pinnately-veined; inflorescence spicate. ....**Sarcostigma**
1. Trees or shrubs.
   3. Cymes axillary; flowers polygamo-dioecious, white. ....**Gomphandra**
   3. Cymes terminal; flowers bisexual, yellow. ....**Nothapodytes**

GOMPHANDRA Wallich ex J.Lindley
Nat. Syst. ed. 2, 439. 1836.

Gomphandra coriacea Wight, Ill. 1 : 103. 1840; Gamble, Fl. Pres. Madras 195. 1915; Manilal, Fl.Silent Valley 52. 1988 ; Mohanan & Sivad., Fl.Agasthyamala 154. 2002. *G. polymorpha* Wight, Ill. 1:103. 1840, p.p., Ic.t. 953-954. 1845; Mast. in Hook.f., Fl. Brit. India 1:586. 1875, p.p. *Kampili-chedi.*

Shrubs. Leaves simple, alternate, elliptic, 6-13 by 3-5.5 cm, base acute, margin subentire, apex acuminate, glaucous below, coriaceous. Inflorescence axillary, 6-10-flowered. Flowers polygamo-dioecious, 4-5- merous; calyx cupular; petals 4-5, connate, tip incurved, hooked; stamens 4-5, filament hariy on the apex at back; ovary 1-locular; ovules 2. Drupe ellipsoid.

Fl. & Fr.: Most of the seasons. *Distr*: Sri Lanka and S. India. Common in semi-evergreen forests. *NAK 230, 723, 777 & 1503* (Ranni, c. 100 m; Konni, c. 200 m).

MIQUELIA C.F. Meissner
Pl. Vasc. Gen. 1:152. 1838. (nom. cons.).

Miquelia dentata Bedd., Madras J. Lit. Sci. Ser. 3, 1:38. 1864; Mast. in Hook. f., Fl. Brit. India 1:593. 1875; Gamble, Fl. Pres. Madras 197. 1915; Ramamoorhty in Sald. & Nicols., Fl. Hassan Dist. 325. 1976; Manilal, Fl. Silent Valley 53. 1988. **(Fig. 21)**

Slender climbers. Leaves alternate, oblong, 8-16 by 4-7 cm, base truncate, margin dentate, apex acuminate, palmately 5-7- veined, membranous. Flowers unisexual, aggregated in subcapitate umbelliform clusteres; calyx-lobes 4-5, minute; petals connate below to form a pedicel like tube in male; free in female flowers; stamens 4-5; ovary 1-locular; ovules 2, pendulous. Drupe oblong, compressed.

Fl. & Fr.: Apr.-Sep. *Distr.*: Southern-Western Ghats. Rare. In evergreen forests usually along stream-sides. *NAK 717, 1709* (Thriveny-Pampa, c.250 m).

NOTHAPODYTES Blume
Mus. Bot. 1:248. 1850

Nothapodytes nimmoniana (Graham) Mabberly in Manilal, Bot. Hist. Hort. Malab. 88. 1980; Manilal, Fl. Silent Valley 53. 1988; Mohanan & Sivad., Fl. Agasthyamala 156. 2002. *Premna nimmnoiana* Graham, Cat. Pl. Bombay 155. 1839. *Stemonurus foetidus* Wight, Ic.t. 955. 1845. *Mappia foetida* (Wight) Miers, Ann. Mag. Nat. Hist. Ser. 2, 9:395. 1852; Mast. in Hook.f., Fl. Brit. India 1:589. 1875; Gamble, Fl. Pres. Madras 196. 1915. *Nothapodytes foetida* (Wight) Sleumer, Notizbl. Bot. Gart. Berlin-Dahlem 15:247. 1940; Ramamoorthy in Sald. & Nicols., Fl. Hassan Dist. 325. 1975.

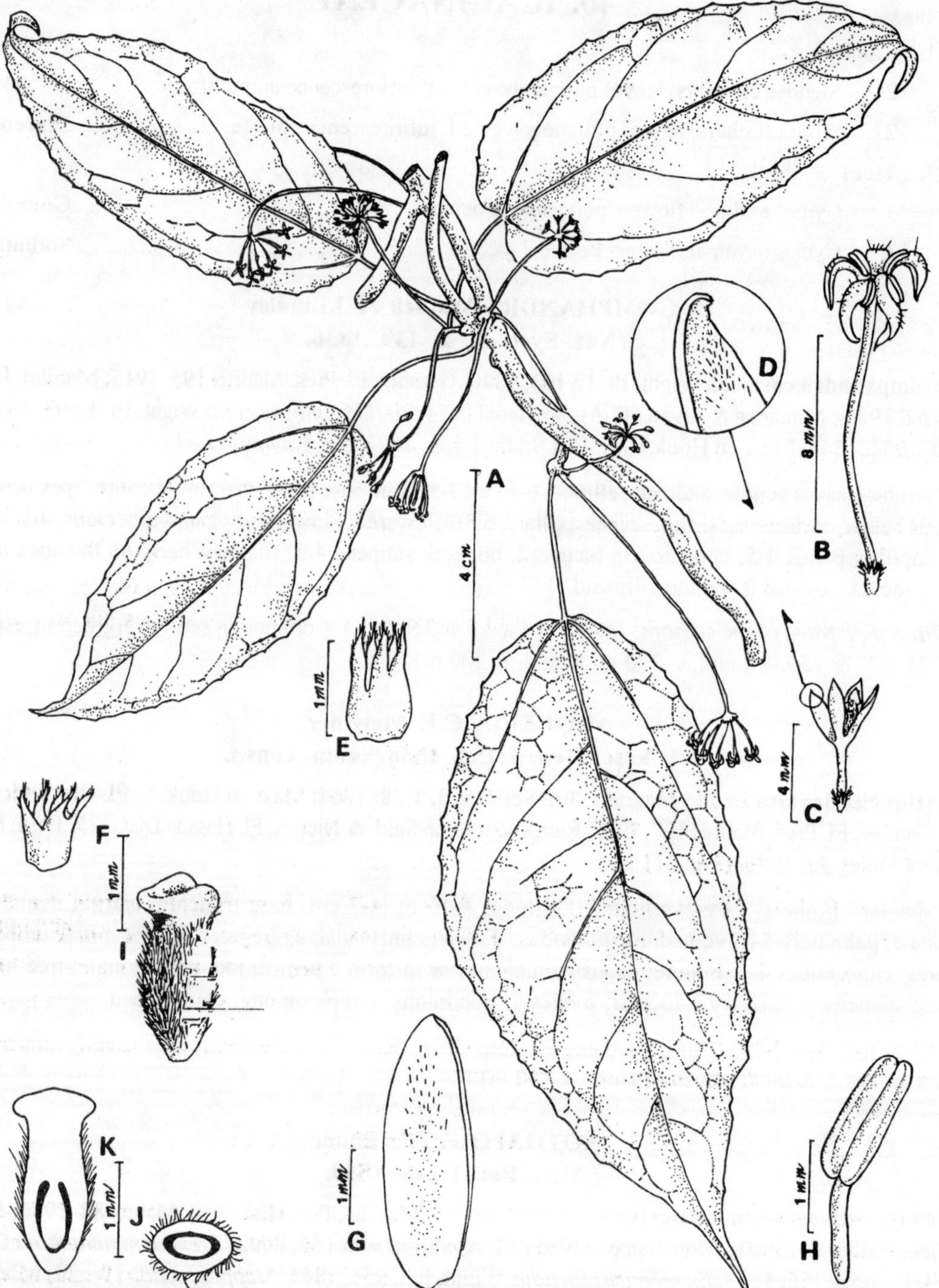

Fig. 21. ***Miquelia dentata*** **Bedd.: A. Twig; B. Male flower; C. Female flower; D. Apex of petal of female flower - enlarged; E. Calyx (Female flower); F. Calyx (Male flower); G. Petals (Male flower); H. Stamen; I. Pistil; J. Ovary - C.S.; K. Pistil - L.S.**

Shrubs or small trees. Leaves ovate-oblong, 10-30 by 7-4 cm, base obliquely rounded, apex shortly acuminate. Cymes terminal, corymbose. Flowers α, dense, bisexual, 5-merous; calyx 5-toothed; petals greenish-yellow, vilous; disc cupular; stamens 5, free; ovary superior, 1-locular; ovules 2, pendulous. Drupe ellipsoid.

Fl. & Fr.: Jun.-Sep. *Distr.*: India, China, Malesia. Rare. Arborescent shrubs in evergreen forests with unpleasent smell-emitting flowers. *BDS 42048* (MH) (Sabarimala slopes, c. 650 m).

SARCOSTIGMA R. Wight et Arnott
Edinburgh New Philos. J. 14:299. 1833.

Sarcostigma kleinii Wight & Arn., Edinburgh New Philos. J. 14:299. 1833; Wight, Ic. t. 1854. 1852; Mast. in Hook.f., Fl. Brit. India 1: 594. 1875; Gamble, Fl. Pres. Madras 199. 1915; Sleumer, Blumea 17:254. 1969; Ramamoorthy in Sald. & Nicols., Fl. Hassan Dist. 326. 1976; ; Mohanan & Sivad., Fl. Agasthyamala 157.2002. *Odal.*

Stout climbers. Leaves oblong or oblong-lanceate, 10-20 by 5-10 cm, base rounded, apex acuminate, shiny, coriaceous. Flowers unisexual, minute, on long pendulous spikes; calyx cupular, faintly 4-5- lobed; petals 5, yellow; stamens 5, free; staminodes 5 in female; ovary superior, 1-locular; ovules 2, pendulous. Drupe ellipsoid.

Fl. & Fr.: Feb.-Jun. *Distr.*: Western Ghats, Malesia. Less common. In evergreen forests and in certain sacred goves. *NAK 2232* (Tazhoor kadavu, c. 100 m). Once it was a common species but now gradually becoming rare.

Fruits edible. Seed-oil used for curing rheumatism, itches, leprosy and piles.

41. CELASTRACEAE

1. Trees or shrubs; flowers bisexual.
    2. Fruit a loculicidal capsule.
        3. Leaves oblong or elliptic-oblong; petals crested; ovule 4 or more per locule. .... **Lophopetalum**
        3. Leaves narrowly elliptic; petals not crested; ovules 2 per locule. .... **Euonymus**
    2. Fruit an indehiscent drupe. .... **Pleurostylia**
1. Climbing shrubs; flower polygamous. .... **Celastrus**

CELASTRUS Linnaeus
Sp. Pl. 196. 1753; Gen. Pl. ed. 5,91.1754.

Celastrus paniculatus Willd., Sp.Pl. 1:1125. 1798; Roxb., Fl. Ind. 1:621. 1832; Wight & Arn., Prodr. 158. 1834; Wight, Ic. t. 158. 1839 & Ill. Ind. Bot. t. 72. 1840; Lawson in Hook. f., Fl. Brit. India 1: 617. 1875; Gamble, Fl. Pres. Madras 208. 1918; Ramamoorthy in Sald. & Nicols., Fl. Hassan Dist. 318. 1976; Matthew, Ill. Fl. Tam. Carnatic t. 142. 1982; Matthew & Britto in Matthew, Fl. Tam. Carnatic 3(1) : 255. 1983. *Kili-thini-panji.*

Climbers. Leaves simple, alternate, ovate or obovate-elliptic, 7-10 by 3-5 cm, base rounded or subacute, apex acuminate, chartaceous or thin-coriaceous. Panicle terminal or axillary. Flowers polygamous, 5-merous; calyx 5-lobed; petals 5, yellow, reflexed; disc cupular, lobed; stamens 5, inserted on the margin of disc; ovary 3-locular; ovules 2 per locule. Capsule globose or ovoid, loculicidal. Seeds enclosed by fleshy orange coloured aril.

Fl. & Fr.: Dec.-May. *Distr.*: India, Myanmar, Indo-China, S. China & Malesia. In evergreen forest borders, commonly climbing on large tree capony. *NAK 15* (Moozhiar, 250 m), *2433* (Appupanthode, c.450 m).

Seeds are used internally and externally in rheumatism, paralysis, leprosy and nervous disorders.

EUONYMUS Linnaeus
Sp. Pl. 197. 1753 ("Evonymus") corr. Linnaeus, Gen. ed. 5.91. 1754.

Euonymus crenulatus Wall. ex Wight & Arn., Prodr. 161. 1834; Wight, Ic. t. 973. 1845; Lawson in Hook.f., Fl. Brit. India 1:608. 1875; Gamble, Fl. Pres. Madras 203. 1918; Mohanan & Sivad., Fl. Agasthyamala 159. 2002. *Malam-kuratha, Dandha-pathri.*

Small trees. Leaves simple, opposite, narrowly elliptic, 5-9 by 1-2.5 cm, base cuneate, margin crenulate towards apex; apex acuminate, thin-coriaceous. Cymes lax, axillary. Flowers bisexual, 5-merous; calyx -lobes 5, spreading; petals 5, brownish, faintly crispate; stamens 5, small; disc large, fleshy, 5-lobed; ovary sunk in the disc, 3-5- locular; ovules 2 per locule. Capsule 5-winged, obovoid, loculicidal. Seeds arillate.

Fl. & Fr.: Sep.-May. *Distr.*: Western Ghats. Rare. An evergreen or shola species. *NAK 2453* (Arampa, c. 500 m).

LOPHOPETALUM Wight ex Arnott
Ann. Nat. Hist. 3:150. 1839.

Lophopetalum wightianum Arn., Ann. Nat. Hist. 3:151. 1839; Wight, Ic.t. 162. 1839; Lawson in Hook.f., Fl. Brit. India 1:615. 1875; Gamble, Fl. Pres. Madras 205. 1918; Ramamoorthy in Sald. & Nicols., Fl. Hassan Dist. 319. 1976 ; Mohanan & Sivad., Fl. Agasthyamala 161. 2002. *Venkotta.*

Large trees. Leaves alternate and/or opposite, elliptic-oblong 6-12 by 3.5-5 cm, base rounded, apex obtusely acuminate, coriaceous. Cymes axillary or terminal. Flowers bisexual, c. 15 mm ϕ, inserted in the disc; ovary 3-locular; ovules 4-6 per locule. Capsule 4-5- angled. Seeds winged.

Fl. & Fr.: Dec.-Jul. *Distr.*: Western Ghats. Common in riverien forests. *NAK 1733, 1839* (Thannithode, c. 70 m), *2332* (Aikadu, c. 50 m).

PLEUROSTYLIA Wight et Arnott
Prodr. 157. 1834.

Pleurostylia opposita (Wall.) Alston in Trimen, Fl. Ceylon 6 (Suppl.) : 48. 1931; Matthew, Fl. Tam. Carnatic 3(1) : 261. 1983; Mohanan & Sivad., Fl. Agasthyamala 162. 2002. *Celastrus opposita* Wall. in Roxb., Fl. Ind. 2:398. 1824. *Pleurostylia wightii* Wight & Arn., Prodr. 157. 1834; Wight, Ic. t. 155. 1839; Lawson in Hook.f., Fl. Brit. India 1:617. 1875; Gamble, Fl. Pres. Madras 211. 1918.

Shrubs. Leaves simple, opposite, elliptic-oblong, to narrowly oblong-lanceate, to 6 by 3 cm, base cuneate, apex obtuse, thin-coriaceous. Cymes few-flowered, axillary. Flowers bisexual, 5-merous; calyx-lobes

5, triangular; petals 5, greenish-yellow; reflexed; disc thick, faintly lobed; stamens 5; ovary 1-2- locular; ovules 2 per locule. Drupe ovoid. Seeds arillate.

Fl. & Fr.: Sep.-Dec. *Distr.*: Sri Lanka, Peninsular India, Madagascar & Mauritius. Rare. In evergreen forest. *CNM 65077* (MH), (Ranni R.F., c. 200 m).

42. HIPPOCRATEACEAE

SALACIA Linnaeus
Mant. 159. 1771.

Climbing shrubs or small trees. Leaves simple, opposite. Flowers in axillary clusters, or in cymes, bisexual, 5-merous, hypogynous; calyx-lobes small; petals yellowish; disc thick; stamens 3, inserted on the margin of disc; ovary 3-locular; ovules 2-8 per locule. Berry globose. Seeds angular.

1. Flowers in axillary cymes; pedicel to 5 mm long; petals orbicular. ....1. **S. fruticosa**
1. Flowers in axillary fasicles; pedicel to 3 cm long; petals elliptic. ....2. **S. malabarica**

1. **Salacia fruticosa** Heyne ex Lawson in Hook.f., Fl. Brit. India 1:628. 1875; Gamble, Fl. Pres. Madras 215. 1918 ; Mohanan & Sivad., Fl. Agasthyamala 164.2002. **(Fig. 22)**

Branchlets profusely muricate. Leaves elliptic-ovate, 4-9 by 2-5 cm, base rounded or cuneate, apex acuminate, coriaceous. Cymes axillary, pedicel to 5 mm long; petals brownish-yellow, orbicular. Berry globose, c. 3 cm ϕ.

Fl. & Fr.: Feb.-May. *Distr.*: Western Ghats (Travancore & Nilgiri hills). Locally abundant but not common. In evergreen forests. *NAK 450 & 2300* (Thekkuthode, c. 70 m), *2363* (Aikadu, 50 m).

2. **Salacia malabarica** Gamble, Kew Bull. 1916 : 133 . 1916 & Fl. Pres. Madras 215. 1918. **(Fig. 23)**

Branchlets faintly muricate. Leaves elliptic-oblong, 9-13 by 4.5-7 cm, base cuneate or rounded, apex acuminate, coriaceous. Flowers in axillary fascicles; pedicel 2-3 cm long; petals greenish-yellow, elliptic. Berry hard, globose, c. 5 cm ϕ.

Fl. & Fr.: Feb.-Apr. *Distr.*: Travancore hills. Very Rare. Along streamsides in evergreen forests. *NAK 2664* (Thannithode-Punnamada dam site, c. 450 m).

43. RHAMNACEAE

1. Plants tendrillar; flowers unisexual; ovary inferior. ....**Gouania**
1. Plants non-tendrillar; flowers bisexual; ovary superior.
    2. Plants armed; leaves palmi-nerved. ....**Ziziphus**
    2. Plants unarmed; leaves pinnate-nerved.
        3. Climbers; style accrescent (form wings on fruit). ....**Smythea**
        3. Trees or shrubs; style deciduous.
            4. Leaves alternate, entire; flowers yellow; ovary 3-locular. ....**Colubrina**
            4. Leaves subopposite, toothed; flowers greenish; ovary 1-locular. ....**Maesopsis**

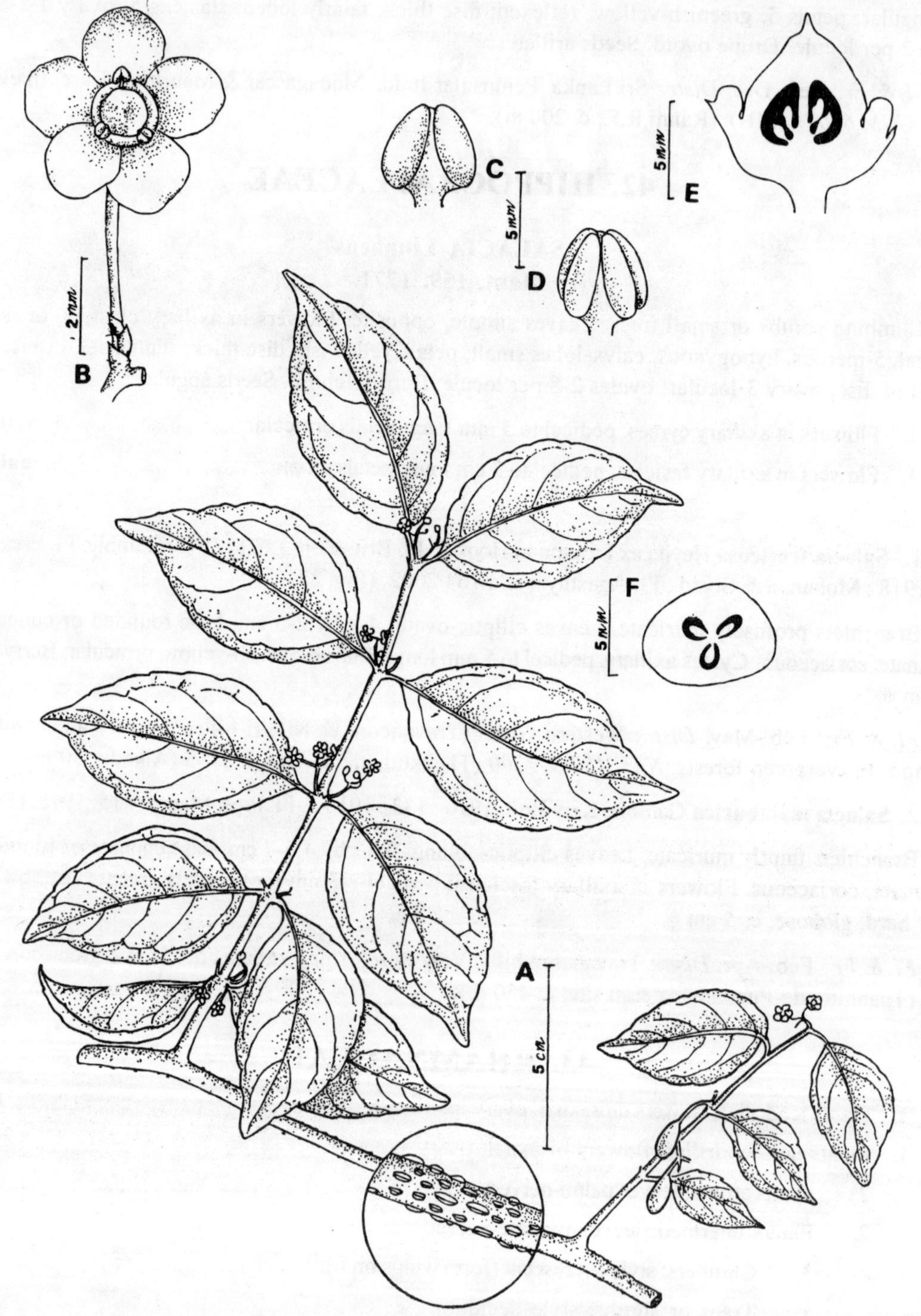

Fig. 22. ***Salacia fruticosa*** **Heyne ex Lawson: A. Twig; B. Flower; C. & D. Stamen- adaxial & abaxial views; E. Ovary - L.S.; F. Ovary - C.S.**

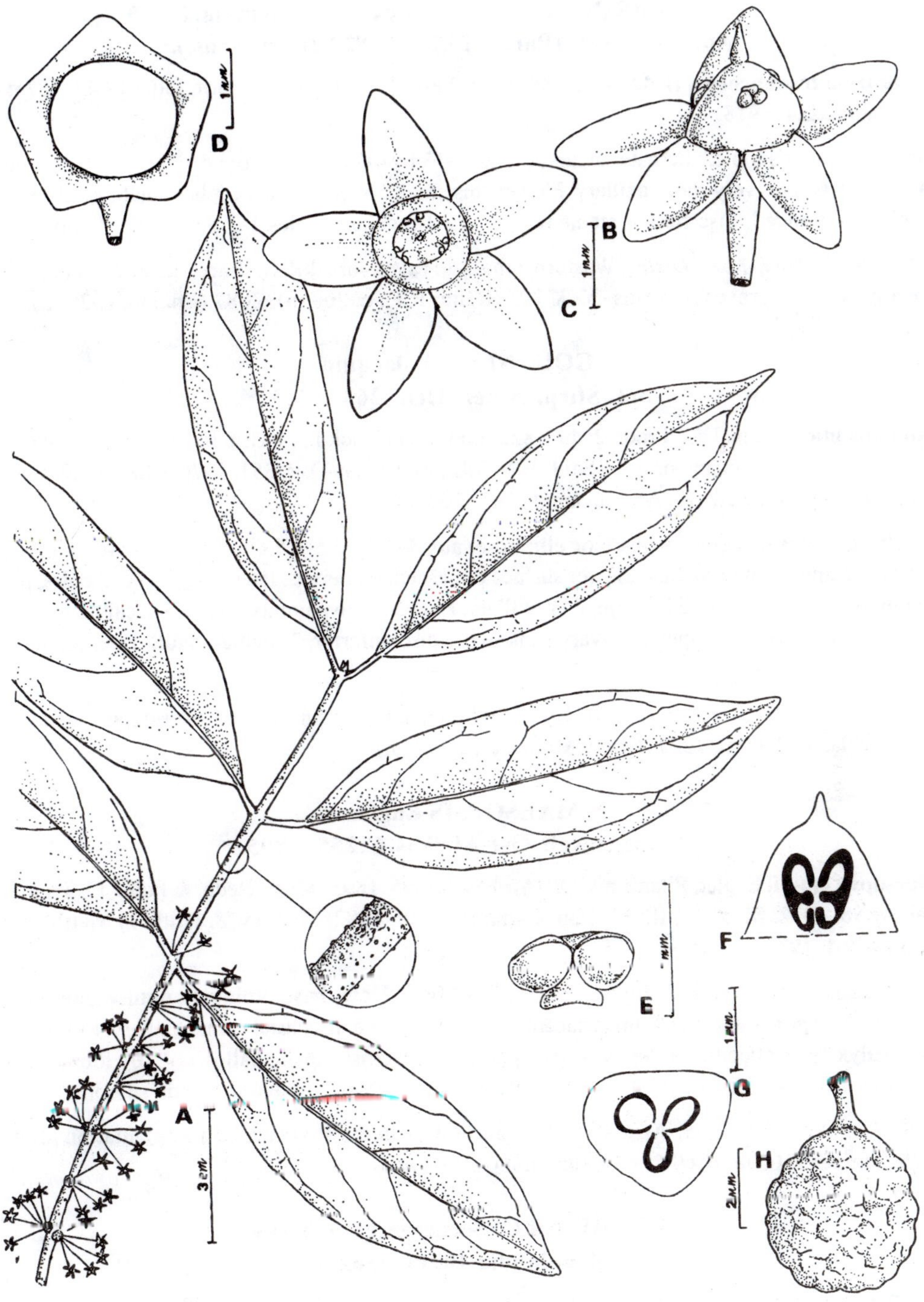

Fig. 23. ***Salacia malabarioa*** **Gamble: A. Twig; B. Flower - side view; C. Flower - top view; D. Calyx; E. Stamen; F. Pistil - L.S.; G. Ovary - C.S.; H. Fruit.**

COLUBRINA L.C. Richard ex A.T. Brongniart
Ann. Sci. Nat. (Paris) 10:368. 1827 (nom. cons.).

Colubrina travancorica Bedd., Ic. t. 188. 1871. Lawson in Hook.f., Fl. Brit. India 1:643. 1875; Gamble, Fl. Pres. Madras 224. 1918.

Shrubs. Leaves simple, alternate, ovate-oblong, 8-13 by 4-7 cm, base rounded, apex obtusely acuminate, thin-coriaceous. Cymes very short, axillary. Flowers bisexual, 5-merous; calyx-lobes small; petals 5, brownish-yellow, clawed, hooded; disc fleshy; stamens 5; ovary 3-locular; ovules 1-2 per locule. Drupe ellipsoid.

Fl. & Fr. : Nov.-Mar. *Distr.*: Western Ghats (Travancore hills). Rare. In evergreen forests and occasionally in dense groves in plains. *NAK 791* (Ranni R.F., c. 200 m), 2087 (Enathu, c. 30 m).

GOUANIA N.J. Jacquin
Sel. Stirp. Amer. Hist. 263. 1763.

Gouania microcarpa DC., Prodr. 2: 40. 1825; Lawson in Hook.f., Fl. Brit. India 1:643. 1875; Gamble, Fl. Prs. Madras 225. 1918; Ramamoorthy in Sald. & Nicols., Fl. Hassan Dsit. 351. 1976; Manilal, Fl. Silent Valley 57. 1988; Mohanan & Sivad., Fl. Agasthyamala 165. 2002.

Tendrillar climbers. Leaves ovate or elliptic-ovate, 4-8 by 2.5-5.5 cm, base cordate, margin crenate-serrate, apex acuminate-mucronulate, lower surface pubescent on nerves, lateral nerves. c. 5 pairs, prominent, chartaceous. Racemes interrupted, terminal or axillary. Flowers polygamous, 5-merous; calyx obconic, lobed; stamens 5, enclosed within the petals; ovary sunk in the disc, inferior, 3-locular, ovule 1 per locule. Capsule, 3-winged. Seeds obovoid.

Fl. & Fr.: Oct.-Mar. *Distr.*: Western Ghats, Sri Lanka. Common. In semi-evergreen forests. *NAK 491* (Lower Moozhiar, c. 250 m), *1223* (Upper Moozhiar, c. 600 m).

MAESOPSIS Engler
Pflanzenwelt Ost-Afrikas C:255. 1895.

Maesopsis eminii Engler, Plfanzenw. Ost-Afrikas, C. 255. 1895; Milne-Redh. & Polh., Fl. Trop. E. Africa Rhamnaceae 38. 1972; Matthew, Ill. Fl. Tam. Carnatic t. 150. 1982; Matthew & Britto in Matthew, Fl. Tam. Carnatic 3(1): 265. 1983.

Trees. Leaves subopposite, elliptic-oblong, 8-12 by 3-4 cm, base obliquely obtuse, margin coarsely toothed or serrate, apex acuminate, thin-coriaceous. Inflorescence a dischasial cyme, axillary. Flowers bisexual, 5-merous; calyx-tube obconic, lobes 5, small; petals 5, greenish; disc shallowely 10- lobed; stamens 5, enclosed within the petals; ovary 1-locular; ovule 1. Drupe oblong-obovoid. Seeds smooth.

Fl. & Fr.: Sep.-Apr. *Distr.*: Africa. An introduced tree in the study are and as plantation on wast area near Kochupampa (1100 m) *NAK 2713* (Kakki hills, c. 100 m).

SMYTHEA B.C. Seemann ex A. Gray
Bonplandia 10:35. 1862.

Smythea bombaiensis (Dalz.) Banerjee & Mukherjee, Ind. For. 96 (3): 214. 1970; Manilal, Fl. Silent Valley 57. 1988. *Ventilago bombaiensis* Dalz. in Hooker's J.Bot. Kew Gard. Misc. 3:36. 1851; Lawson in Hook f., Fl. Brit. India 1:631. 1875; Gamble, Fl. Pres. Madras 218. 1918.

Climbing shrubs. Leaves elliptic-oblong, 3.5-7 by 1.5-2.5 cm, base cuneate, margin glandular, faintly serrate, apex acuminate, chartaceous. Flowers in axillary fascicles, 5-merous, bisexual; calyx-lobes 5, spreading; petals 5, greenish-yellow; disc slightly 5-angled; stamens 5; ovary 2-locular; ovule 1per locule. Nut winged.

Fl. & Fr.: Feb.-Jun. *Distr.*: Western Ghats. Rare. In interior evergreen forests. *NAK* 2402. (Moozhiar, c. 250 m).

ZIZIPHUS P. Miller
Gard. Dict. Abr. ed. 4. 1754.

Erect armed shrubs or shrubby climbers. Leaves palmately 3-5- nerved. Inflorescence an axillary or terminal cyme or a panicle. Flowers bisexual, 5-merous, regular; calyx-lobes 5, keeled within; petals 5 or 0, yellowish or greenish; disc 5-10- lobed; stamens 5, enclosed by petals; ovary immersed in the disc, 2-4-locular; ovule 1per locule. Drupe globose.

1. Erect shrubs; leaves 3-nerved; petals 5.
    2. Leaves white-tomentose beneath, ovate-orbicular; drupe, c. 10 mm ϕ. .... 1. **Z. mauritiana**
    2. Leaves dull-pubescent beneath, ovate-lanceate; drupe, c. 4 mm ϕ. .... 2. **Z. oenoplia**
1. Shrubby climbers; leaves 5-nerved; petals 0. .... 3. **Z. rugosa**

1. **Ziziphus mauritiana** Lam., Encycl. 3:319. 1789; Matthew & Britto in Matthew, Fl. Tam. Carnatic 3(1): 271. 1983. Nicols., Suresh & Mani., An Interpr. Hort. Malab. 214. 1988. *Z. jujuba* (L.) Gaertn., Fruct. Sem. Pl. 1:203. 1788, non Miller, 1768; Roxb., Fl. Ind. 1:608. 1832; Wight, Ic.t. 149. 1871; Lawson in Hook.f., Fl. Brit. India 1:632. 1875; Gamble, Fl. Pres. Madras 219. 1918; *Perin-toddali* Rheede, Hort. Malab. 4:85-86, t.41. 1683. *Ilantha.*

Shrubs or small trees. Leaves ovate-orbicular, 2-3 by 1.5-3 cm, base obliquely subcordate, margin subentire, apex obtuse or rounded, basally 3-nerved, lower surface white-tomentose. Cymes axillary. Flowers c. 5 mm ϕ; petals 5, ovate, greenish. Drupe subglobose, c. 10 mm ϕ.

Fl. & Fr.: Jul.-Dec. *Distr.*: Sri Lanka, India. Planted. Occasionally run Wild. *NAK 2841* (Angamoozhy, c. 250 m).

Fruits edible and stops vomiting and cough.

2. **Ziziphus oenoplia** (L.) Miller, Gard. Dict. (ed. 8), 3.1768; Roxb., Fl. Ind. 1:611. 1832; Lawson in Hook.f., Fl. Brit. India 1:634. 1875; Gamble, Fl. Pres. Madras 220. 1918; Ramamoorthy in Sald. & Nicols., Fl. Hassan Dist. 354. 1976; Matthew & Britto in Matthew, Fl. Tam. Carnatic 3(1) : 272. 1983; Mohanan & Sivad., Fl. Agasthyamala 166.2002.

Shrubs. Leave ovate-lanceate, 4-7 by 2-3 cm, base obliquely rounded, margin faintly serrate, apex acute, basally 3-(4-) nerved,. lower surface dull-pubescent. Cymes axillary. Flowers, c. 3 mm ϕ; petals 5, spathulate, greenish-yellow. Drupe globose, c.4 mm ϕ.

Fl. & Fr.: Nov.-Mar. *Distr.*: Pantropical. Common along wayside thickets in plains and hills. *NAK 175, 209, 1145* (Lower Moozhair, c. 250 m).

The whole plant is astringent and helps digestion. Decoction of bark used to heal wounds.

3. **Ziziphus rugosa** Lam., Encycl. 3:319. 1789; Wight, Ic.t. 339. 1840; Lawson in Hook.f., Fl. Brit. India 1: 636. 1875; Gamble, Fl. Pres. Madras 221. 1918; Ramamoorthy in Sald. & Nicols., Fl. Hassan Dist. 354. 1976; Matthew & Britto in Matthew, Fl. Tam. Carnatic 3(1):273. 1983; Manilal, Fl. Silent Valley 58. 1988; Mohanan & Sivad., & Agasthyamala 166. 2002. *Thodali.*

Stragglers. Leave elliptic-oblong, 7-12 by 3.5-6 cm, base obliquely subcordate, margin faintly serrate, apex shortly acuminate or acute, basally 5-nerved, coriaceous. Cymes in terminal or axillary panicles. Flower c.4 mm ϕ; petals 0. Drupe globose, c. 1 cm ϕ.

Fl. & Fr.: Nov.-May. *Distr.*: Sri Lanka, India. Common in forest borders. *NAK 189, 1549* (Moozhiar, 350 m).

Bark powder mixed with ghee applied to swellings and given to cure ulcers in mouth; fruits edible.

44. VITACEAE

1. Flowers 5-merous; petal-hood 2-fid. .... **Parthenocissus**
1. Flowers 4-merous; petal-hood 0 or 1 (strongly mucronate).
    2. Berry 1-2-seeded; leaves simple (angular or 3- lobed). .... **Cissus**
    2. Berry 2-4-seeded; leaves 3-7-foliolate.
        3. Flowers polygamo-dioecious; stigma 4-lobed; stipule obscure; seeds deeply furrowed. .... **Tetrastigma**
        3. Flowers bisexual; stigma obscure; stipules prominent; seeds deeply furrowed. .... **Cayratia**

CAYRATIA A.L. Jussieu
Dict. Sci. Nat. 10:103. 1818 (nom. cons.).

Climbers; tendrils leaf-opposed. Leaves pedately 3-7- foliolate, alternate; stipules 2. Inflorescence a corymbose cyme, axillary. Flowers bisexual, 4-merous; calyx-lobes obscure; petals 4 cucullate within, disc angular; ovules 2 per locule; stigma obscure. Berry 2-4-seeded. Seeds pyriform, pitted or angular.

1. Leaves pedately (5-) 7-9-foliolate, tomentose, margin subentire. .... 2. **C. pedata**
1. Leaves pedately 5-foliolate, glabrescent, margin serrate.
    2. Seeds pyriform, acute at base. .... 1. **C. japonica**
    2. Seeds sharply angled, rounded at base. .... 3. **C. tenuifolia**

1. **Cayratia japonica** (Thunb.) Gagn. in Lecomte, Notul. Syst. (Paris) 1:349. 1911; Gamble, Fl. Pres. Madras 237. 1918; Ramamoorthy in Sald. & Nicols., Fl. Hassan Dist. 358. 1976. *Vitis japonica* Thunb., Fl. Japan 104. 1784. *V. tenuifolia* Wight & Arn., Prodr. 129. 1834, p.p; Lawson in Hook.f., Fl. Brit. India 1:660. 1875, p.p. **(Fig. 24)**

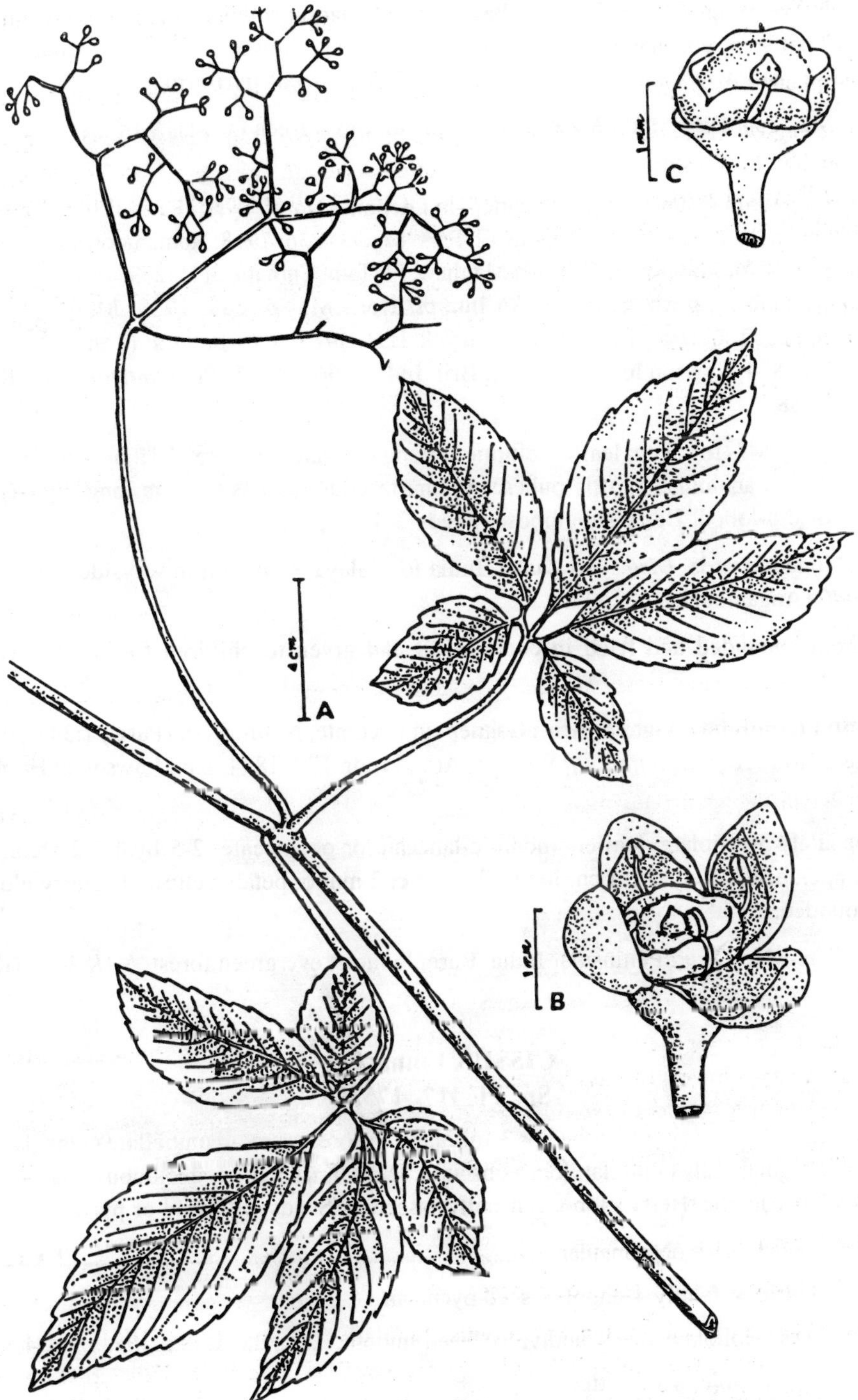

Fig. 24. ***Cayratia japonica*** **(Thunb.) Gagnep. : A. Twig; B. Flower; C. Calyx with disc.**

Climbers. Leaves elliptic, 4-9 by 2-4 cm, base cuneate, margin sharply serrate, apex acute, glabrescent, lateral veins 8-12 pairs, raised, chartaceous. Cymes to 15 cm long. Flowers. c. 3 mm ϕ, petals greenish-white, elliptic-ovate; disc shallowly 4-lobed. Berry globose, c.7 m ϕ. Seeds pyriform.

Fl. & Fr.: Mar.-Jun. *Distr.*: Indo-Malesia. Rare. In moist evergreen forests. *NAK 2411* (Lower Moozhiar, c.300 m).

2. **Cayratia pedata** (Lam.) Juss. ex Gagnep. in Lecomte, Notul. Syst. (Paris 1:346. 1911, var. **pedata** Fyson, Fl.S. India Hill. St. 2:t. 97. 1932; Gamble, Fl. Pres. Madras 236. 1918; Ramamoorthy in Sald. & Nicols., Fl. Hassan Dist. 359. 1976; Matthew & Britto in Matthew, Fl. Tam. Carnatic 3(1): 279. 1983; Manilal, Fl. Silent Valley 59. 1988; Nicols., Suresh & Mani., An Interpr. Hort. Malab. 268. 1988; Mohanan & Sivad., Fl. Agasthyamala 168. 2002. *Cissus pedata* Lam., Encycl. 1:31. 1783. *Vitis pedata* (Lam.) Wall. ex Wight & Arn., Prodr. 1:128. 1834; Lawson in Hook. f., Fl. Brit. India 1:661. 1875. *Belutta tsjori-valli* Rheede, Hort. Malab. 7:19, t.10.1688.

Leaves pedately 5-7- foliolate; leaflets elliptic-oblong or ovate-lanceate, 7-12 by 3-6 cm, base cuneate, margin subentire, apex acuminate, softly pubescent, chartaceous. Cymes to 7 cm long. Flowers c.3 mm ϕ; petals greenish; disc 4-lobed. Berry subglobose. Seeds 2-4.

Fl. & Fr.: June onwards. *Distr.*: Sri Lanka, India to Malaya. Common in wayside thickets, usually in plains. *NAK 2840* (Angamoozhy, c. 250 m).

Young shoots pounded and fried in coconut oil and given to children for worms, vomiting and diarrhoea.

3. **Cayratia tenuifolia** (Wight & Arn.) Gagnep. in Lecomte, Notul. Syst. (Paris) 1:343.1911; Gamble, Fl. Pres. Madras 236. 1918. *Vitis tenuifolia* Wight & Arn., Prodr. 129. 1834, p.p.; Lawson in Hook.f., Fl. Brit. India 1:660. 1875, p.p.

Leaves pedately 5-foliolate; leaflets rhombic-lanceate or oblanceate, 2-5 by 1.2-2.5 cm, base acute, margin serrate, apex acute. Cymes c.4 cm long. Flowers c. 2 mm ϕ; petals yellowish. Berry globose. Seeds angled, base rounded.

Fl. & Fr.: Sep.-Mar. *Distr.*: Peninsular India. Rare. In moist evergreen forest. *NAK* 2466 (Kakkı hill, c. 1100 m).

CISSUS Linnaeus
Sp. Pl. 117. 1753.

Climbing shrubs. Leaves simple, lobed or 3-foliolate. Inflorescence an umbellate cyme, leaf-oppossed. Flowers 4-merous, regular; calyx cupular, lobes obscure; petals 4, triangular, deciduous; disc 4-lobed; ovary 2-locular; ovules 2 per locule. Berry globose, usually 1-seeded. Seeds eillipsoid or pyriform.

1. Stem 4-winged; seeds obtriangular. .... 2. **C. glyptocarpa**
1. Stem cylindric or faintly 4-angular; seed pyriform.
    2. Leaves 3-foliolate, rarely undivided; seed smooth. .... 4. **C. trilobata**
    2. Leaves simple; seed pitted.

3. Branchlets glaucous, cylindric; leaves greenish. .... 1. **C. glauca**

3. Branchlets reddish, faintly angular; leaves variegated. .... 3. **C. javana**

1. **Cissus glauca** Roxb., Fl.Ind. 1:406. 1820; Gamble, Fl. Pres. Madras 234. 1918. *Vitis glauca* (Roxb.) Wight & Arn. in Wight, Cat. 26.1833 & Prodr. 126. 1834; Lawson in Hook.f., Fl. Brit. India 1: 648. 1875; Mohanan & Sivad., Fl. Agasthyamala 169.2002. *Chunnambu-valli.*

Stem glaucous. Leaves broadly ovate, 10-16 by 8-15 cm, base cordate, margin denticulate or serrulate, apex long-acuminate, chartaceous. Cyme c.6 cm long. Flowers c.2 mm f; calyx obscurely 4-lobed; petals greenish-yellow. Berry subglobose. Seed pyriform.

Fl. & Fr.: Sep.-Feb. *Distr.*: Western Ghats. Common in sacred groves and on thickets in plains. *NAK 155* (Adoor, c.40 m), *666* (Pandalam, c.50 m).

2. **Cissus glyptocarpa** (Thw.) Planch. in DC., Monogr. Phan. 5:477. 1887; Gamble, Fl. Pres. Madras 235. 1918. *Vitis glyptocarpa* Thw., Enum. Pl. Zeyl. 62. 1858; Lawson in Hook.f., Fl. Brit. India 1: 645. 1875; Manilal, Fl. Silent Valley 60.1988; Mohanan & Sivad., Fl. Agasthyamala 169. 2002.

Stem 4-winged. Leaves ovate to ovate-lanceate or deltoid, 5-11 by 3-7 cm, base truncate-cordate, margin serrulate, apex long-acuminate, membranous. Cyme 5-7 cm long. Flowers c. 3 mm ϕ; petals brownish, reflexed. Berry globose, reddish-purple. Seeds obtriangular.

Fl. & Fr.: Apr.-Aug. *Distr.*: Sri Lanka, Western Ghats (Travancore hills). Rare. In moist evergreen forests. During full bloom leaves are practically nil. *NAK 2410* (Moozhiar, c. 250 m).

3. **Cissus javana** DC., Prodr. 1:628. 1825; Moniyama, Enum. Flow. Plants of Nepal 2:94. 1979. *Vitis discolor* (Bl.) Dalz. in Hook.f., Fl. Brit. India 1:647.1875. *Cissus discolor* Bl., Bijdr. 181. 1825; Gamble, Fl. Pres. Madras 235. 1918; Ramamoorthy in Sald. & Nicols., Fl. Hassan Dist. 360. 1976; Manilal, Fl. Silent Valley 59.1988.

Branches reddish, fainty angled. Leaves ovate-oblong or ovate-lanceate, base cordate or truncate, margin toothed, apex acuminate, variegated,membranous. Cymes 2-4 cm long. Flowers c.3 mm f; calyx- apex truncate; petals reddish-brown. Berry globose, reddish. Seed obovoid.

Fl. & Fr.: Jul.-Jan. *Distr.*: Indo-Malesia. Common. In evergreen and semi-evergreen forest borders. *NAK 218, 927 & 1149* (Plappally, 400 m).

4. **Cissus trilobata** Lam., Encycl. 1:31. 1783; Gamble, Fl. Pres. Madras 233. 1918; Nicols., Suresh & Mani., An Interpr. Hort. Malab. 170. 1988. *Vitis rheedei* Wight & Arn. in Wight, Cat. 27. 1833 & Prodr. 127 1834; Lawson in Hook.f., Fl. Brit. India 1:653. 1875. *Puria trilobata* (Lam.) N.C. Nair, Biol. Land Pl. 127. 1972 (1974). *Kareta-tsjori-valli* Rheede, Hort. Malab. 7:85, t.45.1688.

Leaves digitately 3-foliolate; leaflets rhombic-lanceate or oblong-lanceate, 5-10 by 2.5-4.5 cm, base very obique, margin toothed, apex acute, chartaceous. Cyme 4 cm long; calyx truncate; petals reddish-brown; disc yellow. Berry globose, white or reddish. Seed ellipsoid.

Fl & Fr.: Jun.-Aug. *Distr.*: Sri Lanka, Western Ghats. Common. On wayside thickets in plains and in sacred groves. *NAK 704, 769* (Plappally, c.400 m).

Cissus quadrangularis L. is seen planted here and there in the district.

PARTHENOCISSUS Planchon in Alph. de Candolle et A.C. de Candolle, Monogr. Phan. 5:447. 1887 (nom. cons.).

Parthonecissus neilgherriensis (Wight) Planch. in DC., Monogr. Phan. 5:2. 1887; Gamble, Fl. Pres. Madras 231. 1918; Manilal, Fl. Silent Valley 60. 1988. *Vitis neilgherriensis* Wight, Ic.t. 965. 1845. *V. himalayana* Brandis, For. Fl.100.1874; Lawson in Hook.f., Fl. Brit. India 1:655. 1875, p.p.

Stout climbers; tendrils branched. Leaves 3-foliolate, rarely terminal ones simple, leaflets ovate-lanceate, 8-14 by 4-10 cm, base obliquely acute, margin serrate-toothed, apex long-acuminate, lateral veins prominent, thin-coriaceous. Inflorescence a large lax cyme, ending in umbellules, terminal or leaf-opposed. Flowers bisexual, 5-merous; calyx cupular, lobes 5, obscure; petals 5, white, hooded at apex with 2 long acumens, disc not distinct; ovary 2-locular; ovules 2 per locule. Berry globose. Seeds 2-4, globose, smooth.

TETRASTIGMA (Miquel) Planchon in Alph. de Candolle et A.C. de Candolle, Monogr. Phan. 5:320, 423.1887.

Stout climbers; branches muricate. Leave pedately 3-5- foliolate. Inflorescence a corymbose axillary or leaf-opposed cyme. Flowers polygamo-dioecious, 4-merous; calyx cupular, nearly truncate at apex; petals 4, caducus, strongly mucronate at apex; stamens 4, around the disc; ovary 2-locular; ovule 2 per locule. Berry globose. Seeds 2-4.

1.Leaf-margin serrate; berries c. 8 mm ϕ; seeds with shallow furrows on face. ....1. **T. leucostaphylum**

1.Leaf margin crenate; berries c.15 mm ϕ; seeds with deep furrows on face. ....2. **T. sulcatum**

1. **Tetrastigma leucostaphylum** (Dennst.) Alston ex Mabberley, Taxon 26:539. 1977; Matthew & Britto in Matthew, Fl. Tam. Carnatic 3(1): 385. 1983; Manilal, Fl. Silent Valley 60.1988; Nicols., Suresh & Mani., An Interpr. Hort. Malab. 270. 1988. *Cissus leucostaphyla* Dennst., Schlues. Hort. Ind. Malab. 17, 19, 33.1818; Forts., Allg. Tentsch. Garten Mag. 3:35, 37, 81. 1818. *Vitis muricata* Wall. ex Wight & Arn., Prodr. 81. 1834. *Tetrastigma muricatum* (Wall. ex Wight & Arn.) Gamble, Fl. Pres. Madras 229. 1918; Ramamoorthy in Sald. & Nicols., Fl. Hassan Dist. 360. 1976. *Vitis lanceolaria* sensu Lawson in Hook.f., Fl.Brit. India 1:660. 1875, p.p., non Wall., 1831-1832. *Vallia-tsjori-valli* Rheede, Hort. Malab. 7:15, t.8. 1688.

Stout climbers. Leaves 3-(5)- foliolate; leaflets oblanceate, 4-10 by 2-5 cm, base cuneate, margin distantly serrulate, apex shortly acuminate, coriaceous. Cymes c.2 cm long. Flowers c.4 mm ϕ; petals greenish-white. Berry globose. Seeds 2-4, oblong.

Fl. & Fr.: May-Aug. *Distr.*: Sri Lanka, Peninsualr & N.E. India to Malesia. Rare. Associated with riverine forest. *NAK* 2537 (Kakki hills, 1100 m).

2. **Tetrastigma sulcatum** (Lawson) Gamble, Fl. Pres. Madras 229.1918; Ramamoorthy in Sald. & Nicols., Fl. Hassan Dist. 361. 1976; Matthew & Britto in Matthew, Fl. Tam. Carnatic 3(1):286. 1983. *Vitis sulcata* Lawson in Hook.f., Fl. Brit. India 1:661. 1875.

Stout climbers; branchlets sulcate, tuberculate. Leaves 3-5-foliolate; leaflets elliptic, 6-15 by 2.5-7 cm, base obliquely rounded or obliquely cuneate, margin crenate-serrate, apex acuminate, coriaceous. Cymes c.3 cm long. Flowers c.3 mm ϕ; petals greenish-white. Berry globose, c.2.5 cm ϕ. Seeds oblong.

Fl. & Fr.: Nov.-Mar. *Distr.*: Peninsular India. Common. In evergreen and semi-evergreen forests. *NAK 387, & 514 & 1350* (Angamoozhy, c. 250 m). A highly variable species with distinct ecotypes.

45. LEEACEAE

LEEA D. Van Royen. ex Linnaeus SYST. NAT. ED. 12, 2: 627. 1767; MANT. 17, 124. 1767 (NOM.CONS.).

Shrubs or small trees. Leaves 2-3-pinnate, alternate; petiole dilated towards base; stipule large, sheathing. Inflorescence a leaf-opposed corymbose cyme. Flowers bisexual, 5-merous, regular; calyx-lobes 5, petals 5 greenish or reddish; stamens 5, connate below to form an urceolate tube; ovary 6-locular; ovule 1 per locule. Berry subglobose. Seeds 3-6, wedge shaped.

1. Flowers bright red; leaves usually 2-pinnate; undershrubs. ....1. **L. guineensis**
1. Flowers greenish-white; leaves 2-3-pinnate; trees. ....2. **L. indica**

1. **Leea guineensis** G. Don, Gen.Syst. 1:712. 1831; Ridsdale, in Blumea 22:92.1974; Sald. in Sald. & Nicols., Fl. Hassan Dist. 356.1976. *L. wightii* Clarke, J. Bot. 19:105. 1881; Gamble, Fl. Pres. Madras 29.1918. **(Fig.25)**

Undershrubs. Leaves 1-(2)- pinnate; leaflets ovate-lanceate or oblong-lanceate, 10-20 by 5-9 cm, base rounded, margin sharply serrate, apex long acuminate, lateral veins c.10 pairs,prominent, thin-coriaceous. Cymes 6-8 cm long. Flowers c.5 mm ϕ; petals lanceate, bright red; staminal tube 5-lobed, lobes 2-toothed. Berry ripening red. Seeds smooth.

Fl. & Fr.: May-Aug. *Distr.*: West Trop. Africa through India to Micronesia. Rare. Among undergrowth in evergreen forests. *NAK 116* (Moozhiar, .250 m), *1817* (Arampa, c.550 m). Reddish inflorescences is very striking.

2. **Leea indica** (Burm.f.) Merr., Philipp. J.Sci. 14:245. 1919; Ridsdale, Blumea 22:95.1974; Sald. in Sald. & Nicols., Fl. Hassan Dist. 356. 1976; Matthew, Ill. Fl. Tam. Carnatic t.160.1982; Matthew & Britto in Matthew, Fl. Tam. Carnatic 3(1):287.1983; Mohanan & Sivad., Fl. Agasthyamala 170. 2002. *Staphylea indica* Burm.f., Fl. Ind. 75,t.24, f.2. 1768. *Leea sambucina* Willd., Sp. Pl. 1:117/1798, nom. illegit; Roxb., Fl. Ind. 1:657. 1832; Lawson in Hook.f., Fl. Brit. India 1:666. 1875; Gamble, Fl. Pres. Madras 240.1918.*Mani porandi.*

Large shrubs or small trees. Leaves 2-3- pinnate; leaflets 3-5(7), oblong-lanceate, 6-16 by 2-5 cm, base cuneate, margin coarsely serrate, apex acuminate, lateral veins c.12 pairs, basally 3-nerved, prominent, chartaceous. Cymes c.8 cm long. Flowers c.5 mm f; petals lanceate, greenish-white; staminal tubes urceolate, 5-lobed. Berry ripening purple. Seeds smooth.

Fl. & Fr.: Most of the seasons. *Distr.*: China, India to Australia. Very common in moist deciduous, evergreen and semi-evergreen forests. *NAK 517 & 825* (Moozhiar, 200-600 m).

Juice of young leaves has digestive properties.

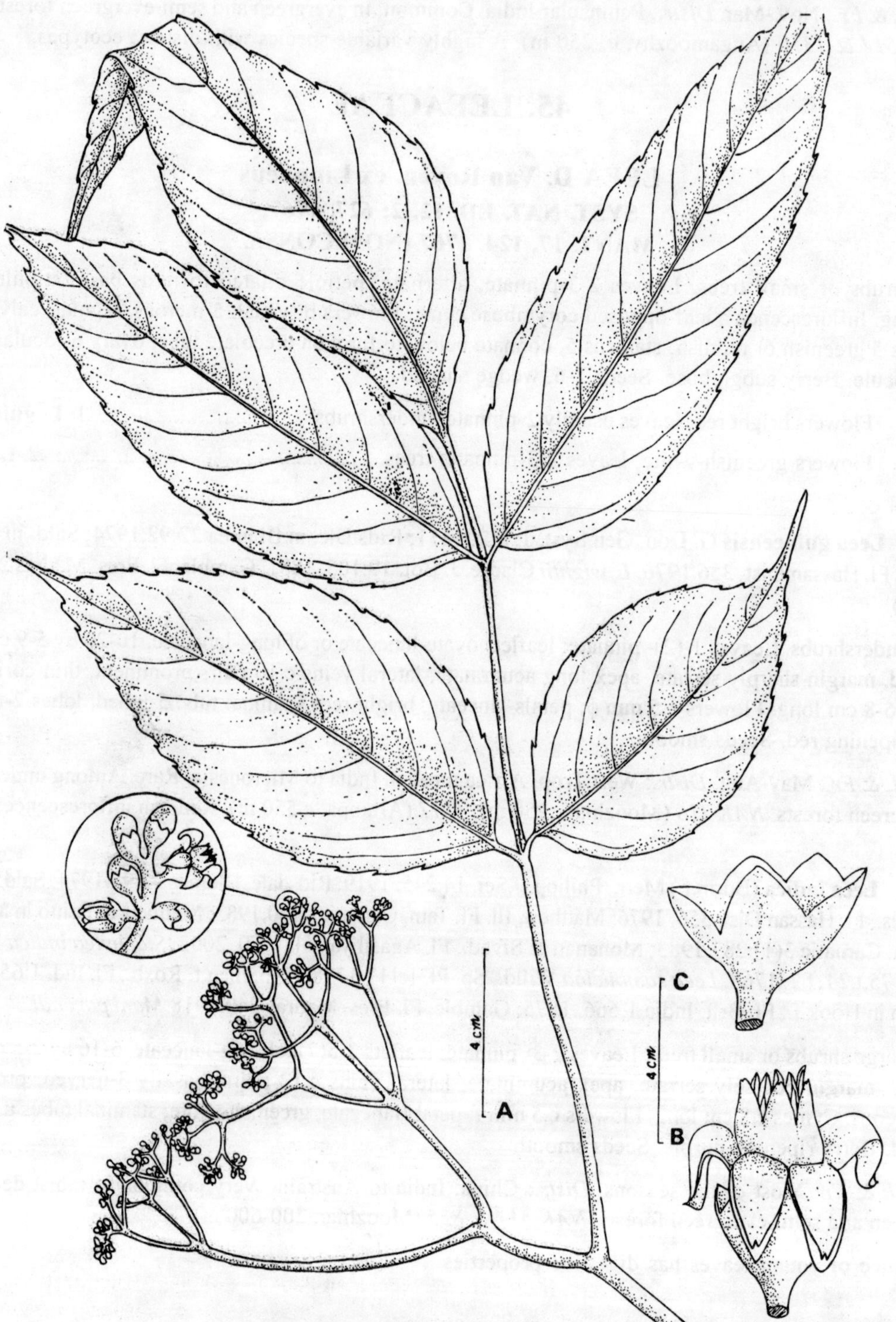

Fig. 25. ***Leea guineensis*** **G. Don: A. Twig; B. Flower; C. Calyx.**

46. STAPHYLEACEAE

TURPINIA Ventenat
Mem. Cl. Sci. Math. Inst. Natl. France 1807(1):3.1807 (nom.cons.).

Turpinia malabarica Gamble, Kew Bull. 1916: 135.1916, Fl. Pres. Madras 241.1918; Merr. & Perry, J. Arnold Arb. 22: 547. 1941; Gandhi in Sald. & Nicols., Fl. Hassan Dist. 361. 1976; Manilal, Fl. Silent Valley 62.1988. *Marali, Kana-kappalam.* **(Fig. 26)**

Trees. Leaves odd-pinnate, opposite; leflets opposite, elliptic to oblanceate, 7-14 by 3.5-7 cm, base cuneate, margin crenate-serrate, apex obtusely acuminate, thin-coriaceous. Inflorescence a large axillary or terminal panicle. Flowers bisexual, 5-merous, regular, c.5 mm ϕ; sepals 5, shortly connate below; petals 5, cream coloured; disc annular, evenly crenate; stamens 5, inserted around the disc; ovary ovoid, sessile, 3-locular; ovules 2 per locule. Berry subglobose. Seeds angular.

Fl. & Fr.: Jan-Apr. *Distr.*: Western Ghats. Not common. In semi-evergreen forests. *NAK 290 & 1464* (Sabarimala slopes, c. 250 m).

47. SAPINDACEAE

1. Climbing herbs; leaves 2-ternate; inflorescence tendrillar. .... **Cardiospermum**
1. Trees/shrubs; leaves 3-foliolate or even-pinnate; Inflorescence non-tendrillar.
    2. Leaves 3-foliolate. .... **Allophylus**
    2. Leaves even-pinnate.
        3. Flowers irregular; disc lobed.. **Lepisanthes**
        3. Flowers regular; disc annular.
            4. Fruits winged; ovules 2 per locule.
            4. Fruits not winged; ovule 1 per locule. .... **Harpullia**
                5. Petals 4-6.
                    6. Drupe tubercled; leaves 4 pairs, glaucous beneath. .... **Dimocarpus**
                    6. Drupe smooth; leaves 2-3 pairs, greenish beneath. .... **Sapindus**
                5. Petals 0.
                    7. Leaflets similar. .... **Schleichera**
                    7. Leaflets dissimilar, lower leaflets stipuliform. .... **Ottonephelium**

ALLOPHYLUS Linnaeus
Sp. Pl. 348. 1753.

Trees. Leaves 3-foliolate, alternate. Inflorescence raceme or panicle with interrupted fasicles. Flowers polygamo-dioecious, 4-merous; sepals 4, in opposite pairs, cucullate; petals 4, obsolete, rarely dorsally scaly; disc unilateral of 4-glands; stames 8, inserted inside the disc; ovary 2-locular; ovule 1per locule. Fruit (Schizocarp) 1-2-lobed. Seeds arillate.

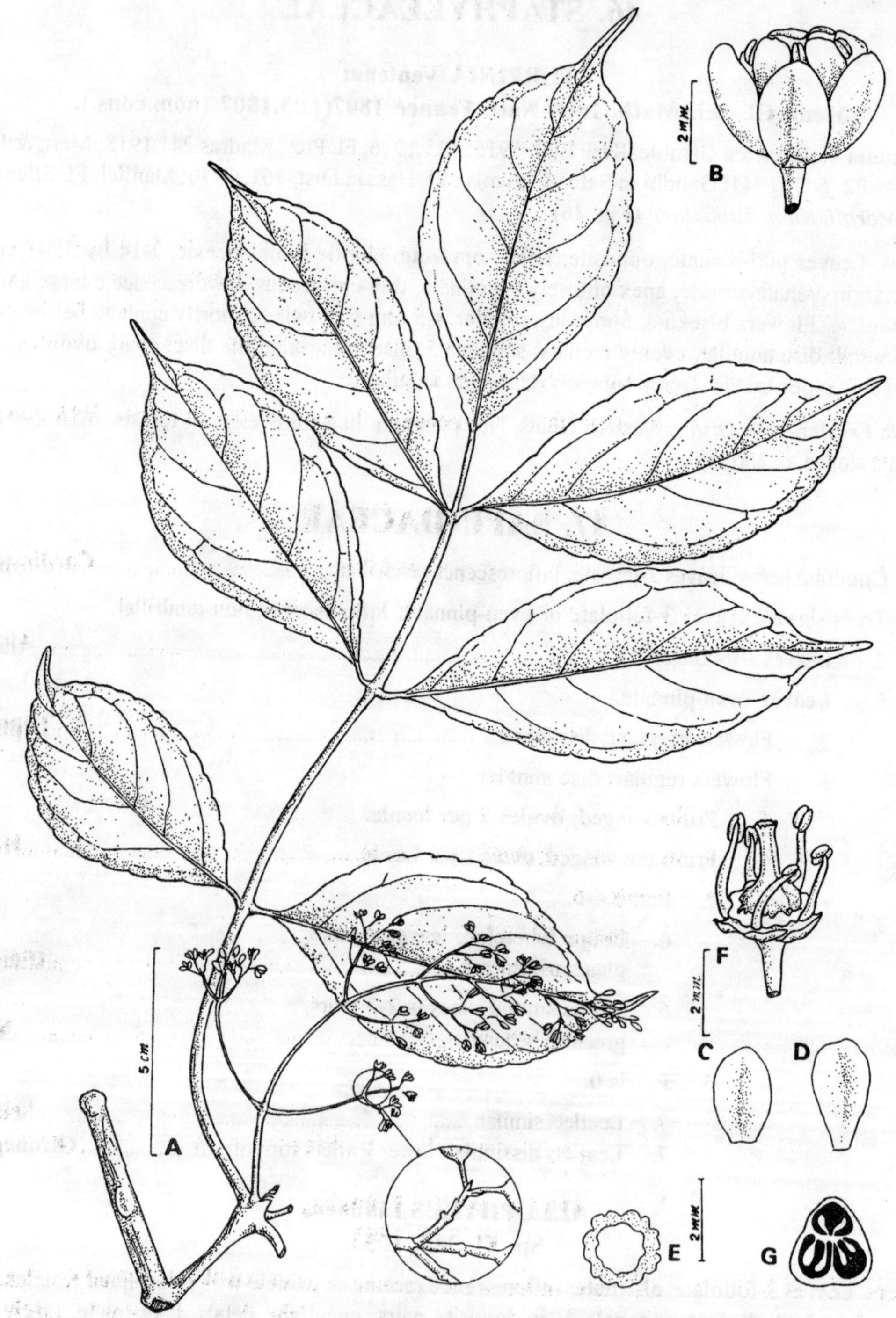

Fig. 26. ***Turpinia malabarica*** **Gamble : A. Twig; B. Flower; C. Sepal; D. Petal; E. Disc-T.S.; F. Flower with petals removed**

1. Panicle much longer than leaves; leaves usually glabrous. ....1.**A. concanicus** var. **lanceolatus**
1. Racemes equaling or slightly longer than leaves;leaves tomentose. ....2. **A. serrulatus**

1. **Allophylus concanicus** Radlk., Sitzungsber. Math.-Phys. Cl. Koenigl. Bayer. Akad. Wiss. Muenchen 20:230. 1890; Mukherjee, J.Econ. Tax.Bot. 1:79.1980, var. **lanceolatus** Gamble, Fl. Pres. Madras 246. 1918.

Leaflets lanceate or ovate-lanceate, 5-9 by 3-5-6.5 cm, base oblique, margin serrate, apex obtusely acuminate, thin-coriaceous. Panicle much branched, exceeding the leaves. Flowers white. Schizocarp 2-lobed, lobes globose.

Fl. & Fr.: Dec.-Apr. *Distr.*: Western Ghats. Rare. In moist evergreen forests. *CNM* 72818 (MH) (Pampa to Thannikudy, c. 1000 m).

2. **Allophylus serrulatus** Radlk., Rec. Bot. Surv. India 3: 341. 1907; Gamble, Fl. Press. Madras 246.1918; Mohanan & Sivad., Fl. Agasthyamala 173.2002.

Leaves tomentose; terminal leaflet obovate or rhombic, laterals oblong-elliptic, 9-16 by 5-10 cm, base cuneate, margin sharply serrate, apex acuminate. Racemes c.15 cm long, unbranched. Flowers pale yellow. Schizocarp 2-lobed.

Fl. & Fr.: May-Dec. *Distr.*: Western Ghats. Rare In moist evergreen and semi-evergreen forests. *NAK 1803* (Angamoozhy, c.250 m).

CARDIOSPERMUM Linnaeus
Sp. Pl. 366. 1753.

Cardiospermum halicacabum L., Sp. Pl. 366. 1753; Wight & Arn., Prodr. 109. 1834; Wight, Ic. t. 508. 1841; Hiern in Hook.f., Fl. Brit. India 1:670.1875; Gamble, Fl. Pres. Madr as 244. 1983; Nicols., Suresh & Mani., An Interpr. Hort. Malab. 237. 1988. *Ulinga* Rheede, Hort. Malab. 8:53-54, t.28. 1688.

Climbers. Leaves 2-ternate, alternate; leaflets ovate-lanceate, 2-4 by 1-2.5 cm, base cuneate or obtuse, margin lobed, apex acute to acuminate, glabrescent, membranous. Inflorescence a corymbose raceme; basal branches tendrillar. Flower polygamo-dioecious, 4-merous; sepal 4, in 2-pairs; petals 4, white, unequal, appendaged; disc of 2-glands; stamens 8, unequal; ovary 3-locular; ovule 1-per locule. Capsule sharply 3-lobed, winged. Seeds with a heart-shaped black marking, arillate.

Fl. & Fr.: Jul.-Dec. *Distr.*: Tropics. Common. Along waysides. *NAK 254, 1628* (Tiruvalla, c.10 m).

Decoction of the whole plant is given as a remedy in colic and hernia.

DIMOCARPUS Loureiro
Fl. Cochinch. 233. 1790.

Dimocarpus longan Lour., Fl. Cochinch. 233. 1790, Matthew, Ill. Fl. Tam. Carnatic t.163. 1982; Matthew & Britto in Matthew, Fl. Tam. Carnatic 3(1) : 293. 1983; Manilal, Fl. Silent Valley 63. 1988; Mohanan & Sivad., Fl. Agasthyamala 174. 2002. *Euphoria longana* Lam., Encycl. 3. 574. 1792; Bedd., Fl. Sylv. . 156. 1871. *Nephelium longana* (Lam.) Cambess., Mem. Mus. Hist. Nat. 18.30.1829; Wight & Arn., Prodr. 113.1834; Hiern in Hook. f., Fl. Brit. India 1: 668.1875; Gamble, Fl. Pres. Madras 252. 1918. *Chempoovam, Pori-punna.*

Trees. Leaves even-pinnate, alternate; leaflets subopposite or alternate, 2-4 pairs, oblong-lanceate to oblanceate, 8-20 (30) by 3-6 (8) cm, base acute or cuneate, apex shortly acuminate, lower surface glaucous, coriaceous. Infloresncne a large panicle, axillary. Flowers polygamo-dioecious, regular; calyx cupular, lobes 4-6, ovate; petals 4-6, cream coloured, reflexed; disc annular; stamens 6-8, inserted within the disc; ovary 2-3- locular; ovule 1per locule. Schizocarp globose, tubercled. Seeds arillate.

Fl. & Fr.: Feb.-May. *Distr.*: Sri Lanka, India, China, Myanmar, Malay Peninsula. Common. In evergreen forests. *NAK 506, 594, 1494* (Kodumudi-Chittar, c.200 m).

Fruits edible.

HARPULLIA Roxburgh
Fl. Indica 2: 441. 1824.

Harpullia arborea (Blanco) Radlk., Sitzungsber Math.-Phys. Cl. Koenigl. Bayer. Akad. Wiss. Muenchen 16: 404. 1890. *Ptelea arborea* Blanco, Fl. Filip. (ed.1) 63.183.*Otonychium imbricatum* Blume, Rumphia 3:180. 1849. *Harpullia imbricata* (Blume) Thw., Enum. Pl. Zeyl.56. 1858; Gamble, Fl. Pres. Madras 253.1918. *Harpullia cupanoides* sensu Hiern in Hook. f., Fl. Brit. India 1:692. 1875, p.p., non Roxb., 1832. *Chittila-madakku.*

Trees. Leaves even-pinnate; leaflets opposite and alternate, elliptic-oblong, 10-20 by 4-6 cm, base cuneate, apex acuminate, thin-coriaceous. Inflorescence a raceme or panicle, axillary or subterminal. Flowers polygamo-dioecious, regular; sepals 4-5; petals 4-5, pale yellow, clawed; scales 0; disc obscure; stamens 5-8; ovary 2-locular; ovules 2 per locule. Capsule inflated, winged. Seeds arillate.

Fl. & Fr.: Jan.-Apr. *Distr.*: Western Ghats. Rare. In moist evergreen forests. Majestic trees with scarlet coloured characteristic fruits. *CNM* 54351 (MH) (Kakki dam area, c. 1100 m).

LEPISANTHES Blume
Bijdr. 237. 1825.

Lepisanthes tetraphylla (Vahl) Radlk., Sitzungsber Math.-Phys; Cl. Koenigl. Bayer. Akad. Wiss. Muenchen 8: 276. 1878; Gamble, Fl. Pres. Madras 247.1918; Gandhi in Sald. & Nicols., Fl. Hassan Dist. 367.1976; Matthew & Birtto in Matthew, Fl. Tam. Carnatic 3(1) : 296. 1983; Manilal, Fl. Silent Valley 64. 1988; Mohanan & Sivad., Fl. Agasthyamala 174. 2002. *Sapindus tetraphylla* Vahl, Symb. Bot. 3: 54.1794. *Molinaea canescens* Roxb., Pl. Corom. t. 60. 1796, Fl. Ind. 2:243. 1832. *Hemigyrosa canescens* (Roxb.) Bl., Rumphia 3:166. 1849; Hiern in Hook. f., Fl. Brit. India 1:671. 1875. *Sapindus deficiens* Wight & Arn., Prodr. 111. 1834. *Hemigyrosa deficiens* (Wight & Arn.) Bedd., Fl. Sylv. t. 231. 1872. *Lepisanthes deficiens* (Wight & Arn.) Radlk., Sitzungsber Math.-Phy. Cl. Koenigl. Bayer. Akad. Wiss. Muenchen 8:276. 1878; Gamble, Fl. Pres. Madras 247. 1918. *Kula-punna.*

Trees. Leaves even-pinnate, alternate; leaflets 4-6(-8) pairs, subopposite, elliptic-oblong, 8-25 by 3-7 cm, base obtuse, apex obtuse, emarginate or acute. Inflorescence lateral cluster of racemes. Flowers polygamo-dioecious, irregular; sepals 5, imbricate; petals 4-5, cream coloured, attached with scales; disc lobed; stamens 8; ovary 3-locular; ovule 1per locule. Drupes 3-gonous, velvetty. Seed 1, non-arillate.

Fl. & Fr.: Mar.-May. *Distr.*: Sri Lanka, India, Myanmar, Malaya Peninsula. Less common. In semi-evergreen forests. *NAK 2005* (Kalleli F.R., c.300 m).

OTONEPHELIUM Radlkofer
Sitzungsber Math.-Phy. Cl. Konigl. Bayer Akad. Wiss. Muenchen 20:253, 288.1890.

Otonephelium stipulaceum (Bedd.) Radlk., Saindc. Hol.-Ind. 71. 1879; Gandhi in Sald. & Nicols., Fl. Hassan Dist. 368. 1976; Mohanan & Sivad., Fl. Agasthyamala 175.2002. *Nephelium stipulaceum* Bedd., Trans. Linn. Soc. London 25: 212. 1865 & Ic. t.103. 1868-1874; Hiern in Hook.f., Fl. Brit. India 1:690. 1875; Gamble, Fl. Pres. Madras 252.1918. *Pori-poovam.*

Trees. Leaves even-pinnate; leaflets alternate, usually 3 pairs, lower pair stipule-like, reniform, 25 by12 mm, upper pairs narrowly obovate or oblanceate, 7-20 by 2.5-8 cm, base cuneate, apex acuminate, lower surface glaucous, thin-coriaceous. Inflorescence an axillary panicle. Flowers c.5 mm f; calyx-lobes 4-6, cream coloured; petals 0; stamens 8; ovary 3-locular; ovule 1-per locule. Fruit ellipsoid, densely echinate. Seeds arillate.

Fl. & Fr.: Feb.-May. *Distr.*: Endemic to Western Ghats. Rare. In moist evergreen and semi-evergreen forests. *NAK 502* (Nilakkal, c.400 m), *1715* (Sabarimala slopes, c.250 m).

SAPINDUS Linnaeus
Sp. Pl. 367. 1753.

Sapindus laurifolia Vahl, Symb. 3:54. 1794; Gamble, Fl. Pres. Madras 250. 1918 ("*laurifolius*"); Gandhi in Sald. & Nicols., Fl, Hassan Dist. 368. 1976; Mohanan & Sivad., Fl. Agasthyamala 176.2002. *Sapindus trifoliata* auct., non L, 1753: Hiern in Hook. f., Fl. Brit. India 1:682. 1875, p.p. "*trifoliatus*". *Pasa-kotta.*

Trees. Leaves even-pinnate, alternate; leaflets subopposite, elliptic, 9-16 by 3-4.5 cm, base oblique, apex acuminate, chartaceous. Inflorescence a terminal panicle. Flowers 5-merous, regular; sepals 5, ovate; petals 5, white, usually scales 0; stamens 8, inserted within in disc; ovary 3-locular; ovule 1 per locule. Drupe ovoid. Seeds globose, black, shiny.

Fl. & Fr.: Nov.-Mar. *Distr.*: India, Sri Lanka. Rare. In evergreen forests. *NAK 1413* (Arampa, 500 m).

Fruits are used as substitute for soap. Seeds yield an oil used in medicine and soap making.

SCHLEICHERA Willdenow
Sp. Pl. 4: 1096. 1806 (nom. cons.).

Schleichera oleosa (Lour.) Oken, Allg. Naturgesh. 3(2): 134. 18411; Merr., Interpr. Herb. Amboin. 337. 1917; Gandhi in Sald. & Nicols., Fl. Hassan Dist. 369. 1976; Matthew, Ill. Fl. Tam Carnatic t.168. 1982; Matthew & Britto in Matthew, Fl. Tam. Carnatic 3(1): 299 1983; Manilal, Fl. Silent Valley 64. 1988; Mohanan & Sivad., Fl. Agasthyamala 176.2002. *Pistacea oleosa* Lour., Fl. Cochinch. 615. 1790. *Schleichera trijuga* Willd., Sp. Pl. 4; 1096.1806; Hiern in Hoo. f., Fl. Brit. India 1:681. 1875; Gamble, Fl Pres. Madras 248. 1918. *Poovam, Poovanam.*

Trees. Leaves even-pinnate; leaflets 3 pairs, subopposite, elliptic-oblong or oblanceate, 6-19 by 2.5-5.5 cm, base cuneate to subcordate, apex acute to obtuse, chartaceous. Inflorescence racemes, clustered towards the branchlet apices, drooping. Flowers polygamo-dioecious; calyx-lobes 4-6, triangular; petals 0; disc flat; stamens 5-8, inserted within the disc; ovary ovoid, 3-locular; ovule 1-per locule. Drupe ovoid. Seed(s) 1 or 2, arillate.

Fl. & Fr.: Mar.-Jun. *Distr.*: Sri Lanka, India, Trop. Himalaya, Indo-China, Malesia. Common. In moist deciduous and semi-evergreen forests. *NAK 539* (Angamoozhy, c.250 m).

Seed-oil is said to be very effective for various forms of skin diseases.

48. SABIACEAE

MELIOSMA Blume
Cat. Buitenzorg 10.1823.

Evergreen trees. Leaves simple or odd-pinnate, alternate; leaflets 9-15, subopposite. Inflorescence a large terminal or axillary panicle. Flowers bisexual; bracteoles 3, sepaloid; petals 3+2, orbicular; stamens 5, fertile ones 2; staminodes 3, filaments cupular at apex, appendaged at base; staminodes 2-fid; ovary 2-locular; ovules 2-per locule. Drupe globose.

1. Leaves odd-pinnate; ovary pubescent; drupe smooth. ....1. **M. pinnata**

1. Leaves simple; ovary glabrous; drupe rugose. ....2. **M. simplicifolia**

1. **Meliosma pinnata** (Roxb.) Walp., Repert. 1:423.1842, ssp. **barbulata** (Cufod.) Beus. in Dassan. & Fosberg, Rev. Handb. Fl. Ceylon 3: 384. 1981; Manilal, Fl. Silent Valley 65.1988; Mohanan & Sivad., Fl. Agasthyamala 177.2002. *M. rohifolia* Maxim. ssp.*barbulata* Cufod., Oest. Bot. Z88. 254. 1939. *Millingtonia arnottiana* Wight, Ill. 1:t.53. 1840. *Meliosma arnottiana* (Wight) Walp., Repert. 1:423. 1842; Hook. f., Fl. Brit. India 2:6. 1876; Gamble, Fl. Pres. Madras 256.1918. *M. pinnata* (Roxb.) Maxim. ssp. *arnottiana* (W.) Beus., Blumea 19:499. 1971; Gandhi in Sald. & Nicols., Fl. Hassan Dist. 69. 1976; Matthew & Britto in Matthew, Fl. Tam. Carnatic 3(1): 301.1983.

Leaves odd-pinnate; leaflets ovate-lanceate or oblong-lanceate, 8-14 by 4-6 cm, base rounded, apex acuminate, rusty-tomentose, chartaceous. Panicle axillary and terminal. Flowers clustered towards the panicle branches; sepals and bracteoles 5, sub orbicular; petals cream coloured. Drupe globose.

Fl. & Fr.: Mar.-Jun. *Distr.*: Sri Lanka, Peninsular India, China, Malaya Peninsula, Malesia. Occasional. In upper hills. *NAK* 2735 (Kakki hills, c.1000 m).

2. **Meliosma simplicifolia** (Roxb.) Walp., Repert. 1:423.1842; Hook.f., Fl. Brit. India 2:5. 1876; Gamble, Fl. Pres. Madras 256. 1918; Beus., Blumea 19: 476. 1971 & in Dassan. & Fosberg, Rev. Handb. Fl. Ceylon 3:381. 1981; Gandhi in Sald. & Nicols., Fl. Hassan Dist. 65. 1976; Manilal, Fl. Silent Valley 65. 1988. *Millingtonia simplicifolia* Roxb., (Hort. Beng. 3. 1814, nom. nud.) Pl. Corom. t.254. 1820. *Cheru-Kallavi.*

Leaves obovate or oblanceate, 15-30 by 5-10 cm, base gradually attenuate, apex acute, lateral veins prominent, chartaceous. Panicle large, terminal or axillary. Flowers small; sepals and bracteoles ± 6, ciliate; petals yellowish or reddish. Drupe subglobose.

Fl. & Fr.: Sep.-Mar. *Distr.*: Sri Lanka, India, Nepal, Malesia. In moist evergreen forests. Occasional. *NAK* 2539 (Kakki hills, c.1100 m).

49. ANACARDIACEAE

1. Leaves simple.
   2. Ovary inferior; petiole often spurred. ....**Holigarna**
   2. Ovary superior; petiole not spurred.
      3. Styles 3 or more.

4. Flowers unisexual; stamens as many as petals; drupe on a fleshy hypocarp. .... **Semecarpus**

4. Flowers bisexual; stamens twice as many as petals; drupe not on a fleshy hypocarp. .... **Buchanania**

3. Style 1.

5. All stamens fertile; style terminal. .... **Nothopegia**

5. Only 1 or 2 stamens fertile; style lateral. .... **Mangifera**

1. Leaves compound. .... **Lannea**

BUCHANANIA K.P.J. Sprengel
J. Bot. (Schrader) 1800 (2): 234. 1802 ('1801').

Trees. Leaves simple, alternate, coriaceous. Inflorescence a panicle, axillary or terminal. Flowers bisexual, 5-merous; calyx-lobes 5, triangular; petals 5, reflexed; disc cuplar, 5-lobed; stamens 10, all fertile; ovary superior, of 5 or 6 free carpels; ovule 1 per carpel. Drupe 1-seeded.

1. Leaves narrowly elliptic-oblong, glabrous beneath; petiolc slender. .... 1. **B. lanceolata**

1. Leaves broadly ovate-oblong, densely pubescent beneath; petiole stout. .... 2. **B. lanzan**

1. **Buchanania lanceolata** Wight, Ic. t.237. 1839; Hook.f., Fl. Brit. India 2: 24.1876; Gamble, Fl. Pres. Madras 259. 1918. *Malamavu.*

Leaves narrowly elliptic-oblong, 8-16 by 3-5 cm, base acute, apex obtusely acuminate, lateral veins 18-20 pairs, prominent, coriacous. Panicle thinly pubescent. Drupe c.12 mm f.

Fl. & Fr.: Nov.-Mar. *Distr.*: Western Ghats (Travancore hills). Rare. In sacred groves. *NAK 2338* (Aikadu sacred grove, Kodumon, c.40 m).

The bark of the tree is medicinal. Kernals of seeds edible.

2. **Buchanania lanzan** Spreng., J.Bot. (Schrader) 2: 234. 1800; Gamble, Fl. Pres. Madras 258. 1918; Ramamoorthy in Sald. & Nicols., Fl. Hassan Dist. 374. 1976; Matthew & Britto in Matthew, Fl. Tam. Carnatic 3(1) : 306.1983; Mohanan & Sivad., Fl. Agasthyamala 178.2002. *B. latifolia* Roxb., Fl. Ind. 2:385. 1832; Hook. f., Fl. Brit. India 2: 23. 1876. *Moonga-pezhu.*

Leaves broadly ovate or oblong-ovate, 13-18 by 6-7 cm, base broadly obtuse, apex obtuse-retuse, lower surface tomentose; petiole stout. Panicle villous. Flowers cream coloured. Drupe subglobose, c.15 mm f.

Fl. & Fr.: Jan.-Apr. *Distr.*: India, Myanmar. Rare. In moist deciduous forests. *NAK* 2337, *CNM 68382* (MH) (Konni R.F., c.250 m).

HOLIGARNA F. (Buchanan) Hamilton ex Roxburgh
Pl. Coromandel 3:79. 1820 ('1819') (nom. cons.).

Trees. Leaves simple, alternate, coriaceous; petiole with deciduous spur-like appendages. Inflorescence a raceme or panicle, axillary and terminal. Flowers polygamo-dioecious; calyx 5-toothed; petals 5; disc obscure; stamens 5, inserted outside the disc; ovary inferior, 1-locular; ovule 1, pendulous. Drupe enclosed fully or partially by a hypocarp.

1. Drupe fully eclosed by the hypocarp; leaves glabrous or thinly pubescent on nerves beneath.
   2. Leaves to 25 cm long, lateral veins 6-9 pairs, apex obtusely acute, brownish when dry. .... 1. **H. arnottiana**
   2. Leaves to 15 cm long, lateral veins 6-9 pairs, apex obtuse or emarginate, nearly black when dry. .... 3. **H. nigra**
1. Drupe partially enclosed by the hypocarp;leaves densely pubescent beneath. ....2. **H. grahamii**

1. **Holigarna arnottiana** Hook.f., Fl. Brit. India 2:36. 1876; Gamble, Fl. Pres. Madras 268. 1918; Nicols., Suresh & Mani., An Interpr. Hort. Malab. 48. 1988; Mohanan & Sivad., Fl. Agasthyamala 179. 2002. *H. longifolia* auct., non Roxb., 1820: Wight & Arn., Prodr. 169. 1834. *Katou - tsjeroe, Cheru* Rheede, Hort. Malab. 4:10-21, t.9.1683.

Leaves oblanceate, 12-14 by 3-7 cm, base attenuate, margin entire- cartilaginous, apex obtusely short-acuminate, lateral veins c.20 pairs, raised on both sides, coriaceous.Racemes terminal, c. 10 cm long. Flowers c.4 mm ϕ; petals pale yellow. Drupe ovoid, resinous.

Fl. & Fr.: Jan.-Jul. *Distr.*: Western Ghats. Common in sacred groves. *NAK 712* (Pampa Valley, c.300 m), 2319 (Aikadu, c. 17 m).

The acrid black juice obtained from the trunk and rind of fruit used as a black varnish.

2. **Holigarna grahamii** (Wight) Kurz, J. Asiat. Soc. Bengal 42: 305. 1872, quoad basionym; Hook. f., Fl. Brit. India 2; 37. 1876; Gamble, Fl. Pres. Madras 268. 1918; Ramamoorthy in Sald. & Nicols., Fl. Hassan Dist. 375. 1976. *Semecarpus grahamii* Wight, Ic. t. 235. 1839. *Holigarna wightii* Balakr., J. Bombay Nat. Hist. Soc. 63: 327. 1966.

Leaves broadly oblanceate, 25-40 by 10-15 cm, base attenuate, apex acumimate, lower surface densely pubescent, lateral veins 20-30 pairs, very prominent beneath, coraceous. Panicle terminal, c.30 cm long. Flowers c.3 mm ϕ, pale yellow. Drupe ovoid, resinous.

Fl. & Fr.: Mar.-Jul. *Distr.*: Western Peninsular India. Rare. In evergreen forests. *BKN* 10433 (CALI) (Moozhiar, c. 250 m).

3. **Holigarna nigra** Bourd., Ind. Forester 30: 95. 1904; Gamble, Fl. Pres. Madras 268. 1988.

Leaves narrowly obovate or oblanceate, 10-15 by 4-6 cm, base narrowed, apex obtuse-emarginate, lateral veins 6-9 pairs, prominent, coriaceous. Racemes terminal. Flowers c.4 mm ϕ; greenish-yellow. Drupe ovoid.

Fl. & Fr.: Mar.-May. *Distr.*: Western Ghats. Rare. In moist evergreen forests. *KV* 66208 (MH)-(Between Pamba Dam & Kakki Dam).

LANNEA A. Richard in J.B.A. Guillemin et al. Fl. Seneg. Tent. 153. 1831 (nom. cons.).

Lannea coromandelica (Houtt.) Merr., J. Arnold. Arbor 19:353. 1938; Ding Hou in Steenis, Fl. Males. 1,8: 478.1978; Gandhi in Sald. & Nicols., Fl. Hassan Dist. 376. 1976; Matthew, Ill. Fl. Tam. Carnatic t. 172. 1982; Matthew & Britto in Matthew, Fl. Tam. Carnatic 3(1): 307. 1983; Nicols., Suresh & Mani., An Interpr.

Hort. Malab. 48. 1988; Mohanan & Sivad., Fl. Agasthyamala 180.2002. *Dialium coromandelicum* Houtt., Nat. Hist. Ser. 2, 2: 39, t.5, f.2. 1774. *Odina wodier* Roxb., Fl. Ind. 2: 293. 1832; Hook.f., Fl. Brit. India 2:29. 1876; Gamble, Fl. Pres. Madras 263. 1918. *Kalesjam* Rheede, Hort. Malab. 4:67-68, t.32. 1683. *Udi.*

Trees. Leaves odd-pinnate; leaflets 4-pairs, opposite, oblong-lanceate, 5-9 by 3-4 cm, base obtuse, apex acuminate, lateral veins c.7 pairs, chartaceous. Racemes clustered, pendulous. Flowers polygamous, 4-merous; calyx-lobes 4; petals 4, greenish-yellow; disc annular, 8-lobed; stamens 8, all fertile; ovary 1-locular. Drupe ovoid or obovoid.

Fl. & Fr.: Jan.-Jun. *Distr.*: Sri Lanka, India, Thailand, China. Common on waysides in plains. *NAK 1352* (Niranam-Tiruvalla, c.10 m).

Decoction of root bark with honey is very good medicine for diarrhoea and dysentery. Wood is suitable for match industry.

MANGIFERA Linnaeus
Sp. Pl. 200. 1753.

Mangifera indica L., Sp. Pl. 200. 1753; Bedd., Fl. Sylv. t. 162. 1871; Hook.f., Fl. Brit. India 2:13. 1876; Gamble, Fl. Pres. Madras 259. 1918; Ramamoorthy in Sald. & Nicols., Fl. Hassan Dist. 376. 1976; Matthew, Ill. Fl. Tam. Carnatic t. 173. 1982; Matthew & Britto in Matthew, Fl. Tam. Carnatic 3(1): 308. 1983; Manilal, Fl. Silent Valley 67. 1988; Mohanan & Sivad., Fl. Agasthyamala 181. 2002. Nicols., Suresh & Mani., An Interpr. Hort. Malab. 48. 1988. *Mavu* Rheede, Hort. Malab. 4:1-4, t.1-2. 1683.

Trees. Leaves simple, oblong or elliptic, 10-20 by 2.5-4 cm, base cuneate, apex acuminate, coriaceous. Panicle terminal. Flowers polygamous, greenish-white; stamens 5, only 1 fertile, 1-locular; style lateral. Drupe large, ovoid or obvoid.

Fl. & Fr.: Mar.-Jun. *Distr.*: India, Myanmar, Thailand. Common. In semi-evergreen and evergreen forests. *NAK 2428* (Kalleli R.F, c.250 m).

Fruits edible, kernel of seeds used in bleeding piles and diarrhoea; and as an anthelmintic.

NOTHOPEGIA Blume
Mus. Bot. 1:203. 1850 (nom. cons.).

Trees. Leaves simple, alternate, coriaceous. Inflorescence a panicle, axillary. Flowers polygamous, 4-merous; calyx small, 4-lobed, petals 4, yellowish; disc annular, 4-lobed; stamens 4, inserted below the disc; ovary 1-locular; ovule 1. Drupe ovoid or subglobose.

1. Branchlets, leaves beneath glabrous; leaves to 14 cm long. ....1. **N. beddomei**
1. Branchlets, leaves beneath softly villous; leaves to 20 cm long. ....2. **N. travancorica**

1. **Nothopegia beddomei** Gamble, Fl. Pres. Madras 265. 1918; Ramamoorthy in Sald. & Nicols., Fl. Hassan Dist. 377. 1976; Matthew, Ill. Fl. Tam. Carnatic t.174. 1982; Matthew & Britto in Matthew, Fl. Tam. Carnatic 3(1) : 309. 1983; Mohanan & Sivad., Fl. Agasthyamala 182. 2002. *N. colebrookeana* sensu Bedd., Fl. Sylv. t.164. 1871, non (Wight) Blume, 1851; Hook. f., Fl. Brit. India 2:40. 1876, p.p.

Leaves elliptic, 7-12 by 1.5-3.5 cm, base cuneate, margin undulate, apex long acuminate, lower surface glaucescent, lateral nerves c.20 pairs, raised below, thin-coriaceous. Racemes axillary, c.2 cm long. Flowers c.4 mm ϕ; calyx-lobes triangular, petals bright yellow. Drupe subglobose.

Fl. & Fr.: Apr.-Jun. *Distr.*: Sri Lanka, Peninsular India. Rare. In moist semi evergreen forests, usually along streamsides. *NAK 2420* (Kalleli R.F., Konni, c.250 m).

2. **Nothopegia travancorica** Bedd. ex Hook. f., Fl. Brit. India 2: 40. 1876; Gamble, Fl. Pres. Madras 265. 1918; Ramamoorthy in Sald. & Nicols., Fl. Hassan Dist. 377. 1976; Mohanan & Sivad., Fl. Agasthyamala 184. 2002.

Leaves oblong-lanceate, 15-22 by 4.5-6 cm, base acute apex acuminate, lower surface softly minute-villous, lateral veins c.25 pairs, coriaceous. Racemes axillary. Flowers small; petals yellow, pubescent; staminal filaments hairy. Drupe ovoid.

Fl. & Fr.: Mar.-May. *Distr.*: Western Ghats. Very Rare. In semi-evergreen forests. *NAK 600* (Thekkuthode, c.70 m).

SEMECARPUS Linnaeus f.
Suppl. Pl. 25, 182. 1782.

Evergreen trees. Leaves simple, alternate, coriaceous. Inflorescence a terminal or axillary panicle. Flowers polygamo-dioecious, regular; calyx 3-5- lobed; petals 3-5; disc annular; stamens 3-5, inserted below the disc; ovary superior, 1-locular. Drupe subglobose or ovoid, seated on a fleshy hypocarp.

1. Leaves, inflorescence and ovary pubescent; hypocarp ± equal to drupe. ....1. **S. anacardium**
1. Leaves, inflorescence and ovary glabrous; hypocarp smaller than drupe.
    2. Leaves oblanceate, c.6 cm wide, base auricled; panicle slender. ....2. **S. auriculatum**
    2. Leaves obovate-oblong, c. 20 cm wide, base cuneate; panicle stout. ......3. **S. travancorica**

1. **Semecarpus anacardium** L.f., Suppl. Pl. 182. 1781; Wight, Prodr. 168. 1834, Ic. t.558. 1842; Hook.f., Fl. Brit. India 2:30. 1876; Gamble, Fl. Pres. Madras 266. 1918; Ramamoorthy in Sald. & Nicols., Fl. Hassan Dist. 378. 1976; Matthew, Ill. Fl. Tam. Carnatic t. 176. 1982; Matthew & Britto in Matthew, Fl. Tam. Carnatic 3(1): 310. 1983; Mohanan & Sivad., Fl. Agasthyamala 185.2002. *Sambiri Then- Kotta, Alakku- Cheru.*

Leaves elliptic-oblong or ovate, 20-35 by 8-15 cm, base obtuse to subcordate, apex retuse or emarginate, lower surface rusty-villous, lateral nerves c.16 pairs, coriaceous. Panicle terminal, to 20 cm long; clayx-lobes 5, triangular; petals 5, greenish. Drupe globose-ovoid.

Fl. & Fr.: May-Mar. *Distr.*: S. Asia, N. Australia. Rare. On hills slopes. *CNM 61217* (MH) (Chuttipara hills-Pathanamthitta, c.100 m).

2. **Semecarpus auriculata** Bedd., Ic. t. 187. 1868-1874; Hook. f., Fl. Brit. India 2:32. 1876; Gamble, Fl. Pres. Madras 267. 1918; Mohanan & Sivad., Fl. Agasthyamala 185. 2002. *Vella- cheru.*

Leaves oblanceate, 17-20 by 4.5-5.5 cm, base attenuate-auricled, apex shortly acumicate, thin coriaceous. Panicle slender, to 25 cm long. Flowers minute; calyx-lobes 5, small; petals 5, greenish.

Fl. & Fr.: Nov.- Mar. *Distr.*: Western Peninsular India. Rare. In evergreen forests. NAK 1198 (Moozhiar, c. 250m), (Appupanthode, c. 400m).

3. **Semecarpus travancorica** Bedd., Fl. Sylv. t.232. 1872; Hook. f., Fl. Brit. India 2:31. 1876; Gamble, Fl. Pres. Madras 267. 1918 ; Mohanan & Sivad., Fl. Agasthyamala 186.2002. *Avukkaram.*

Leaves obovate-oblóng. 14-40 by 8-19 cm, base acute, apex obtuse, lateral nerves c.20 pairs, coriaceous. Panicle stout, to 30 cm long. Flowers small. Drupe subreniform.

Fl. & Fr.: Sep.-Dec. *Distr.*: Western Ghats. Rare. In evergreen forests. *NAK 2553* (Kakki hills, 1100 m).

Note : Extensive plantations of **Anacardium occidentale** L. has replaced a good area of forest in Konni. **Spondias pinnata** (L.f.) Kurz is also planted elsewhere.

50. CONNARACEAE

1. Panicle rusty-pubescent; calyx persistent, clasping the fruit-pedicel. ....**Connarus**
1. Panicle glabrous; calyx accrescent, clasping the fruit-base. ....**Rourea**

CONNARUS Linnaeus
Sp. Pl. 675. 1753.

Scandent shrubs. Leaves odd-pinnate, alternate; leaflets 5, rarely 3, opposite. Inflorescence a branched terminal or axillary panicle. Flowers bisexual, regular; sepals 5, connate at base, persistent; petals 5; stamens 10, unequal, often the ones alternating petals smaller and without anthers; ovaries 5, 4 usually obsolute. Follicle stipitate. Seed arillate.

1. Leaf nervature faintly reticulate; follicle-valves glabrous inside.
    2. Follicle scarlet coloured; leaves elliptic, lateral veins 3-5 pairs. ....1. **C. monocarpus**
    2. Follicle brown or red coloured. leaves elliptic-ovate, lateral veins 7-8 pairs. ....2. **C. wightii**
1. Leaf nervature prominently reticulate; follicle-valves tomentose inside. ....3. **C. sclerocarpus**

1. **Connarus monocarpus** L., Sp. Pl. 675. 1753; Hook. f., Fl. Brit. India 2:50. 1876; Gamble, Fl. Pres. Madras 272. 1918; Nicols., Suresh & Mani., An Interpr. Hort. Malab. 86. 1988. *Kuriel, Perim-kurigil* Rheede, Hort. Malab. 6: 43-44, t.24. 1686. (**Fig. 27**).

Leaflets elliptic-ovate, 5-10 by 2.5-4.5, base rounded, apex acuminate, thin-coriaceous. Flowers c.5 mm ϕ; sepals elliptic-ovate; petals bright-yellow, narrowly obovate; stamens all antheriferous, filaments with glandular appendages; ovary 1-locular; ovules 2. Fruit scarlet coloured within

Fl. & Fr.: Apr-Jun. *Distr.*: Sri Lanka, Peninsular, India. Occasional. In evergreen forests and often in sacred groves. *NAK 2301* (Aikadu sacred grove, c.50 m).

Decoction or paste of roots given in syphilis.

2. **Connarus sclerocarpus** (Wight & Arn.) Schellenb., Candollea 298. 1925; Gamble, Fl. Pres. Madras 273. 1918; Mohanan & Sivad., Fl. Agasthyamala 187.2002. *Rourea? sclerocarpa* Wight & Arn., Prodr. 262. 1864.

Leaflets elliptic, 6-12 by 3-5 cm, base rounded, apex acuminate, nervature prominently reticulate, coriaceous. Panicle densely rusty-tomentose, ovary 1-locular; ovules 2. Fruit red coloured, velvetty within.

Fl. & Fr.: Sep.-Apr. *Distr.*: Western Peninsular India (Travancore hills). Rare. In moist evergreen forest. *CNM 54357* (MH) (Pamba dam area, c. 1100 m).

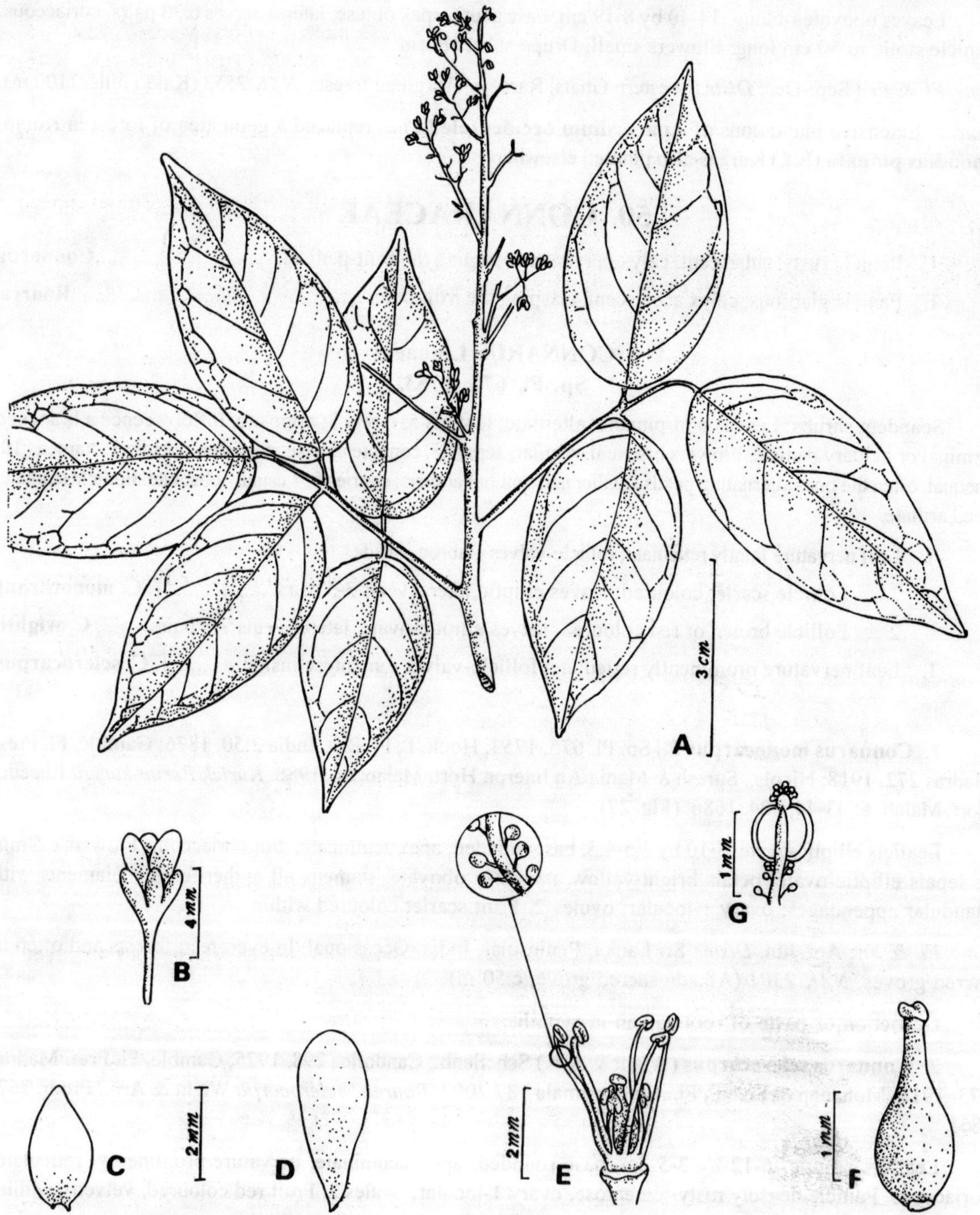

Fig. 27. ***Connarus monocarpus*** **L.: A. Twig; B. Flower; C. Sepal; D. Petal; E. Floral axis with stamens and pistil; F. Pistil; G. Single stamen- apical portion**

3. **Connarus wightii** Hook. f., Fl. Brit. India 2: 51. 1876; Gamble, Fl. Pres. Madras 273. 1918; Ramamoorthy in Sald. & Nicols., Fl. Hassan Dist. 363. 1976. *C. ritchiei* Hook. f., Fl. Brit. India 2:51. 1876; Gamble, Fl. Pres. Madras 273. 1918.

Leaflets elliptic or elliptic-ovate, base rounded, apex acuminate, nervature faint or prominent, lateral veins 5-8 pairs, thin-coriaceous. Fruit pale red within.

Fl. & Fr.: Mar.-May. *Distr.*: Western Peninsular India. Rare. In moist evergreen and semi-evergreen forests. *EV* 80590 (MH), (Ayyappan temple forest, Sabarimala, c.250 m).

ROUREA Aublet
Hist. Pl. Guiane 467. 1775 (non. cons.).

Rourea minor (Gaertn.) Alston in Trimen, Handb. Fl.Ceylon 6: 67. 1931 ('*minus*'); Ramamoorthy in Sald. & Nicols., Fl. Hassan Dist. 363. 1976; Nicols., Suresh & Mani., An Interpr. Hort. Malab. 86. 1988; Mohanan & Sivad., Fl. Agasthyamala 188. 2002. *Aegiceras minus* Gaertn., Fruct. 1:216, t. 46. 1\788. *Connarus santaloides* Vahl, Symb. 4:87. 1794. *Rourea santaloides* (Vahl) Wight & Arn., Prodr. 144. 1844; Gamble, Fl. Pres. Madras 273. 1918.

Scandent shrubs. Leaves odd-pinnate, leaflets 3-5 (9), subopposite or alternate, elliptic or ovate, 5-7 by 2.5-3.5 cm, base rounded, apex acuminate, thin-coriaceous. Panicle axillary. Flowers 5-merous, regular; sepals 5, orbicular; petals 5, white; stamens 10, unequal; carpels 5, 4 usually imperfect. Follicles narrowly ellipsoid. Seeds arillate.

Fl. & Fr.: Sep.-May. *Distr.*: Indo-Malesia. In moist deciduous and semi-evergreen forests. *NAK 1748* (Thannithode, c.70 m).

Decoction of roots prescribed for rheumatism, scurvy and diabetes.

51. FABACEAE (LEGUMINOSAE)

Key to the subfamilies

1. Flowers zygomorphic; perianth segments predominantly imbaricate in bud.
    2. Corolla papilionaceous, aestivation imbricate descending; the posterior petals outermost, the two anterior petals often basally connate. ....**Faboldeae**
    2. Corolla caesalpiniaceous, aestivation imbricate-ascending; the posterior petal innermost the petals typically 5, distinct. ....**Caesalpinioideae**
1. Flowers actinomorphic; perianth segments valvate in bud. ....**Mimosoideae**

51-A. FABOIDEAE (PAPILIONOIDEAE)

1. Twiners, trailers, stragglers or liane.
    2. Leaves even-pinnate, 18-a-foliolate. ....**Abrus**
    2. Leaves odd-pinnate, 3-a-foliolate.
        3. Leaves (3) 5-a-foliolate.

4. Stamens 9 + 1, vexillary stamen adnate to the base of standard petal.; flowers dark purple; liane. .... **Kunstleria**

4. Stamens 9 or 10, monadelphous or isoadelphous, vexillary stamen free at base; flowers white, pink or cream-coloured.

5. Leaflets opposite; ovary sessile or nearly so. .... **Derris**

5. Leaflets alternate, ovary clearly stipitate. .... **Dalbergia**

3. Leaves 3-foliolate.

6. Stamens 10, monadelphous.

7. Stamens alternately sterile; pod to 5 cm long, hooked at apex. ...**Teramnus**

7. Stamens all fertile; pod to 15 cm long, horned at apex. ....**Canavalia**

6. Stamens 9+1, diadelphous (vexillary stamen apically fused in **Pueraria**).

8. Leaves gland-dotted.

9. Pod with oblique or transverse lines between seeds. ....**Atylosia**

9. Pod without oblique or transverse lines between seeds; pod compressed. .... **Rhynchosia**

8. Leaves not gland-dotted.

10. Inflorescence-rachis swollen at nodes.

11. Pod prominently ribbed; standard petal unappendaged. .... **Canavalia**

11. Pod other than ribbed; standard petal non-appendaged; vexillary stamen apically fused. ....**Pueraria**

10. Inflorescence rachis not swollen at nodes.

12. Standard petal unappendaged.

13. Pod turgid or inflated.

14. Pod cylindric with brittle, irritant hairs; climbers. .... **Mucuna**

14. Pod oblong-ellipsoid, without brittle, irritant hairs, transversely nerved; trailers. ....**Pycnospora**

13. Pod compressed, or flattened or cylindric.

15. Calyx apically truncate, obscurely lobed; style dilated above the middle. .... **Dumasia**

15. Calyx apically not truncate, deeply lobed; style uniform.

16. Calyx 2-lipped, upper 2-lobed, lower 3-lobed; flowers in panicles; stragglers. .... **Uraria**

16. Calyx not so; flowers in racemes; climbers.

17. Calyx-lobes setaceous, all free; pod cylindric, obscurely angled, straight. ....**Calopogonium**

17. Calyx-lobes ovate, upper ones connate; pod linear, flat, recurved. .... **Shuteria**

12. Standard petal appendaged.

18. Pod with 4 raised ribs near margins; standard appendaged at back. .... **Centrosema**

18. Pod without raised ribs near margins; standard appendaged within.

19. Style divided into a thin basal part and thick upper part; stipule either peltate or basifixed; pod obscurely septate between seeds. .... **Vigna**

19. Style not divided into a thin basal part and thick upper part; stipule either peltate or basifixed.

20. Pod wing-like; seed one, apical; liane. .... **Butea**

20. Pod not wing-like; seeds more than one, not apical; twiners. **Dunbaria**

1. Herbs, erect or subscandent shrubs or trees.

21. Trees.

22. Leaves 7-9-foliolate; standard petal to 2 by 1 cm.

23. Pod samaroid or with wing-like structures; leaflets alternate.

24. Pod oblong-lanceate, not winged; flowers white; stamens 9, monadelphous. .... **Dalbergia**

24. Pod orbicular with a broad wing; flowers yellow; stamens 10, monadelphous, or (5)+(5) at length. .... **Pterocarpus**

23. Pod woody, obliquely oblong; leaflets opposite. .... **Erythrina**

22. Leaves 3-5-foliolate; standard petal 1 by 0.8 cm. .... **Pongamia**

21. Erect, diffuse, scandent shrubs or herbs.

25. Leaves even-pinnate.

26. Bracts foliaceous; stamens (10) or (9)+1.

27. Bracts orbicular, closely imbricate; stamens 10, monadelphous; pod not glochidiately prickled. .... **Geissaspis**

27. Bracts geminate, not closely imbricate; stamens 9+1; pod glochidiately prickled. .... **Zornia**

26. Bracts not foliaceous stamens (5)+(5). .... **Smithia**

25. Leaves odd-pinnate.

28. Pod 1 or 2-seeded.

29. Bracts dimorphic, clustered as primary and secondary.

30. Flowers yellow; stamens monadelphous; anthers dimorphic; articles of pod 1or 2....... **Stylosanthes**

30. Flowers other than yellow; stamens usually diadelphous; anthers monomorphic; articles of pod 2 or more .... **Desmodium**

29. Bracts unimorphic, not clustered as primary and secondary.

31. Pod with oblique or transverse lines between seeds.

32. Leaves resinous-glandular beneath; flowers usually in pairs; seeds with large strophiole. ....**Atylosia**

32. Leaves densely velvetty-pubescent beneath; flowers in terminal panicle and racemes; seeds with small strophiole. ....**Cajanus**

31. Pod neither obliquely nor transversly lined between seeds.

33. Stamens monadelphous; anthers dimorphic; pod cylindric or globose, inflated. ....**Crotalaria**

33. Stamens diadelphous; anthers unimorphic or monomorphic; pod angular or compressed, not inflated.

34. Keel spurred above claw; indumentum biramous; anthers usually apiculate; pod 4-gonal or terete. ....**Indigofera**

34. Keel other than spurred above claw; indumentum not biramous; anthers not apiculate; pod inflated. ....**Flemingia**

28. Pod 3- or more seeded.

35. Leaves a-foliolate.

36. Pod breaking into 1-seeded segments; articles of pod rectangular; calyx 2-lipped. ....**Aeschynomene**

36. Pod not breaking into 1- seeded segments; articles of pod not rectangular; calyx not 2-lipped. ....**Tephrosia**

35. Leaves 1-3- foliolate.

37. Prostrate herbs; leaves 1-foliolate; calyx glumaceous; corolla not or little exerted. ....**Alysicarpus**

37. Erect woody herbs; leaves 3-foliolate; calyx herbaceous; corolla well exerted. ....**Pseudarthria**

ABRUS Adanson
Fam. 2:327, 511. 1763.

Climbing shrubs. Leaves even-pinnate; leaflets 7-13 pairs, opposite; stipules deciduous. Inflorescence an axillary or terminal pseudoraceme. Flowers zygomorphic; calyx nearly truncate at apex; petals shortly clawed; stamens 9, monadelphous; ovary subsessile; ovules α. Pods flat, septate. Seeds globose or compressed, brightly coloured or black.

1. Seeds globose, brightly red coloured with a black spot; leaflets to 2 cm long. 1. **A. precatorius**
1. Seeds compressed, black coloured; leaflets to 4 cm long. ....2. **A. pulchellus**

1. **Abrus precatorius** L., Syst. Nat. (ed. 12) 2:472. 1767; Roxb., Fl. Ind. 3:257. 1832; Baker in Hook. f., Fl. Brit. India 2: 175. 1876; Gamble, Fl. Pres. Madras 349. 1918; Verdc., Kew Bull. 24:240. 1970; Gandhi in Sald. & Nicols., Fl. Hassan Dist. 235. 1976; Rudd in Dassan. & Fosberg, Rev. Handb. Fl. Ceylon 1:446. 1980; Matthew, Ill. Fl. Tam. Carnatic t. 179. 1982; Britto in Matthew, Fl. Tam. Carnatic 3(1) : 333. 1983. Nicols., Suresh & Mani., An Interpr. Hort. Malab. 119. 1988; Sanjappa, Leg. India 74. 1992. *Kunni* Rheede, Hort. Malab. 8:71-72, t.39. 1688.

Leaflets 10-13 pairs, oblong, 1.5-2.5 by 0.6-0.8 m, base and apex obtuse. Pseudo-racemes to 10 cm long. Flowers c.1 cm ϕ; corolla rose or pink; standard broadly obovate; stamens subequal. Pod cylindric, wrinkled. Seeds subglobose, blood red with a black spot.

Fl. & Fr.: Apr.-Dec. *Distr.*: Tropics. Common. On exposed wayside thickets. *NAK 2105* (Vallicode, c. 100 m).

Leaves and root yield and extract similar to liquorice. Seeds are useful as a medicine in eye disease and rat poison, and also for abortion.

2. **Abrus pulchellus** Wall. ex Thw., Enum. Pl. Zeyl. 91. 1859; Backer in Hook. f., Fl. Brit. India 2.5 17.5 1876; Gamble, Fl. Pres. Madras 350. 1918; Verdcourt, Kew Bull. 24: 246. 1970; Gandhi in Sald. & Nicols., Fl. Hassan Dist. 236. 1976; Sanjappa, Leg. India 74. 1992.

Leaflets 9-12 pairs, oblong, 1-4 by 0.5-1.3 cm, base and apex obtuse. Pseudoracemes to 15 cm long. Flowers c.15 cm ϕ; corolla rose coloured; standard broadly ovate. Pod flat, 4 by 1 cm. Seeds compressed, black.

Fl. & Fr.: Nov.-Mar. *Distr.*: Dist.: Indo-Malesia. Common in forest outskirts, along waysides. *NAK 161* (Adoor, c. 50 m), *1067* (Moozhiar, c.250 m).

AESCHYNOMENE Linnaeus
Sp. Pl. 713. 1753.

Aeshchynomene americana L., Sp. Pl. 713.1753; Rudd, Reinw. 5:25. 1959; Backer & Bakh, Fl. Java 1:599. 1963; Mahesw. & Paul, J. Bombay Nat. Hist. Soc. 72(1): 196. 1975; Chandrabose, Bull. Bot. Surv. India 18 (1-4): 236. 1976; Sanjappa, Leg. India 75.1992.

Subscandent woody herbs. Leaves odd-pinnate; stipules peltate, strigose; leaflets 20-50, subfalcate, sessile. Inflorescence a raceme, axillary. Flowers c. 4 mm ϕ; bracts cordate, strigose along the margins; calyx deeply lobed; corolla orange-red. Pod jointed, slightly curved, puberulent. Seeds 6-8.

Fl. & Fr.: Sep.-Dec. *Distr.*: Mexico to Honduras; now introduced and naturalised in some parts of Peninsular India. Not common. In exposed sunny areas. *NAK 1189* (Upper Moozhiar, 750 m), *2180* (Vayyattupuzha, c. 400 m).

This introduced species is gradually spreading even to interior forests.

ALYSICARPUS Necker ex Desvaux
J. Bot. Agric. 1:120. 1813 (nom. cons.).

Diffuse herbs. Leaves 1-foliolate, alternate; stipule scarious, free or connate. Inflorescence a terminal or axillary raceme. Flowers usually in pairs; bracts scarious; calyx glumaceous, deeply 4-lobed, upper lobe 2-fid; corolla included; petals pink, keeled; stamens 9+1; ovary subsessile; style incurved at apex. Pods subterete or lomentaceous. Seeds subglobose.

1. Pods moniliform, articles subglobose, with hooked hairs. .... 2. **A. monilifer**
1. Pods not moniliform, articles 4-gonous, without hooked hairs.
    2. Racemes to 10 cm long; articles of pod glabrous. .... 1. **A. bupleurifolius**
    2. Racemes to 3 cm long; articles of pod downy pubescent. .... 3. **A. vaginalis**

1. **Alysicarpus bupleurifolius** (L.) DC., Prodr. 2:352. 1825; Wight & Arn., Prodr. 233. 1834; Baker in Hook.f., Fl. Brit. India 2:158. 1876; Gamble, Fl. Pres. Madras 338. 1918; Gandhi in Sald. & Nicols., Fl. Hassan Dist. 276. 1976; Britto in Matthew, Fl. Tam. Carnatic 3(1): 336. 1983; Nicols., Suresh & Mani., An Interpr. Hort. Malab. 121. 1988. *Hedysarum bupleurifolium* L., Sp. Pl. 745. 1753. *Nir-murri* Rheede, Hort. Malab. 9:59, t. 32. 1689.

Leaves narrow-lanceate to oblong-lanceate, 1.5-4 by 0.4-1 cm, base rounded, apex acute, nervature prominent, pubescent below. Racemes 5-13 cm long; standard obovate, wings adherent to keel. Pods stipitate, lower 2 articles closing by persistent calyx.

Fl. & Fr.: Nov.-Jan. *Distr.*: India, Pakistan, Myanmar, China, Malesia. Rare. Along moist-shady cutting in upper hills. *NAK 1340* (Kakki hills, c. 1100 m).

2. **Alysicarpus monilifer** (L.) DC., Prodr. 2:353. 1825; Wight & Arn., Prodr. 232. 1834; Baker in Hook. f., Fl. Brit. India 2:157. 1876; Gamble, Fl. Pres. Madras 338. 1918; Gandhi in Sald. & Nicols., Fl. Hassan Dist. 237. 1976; Britto in Matthew, Fl. Tam. Carnatic 3(1): 336. 1983; Sanjappa, Leg. India 79. 1992. *Hedysarum moniliferum* L., Mant. Pl. 102. 1767; Roxb., Fl. Ind. 3:345. 1932.

Leaves ovate or oblong, 1-3 by 0.5-2 cm, base subcordate, apex obtuse-mucronate. Racemes 2-4 cm long; standards ovate. Pods moniliform, articles subglobose with hooked hairs.

Fl. & Fr.: Nov.-Jan. *Distr.*: India, Pakistan, Ethiopia. Occasional. On exposed way sides. *NAK 2328* (Kodumon, c.40 m).

3. **Alysicarpus vaginalis** (L.) DC., Prodr. 2:253. 1825; Wight & Arn., Prodr. 233. 1834; Baker in Hook. f., Fl. Brit. India 2:158. 1876; Gamble, Fl. Pres. Madras 338. 1918; Britto in Matthew, Fl. Tam. Carnatic 3(1): 339. 1983; Sanjappa, Leg. India 81. 1992. *Hedysaum vaginale* L., Sp. Pl. 746. 153; Roxb., Fl. Ind. 3:345. 1832.

Leaves suborbicular to elliptic oblong or oblong- lanceate, 1-3 by 0.6-1 cm, base obtuse-subcordate, apex obtuse, lower surface sparsely pubescent on nervature. Racemes 3-5 cm long obovate. Pods subterete, articles 4-gonous, strongly nerved.

Fl. & Fr.: Nov.-Jan. *Distr.*: Sri Lanka, Afghanistan, India to Malesia. Common along moist shady and exposed sunny slopes. *NAK 1650*. (Adoor, c.40 m).

Note : This species differs from the previous species by only having moniliform lomentum. In sterile state the differentiation of this is very difficult.

ATYLOSIA R. Wight et Arnott
Prodr. 257. 1834.

Erect shrubs or twiners. Leaves digitately 3-foliolate, alternate, lower surface resinous-glandular. Inflorescence an axillary or terminal raceme, some times flowers in pairs; calyx campanulate, lobes 5, upper lip connate, 2-fid; corolla yellow, often persistent, standard orbicular; stamens 9+1; ovary sessile; style incurved at middle. Pods oblong, compressed, usually transversly lined between the seeds.

1. Erect shrubs; flowers in pairs; pods faintly lined between seeds. .... 1.**A. lineata**
1. Slender climbers; flowers in short racemes;pods clearly lined between seeds. .... 2. **A. scarabaeoides**

1. **Atylosia lineata** Wight & Arn., Prodr. 258, 1834; Baker in Hook. f., Fl. Brit. India 2:213. 1876; Gamble, Fl. Pres. Madras 368. 1918; Gandhi in Sald. & Nicols., Fl. Hassan Dist. 238. 1976; Manilal, Fl. Silent Valley 69. 1988. *A. lawii* Wight, Ic. t.93. 1838. *Glycine lineata* Heyne in Wall, Cat. 5578. 1831-32, nom. nud. *Cajanus lineatus* (Wight & Arn.) van der Maesen, Agric. Univ. Wageningen Pap. 85-4: 143. 1985; Sanjappa, Leg. India 101. 1992. **(Fig. 28)**

Erect shrubs; leaflets obliquely lanceate, 15-25 by 9-12 mm, terminal one oblanceate, 2-4 by 0.7-1.5 cm, base cuneate, apex acute-apiculate, lower surface pubescent with yellow resinous glands. Flowers 2 or 3, axillary and terminal, standard orbicular, keel petals subfalcate. Pods oblong, flat, faintly septate.

Fl. & Fr.: Nov.-Feb. *Distr.*: Western Ghats. Locally abundant. In moist evergreen forests. *NAK 362 & 1170* (Kakki hills, c. 110 m). A close wild relative of *Cajanus cajan* (L.) Millsp.

2. **Atylosia scarabaeoides** (L.) Benth. in Miq., Pl. Jungh. 243. 1825; Baker in Hook.f., Fl. Brit. India 2:215. 1876; Gamble, Fl. Pres. Madras 369. 1918; Britto in Matthew, Fl. Tam. Carnatic 3(1) : 342. 1983; Mohanan & Sivad., Fl. Agasthyamala 191. 2002. *Dolichos scarabaeoides* L., Sp. Pl. 726. 1753. *Cajanus scarabaeoides* (L.) du-Petit -Thou., Dict. Sci. Nat. 6. 617. 1817 (*"Cajan scarabaeoide"*); Sanjappa, Leg. India 101. 1992.

Twiners. Leaflets oblong-ovate or obovate-oblong, 10-25 by 15 mm, base and apex obtuse, finely soft pubescent. Racemes slender, axillary, usually 2-flowered, standard obovate, keel petals subfalcate. Pods oblong, 4-5- septate, strongly margined between seeds.

Fl. & Fr.: Oct.-Dec. *Distr.*: Sri Lanka, India, Pakistan, Nepal, Myanmar, China & Malaya Peninsula. Common. In exposed sunny areas among bushes. *NAK 128, 1094, 2216* (Pathanamthitta, c.100 m).

BUTEA J. Koenig ex Roxburgh
Asiat. Res. 3:469. 1972 (nom. cons.).

Climbing shrubs. Leaves pinnately 3-foliolate; leaflets unequal, terminal one larger; stipules caducous. Inflorescence a large terminal panicle. Flowers fascicled; calyx campanulate, lobes 5, deltoid, upper connate; corolla exerted, petals clawed, standard ovate to orbicular; stamens 9+1, anthers uniform; ovary sessile or stalked, ovules 2. Pods flat, 1- seeded, apically dehiscent .

1. Flowers cream coloured; leaflets 30 by 20 cm; pods tomentose, stalked. .... 1. **B. parviflora**
1. Flowers purple coloured; leaflets 15 by 8 cm; pods subglabrous, sessile............ 2. **B. purpurea**

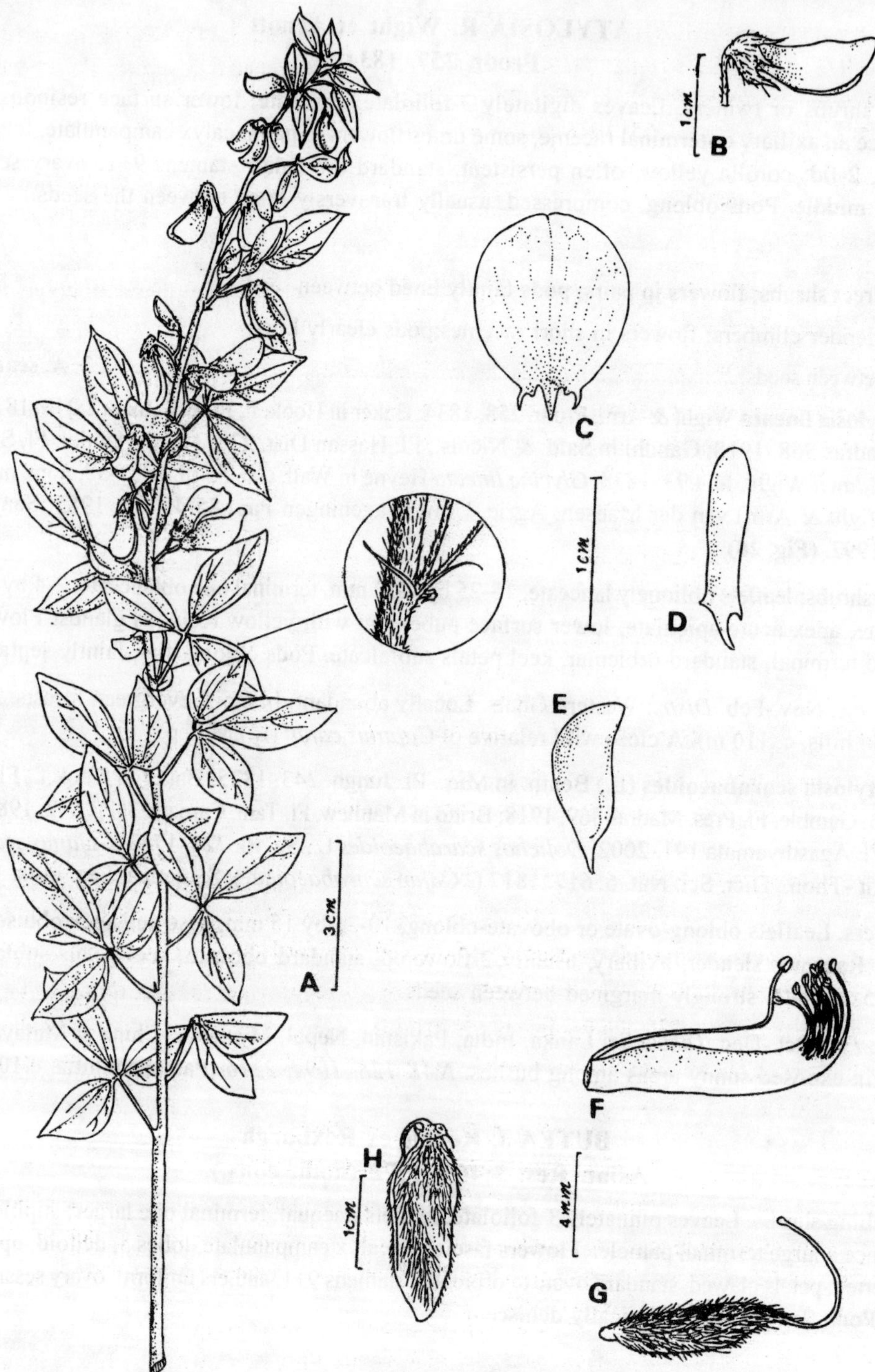

Fig. 28. *Atylosia lineata* Wight & Arn.: A. Twig; B. Flower; C.; D. Wing petal; E. Keel petal; F. Staminal column; H. Pistil; H. Fruit.

1. **Butea parviflora** Roxb., (Hort. Beng. 53. 1814, nom. nud.) Fl. Ind. 3:248. 1832; Britto in Matthew, Fl. Tam. Carnatic 3(1) : 345. 1983; Manilal, Fl. Silent Valley 70. 1988 ; Mohanan & Sivad., Fl. Agasthyamala 192. 2002. *Spatholobus roxburghii* Benth. in Miq., Pl. Jungh. 238. 1852; Baker in Hook. f., Fl. Brit. India 2:193. 1876; Gamble, Fl. Pres. Madras 358. 1918. *Athampu - valli, Chamatha.*

Terminal leaflet ovate-elliptic, 14-25 by 10-15 cm, laterals obovate-oblong, inequilateral, base cuneate, apex subacute-retuse, lower surface greyish-tomentose, thin-coriaceous. Flowers cream coloured. Pods oblong or falcate, 12 by 3 cm, golden-velvetty. Seed 1, apical.

Fl. & Fr.: Aug.-Jan. *Distr.*: Sri Lanka, India. Rare. In semi-evergreen forests along streamsides. *NAK 1422* (Mannarappara, c.250 m).

Bark yields a fibre used for ropes. Ashes of leaves taken with molasses as a remedy in worm infections.

2. **Butea purpurea** (Benth. ex Baker) Blatter, J. Ind. Bot. Soc. 8:137. 1929. *Spatholobus purpureus* Benth. ex Baker in Hook. f., Fl. Brit. India 2:194. 1876; Gamble, Fl. Pres. Madras 358. 1918; Sanjappa, Leg. India 252. 1992. Mohanan & Sivad., Fl. Agasthyamala 192. 2002

Leaflets elliptic-oblong, 12-17 by 5-7 cm, base obtuse-truncate, apex acuminate, thin-coriaceous. Panicle large. Flowers purplish. Pods oblong, flat, 10 by 2 cm. Seed 1, apical.

Fl. & Fr.: Oct.-Feb. *Distr.*; Western Ghats. Rare. In moist evergreen and semi-evergreen forests. *NAK 2628* (Way to valve House, Moozhiar, c. 300 m).

CAJANUS A.P. de Candolle
Cat. Hort.Monspel. 85.1813 (nom. cons.).

Cajanus cajan (L.) Millsp., Publ. Field Columbian Mus. Bot. Ser. 2: 53. 1900; Gandhi in Sald. & Nicolo., Fl. Hassan Dist. 239. 1976; Purseglove, Trop. Crops., Dicots 236. 1977; Britto in Matthew, Fl. Tam. Carnatic 3(1):345. 1983; Nicols., Suresh & Mani., An Interpr. Hort. Malab. 126. 1988; Sanjappa, Leg. India 100. 1992. *Cajanus indicus* Spreng. in L., Syst. Veg. (ed.16) 3:248. 1826, nom. illegit.; Baker in Hook.f., Fl. Brit. India 2:217. 1876; Gamble, Fl. Pres. Madras 369. 1918. *Thora-paeru* Rheede, Hort. Malab. 6:23-24, t. 32.1689.

Erect Shrubs. Leaves 3-foliolate; leaflets elliptic-lanceate, 4-8 by 1-2.5 cm, base cuneate, apex acute-acuminate, tomentose below. Inflorescence a terminal panicle or axillary raceme; corolla exerted, yellow; stamens 9+1; ovary subsessile. Pods linear-oblong, turgid. Seeds 4-6.

Fl & Fr.: Dc. Mar. *Distr.*: Tropics. Planted for edible seeds. Occasionally run wild along way sides.*NAK 2642* (Chittar, c. 250 m).

CALOPOGONIUM Desvaux
Ann. Sci. Nat. (Paris) 9:423.1826.

Calopogonium mucunoides Desv., Ann. Sci. Nat. (Paris) 9:423. 1826; Hepper in Hutch. & Dalz., Fl. West. Trop. Africa 2:563. 1958; Sanjappa, Leg. India 104. 1992.

Densely villous climbers. Leaves 3-foliolate; stipule lanceate, petiole c.6 cm long; leaflets ovate-lanceate, 5-7 by 4-6 cm, base broadly cuneate, apex subacute, mucronate. Racemes lateral, to 10 cm long. Flowers clustered towards apex; calyx-lobes 5, linear; corolla red-pink, petals clawed; stamens 9+1; ovary linear, villous; style short. Pod stout, cylindric, to 3.5 cm long with golden silky hairs.

Fl. & Fr.: Oct.-Jan. *Distr.*: Native of S. America. Now introduced in Tropics. Less Common. Among bushes in plains and hills. *NAK 1004* (Tiruvalla, c.10 m), *1369* (Pampa dam area, c. 1100 m).

Note :Now the distribution of this alien species seems to be rare. But may become a dominant one in future as it looks more competent over native species.

CANAVALIA A.P. de Candolle
Prodr. 2:403. 1825 (nom. cons.).

Canavalia cathartica Thouars, J. Bot. (Desvaux) 1:81. 1813, "Canavali"; Sauer, Brittonia 16:158. 1964; Verdc., Kew Bull. 42 : 659. 1987; Nicols., Suresh & Mani, An Interpr. Hort. Malab. 127. 1988; Sanjappa, Leg. India 107. 1992. *Dolichos virosa* Roxb., Hort. Beng. 55. 1814. *Canavalia virosa* (Roxb.) Wight & Arn., Prodr. 253. 1834; Gamble, Fl.Pres. Madras 359. 1918; Sauer, Brittonia 16:152.1964; Matthew, Ill. Fl. Tam. Carnatic t. 187. 1982; Britto in Matthew, Fl. Tam. Carnatic 3(1): 348.1983. *Dolichos virosus* Roxb., Fl. Ind. 3:301. 1832. *Katu-baramareca* Rheede, Hort. Malab. 8:87-88, t.45. 1688.

Stout climbers. Leaves pinnately 3-foliolate; petiole c. 8 cm long; terminal leaflet broadly ovate, laterals ovate-oblong, 8-12 by 5-9 cm, base broadly cuneate to rounded, apex acuminate, apiculate. Racemes axillary, to 20 cm long. Flowers c.2.5 cm f; calyx 2-lipped, upper 2 connate, lower 3-toothed; corolla lilac, standard large, orbicular; stamens 10, monadelphous, vexillary one shortly free at base; ovary shortly stipitate; style incurved. Pod oblong, compressed, to 15 cm long. Seeds ovate-reniform, compressed.

Fl. & Fr.: Sep.-Dec. *Distr.*: Introduced from S. America; now widely cultivated in India. Occasionally run wild. *NAK 2079* (Enathu, c.40 m).

CENTROSEMA (A.P. de Candolle) Bentham
Commentat. Legum. Gener. 53. 1837 (nom. cons.).

Centrosema virginianum (L.) Benth., Ann. Wien. Mus. 2:210. 1838; Thoth. & Prasad, Curr. Sci. 39:353. 1970; Sanjappa, Leg. India 111. 1992. *Clitoria virginiana* L., Sp. Pl. 753. 1753.

Slender climbers. Leaves 3-foliolate, stipule persistent; leaflets ovate-lanceate, 2-6 by 1-3 cm, base obtuse, apex shortly acuminate, puberulous, thin-coriaceous. Racemes axillary, c.7 cm long; bracts and bracteoles adherent to calyx. Flowers c.2 cm ϕ; calyx faintly spurred; corolla pinkish-white; stamens 9+1. Pods oblong-falcate, apically long-beaked. Seeds oblong.

Fl. & Fr.: Most of the seasons. *Distr.*: Native of tropical America. Very Common. In disturbed forest areas among bushes. *NAK 160* (Adoor, c.40 m).

Extensively used as a covercrop for Rubber plantations.

Note : This species poses a serious threat to native plants; and in a very good indicator of forest destruction.

CROTALARIA Linnaeus
Sp.Pl. 714. 1753.

Erect, diffuse herbs or subshrubs. Leaves 1-3 - foliolate, stipule filiform to foliaceous or 0. Inflorescence usually a raceme or panicle, some times a terminal umbellate-head; calyx-tube short, lobes linear or linear-lanceate; corolla usually yellow, petals clawed, standard orbicular, with a callus above claw; stamens 10, monadelphous, anthers dimorphous, alternately versatile, basifixed; ovary sessile or stalked; ovules 2-α; style sharply incurved or geniculate. Pods inflated or turgid, oblong or globose. Seeds 2-α.

1. Leaves 3-foliolate; petiole long. .... 8. **C. pallida**
1. Leaves 1-foliolate; petiole short or absent.
 2. Racemes few-flowered, lateral or leaf-opposed.
  3. Trailing herbs; flowers c. 1 cm long; pod c. 2 cm long, softly pubescent. .... 4. **C. evolvuloides**
  3. Diffuse or suberect herbs; flowers c. 2.5cm long; pod c. 4 cm long, strigosely pubescent. .... 6. **C. multiflora**
 2. Racemes a-flowered, usually terminal, rarely leaf-opposed.
  4. Pods glabrous.
   5. Pods much exerted from calyx.
    6. Upper leaf surface glabrous.
     7. Stipules and bracts minute, filiform; leaf apex obtuse or retuse. .... 9. **C. retusa**
     7. Stipules and bracts prominent, ovate; leaf apex acute. .... 11. **C. spectabilis**
    6. Upper leaf surface pubescent.
     8. Indumentum dense, golden coloured. .... 10. **C. salicifolia**
     8. Indumentum sparse, grey or pale brownish. .... 1. **C. barbata**
   5. Pods not or slightly exerted from calyx.
    9. Upper calyx-lobes connate.
     10. Umbels ± 3-flowered, often leaf-opposed; pod oblong; seeds c.15. .... 7. **C. nana**
     10. Umbels 6-8- flowered, often terminal; pod globose; seeds c.5. .... 13. **C. umbellata**
    9. Upper calyx-lobes not connate or slightly connate at base; densely silky herbs. .... 2. **C. calycina**
  4. Pods pubescent.
   11. Trailing annual herbs; vegetative parts golden tomentose. .... 3. **C. clarkei**
   11. Erect or subscandent perennial subshrubs.
    12. Stipule semi-lunate.
     13. Erect subshrubs; stem angled; flowers blue; stipule very prominent. .... 14. **C. verrucosa**
     13. Subscandent subshrubs; stem terete; flowers yellow; stipule not so. .... 15. **C. walkeri**

12. Stipule subulate or 0.

14. Flowers in racemes; bracts and bracteoles linear or subulate, deciduous; planted species. ....5. **C. juncea**

14. Flowers in panicles; bracts and bracteoles broadly ovate, persistent; wild species. ....12. **C. subperfoliata**

1. **Crotalaria barbata** Graham ex Wight & Arn., Prodr. 181. 183; Wight, Ic. t. 980. 1845; Baker in Hook. f., Fl. Brit. India 2:76. 1876; Gamble, Fl. Pres. Madras 294. 1918; Sanjappa, Leg. India 116. 1992.

Erect shrubs. Leaves simple; stipule 0; lamina oblanceate, apically linear-lanceate, 4-9 by 1.5-3 cm, base rounded or obliquely subcordate, apex-subacute-mucronate, upper surface thinly pubescent, lower surface densely silky. Racemes terminal, c.30 cm long. Flowers 8-14; calyx golden-tomentose; corolla bright-yellow, slightly exerted. Pods 5 by 3 cm.

Fl. & Fr.: Nov.-Jan. *Distr.*: Western Ghats. Rare. On exposed sunny hill slopes. *NAK 2240* (Mannarappara, c. 200 m).

2. **Crotalaria calycina** Schrank, Fl. Rar. Hort. Monac. t. 12. 1817; Baker in, Hook.f., Fl. Brit. India 2:72. 1876; Gamble, Fl. Pres. Madras 295. 1918; Gandhi in Sald. & Nicols. Fl. Hassan Dist. 241. 1976; Britto in Matthew, Fl. Tam. Carnatic 3(1) : 370. 1983; Sanjappa, Leg. India 118. 1992; Mohanan & Sivad., Fl. Agasthyamala 195. 2002.

Erect herbs. Leaves simple, elliptic or elliptic- lanceate 1.5-4 by 0.8-1.4 cm, base subacute, apex acute-mucronate, upper surface nearly black when dry; corolla pale yellow, not exerted. Pods cylindric, sessile, 2 by 1 cm.

Fl. & Fr.: Aug.-Dec. *Distr.*: Paleotropics. Less common. Among grasses in open hills. *NAK 2757* (Kakki hills, c. 1100 m).

3. **Crotalaria clarkei** Gamble, Kew Bull. 1917:27. 1917 & Fl. Pres. Madras 296. 1918; Sanjappa, Leg. India 118. 1992. **(Fig. 29)**

Trailing woody herbs; branches 4-angled, whole plant golden tomentose. Leaves simple, elliptic-lanceate, 1-3.5 by 0.7-1.3 cm, base subcordate, apex acute or obtuse. Racemes terminal, c. 10 cm long; bracts and bracteoles lanceate. Flowers 2-6 per raceme; corolla bright yellow, standrad ovate-rhombic. Pods cylindric with golden dense spreading h irs. Seeds c.12, smooth.

Fl. & Fr.: Nov.-Jan. *Distr.*: Endemic to Western Ghats (Travancore & Nilgiri hills). Rare. Locally adundant in Kakki hills along exposed moist hill cuttings. *NAK 382, 628, 1329 & 1383* (Kakki hills, c. 1100 m).

4. **Crotalaria evolvuloides** Wight ex Wight & Arn., Prodr. 188. 1834 (excl. syn.); Baker in Hook.f., Brit. India 2:68.1876; Gamble, Fl. Pres. Madras 292. 1918; Britto in Matthew, Fl. Tam. Carnatic 3(1) : 364. 1983; Sanjappa, Leg. India 119. 1992.

Diffuse herbs; branches terete. Leaves simple, ovate to ovate-lanceate, 1-5 by 0.6-2 cm, base rounded, apex obtuse, mucronulate, tomentose. Racemes terminal and lateral, to 4 cm long; bracts ovate-lanceate; corolla yellow, standard broadly ovate. Pods elliptic-oblong, softly pubescent.

Fl. & Fr.: Dec-Feb. *Distr.*: Sri Lanka, Peninsular India. Less common. On exposed hills. *NAK 224* (Athumpumkulam, c.50 m), *2186* (Chittar, c. 350 m).

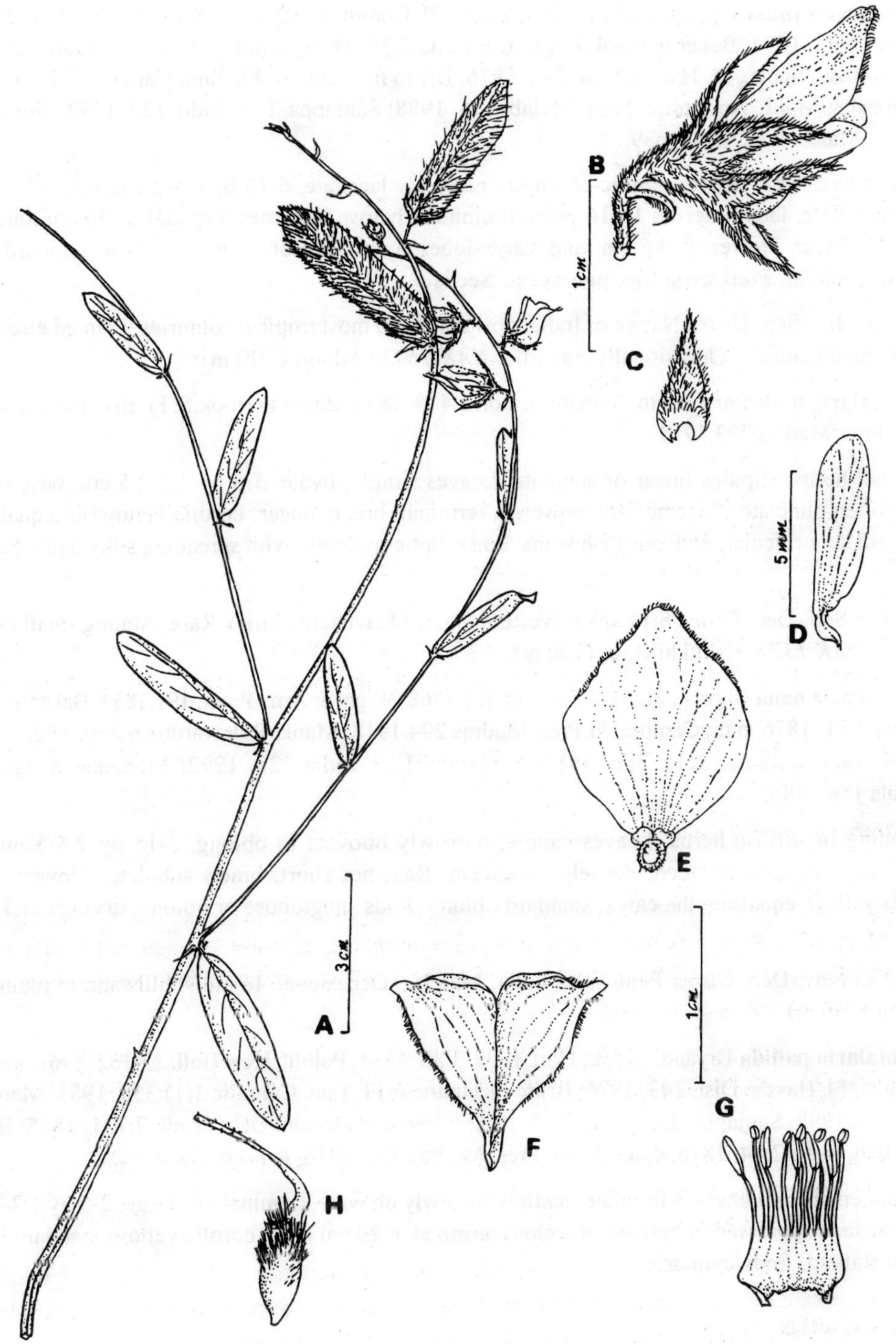

Fig. 29. ***Crotalaria clarkei*** **Gamble: A. Twig; B. Flower; C. Bracteole; D. Wing petal; E. Standard petal; F. Keel petals; G. Staminal column; H. Pistil.**

5. **Crotalaria juncea** L., Sp. Pl. 714. 1753; Roxb., Pl. Corom. t. 193. 1805 & Fl. Ind. 3:259. 1832; Wight & Arn., Prodr. 185. 1834; Baker in Hook.f., Fl. Brit. India 2:79. 1876; Gamble, Fl. Pres. Madras 297. 1918; Gandhi in Sald. & Nicols., Fl. Hassan Dist. 242. 1976; Britto in Matthew, Fl. Tam. Carnatic 3(1): 373. 1983; Nicols., Suresh & Mani., An Interpr. Hort. Malab. 136. 1988; Sanjappa, Leg. India 122. 1992. *Tandale-cotti* Rheede, Hort. Malab. 9:47, t.26. 1689.

Erect undershrubs. Leaves elliptic-oblong to narrowly lanceate, 6-10 by 0.8-2 cm, base obtuse, apex subacute-mucronate, lateral nerves 14-16 pairs, prominent below. Racemes terminal, c. 30 cm long; bracts and bracteoles linear. Flowers c. 3.5 cm long; calyx-lobes equal, lanceate; corolla yellow, standard broadly ovate or suborbicular. Pods cylindric, pubescent. Seeds 5.

Fl. & Fr.: Jul.-Sep. *Distr.*: Native of India; now spread to most tropical countries. Planted elsewhere as a source of green manure. Occasionally run wild. *NAK 1642* (Adoor, c. 20 m).

6. **Crotalaria multiflora** Benth., London J. Bot. 2:478. 1843; Baker in Hook.f., Fl. Brit. India 2:69. 1876; Gamble, Fl. Pres. Madras 293.1918.

Erect tall herbs; stipules linear or subulate. Leaves simple, ovate, 3-5 by 1.5-2.5 cm, base rounded, apex acute, often punctate. Raceme 2-6- flowered, terminal; bracts linear; corolla yellowish, equalling the calyx, standard suborbicular, with purplish veins. Pod elliptic-cylindric with spreading silky hairs. Seeds 10-12.

Fl. & Fr.: Sep.-Dec. *Distr.*: Sri Lanka, Western Ghats (Travancore hills). Rare. Among small bushes in exposed hills. *NAK 1379* (Kakki hills, c. 1100 m).

7. **Crotalaria nana** Burm.f., Fl. Ind. 156, t. 48, f. 2.1768; Wight & Arn., Prodr. 191.1834; Baker in Hook.f., Fl. Brit. India 2:71. 1876, p.p.; Gamble, Fl.Pres. Madras 294.1918; Munk, Reinwardtia 6:210. 1962; Britto in Matthew, Fl. Tam. Carnatic 3(1): 366. 1983; Sanjappa, Leg. India 125. 1992; Mohanan & Sivad., Fl. Agasthyamala 196. 2002.

Ascending or diffuse herbs. Leaves simple, narrowly obovate to oblong, 5-15 by 2.5-5 mm, base rounded or cuneate, apex rounded, densely pubescent. Racemes short; bracts subulate. Flowers c.5 mm long; corolla yellow, equalling the calyx, standard oblong. Pods subglobose or oblong, drying black. Seeds 10-15.

Fl. & Fr.: Nov.-Dec. *Distr.*: Peninsular India, Malesia. Occasional. In moist hills and in plains. *NAK 2080* (Enathu, c.50 m).

8. **Crotalaria pallida** Dryand. in Ait., Hort. Kew. 3:20. 1789; Polhill, Kew Bull. 22:262. 1968; Gandhi in Sald. & Nicols., Fl. Hassan Dist. 243. 1976; Britto in Matthew, Fl. Tam. Carnatic 3(1):358. 1983; Manilal, Fl. Silent Valley 73.1998; Sanjappa, Leg. India 126. 1992. *Crotalaria striata* DC., Prodr. 2:131. 1825; Baker in Hook.f., Fl. Brit. India 2:84. 1876; Gamble, Fl. Pres. Madras 301. 1918. *Kilukki-chedi.*

Erect undershrubs. Leaves 3-foliolate; leaflets narrowly obovate, terminal one large, 2-5 by 1.2-2.5 cm, long; lower surface appressed pubescent. Racemes terminal, c.20 cm long; corolla yellow, standard broadly ovate; ovary stalked. Pods cylindric.

Key to the varieties

1. Leaflets obovate, apex obtuse or retuse. ....**Crotalaria pallida** var. **pallida**
1. Leaflets elliptic, apex acute. ....**Crotalaria pallida** var. **acutifolia**

Crotalaria pallida Dryand., var. **pallida**

Leaflets smaller, apex obtuse or retuse. Flowers lax; standard streaked with pale brown. Pods glabrous.

Fl. & Fr.: Most of the seasons. *Distr.*: Tropics. Most common species of the genus. Along waysides in plains to hills. *NAK 275* (Upper Moozhiar, c.550 m).

Root tied on wrist and neck of patients suffering from dropsy. Bark yields a fibre.

Crotalaria pallida Dryand. var. **acutifolia** Trimen, Handb. Fl. Ceylon 2:19. 1894; Gamble, Fl. Pres. Madras 301. 1918.

Leaflets larger, apex acute-mucronate. Flowers dense; standard streaked with purple. Pods puberulous.

Fl. & Fr.: Sep.-Dec. *Distr.*: Rare. Tropics. In the moist forest clearings. *NAK 276 & 1175* (Upper Moozhiar, c 550 m).

9. **Crotalaria retusa** L., Sp. Pl. 715. 1753; Roxb., Fl. Ind. 3:272. 1832; Wight & Arn., Prodr. 187. 1834; Baker in Hook.f., Fl. Brit. India 2:75. 1876; Gamble, Fl. Pres. Madras 293. 1918; Munk, Reinwardtia 6:212. 1962; Gandhi in Sald. & Nicols., Fl. Hassan Dist. 243. 1976; Britto in Matthew, Fl. Tam. Carnatic 3(1):365. 1983; Nicols., Suresh & Mani., An Interpr. Hort. Malab. 133. 1988; Sanjappa, Leg. India 128. 1992. Mohanan & Sivad., Fl. Agasthyamala 197. 2002. *Tandale-cotti* Rheede, Hort. Malab. 9:45-46, t.25. 1689.

Subshrubs. Leaves simple, lanceate or oblanceate, 1.5-12 by 0.7-4 cm, base cuneate to slightly attenuate, apex obtuse or retuse, mucronulate, lower surface appressed pubescent. Racemes terminal, c.20 cm long; calyx lobes ovate-lanceate; corolla bright yellow, with purplish veins, standard orbicular. Pods narrowly obovoid. Seeds α.

Fl. & Fr.: Oct.-Mar. *Distr.*: Tropics. Common in hills and plains. *NAK 137* (Adoor, c. 40 m) *964* (Sabarimala slopes, c.250 m), *1642* (Arampa, c.500 m).

10. **Crotalaria salicifolia** Heyne ex Wight & Arn., Prodr. 182. 1834; Baker in Hook.f., Fl. Brit. India 2:77. 1876; Gamble, Fl. Pres. Madras 294. 1918; Sanjappa, Leg. India 128. 1992; Mohanan & Sivad., Fl. Agasthyamala 197.2002. **(Fig. 30)**

Tomentose undershrubs. Leaves simple, ovate-elliptic to narrowly linear-lanceate towards branch-apices, 5-12 by 1.5-3 cm, base rounded, apex obtuse, mucronate, thin-coriaceous. Racemes terminal, c. 13 cm long; bracts and bracteoles ovate-lanceate, 5-7 by 2-3 mm; calyx-lobes lanceate; corolla bright yellow, not exceeding the calyx, standard broadly obovate; ovary sessile. Pods ellipsoid. Seeds α.

Fl. & Fr.: Nov.-Feb. *Distr.*: Western Ghats. Rare. In exposed hills among bushes. *NAK 433, 1332 & 1388* (Kakki hills, c. 1100 m).

11. **Crotalaria spectabilis** Roth, Nov. Pl. Sp. 341. 1821; Shah, J. Bombay Nat. Hist.Soc. 69: 445.1972; Sanjappa, Leg. India 130.1992. *Crotalaria sericea* Retz., Obs. Bot. 5: 26.1789, non Burn.f., 1768. Baker in Hook.f., Fl. Brit. India 2:75. 1876; Gamble, Fl. Pres. Madras 293. 1918.

Subshrubs. Leaves simple, oblanceate, 7-12 by 3-6 cm, base cuneate, apex broadly acute or obtuse-mucronulate, lower surface sparsely pubescent, lateral nerves raised on both sides. Racemes termnal, to 20 cm tall; bracts ovate-lanceate. Flowers 1 cm long; corolla bright yellow; ovary sessile. Pods stipitate, ellipsoid.

Fl. & Fr.: Dec.-Mar. *Distr.*: Tropics. Rare. In moist semi-evergreen forests. *NAK 181* (Thekkuthode, c. 70 m).

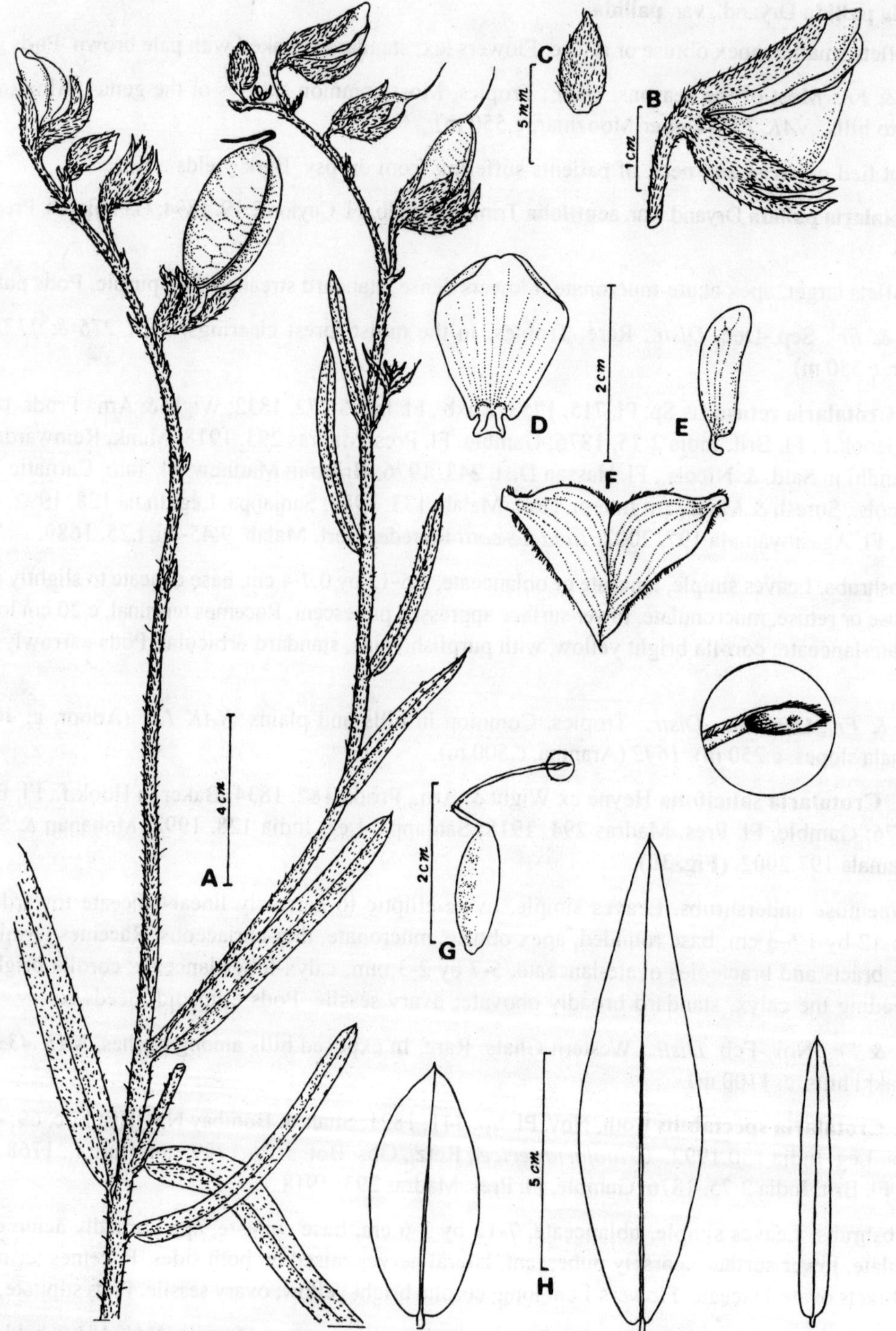

Fig. 30. ***Crotalaria salicifolia*** **Heyne ex Wight & Arn.: A. Twig; B. Flower; C. Bracteole; D. Standard petal; E. Wing petal; F. Keel petals; G. Pistil; H. Various forms of leaf.**

12. **Crotalaria subperfoliata** Wight ex Wight & Arn., Prodr. 184. 1834; Baker in Hook.f., Fl. Brit. India 2:79. 1876; Gamble, Fl. Pres. Madras 298. 1918; Britto in Matthew, Fl. Tam. Carnatic 3(1): 377. 1983; Sanjappa, Leg. India 130. 1992.

Shrubs. Leaves simple, oblanceate, 3-8 by 1.5-2.5 cm, base obtuse, margin slightly reflexed, apex acute-apiculate; upper surface sparsely pubescent, lower surface tomentose, thin-coriaceous. Panicle terminal and lateral, 6-10 cm long; bracts and bracteoles ovate-cordate, persistent; calyx-lobes lanceate; corola yellow, much exerted, standard broadly ovate; ovary stipitate. Pods oblong, puberulous. Seeds α.

Fl. & Fr.: Dec.-Feb. *Distr.*: Peninsular India. Rare. A beautiful shrub in moist evergreen forest borders. *NAK 1276* (Kakki hills, c. 1100 m).

13. **Crotalaria umbellata** Wight ex Wight & Arn., Prodr. 191. 1834; Gamble, Fl. Pres. Madras 294. 1918; Britto in Matthew, Fl. Tam. Carnatic 3(1) : 368. 1983; Hook.f., Fl. Brit. India 2:71. 1876. p.p., non. Burm. f., 1768.

Erect herbs. Leaves simple, elliptic-oblong, 8-25 by 3-7 mm, base rounded, apex subacute- mucronate. Umbels terminal. Flowers 2-6; calyx-lobes lanceate; corolla yellow, hardly exerted, standard ovate-orbicular. Pods globose. Seeds, c. 6, reniform, shiny.

Fl. & Fr.: Dec.-Mar. *Distr.*: Sri Lanka, Peninsular India. Rare. In exposed hill slopes. *NAK 401, 1325* (Kakki hills, c. 1100 m).

14. **Crotalaria verrucosa** L., Sp. Pl. 715. 1753; Wight & An., Prodr. 187. 1834; Wight, Ic. t. 200. 1839; Baker in Hook.f., Fl. Brit. India 2:77. 1876; Gamble, Fl. Pres. Madras 297. 1918; Britto in Matthew, Fl. Tam. Carnatic 3(1) 374. 1983; Nicols., Suresh & Mani., An Interpr. Hort. Malab. 133. 1988; Sanjappa, Leg. India 131. 1992. *Pee-tandale-cotti* Rheede, Hort. Malab. 9:53, t.29. 1689.

Subshrubs. Leaves simple, ovate to rhomboid-deltoid, 6-10 by 4-5 cm, base attenuate or cuneate, apex subacute or rounded; stipule prominent, semi-lunate. Racemes terminal or lateral, c. 15 cm long; Flowers c. 2 cm long; calyx-lobes elongate, lanceate; ovary stipitate; corolla bluish, much exceeding the calyx. Pods oblong-terete, sparsely pubescent. Seeds c.15.

Fl. & Fr.: Most of the seasons. *Distr.*: Pantropical. Rare. Along waysides. *NAK* 2740 (Adoor, c. 23 m).

15. **Crotalaria walkeri** Arn., Nov. Actorum Acad. Caes. Leop. - Carol. Nat. Cur. 18(1):328. 1840; Gamble, Fl. Pres. Madras 297. 1918; Sanjappa, Leg. India 131. 1992. *C. semperflorens* Vent. var. *walkeri* (Arn.) Baker in Hook.f., Fl. Brit. India 2:78. 1876.

Scandent subshrubs. Leaves simple ovate-lanceate or narrowly oblong-lanceate, to elliptic-lanceate, 3-10 by 1-4 cm, base broadly cuneate, apex acute-mucronulate, lower surface appressed hairy. Racemes terminal and lateral, 8-20 cm long; bracts and bracteoles subulate; calyx-lobes linear-lanceate; corolla yellow, much exerted, standard suborbicular. Pods elliptic-oblong, sparsely pubescent. Seeds 4-5.

Fl. & Fr.: Nov.-Feb. *Distr.*: Sri Lanka & Peninsular India. Locally abundant. Along wayside bushes in higher hills *NAK 315, 277, 1169 7 1268,* (Kakki hlls, c. 1100 m).

DALBERGIA Linnaeus f.
Suppl. Pl. 52. 316. 1782 (nom. cons.).

Trees or climbing shrubs. Leaves alternate, odd-pinnate; leaflets alternate. Inflorescence a terminal and/or axillary panicle. Flowers bracteate and bracteolate; calyx - tube campanulate, upper-lobes connate;

corolla exerted; petals clawed; stamens 9 or 10, monadelphous or 5+5, diadelphous. Pod compressed, oblong-lanceate, samaroid. Seeds reniform, compressed.

1. Climbing shrubs; leaflets 3-5; panicles axillary; pods samaroid. ....2. **D. pseudo-sisso**

1. Trees; leaflets 5-7; panicle fascicled, cauline or on the axils of older leaves; pods woody. ....1. **D. latifolia**

1. **Dalbergia latifolia** Roxb., Pl. Corom. t. 113.1799; Baker in Hook.f., Fl. Brit. India 2:231. 1876; Prain, Ann. Roy. Bot. Gard. (Calcutta) 10:t. 62. 1904; Gamble, Fl. Pres. Madras 383. 1918; Gandhi in Sald. & Nicols., Fl. Hassan Dist. 245. 1976; Britto in Matthew, Fl. Tam. Carnatic 3(1):383. 1983; Nicols., Suresh & Mani., An Interpr. Hort. Malab. 134. 1988; Sanjappa, Leg. India 137. 1992; Mohanan & Sivad., Fl. Agasthyamala 200. 2002. *Biti* Rheede, Hort. Malab. 5:115, t. 58.1685.

Deciduous trees; leaflets 7-9, elliptic-orbicular, 3-5 by 2.5-4 cm, base cuneate, apex rounded to emarginate, thin-coriaceous. Inflorescence cauline or on axils of old leaves; corolla cream coloured; petals except standard clawed; stamens 9, monadelphous. Pods slightly thickened, narrowed at both ends.

Fl. & Fr. Apr.-Sep. *Distr.*: Indo-Malesia. Rare. In moist deciduous forests. *NAK 2742* (Ranni, c. 100 m).

2. **Dalbergia pseudo-sisso** Miq., Fl. Ind. Bat. 1:128. 1855; Sanjappa, Leg. India 139. 1992; Mohanan & Sivad., Fl. Agasthyamala 200. 2002. *D. championi* Thw., Enum. Pl. Zeyl. 95. 1858; Baker in Hook.f., Fl. Brit. India 2:231. 1876. *D. rostrata* Graham ex Prain, Ann. Roy. Bot. Gard. (Calcutta) 10:60, t. 36.1905; Gamble, Fl. Pres. Madras 382. 1918.

Scandent shrubs; branchlets lenticellate. Leaves 3-5- foliolate; leaflets obovate to oblanceate, 5-10 by 3-4.5 cm, base cuneate, margin subentire, apex caudate-retuse, thin-coriaceous. Panicles axillary, c. 8 cm long. Flowers c.10 mm long; calyx campanulate; corolla white tinged with pink, petals long-clawed; stamens 9, monadelphous; ovary long-stipitate; style slender; stigma obscure. Pods samaroid.

Fl. & Fr.: Apr.-Sep. *Distr.*: Sri Lanka, Western Ghats. Rare. In moist semi-evergreen forests often along streamsides. *NAK* 544 (Thekkuthode, c. 70 m), *1731* (Thannithode, c. 50 m). This seems to be a new distributional record for Kerala State.

DERRIS Loureiro
Fl. Cochinch. 432. 1790 (nom. cons.).

Climbing shrubs. Leaves odd-pinnate, alternate, leaflets subopposite. Inflorescence axillary or terminal racemes or panicles, sometimes flowers in fascicles; calyx-tube campanulate, truncate, obscurely 5-toothed; corolla exerted; petals clawed; standard suborbicular, non- appendaged; stamens 10, monadelphous, vexillary free at base; ovary subsessile; style incurved. Pod oblong or obliquely orbicular, winged on upper or both sutures. Seeds compressed.

1. Standard petal with two callosites at the base of limb; leaflets 13-19. ....2. **D. canarensis**

1. Standard petal without callosites at the base of limb; leaflets 5-13.

    2. Pod winged along the upper suture; seeds 1-5. ....3. **D. scandens**

    2. Pod winged along upper and lower sutures; seed 1.

3. Pod tomentose; leaflets to 10 cm long. ....1. **D. brevipes**

3. Pod glabrous; leaflets to 7 cm long. ....4. **D. thothathrii**

1. **Derris brevipes** (Benth.) Baker in Hook.f., Fl. Brit. India 2:244. 1878; Gamble, Fl. Pres. Madras 388. 1918; Thothathri, Bull. Bot.Surv. India 3:184. 1961; Manilal, Fl. Silent Valley 75. 1988; Sanjappa, Leg. India 144. 1992. *D. heyneana* Benth., var. *brevipes* Benth., J. Linn. Soc. Bot.4 (Suppl.):110. 1860.

Leaflets 5-7, elliptic-oblong, 6-10 by 3-4 cm, base cuneate, apex obtusely acuminate, thin-coriaceous. Panicle terminal, c. 20 cm long. Flowers pale pink. Pods broadly oblong, silky hairy. Seed 1.

Fl. & Fr.: Mar.-Jul. *Distr.*: Peninsular India. Rare. In evergreen forests. *Deb 30452* (MH) (Pamba-Vandiperiyar, c. 1025 m).

Note : Could not locate this species in the said locality though searched several times during the present study.

2. **Derris canarensis** (Dalz.) Baker in Hook. f., Fl. Brit. India 2:246. 1878; Gamble, Fl. Pres. Madras 387. 1918; Thothathri, Bull. Bot. Surv. India 3:190. 1961; Gandhi in Sald. & Nicols., Fl. Hassan Dist. 247. 1988; Sanjappa, Leg. India 144. 1992. *Pongamia canarensis* Dalz. in Hooker's J.Bot. Kew Gard. Misc. 3:37. 1850. *Derris oblonga* Benth., J. Linn. Soc. Bot. 4 (suppl): 112. 1860. Baker in Hook.f., Fl. Brit. India 2:242. 1878. **(Fig. 31)**

Leaves c.25 cm long; petiole 7-9 cm long; leaflets c.16 pairs, oblong-lanceate or oblanceate, 4-10 by 1-3 cm, base subacute or cuneate, apex obtuse, lower surface pubescent in young leaves, thin-coriaceous. Panicle axillary and/or terminal, c. 15 cm long. Flowers 3-6 per panicle-branches; standard petal suborbicular with two small callosites at base. Pod flat, oblong, upper suture winged. Seed(s) 1 or 2, reniform.

Fl. & Fr.: Jan.-Mar. *Distr.*: Western Ghats. Very Rare. Along streamsides. *NAK* 2367 (Along Achancovil river bank, Naduvathumoozhy, c. 200 m).

3. **Derris scandens** (Roxb.) Benth., J. Linn. Soc. Bot. 4 (Suppl.) : 103. 1860; Baker in Hook.f., Fl. Brit. India 2:240. 1876; Gamble, Fl. pres. Madras 387. 1918; Thothathri, Bull. Bot. Surv. India 3: 1860; Baker in Hook.f., Fl. Brit. India 2:240. 1876; Mathew, Ill. Fl. Tam. Carnatic t. 193. 1982; Britto in Matthew, Fl. Tam. Carnatic 3(1) : 388. 1983; Nicols., Suresh & Mani., An Interpr. Hort. Malab. 135. 1988; Sanjappa, Leg. India 148. 1992. *Noel-valli, Panni-valli* Rheede, Hort. Malab. 6:39, t.22. 1686.

Leaves c.18 cm long; leaflets 9-13, opposite, oblong-obovate, 3-6 by 1.5-3 cm, base obtuse to rounded, apex shortly acuminate, retuse, lower surface finely puberulous, thin-coriaceous. Racemes axillary, to 25 cm long; standard ovate-orbicular. Pods oblong, winged on upper suture, narrowed at both ends. Seeds c.5.

Fl. & Fr.: Jun.-Dec. *Distr.*: Peninsular, Central & Eastern India. Rare. Usually on exposed forest edges along streamsides. *NAK 1951* (Tiruvalla, c.10 m).

Bark yields a coarse fibre.

4. **Derris thothathrii** Bennett, Indian J. For. 1:23. 1978, Sanjappa, Leg. India 148. 1992. Mohanan & Sivad., Fl. Agasthyamala 203.2002. *D. heyneana* auct., non Benth., 1852: Baker in Hook.f., Brit. India 2:244. 1878; Gamble, Fl. Pres. Madras 388. 1918; Thothathri, Bull. Bot. Surv. India 3:186. 1961. **(Fig. 32)**

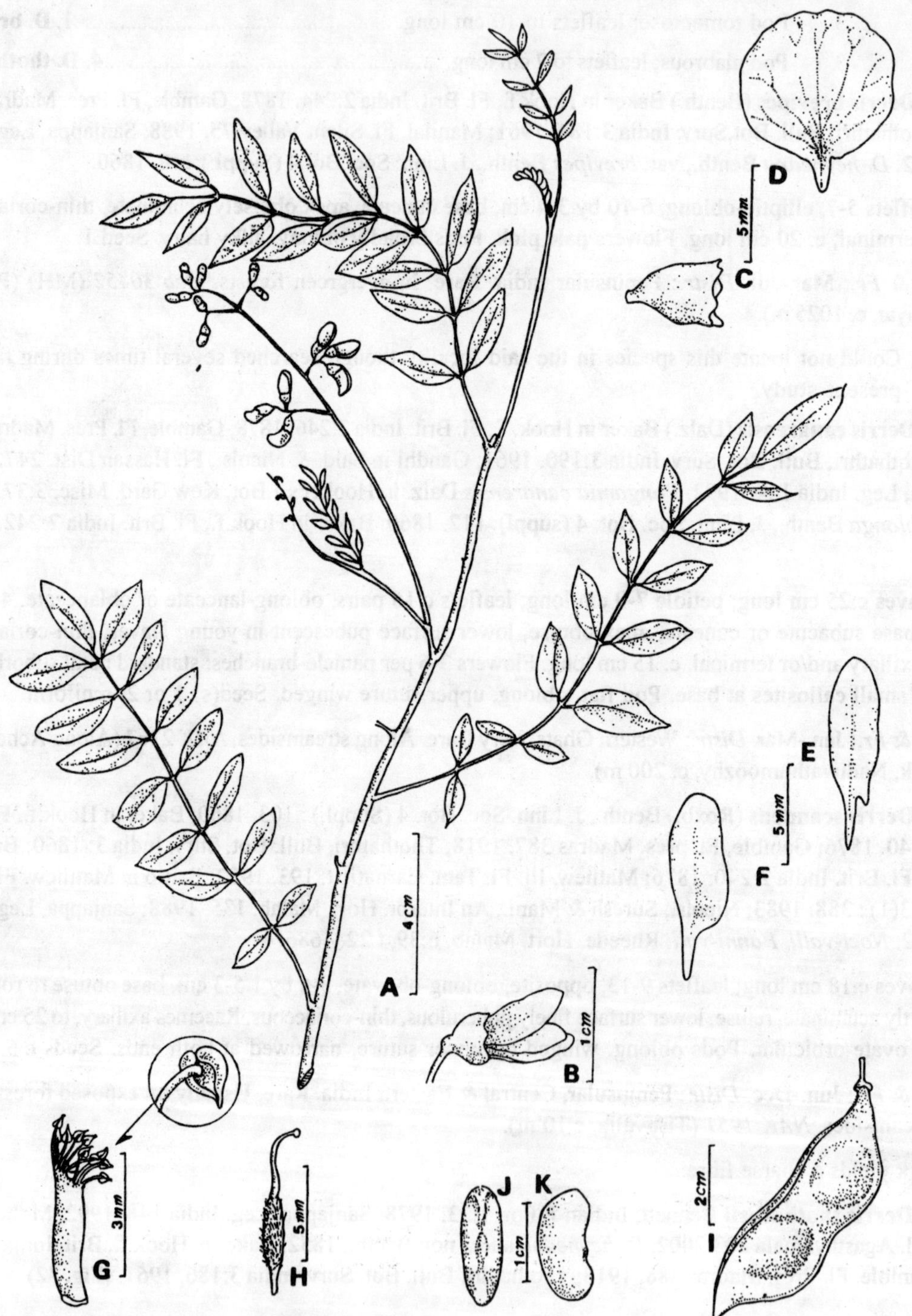

Fig. 31. ***Derris canarensis*** **(Dalz.) Baker: A. Twig; B. Flower; C. Calyx; D. Standard petal; E. Wing petal; F. Keel petal; G. Staminal column; H. Pistil; I. Pod; J & K. Seed - different views.**

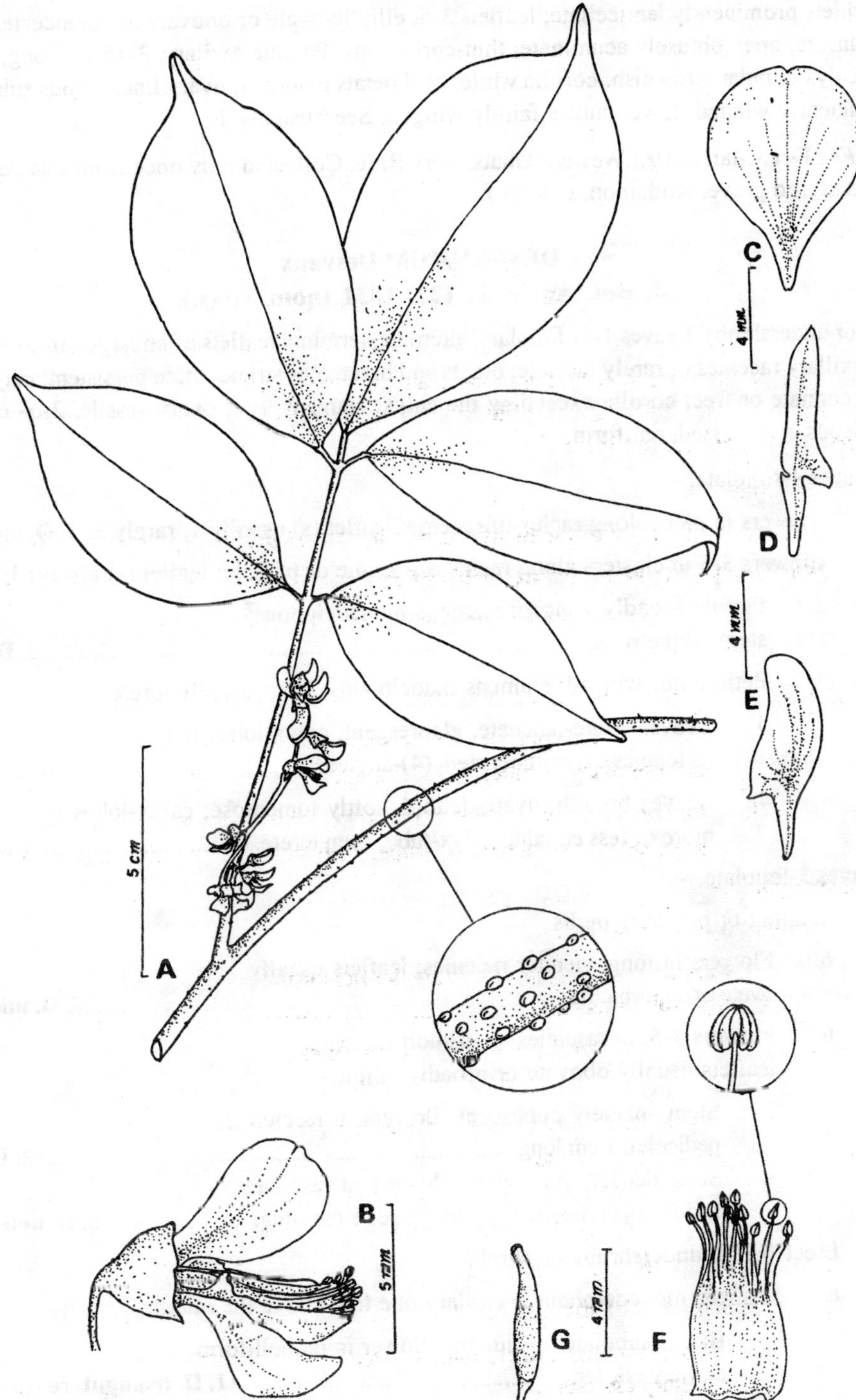

Fig. 32. ***Derris thothathrii*** **Bennett : A. Twig; B. Flower; C. Standard petal; D. Wing petal; E. Keel petal; F. Staminal column; G. Pistil.**

Branchlets prominently lenticellate; leaflets 3-5, elliptic-ovate or obovate to oblanceate, 5-9 by 2. 5-5 cm, base cuneate, apex obtusely acuminate, thin-coriaceous. Panicle axillary, 3-15 cm long, finely golden pubescent; calyx cupular, brownish; corolla white; keel petals glabrous; ovary linear. Pods subfalcate, upper suture prominently winged, lower suture faintly winged. Seed usually 1.

Fl. & Fr.: Dec.-Mar. *Distr.*: Western Ghats. Very Rare. Collected only once from a sacred grove. *NAK 2330* (Aikad sacred grove, Kodumon, c. 40 m).

DESMODIUM Desvaux
J. Bot. Agric. 1: 122. 1831 (nom. cons.).

Herbs or undershrubs. Leaves 1-3- foliolate, alternate, terminal leaflets often larger. Inflorecence usaully terminal or axillary racemes or rarely fascicle; bracts and bracteoles striate, often persistent; calyx 4-5- lobed, upper lobes connate or free; corolla exceeding the calyx; stamens 9+1; ovary sessile, 2-α- ovuled. Pod a lomentum. Seeds compressed, reniform.

1. Leaves 1-foliolate.
    2. Flowers in pairs along rachis of raceme; leaflet (s) usually 1, rarely 3.. 1. **D. alysicarpoides**
    2. Flowers 3-5 in clusters along rachis of raceme or panicle; leaflet (s) always 1.
        3. Petiole broadly winged; stamens monadelphous; stem triquetrous. .... 13. **D. triquetrum**
        3. Petiole not winged; stamens diadelphous; stem usually terete.
            4. Leaves ovate-lanceate, glabrescent, calyx-lobes twice as long as calyx-tube; stem (4)-angled. .... 4. **D. gangeticum**
            4. Leaves broadly ovate-deltoid, softly tomentose; calyx-lobes more or less equal to calyx-tube; stem terete. .... 14. **D. velutinum**
1. Leaves 3-foliolate.
    5. Trailing or prostrate herbs.
        6. Flowers in long, slender racemes; leaflets usually ovate or suborbicular. .... 8. **D. microphyllum**
        6. Flowers 3-5, in fascicles or in short racemes; leaflets usually obovate or broadly elliptic.
            7. Stem sparsely pubescent; flowers in fascicles; pedicel to 1 cm long. .... 12. **D. triflorum**
            7. Stem densely pubescent; flowers in fascicles and in short racemes; pedicel over 1 cm long. .... 6. **D. heterophyllum**
    5. Erect herbs, undershrubs or shrubs.
        8. Stamens monadelphous (vexillary one free above the middle).
            9. Bracts subulate, deciduous; flower in umbelliform axillary clusters. .... 11. **D. triangulare** var. **congestum**
            9. Bracts orbicular, persistent; flowers 5-8, in fascicles on long terminal racemes. .... 10. **D. pulchellum**

8. Stamens diadelphous.

10. Pods with oblong articles, hooked-pubescent. ....7. **D. laxiflorum**

10. Pods with crescent-shaped articles, softly appressed pubescent or glabrous.

11. Lateral leaflets very small, often moving in jerks. ....9. **D. motorium**

11. Lateral leaflets half as long, as long, nearly as long as the terminal.

12. Pedicles shorter than stipules; racemes to 20 cm long; leaves appressed pubescent below. ....5. **D. heterocarpon**

12. Pedicles thrice as long as stipules; racemes to 9 cm long; leaves densely sericeous below. ....2. **D. ferrugineum**

1. **Desmodium alysicarpoides** Van Meeuwen, Reinwardtia 6: 246. 1962; Britto in Matthew, Fl. Tam. Carnatic 3(1) : 392. 1983; Manilal, Fl. Silent Valley 76. 1988; Sanjappa, Leg. India 149. 1992. *Alysicarpus parviflorus* Dalz., J. Bot. Kew Gard. Misc. 3: 211. 1851; Gamble, Fl. Pres. Madras 339. 1918. *Desmodium parviflorum* (Dalz.) Baker in Hook.f., Fl. Brit. India 2: 172. 1876, non Mart. & Galeotti, 1843.

Erect or trailing herbs. Leaves 1-foliolate or 3-foliolate towards branch apices, elliptic to elliptic-lanceate, 1-2 by 0.7-1 cm, base rounded-subcordate, apex acute, lower surface appressed hairy. Racemes terminal, to 20 cm long; primary bracts lanceate, apex setaceous, persistent. Flowers in pairs on rachis; pedicel slender, to 12 mm long; corolla violet to pink; stamens 9+1. Pods with obliquely rounded articles, margins indented.

Fl. & Fr.: Sep.-Dec. *Distr.*: India, Malesia. Rare. In moist open hills among grasses. *NAK 2502* (Kakki hills, c. 1100 m). Highly variable in habit; some are typically trailers while some are typically erect.

2. **Desmodium ferrugineum** Wall. ex Thw., Enum. Pl. Zeyl. 87. 1859; Shindler, Repert. Spec. Nov. Regni Veg. Beih. 49:291. 1928; Ohashi, Ginkgoana 1: 204. 1973; Matthew, Ill.Fl. Tam. Carnatic t. 195. 1982; Britto in Matthew, Fl. Tam. Carnatic 3(1): 393. 1983. *D. rufescens* Wight & Arn., Prodr. 228. 1834, non. DC.: Wight, Ic. t. 984. 1845 & Spicil. Neilgh. t. 57. 1846; Baker in Hook.f., Fl. Brit. India 2: 171. 1876; Gamble, Fl. Pres. Madras 346. 1918.

Key to the subspecies

13. Leaflets elliptic or obovate, apex obtuse-mucronate; pods slightly curved, minutely pubescent. .... **Desmodium ferrugineum** ssp. **ferrugineum**

13. Leaflets lanceate, appex acute or subacute; pods strongly curved, glaborus. ....**Desmodium ferrugineum** ssp. **wynaadense**

Desmodium ferrugineum Wall. ex Thw. ssp. **ferrugineum**

Shrubs; branchlets silky pubescent. Leaves 3-foliolate; leaflets elliptic to obovate, 7-4 by 0.5-3 cm, base rounded, apex rounded, truncate, mucronulate, lower surface silky pubescent, lateral veins faint. Racemes terminal, c.7 cm long. Flowers congested; bracts subulate; corolla lilac; stamens 9+1. Pods c. 3 cm long; sparsely pubescent.

Fl. & Fr.: Nov.-Jan. *Distr.*: Sri Lanka, Peninsular India. Locally abundant. In exposed grassy slopes. *NAK 393, 1378* (Kakki hills, c. 1100 m).

2. **Desmodium ferrugineum** Wall. ex Thw. ssp. **wynaadense** (Bedd. ex Gamble) Ohashi, Ginkgoana 1:206. 1973; Manilal, Fl. Silent Valley 76. 1988. Sanjappa, Leg. India 153. 1992. *D. wynaadense* Bedd. ex Gamble, Fl. Pres. Madras 346. 1918.

Undershrubs, to 2 m high; branches silky pubescent. Leaves 3-foliolate; leaflets elliptic-lanceate, 3-5 (7) by 1-2(3) cm, base rounded, apex acute, mucronate, lower surface silky pubescent, lateral veins prominent. Racemes terminal, c. 20 cm long. Flowers lax; bracts subulate; corolla purple; stamens 9+1. Pods c. 3.5 cm long, curved inwards articles 6-8, reticulate.

Fl. & Fr.: Nov.-Jan. *Distr.*: Western Ghats. Rare. Pretty small shrubs in moist forest borders. *NAK 368 & 1392* (Kakki hills, c. 1100 m).

3. **Desmodium gangeticum** (L.) DC., Prodr. 2:327. 1825; Wight & Arn., Prodr. 225. 1834; Wight, Ic. t. 271. 1839; Baker in Hook.f., Fl. Brit. India 2 : 168. 1876; Gamble, Fl. Pres. Madras 345. 1918; Ohashi, Ginkgoana 1: 184. 1973; Britto in Matthew, Fl. Tam. Carnatic 3 (1) : 394. 1983; Sanjappa, Leg. India 153. 1992; Mohanan & Sivad., Fl. Agasthyamala 205. 2002. *Hedysarum gangeticum* L., Sp. Pl. 746. 1753; Roxb., Fl. Ind. 3:349. 1832. *Moovila, Orila.*

Undershrubs or trailing or erect woody herbs. Leaves 1- foliolate, basal ones rarely 3-foliolate; leaflets ovate-lanceate, 3-10 by 1.5-6 cm, base truncate - cordate, lower surface densely silky pubescent. Racemes terminal, c. 20 cm. long; bracts linear; corolla lilac or white; stamens 9+1. Pods to 2.5 cm long, articles 6-8.

Fl. & Fr.: Nov.-Mar. *Distr.*: Tropics. Common in plains and hills. *NAK 246* (Thekkuthode, c. 70 m), *1471* (Mannarappara, c. 250 m).

The root is crushed and juice is given for vomiting, fever, asthma and dysentery.

4. **Desmodium heterocarpon** (L.) DC., Prodr. 2 : 337. 1825; Van Meeuwen, Reinwardtia 6 : 251. 1962; Ohashi, Ginkgoana 1: 210. 1973; Britto in Matthew, Fl. Tam. Carnatic 3(1) : 396. 1983; Sanjappa, Leg. India 154. 1992; Mohanan & Sivad., Fl. Agasthyamala 205. 2002. *Hedysarum heterocarpon* L., Sp. Pl. 747. 1753. *Hedysarum polycarpum* Poiret in Lam., Encycl. 6:413. 1805. *Desmodium polycarpum* (Poiret) DC., Prodr. 2: 334. 1825; Wight & Arn., Prodr. 227. 1834; Wight, Ic. t. 406. 1840; Baker in Hook.f., Fl. Brit. India 2 : 171. 1876; Gamble, Fl. Pres. Madras 346. 1918.

Suberect or trailing woody herbs. Leaves 3-foliolate; leaflets obovate, terminal one large, 1.5-4 by 1-2.5 cm, base obtuse, apex obtuse or obtusely acute, mucronulate, lower surface greyish pubescent. Racemes terminal or from upper leaf-axils, c. 7 cm long; bracts lanceate, caducous; corolla bluish-purple; stamens 9+1. Pods c. 2 cm long, slightly indented, articles 3-6.

Fl. & Fr.: Oct.-Jan. *Distr.*: India, Japan, China, Malesia. Not common. Among forest undergrowths. *NAK 152, 258, 1143 & 1479* (Arampa, c. 450 m).

5. **Desmodium heterophyllum** (Willd.) DC., Prodr. 2: 334. 1825; Baker in Hook.f., Fl. Brit.India 2 : 173. 1876; Gamble, Fl. Pres. Madras 347. 1918; Gandhi in Sald. & Nicols., Fl. Hassan Dist. 250. 1976; Sanjappa, Leg. India 155. 1992; Mohanan & Sivad., Fl. Agasthyamala 206. 2002. *Hedysarum heterophyllm* Willd., Sp. Pl. 3:1201. 1802.

Prostrate or trailing herbs. Leave 3-foliolate; leaflets ovate-elliptic to obovate, 8-20 by 6-13 mm, base rounded to subcordate, apex obtuse, subacute or slightly retuse, mucronulate, lower surface sparsely hairy on nerves. Flowers 1-3, in a cluster and/or in short axillary racemes; pedicel c. 15 mm long; corolla purplish-blue; stamens 9+1. Pods slightly indented, articles 2-4.

Fl. & Fr.: Sep.-Dec. *Distr.*: Paleotropics. Very common in moist exposed areas in plains and hills. *NAK 695, 906* (Pandalam, c. 50 m), *1066* (Tiruvalla, c. 10 m).

6. **Desmodium laxiflorum** DC., Ann. Sci. Nat. (Paris) 4 : 100. Jan. 1825; Baker in Hook.f., Fl. Brit. India 2 : 164. 1876; Gamble, Fl. Pres. Madras 344. 1918; Ohashi, Ginkgoana 1 : 101. 1973; Britto in Matthew, Fl. Tam. Carnatic 3(1): 396. 1983; Sanjappa, Leg. India 156. 1992.

Undershrubs; branchlets angular. Leaves 3-foliolate; leaflets ovate- lanceate, terminal one rhombic-lanceate, 4-15 by 2-7.5 cm, base acute, apex narrowly acuminate, lower surface appressed silky pubescent, lateral veins c. 10 pairs, prominent. Racemes axillary, c. 10 cm long. Flowers 2-3 in cluster on rachis; corolla purplish-pink or bluish; stamens 9 + 1. Pods linear, to 4 cm long, articlels 6-10, narrow oblong, hooked hairy.

Fl. & Fr.: Sep.-Jan. *Distr.*: India to Formosa, Malesia. Rare. Collected only once. In semi-evergreen forest. *NAK 188* (Thekkuthode, c. 70 m).

7. **Desmodium microphyllum** (Thunb.) DC., Prodr. 2 : 337. 1825; van Meeuwen, Reinwardtia 6 : 254. 1962. *Hedysarum microphyllum* Thunb., Fl. Jap. 284. 1784. *Desmodium parvifolium* DC., Ann. Sci. Nat. (Paris) 4 : 100. 1825 & Prodr. 2 : 234. 1825; Baker in Hook.f., Fl. Brit. India 2 : 174. 1876; Gamble, Fl. Pres. Madras 348, 1918; Sanjappa, Leg. India 157. 1992.

Trailing slender woody herbs. Leaves 3-foliolate, rarely simple; leaflets broadly ovate to suborbicular, 3-6 by 3-5 mm, base cordate, apex rounded, mucronulate, lower surface appressed pubescent. Racemes slender, terminal, c. 8 cm long. Flowers very lax; corolla purplish to pink; stamens 9 + 1. Pods 5-10 mm long, articles 3-5, puberulous.

Fl. & Fr.: Nov.-Jan. *Distr.*: India, China, Malesia. In moist hill cuttings among grasses. Rare, but abundant in Kakki hills, along the way to Ponnambalamedu. *NAK 1413* (Kakki hills, c. 1000 m).

8. **Desmodium motorium** (Houtt.) Merr., J. Arnold Arb. 19 : 345. 1938; Van Meeuwen, Reinwardtia 6 : 254. 1962; Gandhi in Sald. & Nicols., Fl. Hassan Dist. 250. 1976; Manilal, Fl. Silent Valley 78. 1988; Sanjappa, Leg. India 158. 1992; Mohanan & Sivad., Fl. Agasthyamala 207.2002. *Hedysarum motorium* Houtt., Nat. Hist. ed. 2. 10 : 246. 1779. *H. gyrans* L. f., Suppl. Pl. 332. 1781. *Desmodium gyrans* (L.f.) DC., Prodr. 2 : 326. 1825; Wight, Ic. t. 294. 1840; Baker in Hook.f., Fl. Brit. India 2 : 174. 1876; Gamble, Fl. Pres. Madras 348. 1918. *Codariocalyx motorius* (Houtt.) Ohashi, J. Jap. Bot. 40 : 267. 1965, & Ginkgoana 1 : 46. 1973; Britto in Matthew, Fl. Tam. Carnatic 3(1): 351. 1983. *Ramanama-pacha.*

Erect shrubs. Leaves 3-foliolate; lateral leaflets very small, showing jerking movements, terminal one large, elliptic-oblong to ovate-oblong, 4-8 by 1.2-3 cm, base and apex rounded, lower surface appressed-pubescent. Racemes terminal or from upper leaf axils, c. 20 cm long. Flowers very lax; corolla pink; stamens 9 + 1. Pods c. 4 cm long, articles 6-8, pubescent

Fl. & Fr.: Nov.-Feb. *Distr.*: Sri Lanka, India, Himalayas, S.W. Asia, China, Malesia & Austrialia. Rare. In moist hill slopes. *NAK 200, 2140* (Chittar, c. 400 m).

9. **Desmodium pulchellum** (L.) Benth., Fl. Hongk. 83. 1861; Baker in Hook.f., Fl.Brit.India 2 : 162. 1876; Gamble, Fl. Pres. Madras 344.1918; Sanjappa, Leg. India 160. 1992. *Hedysarum pulchellum* L., Sp. Pl. 747.1753.

Dicerma pulchellum (L). DC., Ann. Sci. Nat. (Paris) 4 : 236. 1825 & Prodr. 2 : 339. 1825; Wight, Ic. t. 418 1841. *Phyllodium pulchellum* (L.) Desv., J. Bot. (Desvaux) (Ser. 2) 1 : 124, t. 5, f. 24. 1813; Ohashi, Ginkgoana 1 : 276. 1973; Britto in Matthew, Fl. Tam. Carnatic 3(1) : 440. 1983; Mohanan & Sivad., Fl. Agasthyamala 215. 2002. *Cheru-Pachotti.*

Erect shrubs. Leaves 3-foliolate, terminal one large, elliptic-oblong, 9 by 4.5 cm, laterals ovate, 4 by 3 cm, base rounded, margin repand, apex obtuse, mucronate, lower suface tomentose. Racemes c. 20 cm long, axillary and terminal, primary bracts orbicular, foliaceous, 12 by 12 mm, enclosing 5-8 flowers; corolla white; stamens 9 + 1. Pods c. 7 mm long, articles 2 or 3.

Fl. & Fr.: Nov.-Feb. *Distr.*: Sri Lanka, India, S.E. Asia, Malesia, China & N. Australia. Common in moist deciduous forests. *NAK 470* (Ranni, c. 100 m).

Powder or decoction of bark given with honey in haemorrhage, swellings, diarrhoea, excessive thirst and poisoning.

10. **Desmodium triangulare** (Retz.) Merr., J. Arnold Arb. 23 : 170. 1942, var. **congestum** (Wight & Arn.) Sant., Kew Bull. 276. 1948; Gandhi in Sald. & Nicols., Fl. Hassan Dist. 251. 1976; Manilal, Fl. Silent Valley 78.1988; Sanjappa, Leg. India 160. 1992; Mohanan & Sivad., Fl. Agasthyamala 208. 2002. *Hedysarum triangulare* Retz., Obs. Bot. 3 : 40. 1783. *Desmodium congestum* Wall. ex Wight & Arn., Prodr. 224. 1834; Wight, Ic. t. 209. 1839. *D. cephalotes* (Roxb.) Wight & Arn., var. *congestum* (Wight & Arn.) Prain, J. Asiat. Soc. Bengal 66 : 389. 1898; Gamble, Fl. Pres. Madras 344. 1918.

Erect undershrubs to 2 m tall; branchlets angular; leaves 3-foliolate; leaflets ellipitc-ovate, obovate to oblanceate, 4-10 by 2-5 cm, base cuneate, apex acuminate, lower surface densely silky, glabrescent, thin-coriaceous. Flowers very dense, in umbelliform axillary clusters; corolla creamy-white; ovary sessile. Pods c. 3 cm long articles 4-6, pubescent.

Fl. & Fr.: Nov.-Mar. *Distr.*: Sri Lanka, India, China, Malesia. Very common in moist decidous forest and teak plantations. A good indicator of forest destruction. *NAK 826* (Chelikhuzy-Ranni R.K., c. 200 m). Leaves used a fooder.

11. **Desmodium triflorum** (L.) DC., Prodr. 2 : 334. 1825; Wight & Arn., Prodr. 229.1834, p.p.; Wight, Ic. t. 292. 1839; Baker in Hook.f., Fl. Brit. India 2 : 173. 1876; Gamble, Fl. Pres. Madras 347. 1918; Van Meeuwen, Reinwardtia 6 : 261. 1962; Ohashi, Ginkgoana 1:245. 1973; Gandhi in Sald. & Nicols., Fl. Hassan Dist. 251. 1976; Britto in Matthew, Fl. Tam. Carnatic 3(1) : 399. 1983; Manilal, Fl.Silent Valley 79. 1988; Sanjappa, Leg. India 155. 1992. *Hedysarum triflorum* L., Sp. Pl. 749. 1753. *Cheru-pulladi.*

Slender trailing or prostrate herbs, rooting at lower nodes. Leaves 3-foliolate; leaflets broadly obovate; 4-10 by 3-9 mm, base cuneate, apex rounded-retuse or truncate, lower surface sparsely hairy on nerves. Flowers 1-3, in clusters in leaf axils; corolla pinkish; stamens 9 + 1. Pods linear, slightly indented, articles 3-5, pubescent.

Fl. & Fr.: Aug.-Dec. *Distr.*: Sri Lanka, India, Myanmar, S. China, Malesia to N. Australia. This is the common species of the genus. In exposed areas in plains, usually forming compact masses. *NAK 695 & 867* (Pathanamthitta, c. 50 m).

Decoction of root given to increase mother's milk, and to purify blood, to cure poison ailments, ulcer, leprosy, diarrheoa, epilepsy and insanity.

12. **Desmodium triquetrum** (L.) DC., Prodr. 2 : 326. 1825; Baker in Hook.f., Fl. Brit. India 2 : 163. 1876, p.p.; Gamble, Fl. Pres. Madras 345. 1918; Van Meeuwen, Reinwardtia 6 : 292. 1962; Gandhi in Sald. & Nicols., Fl. Hassan Dist. 252. 1976; Manilal, Fl. Silent Valley 79. 1988 ; Sanjappa, Leg. India 163. 1992. Mohanan & Sivad., Fl. Agasthyamala 208. 2002. *Hedysarum triquetrum* L., Sp. Pl. 1753. *Adekka Panal.*

Undershrubs; branchlets trigonous to triquetrous. Leaves 1-foliolate; petiole broadly winged; lamina oblong-lanceate to ovate-lanceate, 5-15 by 3-5.5 cm; base cordate, margin reflexed, apex acute, glabrous or hairy, chartaceous. Racemes terminal, to 30 cm long; bracts setaceous. Flowers in clusters of 3 or 2 on rachis; corolla bluish-purple; stamens monadelphous. Pod flat, 3 by 0.6 cm, 6-8 jointed, appressed pubescent.

Fl. & Fr.: Sep.-Jan. *Distr.*: Sri Lanka, India, China, Philippines. Common in moist deciduous forests and in teak plantations. A ready colonizer of degrading forests. *NAK 154, 177, 186 & 1360* (Moozhiar, c. 250 m).

Note : Indumentum and shape of leaves are hightly variable.

13. **Desmodium velutinum** (Willd.) DC., Prodr. 2 : 328. 1855; Van Meeuwen, Reinwardtia 6:264. 1962; Britto in Matthew, Fl. Tam. Carnatic 3(1) : 400. 1983; Manilal, Fl. Silent Valley 80. 1988; Sanjappa, Leg. India 165. 1992; Mohanan & Sivad., Fl. Agasthyamala 209. 2002. *Hedysarum velutinum* Willd., Sp. Pl. 3 : 117. 1802. *Hedysarum latifolium* Roxb. ex Ker, Bot. Reg. 5 : t. 355. 1819. *Desmodium latifolium* (Roxb. ex Ker) DC., Prodr. 2 : 328. 1825; Baker in Hook.f., Fl. Brit. India 2 : 168. 1876; Gamble, Fl. Pres. Madras 346. 1918. *Orila.*

Undershrubs. Leaves 1-foliolate, broadly ovate, 5-10 by 4-8 cm, base truncate-obtuse, apex subacute-mucronulate, lower surface densely tomentose, thin-coriaceous. Racemes terminal and axillary, c. 20 cm long; corolla purple to white; stamens 9 + 1; ovary sessile. Pod c. 2 mm long, slightly indented to dorsal side, hooked-hispid pubescent.

Fl. & Fr.: Nov.-Feb. *Distr.*: Africa, Sri Lanka, India, Myanmar, Indo-China, Malesia. Less common. In semi-evergreen and moist deciduous forests. *NAK 191* (Thekkuthode, c. 70 m).

Decoction fo roots given for fever, diarrhoea, burning sensation, vomiting, indigestion, insanity and ulcers.

DOLICHOS Linnaeus
Sp.Pl. 725. 1753.

Dolichos trilobus L., Sp. Pl. 726. 1753; Gandhi in Sald. & Nocols., Fl. Hassan Dist. 252. 1976; Verdc., Taxon 17 : 173. 1968; Matthew, Ill. Fl. Tam. Carnatic t. 196. 1982; Britto in Matthew, Fl. Tam. Carnatic 3(1) : 401. 1983; Manilal, Fl. Silent Valley 80.1988; Sanjappa, Leg. India 167. 1992. *Dolichos falcatus* Klein ex Willd., Sp. Pl. 3 : 1047. 1802; Baker in Hook.f., Fl. Brit. India 2 : 211. 1876, Gamble, Fl. Pres. Madras 366. 1918. *Kattu muthira.*

Twiners. Leaves 3-foliolate; terminal leaflet ovate-rhombold, laterals ovate-lanceate, 5-10 by 3-5 cm, base truncate, apex acute to shortly acuminate, apiculate, sparsely puberulous. Racemes axillary, 5 cm long. Flowers 3-5 per raceme; bracts and bracteoles subulate; calyx campanulate, upper 2 lobes connate; corolla violet, exerted, standard orbicular, auricled and appendaged; stamens 9+1; anthers uniform; ovary subsessile; style incurved, bearded on the inner side. Pods falcate, c.6 cm long. Seeds strophiolate.

Fl. & Fr.: Sep.-Jan. *Distr.*: Africa extending to E. Asia. Rare. On wayside bushes in evergreen forests. *NAK 215, 311 & 1182* (Upper Moozhiar to Anathode, c. 850 m).

Decoction of root given as a remedy in piles, stomachache, constipation, ophthalmia, ulcer, and poison ailments.

DUMASIA A.P. de Candolle
Ann. Sci. Nat. (Paris) 4 : 96. 1825.

Dumasia villosa DC., Ann. Sci. Nat. (Paris) Ser. 1, 4 : 96. 1825; Baker in Hook.f., Fl. Brit. India 2 : 183. 1876; Gamble, Fl. Pres. Madras 351. 1918; Gandhi in Sald. & Nicols., Fl. Hassan Dist. 252. 1976; Matthew, III. Fl. Tam. Carnatic t. 197. 1982; Britto in Matthew, Fl. Tam. Carnatic 3(1) : 402. 1983; Manilal, Fl. Silent Valley 81. 1988Sanjappa, Leg. India 167. 1992. *Dumasia congesta* Graham ex Wight, & Arn., Prodr. 26. 18341; Wight, Ic. t. 445. 1841.

Slender twiners. Leaves 32-foliolate; leaflets ovate-deltoid, 3-7 by 2-4.5 cm, base cuneate, apex obtuse-mucronate, pubescent, basally 3-nerved. Racemes axillary, c. 6 cm long. Flowers c.2 cm long; calyx gibbous at base, lobes 5, obscure; corolla exerted, petals yellow, long clawed, standard obovate, auricled below, keels adherent to wing; stamens 9+1; ovary subessile; style flattened, incurved above. Pods compressed, oblong, torulose, pubescent. Seeds subglobose.

Fl. & Fr.: Nov.-Jan. *Distr.*: Africa, Sri Lanka, India, China, Malesia. Rare. On wayside bushes in evergreen forests. *NAK 1387* (Kakki hills, c. 1100 m).

DUNBARIA R. Wight et Arnott
Prodr. 258. 1834.

Dunbaria heynei Wight & Arn., Prodr. 258. 1834; Baker in Hook.f., Fl. Brit. India 2 : 217. 1876; Gamble, Fl. Pres. Madras 370. 1918; Manilal, Fl. Silent Valley 81. 1988. *Cajanus heynei* (Wight & Arn.) Van der Maesen, Meded. Landbonwhogeschool Wageningen 85-4:129.1985; Sanjappa, Legumes of India 101.1992.

Densely villous, slender twiners. Leaves 3-foliolate; leaflets : terminal one obovate; laterals: oblong-lanceate, inequilateral, 4-9 by 2-5 cm, base subcordate. Flowers lax, c.2 cm long; calyx-tube campanulate, lobes, c.4, lower 2 longer; corolla exerted, petals bright-yellow, clawed, standard broadly obovate, auricled at base; stamens 9+1, anthers uniform; ovary sessile; style long, slender. Pod oblong, compressed, densely viscid-hairy.

Fl. & Fr.: Sep.-Feb. *Distr.*: Sri Lanka, Peninsular India. Rare. On bushes in evergreen forest borders. *NAK 312 & 1375* (Kakki hills, c. 1100 m).

ERYTHRINA Linnaeus
Sp. Pl. 706. 1753.

Erythrina stricta Roxb., (Hort. Beng. 53. 1813, nom. nud.) Fl. Ind. 3 : 251. 1832., Bedd., Fl. Sylv. t. 175. 1872; Baker in Hook.f., Fl. Brit. India 2 : 189. 1876; Gamble, Fl. Pres. Madras 354. 1918; Gandhi in Sald. & Nicols., Fl. Hassan Dist. 253.1976; Matthew, Ill. Fl. Tam. Carnatic t. 200. 1982; Britto in Matthew, Fl. Tam. Carnatic 3(1): 405 1983; Manilal, Fl. Silent Valley 81.1988; Sanjappa, Leg. India 173. 1992; Mohanan & Sivad., Fl. Agasthyamala 210. 2002.

Armed deciduous trees with yellowish prickles. Leaves 3-foliolate; leaflets rhomboid-ovate, 8-12 by 9-20 cm, base truncate, apex subacute. Racemes to 10 cm long. Flowers 2 cm ϕ; calyx spathaceous, apically entire; corolla deep-red; stamens 9 + 1, filaments alternatively longer and shorter; ovary stalked. Pods falcate. Seeds ovoid.

Fl. & Fr.: Mar.-Jun. *Distr.*: India, Nepal, Mayanmar, Thailand, China. Not common. In semi-evergreen forests.

Note : Failed to collect this species but observed and identified some plants in forests of Moozhiar.

FLEMINGIA Roxburgh ex W. Aiton et W.T. Aiton Hortus Kew. ed. 2, 4 : 349. 1812 (nom. cons.)

Woody herbs or undershrubs. Leaves 1-3-foliolate. Inflorescence racemes or panicles, axillary or terminal; bracts often foliaceous and concealing the flowers; calyx-tube short; corolla subexerted, standard obovate, auricled at base; stamens 9 + 1, anthers uniform; ovary subsessile. Pods turgid, oblique. Seeds 2, globose.

1. Leaves 1-foliolate; racemes as long as leaves; flowers white, hidden within the floral bracts.
   2. Floral bracts acute at apex; lateral leaf nerves 8 or more pairs; branchlets terete. .... 4. **F. strobilifera**
   2. Floral bracts usually retuse at apex; lateral leaf nerves 6 or less pairs; branchlets angular. .... 1. **F. bracteata**
1. Leaves 3-foliolate; racemes shorter than leaves; flowers purplish-rose, exposed from floral bracts.
   3. Petiole winged; racemes to 10 cm or more long; flowers lax. .... 3. **F. semialata**
   3. Petiole not or very slightly winged; racemes to 6 cm long; flowers congested. .... 2. **F. macrophylla**

1. **Flemingia bracteata** (Roxb.) Wight, Ic. t. 268. 1840; Gamble, Fl. Pres. Madras 378. 1918. *Hedysarum bracteatum* Roxb., Fl. Ind. 3 : 351. 1832. *Flemingia strobilifera* R. Br. ex Ait., var. *bracteata* (Roxb.) Baker in Hook.f., Fl. Brit. India 2 : 227. 1876. **(Fig. 33)**

Branchlets angular. Leaves 1-foliolate, ovate-lanceate, 5-8 by 2.5-3.5 cm, base rounded, apex acute, lateral nerves 4-6(-7) pairs, gland-dotted below, coriaceous. Racemes usually simple or branched; floral bracts foliaceous, orbicular, pubescent, margin ciliate, apex rounded or retuse; corolla white. Pods oblong.

Fl. & Fr.: Sep.-Dec. *Distr.*: Peninsular & Eastern India. Rare. Among grasses in exposed hills. *NAK 398* (Kakki hills, c. 110 m).

2. **Flemingia macrophylla** (Willd.) Prain ex Merr., Philip. J. Sci. 5 : 130.1910; Gandhi in Sald. & Nicols., Fl. Hassan Dist. 254. 1976; Matthew, Ill. Fl. Tam. Carnatic t. 201. 1982; Britto in Matthew, Fl. Tam. Carnatic 3(1) : 408. 1983; Manilal, Fl. Silent Valley 82. 1988; Sanjappa, Leg. India 176. 1992; Mohanan & Sivad., Fl. Agasthyamala 211.2002. *Crotalaria macrophylla* Willd., Sp. Pl. 3 : 982. 1802. *Flemingia congesta* Roxb. ex Ait., Hort. Kew. (ed 2), 4 : 349. 1812; Baker in Hook.f., Fl. Brit. India 2 : 228. 1876; Gamble, Fl. Pres. Madras 378. 1918.

Stem triquetrous. Leaves 3-foliolate; petiole wingless or very feebly winged; leaflets elliptic-ovate to ovate-lanceate, 8-25 by 4-12 cm, base obtuse-cuneate, apex acuminate, apiculate, pubescent, lower surface reddish-glandular, thin-coriaceous. Racemes 3 cm long, 3-4 clustered together. Flowers congested; bracts oblong-lanceate, to 2 cm long; corolla purplish-rose coloured. Pods oblong, downy pubescent. Seeds 2.

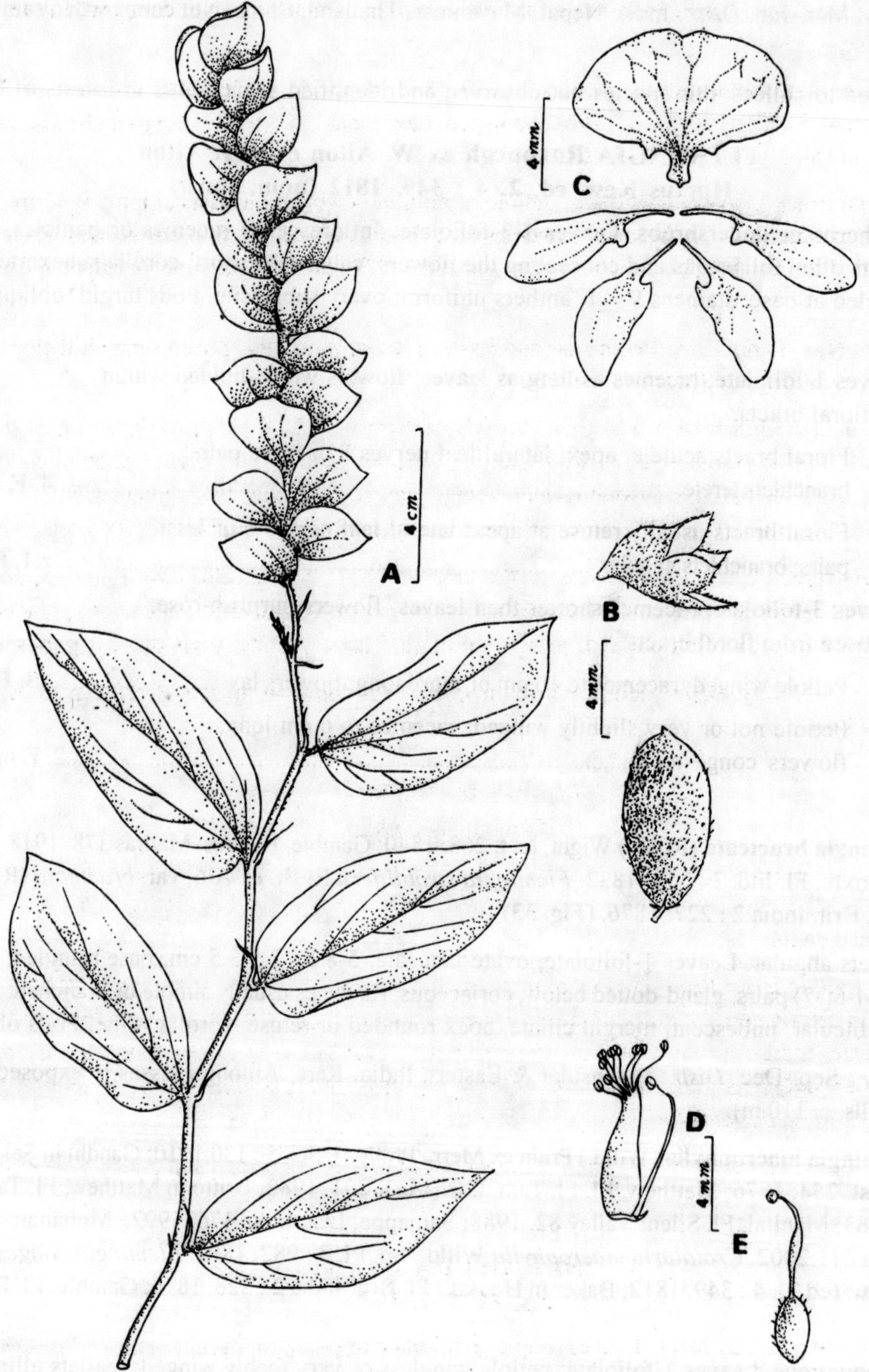

Fig. 33. ***Flemingia bracteata*** **(Roxb.) Wight: A. Twig; B. Calyx; C. Corolla - spread out; D. Staminal column; E. Pistil; F. Pod.**

Fl. & Fr.: Nov.-Jan. *Distr.*: Indo-Malesia. Less common. Among forest undergrowth. *NAK 346, 507 & 1343* (Moozhiar, c. 250 m).

3. **Flemingia semialata** Roxb., Fl. Ind. 3 : 340. 1832; Wight, Ic. t. 326. 1840; Gamble, Fl. Pres. Madras 378. 1918; Manilal, Fl. Silent Valley 82. 1988; Sanjappa, Leg. India 178. 1992. *F. congesta* Roxb., var. *semialata* (Roxb.) Baker in Hook.f., Fl. Brit. India 2 : 229. 1876, p.p.

Stem triquetrous. Leaves 3- foliolate; petiole prominently winged; leaflets elliptic, 9-17 by 2.5- 4.5 cm, base obtuse, apex acuminate, pubescent, lower surface reddish-glandular, thin-coriaceous. Racemes 5-7 cm long, clustered. Flowers lax; bracts triangular-lanceate, c. 7 mm long; corolla purplish. Pods glandular. Seeds 2.

Fl. & Fr.: Nov.-Feb. *Distr.*: Peninsular and Eastern India. Rare. In exposed sunny hill slopes. *NAK 210* (Thekkuthode, c. 70 m), *436* (Maniyaar, c. 100 m).

4. **Flemingia strobilifera** (L.) R. Br. ex Ait., Hort. Kew. (ed. 2), 4 : 350. 1812; Wight, Ic. t. 267. 1840; Baker in Hook.f., Fl. Brit. India 2 : 227. 1876; Gamble, Fl. Pres. Madras 377. 1918; Gandhi in Sald. & Nicols., Fl. Hassan Dist. 255. 1976; Britto in Matthew, Fl. Tam. Carnatic 3(1) : 408. 1983; Manilal, Fl. Silent Valley 83. 1988; Sanjappa, Leg. India 178. 1992. *Hedysarum strobiliferum* L., Sp. Pl. 746. 1753.

Branchlets terete. Leaves 1-foliolate, ovate-lanceate, 5-15 by 3-5 cm, base rounded, apex acute to acuminate, lateral nerves 7-9 pairs, thin-coriaceous. Racemes usually branched, 15 cm long; bracts foliaceous, orbicular, 2 by 3 cm, apex often pointed, margin ciliolate. Flowers white. Pods oblong, pubescent.

Fl. & Fr.: Feb.-May. *Distr.*: Tropics & Subtropics. Less common. In semievergreen forests. *NAK 468* (Maniyaar, c. 75 m).

GEISSASPIS R. Wight et Arnott
Prodr. 217. 1834.

Geissaspis cristata Wight & Arn., Prodr. 218. 1834; Bedd., Ic. t. 293. 1868-1874; Baker in Hook.f., Fl. Brit. India 2 : 141. 1876; Gamble, Fl. Pres. Madras 324. 1918; Sanjappa, Leg. India 181. 1992; Mohanan & Sivad., Fl. Agasthyamala 212. 2002.

Diffuse, trailing herbs. Leaves even-pinnate; leaflets 2 pairs, opposite, obovate, 5-10 by 3-6 mm, base oblique, margin faintly serrulate, apex retuse. Racemes axillary, c. 3 cm long; bracts foliaceous, orbicular, closely imbricate, margin prominently bristly, persistent. Flowers very small, concealed by bracts; calyx hyaline, 2-lipped; stamens 10 in a bundle, anthers uniform; ovary subsessile; style incurved; ovules 2. Pods turgid, 1-2 jointed. Seeds reniform.

Fl. & Fr.: Nov.-Jan. *Distr.*: Sri Lanka, Peninsular India. Rare. In swamps in plains. *NAK 133* (Adoor, c. 40 m).

INDIGOFERA Linnaeus
Sp. Pl. 751. 1753.

Erect or trailing woody herbs. Leaves simple, 3-foliolate or odd-pinnate, alternate. Inflorescence racemes or spikes, axillary; bracts caducous; calyx minute, lobes 5, setaceous; corolla purplish-rose or reddish, caducous, keel petals with a downward spur at base; stamens 9+1, anthers uniform, often apiculate; ovary sessile. Pods oblong-linear, terete, or 4-gonous. Seeds globose-ovoid, 4-gonous or terete.

1. Trailing herbs; leaflets alternate. ....5. **I. spicata**
1. Erect or trailing subshrubs or small trees; leaflets opposite.
    2. Small trees; leaflets c. 10 cm long. ....6. **I. zollingeriana**
    2. Shrubs or trailing subshrubs; leaflets to 5 cm long.
        3. Leaflets 7-13; flower less than 1 cm long;pod not prominently beaked.
            4. Pods terete or faintly angular, slightly curved. ....4. **I. longiracemosa**
            4. Pods 4-gonous, straight, linear or curved.
                5. Leaflets to 5 cm long; pods straight, densely hirsute. ....3. **I. hirsuta**
                5. Leaflets to 2 cm long; pods curved, sparsely hirsute. ....1. **I. constricta**
        3. Leaflets upto 25; flower over 1 cm long; pod prominently beaked ....2. **I. galegoides**

1. **Indigofera constricta** (Thw.) Trimen, Cat. Ceyl. Pl. 23. 1885 & Handb. Fl. Ceylon 2 : 27. 1894; Gamble, Fl. Pres. Madras 312. 1918; Chandrabose, Bull. Bot. Surv. India 18: 327. 1976; Sanjappa, Leg. India 187. 1992. *I. flaccida* Koen. ex Roxb. var. *constricta* Thw., Enum Pl. 411. 1864; Baker in Hook.f., Fl. Brit. India 2:99. 1876.

Erect shrubs. Leaves odd-pinnate, to 12 cm long; leaflets usually 11, opposite, elliptic, 1.5-4 by 0.7-2 cm, base and apex rounded, mucronulate, appressed hairy, membranous. Racmes to 12 cm long, axillary. Flowers α, lax, c.6 mm long; calyx 5-toothed; corolla scarlet coloured, appressed pubescent without; ovary linear. Pods slender, c.4 cm long, curved, constricted between seeds.

Fl. & Fr.: Sep.-Dec. *Dsitr.*: Peninsular and Eastern India. Very rare. Along forest margins in uper hills. *NAK 2612* (Plappally, c. 450 m).

Note : Rapidly vanishing because of the failure to compete with alien species like *Chromolaena odorata, Ageratum conyzoides,* etc.

2. **Indigofera galegoides** DC., Prodr. 225, 1825; Baker in Hook.f., Fl. Brit. India 2 : 100. 1876; Gamble, Fl. Pres. Madras 313. 1918; Rudd in Dassan. & Fosberg, Rev. Handb. Fl. Ceylon 7:130. 1991; Sanjappa, Leg. India 189. 1992.

Erect shrubs. Leaves odd-pinnate, c. 15 cm long; leaflets to 25, elliptic-oblong, 1.5-2.5 by 0.7-1 cm, base cuneate, apex obtuse, mucronate, both surfaces appressed hairy, membranous. Racemes c. 5 cm long; calyx teeth minute; corolla puplish. Pods cylindric, to 7 cm long; straight, prominently beaked. Seeds 8-10.

Fl. & Fr.: Sep.-Jan. *Distr.*: Sri Lanka, India, S. China, Philippines, Malesia. Very rare. In evergreen forest borders. *NAK 1248* (Moozhiar, c. 250 m).

3. **Indigofera hirsuta** L., Sp. Pl. 751. 1753; Baker in Hook.f., Fl. Brit. India 2: 96. 1876; Gamble, Fl. Pres. Madras 312. 1918; J.B. Gillett, Kew Bull. 14:290. 1960; Britto in Matthew, Fl. Tam. Carnatic 3(1) : 420. 1983; Sanjappa, Leg. India 191. 1992; Mohanan & Sivad., Fl. Agasthyamala 213.2002.

Undershrubs. Leaves odd-pinnate, to 15 cm long; leaflets to 11, subopposite, elliptic, 2-4 by 12 cm, base subacute, apex obtuse, apiculate, densely hirsute. Racemes to 20 cm long. Flowers dense, c. 1 cm long; calyx-lobes setaceous; corolla pink to brick-red. Pods straight, 4-gonous. Seeds 5, angular.

Fl. & Fr.: Sep.-Jan. *Distr.*: India, Africa, S. Asia, N. Australia. Less common. In exposed lowlands. *NAK 1908* (Kodumon, c. 70 m).

4. **Indigofera longiracemosa** Boivin ex Baillon, Bull. Mens. Soc. Linn. Paris 1: 399. 1883; Gamble, Fl. Pres. Madras 312. 1918; Britto in Matthew, Fl. Tam. Carnatic 3(1) : 422. 1983; Sanjappa, Leg. India 192. 1992. **(Fig. 34)**

Erect subshrubs; branchlets terete; leaflets 7-9, opposite, obovate, odd one larger, 8-16 by 4-10 mm, base cuneate, apex obtuse, mucronate, both surfaces appressed pubescent, membranous. Racemes axillary, c. 2 cm long. Flowers dense; corolla pinkish-purple. Pods cylindric, straight, shortly beaked. Seeds 6-10, obscurely angular.

Fl. & Fr.: Dec.-Mar. *Distr.*: Peninsular India. Rare. Among bushes in forest borders. *NAK 1511* (Aluvamkudi, c. 400 m).

5. **Indigofera spicata** Forsskal, Fl. Aegypt.-Arab. 138. 1775; J.B. Gillett, Kew Bull. (Add. Ser.) 1:119. 1958; Gandhi in Sald. & Nicols., Fl. Hassan Dist. 258. 1976; Britto in Matthew, Fl. Tam. Carnatic 3(1) : 427. 1983; Sanjappa, Leg. India 195. 1992. Mohanan & Sivad., Fl. Agasthyamala 213.2002. *I. endecaphylla* Jacq., Ic. Pl. Rar. 3(3): 13, t. 570. Mar. 1789 ('*hendecaphylla*'); l.c., 3(16) : 14. 1795 & Collectanea 2 : 358. Apr. 1789; Baker in Hook.f., Fl. Brit. India 2:98. 1876; Gamble, Fl. Pres. Madras 311. 1918.

Trailing woody herbs. Leaves odd-pinnate, c. 2 cm long; leaflets 7-8, obovate, 6-20 by 2.5-8 mm, base cuneate, apex subacute to obtuse, mucronate, lower surface appressed pubescent. Racemes axillary, 5-10 cm long; calyx-lobes setaceous; corolla purplish-rose; ovary sessile. Pods straight, c. 2.5 cm long, faintly 4-gonous, prominently beaked. Seeds 6-8.

Fl. & Fr.: Sep.-Jan. *Distr.*: Sri Lanka, India, Africa, Madagascar. Locally abundant. The common species of the genus of; exposed hill slopes. *NAK 66, 1393 (*Kakki hills, c. 1100 m).

6. **Indigofera zollingeriana** Miq., Fl. Ind. Bat. 1,1:310. 1855; Sanjappa, Leg. India 198. 1992. *I. teysmannii* Miq., Fl. Ind. Bat. 1,1:1083. 1885; Sald. & G. Singh in Sald., Fl. Karnataka 1;474. 1984. *I. benthamiana* Hance, Ann. Sci. Nat. Bot. 4, 18:219. 1862; Ravi, J. Bombay Nat. Hist. Soc. 73:242. 1972.

Small trees. Leaves odd-pinnate, to 30 cm long; leaflets 15-23, opposite, ovate lanceate to oblong-lanceate, 5-10 by 2-3.5 cm, base obtuse, apex gradually acute, pubescent. Racemes to 20 cm long, axillary. Flowers dense, 10 12 mm long; calyx subequally 5-lobed, corolla deep-pink; ovary linear. Pods cylindric, slightly curved, prominently beaked. Seeds 10-15, flat.

Fl. & Fr.: Aug.-Dec. *Distr.*: Native of S. China; now introduced in some parts of Western Ghats. Seen as hedge plants; occasionaly run wild. *CNM 5527* (MH) (Pathanamthitta, c. 50 m).

KUNSTLERIA Prain
in G. King, J. Asiat. Soc. Bengal, Pt. 2, Nat. Hist. 66 (2) : 109. 1897.

Kunstleria keralensis Mohanan *et* Nair, Proc. Ind. Acad. Sci. Sec. B. (Plant Sci.) 90 (3) : 208. 1981; Manilal, Fl. Silent Valley 84. 1988; Sanjappa, Leg. India 199. 1992; Mohanan & Sivad., Fl. Agasthyamala 214.2002. **(Fig. 35)**

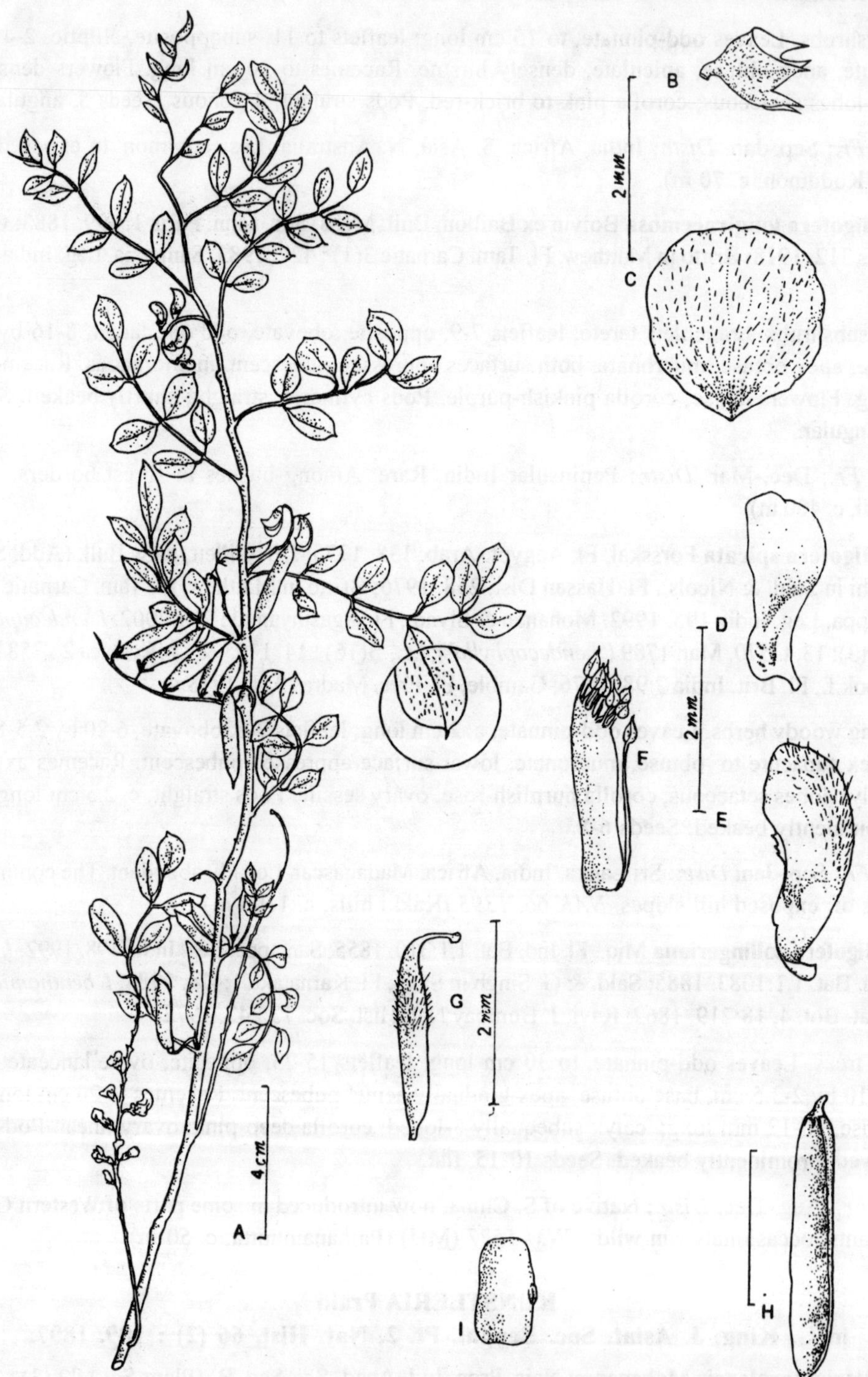

Fig. 34. *Indigofera longiracemosa* Boiv. ex Baill.: A. Twig; B. Calyx; C. Standard petal; D. Wing petal; E. Keel petal; F. Staminal column; G. Pistil; H. Pod; I. Seed.

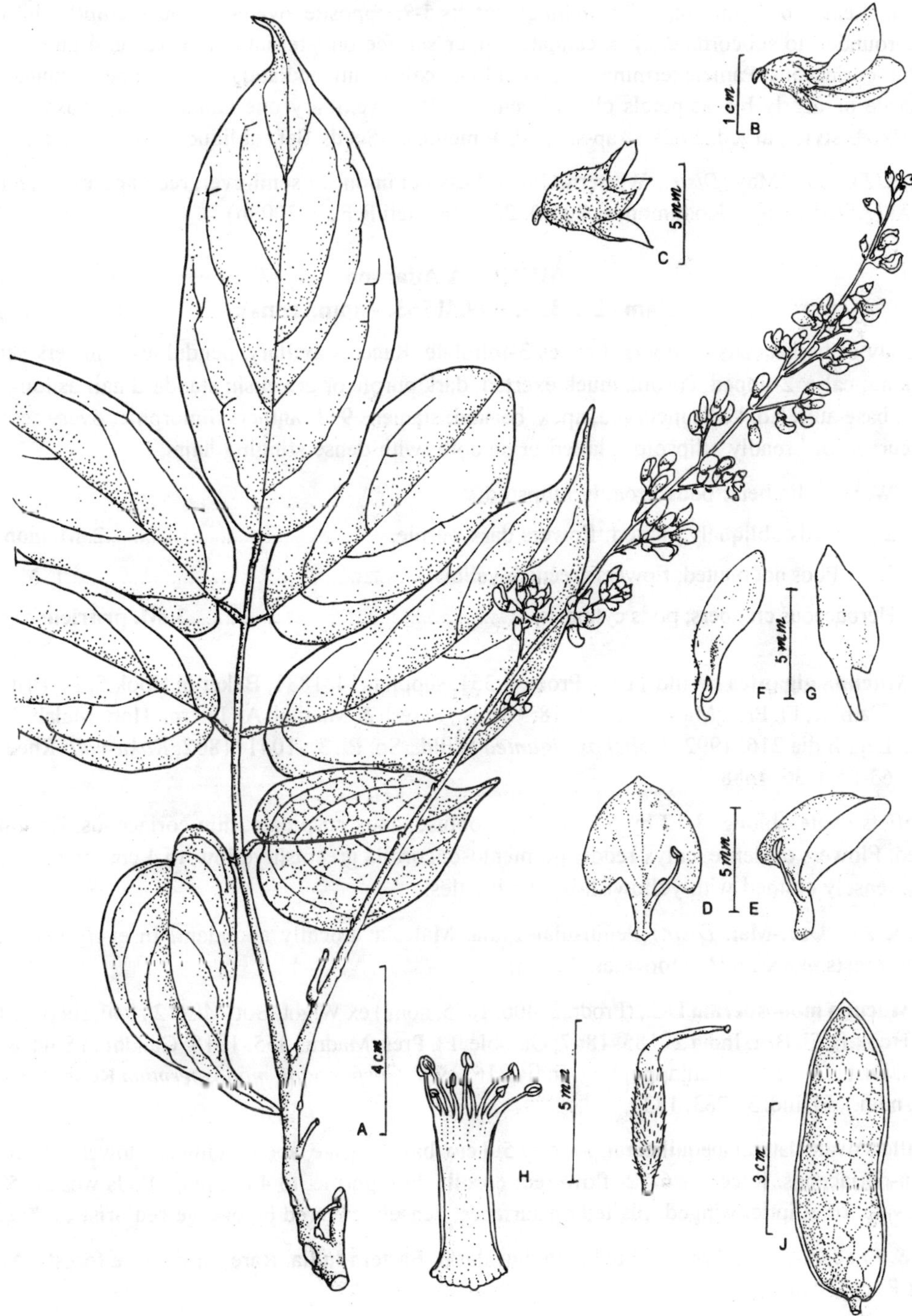

Fig. 35. ***Kunstleria keralensis*** **Mohanan & Nair: A. Twig; B. Flower; C. Calyx; D & E. Standard petal - with vexillary adnate stamen - different views; F. Wing petal; G. Keel petals; H. Staminal column; I. Pistil; J. Pod.**

Liana. Leaves odd-pinnate, c. 30 cm long; leaflets 5-7, opposite, ovate-elliptic to elliptic, 4-15 by 2-5.5 cm, base rounded to subcordate, apex caudate, upper surface bullate, lateral nerves c. 4 pairs, prominent below, thin-coriaceous. Panicle terminal, c. 25 cm long; calyx campanulate, lobes 5, upper 2 connate; corolla dark purple or nearly black; petals clawed; stamens 9+1, vexillary one adnate to the base of standard; ovary stalked; style curved. Pods strap-shaped, tomentose. Seeds flat, obliquely suborbicular.

Fl. & Fr.: Jan.-May. *Distr.*: Western Ghats. Less common. In semi-evergreen and evergreen forests. *NAK 2313* (Aikadu grove, Kodumon, c. 70 m), *2344* (Kalleli R.F., c. 200 m).

MUCUNA Adanson
Fam. 2 : 325, 579. 1763. (nom. cons).

Woody or herbaceous climbers. Leaves 3-foliolate. Racems axillary, pendulous. Flowers very dense, to 6 cm long; calyx 2-lipped; corolla much exerted, dark purple or greenish; standard half as long as wings and keels, base auricled, keels incurved, apex beaked; stamens 9+1, anthers dimorphic; ovary sessile. Pods oblong, curved or broadly ellipsoid, plaited or smooth, with dense stinging hairs.

1. Woody climbers; pods broadly ellipsoid.
   2. Pods obliquely plaited; flowers dark purple. ....2. **M. monosperma**
   2. Pods not plaited; flowers greenish-yellow. ....1. **M. gigantea**
1. Herbaceous climbers; pods cylindric. ....3. **M. pruriens** var. **hirsuta**

1. **Mucuna gigantea** (Willd.) DC., Prodr. 2:351, suppl. t. 14. 1831; Baker in Hook.f., Fl. Brit. India 2 : 186. 1876; Gamble, Fl. Pres. Madras 355. 1918; Nicols., Suresh & Manilal, An Interpr. Hort. Malab. 138. 1988; Sanjappa, Leg. India 216. 1992. *Dolichos gigantea* Willd., Sp. Pl. 3 : 1041. 1802. *Kaku-valli* Rheede, Hort. Malab. 8 : 63-65, t. 36. 1688.

Leaflets ovate-oblong, 8-12 by 5-8 cm, base obtuse, apex cuspidate, thin coriaceous. Racemes long-peduncled. Flowers α, dense, calyx reddish-tomentose; corolla greenish-yellow, to 4 cm long. Pods flat, 12 by 5 cm, densely clothed with yellowish-brown bristles.

Fl. & Fr.: Dec.-Mar. *Distr.*: Peninsular India, Malesia. Locally abundant. In evergreen and semi-evergreen forests. *NAK 2621* (Moozhiar, 250 m).

2. **Mucuna monosperma** DC., (Prodr. 2: 406. 1825, nom.) ex Wight, Bot. Misc. 2 : 346, suppl. t. 12. 1830; Baker in Hook.f., Fl. Brit. India 2: 185. 1867; Gamble, Fl. Pres. Madras 355. 1918; Gandhi in Sald. & Nicols., Fl. Hassan Dist. 259. 1976. Sanjappa, Leg. India 216. 1992.*Carpopogon monospermum* Roxb., (Hort. Beng. 54. 1814, nom.). Fl. Ind. 3 : 283. 1832.

Leaflets ovate, lateral inequilateral, 7-12 by 5-7 cm, base cuneate, apex acuminate, lower surface sparsely hairy, thin-coriaceous. Racemes 6-12- flowered; corolla dark purple, to 4 cm long. Pods woody, 5 by 3cm, broadly ovate or elliptic, winged, plaited on surfaces, densely covered by orange-red bristles. Seed 1, flat.

Fl. & Fr.: Jan.-Apr. *Distr.*: Sri Lanka, Peninsular & Eastern India. Rare. In riverine forests. *NAK 2371* (Kalleli R.F., c. 250 m).

Seeds used as an expectorant and sedative in cough, asthma and afflictions of the tongue.

3. **Mucuna pruriens** (L.) DC., Prodr. 2:405. 1825. *Dolichos pruriens* L. in Stickman, Diss. Herb. Amb. 23. 1754 & Syst. Nat. ed. 10, 1162, 1759, var. **hirsuta** (Wight & Arn.) Wilmot-Dear, Kew Bull. 42:44. 1987;

Sanjappa, Leg. India 218. 1992. *Mucuna hirsuta* Wight & Arn., Prodr. 254. 1834; Baker in Hook.f., Fl.Brit. India 2:187. 1876; Gamble, Fl. Pres. Madras 355. 1918.

Leaflets ovate-elliptic, 8-16 by 4-10 cm, base truncate, apex acute to shortly acuminate, pubescent on both sides, thin-coriaceous. Racemes α-flowered; corolla dark-purple, to 4 cm long. Pod curved at apex, with dense soft silvery brown stinging hairs.

Fl. & Fr.: Nov.-Feb. *Distr.*: India, Sri Lanka, S.E. Asia, Malesia. Common. Among forests undergrowth. *NAK 18* (Upper Moozhiar, c. 500 m), *1165* (Lower Moozhiar, c. 250 m).

PONGAMIA Ventenat
Jard. Malm. t. 28. 1803 (nom. cons.).

Pongamia pinnata (L.) Pierre, Fl. Forest. Cochinch., sub. t. 385. 1899; Thoth., Bull. Bot. Surv. India 3:418. 1962; Matthew, Ill. Fl. Tam. Carnatic t. 212. 1982; Britto in Matthew, Fl. Tam. Carnatic 3 (1) : 441. 1983; . Nicols., Suresh & Manilal, An Interpr. Hort. Malab. 140. 1988; Sanjappa, Leg. India 230. 1992; Mohanan & Sivad., Fl. Agasthyamala 215. 2002. *Cystis pinnatus* L., Sp. Pl. 741. 1753. *Pongamia glabra* Vent., Jard. Malm. 28, t. 28. 1803; Wight & Arn., Prodr. 262. 1835; Wight, Ic. t. 59. 1838; Bedd., Fl. Sylv. S. India. t. 177. 1872; Baker in Hook.f., Fl. Brit. India 2: 240. 1876; Gamble, Fl. Pres. Madras 385. 1918. *Pongam, Minari* Rheede, Hort. Malab. 6:5-6, t. 3. 1686.

Trees. Leaves odd-pinnate; leaflets 3-5 pairs, opposite, ovate or broadly obovate, 6-8 by 4.5-5.5 cm, base cuneate, apex acumiante, thin-coriaceous. Racemes axillary, to 20 cm long. Flowers lax, c. 1 cm long; clayx-tube campanulate, truncate at apex; carolla pinkish-white; stamens 10, monadelphous, vexillary stamen free at base and above; ovary subsessile; ovule 2; style incurved. Pods woody, obliquely oblong. Seed 1, reniform.

Fl. & Fr.: Mar.-Jul. *Distr.*: Sri Lanka, India, Myanmar, Malesia, Polynesia. Common. Along river banks. *NAK 2372* (Thannithode, c. 70 m).

Powder of root bark given with honey for curing rheumatism, leprosy, piles, worm infections and ulcers. Seeds used to cure skin-diseases and abscess.

PSEUDARTHRIA R. Wigth et Arnott
Prodr. 209. 1834.

Pseudarthria viscida (L.) Wight & Arn., Prodr. 209. 1834; Wight, Ic. t. 286. 1840; Baker in Hook.f., Fl. Brit. India 2 : 154. 1876; Gamble, Fl. Pres. Madras 334. 1918; Gandhi in Sald. & Nicols., Fl. Hassan Dist. 261. 1976; Matthew, Ill. Fl. Tam. Carnatic t. 213. 1982; Britto in Matthew, Fl. Tam. Carnatic 3(1) : 442. 1983; Sanjappa, Leg. India 230. 1992; Mohanan & Sivad., Fl. Agasthyamala 216. 2002. *Hedysarum viscidum* L., Sp.Pl. 747. 1753. *Kattu-Moovila.*

Woody herbs, branchlets viscid. Leaves 3-foliolate, terminal one ovate-rhombic, 9 by 7 cm, laterals obliquely-ovate, 4.5 by 3.5 cm, base cuneate-obtuse, margin undulate, ciliate, apex acute, mucronate, lower surface densely pubescent. Racemes axillary and terminal, c. 20 cm long. Flowers lax, in pairs or in fascicles; calyx-tube campanulate, upper 2 lobes connate; corolla exerted, pink; stamens 9+1, anthers uniform; ovary subsessile; style incurved. Pods oblong, viscous pubescent. Seeds subreniform, compressed.

Fl. & Fr.: Oct.-Dec. *Distr.*: Sri Lanka, Peninsular India. Less common. In exposed moist forests borders. *NAK* (Arampa, c. 500 m).

The whole plant is used as a remedy in biliousness, rheumatism, excessive heat, fever, asthma, heart diseases, piles and worm infections.

PTEROCARPUS N.J. Jacquin
Select. Stirp. Amer. Hist. 283. 1763 (nom. cons.).

Pterocarpus marsupium Roxb., Pl. Corom. t. 116. 1799 & Fl. Ind. 3:234. 1832; Wight & Arn., Prodr. 266. 1834; Bedd., Fl. Sylv. t. 21. 1869; Baker in Hook.f., Fl. Brit. India 2 : 239. 1876; Gamble, Fl. Pres. Madras 385. 1918; Gandhi in Sald. & Nicols., Fl. Hassan Dist. 261. 1976; Matthew, Ill. Fl. Tam. Carnatic t. 216. 1982; Britto in Matthew, Fl. Tam. Carnatic 3(1) : 445. 1983; Sanjappa, Leg. India 232. 1992; Mohanan & Sivad., Fl. Agasthyamala 217. 2002. *Venga.*

Medium sized trees. Leaves odd-pinnate; leaflets 5-7, alternate, elliptic-oblong, 7-10 by 4-5 cm, base and apex obtuse, pubescent below, coriaceous. Flowers in terminal or axillary panicles; clayx-tube campanulate, upper 2-lobes larger; corolla golden-yellow; stamens 10, monadelphous, at length split into 2 bundles of 5 each. Pods orbicular, compressed, broadly winged.

Fl. & Fr.: Sep-Dec. *Distr.:* Sri Lanka & Peninsular India. Rare. In semi-evergreen and moist decidous forests. *NAK 631* (Konni, c. 100 m).

Decoction of wood given as a remedy in leprosy, sore-eyes, gonorrhoea, fever, worm infections, rheumatism and constipation.

PUERARIA A.P. de Candolle
Ann. Sci. Nat. (Paris) 4 : 97. 1825.

Pueraria phaseoloides (Roxb.) Benth., J. Linn. Soc. Bot. 9 : 125. 1865; Baker in Hook.f., Fl. Brit. India 2 : 199. 1876. Sanjappa, Leg. India 234. 1992. *Dolichos phaseoloides* Roxb., Fl. Ind. 3: 316. 1832. *Thottapayar.*

Twining herbs. Leaves 3-foliolate; leaflets unlobed, broadly obliquely ovate to rhombic-ovate, 6-10 by 5.5-9 cm, base cuneate, apex acute-apiculate, basally 3-nerved, lower surface appressed silky, chartaceous. Flowers subsessile clustered on long-peduncled racemes; calyx-teeth short, upper 2 connate; corolla reddish-pink; stamens 10, monadelphous; ovary sessile; ovules α. Pods to 10 cm long.

Fl. & Fr.: Dec.-Mar. *Distr.*: India, Myanmar, China, Malesia. Among wayside thickets in hills. *NAK 231* (Athumpumkulam, c. 50 m), *300, 1221* (Moozhiar, c. 250 m).

This species is widely used as a cover crop in rubber plantations.

PYCNOSPORA R. Brown ex R. Wight *et* Arnott
Prodr. 197. 1834.

Pycnospora lutescens (Poir.) Schind., J. Bot. 64: 145. 1926; Gandhi in Sald. & Nicols., Fl. Hassan Dist. 262. 1976; Britto in Matthew, Fl. Tam. Carnatic 3 (1) : 447. 1983; Manilal, Fl. Silent Valley 85, 1988; Nicols., Suresh & Manilal, An Interpr. Hort. Malab. 142. 1988; Sanjappa, Leg. India 236. 1992; Mohanan & Sivad., Fl. Agasthyamala 218. 2002. *Hedysarum leutescens* Poir. in Lam., Encycl. 6 : 417. 1805. *Pycnospora hedysaroides* R. Br. ex Baker in Hook.f., Fl. Brit. India 2 : 153 1876; Gamble, Fl. Pres. Madras 333. 1918. *Nirpullari* Rheede, Hort. Malab. 9 : 67, t. 36. 1689.

Trailing herbs. Leaves 3-foliolate; leaflets obovate, 8-25 by 6-15 mm, base obtuse-subcordate, apex truncate to faintly retuse, pubescent, chartaceous. Racemes terminal, c. 8 cm long. Flowers lax, c. 4 mm ϕ; calyx deeply cleft in one side, lobes 5, lanceate; corolla purplish-blue, exerted; stamens 9+1, anthers uniform; ovary sessile; ovules α. Pods turgid, inflated. Seeds reniform.

Fl. & Fr.: Sep.-Dec. *Distr.*: Tropical Asia, New Guinea & Australia. Common on moist exposed hill cuttings. *NAK 1096 & 2096* (Vallicode, c. 100 m).

RHYNCHOSIA Loureiro
Fl. Cochinch. 425, 460. 1790. (nom. cons.).

Rhynchosia acutissima Thw., Enum. Pl. Zeyl. 413. 1864; Baker in Hook.f., Fl. Brit. India 2:226. 1876; Gamble, Fl. Pres. Madras 375. 1918; Sanjappa, Leg. India 236. 1992.

Twiners. Leaves 3-foliolate; leaflets rhombic-ovate, 7-10 by 6-8 cm, base cuneate, apex acuminate, pubescent beneath, thin-coriaceous. Flowers in short axillary racemes; calyx-tube short, upper 2 lobes connate; corolla yellow with reddish veins, exerted; stamens 9+1, anthers uniform; ovary subsessile; ovules 2. Pods softly pubescent, apex with a curved beak. Seeds subreniform.

Fl. & Fr.: Sep-Dec. *Distr.*: Sri Lanka, Western Ghats (Travancore hills). Rare. In evergreen forests among undergrowths. *NAK 1219* (Upper Moozhiar, c. 500 m).

SHUTERIA R. Wight et Arnott
Prodr. 207. 1834 (nom. cons.).

Shuteria vestita Wight & Arn., Prodr. 207. 1834, var. **vestita.** Wight, Ic. t. 165. 1839; Baker in Hook.f., Fl. Brit. India 2 : 181. 1876; Gamble, Fl. Pres. Madras 350, 1918; Thuan, Adansonia (ser. 2), 12 : 295. 1972; Gandhi in Sald. & Nicols., Fl. Hassan Dist. 263. 1976; Manilal, Fl. Silent Valley 86. 1988; Sanjappa, Leg. India 245. 1992.

Twiners; branchlets densly villous. Leaves 3-foliolate; terminal leaflets ovate-rhombic, 4 by 3 cm, laterals ovate, 3.5 by 2.5 cm, base rounded to cuneate, margin ciliate, apex obtuse-mucronate, sparsely pubescent, thin-coriaceous. Racemes 2-3, from a node or solitary, axillary. Flowers dense, solitary or in pairs on rachis; calyx-lobes upper- 2 connate; corolla purplish, exerted; stamens 9+1; anthers uniform; ovary stipitate; style curved. Pods flat, recurved, tomentose. Seeds reniform.

Fl. & Fr.: Nov-Feb. *Distr.*: Sri Lanka, India (Western Ghats). Less common. On bushes in evergreen forests borders. *NAK 386, 1365* (Kakki hills, c. 1050 m).

SMITHIA W. Aiton
Hort. Kew. 3:496. 1789 (nom. cons.).

Slender trailing or stout erect herbs. Leaves abruptly pinnate; leaflets 1-4 pairs, sensitive to touch; stipules scarious, persistent, appendaged at base. Flowers in condensed 1-sided racemes or in terminal heads; calyx deeply 2-lipped, persistent, upper lobe entire to emarginate, lower often 3-toothed; corolla yellow or cream coloured, exerted, petals clawed, auricled; stamens (5) + (5), anthers uniform; ovary sessile or stalked. Lomentum folded together inside the calyx.

1. Slender diffuse herbs; leaflets often 2 pairs; flowers in racemes. ....1. **S. bigemina**
1. Stout erect herbs; leaflets 4-5 pairs; flowers in panicles. ....2. **S. blanda**

1. **Smithia bigemina** Dalz., Hooker's J. Bot. Kew. Gard. Misc. 3 : 208. 1851; Baker in Hook.f., Fl. Brit. India 2 : 149. 1876; Gamble, Fl. Pres. Madras 329. 1918; Manilal, Fl. Silent Valley 86. 1988; Sanjappa, Leg. India 246. 1992; Mohanan & Sivad., Fl. Agasthyamala 220.2002.

Profusely branched trailing herbs; leaflets 1-2 pairs, narrowly obovate, 6-10 by 2.5-4 mm, base cuneate, rounded, margin ciliate, apex obtuse to retuse, lower surface sparsely hairy, membranous. Racemes axillary and terminal, c. 3 cm long; calyx 5 by 4 mm, upper lip emarginate, lower entire or very faintly 3-lobed, hairy, membranous; corolla yellow. Pods twisted, articles often 2. Seeds reniform.

Fl. & Fr.: Sep.-Feb. *Distr.:* Peninsular India. Locally abundant. Among grasses along wet waysides in hills. *NAK 1246* (Kakki hills, c. 1050 m).

This plant is a good fodder.

2. **Smithia blanda** Wall. ex Wight & Arn., Prodr. 221. 1834; Wight, Ic. t. 986. 1845; Baker in Hook.f., Fl. Brit. India 2: 151. 1876; Gamble, Fl. Pres. Madras 330. 1918; Manilal, Fl. Silent Valley 87. 1988; Sanjappa, Leg. India 246. 1992; Mohanan & Sivad., Fl. Agasthyamala 221. 2002.

Tall erect woody herbs, densely clothed with bulbous based hairs; leaflets 3-4 (-5) pairs, oblong-obovate, 10-24 by 3-6 mm, base oblique, margin serrulate, apex obtuse, lower surface villous hairy. Panicle terminal, corymbiform; calyx c. 6 mm ϕ, upper lip truncate, emarginate, lower lip 3-lobed. Lomentum twisted, articles 4.

Fl. & Fr.: Oct.-Jan. *Distr.:* Peninsular & Eastern India. Rare. Along moist hill cuttings. *NAK 1337* (Kakki hills, c. 1100 m).

STYLOSANTHES O. Swartz
Prodr. 7: 108. 1788.

Stylosanthes fruticosa (Retz.) Alston in Trimen, Fl. Ceylon 6 (Suppl.) : 77. 1931; Nooteb., Reinwardtia 5:449. 1961; Verdc., Kew Bull. 24:59. 1970; Gandhi in Sald. & Nicols., Fl. Hassan Dist. 265. 1976; Matthew, Ill. Fl. Tam. Carnatic t. 224. 1982; Britto in Matthew, Fl. Tam. Carnatic 3 (1) : 463. 1983; Sanjappa, Leg. India 254. 1992; Mohanan & Sivad., Fl. Agasthyamala 222. 2002. *Arachis fruticosa* Retz., Obs. Bot. 5 : 26. 1788. *Stylosanthes mucronata* Willd., Sp. Pl. 3: 1166. 1802, nom. illegit., Wight & Arn., Prodr. 218. 1834; Bedd., Ic. t. 294. 1874; Baker in Hook.f., Fl. Brit. India 2:148. 1876; Gamble, Fl. Pres. Madras 326. 1918.

Woody trailing subshrubs. Leaves 3-foliolate, stipules prominent, adnate to petiole, sheathing; leaflets oblong-elliptic or elliptic, 1.5-3 by 0.4-0.7 cm, base obtuse, apex acute, mucronate, lateral veins prominent, lower surface densely pubescent. Flowers in dense terminal heads, covered by stipule and bracts; calyx-tube slender, long, upper 4 lobes connate; corolla yellow, petals inserted on calyx throat; stamens monadelphous, anthers dimorphous; ovary subsessile; ovules 2-3. Pods compressed, articles 1-2, reticulate, apex hooked. Seeds reniform.

Fl. & Fr.: Sep.-Jan. *Distr.:* Sri Lanka, India, Africa, Madagascar. Rare. Along exposed waysides in hills. *NAK 916* (Way to Plappally, c. 400 m).

TEPHROSIA Persoon
Syn. Pl. 2:328. 1807 (nom. cons.).

Woody herbs or undershrubs. Leaves odd-pinnate; leaflets opposite, closely parallel-nerved. Flowers in terminal or leaf-opposed racemens or solitary or in pairs; calyx-tube campanulate, lobes lanceate, upper

2 connate or free; corolla pink or reddish, petals clawed; stamens diadelphous, vexillary stamen becoming free; ovary sessile; ovules α. Pods linear, flattened. Seed α, reniform, elliptic or ovoid.

1. Flowers small; style glabrous. .... 2. **T. purpurea**
1. Flowers large; style pubescent.
   2. Peduncle equal or slightly exceeding leaflets; leaflets 3-5, rarely 1; pods densely tomentose. .... 1. **T. pulcherrima**
   2. Peduncle twice as long as leaflets; leaflets 7-9(-11); pods sparsely tomentose. .... 3. **T. tinctoria**

1. **Tephrosia pulcherrima** (Baker) Gamble, Fl. Pres. Madras 319.1918; Gandhi in Sald. & Nicols., Fl. Hassan Dist. 266. 1976; Britto in Matthew, Fl. Tam Carnatic 3(1) : 468. 1983; Sanjappa, Leg. India 257. 1992. *T. tinctoria* Pres., var. *pulcherrima* Wight ex Baker in Hook.f., Fl. Brit. India 2 : 112.1876.

Leaflets (1) 3-5-foliolate, elliptic, or oblong-lanceate, 1.5-7 by 1-2.5 cm, base otuse, apex obtuse or subacute, apiculate; lower surface silky-tomentose, thin-coriaceous. Flowers large, congested; calyx velvetty; corolla pinkish red or orange-red. Pods flattened, c. 6 cm long, densely golden tomentose. Seeds strophiolate.

Fl. & Fr.: Oct.-Jan. *Distr.*: Sri Lanka & Peninsular India. Common. Along moist exposed hills among bushes. *NAK 397, 1299* (Kakki hills, c. 1100 m).

2. **Tephrosia purpurea** (L.) Pres., Syn. Pl. 2: 329. 1807; Wight & Arn., Prodr. 213. 1834; Baker in Hook.f., Fl. Brit. India 2 : 112. 1876; Gamble, Fl. Pres. Madras 320.1918; Gandhi in Sald. & Nicols., Fl. Hassan Dist. 266. 1976; Britto in Matthew, Fl. Tam. Carnatic 3(1) : 470. 1983; Nicols., Suresh & Manilal, An Interpr. Hort. Malab. 143. 1988; Sanjappa, Leg. India 285. 1992. *Colingil* Rheede, Hort. Malab. 1 : 103-104, t. 55. 1678.

Leaflets 8-13(-18), narrowly obovate to oblanceate, 10-17 by 5-7 mm, base cuneate, apex obtuse to retuse, mucronate, thin-coriaceous. Racemes axillary or leaf-opposed, 4-8 cm long. Flowers small, α; corolla purplish; ovary linear. Pods slightly falcate, to 4 cm long, puberulous. Seeds ovoid, strophiolate.

Fl. & Fr.: Most of the seasons. *Distr.*: India, Myanmar, Malesia. Common along waysides. *NAK 1077* (Kodumon, c. 70 m), *1641* (Chittar, c. 150 m).

Root pound in milk and boil, applied to leprous wounds. Decoction of whole plant gargled for strengthening gum and teeth.

3. **Tephrosia tinctoria** (L.) Pres., Syn. 2 : 329.1807; Baker in Hook.f., Fl. Brit. India 2:111.1876; Gamble, Fl. Pres. Madras 319. 1918; Britto in Matthew, Fl. Tam. Carnatic 3(1) : 473. 1983; Manilal, Fl. Silent Valley 88. 1988 , Sanjappa, Leg. India 259. 1992; Mohanan & Sivad., Fl. Agasthyamala 222. 2002. *Cracea tinctoria* L., Sp. Pl. 752.1753. *Galega tinctoria* L., Syst. (ed. 10) : 1172. 1759.

Leaflets to 11, oblong-elliptic, 3-8 by 1-1.5 cm, base subacute, apex obtuse to rounded, mucronate, lower surface silky, coriaceous. Racemes to 6 cm long. Flowers large, few; calyx velvetty, corolla orange-red or pink-red. Pod flattened, c. 7.5 cm long; tomentose. Seeds α, elliptic.

Fl. & Fr.: Sep.-Jan. *Distr.*: Sri Lanka, Peninsular India. Rare. On moist hill slopes. *NAK 2141* (Chittar, c. 150 m).

TERAMNUS P. Browne
Cio. Nat. Hist. Jamaica 290. 1756.

Teramnus mollis Benth., J. Linn. Soc. Bot. 8 : 265. 1865; Gamble, Fl. Pres. Madras 353. 1918; Verdc., Kew Bull. 24 : 276. 1970; Gandhi in Sald. & Nicols., Fl. Hassan Dist. 266. 1976. *T. labialis* (L.f.) Spreng., var. *mollis* (Benth.) Baker in Hook.f., Fl. Brit. India 2 : 184.1876; Sanjappa, Leg. India 261. 1992.

Slender twiners. Leaves 3-foliolate; leaflets unequal, terminal one rhombi-lanceate, c. 9 by 4.5 cm, laterals oblong-lanceate, 7 by 3.5 cm, base rounded to subcordate, apex acute, lower surface densely greyish-temontose, thin-coriaceous. Racemes axillary, c. 6 cm long, solitary or 2-3, in clusters. Flowers lax c. 2 mm ϕ; calyx campanulate, lobes ± 4, upper 2 connate; corolla pink, little exerted; petals clawed; stamens 10; monadelphous, anthers alternatively fertile and sterile; ovary stipitate; style curved. Pods compressed, pubescent, apex hooked. Seeds oblong.

Fl. & Fr.: Nov.-Jan. *Distr.*: Tropics. Rare. On bushes in evergreen forests. *NAK 2153, 2283* (Gurunadhan manu, c. 400 m).

The fruits are employed in paralysis, rheumatism and afflictions of the nervous system.

URARIA Desvaux
J. Bot. Agric. 1 : 122. 1813.

Uraria rufescens (DC.) Schindler, Fedde's Repert. Spec. Nov. Regni Veg. 21 : 14. 1925; Van Meeuwen, Reinwardtia 5:453. 1961; Britto in Matthew, Fl. Tam. Carnatic 3 (1) : 476. 1983; Manilal, Fl. Silent Valley 88. 1988; Sanjappa, Leg. India 268. 1992; Mohanan & Sivad., Fl. Agasthyamala 223. 2002. *Desmodium rufescens* DC., Ann. Sci. Nat. (Paris) 4:101. 1825. *Uraria hamosa* (Roxb.) Sw. ex Wight & Arn., Prodr. 222. 1834; Wight, Ic. t. 284. 1840; Baker in Hook.f., Fl. Brit. India 2 : 156. 1876; Gamble, Fl. Pres. Madras 336. 1918.

Undershrubs; branches often straggling. Leaves 3-foliolate; stipules scarious, apex setaceous; leaflets unequal; terminal one elliptic to obovate, 5-9 by 4-4.5 cm, lateral elliptic-oblong, 3-5 by 1.5-2 cm, base rounded, ciliate, apex obtuse to subacute, mucronate, both surfaces pubescent, lateral nerves prominent, chartaceous. Inflorescence a panicle or raceme, terminal or axillary. Flowers 6 mm f; calyx 2-lipped (2+3); corolla pinkish-rose; stamens 9+1; anthers uniform; ovary shortly stipitate. Pods folded up and included within the calyx, sticky pubescent. Seeds orbicular.

Fl. & Fr.: Oct.-Dec. *Distr.*: Sri Lanka, India, S.E. Africa, China & Malesia. *NAK 190* (Thekkuthode, c.70 m), *226 & 1156* (Moozhiar, c. 250 m).

This plant is an ingredient of the dasamoola and used in various ayurvedic preparations.

VIGNA Savi
Nuov. Giorn. Lett. 8 : 113. 1824.

Climbing herbs. Leaves 3-foliolate; stipules peltate or basifixed. Infloresence of axillary racemes, often fascicled on rachis; bracts and bracteoles conspicuous, deciduous; calyx campanulate, 5-lobed, 2-lipped; corolla exerted, keels beaked, often spirally twisted; stamens 9+1, vexillary one free, anthers uniform; ovary sessile; ovules a; style usually incurved, hirsute towards apex. Pods cylindric or flattened, straight or curved. Seeds usually reniform.

1. Glabrescent climbers; leaves to 5 by 4 cm. .... 1. **V. dalzelliana**
1. Hirsute climbers; leaves to 11 by 6 cm.
   2. Leaflets ovate-lanceate, apex acute; stipules linear, minute;flowers pinkish-red. 2. **V. pilosa**
   2. Leaflets broadly ovate, apex acuminate; stipules lanceate, prominent; flowers yellow. .... 3. **V. umbellata**

1. **Vigna dalzelliana** (Kuntze) Verdc., Kew Bull. 24 : 558. 1970. Sanjappa, Leg. India 272. 1992. *Phaseolus dalzellianus* Kuntze, Rev. Gen. Pl. 1 : 202. 1891. *P. pauciflorus* Dalz. in Hookers' J. Bot. Kew Gard. Misc. 3 : 209. 1851, non G. Don, 1832, nec Benth., 1837; Baker in Hook.f., Fl. Brit. India 2: 202. 1876. *P. dalzelli;* Cooke, Fl. Pres. Madras 363. 1918.

Leaflets ovate-lanceate, 3-5 by 2.5-3.5 cm, base cuneate or truncate, apex acuminate, membranous. Racemes few-flowered; bracteoles large; calyx teeth obscure; corolla yellow. Pod narrow-cylindrical, to 5 cm long. Seeds 6-10.

Fl.& Fr.: Dec.-Mar. *Distr.*: Western Ghats. Rare. On bushes in evergreen forests. *NAK 1339* (Pampa dam area, c. 1100 m).

2. **Vigna pilosa** Baker in Hook.f., Fl. Brit. India 2 : 207. 1876; Gamble, Fl. Pres. Madras 365. 1918; Sanjappa, Leg. India 275. 1992.

Leaflets ovate-lanceate, 8-12 by 4-5.5 cm, base subcordate, apex gradually acute-apiculate, pubescent, basally 3-nerved, chartaceous. Racemes short-peduncled. Flowers to 15 mm long; calyx-lobes unequal; corolla pinkish-red, exerted. Pods straight or slightly curved; cylindric, densely pilose, apex beaked. Seeds 8-12, reniform.

Fl. & Fr.: Sep.-Dec. *Distr.*: Peninsular & East India. Rare. In moist semi-evergreen forests. *NAK 185* (Thekkuthode, c. 70 m).

3. **Vigna umbellata** (Thunb.) Ohwi & Ohashi, Jap. J. Bot. 44. : 31. 1969; Verdc., Kew Bull. 24 : 960. 1970; Gandhi in Sald. & Nicols., Fl. Hassan Dist. 267. 1976; Sanjappa, Leg. India 276. 1992; Mohanan & Sivad., Fl. Agasthyamala 224. 2002. *Dolichos umbellatus* Thunb., Trans. Linn. Soc. London 2:339. 1794. *Phaselous calcaratus* Roxb., Fl.Ind. 3 : 289. 1832, Baker in Hook.f., Fl. Brit. India 2 :203. 1876; Gamble, Fl. Pres. Madras 363. 1918.

Leaflets broadly ovate, terminal one rhombic-ovate, 4-9 by 2.5-6 cm, base rounded, apex acuminate, pubescent, membranous. Racemes to 10 cm long; bracteoles linear. Flowers 10-20; calyx-lobes short, deltoid; corolla yellow. Pods cylindric. Seeds 10-12.

Fl. & Fr.: Sep.-Dec. *Distr.*: Sri Lanka, India, Malaya Peninsula. In plains and rarely in hills on bushes. *NAK 162, 358, 2155.* (Mundanpara, c. 200 m).

ZORNIA J.F. Gmelin
Syst. Nat. 2 : 1076, 1096. 1792.

Diffuse herbs. Leaves 2-foliolate; leaflets ovate or lanceate, gland-dotted; stipules foliaceous, peltate. Racemes terminal or axillary, interrupted; bracts foliaceous, enclosing a flower; calyx-tube campanulate, hyaline, 5-toothed; corolla yellow, exerted; petals clawed; stamens monadelphous, filaments alternately long and short, anthers dimorphic; ovary sessile; ovules a. Pods compressed, articles glochidiate-prickled. Seeds subreniform.

1. Leaves anisophyllous, lower oblong-ovate; upper linear-lanceate; lomentum scarcely exerted. .... 1. **Z. gibbosa**
1. Leaves isophyllous, obliquely oblong-ovate; lomentum clearly exerted. ....2. **Z. quilonensis**

1. **Zornia gibbosa** Spanoghe, Linnaea 15 : 192. 1841; Mohl., Webbia16 : 112, ff. 44 & 76. 1961; Wagh, J. Bombay Nat. Hist. Soc. 61 :214. 1964; Gandhi in Sald. & Nicols., Fl. Hassan Dist. 268. 1976; Ravi, Bull. Bot. Surv. India 21(1-4) : 199. 1979; Nicols., Suresh & Manilal, An Interpr. Hort. Malab. 146.1988. Sanjappa, Leg. India 279. 1992. *Z. diphylla* auct., non L., 1753: Gamble, Fl. Pres. Madras 325. 1918. *Nelam-mari* Rheede, Hort. Malab. 9 :161, t. 82, 1689.

Leaves dimorphic, lower leaves obliquely oblong-ovate, upper ones narrowly lanceate, 1.5-3 by 0.3-0.4 cm, base obtuse, apex acute, apiculate, chartaceous. Racemes 3-12-flowered; Lomentum scarcely exerted from bracts; articles 2-5, pubescent with retrorsely scabrid small prickles.

Fl. & Fr.: Sep-Feb. *Distr.*: India, China, Malesia, Australia. Common. On exposed hills and in fallow fields in plains. *NAK 132* (Adoor, c. 40 m), *2215* (Pathanamthitta town, c. 50 m).

2. **Zornia quilonensis** Ravi, J. Bombay Nat. Hist. Soc. 66 (3) : 489. 1970.

Leaves uniform, obliquely oblong-ovate, 1-1.5 by 0.4-0.8 cm, base oblique, apex acute, mucronate, chartaceous. Racemes 3-8- flowered. Lomentum clearly exerted from bracts; articles 2-6, densely puberulous with retrorsely scabrid long prickles.

Fl. & Fr.: Oct.-Jan. *Distr.*: So far reported from different parts of Kerala State only. Rare. *NAK 1995* (Tiruvalla, c. 10 m).

Gliricidia sepium (Jacq.) Kunth ex Walp. is extensively planted in the district as a source of green manure.

51-B. CAESALPINIOIDEAE

1. Leaves deeply 2-lobed; calyx-lobes connate. .... **Bauhinia**
1. Leaves pinnately a-foliolate; calyx-lobes free.
    2. Leaves 2-pinnate; armed stragglers. .... **Caesalpinia**
    2. Leaves 1-pinnate.
        3. Stamens 3, connate. .... **Tamarindus**
        3. Stamens 5-10, free.
            4. Petals 0; calyx-lobes 4. .... **Saraca**
            4. Petals 3-5; calyx-lobes 5.
                5. Stamens 10, anisomorphic, dehiscence poricidal; flowers usually yellow. .... **Cassia**
                5. Stamens 5, isomorphic (alternating with 5 minute staminodes), dehiscence longitudinal; flowers never yellow. .... **Humboldtia**

BAUHINIA Linnaeus
Sp. Pl. 374. 1753.

Bauhinia phoenicea Heyne ex Wight & Arn., Prodr. 296. 1834; Baker in Hook.f., Fl. Brit. India 2 : 283. 1878; Gamble, Fl. Pres. Madras 408. 1919; Gandhi in Sald. & Nicols., Fl. Hassan Dist. 217. 1976; Manilal, Fl. Silent Valley 89. 1988; Sanjappa, Leg. India 4. 1992; Mohanan & Sivad., Fl. Agasthyamala 228. 2002.

Scandent shrubs. Leaves deeply 2-lobed at apex, broadly ovate to suborbicular, 8-15 by 8-13 cm, base truncate cordate, apex obtuse, lower surface thinly pubescent, basally 8-nerved, thin-coriaceous. Panicle axillary and terminal, c. 20 cm long, finely brownish tomentose. Flowers c. 7 cm ϕ; calyx-tube cylindric, lobes 5, spathaceous; corolla reddish, petals long-clawed; stamens 4-5; ovary pubescent. Pods flat, oblanceate, c. 16 cm long, beaked, reddish-tomentose. Seeds compressed, smooth.

Fl. & Fr.: Nov.-Feb. *Distr.*: Western Ghats. Locally abundant. Along forest borders in evergreen types. *NAK 293 & 118* (Moozhiar, c. 250 m).

CAESALPINIA Linnaeus
Sp. Pl. 380. 1753.

Armed stragglers. Leaves 2-pinnate, leaflets opposite, a. Inflorescence a raceme or panicle, axillary or terminal; calyx-tube short, lobes 5, unequal; petals 5, unequal, spreading, clawed; stamens 10, free, subequal; ovary subsessile. Pods oblong, flat or turgid. Seeds ovoid.

1. Pinnae 2-8 pairs; leaflets ovate-elliptic, large; pod flat, winged on upper suture. .... 1. **C. cucullata**
1. Pinnae 10-20 pairs; leaflets oblong, small; pods turgid, not winged on upper suture. .... 2. **C. mimosoides**

1. **Caesalpinia cucullata** Roxb., Fl. Ind. 2 : 358. 1832; Hattink, Reinwardtia 9 : 22. 1974; Britto in Matthew, Fl. Tam. Carnatic 3(1) : 490. 1983; Manilal, Fl. Silent Valley 90. 1988. Sanjappa, Leg. India 10. 1992. *Mezoneuron cucullatum* (Roxb.) Wight & Arn., Prodr. 283. 1834; Baker in Hook.f., Fl. Brit. India 2 : 258. 1878; Gamble, Fl. Pres. Madras 395. 1919, "Mezoneurum". *Kaka-kalingi-valli.*

Branches with hard black prickles; pinnae 2-3 pairs; leaflets usually 3-pairs, ovate to elliptic-ovate, 4-8 by 2.5-3.5 cm, base obtuse, margin subentire, apex acuminate, mucronulate, shiny, thin-coriaceous. Panicle large. Flowers c. 15 mm ϕ; corolla yellow, petals orbicular; stamens exerted; ovules usually 2; style filiform; stigma capitate. Pod flat, papery, broadly winged along upper margin, 10 by 3 cm. Seed(s) 1 or 2, orbicular.

Fl. & Fr.: Jan.-Mar. *Distr.:* India to Malesia. Locally abundant but not common. In evergreen and semi-evergreen forests, usually along streamsides. *NAK 2366* (Kalleli R.F., c. 250 m).

2. **Caesalpinia mimosoides** Lam., Encycl. 1:452. 1785, Wight, Ic. t. 392. 1840; Baker in Hook.f., Fl. Brit. India 2:256. 1878; Gamble, Fl. Pres. Madras 394. 1919; Gandhi in Sald. & Nicols., Fl. Hassan Dist. 220.1976; Britto in Matthew, Fl. Tam. Carnatic 3 (1) : 492. 1983; Manilal, Fl. Silent Valley 90, 1988; Nicols., Suresh & Manilal, An Interpr. Hort. Malab. 126:1988; Sanjappa, Leg. India 12. 1992. *Kal-toddavaddi* Rheede, Hort. Malab. 6:15-16, t.8. 1686.

Branches with reddish glandular hairs and dense stiff prickles. Pinnae c. 20 pairs; leaflets to 15 pairs, sessile, oblong, 10 by 3 mm, base and apex obtuse; membranous. Racemes to 45 cm long, rachis and peduncle prickled. Flowers c. 2.5 cm f; corolla bright yellow; stamens subexerted; ovules c. 4; style slender; stigma simple. Pod turgid, 4 cm long, minutely stiff-bristly. Seeds 2, narrowly oblong.

Fl. & Fr.: Sep.-Jan. *Distr.*: Sri Lanka, Peninsular India. Rare. Massive and dense stragglers (with large beautiful inflorescences) forming impenetrable thickets along forest borders. *NAK 112* (Adoor, c. 40 m).

Root is used for purging, indigestion and rheumatism of joints.

CASSIA Linnaeus
Sp.Pl. 376. 1753.

Trees, shrubs or herbs. Leaves 2-pinnate; leaflets subopposite, often glandular on rachis and/or petiole. Inflorescence racemes or panicles, axillary or terminal. Flowers usually yellow; calyx-tube short, lobes 5; petals 5, clawed, subequal; stamens usualy 10, often 7 fertile and other sterile or all fertile of various sizes or rarely only 5 fertile, anthers dehiscing by apical pores; ovary subsessile or stipitate; ovules a; style incurved. Pod terete or flat, septate between seeds. Seeds usually compressed.

1. Herbs, subshrubs or shrubs.
    2. Leaflets 20-40 pairs, up to 5 mm wide. ....7. **C. mimosoides**
    2. Leaflets less than 20 pairs, 10 mm or more wide.
        3. Viscous herbs; flowers reddish-yellow; leaflets 2 pairs. ....1. **C. absus**
        3. Non-viscous herbs; flowers yellow; leaflets more than 2 pairs.
            4. Leaflets 3-pairs; only certain stamens fertile; erect stout herbs. ....11. **C. tora**
            4. Leaflets 8-10 pairs; all stamens fertile; trailing herbs. ....5. **C. kleinii**
1. Trees or shrubs.
    5. Foliar glands present.
        6. Foliar glands on rachis; leaflets c. 3 pairs. ....11. **C. tora**
        6. Foliar glands on petiole; leaflets more than 3 pairs.
            7. Leaves glabrous.
                8. Leaflet 4-5 pairs, ovate, 2-7 cm long; glands globose; pod compressed. ....8. **C. occidentalis**
                8. Leaflets 5-10 pairs, oblong-lanceate, to 4 cm long; glands clavoid; pod turgid. ....10. **C. sophera**
            7. Leaves hirsute.
                9. Leaflets c. 4. pairs, ovate-lanceate; pod slender, curved, densely hirsute, to 14 cm long. ....4. **C. hirsuta**
                9. Leaflets often less than 4 pairs, elliptic-lanceate; pod stout, straight, sparsely pubescent, to 6 cm long. ....5. **C. intermedia**
    5. Foliar glands absent.
        10. Stipules broad, persistent; racemes spicate; leaflets oblong

to 8 cm f wide. ....2. **C. alata**

10. Stipules linear, deciduous; racemes not spicate.

11. Leaflets c. 5 pairs, ovate-lanceate or elliptic; inflorescene a dropping raceme. ....3. **C. fistula**

11. Leaflets 10 or more pairs, elliptic-oblong; inflorescence an erect corymbose-panicle. ....9. **C. siamea**

1. **Cassia absus** L., Sp. Pl. 378. 1753; Baker in Hook. f., Fl. Brit. India 2:265. 1878; Gamble, Fl. Pres. Madras 403. 1919; de Wit, Webbia 11 : 279 .1955; Matthew, Ill. Fl. Tam. Carnatic t. 231. 1982; Britto in Matthew, Fl. Tam. Carnatic 3(1): 496.1983; Sanjappa, Leg. India 14. 1992.

Stout herbs or subshrubs; branchlets and leaves viscous pubescent; glands on rachis; leaflets 2 pairs, broadly elliptic-ovate, 1.5-3 by 1-2 cm, base very oblique, apex subacute, chartaeous. Racemes leaf-opposed. Flowers c. 15 mm ϕ; calyx-lobes ovate, petals reddish to reddish-yellow; stamens 5, fertile; ovary densely hairy. Pods c. 3.5 cm long, flat, viscous. Seeds ovoid.

Fl. & Fr.: Jun-Jul. *Distr.*: Sri Lanka, Peninsular India, Himalayas. Common. On exposed waste lands. *NAK 2575* (Tiruvalla, c. 10 m).

Seeds powdered and applied beneath the eye-lids in purulent ophthalmia, but not in catarrhal ophthalmia.

2. **Cassia alata** L., Sp. Pl. 378. 1753; Wight & Arn., Prodr. 287. 1834; Wight, Ic.t. 253. 1840; Baker in Hook.f., Fl. Brit. India 2 : 264. 1878; Gamble, Fl. Pres. Madras 404. 1919; de Wit, Webbia 11: 231.1955; Britto in Matthew, Fl. Tam. Carnatic 3 (1): 497. 1983; Sanjappa, Leg. India 14. 1992. *Puzhukkadi-conna.*

Shrubs or small trees. Leaves 30-40 cm long; stipule foliaceous 7 by 6 mm, persistent; leaflets 10 pairs, oblong to slightly obovate, 4-14 by 2.5-7 cm, base obliquely truncate, apex obtuse, mucronulate, chartaceous. Racemes spiciform, dense, to 30 cm long. Flowers c. 2 cm ϕ; calyx-lobes oblong; petals yellow; fertile stamens 7; staminodes 3; ovary angular. Pods oblong, thin, winged on margins, to 10 cm long. Seeds α.

Fl. & Fr.: Sep.-Jan. *Distr.*: Pantropical, originally from S. America. Along streamsides and in waste places. *NAK 1050* (Tiruvalla, c. 10 m).

Leaves used as a remedy for ring worms, and decoction given for veneral afflictions and bites of poisonous insects.

3. **Cassia fistula** L., Sp. Pl. 377. 1753; Baker in Hook.f., Fl.Brit.India 2 : 261. 1878; Gamble, Fl. Pres. Madras 400. 1919; de Wit, Webbia 11:207. 1956; Gandhi in Sald. & Nicols., Fl. Hassan Dist. 219. 1976; Britto in Matthew, Fl. Tam. Carnatic 3 (1) : 498. 1983; Manilal, Fl. Silent Valley 91. 1988; Nicols., Suresh & Manilal, An Interpr. Hort. Malab. 129. 1988; Sanjappa, Leg. India 15. 1992; Mohanan & Sivad., Fl. Agasthyamala 229. 2002. *C. rhombifolia* Roxb., Fl. Ind. 3 : 334. 1832; Wight Ic. t. 269. 1840. *Conna* Rheede, Hort. Malab. 1: 37-38, t. 22. 1678. *Kani-Conna.*

Trees. Leaves 15-40 cm long; leaflets 4-6 pairs, ovate-lanceate, 4-10 by 2-5 cm, base rounded, apex acute,thin-coriaceous. Racemes large, lax, pendulous. Flowers to 4 cm ϕ; calyx-lobes ovate, reflexed; petals yellow; stamens all antheriferous; ovary appressed pubescent. Pods stipitate, cylindric, 40-60 cm long, transversely septate. Seeds α, lenticular.

Fl. & Fr.: Apr.-Jul. *Distr.*: Sri Lanka, India & Myanmar. Common in semi-evergreen and moist deciduous forests and planted as ornamental trees. *NAK 1672* (Arampa, c. 300 m).

Pulp of ripe fruit eaten to cures rheumatism, fever, leprosy, and improves vitality. Decoction of flowers used in stomach affection. Paste of leaves applied to ring worm infections.

4. **Cassia hirsuta** L., Sp. Pl. 378. 1753; Baker in Hook.f., Fl. Brit. India 2 : 263. 1878; de Wit, Webbia 11 : 250. 1956; Gamble, Fl. Pres. Madras 401. 1919; Gandhi in Sald. & Nicols., Fl. Hassan Dist. 220. 1976; Britto in Matthew, Fl. Tam. Carnatic 3(1):501. 1983; Sanjappa, Leg. India 15. 1992; Mohanan & Sivad., Fl. Agasthyamala 230. 2002.

Shrubs; branchlets hirsute. Leaves c. 20 cm long; Leaflets 3-4 pairs, ovate to ovate-lanceate, 2-7 by 1-3.5 cm, base obtuse or rounded, apex acuminate-apiculate, chartaceous. Racemes corymbose, axillary. Flowers c. 2 cm ϕ, sepals broadly ovate; petals yellow; fertile stamens 7; staminodes 3. Pods sessile, flat, c. 14 cm long, subfalcate, densely hirsute. Seeds α.

Fl. & Fr.: Sep-Jan. *Distr.*: Native of tropical America. Now widely distributed in Old World tropics. Along waysides and waste places in plains. *NAK 146* (Adoor, c. 40m), *1043* (Tiruvalla, c. 10 m).

5. **Cassia intermedia** Sharma, Vivek & Rath., Proc. Ind. Acad. Sci. 80 : 301-306, f. 16. 1974; Sanjappa, Leg. India 16. 1992.

Shrubs; branches zig-zag. Leaves to 10 cm long; leaflets 3-4 pairs, elliptic, 1-4 by 0.7-1.5 cm, base obliquely rounded, apex acute, apiculate, both surface tomentose, chartaceous. Racemes terminal and/or from upper leaf-axils. Flowers c. 2 cm ϕ; fertile stamens 7; staminodes 3; ovary sessile. Pods flat, subsessile, straight, to 6 cm long, margins thickened, sparsely pubescent.

Fl. & Fr.: Nov.-Jan. *Distr.*: Hitherto known from Kerala and Tamil Nadu States. Rare. Among wayside thickets in hills. *NAK* 1290 (Kakki hills, c.1000 m).

6. **Cassia kleinii** Wight & Arn., Prodr. 293. 1834; Baker in Hook.f., Fl.Brit. India 2: 266. 1878; Gamble, Fl. Pres. Madras 403. 1919; Britto in Matthew, Fl. Tam. Carnatic 3(1): 501. 1988; Nicols., Suresh & Manilal, An Interpr. Hort. Malab. 130. 1988; Sanjappa, Leg. India 17. 1992; Mohanan & Sivad., Fl. Agasthyamala 230. 2002. *Malamtodda-vadi* Rheede, Hort. Malab. 9:37, t.21. 1689.

Trailing herbs. Leaves to 2 cm; leaflets 8-10 pairs, oblong, subfalcate, 4-8 by 2-3 mm, base obtuse, apex acute, apiculate, chartaceous; gland(s) solitary, peltate. Racemes axillary, 2-flowered. Flowers 1 cm ϕ; calyx-lobes ovate, unequal; petals yellow; stamens 10, all antheriferous, ovary sessile. Pods flat, to 3 cm long; downy tomentose. Seeds flat, ovate.

Fl. & Fr.: Nov.-Mar. *Distr.*: Sri Lanka, Peninsular India. Rare. Among forest undergrowths. *NAK 2652* (Appuppanthode, c. 400 m).

7. **Cassia mimosoides** L., Sp. Pl. 379. 1753; Baker in Hook.f., Fl. Brit. India 2 :266. 1878; Gamble, Fl. Pres. Madras 403. 1919; de Wit, Webbia 11:283. 1955; Gandhi in Sald. & Nicols., Fl.Hassan Dist. 220. 1976; Britto in Matthew, Fl. Tam. Carnatic 3 (1): 502. 1983; Sanjappa, Leg. India 17. 1992; Manilal, Fl. Silent Valley 92. 1988; Mohanan & Sivad., Fl. Agasthyamala 231. 2002.

Diffuse undershrubs. Leaves 7-9 cm long; leaflets 40-60 pairs, obliquely oblong, 8 by 3 mm, base oblique, margin ciliolate, apex subacute, apiculate, chartaceous. Flower(s) solitary, supra-axillary, c. 1 cm ϕ; calyx-lobes lanceate, hairy; petals yellow, ovate; stamens 10, all antheriferous, unequal. Pods flat, stipitate, 4 by 0.5 cm, hairy. Seeds ellipsoid.

Fl. & Fr.: Sep.-Dec. *Distr.*: Old World Tropics. Common in exposed hilly areas. *NAK 62* (Moozhiar, c. 250 m), *914* (Plapally, c. 400 m).

Powdered roots given for spasms in the stomach.

8. **Cassia occidentalis** L., Sp. Pl. 377. 1753; Wight & Arn., Prodr. 290. 1834; Baker in Hook.f., Fl. Brit. India 2:262. 1878; Gamble, Fl. Pres. Madras 401. 1919; de Wit, Webbia 11 : 256. 1955; Gandhi in Sald. & Nicols., Fl. Hassan Dist. 220. 1976; Britto in Matthew, Fl. Tam. Carnatic 3(1) : 505. 1983. Sanjappa, Leg. India 19. 1992. *Senna occidentalis* Roxb., Fl. Ind. 2:343. 1832.

Undershrubs. Leaves to 15 cm long; leaflets 4-5 pairs, elliptic, 2-5 by 1-2 cm, base oblique, margin ciliate, apex acuminate, chartaceous. Racemes terminal and axillary. Flowers 2 cm ϕ; clayx-lobes ovate; petals yellow, oblong-obovate; fertile stamens 7, 3 reduced to staminodes. Pods compressed, c. 10 cm long, shortly stipitate. Seeds ovoid to obovoid, laterally compressed.

Fl. & Fr. : Aug.-Nov. *Distr.*: Pantropical. A common weed of waste places. *NAK 232* (Adoor, c. 20 m).

Leaves are used in foot and mouth disease of cattle. Decoction or powder of root given in honey for leprosy, itches and worms in the intestine.

9. **Cassia siamea** Lam., Encycl. 1:648. 1785; Baker in Hook.f., Fl.Brit. Inda 2 : 264. 1878; Gamble, Fl. Pres. Madras 402. 1919; de Wit, Webbia 11 : 263. 1955; Gandhi in Sald. & Nicols., Fl. Hassan Dist. 221. 1976; Britto in Matthew, Fl. Tam. Carnatic 3 (1) : 508. 1983. Sanjappa, Leg. India 20. 1992. *C. florrida* Vahl, Symb. Bot. 3:57. 1794; Wight & Arn., Prodr. 288. 1834.

Small to medium sized trees. Leaves to 20 cm long; leaflets to 14 pairs, oblong-elliptic, 1.5-3 by 0.7-1.1 cm, base broadly cuneate, margin thickened, apex acute, mucronate, both surfaces appressed hairy, chartaceous; rachis and petiole glandular. Panicle terminal and from upper leaf-axils; calyx-lobes unequal, corolla yellow; stamens 7 antheriferous, 3 become staminodes; ovary sessile, glabrous. Pods oblong, flat, shortly stipitate.

Fl. & Fr.: Oct.-Mar. *Distr.*: South E. Tropical Asia. Widely cultivated as avenue trees. Occasionally seen as wild. *NAK 2603* (Angamoozhy, c. 400 m).

10. **Cassia sophera** L., Sp. Pl. 379. 1753; Baker in Hook.f., Fl. Brit. India 2 : 262.1878; Gamble, Fl. Pres. Madras 401. 1919; Nicols., Suresh & Manilal, An Interpr. Hort. Malab. 130. 1988. Sanjappa, Leg. India 21. 1992. *Punnam-tagera* Rheede, Hort. Malab. 2:101-102, t. 52. 1679.

Under shrubs. Leaves c. 15 cm long; leaflets 5-8 pairs, lanceate to oblong-lanceate, 2-4 by 1-1.3 cm, base oblique, apex acute-apiculate, chartaceous. Racemes umbelliform. Flowers 15 mm ϕ; calyx-lobes ovate; petals yellow; stamens 7, antheriferous, 3 become staminodes. Pods turgid, 8 by 1 cm, septate. Seeds laterally compressed.

Fl. & Fr.: Nov.-Feb. *Distr.*: Sri Lanka, India. Along waysides in hills. *NAK 2101* (Vallicode, c. 150 m).

Juice of bark used in medicine for asthma. Seed-oil used in medicine.

11. **Cassia tora** L., Sp. Pl. 376. 1753; Wight & Arn., Prodr. 240. 1834; Baker in Hook.f., Fl. Brit. India 2:263. 1878, p.p.; Gamble; Fl. Pres. Madras 401. 1919; de Wit, Webbia 11:276. 1955; Brenan, Kew Bull. 13: 248. 1959; Gandhi in Sald. & Nicols., Fl. Hassan Dist. 221. 1976; Britto in Matthew, Fl. Tam. Carnatic 3 (1) : 510. 1983; Nicols., Suresh & Manilal, An Interpr. Hort. Malab. 130. 1988. Sanjappa, Leg. India 30. 1992. *Tagera* Rheede, Hort. Malab. 2:103, t. 53. 1679.

Undershrubs. Leaves c. 4 cm long; rachis with 2 glands at the base of lower leaflets; leaflets 3-pairs, obovate, 1-3 by 0.7-2 cm, base obliquely cuneate, apex obtuse-apiculate, chartaceous. Racemes corymbose, terminal. Flowers c. 15 mm ϕ; calyx-lobes ovate, petals golden-yellow, obovate; fertile stamens 7, staminodes 3, ovary subsessile. Pods flat, c. 10 cm long, shortly stalked. Seeds oblong.

Fl. & Fr.: Oct.-Jan. *Distr.*: India to Polynesia. Common. Forming dense population along waysides especially in waste places. *NAK 942* (Plappally, c. 400 m).

Root rubed into a paste with lime juice and applied in ring worm infections. Leaves and seeds used to cure skin diseases.

HUMBOLDTIA M. Vahl
Symb. Bot. 3 : 106. 1794 (nom. cons.).

Humboldtia vahliana Wight, Ic. tt. 1607 & 1608. 1850; Baker in Hook.f., Fl. Brit. India 2 : 272. 1878; Gamble, Fl. Pres. Madras. 411. 1919; Sanjappa, Leg. India 30. 1992; Mohanan & Sivad., Fl. Agasthyamala 234. 2002.

Trees. Leaves even-pinnate, 15-30 cm long; stipules foliaceous, appendaged, deciduous; rachis flat, slightly winged or wingless; leaflets 3-4 pairs, lanceate to oblong-lanceate, 6-20 by 2.5-5.5 cm, base obtuse-cuneate, apex acuminate, reticulations prominent, coriaceous. Racemes corymbiform, axillary and cauline; bracts foliaceous, persistent. Flowers c. 2.5 cm ϕ; calyx-tube cylindric, lobes 4, broadly elliptic, unequal; petals 4, white, clawed, elliptic-ovate; stamens 5, alternating with 5 staminodes, filaments pink coloured; ovary stipitate. Pods oblanceate, 17 by 5 cm, brownish-velvetty, stipitate, apex beaked. Seeds discoid.

Fl. & Fr.: Mar.-Jun. *Distr.*: Western Ghats. Locally abundant. Common along streamsides. *NAK 622, 1119, 1484 & 1556*. (Ranni, Konni, etc.).

A tree with highly variable leaves.

Powder or decoction of bark given to cure biliousness, impure blood, leprosy, ulcers and epilepsy.

SARACA Linnaeus
Syst. Nat. (ed. 12) 2: 469. 1767.

Saraca asoca (Roxb.) de Wilde, Blumea 15 : 393. 1968; Zuijderhoudt, Blumea 15:422. 1968; Gandhi in Sald. & Nicols., Fl. Hassan Dist. 224. 1976; Nicols., Suresh & Manilal, An Interpr. Hort. Malab. 142. 1988; Sanjappa, Leg. India 35. 1992. *Jonesia asoca* Roxb., Asiat. Res. 4: 365. 1795; Wight, Ic. t. 206. 1839. *Saraca indica* auct., non. L. 1767: Bedd., Fl. Sylv. t. 57. 1870; Baker in Hook.f., Fl.Brit. India 2 : 271. 1878; Gamble, Fl. Pres. Madras 409. 1919. *Asjogam* Rheede, Hort. Malab. 5 : 117, t. 53. 1679.

Trees. Leaves even-pinnate, c. 35 cm long; stipules intrapetiolar, deciduous; leaflets c. 4 pairs, opposite, oblong-elliptic to oblong-lanceate, 8-22 by 2-6 cm, base cuneate, apex acuminate, thin-coriaceous. Panicle corymbose, axillary, c. 10 cm long; bracteoles 2, broadly ovate, embracing the calyx-tube, larger than bract; calyx-lobes 4, petaloid, orange-red, obovate; petals 0; stamens 3-8, free; filaments very slender; anther versatile; ovary stipitate. Pods flat, oblong, woody. Seeds compressed, suboricular.

Fl. & Fr.: Jan.-Mar.*Distr.*: Indo-Malesia. Very rare in wild. In semi evergreen forests. *NAK 2258* (Arampa, c. 500 m). One of the most sacred tree of India.

Powder or decoction of bark given in biliousness, worm infections, ulcers, poisons, pimples, weakness, bowel complaints, haemorrhage and dropsy.

TAMARINDUS Linaeus
Sp. Pl. 34. 1753.

Tamarindus indica L., Sp. Pl. 34. 1753; Wight & Arn., Prodr. 285. 1834; Bedd., Fl. Sylv. t. 184. 1872; Baker in Hook.f., Fl. Brit. India 2 : 273. 1878; Gamble, Fl.Pres. Madras 409. 1919; Gandhi in Sald. & Nicols., Fl. Hassan Dist. 224. 1976; Purseglove, Trop. Crops, Dicots 204. 1977; Anon., (U.S.A.) Trop. Legum.: Resources for future 117. 1979; Britto in Matthew, Fl. Tam. Carnatic 3 (1) : 516. 1983; Nicols., Suresh & Manilal, An Interpr. Hort. Malab. 143. 1988. Sanjappa, Leg. India 36. 1992. *Balam-pulli, Madiram-pulli* Rheede, Hort. Malab. 1:39-40, t. 23. 1678.

Trees. Leaves even-pinnate; leaflets 15-17 pairs, narrowly oblong, 15 by 7 mm, base and apex obtuse, chartaceous. Racemes terminal; bracts and bracteoles coloured. Flowers c. 1 cm ϕ; calyx-tube narrowly turbinate, lobes 4; petals 3, yellow with pink dots; lower pairs scaly; stamens 3, monadelphous; staminodes 2, bristly; ovary stipitate. Pods oblong, subcompressed, mesocarp pulpy, indehiscent. Seeds obovoid-orbicular, compressed, black, shiny.

Fl. & Fr.: Aug.-Dec. *Distr.*; Tropics. Planted for the edible fruits; ocassionally seen as wild in plains. *NAK 132* (Adoor, c. 20 m).

Fruits used in cooking. Pulp of the ripe fruit as well as poultice of the leaves is recommended for application to relieve inflammatory swellings. Pounded seeds cures ulcers and stone in urinary bladder.

Bauhinia pupurea L., **Delonix regia** (Hook). Raf. and **Peltophorum pterocarpum** (DC.) Backer & Heyne are seen in the district planted as avenue trees.

51-C. MIMOSOIDEAE

1. Plants armed.
    2. Flowers 4-merous; stamens 4-8; pods bristly or spinous on margins. .... **Mimosa**
    2. Flowers 5-merous; stamens α; pods glabrous on margins. .... **Acacia**
1. Plants unarmed.
    3. Climbers; pod a lomentum to 1 m long. .... **Entada**
    3. Trees; pod not a lomentum, to 15 cm long.
        4. Stamens 10; pods falcate, woody, anthers gladular at apex. .... **Xylia**
        4. Stamens more than 10; pod straight or spiraly coiled, thin; anthers not glandular at apex.
            5. Pods straight, thin-valvular. .... **Albizia**
            5. Pods spirally coiled; thick-valvular. .... **Archidendron**

ACACIA P. Miller
Gard. Dict. Abr. ed. 4. 1754.

Armed scandent shrubs. Leaves alternate, 2-pinnate, basal pinnae smaller, progressively larger towards apex; leaflets even-pinnate, opposite, often foliar glands present on the rachis or petiole. Inflorescence of terminal and/or axillary globose heads or spikes. Flowers 5-merous, bisexual; calyx-tube 4-5- lobed; petals exerted, connate at base; stamens α, shortly connate at base; ovary stipitate or subsessile. Pods oblong, flat, rarely curved. Seeds ovoid or compressed.

1. Pinnae 6 or 7 pairs; leaflets 13-15 pairs, c.3 mm or more wide. .... **1. A. caesia**
1. Pinnae 10-40 pairs; leaflets 25-60 pairs, less than 3 mm wide.
   2. Petiole-gland conical, leaflets touching together but not overlapping; flowers sessile. .... **3. A. torta**
   2. Petiole-gland not conical, leaflets touching together and overlapping; flowers pedicellate. .... **2. A. pennata**

1. **Acacia caesia** (L.) Willd., Sp. Pl. 4 : 1090. 1806; Gamble, Fl. Pres. Madras. 428. 1919; Kosterm. in Dassan. & Fosb., Rev. Handb. Fl. Ceylon 1: 481. 1980; Britto in Matthew, Fl. Tam. Carnatic 3(1): 522. 1983; Manilal, Fl. Silent Valley 92. 1988; Nicols., Suresh & Manilal, An Interpr. Hort. Malab. 120. 1988. Sanjappa, Leg. India 37. 1992. *Mimosa caesia* L., Sp. Pl. 522. 1753, p.p. *Acacia intsia* var. *casesia* Baker in Hook.f., Fl. Brit. India 2 : 297. 1878. *Instsia* Rheede, Hort. Malab. 6: 7-8, t. 4. 1686.

Branchlets glabrescent, prickles downwards. Leaves c. 13 cm long; pinnae 6-7 (8) pairs; leaflets 13-15 pairs, oblong-falcate, 10 by 3 mm, base subtruncate, apex acute, mucronate, chartaceous; rachis with a hooked gland at base, slender glands between on uppermost 3 pairs or on all pairs. Flower heads c.15 mm f, cream coloured. Flowers sessile, 2 mm ϕ. Pods thin, flat, stipitate, to 15 cm long. Seeds c. 10.

Fl. & Fr.: Oct-Dec. *Distr.*: Sri Lanka, India, Thailand, Malesia. Commonly found as dense impenetrable thickets along forest borders. *NAK 24, 263* (Moozhiar, c. 250 m).

Bark is used in dyeing and as a substitute for soap.

2. **Acacia pennata** (L.) Willd., Sp. Pl. 4:1090. 1806; Baker in Hook.f., Fl. Brit. India 2 : 297. 1878, p.p.; Gamble, Fl. Pres. Madras 429. 1919; Gandhi in Sald. & Nicols., Fl. Hassan Dist. 228. 1976; Britto in Matthew, Fl. Tam. Carnatic 3 (1) : 530. 1983; Manilal, Fl. Silent Valley 92. 1988. Sanjappa, Leg. India 41. 1992. *Mimosa pennata* L., Sp. Pl. 522. 1753.

Branchlets glabrescent. Leaves c. 10 cm long; pinnae 12-15 pairs; leaflets 50-60 pairs, close, overlaping, linear, 6 by 1 mm, base semi-truncate, apex obliquely acute, chartaceous; rachis with glands between upper pinnae. Panicles to 15 cm long; head 1-2 cm ϕ. Flowers pedicelled; corolla white, c. 2 mm ϕ. Pods 15-20 cm long.

Fl. & Fr.: Mar.-May *Distr.*: Paleotropics. Occasionally in semi-evergreen forests. *CNM 63438* (MH) (Ranni R.F., c. 200 m).

3. **Acacia torta** (Roxb.) Craib, Bull. Misc. Inform. 1915 : 410. 1915; Gamble, Fl. Pres. Madras 428. 1919; Gandhi in Sald. & Nicols., Fl. Hassan Dist. 229. 1976; Britto in Matthew, Fl. Tam. Carnatic 3 (1) : 533. 1983. Sanjappa, Leg. India 44. 1992. *Mimosa torta* Roxb., Fl. Ind. 2 : 566. 1832. *Acacia caesia* auct., non. Willd.: Wight & Arn., Prodr. 278. 1834.

Branchlets yellowish pubescent. Leaves to 20 cm long; pinnae 10-14 pairs; leaflets 25-40 pairs, not overlapping, elliptic, to 8 by 2 mm, base truncate, apex sub-acute, chartaceous, rachis with glands between upper 4 pairs of pinnae. Panicles to 20 cm long; heads to 1 cm. ϕ. Flowers c. 2 mm ϕ, sessile; corolla white. Pods flat, to 12 cm long, prominently nerved. Seeds c. 6.

Fl. & Fr.: Feb.-Apr. *Distr.*: Pakistan, India. Rare. On exposed forest borders. *NAK 2657*. (Appuppanthode, c. 400 m).

ALBIZIA Durazzini
Mag. Tosc. 3(4) : 13. 1772.

Trees. Leaves alternate, 2-pinnate; leaflets even-pinnate; petiole and rachis often with glands. Flower -heads in terminal or axillary panicles or in clusters. Flowers 5-merous; calyx-tube 5-lobed; petals 5, basally connate; stamens α, basally connate, filaments coloured; ovary sessile or stalked; ovules α. Pods flat, compressed. Seeds ovoid to orbicular, compressed.

1. Pinnae 6-13 pairs; leaflets 2-8 mm wide; flowers sessile; pods 3-4 cm wide............1. **A. chinensis**
1. Pinnae 2 or 3 pairs; leaflets 1-2 cm wide; flowers pedicellate; pods 4-5 cm wide............2. **A. lebbeck**

1. **Albizia chinensis** (Osb.) Merr., Amer. J. Bot. 3 : 575. 1916; Gandhi in Sald. & Nicols., Fl. Hassan Dist. 230. 1976; Kosterm. in Dassan. & Fosb., Rev. Handb. Fl. Ceylon 1:500. 1980; Britto in Matthew, Fl. Tam. Carnatic 3 (1) : 536. 1983; Sanjappa, Leg. India 54. 1992. Mohanan & Sivad., Fl. Agasthyamala 236. 2002. *Mimosa chinensis* Osbeck, Dagb. Ostind. Resa 233. 1757. *M. marginata* Lam., Encycl. 1: 12. 1783. *Albizia stipulata* (Roxb.) Boivin, Encycl. XIXme Siecle 2 : 33. 1838; Baker in Hook.f., Fl. Brit. India 2 : 300. 1878. *A. marginata* (Lam.) Merr., Philipp. J. Sci. 5 : 23. 1910; Gamble, Fl. Pres. Madras 433. 1919. *Potta-vaga.*

Leaves to 25 cm long; pinnae c. 12 pairs; leaflets 30-40 pairs, oblong-falcate, 6 by 2 mm, base truncate, margin ciliate, apex acute,mid nerve very close to upper margin; petiole with a gland on middle. Flower-heads c. 2.5 cm ϕ; corolla white, filaments slender, pink coloured above the middle. Pods c. 20 by 3 cm.

Fl. & Fr.: Mar.-May. *Distr.*: Sri Lanka, India, Myanmar, Thailand, Indo-china, S. China & Malesia. Common. In disturbed semi-evergreen forests. *NAK 1590* (Pampavalley, c. 1980 m), 1760 (Thannithode, c. 70 m).

2. **Albizia lebbeck** (L.) Willd., Sp. Pl 4 : 1066. 1806; Bedd., Fl. Sylv. t. 53. 1870, Baker in Hook. f., Fl. Brit. India 2: 298. 1878; Gamble, Fl. Pres. Madras 432. 1919; Gandhi in Sald. & Nicols., Fl. Hassan Dist. 230. 1976, Kosterm. in Dassan. & Fosberg, Rev. Handb.Fl. Ceylon 1:502. 1980; Matthew, Ill. Fl. Tam. Carnatic t. 242. 1982; Britto in Matthew, Fl. Tam. Carnatic 3(1) : 538. 1983. Sanjappa, Leg. India 56. 1992. *Mimosa lebbeck* L., Sp. Pl. 516. 1753. *Acacia speciosa* Willd., Sp. Pl. 4: 1066. 1806, Wight & Arn., Prodr. 275. 1834. *Vaga.*

Trees. Leaves to 18 cm long; pinnae 2-3 pairs, leaflets c. 5 pairs, oblong-elliptic, 3-5 by 1.5-2.5 cm, upper pair obovate, 5 by 3 cm, base oblique, apex retuse, glabrescent, chartaceous; glands opposite to two upper most pairs on rachis and on petiolules of all leaflets. Flower- head(s) c. 4 cm ϕ, axillary, solitary or fascicled. Flowers sessile, c. mm ϕ; corolla white; filaments pink coloured. Pods to 24 by 5 cm. Seeds c. 10.

Fl. & Fr.: (Feb.) Mar.-May. *Distr.*: Sri Lanka, India, S.E. Asia & S. China. Often planted; occasionally seen as wild. *NAK 524* (Angamoozhy, c. 450 m).

Wood durable. Leaves used as green manure. Seeds used in the treatment of piles and as an astringent in diarrohoea.

ARCHIDENDRON F.V. Mueller
Fragm. 5: 59. 1865.

Archidendron monadelphum (Roxb.) Nielsen, Adansonia (1991) : 21. 1979; Nicols., Suresh & Manilal, An Interpr. Hort. Malab. 122. 1988; Sanjappa, Leg. India 61. 1992. *Mimosa monadelpha* Roxb., Fl. Ind. 2: 544. 1832. *Pithecellobium bigeminum* sensu Baker in Hook. f., Fl. Brit. India. 2: 303. 1878, non (L.) Mart. ex Benth., 1844; Gamble, Fl. Pres. Madras 435. 1919, "Pithecolobium". *P. monadelphum* (Roxb.) Kosterm., Reinwardtia 3:11. 1954; Gandhi in Sald. & Nicols., Fl.Hassan Dist. 232. 1976; Manilal, Fl. Silent Valley 94. 1988. *Katoukonna* Rheede, Hort. Malab. 6:21, t. 12. 1686.

Trees; branchlets pubescent. Leaves alternate, 2-pinnate, to 25 cm long; pinnae often 2 pairs; leaflets c. 5 pairs, elliptic-oblong, 6-13 by 3.5-4.5 cm, base obliquely cuneate, apex abruptly acuminate, chartaceous, lateral nerves 6-8 pairs forming marginal loopes; glands on the middle and apex of petiole and between leaflets. Flower -heads in large panicles, axillary or terminal. Flowers 4 mm ϕ; calyx 5-lobed; petals 5, white to cream coloured, connate to the middle; stamens α, monadelphous; ovary sessile, minute; ovules α. Pods spirally twisted, dehiscent, orange-red coloured within. Seeds black.

Fl. & Fr.: Jan.-Jul. *Distr.*: Western Ghats, E. Himalayas. Profusely branched trees. Common in semi-evergreen forests. *NAK 516* (Angamoozhy, 400 m), *1487 & 1566* (Arampa, c. 400 m).

The wood is very soft, suitable for match-boxes, planking and battens. Decoction of leaves used externally in leprosy and jaundice and is said to be promoting the growth of hair.

Note : The genus **Pithecellobium** Mart. is mainly American with spinescent stipules and seeds with aril and pleurogram, while the genus **Archidendron** Mueller is mainly Asian - Australian,and the plants are unarmed and seeds without aril and pleurogram.

ENTADA Adanson
Fam. 2: 318.554. 1763 (nom. cons.).

Entada rheedei Sprengel in L., Syst. Veg. (ed. 16) 2 : 325. 1825; Britto in Matthew, Fl. Tam. Carnatic 3(1) : 542. 1483; Panigrahi, Taxon 34: 714. 1985; Nicols., Suresh & Manilal, An Interpr. Hort. Malab. 135. 1988; Sanjappa, Leg. India 65. 1992. Mohanan & Sivad., Fl. Agasthyamala 237. 2002. *E. pursaetha* DC., Prodr. 267.1834; Brenan, Kew Bull. 20: 363. 1967; Gandhi in Sald. & Nicols., Fl. Hassan Dist. 231. 1976; Kosterm. in Dassan. & Fosb., Rev. Handb. Fl. Ceylon 1:464. 1980. *E. scandens* auct., non Benth, 1841: Baker in Hook.f., Fl. Brit. India 2: 287. 1878; Gamble, Fl. Pres. Madras 417. 1919. *Perim-kaku-valli* Rheede, Hort. Malab. 8: 59-60, t. 32-34 1688. *Parand-ka-valli, Makum-ko-ka.*

Extensive climbers. Leaves alternate, bipinnate, c. 12 cm long; pinnae 2 pairs; leaflets 4 pairs, even-pinnate, oblong-obovate, 2-5 by 2-1 cm, base oblique, apex retuse, thin-coriaceous; rachis ends in a forked tendril. Flowers polygamous, dense, small, in long pendulous spikes; calyx-tube 5-toothed; petals 5, cream-colourd; stamens 10, filaments apically dilated, anthers gland-crested; ovary subsessile; ovules c.8. Pods elongate, flat, woody, to 1 m long. Seeds discoid, hard, black-shiny.

Fl. & Fr.: Dec-Jul. *Distr.*; Tropical & S. Africa, Sri Lanka, India to China, Malesia & Australia. Forming dense thickets along streamsides. Locally abundant. *NAK 289* (Upper Moozhiar on way side forest, c. 500 m), *2643* (Thannithode-Along streamside, c. 70 m).

Seeds soaked in water, powdered, dried and taken with sugar in asthma with milk in chest-afflictions and liver complaints.

MIMOSA Linnaeus
Sp. Pl. 516. 1753.

Armed stragglers or spreading herbs. Leaves alternate, 2-pinnate; leaflets even-pinnate, often sensitive to touch; rachis prickly or bristly. Flower heads in terminal racemes or axillary clusters. Flowers sessile, 4-merous, polygamous; calyx minute; petals 4, connate at base; stamens 4-8, much exerted, free; ovary stipitate; style filiform. Pods clustered, stipitate, oblong, flat, spinous or bristly along margins. Seeds ovoid, flat.

1. Straggling subshrubs; pinnae 3-10 pairs; stamens twice as many as petals. .... 1. **M. invisa**
1. Spreading herbs; pinnae 2-3 pairs; stamens as many as petals. .... 2. **M. pudica**

1. **Mimosa invisa** Mart., Herb. Fl. Bras. 121. 1837; N.C. Nair, J. Bombay Nat. Hist. Soc. 61(2) : 470. 1964; Fosberg, Phytologia 15 : 499. 1968; Kosterm. in Dassan. & Fosberg, Rev. Handb. Fl. Ceylon 1: 464. 1980; Manilal, Fl. Silent Valley 94. 1988; Sanjappa, Leg. India 68. 1992.

Key to the varieties

1. Branchlets and pods densely prickled. .... **Mimosa invisa** var. **invisa**
1. Branchlets and pod glabrescent. .... **Mimosa invisa** var. **inermis**

Mimosa invisa Mart. var. **invisa**

Straggling subshrubs; stem 4-angular, with recurved prickles. Leaves to 12 cm long; rachis prickled; pinnae 5-10 pairs; leaflets c. 20 pairs, oblong, 3-7 by 0.75-1 mm, overlapping, base oblique-truncate, apex acute-mucronate. Flowers a, pink. Lomentum flat, margin with recurved prickles.

Fl. & Fr.: Sep.-Dec. *Distr.*: Malesia, India. Occasional. In exposed or cleared forest areas in higher hills. *NAK 326* (Upper Moozhiar, c. 500 m).

Mimosa invisa Mart. var. **inermis** Adelb., Reinwardtia 2(2): 359. 1953; Nayar & Giri, J. Econ. Tax. Bot. 3 : 603. 1982.

Branchlets ribbed, densely hispid, prickles 0. Pinnae 5-6 pairs. Pods not prickled at margin.

Fl. & Fr.: Oct.-Mar. Distr.: *NAK 178* (Adoor, c. 40 m), *311* (Moozhiar, c. 250 m).

An introduced weed and a ready colonizer. An important soil-cover crop and a green manure.

2. **Mimosa pudica** L., Sp. Pl. 518. 1753; Baker in Hook.f., Fl. Brit. India 2 : 291. 1878; Gamble, Fl. Pres. Madras 421. 1919; Gandhi in Sald. & Nicols., Fl. Hassan Dist. 231. 1976; Kosterm. in Dassan. & Fosberg, Rev. Handb. Fl. Ceylon 1: 463. 1980; Matthew, Ill. Fl. Tam. Carnatic t. 246. 1982; Britto in Matthew, Fl.Tam. Carnatic 3(1) : 545. 1983; Manilal, Fl. Silent Valley 94. 1988; Sanjappa, Leg. India 69. 1992. Mohanan & Sivad., Fl. Agasthyamala 238. 2002. *Thotta-vaadi.*

Diffuse or spreading herbs; stem terete with distant, recurved prickles. Leaves to 6 cm long; pinnae 2-3 pairs, palmately arranged; leaflets 14-20 pairs, elliptic-oblong, 4-10 by 1-2 mm, base truncate, margin ciliate, apex acute. Flowers pink, in globose heads. Lomentum flat, slightly recurved, bristly. Seeds ovoid.

Fl. & Fr.: Jun.-Dec. *Distr.*: Pantropical. Frequent in fallow fields, waysides, etc. *NAK 198* (Adoor, c. 20 m).

Leaves made into a paste and applied to hydrocele.

XYLIA Bentham
J. Bot. (Hooker) 4 : 417. 1842.

Xylia xylocarpa (Roxb.) Taub., Bot. Centralbl. 47: 397. 1891; Gamble, Fl. Pres. Madras 417. 1919; Gandhi in Sald. & Nicols., Fl. Hassan Dist. 233. 1976. Sanjappa, Leg. India 73. 1992. *Mimosa xylocarpa* Roxb., Pl. Corom. t. 100. 1798. *Xylia dolabriformis* Benth., J. Bot. (Hooker) 4 : 417. 1842; Bedd., Fl. Sylv.t. 186. 1872; Baker in Hook.f., Fl. Brit. India 2 : 286. 1878. Irul.

Trees. Leaves 2-pinnate, to 20 cm long; pinnae 2; leaflets 4 pairs, elliptic-oblong, 4-10 by 2-3 cm, base cuneate, apex acuminate, thin-coriaceous; petiole with a solitary apical gland; rachis with glands opposite to all pinnae. Flower- heads c. 15 mm ϕ, usually in racemes; calyx 5-toothed; corolla yellowish; petals 5, slightly connate at base; stamens 10; anthers tipped with caducous glands; ovary minute, sessile; ovules α. Pods falcate, woody, flat, 16 by 6 cm. Seeds c. 6.

Fl. & Fr.: Mar.-Dec. *Distr.:* Peninsular India, Myanmar, Malesia, Philippines. Not common. In moist deciduous forests. *NAK 461* (Perunthenaruvi, c. 100 m), *1678* (Thannithode, c. 70 m).

A valuable timber tree. Seeds yield an oil used in rheumatism, ulcers, leprosy, piles and poisons.

Plants of **Acacia auriculiformis** A. Cunn. ex Benth.are seen planted extensively everywhere, and even in grasslands of Kakki Hills. **Samanea saman** (Jacq.) Merr. and **Adenanthera pavonina** L. are planted as avenue trees.

52. ROSACEAE

RUBUS Linnaeus
Sp. Pl. 492.1753.

Armed stragglers. Leaves simple, alternate, 5-α- lobed, basally 5-nerved. Inflorescence a panicle, terminal and/or axillary. Flowers 5-merous, regular, bisexual; calyx-tube broad, lobes 5, persistent; petals 5, white, clawed; stamens α, inserted on the mouth of calyx-tube; carpels α, on a convex receptacle; ovary superior; ovules 2-per ovary, collateral, pendulous. Fruit an aggregate of drupelets, rugose. Seeds pendulous.

1. Bracts shallowly laciniate; leaves 9-or more-lobed, upper surface rugose, tomentum fulvous; fruits reddish. ....1.**R. fairholmianus**
1. Bracts deeply laciniate; leaves 5-(-7)-lobed upper surface not orless rugose, tomentum white; fruits black-purple. ....2. **R. glomeratus**

1. **Rubus fairholmianus** Gard., Calcutta J. Nat. Hist. 8:6.1847; Gamble, Fl. Pres. Madras 441. 1919;Tirvengadum in Dassan. & Fosberg, Rev. Handb. Fl. Ceylon 3:335. 1981; Manilal, Fl. Silent Valley 99. 1988. *R. moluccanus* sensu Hook.f., Fl. Brit. India 2 : 330. 1878, p.p., non L., 1753.

Branchlets fulvous tomentose; prickles slightly recurved. Leaves ovate, 6-11 by 5.5-10.5 cm, 9-12- lobed, lobes acute, base cordate, margin irregulary serrate, upper surface rugose, lower surface white-tomentose, basally 5-nerved, coriaceous. Racemes c. 10 cm long; bracts broadly ovate, shallowly laciniate. Flowers c.2 cm ϕ; calyx cupular, lobes lanceate; petals broadly ovate, margins crispate. Fruits reddish.

Fl. & Fr.: Aug.-Dec. *Distr.*: Sri Lanka, Peninsular & Eastern India, Myanmar, Malesia. Locally abundant. In evergreen forest borders climbing to great heights. *NAK 390 & 1282* (Kakki hills, c. 100 m). Fruits edible.

2. **Rubus glomeratus** Bl., Bijdr. 1111. 1826; Tirvengadum in Dassan. & Fosberg, Rev. Handb. Fl. Ceylon 3:354. 1981; Manilal, Fl. Silent Valley 99. 1988. *Rubus fulvus* Focke, Biblioth. Bot. 17(72) : 81. 1911, nom. illeg., non Sudre, 1902, nec L., 1753; Gamble, Fl. Pres. Madras 441. 1919. *R. molluccanus* sensu Hook.f., Fl. Brit India 2: 230. 1878, p.p., non L., 1753. *Kattu-munthiri.*

Branchlets white-tomentose; stipules and bracts deeply laciniate. Leaves broadly ovate, 3-13 by 2-12 cm, 5-lobed; upper lobe larger, base cordate, margin dentate serrate, apex acute, lower surface white-tomentose. Fruits dark-purple.

Fl. & Fr.: Aug.-Dec. *Distr.*: Peninsular India. Forming dense masses on waysides in higher hills. *NAK 71 & 1283* (Kakki hills, c. 1050 m).

53. CRASSULACEAE

KALANCHOE Adanson
Fam. 2:248. 1763.

Kalanchoe pinnata (Lam.) Pers., Syn. Pl. 1 : 446. 1805; Raym.-Hamet, Bull. Hebr. Boissier (Ser. 2) 8: 21. 1908; Baker in Steenis, Fl. Males. I, 4:199. 1951; Matthew, Fl. Tam. Carnatic 3(1) : 561. 1983. *Cotylendon pinnata* Lam., Encycl. 2 : 141. 1786. *Bryophyllum pinnatum* (Lam.) Oken, Allg. Naturgesch. 3: 1966. [1]841. *B. pinnatum* (Lam.) Kurz, J. Asiat. Soc. Bengal, Pt. 2, Nat. Hist. 40 : 42. 1871; Gamble, Fl. Pres. Madras 451. 1919. *B. calycinum* Salisb., Parad. Lond. t. 3. 1805 & in Curtis, Bot. Mag. t. 1409. 1811; Clarke in Hook.f., Fl. Brit. India 2 : 413. 1878.

Subsucculent tall herbs. Leaves decussate, simple elliptic-ovate, 5-10 by 3-6 cm, base cuneate, margin deeply crenate, apex obtuse to rounded, herbaceous. Inflorescence a cymose panicle. Flowers pendulous; clayx-tube lobed above the middle, lobes 4, triangular; corolla greenish, 8-folded, glandular within; stamens 8, inserted above the constriction of corolla; hypogynous scales 4; carpels 4;ovules α. Follicle enclosed with reddish calyx.

Fl. & Fr: Dec.-Mar. *Distr.*: Pantropical. Locally abundant. Forming localised population on exposed rocks. *NAK 2210* (Pathanamthitta, c. 70 m).

54. DROSERACEAE

DROSERA Linnaeus
Sp.Pl. 218. 1753.

Insectivorous herbs. Leaves rosulate or cauline, spiral, with sticky marginal glands. Racemes terminal or lateral, helicoid. Flowers 4-(5)- merous, bisexual, regular; calyx-tube 5-lobed, lobes persistent; petals 5, fre, spathulate; stamens 5, free; ovay superior, 1-locular; ovules α. Capsule loculicidal. Seeds α.

1. Leaves rosulate; styles 5; stem 0. .... 1.**D. burmannii**
1. Leaves cauline; styles 3; stem developed.
    2. Leaves linear; flowers rose; styles 2-fid to the base. .... 2. **D. indica**
    2. Leaves lunate; flowers white; styles fimbriate. .... 3. **D. peltata**

1. **Drosera burmannii** Vahl, Symb. Bot. 3: 50. 1794; Roxb., Fl. Ind. 2: 113. 1832; Wight & Arn., Prodr. 34. 1834; Clarke in Hook.f., Fl. Brit. India 2:424. 1878; Gamble, Fl. Pres. Madras 452. 1919; Steenis in Steenis, Fl. Males. I, 4: 378.1953; Ramamoorthy in Sald. & Nicols., Fl. Hassan Dist. 159. 1976; Matthew & Britto in Matthew, Fl. Tam. Carnatic 3(1) : 563. 1983; Manilal, Fl. Silent Valley 100. 1988; Mohanan & Sivad., Fl. Agasthyamala 244. 2002.

Acaulescent herbs. Leaves radical, rosulate, spathulate to obovate, 5-12 by 3-9 mm, margin with a reddish glandular tentacles; stipules 3-partite. Flowers c.5 mm f on slender scapes; petals rose colored; ovary ovoid; styles 5, stigma penicellate. Capsule membranous.

Fl. & Fr.: Oct.-Sep. *Distr.*: W. Africa, India, Himalaya, Myanmar to China, Taiwan, Malesia. Rare. On exposed moist rock crevices. *NAK 1914* (Vallicode, c. 150 m).

2. **Drosera indica** L., Sp. Pl.282. 1753; Roxb., Fl.Ind. 2:113. 1983; Wight & Arn., Prodr. 34.1834; Clarke in Hook.f., Fl.Brit. India 2:424. 1878; Gamble, Fl. Pres. Madras 452. 1919; Steenis in Steenis, Fl. Males. I, 4: 379. 1953; Ramamoorthy in Sald. & Nicols., Fl. Hassan Dist. 159. 1976; Matthew, Ill. Fl. Tam. Carnatic t.255. 1982; Matthew & Britto in Matthew, Fl. Tam. Carnatic 3(1) : 564. 1983; Nicols., Suresh & Manilal, An Interpr. Hort. Malab. 102. 1988 ; Mohanan & Sivad., Fl. Agasthyamala 244. 2002. *Akara-puda* Rheede, Hort. Malab. 10 : 39, t. 20.1690.

Caulescent herbs; stipules 0. Leaves cauline, linear, c.3 cm long with glandular tentacles Racemes lateral, α-flowered. Flowers c. 5 mm ϕ; calyx-lobes lanceate, glandular; petals rose-coloured, spathulate; ovary ovoid; styles 2-fid to base. Capsule ovoid. Seeds ellipsoid.

Fl. & Fr.: Sep.-Dec. *Distr.:* Trop. Africa, Madagascar, Sri Lanka, India to Japan, New Guinea and Australia. Rare. On exposed rocks among grasses. *NAK 2134* (Vallicode, c. 150 m).

3. **Drosera peltata** Smith in Willd., Sp. Pl. 1: 1546. 1798; Wight & Arn., Prodr. 34. 1834; Clarke in Hook.f., Fl. Brit. India 2:424. 1878; Gamble, Fl. Pres. Madras 452. 1919; Steenis, Fl. Males. I,4: 380. 1953; Ramamoorthy in Sald. & Nicols., Fl. Hassan Dist. 159. 1976; Matthew & Britto in Matthew, Fl. Tam. Carnatic 3(1): 565.1983; Manilal, Fl. Silent Valley 100. 1988; Mohanan & Sivad., Fl. Agasthyamala 245. 2002.

Slender herbs; rhizome tuberous. Leaves rosulate, cauline, lunate, 2-4 by 1-2.5 mm, peltate, glandular tentacled. Racemes lateral, c. 10-flowered. Flowers c. 10 mm ϕ; sepals 5, ovate-elliptic, fimbriate; petals 5, white, spathulate or narrowly obovate; ovary globose; styles 3, apically fimbriate. Capsule globose. Seeds α.

Fl. & Fr.: Dec.-Mar. *Distr.*: Sri Lanka, India, S.E. Asia, Malesia, China and Japan to Australia. Very rare. On moist shady hill cuttings. *NAK 2553* (Way to Ponnambalamedu, c. 1100 m).

Leaves bruised and mixed with salt and used as a remedy for blister of the skin.

55. RHIZOPHORACEAE

CARALLIA Roxburgh
Pl. Coromandel 3:8, t. 211. 1811 ("1819") (nom. cons.).

Carallia brachiata (Lour.) Merrill, Philipp. J. Sci. Bot. 15:249. 1919; Ding Hou in Steenis, Fl. Males. I, 5 (4) : 485. 1958; Ramamoorthy in Sald. & Nicols., Fl. Hassan Dist. 297. 1976; Nicols., Suresh & Manilal, An Interpr.

Hort. Malab. 216. 1988; ; Mohanan & Sivad., Fl. Agasthyamala 247. 2002. *Diatoma brachiata* Lour., Fl. Cochinch. 296. 1790. *Carallia integerrima* DC., Prodr. 3:33. 1828; Bedd., Fl. Sylv. t. 193. 1872; Henslow in Hook.f., Fl. Brit. India 2:439. 1878; Gamble, Fl. Pres. Madras 459. 1919. *Kare-kandel* Rheede, Hort. Malab. 5:25-26, t. 13. 1685.

Trees; stipules inter-petiolar. Leaves simple, opposite, obovate to elliptic, 10-14 by 5-7.5 cm, base cuneate, apex obtuse, shiny above, coriaceous. Inflorecence axillary, 3-chotomously branched cymes. Flowers bisexual, regular; bracteoles 2; calyx subglobose, 8-lobed, lobes triangular-lanceate; petals 5-8, greenish-yellow, orbicular, clawed, apex 2-fid, margin lacerate; stamens 10-16, inserted on the disc; ovary globose, half-inferior; style subulate; stigma 3-5-lobed; ovules 2 per locule. Fruit globose. Seed 1, subreniform.

Fl. & Fr.: Dec.-Apr. *Distr.*: Continental Asia to N. Australia. In evergreen forests, and also cultivated. *NAK 527* (Angamoozhy, c. 400 m).

Wood useful for furniture. Fruits fried with oil and used to cure ulcers.

56. COMBRETACEAE

1. Trees.
   2. Flowers in capitate heads, all bisexual. .... **Anogeissus**
   2. Flowers in spikes or racemes, basally bisexual, apically male. .... **Terminalia**
1. Climbing shurbs.
   3. Calyx accrescent; petals 0. .... **Calycopteris**
   3. Calyx not accrescent; petals 4 or 5.
      4. Flowers greenish-yellow, 4-merous; hypanthium short; drupe winged. ....**Combretum**
      4. Flowers pinkish or pinkish-white, 5-merous, hypanthium long; drupe wingless. .... **Quisqualis**

ANOGEISSUS (A.P. de Candolle) Guillemin Perrottet et A. Richard, Fl. Seneg. Tent. 1:280, t. 65. 1832.

Anogeissus latifolia (Roxb. ex DC.) Wall. ex Guill. & Perr., Fl. Seneg. Tent. 1:280. 1832; Bedd., Fl. Sylv. t. 15. 1868; Clarke in Hook.f., Fl. Brit. India 2:450. 1878; Gamble, Fl. Pres. Madras 466. 1919; Gandhi in Sald. & Nicols., Fl. Hassan Dist. 292. 1976; A.J. Scott, Kew Bull. 33 (4): 560. 1979; Matthew & Britto in Matthew, Fl. Tam. Carnatic 3(1):575.1983; Manilal, Fl.Silent Valley 101. 1988; Mohanan & Sivad., Fl. Agasthyamala 248. 2002. *Conocarpus latifolia* Roxb. ex DC., Prodr. 3:16. 1828; Wight, Ic. t. 294. 1845.

Trees. Leaves simple, alternate, ablong-elliptic to elliptic-obovate, 3-8 by 2-4 cm, base rounded, apex obtuse, apiculate, thin-coriaceous. Flowers small in cymose clusters aggregated into globose heads; hypanthium cupular; sepals 5, triangular; petals 0; stamens 5+5, exerted; ovary linear, 1-locular; ovules 2, pendulous. Drupes compressed, 2-winged, clustered. Seed 1, flat.

Fl. & Fr.: Sep.-Dec. *Distr.*: Sri Lanka, India, Pakistan. Rare. In dry evergreen forests. *NAK 2023* (Kokkathode, c.400 m).

A gum obtained from bark is useful for cloth-printing and dyeing.

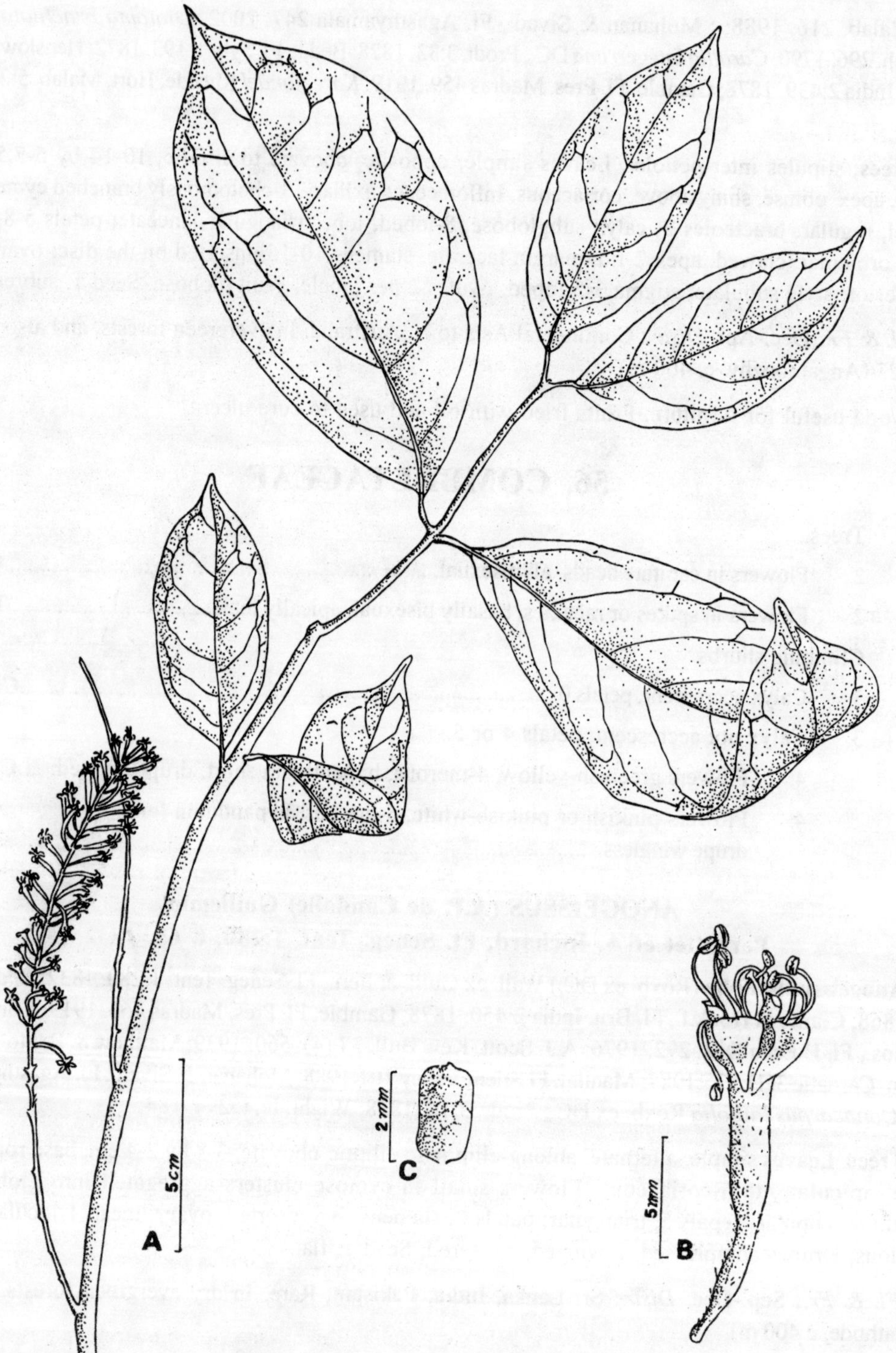

Fig. 36. ***Combretum latifolium*** **Bl.: A. Twig; B. Flower; C. Petal.**

CALYCOPTERIS Lamarck
Tabl. Encycl. t. 357; 2(2):485. 1793.

Calycopteris floribunda (Roxb.) Poiret in Lam., Encycl. (Suppl.) 2:41. 1811 & Tabl. Encycl. 2 : t. 357. 1793; Clarke in Hook.f., Fl. Brit. India 2: 449.1878; Gamble, Fl. Pres. Madras 467. 1919; Exell in Steenis, Fl. Males. I, 4:584. 1954; Gandhi in Sald. & Nicols., Fl. Hassan Dist. 292. 1976; Matthew, Ill. Fl. Tam. Carnatic t. 262. 1982; Matthew & Britto in Matthew, Fl. Tam. Carnatic 3(1): 576. 1983 ; Mohanan & Sivad., Fl. Agasthyamala 248.2002. *Getonia floribunda* Roxb., Pl. Corom.t. 87. 1798 & Fl. Ind. 2:428. 1832; Wight & Arn., Prodr. 315. 1834. *Pullanji.*

Scandent shrubs. Leaves simple, opposite, ovate to elliptic, gradually narrower towards apical portion and forming bracts, 3-11 by 1.3-3 cm, base rounded, apex acuminate, lower surface glandular-tomentose, thin-coriaceous. Racemes dense, axillary, crowded towards the ends of branches; bracts foliaceous; hypanthium c.6 mm long; sepals 5, triangular, accrescent; petals 0; stamens 5 + 5; ovary inferior; 1-locular; ovules 3, pendulous. Drupe with 5 spreading accrescent calyx-lobes. Seed 1.

Fl. & Fr.: Jan.-May. *Distr.*: S.E. Asia, Malesia. Very common. In moist deciduous and degraded forests. A very good indicator of forest destruction. *NAK 223, 1421* (Mannarappara, c. 200 m).

COMBERTUM Loefling
Iter Hispan. 308. 1758(nom. cons.).

Combretum latifolium Bl., Bijdr. 641. 1825, non Don, 1827; Exell in Steenis, Fl. Males. I, 4(5):542. 1954; Gandhi in Sald. & Nicols., Fl. Hassan Dist. 293.1976. *C. extensum* Roxb. ex G. Don, Trans. Linn. Soc. Lond on. 15:414, 422. 1827; Clarke in Hook.f., Fl.Brit. India 2:458.1878; Gamble, Fl. Pres. Madras 469. 1919; Nicols., Suresh & Manilal, An Interpr. Hort. Malab. 85. 1988. *Pee-ula* Rheede, Hort. Malab. 7: 43, t. 23. 1688. **(Fig. 36)**

Climbing shrubs. Leaves simple, opposite, elliptic-obovate to oblong-elliptic, 7-12 by 3.5-6 cm, base cuneate, apex obtusely acuminate. Flowers 4-merous, dense, in pendulou spikes, axillary, polygamo-dioecious; hypanthium narrow-funnelform, c. 6 mm long, constricted below; sepals 4, triangular; petals 4, greenish-yellow; disc lobed, villous; stamens 4 + 4; ovary linear, inferior, 1-locular; ovules 2-3, pendulous. Drupe with 4 membranous wings.

Fl. & Fr.: Feb.-May. *Distr.*: Indo-Malesia. Occasional. In rivereine forests. *NAK 1558* (Velithode, c. 250 m), *2671* (Manniera R.F., c. 150 m).

QUISQUALIS Linnaeus
Sp. Pl. ed. 2, 1:556. 1762.

Quisqualis malabarica Bedd., Ic. t. 155. 1868-1874; Clarke in Hook.f., Fl.Brit.India 2:460. 1878; Gamble, Fl.Pres.Madras 469. 1919.

Climbing shrubs. Leaves simple, opposite to subopposite, elliptic-oblong, 6-12 by 3-5.5 cm, base rounded, apex acuminate, chartaceous. Spikes drooping, few-flowered; bracts subulate; hypanthium c. 1 cm long; sepals 5, subulate; petals 5, pinkish-red, ovate-lanceate; stamens 5+5; ovary linear, inferior, 1-locular; ovules 2-3, pendulous. Drupe linear, 5-angled. Seed 1.

Fl. & Fr.: Feb.-Apr. *Distr.*: Peninsular & Eastern India. Rare. In interior of evergreen forests. *NAK 512* (Nilakkal, c. 400 m).

TERMINALIA Lineaeus
Syst. Nat. ed. 12, 2:674 (err. '638'). 1767;
Mant. Pl. 21, 128. 1767 (nom. cons.)

Trees. Leaves simple, subopposite or alternate, or apically clustered, lateral nerves prominent; usually glands present on the petiole or on the lower part of the midrib beneath. Flowers small in axillary or terminal spikes, 5-merous, polygamo-dioecious, bisexual flowers towards base . Male flowers towards apex; hypanthium apically companulate; sepals 5, triangular; disc 5-lobed; stamens 5 + 5; ovary inferior, 1-locular; ovules 2 or 3, pendulous. Drupe angular or winged, wings unequal. Seed 1.

1. Drupe winged; spikes panicled.
    2. Wings of drupe unequal, one longer, the other two shorter, rusty-tomentose. ....3. **T. paniculata**
    2. Wings of drupe equal, glabrous. ....2. **T. crenulata**
1. Drupe wingless; spikes simple. ....1. **T. bellirica**

1. **Terminalia bellirica** (Gaert.) Roxb., Pl. Corom. t. 198. 1805, ('bellerica') & Fl. Ind. 2: 431. 1832; Wight & Arn., Prodr. 313. 1834, (excl. syn.); Clarke in Hook.f., Fl. Brit. India 2 : 445. 1878; Gamble, Fl. Pres. Madras 463. 1919; Exell in Steenis, Fl. Males. I, 4:569. 1954; Matthew & Britto in Matthew, Fl. Tam. Carnatic 3(1) : 381. 1983; Nicols., Suresh & Manilal, An Interpr. Hort. Malab. 85.1988; Mohanan & Sivad., Fl. Agasthyamala 249. 2002. *Myrobalanus bellirica* Gaert., Sem. Pl. 2 : 90, t. 97. 1791. *Tani* Rheede, Hort. Malab. 4 : 23-24, t. 10.1683.

Leaves crowded towards the ends of branchlets, obovate, 10-16 by 6-8 cm, base cuneate, apex obtusely short-acuminate or emarginate, lower surface glaucescent, lateral nerves c. 6 pairs, thin-coriaceous. Spikes simple, clustered or solitary, axillary, c. 6 cm long; calyx cupular, greenish-yellow. Drupe subglobose, shallowly 5-lobed, wings absent.

Fl. & Fr.: Feb.-Nov. *Distr.*: Sri Lanka, India, Nepal, Myanmar, Thailand, Indo-China, Malesia. Rare. In moist deciduous forests. *NAK 2376* (Vallemthetti, Padam, c. 70 m).

Wood suitable for match boxes and splints and also for paper pulp. Oil from the seeds used as dressing for hair.

2. **Terminalia crenulata** Roth, Nov. Pl. Spec. 380. 1821; Wight & Arn., Prodr. 314. 1834; Gamble, Fl. Pres. Madras 465.1919; Gandhi in Sald. & Nicols., Fl. Hassan Dist. 294. 1976; Matthew & Britto in Matthew, Fl. Tam. Carnatic 3(1) : 584. 1983 ; Mohanan & Sivad., Fl. Agasthyamala 250. 2002. *Thempavu.*

Leaves alternate or subopposite, elliptic-oblong or elliptic-obovate, 7-18 by 5-8 cm, base obliquely cuneate, apex obtuse or subacute, lower surface with a stalked gland near the base, lateral nerves 16-20 pairs, coriaceous. Spikes panicled, terminal and/or axillary, c. 4 cm long; calyx cupular, yellow. Drupe ovoid, broadly equally 5-winged.

Fl. & Fr.: Apr.-Dec. *Distr.*: Sri Lanka, India. Common. In semi-evergreen adn moist deciduous forest. *NAK 892* (Plappally, c. 400 m).

3. **Terminalia paniculata** Roth, Nov. Pl. Spec. 383. 1821; Wight & Arn., Prodr. 315. 1834; Clarke in Hook.f., Fl. Brit. India 2 : 448. 1878; Gamble, Fl. Pres. Madras 465. 1919; Gandhi in Sald. & Nicols., Fl. Hassan

Dist. 294. 1976; Matthew & Britto in Matthew, Fl. Tam. Carnatic 3(1) : 585.1981; Manilal, Fl. Silent Valley 102. 1988; Mohanan & Sivad., Fl. Agasthyamala 251. 2002.

Leaves - upper ones alternate, lower oppoiste, elliptic-oblong, 5-10 by 4-5.5 cm, base turncate to subcordate, apex acute to shallowly obtusely acuminate, lower surface with 2 sessile glands at extreme base, lateral nerves 6-8 pairs, thin-coriaceous. Panicles usually terminal or axillary, to 20 cm long; calyx campanulate. Drupe unequally 3-winged, median wing longer than laterals. Seed 1.

Fl. & Fr.: Jul-Feb. *Distr.:* Peninsular India. Common. In moist deciduous and semi-evergreen forest. *NAK 147 & 454* (Adoor, c.50 m).

Juice of bark or flowers with melted butter and rock-salt applied externally in parotitis. Powder or decoction of root-bark given as a remedy for impure blood, biliousness, leprosy, swellings, syphilis and worm infections.

Trees of **Terminalia catappa** L. are planted as avenue trees.

57. MYRTACEAE

1. Fruit a capsule; adult leaves alternate, young leaves opposite. .... **Eucalyptus**
1. Fruit a berry; all leaves opposite.
   2. Ovary a-locular; seeds α. .... **Psidium**
   2. Ovary 2-locular. Seed(s) 1 or 2.
      3. Calyx-tube produced beyond the ovary; cymes α–flowered. .... **Syzygium**
      3. Calyx-tube not produced beyond the ovary; cymes few-flowered. .... **Eugenia**

EUCALYPTUS L' Heritier de Bruttelle
Sertum Angl. 18. 1789; t. 20. 1792

Trees; bark usually smooth and decorticating. Leaves (juvenile) alternate or opposite, ovate or lanceate; adult ones are alternate, lanceolate or lanceate, pellucid-punctate, coriaceous. Flower(s) solitary or in axillary or terminal panicles or umbels, 4-merous; hypanthium turbinate; operculum hemispherical or conical; stamens α, exerted; ovary inferior, 3-4-locular; ovules α an axile placenta. Fruit globose or obovoid.

Different species of **Eucalyptus** like **E. grandis** Hill ex Maiden and **E. tereticornis** Smith are extensively planted by Department of Social Forestry near Plapally.

EUGENIA Linnaeus
Sp. Pl. 470. 1753.

Trees. Leaves simple, decussate, pellucid-punctate. Flower(s) solitary, in pairs or in short cymes, axillary or terminal, 4-5- merous, bisexual, regular; calyx-tube subglobose, not produced beyond the ovary, lobes 4-5, persistent; petals 4-5, distinct; stamens α, erect or incurved; ovary inferor, 2-locular, often again divided by false septum. Berry globose. Seed(s) 1 or 2.

1. Flowers fascicled in leaf-axils; branchlets and peduncle tomentose. ....1. **E. rottleriana**
1. Flowers in pairs, axillary or lateral; branchlets and peduncle glabrous. ....2. **E. thwaitesii**

1. **Eugenia rottleriana** Wight & Arn., Prodr. 331.1834; Wight, Ic. t. 100.1838; Duthie in Hook.f., Fl.Brit.India 2:502.1879; Gamble, Fl.Pres. Madras 484.1919; Mohanan & Sivad., Fl. Agasthyamala 254. 2002.

Branchlets tomentose. Leaves narrowly lanceate, 2.5-5 by 0.5-5 by 0.5-1.5 cm, base and apex acuminate, thin-coraceous. Cymes 3-4, fascicled, lateral calyx-lobes 4, triangular-ovate; petals 4, white, ciliate on margin. Fruit globose with peristent calyx.

Fl. & Fr.: Jan.-Apr. *Distr.*: Western Ghats. Rare. In evergreen forests. *BKN 10521* (CALI), (Aranamudi - Sabari hills).

2. **Eugenia thwaitesii** Duthie in Hook.f., Fl.Brit. India 2:506. 1879; Ashton in Dassan. & Fosb., Rev. Handb. Fl.Ceylon 2:415. 1981; Manilal, Fl.Silent Valley 103.1988. *E. mooniana* Wight, Ic. t.551.1842; Duthie in Hook. f., Fl.Brit. India 2:505.1879; Gamble, Fl. Pres. Madras 484.1919; Mohanan & Sivad., Fl. Agasthyamala 255. 2002.

Branchlets glabrous. Leaves ovate-lanceate to narrowly or broadly elliptic, 5-8 by 2-3 cm, base cuneate, apex acute to acuminate, shiny, thin-coriaceous. Flowers in pairs; calyx-lobes 4, triangular-ovate; petals 4, pure-white. Fruit subglobose to ovoid with apical persistent ring of calyx.

Fl. & Fr.: Jan.-May. *Distr.*: Sri Lanka, Peninsular India. Very rare. In interior of evergreen forests; *NAK 2741* (Near Ponnambalamedu, c.1100 m).

PSIDIUM Linnaeus
Sp.Pl. 470.1753.

Psidium guajava L., Sp.Pl. 470.1753; Duthie in Hook.f., Fl.Brit. India 2:468.1878, "*guayava*"; Gamble, Fl. Pres. Madras 472.1919; Gandhi in Sald. & Nicols., Fl. Hassan Dist. 280.1976; Matthew & Britto in Matthew, Fl. Tam. Carnatic 3(1) : 593.1983; Manilal, Fl. Silent Valley 103.1988; Nicols., Suresh & Manilal, An Interpr. Hort. Malab. 194. 1988; Mohanan & Sivad., Fl. Agasthyamala 255. 2002. *Pela, Malacka-pela* Rheede, Hort. Malab. 3 : 31-33, t. 34, 35. 1682.

Trees. Leaves elliptic-oblong, 9-12 by 3-5 cm, base rounded to obtuse-cuneate, apex acute-apiculate, lateral nerves prominent, thin-coriaceous. Flower(s) usually solitary, axillary, c. 3 cm ϕ, calyx-tube sub-urceolate, lobes 4, unequally ovate-lanceate; petals 4 or more, white, caducous; ovary globose, α-celled; ovules α. Berry globose crowned by persistent calyx-lobes. Seeds α, subreniform.

Fl. & Fr.: Mar.-May. *Distr.*: Pantropical. Usually planted for its edible fruits. Occasionally run wild. *NAK 1956* (Tiruvalla town, c. 10 m). (Kakki hills, c. 1000 m).

Root bark given to arrest vomiting in Cholera and useful in swollen gums.

SYZYGIUM J. Gaertner
Fruct. 1 : 116. 1788 (nom cons.).

Trees. Leaves simple, decussate or alternate, lateral nerves α, close forming prominent intramarginal nerve. Inflorescence a paniculate cyme, axillary and/or terminal; pedicel often articulated. Flowers 4-merous, regular; calyx turbinate, persistent, adnate, produce above the ovary, lobes 4; petals 4, calyptrate or distinct, caducous; disc present; stamens α, inflexed; ovary 2-locular; ovules α, Berry globose. Seed 1.

1. Calyx-tube with a thickened staminal disc at mouth, lobes prominent; flowers usually large; petals always free.
    2. Leaves 15-25 (30) cm long, oblong or elliptic-oblong, base cordate. ....7. **S. mundagam**
    2. Leaves 5-12 cm long, elliptic, ovate, or linear-lanceate, base cuneate.
        3. Leaves elliptic or ovate.
            4. Calyx-tube funnelform; stamens reddish-pink or yellow; fruit ovoid. ....5. **S. laetum**
            4. Calyx-tube obconic; stamens white; fruit subglobose. ....4. **S. hemisphericum**
        3. Leaves liner-lanceate; lateral nerves faint; clayx tube turbinate. ....8. **S. occidentale**
1. Calyx-tube without a thickened staminal disc at mouth, lobes faint or 0; flowers usually small; petals calyptrate or free.
    5. Petals distinct, caducous.
        6. Calyx-tube turbinate; leaves elliptic, to 12 cm long, base obliquely cuneate. ....3. **S. densiflorum**
        6. Calyx-tube funnelform; leaves ovate-lanceate, to 8 cm long, base obtuse. ....11. **S. zeylanicum**
    5. Petals adherent, calyptrate.
        7. Cymes mostly terminal or from upper leaf-axils.
            8. Branchlets 4-gonous; leaves thick-coriaceous, apex acuminate; calyx-tube turbinate. ....9. **S. rubicundum**
            8. Branchlets terete; leaves thin-coriaceous, apex obtuse; narrow-funnelform, calyx-tube turbinate. ....1. **S. caryophyllatum**
        7. Cymes mostly terminal or from upper leaf-axils, rarely terminal or lateral.
            9. Intramarginal nerve developed, the nerves close and parallel; fruit ellipsoid or globose. ....2. **S. cumini**
            9. Intramarginal nerve not developed; fruits spherical or oblong.
                10. Leaves glaucous beneath; peduncle short; cymes lateral;fruits spherical. ....6. **S. malabaricum**
                10. Leaves greenish beneath; peduncle long; cymes axillary; fruits oblong. ....10. **S. travancoricum**

1. **Syzygium caryophyllatum** (L.) Alston in Trimen, Handb. Fl. Ceylon 6 (suppl.): 116.1931; Gandhi in Sald. & Nicols., Fl. Hassan Dist. 280. 1976; Nicols., Suresh & Manilal, An Interpr. Hort. Malab. 194. 1988;Mohanan & Sivad., Fl. Agasthyamala 260. 2002. *Myrtus caryophyllata* L., Sp. Pl. 472. 1753. *Eugenia caryophyllaea* Wight, Ic. t. 540. 1842; Duthie in Hook.f., Fl. Brit.India 2 : 490. 1878. *Syzygium*

caryophyllaeum auct., non Gaertn., 1788: Gamble, Fl.Pres. Madras 480.1919. *Njara* Rheede, Hort. Malab. 5 : 53, t. 27. 1685.

Leaves deccusate, obovate, 4-10 by 2-5 cm, base cuneate, apex obtuse to rarely emarginate, lateral veins α, slender, forming intramarginal vein, thin-coriaceous. Cymes in terminal panicle, rarely from upper leaf-axils; calyx turbinate, c. 2 mm ϕ; petals white, calyptrate; stamens white. Berry globose.

Fl. & Fr.: Feb.-Jun. *Distr.*: Sri Lanka, Peninsular India. Occasional. In sacred groves. *NAK 2318* (Kodumon, c. 50 m).

Bark ground and taken with milk or ghee as a remedy in diarrhoea, impure blood and phlegm.

2. **Syzygium cumini** (L.) Skeels, U.S. D.A. Bur. Pl. Industr. Bull. 248: 2. 1912; Gandhi in Sald. & Nicols., Fl. Hassan Dist. 281. 1976; Matthew, Ill. Fl. Tam. Carnatic t. 271. 1982; Matthew, Fl. Tam. Carnatic 3(1) : 594. 1983; Nicols., Suresh & Manilal, An Interpr. Hort. Malab. 1294. 1988; Mohanan & Sivad., Fl. Agasthyamala 261. 2002. *Myrtus cumini* L., Sp. Pl. 471. 1753. *Eugenia jambolana* Lam., Encycl. 3 : 198. 1789; Wight, Ic. t. 535. 1842. *Syzygium jambolanum* (Lam.) DC., Prodr. 3 : 259. 1828; Gamble, Fl. Pres. Madras 481. 1919. *Perin-njara* Rheede, Hort. Malab 5 : 57-58, t. 29. 1685.

Leaves ovate-lanceate or elliptic, 6-12 by 3-6 cm, base broadly cuneate, apex shortly acuminate, thin-coriaceous. Cymes in panicles, axillary or terminal. Flowers small; calyx shallowly turbinate, c. 6 mm f, petals white; filaments reddish. Berry globose.

Fl. & Fr.: Feb.-Apr. *Distr.*: Sri Lanka, India, Malesia to Australia. Not common. In evergreen forest. *NAK 1354* (Kakki hills, c. 1100 m).

Wood durable.

3. **Syzygium densiflorum** Wall. ex Wight & Arn., Prodr. 329. 1834. *S. arnottianum* Walp., Repert. 2 : 180. 1843; Gamble, Fl. Pres. Madras 478.1919. *Eugenia arnottiana* (Walp.) Wight, Ic. t. 999. 1845; Duthie in Hook.f., Fl. Brit. India 2 : 483.1878.

Leaves decussate, elliptic, 5-12 by 2.5-5 cm, base obliquely cuneate, apex long acuminate, lateral veins α, close, faint obove, thin-coriaceous. Cymes terminal or from upper leaf-axils. Flowers in dense umbellules; calyx-tube turbinate; petals white. Berry oblong.

Fl. & Fr.: Apr.-Jun. *Distr.*: Western Ghats. Rare. In evergreen and semi-evergreen forests. *NAK 1680* (Sabari hills, c. 250 m).

Fruit astringent, edible. Wood durable.

4. **Syzygium hemisphericum** (Wight) Alston in Trimen, Handb. Fl. Ceylon 6:115.1931; Gandhi in Sald. & Nicols., Fl. Hassan Dist. 281. 1976; Ashton in Dassan. & Fosberg, Rev. Handb. Fl. Ceylon 2 : 430. 1981; Manilal, Fl. Silent Valley 105. 1988. *Eugenia hemispherica* Wight, Ic. t. 525. 1842, Ill. 2 : 14.1850; Duthie in Hook. f., Fl.Brit. India 2:477.1878. *Jambosa hemispherica* (Wight) Walp., Repert. 2 : 191.1842; Gamble, Fl. Pres.Madras 474.1919. *Thol-najaval.*

Leaves decussate, ovate-lanceate to oblanceate, 5-15 by 2-5 cm, base and apex acuminate, lateral veins faint above, coriaceous. Cymes pyramidal, axillary and terminal, solitary or in pairs. Flowers c. 3 cm ϕ, calyx-tube semi-globose; petals white. Berry subglobose to globose.

Fl. & Fr.: Mar.-Jun. *Distr.*: Sri Lanka, Peninsular India. Not common. In evergreen forests. *NAK 2726* (Way to Ponnambalamedu, c. 1100 m).

Berries edible.

5. **Syzygium laetum** (Buch.-Ham.) Gandhi in Sald. & Nicols., Fl. Hassan Dist. Fl. Hassan Dist. 282. 1976; Manilal, Fl. Silent Valley 106.1988; Mohanan & Sivad., Fl Agasthyamala 262. 2002. *Eugenia laeta* Buch.-Ham., Mem. Wern. Nat. Hist. Soc. 5 : 338. 1826. *Jambosa pauciflora* Wight, Ill. 2 : 14. 1841. *Eugenia pauciflora* (Wight) Wight, Ic. t. 526. 1842. *Jambosa laeta* (Buch.-Ham.) Bl., Mus. Bot. 1 : 104. 1849; Gamble, Fl. Pres. Madras 474. 1919. *Eugenia laeta* var. *pauciflora* (Wight) Duthie in Hook. f., Fl. Brit. India 2 : 479. 1878.

Leaves decussate, elliptic, 5-12 by 2-4 cm, base cuneate, apex acuminate, thin coriaceous. Cymes terminal, rarely lateral. Flowers 1-3, large; calyx-tube narrow-conical, lobes 4-5, suborbicular; petals 4-5, caducous; filaments reddish or yellow. Berry ovoid. Seeds angled.

Fl. & Fr.: Jan.-May. *Distr.*: Western Ghats. Common. In evergreen forests. *NAK 976* (Sabari hills, c. 250 m), *1455* (Arampa, c. 450 m).

6. **Syzygium malabaricum** (Bedd.) Gamble, Fl. Pres. Madras 481. 1919. *Eugenia malabarica* Bedd., Fl. Sylv. t. 199. 1972; Duthie in Hook. f., Fl. Brit. India 2 : 497. 1879.

Leaves decussate, obovate-spathulate to obcordate, 5-12 by 2.5-6 cm, base cuneate, apex obtuse to retuse, thin-coriaceous. Cymes lateral, usually cauline. Flowers α, small; calyx-4, minute; petals 4, white, orbicular. Berry spherical.

Fl. & Fr.: Sep.-Jan. *Distr.:* Western Ghats. Very rare. In evergreen forests. *KV 48572 (MH)* Pachakkanam, (Kakki hills, c. 1000 m).

7. **Syzygium mundagam** (Bourd.) Chithra in Nair & Henry, Fl. Tamilnadu 1 : 157.1983; Manilal, Fl. Silent Valley 107. 1988; Mohanan & Sivad., Fl. Agasthyamala 263. 2002. *Eugenia mundagam* Bourd., For. Trees Travancore 182.1908. *Jambosa mundagam* (Bourd.) Gamble, Fl.Pres.Madras 473.1919. *Kaatu-champa, Mundagam.*

Branchlets faintly 4-gonous. Leaves decussate, subsessile, elliptic-oblong, 16-32 by 8-11 cm, base subcordate, apex acute to obtuse, lateral veins and intramarginal vein prominent, pellucid-punctate, coriaceous. Cymes terminal, 3-chotomous. Flowers large; calyx-tube obconic, reddish, lobes 4, suborbicular; petals 4, white; filaments reddish. Berry globose, c. 3 cm. Seeds large, angled.

Fl. & Fr.: Mar.-Jun. *Distr.*: Western Ghats. Rare. In evergreen forests. A small tree with large beautiful flowers. *NAK 2408* (Moozhiar, c. 250 m).

Decoction of bark is useful in nervous disorders, improves complexion and digestion.

8. **Syzygium occidentale** (Bourd.) Gandhi in Sald. & Nicols., Fl. Hassan Dist. 282.1976. *Eugenia occidentalis* Bourd., Ind. For. 30 : 195, t. 3. 1904. *Jambosa occidentalis* (Bourd.) Gamble, Fl. Pres. Madras 474. 1919. *Kari-njara.*

Leaves sub-opposite, linear-lanceate, 4-13 by 0.5-1.7 cm, base cuneate to attenuate, apex gradually acute, lateral veins faint. Cymes terminal, 3-chotomous. Flowers c. 4.5 cm ϕ; calyx-tube conical, lobes 4-5, ovate or orbicular; petals 5, whtie. Berry globose, pale-red, c. 2 cm ϕ. Seeds small, angled.

Fl. & Fr.: Jan.-Apr. *Distr.*: Western Ghats. Rare. Rheophytic trees. *NAK 1620 & 1665* (Pampavalley, c. 100 m).

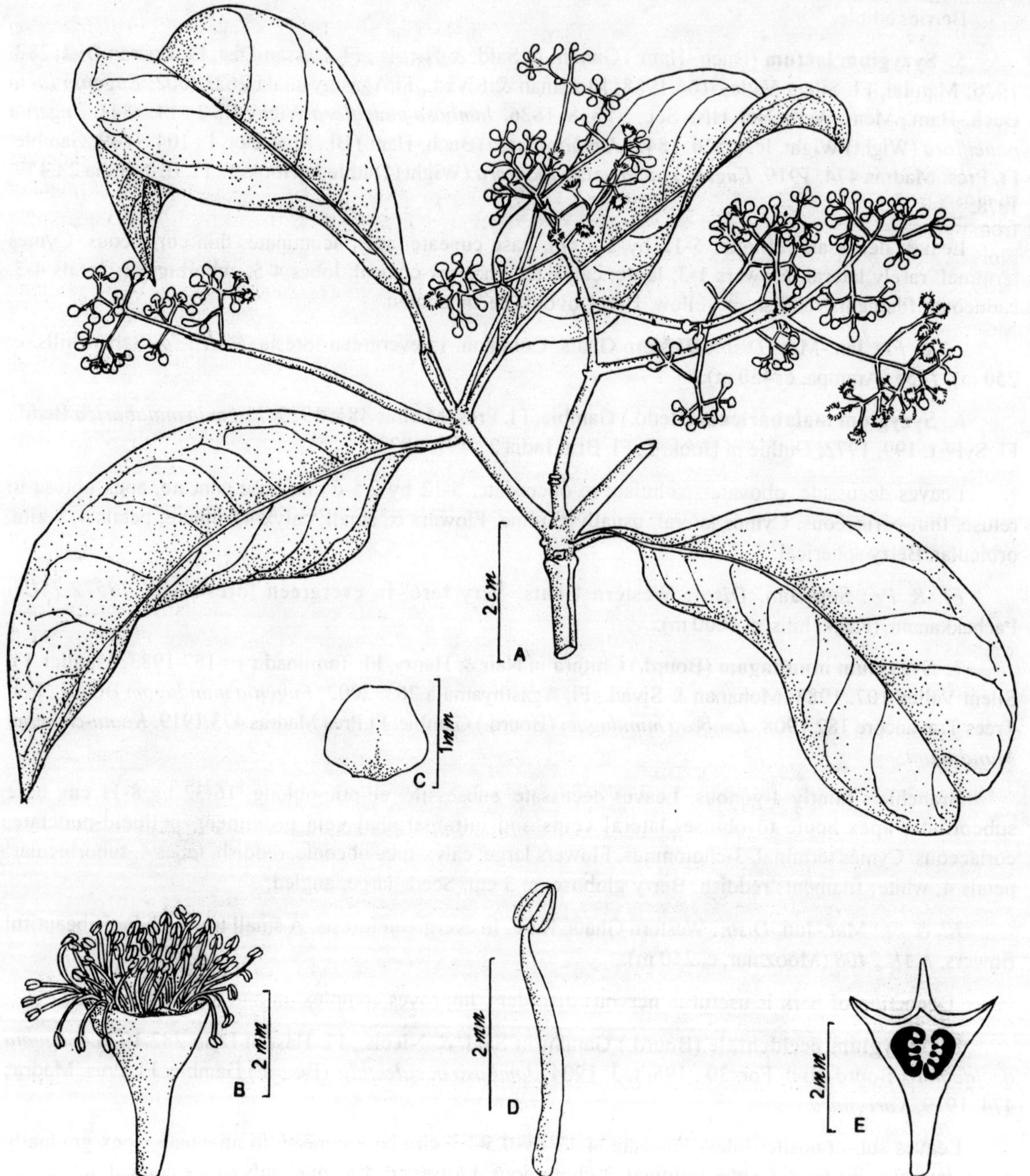

Fig. 37. *Syzygium travancoricum* Gamble: A. Twig; B. Flower; C. Petal; D. Stamen; E. Ovary - L.S.

9. **Syzygium rubicundum** Wight & Arn., Prodr. 330. 1834; Gamble, Fl. Pres. Madras 479. 1919; Mohanan & Sivad., Fl. Agasthyamala 265. 2002. *Eugenia rubicunda* (Wight & Arn.) Wight, Ic. t. 538. 1842; Duthie in Hook.f., Fl. Brit. India 2 : 495. 1878. *Eugenia lissophylla* (Thw.) Bedd., For.Man.Bot. 108.1874; Duthie in Hook.f., Fl. Brit. India 2 : 488. 1878.

Leaves decussate, elliptic-ovate to obovate, 6-8 by 2-3.5 cm, base cuneate, apex caudate, lateral nerves α, close, parallel, forming intramarginal vein, thick-coriaceous, drying greyish. Cymes panicled, terminal or from upper leaf-axils; calyx-tube turbinate, lobes 4; petals 4, white, calyptrate; filaments white. Berry small, globose. Seed solitary.

Fl. & Fr.: Dec.-Apr. *Distr.*: Sri Lanka, Peninsular India. Rare. In evergreen forests. *NAK 367* (Upper Moozhiar, c. 500 m).

10. **Syzygium travancoricum** Gamble, Kew Bull. 1918:240. 1918 & Fl. Pres. Madras 480. 1919; Nair & Mohanan, J. Econ. Tax. Bot. 2 : 233-235. 1981. **(Fig. 37)**

Leaves decussate, ovate-oblong, 10-20 by 4-9 cm, base obliquely rounded to broadly cuneate, apex acute, shiny. Cymes axillary. Flowers small, white. Berry oblong. Seed 1.

Fl. & Fr.: May-Jun. *Distr.*: Western Ghats. An endangered species. In some sacred groves only. *NAK 2732* (Aikkad grove - Kodumon, c. 30 m).

11. **Syzygium zeylanicum** (L.) DC., Prodr. 3 : 260. 1828; Wight, Ic. t. 73. 1838; Gamble, Fl. Pres. Madras 479. 1919; Gandhi in Sald. & Nicols., Fl. Hassan Dist. 283. 1976; Nicols., Suresh & Manilal, An Interpr. Hort. Malab. 196. 1988;Mohanan & Sivad., Fl. Agasthyamala 266.2002. *Myrtus zeylanica* L., Sp. Pl. 472. 1753. *Eugenia spicata* Lam., Encyl. 3 : 201. 1789; Bedd., Fl. Sylv. t. 202.1872.

Leaves decussate, ovate-lanceate, 2-5 by -0.7-1.7 cm, base acute to obtuse, apex acuminate, reticulations prominent below, thick-coriaceous. Cymes of small umbellules, terminal and axillary; calyx-tube small conical, lobes 4; petals 4, white, distinct. Berry globose, white. Seed solitary.

Fl. & Fr.: Jan.-Apr. *Distr.*: Indo-Malesia. Shrubs which are common in sacred groves *NAK 2340* (Aikadu grove, Kodumon, c. 50 m).

Syzygium aromaticum (L.) Merr. & Perry is seen planted elsewhere in the district.

58. BARRINGTONIACEAE

1. Flowers scarlet coloured, c. 1.5 cm ϕ; racemes pendant; berries angular, fibrous. ... **Barringtonia**
1. Flowers white coloured, c. 7 cm α; spikes erect; berries globose,fleshy. ....**Careya**

BARRINGTONIA J.R. Forster et J.G.A. Forster
Charact. Gen. 75, t. 38. 1776 (nom. cons.).

Barringtonia acutangula (L.) Gaertn., Fruct. Sem. Pl. 2 : 97, t. 101. 1790; Roxb., Fl. Ind. 2 : 635. 1832; Clarke in Hook.f., Fl.Brit.India 2 : 508.1879; Gamble, Fl. Pres. Madras 487.1919; Matthew, Ill.Fl.Tam. Carnatic t. 273. 1982; Matthew & Britto in Matthew, Fl. Tam. Carnatic 3(1) : 596. 1983; Nicols., Suresh & Manilal, An Interpr. Hort. Malab. 159. 1988. *Tsjeria-samastravadi* Rheede, Hort. Malab. 4 : 15, t. 7. 1683.

Trees. Leaves simple, alternate, apically clustered, obovate, 7-14 by 3-5 cm, base attenuate, margin serrulate, apex subacute, thin-coriaceous. Racemes axillary, pendant, c. 30 cm long. Flowers, cm. 1 ϕ, bisexual, 4-5 merous; calyx-tube cupular; lobes 4; petals 4, scarlet coloured, basally connate; stamens very a; filaments pink, connate at base; ovary inferior, 2-locular; ovules 2-6-per locule; style long, filiform. Berry angular, fibrous. Seeds ovoid, solitary.

Fl. & Fr.: Mar.-May. *Distr.*: Sri Lanka, Myanmar, Malesia, N. Australia. Less common. Along streamsides. *NAK 1593* (Pampa Valley, c. 100 m).

Rook bark used to poison fish. Seeds powdered and given with milk to reduce enlarged abdomen of children.

CAREYA Roxburgh
Pl. Coromandel 3 : 13.1811 (nom.cons.).

Careya arborea Roxb., Pl. Corom. t. 218. 1811; Wight & Arn., Prodr. 334.1834; Bedd., Fl. Sylv. t. 205. 1872; Clarke in Hook. f., Fl. Brit. India 2 : 511.1879; Gamble, Fl. Pres. Madras 488. 1919; Gandhi in Sald. & Nicols., Fl. Hassan Dist. 157.1976; Matthew, Ill. Fl. Tam. Carnatic 274. 1982; Matthew & Britto in Matthew, Fl. Tam. Carnatic 3(1) : 597.1983; Manilal, Fl. Silent Valley 108.1988; Nicols., Suresh & Manilal, An Interpr. Hort. Malab. 160. 1988; Mohanan & Sivad., Fl. Agasthyamala 266. 2002. *Pelou* Rheede, Hort.Malab. 3 : 35-36, t. 36.1682.

Trees. Leaves simple, alternate, crowded towards branch-apices, obovate, 10-30 by 5-15 cm, base cuneate, margin denticulate to entire, apex acute, thin-coriaceous. Flowers large in erect spikes; calyx-tube campanulate, lobes 4, ovate, persistent; petals 4, white; stamens α, filaments slender, cream coloured; disc annular; ovary inferior, 4-5-locular; ovules α-per locule. Berry large, globose, crowned by persistent calyx. Seeds α.

Fl. & Fr.: Jan.-Apr. *Distr.*: Afghanistan to N. Malaya.Common. In deciduous forests. *NAK 1616* (Chathanthara, c. 70 m).

Infusion of flowers given after child birth to heal ruptures. Juice of fresh bark given with honey as demulcent in cough and cold.

59. MELASTOMATACEAE

1. Epiphytic scandent shrubs; leaves succulent. .... **Medinilla**
1. Terrestrial erect herbs or shrubs; leaves not succulent.
   2. Branches scaly; stamens unequal; fruit-dehiscence irregular. .... **Melastoma**
   2. Branches hairy; stamens equal; fruit-dehiscence apical/and or poricidal.
      3. Flowers 4-5-merous; anther-connective produced; inflorescence fascicled.
         4. Secondary veins parallel; cymes axillary; flowers white. .... **Clidemia**
         4. Secondary veins reticulate; cymes terminal; flowers pink. .... **Osbeckia**
      3. Flowers 3-merous; anther- connective not produced; inflorescence scorpioid cyme. .... **Sonerila**

CLIDEMIA D. Don
Mem. Wern.Nat. Hist. Soc. 4 : 306. 1823.

Clidemia hirta (L.) D. Don, Mem. Wern. Nat. Hist. Soc. 4 : 306. 1823; Nair, Kew Bull. 20 : 155-161. 1966; Bremer in Dassan. & Fosberg, Rev. Handb. Fl. Ceylon 6 : 177. 1987; Mohanan & Sivad., Fl. Agasthyamala 267. 2002 *Melastoma hirta* L., Sp. Pl. 390. 1753.

Profusely branched subshrubs, branchlets terete, strigosely hairy. Leaves simple, opposite, broadly - elliptic ovate or ovate, 6-15 by 3.5-8 cm, base rounded to subcordate, margin crenate-dentate, apex acuminate, densely hairy, basally 5-ribbed, secondary veins parallel, chartaceous. Flowers white, in axillary cymes; calyx-tube campanulate to cupular, lobes 5, linear; petals 5, white, elliptic-oblong; stamens 10; ovary 5-locular, ovules a. Capsule dehiscing apically. Seeds minute.

Fl. & Fr.: Dec.-Mar. *Distr.*: South America, Malesia. Naturalized in Old World. Fairly common along wayside in hills. *NAK 86* (Moozhiar, c. 250 m), *660* (Pampa, c. 260 m).

Note : This introduced species in rapidly flourishing along forest waysides in Kakki hills, Moozhair, Sabarimala, etc. and is posing a serious threat to the indigenous peripheral evergreen species.

MEDINILLA Gaudichaud-Beaupre
in Freycinet, Voyage Mon de
Uranie Physicienne Bot. 484. 1830 ("1826")

Medinilla beddomei Clarke in Hook. f., Fl. Brit. India 2 : 548. 1879; Gamble, Fl. Pres. Madras 496. 1919; Manilal, Fl. Silent Valley 108.1988; Mohanan & Sivad., Fl. Agasthyamala 268. 2002. *M. radicans* sensu Bedd., Ic. t. 184. 1869-1879, non Bl., 1831.

Epiphytic shrubs, rooting at nodes. Leaves simple, opposite, suborbicular, 3-4 by 2.5-3 cm, fleshy, 3-ribbed. Flower(s) solitary, axillary, bisexual, 4-5 merous, epigynous; calyx-tube ovoid, lobes 4-5; petals rose-white, succulent; stamens 8-10, anther curved, acuminate at apex, connective prominent; ovary inferior, 4-6-locular; ovules α per locule. Berry crowned by the limb of calyx.

Fl. & Fr.: Apr.-Jun. *Distr.*: Western Peninsular India. Rare. In evergreen forests. *NAK 2692* (Kakki hills, c. 1050 m.).

MELASTOMA Linnaeus
Sp. Pl. 389. 1753.

Melastoma malabathricum L., Sp. Pl. 390. 1753 ('*malabathrica*'); Clarke in Hook. f., Fl. Brit. India 2 : 523. 1878; Gamble, Fl. Pres. Madras 495. 1919; Ramamoorthy in Sald. & Nicols., Fl. Hassan Dist. 286. 1976; Nicols., Suresh & Manilal, An Interpr. Hort. Malab. 175.1988; Mohanan & Sivad., Fl. Agasthyamala 269. 2002. *Kadali* Rheede, Hort. Malab. 4 : 87-88, t. 42. 1683. *Kalam-potti.*

Shrubs. Leaves simple, opposite, elliptic to elliptic-lancate, 4-11 by 1.5 - 4 cm, base and apex acute, upper surface prominently lineolate, lower surface tomentose, equally 5-veined, thin-coriaceous, drying dull-greenish, Flower(s) solitary or in short panicles, c. 5 cm ϕ; calyx-tube campanulate, lobes 5, ovate, appendages subulate, alternate the calyx-lobes; petals 5, red-purple; stamens 10, alternately long and short, anthers dimorphic; ovary 5-locular; ovules α per locule. Capsule irregularly dehiscent. Seeds α, minute, punctate.

Fl. & Fr.: Most of the seasons. *Distr.*: Throughout India. Fairly common along wet places, near rivulets, etc. *NAK 119* (Adoor, c. 50 m.).

Decoction of flowers given to cool the head, in piles and haemorrhage.

OSBECKIA Linnaeus
Sp. Pl. 345. 1753.

Subshrubs or herbs; branchlets 4-angular. Leaves simple, decussate, basally 3-5-nerved, upper surface lineolate. Cymes terminal, (rarely flower solitary). Flowers 4-5-merous, usually pink; stamens 8 or 10, subequal, anthers usually beaked, dehiscing by apical pores; ovary semi-inferior, 4-or 5-locular; ovules α per locule. Capsule enclosed within the calyx. Seeds α, minutely tubercled.

1. Flowers 5-merous; anthers beaked at apex.
    2. Calyx-lobes obtuse to emarginate at apaex; vegetatative parts densely bristly
        3. Calyx-tube with simple bristles towards base, tufted bristles towards apex; leaves 5-7-nerved. .... 1.**O. aspera**
        3. Calyx-tube with tufted stalked bristles all over, rarely simple bristles towards extreme base; leaves ±5-nerved. .... 2. **O. wightiana**
    2. Calyx-lobes acute; vegetative parts sparsely bristly. .... 3. **O. virgatus**
1. Flowers 4-merous; anther truncate at apex. .... 2. **O. muralis**

1. **Osbeckia aspera** (L.) Bl., Bijdr. Natuurk. Wetensch. 6 : 220.1831. Clarke in Hook. f., Fl. Brit. India 2 : 519. 1879; Gamble, Fl. Pres. Madras 492. 1919; Hansen, Ginkgoana 4 : 82. 1977; Manilal, Fl. Silent Valley 110. 1988; Mohanan & Sivad., Fl. Agasthyamala 278.2002. *Melastoma asperum* L., Sp. Pl. 389. 1753. *Osbeckia kleinii* Wight & Arn., Prodr. 323.1834. *O. courtallensis* Gamble, Kew Bull. 1918: 242.1918, Fl. Pres. Madras 493.1919.

Key to the varieties

1. Leaves elliptic to oblong-lanceate; stem greenish; capsule ribbed. ....**O. aspera** var. **aspera**
1. Leaves lanceate; stem greyish with long retrorse bristles; capsule not ribbed. ....**O. aspera** var. **travancorica**

O. aspera (L.) Bl.var. **aspera**

Leaves elliptic to oblong-lanceate, 3-9 by 1.3-3.5 cm, base acute or obtuse, apex actue, basally 5-7-nerved, outer pair very slender, thin-coriaceous, drying yellowish-green. Flower(s) solitary or 2-3, terminal or from upper leaf-axils; calyx-tube narrowly cupular, lobes 5, oblong; petals 5, pink; stamens 8, anthers beaked at apex; ovary 5-locular. Capsule ribbed.

Fl. & Fr.: Nov.-Dec. *Distr.:* Sri Lanka, Western Ghats. Not common. Among wayside thickets in plains. *NAK 120* (Adoor, c. 30m), *1837* (Tiruvalla, c. 10 m).

Osbeckia aspera (L.) Bl. var. **travancorica** (Bedd. ex Gamble) Hansen in Ginkgoana 4 : 84, t. 13 (B). 1977; Nicols., Suresh & Manilal, An Interpr. Hort. Malab. 176. 1988. *O. travancorica* Bedd. ex Gamble, Kew Bull. 1919: 404. 1919 & Fl. Pres. Madras 493. 1919. *Katou-kadali* Rheede, Hort. Malab. 4 : 91, t. 43. 1683.

Branchlets greyish, with long retrorse bristles. Leaves lanceate, 4-10 by 1.3-3 cm, base obtuse, apex gradually acute or shortly acuminte, lower surface bristly , 5-7-nerved, thin-coriaceous. Flowers c. 4 cm ϕ;

bracteoles; calyx cupular, densely tufted bristly, lobes 5, oblong; petals 5, purple; stamens 8; ovary 5-locular. Capsule subglobose.

Fl. & Fr.: Sep.-Dec. *Distr.*: Western Ghats. Rare. On forest borders especially in exposed hilly areas. *NAK 435, 648, 805* (Ranni, c. 100 m).

2. **Osbeckia muralis** Naud., Ann. Sci. Nat. Bot. Ser. 3. 14: 56. 1850; Hansen, Ginkgoana 4: 41, ff. 2(5) & 9(10), tt. 6. (C & D). 1977. *O. truncata* D.Don ex Wight & Arn., Prodr. 332. 1834, p.p., nom. illegit., Clarke in Hook.f., Fl. Brit. India 2: 514. 1879; Gamble, Fl. Pres. Madras 494. 1919. *Cherkulathi.*

Leaves crowded towards the branchlets, elliptic, 1-4 by 0.6-2 cm, base rounded to obtuse, apex subacute, 3-nerved, chartaceous. Flower 4-merous, c. 15 mm f; calyx cupular, ribbed, long-simple hairy, lobes 4, triangular, stellately bristly at apex; petals 4, purple; stamens 8, anthers short, apex truncate; ovary 4-locular. Capsule small, globose.

Fl. & Fr.: Sep-Jan. *Distr.*: Western Ghats. Slender herbs on exposed areas of hills. *NAK 962* (Sabari hills, c. 250 m).

3. **Osbeckia virgata** D. Don ex Wight & Arn., Prodr. 323. 1834; Hasen, Ginkgoana 4 : 109, ff. 1(8); 2(7) & 35 (5). 1977; Matthew, Ill. Fl. Tam. Carnatic t. 277. 1982; Matthew & Britto in Matthew, Fl. Tam. Carnatic 3(1) : 602. 1983; Manilal, Fl. Silent Valley 111. 1988.

Branchlets slender, sparsely appressed-bristly. Leaves elliptic, 2-6 by 0.6-2 cm, appressed-bristly, base cuneate, margin subentire, apex acute, basally 3-(5)- nerved, chartaceous. Cymes 3-6-flowered, terminal. Flowers 5-merous, c. 2 cm ϕ; calyx-tube subglobose with stalked stellate and simple bristles, lobes 5, lanceate, acute; petals 5, pinkish-purple; stamens 10; ovary 5-locular. Capsule globose.

Fl. & Fr.: Most of the seasons. *Distr.*: Peninsular India, Sri Lanka. The most common species of the genus. In exposed hill cuttings. *NAK 640* (Ranni, c. 70 m).

4. **O. wightiana** Benth. ex Wight & Arn., Prodr. 323. 1834; Wight, Ic. t. 998. 1845; Clarke in Hook. f., Fl. Brit. India 2 : 519; 1879; Gamble, Fl. Pres. Madras 493. 1919; Mohanan & Sivad., Fl. Agasthyamala 280. 2002.*Osbeckia aspera* (L.) Bl., var. *wightiana* (Benth. ex Wight & Arn.) Trimen, Handb. Fl. Ceylon 2 : 196. 1894.

Branchlets reddish with short stiff bristles. Leaves elliptic, 2.5-6 by 1-3.5 cm, base truncate-obtuse, apex acute, lower surface hirsute, 5-nerved, thin coriaceous. Cymes terminal. Flowers c. 4 cm ϕ, calyx globose, very densely tufted hairy, lobes 5, oblong; petals 5, purple; stamens 8; ovary 5-locular. Capsule subglobose, densely villous hairy.

Fl. & Fr.: Nov.-Apr. *Distr.*: Sri Lanka, Western Ghats. Common. Along forest borders of evergreen types. *NAK 363, 435, 1116 & 1317.* (Most of the places)

SONERILA Roxburgh
Fl. Indica 1 : 180 1820 (nom. cons.).

Slender subsucculent herbs, caulescent or acaulescent. Leaves simple, decussate, palmately or pinnately veined. Inflorescence a scorpiod cyme, terminal. Flowers 3-merous; calyx campanulate or long-funnelform, lobes 3, short; petals 3, pink; stamens 3, apex beaked or not; ovary inferior, 3-locular; ovules α. Capsule obconic, ribbed, apically 3-valvular. Seeds α, smooth or minutely tubercled.

1. Leaves nearly palmately veined; stem to 60 cm tall. ....1. **S. brunonis**
1. Leaves pinnately veined; stem to 15 cm tall.
   2. Stem decumbent or creeping; lmeaves opposite or subopposite; anthers beaked. ....2. **S. rheedii**
   2. Stem erect;leaves fascicled or opposite; anthers beaked or not.
      3. Stem very short; leaves fascicle; anthers not beaked. ....3. **S. sahyadrica**
      3. Stem much longer; leaves opposite; anthers beaked. ....4. **S. versicolor**

1. **Sonerila brunonis** Wight & Arn., Prodr. 321. 1834; Wight, Ic. t. 1059. 1846; Clarke in Hook.f., Fl. Brit. India 2:531. 1879; Gamble, Fl. Pres. Madras 499. 1919; Mohanan & Sivad., Fl. Agasthyamala 282. 2002.

Herbs to 60 cm high; stem 4-winged, reddish, rooting higher up from the woody base. Leaves ovate-lanceate, 2.5-7 by 1.2-3.2 cm, base equilateral, rounded to subcordate, margin finely serrulate, apex acute to bluntly acuminate, upper surface sparsely hairy, lower surface glabrous, basaly 5-7 nerved, membranous. Cymes terminal or subterminal, to 14-flowered. Flowers c. 1 cm ϕ; calyx-tube obconic, lobes triangular; petals mauve, ovate; stamens 3, anthers beaked. Capsule obconic, 6-ribbed.

Fl. & Fr.: Oct.-Dec. *Distr.*: Endemic to Western Ghats. Along moist shady hill cuttings in higher hills. *NAK* 1180, 1363 (Kakki hills, c. 1150 m).

2. **Sonerila rheedii** Wight & Arn., Prodr. 321. 1834; Gamble, Fl. Pres. Madras 500. 1919; Ramamoorthy in Sald. & Nicols., Fl. Hassan Dist. 290. 1976; Nicols., Suresh & Manilal, An Interpr. Hort. Malab. 177. 1988; Mohanan & Sivad., Fl. Agasthyamala 283. 2002. *S. wallichii* sensu Clarke in Hook.f., Fl. Brit. India 2 : 538. 1879, p.p., non Benn., 1844. *Soneri-ila* Rheede, Hort. Malab. 9: 127, t. 65 1689.

Decumbent to suberect herbs; stem 2-15 cm long. Leaves opposite to subopposite, elliptic-ovate to oblong-ovate, 3.5-10 by 2.15 cm, base subcordate to obtuse rotund, margin faintly serrulate, apex subacute, puberulous, pinnately veined, membranous, greenish or variegated. Cyme 5-8-flowered; anthers long-beaked. Capsule sparsely glandular hairy.

Fl.& Fr.: Nov.-Dec. *Distr.*: Endemic to Western Ghats. On moist shady hill cuttings. Locally abundant. *NAK 26 & 1126* (Lower Moozhiar, c. 250 m).

3. **Sonerila sahyadrica** Giri & Nayar, Bull. Bot. Surv. India 26 (3 & 4) : 175. 1984.

Small herbs; stem very short, to 2 cm long, bulbous at base. Leaves usually fascicled, ovate, 2-4 by 1.2-2.5 cm, base acute to obtuse, margin serrulate, ciliate, apex acute, pinnately veined, membranous. Cyme few-flowered; anthers without a long beak. Capsule glabrous.

Fl. & Fr.: Nov.-Dec. *Distr.*: Endemic to Western Ghats. Rare. On moist shady hill cuttings. Seen in association with *S. rheedii,* and presumably could be a descendant species. *NAK* 22 (Lower Moozhiar, c. 250 m).

4. **Sonerila versicolor** Wight, Ic. t. 1057. 1846; Clarke in Hook.f., Fl. Brit. India 2 : 535. 1879, p.p; Gamble, Fl. Pres. Madras 499. 1919; Mohanan & Sivad., Fl. Agasthyamala 284. 2002.

Branched herbs. Leaves elliptic, 5-8 by 2-3 cm, base rounded, margin finely serrulate, apex acute, sparsely long hairy on both surfaces, pinnately veined. Flowers few in terminal cymes; petals mauve. Capsules in crowded clusters.

Fl. & Fr.: Jun.-Dec. *Distr.*: Western Ghats. Rare. Along moist shady places. *PR 79288* (MH) (Kakki dam area, c. 1100 m).

60. MEMECYLACEAE

MEMECYLON Linnaeus
Sp. Pl. 349. 1753.

Shrubs or trees. Leave decussate, petioled or sessile, lateral nerves obscure, coriaceous. Flowers 4-merous, axillary or lateral, in condensed or effuse umbels; calyx-tube adnate to ovary, lobes 4, suborbicular; petals 4, blue; disc rayed; stamens 8, subequal, anther- connective spurred at base; ovary inferior, 1-locular; ovules 6-12 per locule. Berry globose. Seeds 1.

1. Leaves sessile, base cordate. ....4. **M. malabaricum**
1. Leaves petioled, base cuneate or rounded.
    2. Leaves drying yellowish.
        3. Trees; leaf apex acute to acuminate; disc-rays faint. ....5. **M. talbotianum**
        3. Shrubs; leaf apex obtuse to retuse; disc-rays conspicous. ....6. **M. umbellatum**
    2. Leaves drying dull-greenish or blackish.
        4. Leaves drying blackish; peduncule and pedicel elongated........1. **M. edule** var. **edule**
        4. Leaves drying dull greenish; peduncle and pedicel very short or 0.
            5. Leaves to 20 cm long, intramarginal nerves and disc-rays very prominent. ....2. **M. heyneanum**
            5. Leaves to 10 cm long, intramarginal nerves and disc-rays very faint. ....3. **M. lawsoni**

1. **Memecylon edule** Roxb., Pl. Corom. t. 82. 1798 & Fl. Ind. 2: 260. 1832; Clarke in Hook.f., Fl. Brit. India 2: 563. 1879, p.p.; Gamble, Fl. Pres. Madras 504. 1919; J.F. Maxwell, Gard. Bull. Straits Settlem. 33: 72. 1980; Matthew, Ill. Fl. Tam. Carnatic t. 275. 1982; Matthew & Britto in Matthew, Fl. Tam. Carnatic 3(1) : 600 1983; Nicols., Suresh & Manilal, An Interpr. Hort. Malab. 175. 1988. *Kurjuvo-murum* Rheede, Hort. Malab. 5:37, t. 19. 1685.

Shrubs. Leaves narrowly elliptic, 5-9 by 2-3 cm, base rounded, revolute, apex acuminate, thin-coriaceous. Umbels to 3 cm long; peduncle c. 5 mm long, stout; pedicel 5-6 mm long, slender; calyx-tube campanulate; petals blue, broadly ovate. Berry globose, blue.

Fl. & Fr.: Mar.-Aug. *Distr.*: Sri Lanka, India, Myanmar, Thailand, Malesia. Common. On forest borders. *NAK* 546 (Thekkuthode, c. 70 m).

2. **Memecylon heyneanum** Benth. ex Wight & Arn., Prodr. 319. 1834; Clarke in Hook.f., Fl. Brit. India 2 : 560. 1879; Gamble, Fl. Pres. Madras 503. 1919; Manilal, Fl. Silent Valley 109. 1988. *M. jambosoides* Wight, Ic. t. 277. 1840.

Large shrubs. Leaves lanceate, 7-18 by 2.5 cm, base acute or broadly cuneate, apex acuminate, lateral veins c. 16 pairs, forming loops with very prominent intramarginal nerves, thin coriaceous. Fascicles c. 3 cm ϕ. Flowers subsessile; calyx-tube campanulate; disc-rays conspicous; petals blue, obovate. Berry globose, dark-purple.

Fl. & Fr.: Feb.-Sep. *Distr.*: Sri Lanka, Peninsular India. Rare. In interior evergreen forests. *NAK 2051* (Way to Devarmalai, c. 600 m).

3. **Memecylon lawsoni** Gamble, Kew Bull. 1919: 226. 1919, Fl. Pres. Madras 503. 1919; Manilal, Fl. Silent Valley 109. 1988.

Trees. Leaves elliptic, 6-10 by 2.5-4 cm, base cuneate, apex shortly acuminate, lateral veins 6-8 pairs, very faint, coriaceous. Fascicles c. 3 cm ϕ. Flowers few, sessile; calyx-tube campanulate; disc-rays inconspicous; petals white to blue, suborbicular. Berry globose, ripening bright yellow.

Fl. & Fr.: Mar.-Aug. *Distr.*: Western Ghats. Rare. In interior of evergreen forests. *NAK* 1810 (Angamoozhy, c. 200 m), *2457* (Arampa, c. 400 m).

4. **Memecylon malabaricum** (Clarke) Cogn. in DC., Monogr. Phan. 7: 1148. 1891; Gamble, Fl. Pres. Madras 505. 1919. *M. amplexicaule* Roxb., var. *malabarica* Clarke in Hook.f., Fl. Brit. India 2: 2559. 1879 p.p., incl. typus; Ramamoorthy in Sald. & Nicols., Fl. Hassan Dist. 288. 1976; Mohanan & Sivad., Fl. Agasthyamala 274. 2002. *Kaikka-thetti.*

Shrubs to small trees; branchlets terete. Leaves lanceate, sessile, 6-12 by 1.5-4 cm, base cordate, apex gradually acute, lateral veins c. 12 pairs, faintly visible, thin coriaceous. Cymes fascicled on lateral tubercles of stem. Berry globose, bluish.

Fl. & Fr.: Feb.-May. *Distr.*: Western Ghats. Not common. In sacred groves. *NAK 1644* (Tiruvalla, c. 10 m).

Decoction of flowers and tender shoots boiled with soil, and used for itches. Leaves boiled in water and in used to bathe children as a remedy for itches.

5. **Memecylon talbotianum** Brandis in Talbot, Bombay List. ed. (2). Append. 1902 & Indian Trees 336. 1906; Gamble, Fl. Pres. Madras 503. 1919; Ramamoorthy in Sald. & Nicols., Fl. Hassan Dist. 288. 1976; Mohanan & Sivad., Fl. Agasthyamala 276. 2002.

Trees. Leaves elliptic-ovate, 4-10 by 2.5-4.5 cm, base cuneate, apex acuminate, lateral veins 6-8 pairs, invisible, drying yellowish. Fascicles axillary and lateral, c. 2 cm ϕ, sessile. Flowers small, sessile; calyx campanulate; disc-rays inconspicous; petals blue. Berry globose.

Fl. & Fr.: Feb.-May. *Distr.*: Western Ghats. Rare. In evergreen forests. *NAK* 1525 (Gurunadhan mannu, c. 300 m).

6. **Memecylon umbellatum** Burm. f., Fl. Ind. 87. 1768; Gamble, Fl. Pres. Madras 504. 1919; Ramamoorthy in Sald. & Nicols., Fl. Hassan Dist. 288. 1976; Matthew, Ill. Fl. Tam. Carnatic t. 276. 1982; Matthew & Britto in Matthew, Fl. Tam. Carnatic 3(1) : 600. 1983; Manilal, Fl. Silent Valley 110. 1988.

Shrubs. Leaves ovate-oblong or elliptic, 2.5-5 by 1.5-3 cm, base cuneate or obtuse, apex obtuse-retuse. Umbels compact; peduncle very short or 0. Flowers sessile. Berry globose.

Fl. & Fr.: Apr.-Nov. *Distr.*: Sri Lanka, Peninsular India. Common species of the genus in the district. In evergreen forest and in sacred groves. *NAK* 176 (Konni, c. 50 m).

61. LYTHRACEAE

1. Herbs; flowers small, solitary, in cymes or racemes.
    2. Flowers in axillary cymes; capsule- dehiscence irregular. ....**Ammania**
    2. Flower(s) solitary or in racemes; capsule dehiscence regular (valvular). ....**Rotala**
1. Trees; flowers large, in panicles. ....**Lagerstroemia**

AMMANIA Linnaeus
Sp. Pl. 119. 1753.

Ammania baccifera L., Sp. Pl. 120. 1753; Clarke in Hook.f., Fl. Brit. India 2: 569. 1879; Gamble, Fl. Pres. Madras 510. 1919; Gandhi in Sald. & Nicols., Fl. Hassan Dist. 272. 1976; Matthew & Britto in Matthew, Fl. Tam. Carnatic 3(1) : 605. 1983.

Erect herbs. Leaves decussate, sessile, oblanceate, 2-8 by 0.5-0.8 cm, base attenuate, apex acute, chartaceous. Cymes dichasial, axillary. Flowers 4-or 5-merous, perigynous; calyx-tube campanulate, lobes 4(5); petals 0; stamens 4(5), subexerted; ovary sessile, 4-5-locular; ovules α; stigma capitate. Capsule subglobose, irregularly dehiscent. Seeds minute, ovoid.

Fl. & Fr.: Sep.-Dec. *Distr.*: Paleotropics. Common. In abandoned paddy fields. *NAK* 1039 (Tiruvalla, c. 10 m).

LAGERSTROEMIA Linnaeus
Syst. Nat. ed. 10, 2: 1068, 1076, 1372. 1759.

Trees. Leaves decussate or upper ones alternate, coriaceous. Panicles terminal, dense. Flowers 6-merous, large, showy; calyx-tube campanulate, ribbed or not, lobes-6; petals 6, clawed, crumpled; stamens α, on the base of calyx-tube; ovary 4-6- locular; ovules a per locule; style curved. Capsule valvular dehiscent. Seeds winged.

1. Calyx-tube ribbed; petals large, mauve. ....1. **L. hirsuta**
1. Calyx-tube smooth; petals small, white or pink. ....2. **L. microcarpa**

1. **Lagerstromia hirsuta** (Lam.) Willd., Sp. Pl. 2:1178. 1799; Nicols., Suresh & Manilal, An Interpr. Hort. Malab. 166. 1988; Mohanan & Sivad., Fl. Agasthyamala 286.2002. *Adambea hirsuta* Lam., Encycl. 1:39. 1783. *Adambea glabra* Lam., Encycl. 1:39. 1783, non *L. glabra* Koechne, 1907. *L. flos-reginae* Retz., Obs. Bot. 5 : 25. 1789 (nom. illegit., incl. type of *Adambea glabra* Lam., 1783). *Lagerstroemia reginae* Roxb., Pl. Corom. t. 65. 1795; Wight, Ic. t. 413. 1843; Bedd., Fl. Sylv. t. 29. 1869. Furtado & Srisuko, Gard. Bull. Straits Settlem. 24 : 261. 1969; Gandhi in Sald. & Nicols., Fl. Hassan Dist. 273. 1976; Matthew & Britto in Matthew, Fl. Tam. Carnatic 3(1) : 606. 1983. Clarke in Hook.f., Fl. Brit. India 2 : 577. 1879; Gamble, Fl. Pres. Madras 513. 1919. *Poo-marthi.*

Leaves oblong-lanceate, 9-20 by 3-6 cm, base cuneate, apex acuminate, lateral veins 14-16 pairs, drying dull-greenish. Panicle dense. Flowers c. 7 cm ϕ; calyx-tube campanulate, ribbed, lobes triangular; petals mauve. Capsule globose.

Fl. & Fr.: Oct.-Apr. *Distr.*: India, Myanmar, New Guinea. Common. Along riversides in midlands. *NAK 1597* (Thannithode, c. 70 m).

Root astringent and seeds narcotic. Often planted as avenue trees.

2. **Lagerstroemia microcarpa** Wight, Ic. t. 109. 1839; Bedd., Fl. Sylv. t. 30. 1869; Furtado & Srisuko, Gard. Bull. Straits Settlem. 24 : 192. 1969; Gandhi in Sald. & Nicols., Fl. Hassan Dist. 273. 1976; Manilal, Fl. Silent Valley 113. 1988; Mohanan & Sivad., Fl. Agasthyamala 286. 2002. *L. lanceolata* Wall. ex Clarke in Hook.f., Fl. Brit. India 2 : 576. 1879; Gamble, Fl. Pres. Madras 513. 1919. *L. thomsonii* Koehne in Engler, Pflanzenr. IV. 216 (17) : 251. 1903; Gamble, Fl. Pres. Madras 513. 1919.

Leaves elliptic-lanceate, 7-10 by 3-45 cm, base broadly cuneate, apex acuminate, lower surface white. Flowers white to pink, in dense panicles; calyx white, smooth, tomentose. Capsule globose.

Fl. & Fr.: Apr.-Jun. *Distr.*: Peninsular and Western India. Less common. In moist semi-evergreen forests. *NAK 932* (Plapally, c. 400 m).

ROTALA Linnaeus
Mant. Pl. 143, 175. 1771.

Annual herbs, terrestrial or emergent aquatics. Leaves decussate or whorled, subsessile, chartaceous. Flower(s) solitry or in axillary racemes or spikes, 4-5-merous; bracteate and bracteolate; calyx-tube campanulate or cupular, lobes alternate usually with appendages; petals minute; stamens 3-5, exerted or not; ovary 2-4-locular; ovules α per locule. Capsule dehiscent by 2-4 valves. Seeds hairy.

1. Flowers 4-merous; capsule 4-valved, seldom longer than calyx.
    2. Flowers in dense spiciform racemes; stamens inserted at the base of calyx-tube. ....4. **R. rotundifolia**
    2. Flower(s) solitary and in axillary spikes.
        3. Calyx-appendages 0; stamens inserted at the middle of calyx-tube. ....1. **R. indica**
        3. Calyx-appendages subulate; stamens inserted below the middle of calyx-tube. ....2. **R. ritchiei**
1. Flowers 5-merous; capsule 3-valved, longer than calyx. ....3. **R. rosea**

1. **Rotala indica** (Willd.) Koehne, Bot. Jahrb. Syst. 1: 172. 1880; Gamble, Fl. Pres. Madras 508. 1919; Van Leeuwen, Blumea 19:54. 1971; Gandhi in Sald. & Ṅicols., Fl. Hassan Dist. 274. 1976; Cook, Boissiera 29 : 108. 1979; Matthew & Britto in Matthew, Fl. Tam. Carnatic 3 (1) : 613. 1983. *Peplis indica* Willd., Sp. Pl. 2 : 244. 1799. *Ammania nana* Roxb., Fl. Ind. 1 : 427. 1832. A. *peploides* Spreng., Syst. 1 : 444, 1825; Clarke in Hook.f., Fl. Brit. India 2 : 566. 1879.

Decumbent or erect marshy herbs; stem 4-angled or terete. Leaves suborbicular, 15 by 10 mm; base cuneate or obtuse, margin thickened, apex obtuse, chartaceous. Flower(s) sessile, 4-merous, axillary, solitary; bracts dimorphic, folicular on main branches, smaller on flowering branches; calyx-tube reddish-violet, lobes triangular ovate, appendages 0; petals 4, persistent; stamens 4, inserted at the middle of calyx-tube. Capsule ellipsoid, dehiscing 2-valvular. Seeds subglobose.

Fl. & Fr.: Nov.-Mar. *Distr.*: Native of S. E. Asia. Introduced and more or less naturalised in Italy, Portugal, Congo and U.S.A. Very common in rice-fields and marshy areas. *NAK 2233, 2325* (Kodumon, c. 30 m).

2. **Rotala ritchiei** (Clarke) Koehne, Bot. Jahrb. 4.386, 1883, & in Engl., Pflanzenr. IV. 216 (17): 36. 1903; Blatt. &Hallb., J. Bombay Nat. Hist. Soc. 25. 709, 1903; Janardhanan, Bull. Bot. Surv. India 21: 230, 1979; Cook, Boissiera 29: 71. 1979. *Ammannia ritchiei* Clarke in Hook.f., Fl. Brit. India 2: 566. 1879; Cooke, Fl. Pres. Bombay 1:577. 1903.

Aquatic or amphibious, slender, annual herbs, to 30 cm tall, with the flowering branch-apices emerging out of water, stem branched, creeping or floating below, simple above. Leaves decussate, ovate-elliptic to obovate, 7-10 by 3-4 mm, base narrowed, apex obtuse or rounded; bracteoles subulate, much shorter than the calyx-tube. Flowers monomorphic, solitary in the axils of bractiform leaves, shortly pedicillate; calyx-tube 2 mm long, tubular-campanulate, lobes-4, triangular, appendages minute; petals 4, 1 mm long, slightly exceeding the calyx-tube; styles very short, persistent; stigma capitate, minutely papillose. Capsule ellipsoid, 3 mm long, opening by 4 valves. Seeds ellipsoid, 0.8 mm long.

Fl. & Fr: Nov.-Dec. *Distr.:* Endemic to Western Ghats. An aquatic plant growing along the shallow margins of fresh water ponds with only tips of flowering branches emerging above water surface. *Ravi 3691* (SNCH). (Ponnambalamedu, c. 1100 m).

3. **Rotala rosea** (Poiret) Cook, Boissiera 29:86. 1979; Matthew, Ill. Fl. Tam. Carnatic t. 283. 1982; Matthew & Britto in Matthew, Fl. Tam. Carnatic 3(1) : 614. 1983. *Ammania rosea* Poiret in Lam., Encycl. (Suppl. 1) 329. 1810. *Rotala leptopetala* (Bl.) Koehne, Bot. Jahrb. Syst. 1 : 162. 1880; Gamble, Fl. Pres. Madras 508. 1919. *Ammania leptopetala* Bl., Mus. Bot. 2 : 134. 1856. *A. pentandra* Roxb., Fl. Ind. 1 : 427. 1832; Wight & Arn., Prodr. 305. 1834; Clarke in Hook.f., Fl. Brit. India 2 : 568. 1879.

Slender herbs. Leaves sessile, linear-lanceate, 5-20 by 2.5 mm, base cuneate, apex obtuse, chartaceous. Flower(s) solitary, axillary, 5-merous, bracteoles linear; calyx-tube campanulate, lobes 5, triangular; appendages subulate; petals 5, pink, obovate, apex-dentate; stamens 5, inserted at or below the calyx-tube. Capsule globose, exceeding the calyx. Seeds brownish.

Fl. & Fr.: Nov.-Dec. *Distr.*: India, S.E. Asia to New Guinea, Philippines, S. China, Korea, Japan. Common. In paddy fields. *NAK 2220* (Pathanamthitta, c. 40 m).

4. **Rotala rotundifolia** (D. Don) Koehne, Bot. Jahrb. 1:175. 1880; Gamble, Fl. Pres. Madras 509. 1919; Gandhi in Sald. & Nicols., Fl. Hassan Dist. 274. 1976; Cook, Boissiera 29 : 49. 1979; Matthew & Britto in Matthew, Fl. Tam. Carnatic 3 (1) : 616. 1983. *Ammannia rotundifolia* Buch.–Ham. ex Roxb., Fl. Ind. 1 : 446. 1820; Wight & Arn., Prodr. 306. 1834, p.p.; Clarke in Hook. f., Fl. Brit. India. 2:566. 1879.

Aquatic to amphibious decumbent annuals or perennials; stem creeping and rooting at lower nodes. Leave subsessile, heteromorphic; submerged ones linear to oblanceate; emerged ones obovate-orbicular, 15 by 10 mm, base cuneate, apex obtuse to rounded, chartaceous. Spikes terminal, often branched; bracts monomorphic, closely imbricating, foliaceous, ovate, 20 by 5 mm, pinkish; bracteoles linear. Flowers 4-merous; calyx-tube campanulate, lobes 4; appendages 0; petals 4, pink; stamens 4, inserted at the base of calyx-tube. Capsule subglobose, 4-valved. Seeds ellipsoid.

Fl. & Fr.: Sep.-Jan. *Distr.*: S. and S. E. Asia from India to Japan. Common. In swamps, shallow pools, ponds, etc. *NAK 1665* (Tiruvalla, c. 10 m).

Lawsonia inermis L., often seen planted as a hedge plant.

62. ONAGRACEAE

LUDWIGIA Linnaeus
Sp. Pl. 118. 1753 ("Ludvigia")
corr. Linnaeus, Gen. ed. 5,55. 1754.

Herbs to subshrubs; branchlets erect, or floating with aerophore. Leaves simple, alternate or decussate. Flower(s) axillary, solitary, 4-or 5-merous; calyx-tube not produced above, lobes 4-5; petals 4-5, spreading, caducous; disc raised with nectaries below the stamens; stamens 4 or 8; ovary inferior, 4-5-locular; ovules α. Capsule terete or angular, ribbed, loculicidal or poricidal. Seeds minute, α, 1-seriate or multiseriate.

1. Aquatic herbs; stems floating with aerophores; flower 5-merous; capsuel thick-walled. .... 1. **L. adscendens**
1. Terrestrial herbs or subshrubs; stems erect without aerophores; flowers 4-merous; capsule thin-walled.
    2. Stamens 4; capsule stout, oblong, to 1 cm long. .... 4. **L. perennis**
    2. Stamens 8; capsule slender, linear, to 2 cm long.
        3. Herbs; flowers, c. 1 cm ϕ; seeds apically 1-seriate, embedded in endocarp. .... 2. **L. hyssopifolia**
        3. Subshrubs; flowers, c. 2 cm ϕ; seeds multiseriate throughout, not embedded in endocarp. .... 4. **L. octovalvis**

1. **Ludwigia adscendens** (L.) H. Hara, J. Jap. Bot. 28 : 291. 1953; Ramamoorthy in Sald. & Nicols., Fl. Hassan Dist. 283. 1976; Raven in Steenis, Fl. Males. I, 8: 104. 1977; Matthew, Ill. Fl. Tam. Carnatic t. 285. 1982; Matthew & Britto in Matthew, Fl. Tam. Carnatic 3(1) : 621. 1983; Nicols., Suresh & Manilal, An Interpr. Hort. Malab. 202. 1988. *Jussiaea adscendens* L., Mant. Pl. 69. 1767. *J. repens* L., Sp. Pl. 388. 1753, non Forster, 1771; Clarke in Hook.f., Fl. Brit. India 2 : 587. 1879; Gamble, Fl. Pres. Madras 516. 19129. *Nir-carambu* Rheede, Hort. Malab. 2:99-100, t. 51. 1679.

Aquatic floating herbs; aerophore spongy, cream coloured, fusiform. Leaves obovate to oblanceate, 15-35 by 5-15 mm ϕ, base attenuate, apex obtuse, lower surface glossy, lateral veins c. 9 pairs, chartaceous. Flower(s) c. 4 cm ; calyx-tube, c.1 cm long; pubescent; lobes 5, narrow-lanceate; petals 5, cream coloured or white with an yellowish-blotch inside; stamens 10; ovary 5-locular. Capsule c. 3 cm long, terete, 10-ribbed, dehiscing by 4-5 valves.

Fl. & Fr.: Sep.-Apr. *Distr.*: Continental Asia, Malesia, Australia. Locally abundant in abandoned paddy fields, shallow pools, ponds, etc. *NAK 1954* (Tiruvalla, c. 10 m).

2. **Ludwigia hyssopifolia** (Don) Exell, Garcia de Orta 5 : 471. 1957; Raven in Steenis, Fl. Males.I,8 : 104. 1977; Matthew & Britto in Matthew, Fl. Tam. Carnatic 3 (1) : 622. 1983. *Jussiaea hyssopifolia* Don, Gen. Hist. 2 : 693. 1832. *J. linifolia* Vahl, Eclog. Amer. 2:32. 1798, non Poiret, 1813; Gamble, Fl. Pres. Madras 1875. 1936. *Neergramp.*

Herbs. Leaves narrow-lanceate, 1.5-5 by 0.5-1.5 cm, base attenuate, apex gradually acute to acuminate, lateral veins c. 12 pairs, chartaceous. Flowers c. 1 cm f, 4-merous; calyx-tube narrow, lobes 4, lanceate;

petals 4, yellow; stamens 8, filaments 4, unequal, 4 very shrot; ovary 4-locular; stigma 4-lobed. Capsule cylindric, c. 2 cm long, thin-walled, 8-ribbed. Seeds ovoid, embedded in endocarp.

Fl. & Fr.: Dec.-Apr. *Distr.*: Tropical Africa, Continental S.E. Asia, Malesia to Micronesia and N. Australia. Fairly common along streamsides, marshes, wet waysides, etc. *NAK 1795, 2254 & 2369* (in plains and hills).

The plant ground into a paste in butter-milk and taken in dysentery. Coconut-oil boiled with bruised plant is applied as a remedy in itches.

3. **Ludwigia octovalvis** (Jacq.) Raven, Kew Bull. 15 : 476. 1962; Ramamoorthy in Sald. & Nicols., Fl. Hassan Dist. 285. 1976; Raven in Steenis, Fl. Males. I,8 : 101. 1977; Matthew & Britto in Matthew, Fl. Tam. Carnatic 3(1) : 622. 1983; Manilal, Fl. Silent Valley 114. 1988; Nicols., Suresh & Manilal, An Interpr. Hort. Malab. 110. 1988; Mohanan & Sivad., Fl. Agasthyamala 287. 2002. *Oenothera octovalvis* Jacq., Enum. Syst. Pl. 19. 1760. *Jussiaea suffruticosa* L., Sp. Pl. 1: 388. 1753; Clarke in Hook.f., Fl. Brit. India 2 : 587. 1879; Gamble, Fl. Pres. Madras 516. 1919. *Cattu-carambu* Rheede, Hort. Malab. 2 : 97, t. 50. 1679.

Subshrubs; branchlets tomentose. Leaves lanceate, oblong or elliptic-oavte, 2-7.5 by 0.5-1.5 cm, base and apex attenuate, chartaceous. Flowers c. 2.5 cm ϕ, 4-merous; calyx-tube elongate, c. 2 cm long, lobes 4, lanceate; petals 4, cream coloured or yellow; stamens 8, unequal, short ones epipetalous. Capsule terete, to 4 cm long, thin-walled, 8-ribbed. Seeds multiseriate.

Fl. & Fr.: Jan.-Mar. *Distr.*: Pantropical. Rare. In mountains. *NAK 2659* (Appuppanthode, c. 450 m).

4. **Ludwigia perennis** L., Sp. Pl. 119. 1753; Ramamoorthy in Sald. & Nicols., Fl. Hassan Dist. 285. 1976; Raven in Steenis, Fl. Males. I,8 : 103. 1977; Matthew, Ill. Fl. Tam. Carnatic t. 286. 1982; Matthew & Britto in Matthew, Fl. Tam. Carnatic 3(1) : 623. 1983; Nicols., Suresh & Manilal, An Interpr. Hort. Malab. 202 : 1988; Mohanan & Sivad., Fl. Agasthyamala 288.2002. *L. parviflora* Roxb., Fl. Ind. 1 : 419. 1832; Clarke in Hook. f., Fl. Brit. India 2 : 588. 1879; Gamble, Fl.Pres. Madras 517. 1919. *Carambu* Rheede, Hort. Malab. 2: 95-96, t. 49. 1679.

Herbs. Leaves elliptic-lanceate, 1-8 by 0.5-1.5 cm, base attenuate, apex acute to acuminate, chartaceous. Flowers, 2 cm ϕ, 4-merous; calyx-tube c. 1 cm long, lobes 4, lanceate; petals 4, yellow; stamens 4, filaments short; ovary 4-locular. Capsule subterete, c. 1 cm long, 4-ribbed. Seeds ellipsoid.

Fl. & Fr.: Most of the seasons. *Distr.*: Trop. Africa, Madagascar, through continental S.E. Asia, and throughout Malesia to Trop. Australia. Common. In marshes, fallow fields, etc. *NAK 1016* (Tiruvalla, c. 10 m).

63. TURNERACEAE

TURNERA Linnaeus
Sp. Pl. 271. 1753.

Turnera subulata J.E. Smith in Rees., Cyclop. 36. 1819; Backer in Steenis, Fl. Males. I,4 : 236. 1951; Vuppulluri, Ind. For. 95(4) : 313. 1969; Madhu. & Rejani, J. Econ. Tax. Bot. 13(1) : 129. 1989. *T. trioniflora* Sims, Bot. Mag. 2106. 1820; *T. elagans* Otto in Nees, Hort. Phys. Berol. 36. 1820. *T. ulmifolia* L., var. *elegans* (Otto) Urban, Monogr. Turn. 139. 1883; Gamble, Fl. Pres. Madras 369. 1935; Matthew & Britto in Matthew, Fl. Tam. Caranatic 3(1) : 627. 1983.

Subshrubs. Leaves simple, alternate; stipules minute. Lamina elliptic-ovate to lanceate, 2.5-6 by 1-3.5 cm, base cuneate-decurrent; margin serrate-dentate, apex obtuse, subacute to acute, sparsely puberulous, gland-dotted, lateral nerves c. 8 pairs, chartaceous; petiole with a pair of glands towards the apex. Flowers bisexual, 5-merous, solitary in upper leaf-axils; bracteoles linear-subulate; pedicel adnate to the petiole throughout; calyx-tube small, lobes 5, linear; petals 5, yellow with deep purple centre; stamens 5, inserted at the base of calyx-tube. Capsule loculicidal, globose. Seeds α, pitted.

Fl. & Fr.: Most of the year. *Distr.*: Native of Trop. America; now introduced in warm countries. *NAK 1994* (Tiruvalla, c. 10 m).

Note : Urban (1883) followed by Gamble (1935) and Matthew (1983) considered this as a polymorphic variety of *T. ulmifolia* L. Here the treatment of Backer (l.c.) is followed.

64. PASSIFLORACEAE

1. Androgynophore 0 or short; flowers unisexual or polygamous; corona obscure; fruit a capsule. .... **Adenia**
1. Androgynophore long; flowers bisexual; corona prominent; fuit a berry. .... **Passiflora**

ADENIA Forsskal
Fl. Aegypt.-Arab. 77. 1775.

Adenia hondala (Gaertn.) de Wilde, Blumea 15 : 1967; Gandhi in Sald. & Nicols., Fl. Hassan Dist. 168. 1976; Nicols., Suresh & Manilal, An Interpr. Hort. Malab. 105. 1988; Mohanan & Sivad., Fl. Agasthyamala 289. 2002. *Granadilla hondala* Gaertn., Fruct. 2 : 480, t. 180, f. 10. 1791. *Bryonia palmata* L., Sp. Pl. 1:1012. 1753 p.p.; *Modecca palmata* Lamk., Encycl. 4:209. 1979; Masters in Hook.f., Fl. Brit. India 2 : 603. 1879; Gamble, Fl. Pres. Madras 525. 1919. *Modecca* Rheede, Hort. Malab. 8 : 39-40 (46), t. 20. 1688. *Palmodecca* Rheede, Hort. Malab. 8 : 41, t. 21. 1688. *Motta-modecca* Rheede, Hort. Malab. 8 : 43, t. 22. 1688. *Orela-modecca* Rheede, Hort. Malab. 8 : 45, t. 23. 1688.

Tuberous-rooted climbers. Leaves simple, entire or palmately 3-5- lobed, deltoid or broadly ovate, 10-14 by 8.5-10 cm; lobes elliptic-lanceate or oblong-lanceate, sinus often with 2 apical glands, base truncate, apex acuminate, basally 5-nerved, chartaceous. Flowers unisexual, 5-merous; calyx-lobes 5, ovate; petals 5, greenish; corona filiform; stamens 5; staminodes 5 in female flowers; ovary globose, superior, 1-locular ovules α in 3 parietal placentae. Capsule loculicidally 3-valvular. Seeds reniform, pitted.

Fl. & Fr.: Aug.-Feb. *Distr.*: Sri Lanka, Western Ghats. Less common. In moist shady evergreen forests. *NAK 770* (Gurunadhan mannu, c. 250 m), *1515* (Moozhiar, c. 250 m).

Tuber is used in medicine for treating alcoholism. Flowers cure biliousness but increase rheumatism.

PASSIFLORA Linnaeus
Sp.Pl. 955. 1753.

Passiflora foetida L., Sp. Pl. 959. 1753; Gamble, Fl. Pres. Madras 524. 1919; Chakrav., Bull. Bot. Soc.Bengal 3(1) : 57. 1949; de Wilde in Steenis, Fl. Males. I, 7: 410. 1972; Gandhi in Sald. & Nicols., Fl.Hassan Dist. 169. 1976; Matthew, Ill. Fl. Tam. Carnatic t. 289. 1982; Matthew & Britto in Matthew, Fl. Tam. Carnatic 3(1) : 630. 1983; Mohanan & Sivad., Fl. Agasthyamala 289. 2002.

Key to the varieties

1. Ovary pubescent; leaf upper surface hispid, margin sparsely glandular-ciliate; berry sparsely pubescent; seeds oblong. ....**Passiflora foetida** var. **foetida**
1. Ovary glabrous; leaf upper surface and margin densely glandular-viscid; berry glabrous; seeds obovate. ....**Passiflora foetida** var. **hispida**

Passiflora foetida L.var. **foetida**. Killip, Bot. Ser. Field Mus. Nat. Hist. 19. (1) : 481. 1938.

Herbaceous vines; stem, leaves and sepals outside hirsute with spreading yellowish hairs, very rarely glabrous or nearly so. Leaves narrowly ovate in outline, shallowly 3-angled to 3-lobed, both surfaces appressed hirsute, never glandular, sparingly glandular-ciliate; ovary hirsute, usually with white stiff-pilose hairs. Berry 3-4 cm ϕ, sparsely hirsute. Seeds oblong.

Fl. & Fr.: Sep.-Feb. onwards. *Distr.*: Native of Trop. & S. America; now naturalized through Tropical Africa, through S.E. Asia to China. Rare. On river-beds, moist shady thickets from plains to hills. *NAK 1089* (Vallicode, c. 200 m).

The plant can be easily recognized in the field by the absence of viscid glands on leaf surfaces (hence they are not sticky to touch) and hirsute berries.

Passiflora foetida L.var. **hispida** (DC.) Killip ex Gleason, Bull. Torrey Bot. Club 58:408. 1931 & Bot. Ser. Field. Mus. Nat. Hist. 19.(1):494. 1938.

Herbaceous vines; stem, leaves and sepals outside hispid with pale yellowish glandular viscid hairs. Leaves broadly ovate in outline, 3-lobed, both surfaces densely or sparsely glandular-hirsute, margin coarsely dentate, glandular-ciliate; ovary glabrous. Berry 2-2.5 cm ϕ. Seeds obovate.

Fl. & Fr.: Nov.-Mar. *Distr.*: Widely distributed in Old World and New World tropics. Common. On river-beds, moist shady waysides, etc. in plains. *NAK 533* (Angamoozhy, c. 250 m).

The plant can be easily recognized in the field by the prsence of viscid glands on lef surfaces (hence they are sticky to touch) and glabrous berries.

65. CUCURBITACEAE

1. Tendrils branched.
    2. Petals fimbriate, white. ....**Trichosanthes**
    2. Petals entire, greenish-yellow.
        3. Flowers to 2.5 cm ϕ; leaves usually deeply lobed; male flowers clustered. ....**Diplocyclos**
        3. Flowers to 4 cm ϕ; leaves usually angular or shallowly lobed; male flowers in racemes. ....**Luffa**
1. Tendrils unbranched.
    4. Calyx-tube enclosed at base by scales. ....**Momordica**
    4. Calyx-tube without scales at base.

5. Male flower(s) solitary. .... **Coccinia**

5. Male flowers in clusters or racemes.

6. All stamens 2-celled; connectives and filaments ciliate. .... **Zehneria**

6. All stamens not 2-celled (one: 1-celled, two: 2-celled); connectives and filaments glabrous.

7. Connective apically crested; trailing herbs. .... **Cucumella**

7. Connective apically not crested; vines.

8. Ovary rostrate at apex. .... **Solena**

8. Ovary not rostrate at apex ....**Mukia**

COCCINIA R. Wight et Arnott
Prodr. 347. 1834.

Coccinia grandis (L.) J. Voigt, Hort. Suburb. Calc. 59. 1845; Matthew, Ill. Fl. Tam. Carnatic t. 293.1982; Matthew & Britto in Matthew, Fl. Tam. Carnatic 3(1) : 637. 1983; Nicols., Suresh & Manilal, An Interpr. Hort. Malab. 95. 1988. *Bryonia grandis* L, Mant. Pl. 126. 1767. *Coccinia indica* Wight & Arn., Prodr. 347. 1834, nom. illegit.; Wight, Ill. Ind. Bot. t. 105 (121). 1850, p.p.; Gamble, Fl. Pres. Madras 537. 1919. *Cephalandra indica* Naudin, Ann. Sci. Nat. Bot. (Ser. 5) : 16. 1866, illegit.; Clarke in Hook.f., Fl. Brit. India 2 : 621. 1879. *Covel* Rheede, Hort. Malab. 8 : 27-28, t. 14. 1688.

Extensive vines; tendrils simple. Leaves alternate, palmately 3-5- lobed or angled, broaldy ovate to orbicular in outline, 3-7 by 3-5 cm, base cordate, margin denticulate, apex acute, upper surface punctate, lower surface glandular. Flower(s) solitary, dioecious. Male flower (s) : calyx campanulate, lobes subulate; corolla white; stamens 3, inserted below the calyx-tube, anthers flexuose, filaments connate. Female flowers: calyx-tube campanulate; corolla white; ovary oblong, glandular pubescent, ovules α , horizontal. Fruit ovoid-oblong. Seeds compressed.

Fl. & Fr.: Dec.-Apr. *Distr.*: Sri Lanka, Peninsular India. Common. On wayside thickets, usually in plains. *NAK 198* (Adoor, c. 20 m).

Juice of tap root used as medicine in diabetes. Green fruits chewed to cure sores in the tongue. Fruits edible.

CUCUMELLA Chiovenda
Fl. Somala 1 : 183. 1929.

Cucumella silentvalleyii Manilal, Sabu *et* P. Mathew, Acta Bot. Indica 13 (2) : 283-284. 1985; Manilal, Fl. Silent Valley 115. 1988.

Slender trailers; stem ribbed, hairy, tendrils simple. Leaves palmately 3-5- angled or shallowly lobed, surborbicular in outline, 3-4.5 by 4-5.5 cm, base cordate, margin denticulate, apex subacute to obtuse, lower surface densely pubescent, basaly 8-10-nerved, membranous. Flowers monoecious but male dominant, 5-7 in axillary clusters; pedicels slender; calyx campanulate, hairy, lobes 5, linear; corolla yellow, fused to middle; stamens 3, subsessile, inserted above the corolla tube, connective producd into a linear hood. Female flower(s) solitary; calyx and corolla as in male; ovary hirsute; ovules α; style with a basal white disc; stigma 3-lobed. Fruits oblong-fusiform, cuspidate at apex, scabrous. Seeds silky.

Fl. & Fr.: Sep.-Mar. *Distr.*: Endemic to Western Ghats. Rare. On moist exposed rocky hills. *NAK 2495* (Way to Pampa dam from Moozhiar, c. 100 m).

This seems to be the first collection of the species from an area other than the type locality.

DIPLOCYCLOS (Endlicher) Post et O. Kuntze Lex. 178. 1903.

Diplocyclos palmatus (L.) C. Jeffrey, Kew Bull. 15 : 352. 1962; Gandhi in Sald. & Nicols., Fl. Hassan Dist. 177. 1976; Matthew, Ill. Fl. Tam. Carnatic t. 297. 1982; Matthew & Britto in Matthew, Fl. Tam. Carnatic 3(1) : 644. 1983; Manilal, Fl. Silent Valley 116. 1988; Nicols., Suresh & Manilal, An Interpr. Hort. Malab. 96. 1988; Mohanan & Sivad., Fl. Agasthyamala 290.2002. *Bryonia palmata* L., Sp. Pl. 1013. 1753, p.p.; Wight, Ic. t. 500. 1841; Clarke in Hook.f., Fl. Brit. India 2 : 622. 1879. *Bryonopsis laciniosa* auct., non (L.), 1753: Naudin. Ann. Sc. Nat. Ser. 4, 12:141. 1859; Gamble, Fl. Pres. Madras 534. 1919; Chakrav., Rec. Bot. Surv. India 17 : 135. 1959; *Nehoemeka* Rheede, Hort. Malab. 8 : 37-38, t. 19. 1688.

Stout vines; tendrils 2-fid. Leaves palmately 5-lobed, 5-7 by 7-9.5 cm, base cordate, margin toothed, apex acute; lobes elliptic-oblong, upper surface scabrid, chartaceous. Flowers monoecious; male and female flowers in axillary clusters; calyx campanulate; corolla greenish-yellow; stamens 3, free, inserted attach middle of calyx-tube, anther cells slightly flexuous; staminodes 3; ovary globose; ovules few, horizontal; stigma 3-fid. Fruit globose with vertical greenish and white patches. Seeds ovoid.

Fl. & Fr.: Oct.-Mar. *Distr.*: N. Trop. Africa, Arabia, Tropical Asia from W. Pakistan to Hainan Malesia and Tropical Australia. Common on thickets and along forest borders. *NAK 54* (Moozhiar, c. 250 m), 464 (Konni, 100 m).

LUFFA P. Miller Gard. Dict. Abr. ed. 4. 1754.

Luffa aegyptiaca Mill., Gard. Dict. (ed. 8) 4 : 500. 1785; Clarke in Hook.f., Fl. Brit. India 2 : 614. 1879; Gamble, Fl. Pres. Madras 533. 1919; Exell, J. Bot. 71 : 352. 1933; Bennet, Fl. Howrah 358. 1979; Nicols., Suresh & Manilal, An Interpr. Hort. Malab. 97. 1988. *Luffa cylindrica* M. Roem., Fam. Nat. Syn. Monogr. 2:63. 1846. *Luffa cylindrica* (L.) M. Roem. ex. T. & H. Durand, Syll. Fl. Cong. 299. 1909, non M. Roem., 1846; Chakrav., Rec. Bot. Surv. India 17:75. 1959; C. Jeffrey, Kew Bull. 15 : 355. 1962; Matthew, Ill. Fl. Tam. Carnatic t. 300. 1982; Matthew & Britto in Matthew, Fl. Tam. Carnatic 3(1) : 646. 1983; Manilal, Fl. Silent Valley 118. 1988. *Cattu-picinna* Rheede, Hort. Malab. 8:15, t. 8. 1688.

Vines; tendrils 3-fid. Leaves 5-(7) lobed, orbicular or broadly ovate in outline 6-12 by 6-11 cm; lobes ovate, base cordate, margin shallowly dentate, apex acuminate,upper surface glandular-punctate, lower surface scabrid, thin-coriaceous. Flowers monoecious, male and female on same axil. Male flowers in racemes, clustered; petals 5 yellow; stamens 5, free, inserted near calyx-tube, anthers, one, 1-celled, others 2-celled. Female flower(s) solitary; ovary oblong, 1-locular; ovules α. Fruit cylindric, to 15 cm long, fibrous within. Seeds α, compressed.

Fl. & Fr.: Feb.-Dec. *Distr.*: Old World tropics. Usually cultivated, occasionally run wild. *NAK* 2242 (Konni, c. 150 m).

Seeds yield an oil and it is said to be emetic and cathartic. Dried fruits are used as brush to flush in baths.

MOMORDICA Linnaeus
Sp. Pl. 1009. 1753 & Gen. Pl ed. 5. 440. 1754.

Momordica dioica Roxb. ex Willd., Sp. Pl. 4 : 605. 1805; Wight & Arn., Prodr. 348. 1834; Clarke in Hook.f., Fl. Brit. India 2 : 617. 1879; Gamble, Fl. Pres. Madras 532. 1919; Chakrav., Rec. Bot. Surv. India 17 : 91. 1959; Gandhi in Sald. & Nicols., Fl. Hassan Dsit. 177. 1976; Matthew & Britto in Matthew, Fl. Tam. Carnatic 3 (1) : 648. 1983; Nicols., Suresh & Manilal, An Interpr. Hort. Malab. 71. 1988. *Erima-pavel* Rheede, Hort. Malab. 8 : 23, t. 12. 1688. *Ben-Pavel* Rheede, Hort. Malab. 8 : 35-36, t. 18. 1688.

Slender vines; tendrils simple. Leaves broadly ovate, deeply 3-5- lobed, 4-9 by 3.5-6 cm, base cordate, margin denticulate, apex acute, membranous. Flowers dioecious, solitary, axillary. Male flowers with long peduncles with foliaceous bracts at apex; petals 5, yellow; stamens 3, inserted at the mouth of calyx-tube; one 1-celled, two 2-celled; staminodes of 2-3 glands; ovary oblong with horizontal ovules. Fruit ovoid, muricate.

Fl. & Fr.: Most of the seasons. *Distr.*: Sri Lanka, India, Myanmar East. to W. China, Malaya. On thickets or in forest clearings of hills. *NAK 46* (Moozhiar, c. 250 m), *948* (Sabarimala, c.250 m).

Seeds fried and eaten as a remedy for cough and pain in the chest. Plaster of roots promotes growth of hairs and prevents its fall. Fruits edible.

Momordica charantia L., is widely cultivated; occasionally run wild. *NAK 2350* (Kalleli, c. 200 m).

MUKIA Arnott
Madras J. Lit. Sci. 17 : 50. 1840.

Mukia maderaspatana (L.) M. Roemer, Syn. Pep. 47. 1846; Gandhi in Sald. & Nicols., Fl. Hassan Dist. 178. 1976; Matthew, Ill. Fl. Tam. Carnatic t. 302. 1982; Matthew & Britto in Matthew, Fl. Tam. Carnatic 3(1) : 650.1983; Nicols., Suresh & Manilal, An Interpr. Hort. Malab. 99. 1988. *Cucumis maderaspatanus* L., Sp. Pl. 1012. 1753. *Bryonia scabrella* L.f., Suppl. Pl. 424. 1781; Roxb., Fl. Ind. 3:724. 1832. *Mukia scabrella* (L.f.) Arn. in Hook., J. Bot. 3:276. 1841; Clarke in Hook.f., Fl. Brit. India 2 : 623. 1879 (excl. syn). *Melothria maderaspatana* (L.) Cogn. in A and C. DC., Monogr. Phan. 3 : 623. 1881; Gamble, Fl. Pres. Madras 539. 1919; Chakrav., Rec. Bot. Surv. India 17: 141. 1959. *Mucca-piri* Rheede, Hort. Malab. 825-26, t. 13. 1688.

Prostrate or climbing vines; branchlets hispid. Tendrils simple. Leaves ovate-deltoid, angular or shallowly 3-5- lobed, 4-6 by 4-5.5 cm, base cordate, margin denticulate, apex acuminate, mucronate, scabrid, chartaceous. Flowers small, monoecious, fascicled; petals 5, yellow; stamens 3, free, inserted at base of calyx-tube. Female flower(s) solitary or in clusters. Berry sessile, ripening red.

Fl. & Fr.: Nov.-Feb. *Distr.*: Africa, India, east to China, Taiwan, Malesia, Australia, New Zealand. Common. In open disturbed areas, on thickets, fences, etc. *NAK 131* (Adoor, c. 30 m).

Root useful in flatulance and when masticated relieves tooth-ache.

SOLENA Loureiro
Fl. Cochinch. 477, 514. 1790.

Solena amplexicaulis (Lam.) Gandhi in Sald. & Nicols., Fl. Hassan Dist. 179. 1976; Matthew, Ill. Fl. Tam. Carnatic t. 303. 1982; Matthew & Britto in Matthew, Fl. Tam. Carnatic 3(1) : 65. 1983; Manilal, Fl. Silent Valley 119. 1988; Nicols., Suresh & Manilal, An Interpr. Hort. Malab. 99. 1988; Mohanan & Sivad., Fl. Agasthyamala 291.2002. *Bryonia amplexicaulis* Lam., Encycl. 1 : 496. 1785. *Solena heterophylla* Lour., Fl.

Cohinch. 1 : 514. 1790. *Zehneria umbelleta* (Klein) Thw., Enum. Pl. Zeyl. 125. 1859; Clarke in Hook.f., Fl.Brit. India 2 : 625. 1879, p.p. *Melothria heterophylla* (Lour.) Cogn. in DC., Monogr. Phan. 3 : 618. 1881; Gamble, Fl. Pres. Madras 539. 1919; Chakrav., Rec.Bot. Surv. India 17 : 146. 1959. *M. amplexicaulis* (Lam.) Cogn. in DC., Monogr. Phan. 3:61. 1881; Gamble, Fl. Pres. Madras 539. 1919; Chakrav., Fasc. Fl. Ind. 11 : 77. 1982. *Karivi-valli* Rheede, Hort. Malab. 8:51, t. 26. 1688.

Vines; tendrils simple. Leaves polymorphic, ovate to oblong, 3-5- angled or lobed, 4-13 by 2.5-7 cm, base cordate, amplexicaule, margin denticulate, apex acuminate, scabrid, chartaceous. Flowers dioecious; Male flowers in racemes; corolla white, campanulate; stamens 3, one 1-celled, two 2-celled, inserted at base of calyx-tube; connective apically produced. Female flower(s) solitary, ovary oblong, apically rostrate. Fruit oblong, ribbed.

Fl. & Fr.: Jul.-Jan. *Distr.*: Western Ghats. Less common. Along forest clearing. *NAK 202* (Pandalam, c.10 m) *442, 675 & 2514* (Ranni, Moozhiar, etc.)

Root- juice given with cumin and sugar in cold milk in spermaetorrhoea. Root fried in ghee, powdered and applied to heal ulcers.

TRICHOSANTHES Linnaeus
Sp. Pl. 1008. 1753.

Vines; tendrils 2-or 3-branched. Leaves deeply lobed or unlobed. Flowers dioecious or monoecious. Male in racemes, bracteate or ebracteate; calyx-tube oblong, apically dilated; corolla white, petals 5, fimbriate; stamens 3, inserted at base of calyx-tube, anthers one 1-celled, two 2-celled. Female solitary; bracts minute; ovary globose; ovules α, horizontal. Fruit globose or ellipsoid, beaked.

1. Leaves 3-5- lobed, fainlty nerved, glandular-pubescent below. .... 1.**T. cucumerina**
1. Leaves unlobed, prominently nerved, glabrous. .... 2. **T. nervifolia**

1. **Trichosanthes cucumerina** L., Sp. Pl. 1008. 1753; Wight & Arn., Prodr. 350. 1834; Clarke in Hook.f., Fl. Brit. India 2 : 609. 1879; Gamle, Fl. Pres. Madras 529. 1919; Chakrav., Rec. Bot.Surv.India 17:31. 1959; Matthew & Britto in Matthew, Fl. Tam. Carnatic 3(1) : 653. 1983; Nicols., Suresh & Manilal, An Interpr. Hort. Malab. 100. 1988. *Puduvulum* Rheede, Hort. Malab. 8 : 29-30, t. 15. 1688.

Leaves orbicular-reniform, 3-5(-7)- lobed, 8-11 by 8-10 cm, base truncate-cordate, margin distantly denticulate, apex acute, lower surface glandular-pubescent, chartaceous. Male racemes ebracteate; calyx turbinate, lobes linear; petals oblong-lanceate, long-fimbriate; ovary ribbed, pubescent. Fruit ovoid-fusiform, beaked.

Fl. & Fr.: Dec.-May. *Distr.*: Sri Lanka, India, Trop. Himalaya, Malesia, Polynesia, N. Australia. Less common. *NAK 77* (Moozhiar, c. 250 m), *466* (Konni, c. 200 m).

Juice of leaves rubbed over the body in remittent fever. Roots act as a powerful cathartic.

2. **Trichosanthes nervifolia** L., Sp. Pl. 1008. 1753; Clarke in Hook.f., Fl. Brit. India 2 : 609. 1879; Gamble, Fl. Pres. Madras 529. 1919; Chakrav., Rec. Bot. Surv. India 17:30, 1959; Gandhi in Sald. & Nicols., Fl. Hassan Dist. 180. 1976; Nicols., Suresh & Manilal, An Interpr. Hort. Malab. 100. 1988. *T. cuspidata* Lam., Encycl. 1 : 190. 1783; Gamble, Fl. Pres. Madras 529. 1919. *Scher-padavalam* Rheede, Hort. Malab. 8 : 31, t. 16. 1688. *Tota-piri* Rheede, Hort. Malab. 8 : 33, t. 17. 1688.

Leaves entire, ovate-lanceate, 6-14 by 5-7 cm, base truncate-cordate, margin minutely denticulate, apex gradually acute, basally 5-nerved, prominent on lowerside, chartaceous. Male few to α , in racemes; ovary ovoid; ovules α. Fruit ovoid-oblong. Seeds semi-ellipsoid, compressed.

Fl. & Fr.: Jan.-Apr. *Distr.*: S. India, Sri Lanka. Rare. On forest clearing. *NAK 277 & 1150* (Moozhiar, c. 250 m).

ZEHNERIA Endlicher
Prodr. Fl. Norfolk. 1: 69. 1833.

Vines; tendrils simple. Leaves broadly ovate-subdeltoid, angular or lobed, membranous. Flowers monoecious or dioecious. Male in umbellate racemes or in clusters; calyx-tube campanulate, lobes 5; corolla campanulate, petals 5, ovate; stamens 3, anthers 2-celled, connective spherical. Female solitary or in clusters; ovary globose, slightly rostrate. Fruit globose, slightly beaked at apex.

1. Leaves brodly ovate or subdeltoid, entire; fruit globose, smooth. .... 1. **Z. maysorensis**
1. Leaves deltoid, angular; fruit ovoid-fusiform, longitudinally 6-costate. .... 2. **Z. thwaitesii**

1. **Zehneria maysorensis** (Wt. & Arn.) Arn. in Hook., J. Bot. 3 : 275. 1841; Jeffrey, Kew Bull. 15 : 366. 1962; Gandhi in Sald. & Nicols., Fl. Hassan Dist. 181.1976; Matthew & Britto in Matthew, Fl. Tam. Carnatic 3(1) : 655. 1983; Manilal, Fl.Silent Valley 122. 1988; Mohanan & Sivad., Fl. Agasthyamala 292. 2002. *Bryonia maysorensis* Wt. & Arn., Prodr. 345. 1834. *Melothria perpusilla* (Bl.) Cogn. var. *subtruncata* Cogn. in DC., Monogr. Phan. 3 : 608. 1881; Gamble, Fl. Pres. Madras 538. 1919; Chakrav., Rec. Bot. Surv. India 17 : 148. 1959. *M. mucronata* sensu Chakrav., Rec. Bot. Surv. India 17 : 148. 1959, non (Bl.) Cogn., 1881. *M. perpusilla* sensu Gamble, Fl. Pres. Madras 538, 1919, p.p., non (Bl.) Cogn., 1881.

Leaves broadly ovate or subdeltoid, 6-8 by 5-6 cm, base subtruncate, margin denticulate, apex acumiante, apiculate, membranous. Male flowers in clusters. Female solitary; corolla white. Fruit globose. Seeds smooth, apiculate.

Fl. & Fr.: Most of the seasons. *Distr.*: Indo-Malesia, Very common on thickets, forest clearing in hills. *NAK 1* (Moozhiar, c. 250 m), *949* (Sabarimala, c. 250 m).

2. **Zehneria thwaitesii** (Schweinf.) Jeffrey, Kew Bull. 15 :371. 1962; Manilal, Fl. Silent Valley 123. 1988. *Melothria thwaitesii* Schweinf., Relique, Kotschy 44, t. 29. 1868. *M. zeylanica* Clarke in Hook.f., Fl. Brit. India 2 : 626. 1879; Gamble, Fl. Pres. Madras 538. 1919.

Leaves deltoid, angular, 2.5-6 by 2-6 cm, base truncate, margin distantly denticulate, apex acuminate hispid on nerves, punctate. Male flowers in axillary fascicles, yellow. Female solitary. Fruits ovoid-fusiform, longitudinally 6-costate. Seeds compressed, ovate-oblong.

Fl. & Fr.: Aug.-Dec. *Distr.*: Sri Lanka, Western Ghats. Rare in evergreen forests. *NAK 12 66* (Upper Moozhiar, c. 550 m), *2529* (Way to Ponnambalamedu c. 1100 m).

66. BEGONIACEAE

BEGONIA Linnaeus
Sp. Pl. 1056. 1753

Succulent herbs or subshrubs; stems jointed, erect or rhizomatous. Leaves simple, alternate, often inequilateral, or basally 3-7- nerved. Flowers monoecious in axillary or terminal cymes. Male flowers with 2-

4 perianth lobes; stamens α , free or basally connate. Females with 2-6 perianth lobes; ovary inferior with 3 persistent wings, 2-4-locular; ovules α; styles 2-4. Capsule ovoid, wings apically broad. Seeds minute, pitted.

1. Stem erect, jointed; leaves ovate-lanceate, inequilateral. ....1.**B. malabarica**

1. Stem rhizomatous; leaves orbicular, equilateral. ....2. **B. subpeltata**

1. **Begonia malabarica** Lam., Encycl. 1 : 393. 1785; Clarke in Hook.f., Fl. Brit. India 2 : 655. 1879; Gamble, Fl. Pres. Madras 546. 1919; Ramamoorthy in Sald. & Nicols., Fl. Hassan Dist. 174. 1976; Matthew & Britto in Matthew, Fl. Tam. Carnatic 3 (1) : 657. 1983; Manilal, Fl. Silent Valley 124. 1988; Nicols., Suresh & Manilal, An Interpr. Hort. Malab. 72. 1988; Mohanan & Sivad., Fl. Agasthyamala 295. 2002. *Tsjieria-narinampuli* Rheede, Hort. Malab. 9:167-163, t. 86. 1689. *Kayyala-puliyan, Janam-kolli.*

Leaves ovate-lanceate, inequilateral, 6-20 by 2.5-10 cm, base very oblique-cordate, margin coarsely denticulate, apex acuminate, glabrous or hairy. Flowers rose-coloured. Capsule broadly winged.

Fl. & Fr.: Most of the seasons. *Distr.*: Western Ghats. Very common on dripping rocky hillsides. *NAK* 192, 356, 1134 *& 1501.* (Most of the hills).

2. **Begonia subpeltata** Wight, Ic. t. 1812. 1852; Clarke in Hook.f., Fl. Brit. India 2:653. 1879; Gamble, Fl. Pres. Madras 546.1919; Manilal, Fl. Silent Valley 124. 1988.

Leaves orbicular, 10-16 by 9-12 cm, base cordate, serratae. Flowers pink in axillary panicles. Capsule elliptic, wings almost equal.

Fl. & Fr.: Sep.-Dec. *Distr.*: Sri Lanka, Western Ghats. Rare. On driping shady rocks. *NAK 1820* (Angamoozhy, c. 250 m).

67. DATISCACEAE

TETRAMELES R. Brown in Denham et Clapperton, Narr. Travels Africa 230. 1826.

Tetrameles nudiflora R. Br. in Benn., Pl. Jav. Rar. 79. t. 17. 1838; Hook.f., Fl. Brit. India 2 : 627. 1879; Gamble, Fl. Pres. Madras 544. 1919; Mohanan & Sivad., Fl. Agasthyamala 293. 2002.

Trees. Leaves simple, broadly ovate, 12-15 by 10-12 cm, base rounded, margin subentire, apex acuminate,lower surface pubescent, thin-coriaceous. Flowers small, dioecious, appearing before the new foliage. Male flowers in panicles; calyx-tube very short, lobes 4, petals 0; stamens 4, inserted on a central disc. Female in racemes, usually clustered towards the branch-apices; ovary inferior, 1-locular; ovules α; styles 4. Capsule ovoid, 4-8- ribbed. Seeds α, minute, ellipsoid.

Fl. & Fr.: Jan. Apr. *Distr.*: Sri Lanka, South India, Andamans. Common. In moist deciduous and semi-evergreen forest. *NAK 2261*(Arampa, c. 450 m).

This tree is characterised by large butress roots and white bark with leaf scars.

68. CACTACEAE

OPUNTIA P. Miller
Gard. Dict. Abr. ed. 4. 1754.

Opuntia vulgaris Miller, Gard. Dict. ed. 8, 1. 1768; Backer & Bakh.f., Fl. Jav. 1 : 316. 1963; Ramamoorthy in Sald. & Nicols., Fl. Hassan Dist.93. 1976; Bennet, Fl. Howrah 253. 1979. *Opuntia monocantha* (Willd.) Haw., Suppl. Pl. Succ. 81. 1819; Burkill, Rec. Bot. Surv. India 4 : 312. 1911; Gamble, Fl. Pres. Madras 548. 1919; Matthew & Britto in Matthew, Fl. Tam. Carnatic 3 (1) : 660. 1983. *Cactus monocanthos* Willd., Enum. Pl. Suppl. 33. 1813. *C. indicus* Roxb., Fl.Ind. 2 : 476, 1832.

Subshrubs; stem flattened, jointed, fleshy, areoles tufted, glochidiate,wooly; spines 1-3 per areole, to 5 cm long. Leaves subulate, caducous. Flower(s) solitary, sessile, bisexual; perianth α, rotate, free, yellow; stamens α, filaments filiform; ovary inferior, 1-locular, glochidiate; ovules α, horizontal. Berry ovoid.

Fl. & Fr.: Aug. onwards. *Distr.*: E.S. America; widely naturalised in India. In exposed dry areas in plains. *NAK 1632* (Tiruvalla, c. 10 m).

69. MOLLUGINACEAE

1. Staminodes present; seeds with a pair of hilar appendages. ....**Glinus**
1. Staminodes absent; seeds without hilar appendages. ....**Mollugo**

GLINUS Linnaeus
Sp. Pl. 463. 1753.

Glinus oppositifolius (L.) A. DC., Bull. Herbs. Boiss. Ser. 2, 1 : 559. 1901; Backer in Steenis, Fl.Males. I,4 : 270. 1951; Pitot, Webbia 19:782. 1965; Ramamoorthy in Sald. & Nicols., Fl. Hassan Dist. 95. 1976; Matthew, Ill. Fl. Tam. Carnatic t.310. 1982; Matthew & Britto in Matthew, Fl.Tam. Carnatic 3(1) : 665. 1983; Sivarajan, J. Taiwan Mus. 41 (2) : 82. 1988; Nicols., Suresh & Manilal, An Interpr. Hort. Malab. 182. 1988. *Mollugo oppositofolia* L., Sp. Pl. 89. 1753; Gamble, Fl. Pres. Madras 552. 1919. *M. spergula* L., Syst. ed. 10, 881. 1759; Clarke in Hook.f., Fl. Brit. India 2 : 662. 1879. *Kaipa-tsjira* Rheede, Hort. Malab. 10 : 47, t. 24. 1690.

Diffuse or prostrate, glabrous or sparsely pubescent herbs. Leaves simple, opposite or in a whorl, oblanceate to elliptic, 7-20 by 3-7 mm, base attenuate, apex obtuse or acute, subsucculent. Inflorescence an axillary fascicle. Flowers 3-10, bisexual, actinomorphic; tepals 5; stamens 3-5, often alternating with linear bifid staminodes; ovary 3-5-locular, superior; ovules α. Capsule globose, loculicidally 3-valvular. Seeds tuberculate, appendaged.

Fl. & Fr.: Jan.-Mar. *Distr.*: Pantropical. Common. In open waste lands, dry riverbeds and in exposed wet sandy areas. *NAK* 1630 (Tiruvalla, c. 10 m).

Crushed leaf juice along with castor oil is a good remedy for ear-ache.

MOLLUGO Linnaeus
Sp. Pl. 49. 1753.

Erect or diffuse annuals; branchlets often angled. Leaves usually in whorls. Inflorescence of axillary or terminal paniculate cymes; tepals 5; stamens 3-5; staminodes 0; ovary 3-locular. Capsule 3-valvular. Seeds α, without hilar appendages.

1. Lower leaves obovate-spathulate; flowers white; seeds areolate. ....1. **M. pentaphylla**

1. Lower leaves elliptic-lanceateor linear; flowers greenish-white; seeds tuberculate. ....2. **M. stricta**

1. **Mollugo pentaphylla** L., Sp. Pl. 89. 1753; Fenzl., Ann. Wien Mus. 1 : 381. 1836; Gamble, Fl. Pres. Madras 553. 1919, p.p; Baker in Steenis, Fl. Males. I , 4: 268. 1951, p.p.; Sivar. & Usha, Taxon 32 : 125. 1983; Matthew & Britto in Matthew, Fl. Tam. Carnatic 3(1) : 668. 1983; Sivar., J. Taiwan Mus. 41 (2) : 86. 1988; Mohanan & Sivad., Fl. Agasthyamala 296. 2002. *M. stricta* auct., non. L., 1762. *M. paniculata* Burm. f., Fl. Ind. 32. 1768.

Leaves (lower ones) in rosette,upper ones in whorls or opposite, obovate-spathulate, 0.7-5 by 05-1.5 cm,base attenuate, apex subacute or obtuse; cauline ones obovate to linear-lanceate. Flowers white; tepals 5; stamens 3(-5). Seeds areolate.

Fl. & Fr.: Jun.-Jul. *Distr.*: India, Malesia, Phillippines, Trop. S. America. Very common in seasonal soils, grasslands and waste places. *NAK 682* (Pandalam, c. 30 m).

Decoction of tender shoots and flowers given in fever.

2. **Mollugo stricta** L., Sp. Pl. : 131. 1762; Burm. f., Fl. Ind. 31, t.5, f.3. 1768; Fenzl, Ann. Wien Mus. 1 : 380. 1836; Clarke in Hook.f., Fl. Brit. India 2 : 663. 1879, p.p.; Sivar. & Usha, Taxon 32 : 125. 1983; Sivar., J. Taiwan Mus. 41 (2) : 87. 1988; Nicols., Suresh & Manilal, An Interpr. Hort.Malab. 182. 1988. *M. pentaphylla* auct. div., non L., 1753. *M. pentaphylla* var. *stricta* (L.) Hochr., Candollea 2: 356. 1925. *Tsjerujonganam* Rheede, Hort. Malab. 10 : 51, t. 26. 1690.

All leaves lanceate to linear-lanceate, 1-4.5 cm by 2-7 mm, base attenuate, apex acute. Flower greenish-white, tepals 5; stamens 3(-5). Capsule 3-valvular. Seeds tuberculate.

Fl. & Fr.: Sep.-Dec. *Distr.*: India, Indo-China, Malesia, Japan & Trop. S. America. Rare. In exposed waste lands in hills & plains. *NAK 1018* (Tiruvalla, c. 10 m).

Whole plant used as an antiseptic.

70. APIACEAE (UMBELLIFERAE)

1. Leaves and umbels simple; prostrate herbs.
    2. Flowers purplish; mericarps 7-9-ribbed. ....**Centella**
    2. Flowers greenish-white; mericarps 3-ribbed. ....**Hydrocotyle**
1. Leaves and umbels compound; erect herbs or subshrubs. ....**Heracle ım**

CENTELLA Linnaeus
Sp. Pl. 236. 1753.

Centella asiatica (L.) Urban in Martius, Fl. Bras. 11 (1) : 287. 1879; Gamble, Fl. Pres. Madras 556. 1919; Buwalda, Blumea 2 : 134. 1936; Gandhi in Sald. & Nicols., Fl.Hassan Dist. 416. 1976; Matthew, Ill. Fl. Tam. Carnatic t. 315. 1982; Matthew & Britto in Matthew, Fl. Tam. Carnatic 3(1) : 676. 1983; Manilal, Fl. Silent Valley 125. 1988; Nicols., Suresh & Manilal, An Interpr. Hort. Malab. 52. 1988; Mohanan & Sivad., Fl. Agasthyamala 297. 2002. *Hydrocotyle asiatica* L., Sp. Pl. 234. 1753; Roxb., Fl. Ind. 2 : 88. 1832; Wight &

Arn., Prodr.366. 1834; Clarke in Hook.f., Fl. Brit. India 2 : 669. 1879. *Codagen* Rheede, Hort. Malab.10 : 91, t. 46. 1690. *Kodangal.*

Prostrate herbs. Leaves simple, alternate or clustered on short shoots; stipule scarious; petiole sheathing at base. Lamina orbicular-reniform, 1.5-3 by 2-3 cm, base cordate, margin crenate-dentate, chartaceous. Inflorescence a simple axillary umbel. Flowers small, bisexual, 5-merous, regular, epigynous; calyx-lobes 5, obscure; petals 5, purplish, minute; stamens 5; ovary inferior, 2-locular; style 2, dilated at base. Cremocarp with 2 laterally compressed 7-9-ribbed mericarps.

Fl. & Fr.: Most of the seasons. *Distr.*: Pantropical. Common along streamsides, marshes, wet places, etc. *NAK 586* (Thannithode, c. 70 m).

HERACLEUM Linnaeus
Sp. Pl. 249. 1753.

Heracleum sprengelianum Wight & Arn., Prodr. 372. 1834; Clarke in Hook.f., Fl. Brit. India 2 : 716. 1879; Gamble, Fl. Pres. Madras 565. 1919; Matthew & Britto in Matthew, Fl. Tam. Carnatic 3 (1) : 680. 1983.

Subshrubs. Leaves 2-or 3-pinnate; leaflets c. 2 pairs, oblong or ovate-lanceate, terminal ones 3-lobed, 5-8 by 3-5 cm, base obtuse, margin serrate, apex acuminate, chartaceous. Umbels compound. Flowers 20-30 per ray; calyx-lobes 5, minute; petals 5, yellow; stamens 5; ovary 2-locular; ovule 1-per locule. Cremocarp winged, ribbed.

Fl. & Fr.: Sept.- Feb. *Distr.*: Peninsular India. Rare. *NCN 50849* (MH). (Moozhiar, c. 1100 m).

HYDROCOTYLE Linnaeus
Sp. Pl. 234. 1753.

Prostrate herbs. Leaves obscurely lobed, base cordate, margin creante-serrate. Umbels simple, usually clustered or solitary, leaf-opposed, sessile or peduncled. Flowers pedicelled; calyx truncate; petals greenish-white. Cremocarp laterally compressed. Mericarps 3-ribbed, vittae 0.

1. Umbel(s) solitary, sessile or subsessile; mericarp-ridges faintly visible; leaves to 4 cm wide. .... 1. **H. conferta**
1. Umbels clustered; long-pedunculed; mericarp-ridges obscure; leaves to 8 cm wide. .... 2. **H. javanica**

1. **Hydrocotyle conferta** Wight, Ic. t. 1002. 1845; Clarke in Hook.f., Fl. Brit. India 2 : 68. 1879; Gamble, Fl. Pres. Madras 556. 1919.

Leaves orbicular-cordate, 2-4 by 2-3.5 cm, umbels subsessile, 5-15-flowered, petiole pubescent. Flowers small. Cremocarp orbicular, subcompressed, obscurely punctate, ridges faintly visible.

Fl. & Fr.: Jan.-Apr. *Distr.*: Western Ghats. Rare. On wet places in hills. *NAK 2754* (Ponnambalamedu, 1100 m).

2. **Hydrocotyle javanica** Thunb., Diss. Hydroc. n. 17, t. 2. 1798; Clarke in Hook. f., Fl. Brit.India 2 : 667. 1879; Gamble, Fl. Pres. Madras 556. 1919; Gandhi in Sald. & Nicols., Fl. Hassan Dist. 416. 1976; Krahulik & Theobald in Dassan. & Fosberg, Rev.Handb. Fl. Ceylon 3 : 481. 1981; Matthew & Britto in Matthew, Fl. Tam. Carnatic 3 (1) : 681. 1983; Manilal, Fl. Silent Valley 125. 1988; Mohanan & Sivad., Fl. Agasthyamala 298. 2002.

Leaves orbicular-reniform, 2-8 by 3-8.5 cm, often palmately lobed, margin coarsely crenate, pubescent. Umbels long-peduncled, dense. Flowers medium sized. Cremocarp suborbicular, ribbed.

Fl. & Fr.: Sep.-Dec. *Distr.*: Trop. Africa, Sri Lanka, India, Malesia. Very common in upper hills among shady forest undergrowth. *NAK 79 & 1186* (Upper Moozhiar, c. 550 m).

71. ARALIACEAE

1. Erect shrubs; leaflets serrate; petals imbricate in bud. .... **Aralia**
1. Scandent shrubs or trees; leaflets entire; petals valvate in bud. .... **Schefflera**

ARALIA Linnaeus
Sp. Pl. 273. 1753.

Aralia malabarica Bedd., Fl. Sylv. t. 15. 1871; Clarke in Hook.f., Fl. Brit. India 2 : 722. 1879; Gamble, Fl. Pres. Madras 569. 1919; Manilal, Fl. Silent Valley 126. 1988.

Large shrubs; stem and petiole prickled. Leaves alternate, 2-pinnate, leaflets lanceate, 8-13 by 2-3.5 cm, base rounded, margin serrate, apex acuminate, pubescent, chartaceous. Inflorescence a large compound terminal umbel. Flowers bisexual; calyx obscruely 5-toothed; petals 5, imbricate; stamens 5; ovary inferior, 2-5-locular; ovule 1-per locule. Berry globose, 5-angled, black when ripe.

Fl. & Fr.: Jan.-Apr. *Distr.*: Western Ghats. Rare. In evergreen forests. *NAK 2446* (Appuppanthode, c. 450 m).

SCHEFFLERA J.R. Forster et J.G.A. Forster
Charact. Gen. 45. 1776 (nom. cons.).

Trees or scandent shrubs. Leaves digitately compound, coriaceous. Inflorescence a large panicled-umbel. Flowers bisexual or polygamous; calyx truncate or faintly 5-toothed; petals 5-6, valvate; stamens 5-6; ovary inferior, 5-6- locular. Drupe 5-6-angled.

1. Trees; leaf base rounded, lateral veins faintly reticulate. .... 1. **S. rostrata** var. **micrantha**
1. Scandent shrubs; leaf base cuneate, lateral veins prominently reticulate. .... 2. **S. venulosa**

1. **Schefflera rostrata** (Wight) Harms in Engl. & Prantil, Pflanzenr. 3(8): 38. 1894. var. **micrantha** (Clarke) Mahesh., Bull. Bot. Surv. India 2 : 76.1960; Manilal, Fl. Silent Valley 127. 1988; Mohanan & Sivad., Fl. Agasthyamala 301.2002. *Heptapleurum rostratum* (Wight) Bedd., var. *micranthum* Clarke in Hook.f., Fl. Brit. India 2 : 729. 1879. *Schefflera micrantha* (Clarke) Gamble, Fl. Pres. Madras 569. 1919.

Trees. Leaves oblong-elliptic, 15-20 by 5-9 cm, base rounded, apex acute, lateral veins faint. Umbels c. 2 cm ϕ, forming terminal panicles. Flowers 8-12; calyx truncate; petals pale green. Drupe angled.

Fl. & Fr.: Apr.-Aug. *Distr.*: Western Ghats. Rare. In evergreen forests. *NAK 2685* (Kakki hills, c. 1100 m).

2. **Schefflera venulosa** (Wight & Arn.) Harms in Engler & Prantl, Pflanzenf. 3(8): 39. 1894; Gamble, Fl. Pres. Madras 570. 1919; Gandhi in Sald. & Nicols., Fl. Hassan Dist. 412. 1976; Manilal, Fl. Silent Valley 128.

1988. *Paratropica venulosa* Wight & Arn., Prodr. 1:377. 1834. *Heptapleurum venulosum* (Wight & Arn.) Seem., J. Bot. 3: 80. 1865; Clarke in Hook.f., Fl. Brit. India 2 : 729. 1879, p.p.

Scandent shrubs; branchlets lenticellate. Leaves elliptic to elliptic-oblong, 7-17 by 3-5.7 cm, base rounded or cuneate, apex acuminate, lateral veins prominent. Umbels in large terminal panicles. Flowers c. 7 mm ϕ; calyx truncate; petals white. Drupe subglobose.

Fr. & Fr.: May.-Aug. *Distr.*: India, Myanmar, Indo-China. Less common. Epiphytic shrubs in evergreen forests. *NAK 1753* (Kalleli, . 20 m), *2474* (Moozhiar, c. 250 m).

72. ALANGIACEAE

ALANGIUM Lamarck

Encycl. Meth. Bot. 1 : 174. 1783 (nom. cons.).

Alangium salviifolium (L.f.) Wangerin in Engl., Pflanzenr. IV. 220 b. 41: 9. 1910; Gamble, Fl. Pres. Madras 572. 1919; Nicols., Suresh & Manilal, An Interpr. Hort. Malab. 42. 1988. *Grewia salviifolia* L.f., Suppl. Pl. 409. 1781. ssp. **hexapetalum** (Lamk.) Wangerin, Pflanzenr. 4 (220 b) : 9. 1910. *Kara-angolam* Rheede, Hort. Malab. 4 : 55-56, t. 26. 1683. *Angolam* Rheede, Hort. Malab. 4: 39-40, t. 17. 1683.

Small trees with drooping branches. Leaves simple alternate, elliptic to oblong-obovate, 5-15 by 3-6 cm, base obtuse-round, aepx caudate, basally 3-5- nerved, lower surface with tufts of hairs in the axils of secondary nerves. Inflorescence an axillary fascicle. Flowers bisexual, regular, epigynous; calyx cupular, lobes 4-6; petals 4-6 (10), white; stamens 4-6(-10), filaments scaly; ovary inferior, 2-locular; ovule 1per locule. Berry oblong, crowned by persistent calyx-lobes. Seeds 2.

Fl. & Fr.: Dec.-Jun. *Distr.*: Sri Lanka, Peninsular India, Andamans, Malesia. Rare. Often subscandent shrubs, along streamsides. *NAK 2364* (Kalleli, c. 250 m).

Powdered bark taken in along with honey is a good remedy in biliousness and impure blood.

73. CORNACEAE

MASTIXIA Blume

Bijdr. 654. 1826.

Mastixia arborea (Wight) Bedd., Fl. Sylv. t. 216; 1872; Clarke in Hook.f., Fl. Brit. India 2 : 745. 1879, p.p.; Gamble, Fl. Pres. Madras 573. 1919; Manilal, Fl.Silent Valley 128. 1988; Mohanan & Sivad., Fl. Agasthyamala 302.2002. *Bursinophyllum arboreum* Wight, Ic.t. 956. 1845. *Mastixia pentandra* auct., non Blume, 1850: Clarke in Hook.f., Fl. Brit. India 2 : 746. 1879; Gamble, Fl. Pres. Madras 574. 1919. ssp. **meziana** (Wangerin) Matthew, Blumea 23:89. 1976. *M. meziana* Wangerin in Feddes Repert. 4 : 336. 1907; Gamble, Fl. Pres. Madras 573. 1919; Ramamoorthy in Sald. & Nicols., Fl. Hassan Dist. 229. 1976. *Nir-kurunthu, Velladambu.*

Trees. Leaves simple, alternate, broadly elliptic to oblong-elliptic, 10-20 by 5-10 cm, base cuneate, apex caudate-acuminate, lateral nerves 6-8 pairs, lower surface glaucous, thin-coriaceous. Inflorescence terminal, cymose panicles, densely ferruginous when young; lower bracts foliaceous, to 5 cm long. Flowers bisexual, epigynous; sepals 5, as long as broad; petals 5, greenish-yellow; stamens 5; ovary inferior, 1-locular, crowned by a fleshy disc. Drupe ellipsoid, to 3.5 cm ϕ, surmounted by calyx and disc.

Fl. & Fr.: Feb.-May. *Distr.*: Western Ghats. Rare. Along streamsides. *NAK 2644* (Thannithode, c. 70 m).

Wood is suitable for match boxes and splints.

74. CAPRIFOLIACEAE

VIBURNUM Linnaeus
Sp. Pl. 267. 1753.

Viburnum punctatum Buch.-Ham. ex D. Don, Prodr. 142. 1825; Clarke in Hook. f., Fl. Brit. India 3:5 : 1880; Kern & Steenis in Steenis, Fl. Males. I, 4 : 186. 1951; Gandhi in Sald. & Nicols., Fl. Hassan Dist. 595. 1976. *V. acuminatum* Wall. ex DC., Prodr. 4 : 325. 1830; Wight, Ic. t. 1021. 1845; Gamble, Fl. Pres. Madras 575. 1919.

Trees. Leaves simple, opposite, elliptic-oblong or obovate, 10-16 by 4-6 cm, base attenuate, apex abruptly acuminate, lower surface with peltate scales. Inflorescence a compound umbel; bracts and bracteoles persistent. Flowers bisexual, regular, epigynous; calyx-lobes 5; corolla white, gamopetalous, campanulate, lobes 5; stamens 5, subexerted, filament inflexed in bud; ovary inferior, 1-locular; ovule 1, pendulous. Drupe ridged, ellipsoid, shortly beaked, punctate.

Fl. & Fr.: Jan.-Aug. *Distr.*: Indo-Malesia. Rare. In evergreen forests. *BDS 40828* (MH) (Pachakkanam, Kakki hills, 1100 m).

75. RUBIACEAE

1. Trees, shrubs or shrubby climbers.
    2. Trees or shrubs.
        3. Flowers aggregated in globose heads; trees.
            4. Calyces of individual flowers connate at base.
                5. Flowers orange-red; ovules a per locule. ....**Ochreinauclea**
                5. Flowers white, ovule 1per locule. ....**Morinda**
            4. Calyces of individual flowers free.
                6. Corolla-lobes imbricate; leaves broadly ovate, 30 cm or more long. ....**Neolamarckia**
                6. Corolla-lobes valvate; leaves oblong or ovate-cordiform.
                    7. Leaves oblong; calyx truncate. ....**Mitragyna**
                    7. Leaves ovate-cordiform; calyx lobed. ....**Haldina**
        3. Flowers in cymes, panicle or in fascicles; shrubs (except *Randia* & *Ixora brachiata*).
            8. Ovule a per locule.
                9. Fruit a loculicidal capsule. ....**Wendlandia**

9. Fruit a berry.

10. Flowers in terminal cymes; berry c. 1 cm ϕ. .... **Tarenna**

10. Flowers solitary, or auxillary fascicles, cymes or in leaf-opposed cymes; berry c. 3 cm ϕ.

11. Stem thorny; flowers solitary or in few flowered auxillary cymes or fascicles. .... **Randia**

11. Stem thornless; flowers in leaf opposed cymes. .... **Aidia**

8. Ovule 1-α per locule.

12. Thorny shrubs. .... **Canthium**

12. Thornless shrubs.

13. Flowers few, clustered.

14. Leaves sessile, unequal. .... **Saprosma**

14. Leaves petioled, equal. .... **Lasianthus**

13. Flowers a in cymes.

15. Flowers 4-merous.

16. Cymes terminal/axillary; stigma entire. .... **Pavetta**

16. Cymes exclusively terminal; stigma divided. .... **Ixora**

15. Flowers 5-merous.

17. Leaves decussate; corolla tube straight; ovary 2-loculed. .... **Psychotria**

17. Leaves not decussate; corolla tube curved; ovary 1-loculed. .... **Chassalia**

2. Shrubby climbers.

18. Calyx-lobes 5, one large, petaloid. .... **Mussaenda**

18. Calyx-lobes 5, all small, tooth-like. .... **Oxyceros**

1. Herbs or undershrubs.

19. Flowers 4-merous.

20. Ovules a per locule.

21. Seeds α, angular.

22. Leaves membranous; fruit a berry. .... **Mycetia**

22. Leaves chartaceous or coriaceous; fruit a capsule. .... **Hedyotis**

21. Seeds few, globose or planoconvex. .... **Neanotis**

20. Ovule 1-per locule.

23. Stem subterete; flowers dimorphic. .... **Knoxia**

23. Stem 4-angular or 4-winged; flowers moromorphic.

24. Capsule dehiscence septicidal. ....**Spermacoce**
24. Capsule dehiscence circumcissile. ....**Mitracarpus**

19. Flowers 5-or 5-7- merous.
25. Prostrate herbs; flower(s) often solitary. ....**Geophila**
25. Woody herbs; flowers in cymes, panicles or racemes.
26. Cymes secund; stipule caducous. ....**Ophiorrhiza**
26. Cymes (or racemes) otherwise; stipule persistent.
27. Cymes subterminal; flowers white; seeds angular. ....**Mycetia**
27. Cymes axillary; flowers rose coloured; seeds globose. ....**Neurocalyx**

AIDIA Loureiro

Fl. Cochinch. 143. 1790.

Aidia gardneri (Thw.)Tirvengadum, Bull. Mus. Hist. Nat. Paris 3e Ser., n° 521. Bot. 3:11. 1978; Mohanan & Sivad., Fl. Agasthyamala 306. 2002. *Griffithia gardneri* Thw., Enum. Pl. Zeyl. 158. 1859; Bedd., Ic. t. 38. 1874. *Randia gardneri* (Thw.) Hook.f., in Benth. & Hook.f., Gen. Pl. 2:88. 1873; Hook.f., Fl. Brit. India 3 : 112. 1881; Gamble, Fl. Pres. Madras 617. 1921.

Small trees; stem thornless. Leaves lanceate, 12-16 by 3.5-5 cm, base cuneate, apex acuminate, shiny black when dry. Cymes leaf-opposed; calyx-lobes obscure; corolla white, mouth villous, lobes oblong. Berry globose.

Fl. & Fr.: Feb.-Aug. *Distr.*: Sri Lanka & Peninsular India. Rare. Evergreen trees with white beautiful flowers. *NAK 2399* (Moozhiar, c. 250 m).

CANTHIUM Lamarck

Encycl. Meth., Bot. 1 : 602. 1785.

Armed or unarmed scandent shrubs or trees. Leaves decussate; stipules interpetiolar, caducous. Inflorescence a lax or dense umbellate cyme. Flowers 4-5 merous; calyx toothed; corolla salver-form; lobes usually reflexed; stamens exerted; ovary 2-locular; ovule 1 per locule. Drupe ellipsoid or compressed. Seeds oblong.

1. Unarmed trees, leaves coriaceous, stigma with an apical slit. ....3. **C. dicoccum**
1. Armed scandent shrubs; leaves chartaceous or thin-coriaceous; stigma without an apical slit.
2. Flowers 4-merous; corolla-lobes obovate; stigma capitate; drupe globose. ....2. **C. coromandelicum**
2. Flowers 5-merous; corolla-lobes lanceate; stigma acute; drupe compressed.
3. Leaves thin-coriaceous, base rounded to subcordate, apex caudate, drying greenish. ....1. **C. angustifolium**
3. Leaves chartaceous, base rounded, apex acuminate, drying black. ....4. **C. rheedei**

1. **Canthium angustifolium** Roxb., Fl. Ind. 2 : 169. 1824; Hook.f., Fl. Brit. India 3 : 135. 1880; Gandhi in Sald. & Nicols., Fl. Hassan Dsit. 574. 1976; Ridsdale in Manilal, Bot.Hist. Hort. Malab. 132. 1980; Nicols., Suresh & Manilal, An Interpr. Hort. Malab. 217. 217. 1988; Mohanan & Sivad., Fl. Agasthyamala 309. 2002. *Plectronia rheedii* DC., var. *angustifolia* (Roxb.) Gamble, Fl. Pres. Madras 625. 1921. *Dondisia leschenaultii* DC., Prodr. 4 : 469. 1830. *Canthium leschenaultii* (DC.) Wight & Arn. in Wight, Cat. 77. 1833 & Prodr. 426. 1834. *Katu-kara-walli* Rheede, Hort. Malab. 7:33, t. 17. 1688.

Profusely armed shrubs. Leaves ovate, 4-8 by 2-4 cm, base rounded, apex caudate-acuminate, thin-coriaceous, shiny, drying greenish. Flowers 5-merous; corolla greenish, villous within. Drupe compressed, obcordiform.

Fl. & Fr.: Mar.-Jun. *Distr.*: Western Ghats. Common usually in plains. *NAK 153, 211, 1063 & 1648* (Adoor, Konni etc.).

Leaves pounded and applied to swellings. Root bark boiled in oil and applied to cure head-ache.

2. **Canthium coromandelicum** (Burm.f.) Alston in Trimen, Handb. Fl. Ceylon 6:152. 1931; Nicols., Suresh & Manilal, An Interpr. Hort. Malab. 219. 1988. *Gmelina coromandelica* Burm.f., Fl. Ind. 132. 1768, excl. syn. *Canthium parviflorum* Lam., 1785; *Webera tetrandra* Willd., Sp. Pl. 1:1224. 1798, nom. illegit. (incl. type of *Canthium parviflorum* Lam.,1795). *Kanden-kara* Rheede, Hort. Malab. 5:71, t. 36. 1685. *Valli-kara* Rheede, Hort. Malab. 7:35-36, t. 18. 1688.

Armed shrubs. Leaves elliptic-ovate to obovate, 2-4 by 1-2.5 cm, base attenuate, apex subacute, chartaceous. Flowers in cymes, 4-merous; corolla greenish, villous within. Drupe globose.

Fl. & Fr.: Apr.-Jun. *Distr.*: Peninsular India. Not common. In evergreen forests. *CNM 55522* (MH)-Konni.

Decoction of the leaves and roots is used in certain stages of flux and the latter has anthelmintic properties. Bark and young shoots are used in dysentery.

3. **Canthium dicoccum** (Gaertn.) Teys. & Binn., Cat. Hort. Bog. 113. 1866 var. **umbellatum** (Wight) Sant. & Merch., Bull. Bot. Surv. India 3:107. 1962; Gandhi in Sald. & Nicols., Fl. Hassan Dist. 575. 1976; Matthew & Rani in Matthew, Fl. Tam. Carnatic 3(2) : 698. 1983; Manilal, Fl. Silent Valley 129. 1988. *C. umbellatum* Wight, Ic. t. 1034. 1845; Hook.f., Fl. Brit. India 3 : 132. 1880. *C. didymum* sensu Bedd., Fl. Sylv. t. 221. 1872, non Gaertn., 1778. *Plectronia didyma* (Gaetrn.) Kurz, var. *umbellata* (Wight) Gamble, Fl. Pres. Madras 624. 1921.

Trees. Leaves elliptic-lanceate, apex shortly acuminate, coriaceous, shiny, 10-15 by 5-7 cm. Flowers in dense umbellate clusters, 5-merous; corolla cream coloured. Drupe ellipsoid or ovoid.

Fl. & Fr.: Nov.-Mar. *Distr.*: Sri Lanka, India, Nepal, Indo-China, Malesia. Not common. In evergreen forests. *CNM 65071* (MH) Ranni. R.F.

4. **Canthium rheedei** DC., Prodr. 4 : 474. 1830; Wight & Arn., Prodr. 426. 1834; Hook.f., Fl. Brit. India 3 : 134. 1880; Matthew & Rani in Matthew, Fl. Tam. Carnatic 3 (2) : 699. 1983, ('rheedii'); Nicols., Suresh & Manilal, An Interpr. Hort. Malab. 219. 1988. *Plectronia rheedei* (DC.) Bedd., Fl. Sylv. 134. 1872; Gamble, Fl. Pres. Madras 625. 1921. *Tsjerou-kara* Rheede, Hort. Malab. 5 : 73-74, t. 37. 1685.

Armed shrubs. Leaves ovate or elliptic-ovate, 4-10 by 2-4 cm, base rounded, apex acuminate, chartaceous, drying black. Flowers in axillary fascicles, 5-merous; corolla greenish, villous within. Drupe compressed, obcordate.

Fl. & Fr.: Aug.-Nov. *Distr.*: Peninsular India & Sri Lanka. Common in semi-evergreen forests. *NAK* 1970 & 2458 (Arampa, c. 400 m).

CHASSALIA Commerson ex Poiret
Encycl. Meth. Bot. Suppl. 2 : 450. 1812.

Chassalia ophioxyloides (Wall.) Craib, Gard. Bull. Straits Settlem. 6 : 474. 1930; Gandhi in Sald. & Nicols., Fl. Hassan Dist. 576. 1976; Manilal, Fl. Silent Valley 131. 1988; Nicols., Suresh & Manilal, An Interpr. Hort. Malab. 221. 1988; Mohanan & Sivad., Fl. Agasthyamala 310. 2002. *Psychotria ophioxyloides* Wall. in Roxb., Fl. Ind. 2: 168. 1824. *Chasalia curviflora* sensu Hook.f., Fl. Brit. India 3:176. 1880, non (Wall.) Thw., 1859; Gamble, Fl. Pres. Madras 643. 1921. *Katou-belluta-ampelpodi* Rheede, Hort. Malab. 5 : 66, t. 33 (on left).1685.

Shrubs. Leaves elliptic to obovate-lanceate to linear lanceate, base rounded, obtuse or cuneate, apex acuminate, chartaceous. Inflorescence a panicled-cyme, terminal. Flowers 5-merous; calyx 5-toothed; corolla white to pink, tube cylindric, curved; stamens 5, usually included; ovary 2-locular, ovule 1-per locule. Drupe globose.

Key to the varieties

1. Leaves elliptic to obovate-lanceate, apex acuminate. **Chassalia ophioxyloides** var. **ophioxyloides**
1. Leaves linear-lanceate, apex acute. ....**Chassalia ophioxyloides** var. **longifolia**

Chassalia ophioxyloides (Wall.) Craib var. **ophioxyloides.**

Shrubs. Leaves elliptic to obovate-lanceate, 6-18 by 2-5 cm, base rounded, obtuse or cuneate, apex acuminate, chartaceous. Flowers white to pink.

Fl. & Fr.: Most of the seasons. *Distr.*: India, Sri Lanka, Myanmar and Philippines. *NAK 27, 143, 783 & 846.* (Almost all places).

Leaf juice together with boiled oil is a good remedy for earache and eye diseases.

Chassalia ophioxyloides (Wall.) Craib var. **longifolia** (Dalz.) Hook.f., Fl. Brit. India 3: 177. 1880; Deb. & Krishna, Bull. Bot. Surv. India 24 : 224. 1983. *Psychotria longifolia* Dalz. in Hook., London J. Bot. 2:133.1845.

Leaves linear-lanceate, apex acute. Flowers pink; calyx short, 5-lobed. Fruit globose, smooth, didymous.

Fl. & Fr.: Jan.-Sep. *Distr.*: India & Malay Peninsula. Rare. In evergreen forests. *NAK 113 & 500* (Kakki hills).

Note : Probably this is the first report of the variety from South India.

GEOPHILA D. Don
Prodr. Fl. Nepal. 136. 1825 (nom. cons.).

Geophila repens (L.) Johnston, Sargentina 8 : 821. 1949; Nicols., Suresh & Manilal, An Interpr. Hort. Malab. 221. 1988. *Geophila herbacea* (Jacq.) K. Schum. in Engler & Prantl, Pflanzenfam. 4 (4) : 99 : 1891; Sant. & Merch., Bull. Bot. Surv. India 3:108. 1961. *Psychotria herbacea* Jacq., Enum. Pl. Carib. 16. 1760. *Geophila reniformis* D. Don, Prodr. 136. 1825, nom. illegit; Hook.f., Fl. Brit. Indfia 3 : 178. 1880; Gamble, Fl. Pres. Madras 643. 1921. *Karinta-kali* Rheede, Hort. Malab. 10 : 41, t. 21. 1690.

Prostrate herbs. Leaves orbicular or broadly ovate to reniform, 15-25 by 10-20 mm, apex rounded or retuse, chartaceous. Flower(s) usually solitary or rarely in umbels, 4-7- merous; calyx-lobes small, persistent; corolla funnel-form, hairy within, 10-15 mm f, lobes 4-7; stamens 4-7, included; ovary 2-locular; ovule 1 per locule. Drupe globose. Seeds plano-convex.

Fl. & Fr.: Sep.-Nov. *Distr.*: Sri Lanka, Peninsular India, Malaya, S. China, Polynesia, Tropical Africa, America. Less common. Pretty herbs with white flowers. In moist shady places in hills. *NAK* 720 (Perunthenaruvi, Ranni, c. 100 m)

HALDINIA Ridsdale
Blumea 24 : 360. 1978.

Haldinia cordifolia (Roxb.) Ridsd., Blumea 24 : 361. 1978; Matthew, Ill. Fl. Tam. Carnatic t. 333. 1982; Matthew & Rani in Matthew, Fl. Tam. Carnatic 3(2) : 707. 1983; Mohanan Sivad., Fl. Agasthyamala 312. 2002. *Nauclea cordifolia* Roxb., Pl. Corom. t. 53. 1796 & Fl. Ind. 1 : 514. 1832; Wight & Arn., Prodr. 391. 1834; Bedd., Fl. Sylv. t. 33. 1869. *Adina cordifolia* (Roxb.) Hook.f. ex Brandis, Forest Fl.N.W. India t. 33. 1874; Hook.f., Fl. Brit. India 3:24. 1880; Gamble, Fl. Pres. Madras 585. 1921; Gandhi in Sald. & Nicols., Fl. Hassan Dist. 472. 1976. *Manjakadambu.*

Trees. Leaves cordiform or suborbicular, 10-15 by 12-17 cm, apex abruptly acuminate, lower surface pubescent, lateral nerves 8-10 pairs, chartaceous. Flowers 5-merous, in axillary globose heads; calyx cupular, obscurely 5-angled, lobes 5, equal; corolla cream coloured, funnel-form; stamens 5, subexerted; ovary 2-locular; ovules 3-5. Capsules clustered.

Fl. & Fr.; Aug.-Nov. *Distr.*: Sri Lanka, India, China to Peninsular Thailand. Rare. Deciduous trees. *CB 49114* (MH) Konni, R.F.

Wood very durable. Bark powdered or its decoction taken as a remedy in biliousness, phlegm, vomiting and various skin diseases; it improves complexion and vitality.

HEDYOTIS Linnaeus
Sp. Pl. 101. 1753.

Erect, prostrate herbs or shrubs. Leaves decussate; stipule interpetioler, often pectinate. Cymes paniculate or umbellate or reduced to a single flower. Flowers small, 4-5-merous; calyx-lobes persistent; corolla white, yellowish or pink, funnel-form or campanulate, lobes valvate; stamens inserted or exerted; ovary 2-locular; ovules many; stigma 2-fid. Capsule septicidal or loculicidal. Seeds α,angular.

1. Erect shrubs or herbs.
    2. Erect shrubs; leaves lanceate, c. 10 by 3 cm; flowers in terminal cymes. ......7. **H. pruinosa**
    2. Erect herbs; leaves linear-lanceate, c. 3 by 0.6 cm; flower(s) solitary or 2-4 per node. ....4. **H. herbacea**
1. Spreading diffuse herbs or undershrubs.
    3. Flowers aggregated in capitate clusters around nodes.
        4. Branchlets (sub-) terete; lateral veins of leaves very prominent; capsule indehiscent. ....1. **H. auricularia**

4. Branchlets 4-gonous; lateral veins of leaves not so; capsule dehiscent.

5. Leaves subsessile, margin scabrulous towards apex, coriaceous; cymes, c. 1 cm ϕ. ....6. **H. nitida**

5. Leaves petioled, margin glabrous, chartaceous; cymes, c. 2.5 cm ϕ. .... 5. **H. membranacea**

3. Flowers distant, 1-4, axillary or many in lax panicle.

6. Flower(s) usually solitary or in pairs, subsessile; peduncle absent. ....2. **H. brachypoda**

6. Flowers in corymbose or umbellate lax cymes, pedicelled; peduncle to 2 cm long. ....3. **H. corymbosa**

1. **Hedyotis auricularia** L., Sp. Pl. 101. 1753; Hook. f., Fl. Brit. India 3:58 1880; Rao & Hemadri, Ind. For. 99 : 375 1973; Manilal, Fl. Silent Valley 131. 1988; Nicols., Suresh & Manilal, An Interpr. Hort. Malab. 222. 1988; Mohanan & Sivad., Fl. Agasthyamala 314. 2002. *Oldenlandia auricularia* (L.) K. Schum .in Engl. & Prantl, Pflanzenfam. 4(4) : 25. 1891; Gamble, Fl. Prcs. Madras 597. 1921; Gandhi in Sald. & Nicols., Fl. Hassan Dist. 585. 1976. *Muriguti* Rheede, Hort. Malab. 10 : 63, t. 32. 1690.

Diffuse woody herbs. Leaves variable, usually elliptic-ovate or elliptic-lanceate to oblong-lanceate, 2-12 by 1-2.5 cm, base rounded or cuneate, apex shortly acuminate, lateral veins prominent. Flowers small; corolla white. Capsule globose, indehiscent.

Fl. & Fr.: Jan.-May. *Distr.*: Sri Lanka, India, Malesia, S. China, Philippines, Australia. Very frequent among forest undergrowth. *NAK 227, 253 & 787* (Almost all places).

Leaf juice quickly heals minor wounds.

2. **Hedyotis brachypoda** (DC.) Sivar. & Biju, Taxon 39 : 672. 1990; Mohanan & Sivad., Fl. Agasthyamala 315.2002. (*H. brachypoda* R.Br. in Wall., Cat. 874. 1829, nom. nud.). *Oldenlandia brachypoda* DC., Prodr. 4: 424. 1830; *Oldenlandia diffusa* sensu auct., non as to type of (Willd.) Roxb., (1820) : Roxb., Fl. Ind. 1 : 444. 1820; Hook.f., Fl. Brit. India 3 : 65. 1880 (incl. var. *diffusa* and var. *extensa*); Trimen, Handb. Fl. Ceylon 2: 315. 1894; Gamble, Fl. Pres. Madras 601. 1921. *Hedyotis diffusa* sensu auct., non Willd., 1798: Bakh. f. in Back. & Bakh. f., Fl. Java 2: 286. 1965; Manilal, Fl. Silent Valley 132. 1988.

Diffuse or prostrate herbs; stem terete. Leaves linear-lanceate, 15-30 by 3-5 mm, base and apex acute, margin faintly revolute. Flower(s) soiltary or in pairs, 4-merous, subsessile; peduncle 0; corolla white, tube much longer than calyx-lobes, glabrous within; stamens inserted near the sinuses of the corolla-lobes. Capsule subglobose, 4-5 mm ϕ.

Fl. & Fr.: Apr.-Aug. *Distr.*: Sri Lanka, India, Bangladesh, China, Japan, Trop.America, Malaya Peninsula, Philippines. Less common. Usually on wet areas, near streamsides, lakes, paddy fields, etc. *NAK* 1789 (Angamoozhy, c. 250 m).

3. **Hedyotis corymbosa** (L.) Lam., Tabl. Encycl. 1: 272. 1972; Rao & Hemadri, Ind. For. 99 : 375. 1973; Dutta, Rev. Ind. Hedyotis 281. 1985; Nicols., Suresh & Manilal, An Interpr. Hort. Malab. 222. 1988; Sivar. & Biju, Taxon 39 : 671. 1990; Mohanan & Sivad., Fl. Agasthyamala 315. 2002. *Oldenlandia corymbosa* L., Sp. Pl. 119. 1753; Hook.f., Fl. Brit. India 3 : 64. 1889; Gamble, Fl. Pres. Madras 600. 1921; Gandhi in Sald. &

Nicols., Fl. Hassan Dist. 586. 1976; Matthew & Rani in Matthew, Fl. Tam. Carnatic 3(2): 727. 1983. *O. corymbosa* var. *corymbosa* Verd., Fl. Trop. E. Africa, Rubiac. 1 : 309. 1976. *Parpadagam* Rheede, Hort. Malab. 10 : 69, t. 35. 1690.

Diffuse or spreading prostrate herbs. Leaves linear-lanceate, 7-30 by 2-4 mm, base attenuate, margin revolute. Flower(s) (1-) 3-8 in corymbose or paniculate cymes, 4- merous; peduncle 2-20 mm long; corolla white, tube with a ring of hairs at throat; stamens 4(-5) inserted at corolla base. Capsule obovoid, c. 3 mm ϕ.

Fl. & Fr.: Apr.-Sep. *Distr.*: Pantropical. A common weed in moist low lands. *NAK 965, 1605 & 1631* (Tiruvalla, Adoor, Sabarimala etc.)

Note : A highly variable, taxonomically puzzling herb with different forms varying according to the habitat.

4. **Hedyotis herbacea** L., Sp. Pl. 102.1753; Rao & Hemadri, Ind. For. 99 : 376 1973; Manilal, Fl. Silent Valley 132. 1988; Nicols., Suresh & Manilal, An Interpr. Hort. Malab. 223. 1988; Mohanan & Sivad., Fl. Agasthyamala 316.2002. *Oldenlandia herbacea* (L.) Roxb., Fl. Ind. 1 : 424. 1820; Gamble, Fl. Pres. Madras 601. 1921; Gandhi in Sald. & Nicols., Fl. Hassan Dist. 1976; Matthew & Rani in Matthew, Fl. Tam. Carnatic 3(2) : 728. 1983. *O. heynei* (R. Br.) G. Don, Gen. Syst. 3 : 531. 1834; Hook.f., Fl. Brit. India 3 : 65. 1880. *Schanganam-pullu* Rheede, Hort. Malab. 10 : 45 t. 23, 1690.

Erect herbs; branchelets sharply 4-angled. Leaves linear-lanceate, 10-30 by 1-3 mm, Flower(s) solitary or in pairs per node, 5-merous; calyx-lobes ovate-lanceate, herbaceous; corolla white; stamens 4, included. Capsule subglobose, c. 2.5 mm ϕ.

Fl. & Fr.: Jun.-Sep. *Distr.*: Sri Lanka, India, Malaya Islands & Trop. Africa. On exposed lands, usually in large populations, from plains to hills. *NAK 678 & 1983.* (Adoor, c. 20 m, Tiruvalla, c. 10 m).

Plants boiled with coconut oil and used against elephentiasis and rheumatism.

5. **Hedyotis membranacea** Thw., Enum. Pl. 143. 1859; Hook.f., Fl. Brit. India 3 : 54. 1880. *Oldenlandia membranacea* (Thw.) O. Kuntze, Rev. Gen. 292-293. 1891; Gamble, Fl. Pres. Madras 600. 1921. **(Fig. 38)**

Diffuse or erect undershrubs; stem stout, 4-angled; stipules glandular pectinate. Leaves oblong-lanceate or elliptic-lanceate, 7-13 by 2-3.5 cm, base and apex acuminate, lateral veins faint, chartaceous. Cymes sessile, c. 2.5 m ϕ. Flowers crowded, very dense; calyx-lobes foliaceous; corolla cream coloured. Capsule obconical, c. 3 mm ϕ. Seeds pitted.

Fl. & Fr.: May-Sep. *Distr.*: Sri Lanka, Southern-Western Ghats, Rare. On moist shady hills. Locally abundant. *NAK 47, 1100 & 1834* (Angamoozhy, c. 250 m).

6. **Hedyotis nitida** Wight & Arn., Prodr. 412. 1834; Hook.f., Fl. Brit.India 3 : 61.1880; Mohanan & Sivad., Fl. Agasthyamala 317.2002. *Oldenlandia nitida* (Wight & Arn.) Gamble, Fl. Pres. Madras 597. 1921; Gandhi in Sald. & Nicols., Fl. Hassan Dist. 586. 1976.

Cherukurane.

Shrubs. Leaves lanceate, 6-13 by 2-5 cm, base broadly acute to rounded, apex acute to shortly acuminate, thin-coriaceous. Cymes corymbiform; bracteoles lanceate; corolla white, tube short, lobes elliptic-lanceate. Berry subglobose, c. 1 cm ϕ.

Fl. & Fr.: Aug.-Nov. *Distr.*: Southern Western Ghats.Rare. In sacred groves. *NAK 2036* (Pathanamthitta, c. 50 m).

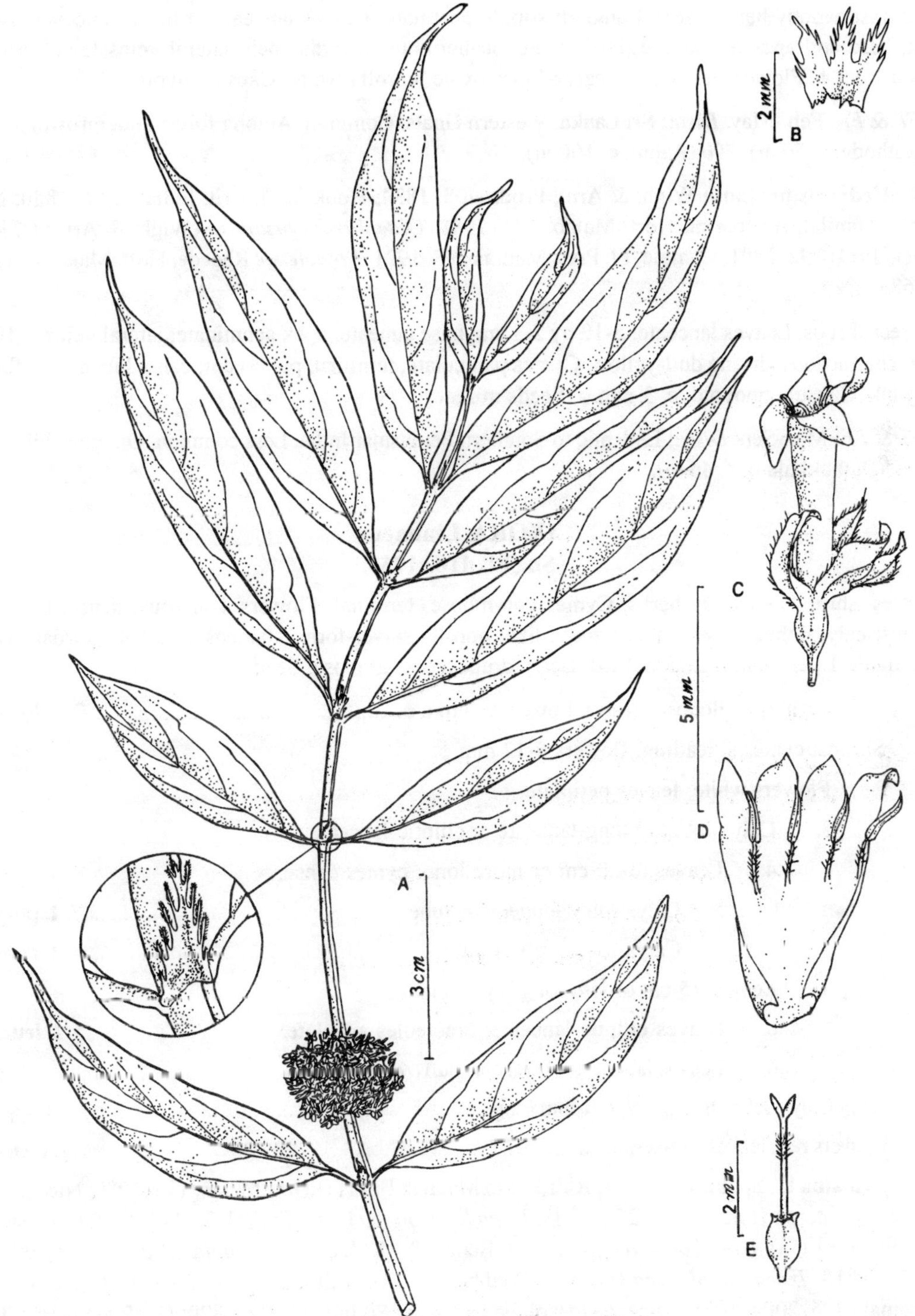

Fig. 38. ***Hedyotis membranacea*** **Thw.: A. Twig; B. Bract; C. Flower; D. Corolla - split open; E. Pistil**

Diffuse woody herbs; stem 4-angled; stipule pectinate. Leaves lanceate to linear lanceate, 3-6 by 1-1.5 cm, base and apex acute, margin revolute, scaberulous towards apex, lateral veins faint, coriaceous. Cymes c.1 cm ϕ. Flowers less dense; calyx-lobes ovate; corolla white. Capsule ovoid.

Fl. & Fr.: Feb.-May. *Distr.*: Sri Lanka, Western Ghats. Common. Among forest undergrowth. *NAK 198* (Thekkuthode, c. 70 m), *790* (Ranni, c. 100 m).

7. **Hedyotis pruinosa** Wight & Arn., Prodr. 408. 1834; Hook.f., Fl. Brit. India 3 : 51. 1880; Nicols., Suresh & Manilal, An Interpr. Hort. Malab. 223. 1988. *Oldenlandia pruinosa* (Wight & Arn.) O. Kuntze, Rev. Gen. Pl. 1: 292. 1891; Gamble, Fl. Pres. Madras 598. 1921. *Poutaletsje* Rheede, Hort. Malab 4 : 117-118, t. 57. 1683.

Erect shrubs. Leaves lanceate, 5-10 by 2-3 cm, base cuneate, apex acuminate, lateral veins c. 10 pairs, slender, chartaceous, drying dull-yellow. Cymes painculate, terminal, pubescent; calyx lobed; corolla white, tinged pink. Capsule obovoid, c. 3 mm ϕ. Seeds minute.

Fl. & Fr.: Mar.-Sep. *Distr.*: Endemic to Southern Peninsular India. Less common. On moist hill cuttings. *NAK 647* (Puthukkulam, c. 40 m).

IXORA Linnaeus
Sp. Pl. 110. 1753.

Trees, shrubs or woody herbs. Cymes corymbose, terminal. Flowers 4-merous, dense; bracts often foliaceous; calyx-lobes 4, lanceate or tooth-like; corolla salver-form; stamens at corolla throat; ovary 2-locular; ovule 1 per locule; stigma 2-fid. Berry drupaceous, globose. Seed 1.

1. Trees; cymes in globose cluster; flowers c. 3 mm ϕ. ....2. **I. brachiata**
1. Shrubs; cymes spreading; flowers 6-12 mm ϕ.
    2. Flowers white; leaves petiolate.
        3. Calyx-lobes oblong-lanceate or elliptic-lanceate.
            4. Leaves to 20 cm or more long; cymes dense.
                5. Calyx-lobes longer than tube. ....7. **I. polyantha**
                5. Calyx-lobes equal to tube. ....4. **I. cuneifolia**
            4. Leaves 15 cm or less long; cymes lax.
                6. Leaves elliptic-lanceate; bracteoles subulate. 5. **I. leucantha**
                6. Leaves lanceate; bracteoles narrowly elliptic. ....1. **I. alba**
        3. Calyx-lobes triangular, c. 1 mm long. ....6. **I. nigricans**
    2. Flowers red; leaves sub-sessile. ....3. **I. coccinea**

1. **Ixora alba** L., Sp. Pl. 110. 1753; Ridsdale in Manilal, Bot.Hist.Hort.Malab. 133. 1980; Nicols., Suresh & Manilal, An Interpr. Hort. Malab. 224. 1988. *Ixora lanceolatea* Lam., Encycl. 3 : 343. 1789 (incl. lectotype of *Ixora alba* L., 1753). *Ixora barbata* Roxb., Hort. Beng. 10. 1814. *Chiococa malabarica* Dennst., Schluessel 12, 19, 37, 1818. *Ixora malabarica* (Dennst.) Mabberley, Taxon 26 : 539. 1977; Mohanan & Sivad., Fl. Agasthyamala 326. 2002. *Ixora lanceolaria* Colebr. ex Roxb., Fl. Ind. 1 : 397. 1820; Hook.f., Fl. Brit. India 3 : 138. 1880; Gamble, Fl. Pres. Madras 630. 1921. *Bem-schetti* Rheede, Hort. Malab. 2 : 19, t. 14. 1679.

2. **Ixora brachiata** Roxb. ex DC., Prodr. 4 : 488. 1830; Wight, Ic. t. 710. 1843; Hook.f., Fl. Brit. India 3 : 142. 1880; Gamble, Fl. Pres. Madras 629. 1921; Mohanan & Sivad., Fl. Agasthyamala 326. 2002.

Trees. Leaves elliptic-oblong or narrowly obovate, 10-20 by 4.5-8 cm, base cuneate, apex obtuse or shortly acuminate, thin-coriaceous. Cymes profusely branched, terminal; bracts stipuliform. Flower buds globose, subsessile, tube slender, c. 4 mm long. Berry globose, c. 7 mm ϕ.

Fl. & Fr.: Jan.-Apr. *Distr.:* Peninsular India. A common evergreen tree. *NAK 451, 1424, 1458 & 2266* (Moozhiar, Armpa, etc.).

Note : In the typical species the inflorscence axis is often reddish and each cymose cluster is dense. But in some collection (*NAK 2267*) they are green and lax respectively.

3. **Ixora coccinea** L., Sp. Pl. 110. 1753; Roxb., Fl. Ind. 1 : 375. 1832; Wight & Arn., Prodr. 427. 1834; Wight, Ic. t. 153. 1839; Hook.f., Fl. Brit. India 3 : 145.1880; Gamble, Fl. Pres. Madras 631. 1921; Matthew & Rani in Matthew, Fl.Tam. Crnatic 3(2) : 711. 1983; Nicols., Suresh & Manilal, An Interpr. Hort. Malab. 225. 1988; Mohanan & Sivad., Fl. Agasthyamala 324. 2002. *'Schetti* Rheede, Hort. Malab. 2 : 17-18, t. 13. 1679.

Woody herbs to small shrubs. Leaves subsessile, oblong to oblanceate, 5-8 by 2.5-3.5 cm, base subcordate, apex subacute. Cymes corymbiform, sessile. Flowers dense; corolla red, tube c. 3.5 cm long. Berry globose, to 1 cm ϕ.

Fl. & Fr.: Most of the seasons. *Distr.*: Sri Lanka, Peninsular India. Common in sacred groves and exposed hills. *NAK 1647* (Tiruvalla, c. 10 m).

Fruits edible. Root and flowers used in medicine for dysentery, fever, gonorrhoea.

4. **Ixora cuneifolia** Roxb. ex DC., Prodr. 4 : 487. 1830; Wight, Ic. t. 709. 1843; Hook.f., Fl. Brit. India 3 : 144. 1880; Gamble, Fl. Pres. Madras 631. 1921; Manilal, Fl. Silent Valley 133. 1988.

Undershrubs. Leaves lanceate or oblong, elliptic-lanceate or oblanceate, 10-20 by 5-7 cm, base cuneate, apex obtusely acuminate, coriaceous. Cyme dense, on stout peduncle; lower bracts often foliaceous. Flowers crowded; corolla white. Berry globose, c. 1 cm ϕ.

Fl. & Fr.: Dec.-Apr. *Distr.*: Peninsular India. Rare. In evergreen or shola forests. *CNM 55788* (MH)- (Pampa dam Area, Kakki hills).

5. **Ixora leucantha** Heyne ex G. Don, Gen. Syst. 3: 572. 1834; Gamble, Fl. Pres. Madras 630. 1921; Mohanan & Sivad., Fl. Agasthyamala 325.2002. *I. lanceolaria* sensu Hook.f., Fl. Brit. India 3 : 138. 1880, p.p., non Colebr. ex Roxb., 1820.

Shrubs. Leaves elliptic to elliptic-lanceate, 3-11 by 2-3 cm base and apex acute, thin-coriaceous. Cymes corymbiform; bracts and bracteoles subulate. Flowers dense; corolla white, tube slender, c. 15 mm long. Berry globose, c. 12 mm ϕ.

Fl. & Fr.: Apr.-Jun. *Distr.*: Endemic to Western Ghats. Not common. Bad smell emitting shrubs in interior of evergreen forests. *NAK 2380 & 2584* (Moozhiar, c. 250 m).

Fl. & Fr.: May-Jul. *Distr.*: Peninsular India. In sacred groves. Occasional. *NAK 121* (Adoor, c. 20 m).

6. **Ixora nigricans** R. Br. ex Wight & Arn., Prodr. 428. 1834; Hook.f., Fl. Brit. India 3 : 148. 1880; Gamble, Fl. Pres. Madras 631. 1921; Gandhi in Sald. & Nicols., Fl. Hassan Dist. 579. 1976; Mohanan & Sivad., Fl. Agasthyamala 326. 2002.

Shrubs. Leaves elliptic-lanceate to oblanceate, 5-15 by 2-4 cm, base cuneate, apex acuminate, chartaceous, shiny, drying black. Cymes c. 12 cm long; bracteoles subulate; calyx-lobe minute; corolla white, tube c. 1 cm long. Berry globose, c. 1 cm ϕ.

Fl. & Fr.: Jan.-Mar. *Distr.*: Indo-Malesia. Very frequent in evergreen forest undergrowth. *NAK 555, 818, 1431, 1563 & 1602*. (Ranni, Konni, Moozhiar, etc.)

7. **Ixora polyantha** Wight, Ic. t. 1066. 1846; Hook.f., Fl. Brit. India 3 : 140. 1880; Gamble, Fl. Pres. Madras 579. 1921; Gandhi in Sald. & Nicols., Fl. Hassan Dist. 579. 1976.

Stout undershrubs. Leaves obovate to broadly oblanceate, 10-20 by 5-8 cm, base cuneate, apex shortly acuminate, thin-coriaceous. Cymes dense, shortly peduncled. Flowers sessile; calyx-lobes elliptic-lanceate to oblong-lanceate, c. 6 mm long; corrola white. Berry globose, c. 15 mm ϕ.

Fl. & Fr.: Mar.-May. *Distr.*: Southern Western Ghats. India. Less common. Among forest undergrowth. *NAK 572* (Maniyaar, c. 70 m), *708* (Ranni, c. 100 m).

Ixora finalaysoniana Wall. ex Don is widely cultivated here and there (*NAK 1542*).

KNOXIA Linnaeus
Sp. Pl. 104. 1753.

Knoxia sumatrensis (Retz.) DC., Prodr. 4: 569. 1830; Gandhi in Sald. & Nicols., Fl. Hassan Dist. 579. 1976; Matthew & Rani in Matthew, Fl. Tam. Carnatic 3 (2) : 714. 1983; Bhatt & Deb, J. Econ. Tax. Bot. 6 : 89. 1985; Manilal, Fl. Silent Valley 134. 1988; Mohanan & Sivad., Fl. Agasthyamala 327.2002. *Spermacoce sumatrensis* Retz., Obs. Bot. 4: 23. 1786; *Knoxia corymbosa* sensu Hook.f., Fl. Brit. India 3 : 128. 1880, non Willd., 1797; Gamble, Fl.Pres. Madras 622. 1921.

Tall perennial slender herbs; branchlets tomentose. Leaves narrowly elliptic-lanceate to lanceate, 3-12 by 1.5-3 cm, base and apex acute, lateral veins c. 10 pairs, prominent, chartaceous. Panicle corymbose, terminal. Flowers small, 4-merous; calyx-lobes 4, small, acute; corolla white, pubescent within near the throat; stamens 4, exerted or inserted; stigma 2-fid. Fruit narrowly ellipsoid. Seeds small.

Fl. & Fr.: Sep.-Jan. *Distr.*: Sri Lanka, India, China, Philippines, Australia. Not common. A variable herb, along moisty grassy hills slopes. *NAK 60, 1091* (Kakki hills, 1100 m).

LASIANTHUS W. Jack
Trans. Linn. Soc. London 14 : 125. 1823. (nom. cons.).

Lasianthus strigillosus Hook.f., Fl. Brit. India 3 : 185. 1880; Gamble, Fl. Pres. Madras 647. 1921.

Erect shrubs; branchlets densely pubescent. Leaves narrowly oblong-lanceate, 5-15 by 1.5-3.5 cm, base obtuse, apex acuminate, lower surface tomentose, lateral veins c. 8 pairs, very prominent. Cymes sessile, few-flowered, axillary. Flowers 5-merous; calyx-lobes triangular-lanceate; corolla white, funnel-form, caducous; stamens 5, included; ovary 4-9- locular; ovule 1per locule. Berry globose, bluish-purple.

MITRACARPUS Zuccarini ex J.A. Schultes et J.H. Schultes
Mant. 3: 210 ('Mitracarpum') 399. 1827.

Mitracarpus hirtus (L.) DC., Prodr. 4:572. 1830; Mohanan & Sivad., Fl. Agasthyamala 331. 2002. *Mitracarpus verticillatus* (Schum. & Thonn.) Vatke, Linnaea 40 : 196. 1876; Sebastine & Ramamoorthy, Bull. Bot. Surv. India 9:291. 1967; Manilal, Fl. Silent Valley 136. 1988. *Staurospermum verticillatus* Schum. & Thonn., Besker. Guin. Pl. 73. 1827. *Mitracarpus scaber* Zucc. in Schultes & Schultes, Mant. 3: 210. 1827.

Erect herbs. Leaves ovate or ovate-lanceate, 4-5 by 1.5-2 cm, base cuneate, apex acute, chartaceous, sessile; stipule pectinate. Flowers small, verticillate in axillary fascicles, 4-merous; calyx-lobes 4, dissimilar; corola white, funnel-form; stamens 4, inserted; ovary 2-locular, ovule 1per locule. Capsule globose, circumcissile in dehiscence.

Fl. & Fr.: Apr.-Sep. *Distr.*: India, Trop. Africa. A common weed on waysides. *NAK 2925* (Adoor, c. 20 m).

This introduced weed is very dominant over the indigenous species and rapidly replace them.

MITRAGYNA P.W. Korthals
Observ. Naucleis Ind. 19. 1839 (nom. cons).

Mitragyna tubulosa (Arn.) Hav., J. Linn. Soc. Bot. 33:71. 1897; Gamble, Fl. Pres. Madras 585. 1921; Mohanan & Sivad., Fl. Agasthyamala 331. 2002. *Nauclea tubulosa* Arn. in Thw., Enum. Pl. Zeyl. 137. 1859. *Stephegyne tubulosa* (Arn.) Benth. & Hook.f., Fl.Brit. India 3 : 25. 1880. *Naikadambu.*

Trees. Leaves elliptic to elliptic-oblong, 8-15 by 3.5-8 cm, base rounded, apex acuminate, chartaceous. Inflorescence of capitate heads arranged in panicles, axillary and/or terminal. Flowers 5-merous; Calyx-tube obconic, truncate; corolla purple; stamens included; ovary 2-locular; ovules a. Capsule in globose heads. Seeds winged.

Fl. & Fr.: Jun.-Sep. *Distr.*: Sri Lanka, Western Peninsular India. Less common. In moist deciduous and semi-evergreen forests. *NAK 804* (Ranni, c. 150 m).

Wood durable.

MORINDA Linnaeus
Sp. Pl. 176. 1753.

Morinda pubescens Sm., Cyclop. 24, n. 3. 1813; Verdc., Kew Bull. 37: 543. 1983; Mohanan & Sivad., Fl. Agasthyamala 332.2002. *M. tomentosa* Roth, Nov. Pl. Sp. 147. 1821; Sant. & Merch., Bull. Bot. Surv. India 3: 108. 1961; Gandhi in Sald. & Nicols., Fl. Hassan Dist. 581. 1976. *M. tinctoria* Roxb., Fl. Ind. (ed. & Carey & Wall.) 1: 543. 1832; Hook.f., Fl. Brit. India 3: 156. 1880; Gamble, Fl. Pres. Madras 651. 1921. *Manjanathi.*

Trees. Leaves broadly or narrowly elliptic, 5-18 by 2-8 cm, base and apex acute, glabrous or pubescent; chartaceous. Cymes few-flowered, terminal or axillary or leaf-opposed. Flowers 5-merous; calyx-tube short, lobes 0; corolla white, lobes 5; stamens 5, included; ovary 2-locular; ovule 1per locule. Fruit a syncarpium of bony pyrenes.

Fl. & Fr.: Most of the seasons. *Distr.*: Sri Lanka, India, Malesia. Highly variable with respect to indumentum. Common. In plains. *NAK 2857* (Adoor, c. 20 m).

Paste of leaves given in diarrhoea, dysentery, and as a febrifuge.

MUSSAENDA Linnaeus
Sp. Pl. 177. 1753.

Mussaenda belilla Ham., Trans. Linn. Soc. London 14:203. 1824.; Ridsdale in Manilal, Bot Hist. Hort. Malab. 131. 1980; Nicols., Suresh & Manilal, An Interpr. Hort. Malab. 226. 1988; Mohanan & Sivad., Fl. Agasthyamala 335.2002. *M.laxa* (Hook.f.) Hutch. ex Gamble, Fl. Pres. Madras 610. 1921; Gandhi in Sald. & Nicols., Fl. Hassan Dist. 582. 1976. *M. frondosa* L. var. *laxa* Hook. f., Fl. Brit. India 3: 89. 1880. *Belilla* Rheede, Hort. Malab. 2: 27-28, t. 18.1679.

Climbing shrubs. Leaves broadly ovate, 6-10 by 4-6 cm, base rounded, apex acuminate, lower surface tomentose, lateral veins c. 12 pairs, prominent. Cymes terminal. Flowers lax, 5-merous; calyx-lobes 5, one petaloid, white, pubescent; corolla orange-red or yellow, funnel-form, villous at throat; stamens 5, included. Berry globose, sparsely pubescent.

Fl. & Fr.: Most of the seasons. *Distr.*: Western Ghats. Very common in all types of forests. *NAK 614* (Thekkuthode, c. 70 m).

Petaloid sepals ground and boiled with castor oil and given for internal swellings and ulcers. Root is also medicinal.

Note : Gamble (1921) distinguished 4 speceis of **Mussaenda** viz. **M. glabrata** Hutch., **M. laxa** (Hook.f.) Hutch. ex Gamb., **M. frondosa** L. and **M. hirsutissima** Hutch. But according to Mayne (1958), "These four speceis form a series characterised by increasing hairiness, increasing density of the inflorescence and decreasing leaf size with increasing elevation. They could be suitably considered as members of an ecoline rather than as distinct species..."

MYCETIA Reinwardt
Syll. Pl. Nov. 2: 9. 1825.

Mycetia acuminata (Wight) Kuntze, Rev. Gen. Pl. 289. 1891; Gamble, Fl. Pres. Madras 612. 1921; Gandhi in Sald. & Nicols., Fl. Hassan Dist. 582. 1976; Manilal, Fl. Silent Valley 137. 1988; Mohanan & Sivad., Fl. Agasthyamala 336.2002. *Lawia acuminata* Wight, Ic. t. 1070. 1846. *Adenosacme lawii* Hook.f., Fl. Brit. India 3:96. 1880.

Woody herbs. Leaves elliptic-lanceate to oblanceate, 15-25 by 4.5-10 cm, base attenuate, apex acuminate, lateral nerves c. 18 pairs, regular, lower surface puberulous, membranous. Cymes terminal. Flowers 4-6-merous; bracts subulate, glandular; calyx-tube globose, lobes small, persistent; corolla yellow; stamens included; ovary 2-6- locular; ovules α. Berry globose. Seeds angular.

Fl. & Fr.: Dec.-Mar. *Distr.*: Southern Western Ghats. Rare. Among forest undergrowth. *NAK 1136* (Moozhiar, c. 250 m), *1441* (Arampa, c. 400 m).

NEANOTIS W.H. Lewis
Ann. Missouri Bot. Gard. 53 : 34. 1966.

Diffuse to procumbent annual or perennial herbs, usually with a foul smell; branchlets terete or angular; stipules broad, pectinate. Leaves decussate. Flowers 4-merous, in corymbose or capitate cymes; corolla funnel-form; stamens on corolla throat; ovary 2-locular, ovules few. Capsule laterally subcompressed. Seeds pitted or reticulate.

1. Cymes corymbose, usually pubescent; flowers dense. ....2. **N. indica** var. **affinis**
1. Cymes otherwise, glabrous; flowers lax. ....1. **N. decipiens**

1. **Neanotis decipiens** (Hook.f.) W.H. Lewis, Ann. Missouri Bot. Gard. 53: 37. 1966; Mohanan & Sivad., Fl. Agasthyamala 337.2002. *Anotis decipiens* Hook.f., Fl. Brit. India 3 : 72. 1880; Gamble, Fl. Pres. Madras 604.1921.

Leaves ovate-lanceate, 2-7 by 1.5-3 cm, base cuneate, decurrent, apex acuminate, sparsely villous hairy, chartaceous. Cymes terminal; calyx-tube ovoid; corolla white, tinged with blue. Capsule 1 by 2 mm. Seeds finely reticulate.

Fl. & Fr.: Sep.-Dec. *Distr.:* Southern Western Ghats. Common. Usually in moist hill slopes. *NAK 331, 1181, 1399* (Pampa dam area, c. 1100 m).

2. **Neanotis indica** (DC.) Lewis, var. **affinis** (Hook.f.) Lewis, Ann. Missouri Bot. Gard. 53 : 39. 1966; Manilal, Fl. Silent Valley 137. 1988. *Putoria indica* DC., Prodr. 3:477.1830. *Anotis leschenaultiana* (Wight & Arn.) Hook.f., var. *affinis* Hook.f., Fl.Brit. India 2: 72.1880; Gamble, Fl. Pres. Madras 604. 1921.

Leaves ovate, 2-5 by 1-3 cm, base rounded or broadly cuneate, decurrent, apex acute, densely villous hairy. Cymes terminal and axillary. Flowers dense; corolla purple. Capsule 3 by 2mm. Seeds pitted.

Fl. & Fr.: Sep.-Dec. *Distr.*: Western Peninsular India. Common. Along moist hill cuttings. *NAK 75 & 2173* (Moozhiar, c. 250 m).

Note : Very variable herbs in habit and indumentum.

NEOLAMARCKIA J. Bosser
Adansonia 6(3) : 245-248. 1984.

Neolamarckia cadamba (Roxb.) Bosser, Bull. Mus. Natl. Hist. Nat. (Paris) Sect. B. Adansonia 6 : 247. 1985 ('1984'); Nicols., Suresh & Mani., An Interpr. Hort. Malab. 227. 1988; Mohanan & Sivad., Fl. Agasthyamala 338.2002. *Nauclea cadamba* Roxb., (Hort. Beng. 17. 1814, nom. nud.) Fl. Ind. 2 : 121.1824. *Anthocephalus cadamba* (Roxb.) Miq., Fl. Ind. Bat. 2 : 135. 1856; Hook.f., Fl. Brit. India 3 : 23. 1880. *A. indicus* Rich., Mem. Soc. Hist. Nat. Paris 5 : 237. 1834; Gamble, Fl. Pres. Madras 583. 1921. *A. chinensis* (Lamk.) Rich. ex Walp., Repert. 2 : 491. 1843; Bakh. f., Taxon 19 : 469. 1970. *Katou tsjaka* Rheede, Hort. Malab. 3 : 29-30, t. 33. 1682.

Trees; branchlets obtusely 4-gonous. Leaves broadly elliptic ovate, 18-20 by 11-18 cm, base truncate, apex acuminate, lower surface tomentose, thin-coriaceous; stipules caducous. Flowers crowded in large globose heads, solitary, terminal, c. 6 cm ϕ; calyx coherent, lobes 5; corolla yellowish-orange, funnel-form; stamens 5 from corolla throat; ovary 4-locular above, 2-locular below. Capsule aggregated in heads.

Fl. & Fr.: May-Sep. *Distr.*: Sri Lanka, India, Malesia. Rare. In riverein forests. *NAK 1838* (Angamoozhy, c. 250 m).

Bark is considered to be a tonic and febrifuge. Tender shoots taken internally to cure dysentery and increase digestion.

NEUROCALYX W.J. Hooker
Icon. Pl. 174. 1837.

Neurocalyx calycinus (R. Br. ex Benn.) Robinson, Proc. Amer. Acad. Arts 45 : 402.1910; Gandhi in Sald. & Nicols., Fl. Hassan Dist. 584. 1976; Manilal, Fl. Silent Valley 138. 1988; Mohanan & Sivad., Fl. Agasthyamala 339.2002. *Argostemma calycinum* R. Br.ex Benn., Pl. Jav. Rar. 97. 1838. *Neurocalyx wightii* Arn., Ann. Nat. Hist. Ser. 1(3) : 22. 1839; Hook.f., Fl. Brit. India 3 : 47. 1880; Gamble, Fl. Pres. Madras 591. 1921.

Suffrutescent herbs. Leaves crowded towards branch-apices, elliptic-lanceate, 6-14 by 3-5 cm, base cuneate, apex obtuse to subacute, lower surface whitish, pubescent, chartaceous; stipule large, bract-like, bifid. Inflorescence an axillary pendulous raceme or a head. Flowers 5-merous; calyx-tube subglobose, lobes petaloid; corolla pink, tube absent, lobes smaller than calyx-lobes; stamens at corolla base, anthers connate to form a conical tube; ovary 2-locular; ovules α. Berry ovoid, dehiscing irregularly.

Fl. & Fr.: May-Aug. *Distr.*: Southern-Western Ghats. Rare. Trees with beautiful globose orange coloured inflorescene. In moist shady forest areas. *NAK 1831* (Angamoozhy, c. 250 m).

OCHREINAUCLEA Ridsdale
Blumea 24 : 332. 1978.

Ochreinauclea missionis (Wall. ex G. Don) Ridsdale, Blumea 24 : 332. 1978. *Nauclea missionis* Wall. ex G. Don, Gen. Hist. 3 : 467. 1834; Hook.f., Fl. Brit. India 3: 27. 1880; Gamble, Fl. Pres. Madras 582. 1921.

Trees. Leaves elliptic-lanceate, 6-14 by 3.5 cm, base cuneate, apex obtuse to subacute, chartaceous; petiole short. Inflorescence globose-cymose heads, solitary, terminal. Flowers 5-merous; calyces fused with fleshy receptacle; corolla orange-red; stamens at corolla throat; ovary 2-locular; ovules α. Capsule in globose heads.

Fl. & Fr.: Mar.-Jun. *Distr.*: Western Peninsular India. Rare. Trees with beautiful globose orar.ge coloured inflorescence, locaily abundant along streamsides. *NAK 625 & 1618* (Thannithode, c. 70 m).

OPHIORRHIZA Linnaeus
Sp. Pl. 150. 1753.

Erect annual or perennial suffrutescent herbs; stipules caducous. Inflorescence a secund-cyme, terminal; bracts and bracteoles in various forms or absent. Flowers 5-merous; corolla funnel-form or tubular; stamens 5, inserted; ovary 2-locular; ovules α. Capsule laterally compressed, obcordate. Seeds minute, angled.

1. Large erect herbs; bracteoles various or absent.
    2. Bracteoles α; corolla-tube more than 7 mm long.
        3. Bracteoles linear, villous; flowers white. ....1. **O. eriantha**
        3. Bracteoles elliptic-lanceate, glabrous; flowers pink. ....3. **O. pectinata**

2. Braceteoles 0; corolla-tube 6-7 mm long. ....2. **O. mungos**

1. Small diffuse herbs; braceteoles subulate, deciduous. ....4. **O. rugosa** var. **prostrata**

1. **Ophiorrhiza eriantha** Wight, Ic. t. 1067. 1846; Hook.f., Fl. Brit. India 3 : 81. 1880; Gamble, Fl. Pres. Madras 608.1921; Mohanan & Sivad., Fl. Agasthyamala 341. 2002.

Branchlets pubescent. Leaves narrowly ovate-lanceate, 7-19 by 1.5-5.5 cm. Cymes densely villous; bracts linear. Flowers subsessile; corolla white, pubescent without. Capsule pubescent.

Fl. & Fr.: Aug.-Oct. *Distr.*: Western Ghats. Rare delicate herbs among forest undergrowth, usually in shades. *NAK 2406 & 2558* (Kakki hills, c. 1100 m).

2. **Ophiorrhiza mungos** L., Sp. Pl. 150. 1753; Hook.f., Fl. Brit. India 3 : 77.1880; Gamble, Fl. Pres. Madras 607. 1921; Manilal, Fl. Silent Valley 139. 1988; Mohanan & Sivad., Fl. Agasthyamala 342.2002. *Amal-poriyan.*

Large herbs. Leaves elliptic or elliptic-lanceate, 10-15 by 6-7 cm, base cuneate attenuate, apex acuminate, chartaceous. Cymes umbelliform; bracts absent; calyx globose; lobes small; corolla white, small. Capsule puberulous.

Fl. & Fr.: Dec.-Mar. *Distr.:* Sri Lanka, Peninsular India, Java & Sumatra. Common. In moist forest undergrowth. Locally very abundant, forming dense populations, especially along wet areas. *NAK 715* (Perunthenaruvi, c. 100 m).

Roots used as an antidote for insect poison and as a remedy for ulcers, leprosy and worm infections in intestine.

3. **Ophiorrhiza pectinata** Arn., Nova Acta Phys.-Med. Acad. Caes. Leop. Carol. Nat. Cur. 18 : 338. 1836; Hook.f., Fl. Brit. India 3: 81. 1880; Gamble, Fl. Pres. Madras 608. 1921; Mohanan & Sivad., Fl. Agasthyamala 343. 2002.

Suffrutescent herbs. Rooting at lower nodes. Leaves lanceate, 6-14 by 2-5.5 cm, glaucescent beneath. Cymes dense; bracts elliptic-lanceate, glabrous; calyx-lobes small; corolla pink. Capsule glabrous.

Fl. & Fr.: Most of the seasons. *Distr.*: Sri Lanka, Peninsular India. Very common in moist hill cuttings. *NAK 64, 565, 716, 991, 1142 & 1396* (in almost all forests).

4. **Ophiorrhiza rugosa** Wall. in Roxb., Fl. Ind. (ed. Carey) 2: 546.1824. var. **prostrata** (D.Don) Deb & Mondal, Bull. Bot. Surv. India 24 : 228. 1983. *O. prostrata* D. Don, Prodr. Fl. Nep. 136. 1825, p.p. (excl. Wall. Cat. 6235); Manilal, Fl. Silent Valley 139. 1988; Mohanan & Sivad., Fl. Agasthyamala 343.2002. *O. harrisiana* Heyne ex G. Don, Gen. Syst. 3 : 523. 1834; Wight, Ic. t. 1162. 1846 "*harrisonii*"; Hook.f., Fl. Brit. India 3 : 78. 1880; Gamble, Fl. Pres. Madras 607. 1921.

Diffuse small soft herbs; branchlets tomentose. Leaves ovate-lanceate, 2-6 by 1-2.5 cm, base rounded, apex acute, puberulous on the nerves below. Cymes small, terminal; bracteoles deciduous; corolla white. Capsule glabrescent.

Fl. & Fr.: Aug.-Dec. *Distr.*: Sri Lanka, India, Nepal, Myanmar. Rare. Among forest undergrowth in shade. *NAK 1144* (Moozhiar, c. 250 m).

OXYCEROS Loureiro
Fl. Cochinch. 96 ('Oxiceros'), 150 ('Oxiceros'), 151. 1790.

Oxyceros rugulosus (Thw.) Tirvengadum, Nordic J. Bot. 3: 466. 1983; Manilal, Fl. Silent Valley 140. 1988; *Griffithia regulosa* Thw., Enum. 159. 1859. *G. speciosa* Bedd., Ic. t. 37. 1874. *Randia rugulosa* (Thw.) Hook.f., Fl. Brit. India 3 : 113. 1880; Gamble, Fl. Pres. Madras 617. 1921; Gandhi in Sald. & Nicols., Fl. Hassan Dist. 590. 1976; Mohanan & Sivad., Fl. Agasthyamala 356.2002. **(Fig. 39)**

Climbing shrubs. Leaves elliptic-lanceate, 10-15 by 3.5-6 cm, base cuneate, apex acuminate, black when dry. Cymes leaf-opposed; bracteoles α; pedicel c. 2 cm long; calyx-lobes obscure; corolla white. Berry globose.

Fl. & Fr.: Feb.-May. *Distr.*: Sri Lanka, Southern Western Ghats. Rare. Large scandent shrubs in evergreen forests. *NAK 1560* (Velithode, Moozhiar, c. 250 m).

PAVETTA Linnaeus
Sp. Pl. 110. 1753.

Shrubs or very small trees; stipules often connate. Leaves decussate. Inflorscence corymbose panicles, bracteoles minute. Flowers 4-merous; calyx toothed; corolla salver-form; stamens inserted at the corolla mouth; ovary 2-locular; ovule 1per locule; style long, much exerted; stigma simple, fusiform. Berry subglobose or globose. Seeds 2.

1. Shrubs to small trees; peduncle pubescent; calyx-lobes minute.
   2. Calyx-lobes 0.2 mm long; branchlets glabrescent. .... 1. **P. indica**
   2. Calyx-lobes 0.6 mm long; branchlets tomentose. .... 2. **P. tomentosa**
1. Shrubs; peduncle glabrous; clayx-lobes triangular-lanceate. .... 3. **P. zeylanica**

1. **Pavetta indica** L., Sp. Pl. 110. 1753; Wight, Ic. t. 148. 1839; Hook.f., Fl. Brit. India 3 : 150. 1880; Gamble, Fl. Pres. Madras 633. 1921; Gandhi in Sald. & Nicols., Fl. Hassan Dist. 587. 1976; Nicols., Suresh & Manilal, An Interpr. Hort. Malab. 228. 1988. *Pavetta* Rheede, Hort. Malab. 5 : 19-20, t. 10. 1685.

Shrubs, branchlets glabrescent. Leaves oblanceate, 7-11 by 3-4 cm, base acute-attenuate, apex acuminate, chartaceous. Corymb terminal; calyx-lobes obsolute, 0.2 mm long; corolla white, tube c. 12 mm long, lobes obovate. Berry globose, c. 6 mm ϕ.

Fl. & Fr.: Apr.-Jul. *Distr.*: South India. Very common in sacred groves and evergreen forests. *NAK 2386* (Appuppanthode, c. 400 m).

Fruits powdered and given in honey for rheumatic fever and dysentery. Root- powder supposed to cure hernia.

2. **Pavetta tomentosa** Roxb. ex Smith in Rees, Cyclop. 26 : n. 2. 1814; Nicols., Suresh & Manilal, An Interpr. Hort. Malab. 228. 1988; Mohanan & Sivad., Fl. Agasthyamala 348.2002. *Ixora tomentosa* (Sm.) Roxb., (Hort. Beng. 10. 1814, nom. nud.) Fl. Ind. 1: 396.120; Gandhi in Sald. & Nicols., Fl. Hassan Dist. 588. 1976; Matthew & Rani in Matthew, Fl. Tam. Carnatic 3(2) : 733. 1983. *P. indica, ssp. tomentosa* (Roxb. ex Smith) Bennet, Fl. Howrah 356. 1979. *Pavetta indica* var. *tomentosa* Hook. f., Fl. Brit. India 3 : 150. 1880; Gamble, Fl. Pres. Madras 633. 1921.

Shrubs to small trees; branchlets tomentose. Leaves broadly to elliptic-lanceate, 10-15 by 5-8 cm, base acute or rounded, tomentose beneath, chartaceous. Cymes profusely branched; calyx-lobes c. 0.6 mm long; corolla white, lobes oblong. Berry 5-8 mm ϕ.

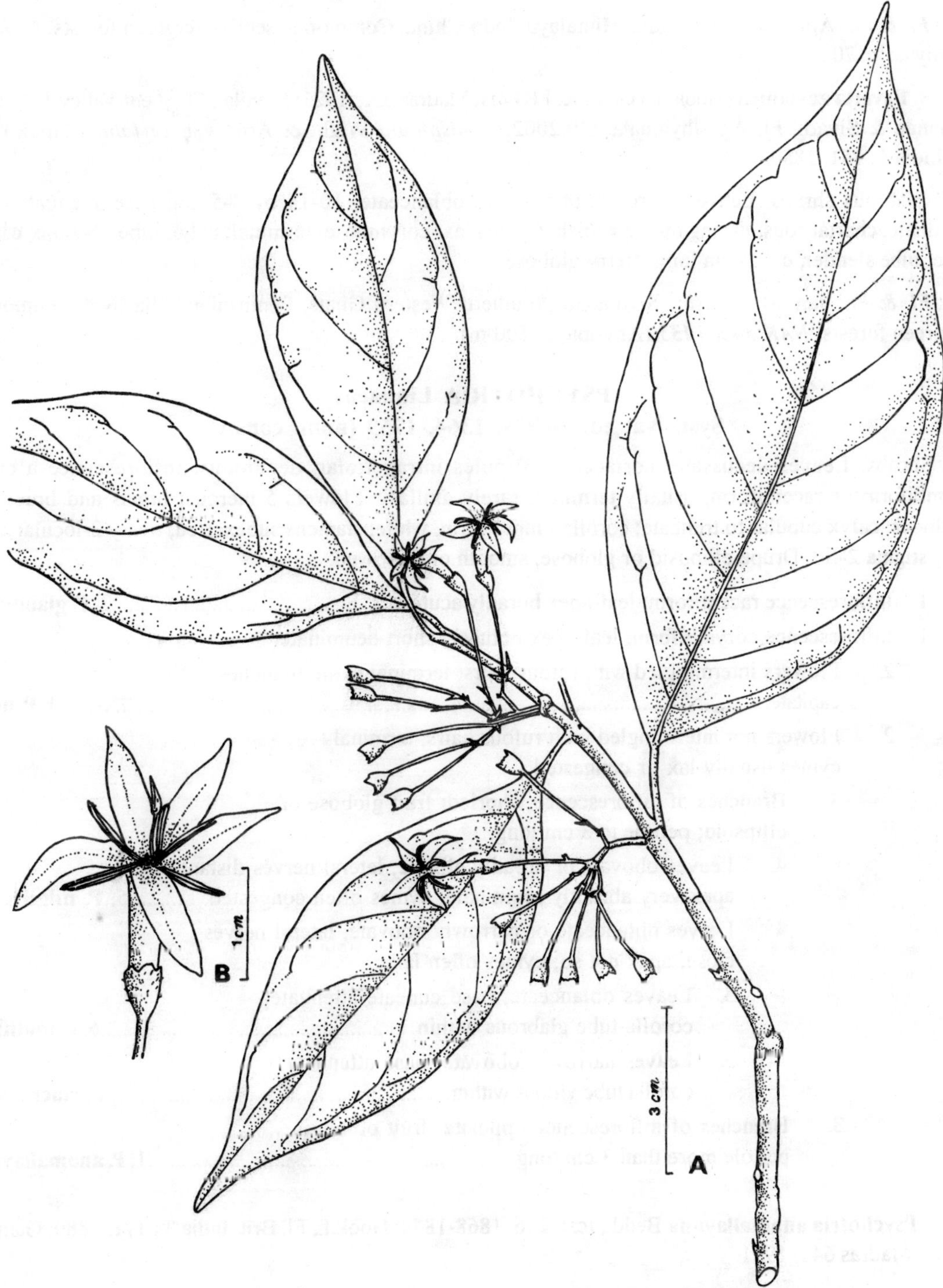

Fig. 39. ***Oxyceros rugulosus*** **(Thw.) Tirveng.: A. Twig; B. Flower.**

Fl. & Fr.: Apr.-Jul. *Distr.*: India, Himalaya, Indo-China. Common in semi- evergreen forests. *NAK 561* (Maniyaar, c. 70 m).

3. **Pavetta zeylanica** (Hook. f.) Gamble, Fl. Pres. Madras 633. 1921; Manilal, Fl. Silent Valley 141. 1988; Mohanan & Sivad., Fl. Agasthyamala 349.2002. *P. hispidula* Wight & Arn., var. *zeylanica* Hook.f., Fl. Brit.India 3 : 151. 1880.

Glabrous shrubs. Leaves narrowly obovate to oblanceate, 10-18 by 3-5 cm, base attenuate, apex acuminate, chartaceous, drying dull-greenish. Cymes lax, corymbose, terminal; calyx-tube obconic; corolla white, tube slender, c. 2.5 cm long. Berry globose.

Fl. & Fr.: Jan.-Mar. *Distr.*: Sri Lanka, Southern Western Ghats, Peninsular India. Not common. In evergreen forests. *NAK 345, 1435* (Arampa, c. 500 m).

PSYCHOTRIA Linnaeus
Syst. Nat ed. 10.929, 1364. 1759 (nom. cons.)

Shrubs. Leaves decussate, coriaceous; stipules intrapetiolar, deciduous. Inflorescence a cyme, corymbiform or racemiform, usually terminal, rarely axillary. Flowers 5-merous; bracts and bracteoles deciduous; calyx cupular to truncate; corolla funnel-form, white; stamens subexerted; ovary 2-locular; style erect; stigma 2-fid. Drupe elliposid or globose, smooth or furrowed.

1. Inflorescence racemiform; leaf-apex boradly acute. .... 2. **P. glandulosa**
1. Inflorescence corymbiform; leaf-apex obtusely short-acuminate.
    2. Flowers intermingled with rufous hairs; terminal cyme-branches capitate. .... 4. **P. nigra**
    2. Flowers not intermingled with rufous hairs; terminal cymes usually lax or congested.
        3. Branches of inflorescence whorled; fruit globose or ellipsoid; petiole to 3 cm long.
            4. Leaves obovate or broadly elliptic, lateral nerves distant, apex very abruptly acuminate; cymes often congested. .... 5. **P. nilgiriensis**
            4. Leaves oblanceate or narrowly obovate, lateral nerves close, apex not so; cymes often lax.
                5. Leaves oblanceate, base cuneate-attenuate; corolla-tube glabrous within. .... 6. **P. nudiflora**
                5. Leaves narrowly obovate, base attenuate; corolla tube villous within. .... 3. **P. macrocarpa**
        3. Branches of inflorescence opposite; fruit oblong; petiole more than 3 cm long. .... 1. **P. anamallayana**

1. **Psychotria anamallayana** Bedd., Ic. t. 236. 1868-1874; Hook.f., Fl. Brit. India 3 : 171. 1880; Gamble, Fl. Pres. Madras 641. 1921.

Leaves oblanceate, 15-25 by 4-7 cm, base cuneate-attenuate, apex acute to abruptly short acuminate; petiole 3-7 cm long. Cymes terminal; branches stout, whorled; bracts and bracteoles conspicuous; calyx broadly obconic; corolla-tube with a ring of hairs within. Fruit oblong or narrowly ovoid.

Fl. & Fr.: Jul.-Sep. *Distr.*: Southern Western Ghats. Rare. In evergreen forests. *NAK 2567* (Pampa, 1100 m).

2. **Psychotria glandulosa** (Dennst.) Suresh in Nicols., Suresh & Manilal, An Interpr. Hort. Malab. 229 : 1988; Mohanan & Sivad., Fl. Agasthyamala 352.2002. *Baldingera glandulosa* Dennst., Schluessel 9, 19, 31.1818. *Grumilea elongata* Wight, Ic. 3(3) : 16, t. 1036. 1845. *Psychotria elongata* (Wight) Hook.f., Fl. Brit. India 3 : 163. 1880; Gamble, Fl. Pres. Madras 641. 1921; Ridsdale in Manilal, Bot.Hist. Hort.Malab. 136. 1980. *Amel-podi* Rheede, Hort.Malab. 5 : 101, t. 51. 1685.

Leaves elliptic or obovate, 7-12 by 3.5-6 cm, base cuneate, apex broadly acute, lateral nerves 10-14 pairs. Cymes racemiform, 5-7 cm long; bracts caducous; corolla-tube short, villous within near throat. Fruit globose.

Fl. & Fr.: Apr.-Aug. *Distr.*: Sri Lanka, Peninsular India. Rare. In shola forest. *NAK 2749* (Ponnambalamedu, c. 1100 m).

3. **Psychotria macrocarpa** Hook.f., Fl. Brit. India 3 : 162. 1880; Gamble, Fl. Pres. Madras 641. 1921; Manilal, Fl. Silent Valley 141. 1988; Mohanan & Sivad., Fl. Agasthyamala 354.2002.

Leaves elliptic or oblanceate, 10-16 by 5-7 cm, apex obtuse or obtusely acuminate. Cymes short, c. 2.5 cm long; bracts caducous. Flowers crowded; corolla white, tube villous within. Fruit ellipsoid.

Fl. & Fr.: Apr.-Aug. *Distr.*: Southern Western Ghats. Rare. In evergreen forests. *NAK 2745* (Kakki hills, c. 1100 m).

4. **Psychotria nigra** (Gaertn.) Alston in Trimen, Handb. Fl. Ceylon 6 : 152. 1931; Gandhi in Sald. & Nicols., Fl. Hassan Dist. 589. 1976; Mohanan & Sivad., Fl. Agasthyamala 354.2002. *Grumilea nigra* Gaertn., Fruct. 1: 138, t. 28, 1788. *Psychotria thwaitesii* Hook.f., Fl. Brit. India 3 : 162. 1880; Gamble, Fl. Pres. Madras 650. 1921.

Leaves elliptic to oblong-elliptic, 10-15 by 4-8 cm, base cuneate, apex obtusely acuminate, lateral nerves, c. 14 pairs, prominent. Cyme-branches often whorled; calyx-lobes persistent; corolla cream coloured. Fruit ellipsoid.

Fl. & Fr.: Mar.-Aug. *Distr.*: Sri Lanka, Southern Western Ghats. Not common. In evergreen forests. *NAK 1686* (Thriveni Pampa, c. 250 m), *2565* (Kakki hills, c. 1100 m)

5. **Psychotria nilgiriensis** Deb & Gang., Taxon 31 : 546. 1982; Mohanan & Sivad., Fl. Agasthyamala 355.2002. *P. congesta* (Wight & Arn.) Hook.f., Fl. Brit. India 3 : 162. 1880, non Spreng. ex DC., 1830; Gamble, Fl. Pres. Madras 640. 1921. *Grumilea congesta* Wight & Arn., Prodr. 432. 1834; Wight, Ic. t. 1037. 1845.

Leaves obovate or oblanceate, 5-12 by 2.5-5 cm, base cuneate, apex very abruptly shortly acuminate, lateral veins 8-10 pairs, distant, greenish when dry. Flowers crowded; corolla white, tube very short, glabrous or villous within. Fruit ellipsoid.

Fl. & Fr.: Sep.-Dec. *Distr.*: Southern Western Ghats. Not common. In semi-evergreen forests. *NAK 562 & 719* (Perunthenaruvi, c. 100 m).

6. **Psychotria nudiflora** Wight & Arn., Prodr. 434. 1823; Hook.f., Fl. Brit. India 2: 175. 1880; Gamble, Fl. Pres. Madras 641. 1921.

Leaves oblanceate, 10-14 by 3.4 cm, base attenuate, apex obtusely acuminate, lateral nerves α, slender, prominent. Cymes c. 5 cm long, branches whorled; corolla creamy-white, tube c. 2 mm long. Fruits ellipsoid.

Fl. & Fr.: Sep.-Dec. *Distr.*: Southern Western Ghats. The most common species of the genus. In evergreen forests. *NAK 815 & 2055* (Angamoozhy, c. 250 m).

RANDIA Linnaeus
Sp. Pl. 1192. 1753.

Trees, shrubs or shrubby climbers; stem often thorny. Leaves often clustered towards branchlet apices. Flowers in short cymes or solitary or in fascicles, axillary or leaf-opposed, 5-merous; calyx cupular, often produced above, lobes imbricate or valvate; corolla salver-form, tube hairy within; stamens exerted; ovary (1-) 2-locular; ovules α. Berry globose or ellipsoid. Seeds often angled.

1. Calyx-tube cupular, lòbes valvate, with intermediate appendages. ....**1. R. brandisii**
1. Calyx-tube turbinate, lobes imbricate, without intermediate appendages. .**2. R. dumetorum**

1. **Randia brandisii** Gamble, Fl. Pres. Madras 616. 1921; Matthew & Rani in Matthew, Fl. Tam. Carnatic 3(2) : 738. 1983. *Randia tomentosa* Wight & Arn., Prodr. 398. 1834, non Bl.

Shrubs to small trees; stem thorny. Leaves obovate, 3-5 by 1-2.5 cm, base cuneate, apex obtuse, lower surface densely pubescent. Flower(s) solitary or in few- flowered axillary cymes; calyx cupular, lobes distant with intermediate appendages; corolla white, turning yellow. Berry subglobose, ribbed.

Fl. & Fr.: Mar.-Jul. *Distr.*: Peninsular India. Common. An evergreen bushy tree. *NAK 656* (Perunthenaruvi, c. 70 m), *1573* (Sabarimala, c. 250 m).

2. **Randia dumetorum** (Retz.) Poiret in Lam., Encycl. Suppl. 2: 829. 1812 & Tabl. Encycl. t. 156. 1791; Wight & Arn., Prodr. 397 1834; Hook.f., Fl. Brit. India 3 : 110. 1880; Gamble, Fl. Pres. Madras 615. 1621; Matthew & Rani in Matthew, Fl. Tam. Carnatic 3 (2) : 738. 1983. *Gardenia dumetorum* Retz., Obs. Bot. 2: 14. 1781; Roxb., Pl. Corom. t. 136. 1800. *Gardenia spinosa* Thunb., Diss. Gard. n. 7, t.2, f.4. 1780; *Randia spinosa* (Thunb.) Poiret in Lam., Encycl. (Suppl.2): 829. 1812. *Catunaregam spinosa* (Thunb.) Tirvengadum, Bull. Mus. Hist. Nat. Paris 35 : 13. 1978.

Small trees; stem thorny. Leaves narrowly obovate, 2-5 by 1-2 cm, lower surface tomentose. Flower(s) solitary or in fascicles, axillary; calyx-tube turbinate, produced above, lobes ovate, 5 by 4 mm, imbricate; corolla yellow. Berry globose, ribbed.

Fl. & Fr.: Jan.-Jun. *Distr.*: India, Sri Lanka, E. Tropical Africa. Not common. Evergreen trees. *NAK 1675* (Sabarimala, c. 250 m).

SPERMACOCE Linnaeus
Sp. Pl. 102. 1753.

Diffuse or erect, usually perennial or annual herbs. Leaves whorled or deccusate. Flowers 4-merous, small, crowded in dense clusters in leaf-axils; bracteoles filiform; calyx-lobes often unequal, persistent; corolla salver-form or funnel-form; stamens subexerted; ovary 2-locular, ovule 1per locule. Capsule globose or compressed.

1. Leaves whorled, 1-veined; erect herbs; stem 4-angled. ....5. **S. pusilla**
1. Leaves decussate; more than 1-veined; usually diffuse or erect herbs; stem 4-winged, 4-angled or terete.
   2. Stem 4-winged.
      3. Calyx-lobes 4; leaf veins prominent. ....3. **S. latifolia**
      3. Calyx-lobes 2; leaf veins faint. ....4. **S. mauritiana**
   2. Stem 4-angled or terete.
      4. Stem 4-angled; corolla-tube much longer than lobes. ....1. **S. articularis**
      4. Stem terete; corolla-tube ± equal to the lobes. ....2. **S. hispida**

1. **Spermacoce articularis** L.f., Suppl. Pl. 119. 1782; Wall. in Roxb., Fl. Ind. 1:378. 1820; Sivar. & Nair, Taxon 35: 366. 1986. *Borreria articularis* (L.f.) Williams, Bull. Herb. Boiss. Ser. 2,5: 956. 1905. *B. articularis*, var. *articularis* Sivar. & Manilal, New Botanist 2:88. 1975. *Spermacoce hispida* auct., mult., non L., 1753: Hook.f., Fl.Brit. India 3:200. 1881; Matthew & Rani in Matthew, Fl. Tam. Carnatic 3(2) : 742 1983. *Borreria hispida* sensu Gamble, Fl. Pres. Madras 654. 1921.

Diffuse herbs; stem 4-angled. Leaves broadly ovate to obovate, 10-15 by 6-10 mm, margin flat, rarely flexuous. Flowers few; corolla funnel-form, pink, tube usually much longer than lobes. Capsule hispid. Seeds oblong, reticulate.

Fl. & Fr.: May.-Aug. *Distr.:* India, China, Malesia, Indo-China, Philippines and Trop. Africa. Less common. In exposed lands in plains. *NAK 634* (Puthukkulam, c. 50 m).

2. **Spermacoce hispida** L., Sp. Pl. 102. 1753; Roxb., Fl. Ind. 1:373. 1820; Sivar. & Nair, Taxon 35: 366. 1986; Nicols., Suresh & Manilal, An Interpr. Hort. Malab. 229. 1988; Mohanan & Sivad., Fl. Agasthyamala 359.2002. *Borreria hispida* (L.) K. Schum. in Engl. & Prantl, Pflanzenfam. 4(4): 144. 1891, non Spruce ex K. Schum., 1888; Gamble, Fl. Pres. Madras 654. 1921; Matthew & Rani in Matthew, Fl. Tam. Carnatic 3(2): 742. 1983, sensu lato. *Taravel* Rheede, Hort. Malab. 9 : 149, t. 76. 1689.

Diffuse to ascending herbs; stem usually terete or subtetragonous, pilose hairy. Leaves ovate to obovate, margins usually flexuous. Flowers sessile; corolla pink, tube short, ± equal to the calyx-lobes. Capsule hispid. Seeds ellipsoid.

Fl. & Fr.: May-Aug. *Distr.*: India, Indo-China, Malesia. Common. In exposed sandy soil. *NAK 2926* (Adoor, c. 20 m).

Roots ressemble sarasaparila in taste. Decoction or powder of root used as remedy in rheumatism, indigestion and biliousness.

3. **Spermacoce latifolia** Aublet, Hist. Pl. Guiane Fr. 1:55, t.19, f 1. 1775; Verdcourt, Fl. Trop. E. Africa, Rubiac. 1:364. 1976; Deb & Dutta, Bull. Bot. Surv. India 18: 216. 1979; Mohanan & Sivad., Fl. Agasthyamala 359,2002. *Borreria latifolia* (Ablet) K. Schum. In Mart., Fl. Bras. 6(6) : 61. 1888. *B. eradii* Ravi, J. Bombay Nat. Hist. Soc. 77:368. 1980. *Tharavu.*

Diffuse herbs; stem greenish, 4-winged, hispid. Leaves broadly ovate to elliptic, 1-8 by 0.6-4.5 cm, base rounded or cuneate, apex acute, lateral veins prominent. Flowers small; corolla white. Capsule densely pubescent. Seeds ellipsoid.

Fl. & Fr.: May.-Aug. *Distr.*: Trop. America, Australia, Trop. Africa and India. Common in exposed lands and waste places. *NAK 635* (Puthukkulam, c. 50 m).

This species is also having the same medicinal properties as that of the previous species.

4. **Spermacoce mauritiana** Gideon ex Verdcourt, Kew Bull. 37: 547. 1983. *Borreria repens* DC., Prodr. 4. 544. 1830, non *Spermacoce repens* Willd. ex Cham. & Schit., 1828. *Borreria ocymoides* auct., non (Burm. f.) DC., 1830: Gamble, Fl. Pres. Madras 654.1921. *S. decandollei* Deb & Dutta, J.Econ.Tax. Bot.5: 044. 1984, nom. superfl.

Diffuse or erect herbs; stem 4-winged, margins minutely pubescent. Leaves ovate to elliptic-ovate, 7-20 by 4-13 mm, base rounded and then attenuate, lateral veins faint. Flowers very small, in axillary and terminal dense clusters; calyx-lobes 2, lanceate. Capsule compressed. Seeds ellipsoid.

Fl. & Fr.: Aug.-Oct. *Distr.*: India, Trop. America, Africa, West Indies, Malesia. Common. Along waysides in hills. *NAK 864* (Maniyaar, c. 70 m).

5. **Spermacoce pusilla** Wall. in Roxb., Fl.Ind. 1:379. 1820; Wall., Cat. 823. 1828; Verdcourt, Fl. Trop. E. Africa 1: 356. 1976, p.p.; Sivar. & Nair, Taxon 35:368. 1986; Manilal, Fl. Silent Valley 144. 1988; Mohanan & Sivad., Fl. Agasthyamala 360.2002. *Borreria pusilla* (Wall.) DC., Prodr. 4:543. 1830. *B. stricta* var. *rosea* Sivar. & Manilal, Bull. Bot. Soc. Univ. Sagar 19:33. 1972. *B. stricta* auct., mult., non L.f. 1791: Gamble, Fl. Pres. Madras 461. 1921.

Erect herbs; stem 4-angled, sparsely pubescent. Leaves linear-lanceate, c. 4 cm long, scabrid,1-veined, lateral veins invisible, margins revolute. Flowers small, in axillary and terminal clusters; corolla pink; capsule glabrescent.

Fl. & Fr.: Aug.-Nov. *Distr.*: India, Myanmar. Rare. On exposed hills. *NAK 2181* (Chittar, c. 200 m).

TARENNA J. Gaertner
Fruct. 1: 139. 1788.

Shrubs to small trees. Stipules imterpetoplar, connate. infloresecnce terminal , corymbose cyme; bracts and bracteoles present. Flowers (4-) -5 merous. Calyx tube shortly lobbed. Corolla lobes contorted. Stamens inserted on the mouth of corolla; anthers exerted. Ovary 2-locular, ovule 1-α ovuled; style hairy. Fruit a berry, usually crowned by calyx limb.

1. Leaves elliptic to elliptic-oblong, base truncate, apex acute. ....**T. asiatica**

1. Leaves obovate to oblanceate, base cuneate-attenuate, apex acuminate. ....**T. canarica**

1. **Tarenna asiatica** (L.) Kuntze ex Schumann, Bot. Tidsskr. 24: 332. 1902; Gandhi in Sald. & Nicols., Fl. Hassan Dist. 591. 1976; Matthew, Ill. Fl. Tam. Carnatic t. 348. 1982; Matthew & Rani in Matthew, Fl. Tam. Carnatic 3(2): 744. 1983; Nicols., Suresh & Manilal, An Interpr. Hort. Malab. 230. 1988; Mohanan & Sivad., Fl. Agasthyamala 362.2002. *Rondeletia asiatica* L., Sp. Pl. 172. 1753. *Chomelia asiatica* (L.) Kuntze, Rev. Gen. P. 1: 278. 1891; Gamble, Fl. Pres. Madras 612. 1921. *Cupi* Rheede, Hort. Malab. 2: 37-38, t. 23. 1679.

Shrubs to small trees. Leaves decussate, elliptic to elliptic-oblong, 8-15 by 3-5 cm, base truncate, apex acute, drying black. Cymes 3-chotomous, corymbose. Flowers 5-merous; calyx-lobes short; corolla white,

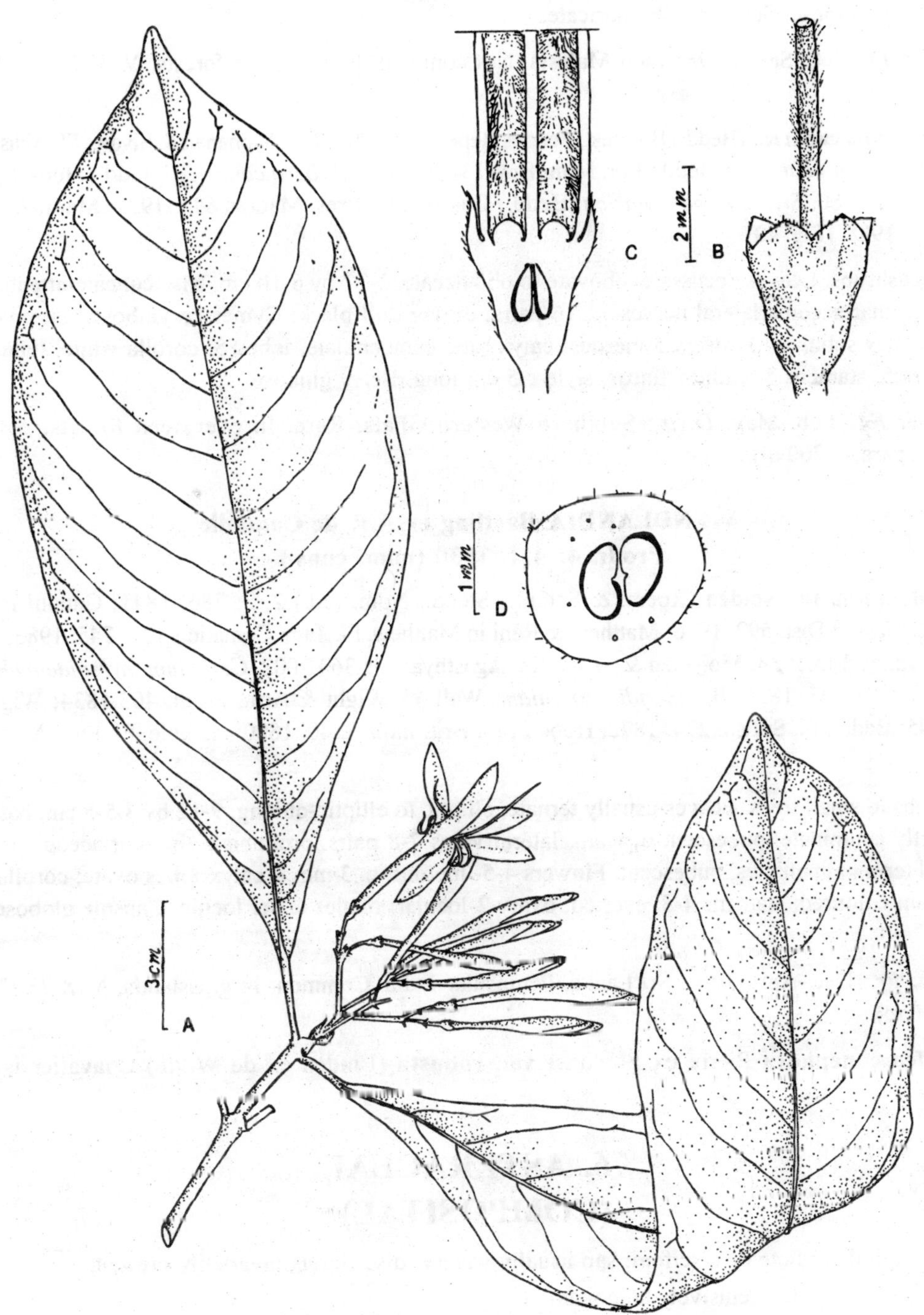

Fig. 40. ***Tarenna canarica*** **(Bedd.) Gamble: A. Twig; B. Calyx and pistil wih basal portion of style; C. Flower - lower portion- L. S.; D. Ovary - C.S.**

salver-form, lobes 5, twisted; stamens 5, exerted, connective produced above; ovary 2-locular; ovules a per locule. Berry globose. Seeds faintly muricate.

Fl. & Fr.: Feb.-Sep. *Distr.*: Indo-Malesia. Not common. In evergreen forests. *NAK 2451* (Moozhiar, c.250 m).

2. **Tarenna canarica** (Bedd.) Bremek., Feddes Repert. 37: 189. 1934; Mohanan & Sivad., Fl. Agasthyamala 363. 2002. *Pavetta canarica* Bedd., For. Man. Bot. 134/8. 1872. *Webera canarica* (Bedd.) Hook.f., Fl. Brit. India 3 : 106. 1880. *Stylocoryne canarica* (Bedd.) Gamble, Fl. Pres. Madras 635. 1921; Manilal, Fl. Silent Valley 144. 1988. **(Fig. 40)**

Erect shrubs. Leaves decussate, obovate to oblanceate, 2-27 by 6-10 cm, base cuneate-attenuate, apex acuminate, chartaceous, lateral nerves 12-16 pairs, drying dull black. Cymes corymbose, terminal; bracts and bracteoles subulate. Flowers 5-merous; calyx-tube campanulate, lobes 5; corolla white, tube c. 3 cm long, lobes 5; stamens 5, anthers linear; style c.5 cm long. Berry globose.

Fl. & Fr.: Feb.-May. *Distr.*: Southern Western Ghats. Rare. In evergreen forests. *NAK 2284* (Mannarappara, c. 200 m).

WENDLANDIA Bartling ex A.P. de Candolle
Prodr. 4: 411. 1830 (nom. cons.).

Wendlandia thyrsoidea (Roem. & Schult.) Steud., Nom. (ed.) 2, 2: 786. 1841; Gandhi in Sald.& Nicols., Fl. Hassan Dist. 592. 1976; Matthew & Rani in Matthew, Fl. Tam. Carnatic 3 (2) : 745. 1983; Manilal, Fl. Silent Valley 145.1988; Mohanan & Sivad., Fl. Agasthyamala 364.2002. *Canthium thyrsoideum* Roem. & Schult., Syst. 6: 207. 1820. *Wendlandia notoniana* Wall. ex Wight & Arn., Prodr. 403.1834; Wight, Ic. t. 1033. 1845; Bedd., Fl. Sylv. t. 224. 1872; Hook.f., Fl. Brit. India 3: 40. 1880; Gamble, Fl. Pres. Madras 588. 1921.

Shrubs to small trees. Leaves usually ternate, elliptic to elliptic-oblong, 9-12 by 3.5-5 cm, base acute, apex shortly acuminate, pubescent beneath, lateral veins 7-8 pairs, prominent, thin-coriaceous. Cymes in pyramidal terminal panicles, pubescent. Flowers 4-5- merous, c. 3 mm f; calyx-lobes ovate; corolla salver-form, cream coloured; stamens 4-5, exerted; ovary 2-locular; ovules α per locule. Capsule globose. Seeds minute.

Fl. & Fr.: Dec.-Apr. *Distr.*: Sri Lanka, Peninsular India. Common. In grasslands. *NAK 1397* (Kakki hills, c. 110 m).

Coffea conephora Pierre ex Frochner var. **robusta** (Linden ex de Wilde) Chavalier is widely cultivated.

76. ASTERACEAE
(COMPOSITAE)

1. Capitula radiate or discifom; sap usually watery; disc florets invariably present.
    2. Capitula exclusively homogamous.
        3. Leaves alternate.
            4. Bracts 2-seriate.

5. Leaves subradical, closely alternate; pappus scales 5, stiff. .... **Elephantopus**
5. Leaves cauline, distantly alternate; pappus hairs not stiff. .... **Mikania**

4. Bracts α-seriate.
6. Capitula in terminal panicles; calyx pappus-like, free.
7. Pappus fugacious; outer bracts foliaceous. .... **Phyllocephalum**
7. Pappus copious, filiform; outer bracts not foliaceous. .... **Vernonia**
6. Capitula in axillary clusters; calyx of 3-spongy, basally connate structures. .... **Struchium**

3. Leaves opposite.
8. Panicles cymose, lax; pappus hairs shorter than corolla-tube; anther-tip truncate; cypsella curved. .... **Adenostemma**
8. Panicles corymbose, dense; pappus hairs as long as corolla-tube; anther-tip produced; cypsella straight.
9. Bracts a-seriate; receptacle epaleaceous; cypsella angled; pappus hairs filiform.
10. Outer bracts lanceate, glandular; ovary glabrous; involucre campanulate. .... **Ageratina**
10. Outer bracts ovate, pubescent, ovary scaly; involucre cylindric. .... **Chromolaena**
9. Bracts (sub-) biseriate; receptacle paleaceous; cypsella ribbed; pappus scales paleaceous. .... **Ageratum**

2. Capitula exclusively heterogamous (except in **Spilanthus,** p.p.).
11. Female florets fewer than bisexual florets.
12. Capitula monoecious; anthers free; filaments 1-2-seriate, radiate, connate; female florets apetalous. .... **Xanthium**
12. Capitula polygamous; anthers connate, filaments free; female florets petalous.
13. Leaves pinnate, or 3-foliolate. .... **Bidens**
13. Leaves simple.
14. Ray florets not bigger than disc florets; rays inconspicuous.
15. Capitula spiny; inner bracts echinate; disc florets sterile. .... **Acanthospermum**
15. Capitula not spiny; inner bracts scarious or herbaceous; disc florets fertile.
16. Capitula homogamous (except in **S. ciliatus**), conical, exclusively terminal; pappus usually absent. .... **Spilanthes**
16. Capitula heterogamous, campanulate, both axillary and terminal; pappus present. .... **Synedrella**

14. Ray florets much larger than disc florets; rays conspicuous.
    17. Leaves 3-nerved from base; pappus of short bristles.
        18. Leaves sparingly serrate; ray florets white; capitula c. 7 mm ϕ. .... **Galinsoga**
        18. Leaves deeply serrate; ray florets yellow; capitula c. 10 mm ϕ. .... **Tridax**
    17. Leaves penni-nerved; pappus 0 of - or scales.
        19. Ray florets yellow; capitula c.15 mm ϕ, paleae vertically folded. .... **Wedelia**
        19. Ray florets white; capitula c.10 mm ϕ; paleae plumose. .... **Eclipta**

11. Female florets (when present) more than bisexual florets, α-seriate, usually filiform.
    20. Capitula homogamous; female florets 0 (except in **Senecio**).
        21. Capitula rayed; corolla of bisexual florets, dilated near middle. .... **Senecio**
        21. Capitula not rayed; corolla of bisexual florets, uniformly cylindric.
            22. Branchlets angled; cypsella ribbed, subsucculent. .... **Gynura**
            22. Branchlets striate; cypsella angled, herbaceous. .... **Emilia**
    20. Capitula heterogamous; female florets present.
        23. Anther-base usually tailed.
            24. Capitula rayed; leaves opposite below, alternate above. .... **Vicoa**
            24. Capitula not rayed; leaves alternate throughout.
                25. Branchlets white-pilose; leaf margin entire; bracts hyaline.
                    26. Inflorescence cymose, laxly spreading; female florets mostly sterile. .... **Anaphalis**
                    26. Inflorescence racemose, densely aggregate; female florets fertile. .... **Gnaphalium**
                25. Branchlets other than white-pilose; leaf margin serrate-dentate; bracts scarious or herbaceous.
                    27. Capitula subsessile, aggregated on a common receptacle; inflorescence globose. pappus 0. .... **Sphaeranthus**
                    27. Capitula stalked, solitary; inflorescence corymbose or paniculate.
                        28. Bracts strongly incurved, scarious; bisexual florets sterile; pappus 0. .... **Epaltes**
                        28. Bracts straight, inner alone scarious; bisexual florets fertile; pappus 1-seriate. .... **Blumea**

23. Anther-base not tailed.
    29. Pappus 0; receptacle inconspicuous; capitula in racemes or panicles; leaves pedately 3-5-fid. .... **Artemisia**
    29. Pappus present; receptacle conspicuous; leaves simple.
        30. Leaves ovate; capitula globose; pappus in female florets 0, in bisexual florets of 2 or 3 slender hairs. .... **Dichrocephala**
        30. Leaves ovate-lanceate; capitula campanulate; pappus present in all flowers, stiff or filiform.
            31. Florets all tubular; leaves fastigiate. .... **Conyza**
            31. Florets both tubular and radiate; leaves spreading,stipule-like leaves present. .... **Erigeron**

1. Capitula exclusively ligulate; sap milky; disc florets 0. .... **Crepis**

Note : This key is mainy adopted from Rani and Matthew (1983)

ACANTHOSPERMUM Schrank
Pl. Rar. Horti. Acad. Monac. 53. 1820 ('1819').

Acanthospermum hispidum DC., Prodr. 5: 522. 1836; Gamble, Fl. Pres. Madras 704. 1921; Ramamoorthy in Sald. & Nicols., Fl. Hassan Dist. 599. 1976; Grierson in Dassan. & Fosberg, Rev. Handb. Fl. Ceylon 1: 206. 1980; Matthew, Ill. Fl. Tam. Carnatic t. 350. 1982; Rani & Matthew, Fl. Tam. Carnatic 3(2) : 758. 1983. *Njeringil.*

Stout herbs. Leaves simple, opposite, obovate to oblanceate, 3-6 by 2-3 cm, base cuneate, margin entire to coarsely dentate-serrate, apex acute, penni-nerved, chartaceous. Capitulum solitary, axillary, subsessile, heterogamous, radiate; bracts 2-seraite, inner bracts prickly with 2 awns; outer florets female; inner bisexual; pappus 0; disc florets yellow; stamens 5, anther-base sagittate; ray florets few. Cypsella of ray florets compressed, smooth, enclosed within echinate inner bracts.

Fl. & Fr.: Jan.-May. *Distr.*: Endemic to S. America. Now introduced into other tropical countries. On exposed lands, chiefly in plains. Locally abundant. *NAK* 2927 (Adoor, c. 20 m).

Note : A noxious weed in plains which is harmful to the local species, especially the ephemerals.

ADENOSTEMMA J.R. Forster et J.G.A. Forster
Charact. Gen. 89, t. 45. 1776.

Adenostemma lavenia (L.) Kuntze, Rev. Gen. Pl. 1: 304. 1891; Gamble, Fl. Pres. Madras 675. 1921; Ramamoorthy in Sald. & Nicols., Fl. Hassan Dist. 600. 1976, Grierson in Dassan. & Fosberg, Rev. Handb. Fl. Ceylon 1: 139. 1980; Rani & Matthew in Matthew, Fl. Tam. Carnatic 3(2): 758. 1983; Manilal, Fl. Silent Valley 146. 1988; Mohanan & Sivad., Fl. Agasthyamala 367.2002. *Verbesina lavenia* L., Sp. Pl. 902. 1753. *Adenostemma viscosum* Forst. f., Char. Gen. Pl. 90, 1776; Hook.f., Fl. Brit. India 3: 342. 1881.

Erect herbs. Leaves simple, opposite, broadly ovate to elliptic-lanceate, 7-20 by 3-13 cm, base subtruncate, decurrent on a long petiole, margin coarsely serrate, apex acute, chartaceous. Capitula in cymose panicles, homogamous, disciform; bracts 2-seriate; florets white, bisexual; pappus of 3-5 subulate hairs; stamens 5; ovary 3-5-gonous. Cypsella curved, angled, papillate.

Fl. & Fr.: Sep.-Jan. *Distr.*: India, China, Malaya, Trop. Africa, Australia. Less common. In moist shady forest floor. *NAK 718* (Pampa valley, c.100 m), *881* (Ranni, c. 100 m).

AGERATINA Spach
Hist. Nat. Veg. Phan. 10:286. 1841.

Ageratina adenophora (Sprengel) King & Robinson, Phytologia 19: 211. 1970; Rani & Matthew in Matthew, Fl. Tam. Carnatic 3(2) : 760. 1983; Manilal, Fl. Silent Valley 146. 1988. *Eupatorium adenophorum* Spreng., Syst. 3: 420. 1826. *E. glandulosum* Kunth, Nov. Gen. Sp. 4: 122. 1820, non Michaux, 1803; Matthew, Rec. Bot. Surv. India 20: 135. 1969.

Viscid undershrubs; stem and young leaves reddish-brown. Leaves opposite, rhombic or deltoid, 4-7 by 3-5 cm, base cuneate, margin serrate, apex subacuminate, chartaceous. Capitula in panicles, homogamous; bracts α; florets white. Cypsella angular.

Fl. & Fr.: Apr.-Jun. *Distr.*: Native of Mexico; now a pantropic weed. Locally abundant. Along waysides in upper hills. *NAK* 2786 (Kakki hills, c. 1100 m).

Note : This introduced weed is slowly and steadily establishing in upper hills, forming dense populations like a monoculture, along the waysides.

AGERATUM Linnaeus
Sp. Pl. 839. 753.

Aromatic herbs. Leaves simple, opposite or upper alternate, broadly ovate, chartaceous. Capitula in dense terminal corymbs, homogamous, disciform; bracts (sub-) biseriate. Florets bisexual; pappus of 5 scales; corolla tubular; stamens 5, anther-base obtuse. Cypsella 5-angled or 5-ribbed.

1. Leaves crenate-serrate; involucral bracts oblong-lanceate, serrulate; capitula usually white. .... 1. **A. conyzoides**
1. Leaves serrate; involucral bracts linear-lanceate, entire; capitula pink or bluish. .... 2. **A. houstonianum**

1. **Ageratum conyzoides** L., Sp. Pl. 839. 1753; Clarke, Compos. Ind. 30. 1876; Hook. f., Fl. Brit. India 3: 243. 1881; Gamble, Fl. Pres. Madras 677. 1921; Ramamoorthy in Sald. & Nicols., Fl.Hassan Dist.601. 1976; Grierson in Dassan. & Fosberg, Rev. Handb. Fl. Ceylon 1: 141. 1980; Matthew, Ill. Fl. Tam. Carnatic t. 352. 1982; Rani & Matthew in Matthew, Fl. Tam. Carnatic 3(2) : 758. 1983; Manilal, Fl. Silent Valley 147. 1988; Mohanan & Sivad., Fl. Agasthyamala 368.2002.

Leaves broadly ovate, 3-8 by 2-5 cm, base truncate to cuneate, margin crenate-serrate, apex acute, basally 3-5- nerved. Capitula white, 6-8 mm ϕ; involucral bracts oblong-lanceate, serrulate. Cypsella sparsely scabrous.

Fl. & Fr.: Most of the seasons. *Distr.*: Pantropical. Another troublesome common weed in hills. *NAK 651* (Puthukkulam, c. 50 m).

Juice of plants boiled with oil and applied in rheumatism.

2. **Ageratum houstonianum** Miller, Gard. Dict. (ed.8), n.2. 1768; Fischer in Gamble, Fl. Pres. Madras 1881. 1936; Grierson in Dassan. & Fosberg, Rev. Handb. Fl. Ceylon 1: 142. 1980; Rani & Matthew in Matthew, Fl. Tam. Carntic 3(2) : 762. 1983.

Leaves broadly ovate to deltoid, 4-8 by 3-6 cm, base truncate, margin serrate, apex acute. Capitula pink or bluish, in dense panicles; involucral bracts linear-lanceate, entire. Cypsella scaberulous.

Fl. & Fr.: Nov.-Jan. *Distr.:* Native of Mexico. Pantropic weed. Common in hill floors; abundant. *NAK 1187* (Kakki hills, c. 1100 m).

This weedy species is dominating over the native species in disturbed upper hilly areas.

ANAPHALIS A.P. de Candolle
Prodr. 6: 271. 1838.

Annual or perennial soft cottony herbs. Leaves simple, alternate, often crowded towards base, sessile, usually 1-5- veined, base subamplexicaul, margin often revolute, apex acute, thick-chartaceous. Capitula in terminal panicles, heterogamous; bracts scarious, white, yellowish or pink, clawed; outer florets female, mostly sterile; inner bisexual; pappus setose, caducous; corolla filiform, often glandular; stamens 5, anther-bases tailed; style 2-fid. Cypsella ellipsoid or angled, papillose or scaly.

Note : Classification of this genus and the allied *Gnaphalium* based on morphological characters is difficult.

1. Robust herbs; leaves elliptic-lanceate, 5-veined. ....2. **A. beddomei**
1. Slender herbs; leaves linear-oblong to linear-lanceate, 1-veined.
   2. Leaves scabrate above, cottony below; bracts pink. ....1. **A. aristata**
   2. Leaves cottony on both sides; bracts yellowish-white. ....3. **A. lawii**

1. **Anaphalis aristata** DC., Prodr. 6: 274. 1838; Wight, Ic. t. 199. 1846; Hook.f., Fl. Brit. India 3: 285. 1881; Gamble, Fl. Pres. Madras 696. 1921; Rani & Matthew in Matthew, Fl.Tam. Carnatic 3(2) : 766. 1983; Manilal, Fl. Silent Valley 147. 1988.

Slender herbs. Leaves linear-oblong to linear-lanceate, 2-6 by 0.2-0.5 cm, base auricled, margin revolute, apex gradually acute, mucronate, upper surface sparsely scabrate, lower surface lanate, 1-veined. Capitula c. 3 mm ϕ; bracts ovate-oblong, pink tinged, apex dentate. Cypsella angled, faintly papillose.

Fl. & Fr.: Dec.-Mar. *Distr.*: Peninsular India. A beautiful herb. Locally very abundant. On exposed hills slopes. *NAK 420 & 1327* (Kakki hills, c. 1100 m).

2. **Anaphalis beddomei** Hook.f., Fl. Brit. India 3: 282. 1881; Gamble, Fl. Pres. Madras 695. 1921; Manilal, Fl. Silent Valley 147. 1988.

Decumbent herbs. Leaves ovate-lanceate or ellitptic-lanceate, 3-7 by 0.7-3 cm, base rounded, then narrowed into subamplexicaul base, margin flat to revolute, 5-(7)-veined. Capitula small; bracts yellowish or white. Cypsella narrowly ellipsoid, scabrous.

Fl. & Fr.: Dec.-Mar. *Distr.*: Peninsular India. On exposed upper hill slopes; abundant. *NAK 400 & 1313* (Kakki hills, c.110 m).

3. **Anaphalis lawii** (Hook.f.) Gamble, Fl. Pres. Madras 696. 1921. *A. oblonga* DC. var. *lawii* Hook.f., Fl. Brit. India 3 : 283. 1881. **(Fig. 41)**

Slender herbs. Leaves linear-lanceate, linear towards stem apex, 5 by 0.3 cm, base subamplexicaul, margin flat or revolute, apex acute-mucronate, both surfaces cottony. Panicle profusely branched. Capitula c. 4 mm ϕ; bracts elliptic-lnceate, white or cream coloured. Cypsella ellipsoid, glandular.

Fl. & Fr.: Oct.-Feb. *Distr.*: Western Peninsular India. A pretty herb on exposed upper hill slopes; abundant. *NAK* 406, *1338 & 1395* (Kakki hills, c. 1100 m).

ARTEMISIA Linnaeus
Sp. Pl. 845. 1753.

Artemisia nilagirica (Clarke) Pamp., Nuov. Giorn. Bot. Ital. 33: 452. 1926; Ramamoorthy in Sald. & Nicols., Fl. Hassan Dist. 602. 1976; Rani & Matthew in Matthew, Fl. & Tam. Carnatic 3(2): 765. 1983; Manilal, Fl. Silent Valley 149.1988; Nicols., Suresh & Manilal, An Interpr. Hort. Malab. 65. 1988. *A vulgaris* L., var.*nilagirica* Clarke, Comp. Ind. 162. 1866. *A. indica* sensu Wight, Ic. t. 1112. 1846, non Willd., 1846. *A. vulgaris* sensu Hook.f., Fl. Brit. India 3: 325. 1881, non L., 1753; Gamble, Fl. Pres. Madras 713. 1921. *Katu-tsjetti-Pu* Rheede, Hort. Malab. 10: 89, t. 45, 1690.

Erect, profusely branched undershrubs. Leaves pedately 3-5-fid, alternate, oblong-lanceate, 3-5 by 0.7-1 cm, base cuneate, apex acute, lower surface white-tomentose. Capitula heterogamous in racemose, axillary and terminal panicles; bracts 2-4- seriate; outer florets female, α-seraite, inner few bisexual; pappus 0; ray flowers glandular without. Cypsella oblong, smooth.

Fl. & Fr.: Feb.-May. *Distr.*: Peninsular India. Rare. *CNM 58468* (MH Kodumon). Garden escape?.

BIDENS Linnaeus
Sp. Pl. 831. 1753.

Bidens biternata (Lour.) Merr. & Sherff ex Sherff, Bot. Gaz. 88: 293. 1929; Ramamoorthy in Sald.& Nicols., Fl. Hassan Dist. 602. 1976; Grierson in Dassan. & Fosberg, Rev. Handb. Fl. Ceylon 1: 229. 1980; Manilal, Fl. Silent Valley 149. 1988; Mohanan & Sivad., Fl. Agasthyamala 369.2002. *Coreopsis biternata* Lour., Fl. Cochinch. 508. 1790. *Bidens pilosa* sensu Hook.f., Fl. Brit. India 3: 309. 1881, non L., 1753; Gamble, Fl. Pres. Madras 709. 1921.

Subshrubs; stem 4-gonous. Leaves pinnately 3-5- foliolate, opposite, leaflets lanceate, 4-7 by 2-3.5 cm, base cuneate, margin sharply serrate, apex acuminate, chartaceous. Capitula heterogamous in corymbose panicles; outer florets female or sterile; inner bisexual, yellow; ray florets white, 1-seriate; pappus of 3-4 awns. Cypsella cuneiform, tetragonous.

Fl. & Fr.: Most of the seasons. *Distr.*: Pantropics. In forest floors and exposed hills. *NAK 658* (Puthukkulm, c. 50 m).

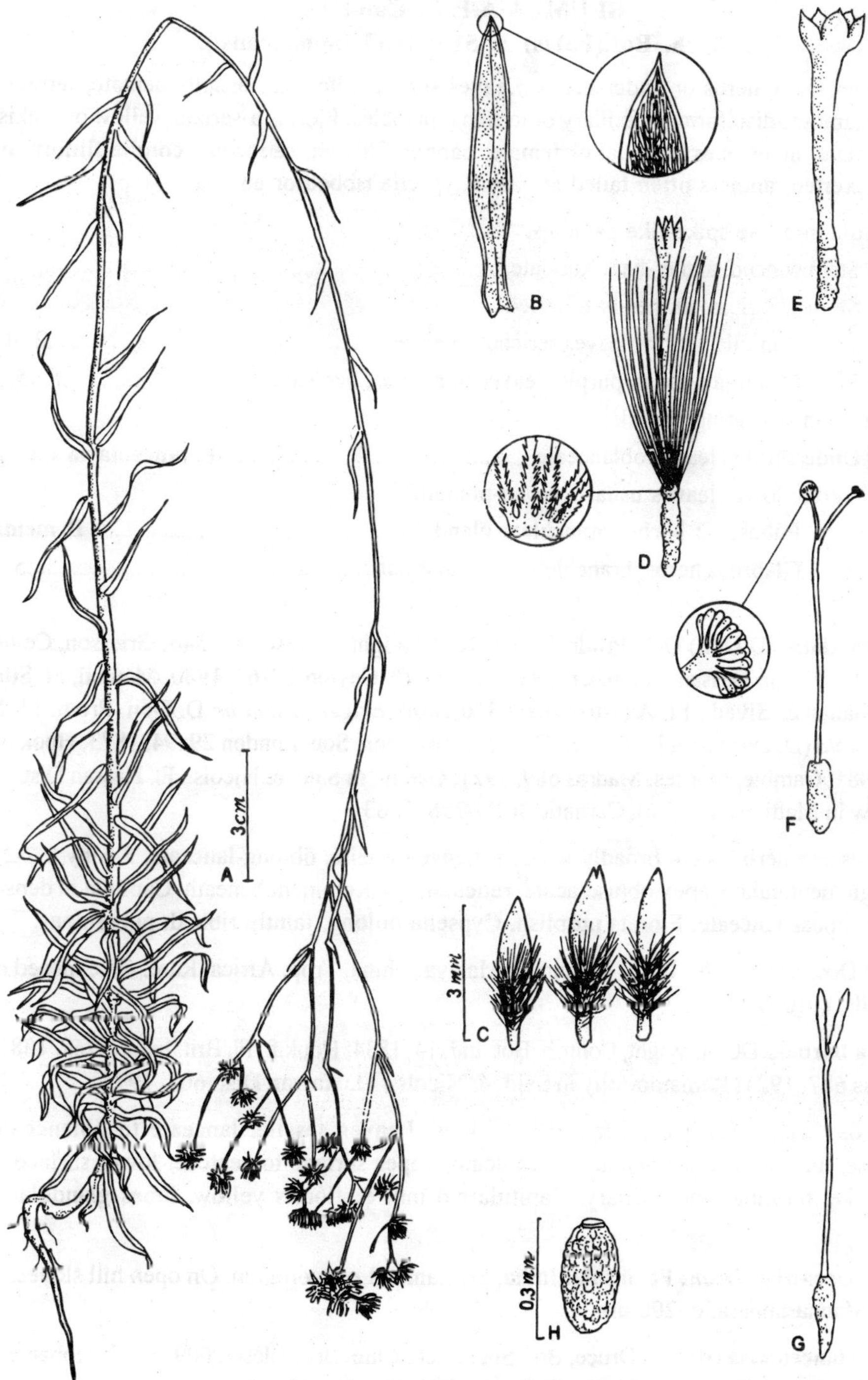

Fig. 41. ***Anaphalis lawii*** **(Hook. f.) Gamble: A. Plant; B. Petal; C. Bracts; D. Outer floret; E. Inner floret; F. Pistil of innner floret; G. Pistil of outer floret; H. Cypsella.**

BLUMEA A.P. de Candolle
Arch. Bot.(Paris) 2: 514. 1833 (nom. cons.).

Perennial aromatic herbs or undershrubs. Leaves simple, alternate, usually dentate, serrate or lyrate. Capitula heterogamous, disciform, in axillary or terminal panicles. Florets a-seriate, yellow or pinkish-purple; outer florets female, inner ones bisexual or female; pappus filiform, caducous; corolla filiform or tubular; stamens 5, subexerted; anthers often tailed at base. Cypsella ribbed or angled.

1. Capitula in dense spike-like panicles.
   2. Stem winged; anther-base sagittate. .... 1. **B. alata**
   2. Stem wingless; anther-base tailed.
      3. Capitula yellow; leaves sericeous beneath. .... 2. **B. barbata**
      3. Capitula pinkish-purple; leaves tomentose beneath. .... 5. **B. mollis**
1. Capitula in spreading panicles.
   4. Undershrubs; leaves oblanceate. .... 3. **B. lanceolaria** var. **spectabilis**
   4. Herbs; lower leaves usually lyrate-pinnatifid.
      5. Pubescent herbs; branchlets glandular. .... 4. **B. membranacea**
      5. Glabrous herbs; branchlets non-glandular. .... 5. **B. virens**

1. **Blumea alata** (D.Don) DC., Prodr. 5: 488. 1836; Wight, Ic. t. 1101. 1846; Grierson, Ceylon J. Sci., Biol. Sci:. 11: 12. 1974, in Dassan. & Fosberg, Rev. Handb. Fl. Ceylon 1: 167. 1980; Manilal, Fl. Silent Valley 150. 1988; Mohanan & Sivad., Fl. Agasthyamala 370.2002. *Erigeron alatum* D. Don, Prodr. Fl. Nep. 171. 1825. *Laggera alata* (D. Don) Schultz-Bip. ex Oliver, Trans. Linn. Soc. London 29: 94. 1873; Hook. f., Fl. Brit. India 3 : 271. 1881; Gamble, Fl. Pres. Madras 689. 1921; Gandhi in Sald. & Nicols., Fl. Hassan Dist. 619. 1976; Rani & Matthew in Matthew, Fl. Tam. Carnatic 3(2) : 758. 1983.

Stout pubescent herbs; stem broadly winged. Leaves sessile, oblong-lanceate, 3-6 by 0.7-2 cm, base decurrent, margin denticulate, apex obtuse-acute, reticulations prominent beneath. Capitula in dense terminal panicles; bracts linear-lanceate. Florets purplish. Cypsella oblong, faintly ribbed, puberulous.

Fl. & Fr.: Dec.-Mar. *Distr.*: India, Myanmar, Malaya, China, Trop. Africa. Rare. On exposed hills. *NAK 1314* (Kakki hills, c. 1000 m).

2. **Blumea barbata** DC. in Wight, Contrib. Bot. Ind. 14. 1834; Hook.f., Fl. Brit. India 3:262. 1881; Gamble, Fl. Pres. Madras 687. 1921; Ramamoorthy in Sald. & Nicols., Fl. Hassan Dist. 604. 1976.

Erect herbs; branchlets brownish, softly villous. Leaves sessile, lanceate to oblanceate, 2.5 by 0.6-1.5 cm, base cuneate, margin dentate, apex acute, upper surface tomentose, lower surface sericeous. Panicle spike-like, terminal and axillary. Capitula c.6 mm ϕ; florets yellow, lobes glandular. Cypsella puberulous.

Fl. & Fr.: Dec.-Mar. *Distr.*: Peninsular India, Sri Lanka. Less common. On open hill slopes. *NAK 126, 1252 & 2109* (Mannarappara, c. 200 m).

3. **Blumea lanceolaria** (Roxb.) Druce, Bot. Soc. Exch. Club. Brit. Isles 4: 609. 1917; Mohanan & Sivad., Fl. Agasthyamala 371.2002.var. **spectabilis** (DC.) Randeria, Blumea 10: 720. 1960; Ramamoorthy in Sald. &

Nicols., Fl. Hassan Dist. 605. 1977; Manilal, Fl. Silent Valley 151. 1988. *B. spectabilis* DC., Prodr. 5: 445. 1836; Hook.f., Fl. Brit. India 3: 269. 1881; Gamble, Fl. Pres. Madras 688. 1921.

Undershrubs. Leaves oblanceate, 10-25 by 2-4 cm, gradually reducing and to form foliaceous bracts, base attenuate, margin denticulate, apex acute to narrowly acuminate, reticulations prominent. Panicle large, pyramidal, c. 40 by 25 c. Capitula c. 15 mm ϕ. Florets yellow. Cypsella pubescent.

Fl. & Fr.: Dec.-Mar. *Distr.*: Sri Lanka, India, Myanmar, Malesia, China. Not common. In forest clearings. *NAK* 353, 1463 (Arampa, c. 450 m).

4. **Blumea membranacea** Wallich ex DC., Prodr. 5 : 440. 1836; Hook.f., Fl. Brit. India 3: 265. 1881; Gamble, Fl. Pres. Madras 687, 1921; Ramamoorthy in Sald. & Nicols., Fl. Hassan Dist. 605. 1976; Manilal, Fl. Silent Valley 152, 1988. *Bhootham-kolli.*

Pubescent herbs. Lower leaves lyrate, upper ones oblanceate, 3-10 by 1-3 cm, base decurrent on a short petiole, margin coarsely serrate-dentate, apex acute, silky pubescent. Panicle lax; florets yellow. Cypsella puberulous.

Fl. & Fr.: Dec.-Feb. *Distr.*: Indo-Malesia. Common weed. On waysides, forest floor, etc. *NAK 518, 1461* (Arampa, c. 450 m).

Note : A highly variable and taxonomically confusing species with different eco lines.

5. **Blumea mollis** (D.Don) Merr., Philipp. J. Sci. 5: 395. 1910; Ramamoorthy in Sald. Nicols., Fl. Hassan Dist. 605. 1976; Grierson in Dassan. & Fosberg, Rev. Handb. Fl. Ceylon 1: 169.1980; Matthew, Ill. Fl. Tam. Carnatic t. 357. 1982; Rani & Matthew in Mathew, Fl. Tam. Carnatic 3(2) : 771. 1983; Manilal, Fl. Silent Valley 152. 1988; Nicols., Suresh & Mani. An Interpr. Hort. Malab. 66. 1988; Mohanan & Sivad., Fl. Agasthyamala 372.2002. *Erigeron molle* D.Don, Prodr. Fl. Nepal. 172. 1825. *Blumea wightiana* DC. in Wight, Contrib. Ind. Bot. 14. 1834; Hook.f., Fl. Brit. India 3: 261. 1881; Gamble, Fl. Pres. Madras 686. 1921. *B. neilgherrensis* Hook.f., Fl. Brit. India 3: 261. 1881; Gamble Fl. Pres. Madras 686. 1921. *Nari-patsja* Rheede, Hort. Malab. 10:123, t. 62. 1690.

Softly pubescent herbs. Leaves oblanceate, 4-10 by 2-4.5 cm, base cuneate, decurrent, margin coarsely serrate, apex acute. Panicle spike-like, dense. Capitula c.5 mm ϕ; florets pinkish-purple, lobes glandular. Cypsella glabrescent.

Fl. & Fr.: Nov.-Feb. *Distr.*: Tropical Africa & Asia. Very common. Along waysides, mostly in plains, usualy after rains. *NAK 235, 1098, 1233 & 1403* (Almost all places).

6. **Blumea virens** Wall. ex DC. in Wight, Contrib. Bot. India 14. 1834; Hook.f., Fl. Brit. India 3 : 264. 1881; Gamble, Fl. Pres. Madras 687. 1921; Randeria, Blumea 10:272. 1960; Grierson in Dassan. & Fosberg, Rev. Handb. Fl. Ceylon 1:173. 1980; Rani & Matthew in Matthew, Fl. Tam. Carnatic 3(2) : 773. 1983.

Tall glabrous herbs. Lower leaves runcinate-lyrate, 18 by 6 cm, upper leaves oblanceate, base decurrent, margin denticulate, apex acute to acuminate, membranous. Panicle axillary and terminal. Capitula c.5 mm ϕ; florets yellow. Cypsella faintly ribbed, puberulous.

Fl. & Fr.: Dec.-Feb. *Distr.*: India, Myanmar, Malesia & China. Not common. Usually in interior evergreen forests. *NAK 352* (Moozhiar, c. 250 m).

CHROMOLAENA A.P. de Candolle
Prodr. 5: 133. 1836.

Chromolaena odorata (L.) King & Robinson, Phytologia 20 : 204. 1970; Matthew, Ill. Fl. Tam. Carnatic t. 369. 1982; Rani & Matthew in Matthew, Fl. Tam. Carnatic 3(2) : 776. 1983; Manilal, Fl.Silent Valley 152. 1988; Mohanan & Sivad., Fl. Agasthyamala 373.2002. *Eupatorium odoratunm* L., Syst. Nat. (ed. 10), 2 : 1205. 1979; Hook. f., Fl. Brit. India 3: 244. 1881. *Communist-pascha.*

Aromatic subshrubs. Leaves simple, opposite, ovate-rhombic, 6-8 by 3-5 cm, base cuneate, acute, margin coarsely serrate, apex acute, basally 3-nerved, chartaceous. Capitula homogamous, disciform in corymbose panicles; florets bluish-white, or pure white, bisexual; pappus 1-seriate, barbellate; corolla tubular; stamens 5, subexerted; style long-exerted, 2-fid. Cypsella 5-anlgled, scabrid.

Fl. & Fr.: Apr.-Aug. *Distr.*: Native of America, now spread in tropical Asia. An aggressive well established weed. *NAK 2803* (Konni, c. 100 m).

Leaf juice quickly heals minor wounds.

CONYZA Lessing
Syn. Gen. Comp. 203. 1832 (nom. cons.).

Conyza japonica (Thunb.) Less., Syn. Gen. Compos. 204. 1832; Hook.f., Fl. Brit. India 3:258. 1881; Gamble, Fl. Pres. Madras 683. 1921; Ramamoorthy in Sald. & Nicols., Fl. Hassan Dist. 608. 1976. *Erigeron japonicum* Thunb., Fl. Jap. 312. 1784.

Pubescent herbs. Leaves simple, alternate, oblanceate, 4-6 by 2-4 cm, base rounded, decurrent, margin crenate, apex obtuse, pinnately-veined, chartaceous. Capitula heterogamous, discoid, in axillary or terminal corymbose panicles; bracts α-seriate. Florets yellow, outer female, inner bisexual; pappus 1-2- seriate, barbellate; corolla filiform-tubular; stamens 5. Cypsella compressed, marginate, pubescent.

Fl. & Fr.: Sep.-Jan. *Distr.*: India, Japan. Very common in exposed upper hills and waysides. *NAK 1237 & 1819* (Kaki hills, c. 1110 m).

CREPIS Linnaeus
Sp. Pl. 805. 1753.

Crepis acaulis (Kurz) Hook.f., Fl. Brit. India 3 : 396. 1881; Gamble, Fl. Pres. Madras 730. 1921.

Scapigerous annual or perennial herbs; stem with milky sap. Leaves rosulate, closely alternate, narrowly oblanceate to spathulate, 4-20 by 0.7-2 cm, base subauricled, margin coarsely denticulate, apex acute, chartaceous. Capitula exclusively ligulate, cylindric, c.2 cm long; disc-florets 0; bracts biseriate, unequal; ray-florets yellow; pappus of a simple barbellate bristles; ligules truncate, 5-toothed; stamens 5, anther-base tailed. Cypsella fusiform, ribbed.

Fl. & Fr.: Apr.-Jul. *Distr.*: India, Myanmar. Rare. On exposed hills of grasslands. *NAK 2766* (Kakki hills, c. 1100 m).

DICHROCEPHALA L' Heritier ex A.P. de Candolle
Arch. Bot. (Paris) 2: 517. 1833.

Dichrocephala integrifolia (L.f.) Kuntze, Rev. Gen. Pl. 1:333. 1891; Grierson in Dassan. & Fosberg, Rev. Handb. Fl. Ceylon 1:149. 1980; Matthew, Ill. Fl. Tam. Carnatic t. 362. 1982; Rani & Matthew in Matthew, Fl. Tam. Carnatic 3(2) : 781. 1983; Manilal, Fl. Silent Valley 53. 1988; Mohanan & Sivad., Fl. Agasthyamala 374. 2002. *Hippia integrifolia* L.f., Suppl. Pl. 389. 1781. *Grangea latifolia* Lam., Tab. Encycl. t. 699, f.1. 1796 & Encycl. (Suppl. 2) : 826. 1812. *Dichrocephala latifolia* (Lam.) DC. in Wight, Contrib. Bot. India 11. 1834; Hook. f., Fl. Brit. India 3: 245. 1881; Gamble, Fl. Pres. Madras 679. 1921.

Erect pubescent herbs, rooting at lower nodes. Leaves deeply lyrate, alternate, terminal lobe broadly ovate, 5 by 3.5 cm, lateral lobes small, margin coarsely serrate, apex acute, chartaceous. Capitula heterogamous, disciform or subradiate, in axillary or terminal panicles; bracts 2-seriate; outer florets yellow, female; inner white, bisexual; pappus 0 or fugacious. Cypsella compressed with a corky margin.

Fl. & Fr.: Dec.-May. *Distr.:* Trop. & Sub-trop. Asia, Africa & Pacific Islands. Open hill slopes above 1000 m. *NAK 1400* (Kakki hills, c. 1100 m).

ECLIPTA Linnaeus
Mant. 2: 157, 286. 1771 (nom. cons.).

Eclipta prostrata (L.) L., Mant.Pl. 286. 1771; Santapau, J. Bombay Nat. Hist. Soc. 54: 475-476. 1957; Grierson in Dassan. & Fosberg, Rev. Handb. Fl. Ceylon 1 : 212. 1980; Matthew, Ill. Fl. Tam. Carnatic t. 365. 1982; Rani & Matthew in Matthew, Fl. Tam. Carnatic 3(2) : 783. 1983; Nicols., Suresh & Manilal, An Interpr. Hort. Malab. 67. 1988. *Verbesina prostrata* L., Sp. Pl. 902. 1753. *V. alba* L., Sp. Pl. 902. 1753. *Eclipta alba* (L.) Hassk., Pl. Jav. Rar. 528. 1848; Hook.f., Fl. Brit. India 3 : 304. 1881; Gamble, Fl. Pres. Madras 795. 1921. *Cajennaem* Rheede, Hort. Malab. 10 : 81-82, t.41. 1690. *Kayyonni.*

Diffuse or ascending herbs. Leaves simple, opposite, sessile, lanceate, 2-5 by 0.5 - 1 cm, base obtuse, margin faintly serrate-dentate, apex acute, both surfaces strigose, thin-coriaceous. Capitula heterogamous, radiate in axillary and terminal clusters; bracts 2-seriate; florets white, outer female, inner bisexual; pappus reduced to 2 teeth; corolla tubular; stamens 5, anther-base obtuse. Cypsella trigonous or compressed.

Fl. & Fr.: Mar.-Jul. *Distr.:* Pantropical. Common on wet marshy places, along paddy field-margins. *NAK 642* (Pandalam, c.50 m).

Juice of plant mixed and boiled with oil, and used against head-ache and dental disorders. Also used by women for growing hairs.

ELEPHANTOPUS Linnaeus
Sp. Pl. 814. 1753.

Elephantopus scaber L., Sp. Pl. 84. 1753; Roxb., Fl. Ind. 3:445. 1832; Hook f., Fl. Brit. India 3 : 242. 1881; Gamble, Fl. Pres. Madras 676. 1921; Grierson in Dassan. & Fosberg, Rev. Handb. Fl. Ceylon 1:135. 1980; Matthew, Ill. Fl. Tam. Carnatic t. 366. 1982; Rani & Matthew in Matthew, Fl. Tam. Carnatic 3(2) : 784. 1983; Nicols., Suresh & Manilal, An Interpr. Hort. Malab. 67. 1988; Mohanan & Sivad., Fl. Agasthyamala 375.2002. *Ana-schovadi* Rheede, Hort. Malab. 10:13-14, t. 7. 1690.

Scapigerous herbs. Leaves simple, radical, closely alternate, oblanceate, 6-18 by 2-5 cm, base attenuate, margin crenate, apex obtuse or broadly acute, strigose above, so on nerves beneath, chartaceous. Capitula homogamous, disciform, sessile, aggregated in dense stalked glomerules, subtended by foliaceous bracts; bracts of capitula 8, biseriate; florets purple, bisexual; pappus of 4-6 stiff bristles; corolla tubular; stamens 5, anther-base auricled; style-arms subulate. Cypsella 10-ribbed.

Fl. & Fr.: Dec.-Apr. *Distr.:* Sri Lanka, India, Malesia, Australia, Trop. Africa. Common. Among forest undergrowth. *NAK 145* (Adoor, c. 40 m).

Decoction of root and leaves mixed with cumin and butter milk, and given in diarrhoea and dysentery. Powder or decoction of roots cures malaria and ulcers.

EMILIA Cassini
Bull. Sci. Soc. Philom. Paris 1871 :68. 1871.

Emilia sonchifolia (L.) DC. in Wight, Contrib. Bot. India 24. 1834; Hook.f., Fl. Brit. India 3:336. 1881; Gamble, Fl. Pres. Madras 716. 1921; Grierson in Dassan. & Fosberg, Rev. Handb. Fl. Ceylon 1:251. 1980; Matthew,Ill. Fl. Tam. Carnatic t. 367. 1982; Rani & Matthew in Matthew, Fl. Tam. Carnatic 3(2) : 787. 1983; Manilal, Fl. Silent Valley 154. 1988; Nicols., Suresh & Manilal, An Interpr. Hort. Malab. 67. 1988; Mohanan & Sivad., Fl. Agasthyamala 376.2002. *Cacalia sonchifolia* L., Sp. Pl. 835. 1753. *Muel-schevi* Rheede, Hort. Malab. 10:135, t. 8.1690.

Profusely branched herbs, glabrous to densely pubescent. Leaves radical and cauline, lyrate-pinnatifid, upper ones undivided, 4-8 by 1.5-5 cm, base acute-auriculate, margin dentate, apex obtuse to subacute. Capitula homgamous, disciform, in terminal, lax corymbs; bracts 2-seriate, linear-lanceate; florets purple, bisexual; papus 1-seriate; stamens 5, exerted. Cypsella angled, pubescent.

Fl. & Fr.: Most of the seasons. *Distr.*: Asia & Africa. Very common. Moist areas of plains and hills. *NAK 67, 693, 913* (Plappally, c. 300 m). A herb with highly variable indumentum.

Juice of leaves dropped into eyes for night blindness and inflammation.

EPALTES Cassini
Bull. Sci. Soc. Philom. Paris 1818 : 139. 1818.

Epaltes divaricata (L.) Cass., Bull. Bot. Sci. Soc. Philom. Paris 1818 : 139. 1818; DC. in Wight, Contrib. Bot. India 6. 1834; Hook.f., Fl. Brit. India 3 : 274. 1881; Gamble, Fl. Pres. Madras 691. 1921; Ramamoorthy in Sald. & Nicols., Fl. Hassan Dist. 614. 1976; Grierson in Dassan. & Fosberg, Rev. Handb. Fl. Ceylon 1:178.1980; Matthew, Ill. Fl. Tam. Carnatic t. 367. 1982; Rani & Matthew in Matthew, Fl. Tam. Carnatic 3(2) : 788. 1983; *Ethulia divaricata* L., Mant. Pl. 10. 1767.

Erect herbs; branchlets winged. Leaves simple, alternate, lanceate to oblong, 3-7 by 0.7-1 cm, base cuneate, decurrent, margin faintly dentate, apex acute, coriaceous. Capitula heterogamous, disciform, 1-3 in terminal clusters; bracts α-seriate, scarious; florets pink to purple, outer florets female, inner bisexual; pappus 0; corolla tubular; stamens 5, included. ovary linearly oblong. Cypsella oblong.

Fl. & Fr.: Dec.-Apr. *Distr.:* India, Myanmar, China, Java. Rare. In low-lying grounds. *CNM 63702* (MH) (Kozencherry).

ERIGERON Linnaeus
Sp. Pl. 863. 1753.

Erigeron canadensis L., Sp. Pl. 863. 1753; Hook.f., Fl. Brit. India 3: 254. 1881; Gamble, Fl. Pres. Madras 681. 1921.

Tall suffrutescent herbs; branchlets slender, pubescent. Leaves simple, narrow linear-lanceate, or oblanceate, 3-10 by 0.3-5 cm, base attenuate, margin coarsely toothed, apex subacute, chartaceous. Capitula heteroganus, radiate, in axillary racemes or panicles; floret yellow, outer few female, ligulate; inner bisexual; pappus filiform; corolla tubular; stamens 5, included., anther-base obtuse. Cypsella compressed, pubescent.

Fl. & Fr.: Sep.-Mar. *Distr.*: Probably a native of North America; now introduced in all tropical countries. On waysides in upper hills. *NAK 1235* (Kakki hills, c. 1100 m).

GALINSOGA Ruiz et Pavon
Prodr. 110, t. 24. 1794.

Galinsoga parviflora Cav., Ic. 3:41, t. 2811. 1796; Hook. f., Fl. Brit. India 3 : 311. 1881; Gamble, Fl. Pres. Madras 711. 1921; Grierson in Dassan. & Fosberg, Rev. Handb. Fl. Ceylon 1:231. 1980; Rani & Matthew in Matthew, Fl. Tam. Carnatic 3(2) : 791. 1983.

Erect slender annuals. Leaves simple, opposite, sessile, lanceate, 1.5-3.5 by 0.7-2 cm, base cuneate or obtuse, margin faintly dentate-serrate, apex shortly acuminate, basally 3-nerved, membranous. Capitula heterogamous, radiate, in terminal and axillary panicles; bracts 1-seriate; bisexual florets yellow, ray florets white; corolla tubular; stamens 5, included, anther-base obtuse. Cypsella of outer florets compressed, inner florets angled, puberulous.

Fl. & Fr.: Dec. Mar. *Distr.:* Native of Trop. S. America. Now in trop. and temperate Asia. In upper hills. *NAK 407* (Kakki hills, c. 1100 m).

GNAPHALIUM Linnaeus
Sp. Pl. 850. 1753:

Gnaphalium polycaulon Pers., Syn. Pl. 2 : 421. 1807; Grierson, Notes Roy. Bot. Gard. Edinburgh 31 : 137. 1972; Ramamoorthy in Sald. & Nicols., Fl. Hassan Dist. 616. 1976; Matthew, Ill. Fl. Tam. Carnatic t. 373. 1982; Rani & Matthew in Matthew, Fl. Tam. Carnatic 3(2) : 795. 1983; Mohanan & Sivad., Fl. Agasthyamala 378. 2002. *G. indicum* auct., non L., 1753: Hook.f., Fl. Brit. India 3 : 289. 1881; Gamble, Fl. Pres. Madras 698. 1921.

Slender erect annuals; vegetative parts white-wooly. Leaves simple, alternate, oblanceate or narrowly obovate, 1-3 by 0.5 - 0.8 cm, base attenuate, apex obtuse, apiculate, chartaceous. Capitula heterogamous, disciform, in dense, leafy, interrupted terminal and axillary racemose clusters; bracts α, linear, wooly; outer florets yellow, female, very α; inner bisexual, 2 or 3; pappus setose; corolla tubular-campanulate; stamens 5, included, anther-bases sagittate. Cypsella oblong, shortly-beaked.

Fl. & Fr.: Jan.-Apr. *Distr.*: A pantropic weed. Along waste places in upper hills only. *NAK 2700* (Kakki hills, c. 1100 m).

GYNURA Cassini
Dict. Sci. Nat. 34 : 391 : 1825 (nom. cons.).

Herbs. Leaves simple, alternate, entire to pinnati-partite, sessile or petiolate. Capitula homogamous, disciform, in terminal long-stalked corymbs; involucre cylindric; bracts (sub) biseriate; pappus filiform; corolla narrowly cylindric; stamens 5, subexerted, anther-base obtuse. Cypsella ribbed.

1. Capitula brick-red; leaves chartaceous, base attenuate to decurrent; lamina (lower) pinnatipartite. .... 1. **G. aurantiaca**
1. Capitula yellowish; leaves subsucculent, base cuneate, auricled or amplexicaul; lamina entire or sinuately lobed.
    2. Stout herbs; capitula 5-7-per corymb. .... 2. **G. nitida**
    2. Slender herbs; capitula 1-5-per corymb. .... 3. **G. pseudo-china**

1. **Gynura aurantiaca** (Bl.) DC., Prodr. 6:300.1838; Backer, Fl.Java 2:425.1965; Matthew, Ill.Fl.Tam. Carnatic t. 361.1982; Rani & Matthew in Matthew, Fl.Tam. Carnatic 3(2) : 798.1983. *Calcalia aurantiaca* Bl., Bijdr. 908.1826. *Crassocephalum crepidioides* auct., non (Benth.) S.Moore, 1912 : Ramamoorthy in Sald. & Nicols., Fl.Hassan Dist. 610.1976.

Subsucculent annuals. Lower leaves pinnatifid, 4-15 by 2.5 cm, base cuneate, decurrent, apex acute. Capitula 5-7 per corymb, brick-red, turbinate; bracts linear-oblong, c.1.5 cm long; corolla tubular. Cypsella prominently ribbed.

Fl. & Fr.: Most of the seasons. *Distr.*: An introduced weed (pantropical). Common along waysides. *NAK 81* (Moozhiar, c. 250 m.). This introduced weed is slowly dominating the indigenous ground-cover species in hills.

2. **Gynura nitida** DC. in Wight, Contrib. Bot. India 24. 1834; Hook.f., Fl. Brit.India 3:333.1881; Gamble, Fl. Pres. Madras 714.1921; Ramamoorthy in Sald. & Nicols., Fl.Hassan Dist. 618. 1976; Matthew, Ill. Fl.Tam. Carnatic t. 876.1982; Rani & Matthew in Matthew, Fl.Tam.Carnatic 3(2) : 800.1983; Manilal, Fl.Silent Valley 154.1988; Mohanan & Sivad., Fl. Agasthyamala 379.2002. **(Fig. 42)**

Stout herbs; branchlets angled, glabrous or puberulous. Leaves oblanceate, 6-8 by 1.5-2.5 cm, base subamplexicaul, margin dentate or faintly serrate, apex acute. Capitula 5-7-per corymb, c. 2 cm m; florets yellow. Cypsella narrowly oblong, prominently ribbed, glabrous or hispid.

Fl. & Fr.: Nov.-Jan. *Distr.*: Western Peninsular India. Along moist shady open hills. Locally abundant. *NAK 74, 13711* (Kakki hills, c. 1100 m).

3. **Gynura pseudo-china** (L.) DC., Prodr. 6 : 299. 1838; Hook. f., Fl.Brit.India 3 : 344. 1881; Gamble, Fl.Pres.Madras 714. 1921. *Senecio pseudo-china* L., Sp.Pl.867. 1753.

Slender tuberous rooted herbs. Leaves rosulate or subradical, oblong-lanceate, 5-14 by 2-3 cm, sinuate or subpinnatifid. Capitula 1-5-per corymb, c. 1.3 cm ϕ; florets yellowish. Cypsella linear, glabrous or puberulous.

Fl. & Fr.: Sep.-Mar. *Distr.*: Sri Lanka, India, Java, China. Rare. In upper hills. *KV 48374* (MH) Pamba, 1075 m.

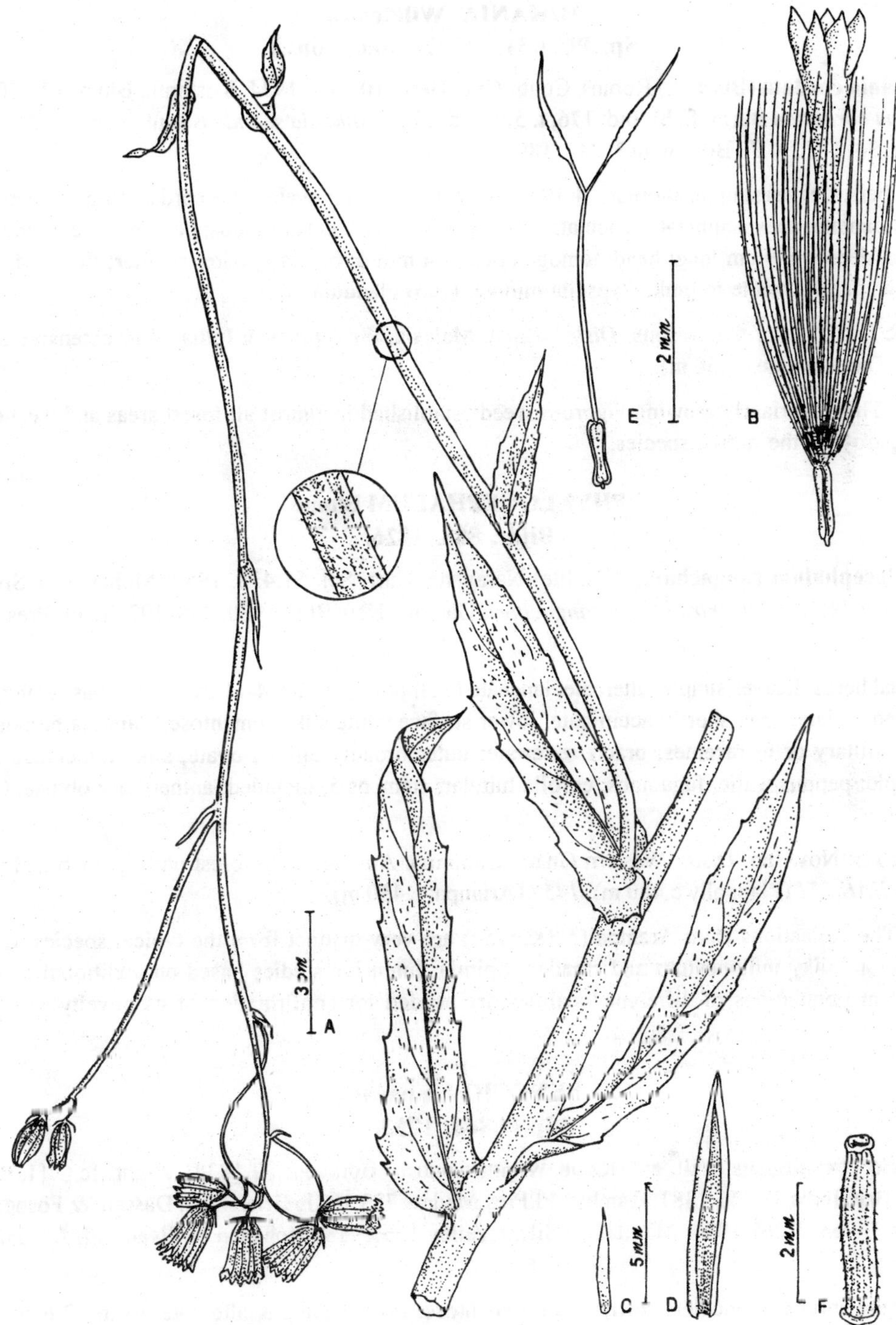

Fig. 42. ***Gynura nitida*** **DC.: A. Twig; B. Flower; C & D. Bracts; E. Pistil; F. Cypsella.**

MIKANIA Willdenow
Sp. Pl. 3(3) : 1742 (nom. cons.)

Mikania cordata (Burm. f.) Robins, Contr. Gray Herb. 104: 65. 1934; Kosterm., Blumea 1 : 504. 1935. *Eupatorium cordatum* Burm. f., Fl. Ind. 176, t. 58, f., d. 1768. *Mikania scandens* acut., non L., 1753: Clarke, Comp. Ind. 34. 1876 & Fl. Brit. India 3: 244. 1882.

Climbers. Leaves simple, alternate, 4-10 by 2-8 cm, triangular ovate, base cordate to subhastate, margin irregularly dentate, apex acuminate-apiculate, basally 5-7-nerved, membranous; petiole c. 7 cm long. Panicle corymbose, dense, c. 9 cm long; head homogamous, 3-4 mm ϕ; bracts 4, outer smaller; florets-4, white ro pale yellow; pappus white to pink. Cypsella minute, (sub) glandular.

Fl. & Fr.: Most of the seasons. *Distr.*: China, Malesia, Philippines & India. Very extensive climbers. *NAK 296* (Thekkuthode, c. 50 m).

Note :This luxuriantly growing vigorous weed established in almost all forest areas and has been very quickly destroying the native species.

PHYLLOCEPHALUM Blume
Bijdr. 888. 1826.

Phyllocephalum rangacharii (Gamble) Narayana, Curr. Sci. 51:438. 1982; Mohanan & Sivad., Fl. Agasthyamala 385.2002. *Centratherum rangacharii* Gamble, Kew Bull. 1920: 338. 1920 & Fl. Pres. Madras 667.1921.

Annual herbs. Leaves simple, alternate, lanceate to elliptic-lanceate, 4-11 by 1-3 cm, base acute, margin distantly denticulate, apex shortly acuminate, lower surface white silky-tomentose. Capitula homogamous, disciform, solitary or in racemes; bracts α-seriate; outer broadly elliptic-ovate, silky-tomentose outside; florets purple; pappus white, fugacious; corolla tubular; stamens 5, included; anther-base obtuse. Cypsella 8-10-ribbed.

Fl. & Fr.: Nov.-Jan. *Distr.*: Western Ghats. Common. Among moist forest undergrowth and exposed hill slopes. *NAK 371* (Plappally c.400 m), *1454* (Arampa, c.450 m).

Note : The collections from Arampa (*NAK 1454*) are very distinct from the typical species by having tall habit, more silky indumentum and smaller capitula. Intensive studies based on additional collections from different populations of different localities are needed for confirmation of its novelty as a distinct taxon.

SENECIO Linnaeus
Sp.Pl.866. 1753.

Senecio corymbosus Wall. ex DC. in Wight, Contrib. Bot. Ind. 22.1834; Wight, Ic.t. 1130. 1846; Hook.f., Fl.Brit. India 3 : 351.1881; Gamble, Fl.Pres.Madras 723.1921; Grierson in Dassan. & Fosberg, Rev. Handb. Fl. Ceylon. 1:261.1980; Manilal, Fl.Silent Valley 156.1988; Mohanan & Sivad., Fl. Agasthyamala 385.2002.

Climbing shrubs; branchlets softly white-tomentose. Leaves simple, alternate, ovate, 2-6 by 1.2-4.5 cm, base cordate, margin denticulate, apex acute or shortly acuminate, lower surface densely white-tomentose, basally 3-5-nerved. Capitula herterogamous, radiate, in axillary corymbose cymes; bracts c.8, uni-seriate, linear-oblong; ray florets female, or neuter; disc-florets bisexual, yellow; stamens 5, included; anther-base obtuse. Cypsella subterete or flat, 5-10-ribbed.

Fl. & Fr.: Jan.-Apr. *Distr.*: Southern Western Ghats & Sri Lanka. Rare. In interior evergreen forests. An extensive climber with beautiful leaves. *NAK 2740* (Kakki hills, c. 1100 m).

SPHAERANTHUS Linnaeus
Sp. Pl. 927. 1753.

Sphaerathus indicus L., Sp. Pl. 92. 1753; Hook.f., Fl. Brit. India 3 : 275. 1881, p.p.; Gamble, Fl. Pres. Madras 692. 1921; Ramamoorthy in Sald. & Nicols., Fl. Hassan Dsit. 623. 1976; Matthew, Ill. Fl. Tam. Carnatic t. 386. 1982; Rani & Matthew in Matthew, Fl. Tam. Carnatic 3(2) : 816. 1983; Nicols., Suresh & Manilal, An Interpr. Hort. Malab. 68, 1988. *Adaca-manjen* Rheede, Hort. Malab. 10:85-96, t. 43. 1690.

Ascending herbs; branchlets winged, wings toothed. Leaves simple, alternate, subsessile, narrowly spathulate, 2-5 by 1-2 cm, base decurrent, margin serrate-dentate, apex obtuse. Capitula heterogamous, disciform, sessile, aggregated on elongate receptacle forming an ovoid or globose glomerule; peduncle stout; florets purple, outer female, inner bisexual; pappus 0; corolla tubular; stamens 5, included; ovary angled or oblong, hirsute. Cypsella angled or smooth.

Fl. & Fr.: Jan.-Apr. Distr.: Sri Lanka, India, Nepal, Myanmar & Malesia to Australia. Occasional. On marshy areas in plains. *NAK 2084* (Enathu, c. 20 m).

The whole plant powdered and gvien with ghee as a remedy in hernia, enlarged spleen, poison, and stone in the urinary bladder.

SPILANTHES N.J. Jacquin
Enum. Pl. Carib. 8, 28, 1760.

Erect, ascending or trailing herbs. Leaves simple, opposite, basally 3-5- nerved, chartaceous. Capitula homogamous or heterogamous, solitary, axillary or terminal, discoid or radiate; florets yellowish or greenish-yellow, outer female, inner bisexual; bracts α-seriate; pappus bristly or 0; orolla tubular. Cypsella laterally compressed or trigonous.

1. Erect herbs; capitula greenish-white. .... **4. S. radicans**
1. Diffuse or trailing herbs; capitula yellowish.
    2. Capitula heterogamous, radiate. .... **2. S. ciliata**
    2. Capitula homogamous, discoid.
        3. Pappus 0; cypsella glabrous on margins. .... **1. S. calva**
        3. Pappus bristly; cypsella ciliate on margins. .... **3. S. paniculata**

1. **Spilanthes calva** DC. in Wight, Contrib. Bot. India 19.1834; Wight, Ic.t. 1109. 1846; Kosterm. & Philip., Blumea 6 : 354. 1950; Ramamoorthy in Sald. & Nicols., Fl. Hassan Dist. 624. 1976; Grierson in Dassan. & Fosberg, Rev. Handb. Fl. Ceylon 1:221. 1980; Rani & Matthew in Matthew, Fl. Tam. Carnatic 3(2) : 817. 1983; Sivar.& Ramesan, J. Econ. Tax. Bot. 10 (1): 143. 1987; Mohanan & Sivad., Fl. Agasthyamala 388.2002. *S. acmella* auct., non (L.) Murray, 1774: Hook. f., Fl. Brit. India 3 : 307. 1881; Gamble, Fl. Pres. Madras 708. 1921.

Trailing herbs. Leaves ovate, 3-6 by 2-3 cm, truncate, margin faintly crenate, apex acute to obtuse, basally 3-5- nerved. Capitula yellowish, terminal and from uper leaf-axils,ovoid, c.12 mm ϕ; peduncle c. 10 cm long; pappus 0. Cypsella glabrous, compressed, apex sinuate.

Fl. & Fr.: Sep.-Dec. *Distr.*: India, Myanmar, Sri Lanka, Java. Common in moist shady localities in plains. *NAK 159 & 1140* (Adoor, c. 20 m).

Flowers are very acrid and used in both tooth-ache and headache.

2. **Spilanthes ciliata** H.B.K., Nov. Gen. et Sp. Pl. 208. 1820; DC., Prodr. 5: 621. 1836; Moore, Proc. Amer. Acad. Arts. & Sci. 42 (20) : 538. 1907; Sivar. & Ramesan, J. Econ. Tax. Bot. 10 (1) : 145. 1987.

Diffuse herbs, rooting at lower nodes. Leaves ovate, 4-7 by 2.5 - 4 cm, base truncate, margin faintly serrate, apex acute. Capitula yellowish, heterogamous, radiate; outer ray florets female, few, ligulate. Cypsella compressed, beaked, margins densely ciliate.

Fl. & Fr.: Jul.-Oct. *Distr.*: Native of Neotropics, now widely naturalised in Peninsular India. Common in marshy areas in plains. *NAK* 707 (Puthukkulam, c. 20 m).

3. **Spilanthes paniculata** Wall. ex DC., Prodr. 5:624. 1836; Kosterm. & Philip., Blumea 6:351. 1950; Ramamoorthy in Sald. & Nicols., Fl. Hassan Dist. 624. 1976; Grierson in Dassan. & Fosberg, Rev. Handb. Fl. Ceylon 1:220. 1980; Sivar. & Ramesan, J.Econ. Tax. Bot. 10(1) : 143. 1987. *S. acmella* auct., non (L.,) Murray, 1774: Moore, Proc. Amer. Acad. Art. & Sci. 42:534. 1907, var. *paniculata* (Wall. ex DC.) Clarke, Comp. Ind. 139. 1876; Hook. f., Fl. Brit. India 3:307. 1881.

Diffuse herbs; rooting at lower nodes. Leaves ovate to lanceate, 3-7 by 2-3.5 cm, base cuneate, margin faintly serrate, apex acute. Capitula yellowish, discoid, solitary or 1-2, axillary appearing as terminal panicles; peduncle 6-10 cm long. Cypsella dimorphic, marginal ones trigonous, inner ones compressed, margin corky, densely ciliate, apex sinuate.

Fl. & Fr.: Sep.-Dec. *Distr.*: Peninsular India, Malesia. An immigrant species. Not common now. Along forest border. *NAK 929* (Plappally, c 400 m), *1495* (Arampa, c.400 m).

4. **Spilanthes radicans** Jacq., Collect. Bot. Chem. Hist. Nat.1(3) : 1714. 1804; Schrad. in DC., Prodr. 5:624. 1836; Sivar. & Ramesan, J. Econ. Tax. Bot. 10 (1) 144. 1987; Manilal, Fl. Silent valley 159. 1988.

Erect stout herbs. Leaves ovate, 3-8 by 1.5-4 cm, base obtuse, margin subentire, apex acute or obtuse, basally 3-nerved. Capitula greenish-yellow, homogamous, discoid, solitary or in pairs; peduncle c.3 cm long; pappus of unequal bristles. Cypsella dimorphic, marginal ones trigonous, inner compressed, margins corky, ciliate, apex sinuate.

Fl. & Fr.: Sep.-Dec. *Distr.:* Native of Central America; now in Peninsular India. Another immigrant species which has been raising serious threat to local species. Common along waysides in hills. *NAK 1228* (Upper Moozhiar, c. 500 m), *2829* (Kalleli, c. 50 m).

STRUCHIUM P. Browne
Civ. Nat. Hist. Jamaica 312. 1756.

Struchium sparganophorum (L.) Kuntze, Rev. Gen.Pl. 1: 366. 1891; Vasudevan Nair, Bull. Bot. Surv. India 8: 202, ff. 1-18. 1977. *Ethulia sparganophora* L., Sp. Pl. ed. 2, 2: 1171.1763.

Erect or ascending herbs; rooting at lower nodes. Leaves simple, alternate, elliptic-lanceate to oblanceate, 7-14 by 3-5 cm, base cuneate, margin crenate-serrate, apex acuminate, chartaceous. Capitula homogamous, discoid, clustered, globose, axillary. Florets white, all bisexual; bracts α-seriate; pappus of 3 basally connate, spongy structures; corolla tubular, stamens 3, included; ovary 2-3- ribbed. Cypsella trigonous, glandular.

Fl. & Fr.: Dec.-Mar. *Distr.*: Native of Trop. America; now in all warmer countries. Common in marshy lowlands. *NAK 1649* (Tiruvalla, c. 10 m).

SYNEDRELLA J. Gaertner
Fruct. 2: 456. 1791 (nom. cons.).

Synedrella nodiflora (L.) Gaert., Fruct. 2: 456, t. 171. 1791; Hook.f., Fl. Brit. India 3 : 308. 1881; Gamble, Fl. Pres. Madras 708. 1921; Ramamoorthy in Sald. & Nicols., Fl. Hassan Dist. 624.1976; Grierson in Dassan. & Fosberg, Rev. Handb. Fl. Ceylon 1: 222. 1980; Matthew, Ill. Fl. Tam. Carnatic t. 388. 1982; Rani & Matthew in Matthew, Fl. Tam. Carnatic 3(2): 818. 1983; Manilal, Fl. Silent Valley 159. 1988; Mohanan & Sivad., Fl. Agasthyamala 389.2002. *Verbesina nodiflora* L., Cent. Pl. 1: 28. 1755.

Erect branched herbs. Leaves simple, opposite, ovate-lanceate, 4-8 by 2-4 cm, base cuneate, decurrent, margin serrate, apex acute, pubescent, basally 3-nerved, thin-coriaceous. Capitula heterogamous, radiate, axillary or at the forks of nodes; bracts 2-seriate; outer foliaceous, inner serecious; florets yellow, outer female, ligulate, inner bisexual; pappus of 2 awns. Cypsella dimorphic, of outer ones compressed with corky margins, inner ones trigonous.

Fl. & Fr.: Most of the seasons. *Distr.*: Native of West Indies; introduced in India, China, Malaya and Polynesia. Common. Along waysides. *NAK 869* (Plappally, c. 400 m), *1028* (Tiruvalla, c. 10 m).

TRIDAX Linnaeus
Sp. Pl. 900. 1753.

Tridax procumbens L., Sp. Pl. 900. 1753; Hook. f., Fl. Brit. India 3: 311. 1881; Gamble, Fl. Pres. Madras 711. 1921; Ramamoorthy in Sald. & Nicols., Fl. Hassan Dist. 626. 1976; Grierson in Dassan. & Fosberg, Rev. Handb. Fl. Ceylon 1: 232. 1980; Matthew, Ill. Fl. Tam. Carnatic t. 391. 1982; Rani & Matthew in Matthew, Fl. Tam. Carnatic 3(2): 821. 1983; Manilal, Fl. Silent Valley 160. 1988.

Procumbent herbs. Leaves ovate, 2-4 by 1-3.5 cm, base and apex acute, margin coarsely serrate, strigose, herbaceous. Capitulum heterogamous, radiate, solitary on long peduncle; bracts α-seriate; florets yellowish white, outer ones female, few, inner bisexual, α; pappus of slender setose hairs; stamens 5, subexerted. Cypsella turbinate, faintly ribbed.

Fl. & Fr.: Most of the seasons. *Distr.*: Native of Central America. Now in Tropics and Subtropics. Very common. A gregarious wide-spread weed. *NAK 2928* (Adoor , c. 20 m).

VERNONIA Schreber
Gen. 2: 541. 1791. (nom. cons.).

Leaves simple, alternate, margin usually dentate or serrate. Capitula homogamous, in terminal and/or axillary corymbose panicles; florets 1-5 or α, bisexual; bracts α-seriate; pappus often 2-seriate, outer short, inner long barbellate; corolla pink or white; anthers linear. Cypsella angled or ribbed.

1. Capitula 1-5- flowered.
    2. Trees; leaves obovate to oblanceate; head 1-flowered. ....1. **V. arborea**
    2. Shrubs; leaves elliptic-lanceate; head c. 5-flowered. ....3. **V. divergens**
1. Capitula 6-15-flowered.
    3. Leaves broadly ovate to suborbicular, lower surface white- tomentose. ....4. **V. indica**
    3. Leaves lanceate, lower surface sparsely pubescent. ....2. **V. cinerea**

1. **Vernonia arborea** Buch.-Ham., Trans. Linn. Soc. London 14: 218. 1824; Hook. f., Fl. Brit. India 3: 239. 1881; Ramamoorthy in Sald. & Nicols., Fl. Hassan Dist. 626. 1976 Grierson in Dassan. & Fosberg, Rev. Handb. Fl. Ceylon 1 : 122. 1980; Rani & Matthew in Matthew, Fl. Tam. Carnatic 3(2) : 826. 1983; Manilal, Fl. Silent Valley 161. 1988; Mohanan & Sivad., Fl. Agasthyamala 391.2002. *Monosis wightiana* DC. in Wight, Contrib. Ind. Bot 1. 1834, Ic. t. 1085. 1845; Bedd., Fl. Sylv. t. 226. 1872. *Vernonia monosis* Benth. ex Clarke, Comp. Ind. 24. 1876, non Schultz.–Bip., 1847; Gamble, Fl. Pres. Madras 672. 1921.

Trees; branchlets, leaves-beneath, peduncle brownish tomentose. Leaves obovate to broadly oblanceate, 8-23 by 4-11 cm, base cuneate, margin coarsely serrate to subentire, apex acuminate, lateral veins prominent below, coriaceous. Panicle profusely branched, c. 25 m long; head 1-flowered, c. 1 cm long; bracts lanceate, receptacle pitted; pappus 2-seriate; corolla pink to purple. Cypsella 10-ribbed, glandular.

Fl. & Fr.: Jan.-Mar. *Distr.*: Sri Lanka, India & Malesia. A common evergreen tree growing at an altitude between 400-100 m. *NAK 349, 1172, 1521 & 1559* (Moozhiar, Arampa, c. 400 m).

2. **Vernonia cinerea** (L.) Less., Linnaea 4: 291. 1829; Clarke, Compos. Ind. 20. 1876; Hook.f., Fl. Brit. India 3: 233. 1881; Gamble, Fl. Pres. Madras 676. 1921; Ramamoorthy in Sald. & Nicols., Fl. Hassan Dist. 627. 1976; Grierson in Dassan. & Fosberg, Rev. Handb. Fl. Ceylon 1: 133. 1980; Rani & Matthew in Matthew, Fl. Tam. Carnatic 3(2) : 827. 1983; Nicols., Suresh & Manilal, An Interpr. Hort. Malab. 69. 1988; Mohanan & Sivad., Fl. Agasthyamala 392.2002. *Conyza cinerea* L., Sp. Pl. 862. 1753. *Puam-curundala* Rheede, Hort. Malab. 10: 127, t. 64. 1690.

Herbs. Leaves lanceate, 3-6 by 1-2.5 cm, base cuneate-attenuate, margin crenate-serrate to toothed, apex subacute-mucronulate, pubescent, chartaceous. Panicle corymbose, lax, terminal bracts linear-lanceate. Capitula c. 15-flowered; papus 2-seriate, soft, white; corolla pink. Cypsella 4-5-angular, villous.

Fl. & Fr.: Most of the seasons.*Distr.:* Tropical Asia, Africa, Australia & New Zealand. A common roadside weed. *NAK 109, 702 & 803* (Addor, Tiruvalla, c. 10 m).

Juice of leaves boiled with coconut oil and applied to cure all fevers and elephantiasis. Seeds used as anthelmintic.

3. **Vernonia divergens** (Roxb.) Edgew., J. Asiat. Soc. Bengal 2: 172. 1853; Hook.f., Fl. Brit. India 3: 234. 1881; Gamble, Fl. Pres. Madras 673. 1921; Ramamoorthy in Sald. & Nicols., Fl. Hassan Dist. 627. 1976; Rani & Matthew, Fl. Tam. Carnatic 3(2): 828. 1938; Manilal, Fl. Silent Valley 161. 1988; Mohanan & Sivad., Fl. Agasthyamala 394.2002. *Eupatorium divergens* Roxb., Fl. Ind. 3: 414. 1832. *Vernonia nilgheryensis* DC., Prodr. 5: 32. 1836; Wight, Ic. t. 1076. 1846.

Shrubs; branchlets striate, pubescent. Leaves elliptic-lanceate, 4-6 by 1-3 cm, gradually smaller towards top, base and apex acute, margin serrate, lower surface greenish, chartaceous. Panicle corymbiform, terminal and axillary; bracts α-seriate, elliptic, glabrescent. Capitula c. 5-flowered; pappus 2-seriate, stiff, cream coloured; corolla pink. Cypsella 10-ribbed.

Fl. & Fr.: Dec.-Mar. *Distr.*: India & Malaya Peninsula. Not common. In interior evergreen forest. *NAK 81* (Moozhiar, c. 300 m).

4. **Vernonia indica** Clarke, Comp. Ind.16. 1876; Gamble, Fl. Pres. Madras 674. 1921; Ramamoorthy in Sald. & Nicols., Fl. Hassan Dist. 628. 1976. *Decaneurum dendigulense* DC. in Wight, Contrib. Bot. Ind. 7. 1834.

Undershrubs; branches ribbed, softly white wooly. Leaves broadly ovate, terminal ones ovate-lanceate, 3-10 by 2-5 cm, base rounded to cuneate, margin serrulate, apex acute, upper surface glabrous, lower surface white-wooly, lateral veins c. 8 pairs; coriaceous. Panicle corymbiform, terminal, c. 15 cm long, tomentose; bracts elliptic oblong. Cypsella, 6-10-ribbed, glandular.

Fl. & Fr.: Nov.-Feb. *Distr.*: Western Peninsular India. In open grassy slopes of upper hills. *NAK 1312* (Kakki hills, c. 1100 m).

VICOA Cassini
Ann. Sci. Nat. (Paris) 17: 418. 1829.

Vicoa indica (L.) DC. in Wight, Contrib. Bot. India 10. 1834; Wight, Ic. t. 1148. 1846; Gamble, Fl. Pres. Madras 701. 1921; Ramamoorthy in Sald. & Nicols., Fl. Hassan Dist. 628. 1976; Grierson in Dassan. & Fosberg, Rev. Handb. Fl. Ceylon 1: 200. 1980; Rani & Matthew in Matthew, Fl. Tam. Carnatic 3(2): 830. 1983; Manilal, Fl. Silent Valley 162. 1988; Mohanan & Sivad., Fl. Agasthyamala 397.2002. *Inula indica* L., Sp. Pl. (ed. 2). 1237. 1763. *Vicoa auriculata* Cass., Ann. Sci. Nat. (Paris) 17: 418. 1829; Hook.f., Fl. Brit. India 3: 297. 1881.

Erect herbs. Leaves oblong-lanceate, 3-9 by 0.5-1.5 cm, sessile, base auricled, margin rarely serrate, apex gradually acute, pubescent, chartaceous. Capitula heterogamous to homogamous, rayed or non-rayed, solitary on long peduncle; bracts α-seraite, pappus 0; corolla yellow. Cypsella terete, pubescent.

Fl. & Fr.: Nov.-Mar. *Distr.*: India, China, Thailand, W. tropical Africa. Common on exposed hill cuttings. *NAK 125, 1059* (Adoor, Tiruvalla, c. 10 m).

WEDELIA N.J. Jacquin
Enum. Pl. Carib. 8, 28. 1760 (nom. cons.).

Wedelia urticifolia DC. in Wight, Contrib. Bot. India 18. 1834; Wight, Ic., t. 1106. 1846; Hook.f., Fl. Brit. India 3: 306. 1881; Gamble, Fl. Pres. Madras 707. 1921; Rani & Matthew in Matthew, Fl. Tam. Carnatic 3(2): 832. 1983, var. **wightii.**

Erect undershrubs; vegetative parts hispid. Leaves ovate-lanceate, 4-8 by 2-3.5 cm, base subtruncate to cuneate, decurrent, margin serrate, apex acuminate, basally 3-nerved. Capitula heterogamous, 1-2, axillary or terminal, bracts 2-seriate; florets yellow, outer female, ligulate, inner tubular. Cypsella cuneate, trigonous, pubescent.

Fl. & Fr.: Nov.-Apr. *Distr.*: Peninsular India. Not common. Among wayside thickets in upper ghats. *NAK 2753* (Kakki hills, c. 1100 m).

XANTHIUM Linnaeus
Sp. Pl. 987. 1753.

Xanthium indicum J. Koenig in Roxb., Fl. Ind. 3:601. 1832; Wight, Ic. t. 1104. 1846; Grierson in Dassan. & Fosberg, Rev. Handb. Fl. Ceylon 1: 209. 1980; Rani & Matthew in Matthew, Fl. Tam. Carnatic 3(2): 833. 1983. *X. strumarium*L., Sp. Pl. 987. 1753, p.p.; Hook.f., Fl. Brit. India 3: 303. 1881, p.p.; Gamble, Fl. Pres. Madras 703. 1921; Ramamoorthy in Sald. & Nicols., Fl. Hassan Dist. 629. 1976.

Erect undershrubs. Leaves broadly ovate, 3-5-lobed, 10-15 by 8-12 cm, base subtruncate, margin coarsely crenate-dentate, scabrid above, 3-nerved basally, chartaceous. Capitula in axillary and terminal clusters, nonoecious. Male 1-flowered, clustered; bracts ovate, stamens 5, exerted. Female head solitary, 2-flowered; corolla and pappus 0. Cypsella 2, enclosed by a prickly glandular-hispid envelope.

Fl. & Fr.: Jan.-Sep. *Distr.*: India, Malaya, Sumatra & Java. Common. In waste places and swamps. *NAK 967* (Sabarimala, c. 200 m).

77. LOBELIACEAE

LOBELIA Linnaeus
Sp. Pl. 2: 929. 1753.

Erect or prostrate herbs or shrubs. Leaves simple, alternate, estipulate. Flower(s) solitary or in racemes, axillary or terminal, bisexual, zygomorphic, epigynous; bracts foliaceous; calyx-lobes 5, connate at base, adnate with ovary; corolla bilabiate, lobes 5, unequal; stamens 5, epipetalous, united to form a staminal-tube, anthers often hairy; ovary 2-locular, inferior; ovules α. Capsule loculicidal. Seeds α, minute.

1. Erect shrubs; flowers in dense racemes; anthers densely hairy on back. .... 3. **L. nicotianifolia**
1. Erect or prostrate herbs; flower(s) solitary; anthers sparsely hairy at tip.
    2. Erect herbs; stem 3-winged; flowers pale blue. .... 2. **L. heyneanus**
    2. Prostrate herbs; stem 3-angled; flowers deep blue. .... 1. **L. alsinoid**

1. **Lobelia alsinoides** Lam., Encycl. 3: 588. 1792; DC., Prodr. 7:378 1839; Ramamoorthy in Sald. & Nicols., Fl. Hassan Dist. 568. 1976; Matthew, Ill. Fl. Tam. Carnatic t. 399. 1982; Matthew & Rani in Matthew, Fl. Tam. Carnatic 3(2): 837. 1983; Mohanan & Sivad., Fl. Agasthyamala 399.2002. *L. trigona* Roxb., Fl. Ind. 2:111. 1824; Wight, Ic. 1178. 1848; Clarke in Hook.f., Fl. Brit. India 3:423. 1881; Gamble, Fl. Pres. Madras 736.1921.

Prostrate herbs; stem 3-angled. Leaves broadly ovate, 1-2 by 1-2 cm, base rounded or truncate, margin serrate, apex acute, sparsely puberulous, chartaceous. Flower(s) solitary, axillary; calyx-lobes lanceate-subulate; corolla deep blue; anthers connivent around stigma, all penicellate at tip. Capsule 5-ribbed. Seeds trigonous.

Fl. & Fr.: Aug.-Dec. *Distr.*: India, Malaya, Sumatra, Java. Not common. In moist areas of plains and hills. *NAK 1373* (Adoor, c.20 m).

2. **Lobelia heyneana** Roth ex Roem. & Schult., Syst. Veg. 5:50. 1819; Santapau, Rec. Bot. Surv. India 16:158.1953; Matthew & Rani in Matthew, Fl. Tam. Carnatic 3(2):838.1983; Manilal, Fl. Silent Valley 162.1988;

Mohanan & Sivad., Fl. Agasthyamala 400.2002. *L. trialata* Buch.–Ham. ex D. Don, Prodr. Fl. Nepal. 157.1825; DC., Prodr. 7:360.1839; Clarke in Hook.f., Fl. Brit. India 3:425.1881; Gamble, Fl. Pres. Madras 736.1921.

Small erect herbs; stem 3-winged. Leaves ovate, to 2 by 1.8 cm, base truncate, decurrent, margin dentate-serrate, apex acute, chartaceous. Flower(s) solitary, axillary; calyx-lobes triangular-subulate; corolla pale blue; only 2 (anterior) anthers, penicellate at tip. Capsule ellipsoid. Seeds compressed.

Fl. & Fr.: Aug.-Dec. *Distr.*: Africa, Sri Lanka, Peninsular India, China, Malesia. Rare. In moist calayey-soils. *NAK 2522* (Kakki hills, c.1100 m).

3. **Lobelia nicot ianifolia** Roth ex Roem. & Schult., Syst. Veg. 5:47.1819 ('*nicotianaefolia*'); Roxb., Fl. Ind. 110.1824; DC., Prodr. 7:381.1839; Wight, Ic. 2:111, t.135.1850; Clarke in Hook. f., Fl. Brit. India 3:428 1881; Gamble, Fl. Pres. Madras 736. 1921; Ramamoorthy in Sald. Nicols., Fl.Hassan Dist. 568.1976; Manilal, Fl.Silent Valley 163. 1988; Mohanan & Sivad., Fl. Agasthyamala 400.2002. *Kattu-pokayila.*

Erect shrubs; stem hollow. Leaves lanceate, 5-30 by 2-6 cm, base acute, margin glandular serrulate, apex acute to acuminate, herbaceous. Flowers in dense racemes; bracts foliaceous; calyx-lobes linear; corolla bluish-white; stamens basally free, anthers densely pilose on back, rarely glabrous. Capsules subglobose. Seeds ellipsoid or ovoid.

Fl. & Fr.: Nov.-Jul. *Distr.*: India, Sri Lanka. Common in evergreen forest borders in upper hills. *NAK 351, 1164* (Upper Moozhiar, c. 500 m).

Infusion of leaves used against asthma, stomach ailments and as an antispasmodic. Roots used as an antidote in scorpion bite.

78. VACCINIACEAE

VACCINIUM Linnaeus
Sp. Pl. 349. 1753.

Vaccinium neilgherrense Wight, Calcutta J. Nat. Hist. 8:173. 1847, Ic.t. 1889.1848 & Spicil. Neilgh.t.129.1851; Clarke in Hook.f., Fl.Brit.India 3:454.1882; Gamble, Fl.Pres Madras 741.1921; Ramamoorthy in Sald. & Nicols., Fl. Hassan Dist. 191.1976; Matthew, Ill. Fl. Tam. Carnatic t.402.1982; Matthew & Rani in Matthew, Fl. Tam. Carnatic 3(2):843.1983; Manilal, Fl. Silent Valley 164.1988.

Small trees. Leaves simple, alternate, estipulate, obovate-lanceate or elliptic-lanceate, 5-10 by 1-2.5 cm, base acute, margin serrulate along upper half, apex shortly acuminate, thin-coriaceous. Flower(s) in racemes or solitary, bisexual actinomorphic, 5-merous; calyx adnate to ovary, lobes 5, triangular-ovate; corolla white, urceolate, 5 toothed; stamens 10, free, in a ring along the disc, anthers produced upwards into 2 tubes; ovary inferior, 5-locular; ovules α, on axile placenta. Berry globose, crowned by persistent calyx-teeth. Seeds α, oblong.

Fl. & Fr.: Feb.-Apr. *Distr.*: Peninsular India. Rare. In shola forests. *NAK 2750* (Way to Ponnambalamedu, c.1100 m).

79. MYRSINACEAE

1. Trees or shrubs.
   2. Flowers in umbels, rose coloured; leaf-margin entire...**Ardisia**

2. Flowers in panicles, white coloured; leaf-marign serrate. .... **Maesa**

1. Climbing shrubs. .... **Embelia**

ARDISIA O. Swartz
Prodr. 3:48.1788 (nom.cons.).

Ardisia pauciflora Heyne in Roxb., Fl. Ind. 2:279.1824; Wight, Ic.t.1214.1848; Clarke in Hook.f., Fl. Brit. India 3:529.1882; Gamble, Fl.Pres Madras 755.1921; Manilal, Fl. Silent Valley 166.1988; Mohanan & Sivad., Fl. Agasthyamala 404.2002.

Shrubs or small trees. Leaves long-lanceate, 6-10 by 1.5-2.5 cm, base and apex acuminate, margin entire or faintly serrate towards apex, thin-coriaceous. Flowers bisexual, 5-merous, actinomorphic, in extra-axillary simple umbels; stamens 5, free, epipetalous; ovary superior, 1-locular; ovules few on free-central placenta. Berry globose. Seed 1, pitted.

Fl. & Fr.: Apr.-Jun. *Distr.*: Sri Lanka, Peninsular India. Rare. In everygreen forests. *NAK 2746* (Kakki hills, c.1100 m).

EMBELIA N.L. Burman
Fl. Ind. 62.1768 (nom.cons.).

Embelia ribes Burm. f., Fl. Ind. 62, t.23.1768; Clarke in Hook.f., Fl.Brit. India 3:513.1882; Gamble, Fl. Pres. Madras 752.1921; Ramamoorthy in Sald. & Nicols., Fl. Hassan Dist. 202.1976; Matthew & Rani in Matthew, Fl. Tam. Carnatic 3(2) : 852.1983; Mohanan & Sivad., Fl. Agasthyamala 408.2002. *E. glandulifera* Wight, Ic.t.1207.1848. *Vizhalari.*

Climbing shurbs; stem tuberculate. Leaves simple, alternate below, opposite above, elliptic-lanceate to oblong-ianceate, 3-7 by 1.5-2.5 cm, base rounded, apex acuminate, lower surface gland-dotted on either side of the midrib, thin-coriaceous. Flowers 4-6-merous, aggregated in axillary or terminal, dense panicles; calyx-lobes ovate, pellucid-dotted; corolla greenish, lobes shortly connate; stamens 5, subexerted; ovary superior, ribbed, 1-locular; ovules few. Drupe globose, white. Seeds globose.

Fl. & Fr.: Jan.-Mar. *Distr.*: Sri Lanka, India, Malaya, S. China. Not common. Extensive stragglers on evergreen trees. *NAK 1286* (Kakki hills, c. 1100 m).

Seeds powdered and used internally in piles and gonorrhoea. Oil boiled with powdered seeds and applied to cure worms in the jaw, head and nostrils.

MAESA Forsskal
Fl. Aegypt.-Arab. 66.1775.

Maesa indica (Roxb.) DC., Trans. Linn. Soc. London 17:134. 1834; Ramamoorthy in Sald. & Nicols., Fl. Hassan Dist. 203. 1976; Matthew, Ill. Fl. Tam. Carnatic t. 409. 1982; Matthew & Rani in Matthew, Fl. Tam. Carnatic 3(2) : 852. 1983; Manilal, Fl. Silent Valley 166. 1988; Mohanan & Sivad., Fl. Agasthyamala 409.2002. *Baeobotrys indica* Roxb., Fl. Ind. 2:230. 1824. *Maesa dubia* (Wall.) DC., Trans. Linn. Soc. London 17:134. 1834; Gamble, Fl. Pres. Madras 749. 1921. *M. perottetiana* A.DC. in DC., Prodr. 8:80. 1844; Gamble, Fl. Pres. Madras 749. 1921. *M. indica,* var. *perottetiana* (A.DC.) Clarke in Hook.f., Fl. Brit. India 3:509. 1882.

Shrubs. Leaves simple, alternate, ovate to elliptic-lanceate, 10-14 by 4-6 cm, base rounded or acute, margin dentate-serrate, apex acute to shortly acuminate. Inflorescene a panicle, axillary or terminal. Flowers small, 5-merous; calyx-lobes 5, ovate; corolla white, lobes 5; stamens 5, subexerted; ovary half-inferior, globular, 1-locular, ovules α, on free-central placenta. Berry globose, white. Seeds angular.

Fl. & Fr.: Most of the seasons. *Distr.*: Sri Lanka, India, Malesia, W. Pakistan. Very common. Along forest clearing. *NAK 1168, 1506* (Chittar, c. 100 m).

80. SAPOTACEAE

1. Flowers yellow; corolla lobes 4; stamens 8. .... **Isonandra**
1. Flowers white or cream coloured; corolla lobes 6-16 (24); stamens (6) 8-24.
    2. Staminodes 0; leavs usually clustered towards branch-apices.
        3. Calyx-lobes 4(5); corolla lobes 6-12. .... **Madhuca**
        3. Calyx-lobes 6; corolla lobes 6. .... **Palaquium**
    2. Staminodes present; leaves alternate throughout; corolla lobes 18 or 24. .... **Mimusops**

ISONANDRA Wight
Icon. 2(1): 4.1840.

Isonandra lanceolata Wight, Ic. t. 359. 1840, forma **lanceolata** Clarke in Hook.f., Fl. Brit. India 3:539. 1882; Gamble, Fl. Pres. Madras 761. 1921; Jeuken, Blumea 6:568. 1952; Manilal, Fl. Silent Valley 167. 1988; Mohanan & Sivad., Fl. Agasthyamala 411.2002.

Shrubs to small trees. Leaves simple, alternate, lanceate or oblanceate, 6-15 by 1.5-4 cm, base cuneate, margin subentire, apex caudate, lateral nerves prominent, coriaceous. Flowers bisexual, in clusters, in leaf-axils; calyx-lobes 4, biseriate; corolla tubular-campanulate, lobes 4, yellowish; stamens 8, attached at base of corolla-tube; ovary superior, 4-locular. Berry ellipsoid. Seeds flattened.

Fl. & Fr.: Dec.-Mar. *Distr.*: Sri Lanka, S. -W. Peninsular India. Not common. In evergreen forests. *NAK 149 & 1792* (Moozhiar & Angamoozhy, c. 250 m).

MADHUCA Hamilton ex J.F. Gmelin
Syst. Nat. 2:773, 799. 1791.

Evergreen or deciduous trees. Leaves simple, alternate, usually crowded towards branch-apices, coriaceous. Flowers in subterminal fascicles; pedicels dilated towards apex; calyx-lobes 4, biseriate; corolla-lobes 6-12; stamens twice as many as corolla-lobes; staminodes 0; ovary ellipsoid or globose, 6-8- locular. Berry fleshy, with milky latex. Seeds ellipsoid.

1. Leaf apex acute; calyx tomentose; anthers 3-toothed at apex. ....1. **M. longifolia** var. **longifolia**
1. Leaf apex obtuse; calyx glabrous; anthers aristate at apex. .... 2. **M. neriifolia**

1. **Madhuca longifolia** (Koen.) Macbr., Contr. Gray Herb. (n.s.) 53:17. 1918,; Royen, Blumea 10:53. 1960; Ramamoorthy in Sald. & Nicols., Fl. Hassan Dist. 193. 1976; Matthew, Ill. Fl. Tam. Carnatic t. 412. 1982; Matthew & Rani in Matthew, Fl. Tam. Carnatic 3(2):858. 1983. *Bassia longifolia* J. Koenig in L., Mant. Pl. 563. 1771; Roxb., Fl. Ind. 2:523. 1832; Wight, Ill. Ind. Bot. t. 147. 1850; Clarke in Hook.f., Fl. Brit. India 3:544. 1882; Gamble, Fl. Pres. Madras 763. 1921, *Aattu-ilippa.*

Leaves elliptic-lanceate or oblanceate, 10-20 by 2.5-6 cm, base acute or obtuse, apex acute; calyx-lobes triangular, rusty-pubescent without; corolla white or cream coloured; stamens, c. 8, in 2 series, anthers 3-toothed at apex; ovary 8-locular. Berry ellipsoid.

Fl. & Fr.: Jan.-Apr. *Distr.*: Sri Lanka, India. Very common. Along streamsides. *NAK* 550, *1562, 1587* & *2308* (Konni & Ranni, R.F.).

Ripe fruit eaten to cure rheumatism, impure blood, tuberculosis, asthma and worms. Seed yield an oil which is useful in skin diseases.

2. **Madhuca neriifolia** (Moon) H.J. Lam, Bull. Bot. Buitenz. Ser. 3, 7:182, 265. 1925; Royen, Blumea 10:98. 1960; Ramamoorthy in Sald. & Nicols., Fl. Hassan Dist. 194. 1976; Mohanan & Sivad., Fl. Agasthyamala 413.2002. *Bassia neriifolia* Moon, Cat. Pl. Ceylon 36. 1824. *B. malabarica* Bedd., For. Man. 140. 1870; Gamble, Fl. Pres. Madras 764. 1921. *Ilippa.*

Leaves elliptic or obovate or linear-oblong, 5-12 by 2-6 cm, base obtuse or acute, pedicels glabrous; calyx-lobes ovate, glabrescent; corolla white, lobes rounded; stamens 16, anthers aristate at apex. Berry ellipsoid.

Fl. & Fr.: Feb.-Apr. *Distr.*: Sri Lanka, India. Not common. Along streamsides. *NAK 447* (Pampavalley, c. 50 m). Known from only one collection.

Decoction of heartwood given to cure ulcers and rheumatism. Flowers soaked in water to extract a sweet liquid which is good for kidney complaints.

MIMUSOPS Linnaeus
Sp. Pl. 349. 1753.

Mimusops elengi L., Sp. Pl. 349. 1753; Wight, Ic. t. 1586. 1850; Clarke in Hook. f., Fl. Brit. India 3:548. 1882; Gamble, Fl. Pres. Madras 765. 1921; Royen, Blumea 6:594. 1952; Ramamoorthy in Sald. & Nicols., Fl. Hassan Dsit. 194. 1976; Matthew, Ill. Fl. Tam. Carnatic t. 414. 1982; Matthew & Rani in Matthew, Fl. Tam. Carnatic 3(2) : 860. 1983; Nicols., Suresh & Manilal, An Interpr. Hort. Malab. 238. 1988; Mohanan & Sivad., Fl. Agasthyamala 414.2002. *Elengi* Rheede, Hort. Malab. 1: 33-34, t. 20. 1678.

Medium to large sized trees. Leaves simple, alternate, elliptic or ovate-lanceate, 7-12 by 3.5-5.5 cm, base acute or attenuate, apex shortly acuminate, lateral nerves α. Flower(s) solitary or very few in axillary clusters: calyx-lobes 8, in 2 series; corolla white, lobes 24, in 3 series of 8 each; stamens 8, alternating with 8 ovate staminodes; ovary 6-8-locular. Berry ovoid, fleshy.

Fl. & Fr.: Apr.-Sep. *Distr.*: Indo-Malesia. Not common in wild. In sacred groves. *NAK 2313* (Kodumon, c. 50 m).

Fruits edible. Pulp of ripe fruit relieves head-ache. Seeds pounded and given internally in human urine in as an antidote in all kinds of animal poison.

PALAQUIUM Blanco
Fl. Filip. 403. 1837.

Palaquium ellipticum (Dalz.) Baill., Traite Bot. Med. Pham. 1500. 1884; Gamble, Fl. & Pres. Madras 764. 1921; Manilal, Fl. Silent Valley 168. 1988; Mohanan & Sivad., Fl. Agasthyamala 414.2002. *Bassia elliptica*

Dalz. in Hook., J. Bot. Kew Gard. Misc. 3:36. 1851; Bedd., Fl. Sylv. t. 43. 1870. *Dichopsis elliptica* Benth. & Hook., Gen. Pl. 2:658. 1865; Clarke in Hook.f., Fl. Brit. India 3:542. 1882.

Evergreen trees with milky latex. Leaves simple, closely alternate, elliptic or elliptic-obovate, 6-9 by 3-5 cm, base cuneate, apex shortly acuminate, coriaceous. Flowers in clusters, lateral or axillary; calyx-lobes 6 in 2 series; corolla white, lobes 6; stamens 12 to 18, uniseriate, connective produced above anthers; staminodes 0; ovary ovoid, 6-locular. Berry fleshy, ellipsoid.

Fl. & Fr.: Feb.-May. *Distr.*: Western and S. W. Peninsular India. Not common. In evergreen forests. *KV 66202* (MH) (Pampa dam area, c. 1000 m).

Manilkara zapota (L.) Royen is seen planted for its fruits.

81. EBENACEAE

DIOSPYROS Linnaeus
Sp. Pl. 1057. 1753.

Evergreen trees. Leaves simple, alternate or subopposite, estipulate, thin-coriaceous. Inflorescence a condensed cyme, usually on old wood or axillary, rarely reduced to a single flower. Flowers regular, hypogynous, dioecious; calyx cupular, lobes 3-5, connate to middle, persistent or accrescent; corolla lobes 3-5, twisted or imbricate; stamens 5-α, distinct, anthers often sagittate; ovary superior, bicarpellary, 2-8-locular, ovules 1-2 per locule; staminodes and rarely pistillodes present in female and male flowers respectively. Berry globose or ovoid. Seed(s) 1-α.

1. Calyx and corolla lobes 5 each; stamens 5-20.
    2. Stamens 5; berry ovoid, ridged. .... 3. **D. hirsuta**
    2. Stamens 10-20; berry broadly ovoid or globose, not ridged.
        3. Stamens 10-16, in pairs.
            4. Leaves elliptic-oblong; stamens 10, filaments glabrous. .... 2. **D. candolleana**
            4. Leaves lanceate or oblong-lanceate; stamens 16, filaments villous. .... 5. **D. nilgirica**
        3. Stamens 20, not in pairs.
            5. Leaves finely reticulate; calyx lobes foliaceous; berry small ovoid. .... 6. **D. paniculata**
            5. Leaves faintly reticulate; calyx lobes not foliaceous; berry large, globose. .... 1. **D. bourdillonii**
1. Calyx and corolla lobes 4 each; stamens 30-40. .... 4. **D. malabarica**

1. **Diospyros bourdillonii** Brandis, Indian Trees 435. 1906; Gamble, Fl. Pres. Madras 777. 1923. **(Fig. 43)**

Leaves oblong or elliptic-oblong, 8-15 by 3-6 cm, base cuneate, apex acuminate, thin-coriaceous. Male flowers in short subumbellate clusters; calyx rusty-tomentose, lobes 5, triangular-ovate; corolla white, lobes 5, orbicular; stamens c. 20, anthers lanceoid. Berry globose, large, c. 12 cm ϕ with a cup like calyx.

Fl. & Fr.: Feb.-Dec. *Distr.*: Southern Western Ghats. (Travancore hills). Rare. In evergreen forests. *NAK 1545* (Lower Moozhiar, c.250 m). Fructescent stage is a wonderful sight.

2. **Diospyros candolleana** Wight, Ic. tt. 1221, 1222. 1848; Clarke in Hook. f., Fl. Brit. India 3:566. 1882; Gamble, Fl. Pres. Madras 773. 1923: Sald. in Sald. & Nicols., Fl. Hassan Dist. 195. 1976. *D. canarica* Bedd., Ic. t. 134. 1874.

Leaves elliptic, 9-15 by 3.5-5.5 cm, acute at both ends, thin-coriaceous. Flowers in fascicles, 5-merous; sepals orbicular in male, lanceate in female; corolla white; stamens 10, usually in pairs; staminodes present in female; ovary globose, 4-locular. Berry depressed-globose.

Fl. & Fr.: Apr.-May. Distr.: Western Peninsular India. Small trees. Not common. In semi-evergreen forests. *CNM 59612* (MH) (Angamoozhy, c. 250 m).

Decoction of root bark taken to cure various swellings and rheumatism.

3. **Diospyros hirsuta** L.f., Suppl. Pl. 440. 1781, A.DC., Prodr. 8:223. 1839; Bedd., Ic.t. 137. 1874; Clarke in Hook.f., Fl. Brit. India 3:565. 1882.

Leaves oblong-lanceate, 12-21 by 4-7 cm, base rounded, margin thick, apex acuminate, thin-coriaceous. Flowers in fascicles. Male flowers densely fascicled; corolla tomentose without; stamens 5. Female flowers 1-3, subsessile; calyx deeply 5-lobed; staminodes 5; ovary c. 10-locular. Berry broadly ellipsoid.

Fl. & Fr.: Mar.-Jun. *Distr.*: Sri Lanka & S. -W. Peninsular India. In evergreen forests. Not Common. *NAK 452* (Thannithode, c. 70 m), *1610* (Chathanthara, c. 100 m).

4. **Diospyros malabarica** (Desr.) Kostel., Allg. Med. Pharm. Fl. 3 : 1099. 1834; Nicols., Suresh & Manilal, An Interpr. Hort.Malab. 102. 1988. *Garcinia malabarica* Desr. in Lam., Encycl. 3:701. 1792. *Diospyros peregrina* auct., non Guerke, 1891: Gamble, Fl. Pres. Madras 777 1921. *D. embryopteris* Pers., Syn. 2:624. 1807; Clarke in Hook.f., Fl. Brit. India 3:556. 1882. *Panitsjika-maram* Rheede, Hort. Malab. 3: 45-47, t. 41. 1682.

Leaves oblong, 10-20 by 2-6 cm, base and apex obtuse, reticulations close, prominent. Flowers 4-merous; male flowers c.8 per cymes; stamens 40 or more, filaments and connective villous; female flowers few; calyx cupular; corolla white, fleshy; staminodes 4; ovary globose. Berry subglobose.

Fl. & Fr.: Mar.-May. *Distr.*: Sri Lanka, India, Malaya. Common in sacred groves. *NAK 2316* (Kodumon, c. 50 m).

Powder of dry fruits applied to cure ulcers. Bark powdered and boiled in oil and applied to cure eczema in children. Seeds yield an oil and used for dysentery and diarrhoea.

5. **Diospyros nilgirica** Bedd., Ic.t. 136. 1874; Clarke in Hook.f., Fl. Brit. India 3:566. 1882; Gamble, Fl. Pres. Madras 755. 1923; Manilal, Fl. Silent Valley 168. 1988. *Karin Chora.*

Leaves oblong, 7-10 by 3-4 cm, base and apex gradually acute, lower surface dense-tomentose, thin-coriaceous. Flowers 5-merous. Male flowers in dense panicles; corolla tubular, tomentose without; stamens 16, in pairs, filaments pubescent. Female flowers 1-2 together; staminodes 8; ovary 8-locular. Berry small, globose.

Fl. & Fr.: Feb.-May. *Distr.*: Peninsular India. Rare. Large trees in evergreen forests. *KV 66213.* (MH) (Pampa Dam area, c. 1000 m).

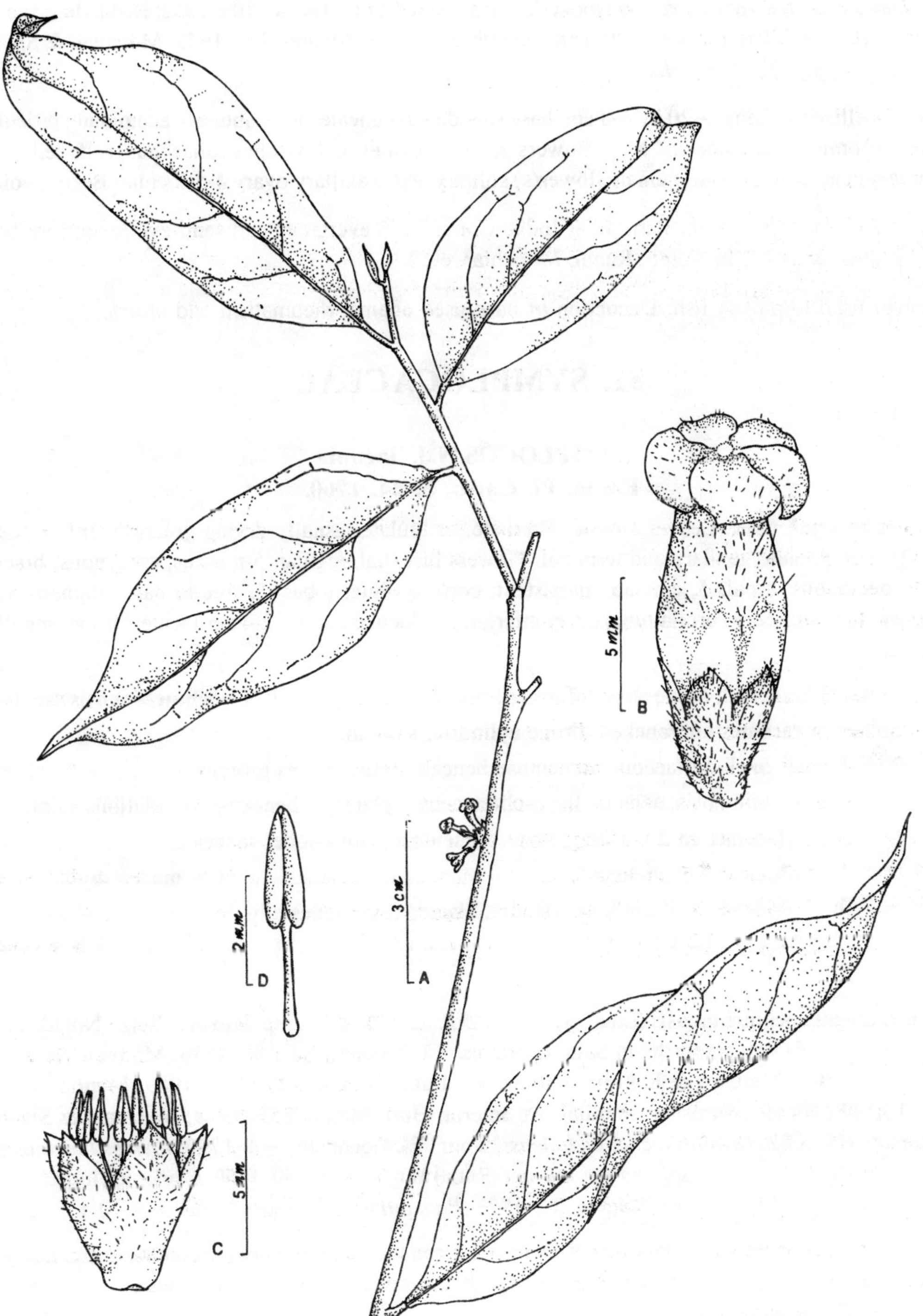

Fig. 43. ***Diospyros bourdillonii*** **Brandis: A. Twig; B. Flower; C. Calyx with stamens; D. Single stamen.**

6. **Diospyros paniculata** Dalz. in Hooker's J. Bot. Kew Gard. Misc. 4 : 109. 1852; Bedd., Ic.t. 125. 1874; Clarke in Hook.f., Fl. Brit. India 3 : 570. 1882; Gamble, Fl. Pres. Madras 775. 1923; Mohanan & Sivad., Fl. Agasthyamala 418.2002. *Kari-vella.*

Leaves elliptic-oblong, 9-20 by 4-8 cm, base rounded to cuneate, apex obtusely acuminate, reticulations very close, prominent, coriaceous. Male flowers in dense cymes; calyx-lobes foliaceous, relfexed; stamens c.20, anthers glandular, beaked. Female flower(s) solitary, extra axillary ovary 4-5-locular. Berry ovoid.

Fl. & Fr.: Jan.-Apr. *Distr.*: Peninsular India. Common in evergreen and semi-evergreen forests. *NAK 497, 1526, 1646 & 1671* (In Konni, Ranni, Moozhiar, etc.)

Leaves used to poison fish. Decoction of bark used against rheumatism and ulcers.

82. SYMPLOCACEAE

SYMPLOCOS N.J. Jacquin
Enum. Pl. Carib. 5: 24. 1760.

Shrubs to small trees. Leaves simple, alternate, estipulate, usually drying greenish. Inflorescence a raceme, spike or panicle, axillary and terminal. Flowers bisexual, regular, 5-merous, epigynous; bracts and bracteoles deciduous; sepals 5, connate, persistent; corolla white, lobes 5, free to base; stamens α, in a-series, adnate to the obscure corolla-tube; ovary inferior, 2-3-locular, ovules 2-4 per locule. Drupe ampulliform or cylindric.

1. Spikes 3-branched; drupe ampulliform, ribbed. ....1. **S. cochinchinensis** ssp. **laurina**
1. Spikes or racemes unbranched; drupe cylindric, smooth.
    2. Leaves thick-coriaceous, tomentose beneath, reticulations prominent. ....3. **S. racemosa**
    2. Leaves thin-coriaceous or thick-chartaceous, glabrous beneath, reticulations faint.
        3. Racemes to 2 cm long, stout, tomentose; calyx-lobes lanceate; drupe, c. 2.5 cm long. ....2. **S. macrophylla** ssp. **rosea**
        3. Racemes to 4 cm long, slender, hispid; calyx-lobes elliptic; drupe c. 1.5 cm long. ....4. **S. wynadense**

1. **Symplocos cochinchinensis** (Lour.) Moore, J. Bot. 52: 148. 1914, ssp. **laurina** (Retz.) Nooteb., Leiden Bot. Ser. 1:156. 1975; Ramamoorthy in Sald. & Nicols., Fl. Hassan Dist. 198. 1976; Matthew, Ill. Fl. Tam. Carnatic t. 417. 1982; Matthew & Rani in Matthew, Fl. Tam. Carnatic 3(2) : 874. 1983; Manilal, Fl. Silent Valley 169. 1988; Nicols., Suresh & Manilal, An Interpr. Hort. Malab. 255. 1988; Mohanan & Sivad., Fl. Agasthyamala 419.2002. *Drupatris cochinchinensis* Lour., Fl. Cochinch. 314. 1790. *Myrtus laurinus* Retz., Obs. Bot. 4:26, 1810. *Symplocos spicata* var. *laurina* (Retz.) Wight, Ill. t. 150. 1850; Clarke in Hook.f., Fl. Brit. India 3:573. 1882; Gamble, Fl. Pres. Madras 782. 1923. *Patsjotti* Rheede, Hort. Malab. 5:9-10, t. 5.1685.

Leaves oblong-lanceate to oblanceate, 9-18 by 3.5-5 cm, base cuneate, margin crenate to crenate-serrate, apex acuminate, thin coriaceous. Spikes often 3-branched; flowers α, dense; bracteoles 3, triangular-ovate; calyx glabrous. Drupe ampulliform, glabrous, ribbed.

Fl. & Fr.: Jan.-Jun. *Distr.*: Sri Lanka, India, Myanmar, Thailand, China, Japan to Malesia..Common along streamsides. *NAK 373* (Moozhiar, c. 250 m), *694* (Thannithode, c. 70 m).

Powder of bark with honey cures biliousness and impure blood. Paste of leaves in oil applied for head diseases.

2. **Symplocos macrophylla** Wall. ex A.DC. in DC., Prodr. 8:257. 1844, ssp. **rosea** (Bedd.) Nooteb., Rev. Symplocac. Old World 229. 1975; Mohanan & Sivad., Fl. Agasthyamala 421.2002. *S. rosea* Bedd., Trans. Linn. Soc. London 25: 219. 1866; Clarke in Hook.f., Fl. Brit. India 3:583. 1882; Gamble, Fl. Pres. Madras 783. 1923. *S. barberi* Gamble, Kew Bull. 1921: 219. 1921 & Fl. Pres. Madras 783. 1923. *Malankuruvi.*

Leaves oblanceate or oblong-lanceate, 8-16 by 2.5-6 cm, base cuneate or rounded, margin serrulate or faintly toothed to entire, thin-coriaceous. Spikes unbranched, c.2 cm long, rose coloured. Flowers few, lax or dense, small; calyx tomentose, lobes lanceate. Drupe ellipsoid-cylindric.

Fl. & Fr.: Jan.-Sep. *Distr.*: S. W. Peninsular India. Not common. In evergreen forests. *NAK 2591 & 2592* (Moozhiar, c. 300 m).

3. **Symplocos racemosa** Roxb., Fl. Ind. 2:539. 1832; Clarke in Hook.f., Fl. Brit. India 3:576. 1882; Ramamoorthy in Sald. & Nicols., Fl. Hassan Dist. 200. 1976. *S. beddomei* Clarke in Hook. f., Fl. Brit. India 3:582. 1882; Gamble, Fl. Pres. Madras 783.1923.

Leaves elliptic or elliptic-lanceate, 7-12 by 3.5-5.5 cm, base cuneate, margin finely serrulate, apex acuminate, lower surface tomentose, lateral veins prominent beneath, thick-coriaceous. Spikes unbranched. Flowers white; calyx tomentose, lobes lanceate. Drupe ellipsoid, smooth.

Fl. & Fr.: Dec.-Feb. *Distr.*: Western Peninsular India. Rare. In evergreen forests and sholas. *NAK 2856* (Kakki hills, c.1100 m).

4. **Symplocos wynadense** (Kuntze) Nooteb., Rev. Symplocac. Old World 293.1975; Mohanan & Sivad., Fl. Agasthyamala 424.2002. *Eugeniodes wynadense* Kuntze, Rev. Gen. Pl. 2:975.1891. *Symplocos acuminata* Bedd., For. Man. Bot. 150.1872 & Ic. 117.1874, non Miq., 1859; Clarke in Hook.f., Fl. Brit. India 3:583.1882; Gamble, Fl. Pres. Madras 783.1923.

Leaves oblanceate to narrowly obovate, 6-15 by 2.5-5 cm, base cuneate, margin serrulate or denticulate, apex acuminate, chartaceous. Spikes unbranched, peduncle slender, hispid. Flowers white. Drupe ovoid or cylindric, smooth.

Fl. & Fr.: Jan.-Sep. *Distr.*: S. W. Peninsular India. Common. Small trees in evergreen and semi-evergreen forests. *NAK 510, 1462 & 2592* (Ranni, Konni, Moozhiar, etc.)

83. OLEACEAE

1. Erect shrubs or small trees; fruit a drupe.
   2. Branchlets smooth or very faintly lenticellate; panicle terminal and/or axillary.
      3. Leaf margin entire; corolla margin incurved. ....**Chionanthus**
      3. Leaf margin usually dentate (or subentire); corolla margins straight. ....**Olea**
   2. Branchlets prominently lenticellate; panicle exclusively terminal. .... **Ligustrum**
1. Climbing shrubs; fruit a berry.

4. Stem subterete; flowers white, c.2.5 cm ϕ...**Jasminum**
4. Stem 4-angled; flowers yellow, c.3 mm ϕ..**Myxopyrum**

CHIONANTHUS Linnaeus
Sp. Pl. 8. 1753.

Small trees. Leaves simple, opposite. Flowers bisexual or male only, 4-merous, in axillary and terminal panicles; calyx very small, 4-toothed; corolla white or cream coloured, lobes 4, free to base, often herbaceous, margin incurved; stamens 2, connective broad; ovary superior, 2-locular; ovules 2 per locule. Drupe ellipsoid or obovoid.

1. Cymes c. 5 cm long; flowers clustered; petals lanceate.
   2. Leaf-base rounded or broadly acute, apex caudate. ....1. **C. linocieroides**
   2. Leaf-base cuneate, apex acuminate...2. **C. mala-elengi**
1. Panicle c.10 cm long; flowers lax; petals elliptic. ....3. **C. ramiflora**

1. **Chionanthus linocieroides** (Wight) Bennet & Raizada, Ind. J. For. 4:68. (May) 1981; Nair & Janardh., J. Bombay Nat. Hist. Soc. 78 : 330 (Aug.) 1981; Mohanan & Sivad., Fl. Agasthyamala 427.2002. *Olea linocieroides* Wight, Ic. t. 1241. 1848. *Linociera wightii* Clarke in Hook.f., Fl. Brit. India 3:608. 1882; Gamble, Fl. Pres. Madras 794. 1923. *Chionanthus wightii* (Cl.) Bahadur & Gaur, Ind. J. For. 4:77. (Aug.) 1981.

Leaves elliptic, 6-8 by 3.5-4.5 cm, base subrounded, apex caudate-acuminate, lateral veins faint, drying greyish. Flowers in small clusters; bracts ovate; calyx-lobes ovate, glabrous, margins ciliate; petals linear-lanceate, coherent at base; ovary glabrous. Drupe ellipsoid.

Fl. & Fr.: Jan.-May. *Distr.*: Peninsular India. Rare. In evergreen forests. *NAK 1430* (Arampa, c. 500 m). Collected only once.

2. **Chionanthus mala-elengi** (Dennst.) Green, Bull. Bot. Surv. India 26:124. 1984; Nicols., Suresh & Manilal, An Interpr. Hort. Malab. 199. 1988. *Forsythia mala-elengi* Dennst., Schluess. Hort. Ind. Malab. 12, 21 & 31. 1818. *Linociera malabarica* Wall. ex G. Don, Syst. 4: 53. 1837-1838; Wight, Ic. t. 1246. 1848; Clarke in Hook.f., Fl. Brit. India 3: 607. 1882; Gamble, Fl. Pres. Madras 794. 1923; Ramamoorthy in Sald. & Nicols., Fl. Hassan Dist. 514. 1976; Matthew & Rani in Matthew, Fl. Tam. Carnatic 3(2): 888. 1983. *Chionanthus malabaricus* (G. Don) Bedd., Fl. Sylv. t. 239 A. 1872. *Mala-elengi* Rheede, Hort. Malab. 5:109-110, t. 55. 1685.

Leaves narrowly elliptic, 5-10 by 2.5 - 4.5 cm, base cuneate, apex obtuse or shortly acuminate, drying greyish. Flowers in small panicles or in clusters; calyx-lobes ovate, pubescent; petals white, basally connate, shortly tubular at base; ovary pubescent. Drupe ellipsoid.

Fl. & Fr.: Feb.-May. *Distr.*: Peninsular India. Common. In riverine forests. *NAK 1207* (Moozhiar, c.250 m) *2063* (Angamoozhy, c. 200 m).

3. **Chionanthus ramiflora** Roxb., Fl. Ind. 1:107. 1832. *Linociera ramiflora* (Roxb.) Wall. ex G. Don, Gen. Syst. 4:52.1838; Ramamoorthy in Sald. & Nicols., Fl. Hassan Dist. 514. 1976; Matthew, Ill.Fl. Tam. Carnatic t. 420. 1982; Matthew & Rani in Matthew, Fl. Tam. Carnatic 3(2): 889. 1983; Manilal, Fl. Silent Valley 172. 1988.

Leaves elliptic or oblong-lanceate, 7-12 by 2-6 cm, base cuneate, margin undulate or serrate, acuminate, drying brownish. Flowers in large panicles, bisexual or male only; calyx 4-toothed; petals cream coloured, oblong, margins incurved, basally connate; ovary glabrous. Drupe ellipsoid.

Fl. & Fr.: Feb.-Apr. *Distr.*: India, Myanmar, S. China, Taiwan, Malesia. Rare. In evergreen shola forests. *NAK 374* (Moozhiar, c. 250 m).

JASMINUM Linnaeus
Sp. Pl. 7.1753.

Climbing shrubs. Leaves simple or 3-foliolate or paripinnate. Inflorescence a corymboid cyme, axillary or terminal, bracteate. Flowers white, fragrant; calyx truncate or campanulate, lobes 5-7; corolla salver-form, lobes 5-7 (-10), imbricate, often biseriate; stamens 2, included, anthers oblong, connective produced; ovary superior, 2-locular; ovule(s) 1per locule. Berry globose or ellipsoid. Seeds non-endospermous.

1. Leaves simple; calyx-lobes linear.
    2. Leaves to 3 cm long, glabrous; bract subulate. ....1. **J. angustifolium**
    2. Leaves to 10 cm long, tomentose; bracts lanceate. ....3. **J. rottlerianum**
1. Leaves 3-foliolate; calyx-lobes toothed. ....2. **J. azoricum**

1. **Jasminum angustifolium** (L.) Willd., Sp.Pl. 1:36.1797; Clarke in Hook.f., Fl. Brit. India 3 : 598.1882; Gamble, Fl. Pres. Madras 790.1923; Matthew & Rani in Matthew, Fl. Tam. Carnatic 3(2) : 879.1983; Nicols., Suresh & Manilal, An Interpr. Hort. Malab. 199. 1988. *Nyctanthes angustifolia* L., Sp. Pl. 6.1753. *Katu-pitsjegam-Mulla* Rheede, Hort. Malab. 6 : 93, t. 53. 1886.

Leaves simple, ovate or lanceate 1-5-3 by 1-1 cm, base rounded or truncate, apex acute-apiculate or subacute, chartaceous. Cymes 1-3-flowered, terminal; bracts linear; calyx-lobes 5-6, linear; corolla 2.5 cm f, lobes 5-9. Berry globose.

Fl. & Fr.: Nov.-Mar. *Distr.:* Sri Lanka, Peninsular India. Common. On exposed slopes and thickets. *NAK 134* (Adoor), *1859* (Tiruvalla, c. 10 m).

Root crushed with rhizome of **Acorus calamus** and mixed lime juice, and applied to cure ringworm infestation.

2. **Jasminum azoricum** L., Sp. Pl. 70. 1753; Ramamoorthy in Sald. & Nicols., Fl. Hassan Dist. 512. 1976, Matthew & Rani in Matthew, Fl. Tam. Carnatic 3(2) : 880.1983; Manilal, Fl. Silent Valley 170. 1988; Mohanan & Sivad., Fl. Agasthyamala 428. 2002. *J. flexile* Vahl, Symb. Bot. 3:1.1794; Clarke in Hook. f., Fl. Brit. India 3:601.1882; Gamble, Fl. Pres. Madras 791. 1923.

Leaves 3-foliolate; leaflets oblong or elliptic-oblong, 5-10 by 2-4 cm, base rounded, or broadly acute, apex acuminate, thin-coriaceous. Cymes dichasial, terminal or axillary; bracts obscure; calyx-lobes 5, triangular; corolla c. 2.5 cm ϕ, lobes 5. Berry ellipsoid.

Fl. & Fr.: Feb.-Apr. *Distr.*: Sri Lanka, Peninsular India. Common. Among thickets of forest borders. *NAK 458, 1245* (Moozhiar, c. 250 m), *1412* (Arampa, c. 400 m).

3. **Jasminum rottlerianum** Wall. ex A.DC. in DC., Prodr. 8:305.1844; Clarke in Hook.f., Fl. Brit. India 3:593.1882; Gamble, Fl. Pres. Madras 789.1923; Ramamoorthy in Sald. & Nicols., Fl. Hassan Dist. 513.1976.

Leaves simple, oblong or elliptic-oblong, 5-10 by 2-4 cm, base truncate or rounded, apex acuminate-apiculate, densely pubescent, chartaceous. Cymes α-flowered, dense, axillary; bracts foliaceous, lanceate; calyx-lobes 5, subulate; corolla c. 2.5 cm ϕ, lobes ± 6. Berry subglobose.

Fl. & Fr.: Most of the seasosn. *Distr.*: S.W. India. Very frequent in deciduous forests and teak plantations. *NAK 504, 1544* (Arampa, c.400 m).

LIGUSTRUM Linnaeus
Sp. Pl. 7. 1753.

Ligustrum robustum (Roxb.) Blume, Mus. Bot. 1:313.1859, ssp. **walkeri** (Decne.) Green, Kew Bull. 40:130.1985. *L. walkeri* Decne., Nouv. Arch. Mus. Hist. Nat. Ser. 2,2:27.1879; Clarke in Hook. f., Fl. Brit. India 3:614.1882; Gamble, Fl. Pres. Madras 797.1923.

Small trees; branchlets prominently lenticellate. Leaves simple, decussate, elliptic lanceate, 4-8 by 1.5 -3 cm, base rounded or acute, apex, acuminate, thin-coriaceous. Inflorescence a large panicle, glabrescent, branches slender. Flowers small, α, bisexual; calyx truncate; corolla funnel-form, lobes 4; stamens 2, included; ovary 2-locular; ovules 2-per locules. Drupe obovoid.

Fl. & Fr.: Nov.-Mar. *Distr.*: Sri Lanka & Peninsular India. Not common. In evergreen and shola forests. Locally abundant. *NAK 69 & 1344* (Pampa dam area, c.1100 m).

MYXOPYRUM Blume
Bijdr. 683. 1825. ('1826')

Myxoprum smilacifolium Bl., Mus. Bot. 1:330.1850; Clarke in Hook. f., Fl. Brit. India 618. 1872; Kiew, Blumea 29:509.1984; Manilal, Fl. Silent Valley 172.1988; Mohanan & Sivad., Fl. Agasthyamala 431.2002. *M. serratulum* A.W. Hill, Kew Bull. 1910:41.1910; Gamble, Fl. Pres. Madras 798.1923. *Chathura-mulla.* **(Fig. 44)**

Climbing shrubs; branches sharply 4-angular. Leaves simple, opposite, broadly elliptic, 8-15 by 3-7 cm, base rounded or obtuse to subcordate, margin serrulate from the middle towards apex, apex acuminate, basally 5-(7)- nerved, coriaceous. Flowers in axillary or terminal 3-chotomous panicles, small, 4-merous, bisexual; calyx-lobes c. 4; corolla yellow, urceolate, lobes 4; stamens 2, included; ovary 2-locular; ovules 1-2 per locule. Berry globose.

Fl. & Fr.: Feb.-Apr. *Distr.*: India & Himalaya. Less common. Extensive climbers in evergreen forests. *NAK 605* (Thannithode, c. 70 m), *1488* (Arampa, c. 400 m).

Leaf-powder taken with ghee is a remedy for asthma, head-ache and ear-diseases.

OLEA Linnaeus
Sp. Pl. 8. 1753.

Olea dioica Roxb., Fl. Ind. 1:105. 1820; Wight, Ill. Ind. Bot. t.151. 1850; Bedd., Fl. Sylv. 3:153.1872; Clarke in Hook. f., Fl. Brit. India. 3:612.1882; Gamble, Fl. Pres. Madras 796. 1823; Ramamoorthy in Sald. & Nicols., Fl. Hassan Dist. 514.1976; Matthew & Rani in Matthew, Fl. Tam. Carnatic 3(2):891.1983; Manilal, Fl. Silent Valley 173.1988; Nicols., Suresh & Manilal, An Interpr. Hort. Malab. 201. 1988; Mohanan & Sivad., Fl. Agasthyamala 432.2002. *Kari-vetti* Rheede, Hort. Malab. 4: 111-112, t.54.1683.

Small trees. Leaves simple, opposite, elliptic or elliptic-oblong, 5-12 by 2-6 cm, base cuneate, slightly decurrent, margin dentate or rarely entire, apex acuminate, chartaceous. Flowers in terminal or axillary panicles, bisexual, 4-merous; calyx 4-toothed; corolla white, lobes 4, triangular-ovate, incurved; stamens 2, attached at base of corolla tube; ovary 2-locular; ovules 2-per locule. Drupe ellipsoid.

Fl. & Fr.: Dec.-Apr. *Distr.*: N.E. India, S. W. India. Common in evergreen forests. *NAK 463, 1377* (Kakki hills, c.1100 m).

Fruits edible.

Note : Leaves very variable from entire to serrate, and membranous to chartaceous.

84. APOCYNACEAE

1. Anthers distinct, free from the stigma.
 2. Flowers in umbellate or panicled cymes.
  3. Leaves whorled, or whorled and opposite.
   4. Corolla lobes twisted to left; fruit a pair fo follicles.
    5. Trees; follicles slender, to 30 cm long; leaves whorled. ....**Alstonia**
    5. Climbing shrubs; follicle linear, to 12 cm long; leaves whorled & opposite. ....**Kamettia**
   4. Corrolla lobes twisted to right; fruit a drupe. ....**Rauvolfia**
  3. Leaves opposite.
   6. Fruit a pair of follicles.
    7. Stamens inserted near the base of corolla-tube; seeds comose. ....**Holarrhena**
    7. Stamens inserted near the throat of corolla; seeds arillate. ....**Tabernaemontana**
   6. Fruit a berry.
    8. Armed erect or subscandent shrubs; ovary 2-locular. ....**Carissa**
    8. Smooth climbing shrubs; ovary 1-locular. ....**Chilocarpus**
 2. Flower(s) solitary or 2-(3-6) in pairs or clusters. ....**Catharanthus**
1. Anthers connivent, adherent to the stigma.
 9. Trees; anthers exerted; follicles connate, at least by their tips. ....**Wrightia**
 9. Climbing shrubs; anthers inserted; follicles free throughout.
  10. Leaves 20 by 15 cm, broadly elliptic to orbicular; flowers yellow, 6-8 cm ϕ. ....**Chonemorpha**
  10. Leaves 6 by 3 cm, elliptic or lanceate; flowers white, 4-6 mm ϕ. ....**Ichnocarpus**

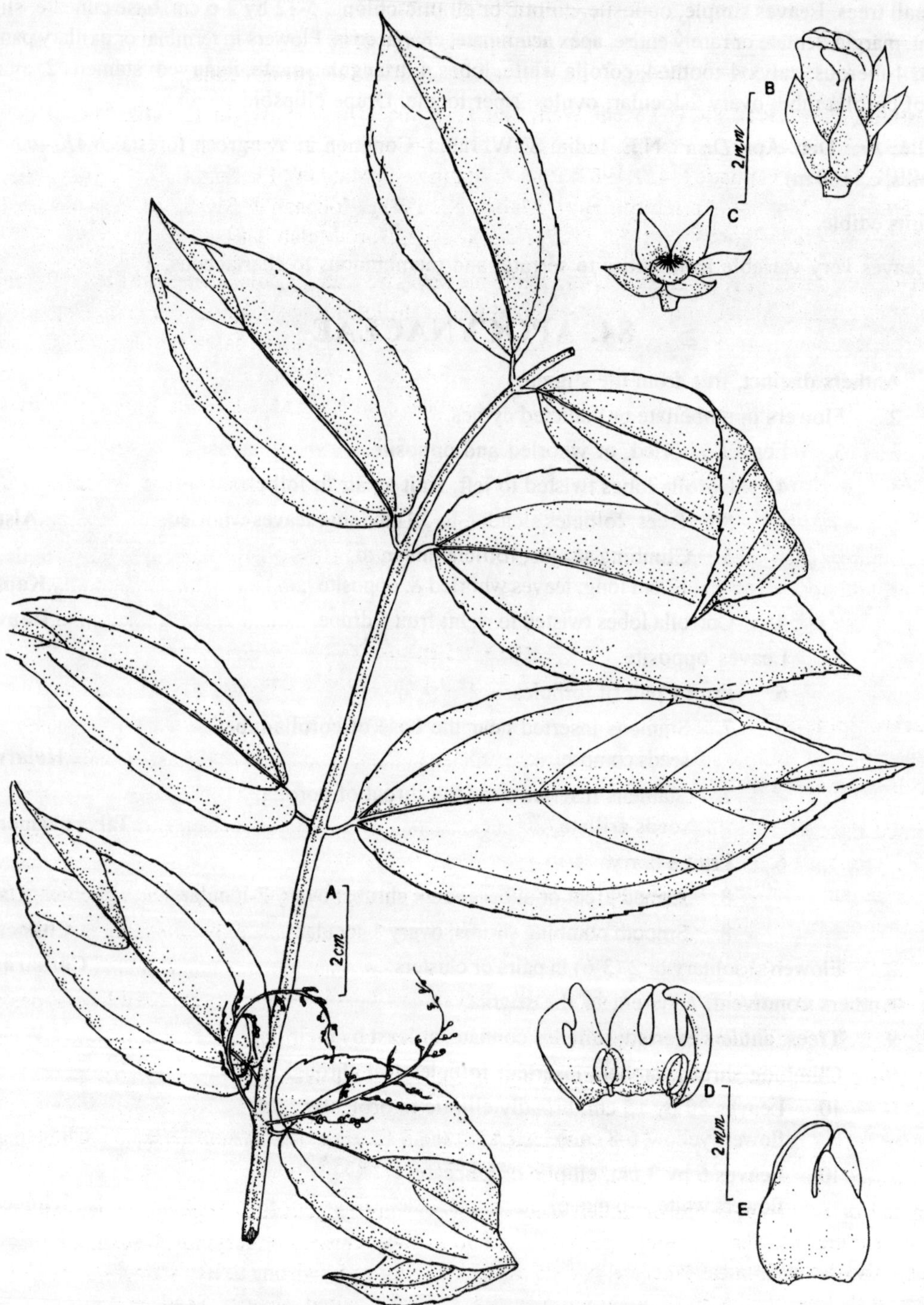

Fig. 44. ***Myxopyrum smilacifolium*** **Bl.: A. Twig; B. Flower; C. Calyx with pistil; D. Corolla - split open; E. Corolla - entire**

ALSTONIA R. Brown
On Asclepiad. 64. 1810;
Mem. Wern. Nat. Hist. Soc. 1:75.1811 (nom. cons.).

Alstonia scholaris (L.) R. Br., Mem. Wern. Nat. Hist. Soc. 1:76.1811; Wight, Ic. t. 422.1843; Hook. f., Fl. Brit. India 3:642.1882; Gamble, Fl. Pres. Madras 810.1923; Stevens in Sald. & Nicols., Fl. Hassan Dist. 429.1976; Matthew, Ill. Fl. Tam. Carnatic t. 427.1982; Rani & Matthew in Matthew, Fl. Tam. Carnatic 3(2) : 903. 1983; Nicols., Suresh & Manilal, An Interpr. Hort. Malab. 52. 1988; Mohanan & Sivad., Fl. Agasthyamala 434. 2002. *Echites scholaris* L., Mant. Pl. 53. 1767. *Pala* Rheede, Hort. Malab. 1:81-82, t. 45. 1678.

Trees. Leaves simple, whorled, 4-7 per node, narrowly obovate or oblanceate, 6-13 by 2.5-5 cm, base cuneate, apex obtuse, lateral veins α, parallel, chartaceous. Inflorescence of superposed umbels, terminating in glomerules, subterminal. Flowers dense, 5-merous, bisexual, hypogynous; calyx cupular, lobes 5, unequal, ovate; corolla white or cream coloured, salver-form, lobes 5, broadly ovate, twisting to left; stamens 5, attached at throat; anthers sagittate; ovaries 2, free; ovules a on marginal placenta. Fruit a pair of slender follicles, to 30 cm long. Seeds compressed, comose.

Fl. & Fr.: Nov.-May. *Distr.*: S. Asia, Malesia to Australia. Common. In sacred groves. *NAK 2673, 2818* (Konni, c.40 m). Flowers highly fragrant.

Bark bitter, used as a remedy in asthma, dropsy and rheumatism. Milky latex of plant boiled with coconut-oil and applied in itches.

CARISSA Linnaeus
Syst. Nat. ed. 12,2: 189.1767 (nom. cons.).

Carissa spinarum L., Mant. Pl.559. 1771; Hook. f., Fl. Brit. India 3:631.1882; Gamble, Fl. Pres. Madras 805.1923; Huber in Abeyw., Rev. Handb. Fl. Ceylon 1(1) : 9.1973; Rani & Matthew in Matthew, Fl. Tam. Carnatic 3(2) : 906.1983. *Cheru-mulli.*

Armed erect or straggling shrubs; branches zig-zag. Leaves simple, decussate, elliptic or suborbicular, 2-4 by 1-3 cm, base rounded, apex obtuse to emarginate, mucronulate, chartaceous. Cymes 5-7-flowered, terminal or axillary. Flowers white, 5-merous, bisexual; calyx-lobes 5, lanceate; corolla lobes 5, oblong, twisting to right; stamens 5, attached the middle of corolla-tube; ovary superior, 1, oblong, 2-locular; ovules 2-per locule. Berry globose, ripening purple.

Fl. & Fr.: Apr.-Jul. *Distr.*: Sri Lanka, India, Myanmar. Less common. In exposed hill slopes and in sacred groves. *NAK 1858* (Tiruvalla, c.10 m).

Fruit is edible; unripe fruit is astringent and useful in biliousness.

CATHARANTHUS G. Don
Gen. Hist. 4:71, 95.1837.

Annual or perennial herbs or undershrubs. Leaves simple, decussate, often glandular, lateral veins regular, ascending. Flower(s) solitary, paired or few in axillary clusters or corymbs, bisexual, hypogynous, 5-merous; calyx-lobes 5, linear-lanceate; corolla salver-form, lobes 5, twisting to left; stamens 5, in the corolla-tube; ovaries 2, free; ovules α, on marginal placenta. Follicles paired, slender. Seeds α.

1. Annual herbs; leaves lanceate; flowers white, c. 1 cm ϕ. .... 1. **C. pusillus**
1. Perennial herbs or undershrubs; leves elliptic-obovate; flowers pink or white. c.3 cm ϕ. .... 2. **C. roseus**

1. **Catharanthus pusillus** (Murray) G. Don, Gen. Hist. 4 : 95.1837; Huber in Abeyw., Rev. Handb. Fl. Ceylon 1(1): 13.1973; Rani & Matthew in Matthew, Fl. Tam. Carnatic 3(2) : 908.1983; Nicols., Suresh & Manilal, An Interpr. Hort. Malab. 53.1988. *Vinca pusilla* Murray, Novi. Comment. Soc. Regiae. Sci.Gott. 3:66,t.2, F.1.1773; Hook.f., Fl.Brit.India 3:640.1882. *Lochnera pusilla* (Murray) Schumann in Engl., Nat. Pflanzenf. 4, 2:145.1895; Gamble, Fl. Pres. Madras 809.1923. *Cupa-vela* Rheede, Hort. Malab. 9:61,t.33.1689.

Annual herbs. Leaves lanceate, 2-6 by 0.8-1.7 cm, base and apex acute, lateral veins 6-10 pairs, chartaceous. Flower(s) solitary or in pairs or 3-5, axillary; calyx-lobes unequal; corolla white, c. 1 cm ϕ. Follicles c. 3 cm long.

Fl. & Fr.: Apr.-Aug. *Distr.*: India, Sri Lanka. On exposed areas. Seasonal. *NAK 636* (Puthukkulam c.20 m). A close relative of famous rose periwinkle. May contain useful alkaloides.

2. **Catharanthus roseus** (L.) G. Don, Gen. Hist. 4:95.1837; Stearn, Lloydia 29:196.1966; Huber in Abeyw., Rev. Handb. Fl. Ceylon 1(1) : 13.1973; Stevens in Sald. & Nocols., Fl. Hassan Dist. 434.1976; Rani & Matthew in Matthew, Fl.Tam. Carnatic 3(2) : 909.1983. *Vinca rosea* L., Syst. Nat. ed. 10, 944. 1759; Hook.f., Fl. Brit. India 3:640.1882. *Lochnera rosea* (L.) Reichenb., Consp. Regni. Veg. 134.1828; Gamble, Fl. Pres. Madras 809.1923.

Perennial woody herbs. Leaves elliptic-obovate, 2-4 by 1-2 cm, base cuneate, apex obtuse or rounded, mucronulate, lateral veins c.10 pairs, chartaceous. Flower(s) solitary or paired; calyx-lobes subequal; corolla pink to white, c.3 cm ϕ. Follicles c. 3 cm long, pubescent.

Fl. & Fr.: Aug.-Jan. *Distr.*: Endemic to Madagascar; naturalized in the tropics. Abundant in wastelands and along waysides. *NAK 1060* (Tiruvalla, c.10 m). The "rose periwinkle".

The whole plant is medicinal and contain alkaloides useful for treatment against cancer.

CHILOCARPUS Blume
Cat. Buitenzorg 22. 1823.

Chilocarpus malabaricus Bedd., Ic. Pl. Ind. Orient. t. 175. 1874; Raghavan, Bull. Bot. Surv. India 6:309.1965; Mohanan & Sivad., Fl. Agasthyamala 436.2002 *Chilocarpus atro-viridis* sensu., non Blume, 1850; Hook.f., Fl. Brit. India 3:626.1882; Gamble, Fl. Pres. Madras 803.1923. **(Fig. 45)**

Climbing shrubs. Leaves simple, decussate, elliptic-oblong, 6-13 by 3-4.5 cm, base cuneate to rounded, apex abruptly acuminate or cuspidate, lateral veins α, close, parallel, forming inframarginal loopes, chartaceous. Cymes axillary. Flowers α, lax, 5-merous, bisexual, hypogynous; calyx-lobes 5; corolla yellow, hypocrateriform, constricted at throat, lobes 5, twisting to right; stamens 5, inserted at corolla-tube; ovary 1, superior, 1-locular; ovules α, on 2 parietal placenate. Berry fleshy, 5-7 cm long.

Fl. & Fr.: Jan-May. *Distr.*: Endemic to Western Ghats. Rare. In evergreen forests. *NAK 1665 & 1857* (Shbari hills, c.200 m).

CHONEMORPHA G. Don
Gen. Hist. 4:69, 76. 1837 (nom.cons.).

Chonemorpha fragrans (Moon) Alston, Ann. Roy. Bot. Gard. (Peradeniya) 11:203. 1929; Stevens in Sald. & Nicols., Fl. Hassan Dist. 433. 1976; Nicols., Suresh & Manilal, An Interpr. Hort. Malab. 54. 1988. *Echites frgrans,* Moon, Cat. Pl. Ceylon 200.1824. *E. macrophylla* Roxb., Fl.Ind. 2:13. 1832. *Chonemorpha macrophylla* (Roxb.) G.Don, Gen. Syst. 4:76.1837; Wight, Ic.t.432.1841; Hook.f., Fl. Brit. India 3:661.1882; Gamble, Fl.Pres. Madras 818.1923. *Belutta-kaka-kodi* Rheede, Hort.Malab.9:7, t.5-6. 1689.

Stout climbers. Leaves simple, decussate, broadly elliptic to suborbicular, 15-20 by 12-15(18) cm, base cordate, apex acuminate, lower surface silky tomentose, coriaceous. Flowers large, 5-merous, bisexual, in paniculate cymes; calyx tubular with a ring of glands at base, lobes 5; corolla yellowish or cream coloured, salver-form, c. 8 cm ϕ, lobes 5, orbicular, twisting to right; stamens 5, included; ovaries 2, free; ovules α, on marginal placenta. Follicle woody, cylindirc, 18 by 2.5 cm. Seeds comose.

Fl. & Fr.: Jul-Nov. *Distr.*: Indo-Malesia. Rare. In evergreen forests and sacred groves. *NAK 1896* (Aikadu, c.20 m), *2145* (Chittar, c.100 m).

Leaf juice boiled with oil is a good medicine for head diseases.

HOLARRHENA R. Brown
On Asclepiad. 51.1810;
Mem. Wern. Nat. Hist. Soc. 1:62.1811.

Holarrhena pubescens (Buch.-Ham.) Wall. ex G. Don, Gen. Hist. 4:78.1837; Rani & Matthew in Matthew, Fl.Tam. Carnatic 3(2):910.1983; Nicols., Suresh & Manilal, An Interpr. Hort. Malab. 55.1988; Mohanan & Sivad., Fl. Agasthyamala 438.2002. *Echites pubescens* Buch.-Ham., Trans. Linn. Soc. London 13:521.1821. *Holarrhena antidysenterica* Wall. ex A.DC. in DC., Prodr. 8:413.1844; Hook.f., Fl. Brit. India 3:644.1882; Gamble, Fl. Pres. Madras 811.1923; Stevens in Sald. & Nicols., Fl. Hassan Dist. 434.1976; Matthew, Ill. Fl. Tam. Carnatic t.430.1982. *H. Codaga* G. Don, Gen. Hist. 4:78. 1837. *Codaga-pala* Rheede, Hort. Malab. 1:85-86, t.47.1678.

Small trees. Leaves simple, decussate, elliptic or broadly ovate to suborbicular, 5-15 by 4.5-8 cm, base rounded, apex acuminate, lower surface densely tomentose. Cymes paniculate or corymbose, terminal or axillary. Flowers 5-merous, bisexual, hypogynous; calyx cupular, lobes 5, unequal, glandular within; corolla white, salver-form, c.3 cm ϕ, lobes 5, oblong, twisting to right; stamens 5, included, attached near the base of corolla tube; ovaries 2, free; ovules α, marginal. Follicles in pairs, distinct, slender, to 25 cm long. Seeds comose.

Fl. & Fr.: Apr.-Aug. *Distr.*: India, Malaya Peninsula. Common. In moist deciduous and semi-evergreen forests. *NAK 657* (Puthukkulam, c.20 m), *2379* (Padom, c.40 m).

Bark bitter, astringent and used as a remedy in piles and dysentery.

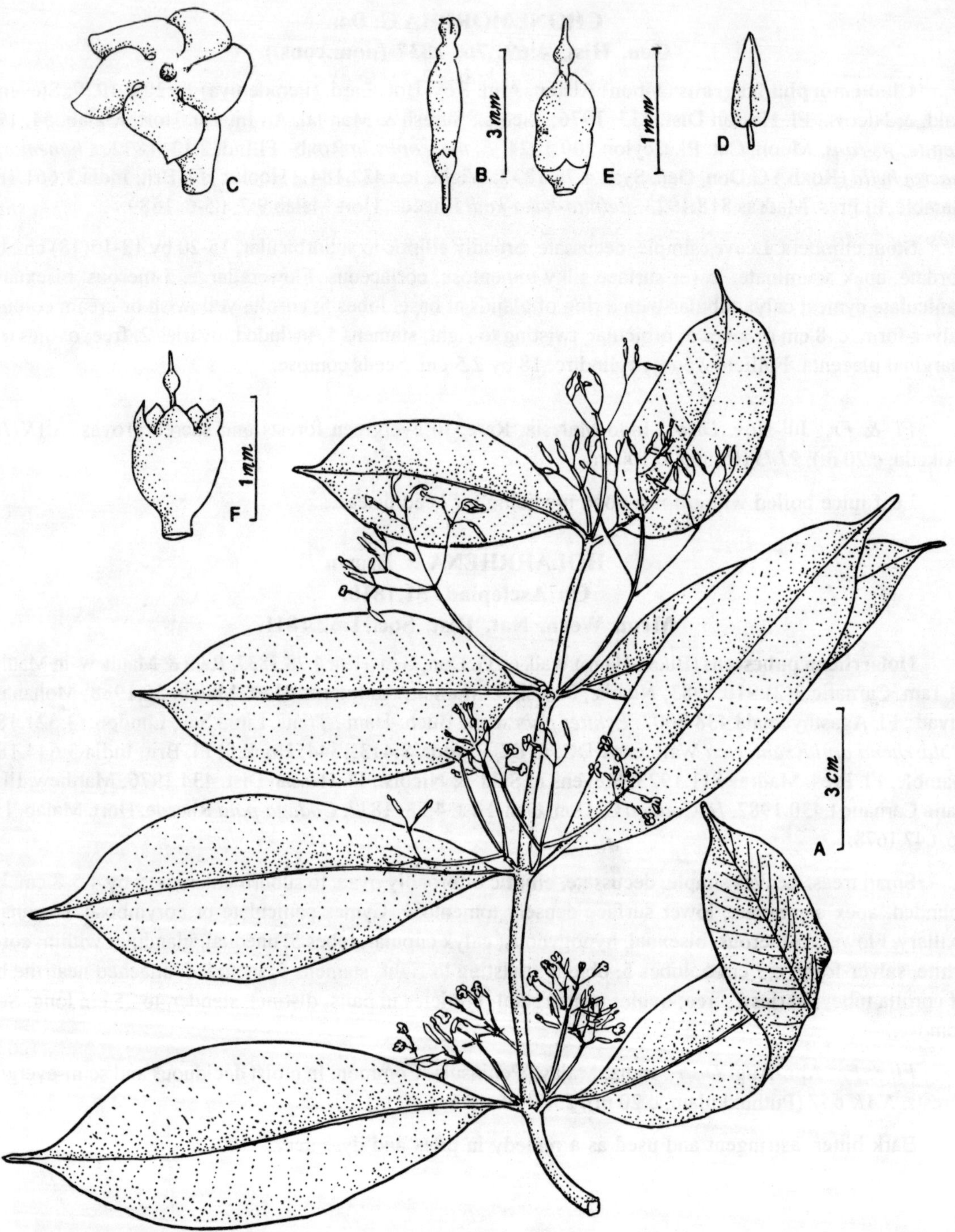

Fig. 45. ***Chilocarpus malabaricus* Bedd.** **: A. Twig; B. Flower bud; C. Corolla; D. Stamen; E. Pistil; F. Calyx with pistil.**

ICHNOCARPUS R. Brown
On Asclepiad. 50.1810;
Mem. Wern. Nat. Hist. Soc. 1:61.1811.

Ichnocarpus frutescens (L.) R.Br. in Aiton, Hort. Kew (ed. 2), 2:69.1811; Hook.f., Fl.Brit. India 3:669.1882; Gamble, Fl. Pres. Madras 820.1923; Stevens in Sald. & Nicols., Fl. Hassan Dist. 435.1976; Matthew, Ill. Fl. Tam. Carnatic t.431.1982; Rani & Matthew in Matthew, Fl.Tam. Carnatic 3(2):912.1983; Nicols., Suresh & Manilal, An Interpr. Hort. Malab. 55.1988; Mohanan & Sivad., Fl. Agasthyamala 439.2002. *Apocynum frutescens* L., Sp. Pl. 213.1753. *Pal-valli* Rheede, Hort. Malab. 9:19, t. 12.1689.

Climbing shrubs. Leaves simple, decussate, elliptic-oblong, 3-6 by 1-3 cm, base acute, apex acute-apiculate, thin-coriaceous. Cymes paniculate, terminal and axillary, rusty pubescent. Flowers 5-merous, bisexual; calyx cupular, lobes 5, subequal, glandular within; corolla white, salver-form, c.6 mm ϕ, lobes 5, twisting to right; stamens 5, included, attached below the middle of corolla-tube; ovaries 2, partly connate; ovules α, on marginal placenta. Follicles slender, 8-12 cm long. Seeds comose.

Fl. & Fr.: Sep.-Mar. *Distr.*: Indo-Malesia, Australia. Common on cleared slopes, on thickets, etc. *NAK 107* (Adoor, c.20 m).

Root powdered and given in milk as a remedy in formation of sugar in urine, impure blood and stone in the urinary bladder.

KAMETTIA Kosteletzky
Allg. Med. Pharm. Fl. 3:1062.1834.

Kamettia caryophyllata (Roxb.) Nicols. & Suresh, Taxon 35:354.1986; Nicols., Suresh & Manilal, An Interpr. Hort. Malab. 5.1988; Mohanan & Sivad., Fl. Agasthyamala 440.2002. *Echites caryophyllata* Roxb., Hort. Beng. 20.1814 ('*coryophyllata*'). *Kamettia malabarica* Kostel., Allg. Med.-Pharm. Fl. 3:1062.1834, nom. illegit (incl. type of *Echites caryophyllata* Roxb. 1814). *Aganosma roxburghii* G. Don, Gen. Syst. 4:77.1837. *Ellertonia rheedei* Wight, Ic. 4(2):2, t. 1295.1848 ('*rheedii*'), nom. illegit. (incl. type of *Echites caryophyllata* Roxb. 1814); Hook. f., Fl. Brit. India 3:641.1882; Gamble, Fl. Pres. Madras 811.1923. *Kametti-valli* Rheede, Hort. Malab. 9: 23-24, t.14-1689. **(Fig. 46)**

Climbing shrubs. Leaves decussate or whorled, elliptic, 7-15 by 3-6.5 cm, base cuneate, apex long-acuminate, lateral nerves 14-16 pairs, parallel, forming inframarginal loops, chartaceous. Cymes subterminal. Flowers 5-merous, bisexual, hypogynous; calyx small, reddish, lobes 5; corolla white, salver-form, c.2 cm ϕ, lobes 5, oblong, twisting to left; stamens 5, at the middle of corolla-tube; ovaries 2, free, ovules α, on marginal palcenta. Follicles slender, cylindric, 8-10 cm long. Seeds winged.

Fl. & Fr.: Sep.-Jan. *Distr.*: Western Ghats. Rare. An extensive climber, usually grows into great height in evergreen forests. *NAK 2607* (Plappally, c.400m).

RAUVOLFIA Linnaeus
Sp.Pl. 208 1753.

Woody herbs or undershrubs. Leaves simple, whorled, chartaceous. Inflorescence of corymbose or umbellate cymes, axillary, or terminal. Flowers 5-merous, bisexual, hypogynous; calyx-lobes 5, free to base, corolla salver-form, tube dilated over stamens, lobes 5, equal, twisting to left; stamens 5, attached above the middle of corolla-tube; ovaries 2, connate; ovules 2-per locule. Drupe of 2 connate pyrenes. Seeds ovoid.

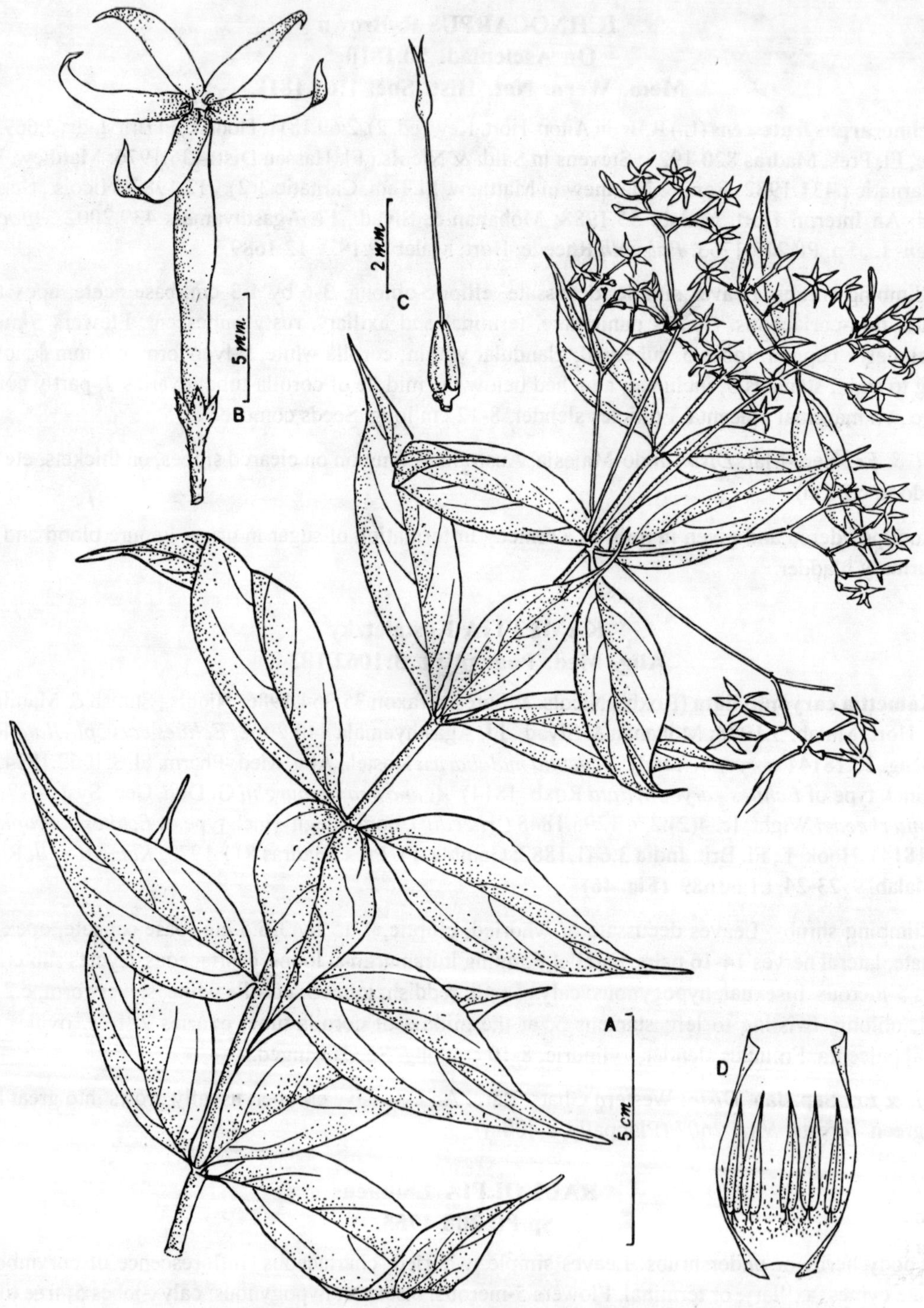

Fig. 46. ***Kamettia caryophyllata*** **(Roxb.) Nicols. & Suresh: A. Twig; B. Flower; C. Pistil; D. corolla tube-split opened showing stamens.**

1. Corolla-tube c.8 mm long, dilated above the middle; calyx greenish. ....1. **R. micrantha**
1. Corolla-tube c.15 mm long, dilated at the middle; calyx reddish. ....2. **R. serpentina**

1. **Rauvolfia micrantha** Hook.f., Fl. Brit. India 3.633.1882; Gamble, Fl. Pres. Madras 808. 1923.

Woody herbs; branchlets lenticellate. Leaves oblanceate to elliptic-lanceate, 6-13 by 2.5-4 cm, base cuneate, apex acuminate. Cymes lax. Flowers few, lax; corolla c.1 cm ϕ, tube straight, dilated above the middle. Drupes obcordiform, c.12 mm ϕ.

Fl. & Fr.: Mar.-May. *Distr.*: Western Ghats. Rare. Among forest undergrowth. *NAK 703, 767, 782, 1824.* (Plapally, c. 300m).

2. **Rauvolfia serpentina** (L.) Benth. ex Kurz, Forest Fl. Burma 2:171.1877; Hook. f., Fl. Brit. India 3:632. 1882; Gamble, Fl. Pres. Madras 807.1923; Huber in Abeyw., Rev. Handb. Fl. Ceylon 1(1):15. 1973; Rani & Matthew in Matthew, Fl. Tam. Carnatic 3(2):917.1983; Nicols., Suresh & Manilal, An Interpr. Hort. Malab. 57.1988; Mohanan & Sivad., Fl. Agasthyamala 443.2002. *Ophioxylon serpentinum* L., Sp. Pl. 1043.1753. *Tsjovanna-ampelodi* Rheede, Hort. Malab. 6:81-82, t. 47.1686.

Undershrubs; branchlets smooth. Leaves oblanceate, 6-15 by 2.5-5.5 cm, base cuneate, apex acuminate. Cymes dense. Flowers c.2 cm ϕ; peduncle, pedicels, calyx reddish; corolla-tube dilated at the middle, slightly curved. Drupe obovoid, c.7 mm ϕ.

Fl. & Fr.: Jun.-Oct. *Distr.*: Tropical Himalayas, India, Sri Lanka, Malaya. Less common. Among forest undergrowth. *NAK 228, 774, 931 & 1768* (Teak plantations).

Roots used against dysentery and bowel complaints; decoction given in snake-bite. The whole plant dried in shade, powdered and given with honey as a remedy in rheumatism, all poisons and epilepsy.

TABERNAEMONTANA Linnaeus
Sp. Pl. 210. 1753.

Shrubs or small trees. Leaves simple, decussate, chartaceous. Inflorescene of umbelliform, terminal or axillary cymes. Flowers 5-merous, bisexual, hypogynous; calyx lobes 5, glandular within; corolla white, salver-form, tube dilated at or above the middle, lobes 5, twisting to left; stamens 5, included; ovaries 2, slightly coherent; ovules α. Follicles in pairs, obliquely ovoid, curved. Seeds with orange red arll.

1. Inflorescence lax; calyx-lobes triangular, glabrous except the marigns, corolla-tube to 1 cm long, dilated near the middle. ....2. **T. gamblei**
1. Inflorescence dense; calyx-lobes suborbicular, hispid, glabrescent; corolla-tube to 3 cm long, dilated near apex. ....1. **T. alternifolia**

1. **Tabernaemontana alternifolia** L., Sp. Pl. 211. 1753; Nicols., Suresh & Manilal, An Interpr. Hort. Malab. 57. 1988; Mohanan & Sivad., Fl. Agasthyamala 445.2002. *T. crispa* Roxb., Hort. Beng. 20.1814, nom. illegit. (incl. type of *T. alternifolia* L., 1753). *T. heyneana* Wall., Bot. Reg. 15:sub t. 1273. 1829. *Ervatamia heyneana* (Wall.) Cooke, Fl. Pres. Bombay 23:134. 1904; Gamble, Fl. Pres. Madras 813. 1923; Stevens in Sald. & Nicols., Fl. Hassan Dist. 433. 1976. *Curutu-pala* Rheede, Hort. Malab. 1:83-84, t. 46. 1678.

Small trees. Leaves elliptic-oblong to oblanceate, 8-15 by 3-6 cm, base cuneate, margin subentire, apex caudate-acuminate, lateral veins 10-14 pairs, close, chartaceous. Panicle dense, slightly pubescent; calyx-lobes suborbicular; corolla-tube, to 2 cm long, dilated at top, lobes crumpled at margins. Follicles ellipsoid, curved.

Fl. & Fr.: Mar.-May. *Distr.*: South-W. Peninsular India. Common. In exposed areas and semi-evergreen forests. *NAK 1609* (Ranni R.F., c. 100 m).

2. **Tabernaemontana gamblei** Subram. & Henr., Bull. Bot. Surv. India 12(1):5. 1970; Mohanan & Sivad., Fl. Agasthyamala 446.2002. *Ervatamia caudata* Gamble, Kew Bull. 1921: 310. 1921 & Fl. Pres. Madras 813. 1923, non *Tabernaemontana caudata* Merr., 1909.

Shrubs. Leaves narrowly elliptic to elliptic-oblanceate, 5-12 by 1.3-3.5 cm, base cuneate, apex caudate-acuminate, lateral veins 6-10 pairs, distant, chartaceous. Panicle lax; calyx-lobes triangular, margin ciliate; corolla-tube c.1.5 cm long, dilated at middle, lobes 5, suborbicular, margin smooth. Follicles not seen.

Fl. & Fr.: Jan. *Distr.*: Endemic to Western Ghats. Common. In evergreen forests. *NAK 1436* (Arampa, c.500 m).

WRIGHTIA R. Brown
On Ascelpiad. 762. 1810.

Small trees. Leaves simple, decussate, chartaceous. Cymes lateral or terminal. Flowers 5-merous, bisexual, hypogynous; calyx cupular, lobs 5, equal, glandular within; corolla white or reddish, salver-form, corona scales within, lobes twisting left; stamens 5, on corolla mouth, exerted; ovaries 2, free or connate. Follicles cylindric, fully or apically connate. Seeds comose.

1. Leaves tomentose beneath; corolla reddish; anthers glabrous at apex; follicles connate throughout. .... 1. **W. arborea**

1. Leaves glabrous beneath; corolla white; anthers ciliate at apex; follicles connate at apex only. .... 2. **W. tinctoria**

1. **Wrightia arborea** (Dennst.) Mabb., Taxon 26:533. 1977; Matthew, Ill. Fl. Tam. Carnatic t.973. 1982; Rani & Matthew in Matthew, Fl. Tam. Carnatic 3(2):920. 1983; Nicols., Suresh & Manilal, An Interpr. Hort. Malab. 59. 1988. *Periploca arborea* Dennst., Schluessel Hort. 13, 23, 35, 35. 1818. *Wrightia tomentosa* Roem. & Schutt. Syst. Veg. 4:414. 1819; Wight, Ic. t. 443. 1841; Hook.f., Fl. Brit. India 3:653. 1882; Gamble, Fl. Pres. Madras 816. 1923. *Nelem-pala* Rheede, Hort. Malab. 9:5, t 3-4. 1689.

Small to medium sized trees. Leaves broadly elliptic, 8-17 by 4.5-9 cm, base rounded or broadly acute, apex acuminate, lower surface tomentose, chartaceous. Cymes terminal, pubescent; calyx-lobes linear-lanceate; corolla reddish, c.2 cm ϕ; corona scales oblong. Follicle stout, c.20 cm long.

Fl. & Fr.: Apr.-Dec. *Distr.*: Not common. In evergreen forests. *NAK 2435* (Appuppanthode, Kalleli, R.F., c. 200 m).

Root used as an antidote against snake bite.

2. **Wrightia tinctoria** (Roxb.) R. Br., Mem. Wern. Nat. Hist. Soc. 1:74. 1811; Hook. f., Fl. Brit. India 3:653. 1882; Gamble, Fl. Pres. Madras 815. 1923; Rani & Matthew in Matthew, Fl. Tam. Carnatic 3(2):921. 1983; Manilal, Fl. Silent Valley 175. 1988; Mohanan & Sivad., Fl. Agasthyamala 446.2002. *Nerium tinctorium* Roxb., Orient. Repert. 1:39, cum tab. 1791. *Thattan-chavana.*

Small trees. Leaves elliptic-oblong, 6-15 by 3-5 cm, base cuneate, apex acuminate, chartaceous. Cymes subterminal; calyx-lobes 5, ovate; corolla white, c.2.5 cm ϕ; corona scales fimbriate. Follicles slender, c.35 cm long.

Fl. & Fr.: Apr.-Jul. *Distr.*: Peninsular & Central India, Myanmar. Less common. In evergreen forests. *NAK 654, 1673* (Sabari hills, c. 200 m).

Fresh leaves chewed as a remedy in tooth ache.

Allamanda cathartica L. (NAK 104) often run wild. **Nerium oleander** L., **Cascabela thevetia** (L.) Lippold, **Plumeria rubra** L. and **Tabernaemontana divaricata** (L.) R. Br. are seen cultivated as garden plants here and there.

85. ASCLEPIADACEAE

1. Erect shrubs; leaf surface farinose. .... **Calotropis**
1. Twining shrubs (except *Ceropegia fimbriifera*); leaf surface not so.
   2. Corolla pitcher-like, zygomorphic, lobes connate by their tips. .... **Ceropegia**
   2. Corolla otherwise, actinomorphic, lobes free.
      3. Leaves thick, fleshy; corona large, stellate. .... **Hoya**
      3. Leaves thin, chartaceous; corona small, spreading
         4. Leaves cordiform.
            5. Corolla greenish-yellow, spotted purple; corona double, scaly. .... **Cosmostigma**
            5. Corolla uniformly greenish; corona single, fleshy. .... **Watakaka**
         4. Leaves oblong or elliptic (rarely cordiform in *Tylophora indica*).
            6. Corolla lobes valvate. .... **Hemidesmus**
            6. Corolla lobes twisted or imbricate.
               7. Corona 5, distinct.
                  8. Corona staminal, closely adnate with staminal column. ..**Tylophora**
                  8. Corona corolline, free. .... **Cryptolepis**
               7. Corona, 1 tubular, plicate, 10-lobed. .... **Cynanchum**

CALOTROPIS R. Brown
On Asclepiad. 28.1810.

Calotropis gigantea (L.) R. Br. in Aiton f., Hort. Kew. 2:78. 1811; Hook.f., Fl. Brit. India 4:17. 1883; Gamble, Fl. Pres. Madras 832. 1923; Stevens in Sald. & Nicols., Fl. Hassan Dist. 440. 1926; Matthew, Ill. Fl. Tam. Carnatic t. 436. 1982; Matthew & Rani in Matthew, Fl. Tam. Carnatic 3(2): 929. 1983; Manilal, Fl. Silent Valley 176. 1988; Nicols., Suresh & Manilal, An Interpr. Hort. Malab. 61. 1988. *Asclepias gigantea* L., Sp. Pl. 214. 1753. *Ericu, Bel-ericu* Rheede, Hort. Malab. 2:53-56, t.31. 1679.

Undershrubs. Leaves subsessile, simple, decussate, elliptic-ovate or oblong-ovate, 10-20 by 5-10 cm, base auricled, apex subacute, farinose, thin-coriaceous. Inflorescence an umbellate panicle, subterminal. Flowers bisexual, actinomorphic; calyx deeply 5-lobed, lobes ovate; corolla pale purple, rotate, lobes 5, ovate, spreading; stamens 5, filaments connate into staminal column, anthers adnate to stigma forming gynostegium; corona staminal, single, laterally compressed, basally incurved, horny, apex 3-fid; ovaries 2, free; ovules α, on marginal placenta. Follicles in pairs, inflated, 8 by 4 cm. Seeds ovate, comose.

Fl. & Fr.: Most of the seasons. *Distr.*: Trop. Asia. Common. Along waysides and exposed waste lands. *NAK 2672* (Tiruvalla, c. 10 m).

Plants yield a fine fibre. Leaves used with pepper for snake bite. Leaf juice boiled with coconut oil and applied as a remedy in itches.

CEROPEGIA Linnaeus
Sp. Pl. 211. 1753.

Twiners or erect herbs. Leaves simple, decussate, linear or ovate or ovate orbicular. Inflorescence an umbellate or recemose cyme or reduced to a single flower, axillary. Flowers 5-merous, bisexual, hypogynous; calyx glandular within, lobes 5, linear, corolla pitcher-like, tube inflated at base, constricted at middle, funnel shaped at apex, lobes 5, apically connate; stamens 5; corona double, outer cupular, 5-10- lobed, inner of 5 clavoid or turbinate processes. Pollinia waxy with pellucid margins. Follicles linear.

1. Erect herbs; leaves linear; flower (s) solitary. ....2. **C. fimbriifera**
1. Twiners; leaves otherwise; flowers in umbellate cymes.
    2. Leaves glabrous; corolla to 4 cm long; follicle 10-12 cm long. ....1. **C. candelabrum**
    2. Leaves puberulous; corola to 8 cm long; follicle c. 30 cm long. ....3. **C. metziana**

1. **Ceropegia candelabrum** L., Sp. Pl. 211. 1753; Hook.f., Fl. Brit. India 4:79. 1883; Gamble, Fl. Pres. Madras 857. 1923; Huber, Mem. Soc. Brot. 12:58. 1957, p.p.; Stevens in Sald. & Nicols., Fl. Hassan Dist. 443. 1976; Manilal, Fl. Silent Valley 176. 1988.

Key to the varieties

1. Flower buds with ± 5 mm long beak. ....**Ceropegia candelabrum** var. **candelabrum**
1. Flower buds with ± 10 mm long beak. ....**Ceropegia candelabrum** var. **biflora**

Ceropegia candelabrum L. var. **candelabrum (Fig. 47)**

Slender twiners. Leaves ovate, 2-4 by 1-2.5 cm, base cordate, margin ciliate, apex acuminate-apiculate, chartaceous. Flowers 2-8, in axillary cymes. Panicle short; calyx lobes equal; corolla c. 2 cm long, tube purplish, lobes yellow towards base and greyish at tip, beak c. 5 mm long, ciliate within. Follicle cylindric.

Fl. & Fr.: Nov.-Jan. *Distr.*: Peninsular India. Not common. Along forest borders. *NAK 2094* (Mannarappara, c.100 m).

Ceropegia candelabrum L. var. **biflora** (L.) M.Y. Ansari, Fasc. Fl. India 16:12.1984. *C. biflora* L., Sp. Pl. 211. 1753. *C. tuberosa* Roxb., Pl. Corom. t.9.1795; Hook.f., Fl. Brit. India 4:70. 1883, p.p.; Gamble, Fl. Pres. Madras 857. 1923; Stevens in Sald. & Nicols., Fl. Hassan Dist. 445. 1976; Matthew & Rani in Matthew, Fl. Tam. Carnatic 3(2): 935. 1983, p.p.; Nicols., Suresh & Manilal, An Interpr. Hort. Malab. 61. 1988. *Njota-njodien-valli* Rheede, Hort. Malab. 9:27-28, t. 16.1689.

Stout twiners. Leaves ovate-lanceate or elliptic-lanceate, 2-5 by 1.3-2.5 cm, base subcordate, apex subacute- apiculate, herbaceous. Flowers 5-15, in axillary cymes. Peduncle long; calyx-lobes linear; corolla greenish-yellow, c.3 cm long, beak c.1 cm long; outer corona membranous, inner herbaceous, suborbicular. Follicles cylindric.

Fl. & Fr.: Jan.-Apr. *Distr.*: Cental & Southern Peninsular India, Sri Lanka. In hills. Occasional. *NAK 2631* (Moozhiar, c.250 m).

2. **Ceropegia fimbriifera** Bedd., Madras Journ. Lit. Sci. 3(1): 53. 1864, Ic. Pl. Ind. Or. 1:38,t.172.1874; Hook.f., Fl. Brit. India 4: 66. 1883; Gamble, Fl. Pres. Madras 856.1923. M. Y. Ansari, Fasc. Fl. India 16:16. 1984.

Erect tuberous rooted herbs; stem terete, glabrous, internodes 2-4 cm long. Leaves simple, linear, to 12 by 0.5 cm, base and apex acute, margin entire, 1-nerved, glabrescent; petiole 0 or very minute. Flower(s) greenish-yellow, solitary; pedicel c. 5 mm long, glabrous; bracts subulate, 1-2 mm long; calyx-lobes 5, linear, c. 3 mm long; corolla-tube ± 2 cm long, inlfated below, tubular portion c. 1 cm long, glabrous inside, lobes 5, lanceate, c.1 cm long, connate at tip, villous hairy inside at base; corona lobes 10, 2-seriate, outer lobes bifid, subulate, divergent, inner lobes spathulate; anther without appendages, pollen masses waxy; ovaries 2, style-apex truncate. Fruits not seen.

Fl.: Sept. *Distr.*: Endemic to South India. Very rare. Among grassy slopes of rocky hills. *NAK* 2017A, 2017B.

3. **Ceropegia metziana** Miq., Ann. Bot. Ind. 3:11.1852; M.Y. Ansari, Fasc. Fl. India 16:25. 1984. *C. brevicollis* Hook.f., Fl. Brit. India 4:74. 1883; Gamble, Fl. Pres. Madras 859. 1923. *C. stocksii* sensu Gamble, Fl. Pres. Madras 859. 1923, non Hook.f., 1883.

Stout twiners. Leaves ovate, elliptic or elliptic-lanceate, 6-14 by 3.6-5 cm, base rounded, apex shortly acuminate, puberulous on both surfaces. Flowers 3-5, in axillary cymes; calyx-lobes linear; corolla purplish with dark blotches, c.8 cm long, tube much inflated, lobes 3.5 cm long. Follicles c. 30 cm long.

Fl. & Fr.: Sep.-Dec. *Distr.*: Western Ghats. Rare. Along forest borders. *NAK 42 & 1366* (Kakki hills, c. 1100 m).

COSMOSTIGMA R. Wight
Contrib. Bot. India 41. 1834.

Cosmostigma racemosum (Roxb.) Wight, Contrib. Bot. India 42. 1834 & Ic.t. 591. 1842; Hook.f., Fl. Brit. India 4:46. 1883; Gamble, Fl. Pres. Madras 846. 1923; Huber in Abeyw., Rev. Handb. Fl. Ceylon 1(1):49.1973; Stevens in Sald. & Nicols., Fl. Hassan Dist. 445. 1976; Matthew, Ill. Fl. Tam. Carnatic tt. 440 & 973 d. 1982; Matthew & Rani in Matthew, Fl. Tam. Carnatic 3(2) : 938. 1983; Nicols., Suresh & Manilal, An Interpr. Hort. Malab. 61. 1988; Mohanan & Sivad., Fl. Agasthyamala 449.2002. *Wattou-valli* Rheede, Hort. Malab. 7:61, t. 32. 1688.

Stout climbers. Leaves simple, decussate, cordiform, 8-10 by 6-15cm, base cordate, apex acuminate, basally 5-7- veined, chartaceous. Cymes racemose, dense, on long peduncles. Flowers bisexual, 5-merous, hypogynous; calyx-lobes 5, ovate, herbaceous; corolla greenish-yellow, spotted purple, rotate; corona staminal, double, bifid at apex; pollinia erect, oblong. Follicles stout, c.15 cm long.

Fl. & Fr.: Jun.-Sep. *Distr.*: Sri Lanka, India. Not common. On hilly slopes of evergreen forests. *NAK 445, 1976* (Moozhiar, c. 200 m).

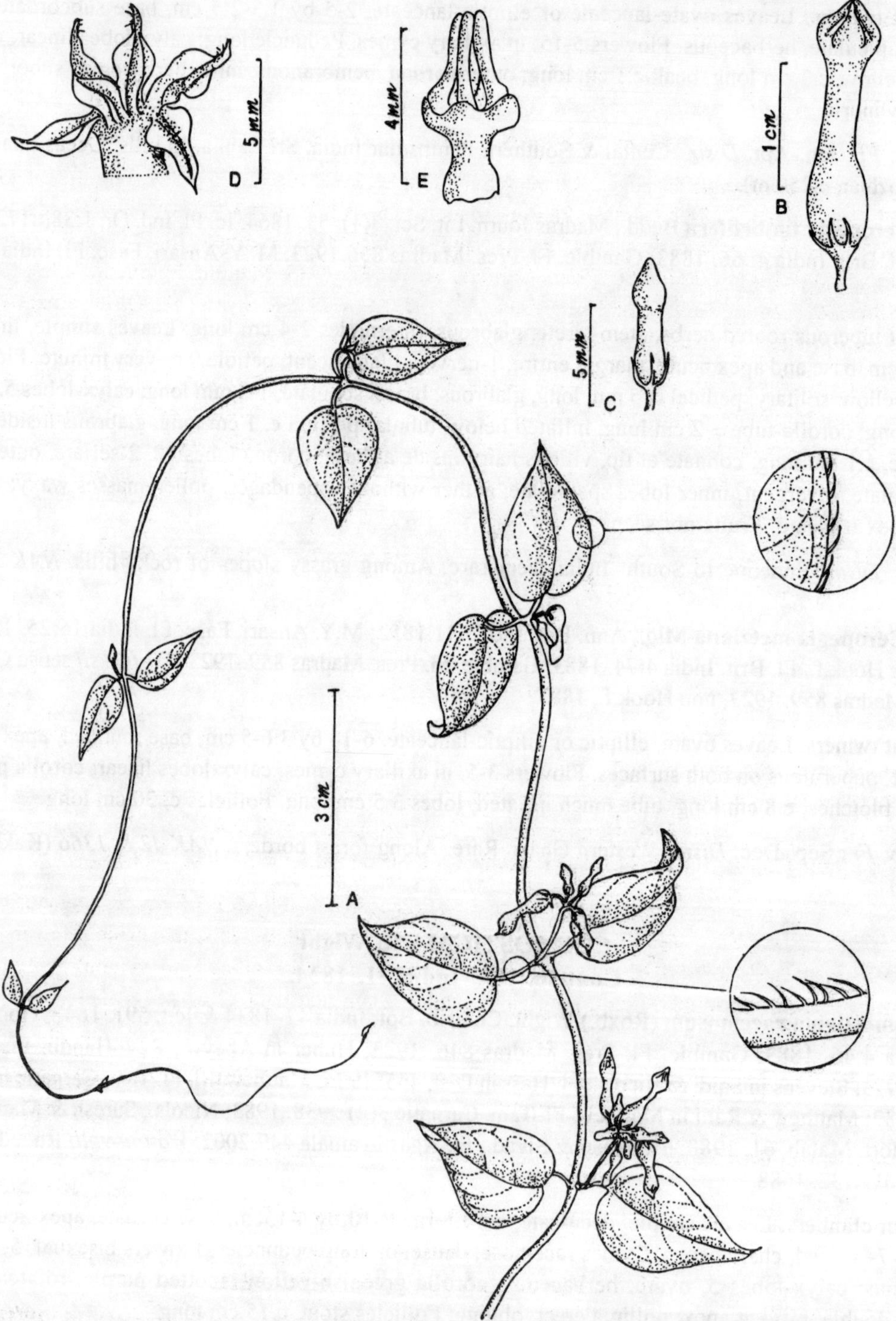

Fig. 47. ***Ceropegia candelabrum*** **L. var.** ***candelabrum*** **: A. Twig; B. Flower; C. Flower Bud; D. Corolla - upper portion; E. Corona.**

Leaves used to cure ulcerous sores.

CRYPTOLEPIS R. Brown
on Asclepiad. 58. 1810;
Mem. Wern. Nat. Hist. Soc. 1:69. 1811.

Cryptolepis buchanani R. Br. ex Roem. & Schult. in L., Syst. Veg. 4:409. 1819; Wight, Ic. t.494. 1841; Hook.f., Fl. Brit. India 4:5. 1883; Gamble, Fl. Pres. Madras 826. 1923. Huber in Abeyw., Rev. Handb. Fl. Ceylon 1(1) : 29.1973; Stevens in Sald. & Nicols., Fl. Hassan Dist. 446. 1976; Matthew, Ill. Fl. Tam. Carnatic t. 441. 1982; Matthew & Rani in Matthew; Fl. Tam. Carnatic 3(2) : 939. 1983; Manilal, Fl. Silent Valley 178. 1988; Nicols., Suresh & Manilal, An Interpr. Hort. Malab. 62. 1988; Mohanan & Sivad., Fl. Agasthyamala 454.2002. *Katu-pal-valli* Rheede, Hort. Malab. 9:17-18, t. 32. 1688.

Stout climbers with milky latex. Leaves simple, decussate, elliptic-oblong, 6-12 by 3-5.5 cm, base acute or cuneate, apex obtuse-cuspidate, lower surface glaucous, chartaceous. Inflorescence an axillary umbellate-raceme. Flowers bisexual, regular, 5-merous; calyx-lobes 5, ovate; corolla white, shortly tubular, lobe imbricate, herbaceous; stamens 5, connivent around stigma; corona single, corolline; ovaries 2, ovules α, on marginal placenta. Follicles lanceoid, c. 7 m long.

Fl. & Fr.: Mar.-Nov. *Distr.*: Sri Lanka, Peninsular India. Less common. On thickets and teak plantations. *NAK 2138, 2346* (Moozhiar, c. 200 m).

Leaf juice used as a medicine to cure rickets in children.

CYNANCHUM Linnaeus
Sp. Pl. 212. 1753.

Cynanchum callialatum Buch.–Ham. ex Wight & Arn. in Wight, Contrib. Ind. Bot. 56. 1834; Wight, Ic.t. 1279. 1848; Hook.f., Fl. Brit. India 4 : 24. 1883; Gamble, Fl. Pres. Madras 836. 1923; Stevens in Sald. & Nicols, Fl. Hassan Dist. 446.1976; Matthew, Ill. Fl. Tam. Carnatic tt. 442 & 974 a. 1982; Matthew & Rani in Matthew, Fl. Tam. Carnatic 3(2) : 942. 1983; Manilal, Fl. Silent Valley 178. 1988 ('*callialata*').

Twiners. Leaves simple, decussate, ovate to ovate-lanceate, 3-10 by 1.5-5.5 cm, base subauricled, apex cuspidate, lower surface glaucous, herbaceous. Inflorescence an axillary umbellate cyme. Flowers bisexual, regular, 5-merous; calyx-lobes 5, equal; corolla white, rotate, lobes 5, spreading; corona tubular, single, plicate, 10-lobed; pollinia pendulous, oblong; ovaries 2; ovules α, on marginal placenta. Follicle ovoid, c. 8 cm long, winged. Seeds comose.

Fl. & Fr.: Sep.-Mar. *Distr.*: Myanmar to South India. Common in forest clearings, and teak plantations. *NAK 179, 926, 987 & 1273* (Almost all hills).

Note : One collection (*NAK 1273*) with linear-lanceate subcordate leaves was made.

HEMIDESMUS R. Brown
on Asclepiad. 45. 1810;
Mem. Wern. Nat. Hist. Soc. 1:56. 1811.

Hemidesmus indicus (L.) R. Br. in Aiton f., Hort. Kew. 2:75. 1811; Wight, 1c.t.594. 1842; Hook. f., Fl. Brit. India 4:5. 1883; Gamble, Fl. Pres. Madras 825. 1923. Huber in Abeyw., Rev. Handb. Fl. Ceylon 1(1): 30.

1973; Stevens in Sald. & Nicols., Fl. Hassan Dist. 448. 1976; Matthew & Rani in Matthew, Fl. Tam. Carnatic 3(2) : 947. 1983; Manilal, Fl. Silent Valley 178 1988; Nicols., Suresh & Manilal, An Interpr. Hort. Malab. 62. 1988; Mohanan & Sivad., Fl. Agasthyamala 455.2002. *Periploca indica* L., Sp. Pl. 211. 1753. *Naru-nindi* Rheede, Hort. Malab. 10:67-68, t. 34. 1690.

Stragglers. Leaves simple, decussate, broadly elliptic to linear, 3-7 by 0.7-3 cm, base rounded, apex gradually acute-apiculate, thin-coriaceous. Inflorescence a lateral subsessile cyme. Flowers bisexual, regular, 5-merous; calyx-lobes 5, broadly ovate; corolla yellow, rotate, lobes 4, fleshy, ovate; corona single, corolline; stamens 5, forming gynostegium. Follicles slender, cylindric, c.10 cm long.

Fl. & Fr.: Jan.-Oct. *Distr.*: Trop. Africa, Sri Lanka, India. Fairly common. In disturbed areas of hills and plains. *NAK 2243* (Mannarappa, c.100 m).

Root is used as substitute for sarasparilla. Fruit pound with milk and given as a remedy for stone in urinary bladder. It improves appetite.

HOYA R. Brown
On Asclepiad. 15. 1810;
Mem. Wern. Nat. Hist. Soc. 1:26. 1811.

Hoya ovalifolia Wight & Arn. in Wight, Contrib. Ind. Bot. 37. 1834; Wight, 1c.t.847.1845; Hook.f., Fl. Brit. India 4:60. 1883; Gamble, Fl. Pres. Madras 849. 1923; Stevens in Sald. & Nicols., Fl. Hassan Dist. 449. 1976; Nicols., Suresh & Manilal, An Interpr. Hort. Malab. 63. 1988. *Mareta-inali* Rheede, Hort. Malab. 11:129, t. 63. 1692. **(Fig. 48)**

Epiphytic subsucculent climbing shrubs, rooting at nodes. Leaves simple, decussate, elliptic-lanceate, 6-9 by 3-4 cm, base cuneate, margin undulate, apex acuminate, thick-fleshy. Flowers in axillary umbels, 5-merous, bisexual; calyx small, deeply 5-lobed; corolla rotate, cream coloured, lobes 5, fleshy; corona scales 5, stellate-spreading, fleshy, concave on upper surface; stamens 5, anthers connivent around stigma; pollen masses globose. Follicles slender, curved, c.10 cm long.

Fl. & Fr.: May-Aug. *Distr.:* Sri Lanka, Peninsular India. Rare. In moist deciduous forests. Extensive climbers on large trees. *NAK 2377* (Padom, c.75 m).

TYLOPHORA R. Brown
On Asclepiad. 17.1810;
Mem. Wern. Nat. Hist. Soc. 1.28.1811.

Twiners. Leaves simple, decussate, chartaceous. Inflorescence of axillary cymes. Flowers bisexual, 5-merous, regular; calyx-lobes 5, herbaceous; corolla rotate, lobes 5, ovate; corona fleshy, staminal, adnate to staminal column; ovaries oblong; ovules α. Follicles paired. Seeds comose.

1. Stout twiners; pedicel thick, pubescent; corolla c.1 cm ϕ. .... 1. **T. indica**
1. Slender twiners; pedicel slender, glabrous; corolla c. 4 mm ϕ. .... 2. **T. pauciflora**

1. **Tylophora indica** (Burm.) Merr., Philipp. J. Sci. Bot. 19:373. 1921; Stevens in Sald. & Nicols., Fl. Hassan Dist. 454. 1976; Matthew & Rani in Matthew, Fl. Tam. Carnatic 3(2) : 962. 1983; Nicols., Suresh & Manilal, An Interpr. Hort. Malab. 64. 1988. *Cynanchum indicum* Burm., Fl. Ind. 70. 1768. *Tylophora asthmatica*

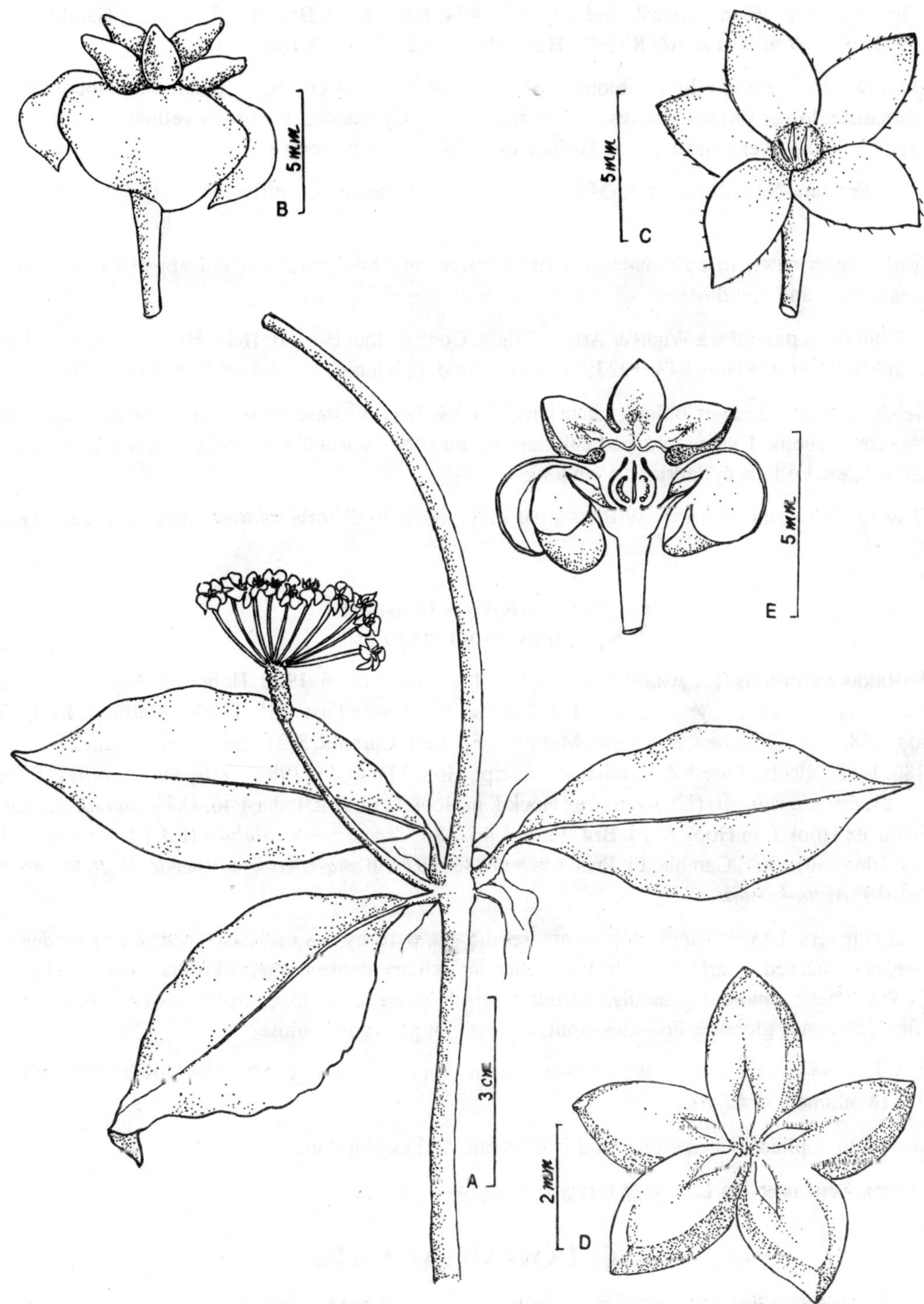

Fig. 48. ***Hoya ovalifolia*** **Wight & Arn.: A. Twig; B. Flower; C. Calyx; D. Corona; E. Flower - L.S.**

(L.f.) Wight & Arn. in Wight, Contrib. Ind. Bot. 51. 1834; Hook.f., Fl. Brit. Ind. 4:44.1883; Gamble, Fl.Pres. Madras 843. 1923. *Nansjera-patsja* Rheede, Hort. Malab. 9:21-22, t. 13.1689.

Stout twiners. Leaves elliptic-oblong to ovate, 5-10 by 2.5-4 cm, base rounded to subcordate, apex obtuse-apiculate, lower surface glaucescent; pedicels stout. Cymes dense; corolla yellow, c.1 cm ϕ; corona free at apex, arching over gynostegium. Follicle cylindric, to 7 cm long.

Fl. & Fr.: Jun.-Nov. *Distr.*: Indo-Malesia. In Plains; occasional. Cultivated? *NAK 124* (Parakode, c. 20 m).

Fresh leaves eaten to cure asthma. Plant bruised and boiled in oil, and applied in rheumatism, biliousness, head and ear diseases.

2. **Tylophora pauciflora** Wight & Arn. in Wight, Contrib. Ind. Bot. 50. 1834; Hook., Fl. Brit. India 4:41. 1883; Gamble, Fl. Pres. Madrs 843. 1923; Stevens in Sald. & Nicols., Fl. Hassan Dist. 454. 1976.

Slender twiners. Leaves oblong-lanceate, 7-11 by 3-5 cm, base truncate or rounded, apex acute-apiculate, chartaceous. Cymes lax; pedicel slender, purplish; corolla 3.7 mm ϕ; corona lobes globose, cuspidate at apex. Follicle cylindric, c.5 cm long.

Fl. & Fr.: Jun.-Aug. *Distr.*: S. W Peninsular India. Rare. In disturbed forest areas. *NAK 2345* (Kalleli, c. 50 m).

WATTAKAKA Hasskarl
Flora 40:99. 1857.

Wattakaka volubilis (L.f.) Stapf in Curtis, Bot. Mag. sub.t. 8976. 1922; Huber in Abeyw., Rev. Handb. Fl. Ceylon 1(1) : 50. 1973; Stevens in Sald. & Nicols., Fl. Hassan Dist. 455. 11976; Matthew, Ill. Fl. Tam. Carnatic t. 458. 1982; Matthew & Rani in Matthew, Fl. Tam. Carnatic 3(2) : 963. 1983; Manilal, Fl. Silent Valley 180. 1988; Nicols., Suresh & Manilal, An Interpr. Hort. Malab. 65. 1988. *Asclepias volubilis* L.f., Suppl. Pl. 170. 1781. *Dregea volubilis* (L.f.) Benth. ex Hook.f. in Hook.f., Fl. Brit. India 4:46. 1883. *Marsdenia volubilis* (L.f.) Benth. ex Hook.f. in Hook.f., Fl. Brit. India 4:46. 1883. *Marsdenia volubilis* (L.f.) T. Cooke, Fl. Pres. Bombay 2:166 (230) 1904; Gamble, Fl. Pres. Madras 846. 1923. *Watta-kaka-kodi* Rheede, Hort. Malab. 9:25-26, t. 15.1689. *Watta-kurinji.*

Stout climbers. Leaves simple, decussate, cordiform, 6-10 by 5-8 cm, base cordate or rounded, apex acuminate, penninerved, chartaceous. Inflorescence an axillary dense umbel. Flowers bisexual, 5-merous, regular; calyx-lobes 5, lanceate, glandular; corolla greenish, rotate, c.15 mm ϕ, lobes 5, ovate; corona staminal, single, fleshy; ovaries globose. Follicles stout, c. 8 cm long. Seeds comose.

Fl. & Fr.: Jun.-Aug. *Distr.*: S. W Peninsular India. Not common. In exposed and disturbed forest areas. *NAK 2345* (Kodumon, c. 40 m).

Root paste applied in snake bite; and it is emetic and expectorant.

Asclepias curassavica L. is seen rarely running wild in plains.

86. LOGANIACEAE

1. Leaves subsucculent, shiny; flowers bright yellow, c. 4 cm ϕ, 5-merous. .... **Fagraea**
1. Leaves chartaceous, not shiny; flowers greenish or white, c. 8 mm ϕ, 4-5-merous.

2. Small erect herbs; flowers 4-merous. ....**Mitrasacme**

2. Large trees or climbing shrubs; flowers 5-merous. ....**Strychnos**

FAGRAEA Thunberg
Kongl. Vetensk. Acad. Nya. Handb. 3:132.1782.

Fagraea ceilanica Thumb., Kongl. Vetensk. Acad. Nya. Handb. 3:132.1782; Clarke in Hook.f., Fl. Brit. India 4:83.1883, '*zeylanica*'; Gamble, Fl. Pres. Madras 865. 1923; Leenh. in Steenis, Fl. Males. Ser.I, 6:315. 1962; Stevens in Sald. & Nicols., Fl. Hassan Dist. 418.1976; Manilal, Fl. Silent Valley 180.1988; Nicols., Suresh & Manilal, An Interpr. Hort. Malab. 162.1988; Mohanan & Sivad., Fl. Agasthyamala 456.2002. *Modogam* Rheede, Hort. Malab. 4:119, t.58.1683.

Epiphytic shrubs. Leaves simple, decussate, obovate, 8-18 by 5-10 cm, base cuneate, apex obtuse, shiny, subsucculent. Inflorescence a raceme or panicle. Flowers 4-or 5-merous, bisexual; calyx deeply 5-lobed; corolla large, bright yellow, funnel-form, c. 4 cm ϕ, lobes 5, broad, twisted to right; stamens 5, adnate at base of corolla-tube, exerted; ovary superior, 2-locular; ovules α. Fruit on ovoid Berry, c. 3 cm ϕ, shiny.

Fl. & Fr.: May-Aug. *Distr.*: Indo-Malesia. Common. In evergreen forests. Scandent or erect small trees; sometimes growing as epiphytes. *NAK 1658, 1822 & 1830* (Angamoozhy, Moozhiar etc., c.200 m).

MITRASACME Labillardiere
Novae Holl. Pl. Spec. 1:35.1805.

Annual erect slender herbs; stem 4-angled. Leaves simple, cauline or sub-radical. Flowers 1-3, axillary or in terminal umbels, 4-merous, bisexual; calyx small, lobes valvate; corolla salver-form, lobes 4, valvate; stamens 4, included; ovary ovoid, superior; styles 2, connate at apex; ovules α. Capsule globose. Seeds α, ellipsoid.

1. Vegetative parts glabrous; leaves linear; flower(s) axillary, solitary or rarely in pairs. ....1. **M. indica**

1. Vegetative parts hispid; leaves ovate; flowers usually in terminal lax umbels. ....2. **M. pygmaea** var. **malaccensis**

1. **Mitrasacme indica** Wight, Ic.t. 1601.1850; Leenh. in Steenis, Fl. Males. I, 6:384. 1962; Matthew, Ill. Fl. Tam. Carnatic t. 460.1982; Matthew & Rani in Matthew, Fl. Tam. Carnatic 3(2): 965.1983. *M. alsinoides* auct., non R.Br., 1810: Clarke in Hook. f., Fl. Brit. India 4:80. 1883; Gamble, Fl. Pres. Madras 864.1923.

Herbs to 8 cm tall. Leaves linear, 5 by 1.5 mm, base and apex acute, subsessile; petiole faintly sheathing at base. Flower(s) solitary or 1-3 in axillary fascicles; calyx-lobes very small; corolla white, lobes ovate; stamens inserted at throat. Capsule subglobose. Seeds reticulate.

Fl. & Fr.: Jun.-Aug. *Distr.*: Sri Lanka, India, Myanmar, Japan, Australia, Malesia. Not common. In exposed grasslands. *NAK 2177* (Chittar, Chootuporuthu para, c.100 m).

2. **Mitrasacme pygmaea** R. Br., Prodr. 453.1810; Ohwi, Fl. Japan 734.1965, var. **malaccensis** (Wight) Hara, J. Jap. Bot. 30:24.1955; Backer & Bakh. f., Fl. Java 2:208.1965; Stevens in Sald. & Nicols., Fl. Hassan Dist. 420.1976. *M. malaccensis* Wight, Ic.t. 1601.1850. *M. polymorpha* auct., non R.Br., 1810: Clarke in Hook.f., Fl. Brit. India 4:80.1883; Gamble, Fl. Pres. Madras 864.1923.

Herbs to 5 cm tall; branches and leaves hispid. Leaves subradical, ovate or ovate-lanceate, 7 by 5 mm, base rounded, apex acute, chartacecous. Flowers few in terminal for subterminal umbels; calyx-lobes small; corolla white, lobes suborbicular. Capsule depressed globose. Seeds ellipsoid, punctate.

Fl. & Fr.: Jun.-Aug. *Distr.:* India to Japan. Rare. On moist grasslands. *NAK 1917* (Vallicode-Kottayam, c.100 m).

STRYCHNOS Linnaeus
Sp.Pl.189.1753.

Trees or scandent shrubs, often tendrillar. Leaves simple, decussate; ovate to orbicular, basally 3-5-ribbed, thin-coriaceous. Flowers in terminal cymes, bisexual, 5-merous; calyx cupular, lobes 5, acute; corolla salver-form, greenish, lobes 5, reflexed; stamens 5, included; ovary superior, depressed globose; ovules few to α. Fruit a large, thick or thin-walled berry. Seeds lenticular or ovate.

1. Trees; flowers greenish; leaves suborbicular or broadly ovate. ....2. **S. nux-vomica**
1. Climbing shrubs; flowers yellow; leaves elliptic or elliptic-ovate.
    2. Corolla-tube and lobes ± equal in length, throat wooly; tendril 2-fid; fruit, c.2 cm ϕ. ....1. **S. minor**
    2. Corolla-tube much longer than lobes, throat sparsely pilose; tendril simple; fruit, c.8 cm ϕ. ....3. **S. wallichiana**

1. **Strychnos minor** Dennst., Schluess., Hort. Malab. 33. 1818; Bisset & Philcox., Taxon 20: 542. 1971. *S. bicirrhosa* Lesch. ex Roxb., Fl. Ind. 2: 267. 1824; Gamble, Fl. Pres. Madras 868. 1923; Nicols., Suresh & Manilal, An Interpr. Hort. Malab. 164. 1988. *S. colubrina sensu* Clarke in Hook. f., Brit. India 4: 87. 1883, p.p., non L., 1753. *Scherukatu-valli-kaniram* Rheede, Hort. Malab. 7: 9, t. 5. 1688.

Climbing shrubs. Tendrils 2-branched. Leaves 6-11 by 3-5 cm, base rounded, apex acuminate, 3(-5)-nerved, coriaceous. Cymes terminal. Flowers small; calyx-lobes elliptic; corolla-tube ± equal to lobes, throat woody. Berry small, brownish, c. 2 cm ϕ.

Fl. & Fr.: Nov.-Feb. *Distr.*: South India. Common. In sacred groves of lowlands. *NAK 2329* (Kodumon, c. 20 m).

2. **Strychnos nux-vomica** L., Sp. Pl. 189.1753; Roxb., Pl. Corom. t.4.1795 & Fl. Ind. 2:261.1824. Bedd., Fl. Sylv. t.243.1872; Clarke in Hook.f., Fl. Brit. India 4:90.1883; Gamble, Fl. Pres. Madras 868.1923; Leenh.in Steenis, Fl. Males. I. 6:349.1962; Matthew, Ill. Fl. Tam. Carnatic t.461.1982; Matthew & Rani in Matthew, Fl. Tam. Carantic 3(2):966.1983; Nicols., Suresh & Manilal, An Interpr. Hort. Malab. 164.1988. *Caniram* Rheede, Hort. Malab. 1:67-68, t.37.1678.

Large trees. Leaves broadly ovate to suborbicular, 5-8 by 4-7 cm, base cuneate, apex obtuse, basally 3-5-nerved, coriaceous. Cymes terminal; Flowers α; calyx-lobes lanceate; corolla-tube longer than lobes; stamens subexerted; ovary depressed globose. Berry large, globose. Seeds lenticular, c. 2 cm ϕ.

Fl. & Fr.: Feb.-Nov. *Distr.*: Sri Lanka, India, Indo-China, Malesia. Common. In sacred groves and evergreen forests. *NAK 2929* (Adoor, c. 20 m).

Juice of fresh bark given in cholera and dysentery. Seeds powdered and given in leprosy, paralysis and snake bites.

3. **Strychnos wallichiana** Steud. ex DC., Prodr. 9:13, 1845; Clarke in Hook. f., Fl. Brit. India 4:90.1883; Bisset & Philcox, Taxon 20:542.1971. *S. colurbrina* L., Sp. Pl. 1:189.1753, auct., nom. confusum. *S. cinnamomifolia* Thw., Enum. 201.1864; Hook.f., Fl. Brit. India 4:89.1883. *S. rheedei* Clarke in Hook.f., Fl. Brit. India 4:87.1883. *S. malaccensis* Clarke in Hoook. f., Fl. Brit. India 4:89.1883. *S. bourdilloni* Brandis, Indian Trees 474.1906. *S. cinnamomifolia,* var. *wightii* Hill, Kew Bull. 1917:194.1917; Gamble, Fl. Pres. Madras 869.1923. *Valli-kanjiram.*

Climbing shrubs. Tendrils unbranched. Leaves elliptic, 4-9 by 2.5-4.5 cm, base rounded or broadly cuneate, apex acuminate, basally 3-nerved, coriaceous. Cymes terminal. Flowers small; calyx-lobes lanceate; corolla-tube longer than lobes, throat sparsely pilose. Berry large, c. 8 cm ϕ, globose. Seeds flat, ovoid.

Fl. & Fr.: Jun.-Sep. *Distr.*: Sri Lanka, India, Malesia. Less common. In evergreen forests. *NAK 501* (Angamoozhy, c.200 m).

Decoction prepared from roots used for rheumatic ulcers, elephantiasis, fever and epilepsy in children.

87. GENTIANACEAE

1. Stamens 4-5, all fertile; ovary 1 or 2- locular.
   2. Ovary 2-locular; style filiform, undivided. .... **Exacum**
   2. Ovary 1-locular; style stout, bilobed. .... **Swertia**
1. Stamens 4, only one fertile; ovary 1-locular.
   3. Flowers pink or white; stigma 2-fid. .... **Canscora**
   3. Flowers yellow, stigma 2-lobed. .... **Hoppea**

CANSCORA Lamarck
Encycl. Meth. Bot. 1:601. 1785.

Annual erect herbs; stem 4-angled or 4-winged. Leaves simple, decussate, basally 3(-5)- nerved, herbaceous or chartaceous. Inflorescence of 2-chotomous terminal or axillary cymes. Flowers 4-merous, bisexual, regular; bracts foliaceous, or subulate, often perfoliate; calyx 4-toothed; corolla tubular, 4-lobed; stamens 4, cupular inserted at middle of corolla-tube, one only fertile; ovary superior, 1-locular; ovules α, on parietal placenta. Capsule oblong-cylindric or ellipsoid. Seeds α, minute.

1. Bracts broadly ovate, perfoliate.
   2. Calyx 4-winged, veins reticulate. .... **2. C. perfoliata**
   2. Calyx wingless, veins stirate. .... **3. C. roxburghii**
1. Bracts linear, not perfoliate. .... **1. C. diffusa**

1. **Canscora diffusa** (Vahl) R. Br., Prodr. 451, (in Obs). 1810; Clarke in Hook.f., Fl. Brit. India 4:103. 1883; Gamble, Fl. Pres. Madras 878. 1923; Ramamoorthy in Sald. & Nicols., Fl. Hassan Dist. 424. 1976; Cramer in Dassan. & Fosberg, Rev. Handb. Fl. Ceylon 3:71.1981; Matthew & Rani in Matthew, Fl. Tam. Carnatic 3(2): 972. 1983; Manilal, Fl. Silent Valley 181.1988. *Gentiana diffusa* Vahl, Symb. Bot. 3:47.1794. *Canscora lawii* Wight, Ic. t. 1327. 1848. *Jeeraka-pullu.*

Slender herbs; stem narrowly 4-winged. Leaves sesile, lower ones ovate-elliptic, 3 by 1.5 cm, upper ones linear-lanceate, base rounded or broadly acute, apex acute, basally 3-nerved, chartaceous. Panicle lax, terminal and axillary; bracts lanceate; calyx-lobes equal, setaceous; corolla pink, lobes spreading. Capsule narrowly ellipsoid. Seeds minutely rugose.

Fl. & Fr.: Nov.-Feb. *Distr.*: Trop. Africa to Indo-Malesia. Common. Among grasses in moist hills. *NAK 118* (Adoor, c. 20 m), *2219* (Pathanamthitta, c. 50 m).

The whole plant used as a medicine in insanity, epilepsy and nervous complaints.

2. **Canscora perfoliata** Lam., Encycl. 1:601. 1785; Clarke in Hook.f., Fl. Brit. India 4:104. 1883; Gamble, Fl. Pres. Madras 879. 1923; Ramamoorthy in Sald. & Nicols., Fl. Hassan Dist. 424. 1976; Manilal, Fl. Silent Valley 182. 1988; Nicols., Suresh & Manilal, An Interpr. Hort. Malab. 148. 1988; Mohanan & Sivad., Fl. Agasthyamala 459.2002. *C. grandiflora* Wight, Ic. t. 1326. 1848. *Cansjan-cora* Rheede, Hort. Malab. 10:103, t.52 1690.

Erect herbs; stem 4-winged. Leaves sessile, elliptic-ovate, 1-2 by 0.6-1 cm, base rounded, apex acute, chartaceous. Flower(s) solitary, terminal and axillary; bracts broadly ovate, foliaceous, perfoliate; calyx winged, lobes lanceate; corolla pale rose, cream- coloured to white, lobes obovate. Capsule cylindric-ellipsoid.

Fl. & Fr.: Nov.-Feb. *Distr.*: S. W India. Less common. Usually in grassy hill slopes. *NAK* 135 (Adoor, c. 20 m), *2108* (Mannarappara, c.100 m).

3. **Cansora roxburghii** Arn. ex Miq., Anal. Bot. Ind. 3:10.1852; Cramer in Dassan. & Fosberg, Rev. Handb. Fl. Ceylon 3:72. 1981; Manilal, Fl. Silent Valley 182. 1988; Mohanan & Sivad., Fl. Agasthyamala 459.2002. *C. wallichii* Clarke in Hook.f., Fl. Brit. India 4:105. 1883; Gamble, Fl. Pres. Madras 879. 1923.

Erect herbs; stem 4-winged. Leaves oblanceate, 1.5-4 by 0.3-1.5 cm, acute at both ends chartaceous. Flower(s) solitary, axillary; calyx not winged, lobes teeth-like; corolla white. Capsule linear-cylindric.

Fl. & Fr.: Dec.-Mar. *Distr.:* S.W Peninsular India. Rare. In grassy hill slopes. *NAK 1242* (Kakki hills, c.1100 m).

EXACUM Linnaeus
Sp. Pl. 112.1753.

Annual or perennial herbs; stem 4-angular. Leaves simple, decussate or whorled, basally 3-7-nerved, chartaceous. Inflorescence of axillary and terminal dichasial cymes. Flowers 4-5-merous, bisexual; calyx winged or not, lobes 4, ovate; corolla tubular-campanulate, lobes-4, broadly obovate, spreading; stamens 4, all fertile, dehiscence porous; ovary globose, 2-locular; ovules α, on bifurcate axile placenta. Capsule globose. Seeds minute, angular.

1. Stem faintly 4-angled or almost terete towards base; leaves succulent; flowers 5-merous. .... 1. **E. courtallense**
1. Stem sharply 4-angled or 4-winged; leaves chartaceous; flowers 4-merous.
   2. Stout herbs; flowers 3-4 cm ϕ; calyx winged. .... 4. **E. tetragonum**
   2. Slender herbs; flowers 8-10 mm ϕ; calyx wingless.
      3. Flowers (sub) sessile, solitary. .... 3. **E. sessile**
      3. Flowers penduncled, in cymes. .... 2. **E. lawii**

1. **Exacum courtallense** Arn., Ann. Sci. Nat. Bot. 11:175. 1839; Wight, Ic. t. 1323. 1848; Clarke in Hook.f., Fl. Brit. India 4:97. 1883; Gamble, Fl. Pres. Madras 873. 1923; Mohanan & Sivad., Fl. Agasthyamala 461.2002.

Key to the varieties

1. Leaves elliptic-lanceate, c.2 cm wide, apex gradually acuminate; flowers few; calyx faintly reticulate. .... **Exacum courtallense** var. **courtallense**

1. Leaves oblong-lanceate, c. 1 cm wide, apex gradually acute; flowers α; calyx prominently reticulate. .... **Exacum courtallense** var. **laxiflorum**

Exacum courtallense Arn.var. **courtallense**

Woody herbs, profusely branched; stem almost terete towards base, apically 4-angled. Leaves elliptic-lanceate, 4-7 by 1.5-2 cm, base rounded, apex gradually acuminate, slightly curved, basally 3-nerved, succulent. Flowes few, solitary or 2-3 in upper leaf-axils, 5-merous; calyx broadly winged, faintly reticulate; corolla deep blue. Capsule ellipsoid.

Fl. & Fr.: Aug.-Nov. *Distr.*: S. W Peninsular India. Rare. On exposed moist or dripping rocks. *NAK 2859* (Thevar mala, c. 250 m),

Exacum courtallense Arn. var. **laxiflorum** Gamble, Fl. Pres. Madras 873. 1923.

Leaves oblong-lanceate, 3-7 by 1.1 cm, apex gradually acute, much curved. Flowers a, very lax; calyx prominently reticulate.

Fl. & Fr.: Sep.-Dec. *Distr.*: S.W Peninsular India. Rare. Among grasses in exposed dripping rocks. *NAK 1086* (Paramada, (Vallicode-Kottayam, c. 100 m).

Note :Mohanan (1984) raised this variety to the rank of species. But inadequency of much striking distinguishing characters prompted us to retain its varietal status.

2. **Exacum lawii** Clarke in Hook.f., Fl. Brit. India 4:98. 1883; Fischer, Rec. Bot. Surv. India 9(1): 119. 1921; Gamble, Fl. Pres. Madras 874. 1923; Manilal, Fl. Silent Valley 182. 1988.

Slender small herbs; stem 4-angled. Leaves elliptic-ovate, 1 by 0.5 cm, base truncate, apex acute, basally faintly 3-nerved. Flowers subsessile, 4-merous, solitary or few in cymes; calyx-lobes without wings; corolla bluish. Capsule globose.

Fl. & Fr.: Dec.-Feb. *Distr.*: S. W Peninsular India. Rare. Among grassy hill slopes. *NAK 1913 & 2118* (Vallicode-Kottayam, c. 100 m).

3. **Exacum sessile** L., Sp. Pl. 112. 1753; Wight, Ic.t. 1324. 1848; Clarke in Hook.f., Fl. Brit. India 4 : 98. 1883; Gamble, Fl. Pres. Madras 874. 1923; Ramamoorthy in Sald. & Nicols., Fl. Hassan Dist. 426. 1976; Cramer in Dassan. & Fosberg, Rev. Handb. Fl. Ceylon 3 : 65. 1981; Matthew & Rani in Matthew, Fl. Tam. Carnatic 3(2) : 975. 1983; Manilal, Fl. Silent Valley 183. 1988; Mohanan & Sivad., Fl. Agasthyamala 461.2002.

Slender small herbs; stem 4-angled. Leaves elliptic-ovate, 1 by 0.5 cm, base truncate, apex acute, basally faintly 3-nerved. Flowers subsessile, 4-merous, solitary, or few in cymes; calyx-lobes without wings; corolla bluish. Capsule globose.

Fl. & Fr.: Dec.-Feb. *Distr.*: Sri Lanka, Peninsular India. Among moist grassy hill slopes. *NAK 2027* (Kokkathode, . 150 m).

4. **Exacum tetragonum** Roxb., Fl. Ind. 1:413. 1820; Klack., Opera Bot. 14:37. 1985; Manilal, Fl. Silent Valley 183. 1988. *E. bicolor* Roxb., Fl. Ind. 1:413. 1820; Wight, Ic. t. 1321. 1848; Clarke in Hook.f., Fl. Brit. India 4:96. 1883; Gamble, Fl. Pres. Madras 873. 1923; Ramamoorthy in Sald. & Nicols., Fl. Hassan Dist. 425. 1976.

Tall stout herbs; stem sharply 4-angled. Leaves ovate to oblong-lanceate, 3-6 by 1-2 cm, base subauriculate, apex acute or very shortly acuminate. Flowers in dense terminal cymes, 4-merous; calyx-lobes cuspidate at apex, broadly winged; corolla bluish-white, c. 3 cm ϕ; anthers, c. 7 mm long, curved. Capsule ovoid. Seeds minute.

Fl. & Fr.: Nov.-Jan. *Distr.*: Peninsular India. In grassy hills slopes. Locally abundant. *NAK 399 & 1330* (Kakki hills, c.1100 m).

The plant has valuable febrifuge properties.

HOPPEA Willdenow
Ges. Naturf. Freunde Berlin Neue Schriften 3:434. 1801.

Hoppea fastigiata (Griseb.) Clarke in Hook. f., Fl. Brit. India 4:100. 1883; Gamble, Fl. Pres. Madras 877. 1923; Ramamoorthy in Sald. & Nicols., Fl. Hassan Dist. 426. 1976; Cramer in Dassan. & Fosberg, Rev. Handb. Fl. Ceylon 3:76. 1981; Matthew & Rani in Matthew, Fl. Tam. Carnatic 3(2): 977. 1983; Manilal, Fl. Silent Valley 184. 1988; Mohanan & Sivad., Fl. Agasthyamala 464.2002. *Cicendia fastigiata* Griseb., Gen. Sp. Gent. 158. 1839.

Erect branched small herbs; branches fastigiate. Leaves broadly ovate, to 7 mm long, base rounded or truncate, apex acute, chartaceous, subsessile. Flowers in 2-3-chotomous terminal and axillary cymes, 4-merous; calyx tubular, lobes 4, lanceate; corolla yellow, tubular, longer than calyx; stamens 4, included, only one fertile; ovary broadly ellipsoid, 1-locular; ovules α, on parietal placenta; stigma 2-lobed. Capsule ellipsoid.

Fl. & Fr.: Nov.-Feb. *Distr.*: Sri Lanka, Central & South India. Not common. On grassy hills. Locally abundant. *NAK 2119* (Vallicode-Kottayam, c. 100 m).

SWERTIA Linnaeus
Sp. Pl. 226. 1753.

Swertia corymbosa (Griseb.) Wight ex Clarke in Hook.f., Fl. Brit. India 4:126. 1883; Gamble, Fl. Pres. Madras 881. 1923; Ramamoorthy in Sald. & Nicols., Fl. Hassan Dist. 427. 1976; Matthew & Rani in Matthew, Fl. Tam. Carnatic 3(2): 977. 1983; Manilal, Fl. Silent Valley 184. 1988. *Ophelia corymbosa* Griseb., Gen. Sp. Gent. 317. 1839; Wight, Ic.t. 1329. 1848.

Herbs, to 25 cm tall; branches 4-angled. Leaves simple, decussate, elliptic-ovate to lanceate, 2-3.5 by 0.6-1.5 cm, base rounded, margin recurved, apex subacute, basally 3-nerved, chartaceous, subsessile. Flowers α, in terminal corymbose cymes, 4-5- merous; calyx deeply 4-5-lobed; corolla pale blue, nearly free, lobes 4-5, elliptic-ovate; stamens 5, inserted near corolla base; ovary 1-locular; ovules α. Capsules failed to collect.

Fl. & Fr.: Oct.-Dec. *Distr.*: S. W India. Rare. On grassy rocks. *NAK 2860* (Kakki hills, c. 1100 m).

88. MENYANTHACEAE

NYMPHOIDES J.Hill
Brit. Herbal 77.1756.

Nymphoides indica (L.) Kuntze, Revis. Gen. Pl. 429. 1891; Cramer in Dassan.& Fosberg, Rev. Handb. Fl. Ceylon 3:207. 1981; Ramamoorthy in Sald. & Nicols., Fl. Hassan Dist. 475. 1976; Matthew, Ill. Fl.Tam. Carnatic t. 974. 1982; Matthew & Rani in Matthew, Fl. Tam. Carnatic 3(2) : 980. 1983; Nicols., Suresh & Manilal, An Interpr. Hort. Malab. 181. 1988. *Menyanthes indica* L., Sp. Pl. 145. 1753; Roxb., Fl. Ind. 2:31. 1824. *Limnanthemum indicum* (L.) Griseb., emend. Thwaites, Enum. Pl. Zeyl. 205. 1860; Clarke in Hook. f., Fl.Brit. India 4:131. 1883; Gamble, Fl. Pres. Madras 883. 1923. *Nedel-ambel* Rheede, Hort. Malab. 11:55-56, t. 28. 1692.

Aquatic submerged herbs, rooting at basal nodes of petiole-like stems. Leaves simple, alternate on the proximal node of stem, peltate, orbicular, 4-5 by 6-8 cm, base deeply cordate, margin entire or faintly sinuate, herbaceous, free floating. Flowers bisexual, 5-merous, regular, long-pedicelled, in umbellate fascicles; calyx-lobes 5, oblong; corolla white with yellowish centre, lobes 5, straight, margin fimbriate; stamens 5, exerted; ovary superior, 1-locular; ovules few on 3 placentae. Capsule ellipsoid. Seeds smooth.

Fl. & Fr.: Sep.-Dec. *Distr.*: Paleotropical. In pools and ponds. Locally abundant. *NAK 1038* (Tiruvalla, c. 10 m).

89. BORAGINACEAE

1. Trees or scandent or trailing shrubs.
   2. Plants riverine; flowers pink; shrubs with trailing branches; flowers solitary or in clusters; stamens subexerted. .... **Rotula**
   2. Plants not riverine; flowers white; scandent shrubs; flowers on scorpiod cyme; stamens included. .... **Tournefortia**
1. Herbs, erect or prostrate.
   3. Style-arms 2, free below, connate above; prostrate herbs. .... **Coldenia**
   3. Style undivided, erect herbs.
      4. Ovary 4-lobed; fruit of 4 nutlets, attached to a carpophore. .... **Trichodesma**
      4. Ovary not lobed; fruit a drupe of 4 pyrenes or of 4 free nutlets. .... **Heliotropium**

COLDENIA Linnaeus
Sp. Pl. 125. 1753.

Coldenia procumbens L., Sp. Pl. 125. 1753; Roxb., Fl. Ind. 1:448. 1832; Clarke in Hook.f., Fl.Brit. India 4:144. 1883; Gamble, Fl. Pres. Madras 892. 1923; Kazmi, J. Arnold Arb. 5:148. 1970; Ramamoorthy in Sald. & Nicols., Fl. Hassan Dist. 479. 1976; Matthew, Ill. Fl. Tam. Carnatic t. 468. 1982; Matthew & Rani in Matthew, Fl. Tam. Carnatic 3(2): 983. 1983; Mohanan & Sivad., Fl. Agasthyamala 464.2002.

Trailing herbs; branchlets scabrid. Leaves simple, alternate, elliptic-ovate, 1-3 by 0.7-1.3 cm, base cuneate, margin coarsely crenate-dentate, apex emarginate, lateral veins impressed above, thick-herbaceous. Flower(s) solitary, axillary, 4-5-merous, bisexual, subsessile; calyx-lobes ovate; corolla cream -coloured, lobes 4 or 5, equal; stamens 4-5, included; ovary 4-lobed, 4-locular; ovule 1 per locule. Drupe of 4 subconnate pyrenes.

Fl. & Fr.: Jan.-Aug. *Distr.*: Sri Lanka, India, Pakistan. On barren paddy fields. Locally abundant. *NAK 2574* (Parakode, c. 20 m).

Roots used in Ayurvedic decoctions to increase vitality and to cure syphilis.

HELIOTROPIUM Linnaeus
Sp. Pl. 130. 1753.

Erect or prostrate herbs. Leaves simple, alternate. Inflorescence a 1-sided spike or raceme, terminal or leaf-opposed. Flowers 5-merous, bisexual, calyx-lobes free; corolla white, very small, campanulate or salver-form; stamens 5, included; ovary imperfectly 4-locular; ovule 1-per locule. Drupe of 4 free or connate pyrenes.

1. Erect herbs; spikes to 20 cm long; nutlets connate in pairs. .... 1. **H. indicum**
1. Prostrate herbs; racemes to 3 cm long; nutlets free. ....2. **H. marifolium**

1. **Heliotropium indicum** L., Sp. Pl. 130. 1753; Roxb., Fl. Ind. 1:454. 1832; Hook.f., Fl. Brit. India 4:152. 1883; Gamble, Fl. Pres. Madras 896. 1923; Ramamoorthy in Sald. & Nicols., Fl. Hassan Dist. 481. 1976; Matthew, Ill. Fl. Tam. Carnatic t. 473. 1982; Matthew & Rani in Matthew, Fl. Tam. Carnatic 3(2): 988. 1983; Nicols., Suresh & Manilal, An Interpr. Hort. Malab. 75. 1988. *Ben-patsja* Rheede, Hort. Malab. 10:95-95, t. 48. 1690.

Erect herbs. Leaves subdeltoid or broadly ovate, base truncate, margin undulate, apex acute, chartaceous; petiole 3-6 cm long. Spikes terminal or leaf-opposed, curved, to 20 cm long; calyx-lobes linear; corolla salver-form, lobes ovate. Nutlets connate in pairs, ribbed.

Fl. & Fr.: Sep.-Mar. *Distr.*: Pantropical. Usually in wet places, along streamsides, paddy fields, etc. *NAK 1656, 1929* (Vallicode-Kottayam, c. 100 m).

Juice of leaves applied to wounds, sores, ophthalmia and scorpion stings. Decoction or powder of the whole plant given to fatten the body.

2. **Heliotropium marifolium** Retz., Obs. Bot. 2:8. 1781; Clarke in Hook. f., Fl. Birt. India 4:152. 1883. p.p.; Backer & Bakh. f., Fl. Java 2:462. 1965. *H. scabrum* Retz., Obs. Bot. 2:8.1781; Wight, Ic.t. 1389. 1848; Clarke in Hook.f., Fl.Brit. India 4:152. 1883; Gamble, Fl. Pres. Madras 897. 1923; Matthew & Rani in Matthew, Fl. Tam. Carnatic 3(2):989. 1983.

Spreading prostrate herbs; vegetative parts softly hirsute. Leaves elliptic or elliptic-oblong, 6-10 by 3.5 mm, base and apex acute, 1-nerved, thin-coriaceous; petiole c. 4 mm long. Racemes subterminal, c. 3 cm long; calyx-lobes triangular; corolla salver-form. Nutlets free.

Fl. & Fr.: Jan.-May. *Distr.*: Sri Lanka, Peninsular India. Less common. In exposed hill slopes and plains. *NAK 1641* (Tiruvalla, c. 10 m).

ROTULA Loureiro
Fl. Cochinch. 121. 1790.

Rotula aquatica Lour., Fl. Cochinch. 121. 1790; Robinson, Philipp. J. Sci. 4:693. 1909; Gamble, Fl. Pres. Madras 893. 1923; Ramamoorthy in Sald. & Nicols., Fl. Hassan Dist. 482. 1976; Matthew, Ill. Fl. Tam. Carnatic t. 471. 1982; Matthew & Rani in Matthew, Fl. Tam. Carnatic 3(2) : 1003. 1983; Manilal, Fl. Silent Valley 186. 1988. *Rhabdia lycioides* Mart., Nov. Gen. Sp. Pl. 2:137, t.195. 1827; Clarke in Hook. f., Fl. Brit. India 4:145. 1883. *Ehretia cuneata* Wight, Ic. t. 1385. 1848. *Kallur-vanchi.*

Shrubs with long trailing branches. Leaves simple, closely alternate, narrowly obovate or elliptic-oblong, 1-2 by 0.3-0.7 cm, base cuneate, margin ciliolate, apex obtuse, thin-coriaceous. Flower(s) solitary or in clusters, 5-merous, bisexual, sessile; calyx lobes 5, free, lanceate; corolla pink, salver-form, c.10 mm ϕ, lobes 5, broadly elliptic, spreading; stamens 5, subexerted; ovary 4-locular; ovule 1-per locule. Drupe of 4 pyrenes.

Fl. & Fr.: Nov.-Mar. *Distr.*: India, S.E. Asia. In rock crevices of streams. Locally abundant. *NAK 1195* (Moozhiar, c.250 m), *1423* (Mannarappara, c. 200 m).

Roots used in Ayurvedic preparation and is a best medicine for stone in urinary bladder, piles and veneral diseases.

TOURNEFORTIA Linnaeus
Sp. Pl. 140. 1753.

Tournefortia heyneana Wall. ex Clarke in Hook. f., Fl. Brit. India 4:145. 1883; Gamble, Fl. Pres. Madras 893. 1923; Manilal, Fl. Silent Valley 186. 1988.

Subscandent shrubs. Leaves simple, alternate, oblong-lanceate, 8-16 by 2-5.5 cm, base obliquely truncate or broadly cuneate, apex acuminate, lateral veins 6-7 pairs, impressed above, chartaceous. Cymes scorpioid, dense, dichotomous. Flowers 5-merous, bisexual; calyx lobes lanceate; corolla white, c. 3 mm ϕ, tube bottle-form, c. 1 cm long, lobes small; stamens 5, included; ovary 4-locular; style obscurely 2-lobed. Drupe subglobose.

Fl. & Fr.: Sep.-Mar. *Distr.:* Peninsular India. Rare. In evergreen forests. *NAK 2515 & 2678* (Kakki hills, c.1100 m).

TRICHODESMA R. Brown
Prodr. 496. 1810 (nom. cons.).

Trichodesma zeylanicum (Burm.f.) R. Br., Prodr. 496. 1810 ('*zeylanica*'); Clarke in Hook. f., Fl. Brit. India 4:154. 1883; Gamble, Fl. Pres. Madras 899. 1923; Ramamoorthy in Sald. & Nicols., Fl. Hassan Dist. 483. 1976; Matthew & Rani in Matthew, Fl. Tam. Carnatic 3(2): 994. 1983. *Borago zeylanica* Burm. f., Fl. Ind. 41. 1768.

Erect hispid herbs. Leaves simple, opposite, elliptic-lanceate, 5-8 by 1-2.5 cm, base acute, apex gradually tapering, lateral nerves 4-5 pairs, white-hirsute below, scabrous above, chartaceous. Flower(s) solitary, axillary; calyx cupular, lobes 5; corolla bluish, salver-form, lobes 5, broadly ovate, abruptly acuminate at apex; stamens 5, connivent into a cone, exerted; ovary conical, 4-lobed, 4-locular; ovule 1 per locule. Drupe of 4 free nutlets.

Fl. & Fr.: Nov.-Jan. *Distr.*: Sri Lanka, Peninsular India, Malaya, Australia. Occasional. On exposed slopes. *NAK 2862* (Kakki hills, c. 1100 m).

90. CORDIACEAE

1. Ovary 4-locular; ovule 1-per locule. .... **Cordia**
1. Ovary 2-locular; ovules 2-per locule. .... **Ehretia**

CORDIA Linnaeus
Sp. Pl. 190.1753.

Trees or shrubs. Leaves simple, alternate or subopposite. Inflorescence of corymbose or scorpioid cymes. Flowers 5-merous, bisexual or functionally unisexual; calyx accrescent, lobes 5; corolla funnel-form, lobes 5, spreading; stamens 5-8, exerted; ovary 4-locular; ovule 1-per locule. Drupe ovoid or ellipsoid.

1. Shrubs; leaves penninerved; inflorescence a scorpiod cyme; stamens 5. .... 1. **C. cylindristachya**
1. Trees; leaves basally 3-nerved; inflorescence a lax corymbose cyme; stamens 8. .... 2. **C. octandra**

1. **Cordia cylindristachya** Roem. & Schult., Syst. 4:459. 1819; Ravi & Murali., Bull. Bot. Surv. India 30(1):171. 1990.

Profusely branched densely foliaceous shrubs. Leaves elliptic-ovate to broadly elliptic, 8-14 by 6-8 cm, base cuneate, margin serrate, apex acuminate, scabrid, thin-coriaceous. Cymes scorpiod, terminal. Flowers dense; calyx campanulate; corolla white, c. 5 mm ϕ. Drupe globose.

Fl. & Fr.: Most of the seasons. *Distr.*: Originally of Trop. America, now in all warmer countries. Often planted for fencing. *NAK 2930* (Parakkode, c. 20 m).

2. **Cordia octandra** A. DC., Prodr. 9:477. 1845; Clarke in Hook. f., Fl. Brit. India 4:140. 1883; Gamble, Fl. Pres. Madras 889. 1923.

Medium-sized trees. Leaves elliptic-ovate to broadly elliptic, 8-14 by 6-8 cm, base cuneate, margin serrate, apex acuminate, lateral veins c. 5 pairs, basally 3-nerved, pubescent, chartaceous. Cymes corymbose, subterminal. Flowers lax; calyx campanulate, c. 1 cm ϕ; corolla white, c. 2 cm ϕ, fragrant; stamens 8, exerted. Drupes could not be collected.

Fl. & Fr.: Sep.-Jan. *Distr.*: Endemic to Tranvancore hills. Very rare. In semi-evergreen forests. *NAK 943* (Near Ganapathy temple, Pampa-Sabarimala, c. 200 m).

Known from only one collection by a single tree!

EHRETIA P. Browne
Civ. Nat. Hist. Jamaica 168. 1756.

Ehretia canarensis (Clarke) Gamble, Fl. Pres. Madras 891. 1923; Ramamoorthy in Sald. & Nicols., Fl. Hassan Dist.481. 1976; Mohanan & Sivad., Fl. Agasthyamala 465.2002. *E. laevis* Roxb., var. *canarensis* Clarke in Hook.f., Fl. Brit. India 4:142. 1883. **(Fig. 49)**

Trees. Leaves simple, alterante, elliptic, to oblanceate, 6-10 by 3-4 cm, base cuneate, apex acuminate, lateral veins 6-8 pairs, thin-coriaceous. Flowers in axillary unilateral corymbose cymes, bisexual, 5-merous; calyx small, campanulate, lobes 5; corolla white, c. 5 mm ϕ, lobes 5, spreading; stamens 5, attached at middle of corolla-tube, exerted; ovary superior, 2-locular; style bifid; ovules 2 per locule. Drupe ellipsoid.

Fl. & Fr.: May-Aug. *Distr.*: S.W India. In evergreen forests. Not common. *NAK 232, 1807* (Angamoozhy, c. 200 m).

91. CONVOLVULACEAE

1. Leafless, parasitic twiners; corolla with 5 epipetalous fimbriate scales inside. ....**Cuscuta**
1. Leafy, non-parasitic twiners (except prostrate or erect *Evolvulus*); corolla without epipetalous fimbriate scales inside.
    2. Style 1 or 0.
        3. Pollen grains spinulose.
            4. Corolla urceolate; stamens with dorsal scales. ....**Lepistemon**
            4. Corolla funnel-form or salver-form, or subrotate; stamens without dorsal scales .
                5. Fruit Indehiscent, epicarp fleshy; seeds embedded in mealy pulp. ....**Ipomoea**
                5. Fruit dehiscent, epicarp thin, fragile; seeds not embedded in mealy pulp. ....**Argyreia**
        3. Pollen grains smooth.
            6. Style 1; corolla funnel-form or salver-form, lobes entire or faintly lobed.
                7. Ovary imperfectly 2-locular (often 1-locular); stigma-lobes oblong. ....**Hewittia**
                7. Ovary perfectly 2-or 4-locular; stigma-lobes capitate or globose.
                    8. Calyx-lobes unequal; outer sepals decurrent on pedicel. ....**Aniseia**
                    8. Calyx-lobes subequal; outer sepals not decurrent on pedicel. .....**Merremia**
            6. Style 0; corolla subrotate, lobes deeply 5-fid. ....**Erycibe**
    2. Styles 2 or more.
        9. Prostrate or ascending herbs; styles free;leaves small, elliptic. ....**Evolvulus**
        9. Woody twiners; styles basally connate, unbranched; leaves large cordiform. ....**Bonamia**

ANISEIA J.D. Choisy

Mem. Soc. Physc. Geneve 6 : 481. 1834.

Aniseia martinicensis (Jacq.) Choisy, Mem. Soc. Phys. Hist. Nat. Geneve 6 : 483. 1838; Ooststr., Blumea 3:279. 1939 & in Steenis, Fl. Males. I,4 : 435. 1935; Austin in Dassan. & Fosberg, Rev. Handb. Fl. Ceylon 1. 291. 1980; Nicols., Suresh & Manilal, An Interpr. Hort. Malab. 86. 1988; Mohanan & Sivad., Fl. Agasthyamala 466.2002. *Convolvulus martinicensis* Jacq., Select. Strip. Amer. 26. t. 17. 1763. *C. uniflorus* N. Burm., Fl. Ind. 47, t. 21. f.2. 1768. *Convolvulus emarginatus* Vahl, Symb. Bot. 3:2. 1794. *Ipomoea uniflora* (Burm. f). Roem.

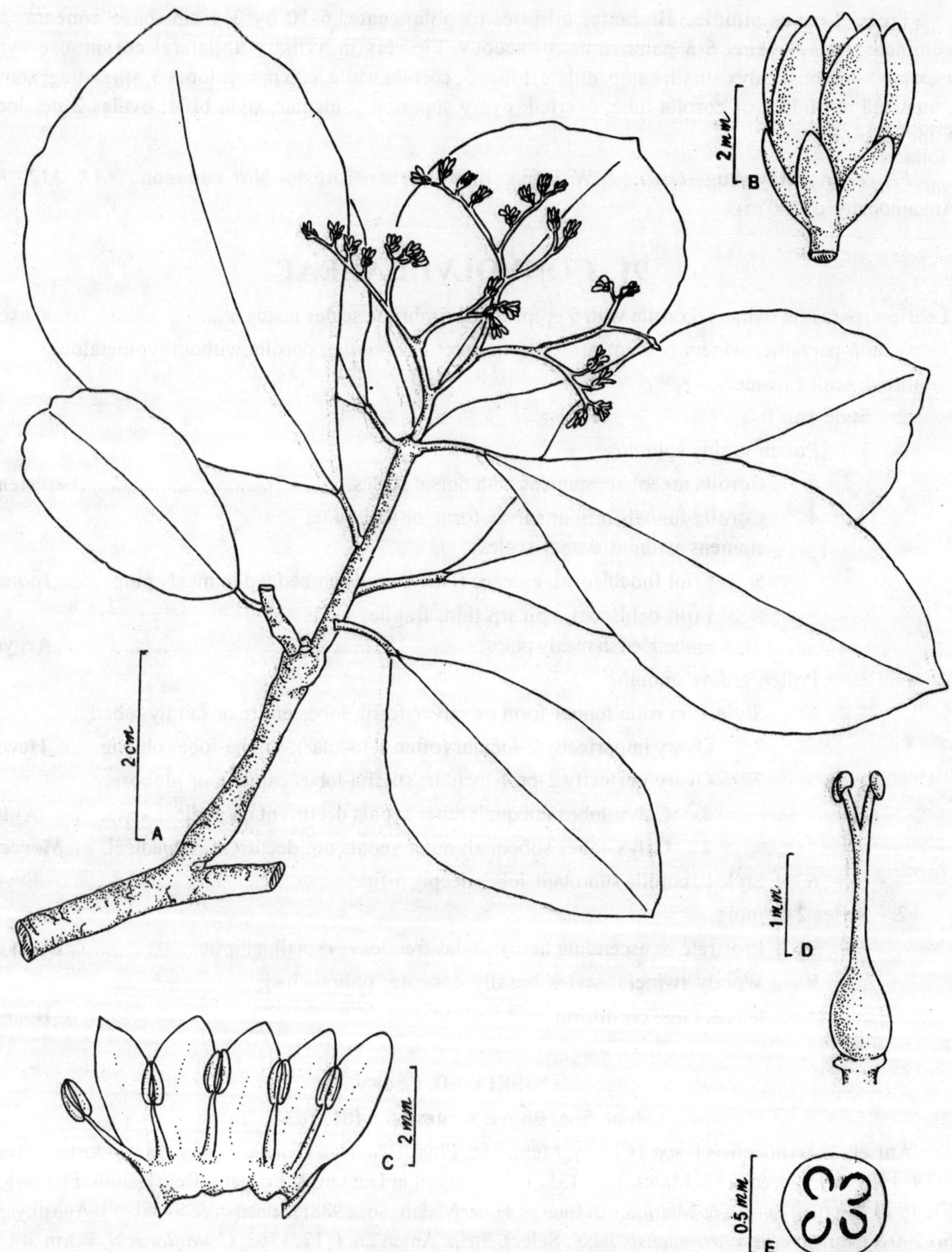

Fig. 49. ***Ehretia canarensis*** **(Clarke) Gamble: A. Twig; B. Flower; C. Corolla - split open end showing stamens; D. Pistil; E. Ovary - C.S.**

& Schult., Syst. Veg. 4:247. 1819; Clarke in Hook.f., Fl. Brit. India 4:201. 1883. *Ben-tiru-tali* Rheede, Hort. Malab. 11:111, t. 54. 1692. *Venthiruthali.*

Twiners. Leaves oblong to elliptic-oblong, 3-6 by 1-2 cm, base acute, apex emarginate, mucronulate, petiole 5-10 mm long. Flower(s) usually solitary or rarely 2-3 in axillary cymes; calyx-lobes 5, unequal, outer 3 foliaceous, ovate, inner 2 lanceate; corolla white, c. 2 cm ϕ, campanulate or funnel-form; stamens included; ovary 2-locular; ovules 2-per locule; style filiform. Capsule ovoid. Seed glabrous.

Fl. & Fr.: Jun.-Dec. *Distr.*: Sri Lanka, India, Nepal, Trop-Africa, Malesia. Not common. Usually among thickets near rivulets. *NAK 1068* (Tiruvalla, c. 10 m).

ARGYREIA Loureiro
Fl. Cochinch. 95, 134. 1790.

Scandant, tomentose shrubs. Leaves cordiform or broadly elliptic, densely villous hairy. Cymes a-flowered, axillary; bracts often foliaceous, persistent or caducous. Flowers 5-merous; calyx-lobes large, accrescent; corolla funnel-form; stamens included, pollen grains spinulose; ovary 2-4-locular; ovule(s) 1-or 2-per locule. Berry globose. Seeds glabrous.

1. Leaves broadly elliptic; bracts linear, caducous; ovary 2-locular. .... 1. **A. elliptica**
1. Leaves cordiform; bracts foliaceous, persistent; ovary 4-locular. .... 2. **A. hirsuta**

1. **Argyreia elliptica** (Roth) Choisy, Mem. Soc. Phys. Geneve 6:417. 1834; Sant., Fl. Khandala ed. 3, 172. 1967; Austin in Dassan. & Fosberg, Rev. Handb. Fl. Ceylon 1:294. 1980; Gandhi in Sald. & Nicols., Fl. Hassan Dist. 466. 1976; Rani & Matthew in Matthew, Fl. Tam. Carnatic 3(2): 1010. 1983. *Ipomoea elliptica* Roth, Nov. Pl. Sp. 113. 1821. *Lettsomia elliptica* (Roth) Wight ex Clarke in Hook.f., Fl. Birt. India 4:192. 1883; Gamble, Fl. Pres. Madras 911. 1923.

Tomentose climbers. Leaves broadly elliptic or elliptic-ovate, 6-14 by 3-10 cm, base rounded, apex subacute, both surfaces tomentose, chartaceous; petiole 2-6 cm long. Cymes corymbose; bracts caducous; calyx-lobes subequal, suborbicular; corolla pink, c. 3 cm ϕ; ovary 2-locular, ovules 2 per locule. Berry globose, ripening orange coloured. Seeds glabrous.

Fl. & Fr.: Nov.-Mar. *Distr.*: Sri Lanka, Peninsular India. Not common. In hills. *NAK 48, 2137* & *2667* (Moozhiar, Chelikkal, c. 500 m).

2. **Argyreia hirsuta** Wight & Arn., Nova Acta. Phys.-Med. Acad. Caes. Leop.-Carol. Nat. Cur. 18:356. 1836; Clarke in Hook.f., Fl. Brit. India 4:189.1883; Gamble, Fl.Pres. Madras 909. 1923. *Rivea hirsuta* (Wight & Arn.) Wight, Ic. t. 891. 1845.

Large climbers. Leaves cordiform or broadly ovate, 10-22 by 7-19 cm, base truncate, rounded to cordate, apex acuminate, chartaceous. Cymes dense, corymbose; peduncle c. 10 cm long; bract foliaceous, linear to elliptic-lanceate, 2-4 by 0.1-1 cm, tomentose without; corolla pink, c. 6 cm ϕ, pubescent without; ovary 4-locular; ovule 1 per locule. Capsule globose. Seeds glabrous.

Fl. & Fr.: Most of the seasons. *Distr.*: Western Peninsular India. Very common in hills. *NAK 5, 832, 1167* (Moozhiar, Arampa etc.)

BONAMIA Du Petit- Thouars
Pl. Isles Afrique Austr. 33 t. 8. 1804 (nom. cons.).

Bonamia semidigyna (Roxb.) Hallier f., Bot. Jahrb. 16:528. 1893; Ooststr., Blumea 3:76. 1939 & in Steenis, Fl. Males. I, 4: 398. 1953. *Convolvulus semidigynus* Roxb., Hort. Beng. 13. 1814. (nom. nud.) & Fl. Ind. 2:47. 1824; *Breweria cordata* Bl., Bijdr. 722. 1825; Clarke in Hook. f., Fl. Brit. India 4: 223. 1883. *B. roxburghii* Choisy, Mem. Soc. Phys. Geneve 6: 493. 1833.

Tomentose climbers. Leaves ovate-coradate, 4-9 by 3-6.5 cm, base cordate, apex acuminate- apiculate, lateral nerves 5-6 pairs, softly tomentose; petiole 2-4 cm long. Flowers 2-5, in umbellate cymes, 5-merous; bracts oblanceate, peduncle c. 9 cm long; sepals subequal, elliptic-ovate to ovate-oblong; corolla white, c. 4 cm ϕ, funnel-form, pilose without; stamens included; ovary 2-locular, ovules 2 per locule; style 2-fid. capsule subglobose. Seed black, glabrous.

Fl. & Fr.: Nov.-Mar. *Distr.*: India, Indo-China, Malesia. Rare. On thickets along streamsides. *NAK 116, 456, 2295* (Adoor & Moozhiar, c. 20 to 400 m).

CUSCUTA Linnaeus
Sp. Pl. 124. 1753.

Parasitic twiners; branchlets yellowish; slender or stout. Leaves 0. Flowers in fascicles or in racemes, 4-5- merous; calyx-lobes nearly free, subequal; corolla globose or funnel-form, often scaley-fimbriate structures present inside corolla-tube; stamens exerted, pollen grains smooth; ovary 2-locular; ovules 2 per locule; style(s) 1 or 2. Capsule subglobose. Seeds glabrous.

1. Stem slender; flowers in fascicles; capsule dry. ....1. **C. chinensis**
1. Stem stout; flowers in racemes; capsule succulent. ....2. **C. reflexa**

1. **Cuscuta chinensis** Lam., Encycl. 2:229. 1786; Wight, Ic. t. 1373. 1848; Clarke in Hook.f., Fl. Brit. India 4:226. 1883; Gamble, Fl. Pres. Madras 931. 1923; Austin in Dassan. & Fosberg., Rev. Handb. Fl. Ceylon 1:305. 1980; Rani & Matthew in Matthew, Fl. Tam. Carnatic 3(2) : 1013. 1983.

Slender twiners. Flowers 5-7, in lateral fascicles; calyx-lobes scarious, broadly ovate; corolla pale yellow, c. 5 mm ϕ, widely funnel-form; stamens 5, scales 4; stigma fimbriate. Capsule dry, irregularly dehiscent.

Fl. & Fr.: Nov.-Mar. *Distr.*: Sri Lanka, India, Australia and China. Common. Usually parasitic upon undershrubs. *NAK 1351* (Moozhiar, c. 250 m).

2. **Cuscuta reflexa** Roxb., Pl. Corom. t. 104. 1799 & Fl. Ind. 1:446. 1832; Clarke in Hook.f., Fl. Brit. India 4:225. 1883; Gamble, Fl. Pres. Madras 931. 1923; Ooststr., Blumea 3: 70. 1938 & in Steenis, Fl. Males. I,4 : 393. 1953; Austin in Dassan. & Fosberg, Rev. Handb. Fl. Ceylon 1:306. 1980; Rani & Matthew in Matthew, Fl. Tam. Carnatic 3(2) : 1014. 1983.

Stout succulent twiners. Flowers in lateral racemes; calyx cupular, lobes herbaceous, suborbicular; corolla-lobes cream- coloured or white, c. 8 mm ϕ, funnel-form; scales 5; stigma acute. Capsules succulent, cirumcissile. Seeds black.

Fl. & Fr.: Nov.-Mar. *Distr.*: Sri Lanka, India, China and Malesia. Parasitic upon trees like *Coffea, Croton, Adhatoda*. *NAK 2650*. (Thannithode, c. 100 m).

ERYCIBE Roxburgh
Pl. Coromandel 2:31.1802.

Erycibe paniculata Roxb., Pl. Corom. 2:31, t. 159. 1798; Clarke in Hook.f., Fl. Brit. India 4:180. 1883; Gamble, Fl. Pres. Madras 930. 1923; Hoogland, Blumea 7:352. 1953; Gandhi in Sald. & Nicols., Fl. Hassan Dist. 467. 1976; Austin in Dassan. & Fosberg, Rev. Handb. Fl. Ceylon 1:308. 1980; Nicols., Suresh & Manilal, An Interpr. Hort. Malab. 87. 1988; Mohanan & Sivad., Fl. Agasthyamala 469.2002. *E. paniculata* var. *wightiana* (Grah.) Clarke in Hook. f., Fl. Brit. India 4:181. 1883. *Erima-tali* Rheede, Hort. Malab. 7:73, t. 39. 1688.

Stout climbers. Leaves simple, broadly elliptic to elliptic-lanceate, 5-10 by 2-6 cm, base rounded, apex obtusely acuminate, coriaceous; petiole c. 10 cm long. Flowers in large axillary panicles, 5-merous; calyx and outside corolla rusty tomentose; calyx-lobes suborbicular; corolla white, c. 1 cm ϕ, lobes deeply bilobed; stamens exerted, pollens smooth; ovary 1-locular; ovule 4, basal; style 0; stigma subglobose. Berry ellipsoid. Seed 1.

Fl. & Fr.: Jul.-Dec. *Distr.*: India, Himalayas, Andaman Islands. Not common. In sacred groves and along streamsides. *NAK 1876* (Kalleli, c. 200 m).

EVOLVULUS Linnaeus
Sp. Pl. ed. 2:391. 1762.

Prostrate or ascending herbs. Leaves small. Flower(s) solitary or few in axillary cymes, 5-merous; sepals subequal, not enlarged in fruit; corolla rotate or funnel-form, blue or white; stamens included or exerted, pollen smooth; ovary 2-locular; ovules 2-per locule; styles 2. Capsule globose or ovoid. Seeds smooth or faintly verrucose.

1. Leaves elliptic or elliptic-lanceate; flowers blue; capsule 2-locular. .... 1. **E. alsinoides**
1. Leaves oblong-orbicular; flowers white; capsule 1-locular. .... 2. **E. nummularius**

1. **Evolvulus alsinoides** (L.) L., Sp. Pl. (ed. 2) 392. 1762; Clarke in Hook. f., Fl. Brit. India 4 : 220. 1883; Gamble, Fl. Pres. Madras 923. 1923; Ooststr., Mon. Evolv. 26. 1934, Blumea 3: 74. 1938 & in Steenis, Fl. Males. I,. 4:395. 1953; Stearn, Taxon 21 : 649. 1972; Gandhi in Sald. & Nicols., Fl. Hassan. Dist. 468. 1976; Austin in Dassan. & Fosberg, Rev. Handb. Fl. Ceylon 1:309. 1980, Ranl & Matthew in Matthew, Fl. Tam. Carnatic 3(2) : 1015. 1983; Nicols., Suresh & Manilal, An Interpr. Hort. Malab. 88. 1988; Mohanan & Sivad., Fl. Agasthyamala 470.2002. *Convolvulus alsinoides* L., Sp. Pl. 157. 1753. *Vistnu-clandi* Rheede, Hort. Malab. 11 : 131-132, t. 64. 1692.

Key to the varieties.

1. Prostrate herb; internodes, c. 5 mm long; leaves usually in 2 distinct rows. .... **Evolvulus alsinoides** var. **hirsutus**
1. Ascending or erect herbs; internodes 10-15 mm long; leaves not in two distinct rows. .... **Evolvulus alsinoides** var. **decumbens**

Evolvulus alsinoides (L.) L var. **decumbens** (R.Br.) Ooststr., Mon. Evov. 38. 1934 & in Steenis, Fl. Males. I, 4 : 396. 1953.

Ascending to erect slender villous herbs; branchlets terete; internodes 10-15 mm long. Leaves lanceate, 7-20 by 3-7 mm, base rounded or acute, apex acute-mucronulate, subsessile. Flower(s) 1-2 or 3 in axillary cymes; peduncle slender, c. 2.5 cm long; bracts linear; pedicel c. 5 mm long; calyx-lobes linear-lanceate, c. 3 mm long; corolla pale blue, c. 7 mm ϕ. Capsule ovoid, 2-locular. Seeds smooth.

Fl. & Fr.: Sep.-Dec. *Distr.*: India, S. China, Indo-China to Australia, Pacific Islands, and Malesia. Not common. Usually among grasses in dripping rocks in hills. *NAK 2028*. (Kalleli, c. 400 m).

Evolvulus alsinoides(L.) L. var. **hirsutus** (Lamk.) Ooststr., Mon. Evolv. 38. 1934 & in Steenis, Fl. Males. 1,4 : 396. 1953.

Prostrate, densely villous herbs. Leaves more or less distinctly in 2 rows, broadly elliptic to elliptic-lanceate, 6-12 by 4-7 mm, base rounded or acute, apex subacute or obtuse. Flower(s) solitary or 2-1 in axillary cymes; peduncle c. 1 cm long; calyx-lobes lanceate. Capsule ovoid.

Fl. & Fr.: Sep.-Dec. *Distr.*: India, Malesia. Common in open wet areas in lowlands. *NAK 2901* (Adoor, c. 20 m).

2. **Evolvulus nummularius** (L.) L., Sp. Pl. (ed. 2) 391. 1762; Ooststr., Mon. Gen. Evolv. 114. 1934 & in Steenis, Fl. Males. I, 5 : 558. 1958; Stearn, Taxon 21 : 649. 1972; Austin in Dassan. & Fosberg, Rev. Handb. Fl. Ceylon 1:311. 1980; Rani & Matthew in Matthew, Fl. Tam. Carnatic 3(2): 1016. 1983; Mohanan & Sivad., Fl. Agasthyamala 470.2002. *Convolulus nummularius* L., Sp. Pl. 157. 1753. *Volvulopsis nummularius* (L.) Roberty, Candollea 14 : 28.1952.

Prostrate herbs. Leaves oblong-orbicular, 10-20 by 8-18 cm, base turncate to subcordate, apex obtuse, chartaceous; petiole 3-5 mm long. Flowers 1-2, in leaf-axils; calyx-lobes ciliate; corolla white, c. 6 mm ϕ; stamens sub-exerted. Capsule globose. Seeds brown to black.

Fl. & Fr.: Jul.-Dec. *Distr.*: Native of Brasil; now naturalised in Trop. Africa, Madagascar, India and Nepal. A common weed of roadsides and wastelands. *NAK 688* (Pandalam, c. 20 m).

HEWITTIA R. Wigth et Arnott
Madras J. Lit. Sci. 5:17, 22. 1837.

Hewittia malabarica (L.) Suresh in Nicols., Suresh & Manilal, An Interpr. Hort. Malab. 88. 1988. *Convolvulus malabaricus* L., Sp. Pl. 155. 1753; *C. sublobatus* L.f., Suppl. Pl. 135. 1782. *C. bicolor* Vahl, Symb. 3:25. 1794, non Desr., 1792. *Ipomoea malabarica* (L.) Roem. & Schult., Syst. 4:235.1819. *Argyreia malabarica* (L.) Choisy, Mem. Soc. Phys. Hist. Nat. Geneve 6:420. 1834. *Hewittia bicolor* Wight & Arn., Madras J. Sci. 1(5) : 22. 1837. *H. sublobata* (L.f.) Kuntze, Rev. Gen. Pl. 2:441. 1891; Ooststr. in Steenis, Fl. Males. I, 4 : 438. 1953; Austin in Dassan. & Fosberg, Rev. Handb. Fl. Ceylon 1:312. 1980; Rani & Matthew, in Matthew Fl. Tam. Carnatic 3(2) : 1016. 1983. *Kattu-kelengu* Rheede, Hort. Malab. 11:105, t. 51. 1692.

Twiners. Leaves ovate, entire or angled, 4-8 by 3-5 cm, basally 3-nerved, apex acute, chartaceous. Flower(s) solitry, axillary; bracts linear-lanceate; calyx-lobes unequal; corolla yellowish or cream- coloured, c. 3.5 cm ϕ; stamens included, pollen grains smooth; ovary imperfectly 2-locular; ovules 2-per locule; stigmas 2. Capsule, 4-valved. Seed subtrigonous.

Fl. & Fr.: Dec.-Mar. *Distr.*: Trop. Africa and Asia, Malesia, Polynesia. Very common. In midlands to hills. *NAK* 2912 (Adoor, c. 20 m).

IPOMOEA Linnaeus
Sp. Pl. 159. 1753.

Herbaceous twiners, rarely subscandent shrubs. Leaves alternate, entire or palmately 3-7- lobed. Flower(s) in umbellate cymes or panicles, rarely solitary, axillary, 5-merous, showy; calyx-lobes often accrescent; corolla funnel-form or salver-form, lobes plicate; stamens usually included; ovary 2-4- locular. Capsule thin- walled, dehiscent. Seeds glabrous or pubescent.

1. Subscandent (rarely erect) shrubs. ....4. **I. carnea** ssp. **fistulosa**
1. Twining or creeping herbs.
    2. Creeping aquatic herbs. ....2. **I. aquatica**
    2. Twining terrestrial herbs.
        3. Leaves entire (rarely 3-lobed in *I. hederifolia*).
            4. Corolla white or pinkish white.
                5. Leaves broadly cordiform; corolla-tube 8-10 cm long. ....1. **I. alba**
                5. Leaves triangular-cordiform or transversely elliptic-oblong; corolla-tube oblong, c. 3 cm long.
                    6. Leave transversely elliptic-oblong; flowers pink-purple to white, c.5 cm ϕ. ....13. **I. pes-caprae** ssp. **brasiliensis**
                    6. Leaves triangular cordiform; flowers pinkish-white, c. 3 cm ϕ. ....9. **I. marginata**
            4. Corolla red or pink.
                7. Bracts large, perfoliate; seeds glabrous. ....15. **I. pileata**
                7. Bracts obscure or 0; seeds pubescent or pitted.
                    8. Seeds pubescent; corolla 2-3 cm ϕ.
                        9. Leaves entire to obscurely 3-lobed; corolla red. .....7. **I. hederifolia**
                        9. Leaves perfectly entire; corolla yellow. ....12. **I. obscura**
                    8. Seeds glabrous, pitted; corolla, c. 1 cm ϕ. ....6. **I. eriocarpa**
        3. Leaves plamately (2) 3-7-lobed (rarely entire).
            10. Densely hirsute plants, flowers in pedunculate capitate heads.
                11. Bracts broadly ovate-cordate, c. 2 cm long. ....5. **I. deccana**
                11. Bracts ovate-oblong, c. 2.5 cm long. ....14. **I. pes-tigridis**
            10. Glabrous or glabrescent herbs (except *I. indica*); flowers in corymbose or umbellate cymes or panicles.
                12. Seeds glabrous except at margins.
                    13. Calyx-lobes linear-lanceate.

14. Inflorescence a dense axillary cluster; corolla violet- blue. ....8. **I. indica**

14. Inflorescence a cyme or flower solitary; corolla bluish white to pink. .... 11. **I. nil**

13. Calyx-lobes elliptic to elliptic- ovate. ....16. **I. triloba**

12. Seeds pubescent.

15. Leaves pedately 5-9- foliolate; stipular leaves present; corolla c. 5 cm long. ....3. **I. cairica**

15. Leaves palmately 5-7-lobed; stipular leaves absent; corolla c. 8 cm long. ....10. **I. mauritiana**

1. **Ipomoea alba** L., Sp. Pl. 161. 1753; Ooststr. in Steenis, Fl. Males. I, 4 : 480. 1953; Gandhi in Sald. & Nicols., Fl. Hassan Dist. 469. 1976; Austin in Dassan. & Fosberg, Rev. Handb. Fl. Ceylon 1:317. 1980; Rani & Matthew in Matthew, Fl. Tam. Carnatic 3(2) : 1020. 1983; Manilal, Fl. Silent Valley 188. 1988; Nicols., Suresh & Manilal, An Interpr. Hort. Malab. 89. 1988. *I. bona-nox* L., Sp. Pl. (ed. 2) 228. 1762; Clarke in Hook.f., Fl. Brit. India 4 : 197. 1883. *Calonyction bona-nox* (L.) Boger, Hort. Maurit. 227. 1837; Gamble, Fl. Pres. Madras 920. 1923. *Munda-valli* Rheede, Hort. Malab. 11:103, t. 50. 1692.

Extensive twiners. Leaves cordiform, 6-15 by 4-12 cm, base cordate, apex acuminate; petiole 5-17 cm long. Flowers white, solitary or few in cymes; calyx-lobes subequal; corolla salver-form, c. 10 cm ϕ, tube c.10 cm long; stamens subexerted. Capsule subglobose. Seeds pubescent.

Fl. & Fr.: Dec.-Mar. *Distr.*: Native of Trop. America, now widely naturalised in subtropics. Common. *NAK 286, 1419* (Moozhiar, c. 250 m).

Leaves and roots used as remedies against snake bite.

2. **Ipomoea aquatica** Forsk., Fl. Aegypt.-Arab. 44. 1775; Ooststr. in Steenis, Fl. Males. I, 4:473. 1953; Gandhi in Sald. & Nicols., Fl. Hassan Dist. 469. 1976; Austin in Dassan. & Fosberg, Rev. Handb. Fl. Ceylon 1:318. 1980; Rani & Matthew in Matthew, Fl. Tam. Carnatic 3(2) : 1021. 1983; Nicols., Suresh & Manilal, An Interpr. Hort. Malab. 89. 1988. *I. reptans* Poiret in Lam., Encycl. (Suppl. 3) : 460. 1814, non L., 1753; Gamble, Fl. Pres. Madras 916. 1923. *Ballel* Rheede, Hort. Malab. 11:107-108, t. 52. 1692.

Creeping or floating aquatic herbs. Leaves varies in form, usually oblong-lanceate or narrowly triangular, 5-10 by 2-6 cm, base hastate, apex acute; petiole 6-10 cm long. Flowers purplish-white, solitary or few in cymes; sepals subequal; corolla funnel-form, c. 5 cm ϕ, tube to 2 cm long; stamens included. Capsule globose. Seeds glabrous.

Fl. & Fr.: Sept.-Jan. *Distr.*: Tropics. Common in fresh water ponds and fallow fields. *NAK 2909* (Tiruvalla, c. 10 m).

3. **Ipomoea cairica** (L.) Sweet, Hort. Brit. 287. 1827; Gamble, Fl. Pres. Madras 918. 1923; Ooststr. in Steenis, Fl. Males. I, 4: 478. 1953; Gandhi in Sald. & Nicols., Fl. Hassan Dist. 470. 1976; Austin in Dassan. & Fosberg, Rev. Handb. Fl. Ceylon 1:322. 1980; Rani & Matthew in Matthew, Fl. Tam. Carnatic 3(2) :1024. 1983. *Convolvulus cairicus* L., Syst. Nat. (ed. 10) 922. 1759. *Ipomoea pulchella* Wight, Ic. t. 156. 1839, non Roth, 1821.

Extensive climbers. Leaves pedately 5-7-foliolate, orbicular in outline, 3-8 by 4-10 cm, lobes elliptic ovate, apex obtuse to emarginate. Flower(s) solitary or in short racemes, c. 4 cm ϕ; calyx-lobes subequal; corolla pink. Capsule globose. Seeds pubescent.

Fl. & Fr.: Sep.-Jan. (March). *Distr.*: Paleotropics. Garden escape; common on wayside thickets. *NAK* 1030 (Tiruvalla, c. 10 m).

4. **Ipomoea carnea** Jacq., Enum. Syst. Pl. 13. 1760; Gamble, Fl. Pres. Madras 919. 1923; Bor & Raiz., Beautiful Ind. Climbers shrubs 6. 1954, ssp. **fistulosa** (Choisy) D. Austin, Taxon 26 : 237. 1977 & in Dassan. Fosberg, Rev. Handb. Fl. Tam. Carnatic 3(2) : 1024. 1983. *I. fistulosa* Mart. ex Choisy in DC., Prodr. 9:349. 1845; Bhattacharya, J. Bombay Nat. Hist. Soc. 73: 318. 1976.

Erect to subscandent shrubs; stems fistulose at maturity, tomentose. Leaves ovate-lanceate, 4-14 by 2.5-9 cm, base cordate, apex gradually acuminate, lateral veins 8-10 pairs, prominent. Cymes α-flowered, axillary or subterminal; peduncle c. 10 cm long; calyx-lobes subequal; corolla pale pink, c. 9 cm long. Capsule ovoid. Seeds pubescent.

Fl. & Fr.: Jun.-Dec. *Distr.*: Circumtropics. Common in wastelands and gardens. A rapidly spreading troublesome weed in lowlands. *NAK 623, 1033 & 1577* (Adoor, Pandalam & Tiruvalla).

5. **Ipomoea deccana** Austin in Dassan.& Fosberg, Rev. Handb. Fl. Ceylon 1:324. 1980; Manilal, Fl. Silent Valley 188. 1988. *I. bracteata* Wight, Ic. 4(2) : 14, t. 1374. 1850; Clarke in Hook.f., Fl. Brit. India 4:203. 1883; Gamble, Fl. Pres. Madras 918. 1923.

Twining herbs. Leaves broadly ovate in outline, 3-5-lobed, 2-6 by 3-6 cm, base cordate, apex acute, basally 5-nerved, densely villous, petiole 2-5 cm long. Flowers in capitate cymose clusters; bracts broadly ovate, foliaceous 2.5 by 1.5 cm, base cordate, apex obtuse; sepals unequal-lanceate, linear; corolla pink, c.15 mm ϕ. Capsule subglobose. Seeds pubescent.

Fl. & Fr.: Nov.-Mar. *Distr.*: Peninsular India. Common in moist rocky hills usually of streamsides. *NAK 1307, 2151* (Moozhiar & Chittar).

6. **Ipomoea eriocarpa** R.Br., Prodr. 484. 1810; Clarke in Hook.f., Fl. Brit. India 4:204. 1883; Ooststr. in Steenis, Fl. Males. I, 4: 462 : 1953; Gandhi in Sald. & Nicols., Fl. Hassan Dist. 470. 1976; Austin in Dassan. & Fosberg, Rev. Handb. Fl.Ceylon 1:325. 1980; Rani & Matthew in Matthew, Fl. Tam. Carnatic 3(2) : 1026. 1983; Manilal, Fl. Silent Valley 188. 1988. *I. hispida* (Vahl) Roem. & Schult., Syst. Veg. 4 . 238. 1819; Gamble, Fl Pres. Madras 915. 1923. *I. sessiliflora* Roth, Nov.Pl. Sp. 116. 1821; Wight, Ic. t. 169. 1839.

Slender hispid twiners. Leaves triangular-ovate to lanceate, 3-5 by 1.2-3 cm, base cordate to sub-hastate, apex gradually acute-apiculate, chartaceous. Flower(s) solitary or aggregated in subsessile clusters, axillary; calyx-lobes linear-lanceate; corolla pink, c. 1 cm long. Capsule small, subglobose. Seeds pitted.

Fl. & Fr. Nov.-Mar. *Distr.*: Paleotropics. Rare. In rocky grasslands. *NAK 2116 & 2142* (Chittar).

Leaves bruised and used to cure epilepsy, fever, ulcers and poisons.

7. **Ipomoea hederifolia** L., Syst. Nat. (ed. 10) 925. 1759; O'Donell, Lilloa 29:45. 1959; Gandhi in Sald. & Nicols., Fl. Hassan Dist. 471. 1976; Austin in Dassan. & Fosberg, Rev. Handb. Fl. Ceylon 1:325. 1980; Rani & Matthew in Matthew, Fl. Tam. Carnatic 3(2) : 1027. 1983; Mohanan & Sivad., Fl. Agasthyamala 472.2002. *I. angulata* Lam., Tabl. Encycl. 1:464. 1793; Ooststr. in Steenis, Fl. Males. I, 4 : 481. 1953. *I. phoenicea* (Roxb), Fl. Ind. 2:92.1824.*Quamoclit phoenicea* (Roxb). Choisy, Mem. Soc. Phys. Geneve 6:433. 1834;

Gamble, Fl. Pres. Madras 919. 1923. *I. coccinea* auct, non. L., 1753: Clarke in Hook.f., Fl. Brit. India 4 : 199. 1883.

Herbaceous twiners. Leaves broadly ovate to rarely shallowly 3-lobed, 4-8 by 3.5-7.5 cm, base cordate, apex acuminate; petiole to 6 cm long. Cymes few- flowered, axillary; calyx-lobes subequal; corolla red, salver-form; stamens exerted; ovary conical. Capsule globose. Seeds puberulous.

Fl. & Fr.: Sep.-Mar. *Distr.*: Native of Trop. America; now naturalised in the tropics. A garden escape, occasionally found on wayside thickets. *NAK 1123* (Moozhiar, c. 250 m).

8. **Ipomoea indica** (Burm. f.) Merr., Interpr. Herb. Amboin. 445. 1917; Fosberg, Micronesica 2:151. 1967 & Bot. Not. 129 : 37. 1976; Austin in Dassan. & Fosberg, Rev. Handb. Fl. Ceylon 1:327. 1980; Rani & Matthew in Matthew, Fl. Tam. Carnatic 3(2) : 1028. 1983. *Convolvulus indicus* Burm. f. in Rumph., Herb. Amboin. 7:(6) 1755. *Ipomoea congesta* R.Br., Prodr. 485. 1810; Ooststr. in Steenis, Fl. Males. I, 4: 465. 1953; Matthew, Rec. Bot. Surv. India 20 : 158. 1969.

Hispid twiners. Leaves broadly ovate in outline, 3-lobed, 5-8 by 6-8 cm, base cordate, apex acute; petiole 3-6 cm long. Cymes umbellate, dense, axillary; calyx-lobes linear-lanceate; corolla bluish-violet, 6-7 cm ϕ; stamens included. Capsule ovoid. Seeds glabrous.

Fl. & Fr.: Sep.-Mar. *Distr.*: Native of S. America, widely naturalised in the tropics. A garden escape. *NAK 1292* (Adoor, c. 20 m).

9. **Ipomoea marginata** (Desr.) Verdc., Kew Bull. 42 : 658. 1987; Nicols., Suresh & Manilal, An Interpr. Hort. Malab. 91. 1988. *Convolvulus marginatus* Desr. in Lam., Encycl. 3:558. 1792. *Ipomoea sepiaria* Koeing ex Roxb., Hort. Beng. 14. nom. illegit. (incl. type of *Convolvulus marginatus* Desr., 1792); Roxb., Fl. Ind. 2:90. 1824; Clarke in Hook.f., Fl. Brit. India 4 : 209. 1883; Gamble, Fl. Pres. Madras 916. 1923; Verdc., Fl. Trop. Africa, Convolvulac. 117. 1963; Austin in Dassan. & Fosberg, Rev. Handb. Fl. Ceylon 1:338. 1980; Rani & Matthew in Matthew, Fl. Tam. Carnatic 3(2) : 1032. 1983. *I. maxima* auct., non (L.) Sweet, 1830: Ooststr. in Steenis, Fl. Males. I, 4: 472. 1953. *Tiru-tali* Rheede, Hort. Malab. 11:109-110, t. 53. 1692.

Extensive twiners. Leaves triangular-cordiform, 5-8 by 3-5 cm, base cordate, apex acute-apiculate, thinly pubescent, petiole 1-2 cm long. Cymes sub-umbellate; calyx-lobes sub-equal, obovate; corolla pinkish-white, c. 3 cm ϕ, salver-form, stamens included. Capsule globose. Seeds pubescent.

Fl. & Fr.: Dec.-Mar. *Distr.:* Trop. Africa, Trop. Asia, Malesia, Queensland, Australia. Common in exposed wastelands. *NAK 2903* (Adoor, c. 20 m).

10. **Ipomoea mauritiana** Jacq., Collect. 4:216. 1791; Jacq., Hort. Schoenbr. 2:39, t. 200. 1797; Verdc. in Fl. Trop. E. Africa, Convolvulac. 135. 1963; Austin in Dassan. & Fosberg, Rev. Handb. Fl. Ceylon 1:331. 1980; Nicols., Suresh & Manilal, An Interpr. Hort. Malab. 92. 1988; Mohanan & Sivad., Fl. Agasthyamala 472. 2002. *Convolvulus paniculatus* L., Sp. Pl. 156. 1753; *Ipomoea paniculata* (L.) R. Br., Prodr. 486. 1810, non N.Burm., 1768: Gamble, Fl. Pres. Madras 918. 1923. *Batatas paniculata* (L.) Choisy, Mem. Soc. Phys. Hist. Nat. Geneve 6:436. 1834. *Ipomoea digitata* auct., non L., 1759: Clarke in Hook.f., Fl. Brit. India 4:202. 1883. *Pal-Modecca* Rheede, Hort. Malab. 11:101-102, t. 49. 1692.

Stout twiners. Leaves 5-7-lobed, 8-14 by 9-12 cm, lobes lanceate, base cordate, apex acuminate; petiole 5-10 cm long. Flowers in corymbose panicles, c. 4 cm ϕ; bracts caducous; peduncle 8-20 cm long; calyx-lobes subequal, suborbicular; corolla pink, widely campanulate, 5-7 cm long; stamens included; ovary clearly 4-locular. Capsule ovoid. Seeds wooly.

Fl. & Fr.: Jul.-Dec. *Distr.*: Sri Lanka, India, Trop. Asia, Africa, America & Australia. Common. *NAK 621 & 1898* (Pandalam & Tiruvalla).

Powdered root stock along with honey or wine increase secretion of milk.

11. **Ipomoea nil** (L.) Roth, Catal. Bot. 1:36. 1797; Ooststr. in Steenis, Fl. Males. I,4:465. 1953; Shinn., Taxon 14:231. 1965; Gandhi in Sald. & Nicols., Fl. Hassan Dist. 472. 1976; Austin in Dassan. & Fosberg, Rev. Handb. Fl. Ceylon 1:332. 1980; Rani & Matthew in Matthew, Fl. Tam. Carantic 3(2) : 1028. 1983. *Convolvulus nil* L., Sp. Pl. (ed. 2) 219. 1762. *Ipomoea hederacea* auct., non (L.) Jacq, 1786 Hook.f., Fl. Brit. India 4: 199. 1883; Gamble, Fl. Pres. Madras 917. 1923.

Pubescent twiners. Leaves ovate in outline, 3-lobed, 6-8 by 5-7 cm, base cordate, lobes acute to acuminate at apex. Cymes umbellate, axillary, few- flowered; bracts linear; calyx-lobes subequal, linear-lanceate; corolla bluish, c. 4 cm ϕ, funnel-form. Capsule subglobose. Seeds pubescent.

Fl. & Fr.: Dec.-Apr. *Distr.*: Native of N. America; now naturalised in tropics. Less common. In lowlands. *NAK 2906* (Adoor, c. 20 m).

12. **Ipomoea obscura** (L.) Ker Gawler, Bot. Reg. 3:t. 239. 1817; Clarke in Hook.f., Fl. Brit. India 4:207. 1883; Gamble, Fl. Pres. Madras 916. 1923; Ooststr. in Steenis, Fl. Males. I,4: 471. 1953 Gandhi in Sald. & Nicols., Fl. Hassan Dist. 472. 1976; Austin in Dassan. & Fosberg, Rev. Handb. Fl. Ceylon 1:333. 1980; Rani & Matthew in Matthew, Fl. Tam. Carnatic 3(2) : 1030. 1983. *Convolvulus obscurus* L., Sp. Pl. (ed. 2) 220. 1762; Roxb., Fl. Ind. 2:52. 1824.

Slender twiners. Leaves ovate-cordiform, 3-6 by 2.5-4 cm, base cordate, apex acute to acuminate. Flower(s) solitary or few in axillary cymes; bracts linear-lanceate; cayx-lobes ovate-lanceate, sub-equal, corolla yellow with a purple throat, c.15 mm ϕ, funnel-form. Capsule ovoid. Seeds tomentose.

Fl. & Fr.: Sep.-Mar. *Distr.*: Sri Lanka, India, Malesia & E. Africa. Common. Among wayside thickets. *NAK 2913* (Adoor, c. 20 m).

13. **Ipomoea pes-caprae** (L.) R. Br. in Turkey, Narr. Exped. R. Zaire 477. Mar. 1818; Sweet, Hort. Suburb. Lond. 35. 1818; Gamble, Fl. Pres. Madras 917. 1923; Austin in Dassan. & Fosberg, Rev. Handb. Fl. Ceylon 1:344. 1980; Rani & Matthew in Matthew, Fl. Tam. Carnatic 3(2) : 1030. 1983, ssp. **brasiliensis** Ooststr., Blumea 3: 533. 1940 & in Steenis, Fl. Males. I, 4:475. 1953; Ravi, Bull. Bot. Surv. India 17: 197. 1975. *Adumbuvalil.*

Stout creepers. Leaves shallowly 2-lobed, 2-5 x 3-6 cm, base truncate, lateral nerves c. 6 pairs, coriaceous; petiole, 2-4 cm long. Flower(s) solitary or few in axillary panciles; calyx-lobes unequal, ovate, acuminate, outer surface wrinkled; corolla pink to white, c. 5 cm, ϕ; ovary 4-6- locular. Capsule subglobose. Seeds thinly pubescent.

Fl. & Fr.: Nov.-Mar. *Distr.*: Patropical. Common. On exposed grounds in plains. *NAK 113* (Adoor, c. 20 m).

Powdered leaves purify water and are used against rheumatism.

14. **Ipomoea pes-tigridis L.,** Sp. Pl. 162. 1753; Roxb., Fl. Ind. 2:93. 1824; Wight, Ic. t. 836. 1844-45; Clarke in Hook. f., Fl. Brit. India 4; 204. 1883; Gamble, Fl. Pres. Madras 918. 1923; Ooststr. in Steenis, Fl. Males. I, 4:467. 1953; Austin in Dassan. & Fosberg, Rev. Handb. Fl. Ceylon 1:336. 1980; Rani & Matthew in Matthew, Fl. Tam. Carnatic 3(2): 1031. 1983; Nicols., Suresh & Manilal, An Interpr. Hort. Malab. 92. 1988. *Pulli-schovadi* Rheede, Hort. Malab. 11:121, t. 59. 1962.

Hispid twiners. Leaves 5-6-lobed, lobes obovate, base decurrent, apex acute to acuminate, sericeous; petiole 5-8 cm long. Flowers aggregated in capitate clusters; penduncle c. 7 cm long; calyx-lobes unequal; corolla white, funnel-shaped, c. 4 cm, ϕ. Capsule globose. Seeds pubescent.

Fl. & Fr.: Sept. onwards. *Distr.*: Tropical Asia and Africa. Less common. On laterite soil. *NAK 2908* (Adoor, c. 20 m).

15. **Ipomoea pileata** Roxb., Fl. Ind. 2:94. 1824; Wight Ic. t. 1363. 1848; Clarke in Hook. f., Fl. Brit. India 203. 1883; Gamble, Fl. Pres. Madras 916. 1923.

Hispid twiners. Leaves ovate, 3-8 × 2.5-5.5 cm, base cordate, apex acuminate, apiculate; chartaceous; petiole 2-7 cm long. Flowers aggregated in capitate clusters; peduncle c. 7 cm long; bracts perfoliate, foliaceous, cucullate; calyx-lobes linear-lanceolate; corolla pinkish purple, c. 2.5 cm, ϕ, funnel-shaped. Capsule subglobose. Seeds glabrous.

Fl. & Fr.: Nov.-Jan. *Distr.*: Sri Lanka, India. Rare. Known only from one collection. On exposed hills. *NAK 2189* (Chittar, c. 250 m).

16. **Ipomoea triloba** L., Sp. Pl. 161. 1753; Ooststr., Blumea 3:509. 1940 & in Steenis, Fl. Males. I, 4: 468. 1953. *Convolvulus dentatus* Blanco, Fl. Filip. ed. 1, 89. 1837, non Vahl, 1794. *Ipomoea blancoi* Choisy in DC., Prodr. 9. 389. 1845. **(Fig. 50)**

Twiners. Leaves broadly ovate, entire to deeply 3-lobed, 4-10 x 3-8 cm, base cordate, margin often deeply dentate, apex acuminate-apiculate, chartaceous; petiole 1-10 cm long. Flowers aggregated in umbellate cymes; peduncle c. 10 cm long; bracts minute; calyx-lobes sub-equal, elliptic-ovate, densely hairy without; corolla pinkish purple, c. 23 cm ϕ, funnel-shaped. Capsule subgolose, densely hispid. Seeds glabrous except at one margin.

Fl. & Fr.: Sep.-Mar. *Distr.*: Tropics. Not common. An introduced weed. *NAK 954* (Sabarimala, c. 250 m).

LEPISTEMON Blume
Bijdr. 722. 1926.

Herbaceous twiners; stem velutinous. Leaves alterante, long petioled; petiole to 8 cm long, lamina entire to 3-lobed, brown tomentose on both sides. Flowers in axillary cymes; calyx 5-partite;lobes glabrous or pubsecent, mottled or not; corolla urceolate, inflated at base, strongly plaited, cream or yellow; disc of 5 obovoid or sub-orbicular scales; stamens 5, included, inserted at the base of the disc-scales; ovary conical, 2-celled, or 4-celled, 4-ovuled; capsule glabrous, dorsiventrally compressed, deeply 4-lobed at the apex or entire; seed 4, ovoid, tomentose or glabrous.

1. Capsule deeply 4-lobed at the apex; seeds pubescent. ....2. **L. verdcourtii**
1. Capsule entire at the apex; seeds glabrous. ....1. **L. leiocalyx**

1. **Lepistemon leiocalyx** Stapf, Kew Bull. 1895: 113. 1895; Gamble, Fl. Pres. Madras 921. 1923.

Leaves broadly ovate, often 3-lobed, 8-10 by 4-7 cm, base deeply cordate, apex gradually acuminate, mucronate, densely soft villous, chartaceous. Flowers in dense axillary cymes, bisexual, 5-merous; calyx-lobes 5, subequal; corolla yellow, pollen grains spinulose, disc cupular, ovary 2-locular; ovules 2 per locule. Capsule 4-valved. Seeds glabrous.

Fl. & Fr.: Jan.-Mar. *Distr.*: Western Ghats. Rare. On forest borders. *NAK 1518* (Gurunadhan mannu, c. 300 m).

2. **Lepistemon verdcortii** Mathew *et* Biju, Kew Bull. 46 (3) : 559-562. 1991.

Leaves broadly ovate, 4-9 by 3-5 cm, base deeply cordate, apex gradually acuminate, mucronate, densely soft villous, chartaceous; corolla yellow, urceolate, short; stamens 5, included, arising from the ventral sides of 5 scales; pollen grains spinulose, disc cupular, Capsule 4-valved, deeply notched at the apex. Seeds pubescent.

Fl. & Fr.: Jan.-Mar. *Distr.*: Konni. Rare. On forest borders. *Biju 42198* (Velithode, c. 250 m).

MERREMIA Dennstedt ex Endlicher Gen. 1403. 1841 (nom. cons.).

Twiners or creepers; branchelets terete or winged. Leaves simple, to 3-7-lobed, base cordate, truncate, auriculate or hastate; petiole 0 to 10 cm long. Flower (s) solitary or in corymbose or subumbellate-cymes; calyx-lobes rarely coriaceous; corolla white to yellow, campanulate or funnel-form; stamens unequal; ovary 2-4-locular. Capsule globose or conical. Seeds pubescent or glabrous.

1. Leaves simple (rarely 3-lobed in *M. hederacea*)
    2. Branchlets winged; calyx-lobes thick-coriaceous. ....4. **M. turpethum**
    2. Brachlets not winged; calyx-lobes herbaceous.
        3. Leaf base truncate or cordate; seeds hairy.
            4. Leaves cordiform; flowers yellow. ....2. **M. hederacea**
            4. Leaves elliptic-oblong; flowers white. ....5. **M. umbellatum**
        3. Leaf base auriculate or hastate; seeds glabrous. ....3. **M. tridentata**
1. Leaves 3-7-lobed.
    5. Leaves divided to below the middle; flowers white. ....1. **M. dissecta**
    5. Leaves divided to above the middle; flowers yellow. ....6. **M. vitifolia**

1. **Merremia dissecta** (Jacq.) Hallier f., Bot. Jahrb. Syst. 16:552. 1893; Gamble, Fl. Pres. Madras 928. 1923; Ooststr. in Steenis, Fl. Males. I,4 : 448. 1953; Austin in Dassan. & Fosberg, Rev. Hanb. Fl. Ceylon 1:349. 1980, Rani & Matthew in Matthew, Fl. Tam. Carnatic 3(2) : 1036. 1983. *Convolvulus dissectus* Jacq., Observ. Bot. 2:4, t. 28. 1767. *Ipomoea sinuata* Ortega, Nov. Pl. Deser. Dec. 84. 1798; Clarke in Hook.f., Fl. Brit. India 4:214. 1883.

Hirsute or glabrous twiners. Leaves dissected, lobes 5-7, lanceolate, 4-7 by 1-3 cm, margin irregularly lobed, apex acuminate; petiole c. 8 cm long. Flowers in axillary cymes; calyx-lobes unequal; corolla white with pinkish throat, c. 4 cm ϕ, campanulate. Capsule globose. Seeds glabrous.

Fl. & Fr.: Sep.-Jan. *Distr.*: Native of Trop. America; now in Sri Lanka, India, Malesia. *NAK 2915* (Adoor, c. 20 m).

2. **Merremia hederacea** (Burm. f.) Hallier f., Bot. Jahrb. 18:118. 1893; Ooststr. in Steenis, Fl. Males. I,4 : 441. 1953; Verdc., Fl. Trop. E. Africa, Convolvulac. 54. 1963; Austin in Dassan. & Fosberg, Rev. Handb. Fl. Ceylon 1:350. 1980; Rani & Matthew in Matthew, Fl. Tam. Carnatic 3(2) : 1037. 1983; Nicols., Suresh &

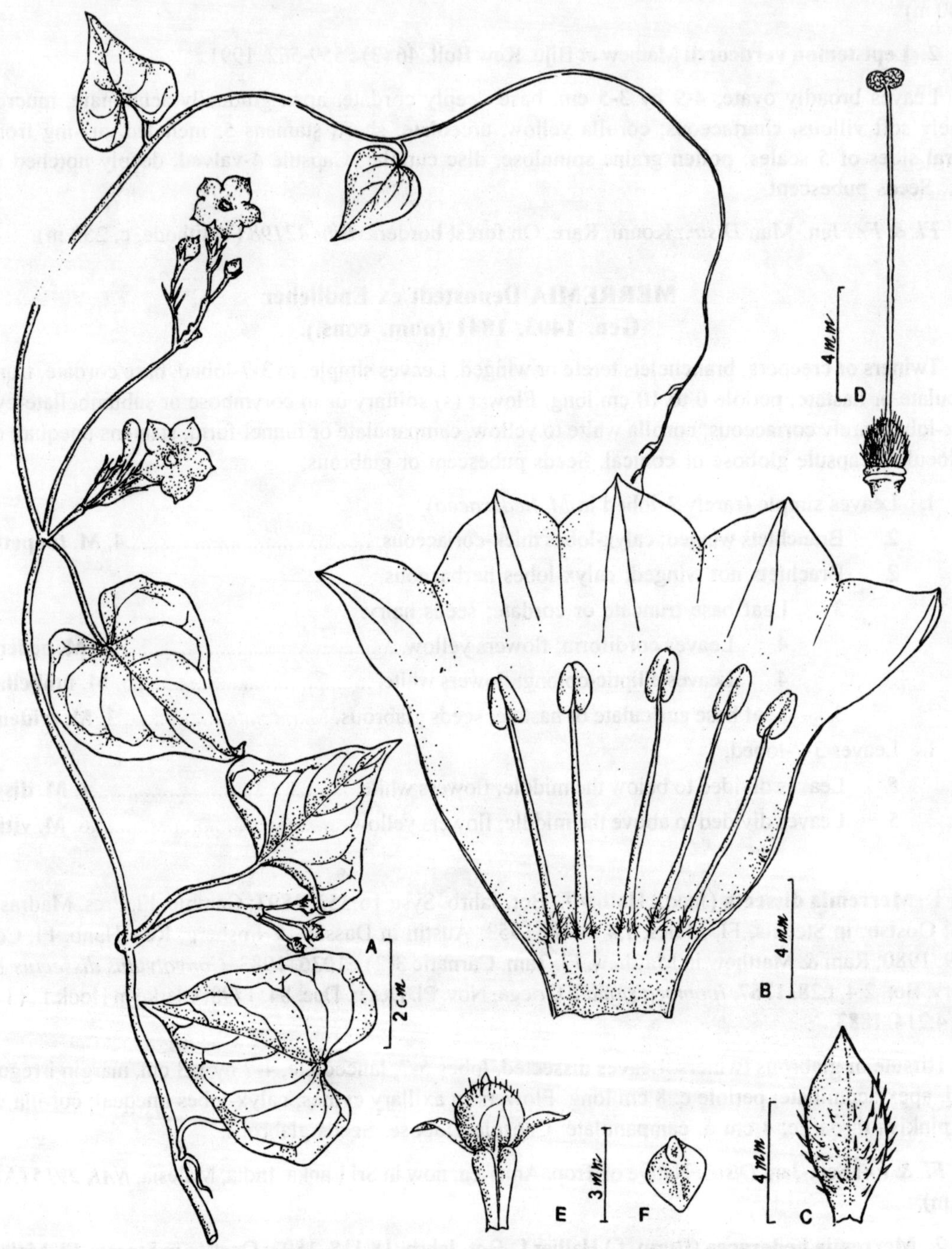

Fig. 50. ***Ipomoea triloba*** **L.: A. Twig; B. Corolla split-open end showing stamens; C. Calyx lobe; D. Pistil; E. Fruit; F. Seed.**

Manilal, An Interpr. Hort. Malab. 92. 1988. *Evolvulus hederaceus* Burm.f., Fl. Ind. 77, t. 30, f. 2. 1768. *Merremia convolvulacea* Dennst. ex Hallier f., Bot. Jahrb. 16:552. 1983. *Ipomoea chryseides* Ker Gawler, Bot. Reg. 4:t. 270. 1818; Wight, Ic. t. 157. 1839; Clarke in Hook.f., Fl. Brit.India 4:206. 1883. *Merremia chryseides* H. Hallier, Bot. Jahrb. Syst. 16:552.1893; Gamble, Fl. Pres. Madras 929. 1923. *Kudici-valli.* Rheede, Hort. Malab. 8: 52 (51), t. 27. 1688.

Slender twiners. Leaves ovate-cordiform, rarely 3-lobed, 2-4 by 1.3-3 cm, base cordate, margin dentate, apex acuminate-apiculate, thinly pubescent; petiole 1-3 cm long. Cymes α-flowered, axillary; calyx-lobes unequal; corolla yellow, c. 15 mm ϕ, funnel-form. Capsule ovoid, obscurely angular. Seeds puberulous.

Fl. & Fr.: Dec.-Mar. *Distr.:* Sri Lanka, India, Malesia, S. China, Australia, Trop. Africa. *NAK 1416* (Arampa, c. 400 m).

3. **Merremia tridentata** (L.) Hallier f., Bot. Jahrb. Syst. 16.552. 1893; Gamble, Fl. Pres. Madras 928. 1923; Austin in Dassan. & Fosberg, Rev. Handb. Fl. Ceylon 1:351. 1980; Ooststr., Blumea 3:315. 1939 & in Steenis, Fl. Males. I, 4:445. 1953; Nicols., Suresh & Manilal, An Interpr. Hort. Malab. 93. 1988; Mohanan & Sivad., Fl. Agasthyamala 474.2002. *Convolvulus tridentatus* L., Sp. 157. 1753. *Ipomoea tridentata* (L.) Roth, Arch. Bot. 1(2):38. 1798. *Sendera-clanti* Rheede, Hort. Malab. 11:133, t. 65. 1692.

Key to the subspecies

1. Leaf base auricled, apex dentate; peducle upto 2mm long. ....**Merremia tridentata** ssp. **tridentata**

1. Leaf base hastate, apex acute-mucronate; peduncle upto 8 mm long. ....**Merremia tridentata** ssp. **hastata**

Merremia tridentata (L.) Hallier f. ssp. **tridentata**

Prostrate or trailing herbs. Leaves subsessile to sessile, linear-oblong, 5-20 by 3-4 mm, base auricled, margin and apex dentate. Flower(s) solitary or very few in axillary cymes; peduncle to 2 cm long; calyx-lobes subequal; corolla cream-coloured or yellow, throat purplish, c.15 mm ϕ. Capsule globose. Seeds glabrous.

Fl. & Fr.: Sep.-Jan. *Distr.*: Trop. Asia, Trop. Africa, Malesia, Australia. Common. On exposed grounds. *NAK 1654* (Tiruvalla, c. 10 m).

Merremia tridentata (L.) Hallier f. ssp. **hastata** (Hallier f.) Ooststr., Blumea 3:317. 1939 & in Steenis, Fl. Males. I, 4: 445. 1953; Gandhi in Sald. & Nicols., Fl. Hassan Dist. 473. 1976; Nicols., Suresh & Manilal, An Interpr. Hort. Malab. 93. 1988. *Convolvulus hastatus* Desr. in Lam., Encycl. 3:542. 1789, non Forsskal, 1775; *Merremia hastata* (Desr.) Hallier f., Bot. Jahrb. Syst. 16: 552. 1893; Gamble, Fl. Pres. Madras 652. 1923. *Tala-neli* Rheede, Hort. Malab. 11:113, t. 55. 1692.

Trailing herbs. Leaves subsessile, oblong to oblong-lanceate, 4-10 by 1-2 cm; base hastate dentate, apex acute-mucronate; peduncle 5-8 cm long; calyx-lobes ovate-lanceate; corolla pale yellow to white, campanulate. Capsule globose. Seeds glabrous.

Fl. & Fr.: Sep.-Mar. *Distr.*: Trop. Asia & Africa, Malesia, Australia. Common. *NAK 873* (Pandalam, c. 20 m).

4. **Merremia turpethum** (L.) Shah & Bhatt, J. Bombay Nat. Hist. Soc. 74:567. 1978; Rani & Matthew in Matthew, Fl. Tam. Carnatic 3(2) : 1039. 1983. *Convolvulus turpethum* L., Sp. Pl. 155. 1753; Roxb., Fl. Ind.

2:57. 1824. *Ipomoea turpethum* (L.) R. Br., Prodr. 485. 1810; Clarke in Hook.f., Fl. Brit. India 4:212. 1883. *Thrikolpa konna.Operculina turpethum* (L.) Silva Manso, Enum. Subst. Braz. 16: 49. 1836; Gamble, Fl. Pres. Madras 929 (653). 1923; Ooststr. in Steenis, Fl. Males. I,4:56. 1953; Austin in Dassan.& Fosberg, Rev. Handb. Fl. Ceylon 1:356. 1980.

Stout twiners, branchlets winged. Leaves broadly ovate, 5-10 by 4-8.5 cm, base cordate, apex acute or obtuse, mucronate, thinly pubescent; petiole 2-7 cm long. Flower(s) solitary or few in axillary cymes; calyx-lobes ovate, outer larger, 2 by 1.5 cm, thick coriaceous at age; corolla white, c. 3 cm ϕ; ovary 2-locular.

Capsule ovoid. Seeds glabrous.

Fl. & Fr.: Jul.-Dec. *Distr.*: Old World Tropics and Polynesia. Rare. Known from only one collection in fruits. Evergreen forest clearings. *NAK 14* (Moozhiar, c. 250 m).

5. **Merremia umbellata** (L.) Hallier f., Bot. Jahrb. Syst. 16:552. 1893; Gamble, Fl. Pres. Madras 928. 1923; Ooststr. in Steenis, Fl. Males. I, 4.449. 1953; Gandhi in Sald. & Nicols., Fl,. Hassan Dist. 473. 1976; Austin in Dassan. & Fosberg, Rev. Handb. Fl. Ceylon 1:354. 1980; Rani & Matthew in Matthew, Fl. Tam. Carnatic 3(2): 1040. 1983; Mohanan & Sivad., Fl. Agasthyamala 474. 2002. *Convolvulus umbellatus* L., Sp. Pl. 155. 1753. *Convolvulus cymosus* Desr. in Lam., Encycl. 3: 556. 1752. *Ipomoea cymosa* (Desr.) Roem. & Schult., Syst. Nat. 4:241. 1819; Clarke in Hook.f., Fl. Brit. India 4:211. 1883.

Twiners. Leaves elliptic-oblong to oblong-lanceate, 4-9 by 1.5-5 cm, base truncate to auriculate, apex acuminate, mucronate; thinly pubescent to glabrous; petiole 1-4 cm long. Flowers in subumbellate axillary cymes; peduncle c. 4 cm long; calyx-lobes subequal; corolla white, 3-4 cm ϕ, pubescent without. Capsule globose. Seeds long hairy.

Fl. & Fr.: Most of the seasons. *Distr.:* Trop. Asia & Africa. A very variable common species of the genus. *NAK 532, 1352, 1512 & 1519* (Almost all forests).

6. **Merremia vitifolia** (Burm. f.) Hallier f., Bot. Jahrb. Syst. 16:552. 1893; Gamble, Fl. Pres. Madras 928. 1923; Ooststr. in Steenis, Fl. Males. I, 4 : 448. 1953; Gandhi in Sald. & Nicols., Fl. Hassan Dist. 474. 1976; Austin in Dassan. & Fosberg, Rev. Handb. Fl. Ceylon 1:355. 1980; Rani & Matthew in Matthew, Fl. Tam. Carnatic 3(2) : 1040. 1983; Manilal, Fl. Silent Valley 190. 1988; Mohanan & Sivad., Fl. Agasthyamala 475.2002. *Convolvulus vitifolius* Burm.f., Fl. Ind. 45, t. 18, f. 1. 1768. *Ipomoea vitifolia* (Burm.f.) Bl., Bijdr. 709. 1825; Clarke in Hook.f., Fl. Brit. India 4:213. 1883.

Extensive twiners. Leaves 3-5-lobed, 6-12 by 5-12 cm, base cordate, margin dentate-serrate, basally 7-ribbed; petiole 3-10 cm long. Flowers in axillary cymes; peduncle c. 3 cm long; calyx-lobes unequal, corolla yellow, c. 7 cm ϕ. Capsule ovoid. Seeds glabrous.

Fl. & Fr.: Most of the seasons. *Distr.*: Sri Lanka, India, Nepal, Indo-China, Malesia. Very common in hills. *NAK 203, 268* (Almost all forests).

92. SOLANACEAE

1. Corolla trumpet-shaped, lobes shorter than tube; fruit a capsule. ....**Datura**
1. Corolla stellate (rotate) or campanulate, lobes longer than tube; fruit a berry.
    2. Anthers connivent in a cone, dehiscence porous; flowers in fascicles or cymes.
        3. Stragglers; flowers in fascicles; calyx cupular or saucer-form. ....**Lycianthes**

3. Erect subshrubs; flowers in cymes; calyx campanulate. ....**Solanum**

2. Anthers not connivent in a cone, dehiscence longitudinal; flower(s) solitary. ....**Physalis**

DATURA Linnaeus
Sp. Pl. 179. 1753.

Subshrubs or shrubs. Leaves simple, alternate, entire or lobed. Flower(s) solitary, axillary, pedicellate, 5-merous, bisexual, tubular, herbaceous, lower part persistent; corolla trumpet-form, plicate, tube gradually widened from middle; stamens 5, near the middle of tube, included; ovary superior, conical, obliquely placed, echinate, 2-locular; ovules α, on bifurcate placenta. Capsule globose, densely softly prickly. Seeds α, compressed.

1. Large shrubs; leaves entire; calyx to 12 cm long; corolla to 25 cm long. ....1. **D. arborea**

1. Undershrubs; leaves usualy sinuate; calyx to 3 cm long; corolla to 16 cm long. ....2. **D. stramonium**

1. **Datura arborea** L., Sp. Pl. 179. 1753; Fyson, Fl. South Indian Hill St. 2:t. 363. 1932; Gamble, Fl. Pres. Madras 941. 1923.

Large shrubs. Leaves broadly elliptic-ovate, 15-18 by 8-11 cm, base obliquely rounded, apex acute. Flowers pendulous; calyx to 12 cm long; slightly inflated; corolla white, to 25 cm long. Capsule globose.

Fl. & Fr.: Dec.-Mar. *Distri.:* Along waysides. Garden escape. *NAK 1291* (Near I.B., Kakki hills, c. 1100 m).

2. **Datura stramonium** L., Sp. Pl. 179. 1753; Clarke in Hook.f., Fl.Brit. India 4:242. 1883; Gamble, Fl. Pres. Madras 941. 1923; Nicols., Suresh & Manilal, An Interpr. Hort. Malab. 248. 1988; *Hummatu* Rheede, Hort. Malab. 2:47-48, t. 28. 1679.

Undershrubs. Leaves broadly elliptic or elliptic-ovate, usually sinuate, 6-13 by 3.5-9.5 cm, base obliquely cuneate, apex shortly acuminate, slightly pubsecent. Flowers not pendulous; calyx to 3 cm long, not inflated; corolla white, to 16 cm long. Capsule globose, c. 5 cm ϕ.

Fl. & Fr.: Jul.-Sep. *Distr*: Tropics. Less common. On waysides in plains. *NAK 632* (Malayalapuzha, c. 40 m).

Juice of fruits with opium and oil applied in parasitic skin diseases. Smoking of dried leaves and stem relieves asthma. Seed powder is a good remedy for tooth-ache.

LYCIANTHES (Dunal) Hassler
Annualre Conserv. Jard. Bot. Geneve 20:180. 1917.

Lycianthes laevis (Dunal) Bitter, Abh. Nat. Ver. Bramen 24:484. 1920; Backer & Bakh.f., Fl. Java 2:476. 1965 ('*levis*'); Gandhi in Sald. & Nicols., Fl. Hassan Dist. 59. 1976; Manilal, Fl. Silent Valley 191. 1988; Mohanan & Sivad., Fl. Agasthyamala 476.2002. *Solanum laeve* Dunal, Solan. Synop. 22. 1816; Clarke in Hook.f., Fl. Brit. India 4:231. 1883; Gamble, Fl. Pres. Madras 936. 1923.

Unarmed straggling shrubs. Leaves elliptic-lanceate, 10-20 by 3-6 cm, base and apex acuminate, upper surface sparsely pubsecent, membranous. Flowers in fascicles on upper leaf-axils, 5-merous, bisexual; calyx cupular or saucer-form, 5-dentate; corolla white with purple tinge, stellate (rotate). lobes 5, ovate; stamens 5, inserted at base, anthers longer than filaments connivent into a cone, dehiscence porous; ovary superior; ovules α. Fruit a berry, crimson red when ripe.

Fl. & Fr.: Dec.-Feb. *Distr.*: South India. Rare. Among forest undergrowth in upper hills. *NAK 2589* (Arampa, c. 400 m).

PHYSALIS Linnaeus
Sp.Pl. 182. 1752.

Herbs or subshrubs. Leaves alternate or opposite, entire to irregularly toothed, glabrous to strigose pubescent. Flower(s) solitary, axillary, 5-merous; calyx campanulate, accrescent, inflated; corolla broadly campanulate, lobes 5, plicate; stamens 5, exerted, anthers distinct, dehiscence longitudinal; ovary 2-locular. Berry globose, overtopped by inflated calyx. Seeds smooth or ruguoe.

1. Glabrous herbs, to 75 cm tall; fruiting calyx to 2 cm long; corolla c. 6 mm ϕ. .... 1. **P. angulata**
1. Pubescent herbs to 1 m tall; fruiting calyx to 3 cm long; corolla c. 1 cm ϕ. .... 2. **P. peruviana**

1. **Physalis angulata** L., Sp. Pl. 183. 1753; Schult. in Urban, Symbol. Antill. 4:141. 1909; Heine in Hepper, Fl.W. Trop. Africa (ed. 2), 2:329. 1963; Waterfall, Rhodora 69:216.1967; Hepper in Dassan. & Fosberg, Rev. Handb. Fl. Ceylon 6:394. 1988; Nicols., Suresh & Manilal, An Interpr. Hort. Malab. 248.1988. *P. minima* L., Sp. Pl. 183. 1753; Clarke in Hook.f., Brit. India 4:238. 1883; Gamble, Fl. Pres. Madras 939. 1923; Gandhi in Sald. & Nicols., Fl. Hassan Dist. 460. 1976; Matthew, Ill. Fl. Tam. Carnatic t. 490. 1982; Matthew & Rani in Matthew, Fl. Tam. Carnatic 3(2): 1056. 1983; Manilal, Fl. Silent Valley 192. 1988. *Inota-inodien* Rheede, Hort Malab. 10:139, t. 70. 1690. *Pee-jnota-jnodien* Rheede, Hort. Malab. 10:141, t. 71.1690.

Slender herbs. Leaves elliptic or elliptic-lanceate, 3-7 by 1.5-3 cm, base acute or obliquely rounded, margin coarsely toothed, apex acute, glabrous or sparsely pubescent; calyx-lobes triangular-ovate; corolla uniformly yellow or with a central purple spot, c. 6 mm ϕ. Berry, c. 7 mm ϕ, accrescent calyx, to 2 cm long.

Fl. & Fr.: Jul.-Dec. *Distr.*: Tropical Asia, Africa, Australia. Common. A weed of moist places. *NAK 963* (Sabari hills, c. 250 m), *1056* (Kodumon, c. 40 m).

Fruit is edible and is said to be a tonic, diuretic and purgative. Decoction of roots given to cure ulcers, fits, eczema and colic in children.

Note : This species is morphologically highly variable.

2. **Physalis peruviana** L., Sp. Pl. (ed. 2) 1670. 1763; Roxb., Fl. Ind. 1:562. 1832; Clarke in Hook.f., Fl. Brit. India 4:238. 1883; Gamble, Fl. Pres. Madras 939. 1923; Matthew & Rani in Matthew, Fl. Tam. Carnatic 3(2) : 1057. 1983.

Stout herbs to 1 m tall. Leaves cordiform, 5-9 by 3.5-7 cm, usually entire or sinuate-angular, apex acute, velvetty pubescent; calyx campanulate, lobes ovate; corolla yellowish with 5 blotches near the throat, c. 1 cm ϕ. Berry c. 1.5 cm ϕ; accrescent calyx to 3.5 cm long.

Fl. & Fr. : Most of the seasons. *Distr.*: Trop. S. America. Naturalised elsewhere. Seen in upper hills only. *NAK 379* (Kakki dam area, c. 1100 m).

Juice of leaves is a medicine for worm infestation in children.

SOLANUM Linnaeus
Sp. Pl. 184. 1753.

Armed or unarmed (sub) shrubs. Leaves opposite, unequal, entire, sinuate or pinnatifid. Flowers bisexual, 4-merous, in axillary or extra axillary or terminal cymes, corymbs or umbels; calyx-lobes 5, basally connate, accrescent; corolla stellate (rotate); stamens 5, anthers connivent into a cone, dehiscence porous; ovary globose, 2-locular, ovules α, on swollen placenta. Berry globose, pulpy. Seeds compressed, triangular or circular.

1. Leaves with prickles.
    2. Calyx smooth; berry densely hispid; leaves with pinnate lobes, coriaceous, velvetty. .... 3. **S. lasiocarpum**
    2. Calyx prickly; berry glabrous; leaves not so, chartaceous, tomentose or villous hairy.
        3. Stem and leaves with dense slender, prickles of c. 12 mm long; leaves chartaceous, villous hairy; berry c.2 cm α, ripening red. .... 1. **S. capsicoides**
        3. Stem and leaves with sparse stout, prickles of c. 5 mm long; leaves thick-chartaceous, softly tomentose; berry c. 8 mm ϕ, ripening yellow. .... 5. **S. violaceum**
1. Leaves without prickles.
    4. Shrubs to 3 m tall; flowers white; leaves sinuate, apex acute. .... 4. **S. torvum**
    4. Shrubs to 4 m tall; flowers violet; leaves entire, apex acuminate. .... 2. **S. giganteum**

1. **Solanum capsicoides** All., Melanges Philos. Mat. Soc. Roy. Turin 12:64. 1773; Symon, J. Adelaide Bot. Gard. 4:101. 1981; Hepper in Dassan. & Fosberg, Rev. Handb. Fl. Ceylon 6 : 375. 1987. *Solanum aculeatissimum* Jacq., Ic. Pl. Rar. 1: 41. 1781 & Coll. Bot. 1:100. 1787; Clarke in Hook.f., Fl. Brit. India 4:237. 1883; Gamble, Fl. Pres. Madras 939. 1923.

Densely prickly subshrubs. Leaves sinuately lobed, 6-9 by 4-8 cm, base obliquely cordate, lobes acuminate, sparsely villous hairy, chartaceous; calyx prickly; corolla white. Berry c. 2 cm ϕ, ripening crimson-red. Seeds punctate, winged.

Fl. & Fr.: Sep.-Jan. *Distr* : Indigenous to Malaya now naturalised in other parts of tropics. On waysides in upper hills; localised. *NAK 1386* (Pampa dam area, c. 1100 m).

2. **Solanum giganteum** Jacq., Collect. 4:125. 1791; Wight, Ic. t.893. 1844-1845; Clarke in Hook.f., Fl.Brit. India 4:233. 1883; Gamble Fl. Pres. Madras 937. 1923; Matthew & Rani in Matthew, Fl. Tam. Carnatic 3(2): 1060. 1983.

Large shrubs; prickly only along the branchlets. Leaves broadly oblong-lanceate, 15-25 by 4-8 cm, base attenuate, apex gradually acuminate, lower surface fulvous tomentose, thick-chartaceous; calyx without prickles; corolla violet. Berry c. 7 mm ϕ, glabrous. Seeds faintly dotted.

Fl. & Fr.: Jun.-Mar. *Distr.*: Sri Lanka, Peninsular India. Rare. Along forest borders. *Deb 30350* (MH) (Pampa, c. 1025 m).

3. **Solanum lasiocarpum** Dunal, Hist. des Solanum 223. 1813; Whalen *et al.,* Gentes Herb. 12:100.1981; Manilal, Fl. Silent Valley 193. 1988; Nicols., Suresh & Manilal, An Interpr. Hort. Malab. 250. 1988; Mohanan & Sivad., Fl. Agasthyamala 478.2002. *S. stramonifolium* Jacq., Misc. 2:298. 1781; Gandhi in Sald. & Nicols., Fl. Hassan Dist. 462. 1976. *S. ferox* sensu Clarke in Hook.f., Fl. Brit. India 4:235. 1883, non L., 1753; Gamble, Fl. Pres. Madras 937.1923, var. *majus* Wight, Ic.t. 1399. 1848. *Ana-schunda* Rheede, Hort. Malab. 2:65-66, t.35.1679.

Stout prickly shrubs; stem and leaves stellate velvetty. Leaves elliptic-lanceate, with shallow coarse triangular lobes, 16-40 by 10-25 cm, base shallowly cordate, apex acute, coriaceous; calyx without prickles; corolla white, Berry globose, golden yellow, densely hispid.

Fl. & Fr.: Dec.-Feb. *Distr.*: S.E. Asia, Indo-Malesia. Rare. On forest borders. *NAK 1439* (Arampa, c. 400 m).

Roots used in Ayurvedic decoctions for rheumatic fever.

4. **Solanum torvum** Sw., Nov. Gen. Sp. 47. 1788; Wight, Ic. t. 345. 1840; Clarke in Hook.f., Fl. Brit. India 4: 234. 1883; Gamble, Fl. Pres. Madras 937. 1923; Gandhi in Sald. & Nicols., Fl. Hassan Dist. 62. 1976; Matthew & Rani in Matthew, Fl. Tam. Carnatic 3(2) : 1064. 1983; Mohanan & Sivad., Fl. Agasthyamala 478.2002.

Prickled shrubs, prickles few, scattered. Leaves sinuate, 6-15 by 5-12 cm, base obliquely subcordate, apex acute, lower surface stellate-pubescent, chartaceous. Cymes corymbose, extra-axillary. Flowers α, dense; calyx cupular; corolla white, c. 2.5 cm ϕ. Berry globose, c. 1 cm ϕ. Seeds smooth.

Fl. & Fr.: Jan.-Apr. Distr.: W. Indies; widely naturalised in other parts of tropics. Common. In plainsand hills. *NAK 945* (Sabari hills, c. 250 m), *1444* (Arampa, c. 400 m).

5. **Solanum violaceum** Ortega, Nov. Pl. Descr. Dec. 56. 1798; Nicols., Suresh & Manilal, An. Interpr. Hort. Malab. 252. 1988. *S. anguivi* sensu Matthew & Rani in Matthew Fl. Tam. Carnatic 3(2) : 1058. 1983, non Lam., 1754. *S. indicum* auct., non., 1753: Hook.f., Fl. Brit. India 4:234. 1883; Gamble, Fl. Pres. Madras 938. 1923. *Scheru - schunda* Rheede, Hort. Malab. 2:67-68, t. 36. 1679.

Prickled shrubs. Leaves sinuate, 8-12 by 5-8 cm, base obliquely truncate, apex subacute, prickly along midrib, thick chartaceous. Racemes 8-10- flowered, extra-axillary; calyx prickly; corolla bluish, c. 2.5 cm ϕ. Berry globose, c. 8 mm ϕ. Seeds pitted.

Fl. & Fr.: Jan.-Jul. *Distr.*: India, Malaya, Indo-China, Philippines. Common. On waysides and waste places in plains. *NAK 673* (Pandalam, c. 20 m).

Seeds used as a remedy for tooth-ache. Fruit eaten as vegetable. Decoction of roots given to cure diarrhoea, dysentery, typhoid and other fevers.

93. SCROPHULARIACEAE

1. Leaves both alternate and decussate in the same plant.
    2. Diffuse herbs; lower leaves often rosulate; calyx-lobes foliaceous; corolla throat with a 2-lobed palate. .... **Mazus**

2. Erect herbs; lower leaves not so; calyx-lobes not foliaceous; corolla throat often without a 2-lobed palate.
   3. Corolla-tube pouched; stamens attached at the base of mouth. .... **Angelonia**
   3. Corolla-tube cylindric; stamens attached at throat.
      4. Calyx spathaceous, split on one side.
         5. Anthers with 1 fertile cell; corolla-lobes subequal; leaves 4 by 0.8 cm; scabrid herbs. .... **Centranthera**
         5. Anthers with 2-fertile cells; corolla 2-lipped.; leaves 6 by 1.5 cm; glabrous herbs. .... **Pedicularis**
      4. Calyx tubular, not split on one side.
         6. Leaves simple; flowers yellow.
            7. Corolla-tube straight; plants drying black. .... **Buchnera**
            7. Corolla-tube bent; plants drying brownish. .... **Striga**
         6. Leaves trifid or pinnatisect; flowers pinkish-purple. .... **Sopubia**

1. Leaves decussate or whorled.
   8. Corolla rotate, lobes 4, subequal. .... **Scoparia**
   8. Corolla tubular, 2-lipped, unequal.
      9. Erect or ascending herbs.
         10. Stamens 2; staminodes 2.
             11. Leaves subradical, upper ones distant; staminodes minute. .... **Dopatrium**
             11. Leaves cauline; staminodes linear, 2-fid. .... **Lindernia** p.p
         10. Stamens 4; staminodes 0.
             12. Anthers 2-celled; calyx-lobes equal.
                 13. Stout tall herbs; stem 4-gonous. .... **Artanema**
                 13. Slender small herbs; stem terete. .... **Limnophila**
             12. Anthers 1-celled; calyx-lobes unequal. .... **Adenosma** *(***A. indiana** p.p.*)*
      9 **Prostrate, diffuse or trailing herbs.**
         14. Calyx-lobes unequal. .... **Adenosma** *(***A. subrepens** p.p.*)*
         14. Calyx-lobes equal.
             15. Stamens 2. .... **Microcarpaea**
             15. Stamens 4.
                 16. Calyx-lobes foliaceous or winged.
                     17. Calyx-lobes free, foliaceous; flowers small, white (anther cells contigous). .... **Bacopa**
                     17. Calyx-lobes basally connate, winged or keeled; flowers large, purple. .... **Torenia**
                 16. Calyx-lobes not foliaceous or winged. .... **Lindernia** p.p.

ADENOSMA R. Brown
Prodr. 442. 1810.

Erect or trailing herbs. Leaves simple, opposite or whorled. Flowers axillary or in dense capitate terminal clusters. Flowers bisexual, 5-merous, zygomorphic; calyx-lobes 5, unequal; corolla tubular, 2-lipped, upper entire or notched, lower 3-lobed; stamens 4, didymous, anthers cells of all or 2 are imperfect; ovary superior, 2-locular; ovules α. Capsule ovoid or ellipsoid. Seeds α, punctate or pitted.

1. Erect herbs; flowers in terminal capital clusters or spikes. ....1. **A. indiana**
1. Trailing herbs; flower(s) axillary, solitary. ....2. **A. subrepens**

1. **Adenosma indiana** (Lour.) Merr., Trans. Amer. Philos. Soc. n.s. 24: 351. 1935 ("indianum'); Cramer in Dassan. & Fosberg, Rev. Handb. Fl. Ceylon 3:448. 1981; Nicols., Suresh & Manilal, An Interpr. Hort. Malab. 238. 1988. *Manulea indiana* Lour., Fl. Cochinch. 2:386. 1790. *Adenosma capitatum* (Benth.) Hance, J. Linn. Soc. Bot. 13:114.1873; Hook.f., Fl. Brit. India 4:264. 1884; Gamble, Fl. Pres. Madras 949. 1923. *Carimtumba* Rheede, Hort. Malab. 10:185, t. 93. 1690.

Erect herbs. Leaves ovate to elliptic-ovate, 1.5-2.5 by 0.7-1 cm, base rounded, margin serrate, apex acute, thick-chartaceous. Flowers dense in terminal cylindric spikes or clusters; bracts linear; calyx-lobes lanceate, upper largest; corolla bluish or violet. Capsule ellipsoid. Seeds punctate.

Fl. & Fr.: Most of the seasons. *Distr.*: Sri Lanka, Peninsular and Eastern India, Malaya. Common. On exposed areas, near paddy fields, on waysides, etc. *NAK 117 & 1064* (Adoor, c. 20 m).

2. **Adenosma subrepens** Benth., Gen. Pl. 2:949. 1883; Hook.f., Fl. Brit. India 4:263. 1884; Gamble, Fl. Pres. Madras 949. 1923; Mohanan & Sivad., Fl. Agasthyamala 480.2002.

Trailing herbs with ascending branches, vegetative parts villous hirsute. Leaves broadly ovate, 2-3.5 by 1.5-2.5 cm, base rounded, margin crenate, apex acute, chartaceous. Flower(s) axillary, solitary, subsessile; sepals unequal, 3 outer sepals in fruit very large, 2 inner small; corolla bluish-purple. Capsule lanceoid. Seed pitted.

Fl. & Fr.: Sep.-Jan. *Distr.*: Sri Lanka, Peninsular India. Rare. On marshes in evergreen forests. *NAK 2599* (Angamoozhy, c.250 m).

ANGELONIA Humboldt et Bonpland
Pl. Aequin. 2:92, t. 108. 1812 ('1809').

Angelonia biflora Benth. in DC., Prodr. 10:254. 1846; Matthew, Ill. Fl. Tam. Carnatic t. 493. 1982; Rani & Matthew in Matthew, Fl. Tam. Carnatic 3(2) : 1070. 1983.

Erect viscid herbs. Leaves simple, subopposite, sessile, lanceate or oblong-lanceate, base acute, or gradually attenuate, margin serrulate, apex acute, lateral veins faint,chartaceous. Flowers 2-4 per node; pedicels slender, c. 12 mm long, often curved; bracts and bracteoles 0; calyx-lobes 5, subequal, ovate lanceate, glandular-pubescent outside; corolla bluish-violet, tube obliquely pouched, with an apical small hooded lobe, wide-mouthed, glandular within; lobes 5, obscurely 2-lipped; stamens 4, equal, exerted, filaments stout; ovary globose, glandular hairy, 2-locular; ovules α, on swollen axile placenta. Capsule globose, c. 8 mm ϕ. Seeds minute.

Fl. & Fr.: Sep. *Distr.*: S. America, now naturalized in tropics. In moist exposed areas. *NAK 1579* (Vetchuchira, c. 100 m).

ARTANEMA D. Don
in Sweet, Brit. Fl. Gard. Ser. 2, 3: t.234. 1834 (nom. cons.).

Artanema longifolium (L.) Vatke, Linnaea 43: 207. 1882; Cramer in Dassan. & Fosberg, Rev. Handb. Fl. Ceylon 3:404. 1981; Nicols., Suresh & Manilal, An Interpr. Hort. Malab. 228. 1988; Mohanan & Sivad., Fl. Agasthyamala 480. 2002. *Columnea longifolia* L., Mant. Pl. 90. 1767. *Archimonis sesamoides* Vahl, Symb. Bot. 2:71.1791. *Artanema sesamoides* (Vahl) Benth., Scroph. Ind. 39. 1835; Wight, Ic. t. 1410. 1849; Hook.f., Fl. Brit. India 4:274. 1884; Gamble, Fl. Pres. Madras 955. 1923. *Bahel-tsjulli* Rheede, Hort. Malab. 9:169-170,t.87. 1689.

Erect stout tall herbs; stem 4-gonous, angles scabrid. Leaves simple, opposite, lanceate, 5-10 by 1.5-2.5 cm, base acute, margin usually serrulate, apex acuminate, chartaceous. Flowers in terminal racemes; bracts foliaceous; bracteoles 0; calyx-lobes 5, nearly free, ovate, acuminate; corolla violet-purple, 2-lipped, tube broad, upper lobe emarginate, lower 3-lobed; stamens 4, didymous, 2 posterior included, 2 anterior with long filaments, anther-cells confluent at their tips; ovary globose; ovules α; style slender; stigma 2. Capsule globose. Seeds α, minute, papillose.

Fl. & Fr.: Jul.-Nov. *Distr.*: Sri Lanka, Peninsular India, Trop. Africa, Sumatra, Java & Malaya. Not common. Along streamlet sides in interior evergreen forests. *NAK 2042* (Naduvathumoozhy, c.200 m), *2690* (Kakki hills, c. 1100 m).

Decoction of roots administered in rheumatism, diarrhoea, indigestion, stone in the urinary bladder, syphilis and swellings. Seeds powdered and taken with ghee to cure biliousness and impure blood.

BACOPA Aublet
Hist. Pl. Guiane 128, t. 48. 1775 (nom. cons.).

Bacopa monnieri (L.) Pennell, Proc. Acad. Nat. Sci. Philadelphia 98:94. 1946; Cramer in Dassan. & Fosberg, Rev. Handb. Fl. Ceylon 3: 420. 1981; Matthew, Ill. Fl. Tam. Carnatic t. 494. 1982; Rani & Matthew in Matthew, Fl. Tam. Carnatic 3(2) : 1071. 1983; Nicols., Suresh & Manilal, An Interpr. Hort. Malab. 239. 1988. *Lysimachia monnieri* L., Cent. Pl. 2:9. 1756. *Moniera cuneifolia* Michaux., Fl. Bor. Amer. 2:22. 1803; Gamble, Fl. Pres. Madras 953. 1923. *Herpestis monnieria* (L.) Kunth, Nov. Gen. Sp. 2:366. 1818, nom. illegit.; Hook.f., Fl. Brit. India 4: 272. 1884. *Brami* Rheede, Hort. Malab. 10:27, t. 14. 1690.

Prostrate subsucculent herbs. Leaves simple, subsessile, decussate, oblong, 10 by 4 mm, base and apex obtuse, herbaceous. Flower(s) axillary, solitary, ebracteate; bracteoles 2; calyx-lobes 5, unequal; corolla white with violet and green bands in throat, lobes 5, faintly 2-lipped; stamens 4, didymous; ovary oblong-globose; style slightly deflexed. Capsule oblong-globose. Seeds oblong, striate.

Fl. & Fr.: May.-Aug. *Distr.*: Paleotropical. Not common. In marshy areas, usually in plains. *CNM 65076* (MH) (Ranni, c. 100 m).

Leaves and roots are diuretic. The whole plant used in medicine as nervine tonic.

BUCHNERA Linnaeus
Sp. Pl. 630. 1753.

Buchnera hispida Buch.-Ham. ex G. Don, Prodr. Fl. Nepal. 91. 1825; Wight, Ic. t. 1413. 1849; Hook.f., Fl. Brit. India 4:298. 1884; Gamble, Fl. Pres. Madras 966. 1924; Sald. & Nicols., Fl.Hassan Dist. 517. 1976; Manilal, Fl. Silent Valley 194. 1988.

Hispid annuals, to 60 cm tall. Leaves simple, decussate, upper ones alternate, subsesile, oblanceate or oblong-lanceate 5-7 by 1-1.5 cm, base cuneate, margin coarsely toothed, apex acute, lateral nerves 4 pairs, 3-nerved from base. Upper leaves linear-lanceate, base attenuate, margin entire, apex acute, chartaceous. Flowers in terminal spikes; bracts lanceate; bracteoles linear; calyx-tubular, lobes 5; fruiting calyx 10-ribbed; corolla bluish-puple, tube slender, straight, villous within, lobes 5, subequal, spreading; stamens 4, didymous, anthers 1-celled, filament produced above. Capsule globose, loculicidal. Seeds minute.

Fl. & Fr.: Oct.-Dec. *Distr.*: India, Madagascar, Trop. Africa. Not common. In exposed upper hills. *NAK 1372* (Kakki hills, c. 1100 m).

Note : Whole plant become black when dry.

CENTRANTHERA R. Brown
Prodr. 438. 1810.

Centranthera indica (L.) Gamble, Fl. Pres. Madras 971. 1924; Sald. & Nicols., Fl. Hassan Dist. 518. 1976; Cramer in Dassan. & Fosberg, Rev. Handb. Fl. Ceylon 3: 396. 1981; Manilal, Fl. Silent Valley 195. 1988; Nicols., Suresh & Manilal, An Interpr. Hort. Malab. 240. 1988. *Rhinanthus indica* L., Sp. Pl. 603. 1753. *Centranthera hispida* R. Br., Prodr. 438. 1810; Hook.f., Fl. Brit. India 4:301.1883; Gamble, Fl. Pres.Madras 971. 1923. *C. procumbens* Benth. in DC., Prodr. 10:525. 1846; Hook.f., Fl. Brit. India 4:301. 1884. *Corosinam* Rheede, Hort. Malab. 9:133, t. 68. 1689.

Scabrid annuals. Leaves simple, basally opposite, upper ones alternate, elliptic, oblong or linear-oblong, 1.5-4 by 0.5-0.8 cm, base rounded, apex acute, scabrid, thin-coriaceous. Flowers subsessile, axillary, solitary; bracteoles 2; calyx spathaceous, split on one side, faintly 3-5-lobed; corolla rose coloured or white, tube long, incurved, dilated above, lobes 5, subequal, spreading, stamens 4, didymous, included, anthers transverse, base mucronate; ovary globose; style simple. Capsule ovoid, loculicidal. Seeds α, conical, spirally reticulate.

Fl. & Fr.: Aug.-Nov. *Distr.*: Paleotropical. Not common. On exposed grassy hills tops. *NAK 1080* (Vallicode-Kottayam, c. 100 m), *2018* (Kokkathode, c. 250 m).

DOPATRIUM F. Hamilton ex Bentham
in Lindley, Edward's Bot. Reg. ad t. 1770 (4) 1835.

Dopatrium junceum (Roxb.) Buch.-Ham. ex Benth., Scorph. Ind. 31. 1835; Hook. f., Fl. Brit. India 4: 274: 1884; Gamble, Fl. Pres. Madras 954. 1923; Cramer in Dassan. & Fosberg, Rev. Handb. Fl. Ceylon 3: 424. 1981; Rani & Matthew in Matthew, Fl. Tam. Carnatic 3(2): 1074. 1983. *Gratiola juncea* Roxb., Pl. Corom. t. 129. 1800 & Fl. Ind. 1:142. 1832.

Slender annuals. Leaves simple, decussate, subrosulate and cauline, oblong, 2 by 0.5 cm, base and apex acute, basally 3-nerved, sessile, decurrent on stem, fleshy. Flowers axillary, in decussate pairs, lower ones subsessile, upper ones pedicellate; bracts linear, bracteoles 0; calyx-lobes 5, obtuse; corolla bluish-pink, lobes 5, upper lip covering the lower lip in bud; stamens 2, included, anthers 2-celled, bearded; ovary globose; stigma 2-lamellate. Capsule globose, on erect pedicel. Seeds minute, pitted.

Fl. & Fr.: Dec.-Mar. *Distr.*: Eastern tropics. In paddy fields. Locally abundant; seasonal. *NAK 1869* (Adoor, c. 20 m).

LIMNOPHILA R. Brown
Prodr. 442. 1810 (nom. cons.).

Partly submerged erect herbs. Leaves decussate or ternate, simple or much dissected, chartaceous. Flower(s) solitary, axillary or in terminal, lax racemes; calyx-lobes 5, subequal or unequal; corolla tubular, lobes 5, unequal, 2+3; stamens 4, didymous, anthers disjunct, horizontal, stipitate, deflexed. Capsule loculicidal/septicidal. Seeds angular.

1. **Leaves monomorphic, pinnately veined; calyx-nerves often striate.**
    2. Leaves opposite; flower(s) subsessile, solitary, axillary, or in racemes. ....4. **L. repens**
    2. Leaves opposite or 3- ternate; flowers long pedicelled, solitary or in racemes.
        3. Leaf-base subamplexicaul; flowers in axillary or terminal racemes. ....1. **L. aromatica**
        3. Leaf-base cuneate or attenuate; flower(s) solitary or 2 or 3-per node. ....2 **L. chinensis**
1. Leaves dimorphic, 3-nerved from base; calyx-nerves not striate. ....3. **L. indica**

1. Limnophila aromatica (Lam.) Merr., Interpr. Rumph. Herb. Amboin. 466-1917; Nicols., Suresh & Manilal, An Interpr. Hort. Malab. 240. 1988; Mohanan & Sivad., Fl. Agasthyamala 481.2002. *Ambulia aromatica* Lam., Encycl. 1:128 1783. *Limnophila gratissima* Blume, Bijdr. 749. 1886; Hook.f., Fl. Brit. India 4:268. 1884; Gamble, Fl. Pres. Madras 951. 1923. *Manga-nari* Rheede, Hort. Malab. 10: 11-12, t. 6. 1690.

Erect glabrous herbs, to 60 cm tall. Leaves often 3-5, oblong, lanceate, 2.5-5 by 0.7-1.3 cm, base subamplexicaul, margin serrulate, apex subacute, penninerved, chartaceous. Flowers in terminal or rarely axillary racemes; pedicel c. 2 cm long, glandular puberulous; calyx-lobes lanceate, striate with age; corolla purple, c. 1 cm ϕ; stamens included; ovary ovoid or globose. Capsule oblong. Seeds minute.

Fl. & Fr.: Sep.-Dec. *Distr.*: Sri Lanka, India, Malesia, Philippines, China, Japan to N. Australia. Flowering usually after rains. Common in water logged areas. *NAK 140, 765, 1635* (Moozhiar, c. 250 m).

The juice of leaves acts as a mild purgative and is supposed to purify blood.

2. **Limnophila chinensis** (Osbeck) Merr., Amer. J. Bot. 3: 581. 1916; Philcox, Kew Bull. 24: 151. 1970; Cramer in Dassan. & Fosberg, Rev. Handb. Fl. Ceylon 3:428. 1981; Matthew, Ill. Fl. Tam. Carnatic t. 497. 1982; Rani & Matthew in Matthew, Fl. Tam. Carnatic 3(2) : 1079. 1983. *Columnea chinensis* Osbeck, Dagb Ostind. resa 230. 1757. *Stemodia hirsuta* Heyne ex Benth., Edward's Bot. Reg. 17: sub t. 1470. 1832. *Limnophila hirsuta* (Heyne ex Benth.) Benth. in DC., Prodr. 10: 388. 1846; Hook.f., Fl. Brit. India 4:268. 1884; Gamble, Fl. Pres. Madras 951. 1923.

Spreading hirsute herbs. Leaves decussate to ternate, obovate-spathulate, 3-5 by 0.9-15 cm, base cuneate or attenuate, margin serrulate, apex acute, penninerved, chartaceous. Flower(s) solitary, axillary or 2-or 3-per node; calyx-lobes lanceate, striate with age; corolla pinkish-purple, c. 6 mm ϕ; stamens included. Capsule lanceoid. Seeds minute.

Fl. & Fr.: Feb.-May-Apr. *Distr.*: Paleotropics. Less common. Along the marshy areas. *NAK 444, 471, 1524, 2600* (Plains & hills of different localities).

3. **Limnophila indica** (L.) Druce, Bot. Exch. Club Soc. Brit. Isls. 3: 420. 1914; Philcox, Kew Bull. 24: 115. 1970; Cramer in Dassan. & Fosberg, Rev. Handb. Fl. Ceylon 3: 432. 1981; Rani & Matthew in Matthew,

Fl. Tam. Carnatic 3(2) : 1080. 1983; Nicols., Suresh & Manilal, An Interpr. Hort. Malab. 240.1988. *Hottonia indica* L., Syst. Nat. (ed. 10) 919. 1759. *Limnophila gratioloides* R. Br., Prodr. 442. 1810; Hook. f., Fl. Brit. India 4: 271. 1884; Gamble, Fl. Pres. Madras 952. 1923. *Tsjeri-manga-nari* Rheede, Hort. Malab. 9:165, t. 85. 1689. *Tsjudan-tsjera* Rheede, Hort. Malab. 12:71, t. 36. 1693.

Aquatic ascending herbs. Leaves dimorphic; submerged leaves deeply dissected; upper leaves decussate to whorled, 1.5-2.5 by 0.5-0.8 cm, base cuneate, decurrent, margin serrulate, apex acute, basally 3-nerved, chartaceous, subsessile. Flower (s) axillary, solitary; pedice l 0.8 mm long; calyx-lobes linear-lanceate; ovary faintly 2-lobed. Capsule globose.

Fl. & Fr.: Sep.-Dec. *Distr.:* Paleotropics. Aquatic herbs. Common. *NAK* 139, 1036 (Adoor, Tiruvalla, c. 10 m). An extemely variable plant with many forms.

Juice of leaves along with ginger and cumin administered in dysentery.

4. **Limnophila repens** (Benth.) Benth. in DC., Prodr. 10:387. 1846; Philcox, Kew Bull. 23:154. 1970; Rani & Matthew in Matthew, Fl. Tam. Carnatic 3(2) : 1081. 1983. *Stemodia repens* Benth., Edwards'Bot. Reg. 17: sub. t. 1470. 1832 & Scroph. Ind. 24. 1835. *Limnophila conferta* Benth. in DC., Prodr. 10:387.1846; Hook.f., Fl. Brit. India 4:266. 1884; Gamble, Fl. Pres. Madras 951. 1923.

Ascending herbs. Leaves decussate, obovate, 1-2 by 0.4-0.8 cm, base cuneate, margin serrate, apex acute, glabrescent, penninerved, chartaceous. Flower(s) axillary, solitary or in short racemes; pedicel c.1 mm long;calyx striate, strigose without, lobes linear; corolla pink, c. 5 mm ϕ; stamens included; ovary obovoid. Capsule ellipsoid.

Fl. & Fr.: Sep.-Dec. *Distr.*: India, Trop. Himalayas, Myanmar, East to S. China, Malesia. Rare. In forest swamps. *NAK 2239* (Mannarappara, c.150 m).

LINDERNIA Allioni

Melanges Philos. Math. Soc. Roy. Turin (Misc. Taur.) 3(1): 178. 1766.

Erect or prostrate herbs. Leaves simple, decussate, 3-nerved or penninerved, chartaceous, subsessile or petiolate. Flower(s) solitary, axillary, leaf-opposed or in racemes; bracts present or not; bracteoles 0, calyx 5-toothed or 5-lobed; corolla caducous, lobes 5, 2-lipped, upper lip entire or 2-lobed, lower lip 3-lobed; stamens 2 or 4, included or not, anthers divaricate, all perfect or lower pair reduced to staminodes; ovary oblong-globose or globose; stigma 2-lamellate. Capsule globose to linear-lanceoid, septicidal. Seeds minute, rugose.

1. Perfect stamens 2; staminodes 2.
   2. Prostrate or spreading herbs.
      3. Capsule linear-lanceoid, much longer than calyx.
         4. Leaves petioled; calyx margin entire; staminodes pubescent. .... **9. L. ruellioides**
         4. Leaves sessile; calyx margin serrulate; staminodes glabrescent. .... **2. L. antipoda**
      3. Capsule subglobose or ovoid. .... **8. L. rotundifolia**
   2. Erect or ascending herbs.
      5. Leaf 3-nerved from base, margin entire; staminodes 2-fid.

6. Fruiting pedicel deflexed; corolla to 7 mm f, deep blue with whitish-lobes. .... 6. **L. hyssopioides**

6. Fruiting pedicel erect; corolla to 3 mm f, white with two yellow dots at the throat. .... 7. **L. parviflora**

5. Leaf penninerved, margin serrate or dentate; staminodes simple. .... 4. **L. ciliata**

1. Perfect stamens 4; staminodes 0.

7. Capsules equalling or shorter than calyx.

8. Flower(s) in terminal racemes, also axillary, solitary.

9. Calyx shortly toothed. .... 5. **L. crustacea**

9. Calyx deeply divided. 10. **L. viscosa**

8. Flower(s) solitary, always axillary, only. .... 3. **L. caespitosa**

7. Capsules much exceeding the calyx; calyx deeply divided; anthers spurred. .... 1. **L. anagallis**

1. **Lindernia anagallis** (Burm. f.) Pennell, J. Arnold Arb.24:252. 1943; Mukherjee, J. Ind.Bot. Soc. 24:133. 1945; Philcox, Kew Bull. 22:45. 1968; Cramer in Dassan. & Fosberg, Rev. Handb. Fl. Ceylon 3:410. 1981; Rani & Matthew in Matthew, Fl. Tam. Carnatic 3(2) : 1084. 1983; Manilal, Fl. Silent Valley 195. 1988. *Ruellia anagallis* Burm. f., Fl. Ind. 135. 1768. *Vandellia pedunculata* Benth., Scroph. Ind.17. 1835; Hook.f., Fl. Brit. India 4:282. 1884; Gamble, Fl. Pres. Madras 959. 1923. *V. angustifolia* Benth., Scroph. Ind. 37. 1835. *Lindernia angustifolia* (Benth.) Wettst. in Engler & Prantl, Natur. Pflanzenfam. 4, 36:79.1891; Mukherjee, J. Ind. Bot. Soc. 24:132. 1945.

Prostrate or spreading herbs. Leaves broadly ovate to elliptic or linear (very variable), 6-12 by 2.5-8 mm, base truncate, margin subentire, apex subacute, sessile. Flower(s) solitary or rarely 2-per node; calyx-lobes 5, divided to the base, lanceate; corolla white with rose blotches or pinkish-violet; stamens with a rounded appendage. Capsule lanceoid, much exceeding the calyx.

Fl. & Fr.: Nov.-Jan. *Distr.*: Sri Lanka, India, Malesia. Less common. Much variable herbs; in marshy lowlands and paddy fields. *NAK 2222* (Pathanamthitta, c. 50 m), *2303 & 2349* (Kodumon, c.40 m).

2. **Lindernia antipoda** (L.) Alston in Trimen, Fl. Ceylon 6 (Suppl.): 214. 1931; Philcox, Kew Bull. 22:57. 1968; Cramer in Dassan. & Fosberg, Rev. Handb. Fl. CEylon 3:413. 1981; Rani & Matthew in Matthew, Fl. Tam. Carnatic 3(2) : 1085. 1983. *Ruellia antipoda* L., Sp. Pl. 635. 1753. *Gratiola veronicaefolia* Retz., Obs. Bot. 4:3. 1786-87. *G. grandiflora* Roxb., Pl. Corom. t. 179. 1805. *Bonnaya veronicifolia* (Retz.) Sprengel in L., Syst. Veg. (ed. 16) 1:41. 1824; Hook.f., Fl. Brit. India 4: 285. 1884. *Ilysanthes veronicifolia* (Retz.) Urban, Ber. Deutsch. Bot. Ges. 2:436. 1884; Gamble, Fl. Pres. Madras 962. 1923. *Vandellia angustifolia* Mukherjee, J. Ind. Bot. Soc. 24: 133. 1945.

Prostrate or diffuse herbs. Leaves variable, elliptic-obovate, 15-25 by 5-10 mm, penninerved, base cuneate, margin subentire to serrate. Flower(s) axillary, solitary or in lax terminal racemes; bracts leafy; calyx-lobes 5, free, lanceate, margin serrulate; corolla bluish with an yellow mouth; stamens 2; staminodes 2, hooked. Capsule linear-lanceoid, twice as long as the calyx.

Fl. & Fr.: Nov.-Feb. *Distr.*: Trop. & Subtrop. Asia & N.Australia. Less common. In swamps, grasslands, wastelands and as a weed in cultivated lands. *NAK 1626* (Pampavalley, c. 100 m).

3. **Lindernia caespitosa** (Bl.) Panigrahi, Taxon 33:320. 1984; Nicols., Suresh & Manilal, An Interpr. Hort. Malab. 241. 1988. *Diceros caespitosus* Bl., Bijdr. 753. 1826. *Gratiola pusilla* Willd., Sp. Pl. 1:105. 1797. *Torenia hirta* Cham. & Schlech., Linnaea 2:571. 1827. *Vandelia scabra* Benth., Scroph. Ind. 36. 1835, nom. illegit. (incl. type of *Torenia hirta* Cham. & Schlech., 1827). *Vandelia caespitosa* (Bl.) Benth. in DC., Prodr. 10:415. 1846. *Lindernia pusilla* (Willd.) Boldingh, Zakfl. Landbouwstr. Java 165. (Oct.)1916. *L. pusilla* (Thunb.) Merr., Philipp. J. Sci. Bot. 11:3112. 1916. Philcox, Kew Bull 24:33. 1968 [as '(Thunb.) Boldingh'] non Boldingh, 1916. *Nanschera-canschabu* Rheede, Hort. Malab. 10:99, t.50.1690.

Prostrate or diffuse herbs, rooting at nodes, hirsute. Leaves ovate to orbicular, 10-15 by 4-8 mm, penninerved,base truncate, margin crenate-serrate, apex acute, sessile. Flowers 2-per node; pedicel to 15 mm long; calyx-lobes 5, divided to the base, lanceate, glandular pubescent; corolla white with an yellow throat; stamens 4, lower pair appendaged. Capsule subglobose, equalling or shorter than fruiting calyx.

Fl. & Fr.: Sep.-Dec. *Distr.*: Africa. Sri Lanka, India, Himalaya, Myanmar, S. China, Malesia, Polynesia. Less common. In marshy lowlands to lower ghats. *NAK 2008, 2115* (Naduvathumoozhy & Vallicode-Kottayam, c.100 m).

4. **Lindernia ciliata** (Colsm.) Pennell, Brittonia 2:182. 1936; Mukherjee, J. Ind. Bot. Soc. 24:133.1945; Philcox, Kew Bull. 22:51. 1968; Cramer in Dassan. & Fosberg, Rev. Handb. Fl. Ceylon 3:412. 1981; Rani & Matthew in Matthew, Fl. Tam. Carnatic 3(2) : 1086. 1983; Nicols., Suresh & Manilal, An Interpr. Hort. Malab. 242. 1988; Mohanan & Sivad., Fl. Agasthyamala 482.2002. *Gratiola ciliata* Colsm., Prodr. Descr. Gratiol. 14.1793. *G. serrata* Roxb., Fl. Ind. 1:140. 1820. *Bonnaya brachiata* Link & Otto, Icon. Pl. Select. 25, t.11. 1820; Hook.f., Fl. Brit. India 4: 284.1884. *Ilysanthes serrata* (Roxb.) Urban, Ber. Deutsch. Bot. Ges. 2:436. 1884; Gamble, Fl. Pres. Madras. 962. 1923. *Pee-tsjnga-puspam* Rheede, Hort. Malab. 9:115, t.59. 1689.

Erect annuals. Leaves elliptic-oblong, 1-3 by 0.6-1 cm, base rounded, margin aristate-serrulate, apex obtuse or acute, penninervd. Racemes terminal, to 10 cm long; bracts scarious; calyx-lobes 5, linear; corolla bluish or pink, rarely white; stamens 2; staminodes 2, divergent. Capsule linear-lanceoid, exceeding the fruiting calyx.

Fl. & Fr.: Most of the seasons. *Distr.*: Sri Lanka, India, Malaya. Very common. Usually after monsoon. *NAK 686, 874, 901* (Plains & Hills).

5. **Lindernia crustacea** (L.) F. Muell., Syst. Cens. Austral. Pl. 97. 1882; Mukherjee, J. Ind. Bot. Soc. 24 : 130.1945; Philcox, Kew Bull. 22:17.1968; Cramer in Dassan. & Fosberg, Rev. Hanb. Fl. Ceylon 3:407. 1981; Rani & Matthew in Matthew, Fl. Tam. Carnatic 3(2) : 1088. 1983; Nicols., Suresh & Manilal, An Interpr. Hort. Malab. 243. 1988; Mohanan & Sivad., Fl. Agasthyamala 483.2002. *Capraria crustacea* L., Mant. Pl. 87. 1767. *Vandellia crustacea* (L.) Benth., Scroph. Ind. 35. 1835; Wight, Ic.t. 863. 1844-45; Hook.f., Fl. Brit. India 4:279. 1884; Gamble, Fl. Pres. Madras 959. 1923. *Katu-pee-tsjanga-puspam* Rheede, Hort. Malab. 9:113, t.58. 1689.

Erect or diffuse herbs. Leaves broadly ovate, 0.6-1.5 by 0.5-1.5 cm, base truncate, margin serrate, apex subacute, penninerved. Flower(s) solitary or 2-per node on terminal racemes; pedicel c.1.5 cm long; calyx campanulate, lobes 5, small; corolla rose to purple or white with pink throat; stamens 4, lower pairs appendaged. Capsule obovoid, equalling or smaller than the fruiting calyx.

Fl. & Fr.: Most of the season. *Distr.*: Trop. & Subtrop. Asia, Australia, Trop. Africa & America. Very common in almost all types of habitats. *NAK 664, 687, 807 & 2777* (Plains and Hills).

6. **Lindernia hyssopioides** (L.) Haines, Bot. Bihar & Orissa 635. 1922; Mukherjee, J. Ind. Bot. Soc. 24:132. 1945; Philcox, Kew Bull. 22:50. 1968; Sald. & Nicols; Fl. Hassan Dist. 522. 1976; Cramer in Dassan. & Fosberg, Rev. Handb. Fl. Cylon 3:414. 1981; Rani & Matthew in Matthew, Fl. Tam. Carnatic 3(2) : 1090. 1983; Manilal, Fl. Silent Valley 195. 1988. *Gratiola hyssopioides* L., Mant. Pl. 174. 1771. *Ilysanthes hyssopioides* (L.) Benth. in DC., Prodr. 10:419. 1846; Wight, Ic. t. 857 1844-45; Hook.f., Fl. Brit. India 4:283. 1884; Gamble, Fl. Pres. Madras 916. 1923.

Slender erect herbs; stem 4-angled. Leaves linear-lanceate, basal ones elliptic-ovate, 1-3 by 0.3-0.5 cm, base cuneate or truncate, apex acute, faintly 3-nerved from base. Flower(s) axillary, solitary; bracts linear; pedicel 2-4 cm long; calyx-lobes 4, linear-lanceate; corolla white, with linear, hairy, basal appendages. Capsule obliquely ovoid twice as long as the fruiting calyx.

Fl. & Fr.: Sep.-Dec. (May). *Distr.*: S.E. Asia, Malesia, China. Common. In wet lowlands, paddy fields, etc. *NAK 1967* (Tiruvalla, c.10 m), *2322* (Kodumon, c. 40 m).

7. **Lindernia parviflora** (Roxb.) Haines, Bot. Bihar & Orissa 2:635. 1922; Mukherjee, J. Ind. Bot. Soc. 24: 132. 1945; Rani & Matthew in Matthew, Fl. Tam. Carnatic 3(2):1091. 1983. *Gratiola parviflora* Roxb., Pl. Corom. t.203. 1811 & Fl. Ind. 1:141.1820. *Ilysanthes parviflora* (Roxb.) Benth. in DC., Prodr. 10:419. 1846; Hook.f., Fl. Brit. India 4:283. 1884; Gamble, Fl. Pres. Madras 961. 1923.

Erect branched or unbranched annuals, to 20 cm tall. Leaves ovate to oblong-elliptic, 3-7 by 2.4 mm, gradually smaller towards branch-apices, base rounded, apex acute, basally 3-nerved. Flower(s) axillary, solitary; bracts leafy; pedicel c.4 mm long; calyx-lobes 5, free, lanceate; corolla pale violet; stamens 2; staminodes 2-fid, glabrescent. Capsule globose or ellipsoid, exceeding the calyx.

Fl. & Fr.: Dec.-Feb. *Distr.*: Trop. Africa, India, Himalayas, Myanmar, Indo-China. Not common. Come up usually after rains as a weed in cultivated lands. *NAK 2302* (Kodumon, c.40 m).

8. **Lindernia rotundifolia** (L.) Alston in Trimen, Handb. Fl. Ceylon 6:214. 1931; Mukherjee, J. Ind.Bot. Soc. 24:132. 1945; Nicols., Suresh & Manilal, An Interpr. Hort. Malab. 243. 1988. *Gratiola rotundifolia* L., Mant. Pl. 274. 1767. *Ilysanthes rotundifolia* (L.) Benth. in DC., Prodr. 10:420. 1846; Hook.f., Fl. Brit. India 4:254. 18841; Gamble, Fl. Pres. Madras 962. 1923. *Tsjanga-puspam* Rheede, Hort. Malab. 9:111, t. 57. 1689.

Prostrate herbs. Leaves ovate or elliptic-ovate, 0.8-2 by 0.5-1 cm, base and apex rounded, puberulous or glabrous, chartaceous. Flowers solitary or rarely 2-per node; calyx-lobes 5, divided to base, lanceate, corolla white with violet-tinge; perfect stamens 2; staminodes 2. Capsule subglobose or ovoid.

Fl. & Fr.: Sep.-Dec. *Distr.*: Tropics. Common. Along moist shady wet places in plains. *NAK 2856* (Tiruvalla, c. 10 m).

9. **Lindernia ruellioides** (Colsm.) Pennell, Brittonia 2:182. 1936 & J. Arnold. Arb. 20:81. 1939; Mukherjee, J. Ind. Bot. Soc. 24:133. 1945; Philcox, Kew Bull. 22: 54. 1968; Rani & Matthew in Matthew, Fl. Tam. Carnatic 3(2) . 1092. 1983; Mohanan & Sivad., Fl. Agasthyamala 483.2002. *Gratiola ruellioides* Colsm., Prodr. Descr. Gratiol. 12. 1793. *G. reptans* Roxb., Fl. Ind. 1:140. 1820. *Bonnaya reptans* (Roxb.) Spreng. in L., Syst. Veg. (ed. 16) 1:41. 1824; Hook.f., Fl. Brit. India 4: 284. 1884. *Ilysnthes reptans* (Roxb.) Urban, Ber. Deutsch Bot. Ges. 2: 436. 1884; Gamble, Fl. Pres. Madras 962. 1923.

Trailing herbs, rooting at nodes. Leaves elliptic-obovate, 2-5 by 1-2 cm, base attenuate, margin serrate, apex acute, penninerved. Racemes terminal. Flowers 2 per node; bracts linear; cayx-lobes 5, lanceate; corolla pinkish; stamens 2; staminodes 2, hooked. Capsule linear-lanceoid, deflexed.

Fl. & Fr.: Sep.-Dec. *Distr.*: Trop. & Subtrop. Asia. Common. On exposed marshes. *NAK 1139, 1691, 2010, 2016* (in hills).

10. **Lindernia viscosa** (Hornem.) Merr., Sp. Blanc. 14. 1918 & Enum. Philipp. Pl. 3:349. 1923; Philcox, Kew Bull. 22:38. 1968. *Gratiola viscosa* Hornem. Enum. Pl. Hort. Hafn. 19:1807. *Vandellia hirsuta* Benth., Scroph. Ind. 36. 1835; Hook.f., Fl. Brit. India 4:280. 1884. Gamble, Pres. Madras 959. 1923. *Lindernia hirsuta* (Benth.) Wettst. in Engl. & Prantl., Pflanzenfam. 4, 36 : 79. 1891; Mukherjee, J. Ind. Bot. Soc.24:131. 1945.

Erect or diffuse hirsute herbs. Leaves broadly ovate, 1.5-4 by 1-3 cm, base rounded to truncate, margin crenate-serrate, apex subacute, penninerved. Flowers axillary and in terminal racemes, 2 per node, c. 7 mm long; calyx-lobes 5, linear-lanceate; corolla white with yellow throat; stamens 4. Capsule subglobose, almost equalling the fruiting calyx.

Fl. & Fr.: Sep.-Dec. *Distr.*: Sri Lanka, India, Malesia, Phillipines & China. Less common. On exposed moist shady hills. *NAK 792 & 2773* (Ranni, c. 150 m).

MAZUS Loureiro
Fl. Cochinch. 385. 1790.

Mazus pumilus (Burm. f.) Steenis, Nova Guinea n.s. 9:31. 1958; Shah, J. Bombay Nat. Hist. Soc. 66:233. 1969. *Lobelia pumila* Burm. f., Fl. Ind. 186, t. 60, f.3. 1768. *Mazus rugosus* Lour., Fl. Cochinch. 385. 1790; Hook.f., Fl. Brit. India 4:259. 1884; Gamble, Fl. Pres. Madras 947. 1923. *Lindernia japonica* Tnunb., Fl. Jap. 253. 1784. *M. japonicus* (Thunb.) O. Ktze., Rev. Gen. 462. 1891; Santapau, J. Bombay Nat. Hist. Soc. 49:48. 1950.

Diffuse herbs. Leaves opposite, rosulate at base, obovate-spathulate, 2-5 by 1-2.5 cm, base obtuse, margin crenate-serrate, apex rounded, penninerved. Flowers in terminal scapose racemes; bracts minute; calyx campanulate, lobes 5, foliaceous; corolla pale blue, 2-lipped, upper 2-lobed, lower large 3-lobed, throat with 2-lobed palate; stamens 4, didymous, anthers divaricate; ovary ellipsoid; ovules α; stigma 2-lamellate. Capsule globose, loculicidal. Seeds α, rugose.

Fl. & Fr.: Sep.-Dec. *Distr.*: Afghanistan, India, Java, China, Japan, Phillipines. Rare. On exposed moist hills. *NAK 1405* (Kakki hills, c. 1100 m).

MICROCARPAEA R. Brown
Prodr. 4:35. 1810.

Microcarpaea minima (J. Koenig) Merr., Philipp. J. Sci. 7:100. 1912; Cramer in Dassan. & Fosberg, Rev. Handb. Fl. Ceylon 3(2): 1098. 1983. *Paederota minima* J. Koenig in Retz., Obs. Bot. 5:10.D89. *Microcarpaea muscosa* R. Br., Prodr. 436. 1810; Hook.f., Fl. Brit. India 4:287. 1884; Gamble, Fl. Pres. Madras. 963. 1923.

Prostrate herbs. Leaves simple, decussate, minute, narrowly elliptic-oblong, 2-4 by 0.5-1 mm, base decurrent, apex obtuse, 1-nerved, herbaceous. Flower(s) axillary, solitary, sessile; bracts and bracteoles 0; calyx tubular, lobes 5, lanceate; corolla purplish, minute, caducous, lobes 5, 2+3, subequal; stamens 2, exerted, anther 1-celled, cofluent; ovary cylindric. Capsule ellipsoid, enclosed within the calyx-base. Seeds ellipsoid.

Fl. & Fr.: Nov.-Feb. *Distr.*: Sri Lanka, India, Nepal, East to S. China, Malesia, Australia. Not common. On moist clayey soil in hills. *NAK 2158* (Vayyattupuzha, c.250 m).

PEDICULARIS Linnaeus
Sp. Pl. 607. 1753.

Pedicularis zeylanica Benth., Scroph. Ind. 54. 1835; Wight, Ic. t. 1419. 1849; Hook.f., Fl. Brit. India 4:317. 1884; Gamble, Fl. Pres. Madras 972.1923.

Erect annuals. Leaves pinnatifid, decussate, alternate or whorled, 2-3.5 by 0.8-1.5 cm, lobes rounded, margin crenulate, chartaceous; petiole c. 3 mm long. Flowers in terminal dense racemes; bracts pinnatifid; bracteoles 0; calyx tubular, lobes 5, nearly free, crested; corola rose-pink, 2-lipped, upper lip hooded, lower 3-lobed; stamens 4, didymous, anthers of upper in pairs, lobes 2, parallel; ovary ovoid; style slender. Capsule obliquely ovoid-lanceoid, loculicidal. Seeds α, ellipsoid, punctate.

Fl. & Fr.: Sep.-Dec. *Distr.*: Sri Lanka, Peninsular India. Rare. In evergreen forests and grasslands. *BDS 42006* (MH) (Uppupara, c. 1000 m).

SCOPARIA Linnaeus
Sp. Pl. 116. 1753.

Scoparia dulcis L., Sp. Pl. 116. 1753; Hook.f., Fl. Brit. India 4:289. 1884; Gamble, Fl. Pres. Madras 964. 1924; Sald. & Nicols., Fl. Hassan Dist. 525. 1976; Cramer in Dassan. & Fosberg, Rev. Handb. Fl. Ceylon 3:439. 1981; Matthew, Ill. Fl. Tam. Carnatic t. 504. 1982; Rani & Matthew in Matthew, Fl. Tam. Carnatic 3(2): 1100. 1983; Manilal, Fl. Silent Valley 196. 1988 ; Mohanan & Sivad., Fl. Agasthyamala 484.2002. *Kallurukki.*

Erect stout herbs. Leaves simple, decussate to whorled, 1-4 by 0.5-2 cm, base cuneate, margin serrate, apex acute, penninerved, punctate, chartaceous. Flower(s) axillary, solitary or 2; bracts and bracteoles 0; calyx-lobes 4, free, oblong; corolla white, rotate, lobes 4; stamens 4, equal, exerted, anthers 2-celled; ovary globose; stigma truncate. Capsule ellipsoid-globose, septicidal. Seeds 4-angled, reticulate.

Fl. & Fr.: Most of the seasons. *Distr.*: Native of Trop. America, now pantropical. A very common weed. *NAK 633* (Malayalapuzha, c. 30 m).

Roots are used in Ayurvedic medicines for stone in urinary bladder.

SOPUBIA F. Hamilton ex D. Don
Prodr. Fl. Nepal. 88. 1825.

Erect herbs. Leaves linear to pinnatifid, alternate below, decussate to whorled above, 1-nerved, subpetiolate. Flower(s) axillary, solitary or 2-3 per node or in lax terminal racemes; bracteoles 2, linear; calyx-lobes 5, connate below; corolla tubular or rotate, lobes 5, equal or unequal; stamens 4, equal or didymous, anthers 2-celled, 1-cell aborted, fertile cells connivent in pairs; ovary cylindric; stigma 2-lamellate. Capsule loculicidal. Seeds striate.

1. Flowers pinkish-purple; lower leaves pinnatisect; capsule oblong. ....1. **S. delphinifolia**
1. Flowers yellow with a purple centre, lower leaves trifid, capsule subglobose. ....2. **S. trifida**

1. **Sopubia delphinifolia** (L.) G. Don, Gen. Syst. 4:560. 1837-1838; Hook.f., Fl. Brit. India 4:302. 1884; Gamble, Fl. Pres. Madras 970. 1924; Sald. & Nicols., Fl. Hassan Dist. 525. 1976; Cramer in Dassan. & Fosberg, Rev. Handb. Fl. Ceylon 3:394. 1981; Rani & Matthew, Fl. Tam. Carnatic 3(2) : 1101. 1983; Manilal, Fl. Silent Valley 196. 1988; Mohanan & Sivad., Fl. Agasthyamala 485.2002. *Gerardia delphinifolia* L. in Juslenius, Cent. Pl. 2: 21. 1756; Roxb., Pl. Corom. t. 90. 1798.

Erect branched stout herbs. Leaves pinnatisect, segments filiform, upper ones 3-fid. Flowers(s) axillary, solitary; calyx 5-ribbed, lobes 5, subulate; corolla pinkish-purple, tubular, lobes 5, broadly ovate; ovary 2-lobed. Capsule oblong, 2-4-valvular. Seeds truncate, longitudinally striate.

Fl. & Fr.: Oct.-Dec. Distr.: Sri Lanka, India. Rare. On grasslands. Locally abundant. *NAK 2113* (Vallicode-Kottayam, c. 200 m).

2. **Sopubia trifida** Buch.-Ham. ex D. Don, Prodr. Fl. Nepal. 88. 1825; Hook.f., Fl. Brit. India 4:302. 1884; Gamble, Fl. Pres. Madras 970. 1924; Sald. & Nicols., Fl. Hassan Dist. 525. 1976; Cramer in Dassan. & Fosberg, Rev. Handb. Fl. Ceylon 3:394. 1981; Matthew, Ill. Fl. Tam. Carnatic t. 505. 1982; Rani & Matthew in Matthew, Fl. Tam. Carnatic 3(2) : 1102. 1983; Mohanan & Sivad., Fl. Agasthyamala 485.2002.

Stout branched herbs. Leaves 3-fid below, linear above. Flower(s) axillary, solitary; calyx campanulate, lobes triangular, pilose within; corolla yellow with a purple centre, rotate, lobes suborbicular; fertile stamens bearded above. Capsule subglobose, 2-valvular. Seeds oblong, spirally striate.

Fl. & Fr.: Sep.-Dec. *Distr.*: Sri Lanka, India, Himalayas, Indo-China, W. & C. China, Malesia. Rare. in Grassy slopes. *NAK 1334 & 2532* (Pampa dam area, c.1100 m).

STRIGA Loureiro
Fl. Cochinch. 17. 1790.

Striga asiatica (L.) Kuntze, Rev. Gen. Pl. 2:466. 1891; Sald. & Nicols., Fl. Hassan Dist. 526. 1976; Matthew, Ill. Fl. Tam. Carnatic t.507. 1982; Rani & Matthew in Matthew, Fl. Tam. Carnatic 3(2) : 1104. 1983; Manilal, Fl. Silent Valley 197. 1988; Nicols., Suresh & Manilal, An Interpr. Hort. Malab. 244. 1988; Mohanan & Sivad., Fl. Agasthyamala 486.2002. *Buchnera asiatica* L., Sp. Pl. 630. 1753. *Striga lutea* Lour., Fl. Cochinch. 22. 1790; Hook.f., Fl. Brit. India 4:299. 1884; Gamble, Fl. Pres. Madras 968. 1924; Cramer in Dassan. & Fosberg, Rev. Handb. Fl. Ceylon 3:400. 1981. *Kalu-polapen* Rheede, Hort. Malab. 9:129, t. 66. 1689.

Erect scabrid herbs. Leaves simple, decussate below, alternate above, narrowly oblong, 1.5-3.5 by 0.2-0.3 cm, base decurrent, apex acute, 1-nerved, thick, subsessile. Flower(s) subessile in lax spikes; bracts and bracteoles linear; calyx 15-ribbed, lobes 4, linear; corolla yellow, tubular; curved, lobes 5, unequal, 2+3; stamens 4 in pairs; ovary globose; style 2-lamellate. Capsule ovoid. Seeds oblong.

Fl. & Fr.: Jul.-Sep. *Distr.*: Trop. & S. Africa, India, Sri Lanka, S.E. Asia. Locally abundant but generally rare. Open grasslands. *NAK 2021*(Naduvathumoozhy, c.150 m) *2521* (Kakki hills, c.1100 m).

TORENIA Linnaeus
Sp. Pl. 619. 1753.

Trailing herbs; rooting at nodes. Leaves simple, decussate, Flower(s) axillary or lateral umbels or racemes; bracteoles 0; calyx faintly 2-lipped, tube winged or keeled, lobes 5; corolla 2-lipped, tube cylindric, curved, upper lip emarginate or 2-lobed, lower 3-lobed, spreading; stamens 4, didymous, lower pair arched, appendaged, anthers connate in pairs; ovary oblong-cylindric; ovules α; stigma 2-lamellate. Capsule septicidal. Seeds α, rugose.

1. Calyx keeled; leaves deltoid, margin serrate. .... 1. **T. bicolor**
1. Calyx winged; leaves ovate-lanceate, margin crenate-serrate. .... 2. **T. tranvancorica**

1. **Torenia bicolor** Dalz. in Hooker's J.Bot. Kew Gard. Misc. 3:38. 1851; Hook.f., Fl. Brit. India 4:278. 1884; Gamble, Fl. Pres. Madras 957. 1924. Sald., Bull. Bot. Surv. India 8:129. 1966; Nicols., Suresh & Manilal, An Interpr. Hort. Malab. 244. 1988; Mohanan & Sivad., Fl. Agasthyamala 487.2002. *Kaka-pu* Rheede, Hort. Malab. 9:103, t.53. 1689.

Decumbent annuals; stem 4-angled. Leaves deltoid, 1-2 cm, base truncate to subcordate, margin serrate, apex acute, sparsely pubescent. Flower(s) axillary, solitary or in pairs, pedicel to 3.5 cm long; calyx strongly keeled, decurrent on pedicel; corolla deep purple, lobes yellowish or whitish. Capsule elongated. Seeds subglobose, rugose.

Fl. & Fr.: Sep.-Dec. *Distr.*: Peninsular India. Not common. Peak of flowering usually after monsoon. In moist shady hills. *NAK 882* (Lahai Estate, c.300 m), *2011* (Kokkathode, c.250 m).

2. **Torenia travancorica** Gamble, Fl. Pres. Madras 957. 1924; Hill, Bot. Mag. t. 9615. 1942; Stearn in Chittenden, Dict. Gard. 2127. 1951; Sald., Bull. Bot. Surv. India 8:130. 1966; Mohanan & Sivad., Fl. Agasthyamala 488.2002. *Torenia asiatica* sensu Hook.f., Fl. Brit. India 4:277. 1884, p.p. non L., 1753, nec Benth., 1835.

Trailing herbs; stem 4-angled, glabrous. Leaves ovate-lanceate, 2-5 by 1-2.5 cm, base truncate, margin crenate-serrate. Flowers 3-5, in clusters, axillary; pedicel c. 4 m long;calyx winged, decurrent on pedicel; corolla, c. 4 cm long, tube yellow, upper lobes purple, lower lobes white with bluish-purple blotches. Capsule narrowly elongated. Seeds pitted.

Fl. & Fr.: After rains. *Distr.*: Peninsular India. Common on moist exposed hills. *NAK 650, 734, 1491* (In lower and upper ghats).

94. OROBANCHACEAE

1. Calyx spathaceous, split at anterior side. .... **Aeginetia**
1. Calyx tubular-campanulate, faintly 2-lipped. .... **Christisonia**

AEGINETIA Linnaeus
Sp. Pl. 632. 1753.

Aeginetia indica L., Sp. Pl. 632. 1753; Roxb., Pl. Corom. 91. 1798; Wight, Ic. 895. 1844-1845; Hook.f., Fl. Brit. India 4: 320. 1884; Gamble, Fl. Pres. Madras 974. 1924; Sald. & Nicols., Fl. Hassan Dist. 532. 1976; Matthew, Ill. Fl. Tam. Carnatic t. 508. 1982; Matthew & Rani in Matthew, Fl. Tam. Carnatic 3(2) : 1108. 1983; Manilal, Fl. Silent Valley 198. 1988; Nicols., Suresh & Manilal, An Interpr. Hort. Malab. 204. 1988; Mohanan & Sivad., Fl. Agasthyamala 489.2002. *Tsjomcumulu* Rheede, Hort. Malab. 11:97, t. 47. 1692.

Leafless, achlorophyllous, stoloniferous root parasites. Flowers on an elongate scape, zygomorphic, bisexual; bracteoles 0; calyx spathaceous, split at anterior side, corola purple, tubular, twisted, curved, lobes 5, subequal, obscurely 2-lipped, 3+2; stamens 4, didymous, coherent in pairs, clavate, deflexed into a long massive spur in lower pair; ovary irregularly lobed, 1-locular; ovules α, on two parietal multi-fid, placentae; style fleshy; stigma massive, capitate-peltate. Capsule 2-valvular, enclosed by the calyx. Seeds minute.

Fl. & Fr.: Dec.-Feb. *Distr.*: Sri Lanka, India, Himalayas, Myanmar, Indo-China, China, Japan, Philippines & Malaya. Rare. Along moist shady forest borders. *NAK 2863* (Kakki hills, c. 1100 m).

CHRISTINSONIA G. Gardner
Calcutta J. Nat. Hist. 8: 153. 1897.

Simple or tufted subsucculent, achlorophyllous, parasitic herbs; stem bearing scales. Flowers zygomorphic, bisexual; calyx tubular, obscurely 2-lipped, lobes 5; corolla tubular or funnel-form, slightly curved or straight, lobes 5, subequal, ovary 1-locular, superior; ovules α, on large 2 or more lobed intruded placenta(e); stigma peltate. Capsule 2-valvular. Seeds α, reticulate.

1. Herbs, to 20 cm tall; stem glandular hairy; flowers yellow; calyx orange coloured. ....1. **C. bicolor**
1. Herbs, to 30 cm tall; stem glabrous; flowers purplish-pink, except the yellow lower lip; calyx white coloured. ....2. **C. tubulosa**

1. **Christisonia bicolor** Gard., Calcutta J. Nat. Hist. 8: 159. 1897; Hook.f., Fl. Brit. India 4: 322. 1884; Gamble, Fl. Pres. Madras 976. 1924. *C. aurantiaca* Wight, Ic. 1486. 1849.

Small herbs. Scape brownish-yellow; scales α, glandular-pubescent. Flowers about 10 on corymbs or racemes; bracteoles 2, foliaceous; calyx orange-coloured; corolla yellow, tubular, lobes equal. Capsule ovoid.

Fl. & Fr.: Sep.-Dec. *Distr.*: Sri Lanka, Peninsular India. Rare. In shola forests. *PR 79299* (MH) (Anathode, c. 1000 m).

2. **Christisonia tubulosa** (Wight) Benth. ex Hook.f., Fl. Brit. India 4: 321. 1884; Gamble, Fl. Pres. Madras 975. 1924; Mohanan & Sivad., Fl. Agasthyamala 490.2002. *Oligopholis tubulosa* Wight, Ic.t. 1422. 1849.

Tall herbs; scape glabrous; scales few. Flowers few in racemes; bracteoles 0; calyx tubular, white-coloured, lobes triangular; corolla purplish-pink, except the yellow lowerlip, lobes orbicular, subequal. Capsule ovoid.

Fl. & Fr.: Nov.-Jan. *Distr.*: Peninsular India. Rare. On moist shady evergreen forest floor. *NAK 1197* (Lower Moozhiar, c. 250 m).

95. LENTIBULARIACEAE

UTRICULARIA Linnaeus
Sp. Pl. 18. 1753.

Carnivorous erect or free-floating, scapigerous herbs with bladder-like traps; roots and leaves replaced by variously modified stem; traps intermingled with rhizoids and foliar organs. Inflorescence a raceme, on a terminal erect or twining scape; bracts base-/medifixed; bracteloes present or 0. Flowers subsessile, zygomorphic, bisexual; calyx-lobes 2, often accrescent; corolla 2-lipped, lower deflexed, produced into a spur below; stamens 2, included; ovary superior, 1-locular; ovules 2-α, on a free basal placenta. Capsule globose, circumcissile or irregularly repturing. Seeds lenticular or ellipsoid.

1. Aquatic sebmrged herbs; foliar organs finely dissected; bracteoles absent. ....1. **U. aurea**
1. Terrestrial, erect or trailing herbs; foliar organs entire; bractolees present.
    2. Foliar organs linear-oblanceate; seeds smooth.
        3. Bracts basifixed.
            4. Flowers blue or bluish-purple or violet.

5. Herbs to 7 cm high; stem erect; flowers small, bluish-purple. ....8. **U. uliginosa**

5. Herbs to 20 cm long; stem twining; flowers large, bluish violet. ....4. **U. reticulata**

4. Flowers yellow; erect slender herbs; spur curved upwards.

6. Lower lip of the corolla ligulate; fruiting pedicel erect. ....2.**U. bifida**

6. Lower lips of the corolla eligulate; fruiting pedicel recurved. ....7. **U. subramanii**

3. Bracts medifixed.

7. Spur longer than lower lip of corolla, curved upwards, apex acute. ....3. **U. caerulea**

7. Spur shorter than lower lip of corolla, straight, apex obtuse. ....5. **U. roseo-purpurea**

2. Foliar organs reniform; seeds glochidiate. ....6. **U. striatula**

1. **Utricularia aurea** Lour., Fl. Cochinch. 26. 1790; Taylor in Steenis, Fl. Males. I,8:296. 1977; Rani & Matthew in Matthew, Fl. Tam. Carnatic 3(2) :1111. 1983. *U.flexuosa* Vahl, Enum. Pl. 1:198. 1804; Clarke in Hook.f., Fl. Brit. India 4:329. 1884; Gamble, Fl. Pres. Madras 980. 1924. *U.fasciculata* Roxb., Fl. Ind. 1:143. 1820; Wight, Ic.t. 1568. 1850.

Aquatic herbs; stolons much branched. Foliar organs much dissected, dichotomously branched;traps lateral or at the point of forking; scape 5-7- flowered; scale and bracteoles 0; bracts lanceate; calyx-lobes subequal; corolla yellow, upper lip suborbicular, lower lip orbicular, spur slightly curved, facing downwards; ovary globose. Capsule enclosed by persistent calyx. Seeds faintly reticulate.

2. **Utricularia bifida** L., Sp. Pl. 18.1753; Clarke in Hook.f., Fl. Brit. India 4: 332. 1884; Gamble, Fl. Pres. Madras 982. 1924; Taylor in Steenis, Fl. Males. I, 8 : 281. 1977. Matthew & Rani in Matthew, Fl. Tam. Carnatic 3(2) : 1112.1983.

Foliar organs linear-subulate, base cuneate, traps reniform; scape 2-5- flowered; scale lanceate; bracts ovate-lanceate, basifixed; bracteoles linear-filiform; calyx-lobes unequal; corolla yellow, upper lip obtuse, lower lip broadly ovate, orbicular, spur facing upwards; ovary globose. Capsule ovoid-globose. Seeds ovoid or globose, reticulate.

Fl. & Fr.: Jul.-Sep. *Distr.*: India, China, Japan, Indo-China, Malesia to N. Australia. Rare. Along marshy grounds. *NAK 2137* (V. Kottayam, c. 100 m).

3. **Utricularia caerulea** L., Sp. Pl. 18. 1753; Wight, Ic. t. 1583. 1850; Gamble, Fl. Pres. Madras 983. 1924; Taylor in Steenis, Fl. Males I, 8: 287. 1977; Matthew & Rani in Matthew, Fl. Tam. Carnatic 3(2): 112. 1983; Mohanan & Sivad., Fl. Agasthyamala 491.2002 *U. racemosa* Wall. ex Walp. in Meyen, Observ. Bot. 19:401. 1843; Wight, Ic. t. 1584, f. 1. 1850; Clarke in Hook.f., Fl. Brit. India 4:333. 1884. *U. nivea* Vahl, Enum. Pl. 1:203. 1804; Roxb., Fl. Ind. 1: 144. 1820.

Foliar organs linear-oblanceate; traps minute, ovoid or obovoid; scape 5-7-flowered; scales and bracts lanceate, medifixed; bracteoles linear-lanceate; calyx-lobes subequal; corolla cream or purple, upper lip 2-

lobed, lower lip flat, broadly suborbicular-ovate; spur straight, horizontal, slightly curved forward. Seeds ellipsoid, faintly reticulate.

Fl. & Fr.: Sep.-Dec. *Distr.*: Madagascar, India, China, Japan, Australia, Malesia. Common. *NAK 1082 & 2133* (Vallicode-Kottayam, c. 150 m).

4. **Utricularia reticulata** J.E. Smith, Exot. Bot. t. 119. 1805; Wight, Ic.t. 1571, f.2.1850; Clarke in Hook.f., Fl. Brit. India 4: 331. 1884, excl. var. *uliginosa;* Gamble, Fl. Pres. Madras 982. 1924; Nicols., Suresh & Manilal, An Interpr. Hort. Malab. 162. 1988. *Nelipu* Rheede, Hort. Malab. 9: 137, t. 70. 1689.

Terrestrial herbs; traps minute, ovoid; scape twining, 3-5- flowered; scales and bracts minute, basifixed; cayx-lobes ovate, fairly enlarged in the fruit, decurrent; corolla bluish, large, upper lip broadly obovate, lower lip suborbicular, c. 2 cm ϕ; spur slightly curved. Capsule globose. Seeds rhomboid, reticulate.

Fl. & Fr.: Jan.-Mar. *Distr.:* Sri Lanka, Peninsular India. Common. In paddy fields. Locally abundant. *NAK 2334* (Kodumon, c. 30 m).

5. **Utricularia roseo-purpurea** Stapf ex Gamble, Fl. Pres. Madras 983. 1924; Subram. & Banerjee, Bull. Bot. Surv. India 10:103. 1968. *U. racemosa* sensu Wight, Ic. t. 1584, f. 1.1850, non Wall. ex Walp., 1843. *U. rosea* sensu Clarke in Hook.f., Fl. Brit. India 4: 333. 1884, non Edgew., 1847.

Terrestrial herbs; foliar organs rosulate, spathulate, 3-5 by 1-2 mm, 1-nerved, base attenuate, decurrent on a pseudo- petiole, apex rounded; traps ovoid; scape 6-20 cm high; scales and bracts medifixed; calyx-lobes unequal, broadly ovate or triangular, ovate; corolla pale violet to pinkish with an yellow spot on palate, upper lip broadly obovate or obcordate, sparsely reticulately veined, lower broadly orbicular; spur subulate, straight shorter than lower lip, apex obtuse; ovary globose. Capsule globose. Seeds ellipsoid or obovoid, reticulate, minutely tuberculate.

Fl. & Fr.: Sep.-Dec. *Distr.*: Peninsular India. Rare. On dripping or moist rocks. Locally abundant. *NAK 2024* (Kakkathode, c. 200 m).

Note :Closely related to **U. caerulea** L., but can be distinguished by having the following characters like, large flowers, straight, obtuse short spur, reticulate and minutely turberculate seeds.

6. **Utricularia striatula** Smith in Rees, Cyclop. 37, n. 17. 1819; Gamble, Fl. Pres. Madras 983. 1924; Taylor in Steenis, Fl. Males. I, 8: 289. 1977; Matthew & Rani in Matthew , Fl. Tam. Carnatic 3(2): 1123.1983; Mohanan & Sivad., Fl. Agasthyamala 493.2002. *U. orbiculata* Wall. ex A.DC. in DC., Prodr. 8:18. 1844; Clarke in Hook.f., Fl. Brit. India 4:334. 1884.

Terrestrial, often lithophytic herbs; foliar organs simple, orbicular to reniform, obtuse; traps rotund, minute; bracts and scales linear, medifixed; calyx-lobes unequal; corolla pink or pinkish-white, upper lip 5-angled, spur curved upwards. Capsule globose, persistent, shorter than calyx. Seeds obovoid, glochidiate.

Fl. & Fr.: Sep.-Dec. *Distr.*: India to China, Indo-China, Malesia, Trop. Africa. Common species of the genus in the district. On dripping vertical rocks. *NAK 2132* (Vallicode-Kottayam, c. 150 m).

7. **Utricularia subramanii**. Janarth. & A.N. Henry, J. Bombay Nat. His. Soc. 87 (3): 441-442. 1990.

Terrestrial herbs. Foliar organs linear-subulate, base cuneate, decurrent to a pseudo-petiole; traps reniform; scape 2-5- flowered; scale lanceate; bracts ovate-lanceate, basifixed; bracteoles linear-filiform; calyx-lobes unequal, acute to acuminate or dentate at apex; corolla yellow, upper lip obtuse, lower lip broadly ovate, orbicular, ligulate at middle; ovary globose. Capsule ovoid-globose; fruiting pedicel erect. Seeds oblongoid; testa not striated within.

Fl. & Fr.: Jul.-Sep. *Distr.*: Endemic to Western Ghats. Rare. Mountain marshes. *CNM 58341* (MH) (Ranni, c. 100 m).

8. **Utricularia uliginosa** Vahl, Enum. Pl. 1:203. 1804; Gamble, Fl, Pres. Madras 981. 1924; Mohanan & Sivad., Fl. Agasthyamala 493.2002. *U. affinis* Wight in Hooker's J. Bot. Kew Gard. Misc. 1:373. 1849 & Ic.t. 1580, f.1.1850; Clarke in Hook.f., Fl. Brit. India 4:330. 1884.

Terrestrial herbs. Foliar organs rosulate, spathulate, 1-nerved, base cuneate and decurrent on a pseudo-petiole; scape erect, straight, 3-6-flowered; scales and bracts lanceate, basifixed; calyx-lobes subequal, ovate, acute; corolla bluish-purple, upper lip obovate or obcordate, lower orbicular, spur straight, conical, slightly curved forward. Capsule globose. Seeds subglobose, aerolate.

Fl. & Fr.: Sep.-Dec. *Distr.*: Peninsular India. Locally abundant and common. On moist rocky hills. *NAK 2120 & 2176* (Vallicode-Kottayam, c. 150 m).

96. GESNERIACEAE

1. Epiphytic pendulous subshrubs; seeds with one or more hairs near the hilum. .... **Aeschynanthus**
1. Terrestrial erect herbs (succulent/woody); seeds without hairs near the hilum.
   2. Fruit a berry; flowers white. .... **Rhynchotechum**
   2. Fruit a loculicidal capsule; flowers other than white.
      3. Scapigerous herbs; capsule linear; flowers pale lilac or bluish-yellow; leaf base rounded or subcordate. .... **Didymocarpus**
      3. Non- scapigerous herbs; capsule ellipsoid; flowers bluish; leaf base oblique. .... **Rhynchoglossum**

AESCHYNANTHUS Jack

Trans. Linn. Soc. London 14:42, t.2, f.3. 1823 (nom. cons.).

Aeschynanthus perrottettii A.DC. in DC., Prodr. 9:261. 1845; Clarke in Hook.f., Fl. Brit. India 4:339. 1884, (incl. vars.); Gamble, Fl. Pres. Madras 985. 1924; Subram. & Henry, Bull. Bot. Surv. India 12:1. 1970; Mohanan & Sivad., Fl. Agasthyamala 494.2002. *A. ceylanica* sensu Wight, Ic. t. 1347. 1848, non Gard., 1846. A. *perrottettii* A.DC. var. *planiculmis* Clarke in Hook.f., Fl. Brit. India 4: 340. 1884, "*platyculmis*". A. *planiculmis* (Clarke) Gamble, Fl. Pres. Madras 985. 1924.

Epiphytic pendulous subshrubs; stem rooting at nodes. Leaves simple, decussate, elliptic-lanceate or oblong-lanceate, 3-6 by 1-2 (1) cm, base cuneate, apex acuminate, subsucculent. Flowers in axillary or terminal fascicles, rarely solitary; bracts caducous; bracteoles small; calyx 5 fid; corolla purplish-red, tubular, c. 5 cm long, ventricose, curved, 2-lipped, upper small, 2-lobed, lower large, 3-lobed; stamens 4, perfect, often a staminode present; ovary superior; ovules parietal. Capsule linear-cylindric, loculicidal, 2-valved. Seeds α, minute, with one or more hairs at hilum.

Fl. & Fr.: Dec.-Mar. *Distr.*: Western Ghats (Malabar Coast). Frequent in evergreen forests. Flowers beautiful, caducous. *NAK 2148* (Moozhiar, c. 250 m).

DIDYMOCARPUS Wallich
Edinburgh Philos. J. 1:378.1819 (non. cons.).

Didymocarpus humboldtianus Gardn., Calcutta J. Nat. Hist. 6:477. 1846; Clarke in Hook.f., Fl. Brit. India 4:353. 1884; Gamble, Fl. Pres. Madras 989. 1924, "*humboldtiana*" Theobald & Grupe in Dassan. & Fosberg, Rev. Handb. Fl. Ceylon 3:84. 1981; Manilal, Fl. Silent Valley 200. 1988.

Terrestrial scapigerous herbs. Leaves simple, in a basal rosette, broadly ovate, 4-8 by 2.5-6.5 cm, base broadly decurrent on petiole margin coarsely crenate-serrate, apex rounded, sparsely villous-pubescent on both surfaces. Flowers in several subumbellate, scapose fascicles; bracts minute; calyx 5-lobed; corolla white, tube short-campanulate, lobes 5, rounded, unequal; stamens 2 with 2 or 3 staminodes, included, filaments curved; anthers coherent, bilocular; ovary 1-locular, oblique. Capsules linear-cylindrical. Seeds α, reticulate.

Fl. & Fr.: Sep.-Nov. *Distr.*: Sri Lanka, Peninsular India (Western Ghats). Rare. On dripping rocks. Locally abundant. *NAK 2554* (way to Ponnambalamedu, c. 1200 m).

RHYNCHOGLOSSUM Blume
Fl. Jav. Praef. 2:18. 1828.

Rhynchoglossum notonianum (Wall.) Burtt, Notes R. Bot. Gard. Edinburgh 24: 170. 1962; Sald. in Sald. & Nicols., Fl. Hassan Dist. 531. 1976; Theobald & Grupe in Dassan. & Fosberg, Rev. Handb. Fl. Ceylon 3: 99.1981; Manilal, Fl. Silent Valley 200. 1988; Mohanan & Sivad., Fl. Agasthyamala 496.2002. *Wulfenia notoniana* Wall., Tent. Fl. Nepal. 46. 1826. *Klugia notoniana* (Wall.) A. DC., Prodr. 9: 276. 1845; Wight, Ic. 1353. 1848; Clarke in Hook.f., Fl. Brit. India 4: 366. 1884; Gamble, Fl. Pres. Madras 990. 1924.

Succulent annuals. Leaves simple, alternate, ovate-oblong, 10-20 by 5-8 cm, base very oblique, apex acute, lateral nerves ϕ, parallel, prominent, membranous. Flowers in axillary, leaf-opposed or terminal racemes; bracts minute; bracteoles linear; calyx campanulate, 5-angled, one angle often winged; corolla bluish, 2-lipped; upper lip very small, lower large, obtuse or faintly 3-lobed, with 2 hooded depressions at base; stamens 4, included, anthers perfect in pairs; disc fleshy, annular; ovary ovoid; ovules α, on parietal placenta. Capsule ovoid, loculicidal, 2-valved. Seeds ellipsoid, reticulate.

Fl. & Fr.: Aug.-Dec. *Distr.*: Peninsular India, Sri Lanka. On dripping rocks at forest outskirts. Locally abundant and common. *NAK 776* (Plappally, c. 400 m).

RHYNCHOTECHUM Blume
Bijdr. 775. 1826.

Rhynchotechum permolle (Nees) Burtt, Notes R. Bot. Gard. Edinburgh 24: 39. 1962; Sald. in Sald. & Nicols., Fl. Hassan Dist. 531. 1976; Theobald & Grupe in Dassan. & Fosberg, Rev. Handb. Fl. Ceylon 3: 105. 1981; Manilal, Fl. Silent Valley 201. 1988; Mohanan & Sivad., Fl. Agasthyamala 497.2002. *Isanthera permollis* Nees, Trans. Linn. Soc. London 17: 82. 1934; Wight, Ic. t. 1355. 1848; Clarke in Hook.f., Fl. Brit. India 4: 372. 1884; Gamble, Fl. Pres. Madras 992. 1924.

Subshrubs; branchlets tawny-tomentose, glabrescent. Leaves simple, alternate, oblanceate 7-25 by 3-9 cm, base cuneate, margin serrulate, apex acuminate, lateral veins, 16-29 pairs, prominent, lower surface sparsely tomentose, chartaceous. Flowers in axillary dense cymes; calyx densely woolly-tomentose, lobes 5; corolla white, obscurely 2-lipped, lobes 5, ovate; stamens 4, adnate at corolla base, anthers all perfect, 1-celled; disc small, annular; ovary ovoid. Fruit an ovoid berry. Seeds minute, ellipsoid.

Fl. & Fr.: May-Sep. *Distr.*: Sri Lanka, Peninsular India (Western Ghats). In moist shady areas of evergreen forests. Locally abundant but not common. *NAK 2047* (Angamoozhy, c.250 m), *2593* (Arampa, c. 300 m).

97. BIGNONIACEAE

1. Semi-evergreen trees; flowers bright yellow with reddish veins; capsule slender, faintly spiral, smooth, 4-angled. .... **Stereospermum**
1. Deciduous trees; flowes white with yellow or pinkish tinge; capsule stout, straight, coarsely tubercled, cylindric. .... **Radermachera**

RADERMACHERA Zollinger et Moritzi in Zollinger, Syst. Verzeichniss Ind. Archipel. 3: 53.1855.

Radermachera xylocarpa (Roxb.) K. Schum. in Engler & Prantl, Pflanzenf. 4(36):243. 1895; Gamble, Fl. Pres. Madras 999. 1924; Mohanan & Sivad., Fl. Agasthyamala 499.2002. *Bingonia xylocarpa* Roxb., Fl. Ind. 3:108. 1832. *Stereospermum xylocarpum* (Roxb.) Wight, Ic. t. 1335 & 1336. 1848; Hook.f., Fl Brit. India 4: 383. 1889. *Mala-muringa, Edangkorana.*

Deciduous trees. Leaves 2-pinnate, opposite; leaflets eliptic or elliptic-ovate, 5-8 by 2-4 cm, base acute, apex acuminate, chartaceous. Flowers in terminal racemes or corymbs; calyx campanulate, 5-lobed; corolla white with yellowish or pinkish veins, c. 5 cm long, funnel-form, obscurely 2-lipped, lobes 5, rounded, spreading; stamens 4, didymous; staminode 1, anthers divaricate. Capsule stout, coarsely tubercled to 90 cm long. Seeds flat, both ends winged.

Fl. & Fr.: Jan.-Mar. *Distr.:* South India. In deciduous forests. Rare. *NAK 2649* (Appuppanthode, c. 150 m), Fruit is very characteristic.

Resin from wood used as a remey for skin eruptions.

STEREOSPERMUM Chamisso Linnaea 7: 720. 1833 ('1832').

Stereopermum colais (Buch.-Ham. ex Dillwyn) Mabb., Taxon 27: 553. 1978; Manilal, Fl. Silent valley 201. 1988; Nicols., Suresh & Manilal, An Interpr. Hort. Malab. 73. 1988; Mohanan & Sivad., Fl. Agasthyamala 499.2002. *Bignonia colais* Buch.-Ham. ex Dillwyn, Rev. Hort. Malab. 28. 1839. *Dipterospermum personatum* Hassk., Fl. 25(2) Beibl.: 28. 1842. *Stereospermum tetragonum* DC., Prodr. 9: 210. 1845; Gamble, Fl. Pres. Madras 998. 1924. *S. chelonoides* sensu Wight, Ic. 1341. 1845, non (L.f.) DC., 1838; Clarke in Hook. f., Fl. Brit. India 4. 382. 1884. *S. personatum* (Hassk.) Chatterjee, Bull. Bot. Soc. Bengal 2(1): 70. 1948; Santisuk, Kew Bull. 28: 178. 1973; Matthew & Rani in Matthew, Fl. Tam. Carnatic 3(2): 1138, 1983. *Padiri* Rheede, Hort. Malab. 6: 47-48, t. 26.1686.

Trees. Leaves 1-pinnate, c. 30 long; leaflets 9-11, elliptic-ovate, 6-12 by 3-5 cm, base obliquely cuneate or rounded, margin subentire, apex caudate, chartaceous. Flowers in terminal large panicles; calyx campanulate, lobes 5, unequal; corolla bright yellow, curved, 2-lipped, upper 2-lobed, lower 3-lobed, lobes orbicular, margin crispate, villous within; stamens 4, didymous, filaments woolly at base; staminode 1, disc cupular, fleshy. Capsule to 60 cm long, slender, spiral, 4-angled. Seeds trigonous.

Fl. & Fr.: Mar.-May. *Distr.*: Sri Lanka, India, Trop. Himalayas, Indo-china, Malesia. Not common. In semi-evergreen and evergreen forests. *NAK 556* (Thekkuthode, c. 50 m), *1674* (Sabari hills, c. 250 m).

Flowers and fruits eaten as a remedy in biliousness diarrhoea, poison and impure blood. Decoction of roots given as a remedy in asthma, cough, swellings, vomiting and excessive thirst.

Oroxylum indicum (L.) Vent., **Spathodea campanulata** P. Beauv. and **Pajanelia longifolia** (Willd.) Schum. are seen planted in the district.

98. PEDALIACEAE

SESAMUM Linnaeus
Sp. Pl. 634. 1753.

Sesamum mulayanum N.C. Nair, Bull. Bot. Surv. India 5: 251-253. 1964.

Annuals to 2 m tall; stem 4-angled. Leaves often heteromorphic, upper ones usually alternate, ovate, basal opposite, 3-lobed or partite, 4-8 by 2-4 cm, base cuneate, margin coarsely toothed, apex acute, pubescent, thick-chartaceous; petiole c. 2.5 cm long, gradually shortening upwards. Flower(s) solitary, axillary; pedicel short, with 2 yellow sessile glands at base; calyx persistent, lobes 5, lanceate; corolla pinkish, to 3.5 cm long, obliquely campanulate; stamens 4, didymous, connectives prolonged and terminating in a globose gland; disc annual; ovary cylindric, pilose, 4-locular due to false septum. Capsule erect, oblong, 4-angled, 4-grooved, apex beaked. Seeds brownish, broadly ovate, margin thick, broad, reticulate.

Fl. & Fr.: May.-Jul. *Distr.:* Eastern & Penin. India. On open places. *NAK 637* (Puthukkulam, c. 50 m). *Note :* This species was first repoted from Mahendragarh district in Punjab. Now, they are seen in waste places of plains in Quilon and Pathanamthitta districts. The seeds are larger and stout and would be useful for breeding purposes.

Sesamum orientale L. is cultivated and often run wild. *NAK 690* (Pandalam, c. 20 m).

99. ACANTHACEAE

1. Climbers; calyx teeth a or absent. .... **Thunbergia**
1. Erect, straggling or prostrate herbs; calyx teeth 4-5.
 2. Seeds supported on retinacula.
  3. Corolla-lobes twisted to left.
   4. Ovules more than 2 per locule; seeds 6 or more-per capsule.
    5. Corolla 2-lipped; flowers in axillary whorls; capsule enclosing seeds throughout. .... **Hygrophila**
    5. Corolla subequally 5-lobed; flower(s) solitary or in cymes; capsule enclosing seeds at base only. .... **Dipteracanthus**
   4. Ovules 2-per locule; seeds 4 or few per capsule.
    6. Spreading herbs; spikes 1-sided; bracts reniform, closely imbricating. .... **Phaulopsis**
    6. Erect shrubs or subshrubs; spikes not so; bracts otherwise.
     7. Fertile stamens 2.
      8. Corolla-tube twice as long as lobes; leaves glabrous. .... **Eranthemum**

8. Corolla-tube as long as or shorter than lobes; leaves wooly beneath. ....**Phlebophyllum**

7. Fertile stamens 4; corolla-tube short, ventricose above..**Nilgirianthus**

3. Corolla-lobes imbricate.

9. Ovules 3-10 per locule; seeds 4-6 or more per capsule.

10. Capsule compressed; seeds 4-8, ovoid.

11. Seeds 4 per capsule; flowers 1-sided on the peduncle. ....**Indoneesiella**

11. Seeds c. 8 per capsule; flowers on all sides of the peduncle. ....**Andrographis**

10. Capsule cylindric or subterete; seeds 16-24, flattened. ....**Gymnostachyum**

9. Ovules 2 or 1 per locule; seed 1-2 per capsule.

12. Corolla-lobes 5, subequal.

13. Calyx-lobes 4; outer pair larger, enclosing the inner. ....**Barleria**

13. Calyx-lobes 5; all equal or subequal.

14. Anther-cells, basally acute or mucronate; seeds 4 per capsule. ....**Asystasia**

14. Anther-cells muticous; seeds 8 or more per capsule. ....**Dipteracanthus**

12. Corolla-lobes 5, 2-lipped.

15. Fertile stamens 4; bracts α, spine-tipped. ....**Lepidagathis**

15. Fertile stamens 2; bracts few, non spine-tipped.

16. Anther-cells divergent or superposed; connective broad; seeds 4.

17. Anther-cells unequal, divergent, appendaged; stamens subexerted.

18. Both the anther cells basally long appendaged. ....**Justicia**

18. Only lower anther cell basally shortly appendaged. ....**Rungia**

17. Anther-cells equal, superposed, both muticous; stamens well exerted.

19. Corolla white; upper lip much narrower than the other. ....**Rhinacanthus**

19. Corolla pink; upper lip nearly as broad as the lower.

20. Anther cells rotund, decurrent; seeds papillose. ....**Peristrophe**

20. Anther cells reniform, distinct; seeds glochidiate. ....**Dicliptera**

16. Anther-cells parallel, connective narrow; seeds 2. ....**Ecbolium**

2. Seeds not supported on retinacula.

21. Calyx-lobes 4; stamens 2; bracts foliaceous. ....**Nelsonia**

21. Calyx-lobes 5; stamens 4; bracts non-foliaceous. ....**Staurogyne**

ANDROGRAPHIS Wallich ex C. G.D. Nees in Wallich, Pl. Asiat. Rar. 3: 77, 116. 1832.

Herbs or subshrubs; branchlets often 4-gonous. Leaves simple, decussate, entire, surface sometimes lineolate, chartaceous. Flowers in terminal and axillary dense/lax racemose panicles, bisexual, zygomorphic, pedicellate; bracts lanceate; bracteoles minute or 0; calyx-lobes 5, subequal, shortly connate below, glandular hairy; corolla tubular, 2-lipped, 2+3, imbricate, upper 2-lobed, lower lip deflexed, 3-lobed, broadly elliptic-oblong; stamens 2, exerted, anthers unequal, often bearded; ovary superior, 2-locular; ovules 2-6 per locule; style filiform; stigma 2-fid. Capsule oblong or ellipsoid. Seeds pitted, glabrous, supported by spoon-shaped or hooked retinacula.

1. Diffuse herbs; flowers in axillary lax simple racemes; anther cells bearded.

2. Stem and racemes slender; leaves lanceate, to 6 by 2 cm; flowers 4-6 per raceme. ....1.**A. atropurpurea**

2. Stem and racemes stout; leaves elliptic-lanceate or ovate-lanceate, 8 by 3.5 cm; flowers 4-16 per raceme. ....3. **A. macrobotrys**

1. Stout herbs or subshrubs; flowers in terminal/axillary, dense paniculate racemes; anther cells glabrous.

3. Leaves elliptic-lanceate, base acute or attenuate, thin-chartaceous; petiole 2-10 mm long; corolla prominently ventricosed. ....4. **A. producta**

3. Leaves oblong-lanceate, base obtuse or rounded, thick-chartaceous; petiole less than 2 mm or 0; corolla not prominently ventricosed. ....2. **A. lineata**

1. **Andrographis atropurpurea** (Dennst.) Alston ex Mabb., Taxon 26 : 539.1977; Nicols., Suresh & Manilal, An Interpr. Hort. Malab. 35. 1988. *Justicia atropurpurea* Dennst., Schluess. Hort. Ind. Malab. 12, 22 & 35. 1818 & in Fortsetz., Allg. Tentsch. Gart-Mag. 3:31, 40 & 83. 1818. *Andrographis wightiana* Arn. ex Nees in DC., Prodr. 11 : 517. 1847; Wight, Ic. t. 1558. 1850; Clarke in Hook.f., Fl. Brit. India 4: 503. 1884; Gamble, Fl. Pres. Madras 1048. 1924. *Katu-karivi* Rheede, Hort. Malab. 9:83, t. 44. 1689.

Slender diffuse herbs with trailing branches. Leaves lanceate, 4-6 by 1-2 cm, base acute or cuneate, decurrent on petiole, apex acuminate, chartaceous. Flowers 4-6 (8) per raceme, very lax; calyx-lobes linear, glabrescent, often lineolate; corolla white with a pinkish lower lip. Capsule linear, oblanceoid, puberulous. Seeds c.10 small.

Fl. & Fr.: Jul.-Sep. *Distr.:* South India (Western Ghats). Common. Among forest undergrowth. *NAK 228, 768 & 932* (Plappally, c. 400 m).

2. **Andrographis lineata** Wall. ex Nees in Wall., Pl. Asiat. Rar. 3 : 116. 1832; Clarke in Hook.f., Fl. Brit. India 4: 50. 1884; Gamble, Fl. Pres. Madras 1049. 1924.

Stout erect herbs. Leaves oblong-lanceate, 4-7 by 1-1.3 cm, base rounded or obtuse, margin stiff ciliolate, apex acute, hispid, thick-chartaceous; petiole 0 or very short. Flowers dense in terminal panicles, glandular hirsute; calyx-lobes lanceate, pale purple, slightly ventricose. Capsule oblong-cylindric. Seeds small.

Fl. & Fr.: Sep.-Dec. *Distr.*: South India (Western Ghats). Not common. On exposed grassy hills. *NAK 1366* (Kakki hills, c. 1000 m).

3. **Andrographis macrobotrys** Nees in DC., Prodr. 11 : 516. 1847, in Wall., Pl. Asiat. Rar. 3:116. 1832; Clarke in Hook.f., Fl. Brit. India 4: 503. 1884; Gamble, Fl. Pres. Madras 1048. 1924.

Stout diffuse subshrubs. Leaves elliptic-lanceate or ovate-lanceate, 4-9 by 2.5-3.5 cm, base rounded, apex acute or acuminate, chartaceous. Flower up to 16 per node; very lax; calyx-lobes, linear-subulate, glabrescent; corolla pink with red purple lower lip. Capsule oblong-lanceoid, glabrous, Seeds c. 8, small.

Fl. & Fr.: Jul.-Sep. *Distr.*: Sri Lanka, S. India. Not common. On disturbed forest areas. *NAK 797* (Ranni, c. 100 m), *891* (Plappally, c. 400 m).

4. **Andrographis producta** Gamble, Fl. Pres. Madras 1049. 1924. *A. neesiana* var. *producta* Clarke in Hook.f., Fl. Brit. India 4: 504. 1884.

Erect subshrubs. Leaves elliptic-lanceate, 7-13 by 1.5-4 cm, base attenuate, decurrent on petiole, apex gradually acute or acuminate, lateral veins prominent, thin-chartaceous; petiole c. 1 cm long. Flowers subdense in axillary and terminal panicles, glandular hispid; corolla white with pinkish blotches. Capsule cylindric, oblanceoid, hispid. Seeds 4-6, black.

Fl. & Fr.: Sep.-Dec. *Distr.*: Western Ghats. Locally abundant but generally rare. Among forest undergrowth in upper hills. *NAK 68, 389, 1296 & 1368* (Kakki hills, c. 1200 m).

ASYSTASIA Blume
Bijdr. 796. 1826.

Subscandent subshrubs. Leaves decussate, petiolate. Inflorescence a secund raceme, terminal; bracts lanceate; bracteoles minute. Flowers subsessile, bisexual; calyx-lobes 5, equal, shortly connate, linear or lanceate; corolla funnel-form, lobes 5, spreading, suborbicular; stamens 4, didynamous, included, connate in pairs, anther cells unequal; ovary 4-ovuled. Capsule elliptic-oblong, beaked. Seeds suborbicular or angled, rugose or tuberculate.

1. Flowers yellow; leaves broadly ovate, base subtruncate or broadly rounded, thick-chartaceous. .... 2. **A. dalzelliana**
1. Flowers violet; leaves ovate-lanceate, (elliptic) oblong-lanceate or linear-oblong, base rounded or acute, chartaceous.
    2. Leaves (elliptic) oblong-lanceate to linear oblong, to 14 cm long, base acute or cuneate. .... 1. **A. chelonoides**
    2. Leaves ovate-lanceate to 10 cm long, base rounded or broadly acute. ....3. **A. gangetica**

1. **Asystasia chelonoides** Nees in Wall., Pl. Asiat. Rar. 3: 89. 1832; Clarke in Hook.f., Fl.Brit. India 4: 493. 1884; Manilal, Fl. Silent Valley 202. 1988; Mohanan & Sivad., Fl. Agasthyamala 504. 2002. *A. chelonoides* var. *qudrangularis* Clarke in Hook.f., Fl. Brit. India 4: 494. 1884; Gamble, Fl. Pres. Madras 1063. 1924.

Stragglers. Leaves elliptic-lanceate, oblong-lanceate or linear-oblong, 6-14 by 1.5 3.5 (1.5) cm, base acute, apex gradually acuminate, sparsely hairy, thin-chartaceous. Racemes terminal or from upper leaf-axils, simple, or sparingly branched. Flowers 8-10-per raceme; calyx-lobes oblong; corolla purplish-violet. Capsule c. 3 cm long. Seeds 4, angular.

Fl. & Fr.: Jan.-Mar. *Distr.*: Sri Lanka, Peninsular India. Less common. *NAK 1450 & 1498* (Arampa, c. 400 m).

2. **Asystasia dalzelliana** Santapau, Kew Bull. 1948: 276. 1948; Rani & Matthew in Matthew, Fl. Tam. Carnatic 3(2): 1156. 1983. *A. violacea* Dalz. ex Clarke in Hook.f., Fl. Brit. India 4 : 494. 1884, non Dalz., 1850; Gamble, Fl. Pres. Madras 1063. 1924.

Erect or diffsue herbs. Leaves broadly ovate, 2-3.5 by 1.8-2.5 cm, base subtruncate, apex acuminate, lateral nerves close, densely lineolate, thick-chartaceous. Racemes terminal, simple. Flower about 6 per raceme; calyx-lobes lanceate; corolla yellow. Capsule to 1.5 cm long, pubescent. Seeds 2-4.

Fl. & Fr.: Sep.-Dec. *Distr.*: Peninsular India. Often in gardens, also run wild. *NAK1048* (Tiruvalla, c. 10 m).

3. **Asystasia gangetica** (L.) T. And. in Thw., Enum. Pl. Zeyl. 235. 1860; Gamble, Fl. Pres. Madras 1063. 1924; Santapau, Bot. Mem. Univ. Bombay 2: 68. 1951; Matthew, Ill. Fl. Tam. Carnatic t. 521. 1982; Rani & Matthew in Matthew, Fl. Tam. Carnatic 3(2): 1157. 1983; Nicols., Suresh & Manilal, An Interpr. Hort. Malab. 36. 1988. *Justicia gangetica* L., Cent. Pl. 2 : 3. 1756. *Asytasia coromandeliana* Wight ex Nees in Wall., Pl. Asiat. Rar. 3:89. 1832; Wight Ic. t. 1506. 1850; Clarke in Hook.f., Fl. Brit. India 4:493. 1884. *Valli-upu-dali* Rheede, Hort. Malab. 9:85, t. 45. 1689.

Straggling or diffuse perennial herbs. Leaves ovate-lanceate, 3-10 (12) by 1.5-5.5 cm, base rounded, apex acute or shortly acuminate, sparsely hairy, thin-chartaceous. Racemes terminal and upper leaf axillary simple. Flowers 10-16-per raceme; calyx-lobes oblong-lanceate; corolla usually violet, c. 3 cm long. Capsule c. 3 cm long, pubescent. Seeds, c.4, angular.

Fl. & Fr.: Most of the seasons. *Distr.*: Arabia, Africa, Sri Lanka, Peninsular India, Malaya. Common. Flowers usually violet, said to be variable. *NAK 21, 144 & 318* (Plains and hills).

BARLERIA Linnaeus
Sp. Pl. 636. 1753.

Armed or unarmed subshrubs. Leaves decussate, petiolate or sessile. Flower(s) solitary or in racemes, axillary; calyx-lobes 4, free, unequal, outer pair larger, enclosing the inner; corolla funnel-form or salver-form, lobes 4, spreading. Fertile stamens 2, subexerted; staminodes 0 or 2; ovary 2-or 4-ovuled; disc cupular, half embracing the ovary. Seeds 2 or 4, compressed.

1. Plants armed; leaf apex spine-tipped; flowers yellow. .... 3. **B. prionitis**
1. Plants unarmed; leaf apex not spine-tipped; flowers other than yellow.
    2. Flowers 1-4, in axillary lax cymes; outer calyx-lobes spinescent on margins. .... 2. **B. cristata**
    2. Flowers 8-16, in subterminal, dense racemes; outer calyx-lobes not spinescent on margins. .... 1. **B. courtallica**

1. **Barleria courtallica** Nees in DC., Prodr. 11:226.1847; Wight, Ic.t. 1529. 1850; Clarke in Hook.f., Fl. Brit. India 4:489. 1884; Gamble, Fl. Pres. Madras 1060. 1924; Rani & Matthew in Matthew, Fl. Tam. Carnatic 3(2): 1160. 1983; Mohanan & Sivad., Fl. Agasthyamala 505.2002. *Ven-kurinji.*

Subshrubs. Leaves elliptic to broadly elliptic-lanceate, 4-20 by 1.5-7 cm, base acute to attenuate, apex acuminate, thick-chartaceous. Racemes subterminal; bracts linear-lanceate; calyx-lobes 4, glandular pubescent, outer 2 elliptic-obovate, prominently 3-5-nerved, apex shortly 2-fid; corolla bluish, funnel-form, c. 3.5 cm ϕ; fertile stamens 2; ovary sericeous; ovules 4. Capsule elliptic, 1.5 by 0.7 cm. Seeds 4, orbicular, pubescent.

Fl. & Fr.: Dec.-Feb. *Distr.*: Peninsular India. Among moist evergreen forest undergrowth. Locally abundant, but generally not common. *NAK 310* (Upper Moozhiar, c. 550 m), *1457* (Armpa, c. 450 m).

Decoction of root administered to cure rheumatism and pneumonia. Leaf juice boiled with oil and applied to cure ear and head diseases.

2. **Barleria cristata** L., Sp.Pl. 636. 1753; Roxb., Fl. Ind. 3: 37. 1832; Wight, Ic. t. 453. 1841; Clarke in Hook.f., Fl. Brit. India 4: 488. 1884; Gamble, Fl. Pres. Madras 1060. 1924; Rani & Matthew in Matthew, Fl. Tam. Carnatic 3(2) : 1162. 1983.

Subshrubs. Leaves elliptic-lanceate, 4-10 by 2-4 cm, base cuneate-decurrent, apex acuminate, closely lineolate, chartaceous. Flowers 1-4, in axillary lax cymes; bracts lanceate, 1-ribbed; calyx-lobes scarious, outer 2-obovate, prominently nerved, margin spinous; corolla violet or white, funnel-form, c. 4 cm ϕ; stamens 5; ovary oblong-conic. Capsule oblong, appressed hairy. Seeds 4.

Fl. & Fr.: Dec.-Feb. *Distr.*: India, Myanmar, Indo-China, S. China. Philippines. Occasionally run wild. Grown in gardens. *NAK 114* (Adoor, c. 30 m).

3. **Barleria prionitis** L., Sp. Pl. 636. 1753; Roxb., Fl. Ind. 3:36. 1832; Wight, Ic.t. 452. 1841; Clarke in Hook.f., Fl. Brit. India 4:482. 1884; Gamble, Fl. Pres. Madras 1058. 1924; Rani & Matthew in Matthew, Fl. Tam. Carnatic 3(2): 1166. 1983; Nicols., Suresh & Manilal, An Interpr. Hort. Malab. 37. 1988. *Coletta-veeta* Rheede, Hort. Malab. 9:77-78, t.41.1689. *Chen-mulli.*

Armed subshrubs. Leaves ovate-lanceate, 4-9 by 1-3.5, base attenuate, margin subentire,apex acuminate, spine-tipped, chartaceous. Flowers in cymose clusters or solitary, axillary; calyx-lobes: outer 2 obovate, spine-tipped; corolla golden yellow or orange- coloured; stamens 4, staminodes villous; ovary ovoid; ovules 2. Capsule ovoid, beaked. Seeds 2, suborbicular.

Fl. & Fr.: Nov.-Jan. *Distr.:* Pantropical. Grown in gardens, occasionally run wild. *NAK 1054* (Tiruvalla, c. 10 m).

Ash of burnt plant mixed with water and boiled with rice into *canjee* (Porridge) and given as a remedy in dropsy and cough.

DICLIPTERA A.L. Jussieu
Ann. Mus. Natl. Hist. Nat. 9: 267.1807 (nom. cons.).

Dicliptera cuneata Nees in Wall., Pl. Asiat. Rar. 3:111, 1832; Wight, Ic.t. 1552. 1850; Clarke in Hook.f., Fl. Brit. India 4: 552. 1885; Gamble, Fl. Pres. Madras 1073. 1924; Matthew, Ill. Fl. Tam. Carnatic t. 524 1982; Rani & Matthew in Matthew, Fl. Tam. Carnatic 3(2) : 1170. 1983.

Woody herbs. Leaves broadly elliptic-ovate, base cuneate, apex acuminate, lateral veins c.10 pairs, prominent, chartaceous. Flowers in cymes, axillary; bracts of the cymes linear; floral bracts elliptic-ovate, 2 by 1 cm; bracteoles linear; calyx-lobes subequal, shortly connate; corolla pink, bilipped; stamens 2, exerted; ovary oblong; ovules 4. Capsule ellipsoid. Seeds 4, orbicular-ovate.

Fl. & Fr.: Nov.-Feb. *Distr.*: Peninsular India. Less common. Among forest undergrowth of upper hills. *CNM 59580* (Upper Moozhiar, c. 500 m).

DIPTERACANTHUS C.G.D. Nees in Wallich, Pl. Asiat. Rar. 3:75, 81. 1832.

Dipteracanthus prostatus (Poiret) Nees in Wall., Pl. Asiat. Rar. 3:81. 1832; Matthew, Ill. Fl. Tam. Carnatic t. 525. 1982; Rani & Matthew, Fl. Tam. Carnatic 3(2) 1171. 1983. *Ruellia prostrata* Poiret in Lam., Encycl. 6:349. 1804; Bedd., Ic. t. 282. 1872; Clarke in Hook.f., Fl. Brit. India 4: 411. 1884; Gamble, Fl. Pres. Madras 1017. 1924.

Diffuse woody herbs. Leaves ovate-elliptic to deltoid, 2-6 by 1.5-3 cm, base truncate or acute, apex subacute,strigose, thick-chartaceous. Flower(s) solitary, axillary; calyx-lobes 5, linear-subulate; corolla bluish, c. 15 mm ϕ, funnel-form, lobes 5, subequal, spreading; stamens 4, didymous, included; ovary pubescent; ovules 12-16. Capsule c. 1.5 cm long, oblong cylindric. Seeds orbicular, compressed.

Fl. & Fr.: Most of the seasons. *Distr.*: India. Common. In plains. *NAK 1779* (Thekkuthode, c. 50 m).

Juice of leaves boiled with copal and used to cure gonorrhoea.

ECBOLIUM S. Kurz J. Asiat. Soc. Bengal, Pt. 2, Nat. Hist. 40: 755. 1875.

Ecbolium viride (Forssk.) Alston in Trim., Handb. Fl. Ceylon 6. 229. 1931; Matthew, Ill. Fl. Tam. Carnatic t. 527. 1982; Rani & Matthew in Matthew, Fl. Tam. Carnatic 3(2) : 1174. 1983; Manilal, Fl. Silent Valley 203. 1988; Nicols.,Suresh & Manilal, An Interpr. Hort. Malab. 38. 1988. *Justicia viridis* Forsskal, Fl. Aegypt. Arab. 5. 1774. *Ecbolium linneanum* Kurz, J. Asiat. Soc. Bengal. 2:75. 1871; Wight, Ic. t. 463. 1841; Clarke in Hook.f., Fl. Brit. India 4 : 544. 1885; Gamble, Fl. Pres. Madras 1074. 1924. *Carim curini* Rheede, Hort. Malab. 2:31-32, t. 20. 1679. *Odiya-Madantha.*

Subshrubs. Leaves elliptic-lanceate to elliptic-ovate, 6-14 by 1.5-3.5 cm, base attenuate, apex gradually acute, lateral veins 6-8 pairs, curved-ascending, glabrous, thick-chartaeous. Flowers in dense terminal and axillary spikes; bracts closely imbricate, lanceate, 2 by 0.8 cm, margin ciliolate; calyx-lobes 5, subequal; corolla bluish-green, tube c. 3 cm long, lobes 5, 2-lipped; lower lip spreading with 2 wing-like lateral lobes; stamens 2, exerted; ovary subglobose; ovules 2. Capsule ovoid, compressed. Seeds 2, orbicular-ovate, compressed, tuberculate.

Fl. & Fr.: Nov.-Feb. *Distr.*: Africa, Sri Lanka, India, Malaya, Nepal. Less common. *NAK 1527* (Aluvamkudi, c. 250 m).

Roots used to cure rheumatism and jaundice. Decoction of leaves used for stricture.

ERANTHEMUM Linnaeus Sp. Pl. 9. 1753.

Eranthemum capense L., Sp. Pl. 9. 1753; Matthew, Ill. Fl. Tam. Carnatic t. 62. 1982; Rani & Matthew, Fl. Tam. Carnatic 3(2): 1176. 1983. *E. montanum* Roxb., Fl. Ind. 1:110. 1820; Wight, Ic. t. 466. 1841; Gamble, Fl. Pres. Madras 1025. 1924. *Justicia montana* Roxb., Pl. Corom. 176. 1805. *Daedalacanthus montanus* T. And. in Thw., Enum. Pl. Zeyl. 229. 1860; Clarke in Hook.f., Fl. Brit. India 4: 421. 1884.

Subshrubs. Leave elliptic-lanceate, 6-16 by 2.5-5.5 cm, base attenuate, margin subentire, apex acuminate, densely lineolate, chartaceous. Flowers in dense terminal, rarely axillary spikes, glandular pubescent; bracts and bracteoles linear-lanceate; calyx-lobes 5, equal, shortly connate, glandular-hispid; corolla bluish, 1.5 cm ϕ, salver-form, lobes 5, subequal; stamens 2, exerted; staminodes 2, basally connate to the filament; ovary cylindric; style sparsely hairy; ovules 4. Capsule cylindric, c.15 mm long. Seeds 4, discoid.

Fl. & Fr.: Jan-Mar. *Distr.:* Sri Lanka, Peninsular India. Locally abundant. Among forest under growth of upper hills. *NAK 267* (Upper Moozhiar), *1411* (Mannarappara, c. 250 m) & *2271* (Moozhiar, c. 250 m).

GYMNNOSTACHYUM C.G.D. Nees in Wallich, Pl. Asiat. Rar. 3: 76, 106. 1832.

Scapigerous herbs or subshrubs. Leaves entire. Flowers subsessile in axillary cymes or in terminal scapose panicle; calyx-lobes 5, basally connate; corolla tubular or campanulate, 2-lipped, lobes 5, subequal; stamens 2, included; ovary 2-locular; ovules usually 2 rows per locule. Capsule linear. Seeds 16-24, retinacula long, curved.

1. Scapigerous herbs; cymes scapose; leaves radical. ....2. **G. polyanthum**
1. Non- scapigerous subshrubs; cymes mostly axillary; leaves cauline. ....1. **G. canescens**

1. **Gymnostachyum canescens** (Nees) T. And., J. Linn. Soc. Bot. 9:505. 1867; Clarke in Hook.f., Fl. Brit. India 4: 509. 1884; Gamble, Fl. Pres. Madras 1053. 1924. *Cryptophragmium canescens* Nees in Wall., Pl.Asiat. Rar. 3:100. 1832; Wight, Ic. t. 1495. 1849.

Subshrubs; branchlets glandular-hairy. Leaves oblanceate or elliptic-obovate, 9-15 (20) by 4.5-6 cm, base attenuate, decurrent, margin denticulate, apex acute to shortly acuminate, lateral veins 10-12 pairs, parallel, prominent, thick-chartaceous. Cymes 3-branched, axillary and terminal, to 6 cm long, glandular pubescent. Flowers dense; corolla yellowish. Capsule not collected.

Fl.: December. *Distr.*: Western Ghats. Very rare. Among forest undergrowth. *NAK 287* (Upper Moozhiar, c. 550 m).

2. **Gymnostachyum polyanthum** Wight, Ic. t. 1494. 1849; Clarke in Hook.f., Fl. Brit. India 4: 508. 1884; Gamble, Fl. Pres. Madras 1053. 1924.

Scapigerous herbs; stem very short, puberulous. Leaves elliptic-ovate, 4-12 by 2-5.5 cm, base sublyrate, decurrent, margin subentire, apex subacute, non lineolate, chartaceous; petiole c. 3 cm long, scape to 25 cm long, sharply 4-angled (4-winged). Cymes condensed, forming false raceme; corolla pale purple. Capsule oblong, cylindric, c. 2 cm long. Seeds c. 14, punctate.

Fl. & Fr.: Apr. Jul. *Distr.*: Sri Lanka, India. Common. *NAK 1640* (Tiruvalla, c. 10 m).

Leaf juice used as a remedy in flatulence, diarrhoea and fever.

HYGROPHILA R. Brown Prodr. 479. 1810.

Perennial herbs, often armed; branchlets 4-gonous. Leaves decussate, petiolate or subsesile. Flowers in axillary sessile whorls around the nodes; bracts and bracteoles lanceate; calyx-lobes 5, subequal, partly connate; corolla shortly tubular, 2-lipped, lobes 5, upper 2-lobed, lower lip 3-lobed, deflexed, palate, crested; stamens 4, exerted; ovary cylindric-oblong; ovules 4 or more per locule. Capsule cylindric, oblong. Seeds orbicular, flattened.

1. Plants armed; calyx-lobes 4; ovules 8. ....2. **H. schulli**
1. Plants unarmed; calyx-lobes 5; ovules 16 or more. ....1. **H. ringens**

1. **Hygrophila ringens** (L.) R. Br. ex Steud., Nom. 1:418. 1821; Nicols., Suresh & Manilal, An Interpr. Hort. Malab. 39. 1988. *Ruellia ringens* L., Sp.Pl. 635. 1753. *Ruellia salicifolia* Vahl, Sym. Bot. 3;84.1794. *Hygrophila salicifolia* (Vahl) Nees in Wall., Pl. Asiat. Rar. 3: 81. 1832; Wight, Ic. t. 1490. 1849; Clarke in Hook.f., Fl. Brit. India 4: 407. 1884; Rani & Matthew in Matthew, Fl. Tam. Carnatic 3(2) : 1180. 1983. Roxb., Fl. Ind. 3: 50.1832. *Hygrophila angustifolia* auct., non R. Br., 1810: Gamble, Fl. Pres. Madras 1016. 1924. *Nir-schuli* Rheede, Hort. Malab. 2:89-90, t. 46. 1679.

Herbs, often branched, rooting at lower nodes. Leaves narrowly oblong to oblong-lanceate, 4-9 by 0.7-1 cm, base ± attenuate, apex gradually acute, chartaceous; calyx-lobes unequal; corolla white with pink lines, glandular, pubescent without; ovules 8-12 per locule. Capsule c. 15 mm long. Seeds 8-12.

Fl. & Fr.: Jan.-Mar. *Distr.*: Sri Lanka, India, China, Indo-China, Japan, Malaya. Common. Near rivulets, on marshy areas. *NAK 430 & 4479* (Maniyaar, c. 70 m).

2. **Hygrophila schulli** (Ham.) M.R. & S.M. Almeida, J. Bombay Nat. Hist. Soc. 83 (Suppl.) : 221.1986; Nicols., Suresh & Manilal, An Interpr. Hort. Malab. 40. 1988. *Bahel suchulli* Ham., Trans. Linn. Soc. London 14 : 289.1825. *Barleria longifolia* L. in Torner, Cent. Pl. II:22.1756, non *Hygrophila longifolia* Nees, 1847. *Asteracantha longifolia* (L.) Nees in Wall., Pl. Asiat. Rar. 3:90.1832. *Hygrophila spinosa* T. And. in Thw., Enum. Pl. Zeyl. 225.1860; Clarke in Hook.f., Fl. Brit. India 4:408.1884; Matthew, Ill. Fl.Tam. Carnatic t.530. 1982; Rani & Matthew in Matthew, Fl. Tam. Carnatic 3(2) : 1178. 1983. *Bahel-schulli* Rheede, Hort. Malab. 2 : 87-88, t. 45.1679.

Armed wooly herbs. Leaves appear whorled, outer pair larger, lanceate, 5-10 by 1.5-2.5 cm, base cuneate, margin subentire, apex acute, scabrous; subsessile, thorns 2-3 cm long, straight or curved; calyx-lobes unequal; corolla purple; lower lip with an yellow plate; ovules 4 per locule. Capsule c. 1 cm long. Seeds 4, orbicular.

Fl. & Fr.: Dec.-Apr. *Distr.*: India, Myanmar, Indo-China, Malaya. Locally abundant. Gregarious herbs forming dense populations in marshy wet lands. *NAK 2868* (Parakkode, c. 20 m).

Decoction of the whole plant used as a remedy in rheumatism, diseases of the genito-urinary tracts.

INDONEESIELLA Sreemadhavan
Phytologia 16:466.1968.

Indoneesiella echioides (L.) Sreemadh., Phytologia 16:466.1968; Rani & Matthew in Matthew, Fl. Tam. Carnatic 3(2) : 1181.1983; Nicols., Suresh & Manilal, An Interpr. Hort. Malab. 40. 1988; Mohanan & Sivad., Fl. Agasthyamala 508.2002. *Justicia echioides* L., Sp. Pl. 16. 1753; Roxb., Fl. Ind. 1:119.1820. *Andrographis echioides* (L.) Nees in Wall., Pl. Asiat. Rar. 3:117.1832, p.p.; Wight, Ic. t. 467.1841; Clarke in Hook.f., Fl. Brit. India 4:505.1884; Gamble, Fl. Pres. Madras 1051. 1924. *Neesiella echioides* (L.) Sreemadh., Phytologia 15:271.1967. *Pee-tumba* Rheede, Hort. Malab. 9 : 87, t. 45. 1689. *Gopuram-thangi.*

Branched erect herbs. Leaves oblong to oblanceate, 1-4 by 0.5-1 cm, base narrowed or rounded, apex subacute, pubescent, subsessile, chartaceous, drying dark greenish. Racemes usually unbranched, unilateral. Flowers subsessile; calyx-lobes 5, subequal, shortly connate; corolla white with brown tinge, tubular, slightly ventricose, 2-lipped, lobes 5, unequal, 2+3; stamens 2, exerted; anther cells unequal, 1 cell bearded below; ovary ellipsoid; ovules 4. Capsule ellipsoid. Seeds 4, ovoid.

Fl. & Fr.: Apr.-Jul. *Distr.*: Sri Lanka, India. Common on exposed sunny areas in plains. *NAK 1640* (Tiruvalla, c. 10 m).

JUSTICIA Linnaeus
Sp. pl. 15. 1753.

Diffuse herbs or subshrubs. Leaves entire, often lineolate, petiolate. Flowers aggregated in dense, linear or stout spikes; bracts various; bracteoles 2 or 0, linear-lanceate; calyx-lobes 4 or 5, equal or unequal, nearly free, margin scarious; corolla pinkish or white with purple streaks, ± tubular, lobes 5, 2-lipped, 2+3; stamens 2, subexerted; anthers unequal, appendaged; ovary oblong-cylindric; ovules 4. Capsule with a basal solid beak. Seeds 4; ovate, orbicular or subtrigonous.

1. Shrubs or subshrubs.
  2. Bracts scarious with green nerves, foliaceous. .... 1. **J. betonica**
  2. Bracts otherwise, not foliaceous.
    3. Erect shrubs; branches stout; flowers c. 3 cm long, in false verticels on peduncle. .... 6. **J. santapaui**
    3. Suberect subshrubs; branches slender; flowers c. 1 cm long, in pairs on peduncle. .... 7. **J. wynaadensis**
1. Herbs (woody or slender).
  4. Bracteoles shorter than calyx-lobes.
    5. Bracts ovate; calyx-lobes nearly glabrous; erect woody herbs. .... 2. **J. diffusa**
    5. Bracts lanceate; calyx-lobes conspicuously ciliate; prostrate herbs. .... 5. **J. prostrata**
  4. Bracteoles as long as calyx-lobes.
    6. Bracts lanceate; indumentum simple; of spikes short and dense. .... 4. **J. procumbens**
    6. Bracts obovate or oblanceate, indumentum (hairs) jointed; of spikes long and lax. .... 3. **J. japonica**

1. **Justicia betonica** L., Sp. Pl. 15.1753; Roxb., Fl. Ind. 1:29.1820; Clarke in Hook.f., Fl. Brit. India 4:525.1885; Gamble, Fl. Pres. Madras 1078.1924; Matthew, Ill. Fl.Tam. Carnatic t.534.1982, Manilal, Fl.Silent Valley 205.1988; Rani & Matthew, Fl.Tam. Carnatic 3(2): 1184.1983; Nicols., Suresh & Manilal, An Interpr. Hort. Malab.41.1988. *Bem-curini* Rheede, Hort. Malab.2:33-34, t. 21.1679. *Valla-kurinji.*

Shrubs. Leaves ovate-lanceate, 8-15 by 3.5-6 cm, base and apex acuminate, margin subentire, lateral veins c. 8 pairs, prominent. Spikes terminal or subterminal; bracts and bracteoles leafy, scarious with green nerves; corolla white with purple lines; stamens 5, calyx-lobes 5; corolla white with purple lines; stamens curved. Capsule c. 1.5 cm long, pubescent.

Fl. & Fr.: Jan.-Jul.- (April) September. *Distr.*: Paleotropics. Common. Among thickets along forest borders. *NAK 247* (Konni, c.100 m), *940* (Plapally, c.450 m).

2. **Justicia diffusa** Willd., Sp. Pl. 1:87.1797; Clarke in Hook. f., Fl. Brit. India 4:538.1855; Gamble, Fl. Pres. Madras 1081.1924; Rani & Matthew, Fl. Tam. Carnatic 3(2): 1185.1983; Manilal, Fl.Silent Valley 205.1988.

Woody herbs. Leaves elliptic-ovate, 2-6 by 1.5-3.5 cm, base subrounded or cuneate, decurrent, apex acute to very shortly acuminate, irregularly lineolate; petiole to 3 cm long. Spike terminal and axillary, linear; bracts and bracteoles ovate, scarious; calyx-lobes 4; corolla pale pink. Capsule minutely tapering towards apex. Seeds minutely papillose.

Fl. & Fr.: Aug.-Oct. *Distr.*: Sri Lanka, India, Myanmar, Thailand. Less common. On bare hill slopes, streamsides, etc. *NAK 2249* (Mannarappara, c.250 m).

3. **Justicia japonica** Thunb., Fl. Jap..20.1784. *J. simplex* D. Don, Prodr. 118.1825; Clarke in Hook. f., Fl. Brit. India 4:539.1885; Gamble, Fl.Pres. Madras 1080.1924; Rani & Matthew in Matthew, Fl. Tam. Carnatic 3(2): 1189.1983. *Rostellularia japonica* (Thunb.) Ellis, Bull. Bot. Surv. India 22:196.1980 (82).

Subdecumbent, diffuse or erect herbs. Leaves ovate or elliptic-ovate, 8-2.5 by 0.16-1.5 cm, base rounded, decurrent, apex acute, lateral veins c. 4 pairs, subparallel, transversely lineolate; bracts ovate, prominently green-veined, closely short ciliate; clayx lobes 4; corolla pink. Capsule c. 4 mm long, apex hairy. Seeds ovate.

Fl. & Fr.: Most of the seasons. *Distr.*: Africa, Sri Lanka, India, Myanmar, Thailand, Malaya. On bare slopes in hills. *NAK 773, 806 1982* (Plappally, Ranni).

4. **Justicia procumbens** L., Sp. Pl. 15.1753; Clarke in Hook.f., Fl. Brit. India 4:539.1885; Gamble, Fl. Pres. Madras 1080.1924; Rani & Matthew in Matthew, Fl. Tam. Carnatic 3(2): 1188.1983; Nicols., Suresh & Manilal, An Interpr. Hort. Malab.41.1988 Mohanan & Sivad., Fl. Agasthyamala 509.2002. *Rostellularia procumbens* (L.) Nees in Wall., Pl. Asiat.Rar. 3:101.1832, non Gaertn.,1807. *Tsjerutardevel* Rheede, Hort. Malab. 10:187, t. 94.1690.

Diffuse or erect herbs. Leaves linear-lanceate, 2.5-7 by 0.5-1 cm, base and apex acute, strigose or glabrous. Spikes terminal, 1-2.5 cm long; bracts lanceate or elliptic-ovate; bracteoles linear-lanceate; calyx-lobes 4; corolla pink. Capsule oblong, 5 mm long, sparsely pubescent. Seeds orbicular.

Fl. & Fr.: Sep.-Mar. *Distr.*: Sri Lanka, India, Malaya, Australia. Common. On bare slopes of hills. *NAK 38* (Kakki hills, c.1000 m), *820* (Ranni, c.250 m).

Juice of leaves dropped into eyes as a remedy in ophthalmia.

5. **Justicia prostrata** (C.B. Clarke) Gamble, Fl. Pres. Madras 1081.1924; Rani & Matthew in Matthew, Fl. Tam. Carnatic 3(2): 1188.1983. *J. diffusa* Willd. var. *prostrata* C.B. Clarke in Hook. f., Fl. Brit. India 4:538.1885.

Prostrate or diffuse herbs. Leaves ovate to suborbicular, 1-2 by 0.5-1 cm, base rounded, apex acute or obtuse. Spikes linear, terminal or subterminal; bracts and bracteoles lanceate; calyx-lobes 4; corolla pink. Capsule oblong, c. 4 mm long. Seeds papillose.

Fl. & Fr.: Sep.-Jan. *Distr.*: Sri Lanka, India. Common. On exposed areas of plains. *NAK 1996* (Tiruvalla, c.10 m).

6. **Justicia santapaui** Bennet, J. Bombay Nat. Hist. Soc. 67:358.1970. *J. montana* (Nees) Wall. ex T. And., J. Linn. Soc. Bot. 9:509.1867, non Roxb., 1805; Clarke in Hook.f., Fl. Brit. India 4:525.1885; Gamble, Fl. Pres. Madras 1078. 1924. *Hemichoriste montana* Nees in Wall., Pl.Asiat. Rar. 3:102.1832; Wight, Ic.t. 1538.1850.

Justicia andersonii Ramamoorthy in Sald. & Nicols., Fl. Hassan Dist. 551.1976; Manilal, Fl. Silent Valley 204.1988.

Erect stout bushy herbs. Leaves ovate-lanceate, 15-25 by 6-12 cm, base attenuate, apex acuminate, lateral veins c. 8 pairs. Racemes terminal, stout, often branched, to 30 cm long; bracts and bracteoles elliptic-lanceate. Flowers in false verticels; calyx-lobes 5; corolla white with purple spots, c. 3 cm long. Capsule clavoid, c. 2.5 cm long. Seeds 4.

Fl. & Fr.: Jan.-Apr. *Distr.*: Peninsular India. Not common. Along moist forest floors in hills. *NAK 519* (Plappally, c.450 m), *NAK 2251* (Arampa, c. 500m).

7. **Justicia wynaadensis** (Nees) Heyne ex T.And., J. Linn. Soc. Bot. 9:515.1867; Clarke in Hook.f., Fl.Brit.India 4:533.1885; Gamble, Fl. Pres. Madras 1079.1924. *Gendarussa wynaadensis* Nees in Wall., Pl. Asiat. Rar. 3:105.1832. *Adhatoda wyanaadensis* (Nees) Nees in DC., Prodr. 11:406.1847; Wight, Ic.t. 1545.1850.

Erect slender bushy herbs with weak branches. Leaves elliptic-lanceate, 8-14 by 2-5 cm, base and apex acuminate, lateral veins 8-10 pairs. Racemes axillary and terminal, slender; bracts and bracteoles linear lanceate. Flowers in pairs; calyx-lobes 5; corolla white with pink blotches. Capsule clavoid, 4-gonous, c.1 cm long, villous. Seeds tuberculate.

Fl. & Fr.: Sep.-Mar. *Distr.*: Peninsular India. Not common. On forest floors. *NAK 2627* (Upper Moozhiar, c.500 m).

Note : **Justicia adhatoda** L. is cultivated and rarely run wild. *NAK 2805* (Kalleli, c.100 m).

LEPIDAGATHIS Willdenow
Sp. Pl. 3(1) : 400.1800

Lepidagathis incurva Buch.-Ham. ex D.Don, Prodr. Fl. Nepal. 119.1825, var. **mucronata** (Nees) Clarke ex Cooke, Fl. Pres. Bombay 2:397.1904. *L. mucronata* Nees in Wall., Pl. Asiat. Rar. 3:95. 1832. *L. hyalina* Nees var. *mucronata* (Nees) Clarke in Hook.f., Fl. Brit. India 4:521.1885; Gamble, Fl. Pres. Madras 1068. 1924; Manilal, Fl. Silent Valley 206.1988.

Erect herbs. Leaves elliptic-lanceate, oblanceate or ovate- lanceate, 4-10 by 2-3.5 cm, base rounded and decurrent or attenuate, margin faintly crenate, apex acute, lateral veins 8-10 pairs, prominent, hispid or glabrescent, chartaceous; petiole 1 4 cm long. Flowers sessile, aggregated in condensed terminal and axillary villous spikes; bracts elliptic lanceate, long mucronate, 1-nerved, coriaceous, scarlous; calyx lobes 5, uncqual; corolla white with purple or brown spots, 2-lipped, 2+3; stamens 4, included, anther-cells unequal, minutely ciliate; ovary globose; ovules 2 per locule. Capsule ellipsoid, apically ciliate. Seeds flattened, hairy.

Fl. & Fr.: Sep.-Mar. *Distr.*: India, Myanmar, S. China. Locally abundant. On waysides in upper hills. *NAK 269, 445, 1269, 1474* (Kakki hills, c. 1100 m).

Note : Specimens of one collection (*NAK 2307* - Aikadu, c. 30 m) were diffuse plants and are rooting at lower nodes and with all ovate, shortly petiolate leaves. They are collected from an open barren land and further collections are needed for knowing the range of variation.

NELSONIA R. Brown
Prodr. 480.1810.

Nelsonia canescens (Lam.) Spreng., Syst. 1:42.1842; Bremek., Reinwardtia 3:248.1955; Santapau & Wagh, Bull. Bot. Surv. India 5: 107.1964. *Justicia canescens* Lam., Encyl. 1:14.1783. *Nelsonia campestris* R. Br., Prodr. 1:481.1810; Clarke in Hook.f., Fl. Brit. India 4:394. 1985; Gamble, Fl. Pres. Madras 1010.1924.

Diffuse herbs with trailing branches; young parts softly villous. Leaves in whorls of 3, or upper decussate, oblanceate, 6-10 by 3-4.5 cm, base cuneate, margin subentire, apex obtuse, lateral veins c. 8 pairs, puberulous, thin-chartaceous. Flowers in dense broad spikes, axillary and terminal; bracts ovate, closely imbricate, densely villous; calyx-lobes 4, unequal; corolla bluish, lobes 5, 2-lipped; stamens 2, exerted, anther-cells 2, divergent; ovary 2-locular; ovules 8-10 per locule. Capsule oblong. Seeds subglobose, granular.

Fl. & Fr.: Feb.-Apr. *Distr.*: Africa, Sri Lanka, India, China, Australia. Rare. In moist deciduous forests. *NAK 476* (Maniyaar, c. 70 m).

NILGIRIANTHUS Bremekamp
Verh. Kon. Ned. Akad. Wetensch. Afd. Natuurk., Tweede Sect. 41 (1) : 171.1944.

Subshrubs or shrubs. Leaves thick-chartaceous to coriaceous, penninerved, densely lineolate. Inflorescence axillary and/or terminal dense spikes or heads, often subtended by leafy bracts; floral bracts and bracteoles various; calyx-lobes 5, subequal, connate at base; corolla tubular, ventricose above, lobes 5, subequal; stamens 4, monadelphous, included or exerted; ovary 2 or 4-ovuled; style filiform. Capsule oblong. Seeds orbicular ellipsoid.

1. Flowers in heads, or slender elongate or strobiloid spikes; stem usually terete.
   2. Flowers in heads; vegetative parts and inflorescence fulvous hairy; peduncle flattened. .... 5. **N. punctatus**
   2. Flowers in slender elongate or strobiloid spikes; vegetative parts and inflroescence not so; peduncle terete.
      3. Flowers in slender elongate unbranched spikes.
         4. Leaves narrowly lanceate, lateral veins faint. .... 2. **N. beddomei**
         4. Leaves elliptic-lanceate, lateral veins prominent. .... 3. **N. ciliatus**
      3. Flowers in strobiloid branched spikes;peduncle often with 2-3 pairs of bracts. .... 4. **N. heyneanus**
1. Flowers in dense, broad spikes; stem 4-gonous (4-winged), grooved. .... 1. **N. barbatus**

1. **Nilgirianthus barbatus** (Nees) Bremek., Mat. Mon. Strob. 172.1944; Santapau, Bot. Mem. Univ. Bombay 2:39. 1951; Sald. & Nicols., Fl. Hassan Dist. 555.1976; Manilal, Fl. Silent Valley 208.1988. *Strobilanthes barbatus* Nees in Wall., Pl.Asiat. Rar. 3:85.1832; Bedd., Ic.t.212.1874; Clarke in Hook.f., Fl. Brit. India 4:437.1885; Gamble, Fl. Pres. Madras 1037.1924.

Shrubs; stem 4-angled, often 4-winged. Leaves elliptic-lanceate, 12-18 by 5-8 cm, base acute, decurrent, margin crenulate, apex acuminate, lateral veins c.12 pairs; petiole auricled at base. Flowers in dense capitate spikes; bracts ovate, acuminate, ciliate; bracteoles linear; calyx divided near to base; corolla white. Capsule c.1 cm long. Seeds 2, ellipsoid.

Fl. & Fr.: Sep.-Dec. *Distr.:* Peninsular India. Locally abundant. On upper hills. Flowering is said to be once in about 7 years. *EV 80531* (MH) (Kadambupara), Konni R.F.

2. **Nilgirianthus beddomei** Bremek., Mat. Mon. Strob. 172.1944. *Srobilanthes adenophorus* sensu Bedd., Ic. t. 225.1868-1874, non Nees, 1847; Clarke in Hook.f., Fl. Brit. India 4:440.1885; Gamble, Fl. Pres. Madras 1040.1924.

Shrubs. Leaves narrowly lanceate, 10-16 by 2.5-4 cm, base and apex acuminate, marign subentire, lateral veins c. 6 pairs, faint; bracts oblong-ovate, obtuse; bracteoles ligule- like; calyx-lobes connate to middle; corolla pale blue. Capsule not seen.

Fl.: September. *Distr.*: Western Ghats. Rare. In shola forests. *KV 48373* (MH) (Pamba, 1075 m). Fruiting specimens were not available.

Note : This species closely ressembles with **N. ciliatus** (Nees) Bremek., and also flowers annually.

3. **Nilgirianthus ciliatus** (Nees) Bremek., Mat. Mon. Strob. 172. 1944. *Strobilanthes ciliatus* Nees in Wall., Pl. Asiat. Rar. 3:85.1832; Bedd., Ic. t. 211. 1868-1874; Clarke in Hook.f., Fl. Brit. India 4:439.1885; Gamble, Fl. Pres. Madras 1039.1924.

Subshrubs. Leaves elliptic-lanceate, 7-15 by 3.5-5.5 cm, base attenuate, margin dentate-serrate, apex long-acuminate, lateral veins c. 8 pairs, prominent; bracts obovate, subacute; bracteoles oblanceate; calyx-lobes connate near to middle; corolla white with pink blotches on lower lip. Capsule narrowly obovoid, c. 1 cm long, pubescent. Seeds orbicular, hairy.

Fl. & Fr.: Nov.-Feb. *Distr.*: Peninsular India. Common. In evergreen forests, usually along streamsides. *NAK 1410* (Mannarappara, R.F., c. 250 m).

4. **Nilgirianthus heyneanus** (Nees) Bremek., Mat. Mon. Strob. 173.1944. *Strobilanthes heyneanus* Nees in Wall., Pl. Asiat. Rar. 3:85.1832; Clarke in Hook.f., Fl. Brit.India 4:443.1885; Gamble, Fl. Pres. Madras 1041.1924. *S. rugosus* Wight, Ic.t. 1619.1850.

Woody herbs or subshurbs. Leaves elliptic-lanceate or ovate-lanceate, 5-13 by 2.5-6 cm, base cuneate, or acute, decurrent, margin crenate-serrate, apex acuminate, lateral veins c. 8 pairs, prominent. Flowers in strobiloid spikes; peduncle often subtended by 2-3 pairs of sterile bracts; calyx lobes linear-lanceate; corolla yellow. Capsule oblong cylindric, c. 1 cm long. Seeds 4, Compressed.

Fl. & Fr.: Nov.-Feb. *Distr.*: Peninsular India. The common species of the genus in the district. On moisty forest floors. *NAK 291* (Moozhiar), *1477* (Arampa, c. 400 m).

5. **Nilgirianthus punctatus** (Nees) Bremek., Mat. Mon. Strob. 173. 1944; Manilal, Fl. Silent Valley 209.1988. *Ruellia punctata* Nees in DC., Prodr. 11:147. 1847; Wight, Ic.t. 1563. 1850. *Strobilanthes anceps* Nees in Hook., Comp. Bot. Mag. 2:312.1836; Bedd., Ic.t. 204.1874; Clarke in Hook.f., Fl. Brit. India 4:442,1885; Gamble, Fl. Pres. Madras 1038.1924.

Shrubs, vegetative parts and inflorescence glandular viscid. Leaves ovate-lanceate, 8-12 by 3-4 cm, base obtuse, decurrent, margin subentire, apex long acuminate, lateral veins c. 6 pairs, flowers in capitate heads or condensed spikes; bracts orbicular; calyx small, free to base; corolla white. Capsule oblong, pubescent. Seeds 4.

Fl. & Fr.: Mar.-May. *Distr.*: Sri Lanka, S.W. Peninsular India. Vary rare. In evergreen forests. *CB 4926* (MH) (Plapally to Chalakkayam).

PERISTROPHE C.G.D. Nees in Wallich, Pl. Asiat. Rar. 3:77, 112.1832.

Peristrophe montana Nees in Wall., Pl. Asiat. Rar. 3:113. 1832; Wight, Ic. 1553.1850; Clarke in Hook.f., Fl. Brit. India 4:556.1885; Gamble, Fl. Pres. Madras 1084.1924; Manilal, Fl. Silent Valley 210.1988.

Erect subshrubs. Leaves elliptic-lanceate, 7-16 by 3-7 cm, base cuneate, margin entire, apex acuminate, lateral veins c.8 pairs, thinly lineolate, chartaceous. Flowers in terminal and axillary clusters; bracts decussate, foliaceous; calyx free to base, lobes 5, 2-lipped; stamens 2, anther-cells 2, lincar; ovary 2-locular; ovules 2 per locule. Capsule ellipsoid. Seeds 4, compressed, papillose.

Fl. & Fr.: Dec.-Mar. *Distr.:* Sri Lanka, Peninsular India. Locally abundant but not common in the district. Among forest undergrowth in upper hills. *NAK 2626* (Upper Moozhiar, c.550 m).

PHAULOPSIS Willdenow
Sp.Pl. 3:4, 342.1799-1800 ('Phaylopsis'); corr. K.P.J. Sprengel, Anleit. ed. 2.2:422.1817 (nom.et orth. cons.)

Phaulopsis imbricata (Forssk.) Sw., Hort. Brit. 327.1827; Rani & Matthew in Matthew, Fl. Tam. Carnatic 3(2):1194., Pl. 91.1983; Manilal, Fl. Silent Valley 210.1988; Mohanan & Sivad., Fl. Agasthyamala 514.2002. *Ruellia imbricata.* Forssk., Fl. Aeg.-Arab.113.1775. *Ruellia dorsiflora* Retz., Obs. Bot. 6:31.1791. *Micranthus oppositifolius* Wendl., Bot. Beobacht. t.39.1798; Gamble, Fl.Pres. Madras 1022.1924. *Phaulopsis parviflora* Willd., Sp. Pl. 3:342.1800; Clarke in Hook.f., Fl. Brit. India 4:417.1885. *Phaulopsis dorsiflora* (Retz.) Santapau, Kew Bull. 3:276.1948 & Bot. Mem.Univ. Bombay 2:30. 1951.

Erect or diffuse herbs. Leaves elliptic-lanceate or obliquely ovate, 3-7 by 1.5-4 cm, base obliquely acute, margin subentire, apex acute, pubescent, chartaceous. Flowers in dense 1-sided spikes, terminal and axillary; bracts foliaceous, orbicular, closely imbricate enclosing c. 3 flowers; calyx-lobes 5, unequal, free; corolla white, lobes 5, subequal, 2-lipped, 2+3; stamens 4, exerted, anther-cells divergent; ovary oblong; ovules 4. Capsule oblong-cylindric, hairy. Seeds 4, orbicular, hairy.

Fl. & Fr.: Mar.-May. *Distr.*: Peninsular India. Common. Moist hill slopes. *NAK 196* (Konni, c.100 m), *242* (Athumpumkulam, c. 50 m).

PHLEBOPHYLLUM C.G.D. Nees
in Wallich, Pl. Asiat. Rar. 3:75, 83.1832.

Phlebophyllum lawsonii (Gamble) Bremek., Mat. Mon. Strob. 279.1944; Ramach. *et al.,* J. Econ. Tax. Bot. 1:96.1980; Mohanan & Sivad., Fl. Agasthyamala 515.2002. *Strobilanthes lawsonii* Gamble, Kew Bull. 1923:374.1923 & Fl. Pres. Madras 1037.1924.

Erect shrubs, young parts and leaves beneath softly-woolly. Leaves ovate-lanceate, 8-15 by 3-5 cm, base rounded, apex long-acuminate, lateral nerves c. 10 pairs, parallel, prominent, thin-coriaceous. Flowers in interrupted slender spikes, appressed wooly, glandular hispid in fruiting stage; bracts ovate-lanceate; bracteoles small; calyx-lobes subequal, accrescent; corolla pale blue, lobes 5, spreading; stamens 2, anthers 2-celled; ovary 2-locular; ovules 2 per locule. Capsule ellipsoid. Seeds 4, hairy.

Fl. & Fr.: Sep.-Dec. *Distr.*: S.W. Peninsular India. Rare. In shola forests. *PR 79296* (MH) *NAK 2744* (Ponnambalamedu, c. 1200 m).

PSEUDERANTHEMUM Radlkofer
Sitzungsber. Math. Phys. C.L. Konigl. Bayer. Akad. Wiss. Munchen 13:282. 1883.

Pseuderanthemum malabaricum (Clarke) Gamble, Fl. Pres. Madras 1064.1924. *Eranthemum malabaricum* Clarke in Hook.f., Fl. Brit. India 4:497.1885.

Subshrubs. Leaves decussate, elliptic-lanceate, 8-12 by 2.5-4.5 cm, base cuneate, margin subentire, recurved, apex acuminate, prominently lineolate, thin-coriaceous. Flowers in terminal, often in unbranched spikes; bracteoles linear; calyx-lobes 5, nearly free; corolla white, tubular, lobes 5, unequal, spreading; fertile stamens 2; staminodes 2, anther 2-celled; disc annular; ovary 2-locular; ovules 2 per locule. Capsule long-stalked, pubescent. Seeds 4, compressed, orbicular.

Fl. & Fr.: Jan.-Mar. *Distr.*: Sri Lanka, Peninsular India. Rare. In forest undergrowth. *NAK 1476* (Kodumudi, Chittar, c. 350 m).

RHINACANTHUS C.G.D. Nees in Wallich, Pl. Asiat. Rar. 3:76, 108.1832.

Rhinacanthus nasuta (L.) Kurz, J. Asiat. Soc. Bengal 39:79. 1870; Matthew, Ill. Fl. Tam. Carnatic t.539. 1982; Rani & Matthew in Matthew, Fl. Tam. Carnatic 3(2): 1196.1983; Nicols., Suresh & Manilal, An Interpr. Hort. Malab. 42.1988. *Justicia nasuta* L., Sp. Pl. 16.1753; Roxb., Fl. Ind. 1:121.1820. *Rhinacanthus communis* Nees in Wall., Pl. Asiat. Rar. 3:109.1832; Wight, Ic.t. 464.1841; Clarke in Hook.f., Fl. Brit. India 4:541.1885; Gamble, Fl. Pres. Madras 1083.1924. *Pul-colli* Rheede, Hort. Malab. 9:135, t. 69.1689.

Subshrubs. Leaves elliptic-lanceate ot oblanceate, 7-14 by 2-6 cm, base cuneate, apex acute, lateral veins c. 8 pairs, chartaceous. Flowers subesssile, in terminal or axillary divariate panicles; bracts and bracteoles lanceate; calyx-lobes 5, basally connate; corolla white, tube slender, long, lobes 5, 2-lipped, upper linear, lower broad, 3-lobed; stamens 2, exerted, anther-cells 2, superposed; ovary 2-locular; ovules 2 per locule. Capsule linear ellipsoid. Seeds 4, orbicular, rugose.

Fl. & Fr.: Dec.-Mar. *Distr.*: Peninsular India, Sri Lanka. Locally abundant. Often in teak plantations. *NAK 1415* (Mannarappara, c. 200 m).

Leaves and root used as antidote to snake poison. Seeds used against ringworm infestation.

RUNGIA C.G.D. Nees in Wallich, Pl. Asiat. Rar. 3:77, 109, 1832.

Erect or diffuse herbs. Leaves lineolate. Flowers in dense 1-sided spikes, terminal or axillary; bracts closely imbricate in 2-4 rows, , 2-rows sterile, often dissimilar, margin scarious, bracteoles small; calyx-lobes 5, connate at base; corolla pinkish or white, lobes 5, 2-lipped, upper 2, connate, slightly emarginate, lower 3-lobed; stamens 2, subexerted, anther-cells 2, superposed, lower appendaged; ovary 2-locular; ovules 2 per locule. Capsule ovoid. Seeds 4, compressed, orbicular, tuberculate.

1. Diffuse herbs; leaves elliptic or obovate, 2 by 1 cm;
   spikes, c. 15 mm long. .... 1. **R. parviflora**
1. Erect herbs; leaves lanceate, 6 by 2 cm;
   spikes usually more than 15 mm long (rarely shorter). .... 2. **R. pectinata**

1. **Rungia parviflora** (Retz.) Nees in Wall., Pl. Asiat. Rar. 3:110.1832; Clarke in Hook.f., Fl. Brit. India 4:550.1885; Gamble, Fl. Pres. Madras 1071.1924. *Justicia parviflora* Retz., Obs. Bot. 5:9.1788. *Rungia longifolia* sensu Bedd., Ic. t. 266.1868-1874, non Nees, 1832.

Diffuse or ascending herbs. Leaves elliptic or elliptic-ovate, 2 by 1 cm, base and apex acute, lateral veins c. 3 pairs. Spikes c. 15 mm long; floral bracts glabrescent; corolla pinkish-white. Capsule ellipsoid.

Fl. & Fr.: Most of the seasons. *Distr.*: Sri Lanka, Peninsular India. Common. On exposed hill slopes. *NAK 475* (Maniyaar, c. 70 m), *2623* (Moozhiar, c. 250 m).

2. **Rungia pectinata** (L.) Nees in DC., Prodr. 11 : 459.1847, p.p.; Wight, Ic.t. 1547; Sald. & Nicols., Fl. Hassan Dist. 559.1976. 1850; Manilal, Fl. Silent Valley, 211.1988; Mohanan & Sivad., Fl. Agasthyamala 516.2002. *Justicia pectinata* L., Amoen. Acad. 4:299.1759. *Rungia parviflora* (Retz.) Nees var. *pectinata* (L.) Clarke in Hook.f., Fl. Brit. India 4:550.1885; Gamble, Fl.Pres. Madras 1071.1924.

Erect, profusely branched subshrubs. Leaves lanceate, 3-7 by 1-2.5 cm, base and apex acute, lateral veins c. 5 pairs. Spikes usually more than 15 mm long; floral bracts villous; corolla white. Capsule ellipsoid.

Fl. & Fr.: Nov.-Feb. *Distr.*: Sri Lanka, Peninsular India. Common. Along streamsides in hills. *NAK 2187* (Vayyattupuzha, c. 400 m).

STAUROGYNE Wallich
Pl. Asiat. Rar. 2:80, t.186.1831.

Erect or diffuse villous herbs. Leaves opposite or alternate, often with smaller leaves in axils. Flowers in branched or unbranched spikes; bracts and bracteoles present; calyx-lobes 5, nearly free; corolla funnel-form, lobes5, 2-lipped; stamens 4, didynamous, included; ovary 2-locular; ovules α per locule. Capsule oblong. Seeds α, smooth or pitted; retinacula.

1. Erect herbs; upper leaves alternate; seeds smooth. .... 1. **S. glauca**
1. Diffuse herbs; all leaves opposite; seeds pitted. ....2. **S. zeylanica**

1. **Staurogyne glauca** (Nees) Kuntze, Rev. Gen. Pl. 1:497.1891; Gamble, Fl. Pres. Madras 1011. 1924. *Ebermaiera glauca* Nees in DC., Prodr. 11:73.1847; Wight, Ic..t. 1488. 1849; Clarke in Hook.f., Fl. Brit. India 4:395.1885.

Erect small herbs. Leaves alternate, margin dentate, apex obtuse, submembraneous; bracts oblanceate; bracteoles linear; corolla pink. Capsule c. 6 mm long. Seeds smooth.

Fl. & Fr.: Jan.-Mar. *Distr.*: Eastern & Peninsular India. Rare. Among forest undergrowth. *NAK 548* (Thekkuthode, c. 50 m).

2. **Staurogyne zeylanica** (Nees) Kuntze, Rev. Gen. Pl. 1:497.1891; Gamble, Fl. Pres. Madras 1011.1924; Mohanan & Sivad., Fl. Agasthyamala 517.2002. *Ebermaiera zeylanica* Nees in DC., Prodr. 11:74.1847; Clarke in Hook.f., Fl. Brit. India 4:397.1885.

Diffuse herbs with trailing branches. Leaves opposite, long-petioled, ovate or obovate to oblanceate, upper leaves smaller, 1.5-8 by 1-1.35 cm, base attenuate, apex obtuse, thin-chartaceous, bracts narrowly obovate, bracteoles linear; corolla pale violet. Capsule lanceoiod, c. 7 mm long. Seeds pitted.

Fl. & Fr.: Jan.-Mar. *Distr.*: Sri Lanka, Peninsular India. Rare. Among moist forest undergrowth. *NAK 2255* (Mannarappara, c. 200 m), *2590* (Arampa, c. 400 m).

THUNBERGIA Retzius
Physiogr. Salsk. Handl. 1(3) : 163.1780 ("1776") (nom.cons.).

Climbers. Leaves decussate, basally 3-5-nerved, chartaceous. Flower(s) solitary or in pendulous racemes, axillary or terminal; bracts 2; bracteoles spathaceous, enclosing calyx, persistent; calyx with 12-16 teeth, decidous; corolla funnel-form, lobes 5; stamens 4, didynamous, anther - pairs unequal; ovary globose, 2 locular; ovules 2-per locule. Capsule subglobse, apically beaked. Seeds 4, globose.

1. Leaves pinnately nerved; flower(s) white, solitary. ....1. **T. fragrans**
1. Leaves palmately nerved; flowers yellow, in pendent racemes. ....2. **T. mysorensis**

1. **Thunbergia fragrans** Roxb., Pl. Corom. t. 67.1796 & Fl. Ind. 3:33.1832; Clarke in Hook.f., Fl. Brit. India 4:390. 1885; Gamble, Fl. Pres. Madras 1007.1924; Matthew, Ill. Fl. Tam. Carnatic t.547.1982; Rani & Matthew in Matthew, Fl. Tam. Carnatic 3(2) : 1211.1983.

Slender twiners. Leaves triangular- ovate, 5-7 by 2.5-3.5 cm, base cordate, margin coarsely dentate towards base, apex obtuse, apiculate, membranous. Flower(s) solitary; calyx-lobes c. 15, subulate; corolla white. Capsule scabrid. Seeds reticulate.

Fl. & Fr.: Dec.-Mar. *Distr.*: Sri Lanka, India, Nepal, Myanmar, Indo-China, W. China. *NAK 1361* (Pampa, c. 1100 m). A common climbing or twining herb with beautiful non-fragrant white flowers.

2. **Thunbergia mysorensis** (Wight) T. And., J. Linn. Soc. Bot. London 9 : 448.1867; Clarke in Hook.f., Fl. Brit. India 4: 393.1885; Gamble, Fl. Pres. Madras 1008. 1924; Rani & Matthew in Matthew, Fl. Tam. Carnatic 3(3): 1212.1983; Mohanan & Sivad., Fl. Agasthyamala 520.2002. *Hexacentris mysorensis* Wight, Ic.t. 871. 1844-45.

Stout twiners. Leaves lanceate, 8-16 by 3.5-5.5. cm, base cuneate, margin coarsely dentate, apex acuminate, thin-coriaceous. Flowers in terminal, pendant racemes; calxy-lobes reduced to a circular rim; corolla yellow. Capsule c. 3 cm long.

Fl. & Fr.: Nov.-Mar. *Distr.*: Peninsular India. Not common. In evergreen forests. *NAK 1176* (Moozhiar, c. 250 m), *1530* (Aluvamkudi, c. 400 m). An extensive climber with very showy flowering racemes.

Note : Recent trend is to treat the genus **Thunbergia** under a separate family **Thunbergiaceae. Thunbergia erecta**. (Benth.) And. and **T. grandiflora** (Rottl.) Roxb., are seen planted in gardens; sometimes the former runs wild.

100. VERBENACEAE

1. Stragglers; bracts forming an involucre. ....**Sphenodesme**
1. Non-stragglers (except **Premna glaberrima**); bracts not forming an involucre.
   2. Leaves digitiately 3-7-foliolate. ....**Vitex**
   2. Leaves simple.
      3. Inforescence a spike; fruiting bracts conspicuous.
         4. Calyx truncate, smooth; stigma lateral, spikes condensed. ....**Lantana**
         4. Calyx tubular, ribbed; stigma terminal; spikes elongate. ....**Stachytarphetta**
      3. Inflorescence a raceme or cyme; fruiting bracts inconspicuous or 0.
         5. Corolla regular, lobes either 4 or 6-8.
            6. Flowers purple, 4-merous; drupe (Pyrenes 4) 1-loculed. ....**Callicarpa**
            6. Flowers cream- coloured or yellowish, 6-8-merous; drupe 1,4-loculed. ....**Tectona**
         5. Corolla bilabiate, lobes 5.
            7. Stamens inserted below the tube, later exerted. ....**Clerodendrum**
            7. Stamens inserted near the middle of tube, included/subexerted.

8. Corolla-tube inflated, lobes clearly 2-lipped.
 9. Recemes paniculate; corolla yellow, c. 2 cm ϕ. ....**Gmelina**
 9. Racemes corymbose; corolla cream-coloured or pinkish-white, c. 7 mm ϕ. ....**Premna**
8. Corolla-tube cylindric, lobes obscurely 2-lipped. ....**Citharexylum**

CALLICARPA Linnaeus
Sp. Pl. 111. 1753.

Callicarpa tomentosa (L.) Murray, Syst. Veg., ed. 13, 130. 1774; Moldenke, Fedd. Repert. 40 : 106.1936 & in Dassan. & Fosberg, Rev. Handb. Fl. Ceylon 4:299.1983; Sald. & Nicols., Fl. Hassan Dist. 486.1976; Matthew, Ill. Fl. Tam. Carnatic tt. 548 & 975. 1982; Matthew & Rani in Matthew, Fl. Tam. Carnatic 3(2):1217. 1983; Manilal, Fl. Silent Valley 213.1988; Nicols., Suresh & Mani. An Interpr. Hort. Malab. 259.1988; Mohanan & Sivad., Fl. Agasthyamala 521.2002. *Tomex tomentosa* L., Sp. Pl. 118.1753. *Callicarpa lanata* L., Mant. Pl. 2:331.1771; Clarke in Hook. f., Fl. Brit. India 4:567.1885; Gamble, Fl. Pres. Madras 1092.1924. *Tondi-teregam* Rheede, Hort. Malab. 4:123-124, t. 60.1683. *Cheru-thekku, Nai-kumbil.*

Small trees; branchlets, petiole, leaves beneath and peduncle fulvous-tomentose. Leaves decussate, broadly elliptic-ovate to triangular ovate, 12-30 by 5-16 cm, base rounded, acute or truncate, margin denticulate or entire, apex acuminate, lateral veins 8-10 pairs, pubescent on midrib above, coriaceous. Flowers α, 4-merous in axillary and terminal umbellate cymes; bracts linear, deciduous; calyx cupular, lobes 4, obscure; corolla purple, c. 5 mm ϕ, lobes 4; stamens 4, exerted; ovary globose, 2-locular; ovules 2, basally axile. Drupe globose, of 4, 1-loculed pyrenes.

Fl. & Fr.: Jan.-Apr. *Distr.*: Sri Lanka, Peninsular India. Common. Along forest edges. *NAK 536, 1285, 1576 & 2436.* (Almost all hills). Morphology of leaf is highly variable.

Leaves, root and bark used in skin diseases. Decoction of roots given for treating rheumatism, fever, cough, asthma, swellings, worms, ulcers, and piles.

CITHAREXYLUM Linnaeus
Sp. Pl. 625. 1753.

Citharexylum subserratum Sw., Nov. Gen. Pl. 91.1788; Matthew & Rani in Matthew, Fl. Tam. Carnatic 3(2): 1218.1983. *C. fruticosum* L. var. *subserratum* (Sw.) Mold., Phytologia 1:17.1933 & 6:383.1958.

Medium- sized trees. Leaves elliptic-lanceate, 5-15 by 4.5-6 cm, base acute, apex obtuse, veins finely reticulate, chartaceous. Flowers in pendulous racemes; calyx cupular, 4-toothed; corolla white, lobes 5, faintly 2-lipped; stamens 4, included; ovary globose. Fruits not seen.

Fl. : Apr. *Distr.*: Argentina northwards to southern U.S.A. Planted, occasionally run wild. Rare. *NAK 2421* (Kalleli, c. 100 m).

CLERODENDRUM Linnaeus
Sp. Pl. 637. 1753

Shrubs. Leaves decussate or whorled. Flowers zygomorphic, 5-merous, aggregated in axillary and/ or terminal cymose-panicles; bracts deciduous; calyx cupular or campanulate, lobes 5, accrescent; corolla salver-form, obscrurely or distinctly 2-lipped, lobes 5, unequal; stamens 4, involute in bud, exerted later, didynamous; ovary obscurely 4-locular; ovule 1-per locule. Drupe globose or 4-lobed; pyrenes 3, supressed. Seeds oblong-pyriform.

1. Leaves cordiform, margin entire, tomentose. ....4. **C. viscosum**
1. Leaves otherwise, margin serrate or denticulate, glabrous or pubescent.
    2. Inflorescence a lax panicle; corolla single, red or bluish.
        3. Flowers reddish; leaves membranous, margins denticulate. ....1. **C. paniculatum**
        3. Flowers bluish; leaves thin-coriaceous, margin coarsely serrate. ....3. **C. serratum**
    2. Inflorescence a condensed dense panicle; corolla double, white. ....2. **C. philippinum**

1. **Clerodendrum paniculatum** L., Mant. Pl. 90. 1767 (*'paniculata'*); Clarke in Hook.f., Fl. Brit. India 4: 593.1885.

Erect subshrubs. Leaves orbicular, 7-18 by 8-19 cm, base cordate, margin faintly lobed, denticulate, apex acute, membraneous. Inflorescence a large terminal panicle; calyx deeply 5-lobed; corolla reddish. Drupe globose.

Fl. & Fr.: Sep.-Dec. *Distr.*: India, S.W. China, Java. Occasional. Along waysides. A garden species, now running wild. *NAK 729* (Perunthenaruvi, c. 100 m).

2. **Clerodendrum philippinum** Schauer in DC., Prodr. 11 : 667.1847 ("*Clerodendron*"); Howard & Powell, Taxon 17:54. 1968; Sald. & Nicols., Fl. Hassan Dist. 487.1976; Matthew & Rani in Matthew, Fl. Tam. Carnatic 3(2):1219. 1983. *Vokameria fragrans* Vent., Jard. Malm. t. 70.1804. *Clerodendrum fragrans* Willd., Enum. Hort. Berol. 659.1809; Gamble, Fl. Pres. Madras 1100.1924.

Erect subshrubs. Leaves deltoid, 8-17 by 6.5-14 cm, base truncte, margin coarsely dentate, apex acuminate, basally 3-veined, pubescent, chartaceous. Flowers in terminal condensed cymes; calyx tubular, lobes 5; corolla white, double. Fruits could not be collected.

Fl.: Feb. *Distr.*: Native of China. Occasional. On waysides in hills. A garden species, now running wild. *NAK 1517* (Gurunadhan mannu, c. 500 m).

3. **Clerodendrum serratum** (L.) Moon, Cat. Pl. Ceylon 46, No. 382.1824; Wight, Ic.t. 1472.1849; Clarke in Hook. f., Fl. Brit. India 4:592.1885; Gamble, Fl. Pres. Madras 1100. 1924 ("*Clerodendron*"); Sald. & Nicols., Fl. Hassan Dist. 487. 1976; Moldenke in Dassan. & Fosberg., Rev. Handb. Fl. Ceylon 4:417.1983; Matthew & Rani in Matthew, Fl. Tam. Carnatic 3(2):1220. 1983; Manilal, Fl. Silent Valley 214. 1988; Nicols., Suresh & Mani., An Interpr. Hort. Malab. 260.1988; Mohanan & Sivad., Fl. Agasthyamala 522.2002. *Volkameria serrata* L., Mant. Pl. 90.1767. *Tsjerou-theka* Rheede, Hort. Malab. 4:61-62, t. 29.1683.

Shrubs. Leaves decussate or in whorls, broadly elliptic-ovate, 8-25 by 5-15 cm, base oblique, acute or rounded, margin coarsely serrate, apex acuminate, thin-coriaceous. Flowers in large terminal panicles; calyx-lobes triangular; corolla bluish, 2-lipped; stamens 4. Drupe 4-lobed.

Fl. & Fr.: Jul.-Dec. *Distr.*: Sri Lanka, India, Himalaya, Myanmar. Occasional. In grasslands. Very beautiful plant, worth introducing in gardens. *NAK 2542* (Ponnambalamedu, c. 1200 m).

4. **Clerodendrum viscosum** Vent., Jard. Malm. 1:t.25. 1803; Sald. & Nicols., Fl. Hassan Dist. 488. 1976; Moldenke in Dassan. & Fosberg, Rev. Handb. Fl. Ceylon 4:473. 1983; Manilal, Fl. Silent Valley 214.1988; Nicols., Suresh & Mani., An Interpr. Hort. Malab. 260. 1988; Mohanan & Sivad., Fl. Agasthyamala 523.2002.

C. infortunatum sensu Wight, Ic. t. 1471.1849 non L., 1753; Clarke in Hook. f., Fl. Brit. India 4: 594.1885; Gamble, Fl. Pres. Madras 1100. 1924. *Peragu* Rheede, Hort. Malab. 2:41-42, t. 25. 1679.

Woody herbs to small trees. Leaves cordiform, 10-25 by 9.5-20 cm, base cordate, apex acuminate, basally 5-(7-) nerved, tomentose, thin-coriaceous. Panicle terminal; calyx lobes ovate, large; corolla white. Drupe black, 4-lobed.

Fl. & Fr.: Dec.-Feb. *Distr.*: Sri Lanka, India, Malaya. Very abundant, appearing in thickets along disturbed forest areas. *NAK 2527* (Kochupampa, c. 1200 m).

Decoction of root-bark cures diarrhoea, dysentery and bowel-complaints, poisons and gonorrhoea.

GMELINA Linnaeus
Sp. Pl. 626. 1753.

Gmelina arborea Roxb., Pl. Corom. t. 246. 1815 & Fl. Ind. 3: 84. 1832; Wight, Ic. t. 1470. 1849 & Spicil. Neilgh. t. 193.1851; Clarke in Hook.f., Fl. Brit. India 4: 581. 1885; Gamble, Fl. Pres. Madras 1097. 1924; Matthew, Ill. Fl. Tam. Carnatic t. 550. 1982; Matthew & Rani in Matthew, Fl. Tam. Carnatic 3(2) : 1222. 1983; Nicols., Suresh & Mani., An Interpr. Hort. Malab. 261. 1988; Mohanan & Sivad., Fl. Agasthyamala 523.2002. *Cumbulu* Rheede, Hort. Malab. 1: 75-76, t. 41. 1678.

Medium- sized trees. Leaves broadly ovate, 10-18 by 9-13 cm, base truncate or subcordate, apex acute, lateral nerves 5-7 pairs, lower pair forking from the base, upper surface glabrous, 2-glandular at extreme base, lower surface fulvous-tomentose. Flowers 5-merous, in axillary or terminal panicles; calyx cupular, unequally 4-5- lobed; corolla yellow, personate, upper lip 3-loboed; stamens 4, subexerted; ovary globose, 4-locular; ovule 1 per locule. Drupe fleshy. Seeds 2-4, oblong.

Fl. & Fr.: Feb.-Apr. *Distr.:* Sri Lanka, India, Himalaya, Philippines. Not common. In open forests. *NAK 535* (Plappally, c. 350 m).

Decoction of roots given as a remedy in rheumatism, consumption, weakness, piles, impotency and to improve digestion. Wood suitable for match industry.

LANTANA Linnaeus
Sp. Pl. 626. 1753.

Lantana camara L., Sp. Pl. 627. 1753 var. **aculeata** (L.) Moldenke, Torreya 34: 9. 1934 & in Dassan. & Fosberg, Rev. Handb. Fl.Ceylon 4: 225. 1983; Sald. & Nicols., Fl. Hassan Dist.489. 1976; Matthew & Rani in Matthew, Fl. Tam. Carnatic 3(2): 1224. 1983; Manilal, Fl. Silent Valley 214. 1988. *Lantana aculeata* L., Sp. Pl. 627.1753; Clarke in Hook.f., Fl. Brit. India 4: 562. 1885; Gamble, Fl. Pres. Madras 1087. 1924.

Straggling armed shrubs. Leaves elliptic-ovate, 4-8 by 2.5-5.5 cm, base subcordate or truncate, margin crenate- serrate, apex acute to shortly acuminate, veins impressed above, rugose. Flowers aggregated in condensed axillary umbellate-spikes; bracts closely imbricating; calyx truncate; corolla pinkish- red, yellow or mixed of these, lobes 5, obscurely 2-lipped; stamens, 4, included; ovary 2-locular; ovule 1 per locule, axile. Drupe fleshy, globose. Seeds reticulate.

Fl. & Fr.: Most of the seasons. *Distr.*: Native of Tropic. America, now in tropics and subtropics. Very common. Most aggressive weed of distrubed areas. On upper hills, they form extensive and impenetrable thickets. Flowers are with different shades. *NAK 677* (Pandalam, c. 30 m), *2505* (Kakki hills, c. 1100 m).

PREMNA Linnaeus
Mant. Pl. 154. 252. 1771 (nom. cons.).

Erect or straggling shrubs to small trees. Leaves decussate, petiolate. Flowers zygomorphic, 5-merous, aggregated in terminal corymbs; bracts linear, deciduous; calyx truncate or toothed; corolla tubular-bilabiate, villous within, upper lip shorter, 2-lobed, lower longer, 3-lobed; stamens 4, included, rarely exerted; ovary globose, 4-locular, ovule 1 per locule. Drupe globose or obovoid. Seeds oblong.

1. Trees; leaves coriaceous; vein-reticulations prominent beneath; petiole to 4 cm long; panicle dense, pink. .... 1. **P. coriacea**
1. Stragglers; leaves chartaceous; vein-reticulatious not prominent; petiole to 2 cm long; panicle subdense, green. .... 2. **P. glaberrima**

1. **Premna coriacea** Clarke in Hook.f., Fl. Brit. India 4: 573. 1885; Gamble, Fl. Pres. Madras 1095. 1924.

Small trees or large shrubs. Leaves elliptic-oblong, 9-15 by 4.5-7 cm, base truncate or rounded, apex long-acuminate, lateral veins c. 4 pairs, prominent beneath, glabrous, coriaceous. Panicle dense, to 12 cm ϕ, pink; calyx truncate; corolla pinkish-white. Drupe obovoid.

Fl. & Fr.: Apr.-Jul. *Distr.*: Eastern India & S. W. Peninsular India. Very rare. In shola forests. *NAK 2732* (Way to Ponnambalamedu, c. 1200 m).

2. **Premna glaberrima** Wight, Ic. t. 1484. 1849; Clarke in Hook.f., Fl. Brit. India 4: 577.1885 ("*integerrima*"); Gamble, Fl. Pres. Madras 1096. 1924; Mohanan *et al.*, Bull. Bot. Surv. India 22: 107. 1980; Manilal, Fl. Silent Valley 215. 1988; Mohanan & Sivad., Fl. Agasthyamala 525.2002.

Stragglers, rarely small trees. Leaves oblong or elliptic-oblong, 8-14 by 3.5-7 cm, base acute, apex acuminate, lateral veins c. 4 pairs, chartaceous. Panicle corymbose, greenish, c. 8 cm ϕ; calyx obscurely toothed; corolla white. Drupe globose.

Fl. & Fr.: Mar.-May. *Distr.*: Endemic to Tamil Nadu & Kerala. Less common. In evergreen forests. *NAK 564* (Maniyaar, c. 70 m, *1660* (Thriveni pampa, c. 250 m) *1802* (Angamoozhy, c. 273 m).

Note : Henry *et al.* (1979) listed this species as very rare/endangered.

SPHENODESME W. Jack
Malayan Misc. 1(1): 19. 1820.

Sphenodesme involucrata (Presl) Roxb., Proc. Amer. Acad. 51: 531. 1916, var. **paniculata** (Cl.) Munir, Gard. Bull. Singapore 21: 338. 1966. *S. paniculata* Clarke in Hook. f., Fl. Brit. India 4: 600. 1885; Gamble, Fl. Pres. Madras 1104. 1924, p.p.

Scandent tomentose shrubs. Leaves decussate, broadly ovate or elliptic, 7-12 by 5-8 cm, base rounded, apex subacute, golden tomentose, thin-coriaceous. Flowers 3-7, in capitate cymes, subtended by an involucre of 6 foliaceous bracts; calyx campanulate, lobes 4-5; corolla white, funnel-form, lobes 4-6; stamens 4-6, included or exerted; ovary 2-locular; ovules 2 per locule. Drupe obovoid. Seed 1, globose.

Fl. & Fr.: Jan.-Apr. *Distr.*: Southern Western Ghats. Rare. In sacred groves. *NAK 2311* (Parakkode, c. 20 m).

STACHYTARPHETA Vahl
Enum. Pl.1:205. 1804 (nom. cons.).

Erect subshrubs; branchlets 4-gonous. Leaves decussate, crenate- serrate, petiolate. Flowers zygomorphic, partly embedded in the excavations of cylindric spikes, bracts lanceate; calyx-tubular, 4-ribbed, teeth 4, unequal; corolla salver-form, curved, lobes 5, subequal; stemens 2, included; ovary 2-locular, ovate 1 per locule, basal. Fruit cylindric. Seeds oblong.

1. Calyx-teeth 5, 4 (sub)equal, 5th one very short; flowers white. .... 1. **S. cayennensis**
1. Calyx-teeth 4, all equal or nearly so; flowers bluish-pink.
    2. Leaves pale green, margin coarsely crenate-serrate. .... 2. **S. jamaicensis**
    2. Leaves bright green, margin finely serrate. .... 3. **S. urticaefolia**

1. **Stachytarpheta cayennensis** (L.C. Rich.) Schou. in DC., Prodr. 11: 562. 1847; Danser, Ann. Jard. Bot. Buitenz. 40: 2. 1929; Brenan, Kew Bull. 5: 223. 1950; Nair *et al.*, J. Bombay Nat. Hist. Soc. 79(1) :230. 1982.

Subshrubs. Leaves elliptic-lanceate or ovate-elliptic, 2-5 by 1-3 cm, base cuneate, margin crenate-serrate, apex acute or obtuse, drying blackish. Spike drooping, 12-30 cm long; bracts linear-lanceate; calyx-teeth 5, anterior one very small; corolla white. Fruits c. 3.5 mm long.

Fl. & Fr.: Dec.-Mar. *Distr.*: Native of Trop. America, now introduced into tropics and subtropics Not Common. Along waysides in hills. *NAK 763* (Lahai, c. 300 m), *1270* (Pampa colony, c. 1000 m).

2. **Stachytarpheta jamaicensis** (L.) Vahl, Enum. Pl. 1: 206. 1804; Matthew & Rani in Matthew, Fl. Tam. Carnatic 3(2): 1232. 1983 *Verbena jamaicensis* L., Sp. Pl. 19. 1753. *Stachytarphetta indica* auct., non (L.) Vahl, 1804: Clarke in Hook.f., Fl. Brit. India 4: 564. 1885; Gamble, Fl. Pres. Madras 1090. 1924; Matthew, Ill. Fl. Tam. Carnatic t. 556. 1982.

Subshrubs. Leaves elliptic-lanceate, 5-10 by 3.5-4.5 cm, base cuneate, decurrent, margin coarsely creante-serrate, apex obtuse or rounded, pale-green. Spike 10-25 cm long; calyx-lobes, 4, subequal; corolla bluish-pink. Fruits oblong.

Fl. & Fr.: Jun.-Aug. *Distr.*: Trop. Asia, America. Common along disturbed grounds. *NAK 624* (Thannithode, c. 70 m), *1009* (Tiruvalla, c. 10 m).

3. **Stachytarpheta urticaefolia** (Salisb.) Sims, Bot. Mag. 43: Pl. 1848. 1816; Moldenke in Dassan. & Fosberg, Rev. Handb. Fl. Ceylon 4: 257. 1983; Mohanan & Sivad., Fl. Agasthyamala 525.2002. *Cymburus urticaefolius* Salisb., Parad. Lond. Pl. 53. 1806. *Stachytarphetta indica* auct., non Vahl 1804: Calrke in Hook.f., Fl. Brit. India 4: 564. 1885, p.p.; Gamble, Fl. Pres. Madras 1090. 1924, p.p. *Kadaladi.*

Subshrubs or woody herbs. Leaves elliptic-ovate, 4-7 by 3-4 cm, base rounded, abruptly decurrent, margin finely serrate, apex acute, bright-green. Spike 10-20 cm long, calyx-lobes 4, subequal; corolla bluish-violet. Fruits oblong.

Fl. & Fr.: Sep.-Dec. *Distr.*: Tropics. Common. Along waysides, disturbed areas in plains and hills. *NAK 136* (Adoor, c. 20 m).

TECTONA Linnaeus f.
Suppl. Pl. 20 ("Thecktona"), 151. 1782 (nom. cons.).

Tectona grandis L.f., Suppl. Pl. 151. 1781; Roxb., Pl. Corom. t. 6. 1785 & Fl. Ind. 1 : 600. 1832; Clarke in Hook.f., Fl. Brit. India 4: 570. 1885; Gamble, Fl. Pres. Madras 1092. 1924; Matthew, Ill. Fl. Tam. Carnatic t. 558. 1982; Matthew & Rani in Matthew, Fl. Tam. Carnatic 3(2): 1236. 1983; Nicols., Suresh & Mani., An Interpr. Hort. Malab. 265. 1988. *Theka* Rheede, Hort. Malab. 4: 57-58, t. 27. 1683.

Deciduous trees. Leaves broadly elliptic, 20-40 by 15-30 cm, base rounded, apex obtuse to acute, stellate-pubescent below, thin-coriaceous. Flowers in terminal and/or axillary cymose panicles; calyx 5-7-lobed; corolla cream -coloured, lobes 7; stamens c. 7, exerted; ovary 4-locular; ovules 1 per locule. Drupe globose. Seeds oblong.

Fl. & Fr.: Jun.-Sep. *Distr.*: Native of Trop. S. Asia & Malesia. *NAK 2869* (Kalleli, c. 100 m).

Extensively planted for timber. Flowers powdered and taken to cure phlegm, bile and syphilis. Bark powdered and applied to cure ulcers.

Note : Most of the forest areas in the district are replaced by monoculture of this species.

VITEX Linnaeus
Sp. Pl. 638 ('938'). 1753.

Trees. Leaves digitately, 3-7- foliolate. Flowers 5-merous, aggregated in terminal or axillary corymbose racemose-panicles; calyx cupular or campanulate, truncate or faintly 5-toothed; corolla personnate; stamens 4, didynamous, exerted; ovary 2 or imperfectly 4-locular; ovules 4. Drupe ovoid or ellipsoid. Seeds ovate-oblong.

1. Leaflets petiolate.
    2. Leaflets narrowly elliptic-lanceate; bracts minute. .... 1. **V. altissima**
    2. Leaflets broadly elliptic-oblanceate; bracts foliaceous. .... 4. **V. pinnata**
1. Leaflets subsessile.
    3. Leaflets 7; panicles dense, c. 10 cm long. .... 2. **V. leucoxylon**
    3. Leaflets 3; panicle lax, c. 25 cm long. .... 3. **V. peduncularis**

1. **Vitex altissima** L.f., Suppl. Pl. 294. 1781; Roxb., Fl. Ind. 3: 71. 1832; Wight, Ic. t.1466. 1849; Clarke in Hook.f., Fl. Brit. India 4: 584. 1885; Gamble, Fl. Pres. Madras 1102. 1924; Matthew & Rani in Matthew, Fl. Tam. Carnatic 3(2): 1239. 1983; Nicols., Suresh & Mani., An Interpr. Hort. Malab. 265. 1988; Mohanan & Sivad., Fl. Agasthyamala 526.2002. *Mail-clou* Rheede, Hort. Malab. 5:1-2, 6.1. 1685.

Trees. Leaves 3-foliolate; leaflets subsessile, narrowly elliptic-lanceate, 5-10 by 1.5-2.5 cm, base acute, apex gradually acuminate, lower surface tomentose. Cymose-panicle terminal or from upper leaf-axils; calyx cupular; corolla bluish-purple. Drupe globose, c. 5 mm ϕ.

Fl. & Fr.: Mar.-May. *Distr.*: Peninsular India. Less common. Evergreen forest slopes. *NAK 1657* (Sabari hills, c.250 m).

2. **Vitex leucoxylon** L. f., Suppl. Pl. 293. 1781; Roxb., Fl. Ind. 3: 74. 1832; Clarke in Hook. f., Fl. Brit. India 4: 587. 1885; Gamble, Fl. Pres. Madras 1103. 1924; Matthew, Ill. Fl. Tam. Carnatic t. 559. 1982; Matthew

& Rani in Matthew, Fl. Tam. Carnatic 3(2): 1239. 1983; Nicols., Suresh & Mani., An Interpr. Hort. Malab. 266. 1988. *Kariil* Rheede, Hort. Malab. 4:75-76, t. 36. 1683. *Nir-Notchi.*

Trees. Leaves 5-foliolate; leaflets petiolulate, elliptic-oblong, lateral 2 leaflets small, 4-10 by 1.5-3.5 cm, base acute, apex acuminate, lower surface thinly pubescent. Cymes corymbose, axillary; calyx cupular; corolla white with purplish throat. Drupe ellipsoid.

Fl. & Fr.: Mar.-May. *Distr.*: Sri Lanka, Peninsular India. Common. Along steam sides. *NAK 1578, 1594 & 1667* (Thannithode,Thekkuthode etc.).

Decoction of root is given in intermittent fever. Timber is hard and strong.

3. **Vitex peduncularis** Wall. ex Schauer in DC., Prodr. 11: 687. 1847; Clarke in Hook.f., Fl. Brit. India 4: 587. 1885; Gamble, Fl. Pres. Madras 1103. 1924.

Trees. Leaves 3-foliolate; leaflets petiolulate, oblong-lanceate, 10-16 by 2.5-4 cm, base acute, apex acuminate, glabrous. Cymose-panicle very lax, c. 25 cm long; calyx cupular; corolla white with yellow throat. Drupes subglobose.

Fl. & Fr.: Feb.-Apr. *Distr.*: India. Less common. In moist deciduous forests. *NAK 2425* (Kalleli, c. 100 m).

Note : In young plants, leaves are with winged petioles and serrate margins.

4. **Vitex pinnata** L., Sp. Pl. 638. 1753; Nicols., Suresh & Mani., An Interpr. Hort. Malab. 266. 1988; Mohanan & Sivad., Fl. Agasthyamala 527.2002. *V. pubescens* Vahl, Symb. Bot. 3: 85. 1794; Clarke in Hook.f., Fl. Brit. India 4: 585. 1885; Gamble, Fl. Pres. Madras 1103. 1924; *V. arborea* Roxb., Fl. Ind.3: 73. 1832; Wight, Ic. t. 1465. 1849.

Trees. Leaves 3-foliolate; leaflets elliptic-oblanceate, or ovate-lanceate, 7-15 by 4.5-6.5 cm, base acute or cuneate, apex shortly acuminate, lateral veins prominent beneath, pubescent. Cymose-panicle subterminal; bracts foliaceous, persistent; calyx broadly cupular; corolla bluish. Drupe globose.

Fl. & Fr.: Mar.-May. *Distr.*: Sri Lanka, Peninsular India. Common in sacred groves. *NAK 701* (Pandalam, c. 40 m).

101. LAMIACEAE (LABIATAE)

1. Calyx-teeth 8-10; upper lip of corolla hooded. .... **Leucas**
1. Calyx-lobes 5; upper lip of corolla straight or spreading.
    2. Calyx-lobes equal, not 2-lipped.
        3. Calyx campanulate, 10-ribbed; corolla cream coloured. .... **Gomphostemma**
        3. Calyx tubular- campanulate, smooth; corolla purplish.
            4. Calyx c. 5 mm long; corolla with lobes 4. .... **Pogostemon**
            4. Calyx c. 8 mm long; corolla with lobes 5.
                5. Calyx-lobes lanceate; corolla-lower lip longer than upper lip. ...**Anisomeles**
                5. Calyx-lobes linear; corolla-lower lip shorter than upper lip. .... **Hyptis**

2. Calyx-lobes unequal, 2-lipped.
   6. Flowers minute in capitate heads. .... **Acrocephalus**
   6. Flowers not so, other than capitate heads.
      7. Calyx oblique, lower lip truncate. .... **Anisochilus**
      7. Calyx equally 2-lipped, lower lip 2-4-toothed.
         8. Calyx as long as corolla-tube; anthers 1-celled. .... **Ocimum**
         8. Calyx shorter than corolla tube; anthers 2-celled.
            9. Staminal filaments free. .... **Plectranthus**
            9. Staminal filaments connate. .... **Coleus**

ACROCEPHALUS Bentham in Edward's, Bot. Reg. 15: t. 1282. 1829.

Acrocephalus hispidus (L.) Nicols. & Sivad., Taxon 29: 324. 1980; Manilal, Fl. Silent Valley 215. 1988; Nicols., Suresh & Mani., An Interpr. Hort. Malab. 151. 1988; Mohanan & Sivad., Fl. Agasthyamala 528.2002. *Gomphrena hispida* L., Sp. Pl. ed. 2, 1: 326. 1762. *Prunella indica* Burm.f., Fl. Ind. 130. 1768, ("*Brunella*"). *Ocimum capitatum* Roth, Nov. Pl. Sp. 276. 1821. *Acrocephalus capitutus* (Roth) Benth., Bot. Reg. 15: Sub t. 1300. 1830; Hook.f., Fl. Brit. India 4: 611. 1885. *Acrocephalus indicus* (Burm. f.) Kuntze, Rev. Gen. Pl. 2: 511. 1891; Gamble, Fl. Pres. Madras 1115. 1924. *Minangani* Rheede, Hort. Malab. 9: 141, t. 72. 1689.

Erect annuals. Leaves decussate, ovate-lanceate or elliptic-lanceate, 2.3 by 1-1.2 cm, base attentuate, margin coarsely serrate, apex acute, thinly pubescent on nerves beneath, chartaceous. Flowers minute, aggregated in capitate, axillary or terminal, clusters. Capitula c. 8 mm ϕ, subtended by a pair of floral leaves and many bracts; calyx tubular, 2-lipped, upper entire, lower usually 4-toothed; corolla white, inconspicuously 2-lipped, 4+1; stamens 4, didynamous, declinate, anthers 1-celled; disc annular; ovary 4-partite; style gynobasic, shortly bifid at apex. Fruit a Carcerulous; nutlets smooth.

Fl. & Fr.: Jul.-Sep. *Distr.*: India, Myanmar, Sumatra & Java. Less common. On exposed waste places. In plains. *NAK 1907* (Kodumon, c. 50 m).

ANISOCHILUS Wallich ex Bentham in Lindley, Edward's Bot. Reg. 15: t. 1300. 1830.

Stout herbs. Leaves decussate or whorled. Flowers in cylindric, dense terminal spikes; bracts caducous; calyx oblique, lobes 5, 1+4; corolla tubular, 2-lipped, inflexed above, decurved, lower lip elongate, concave; stamens 4, didynamous, exerted, anthers 2-celled; ovary 4-lobed; style gynobasic; disc anteriorly developed. Nutlets orbicular.

1. Leaves decussate, broadly ovate; peduncle branched; spike to 4 cm long. .... 1. **A. carnosus**
1. Leaves whorled, oblong-lanceate; peduncle often unbranched; spike to 12 cm long. .... 2. **A. verticillatus**

1. **Anisochilus carnosus** (L.f.) Wall., Pl. Asiat. Rar. 2: 18. 1831; Wight, Ill. Bot. t. 176. 1850; Hook.f., Fl.Brit. India 4: 627. 1885; Gamble, Fl. Pres. Madras 1126. 1924; Cramer in Dassan. & Fosberg, Rev. Handb. Fl. Ceylon 3: 152. 1981; Matthew, Ill. Fl.Tam. Carnatic t. 561. 1982; Matthew & Rani in Matthew, Fl. Tam. Carnatic 3(2): 1248. 1983; Nicols., Suresh & Mani., An Interpr. Hort. Malab. 151. 1988. *Lavendula carnosa* L. f., Suppl. Pl. 273. 1781. *Katu-kurka* Rheede, Hort. Malab. 10: 179, t. 90. 1690.

Leaves broadly ovate, 2-5 by 2.3-4 cm, base cordate, margin crenate, apex acute or obtuse, verrucose above, pubescent below, subsucculent; corolla purple, upper lip entire, lower lip 4-lobed. Nutlets smooth.

Fl. & Fr.: Sep.-Dec. *Distr.*: Sri Lanka, India, Myanmar. Common. Locally abundant. On exposed moist grassy slopes. *NAK 1087* (Vallicode, c. 150 m).

Fresh juice of leaves with sugar given to children to cure coughs. Decoction of whole plant given to remove phlegm.

2. **Anisochilus verticillatus** Hook.f., Fl. Brit. India 4: 629. 1885; Gamble, Fl. Pres. Madras 1127. 1924.

Stout herbs; stem longitudinally ridged. Leaves whorled, oblong-lanceate to oblanceate, 2-8 by 0.3-0.9 cm, base narrowed, margin entire to faintly crenate towards apex, apex acute, lateral veins c. 4 pairs, almost parallel, prominent on lower surface. Spikes to 12 cm long; bracts lanceate, c. 5 mm long; calyx short, tubular; corolla pale purple. Nutlets erect.

Fl. & Fr.: Sep.-Dec. *Distr.*: Endemic to Western Ghats. Very rare. On moist exposed grassy hill slopes. *NAK 1088* (Vallicode, c. 150 m).

Note : No representation in MH. Probably this is teh first collection from Travancore hills.

ANISOMELES R. Brown
Prodr. 503. 1810.

Anisomeles indica (L.) Kuntze, Rev. Gen. Pl. 2: 512. 1891; Gamble, Fl. Pres. Madras 1140.1924; Sald. & Nicols., Fl. Hassan Dist. 498. 1976; H. Keng in Steenis, Fl.Males. I, 8: 329. 1978; Cramer in Dassan . & Fosberg, Rev. Handb. Fl. Ceylon 3: 176. 1981; Matthew & Rani in Matthew, Fl. Tam. Carnatic 3(2): 1250. 1983; Manilal, Fl. Silent Valey 216. 1988; Nicols., Suresh & Mani., An Interpr. Hort. Malab. 152. 1988; Mohanan & Sivad., Fl. Agasthyamala 530.2002. *Nepeta indica* L., Sp. Pl.571. 1753. *Nepeta indica* L., Sp. Pl. 571. 1753. *Anisomeles ovata* Aiton f., Hortus Kew. 3: 364. 1811; Wight,Ic. t. 865. 1844-45; Hook.f., Fl. Brit. India 4: 672. 1885. *Tsjadaen* Rheede, Hort. Malab. 10: 175, t. 88. 1690.

Shrubs; branchlets and leaves densely pubescent Leaves decussate, broadly ovate to obovate, 4-8 by 3-6 cm, base cuneate, margin coarsely crenate-serrate, apex acute, chartaceous. Flowers in axillary whorls of short spikes; bracts 2; calyx ovoid, reticulate, lobes 5, equal; corolla pale pink, tubular, 2-lipped, lobes 2+3; stamens 4, didynamous, anthers of upper pair 2-celled, of lower 1-celled; ovary 4-lobed; disc subentire. Nutlets 4, lenticular.

Fl. & Fr.: Most of the seasons. *Distr.*: Sri Lanka, India, China, Malesia. Common in wastelands, waysides, etc. *NAK 1032* (Tiruvalla, c. 10 m), *1222* (Upper Moozhiar, c. 600 m).

Leaves boiled in water for children to bathe, and as a remedy in fever and itches.

COLEUS Loureiro
Fl. Cochinch. 358. 1790.

Coleus malabaricus Benth. in Wall., Pl. Asiat. Rar. 2: 16. 1839; Hook.f., Fl. Brit. India 4: 626. 1885; Gamble, Fl. Pres. Madras 1124. 1924; Cramer in Dassan. & Fosberg, Rev. Handb. Fl. Ceylon 3: 145. 1981; Mohanan & Sivad., Fl. Agasthyamala 531.2002.

Subsucculent herbs. Leaves decussate, broadly ovate, 6-20 by 5-13 cm, base cordate, margin crenate, apex acuminate, lateral nerves c. 6 pairs, lower surface red-glandular, chartaceous. Verticils in 25-30 cm long racemes; floral leaves, bracts caducous; calyx 2-lipped, 1+4, glabrous within; corolla rose-purple, 2-lipped, 4+1, tube decurved, acuminate; lower lip long, entire; stamens 4, didynamous, connate below in a sheath around the style; ovary 4-lobed; disc anteriorly developed. Nutlets ovoid, trigonous, smooth.

Fl. & Fr.: Sep.-Dec. *Distr.*: Sri Lanka, South India. Not common. On moist shady places of evergreen forests. Locally abundant. *NAK 90, 1320, 2614* (Kakkidam area, c. 1100 m).

Note :The circumscription of the genus **Coleus** Lour. is confusing. Some authors merged it with **Plectranthus**. But here the treatment of Cramer (Kew Bull. 32: 551-56:1977) who disagreed with the merger is followed.

GOMPHOSTEMMA Wallich ex Bentham in Wallich, Pl. Asiat. Rar. 2: 12. 1831.

Subshrubs, stellate-tomentose. Leaves decussate, dentate or serrate, closely reticulate beneath, chartaceous. Flowers in whorls of short cymes, axillary, or in terminal spikes; bracts α, linear; calyx-lobes 5, 10-ribbed; corolla yellow or cream-coloured, 2 lipped, upper lip hooded, lower broadly 3-lobed; stamens 4, didynamous; anthers 2-celled; ovary 4-lobed; disc swollen. Nutlets corky.

1. Cymes axillary; calyx-lobes linear. .... 1. **G. eriocarpon**
1. Cymes terminal; calyx-lobes lanceate. .... 2. **G. heyneanum**

1. **Gomphostemma eriocarpon** Benth. in Wall., Pl. Asiat. Rar. 2: 12. 1830-1831; Hook.f., Fl. Brit. India 4: 698. 1885; Gamble, Fl. Pres. Madras 1157. 1924; Mohanan & Sivad., Fl. Agasthyamala 532.2002. *G. oblongum* Wight, Ic.t. 1457. 1849, non Wall. ex Benth., 1830-1831.

Leaves elliptic-lanceate or oblanceate, 7-12 by 2.5-4.5 cm, base cuneate, margin dentate, apex acute, upper surface sparsely simple, pubescent. Cymes sessile; calyx-lobes linear; corolla cream- coloured. Nutlets corky, pubescent.

Fl. & Fr.: Aug.-Dec. *Distr.*: Peninsular India. Less common. Among shady forest undergrowth in evergreen forests. *NAK 2559* (Kakki hills, c. 1100 m), *2052* (Angamoozhy, c. 250 m).

2. **Gomphostemma heyneanum** Wall. ex Benth. in Wall., Pl. Asiat. Rar. 2: 12. 1831; Wight, Ic. t. 1456.1849; Gamble, Fl. Pres.Madras 1167. 1924; Sald. & Nicols., Fl. Hassan Dist. 501. 1976; Manilal., Fl. Silent Valley 218. 1988. *G. strobilinum* Wall., var. *heyneana* Hook.f., Fl. Brit. India 4: 696. 1885.

Leaves elliptic-ovate, 10-20 by 5-10 cm, base rounded, margin serrate, apex subacute, lower surface woolly. Flowers in terminal racemes; calyx-lobes ovate-lanceate; corolla pale yellow. Nutlets glabrous.

Fl. & Fr.: Sep.-Nov. *Distr.*: Peninsular India. Occasional in shady forest areas of upper hills. *CNM 12834* (MH).

HYPTIS N. J. Jacquin Collectanea 1: 101, 103. 1787 (nom. cons.).

Subshrubs, erect or subscandent. Leaves decussate. Flowers in cymes or capitate, clusters terminal and axillary; calyx-lobes 5, subequal; corolla lobes 5, 2-lipped, 2+3; lower lip deflexed saccate; stamens 4, didynamous, sub-exerted; ovary 4-lobed, disc anteriorly developed. Nutlets 2, ovoid.

1. Subscandent subshrubs; flowers white, in capitate heads. .... 1. **H. rhomboidea**
1. Erect subshrubs; flowers blue, in cymose racemes. .... 2. **H. suaveolens**

1. **Hyptis rhomboidea** Mart. & Gal., Bull. Acad. Bru. 11(2): 188. 1844; Keng, Garden Bull. Singapore 24: 92. 1969. *H. capitata* auct., non Jacq., 1781-1786: Mukherjee, Rec. Bot. Surv. India 14(1): 63. 1940; Varghese, J. Bombay Nat. Hist. Soc. 76:200. 1980.

Subscandent woody herbs.Leaves ovate-lanceate, 4-25 by 2-9 cm, base cuneate-decurrent, margin coarsely serrate, apex acute, pubescent, Flowers white in capitate heads; innvolucral bracts foliaceous. Nutlets ovate-oblong, subtrigonous, blackish.

Fl. & Fr.: Apr.-Dec. *Distr.*: On disturbed forest areas. A recently introduced alarming weed. *NAK 1045* (Tiruvalla, c. 10 m).

2. **Hyptis suaveolens** (L.) Poit., Ann. Mus. Natl. Hist. Nat. 7: 472, t. 29, f. 2. 1806; Hook.f., Fl. Brit. India 4: 630.1885; Gamble, Fl. Pres. Madras 1129. 1924; Keng in Steenis, Fl. Males I, 8: 37. 1978; Cramer in Dassan. & Fosberg, Rev. Handb. Fl. Ceylon 3: 155. 1982; Matthew, Ill. Fl. Tam. Carnatic t. 567. 1982; Matthew & Rani in Matthew, Fl. Tam. Carnatic 3(2): 1256. 1983; Mohanan & Sivad., Fl. Agasthyamala 532.2002. *Ballota suaveolens* L., Syst.Nat.ed. 10,1100.1759.

Erect subshrubs or woody herbs. Leaves long-petioled, broadly elliptic-ovate, 2-7 by 1-4 cm, base obliquely truncate or acute, margin coarsely serrulate, apex acute, chartaceous. Flowers blue, in short stalked axillary cymes. Nutlets ovoid.

Fl. & Fr.: Sep.-Oct. *Distr.*: Native of C. America now naturalized in tropics. A common weed. On wastelands, waysides, etc. *NAK 858* (Plappally, c. 400 m).

LEUCAS R. Brown
Prodr. 504. 1810.

Erect or diffuse herbs or subshrubs; stem 4-angled. Leaves deccusate, petioled. Flowers in paired cymes, or aggregated into terminal or axillary pseudo-verticils or in pairs; calyx tubular, turbinate or campanulate, mouth straight or oblique, 8-10- ribbed, 8-10-toothed; corolla-lobes 5, 2-lipped, 1+4, upper hooded, lower broad, flat, deflexed; stamens 4, included or sub-exerted; ovary 4-locular; disc 4-lobed. Nutlets 4, subtrigonous.

1. Flowers 2-4 per node (not aggregated in verticils); bracts linear. ....1. **L. biflora**
1. Flowers 6 or more, aggregated in verticils.
   2. Mouth of the calyx-tube oblique.
      3. Calyx-teeth 7-10, very unequal, the posterior one much larger than the others, mouth often constricted. ....3. **L. indica**
      3. Calyx-teeth 8, subequal, the posterior one only slightly larger, mouth open. ....5. **L. zeylanica**
   2. Mouth of the calyx-tube, straight.
      4. Upper lip of corolla rufous; leaves hirsute below. ....4. **L. vestita**
      4. Upper lip of corolla villous; leaves tawny pubescent below. ....2. **L. hirta**

1. **Leucas biflora** (Vahl) R. Br., Prodr. 504. 1810; Wight, Ic. t. 866. 1844-45; Hook.f., Fl. Brit. India 4: 683.1885; Gamble, Fl. Pres. Madras 1150. 1924; Cramer in Dassan. & Fosberg, Rev. Handb. Fl. Ceylon 3: 186. 1982; Matthew & Rani in Matthew, Fl. Tam. Carnatic 3(2): 1260. 1983; Manilal, Fl. Silent Valley 219. 1988. *Phlomis biflora* Vahl, Symb. Bot. 3: 77. 1794.

Diffuse herbs. Leave ovate, 2-5 by 1-2.5 cm, base cuneate, margin crenate, apex acute, glabrescent, chartaceous. Flowers 2-4, in lax verticils; calyx 10-ribbed, teeth 10, subequal; corolla white. Nutlets ovoid.

Fl. & Fr.: Jan.-Jul. *Distr.*: Sri Lanka, Peninsular India. Not common. On moist shady embankments. *NAK 446* (Maniyaar, c. 50 m), *2287* (Mannarappara, c. 200 m).

2. **Leucas hirta** (Roth) Spreng. in L., Syst.2: 743. 1825; Hook.f., Fl.Brit. India 4: 687. 1885; Gamble, Fl. Pres. Madras 1153. 1924; Sald. & Nicols., Fl. Hassan Dist. 503. 1976; Matthew & Rani in Matthew, Fl. Tam. Carnatic 3(2):1261. 1983 *Phlomis hirta* Heyne ex Roth, Nov. Pl. Sp. 264. 1824.

Erect herbs, tawny pubescent. Leaves elliptic-oblong, 2-3.5 by 1-1.5 cm, base and apex obtuse, margin crenate, silky tomentose, coriaceous. Flowers in terminal verticils. bracts linear, calyx-throat villous within, teeth 10, alternately short and long; corolla white. Nutlets ovoid.

Fl. & Fr.: Nov.-Jan. *Distr.*: Peninsular India. Not common. On dry hills slopes. *NAK* 1357 (Pampadam area, 110 m).

3. **Leucas indica** (L.) R. Br. ex Vatke, Oester. Bot. Z. 25: 95. 1875; Matthew, Ill. Fl. Tam. Carnatic t. 569. 1982; Matthew & Rani in Matthew, Fl. Tam. Carnatic 3(2): 1262. 1983; Mohanan & Sivad., Fl. Agasthyamala 534.2002. *Leonurus indicus* L., Syst. Nat. (ed. 10) 1101. 1760; Burm. f., Fl. Ind. 127. 1768. *Phlomis linifolia* Roth, Nov. Pl. Spec. 260. 1821. *Leucas lavandulifolia* Sm. in Rees, Cyclop. 20. n. 2. 1812; Keng in Steenis, Fl. Males. I, 8: 338. 1978. *L. linifolia* (Roth) Spreng. in L., Syst. Veg. (ed. 16) 2: 743. 1825; Hook.f., Fl. Brit. India 4: 690. 1885; Gamble, Fl. Pres. Madras 1149. 1924.

Erect herbs. Leaves linear-lanceate, 4-8 by 0.4-0.7 cm, base and apex cuneate, margin subentire thick-chartaceous. Verticils axillary; calyx-tube sightly curved, mouth oblique, teeth 8, posterior much longer than rest; corolla white. Nuts ovoid.

Fl. & Fr.: Most of the seasons. *Distr.*: Sri Lanka, India, China, Malesia to Japan. Less common. A weed of wasteplaces. *NAK 151* (Adoor, c. 40 m).

4. **Leucas vestita** Benth. in Wall., Pl. Asiat. Rar. 1: 61. 1830; Wight, Ic. t. 338. 1840; Hook.f., Fl. Brit. India 4: 686. 1885; Gamble, Fl. Pres. Madras 1153. 1924; Matthew & Rani in Matthew, Fl. Tam. Carnatic 3(2): 1266. 1983.

Diffuse herbs, spreading tomentose. Leaves lanceate, 3-7 by 1.2-2 cm, base obtuse, margin crenate-dentate, apex acute, chartaceous. Verticils terminal; calyx-mouth straight, villous within; corolla white with rufous upper lip. Nutlets brown.

Fl. & Fr.: Nov.-Jan. *Distr.*: Peninsular India. Common. On exposed grasslands. *NAK* 396, 1305 (Kakki hills, c. 1200 m).

5. **Lecuas zeylanica** (L.) R. Br., Prodr. 504. 1810; Wight, Ill. 2: t. 176.1850; Hook. f., Fl. Brit. India 4: 689. 1885; Gamble, Fl. Pres. Madras 1150. 1924; Cramer in Dassan. & Fosberg, Rev. Handb. Fl. Ceylon 3: 183. 1981; Manilal, Fl. Silent Valley 220. 1988. *Phlomis zeylanica* L., Pl. 586. 1753.

Erect hispid herbs. Leaves lanceate, 5-8 by 1-1.5 cm, base and apex acute, margin subentire, chartaceous. Verticils terminal and subterminal; calyx mouth oblique, teeth 8, subequal, the posterior one slightly larger; corolla white. Nutlets smooth, brownish.

Fl. & Fr.: May-Jul. *Distr.*: Throughout Southern and S.E. Asia. Common. A weed in waste places and along roadsides. *NAK 596* (Thekkuthode,c. 50 m).

OCIMUM Linnaeus
Sp. Pl. 597. 1753.

Ocimum americanum L., Cent. Pl. 1: 15. 1755 Keng in Steenis, Fl. Males. I, 8: 376. 1978. *O. canum* Sims, Bot. Mag.t. 2452. 1823; Hook.f., Fl. Brit. India 4: 607. 1885; Mukherjee, Rec. Bot. Surv. India 14: 17. 1940; Gamble, Fl. Pres. Madras 1111. 1924. Cramer in Dassan. & Fosberg, Rev. Handb. Fl. Ceylon 3: 114. 1981; Matthew & Rani in Matthew, Fl. Tam. Carnatic 3(2) : 1269. 1983. *Katu-thulasi.*

Herbs; branchlets terete to sub 4-angular. Leaves elliptic-oblong, 1-3.5 by 0.5-2 cm, base truncate, margin sparingly serrate, apex acute, chartaceous. Flowers in verticils, aggregated into a panicle; bracts ovate; calyx-lobes 5, 2-lipped, 1+4; corolla white, bilabiate, 1+4; stamens 4, didynamous; anthers 1-celled; styles gynobasic, equally 2-fid at apex; disc 4-lobed. Nuts basilar.

Fl. & Fr.: Jul.-Dec. *Distr.*: Paleotropics. Common. On wastelands. *NAK 1856* (Kodumon, c. 40 m).

PLECTRANTHUS L' Heritier
Stirp. Novae 84. 1788 (nom. cons.).

Plectranthus wightii Benth., Labiat. Gen. Spec. 41. 1832; Wight, Ic. t. 1429. 1849; Hook.f., Fl. Brit. India 4: 619. 1885; Gamble, Fl. Pres. Madras 1120. 1924; Rani & Matthew in Matthew, Fl. Tam. Carnatic 3(2): 1276. 1983; Mohanan & Sivad., Fl. Agasthyamala 536.2002.

Tall herbs. Leaves broadly ovate, 2-10 by 1.5-8.5 cm, base truncate, margin deeply crenate, apex acute, both surfaces strigose; cymes lax, branched to 25 cm long; calyx-lobes 5, subequal, 3+2, obscurely 2-lipped; corolla white, lobes 5, 2-lipped, 4+1, upper lip recurve stamens 4, free, didynamous, anthers 2-celled; style gynobasic; disc hooded. Nutlets subtrigonous.

Fl. & Fr.: Nov.-Jan. *Distr.*; Peninsular India. Less common. On moist shady wet areas of higher hills. *NAK 1281, 1319* (Kakki hills, c. 1100 m).

POGOSTEMON Desfontaines
Mem. Mus. Hist. Nat. 2: 154. 1815.

Erect herbs or shrubs. Leaves decussate, petiolate. Flowers in decussate or verticillate cymules aggregated into dense, cylindric or secund spicate-racemes; bracts and bracteoles foliaceous or linear-lanceate; calyx tubulər, lobes 5, subequal; corolla tubular, lobes 4, bilabiate; stamens 4, unequal or subequal, exerted, filaments bearded, anthers 2-celled; style gynobasic, stigma 2-fid; disc entire. Nutlets erect, basilar.

1. Spicate racemes paniculate.
    2. Inflorescence 1-sided; flower-clusters secund; leaf-pair unequal. .... 3. **P. paniculatus**
    2. Inflorescence even; flower- clusters decussate or verticillate; leaf-pair equal.
        3. Flower-clusters interrupted; bracts lanceate. .... 2. **P. heyneanus**
        3. Flower-clusters continous; bracts broadly ovate, foliaceous. .... 1. **P. benghalensis**
1. Spicate racemes not paniculate (Unbranched). .... 4. **P. travancoricus**

1. **Pogostemon benghalensis** (Burm. f.) Kuntze, Rev. Gen. Pl. 2: 517. 1891; Matthew, Ill. Fl. Tam. Carnatic. t. 574. 1982; Matthew & Rani in Matthew, Fl. Tam. Carnatic 3(2): 1278. 1983; Mohanan & Sivad., Fl. Agasthyamala 537.2002. *Origanum benghalense* Burm.f., Fl. Ind. 128, t. 38, f. 3. 1768. *Pogostemon plectranthoides* Desf., Mem. Mus. Hist. Nat. 2: 155, t. 6. 1815; Hook.f., Fl.Brit. India 4: 632. 1885; Gamble, Fl. Pres. Madras 1133. 1924.

Subshrubs, to 2 m tall. Leaves ovate to ovate-lanceate, 3-10 by 2-5cm, base cuneate, margin doubly-crenate, apex acute to acuminate. Spicate-racemes dense, branched, brownish; calyx-lobes subequal; corolla pink-violet, c. 4 mm ϕ, obscurely 2- lipped, 3+1.

Fl. & Fr.: Nov.-Feb. *Distr.*: India, Himalayas. Rare. On forest floors in shades. *NAK 384* (Kakki hills, c.1100 m).

2. **Pogostemon heyneanus** Benth. in Wall., Pl. Asiat. Rar. 1: 31.1830; Wight, Ic. t. 1440. 1849; Gamble, Fl. Pres. Madras 1133. 1924; Mohanan & Sivad., Fl. Agasthyamala 538.2002. *P. petchouly* sensu Hook.f., Fl. Brit. India 4: 633. 1885 ("*patchouli*"), non Pellit., 1844, excl. var. *suavis;* Nicols. Suresh & Mani., An Interpr. Hort. Malab. 156. 1988. *Cottam* Rheede, Hort. Malab. 10:153, t. 77. 1690.

Shrubs, to 2 m tall. Leave ovate-lanceate, 3-7 by 2-4 cm, base acute, shortly decurrent, margin irregularly serrate, apex acute, (sub) fleshy. Spicate racemes interrupted, branched to 9 cm long; calyx-lobes subequal; corolla white, c. 2 cm ϕ, lobes obscurely 2-lipped, lobes 3+1.

Fl. & Fr.: Nov.-Feb. *Distr.*: Sri Lanka, Continental SE. Asia. Less common. Along streamsides. *NAK 1468* (Arampa, c. 400 m), *2248* (Mannarappara, c. 250 m).

3. **Pogostemon paniculatus** (Willd.) Benth. in Wall., Pl. Asiat. Rar. 1 : 30; Hook. f., Fl. Brit. India 4: 631. 1885; Gamble, Fl. Pres. Madras 1132. 1924; Sald. & Nicols., Fl. Hassan Dist. 507.1976; Matthew & Rani in Matthew, Fl. Tam. Carnatic 3(2) : 1280. 1983; Manilal, Fl. Silent Valley 222. 1988; Nicols., Suresh & Mani., An Interpr. Hort. Malab. 156. 1988; Mohanan & Sivad., Fl. Agasthyamala 538.2002. *Elsholtzia paniculata* Willd., Sp. Pl. 3 : 59. 1800. *Manam-podam* Rheede, Hort. Malab. 10 : 129, t. 65. 1690.

Herbs to 1 m tall. Leaves unequal, ovate to oblanceate, 4-10 by 2-5 cm, base cuneate, margin coarsely serrate, apex acute, pubescent. Panicle unilateral, 15 cm long; spikes secund, flattened subtended by oblique bracts; calyx-lobes subequal; corolla violet, 2.5 mm ϕ, obscurely 2-lipped, 3+1.

Fl. & Fr.: Sep.-Dec. *Distr.*: Peninsular India, Mynamar. Common. On forest floors and plains. *NAK 1055* (Kodumon, c. 40 m), *1192* (Upper Moozhiar), *2807* (Konni, c. 100 m).

4. **Pogostemon travancoricus** Bedd., Ic. t. 159. 1868-74; Hook.f., Fl. Brit. India 4: 637. 1885; Gamble, Fl. Pres. Madras 1135. 1924.

Subshrubs, to 1 m tall. Leaves ovate, 3-5 by 2-4 cm, base rounded or cuneate, margin doubly serrate, apex acute. Spicate-racemes unbranched; bracts minute; calyx-lobes triangular; corolla violet.

Fl. & Fr.: Feb.-Apr. *Distr.*: Peninsular India. Very rare. In mountain forests. *CNM 73446* (MH) (Chokkampatti hills).

Salvia officinalis L., occasionally seen as a garden escape in hills (Kakki hills).

102. NYCTAGINACEAE

BOERHAVIA Linnaeus
Sp. Pl. 3. 1753.

Erect or diffuse herbs. Leaves opposite in unequal pairs. Inflorescence a panicle of umbellate capitate clusters. Flowers actinomorphic, 5-merous, bisexual; perianth funnel-form, basally sepaloid, constricted, 5-ribbed; upper part deep pink, or pinkish-white, expanded; stamens 1-3, included; ovary superior, 1-locular; ovule 1, basal; stigma peltate. Fruit (Anthocarp) 5-ribbed.

1. Erect herbs; fruit (anthocarp) obconic, apex truncate. ....1. **B. erecta**
1. Diffuse herbs; fruit (anthocarp) club-shaped, apex obtuse. ....2. **B. repens**

1. **Boerhavia erecta** L., Sp. Pl. 3. 1753; Stemm. in Steenis, Fl. Males. I, 6.:454. 1964; Matthew & Rani in Matthew, Fl. Tam. Carantic 3(2): 1288. 1983; Mohanan & Henry, Fl. Thiruvananthapuram 374.1994. *B. punarnava* Shah & Krishnamoorthy, J.Sci. Industr. Res. 210:254. 1961; Nair, Bull. Bot. Surv. India 9:288. 1967.

Erect herbs. Leaves elliptic-ovate to ovate-lanceate 2-5 by 1.2-3.5 cm, base cuneate or subacute, margin repand, apex acute, subsucculent. Panicle 2-4 cm long; perianth white with pink stripes, c. 3 mm ϕ; stamen 1 or 3. Anthocarp obconic, apex truncate.

Fl. & Fr.: Aug.-Dec. *Distr.*: Pantropics. Common in waste lowlands. *NAK 1000* (Tiruvalla, c.10 m).

2. **Boerhavia repens** L., Sp. Pl. 3. 1753; Fosberg, Smiths. Contrib. Bot. 39:8. 1978; Nicols., Suresh & Mani., An Interpr. Hort. Malab. 40. 1988. *B. diffusa* sensu Gamble, Fl. Pres. Madras 1162. 1925; Gandhi in Sald. & Nicols., Fl. Hassan Dist. 92. 1976; Matthew, Ill. Fl. Tam. Carnatic t. 578. 1982; Matthew & Rani in Matthew, Fl. Tam. Carnatic 3(2) : 1287. 1983, non L., 1753. *Talu-dama* Rheede, Hort. Malab. 7:105, t.56. 1988.

Diffuse herbs. Leaves broadly ovate, 2-4.5 by 2-4.5 cm, base truncate,apex obtuse, subsucculent. Umbellate cymes mostly axillary, 5-6 cm long; perianth deep pink, c. 4 mm ϕ; stamens 2 or 3. Anthocarp club-shaped, apex rounded, glandular.

Fl. & Fr.: Most of the season. *Distr.*: Pantropical. Very common. On waysides, wastelands, etc. *NAK 998* (Tiruvalla, c. 10 m).

Decoction of root is used as a remedy for rheumatism, cough, asthma, chest-pain, piles and swellings.

Note: This common species generally has been misidentified as *Boerhavia diffusa* L., where inflorescence is paniculate and strictly terminal.

Mirabilis jalapa L. occasionally run wild. **Bougainvillea** spp. with different varieties are seen cultivated in gardens.

103. AMARANTHACEAE

1. Leaves decussate.
   2. Anthers 1-celled. ....**Alternanthera**
   2. Anthers 2-celled.
      3. Flower-cluster with only 1 or 3 flower(s) perfect, others sterile. ....**Cyathula**

3. Flower-cluster with all flowers perfect.
    4. Spikes elongate; bracts & bracteoles spine-tipped; staminodes 5. .... **Achyranthes**
    4. Spikes globose to elongate; bracts and bracteoles not spine-tipped; staminodes 0. .... **Gomphrena**

1. Leaves alternate.
    5. Flowers bisexual.
        6. Inflorescence a dense spike; staminodes 5. .... **Aerva**
        6. Inflorescence either a panicle or a globose head; staminodes 0.
            7. Tall erect berbs; inflorescence a panicle. .... **Indobanalia**
            7. Diffuse herbs; inflorescence a globose head. .... **Allmania**
    5. Flowers unisexual. .... **Amaranthus**

ACHYRANTHES Linnaeus
Sp. Pl. 204. 1753.

Achyranthes aspera L., Sp. Pl. 204. 1753 var. **porphyristachya** (Wall. ex Moq.) Hook.f., Fl. Brit. India 4: 730. 1885; Gamble, Fl. Pres. Madras 1176. 1925. *A. porphyristachya* Wall. ex Moq. in DC., Prodr. 13(2):316. 1849.

Straggling subshrubs. Leaves decussate, oblanceate, 3-6 by 2-3 cm, base cuneate, apex acuminate, pubescent, chartaceous. Flowers in terminal and/or axillary spikes, bisexual; bracts and bracteoles spine-tipped; tepals 5, unequal, 1-3- nerved; stamens 5, connate, alternate with fimbriate staminodes; ovary superior, 1-locular, ovule 1, basal. Fruit a dehiscent utricle. Seed 1.

Fl. & Fr.: Sep.-Jan. *Distr.*: Sri Lanka, Peninsular India. Common along wayside thickets in upper hills. *NAK 1263* (Kakki hills, c. 1100 m).

AERVA Forsskal
Fl. Aegypt.-Arab. 170. 1775 (non. cons.).

Aerva lanata (L.) Juss., Ann. Mus. Natl. Hist. Nat. 2:131.1803; Schult. in L., Syst. Veg. 5:564. 1819; Wight, Ic.t. 723. 1840; Hook.f., Fl. Brit. India 4:728. 1885, Gamble, Fl. Pres. Madrs 1178. 1925; Ramamoorhty in Sald. & Nicols., Fl. Hassan Dist. 104. 1976; Backer in Steenis, Fl. Males. I, 4:84. 1949; Towns. in Dassan. & Fosberg, Rev. Handb. Fl. Ceylon 1:32:1980; Matthew & Rani in Matthew, Fl. Tam. Carnatic 3(2): 1299. 1983; Nicols., Suresh & Mani., An Interpr. Hort. Malab. 44. 1988. *Achyranthes lanata* L., Sp. Pl. 204. 1753. *Scheru-bula* Rheede, Hort. Malab. 10:57, t.29. 1690.

Erect or decumbent herbs. Leves closely alternate, ovate or orbicular, or broadly obovate, 1.5-3.5 by 1-2.5 cm, base cuneate, apex obtuse, chartaceous. Flowers minute, bisexual, in dense terminal and axillary spikes; bracts and bracteoles membranous; tepals 5, unequal; stamens 5, shortly connate below, alternate with subulate staminodes; ovary 1-ovuled. Utricle irregularly rupturing.

Fl. & Fr.: Sep.-Jan. *Distr.*: Tropics & Subtropics. Common on waysides in lowlands. *NAK 1005* (Tiruvalla, c. 10 m).

The plant made into a paste and taken along with milk to cure diabetes.

ALLMANIA R. Brown ex R. Wight
J. Bot. (Hooker) 1:225. 1834.

Allmania nodiflora (L.) R. Br. ex Wight in Hook., J. Bot. 1:227, t.128. 1834; Hook.f., Fl. Brit. India 4:716. 1885; Gamble, Fl. Pres. Madras 1167. 1925; Backer in Steenis, Fl. Males. I, 4:74:1949; Ramamoorthy in Sald. & Nicols., Fl. Hassan Dist. 105. 1976; Towns. in Dassan. & Fosberg, Rev. Hand. Fl. Ceylon 1:7. 1980; Matthew, Ill. Fl. Tam. Carnatic t.583. 1982; Matthew & Rani in Matthew, Fl. Tam. Carnatic 3(2): 1301. 1983; Manilal, Fl. Silent Valley 225. 1988; Nicols., Suresh & Mani., An Interpr. Hort. Malab. 44. 1988; Mohanan & Sivad., Fl. Agasthyamala 540.2002. *Celosia nodiflora* L., Sp. Pl. 205. 1753. *Pee-tardavel* Rheede, Hort. Malab. 9:153, t. 79. 1689. *Pee-coipa* Rheede, Hort. Malab. 10:133, t. 67. 1690.

Ascending herbs. Leaves alternate, oblong-spathulate, or elliptic-oblong, 2-6 by 0.5-2 cm, base attenuate, apex obtuse to acute, chartaceous. Flowers bisexual, in globose spikes; bracts aristate; tepals 5, silvery-white, 1-nerved, persistent; stamens 5, unequal; ovary globose, 1-ovuled. Utricle circumcissile. Seeds compressed, arillate.

Fl. & Fr.: Jun.-Aug. *Distr.*: Tropics. Less common. On exposed bare slopes. *CNM 2475* (Kalleli, c. 100 m).

ALTERNANTHERA Forsskal
Fl. Aegypt.-Arab. 28. 1775.

Prostrate herbs. Leaves decussate or clustered. Spikes axillary, globose to elongate. Flowers bisexual; bracteole 2; tepals 5, unequal; stamens 3 or 5, basally connate, alterante with subulate staminodes, anthers 1-celled; ovary 1-ovuled. Utricle indehiscent.

1. Bracts and bracteoles glabrous; fertile anthers 3. .... 1. **A. sessilis**
1. Bracts and bracteoles barbellate; fertile anthers 5. .... 2. **A. tenella**

1. **Alternanthera sessilis** (L.) R. Br. ex DC., Cat. Pl. Hort. Monsp. 4:77. 1813; Wight, Ic.t. 727. 1843; Hook.f., Fl. Brit. India 4: 731. 1885; Backer in Steenis, Fl. Males. I, 4:92. 1949; Ramamoorthy in Sald.& Nicols., Fl. Hassan Dist. 106. 1976; Towns. in Dassan. & Fosberg, Rev. Handb. Fl. Ceylon 1:49. 1980; Matthew, Ill. Fl. Tam. Carnatic t. 584. 1982; Matthew & Rani in Matthew, Fl. Tam. Carnatic 3(2): 1304. 1983; Nicols., Suresh & Mani., An Interpr. Hort. Malab. 45. 1988. *Gomphrena sessilis* L., Sp. Pl. 225. 1753. *Alternanthera triandra* Lam., Encyl. 1:95. 1783; Gamble, Fl. Pres. Madras 1179. 1925. *Coluppa* Rheede, Hort. Malab. 10:21-22, t.11.1960.

Prostrate herbs. Leaves lanceate to spathulate, 1.5-5 by 0.5-1 cm, base cuneate, apex acute or obstuse, chartaceous. Spikes white, globose; tepals 5, subequal; stamens, 4; staminodes 2, subulate. Utricle cordiform, strongly compressed. Seeds orbicular.

Fl. & Fr.: Most of the seasons. *Distr.:* Tropics & Subtropics. In moist clayey soil in lowlands. *NAK 1021* (Tiruvalla, c. 10 m).

2. **Alternanthera tenella** Colla, Mem. Acad. Sci. Terine Turin 33:131, t.9. 1820; Mears, Proc. Acad. Nat. Sci. Philadelphia 129: 19.1977; Sivar. & Matthew., Ind. J. For. 7(1): 49. 1984. *A. polygonoides* var. *erecta* Mart. Mears, l.c. in syn., non sensu Shrivastava & Santapau, 1955.

Prostrate herbs. Leaves elliptic to oblanceate, 3-4.5 by 1-2 cm, base cuneate, decurrent, apex subacue, chartaceous. Spikes white ovoid; bracts and bracteoles barbellate; tepals 5, spine-tipped; stamens 5, staminodes 3-4- toothed. Utricle cordiform. Seeds discoid, faintly reticulate.

Fl. & Fr.: Jul.-Dec. *Distr.*: Tropics & Subtropics. In moist clayey soil in lowlands. *NAK 1937* (Tiruvalla, c. 10 m).

AMARANTHUS Linnaeus
Sp. Pl. 989. 1753.

Monoecious herbs; stem subsucculent. Leaves alterante, apparently clustered with small axillary leaves. Flowers unisexual; male and female flowers mixed, in axillary or terminal spiciform racemes or panicles; bracteoles 2, persistent; tepals 3 or 5, hyaline, free in males, shortly connate in female; stamens 3 or 5, free; ovary 1-ovuled. Utricle circumcissile or indehiscent.

1. Spinescent herbs; utricle dehiscent (circumcissile)..1. **A. spinosus**
1. Non-spinescent herbs; utricle indehiscent..2. **A. viridis**

1. **Amaranthus spinosus** L., Sp. Pl. 991. 1753; Wight, Ic. t. 513. 1841; Hook.f., Fl. Brit. India 4:718. 1885; Gamble, Fl. Pres. Madras 1170. 1925; Sald. & Nicols., Fl. Hassan Dist. 106. 1976; Backer in Steenis, Fl. Males. I, 4:78.1949; Ramamoorhty in Sald. & Nicols., Fl. Hassan Dist. 106. 1976; Towns. in Dassan. & Fosberg, Rev. Handb. Fl. Ceylon 1:9. 1980; Matthew & Rani in Matthew, Fl. Tam. Carnatic 3(2): 1307. 1983; Manilal, Fl. Silent Valley 226. 1988; Mohanan & Sivad., Fl. Agasthyamala 541.2002. *Mullen-Keera.*

Spinescent herbs. Leaves elliptic-lanceate, 3-7 by 2-3.5 cm, base cuneate, apex obtuse or subacute. Flowers 1.5 mm ϕ; bracts deltoid; tepals 5, unequal; stamens 5. Utricle circumcissile.

Fl. & Fr.: Jan.-May. *Distr.:* Tropics. Common in waste lands, roadsides, etc. *NAK 1017* (Tiruvalla, c. 10 m).

Whole plant in decoction given to cure swellings and to clean kidneys. Root is a specific medicine in colic and gonorrhoea.

2. **Amaranthus viridis** L., Sp. Pl. (ed. 2) 1405. 1763; Roxb., Fl. Ind. 3:605. 1832; Hook.f., Fl. Brit. India 4:720. 1885; Gamble, Fl. Pres. Madrs 1171. 1925; Towns. in Dassan.& Fosberg, Rev. Handb. Fl. Ceylon 1:19. 1980; Matthew & Rani in Matthew, Fl.Tam Carnatic 3(2): 1308. 1983. *Kuppa-keera.*

Non-spinescent herbs. Leaves broaldy ovate, 1.5-5 by 1.2-3 cm, base cuneate, apex obtuse. Panicles terminal and axillary. Flowers 2 mm ϕ; tepals 2; stamens 3. Utricle indehiscent.

Fl. & Fr.: Jul.-Dec. *Distr.*: Tropics and Subtropics. Common. Along waysides and as a weed of cultivated plants. *NAK 1029* (Tiruvalla, c. 10 m), *1401* (Pampa dam area, c. 1100 m)

CYATHULA Blume
Bijdr. 548. 1826 (nom. cons).

Cyathula prostrata (L.) Blume, Bijdr. 549. 1826; Hook.f., Fl.Brit. India 4: 723. 1885; Gamble, Fl. Pres. Madras 1172. 1925; Ramamoorthy in Sald. & Nicols., Fl.Hassan Dist. 107. 1976; Nicols., Suresh & Mani., An Interpr. Hort. Malab. 47. 1988; Mohanan & Sivad., Fl. Agasthyamala 541.2002. *Achyranthes prostrata* L., Sp. Pl. (ed 2) 1:296. 1762. *Scheru-cadelari* Rheede, Hort. Malab. 10: 157, t. 79. 1690.

Decumbent herbs. Leaves opposite, rhombic, 3-8 by 2-4 cm, base cuneate, apex acute-apiculate, pubescent, chartaceous. Flowers in clusters, 1-2 perfect, bisexual, other sterile; tepals of bisexual flowers 5-lobed; of sterile reduced to hooked spines; stamens 6, connate below; staminodes lacerate; ovary 1-ovuled. Utricle circumcissile.

Fl. & Fr.: Sep.-Dec. *Distr.:* Tropics. Common. Among moist shady forest undergrowth. *NAK 114* (Lower Moozhiar, c. 250 m).

GOMPHRENA Linnaeus
Sp. Pl. 224. 1753.

Diffuse or erect herbs. Leaves decussate. Spikes terminal and/or axillary, subtended by basal leaves. Flowers bisexual, 2-bracteolate; tepals 5, unequal; stamens 5, connate below; ovary 1-ovuled. Utricle irregularly rupturing.

1. Spikes cylindric, silvery-white; ascending or diffuse herbs. ....1. **G. celosioides**
1. Spikes globose, pink; erect herbs. ....2. **G. globosa**

1. **Gomphrena celosioides** Martius, Beitr. Amarantac. 93. 1825 & Nova Acta Phys. Med. Acad. Caes. Leop.-Carol. Nat. Cur.13 : 30. 1826; Backer in Steenis, Fl. Males. I, 4: 96. 1949; Ramamoorthy in Sald. & Nicols., Fl. Hassan Dist. 108. 1976; Towns. in Dassan. & Fosberg, Rev. Handb. Fl. Ceylon 1:53. 1980; Matthew, Ill. Fl. Tam. Carnatic t.588. 1982; Matthew & Rani in Matthew, Fl. Tam. Carnatic 3(2) : 1312. 1983. *G. decumbens* auct., non Jacq., 1805: Gamble, Fl. Pres. Madras 1179. 1925.

Diffuse or ascending herbs.Leave oblanceate, 2-5 by 0.8-1.5 cm, base cuneate, apex subacute, hairy. Spikes elongate or globose, silvery-white; tepals fainty 1-nerved. Seeds faintly reticulate.

Fl. & Fr.: Most of the seasons. *Distr.*: Tropics. Common. In open lowlands and occasionally in hills. *NAK* 1780 (Thekkuthode, c. 70 m).

2. **Gomphrena globosa** L., Sp. Pl. 224. 1753; Roxb., Fl. Ind. 2: 63. 1832; Wight, Ic.t. 1784. 1862; Hook.f., Fl. Brit.India 4: 732. 1885; Gamble, Fl. Pres. Madras 1179. 1925; Backer in Steenis, Fl. Males. I, 4 : 95. 1949; Towns. in Dassan. & Fosberg, Rev. Handb. Fl. Ceylon 1:55. 1980; Matthew & Rani in Matthew, Fl. Tam. Carnatic 3(2) : 1313. 1983; Nicol., Suresh & Mani., An Interpr. Hort.Malab. 46. 1988. *Wadapu* Rheede, 10:73-74, t. 37. 1690.

Erect herbs. Leaves obovate-lanceate, 5-10 by 1.5-4 cm, base cuneate, apex subacute. Spikes globose, elongate with age, pink. Flowers c. 4 mm ϕ; tepals prominently 1-nerved. Seeds shiny.

Fl. & Fr.: Most of the seasons. *Distr.*: Tropics. A garden escape, now naturalised. *NAK 520* (Plappally, c. 400 m).

INDOBANALIA A.N. Henry et B. Roy
Bull. Bot. Surv. India 10:274. 1969.

Indobanalia thyrsiflora (Moq.) Henry & Roy, Bull. Bot. Surv. India 10 : 274. 1969; Ramamoorthy in Sald. & Nicols., Fl. Hassan Dist.108. 1976; Manilal, Fl. Silent Valley 226. 1988. *Banalia thyrsiflora* Moq. in DC., Prodr. 13(2):278. 1849; Wight, Ic. t. 1774. 1852; Hook.f., Fl. Brit. India 4:716. 1885; Gamble, Fl. Pres.Madras 1167. 1925.

Tall herbs. Leaves alternate, ovate-lanceate, 4-17 by 2.5-8.5 cm, base acute, apex acuminate, membranous. Flowers bisexual, in clusters on axillary or terminal panicled spikes; bracts and bracteoles hyaline; tepals 5, 3-nerved; stamens 5, connate below; ovary ovoid, compressd, 1-ovuled. Utricle subglobose, indehiscent.

Fl. & Fr.: Sep.-Feb. *Distr.:* Western Ghats. Rare. On moist shady forest floors in higher hills. *NAK 1433* (Arampa, c. 800 m), *2680 & 2693* (Kakki hills, c. 1100 m).

104. CHENOPODIACEAE

CHENOPODIUM Linnaeus

Sp. Pl. 218. 1753.

Chenopodium ambrosioides L., Sp. Pl. 219. 1753; Wight, Ic. t. 1786. 1852; Hook.f., Fl. Brit. India 5:4. 1886; Gamble, Fl. Pres. Madras 1181. 1925; Backer in Steenis, Fl.Males I, 5 : 101. 1949; Ramamoorthy in Sald. & Nicols., Fl. Hassan Dist. 101. 1976; Matthew & Rani in Matthew, Fl. Tam. Carnatic 3(2): 1322. 1983.

Woody herbs. Leaves alternate, oblong-lanceate or oblonaceate, 5-10 by 1-2.5 cm, base cuneate, apex subacute, chartaceous. Flowers small, bisexual, 5-merous, aggregated in terminal or axillary slender panicle; tepals 5; stamens 5; ovary tiped with glandular vesicles. Utricle enclosed in herbaceous tepals, dehiscent. Seeds horizontal.

Fl. & Fr.: Most of the seasons. *Distr.*: Tropics. Not common. Along moist waysides in upper hills. *NAK 2551* (Gavi-Pampa, c. 1200 m).

105. POLYGONACEAE

POLYGONUM Linnaeus

Sp. Pl. 359. 1753.

Herbs or shrubs. Leaves alternate; stipules connate into a tubular membranous ocrea. Infloresence usually a spike or panicles of capitate clusters; bracts and bracteoles ocreaceous. Flowers bisexual, regular; pedicels articulate below the periath. Perianth 4-5 lobbed, outer two lobes usually smaller. Stamens 5-8, filament often dialated at base, anthers dorsifixed. Ovary compressed or 3-gonous, 1-locular with one basal ovule; styles 2 or 3 branched, stigma capitate. Nutlet trigonous; covered by the persistent perianth.

1. Stragglers or prostrate herbs.
    2. Stragglers; leaves more than 3 cm broad; inflorescence of capitate clusters. .... 2. **P. chinense**
    2. Prostrate herbs; leaves less than 0. 5 mm broad; infloresence of axillary fasicles. .... 4. **P. plebium**
1. Erect subshrubs; inflorescence of racemes.
    3. Ocrea smooth at mouth; nut biconvex. .... 3. **P. glabrum**
    3. Ocrea barbellate at mouth; nut trigonous. .... 1. **P. barbatum**

1. **Polygonum barbatum** L., Sp. Pl. 362. 1753; Wight, Ic. t. 1798. 1852; Hook.f., Fl. Brit. India 5:37. 1886; Gamble, Fl. Pres. Madras 1189. 1925; Gandhi in Sald. & Nicols., Fl. Hassan Dist. 109. 1976; Nicols., Suresh & Mani., An Interpr. Hort. Malab. 211. 1988; Mohanan & Sivad., Fl. Agasthyamala 542.2002. *Persicaria barbata* (L.) Hara, Fl. E. Himal. 1:70. 1966; Matthew & Rani in Matthew, Fl. Tam. Carnatic 3(2) : 1334. 1983. *Belutta-modela-muccu* Rheede, Hort. Malab. 10:159, t.80. 1690. *Velutta-modela-muccu* Rheede, Hort. Malab. 12 : 145, t. 76.1693.

Stout herbs. Leaves lanceate, 7-15 by 2-3 cm, base acute to cuneate, apex gradually acute, thinely strigose, chartaceous; ocrea tubular, to 4 cm long, mouth long-barbellate. Flowers in terminal spikes; perianth white; style-arms 3. Nutlets 3-gonus.

Fl. & Fr.: Nov.-Feb. *Distr.*: Africa, Asia, Australia. Not common. Along streamsides in hills. Locally abundant. *NAK 1408* (Mannarappara, c. 250 m).

2. **Polygonum chinense** L., Sp. Pl. 363. 1753; Wight, Ic. t. 1806. 1852; Hook.f., Fl. Brit. India 5 : 44, 1886; Gamble, Fl. Pres. Madras 1190. 1925; Gandhi in Sald. & Nicols., Fl. Hassan Dist. 109. 1976; Nicols., Suresh & Mani., An Interpr. Hort. Malab. 212. 1988; Mohanan & Sivad., Fl. Agasthyamala 543.2002. *Persicaria chinensis* (L.) Gross, Bot. Jahrb. Syst. 49 : 269, 277, 315. 1913; Matthew & Rani in Matthew, Fl. Tam. Carnatic 3(2) : 1334. 1983; Manilal, Fl. Silent Valley 226. 1988. *Piripu* Rheede, Hort. Malab. 7: 101, t. 54. 1688.

Stragglers. Leaves elliptic-oblong, 7-15 by 3-7 cm, base obtusely truncate, apex shortly acuminate, chartaceous; ocrea 4-7 cm long, mouth obiquely cleft. Flowers in stalked capitate clusters aggregated in terminal corymbose-panicles; perianth white; style-arms 3. Nutlets 3-gonous.

Fl. & Fr.: Nov.-Feb. *Distr.*: Indo-Malesia. Common along wayside forest floors in hills. *NAK 316, 1533* (Kakki hills, c. 1100 m).

3. **Polygonum glabrum** Willd., Sp. Pl. 2 : 447. 1799; Wight, Ic. t. 1797. 1852; Hook.f., Fl. Brit. India 5:34. 1886; Gamble, Fl. Pres. Madras 1189. 1925; Gandhi in Sald. & Nicols., Fl. Hassan Dist. 109. 1976; Manilal, Fl. Silent Valley 227. 1988; Nicols., Suresh & Mani., An Interpr. Hort. Malab. 212. 1988. *Persicaria glabra* (Willd). M. Gomez, Ann. Inst. Segunda Enself. Habana 2: 278. 1896; Matthew, Ill. Fl. Tam. Carnatic t. 89.1982; Matthew & Rani in Matthew, Fl. Tam. Carnatic 3(2) : 1336. 1983. *Schovanna-modela-muccu* Rheede, Hort. Malab. 12 : 147, t. 77. 1693.

Stout herbs. Leaves narrowly lanceate, 10- 21 by 1.5-2.5 cm, base acute or cuneate, apex gradually acute; ocrea tubular, to 4 cm long, mouth truncate, smooth; perianth rose or white; style-arms 3. Nutlets 3-gonous.

Fl. & Fr.: Dec.-Feb. *Distr.*: Africa, Indo-Malesia. Very common. On marshes in hills and lowlands. *NAK 993* (Sabarimala, c. 250 m), *1288 & 1608* (Kakki hills, c. 1100 m).

4. **Polygonum plebeium** R. Br., Prodr. 420. 1810 ("*plebejum*"); Hook.f., Fl. Brit. India 5:27. 1886; Gamble, Fl. Pres. Madras 1188. 1925; Steward, Contr. Gray Herb. 5(88) : 24. 1930; Gandhi in Sald. & Nicols., Fl. Hassan Dist. 111. 1976; Matthew & Rani in Matthew, Fl.Tam. Carnatic 3(2) : 1339. 1983. *P. indicum* Heyne in Roth, Nov. Pl. Spec. 208.1821; Wight, Ic.t. 1808. 1852.

Prostrate herbs. Leaves oblanceate or narrowly obovate, 5-12 by 2-4 mm, base attenuate, shortly decurrent, apex obtuse or subacute, thick-chartaceous, subsessile; ocrea hyaline, scarious, obliquely tubular, lacerate. Flowers bisexual, in axillary fascicles; bracts hayline; perianth rose-coloured, 2 mm ϕ, lobes 5, unequal; stamens 5, included; disc-lobes glandular; ovary 3-gonous, 1-locular; ovule 1, basal. Nutlets 3-gonous.

Fl. & Fr.: Mar.-Apr. *Distr.*: Africa, Asia, Indo-Malesia, Australia. Rare. On fallow fields in upper hills. *NAK 2735* (Ponnambalamedu, c. 1100 m).

106. PODOSTEMACEAE

1. Flowers actionomorphic; tepals 3.
   2. Stiff, branched herbs; thallus basally attached; roots filiform. .... **Indotristicha**
   2. Succulent scarcely branched berbs; thallus attached throughout; roots thalloid. .... **Dalzellia**
1. Flowers zygomorphic; tepals 0.
   3. Capsule isolobus; thallus usually free-floating. .... **Polypleurum**
   3. Capsule anisolobus; thallus usually attached.
      4. Capsule smooth.
         5. Thallus fucoid; leaves fleshy, distichous. .... **Cladopus**
         5. Thallus crustaceous; leaves not fleshy, 4-ranked. .... **Willisia**
      4. Capsule ribbed. .... **Zeylanidium**

CLADOPUS H. Muller
Ann. Jard. Bot. Buitenzorg, ser.2, 1:115. 1899.

Cladopus hookerianus (Tul.) Cusset, Fl. Cambodge, Laos, Viet-Nam 14: 71.1973, Adansonia 1:24. 1992; Nag. & Arekd, Bull. Bot. Surv. India 23: 231. 1981. *Mniopsis hookeriana* Tul., Ann. Sc. Nat. ser. 3, 2:105. 1849. *Podostemon hookerianus* (Tul.) Wedd. in DC., Prodr. 17 : 74. 1873; Hook.f., Fl. Brit. India 5 : 65. 1886. *Griffithella hookeriana* (Tul.) Warm., Danske Vid.-Selsk. Skr. ser. 2 : 13. 1901; Willis, Ann. Roy. Bot. Gard. (Peradeniya) 1(3) : 233. 1902; Gamble, Fl. Pres. Madras 1197. 1197. 1925.

Lithophytic herbs. Thallus variable, thread-like or disc-like, attached firmly throughout the length, bearing leafy secondary shoots. Leave sheathing, distichous. Flowers minute, zygomorphic, bisexual, naked, emerging from a spathe; stamens 20, monadelphous; staminodes 2, linear; ovary superior, oblique, 2-locular: ovules α, on a massive placenta. Capsule nearly globose, splitting into 2 unequal lobes.

Fl. & Fr.: Dec.-Feb. *Distr.*: Western Ghats. Rare. On rocks in running water. *NAK 486* (Maniyaar, Kallar river, c. 70 m).

INDOTRISTICHA Royen
Acta Bot. Neerl. 8. 474. 1959.

Indotristicha ramosissima (Wight) Royen, Acta Bot. Neerl. 8: 474. 1959; Mohanan & Sivad., Fl. Agasthyamala 544.2002. *Dalzellia ramosissima* Wight, Ic. t. 1920, f. 1. 1852. *Terniola ramosissima* (Wight) Wedd. in DC., Prodr. 17 : 47. 1873; Hook.f., Fl. Brit. India 5 : 63. 1886. *Tristicha ramosissima* (Wight) Willis, Ann. Roy. Bot. Gard. (Peradeniya) 1(3) : 208, 293, 306, tt. 5-9, f. 29. 1902; Gamble, Fl. Pres. Madras 1194. 1925. **(Fig. 51)**

Submerged herbs, basally attached, roots creeping; secondary shoots usually in pairs, floating, frequently branched; branches of two kinds, long ones with the structure of main axis and short ones with delicate axis and a small tristichous leaves. Flower regular, terminal, basally subtended by 2-3 imbricate leaves; tepals 3, basally connate; stamens 3, hairy; ovary 3-locular; style-arms 3. Capsule 3-ribbed. Seeds minute, α.

Fl. & Fr.: Feb-Mar. *Distr.*: Western Ghats. Sri Lanka, Common. On submerged rocks in streams. *NAK 481* (Kallar river, Maniyaar, c. 70 m).

POLYPLEURUM (Tulsane) Warming
Kongel. Danske Vidensk-Selsk. Skr. Ser. 6, II:64.1901

Polyplerum stylosum (Wight) Hall, Kew Bull. 26:131.1971. *Dicraea stylosa* Wight, Ic. t. 1917, f.2. 1852; Gamble, Fl. Pres. Madras 1196. 1925; Manilal, Fl. Silent Valley 228.1988. *Podostemon stylosus* (Wight) Benth. ex Hook.f., Fl. Brit. India 5:64.1886. *Dicraea algaeformis* Bedd., Trans. Linn. Soc. London 25 : 223, t.24. 1865. *P. algaeformis* (Bedd.) Benth. ex Hook.f., Fl. Brit. India 5:65.1886.

Thallus basally attached with free floating branches, rarely attached throughout the length; secondary shoots marginal. Leaves subulate. Flowers zygomorphic, basally enclosed by sheathing spathes and fleshy bracts; pedicel c. 2 cm long; tepals 0; stamen(s) usually 1 or 2; staminodes 2, linear; ovary ellipsoid, 1-locular. Capsule 8-10-ribbed, isolobous.

Fl. & Fr.: Dec.-Feb. *Distr.*: Western Ghats. Very common. On rocks in flowing streams. *NAK 486* (Kallar river.-Completely attached forms), *NAK 2668* (Punnamada river, Thannithode.–The common form with very long floating branches).

DALZELLIA Wight.
Icon. Pl. Ind. Or. 5:34.1852.

Dalzellia zeylanica (Gard.) Wight, Ic., t.5. Pl. Ind. Or. 5:34, t.1919.1852; Brink, Taxon 18:598. 1969; Mohanan & Sivad., Fl. Agasthyamala 544.2002. *Tristicha zeylanica* Gard., Calcutta J. Nat. Hist. 7:177.1846. *Lawia zeylanica* (Gard.) Tul., Ann. Sc. Nat. Ser. 3, 2:112.1849, *emend* Willis, Ann. Roy. Bot. Gard. (Peradeniya 1(3) : 213.1902; Gamble, Fl. Pres. Madras 1195.1925. *Terniola zeylanica* (Gard.) Tul., Archives Mus. Hist. Nat. Paris 6:190-192, t. 13, f.3. 1852; Hook.f., Fl. Brit. India 5:62.1886; Ramamoorthy in Sald. & Nicols., Fl. Hassan Dist. 269.1976.

Thallus firmly attached throughout the length on rocks, suborbicular or stellate. Leaves lanceoid, acute usually reddish-brown. Flowers regular, bisexual, on margins or upper surface of thallus, enclosed basally by sheathing cupules; tepals 3; stamens 3; ovary 3-locular. Capsule ellipsoid or obovoid, 9-ribbed, 3-locular. Seeds minute.

Fl. & Fr.: Dec.-jan. *Distr.*: Sri Lanka, Western Ghats. Common. On exposed rocks of streams *NAK 2234 & 2235* (Perunthenaruvi, Ranni, c.100 m).

WILLISIA Warming
Kongel. Danske Vidensk.-Skr. ser. 6, II:58, 65.1901.

Willisia selaginoides (Bedd.) Warm. ex Willis, Ann.Roy.Bot. Gard. (Peradeniya) 1(3) : 255.1902; Gamble, Fl. Pres. Madras 1198. 1925; Manilal, Fl. Silent Valley 228.1988; *Mniopsis selaginoides* Bedd., Madras J. Sci. Ser. 3(1) : 54.1864. *Podostemon selaginoides* Benth. in Benth. & Hook.f., Gen. Pl. 3:113. 1880; Hook.f., Fl. Brit. India 5:68. 1886.

Thallus crustose, tufted, firmly attached on rocks; secondary shoots erect, tufted with scaly leaves, triquterous. Flowers zygomorphic, naked, sessile, basally enclosed by spathes; stamens 2, monadelphous; staminodes 2, linear; ovary ellipsoid. Capsule smooth or faintly 2-ribbed.

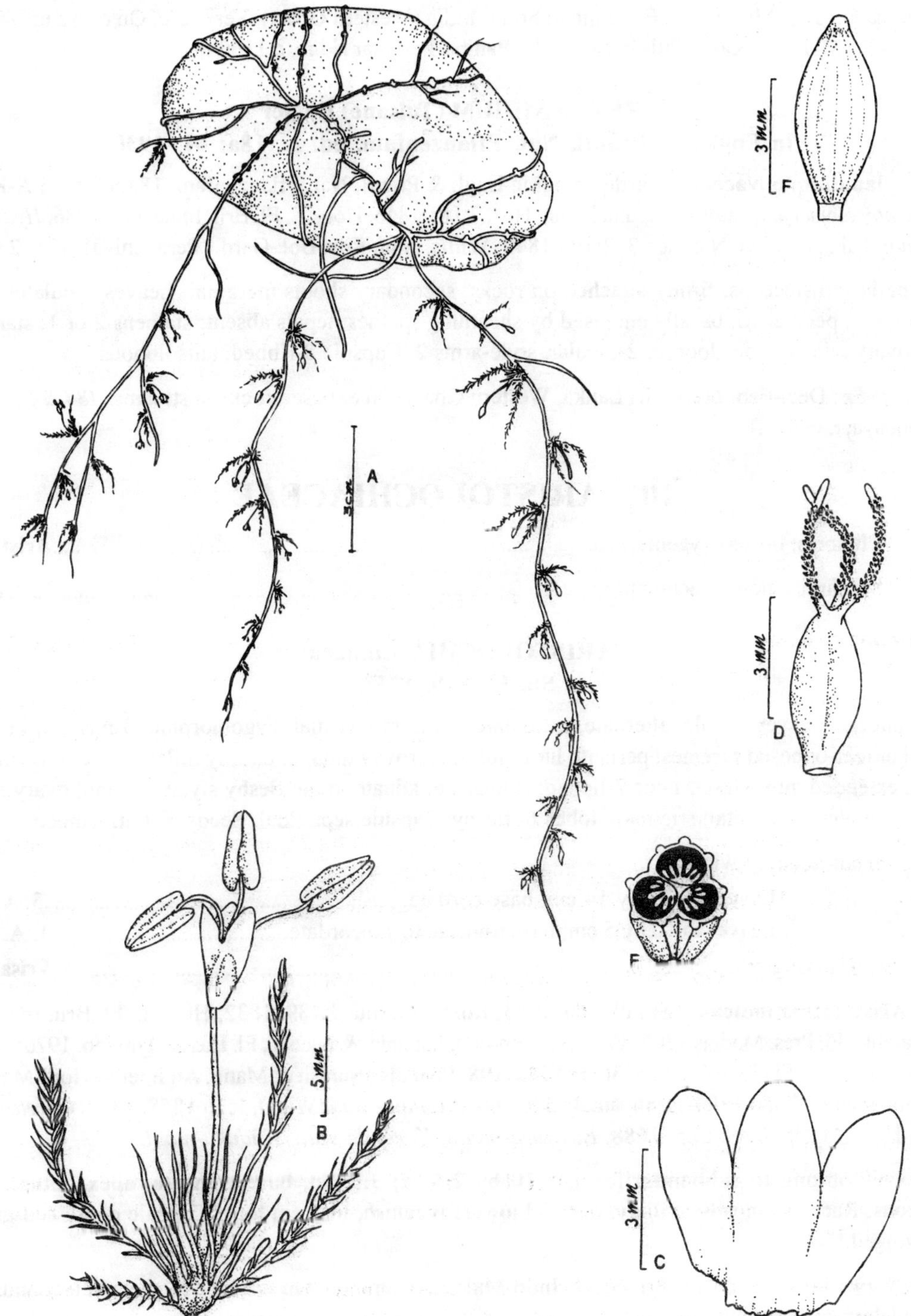

Fig. 51. ***Indotristicha ramosissima*** **(Wight) Royen: A. Plant with substratum; B. Flower; C. Perianth; D. Pistil; E. Ovary - C.S.; F. Fruit.**

Fl. & Fr.: Jan.-Mar. *Distr.*: Endemic to South India (Western Ghats). Very rare. On rocks in water falls. *CNM 72828* (MH), (In Kakki falls between the Pamba-Periyaar ridge, c.1100 m).

ZEYLANIDIUM (Tulsane) Engler in Engler et Prantl, Nat. Pflanzenfam. ed. 2., 18a: 61. 1930.

Zeylanidium olivaceum (Gardn.) Engl.in Engl. & Prantl, Nat. Pflanzenfam. 18a:62, f. 55 A-K. 1930. *Podostemon olivaceum* Gardn., Calcutta J. Nat. Hist. 7:181.1846; Hook.f., Fl. Brit. India 5:66.1886. *Hydrobryum olivaceum* Tul., Ann. Sci. Nat. Ser. 3, 2:104.1849; Willis, Ann. Roy. Bot. Gard. (Peradeniya) 1(3) : 239.1902.

Thallus crustaceous, firmly attached on rocks; secondary shoots marginal. Leaves subulate. Flowers zygomorphic, pedicelled, basally enclosed by sheathing spathes; tepals absent; stamens 2 or 1; staminodes linear; ovary ellipsoid or globose, 2-locular, style-arms 2. Capsule 8-ribbed, anisolobous.

Fl. & Fr.: Dec.-Feb. *Distr.*: Sri Lanka, Western Ghats. On exposed rocks in streams. *484 & 560* (Kallar river, Maniyaar, c.70 m).

107. ARISTOLOCHIACEAE

1. Climbers; flowers zygomorphic. .... **Aristolochia**
1. Subshrubs; flowers actinomorphic. .... **Thottea**

ARISTOLOCHIA Linnaeus Sp. Pl. 960. 1753.

Climbers. Leaves simple, alternate, estipulate. Flowers bisexual, zygomorphic, 3-merous, in axillary terminal or leaf-opposed racemes; perianth lurid, lobes narrowly tubular, basally inflated, enclosing the stylar column, extended into a limb, 1- or 2-lipped; stamens 6, adnate to the fleshy stylar column; ovary inferior, 6-locular; ovules α, parietal; stigmatic lobes 6, fleshy. Capsule septicidal. Seeds α, flat, winged or not.

1. Seeds broadly winged.
    2. Leaves to 25 by 14 cm, base cordate. .... 3. **A. tagala**
    2. Leaves to 10 by 5 cm, base truncate to subcordate. .... 1. **A. indica**
1. Seeds wingless. .... 2. **A. krisagathra**

1. **Aristolochia indica** L., Sp. Pl. 960. 1753; Roxb., Fl. Ind. 3:489. 1832; Hook.f., Fl. Brit. India 5: 75. 1886; Gamble, Fl. Pres. Madras 1202. 1925; Ramamoorthy in Sald. & Nicols., Fl. Hassan Dist. 56. 1976; Matthew & Rani in Matthew, Fl. Tam. Carnatic 3(2) : 1345. 1983; Nicols., Suresh & Mani., An Interpr. Hort. Malab. 60. 1988; Mohanan & Sivad., Fl. Agasthyamala 546.2002. *A. lanceolata* Wight, Ic. t. 1858. 1852. *Careloe-vegon* Rheede, Hort. Malab. 8:49, t.25. 1688. *Easwara-mulla, Karenda-valli, Kuduka-mooli.*

Leaves oblong to subpanduriform, 5-10 by 2.5-(5) 3.5 cm, base truncate, apex obtuse-retuse, chartaceous. Racemes mostly axillary, dense. Flowers greenish, to 5 cm long. Capsule depressed-globose. Seeds winged.

Fl. & Fr.: Jul.-Nov. *Distr.*: Sri Lanka, Indo-Malesia. Common. On wayside thickets in lowlands. *NAK 685* (Pandalam, c.20 m).

Juice of leaves boiled in oil and is applied to swellings and eczema. The root used as an antidote against snake-bite, bites of poisonous insects, etc.

2. **Aristolochia krisagathra** Sivar. & Pradeep, Pl. Syst. Evol. 163:31-34. 1989; Mohanan & Sivad., Fl. Agasthyamala 546.2002.

Leaves oblong-lanceate, 6-20 by 2-6 cm, base truncate to subcordate, apex accumiate-apiculate, chartaceous. Racemes axillary and terminal. Flowers dark-purple. Capsule subglobose or ovoid. Seeds çordiform or triangular, wingless, muricate.

Fl. & Fr.: Sep.-Jan. *Distr.*: Endemic to Travancore. Common. On thickets and bushes along waysides in evergreen forests. *NAK 264 & 1200* (Moozhair, c.250 m), *2114* (Vallicode-Kottayam, c.150 m).

3. **Aristolochia tagala** Cham., Linnaea 7:207, t.5, f.3. 1832; Gamble, Fl. Pres. Maras 1202. 1925; Ramamoorthy in Sald. & Nicols., Fl. Hassan Dist. 57.1976; Matthew & Rani in Matthew, Fl. Tam. Carnatic 392: 1345. 1983; Manilal, Fl. Silent Valley 229.1988; Mohanan & Sivad., Fl. Agasthyamala 547.2002. *A. acuminata* Roxb., Fl. Ind. 3:489.1832, non Lam., 1783; Wight, Ic.t. 771. 1844. *A roxburghiana* Klotzsch, Monatsber. Deutsch. Akad. Wiss. Berlin. 596. 1859; Hook.f., Fl. Brit. India 5:75.1886.

Leaves broadly ovate, 8-25 by 5-14 cm, base deeply cordate, apex acuminate, chartaceous. Racemes axillary, 4-6 -flowered. Flowers to 7 cm long, dark-purple, limb entire, rolled back, 1-lipped. Capsule subglobose, c. 3.5 cm ϕ. Seeds cordiform, winged all around.

Fl. & Fr.: Sep.-Feb. *Distr.*: Sri Lanka, Indo-Malesia. Not common. On forest borders in hills. *NAK 924* (Plappally, c.400 m), *NAK 1212* (Upper Moozhiar, c.550 m).

THOTTEA Rottboell
Nye Saml. Kongel. Danske Vidensk. Selsk. Skr. 2:529.1783.

Subshrubs or woody herbs. Leaves alternate, 3-5-nerved from base, reticulations prominent beneath, coriaceous. Flowers actinomorphic few to ϕ, in axillary or subradical cymes; perianth campanulate, 3-lobed or 3-fid, deep purple or creamy; stamens 9-15, 1-or 2-seriate, triadelphous or free; ovary inferior, 4-lobed; stylar column short or none; stigmatic lobes 4-9, linear. Capsule 4-angled, pubescent to glabrescent. Seeds angled.

1. Cymes subradical; flowers 8-9 per rachis; stamens 15, free in 2 whorls. .... 1. **T. dinghoui**
1. Cymes axillary; flower(s) 1-3 per axils; stamens 9, connate, triadelphous.
    2. Perianth-lobes broadly elliptic, divided to base. ....3. **T. siliquosa**
    2. Perianth lobes brodly ovate, divided to middle. ....2. **T. duchartrei**

1. **Thottea dinghoui** Swarup., Blumea 28: 407-411.1983.

Ceaspitose woody herbs. Leaves lanceate or elliptic-lanceate, 12-18 by 6.5-7 cm, base obtuse, apex acute, basally 3-nerved, lower surface tomentose. Cyme subradical, 10-20 cm long. Flowers cream, coloured, 8-12 per rachis; perianth lobes free; stamens 15, in 2 whorls, outer 9 in 3 groups of 2 each; stigma 4-5, linear. Capsule puberulent. Seeds trigonous.

Fl. & Fr.: May-Nov. *Distr.*: Endemic to Western Ghats (Travancore hills). Very rare. Among undergrowth in evergreen forests. *NAK 1823* (Angamoozhy, c.250 m).

2. **Thottea duchartrei** Sivar. & Babu , Ind.J. For.267.1985. *Kodassari.*

Subshrubs. Leaves oblanceate or narrowly obovte, 10-20 by 5-8 cm, base obtuse or broadly acute, apex obtusely acuminate. Flower(s) 13-per axil; perianth deep purple, lobes 3, broadly ovate, divided to middle. Capsule greenish, glabrescent.

Fl. & Fr.: Most of the season. *Distr.*: Peninsular India (Travancore hills). Most common. On thickets, bushes, in plantations of midlands. *NAK 220, 1473 & 1600 2268, 2473* (Most of the midlands and hills).

Root powdered and given in hot water as an antidote to poison; and externally applied in swellings.

3. **Thottea siliquosa** (Lam.) Ding Hou, Blumea 27(2) : 327.1981; Manilal, Fl. Silent Valley 229.1988; Nicols., Suresh & Mani., An Interpr. Hort. Malab. 28. 1982.; Mohanan & Sivad., Fl. Agasthyamala 548.2002. *Apama siliquosa* Lam., Encycl. 1:91. 1783; Gamble, Fl. Pres Madras 1200. 1925; Ramamoorthy in Sald. & Nicols., Fl. Hassan Dist. 56. 1976. *Bragantia wallichii* R. Br. ex Wight, Ic. t.520. 1841; Hook.f., Fl. Brit. India 5:73.1886. *Alpam* Rheede, Hort. Malab. 6:51-52, t.28. 1686.

Subshrubs. Leaves oblong-lanceate, 12-22 by 4.5.5 cm, base acute, apex gradually acuminate. Cymes axillary, peduncle short or 0. Flower(s) 1-3 per axils; perianth deep purple, lobes 3, broadly elliptic, divided to base. Capsule 4-angled, pubescent.

Fl. & Fr.: Nov.-Mar. *Distr.*: Sri Lanka, Peninsular India. Not common. On forest edges in upper hills. *NAK 1385* (Kakki hills, c.1100 m).

108. PIPERACEAE

1. Root- climbing shrubs; flowers unisexual. .... **Piper**
1. Erect shrubs or herbs; flowers bisexual.
    2. Erect shrubs; spikes stout; leaves to 30 cm long. .... **Lepianthes**
    2. Erect herbs; spikes slender; leaves to 3 cm long. .... **Peperomia**

LEPIANTHES Rafinesque
Sylva Tellur. 84. 1838.

Lepianthes umbellata (L.) Raf., Sylva Tellur. 84. 1838; Ramamoorthy in Sald. & Nicols., Fl. Hassan Dist. 52. 1976; Manilal, Fl. Silent Valley 230. 1988; Mohanan & Sivad., Fl. Agasthyamala 549.2002. *Piper umbellatum* L., Sp. Pl. 30. 1753. *P. subpeltatum* Willd., Sp. Pl. 1:166. 1798; Hook.f., Fl. Brit. India 5:95 1886. *Heckeria subpeltata* (Willd.) Kunth, Linnaea 13:171. 1839; Gamble, Fl. Pres. Madras 1208. 1925. *Pothomorphe subpeltata* (Willd.) Miq., Syst. 213. 1843; Wight, Ic. t. 1925. 1853.

Shrubs; branches succulent. Leaves simple, alternate, orbicular-ovate, 20-30 by 15-25 cm, base cordate, apex acute to shortly acuminate, puberulous, chartaceous, peltate, long-petioled. Flowers minute, sessile,bisexual, closely aggregated in cylindric, umbellate, fleshy, axillary spikes; bracts peltate; perianth 0; stamens 2 or 3, anthers small, 4-locular; ovary superior, 1-locular; ovule 1; stigma sessile. Fruit a dry trigonous berry.

Fl. & Fr.: Dec.-Feb. *Distr.*: Sri Lanka, Trop. Africa, India, Malaya. Less common. Along moist shady forest floors in upper hills. Locally abundant. *NAK 10 & 113* (Moozhiar, c. 250 m).

PEPEROMIA Ruiz et Pavon
Prodr. 8. 1794.

Succulent, slender herbs. Leaves decussate or alternate, 3-5(-7)-nerved from base, pellucid-punctate; bracts peltately or basally attached. Flowers bisexual; stamens 2, anther-lobes confluent; stigma sessile. Berry thin-walled, rugose or smooth, partly embedded in spikes.

1. Terrestrial, occasionally epiphytic herbs; leaves cordiform. ....1. **P. pellucida**

1. Epiphytic herbs; leaves obovate. ....2. **P. portulacoides**

1. **Peperomia pellucida** (L.) H.B.K., Nov. Gen. Sp. 1:64. 1816; Gamble, Fl. Pres. Madras 1210. 1925; Mohanan & Sivad., Fl. Agasthyamala 551.2002. *Piper pellucidum* L., Sp. Pl. 30. 1753.

Slender succulent herbs. Leaves decussate and alternate, cordiform, 2-3 by 1.5 2.5 cm, base cordate, apex shortly acuminate, 5-7- nerved, membranous when dry. Spikes slender, leaf-opposed, c. 5 cm long. Berry ovoid, ribbed, reticulate.

Fl. & Fr.: Sep.-Dec. *Distr.*: Native of S. America, now introduced into tropics. Common. On moist walls and shady areas of lowlands. *NAK 871* (Pathanamthitta, c.50 m).

2. **Peperomia portulacoides** (Lam.) Dietr., Sp. Pl. ed. 6,1:172. 1831; Wight, Ic. t. 1922, f.2. 1853; Hook.f., Fl. Brit. India 5:98. 1886; Gamble, Fl. Pres. Madras 1210. 1925; Ramamoorthy in Sald. & Nicols., Fl. Hassan Dist. 53. 1976; Mohanan & Sivad., Fl. Agasthyamala 552.2002. *Piper portulacoides* Lam., Tabl. Encycl. 1:82. 1791.

Subsucculent herbs. Leaves decussate and alternate, obovate, 0.6-2.5 by 0.5-2 cm, base cuneate, apex obtuse, basally 3-nerved, fleshy. Spikes cylindric, axillary leaf-opposed or terminal, to 5 cm long. Berry ovoid, scaly.

Fl. & Fr.: Sep.-Dec. *Distr.*: Peninsular India. Madagascar & Mauritius. Rare epiphytic herbs in evergreen forests. *NAK 961* (Sabari hills, c. 250 m), *2547* (Kakki hills, c. 1100 m).

PIPER Linnaeus
Sp. Pl. 28. 1753.

Climbers; nodes swollen. Leaves alternate, 5-9-nerved from base, membranous to coriaceous. Flowers unisexual, sessile in axillary pendant or erect spikes, partly embedded in fleshy peduncles, enclosed within bracts; bracteoles present or 0; stamens 2 or 3, anthers 2-celled, convergent; ovary ovoid or ellipsoid; style 2-4, spreading. Berry globose to ellipsoid.

1. Fruiting spikes globose.
   2. Flowers enclosed in globose sacs of connate bracts. ....6. **P. trichostachyon**
   2. Flowers not enclosed in globose sacs of connate bracts. ....4. **P. mullesua**
1. Fruiting spikes not globose.
   3. Trailors; fruiting spikes thick. ....3. **P. longum**
   3. Climbers; fruiting spikes slender.
      4. Leaves usually broadly ovate; bracts forming an involucral-cup below the flowers. ....5. **P. nigrum**
      4. Leaves usually elliptic or elliptic-ovate; bracts adnate medianly to rachis with free margins.
         5. Pubescent climbers. ....2. **P. hymenophyllum**
         5. Glabrous climbers.
            6. Female spikes to 25 cm long; leaves glabrous beneath. ....7. **P. trioicum**

6. Female spikes to 15 cm long; leaves silvery-scaly beneath. .... 1. **P. argyrophyllum**

1. **Piper argyrophyllum** Miq., Syst. Piper. 330. 1844; Wight, Ic.t. 1941. 1853; Hook.f., Fl. Brit. India 5:93. 1886; Gamble, Fl. Pres. Madras 1207. 1925; Ramamoorthy in Sald. & Nicols., Fl. Hassan Dist. 53. 1976; Nicols., Suresh & Mani., An Interpr. Hort. Malab. 207. 1988. *Amolago* Rheede, Hort. Malab. 7:31, t. 16. 1688.

Slender climbers. Leaves broadly or narrowly elliptic, 6-12 by 2.5-6 cm, base acute or oblique, apex acuminate, basally 5-nerved, lower surface often with silvery scales, thin-chartaceous. Male spikes slender, 10-15 cm long. Female-spikes to 15 cm long. Berries globose.

Fl. & Fr.: Jan.-Mar. *Distr.*: Sri Lanka, Western Ghats. Less common. In evergreen forests. *NAK* 606 (Thekkuthode, c. 70 m), *NAK 1531* (Arampa, c. 400 m).

2. **Piper hymenophyllum** Miq. in Hook., London J. Bot. 5:554. 1846; Wight, Ic. t. 1942. 1853; Hook.f., Fl. Brit. Inida 5:93. 1886; Gamble, Fl. Pres. Madras 1207. 1925; Matthew, Ill. Fl. Tam. Carnatic t. 603. 1982; Ramamoorthy in Sald. & Nicols., Fl. Hassan Dist. 53. 1976; Matthew & Rani in Matthew, Fl. Tam. Carnatic 3 (2): 1351. 1983; Manilal, Fl. Silent Valley 231. 1988; Mohanan & Sivad., Fl. Agasthyamala 555.2002.

Slender hispid climbers. Leaves ovate to ovate-elliptic, 8-13 by 4.5-8 cm, base rounded apex acuminate, 7-nerved from base, chartaceous. Male spikes c. 10 cm long. Female-spikes to c. 9 cm long. Berries globose.

Fl. & Fr. Jan-Apr. *Distr.*: Western Ghats. Not common. In evergreen forests! *NAK 1109, 2398 & 2462* (Moozhiar, c. 250 m).

3. **Piper longum** L., Sp. Pl. 29. 1753; Hook.f., Fl. Brit. India 5:83 1886; Gamble, Fl. Pres. Madras 1205. 1925; Nicols., Suresh & Mani., An Interpr. Hort. Malab. 208. 1988. *Chavica roxburghii* Miq., Syst. Piperac. 239. 1843; Wight, Ic.t. 1928. 1853. *Cattu-tirpali* Rheede, Hort. Malab. 7: (27) - 28, t. 14. 1688.

Creeping stout subshrubs, Leaves: upper elliptic-ovate, 6-12 by 3-5.5 cm; lower ovate, base obliquely cordate, apex acuminate, basally 5-nerved, chartaceous. Male-spikes 3-6 cm long. Female spikes (fruiting) 4 by 0.7 cm. Berries small, ellipsoid.

Fl. & Fr.: Sep.-Dec. *Distr.*: Sri Lanka, India, Malesia, Common. Among undergrowth in moist deciduous and semi-evergreen forests. *NAK* 610, 721 & 785, 1409 (Almost all hills).

Decoction of fruits in honey given for cough and throat pain, and fruit used as a spice.

4. **Piper mullesua** Buch.-Ham.ex D. Don, Prodr. Fl. Nepal 20. 1825; Mohanan & Sivad., Fl. Agasthyamala 556.2002. *P. brachystachyum* Wall. ex Hook.f., Fl. Brit. India 5:87. 1986; Gamble, Fl. Pres. Madras 1206. 1925; Manilal, Fl. Silent Valley 231. 1988. *Chavica sphaerostachya* Miq., Syst. Piperac. 279. 1843; Wight, Ic. t. 1931. 1853.

Creeping or sout climbers. Leaves broadly elliptic-ovate, 6-9 by 3.5-6.5 cm, base rouned, apex acuminate, basally 5-(7)-nerved chartaceous. Male spikes slender, c. 5 cm long; female spikes (fruiting) globose. Berries small, ellipsoid.

Fl. & Fr.: Jan.-Apr. *Distr.*: Peninsular & Eastern India. Not common. In semievergreen and evergreen forests. *NAK 595* (Maniyaar, c.70 m), *NAK 2687* (Kakki hills, c. 1100 m).

5. **Piper nigrum** L., Sp. Pl. 28. 1753; Hook.f., Fl. Brit. India 5:90. 1886; Gamble, Fl. Pres. Madras 1206. 1925; Ramamoorthy in Sald. Nicols., Fl. Hassan Dist. 54. 1976; Matthew & Rani in Matthew, Fl. Tam. Carnatic

3 (2) : 1352. 1983; Manilal, Fl. Silent Valley 231. 1988; Nicols., Suresh & Mani., An Interpr. Hort. Malab. 209. 1988. *Molago-codi* Rheede, Hort. Malab. 7:23-24, t. 12. 1688. *Cattu-molago* Rheede, Hort. Malab. 7:25, t. 13. 1688.

Stout climbers. Leaves broadly ovate to elliptic-ovate, 10-15 by 6-9 cm, base cuneate, apex acuminate, basally 3-(5)- nerved, 2-pairs rather higher up. Spikes stout, to 6 cm long. Berries globose, ripening crimson-red.

Fl. & Fr.: Jan.-Apr. *Distr.*: Sri Lanka, Peninsular India. Less common in wild. In semi-evergreen and evergreen forests. *NAK 540* (Angamoozhy, c. 200 m). Some collections are with considerably large fruits.

Fruit is the well known spice, black pepper. Root powder or decoction taken with honey cures cough, asthma and rheumatism.

6. **Piper trichostachyon** (Miq.) C.DC., Prodr. 16 (1): 242. 1869; Hook.f., Fl. Brit. India 5:80. 1886; Gamble, Fl. Pres. Madras 1205. 1925. 1925; Ramamoorthy in Sald. & Nicols., Fl. Hassan Dist. 54. 1976; Manilal, Fl. Silent Valley 232. 1988. *Muldera trichostachya* Miq. in Hook., London J. Bot. 5:556. 1846; Wight, Ic. t. 1944. 1852.

Stout climbers, nodes very much swollen. Leaves narrowly elliptic, 6-12 by 2-4 cm, base, acute, apex shortly acuminate, basally 3-nerved, a pair rather higher up, coriaceous. Spikes to 10 cm long, glabrescent; bracts connate. Berry globose, yellow when ripe.

Fl. & Fr.: Mar.-Apr. *Distr.*: Peninsular India. Rare. In evergreen forests. *NAK 1297 & 2463* (Moozhiar, c.250 m).

7. **Piper trioicum** Roxb., Fl.Ind.1:151.1820; Wight, Ic.tt. 1935 & 1936. 1853; Subram. & Henry, Bull. Bot.Surv. India 12:4.1976; Ramamoorthy in Sald. & Nicols., Fl. Hassan Dist. 54.1976. *P. attenuatum* Buch.-Ham. ex Miq., Syst. Piperac. 306.1844; Hook.f., Fl. Brit. India 5:92.1886; Gamble, Fl.Pres.Madras 1207.1925; Matthew & Rani in Matthew, Fl. Tam. Carnatic 3(2):1351.1983. *P. sylvestre* sensu Wight, Ic.t. 1937.1853, non Lour., 1790; Hook.f., Fl. Brit. India 5:93.1886.

Slender climbers. Leaves broadly ovate to elliptic-ovate, 9-13 by 5-9 cm, base rounded or oblique, apex acuminate, clearly 7-nerved from base, thick-chartaceous. Male spikes slender, to 18 cm long; female spikes to 20 cm long. Berries ellipsoid, ripening yellow.

Fl. & Fr.: Most of the seasons. *Distr.*: India to Java. Most common specis of the genus. In semi-evergreen forests. *NAK 207 618, 788, 930 & 2050* (Almost all forests).

Leaves rather variable in shape.

Note : **Piper betle** L. is widely cultivated for its leaves. *NAK 2586* (Arampa, c.400 m).

109. CHLORANTHACEAE

SARCANDRA Gardner
Calcutta J.Nat.Hist.6:348.1846.

Sarcandra chloranthoides Gard., Calcutta.J.Nat.Hist. 6:348. 1846; Wight, Ic.t.1946. 1853; Verdc., Kew Bull.39:66. 1983; Mohanan & Sivad., Fl. Agasthyamala 557.2002. *Chloranthus brachystachya* sensu Hook.f., Fl. Brit. India 5:100.1886, p.p. (under undeterminable and excluded species) non Blume, 1829; Gamble, Fl. Pres.Madras 1211.1925. *Sarcandra irvingbaileyi* Swamy, Proc. Natl. Inst.Sci.India Pt.B. 19:303, ff.1-7, t.11/

8-11 & t. 12/12-13.1953. *Chloranthus grandifolius* Miq., Fl.Ind.Bat. 1(1): 802.1856. *S. grandifolia* (Miq.) Subram. & Henry, Bull. Bot. Surv. India 12:5. 1970; Manilal, Fl.Silent Valley 232.1988.

Erect shurbs. Leaves simple, decussate, elliptic-lanceate, 10-20 by 5.5-7.5 cm, base acute, margin spinous-serrate, chartaceous; petioles connate in a sheath. Flowers greenish-brown, unisexual, connate on a bract and in terminal or subterminal spikes or panicles; perianth 0; stamens 1-3, anther 1-2-celled; ovary 1-locular; ovule 1, pendulous. Drupe globose, black. Seed endospermous.

Fl. & Fr.: Dec.-May. *Distr.*: Sri Lanka, Peninsular & Eastern India, China, Philippines, Japan. Less Common. In evergreen forest floor. *NAK 1362 & 2701* (Kakki Hills, c. 1000 m), *2469* (Arampa, c.400 m). A vesselless densly foliaceous shrub.

110. MYRISTICACEAE

1. Anther-column stipitate.
    2. Anthers connate by their back. .... **Myristica**
    2. Anthers connate by their base. .... **Knema**
1. Anther-column sessile. .... **Gymnacranthera**

GYMNACRANTHERA Warburg
Ber. Deutsch Bot.Ges. 13:90, 91, 94.1896.

Gymnacranthera eugeniifolia (A.DC.) Sinclair, Gard.Bull.Singapore 16:447.1958 & 17:113. 1958. *Myristica eugeniaefolia* A.DC., Ann. Sc. Nat. 4(4):29.1855; Hook.f., Fl. Brit.India 5:113.1886. *M. farquhariana* Wall. ex Hook.f. & Thoms., Fl.Ind. 162.1855; Hook.f., Fl. Brit. India 5:108. 1886. *Gymnacranthera farquhariana* (Wall.ex Hook.f. & Thoms). Warb., Nov. Act. Nat. Cur. 68: 365.1897. *G. canarica* Warb., Monogr. Myrist. 368. 1897. *Unda-payin.*

Large trees. Leaves simple, alternate, oblong-lanceate, 12-22 by 4-6.5 cm, base obtuse, apex acuminate, lower surface glaucous, coriaceous. Flowers unisexual, deep yellow, male clusters in axillary panicles; stamens 6-12, connate by their back, free above; ovary ovoid, superior, 1-locular; ovule 1, basal. Capsule ovoid. Seeds arillate, ruminate endospermous.

Fl. & Fr.: Feb.-Apr. *Distr.*: India, Sumatra, Malaya Peninsula. Rare. In sacred groves. *NAK 2326* (Aikadu grove, Kodumon, c.70 m).

KNEMA Loureiro
Fl. Cochinch. 604. 1970.

Knema attenuata (Wall.ex Hook. f. & Thoms.) Warb., Monogr. Myrist. 590.1897; Gamble, Fl. Pres. Madras 1215. 1925; Gandhi in Sald. & Nicols., Fl. Hassan Dist. 42. 1976; Mohanan & Sivad., Fl. Agasthyamala 558.2002. *Myristica attenuata* Wall. ex Hook.f. & Thoms., Fl. Ind. 157. 1885 & in Hook.f., Fl. Brit. India 5:110. 1886. *M. corticosa* Bedd., Fl. Sylv.t. 278.1872. *Chora-payin, Chennelli.*

Trees. Leaves alternate, oblong-lanceate, 10-20 by 3.5-6.5 cm, base acute, apex acuminate, lateral veins c.16 pairs, prominent, parallel, lower surface glaucous, coriaceous. Flowers dioecious. Male and female in axillary or lateral fascicles, rusty pubescent; perianth 3-lobed; stamens 8-20, connate by their bases, form a peltate disc, stalked; ovary ovoid, superior, 1-locular; ovule 1, basal. Capsule ellipsoid, rusty pubescent. Seeds arillate.

Fl. & Fr.: Dec.-Mar. *Distr.*: Western Ghats. Common. In evergreen forests. *NAK 280, 534, 1612* (Almost all hills).

MYRISTICA Gronovius
Fl. Orient. 141.1755 (nom.cons.).

Evergreen trees. Leaves alternate, pellucid-punctate, estipulate. Flowers dioecious in axillary fascicles or umbels, raised on a stout short shoot; perianth-lobes 3; stamens adnate to form a fleshy cylindric column; anthers attached by their back; ovary superior, 1-locular; ovule 1, basal; style 2-fid. Drupe fleshy. Seeds arillate, endosperm ruminate.

1. Leaves coriaceous, lower surface glaucous; peduncle stout. .... 1. **M. dactyloides**
1. Leaves chartaceous, lower surface brownish; peduncle slender. .... 2. **M. malabarica**

1. **Myristica dactyloides** Gaertn., Fruct. 1:195, t.41,f.2a-d. 1788; Sincl., Gard.Bull. Singapore 23:445.1968; Gandhi in Sald. & Nicols., Fl. Hassan Dist. 42.1976; Matthew, Ill. Fl. Tam. Carnatic t. 604.1982; Matthew & Rani in Matthew, Fl. Tam.Carnatic 3(2): 1353.1983; Manilal, Fl. Silent Valley 233. 1988; Mohanan & Sivad., Fl. Agasthyamala 559.2002. *M.laurifolia* Hook.f. & Thoms., Fl. Ind.163.1855; Hook.f., Fl.Brit. India 5:103.1886, Ind., incl. var. *lanceolata. M.beddomei* King, Ann.Roy Bot. Gard.(Calcutta) 2:291, t.118,f. 1-8.1891; Gamble, Fl.Pres.Madras 1214.1925. *M. contorta* Ward., Monogr. Myrist. 507.1897; Gamble, Fl. Pres. Madras 1214. 1925. *Adekka-payin.*

Trees; bark grey. Leaves oblong-lanceate or elliptic-lanceate, 20-30 by 8-10 cm, base acute or obtuse, lower surface glaucous, lateral veins α, prominent beneath, coriaceous. Flowers small, ovoid. Male flowers in short umbels on axillary shoots; perianth greenish-yellow; staminal column, c. 5 mm long. Female flowers few, subsessile. Drupe broadly ovoid or ellipsoid, c. 5 cm ϕ, aril reddish, scarcely meeting the tip of seed.

Fl. & Fr.: Jan.-Apr. *Distr.*: Sri Lanka, Peninsular India. Common. In evergreen forests. *NAK 2152* (Mundanpara, c. 100 m), *2586* (Arampa, c.550 m).

Aril commercially used for making certain dyes.

2. **Myristica malabarica** Lam., Hist.Acad.Roy.Sci.Mem.Math.Phys. (Paris) 162.1791; Bedd., Fl.Sylv.t. 269.1872; Hook.f. & Thoms., Fl. Ind. 163.1855; Hook.f., Fl.Brit.India 5;103 1886; Gamble, Fl.Pres.Madras 1213.1925; Sincl., Gard. Bull Singapore 23.168, t.9. 1968; Manilal, Fl.Silent Valley 233.1988; Nicols., Suresh & Mani., An Interpr. Hort. Malab. 190. 1988; Mohanan & Sivad., Fl. Agasthyamala 560.2002. *Panam-palca* Rheede, Hort. Malab. 4:9-10, t.5.1683.

Trees; bark greyish-black. Leaves elliptic-oblong, 12-20 by 5-6 cm, base obliquely acute, apex acuminate, lower surface brownish, lateral veins 6-8 pairs, prominent beneath, thick chartaceous. Flowers large, globose. Male and female in short sessile axillary umbels, rusty tomentose. Drupe cylindrical, tomentose, aril golden yellow, completely covering the seed.

Fl. & Fr.: Sep.-Nov. *Distr.*: Western Ghats. In evergreen forests. Very rare. *NAK 2355* (Appupanthode, Kalleli R.F., c. 200 m). Known from only one collection.

Note : **Myristica fragrans** Houtt. is seen planted in the district.

111. LAURACEAE

1. Leaves 3-(-5)-nerved from base.
   2. Flowers unisexual, in axillary umbels; tepals 4. .... **Neolitsea**
   2. Flowers bisexual, in subterminal panicles; tepals 6. .... **Cinnamomum**
1. Leaves penni-nerved.
   3. Inflorescence an umbel.
      4. Leaves alternate or subopposite; involucral bracts subpersistent. .... **Litsea**
      4. Leaves subverticillate, rarely opposite; involucral bracts deciduous. .... **Actinodaphne**
   3. Inflorescence a panicle or raceme.
      5. Berry ellipsoid; staminal-glands subsessile. .... **Alseodaphne**
      5. Berry globose; staminal-glands stipitate.
         6. Anthers 2-celled; stigma capitate. .... **Beilschmiedia**
         6. Anthers 4-celled; stigma peltate. .... **Persea**

ACTINODAPHNE C.G.D.Nees in Wallich, Pl.Asiat.Rar. 2:61, 68.1831.

Trees. Leaves simple, subverticillate, 3-(-5)-nerved from base or penni-nerved, coriaceous. Flowers dioecious, actinomorphic, 3-merous, in axillary or lateral racemes or umbellate-clusters. Male flowers : tepals subequal; stamens 9, in 3 series, outer 2 series opposite to petals, all filaments with a pair of glands at base, anther 4-celled, valvular, introrse. Female flowers: tepals 6; staminodes 9, outer 2 series spathulate or linear, eglandular, inner row linear, glandular at base; ovary superior, ovoid, 1-locular, ovule 1, pendulous; style terminal, stigma dilated. Berries subtended by persistent perianth lobes. Seeds subglobose.

1. Leaves elliptic, to 7 cm long, apex obtuse. .... 1. **A. campanulata** var. **obtusa**
1. Leaves elliptic-lanceate, to 17 cm long, apex acute to acuminate. .... 2. **A. malabarica**

1. **Actinodaphne campanulata** Hook.f., Fl.Brit. India 5:148.1886 var. **obtusa** Gamble, Fl.Pres.Madras 1230. 1925. ; Mohanan & Sivad., Fl. Agasthyamala 563.2002.

Small trees or large shrubs. Leaves 4-8 in a whorl, elliptic, 5-7 by 2-3.5 cm, base and apex obtuse. Male flowers c. 3 mm ϕ, subglobose, in slender axillary fascicles; stamens glabrous. Female flowers c.12 mm long, silky; style slender, stigmas 3. Berries globose.

Fl. & Fr.: Feb.-Apr. *Distr*: Endemic to Western Ghats. Very rare. In evergreen forests. *CNM 72807* (MH) (Chokkampatti hills, c. 1000 m).

2. **Actinodaphne malabarica** Balakr., J. Bombay Nat.Hist.Soc. 63:329.1967. *A. hirsuta* Hook.f., Fl. Brit. India 5: 152. 1886, non Blume, 1851; Gamble, Fl. Pres.Madras 1231.1925; Gandhi in Sald. & Nicols., Fl. Hassan Dist. 45.1976. *A.hookeri* sensu Bedd., Fl.Sylv.t.296.1873, p.p., non Meisner, 1864.

Large trees. Leaves ± 7 per whorl, elliptic-lanceate, 10-15 by 4-6 cm, base attenuate, apex gradually acuminate, upper surface shiny, lower surface densely hirsute- Male and female flowers long-pedicelled, densely rusty tomentose. Berry globose, seated on cup-shaped entire perianth-tube.

Fl. & Fr.: Sep.-Dec. *Distr.*: Endemic to Western Ghats. Rare. *NAK 2998* (Kakki hills, c. 1100 m).

ALSEODAPHNE C.G.D. Nees in Wallich, Pl.Asiat. Rar. 2:61,71. 1831.

Alseodaphne semecarpifolia Nees in Wall., Pl.Asiat. Rar. 2:72.1831; Wight, Ic.t. 1826 & 1827. 1852; Hook.f., Fl.Brit. India 5:144.1886; Gamble, Fl.Pres. Madras 1226.1925; Matthew, Ill.Fl.Tam. Carnatic t.605.1982; Matthew & Rani in Matthew, Fl.Tam. Carnatic 3(2): 1356.1983.

Trees. Leaves alternate or clustered towards branch-apices, obovate to oblanceate, 6-11 by 2-6 cm, base acute to cuneate, apex acute, thin-coriaceous. Flowers bisexual, in branched sub-terminal panicles; *tepals* 6, equal, 3+3, ovate; stamens 9, glandular; staminodes 3; ovary ellipsoid. Berry ellipsoid. Seeds subglobose.

Fl. & Fr.: Feb.-Apr. *Distr.*: Sri Lanka, Peninsular India. Rare. In shola forests. *CNM 73427* (MH) (Chokkampatti hills, c. 1000 m).

BEILSCHMIEDIA C.G.D.Nees in Wallich, Pl.Asiat Rar.2:61.69.1831.

Beilschmiedia bourdillonii Brandis, Ind. Trees 528. 1907; Gamble, Fl.Pres.Madras 1221.1925; Matthew, Ill.Fl. Tam. Carnatic t.606.1982; Matthew & Rani in Matthew, Fl. Tam. Carantic 3(2) : 1357.1983. *B. fagifolia* Bedd., Fl.Sylv. t. 263.1872, non Nees, 1830.

Trees. Leaves alternate to subopposite, elliptic to obovate, 8-12 by 3-4 cm, base and apex acute, lateral nerves 10-12 pairs. Flowers bisexual, in subterminal panicles; tepals 6, subequal, 3+3; stamens 9, filaments bearded, glands stipitate, anthers 2-celled; staminodes 3; ovary oblong-ovoid. Berry purplish, cylindric.

Fl. & Fr.: Feb.-Apr. *Distr.*: Peninsular India. Very rare. In semi-evergreen forests. *Bourdillon 45* (UCT) (Konni R.F., c. 200 m). Could not locate this species during the present study, and the description given is based on the literature.

CINNAMOMUM Schaeffer Bot.Exped. 74. 1760 (nom. cons.).

Trees, bark often fragrant. Leaves opposite, basally 3-nerved, coriaceous. Flowers bisexual in lax panicles, pseudo terminal; tepals 6, 3+3, basally connate to form a campanulate tube; stamens 9, in 3 series, inner series glandular at base; anthers 4-celled in outer 2 series, introrse; in the inner series 2-celled; staminodes 3; ovary superior, 1-locular; ovule 1, penulous. Berry enclosed by persistent perianth-tube.

1. Leaf-undersurface glabrous or microscopically sparsely appressed puberulent. .... 1. **C. malabatrum**
1. Leaf-under surface densely minutely tomentose. .... 2. **C. sulphuratum**

1. **Cinnamomum malabatrum** (Burm.f.) Blume, Rumphia 1:38, t.13, f.3-4. 1836, Kosterm., Bull. Bot. Surv. India 25:102.1983; Manilal, Fl.Silent Valley 235.1988; Nicols., Suresh & Mani., An Interpr. Hort. Malab. 157.1988. *Laurus malabatrum* Burm.f., Fl. Ind. 92.1768, p.p., (*quoad cit. Katou karua* Rheede, Hort. Malab. 5.t.53. 1685). *Cinnamomum iners* auct., non Reinw. ex Blume: 1826: Hook.f., Fl. Brit. India 5:130. 1886; Gamble, Fl. Pres. Madras 1224. 1925. *Katou-karua* Rheede, Hort. Malab. 5:105-106, t 53.1685.

Trees, to 20 m tall. Leaves mostly oblong to elliptic-oblong, 15-30 by 5.5-8 cm, base acute, apex acuminte or obtuse, lower surface galbrous or microscopically puberulous. Panicle to 25 cm long, minutely pilose. Flowers yellowish; tepals fleshy, ovate. Berry ellipsoid, 8 by 10 mm, on a deep perianth-cup.

Fl. & Fr.: Feb.-May. *Distr.*: Western Ghats. Common. In lowlands and hills. *NAK 426, 472, 1509* (Maniyaar, Moozhiar etc.).

Bark used to adulterate cinnamon, or as a substitue for it.

2. **Cinnamomum sulphuratum** Nees in Wall., Pl. Asiat. Rar. 2:74.1831; Hook. f., Fl.Brit. India 5:132. 1886; Gamble, Fl.Pres. Madras 1225. 1925; Gandhi in Sald. & Nicols., Fl. Hassan Dist. 47.1976; Kosterm., Bull. Bot. Surv. India 25:114.1983; Manilal, Fl. Silent Valley 235.1988; Mohanan & Sivad., Fl. Agasthyamala 567.2002.

Trees to 10 m tall. Leaves ovate-elliptic to elliptic-lanceate, 7-12 by 3-5.5 cm, base acute, apex obtusely acuminate, lower surface densely tomentellose, glabrescent. Panicle to 15 cm long, minutely tomentellose. Flowers cream- coloured; tepal thick, ovate. Berry ellipsoid, 1 by 1.5 cm, on cup-shaped perianth.

Fl. & Fr.: Feb.-May. *Distr.*: South India. Not common. In evergreen and semievergreen forests. *NAK 2696* (Kakki hills, c.1100 m).

LITSEA Lamarck
Encycl.Meth. Bot. 3:574.1792 (nom.cons.).

Trees. Leaves alternate, coriaceous. Flowers dioecious, in extra-axillary umbels; involucral bracts 2+2, unequal. Male: tepals present or not; stamens 9 or 12, in 3-4 series, filaments bearded, anthers 4-celled, introrse; pistillode globose. Female : tepals normally 6, caducous, staminodes 3-4 series, linear, inner series glandular; ovary globose; stigma 2-lobed. Berry globose or ellipsoid with a basal rim of perianth.

1. Leaf - lower surface glabrous.
    2. Weak shrubs; male umbellule solitary; female few in racemes, both peduncled; leaves chartaceous. .... 7. **L. venulosa**
    2. Trees; male & female umbellules in clusters or in racemes; leaves coriaceous.
        3. Umbellules in clusters. .... 2. **L. coriacea**
        3. Umbellules in racemes. .... 5. **L. oleoides**
1. Leaf - lower surface pubescent.
    4. Umbellules in clusters.
        5. Leaves all alternate, coriaceous.
            6. Leaves thick-coriaceous, lateral nerves 10-12 pairs; berry ellipsoid. .... 4. **L. insignis**
            6. Leaves thin-coriaceous, lateral nerves 12-18 pairs; berry cylindric. .... 1. **L. bourdillonii**
        5. Leaves, upper ones opposite, chartaceous. .... 6. **L. travancorica**
    4. Umbellules in racemes. .... 3. **L. floribunda**

1. **Litsea bourdillonii** Gamble, Kew Bull. 1925: 131. 1925 & Fl. Pres. Madras 1237.1925.

Trees. Leaves obovate, 12-20 by 5-9 cm, base cuneate, apex shortly acuminate, lower surface minutely fulvous pubescent. Male & female flowers in dense lateral or extra-axillary clusters. Berry cylindric, c.15 mm long seated on cupular, 3-4-lobed perianth-tube.

Fl. & Fr.: Dec.-Apr. *Distr.:* Western Ghats. Rare. In evergreen forests. *BDS 43924* (MH) (Pamba, c. 1000 m).

2. **Litsea coriacea** (Heyne ex Meisner) Hook.f., Fl. Brit. India 5:166. 1886; Gamble, Fl.Pres. Madras 1236. 1925; Manilal, Fl. Silent Valley 236. 1988; Mohanan & Sivad., Fl. Agasthyamala 569.2002. *Tetranthera coriacea* Heyne ex Meisner in DC., Prodr. 15(1) : 186.1864. *Mara vetti-thali.*

Small trees. Leaves elliptic-lanceate or oblong-lanceate, 10-15 by 4-6 cm, base acute, apex acuminate, coriaceous. Male & female flowers in subsessile dense umbellules. Berry ovoid, seated on cupular, entire perianth-tube.

Fl. & Fr.: Most of the seasons. *Distr.*: Peninsular India. Most common species of the genus. *NAK 434, 538, 889, 995 & 1194* (In almost all forests).

3. **Litsea floribunda** (Bl.) Gamble, Fl.Pres. Madras 1238. 1925; Gandhi in Sald. & Nicols., Fl.Hassan Dist. 48. 1976; Manilal, Fl. Silent Valley 236.1988; Mohanan & Sivad., Fl. Agasthyamala 569.2002. *Cylicodaphne floribunda* Bl., Mus.Bot. 1:387. 1857. *Tetranthera wightiana* Bedd., Fl. Sylv. t. 293.1873. *Litsea wightiana* (Bedd.) Wall. ex Hook.f., Fl.Brit. India 5:177.1886, p.p.

Trees. Leaves usually broadly oblanceate to oblong-lanceate or elliptic, 12-25 by 7-10.5 cm, base acute to obtuse, apex acuminate, lower surface densly soft tomentose, coriaceous. Umbellules in axillary panicles, rusty tomentose. Berry small, almost enclosed by cupular perianth-tube.

Fl. & Fr.: Sep.-Apr. *Distr.:* Western Ghats (Travancore Hills). Very common. In evergreen forests. *NAK 304, 1135, 1294, 2442, 2625 & 2660.* (Almost all forests).

4. **Litsea insignis** Gamble, Kew Bull. 1925:130.1925 & Fl. Pres.Madras 1237.1925; Mohanan & Sivad., Fl. Agasthyamala 570.2002.

Tall trees. Leaves elliptic-oblong, 12-23 by 6-12 cm, base acute, apex obtuse, lower surface fulvous pubescent, lateral nerves 10-12 pairs, nervature parallel, conspicuous, coriaceous. Umbellules in axillary or lateral clusters. Berry ellipsoid, c. 25 cm long, enclosed by deep obconic perianth-tube.

Fl. & Fr.: Sep.-Mar. *Distr.*: Western Ghats (Travancore Hills). Rare. In evergreen forests. *BDS 42474* (MH) (Thriveny - Pamba, c. 250 m).

5. **Litsea oleoides** (Meisner) Hook.f., Fl.Brit. India 5:175.1886; Gamble, Fl.Pres.Madras 1236.1925; Mohanan & Sivad., Fl. Agasthyamala 571.2002. *Tetranthera oleoides* Meisner in DC., Prodr. 15(1) : 195.1864.

Medium to large-sized trees. Leaves elliptic-oblong, 10-18 by 4.5-8 cm, base obtuse or obliquely acute, apex acuminate, lateral veins faint, thick-coriaceous. Umbellules in axillary or lateral racemes, pubescent. Berry subglobose, seated on short perianth-tube.

Fl. & Fr.: Sep.-Mar. *Distr.*: Western Ghats. Rare. In evergreen forest of upper hills. *NAK 2516 & 2517* (Kakki hills, c. 1000 m).

6. **Litsea travancorica** Gamble, Kew Bull. 1925:132.1925 & Fl.Pres. Madras 1237.1925.

Small trees. Leaves elliptic-lanceate, 10-16 by 4-7 cm, base rounded, apex acuminate, lower surface pinkish, pubescent on nerves, lateral nerves 14-16 pairs, prominent, forming marginal loops, thin-coriaceous, or chartaceous. Umbellules few, in axillary clusters. Berry obovoid, seated on lobed perianth-tube.

Fl. & Fr.: Feb.-May. *Distr.*: Western Ghats. (Travancore hills). Very rare. In evergreen forests. *NAK 521 & 1787* (Angamoozhy, c.250 m).

7. **Litsea venulosa** (Meisner) Hook.f., Fl. Brit. India 5:161.1886; Gamble, Fl. Pres. Madras 1235.1925; Mohanan & Sivad., Fl. Agasthyamala 572.2002. *Tetranthera venulosa* Meisner in DC., Prodr. 15(1): 187.1864.

Subscandent shrubs. Leaves elliptic, 8-14 by 3.5 cm, base acute, margin slightly recurved, apex caudate, lateral veins c. 8 pairs, irregular, lower surface glaucous, thin coriaceous. Male umbellule solitary, female few in racemes, both peduncled. Berry subglobose.

Fl. & Fr.: May-Aug. *Distr.*: Western Ghats. Rare. In evergreen forests. *NAK 2465* (Arampa, c.430 m).

NEOLITSEA (Bentham) Merrill
Philipp. J. Sci.1 Suppl.56.1906 (nom. cons.).

Neolitsea scrobiculata (Meisner) Gamble, Fl.Pres.Madras 1240. 1925; Gandhi in Sald. & Nicols., Fl. Hassan Dist. 49.1976; Matthew & Rani in Matthew, Fl. Tam. Carnatic 3(2): 1362. 1983; Manilal, Fl.Silent Valley 237. 1988. *Litsea scrobiculata* Meisner in DC., Prodr. 15(1): 223.1864. *L. zeylanica* sensu Hook.f., Fl.Brit.India 5:178.1886, p.p., non Nees, 1823.

Small trees. Leaves alternate, or subopposite towards apices, 7-11 by 3.5-5 cm, basally 3-ribbed, base acute, apex acuminate, coriaceous. Flowers dioecious, 3-7 in umbellules, in clusters or short racemes, towards upper leaf-axils; involucral bracts 4; tepals 4. Male flowers : stamens 6 in 2-series, outer eglandular, anthers 2-celled. Female flowrs : staminodes 6, linear; ovary ovoid. Berry ellipsoid, seated on perianth-tube.

Fl. & Fr.: May-Aug. *Distr.*: Peninsular India. Rare. In shola forests. *NAK 2518* (Kakki hills, c.1100 m).

PERSEA P. Miller
Gard. Dict. Abr. ed. 4. 1754 (nom.cons.)

Persea macrantha (Nees) Kosterm., Reinwardtia 6: 193. 1962 & 7:537.1969; Gandhi in Sald. & Nicols., Fl. Hassan Dist. 49. 1976; Matthew, Ill. Fl. Tam. Carnatic t. 611. 1982; Matthew & Rani in Matthew, Fl.Tam. Carnatic 3(2) : 1363. 1983; Mohanan & Sivad., Fl. Agasthyamala 573.2002. *Machilus macrantha* Nees in Wall., Pl. Asiat. Rar. 2:70. 1831 & 3:31.1831; Wight, Ic.t. 1824.1852; Bedd., Fl.Sylv. t. 264.1872; Hook.f., Fl.Brit. India 5:140.1886; Gamble, Fl.Pres. Madras 1227.1925.

Trees. Leaves alternate, elliptic-oblong, 9-16 by 3.5-6 cm, base obliquely acute, apex obtuse or obtusely short acuminate, coriaceous. Flowers bisexual in subterminal lax panicles; tepals 6, subequal 3+3; stamens 9 in 3-series, glands stipitate, anthers 4-celled; staminodes 3; ovary superior, obovoid. Berry globose, seated on persistent perianth.

Fl. & Fr.: Dec.-Apr. *Distr.*: Sri Lanka, Peninsular India. Less common. In evergreen forests in upper hills. *NAK 375, 630, 1254 & 1384* (Kakki hills, c.1100 m).

Note : **Persea americana** Miller is seen planted near Moozhiar (*NAK 262*).

112. ELAEAGNACEAE

ELAEAGNUS Linnaeus
Sp. Pl. 121.1753.

Straggling shrubs; young parts lepidote. Leaves simple, alternate. Flowers bisexual, regular, 4-merous, aggregated in axillary fascicles. Perianth tubular-urceolate, lobes 4, spreading or incurved; stamens 4, subexerted; ovary superior, 4-locular, ovule 1, basal. Fruit a drupaceous nut, closely covered by persistent perianth.

1. Leaf - lower surface coppery; nut winged...1. **E. conferta**
1. Leaf - lower surface silvery; nut wingless. ....2. **E. indica**

1. **Elaeagnus conferta** Roxb., Fl.Ind. 1:460.1820; Gamble, Fl.Pres. Madras 1246. 1925; Gandhi in Sald. & Nicols., Fl.Hassan Dist. 295. 1976. *E.latifolia* auct., non L., 1753.: Bedd., Fl. Sylv.t.180.1872; Hook.f., Fl.Brit. India 5:202. 1886, p.p.

Leaves ellptic to elliptic-obovate, 4-9 by 2-5 cm, base obtuse, apex acuminate, lower surface coppery, lateral nerves c. 4 pairs, chartaceous. Fascicle few-flowered; perianth-tuber urceolate, c. 10 mm long. Nut oblong, ribbed.

Fl. & Fr.: Dec.-Mar. *Distr.*: Sri Lanka, Peninsular & Eastern India, Myanmar, Malesia, China. Rare. In Shola forest edges. *NAK 1293* (Kakki Hills, c.1100 m).

2. **Elaeagnus indica** Servettaz, Bull. Herb. Boissier Ser. 2, 8:393. 1908; Gamble, Fl.Pres. Madras 1246.1925; Matthew, Ill.Fl.Tam. Carnatic t.615. 1982; Matthew & Rani in Matthew, Fl.Tam. Carnatic 3(2) : 1369. 1983.

Leaves broadly elliptic to suborbicular, 8-10 by 5.5-6.5 cm, base rounded or acute, apex shortly acuminate, lower surface silvery, lateral veins c. 4 pairs, thick-chartaceous. Fascicles 3-4-flowered; perianth tubular-urceolate. Nut obovoid, ribbed.

Fl. & Fr.: Dec.-Apr. *Distr.*: Peninsular India. Rare. Along streamside forests. *NAK 2226* (Thazhoor Kadavu, Pathanamthitta, c. 1100 m).

113. LORANTHACEAE

1. Leaves lanceate; bracteoles 2. ....**Macrosolen**
1. Leaves otherwise; bracteoles 0.
    2. Corolla lobes free; floral buds angular. ....**Helixanthera**
    2. Corolla lobes connate towards base; floral buds terete.
        3. Corolla lobes 4 ....**Scurrula**
        3. Corolla lobes 5.
            4. Corolla densely tomentose, lobes straight. ....**Taxillus**
            4. Corolla glabrous, lobes spirally twisted. ....**Dendrophthoe**

DENDROPHTHOE C. F. P. Martius
Flora 13:109.1830.

Dendrophthoe elasticus (Desr.) Danser, Bull. Bot. Buitenzorg Ser. 3, 10: 308.1929; Nicols., Suresh & Mani., An Interpr. Hort. Malab. 165.1988. *Loranthus elasticus* Desr. in Lam., Encycl. 3: 599. 1792; Wight, Ic.t.343. 1890; Hook.f., Fl.Brit. India 5:216.1886; Gamble., Fl.Pres. Madras 1254. 1925. *Helicanthes elastica* (Desr.) Danser, Verh. Kon. Ned.Inst. Wetensch., Afd. Natuurk., Tweede Sect. 29 (6):.55.1933; Matthew, Ill.Fl.Tam. Carnatic t.617. 1982; Matthew & Rani in Matthew, Fl.Tam.Carnatic 3(2): 1373. 1983; Manilal, Fl.Silent Valley 239.1988. *Velutta-itti-canni* Rheede, Hort. Malab. 10:7 (not 51), t. 3. 1690.

Semi- parasitic subshrubs. Leaves simple, decussate, obovate to elliptic, 5-10 by 2.5-6 cm, base acute, apex obtuse or rounded, basally 3-nerved, coriaceous. Flowers bisexual, aggregated in short axillary fascicles; bracts partly enclosing calyx-base, bracteoles 0; outer perianth sepaloid (calyx), flask-shaped, marign truncate, inner perianth petaloid (corolla), tubular, tube terete, split lengthwise towards one side, lobes 5, spirally twisted; stames 5, exerted, anthers 2-celled; ovary inferior, adnate to calyx-base, 1-locular; ovule 1, basal style slender, exerted. Berry obovoid.

Fl. & Fr.: Dec.-Apr. *Distr.*: Peninsular India. Parasitic, commonly on **Mangifera indica** L. *NAK 2870* (Kodumon).

HELIXANTHERA Loureiro
Fl. Cochinch. 142. 1790.

Semi-parasitic shrubs. Leaves opposite, penninerved, coriaceous. Flowers aggregated in axillary dense racemes; corolla curved, angular, lobes 4, free; calyx cupular, tomentose; stamens 4, filaments flat; style slender, stigma globose. Berry ovoid.

1. Inflorescence pubescent; flowers large, purple. ....1. **H. obtusata**
1. Inforescence glabrous; flowers small brick-red. ....2. **H. wallichiana**

1. **Helixanthera obtusata** (Schult.) Danser, Bull. Jard.Bot. Buitenz. Ser.3, 10 : 317. 1929; Manilal, Fl.Silent Valley 240. 1988. *Loranthus obtusatus* Schult., Syst. 7 (2): 1650.1830; Hookf., Fl.Brit. India 5: 205.1886; Gamble, Fl.Pres. Madras 1251.1925.

Stout parasitic shrubs; branches lenticellate. Leaves ovate to ovate-oblong, 8-12 by 5.7.5, base rounded, apex obtusely acute, coriaceous. Racemes 6-10 cm long, α-flowered, pubescent; corolla orange-red or purple, c. 15 mm long. Berry ovoid.

Fl. & Fr.: Apr.-Jun. *Distr.*: Peninsular India. Less common. In evergreen forests. *NAK 2733* (Kakki hills, c. 1000 m). Parasitic on *Maesa indica.*

2. **Helixanthera wallichiana** (Schult.) Danser, Bull. Jard. Bot. Buitenz. Ser. 3,13:317.1929; Ramamoorthy in Sald. & Nicols., Fl. Hassan Dist. 307.1976; Manilal, Fl. Silent Valley 240.1988; Nicols., Suresh & Mani., An Interpr. Hort. Malab. 165.1988; Mohanan & Sivad., Fl. Agasthyamala 581.2002. *Loranthus wallichianus* Schult., Syst. 7:100.1820; Wight, Ic.t. 143.1839; Hook.f., Fl.Brit. India 5:204.1886; Gamble, Fl.Pres.Madras 1251.1925. *Kanneli-itti-canni* Rheede, Hort. Malab. 10:9, t.5.1690.

Parasitic subshrubs. Leaves ovate to elliptic-ovate, 4-10 by 3-5 cm, base and apex acute, coriaceous. Racemes 2-4 cm long, glabrous. Flowers lax; corolla brick-red, c. 5 mm long. Berry ovoid.

Fl. & Fr.: Apr.-Jun. *Distr.:* Peninsular India. Common in semi-evergreen forests and Teak plantations. *NAK 613, 916 & 1813* (Thannithode & Angamoozhy). Parasitic on **Helecteres isora.**

MACROSOLEN (Blume) Blume in J.A. Schultes et J.H. Schultes in J.J. Roemer et J.A. Schultes, Syst. Veg. 7:1731. 1830.

Macrosolen parasiticus (L.) Danser, Blumea 4:36. 1936; Ramamoorthy in Sald. & Nicols., Fl. Hassan Dist. 308. 1976; Manilal, Fl. Silent Valley 240.1988; Nicols., Suresh & Mani., An Interpr. Hort. Malab. 166.1988; Mohanan & Sivad., Fl. Agasthyamala 582.2002. *Lonecera parasitica* L., Sp.Pl.175.1753. *Loranthus loniceroides* L., Sp.Pl. (ed.2), 473.1762; Wight, Ic.t.203.1839; Hook.f., Fl. Brit. India 5:221. 1886. *Elytranthe loniceroides* (L.) G. Don, Gen.Syst. 3:427. 1834; Gamble, Fl.Pres.Madras 125.5.1925. *Itti-canni* Rheede, Hort. Malab. 7:55, t.29.1688.

Stout parasites. Leaves opposite, lanceate, 6-10 by 2.5-3 cm, base acute, apex acuminate, coriaceous. Flowers in condensed peduncled-spikes; bract 1; bracteoles 2, subequal, broadly ovate to orbicular; calyx tubular; corolla reddish towards base, greenish above, divided near half-way, curved, lobes 5, equal, reflexed; stamens 5, anthers 2-celled, usually septate; ovary ovoid, enclosed in the calyx-tube. Berry globose.

Fl. & Fr.: Apr.-Jul. *Distr.:* Sri Lanka, Peninsular India. Common. In evergreen & moist deciduous forests. *NAK 2429* (Appupanthode, c. 450 m).

SCURRULA Linnaeus Sp.Pl. 110.1753.

Scurrula parasitica L., Sp.Pl.110.1753; Wiens in Abeyw., Rev.Handb.Fl.Ceylon 1:73 1973; Ramamoorthy in Sald. & Nicols., Fl.Hassan Dist. 309. 1976; Matthew, Ill. Fl.Tam. Carnatic t. 618. 1982; Matthew & Rani in Matthew, Fl.Tam. Carnatic 3(2): 1374.1983; Manilal, Fl.Silent Valley 241. 1988; Mohanan & Sivad., Fl. Agasthyamala 583.2002. *Loranthus buddlejoides* Desr. in Lam., Encycl. 3:600. 1792; Wight & Arn., Prodr. 383:1834; Gamble, Fl.Pres. Madras 1251:1925, ("*buddleioides*"). *L.scurrula* (L.) L., Sp.Pl. (ed. 2) 472. 1762.

Parasitic subshrubs; branchlets tawny tomentose. Leaves subopposite, broadly elliptic to ovate, 5-10 by 3-5 cm, base and apex rounded or obtuse, rusty tomentose, glabrescent, thin-coriaceous. Flowers in short axillary racemes; calyx truncate; corolla pale orange-red, split lengthwise towards one side, lobes 4; stamens 4; ovary ovoid. Berry elliptic.

Fl. & Fr.: Dec.-Apr. *Distr.*: South Asia, Indo-Malaysia. Common in evergreen forests. *NAK 183 & 299* (Moozhiar, c. 250 m).

TAXILLUS van Tieghem Bull. Soc. Bot. France 42: 256.1895.

Taxillus tomentosus (Heyne ex Roth) van Tieghem, Bull. Soc.Bot. France 42:256.1895; Mohanan & Sivad., Fl. Agasthyamala 584.2002. *Loranthus tomentosus* Heyne ex Roth, Nov.Pl.Sp.191.1821; Hook.f., Fl.Brit.India 5:212. 1886; Gamble, Fl.Pres. Madras 1252.1925.

Parasitic subshrubs. Leaves alternate, obovate to obovate-oblong, to 4 by 2 cm, base cuneate, apex rounded, lower surface rusty-tomentose, penninerved, coriaceous. Flowers in axillary fascicles; bract 1; calyx obscurely 5-toothed; corolla yellow, curved, split lengthwise at one side, lobes 5; stamens 5, filaments short; ovary cylindric. Berry ellipsoid.

Fl. & Fr.: Sep.-Dec. *Distr.*: Sri Lanka, Peninsular India. Rare. In evergreen forests. *NAK 2493* (Moozhiar, c. 250 m).

114. VISCACEAE

VISCUM Linnaeus
Sp.Pl.1023.1753.

Viscum orientale Willd., Sp.Pl. 4:737.1806; Wight & Arn., Prodr. 379.1834; Hook.f., Fl.Brit.India 5:224. 1886; Gamble, Fl.Pres. Madras 1258.1925; Danser, Blumea 4:299. 1941; Wiens in Abeyw., Rev.Handb. Fl.Ceylon 1:61.1973; Matthew, Ill. Fl.Tam. Carnatic t.620.1982; Matthew & Rani in Matthew, Fl.Tam. Carnatic 3(2):1378.1983.

Parasitic subshrubs; branchlets 2-3-chotomous; internodes terete. Leaves decussate, lanceate to spathulate, 2.5-5 by 1-2 cm, base cuneate, apex acute to obtuse, basally 3-nerved, coriaceous. Flowers unisexual, white to yellow in triads, axillary clusters. Male flowers : subsessile; tepals 4, ovate; stamens 4, sessile, dehiscing poricidally. Female flowers: raised by a thick cylindric stalk; stigma rounded, sessile. Berry obovoid.

Fl. & Fr.: Feb.-Apr. *Distr.:* Sri Lanka, India, Bangladesh, China, Malaya, New Guinea, Australia. Rare. In semi-evergreen forests. *NAK 1726* (Thannithode, c.70 m), *2265* (Mannarappara, c.250 m).

115. SANTALACEAE

SCLEROPYRUM Arnott
Mag. Zool. Bot. 2: 549. 1838 (nom. cons.).

Scleropyrum pentandrum (Dennst.) Mabberley, Taxon 26:533.1977; Nicols., Suresh & Mani., An Interpr. Hort. Malab 236.1988; Mohanan & Sivad., Fl. Agasthyamala 587. 2002. *Pothos pentandrus* Dennst., Schuless. Hort. Ind. Malab. 16, 24, 33.1818. *Sphaerocarya wallichiana* Wight & Arn., Edinburgh New Philos. J. 15:180.1833. *Scleropyrum wallichianum* (Wight & Arn.) Arn., Mag. Zool. Bot. 2: 550.1838; Wight, Ic.t.241.1839; Hook.f., Fl. Brit. India 5:234.1886; Gamble, Fl.Pres.Madras 1262.1925; Gandhi in Sald. & Nicols., Fl. Hassan Dist. 304. 1976. *Idou-moulli* Rheede, Hort. Malab. 4:41-42, t.18.1683. *Tiri-itti-canni* Rheede, Hort. Malab. 7:57, t.30.1688.

Small trees with axillary spines. Leaves simple, alternate, ovate-oblong, 10-17 by 4-7.5 cm, 3-nerved from base, base rounded to subcordate, apex obtuse, coriaceous. Flowers polygamous, in pendulous spikes, perianth-lobes 4-5; stamens 4-5, inserted at the base of lobes, filaments 2-fid, anther-cells separate, dehiscing transversely; ovary inferior, 1-locular; ovules 3, pendulous; style short; stigma large, peltate. Drupe crowned by remains of perianth. Seeds globose.

Fl. & Fr.: Sep.-Feb. *Distr.*: Sri Lanka, Peninsular India. Rare. In evergreen forests. *BDS 40830* (MH) (Pachakanam, c.850 m), *CB 49197* (MH) (Perundenaruvi, c. 100 m).

116. BALANOPHORACEAE

BALANOPHORA J. R. Forester et J. G. A. Forster
Charact. Gen.99, 100.1776.

Parasitic, achlorophyllous rootless rhizomatous fleshy herbs. Rhizome massive, warted; stem stout, closely covered by imbricate scaley leaves. Flowers unisexual, monoecious or dioecious. Male : sessile or pedicellate, fully covering the exposed part of heads or basal to female flowers; tepals 4, valvate; stamens 4, connate into a synandrium, anthers connate by their back, dehiscence extrorse. Female flowers : minute, sessile, embedded, intermingled with spadicles; tepals 0; ovary 1-locular; ovule 1, pendulous; style filiform. Fruit nut-like.

1. Monoecious herbs; male flowers 20-40, basal to female heads. ....1. **B. abbreviata**

1. Dioecious herbs; male flowers very a, on separate heads. ....2. **B. fungosa** ssp. **indica**

1. **Balanophora abbreviata** Bl., Enum. Pl. Jav. 1: 98. 1827; Govindappa *et al.,* Current Sci. 44(3) :97. 1975. *B. polyandra* Haines, Bot. Bihar & Orissa 3:806. 1924. *Acroblastom ambavanense* Venkatareddi, Willdenowia 5:389.1969. **(Fig. 52)**

Monoecious herbs. Rhizome finely warted with lenticels, cream coloured. (Scale-) Leaves 4-6, ovate, acute, stiff. Heads ovoid or lanceoid. Male flowers 20-40, sessile, synandrium with 15-25 anther-loculi. Female flowers fully covering the exposed part of head. Spadicles obovoid with a narrow basal and upper portion. Fruits could not be collected.

Fl. & Fr.: Sep.-Dec. *Distr.*: India, Africa, Madagascar, Indo-China, Malesia & Pacific Islands. Rare. In moist deciduous forests. *NAK 2022* (Kalleli-moozhy, Konni c.150 m).

2. **Balanophora fungosa** Forst. & Forst. f., Char. Gen. Pl.100,t.50.1776, ssp. **indica** (Arn.) Hansen, Dansk Bot. Ark.28:100.1972; Ramamoorthy in Sald. & Nicols., Fl.Hassan Dist. 313.1976; Matthew, Ill. Fl.Tam.Carnatic t.623.1982; Matthew & Rani in Matthew, Fl. Tam. Carnatic 3 (2): 1483. 1983; Manilal, Fl.Silent Valley 242.1988; Mohanan & Sivad., Fl. Agasthyamala 587.2002. *Longsdorffia indica* Arn., Ann. Nat.Hist. 2:37.1838. *Balanophora indica* (Arn.) Wallich ex Griffith, Trans. Linn. Soc. London 20:95.1846; Hook.f., Fl.Brit. India 5:237.1886; Gamble, Fl.Pres.Madras 1263.1925.

Dioecious herbs. Rhizome cream-coloured. Leaves closely spiral, suborbicular, 1.5-2.5 by 1.5-2 cm. Male flowers : very α, in ovoid to ellipsoid heads, cream- coloured or yellow, c. 5 cm ϕ; synandrium oblong-cylindric, raised on a short torus. Female flowers: very α, in globose rusty brown heads, c. 4.5 cm ϕ; spadicles club-shaped. Nut obovoid.

Fl. & Fr.: Jan.-Mar. *Distr.*: Sri Lanka, India, China, Malesia, Sumatra, Australia. Rare. In evergreen hill floors. *NAK 2623* (Sabari hills, c. 250 m).

117. EUPHORBIACEAE

1. Inflorescence a cyathium.
    2. Cyathia actinomorphic; branchlets straight. .... **Euphorbia**
    2. Cyathia zygomorphic; branchlets zigzag. ....**Pedilanthus**
1. Inflorescence other than cyathium.

3. Ovules 2-per locule; stamens opposite the tepals.
4. Tepals 2-seriate (inner often smaller).
5. Tepals valvate; stamens connate. .... **Bridelia**
5. Tepals imbricate; stamens free. .... **Actephila**
4. Tepals 1-seriate.
6. Disc present (usually glandular).
7. Flowers in axillary clusters; ovary 3-locular.
8. Disc villous; stamens α; leaves large (± 20 cm long). .... **Drypetes**
8. Disc glabrous; stamens definite (2-5); leaves small (to 10 cm long).
9. Dioecious shrubs; leaves obovate; capsule fleshy. .... **Securinega**
9. Monoecious herbs to trees; leaves other than obovate; capsule dry. .... **Phyllanthus**
7. Flowers in spikes or racemes; ovary 1-locular. .... **Antidesma**
6. Disc absent.
10. Ovary half inferior; female flowers pendulous. .... **Breynia**
10. Ovary superior; female flowers not so.
11. Stamens connate.
12. Shrubs; stamens 3, connate to form a trigonous column, anthers sessile. .... **Sauropus**
12. Trees; stamens 3-8, connate by their connectives, anther subsessile. .... **Glochidion**
11. Stamens free.
13. Leaves simple; ovary 3-locular.
14. Both male and female flowers in dense cauline spikes. .... **Baccaurea**
14. Male flowers in axillary catkin like spikes; female in clusters. .... **Aporusa**
13. Leaves 3-foliolate; ovary 2-locular. .... **Bischoffia**
3. Ovule(s) 1-per locule; stamens alternating the tepals.
15. Tepals 2-seriate.
16. Calyx accrescent. .... **Dimorphocalyx**
16. Calyx not accrescent.
17. Stamens α, latex absent.
18. Stamens monadelphous. .... **Aleurites**
18. Stamens free. .... **Croton**
17. Stamens definite; latex present. .... **Jatropha**
15. Tepals 1-seriate.

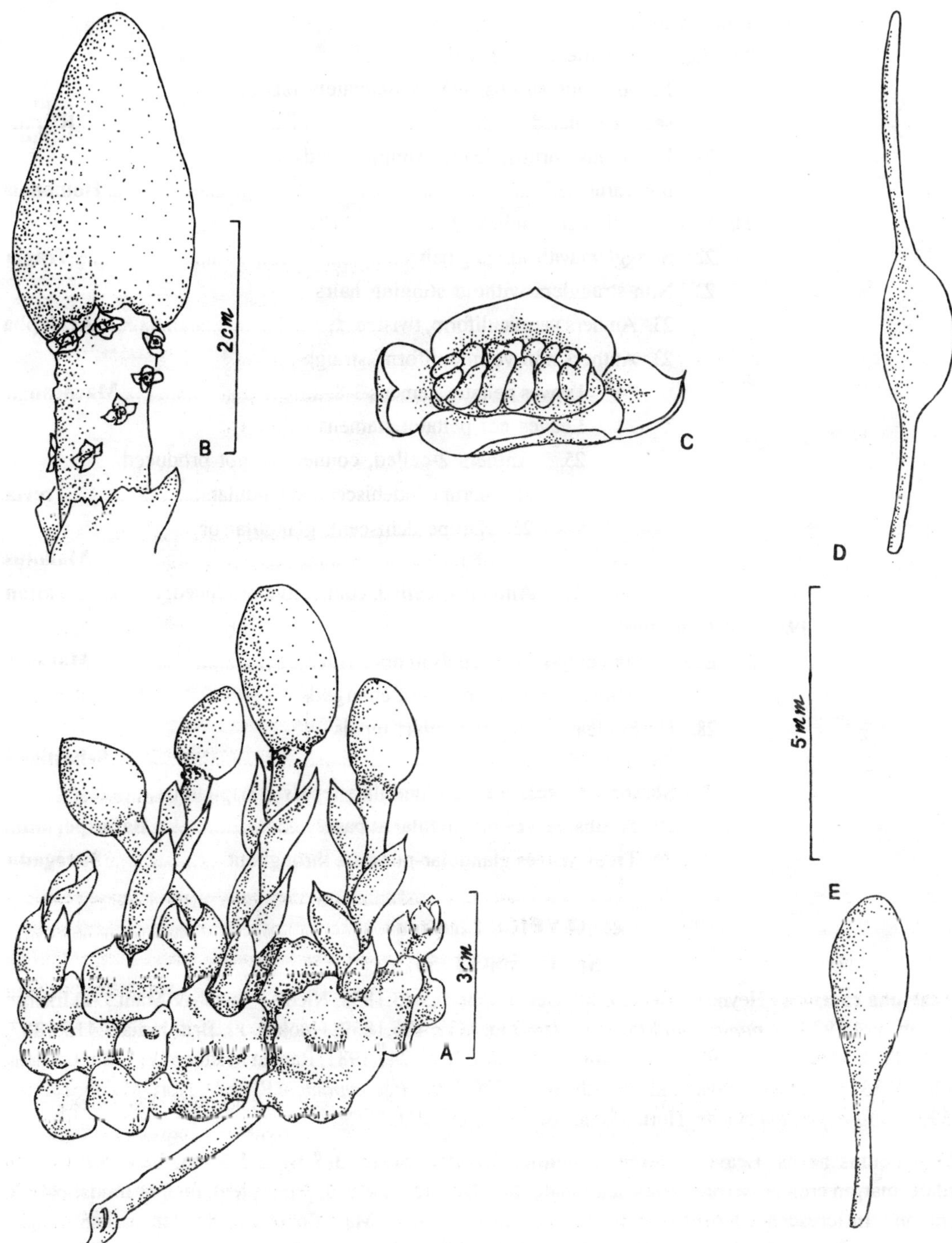

Fig. 52 . *Balanophora abbreviata* Bl.: A. Habit; B. Inflorescence; C. Male flower; D. Female flower; E. Spadicle.

19. Tepals valvate.
   20. Staminal filaments branched.
      21. Monoecious shrubs; leaves palmately-lobed; seeds caruncled. .... **Ricinus**
      21. Dioecious shrubs; leaves simple; seeds non-caruncled. .... **Homonoia**
   20. Staminal filaments unbranched.
      22. Stragglers with stinging hairs. .... **Tragia**
      22. Non-stragglers without stinging hairs.
         23. Anthers vermiculiform, twisted. .... **Acalypha**
         23. Anthers non-vermiculiform, straight.
            24. Leaves peltate; stamens 3-8. .... **Macaranga**
            24. Leaves not peltate; stamens α.
               25. Anthers 2-celled, connective not produced.
                  26. Drupe indehiscent, eglandular. .... **Trevia**
                  26. Drupe dehiscent, glandular or echinate. .... **Mallotus**
               25. Anthers 4-celled, connective produced. .... **Cleidion**
19. Tepals imbricate.
   27. Leaves palmately-lobed; tepals in male enlarged. .... **Manihot**
   27. Leaves simple; tepals in male not enlarged.
      28. Herbs; leaves non-glandular; tepals inconspicuous. .... **Sebastiana**
      28. Shrubs or trees; leaves glandular; tepa*ls* conspicuous.
         29. Shrubs; leaves biglandular at base. .... **Baliospermum**
         29. Trees; leaves glandular-punctate throughout........ **Suregada**

ACALYPHA Linnaeus
Sp. Pl. 1003.1753

Acalypha racemosa Heyne ex Baill., Etud. Gen. Euphorb. 443.1858; Nicols., Suresh & Mani., An Interpr. Hort. Malab. 104.1988. *A. paniculata* Miq., Fl. Med. Ind. 1(2): 406.1859; Hook.f., Fl. Brit. India 5:415.1887; Gamble, Fl. Pres. Madras 1330.1925; Airy Shaw, Kew Bull. 32(2) : 247.1981; Rani in Matthew, Fl. Tam. Carnatic 3(2) : 1409.1983; Mohanan & Sivad., Fl. Agasthyamala 590.2002. *Pee-cupameni* Rheede, Hort. Malab. 10:163, t. 82.1690. *Welia-cupameni* Rhede, Hort. Malab. 10:165, t.83.1690.

Monoecious herbs. Leaves alternate, simple, broadly, ovate, 3-8 by 2-5.5 cm, base truncate to subcordate, margin crenate-serrate, apex acuminate, basally 3-(5-) nerved, puberulent, membranous; petiole to 10 cm long. Inflorescence terminal or axillary panicles or spikes. Male flowers in slender axillary apikes; tepals 4; stamens 8 on a convex receptacle, anthers vermiculiform. Female flowers in terminal panicles; bracts minute; tepals 3-6; ovary superior, 3-lobed, 3-locular; ovule 1 per locule; style filiform in 3 groups of 3 each. Capsule a 2-valved coccus. Seeds globose.

Fl. & Fr.: Jul.-Dec. *Distr.*: Trop. Africa, Sri Lanka, Peninsular India, Java. Common. In moist shady waysides in hills. *NAK 217, 361, 1467* (Puthukkulam, c. 100 m & Arampa, c.500 m).

Juice with sesamum oil used in inflammation and pain in the belly.

ACTEPHILA Blume
Bijdr. 581.1825-1826

Actephila excelsa (Dalz.) Muell.-Arg., Linnaea 32:78.1863; Hook.f., Fl.Brit.India 5:282.1887; Gamble, Fl.Pres. Madras 1283.1925; Mohanan & Sivad., Fl. Agasthyamala 591.2002. *Anomospermum excelsum* Dalz. in Hooker's J. Bot. Kew Gard. Misc. 3:228.1851. *Actephila neilgherrensis* Wight, Ic.t.1910.1852.

Stout shrubs. Leaves alternate, oblong-lanceate to oblanceate, 10-18 by 5-8 cm, base acute, apex shortly acuminate, coriaceous. Flowers monoecious or dioecious. Male flowers : short- pedicelled; tepals 5+5, outer ovate-orbicular, inner smaller, spathulate; disc broad, lobed; stamens 5, inserted on the disc, filaments connate at base and enclosing the pistillode. Female flowers : solitary or very few in axillary clusters, long-pedicelled; ovary sessile, 3-locular; ovules 2 per locule; style 3, 2-fid. Fruits could not be collected.

Fl.: Apr.-May. *Distr.*: Sri Lanka, Peninsular India, Java. Rare. In evergreen forests. *NAK 2387* (Lower Moozhiar, c.250 m).

ALEURITES J. R. Forster et J. G. A. Forster
Charact. Gen. 111. 1776.

Aleurites moluccana (L.) Willd., Sp. Pl.4:590.1805; Bedd., Fl. Sylv. t.276.1872; Hook.f., Fl. Brit. India 5:384.1887; Gamble, Fl. Pres. Madras 1218.1925; Airy Shaw, Euphorb. Borneo 29.1975; Matthew, Ill. Fl. Tam. Carnatic t. 626.1982; Rani in Matthew, Fl. Tam. Carnatic 3(2):1411.1983. *Jatropha moluccana* L., Sp. Pl.1006.1753.

Monoecious or dioecious trees; tender parts and inflorescence lepidotous. Leaves broadly elliptic to deltoid, 10-15 by 8-13 cm, base cuneate, apex acuminate, 3-5-nerved from base, thin-coriaceous; petiole apically 2-glandular. Panicles terminal; tepals 2-seriate, outer 2 or 3, inner 5, petaloid; stamens α, monadelphous on a conical receptacle; disc-glands 5; ovary 2-locular, ovule 1 per locule; disc obscure. Drupe ellipsoid, lepidote.

Fl. & Fr.: Dec.-Feb. *Distr.*: India, China to Polynesia, New Zealand. Planted *NAK 572* (Moozhiar, near K.S.E.B. Staff quarters, c. 250 m). The silvery nature of leaves is striking even from a long distance.

ANTIDESMA Linnaeus
Sp. Pl. 1027.1753.

Small, dioecious trees. Leaves alternate. Flowers in catkin-like spikes, axillary or terminal; tepals 3-5; stamens 2-5, inserted on or within the disc; ovary 1-locular, ovules 2; stigmas 2-4. Fruit a Drupe.

1. Leaves pubescnet beneath; stamens 4. .... 2. **A. ghaesembilla**
1. Leaves glabrous beneath; stamens 2, 3 or 5.
    2. Stamens 2; pistillode 0. .... 1. **A. acidum**
    2. Stamens 3 or 5; pistillode present.
        3. Spikes to 4 cm long; flowers sessile. .... 4. **A. zeylanicum**
        3. Spikes to 10 cm long; flowers pedicellate. .... 3. **A. menasu**

1. **Antidesma acidum** Retz., Obs. Bot. 5 : 30.1788; Rani in Matthew, Fl. Tam. Carnatic 3(2): 1414.1983; Mohanan & Sivad., Fl. Agasthyamala 593.2002. *Stilago diandra* Roxb., Pl. Corom. t.166.1802 & Fl. Ind. 3:759.1832. *Antidesma diandrum* (Roxb.) Roth, Nov. Pl.Spec. 369.1821; Hook f., Fl.Brit. India 5:361.1887, p.p.; Gamble, Fl. Pres. Madras 1298.1925.

Leaves obovate, 3-9 by 2-4 cm, base acute, apex shortly acuminate, lateral nerves c. 6 pairs, chartaceous. Spikes 3-5 cm long. Flowers pedicellate; tepals 5, ovate; stamens 2, inserted within the disc; pistillode 0; ovary obovoid. Drupe ellipsoid.

Fl. & Fr.: Jan.-Apr. *Distr.*: India, Myanmar, S.China, Indo-China, Java. Less common. Along streamsides. *NAK 584 & 1724* (Thannithode, c.70 m).

2. **Antidesma ghaesembilla** Gaertn., Fruct. 1:189.1788; Hook. f., Fl. Brit. India 5:357.1887; Gamble, Fl.Pres. Madras 1298.1925; Rani in Matthew, Fl. Tam. Carnatic 3(2): 1413.1983. Mohanan & Sivad., Fl. Agasthyamal 596. 2002. *Antidesma pubescens* Roxb., Pl. Corom. t. 167. 1802 & Fl. Ind. 1806; Almeida, J. Bombay Nat. Hist. Soc. 84:492-493.1987. *A.paniculatum* Roxb. ex Willd., Sp. Pl. 4:764.1806; Wight, Ic.t. 870.1844-1845.

Leaves broadly oblong-elliptic, 4-9 by 2.5-5 cm, base rounded, apex obtuse to obtusely short acuminate, lower surface densly pubescent. Flowers: female in tomentose panicles, male in spikes; tepals 5, basally connate; stamens 4, inserted within the disc; pistillode linear; ovary globose. Drupe broadly ovoid.

Fl. & Fr.: Jan.-Apr. *Distr.*: India, Sri Lanka, S.China, Malesia, Australia. Rare. In semi-evergreen forests. *NAK 566* (Maniyaar, c.70 m).

3. **Antidesma menasu** Miq. ex Tul., Ann. Sci. Nat. Bot. Ser. 3,15:215.1851; Hook.f., Fl.Brit.India 5:364.1887; Gamble, Fl. Pres. Madras 1297.1925; Ramamoorthy in Sald. & Nicols., Fl. Hassan Dist. 331.1976; Matthew, Ill. Fl. Tam. Carnatic t. 627.1982; Rani in Matthew, Fl. Tam. Carnatic 3(2):1413.1983; Nicols., Suresh & Manilal, An Interpr. Hort. Malab. 105.1988; Mohanan & Sivad., Fl. Agasthyamala 596.2002. *Noeli-tali, Nuli-tali* Rheede, Hort. Malab. 4:115-116, t. 56.1683.

Leaves oblong-lanceate, 7-16 by 3-5 cm, base broadly acute to obtuse, apex acuminate, lateral nerves c.6 pairs, chartaceous. Flowers in racemes or panicles, to 10 cm long; pedicellate; tepals 3, free; stamens 5, inserted in the cavity of disc; pistillode present; ovary globose. Drupe globose, ripening red.

Fl. & Fr.: Mar.-May. *Distr.:* Peninsular India. Very common in semi-evergreen forest. *NAK 1643, 1728, 1592, 1758 & 1730* (Almost all forests).

4. **Antidesma zeylanicum** Lam., Encycl. 1:207.1782; Gamble, Fl. Pres. Madras 1297.1925; Rani in Matthew, Fl. Tam. Carnatic 3(2) : 1414.1983; Mohanan & Sivad., Fl. Agasthyamala 597.2002. *A. alexiteria* L., Sp. Pl. 1027.1753, p.p.; Hook.f., Fl.Brit. India 5: 359.1887.

Leaves obovate, 4-7 by 2-3 cm, base acute, apex acute-apiculate, lateral nerves 6-8 pairs, thin-coriaceous. Spikes terminal and axillary, to 4 cm long. Flowers sessile; tepals 3, free; stames 3, inserted in the cavity of disc; ovary globose. Drupe subglobose.

Fl. & Fr.: Mar.-May. *Distr.:* Sri Lanka, Peninsular India. Rare In evergreen forests. *NAK 2472* (Arampa, c.500 m).

APOROSA Blume
Bijdr. 514. 1825-1826.

Dioecious trees. Leaves alternate, chartaceous or coriaceous. Flowers minute, aggregated in catkin-like axillary solitary or clustered spikes; tepals 4; stamens 1-5, central; disc 0; pistillode minute or 0; ovary 2-(3) locular; ovules 2 per locule. Fruit a capsule.

1. Male spikes c. 1 cm long; leaves elliptic, long acuminate, chartaceous. .... 1. **A. acuminata**
1. Male spikes c. 5 cm long; leaves elliptic-oblong shortly acuminate, coriaceous. .... 2. **A. lindleyana**

1. **Aporosa acuminata** Thw., Enum. Pl. Zeyl. 288.1861; Hook.f., Fl. Brit.India 5:348.1887; Gamble, Fl. Pres.Madras 1309.1925; Mohanan & Sivad., Fl. Agasthyamala 598.2002.

Leaves elliptic, 6-11 by 2.5-4 cm, base acute, apex acuminate, lateral nerves c. 6 pairs, chartaceous. Male spikes short, c. 1 cm long; villous; stamens 2. Female flowers seesile; ovary hirsute, 3-locular; stigmes 3,2-fid, recurved. Capsule gubglobose, pubescent.

Fl. & Fr.: Jan.-Mar. *Distr.*: Sri Lanka, S. W Peninsular India. Less common. In semi-evergreenand evergreen forests. *NAK 492* (Maniyaar, c.70 m), *1127, 1432* (Arampa, c.500 m).

2. **Aporosa lindleyana** (Wight) Baill., Etud. Gen. Euphorb. 645.1858; Hook.f., Fl. Brit. India 5: 349.1887; Gamble, Fl.Pres. Madras 1309.1925; Ramamoorthy in Sald. & Nicols., Fl. Hassan Dist. 331. 1976; Nicols., Suresh & Manilal, An Interpr. Hort. Malab. 106.1988; Mohanan & Sivad., Fl. Agasthyamala 598.2002. *Scepa lindleyana* Wight, Ic.t.361.1840. *Amvetti, Vetti-tali* Rheede, Hort. Malab. 5: 107-108, t.54.1685.

Leaves elliptic-oblong, 7-14 by 3-6.5 cm, base rounded, apex shorlty acuminate, lateral nerves 6-8 pairs, coriaceous. Male spikes c. 4 cm long; bracts ciliate; stamens 2-3. Female flowers pedicellate; ovary fusiform, stipitate, 2-3-locular; style short. Capsule fusiform.

Fl. & Fr.: Jan.-Apr. *Distr.*: Sri Lanka, Peninsular India. Common. In moist deciduous and semi-evergreen forests. *NAK 424, 438, 1532, 1553, 1592* (Maniyaar, Thekkuthode, c. 70 m).

BACCAUREA Loureiro
Fl. Cochinch. 641, 661.1790.

Baccaurea courtallensis (Wight) Muell. Arg. in DC., Prodr. 15(2) : 1125.1866; Gamble, Fl. Pres. Madras 110.1925; Mohanan & Sivad., Fl. Agasthyamala 599.2002. *Pierardia courtallensis* Wight, Ic. 5(2) : 30, t.1912.1852, (Plate named as *P. macrostachys*). *Mootti-pazham, Mootti-thoori.*

Small dioecious trees. Leaves alternate, clustered towards branchlets, obovate to oblanceate, 6-13 by 2.5-5.5 cm, base cuneate, apex acuminate, chartaceous. Flowers small in clusters on spike which are in tufts on tubercles on the trunk and branches. Male : tepals 4, stamens 4-6 free; pistillode present, disc 0. Female : tepals 5; ovary globose, 2-5- locular; stigmas 3, sessile. Capsule globose or ellipsoid, crimson coloured.

Fl. & Fr.: Feb.-Apr. *Distr.*: Western Ghats. Common. In evergreen forests. *NAK 495 & 503* (Nilakkal, c. 400 m), *1502* (Kodumudi-Chittar, c.400 m). Trees in full bloom is a wonderful sight. Inflorescence reddish or yellow.

BALIOSPERMUM Blume
Bijdr. 604.1826

Baliospermum solanifolium (J.Burm.) Suresh in Nicols., Suresh & Manilal, An Interpr. Hort. Malab. 106.1988. *Croton solanifolius* J. Burm., Fl. Malab. 6. 1769 ("*solanifolium*"). *Jatropha montana* Willd., Sp.Pl. 4:563.1805. *Baliospermum axillara* Bl.,Bijdr. 604.1826; Hook.f., Fl.Brit. India 5:461.1887. *B. polyandrum* Wight, Ic. t. 1885.1852. *Baliospermum montanum* (Willd.) Muell.-Arg. in DC., Prodr. 15(2):1125.1866; Gamble, Fl.Pres. Madras 1342.1925; Airy Shaw, Kew Bull. 26(2):222.1972; Manilal, Fl.Silent Valley 245.1988. *Naga-danti* Rheede, Hort. Malab.10: 151, t.76.1690.

Monecious shrubs. Leaves alternate, ovate-oblong, lancete towards apices, 8-15 by 4-8 cm, entire to palmately 3-5-angled, base cuneate, 2-glandular, apex shortly acuminate, thin-coriaceous, long petioled. Flowers small, in axillary racemes. Male flowers c. 5 mm f; tepals 4-5, orbicular; disc glands lobulate; stamens 15-20. Female flowers: c. 6 mm ϕ; tepals 5, ovate, dentate; disc annular or cupular; ovary 3-locular; ovule 1 per locule; style 2-fid. Capsule obovoid. Seeds ovoid.

Fl. & Fr.: Nov.-Jan. *Distr.*: India, Java. Rare. In evergreen forests. *NAK 2848* (Karippanthode, Kallei R.F., c.100 m).

Seeds and roots used as purgative.

BISCHOFIA Blume
Bijdr. 1168. 1826-1827.

Bischofia javanica Bl., Bijdr. 1168.1827; Bedd., Fl. Sylv. t. 259.1872; Hook.f., Fl. Brit. India 5:345.1888; Gamble, Fl. Pres. Madras 1312.1925; Ramamoorthy in Sald. & Nicols., Fl. Hassan Dist. 332. 1976; Matthew, Ill. Fl. Tam. Carnatic tt.663 & 976 d.1982; Matthew & Rani in Matthew, Fl. Tam. Carnatic 3(2): 1485.1983; Manilal, Fl. Silent Valley 246.1988; Mohanan & Sivad., Fl. Agasthyamala 599.2002. *Microelus roeperianus* Wight & Arn., Edinburgh New Philos. J. 14: 298.1833; Wight, Ic.t. 1880.1852.

Dioecious trees. Leaves 3-foliolate, alternate; leaflets elliptic-ovate, 7-15 by 4.5-8.5 cm, base acute, margin crenate-dentate, apex acuminate, coriaceous. Flowers in axillary lax panicles. Male flowers : tepals 5, ovate, obtuse; stamens 5, anthers large; disc 0; pistillode disciform, stalked. Male flowers : tepals 5; ovary 2-locular; ovules 2 per locule; styles 3, long. Berry Globose. Seeds 3-gonous.

Fl. & Fr.: Feb.-Apr. *Distr.*: India, Malaya & Pacific Islands. Not common. In evergreen forests. *NAK 2549 & 2730* (Kakki hills, c.1100 m).

Note : Some authors treated *Bischofia* under a separate family Bischofiaceae.

BREYNIA J.G. Forster et J.G.A. Forster
Charact. Gen. 145.1776 (nom.cons.)

Monoecious shrubs. Leaves alternate-distichous. Flowers axillary. Male: pendulous, perianth-lobes 6, connate into a turbinate tube; stamens 3, included. Female: perianth lobes 6, shortly connate; disc 0; ovary half-inferior, truncate or cupular, 3-loculed; ovules 2 per locule. Capsule globose, with accrescent perianth.

1. Leaves glaucous beneath; fruiting perianth enlarged. ....1. **B. retusa**
1. Leaves greenish beneath; fruiting perianth hardly enlarged. ....2. **B. vitis-idaea**

1. **Breynia retusa** (Dennst.) Alston, Ann. Roy. Bot. Gard. (Peradeniya) 11: 204. 1929; Airy Shaw, Kew Bull. 26(2) : 227. 1927; Ramamoorthy in Sald. & Nicols., Fl. Hassan Dist. 333. 1976. Matthew, Ill. Fl. Tam. Carnatic t.628. 1982; Rani in Mathew, Fl. Tam. Carnatic 3(2) : 1415. 1983; Manilal, Fl. Silent Valley 246. 1988; Nicols., Suresh & Manilal, An Interpr. Hort. Malab. 106. 1988; Mohanan & Sivad., Fl. Agasthyamala 600.2002. *Phyllanthus retusus* Dennst., Schluess. Hort. Malab. 24. 1818. *P. patens* Roxb., Fl. Ind. 3:667. 1832. *Melanthesa turbinata* Wight, Ic.t.1897. 1852. *M. obliqua* Wight, Ic. t. 1898. 1852. *Breynia patens* (Roxb.) Rolfe, J. Bot. 11:359. 1882; Hook.f., Fl. Brit. India 5:329. 1887; Gamble, Fl. Pres. Madras 1304. 1925. *Perin-nirouri, Ma-nirouri* Rheede, Hort. Malab. 5:85-86, t. 43. 1685.

Leaves elliptic-ovate, 1-2 by 0.7-11.3 cm, base aucte, apex obtuse, lower surface glaucous. Flowers drooping, yellow, male 4 mm ϕ; female 6 mm ϕ. Capsule subglobose, c. 12 mm ϕ, orange red; fruiting perianth enlarged.

Fl. & Fr.: Feb.-Apr. *Distr.*: Sri Lanka, India, Nepal, Myanmar. Common. In evergreen forest borders and in degraded forests. *NAK* 516 (Velithode-Moozhiar. c. 250 m), *1745* (Thannithode, c. 70 m).

2. **Breynia vitis-idaea** (Burm. f.) Fischer, Bull. Misc. Inform. 1932:65. 1932; Rani in Matthew, Fl. Tam. Carnatic 3(2) : 1416.1983; Nicols., Suresh & Manilal, An Interpr. Hort. Malab. 106. 1988; Mohanan & Sivad., Fl. Agasthyamala 601.2002. *Rhamnus vitis-idaea* Burm. f., Fl. Ind. 61. 1768. *Phyllanthus rhamnoides* Retz., Obs. Bot. 5:30. 1788. *Melanthesa rhamnoides* (Retz.) Blume, Bijdr. 591. 1826; Wight, Ic. t. 1898, f. 1. 1852. *Breynia rhamnoides* (Retz.) Muell.-Arg. in DC., Prodr. 15(2) : 440. 1866; Hook.f., Fl. Brit. India 5:330. 1887; Gamble, Fl. Pres. Madras 1304. 1925. *Niruri* Rheede, Hort. Malab. 2:45-46, t. 27. 1669.

Leaves broadly ovate to elliptic-ovate, 2-3 by 1.5-2.5 cm, base acute, apex subacute, lower surface greenish. Flowers drooping. Male yellow, 2.5 mm ϕ; anthers linear. Female reddish, c.1.5 mm ϕ; ovary truncate. Capsule subglobose, c. 5 mm ϕ; fruiting perianth hardly enlarged.

Fl. & Fr.: Sep.-Dec. *Distr.*: Sri Lanka, India, Nepal, SE. Asia, China, Malesia. Not common. In evergreen forest borders. *NAK 1193* (Moozhiar, c. 250 m).

BRIDELIA Willdenow
Sp. Pl. 4(2) : 978.1806 ("1805").

Scandent shrubs or small trees. Leaves alternate, nervature prominent, coriaceous. Flowers monoecious or dioecious, in axillary spicate clusters; tepals 10, 5 + 5; male : flowers pedicelled, outer *tepals* cucullate, dentate; stamens 5 on an erect gonophore; disc flat, annular; pistillode linear. Female flowers sessile; inner tepals spathulate; ovary 2-locular; ovules 2 per locule, styles 2; disc corona-like. Drupe globose or subglobose, black.

1. Trees; drupe globose. .... 1. **B. retusa**
1. Scandent shrubs; drupe subglobose. .... 2. **B. scandens**

1. **Bridelia retusa** (L.) Spreng., Syst. Veg. 3:48. 1826; Bedd., Fl. Sylv. t. 260. 1872; Hook.f., Fl.Brit. India 5:268. 1887, p.p.; Gamble, Fl. Pres. Madras 1280. 1925; Mohanan & Sivad., Fl. Agasthyamala 602.2002. *Clutia retusa* L., Sp. Pl. 1042. 1753, ('*Cluytia*'). *Mullu venga.*

Small trees; stem often thorny. Leaves elliptic-oblong or obovate, 8-15 by 4-8 cm, base obtuse, apex acute to retuse, lateral veins strong, branched near margin, thin-coriaceous. Flowers monoecious, in clusters in spikes.

Fl. & Fr.: Mar.-Apr. *Distr.*: Sri Lanka, India, Myanmar, Malaya. Occasional. *NAK 1173* (Upper Moozhiar, c. 550 m).

2. **Bridelia scandens** (Roxb.) Willd., Sp. Pl. 4 : 979. 1806; Gamble, Fl. Pres. Madras 1281. 1925; Ramamoorthy in Sald. & Nicols., Fl. Hassan Dist. 334. 1976 Manilal, Fl. Silent Valley 247. 1988; Nicols., Suresh & Manilal, An Interpr. Hort. Malab. 106. 1988; Mohanan & Sivad., Fl. Agasthyamala 603.2002. *Clutia scandens* Roxb., Pl. Corom. t. 173. 1802, ('*Cluytia*'). *Scherunam-cottam* Rheede, Hort. Malab. 2:23-24, t.16. 1679.

Climbing shrubs. Leaves elliptic-oblong, 3-6 by 2-6 cm, base subcordate, apex subacute to rounded to retuse, lateral veins c. 8 pairs, unbranched. Flowers greenish-yellow, in axillary fascicles. Drupe globose.

Fl. & Fr.:Nov.-Mar. *Distr.*: Sri Lanka, Eastern & Peninsular India, Malaya Islands, Philippines, Trop. Africa. Very common. In evergreen forest borders. *NAK 261, 515, 2256* (Moozhiar & other forests).

CLEIDION Blume
Bijdr. 613. 1826.

Cleidion spiciflorum (Burm. f.) Merr., Interpr. Rumph. Amboin. 322, (in nota) 1917. *Acalypha spiciflora* Burm. f., Fl. Ind. 203, t. 61, f. 2. 1768. *Cleidion javanicum* Blume, Bijdr. 613.1826; Bedd., Fl. Sylv. t. 272. 1872; Hook.f., Fl. Brit India 5:444. 1887; Gamble, Fl. Pres. Madras 1325. 1925.

Trees. Leaves alternate, elliptic-lanceate, 10-18 by 4-10 cm, base acute, margin entire or toothed, apex acuminate, chartaceous. Flowers dioecious or monoecious. Male in long axillary racemes; perianth globose, lobes 3-4; stamens α, on a conical receptacle; pistillode 0. Female solitary, axillary; pedicel to 4 cm long, thickened towards apex; perianth-lobes minute, 3-5; ovary 3-locular; ovule 1 per locule. Capsule c. 2.5 cm f, subglobose.

Fl. & Fr.: Nov.-Mar. *Distr.*: Sri Lanka, Peninsular India, Java. Not common. In evergreen forests. *NAK 2280 & 2285* (Mannarappara, c. 250 m).

CROTON Linnaeus
Sp. Pl. 1004. 1753.

Monoeious trees, climbing or erect shrubs. Leaves alternate, base 2-glandular, lateral veins c. 6 pairs, basal pair prominent. Flowers clustered and form terminal racemes; male: tepals 10, biseriate; stamens, c. 20, free on raised receptacle; disc-glands 5. Female : tepals 5 or 10 in 2 series; ovary 3-locular; ovule 1 per locule; styles 3, once or twice forked; disc-glands 5. Capsule subglobose or obovoid.

1. Climbing shrubs .... 2. **C. caudatus**
1. Trees, herbs or subshrubs.
    2. Trees.
        3. Leaves silvery beneath.
            4. Leaves elliptic ovate to oblong-lanceate; stamens α. .... 7. **C. zeylanicus**
            4. Leaves ovate; stamens 10-12. .... 5. **C. malabaricus**
        3. Leaves not silvery beneath.
            5. Inflorescence stellate-tomentose. .... **C. laccifer**
            5. Inflorescence glabrous.

6. Basal pairs of leaf-nerves inconspicous; petiole to 2.5 cm long. ....3. **C. klotzschianus**

6. Basal pair of leaf-nerves conspicous; petiole to 5 cm long. ....6. **C. tiglium**

2. Herbs or subshrubs. ....1. **C. bonplandianus**

1. **Croton bonpalndianus** Baillon, Adansonia 4:339. 1864; Croizat, J. Bombay Nat. Hist. Soc. 41:573. 1940, ("*bonplandianum*"); Ramamoorthy in Sald. & Nicols., Fl. Hassan Dist. 335. 1976; Rani in Matthew, Fl. Tam. Carnatic 3(2) : 1420. 1983; Mohanan & Sivad., Fl. Agasthyamala 604.2002. *C. sparsiflorus* Morong. Ann.New York Acad. Sci. 7: 221. 1893; Gamble, Fl. Pres. Madras 1316. 1925.

Herbs to subshrubs. Leaves ovate-lanceate, 2-5 by 1-2 cm, base acute, margin faintly serrulate, apex gradually acute, chartaceous. Racemes c. 10 cm long. Flowers greenish-yellow; tepals 5 in female. Capsule 4 mm ϕ, warty.

Fl. & Fr.: Most of the seasons. *Dsitr.*: Native of S. America, naturalized in Subtropics. This introduced noxious weedy species has occupied almost all arable and exposed lands. *NAK 2913* (Adoor, c. 20 m).

2. **Croton caudatus** Geiseler, Crot. Monogr. 73. 1807; Hook.f., Fl. Brit. India 5:388. 1887; Gamble, Fl. Pres. Madras 1315. 1925. **(Fig. 53)**

Scandent shrubs; young parts with stellate tomentum. Leaves ovate to elliptic-ovate, 4-9 by 3-5 cm, base cordate, margin coarsly serrulate, apex acuminate, stellate tomentose, chartaceous. Racemes terminal, to 20 cm long. Male : tepals 10, 5+5, subequal, woolly; disc-glands minute; stamens 18-30, well exerted. Female : tepals 5+5, inner very minute. Capsule subglobose. Seeds compressed.

Fl. & Fr.: Mar.-May. *Distr.*: Sri Lanka, India, Java, Philippines. Rare. In evergreen forests. A very variable plant. *NAK* 1684, 2401 (Sabari Hills, c. 250 m).

3. **Croton klotzschianus** (Wight) Thw., Enum. Pl. Zeyl. 276. 1861; Hook.f., Fl. Brit. India 5:392. 1887; Gamble, Fl. Pres. Madras 1315. 1925.

Shrubs, often straggling. Leaves elliptic-ovate or oblong, 7-9 by 2.5-4 cm, base and apex acute, margin serrulate, membranous. Racemes 3-6 cm long; male flowers: pedicel long; tepals gland-dotted; disc-glands 5, large. Female flowers : pedicel short, outer tepals ovate, inner ones subulate. Capsule subglobose.

Fl. & Fr.: Feb.-Apr. *Distr.*: Sri Lanka, Peninsular India. Not common. In semi-evergreen forests. *CNM* 61208 (MH) (Placherry R.F., c. 300 m).

4. **Croton laccifer** L., Sp. Pl. 1005. 1753; Wight, Ic. t. 1915. 1852; Gamble, Fl. Pres. Madras 1315. 1925; Airy Shaw, Kew Bull. 26 : 248. 1972; Rani in Matthew, Fl. Tam. Carnatic 3(2) : 1420. 1983; Manilal, Fl. Silent Valley 217. 1988. *C. aromaticus* auct., non L., 1753: Hook.f., Fl. Brit. India 5:388. 1887, p.p.

Shrubs to small trees. Leaves ovate or lanceate, 5-10 by 2.5-6 cm, base cordate or rounded, margin denticulate, apex acuminate, basally 3-nerved, membranous. Racemes 10-15 cm long, stellate-tomentose. Male flowers : tepals woolly; stamens c. 20; disc-glands small. Female flowers: pedicel stout; outer tepals broadly ovate; inner ones minute, filiform. Capsule subglobose.

Fl. & Fr.: Jun.-Sep. *Distr.:* Sri Lanka, Peninsular India. Rare. In evergreen forests. *Deb 30443* (MH) (Pamba, c. 1025 m).

5. **Croton malabaricus** Bedd., Ic.t. 181. 1868-1874; Hook.f., Fl. Brit. India 5:386. 1887; Gamble, Fl. Pres. Madras 1314. 1925; Ramamoorthy in Sald. & Nicols., Fl. Hassan Dist. 335. 1976; Mohanan & Sivad., Fl. Agasthyamala 605.2002.

Trees. Leaves ovate or rhombic-ovate, 8-15 by 5-8 cm, base cuneate, apex acuminate, lower surface silvery, chartaceous. Racemes few-flowered. Male : tepals 10, woolly, stamens 10-1, villous. Female : *tepals* 5, ovate-oblong; ovary stellate-tomentose; style- arms long, slender, 2-fid. Capsule obovoid, lepidote.

Fl. & Fr.: Apr.-May. *Distr.*: Western Ghats. Rare. In evergreen forests. *NAK 1794* (Angamoozhy, c. 250 m). The silvery nature of leaves is striking from a long distance.

6. **Croton tiglium** L., Sp. Pl. 1004. 1753; Hook. f., Fl. Brit. India 5 : 393. 1887; Gamble, Fl. Pres. Madras 1315. 1925; Airy Shaw, Kew Bull. 26(2): 250. 1972; Nicols., Suresh & Manilal, An Interpr. Hort. Malab. 108. 1988. *Cadel-avanacu* Rheede, Hort. Malab. 2 : 61-62, t.33-1679.

Small trees. Leaves ovate-lanceate to oblong-lanceate, 10-15 by 5.5-7.5 cm, base cuneate, apex acuminate, basally 3-(5)-nerved, sparsely stellate-tomentose, membranous, drying yellowish. Racemes 3-7 cm long. Male flowers : tepals 10, linear, inner woolly at margins; stamens 15-20, disc-glands small. Female flowers : tepals 5, villous within at base. Capsule obovoid. Seeds 3-gonous.

Fl. & Fr.: Nov.-Jan. *Distr.:* Sri Lanka, India, Myanmar, China, Malesia. Cultivated, also run wild *NAK* 257 (Athumpumkulam, c. 70 m).

Dried seeds and leaves ground and applied to snake bites. Leaves and fruits ground and used for fish poisoning.

7. **Croton zeylanicus** Muell.-Arg., Linnaea 34 : 107. 1865 & in DC., Prodr. 15(2): 581. 1866; Manilal, Fl. Silent Valley 247. 1988; Mohanan & Sivad., Fl. Agasthyamala 605.2002. *C. reticulatus* Heyne ex Muell.-Arg. in DC., Prodr. 15(2) : 580. 1866; Hook.f., Fl. Brit. India 5: 386. 1887; Gamble, Fl. Pres. Madras 1314. 1825.

Shrubs to very small trees. Leaves elliptic-ovate to oblong lanceate, 6-16 by 1.5-4 cm, base rounded, apex acuminate, lower surface silvery, drying-yellowish brown. Racemes to 8 cm long, few-flowered. Male : tepals 10, woolly; stamens 15-18, glabrous except at base. Female : tepals 5, linear-oblong; ovary globose, stellate. Capsule narrowly ellipsoid.

Fl. & Fr.: Aug.-Nov. *Distr.:* Sri Lanka, Peninsular India. Not common. In evergreen forest. *NAK 2560* & *2694* (Kakki hills, c. 1100 m).

DIMORPHOCALYX Thwaites
Enum. Pl. Zeyl. 278. 1861.

Dimorphocalyx lawianus (Muell.-Arg.) Hook. f., Fl. Brit. India 5 : 404. 1887; Gamble, Fl. Pres. Madras 1337. 1925. *Trigonostemon lawianus* Muell.-Arg., Linnaea 34: 212. 1865, p.p.

Small dioecious trees. Leaves alternate, broadly elliptic or eliptic-oblong, 7-15 by 5-8 cm, base rounded, margin faintly crenate, apex obtusely acuminate, chartaceous, drying black. Male flower(s): in axillary cymes or racemes or solitary; tepals 10, 5+5, outer connate, cupular; stamens 10-20, biseriate; disc-glands 5. Female flower(s): solitary or 2-3 in terminal or leaf-opposed clusters; tepals 10, 5+5, outer foliaceous, accrescent; ovary 3-locular, ovule 1 per locule; style 3, connate at base 2-fid; disc annular. Capsule enclosed by accrescent calyx.

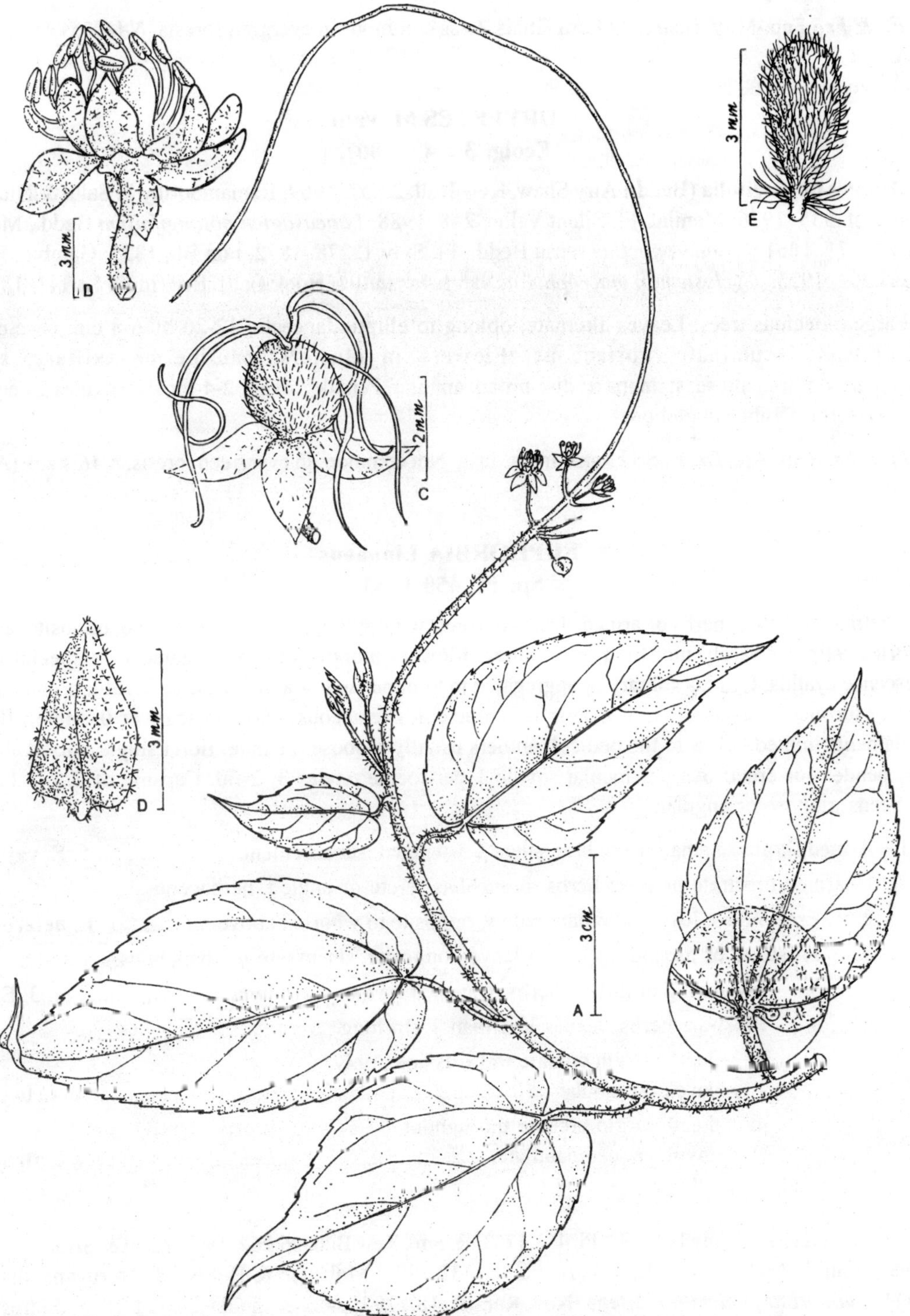

Fig. 53. *Croton caudatus* Geiseler: A. Twig; B. Male flower; C. Female flower; D. Outer Petal; E. Inner petal.

Fl. & Fr.: Feb.-May. *Distr.*: Western Ghats. Less common. In evergren forests. *NAK 2594* (Arampa, c. 550 m).

DRYPETES M. Vahl
Ecolg. 3 : 49. 1807.

Dryptes oblongifolia (Bedd.) Airy Shaw, Kew Bull. 23:57. 1969; Ramamoorthy in Sald. & Nicols., Fl. Hassan Dist. 335. 1976; Manilal, Fl. Silent Valley 248. 1988. *Laneasagum oblongifolium* Bedd., Madras J. Sci. n.s. 6 : 71. 1861. *C. macrophyllus* sensu Bedd., Fl. Sylv. t. 278. 1872, non Bl., 1825; Gamble, Fl. Pres. Madras 1302. 1925. *Cyclostemon macrophyllus* var. *peninsularis* Hook.f., Fl. Brit. India 5 : 341. 1887.

Large dioecious trees. Leaves alternate, oblong to elliptic-lanceate, 12-20 by 6-8 cm, base oblique, apex obtusely acuminate, coriaceous. Flowers in clusters, cauline or axillary, sessile. Male : tepals 4-5, tomentose; stamens α; disc broad, annular. Female : ovary 2-4-locular; ovules 2 per locule; stigma reniform. Drupe subglobose.

Fl. & Fr.: Feb.-Apr. *Distr.*: Sri Lanka, India, Java. Not common. In evergreen forests. *NAK 1449* (Arampa, c.550 m).

EUPHORBIA Linnaeus
Sp. Pl. 450 1753.

Prostrate or diffuse herbs or armed shrubs to trees, with milky latex. Leaves simple, opposite, alternate or whorled; stipules rarely modified into spines. Flowers monoecious, aggregated in a special type of inflorescence cyathia. Cyathia solitary or aggregated in terminal and/or axillary cyme, involucre campanulate, cupular or turbinate; bracts often 5, glands 1-5; bracteoles setaceous; perianth scaly or 0. Male : florets 1 to α, filaments jointed below to the pedicel; anthers usually globose. Female: floret single, raised above on a stalk, pendent or erect; ovary 3-locular; ovule 1 per locule; styles 3, 2-fid. Capsule of 2 to 3 bivalved cocci. Seeds globose or angular.

1. Armed shrubs or small trees; branchlets 3-5-winged, subsucculent. ....5. **E. vajravelui**
1. Unarmed prostrate or erect herbs; branchlets terete or angled, herbaceous.
    2. Erect herbs; leaves alternate below, opposite to whorled above. ....1. **E. heterophylla**
    2. Prostrate or ascending herbs; leaves opposite, decussate or distichous.
        3. Ascending or diffuse herbs; leaves more than 1 cm long. ....3. **E. hirta**
        3. Prostrate herbs; leaves less than 1 cm long.
            4. Leaf-margin crenate towards apex only; involucre tubular. ....**2. E. heyneana**
            4. Leaf-margin crenate throughout; involucre campanulate. ....**4. E. thymifolia**

1. **Euphortia heterophylla** L., Sp. Pl. 453.1753; R. Sm., Kew Bull. 26:264. 1972; Mani & Sivar., Fl. Calicut 257. 1982; Rani in Matthew, Fl. Tam. Carnatic 3: 1432.1983; Mohanan & Henry, Fl. Thiruvananthapuram 414.1994. *Euphorbia geniculata* Ortega, Nov. Rar. Pl. 18.1797; Fischer in Gamble, Fl. Pres Madras 1884. 1936. *E. prunifolia* Jacq., Hort. Schoenbr. 3:15, t. 277. 1798; Hook.f., Fl. Brit. India 5 : 266. 1887, *in adnot.*

Erect herbs. Leaves alternate or opposite below, rarely whorled above, obovate to oblanceate, 5-9 by 2-5 cm, base and apex acute, chartaceous. Cyathia in terminal clusters; involucre campanulate, gland 1; appendages 0. Male florets 12-15, bracteolate; female pendulous; styles 3, erect, 2-fid. Capsule 3-gonous. Seeds angled.

Fl. & Fr.: Most of the seasons. *Dsitr.*: Native of C. America, now a pantropical weed. Not common. Along waysides in waste places. *NAK 1879* (Adoor, c. 50 m).

2. **Euphorbia heyneana** Sprengel in L., Syst. Veg. (ed. 16) 3: 791. 1826; Panigr., Kew Bull. 29:695. 1974; Rani in Matthew, Fl. Tam.Carnatic 3(2) : 1433. 1983. *E. microphylla* Heyne ex Roth, Nov. Pl. Spec. 29. 1821, non Lam., 1788; Hook.f., Fl. Brit. India 5: 252. 1887; Gamble, Fl. Pres. Madras 1276. 1925.

Prostrate or spreading herbs. Leaves opposite, distichous, elliptic-oblong, 3-9 by 2-5 mm, base obliquely subcordate, margin crenate towards apex, apex rounded, faintly 3-nerved, herbaceous. Cyathia few, axillary; involucre tubular, glands 4; appendages inconspicuous. Male florets 1-4, ebracteolate. Female pendulous; style 2-fid. Capsule obscurely keeled. Seed 4-angled, furrowed.

Fl. & Fr.: Most of the seaosns. *Distr.*: Pakistan, India, Bangladesh, Mynmar. Common on exposed waysides. *NAK 1051* (Kodumon, c. 70 m).

3. **Euphorbia hirta** L., Sp. Pl. 454. 1753; Gamble, Fl. Pres. Madras 1275. 1925; Airy Shaw, Kew Bull. 26(2) : 264. 1972; Ramamoorthy in Sald. & Nicols., Fl. Hassan Dist. 338. 1976; Rani in Matthew, Fl. Tam. Carnatic 3 (2) : 1434. 1983; Manilal, Fl. Silent Valley 249. 1988; Mohanan & Sivad., Fl. Agasthyamala 608.2002. *E. pilulifera* L., Sp. Pl. 454.1753; Hook.f., Fl.Brit. India 5:250. 1887.

Ascending or diffuse herbs. Leaves desuccate, broadly oblong to elliptic-lanceate, 1-2.5 - 0.7-1.5 cm, base oblique, margin serrulate, apex acute, basally 3-nerved, chartaceous. Cyathia very small, α, aggregated in single or paired axillary clusters; involucre campanulate, glands 5, minute. Male florets 4-6, ebracteolate. Female florets laterally pendulous; styles 2-fid from base. Capsule pubescent. Seeds 4-anlged, fainty furrowed.

Fl. & Fr.: Most of the seaons. *Distr.*: Pantropical. Common on waysides and wastelands in lowlands. *NAK* 2871 (Adoor, c.20 m).

The whole plant crushed and cooked with rice and made into 'canji' and taken against worms in children, bowel-complaints, cough and gonorrhoea.

4. **Euphorbia thymifolia** L., Sp. Pl. 454. 1753; Roxb., Fl. Ind. 2:473. 1832; Hook. f., Fl. Brit. India 5 : 252. 1887; Gamble, Fl. Pres. Madras 1276. 1925; Ramamoorhty in Sald. & Nicols., Fl. Hassan Dist. 339. 1976; Rani in Matthew, Fl. Tam. Carnatic 3(2) : 1439. 1983; Nicols., Suresh & Manilal, An Interpr. Hort. Malab. 110. 1988. *Culcotten-pala* Rheede, Hort. Malab. 10:65, t.33. 1690.

Prostrate herbs; branchlets hispid. Leaves opposite, distichous, oblong, 2-7 by 2-4 mm, base obliquely truncate, margin serrulate, apex obtuse, 3-nerved, subsessile. Cyathia in axillary, clusters. involucre campanulate,; glands 4. Male florets 1-4, ebracteolate. Female laterally pendulous; style 3-fid from base. Capsule obtusely angled, glabrescent. Seeds 4-angular.

Fl. & Fr.: Feb.-May. *Distr.*: Tropical Asia. Common. On exposed waysides. *NAK* 1505 (Chittar, c. 400 m).

5. **Euphorbia vajravelui** Binojk. & Balakr., Cact. Succ. J. 63: 229. 1992.

Armed shrubs or small trees; branchlets 3-5-winged, subsucculent. Leaves caducous. Cyathia in axillary cymes; involucre cupular; glands 5, thick. Male florets in 5 groups. Female erect; styles 3, 2-fid. Capsule 3-gonous. Seeds globose.

Fl. & Fr.: Jan.-Apr. *Distr.*: Sri Lanka, India. Rare, but locally very abundant. Profusely branched small trees, on exposed rocky hills. *NAK 2839* (Kattathipara, Allungal - Kokkathode, c. 150 m).

Root is used against snake bites; latex used in dropsy and chronic affections of ears and eyes.

Note : This species was originally reported from Kalakkad forest in Tirunelveli distric of Tamil Nadu. The present collection is showing the extended distribution of this species.

Euphorbia nivulea Buch.-Ham. and **E. tirucalli** L. are seein occasionally running wild. **E. pulcherrima** Willd. ex Klotzsch. is seen planted in gardens.

GLOCHIDION J. R. Forster et J. G. A. Forster Charat. Gen. 113, t. 57. 1776 (nom. cons.).

Monoecious or dioecious trees. Leaves simple, alternate, distichous. Flowers in axillary clusters; male : pedicellate, tepals 6, basally connate or free, biseriate; stamens 3-8, exerted, connate by their connectives, anther dehiscence longitudinal; pistillode 0. Female: subsessile; tepals 6, connate; ovary 3-5- locular; ovules 2-per locule; disc 0. Capsule 3-5, bivalved cocci. Seeds subtrigonous.

1. Leaves and branchlets pubescent. .... 3. **G. tomentosum**
1. Leaves and branchlets glabrous.
 2. Dioecious trees; stamens 4 or more. .... 4. **G. zeylanicum**
 2. Monoecious trees; stamens 3.
  3. Leaf-base unequal. .... 1. **G. ellipticum**
  3. Leaf-base (sub) equal. .... 2. **G. malabaricum**

1. **Glochidion ellipticum** Wight, Ic. t. 1906. 1852; Hook.f., Fl. Brit. India 5:321. 1887; Gamble, Fl. Pres. Madras 1308. 1925; Ramamoorthy in Sald. & Nicols., Fl. Hassan Dist. 341. 1976; Matthew, Ill. Fl. Tam. Carnatic t. 640. 1982; Rani in Matthew, Fl. Tam. Carnatic 3(2) : 1445. 1983; Mohanan & Sivad., Fl. Agasthyamala 609.2002.

Monoecious trees. Leaves oblong-lanceate, 6-12 by 2-4 cm, base very unequal, apex sharply acuminate; lateral nerves 6 pairs, thin-coriaceous. Male and female flowers greenish on same node. Male : pedicel c. 1 cm long; tepals linear-oblong; stamens 3. Female: sessile, oblong, 6-toothed; ovary 4-6-locular. Capsule subglobose.

Fl. & Fr.: Sep.-Dec. *Distr.*: Peninsular India. Rare. In evergreen forests. *NAK 2555* (Kakki hills, c. 1100 m).

2. **Glochidion malabaricum** Bedd., For. Man. Bot. 194. 1873; Hook.f., Fl. Brit. India 5:319. 1887; Gamble, Fl. Pres. Madras 1308. 1925; Mohanan & Sivad., Fl. Agasthyamala 609.2002.

Monoecious trees. Leaves oblong-lanceate, 9-13 by 5-5.5 cm, base (sub) equal, apex obtusely short-acuminate, lateral nerves arching, chartaceous. Male and female flowers in same node. Male : pedicel slender, c. 8 mm long; tepals lanceate; stamens 3. Female : sessile, minute; tepals linear; ovary 3-5-locular; style conical. Capsule subglobose.

Fl. & Fr.: Fb.-May. *Distr.*: Western Ghats. Not common. In evergreen forests. *NAK 522 & 1242* (Angamoozhy, c. 250 m).

3. **Glochidion tomentosum** Dalz. in Hooker's J. Bot. Kew Gard. Misc. 2:38. 1850; Hook.f., Fl. Brit. India 5:309. 1887; Gamble, Fl. Pres. Madras 1306. 1925; Ramamoorthy in Sald. & Nicols., Fl. Hassan Dist. 342. 1976. *Nellikka-puli.*

Dioecious trees. Leaves broadly ovate to ovate-oblong, 4-13 by 3.5-7 cm, base rounded, apex obtusely short-acuminate, midrib above and lower surface finely soft tomentose, coriaceous. Flowers yellow in axillary clusters; tepals ovate, pubescent, stamens 5-7; ovary 4-5-locular; style stout, conical 4-5-fid. Capsule subglobose, beaked.

Fl. & Fr.: Mar.-May. *Distr.*: Western Ghats. Rare. In riparian forests. Known from only one collection. *NAK 1734* (Thannithode, c. 70 m).

4. **Glochidion zeylanicum** (Gaertner) A. Juss., Euphorb. Gen. 107. 1824; Hook. f., Fl. Brit India 5:310. 1887; Gamble, Fl. Pres. Madras 1306. 1925; Ramamoorthy in Sald. & Nicols., Fl. Hassan Dist. 342. 1976; Rani in Matthew, Fl. Tam. Carnatic 3(2): 1446. 983; Nicols, Suresh & Manilal, An Interpr. Hort. Malab.112. 1988; Mohanan & Sivad., Fl. Agasthyamala 610.2002. *Bridelia zeylanica* Gaertner, Fruct. 2:128, t.109. 1790. *Pevetti, Pee-vetti* Rheede, Hort. Malab. 4: 113, t.55. 1683. *Neer-vetti.*

Dioecious trees. Leaves broadly oblong-elliptic, 6-12 by 3.5-6 cm, base obliquely cunete or truncate, apex obtusely short-acuminate, lateral nerves c. 6 pairs, coriaceous. Flowers yellow in clusters. Male: tepals 6; stamens 4 or more. Female: tepals 6, almost free; ovary subglobose, 4-5- locular; style beaked, thick. Capsule subglobose, in clusters, beaked.

Fl. & Fr.: Jan.-Apr. *Distr.*: Sri Lanka, India, Malaya Peninsula. Most common species of the genus. In evergreen and semi-evergreen forests and in lowlands. *NAK 431, 469 & 1625* (Almost all places). A common tree with highly variable leaves.

Seed-oil is a good medicine for rheumatism.

HOMONOIA Loureiro
Fl. Cochinch. 601, 636. 1790.

Homonoia riparia Lour., Fl. Cochinch. 637. 1790, Hook.f., Fl. Brit. India 5:455. 1887; Gamble, Fl. Pres. Madras 133. 1925; Airy Shaw, Kew Bull. 26(2):282. 1972, Ramamoorthy in Sald. & Nicols., Fl. Hassan Dist. 342. 1976, Matthew,Ill. Fl. Tam. Carnatic t. 641.1982; Rani in Matthew, Fl. Tam. Carnatic 3(2) : 1447. 1983; Manilal, Fl. Silent Valley 251. 1988. *Aattu-vanchi.*

Dioecious shrubs. Leaves alternate, linear-lanceate, 10-20 by 1.5-1.8 cm, base and apex acute, lower surface glandular-scaly, coriaceous. Flowers in axillary short spikes. Male : perianth cupular, lobes 3; stamens α, exerted, filametns branched, anthers broadly ellipsoid. Female : tepals 5, free, ovate; ovary 3-lobed, 3-locular; ovule 1 per locule; styles 3, papillose. Capsule subglobose of 3-valved cocci.

Fl. & Fr.: Apr.-Aug. *Distr.*: India, SE. Asia, Formosa, Malesia to E. New Guinea. Common. In rocky crevices on river beds. *NAK 557, 1417* (Thekkuthode, Mannarappara, etc.)

Decoction of root cures piles, stone in the urinary bladder, chest-pain, disorder in the vagina, gonorrhoea, syphilis and thirst. It accelerates the flow of urine.

JATROPHA Linnaeus
Sp. Pl. 1006. 1753.

Undershrubs. Leaves lobed, alternate; stipules glandular, usually persistent. Flowers unisexual or polygamous in terminal and/or axillary dichasia; perianth 2-seriate, 5+5, outer sepaloid, inner ones petaloid, free or connate; stamens 8-10, monadelphous or diadelphous, exerted; ovary globose, 3-locular; ovule 1 per locule; styles 3; disc-glands 5, free. Capsule smooth or ribbed, 3-valved. Seeds ovoid to 3-gonous.

1. Leaf-margin smooth; inner tepals connate near half way. .... 1. **J. curcas**
1. Leaf-margin glandular-ciliate; inner tepals free. .... 2. **J. gossypifolia**

1. **Jatropha curcas** L., Sp. Pl. 1006. 1753; Roxb., Fl. Ind. 3:686. 1832; Hook.f., Fl. Brit. India 5:383. 1887; Pax in Engl., Pflanzenr. IV. 147:77. 1910; Gamble, Fl. Pres. Madras 1340. 1925; Airy Shaw, Euphorb. Borneo 137. 1975; Ramamoorthy in Sald. & Nicols., Fl. Hassan Dist.343. 1976; Dehgan & Webster, Univ.California Publ. 74:52. 1979; Rani in Matthew, Fl. Tam. Carnatic 3(2) : 1449. 1983. *Kadalavanakku.*

Leaves shallowly 5-lobed or entire, cordiform, 8-15 by 6-12 cm, base truncate to cordate, apex acute, basally 5-nerved, chartaceous. Cymes to 12 cm long. Flowers yellowish-green, c. 8 mm ϕ; outer tepals equal in male, unequal in female; stamens 5+5. Capsule globose.

Fl. & Fr.: May-Aug. *Distr.*: New World tropics. Common. Planted as hedge plants, occasionally run wild. *NAK 1949* (Tiruvalla, c. 10 m).

Juice of leaves applied to cure piles. Decoction of bark given to cure rheumatism and leprosy.

2. **Jatropha gossypifolia** L., Sp. Pl. 1006. 1753; Hook.f., Fl. Brit. India 5:383. 1887; Pax in Engl., Pflanzenr. IV. 147:26. 1910; Gamble, Fl. Pres. Madras 1340. 1925; Rani in Matthew, Fl. Tam. Carnatic 3(2) : 1450. 1983.

Leaves in close spiral, deeply 3-5-lobed, 7-11 by 8-12 cm, lobes obtusely acuminate, base cordate, margins glandular-ciliate, apex obtusely acuminate, 5-nerved, chartaceous. Cymes c. 6 cm long. Flowers yellowish green, unisexual; outer tepals glandular-hairy, inner tepals reddish; stamens 8, biseriate, 5+3. Capsule 3-lobed.

Fl. & Fr.: Jul.-Sep. *Distr.*: New-World Tropics. Planted elsewhere in the tropics, occasionally run wild. *NAK 663* (Pandalam, c. 40 m).

Juice of leaves dropped into the eyes to cure ophthalmia and other eye-diseases.

MACARANGA Du Petit-Thouars
Gen. Nova Madag. 26. 1806.

Dioecious trees. Leaves alternate, peltate; stipules large, deciduous. Flowers in large axillary panicles Male : α; bracts with or without a glandular appendage; tepals 3, spreading,concave; stamens 3 or 6-8, free or basally connate, exerted. Female : tepals 2-4, basally connate, ovary globose, echinate, 1-locular; ovules 3; style lateral. Capsule globose, pubescent or echinate. Seed 1, globose.

1. Stamens 6-8; leaf margin clearly toothed. .... 1. **M. indica**
1. Stamens 3; leaf margin entire or faintly toothed. .... 2. **M. peltata**

1. **Macaranga indica** Wight, Ic.t. 1883. 1852 & t. 1949. 1853; Bedd., Fl. Sylv. t. 287 A. 1872; Hook.f., Fl. Brit. India 5:446. 1887; Gamble, Fl. Pres. Madras 1326. 1625; Airy Shaw, Kew Bull. 23:93. 1969; Whitmore, Gard. Bull. Straits Stettlem. 31:52. 1978; Rani in Matthew, Fl. Tam. Carnatic 3(2) : 1453. 1983; Manilal, Fl. Silent Valley 252. 1988.

Leaves broadly ovate to suborbicular, 11-17 by 9.5-15 cm, margin toothed, apex acuminate, lower surface tawny pubescent, yellow glandular. Panicle branches zigzag; bracts spathulate with a large glandular appendage; stamens 6-8, shortly connate below. Capsule pubescent when young.

Fl. & Fr.: Oct.-Dec. *Distr.*: Sri Lanka, Peninsular India, S.E. Asia. Less common. In evergreen forests. *CNM 59621* (MH) (Angamoozhy, c. 250 m).

2. **Macaranga peltata** (Roxb.) Muell.-Arg. in DC., Prodr. 15(2): 1010. 1866; Gamble, Fl. Pres. Madras 1326. 1925; Ramamoorthy in Sald. & Nicols., Fl. Hassan Dist. 343. 1976; Whitmore., Gard. Bull. Straits Settlem. 31:55. 1978; Matthew, Ill. Fl. Tam. Carnatic t. 644. 1982; Rani in Matthew, Fl. Tam. Carnatic 3(2) : 1453. 1983. *Osyris peltata* Roxb.,Fl. Ind. 3:755. 1832. *Macaranga roxburghii* Wight, Ic.t. 1949, f.4. 1853; Hook.f., Fl. Brit. India 5:448. 1887.

Leaves suborbicular, 12-20 by 10-15 cm, margin entire or faintly crenate, apex acuminate, lower surface yellow-glandular, puberulous. Panicle branches not zigzag; bracts ovate; stamens 3, free. Capsule echinate when young.

Fl. & Fr.: Feb.-Apr. *Distr.*: Sri Lanka, India, Andaman Isls. Common in lowlands and hills. *NAK 523* (Angamoozhy, c. 250 m), *1550* (Moozhiar, c. 250 m). A fast growing good shady tree.

Leaves used as a green manure. Leaves and bark boiled in water and used to cure ulcers.

MALLOTUS Louriero
Fl. Cochinch. 601, 635. 1790.

Dioecious trees. Leaves simple, opposite or alternate, lower surface often glandular, thin-coriaceous. Flowers in paniculate or simple, axillary or terminal racemes. Male : subsessile, often clustered in the axils of bracts; tepals 4; stamens c. 20, free, exerted. Female : flowers lax; tepals 3-6, free or connate; ovary 3-or 4-locular; ovule 1 per locule; styles 3, plumose or papillose. Capsule of 3 or 4 cocci, epicarps smooth or echinate. Seeds globose or angular

1. Leaves broadly ovate to deltoid, lower surface white-tomentose. .... 4. **M. tetracoccus**
1. Leaves otherwise, lower surface yellow or reddish-glandular pubescent.
    2. Leaves opposite. .... 2. **M. resinosus**
    2. Leaves opposite and/or alternate.
        3. Leaves alternate, lower surface red-glandular. .... 1. **M. philippensis**
        3. Leaves opposite and alternate, lower surface yellow-glandular. .... 3. **M. rhamnifolius**

1. **Mallotus philippensis** (Lam.) Muell.-Arg., Linnaea 31:196. 1865; Hook.f., Fl. Brit. India 5 : 442. 1887 ("*philippinensis*"); Gamble, Fl. Pres. Madras 1322. 1925; Airy Shaw, Euphorb. Borneo 168. 1975; Ramamoorthy in Sald. & Nicols., Fl. Hassan Dist. 344. 1976; Matthew, Ill. Fl. Tam. Carnatic t. 645. 1982; Rani in Matthew, Fl. Tam. Carnatic 3(2): 1456. 1983; Nicol., Suresh & Manilal, An Interpr. Hort. Malab. 112. 1988;

Mohanan & Sivad., Fl. Agasthyamala 612.2002. *Ponnagam* Rheede, Hort. Malab. 5:41-42, t. 21. 1685. *Tsjerou-ponnagam* Rheede, Hort. Malab. 5:53, t. 22. 1685. *Pee-ponnagam* Rheede, Hort. Malab.5:37, t. 24. 1685.

Key to the varieties

1. Lower surface of leaf glandular only. ....**Mallotus philippensis** var. **philippensis**

1. Lower surface of leaf glandular with soft stellate tomentum. ....**Mallotus philippensis** var. **tomentosa**

Mallotus philippensis (Lam.) Muell.-Arg. var. **philippensis**

Leaves alternate, ovate-lanceate, 7-27 by 3.5-9 cm, base rouneded or acute, margin subentire, apex gradually acuminate, basally 3-ribbed, lower surface red-glandular. Panicles subterminal. Male : tepals 4, free. Female: tepals 3, bifid; ovary globose, pubescent, 3-locular. Capsule smooth, red-gladular.

Fl. & Fr.: Nov.-Mar. *Distr.*: Sri Lanka, India, Malesia to Australia. Common. In evergreen forests. *NAK 294 & 295* (Moozhiar, c. 250 m), *453* (Konni, c. 150 m), *525* (Angamoozhy, c. 250 m).

Capsule and seeds are the source of valuable dye 'kamala' and their powder used medicinally as a remedy for tape-worm infestation.

Mallotus philippensis (Lam.) Muell.-Arg. var. **tomentosus** Gamble, Fl. Pres. Madras 1322. 1925. *Sindooram.*

Leaves elliptic-oblong, 5-9 by 3.5-5.5 cm, base rounded, apex acute to cuspidate, lower surface softly stellate-tomentose. Capsule glandular.

Fl. & Fr.: Nov.-Feb. *Distr.*: Western Ghats. Less common. In evergreen forests. *NAK 1414* (Achancoil-Mannarappara, c. 250 m).

2. **Mallotus resinosus** (Blanco) Merr., Sp. Blancoanae 222. 1918 & Enum .Philip. Fl. Pl. 2:432. 1923; Airy Shaw, Kew Bull. 26:294. 1972, 31:392. 1976, 35:656. 1980, 36:326. 1981; Kew Bull. Add. Ser. IV:161. 1976, VII:171. 1980; Rammoorthy in Sald. & Nicols., Fl. Hassan Dist. 344. 1976; Chakrab., J. Econ. Tax. Bot. 11:22. 1987; Balakr. & Chakrab., Rheedea 1:37. 1991; Mohanan & Sivad., Fl. Agasthyamala 612.2002. *Adelia resinosa* Blanco, Fl. Philip. (ed. 2), 562. 1845; Muell.-Arg. in DC., Prodr. 15(2):731. 1866. *Claoxylon muricatum* Wight, Ic.t. 1886. 1852. *Mallotus intermedius* (Baill.) Balakr., Bull. Bot. Surv. India 10:245. 1968. *M. walkerae* Hook.f., Fl. Brit. India 5:346. 1887. *M. muricatus* (Wight) Muell.-Arg. var. *walkerae* (Hook.f.) Pax & Hoffm. in Engl., Pflanzenr. IV.147 (7):190.1914.

Key to the varieties

1. Spines of capsule long, slender. ....**Mallotus resinosus** var. **resinosus**

1. Spines of capsule short, conical. ....**Mallotus resinosus** var. **muricatus**

Mallotus resinosus (Blanco) Merr. var. **resinosus**

Leaves opposite, unequal, ovate-lanceate or elliptic-lanceate, 7-20 (30) by 3-9(12) cm, base acute, margin repand-dentate, apex acuminate, penni-nerved, nervature prominent, yellow-glandular beneath. Spikes axillary, to 15 cm long. Male : tepals 4, obovate. Female : tepals 4-6, lanceate. Capsules with slender glabrous spines.

Fl. & Fr.: Dec.-Mar. *Distr.*: Sri Lanka, S. India, Andaman Isls., Myanmar, Indo-China, Thailand, Malesia, Philippines, Java to Australia. Less common. In evergreen forest. *NAK 1440* (Arampa, c. 800 m).

Mallotus resinosus (Blanco) Merr. var. **muricatus** (Wight) Balakr. & Chakrab., Rheedea 1:39.1991. *Claoxylon muricatum* Wight, Ic.t. 1886. 1852.

Shrubs to very small trees. Capsule with short conical spines.

Fl. & Fr.: Dec.-Apr. *Distr.*: Endemic to Peninsular India. Rare. In evergreen forests. *NAK 2289* (Mannarappara, c. 250 m).

3. **Mallotus rhamnifolius** Muell.-Arg., Linnaea 34:196. 1865; Hook.f., Fl. Brit. India 5:440. 1887; Gamble, Fl. Pres. Madras 1322. 1925; Nicols., Suresh & Manilal, An Interpr. Hort. Malab. 132. 1988; Mohanan & Sivad., Fl. Agasthyamala 613.2002. *Pee-tsjerou-ponnagam* Rheede, Hort. Malab. 5:45, t.23. 1685.

Leaves opposite and alternate, elliptic-lanceate or rhombic-lanceate, 7-16 by 3.5-6.5 cm, base cuneate or obtuse, apex acuminate-apiculate, basally 3-ribbed, lower surface yellow-glandular, nervature with sparse simple tomentum. Spikes axillary, to 15 cm long. Male : tepals 4, broadly ovate. Female : tepals 5, lanceate. Capsule 3-valvular, stellate-tomentose.

Fl. & Fr.: Aug.-Dec. *Distr.*: Sri Lanka, Peninsular India. Less common. In midlands usually along streamsides. *NAK 698* (Pandalam, c. 50 m).

4. **Mallotus tetracoccus** (Roxb.) Kurz, J.Asiat. Soc. Bengal 16:245.1873; Airy Shaw, Kew Bull. 26(2):298. 1972; Ramamoorthy in Sald. & Nicols., Fl. Hassan Dist. 345. 1976; Rani in Matthew, Fl. Tam. Carnatic 3(2): 1457. 1983; Manilal, Fl. Silent Valley 253. 1988; Mohanan & Sivad., Fl. Agasthyamala 614.2002. *Rottlera tetracocca* Roxb., Fl. Ind. 3:826. 1832. *Mallotus albus* auct., non Muell.-Arg. 1865: Pax in Engl., Pflanzenr. IV: 147(7). 168. 1914, excl. basionym. *M. albus* var. *occidentalis* Hook.f., Fl. Brit. India 5:429. 1887; Gamble, Fl. Pres. Madras 1321. 1925. *Vatta-kumbil, Pori-vatta.*

Leaves broadly ovate to deltoid, subpeltate, 5-16 by 3-12.5 cm, base broadly cuneate, or subtruncate, apex acuminate, lower surface white-tomentose. Female flowers in stout panicles, to 10 cm long. Male flowers in slender panicles, to 25 cm; Male: tepals 4, ovate; stamens on a fleshy receptacle. Female flowers : tepals 4, connate at base, ovary 4-locular. Capsule 3-or 4-valved, echinate with soft woolly spines. Seeds angular.

Fl. & Fr.: Most of the seasons. *Distr.*: Sri Lanka, India, SW. China. Very common. In evergreen forests. *NAK 385* (Moozhiar, c. 250 m), *740* (Pampa Valley, c. 100 m).

Wood suitable for making match boxes.

PEDILANTHUS Necker ex Poiteau
Ann. Mus. Natl. Hist. Nat. 19:388. 1812.

Pedilanthus tithymaloides (L.) Poit., Ann. Mus. Natl. Hist. Nat. 19:390, t. 19.1812; Hook.f., Fl. Brit. India 5: 239. 1886; Gamble, Fl. Pres. Madras 1346. 1925; Rani in Matthew, Fl.Tam. Carnatic 3(2): 1461. 1983. *Euphorbia tithymaloides* L., Sp. Pl. 453. 1753.

Erect subshrubs; branchlets zigzag, latex milky. Leaves alternate, elliptic-ovate, variegated, deciduous. Cyathia terminal, cymose, subtended by petaloid bracts; glands 4, unequal; appendages and perianth absent. Male florets c. 25, female single; ovary 3-locular; ovule 1 per locule. Capsule of 3 bivalved cocci.

Fl. & Fr.: Apr.-Aug. *Distr.*: Native of W. Indies. Planted as hedge plants. Occasionally run wild and getting naturalized. *NAK 1606* (Ranni, c. 100 m).

PHYLLANTHUS Linnaeus
Sp. Pl. 981. 1753.

Monoecious erect herbs, subshrubs or trees. Leaves alternate, distichous; branchlets resemble pinnate leaves. Flower(s) solitary or clustered in fascicles, axillary or terminal, tepals 4, 5 or 6, persistent; stamens 3-6, free or connate; ovary 3-locular; ovules 2 per locule, or 6-locular due to false septum; styles 3; disc-glands 4, 5 or 6, free, alternate with tepals in male, usually annular in female. Capsule of 3 bivalved cocci, smooth or verrucose, rarely a fleshy and indehiscent drupe. Seeds 6, triquetrous.

1. Shrubs or trees.
    2. Shrubs or scandent shrubs; stamens 4 or 5; fruits berries or capsules.
        3. Shrub; leaf base rounded; stamens 4; friut a capsule. ....2. **P. baillonianа**
        3. Scandent shrub, leafbase acute; stamens 5; fruit a berry. ....8. **P. reticulatus**
    2. Trees; stamens 3; drupe greenish-yellow.. ....4. **P. emblica**
1. Herbs or subshrubs.
    3. Tepals 5; disc-glands 5.
        4. Leaves oblong or oblong elliptic, disc of female flowers ovate, crenate..1. **P. amarus**
        4. Leaves elliptic to oblanceate, disc of female flowers linear. ....7. **P. rheedii**
    3. Tepals 6; disc-glands 5-6.
        5. Subshrubs; leaves elliptic or obliquely ovate, to 2-4 cm long.
            6. Leaves elliptic; disc-glands 6. ....6. **P. gardnerianus**
            6. Leaves obliquely ovate; disc-glands 4. ....5. **P. gageana**
        5. Herbs; leaves otherwise, to 3 cm long.
            7. Leaves elliptic-ovate to linear oblong; disc of male star-shaped. ....3. **P. debilis**
            7. Leaves oblong-lanceate; disc of male lobular.
                8. Stamens free; seeds minutely tubercled. ....10. **P. virgatus**
                8. Stamens connate; seeds transversely ridged. .... 9. **P. urinaria**

1. **Phyllanthus amarus** Schum. & Thonn., Kongl. Danske Vidensk.- Selsk. Skr. 4:195. 1829; Airy Shaw, Euphorb. Borneo 182. 1975; Matthew, Ill. Fl. Tam. Carnatic t. 649. 1982; Rani in Matthew, Fl. Tam. Carnatic 3(2): 1464. 1983; Mitra & Jain, Bull. Bot. Surv. India 27:164. 1985; Mohanan & Sivad., Fl. Agasthyamala 616.2002. *P. fraternus* Webster, Contr. Gray Herb. No. 176:53. 1955. *P. niruri* auct., non L., 1753: Wight, Ic. t. 1894. 1882; Hook.f., Fl. Brit. India 5:298. 1887; Gamble, Fl. Pres. Madras 1290. 1925. *Keezhar-nelli.*

Herbs. Leaves oblong to oblong-elliptic, 8-10 by 3-4 mm,base and apex obtuse, lower surface glaucous; stipules lanceate. Cymules (almost all) bisexual or male, few, male towards terminal and female more towards base. Male flowers: tepals 5, ovate, stamens 3, connate below, dehiscence transverse; disc-glands 5. Female flowers : tepals 5, oblong; style erect, recurved. disc ovate, acute. Capsule c. 2 mm ϕ. Seeds muriculate.

Fl. & Fr.: Most of the seasons. *Distr.*: Probably native to America, now pantropical. Along waste places in lowlands. *NAK 1637* (Tiruvalla, c. 10 m).

Juice of whole plant is a powerful remedy against jaundice.

2. **Phyllanthus baillonianus** Muell.-Arg., Linnaea 32:47.1863; Hook.f., Fl. Brit. India 5:300. 1887. *Reidia bailloniana* Gamble, Fl. Pres. Madras 1293. 1925.

Shrubs. Leaves ovate to ovate-lanceate, 2-5 by 1.3-3 cm, base rounded, apex acute, apiculate, lower surface glaucous, lateral veins c. 6 pairs, forming marginal loops, thick-chartaceous. Male flowers : towards lower leaf-axils, pedicel slender; tepals 4, ovate, margin scarious, acute; stamens 4, anther dehiscence transverse; disc-glands 4, large, tubercled. Female flowers : axillary and terminal, or in drooping racemes; pedicel c. 2 cm long; tepals 6, ovate; disc annular, shallowly 6-lobed; ovary 6-chambered due to false septem; ovule 1 per locule. Capsule subtended by enlarged tepals.

Fl. & Fr.: Sep.-Dec. *Distr.*: Sri Lanka, Peninsular India. Rare. In evergreen forests. *NAK 2494* (Kakki hills, c. 1100 m).

3. **Phyllanthus debilis** Klein ex Willd., Sp. Pl. 4:582. 1804; Webster, J. Arnold Arb. 38:307, t.19, f. C & D. 1957, excl. syn. *P. debilis* Herb. Ham. ex Hook.f., Fl. Brit. India 5:299:1887; Gamble, Fl. Pres. Madras 1290. 1925; Mitra & Jain, Bull. Bot. Surv. India 27:168. 1985; Mohanan & Sivad., Fl. Agasthyamala 617.2002. *P. niruri* var. *javanicus* Muell.-Arg., Linnaea 32:43. 1863, & in DC., Prodr. 15(2):407. 1866. *P. niruri* var. *debilis* (Willd.) Muell.-Arg., Linnaea 32:43. 1863, & in DC., Prodr. 15(2):407. 1866. *P. mukerjeeanus* Mitra & Bennet, Bull. Bot. Soc. Bengal 19:145, f.1, t.1.1966. *P. niruri* auct., non L., 1753: Hook.f., Fl. Brit. India 5:298. 1887, p.p.

Herbs; branchlets with conspicuous normal leaves in lower nodes when young, 4-gonous. Leaves elliptic-lanceate to linear-lanceate, 10-35 by 2-5 mm, base attenuate or broadly cuneate, lower surface subglaucous. Flowers in proximal 3-5 axils with 3-4 male flowers; succeeding axils with solitary female flower(s). Male : tepals 6, obovate, hyaline, disc-glands 6, roundish; stamens 3, connate below into a column, anther dehiscence transverse. Female : tepals 6, elliptic-oblong, margin hyaline; disc saucer-form, faintly 6-lobed; style-arms recurved. Capsule c. 2.5 mm ϕ. Seeds 3-gonous, 7-8- longitudinally ribbed, transversely striate.

Fl. & Fr.: May-Aug. *Distr.*: Sri Lanka, India, Myanmar, Indonesia, Pacific islands and West-Indies. Common. On moist lands, as a weed in rice fields, etc. *NAK 645* (Puthukkulam, c. 100 m).

Note :*P. debilis* Herb. Ham. ex Hook.f. (1887) being a later homonym is replaced by *P. airy-shawi* Brunel & Roux (Nord. J.Bot. 4: 47. 1984).

4. **Phyllanthus emblica** L., Sp. Pl. 982. 1753; Roxb., Fl. Ind.3:671. 1832; Hook.f., Fl. Brit. India 5:289.1887; Airy Shaw, Euphorb. Borneo 183. 1975; Matthew, Ill. Fl. Tam. Carnatic t. 648. 1982; Rani in Matthew, Fl. Tam. Carnatic 3(2):1466. 1983; Manilal, Fl. Silent Valley 253. 1988; Nicols., Suresh & Manilal, An Interpr. Hort. Malab. 114. 1988; Mohanan & Sivad., Fl. Agasthyamala 617.2002. *Emblica officinalis* Gaertn., Fruct. 2:122. 1791; Wight, Ic. t. 1896. 1852; Gamble, Fl. Pres. Madras 1295. 1925. *Nilicamaram* Rheede, Hort. Malab. 1:69-70., t.38. 1678. *Nelli.*

Trees; branchlets reddish brown, pubescent. Leaves oblong-elliptic, 10-12 by 2-4 mm, base truncate, apex apiculate. Flowers: male and female mixed in fascicles or more usually the upper male and lower ones female. Male flowers : tepals 6, oblanceate; stamens 3, connate, dehiscence vertical; disc-glands 6. Female flowers : tepals 6, oblanceate, styles recurved, fimbriate. Drupe c. 3 cm ϕ, depressed-globose, fleshy, juicy. Seeds 3-gonous, crustaceous.

Fl. & Fr.: Jan.-Mar. *Distr.*: India, Myanmar, S. China, Malesia. Less common. On grassalands. *NAK 2043* (Pappinippara, Kalleli R.F., c. 200 m).

Fruits edible. Dried fruits powdered or decoction given to cure vomiting, rheumatism, giddiness and syphilis.

5. **Phyllanthus gageana** (Gamble) Mohanan, J. Econ. Tax. Bot. 6:480.1985. *Reidia gageana* Gamble, Kew Bull. 1925: 331. 1925 & Fl. Pres. Madras 1292. 1925.

Subshrubs. Leaves obliquely ovate, 2-4 by 1-2 cm, base and apex acute, lower surface glaucous, chartaceous. Male flowers : 3-5, subsessile, towards lower leaf-axils; bracts α, lanceate; tepals 4, lacerate. Female flowers towards upper leaf-axils; ovary 3-locular. Capsule globose.

Fl. & Fr.: Oct.-Dec. *Distr.*: Peninsular India. Rare. On moist forest floors. *NAK 1569 & 2613* (Angamoozhy, c. 250 m).

6. **Phyllanthus gardnerianus** (Wight) Baill., Stud. Gen. Euphorb. 628. 1858; Gamble, Fl. Pres. Madras 1290. 1925; Mohanan & Sivad., Fl. Agasthyamala 618.2002. *Macraea gardneriana* Wight, Ic. t. 1902, f.3.1852. *Phyllanthus simplex* Retz., var. *gardneriana* (Wight) Muell. Arg., Linnaea 32: 32.1863; Hook.f., Fl. Brit. India 5:295. 887.

Slender subshrubs with weak branches; stipules lanceate, peltate. Leaves elliptic, 3-3.5 by 1.1-1.2 cm, base cuneate or obtuse, apex obtuse, lower surface glaucous. Male flowers: α, in axillary fascicles; pedicels slender, c. 10 mm long; tepals 6; stamens 3, free; disc-glands 6, brownish. Female flower(s) solitary; pedicel slender, to 15 mm long; tepals 6; disc annular, faintly 6-lobed; style recurved. Capsule c. 5 mm ϕ, faintly verruculose. Seeds faintly striate.

Fl. & Fr.: Jul.-Nov. *Distr.*: Sri Lanka, Peninsular India. Not common. In hills. *NAK 865* (Ranni R.F., c. 200 m), *973* (Sabarimala, c. 250 m).

7. **Phyllanthus reticulatus** Poiret in Lam., Encycl. 5:298. 1804; Hook.f., Brit. India 5:288. 1887; Airy Shaw, Euphorb. Borneo 185. 1975; Matthew, Ill. Fl. Tam. Carnatic t. 643. 1982; Rani in Matthew, Fl. Tam. Carnatic 3(2):1468. 1983; Nicols., Suresh & Manilal, An Interpr. Hort. Malab. 115. 1988. *Kirganelia reticulata* (Poiret) Baillon, Etude Euphorb. 613. 1874; Gamble, Fl. Pres. Madras 1294. 1924. 1925. *Katou-nirouri* Rheede, Hort. Malab. 5:87, t. 44. 1685.

Scandent shrubs. Leaves oblong-elliptic, 2-3 by 1-1.5 cm, base and apex acute; stipules lanceate. Male flowers : 5-7 in fascicles; tepals 5, unequal, obovate; stamens 5, inner 2 or 3 connate, others free; disc-glands 5. Female flowers: fewer, towards lower side; tepals 5, shortly connate; ovary globose, 5-12-locular; disc-glands 5. Berry dark blue, globose, c. 7 mm ϕ. Seeds 3-gonous.

Fl. & Fr.: Most of the seasons. *Distr.*: Trop. Africa, India, Himalayas, Sri Lanka. *NAK 1633* (Tiruvalla, c. 10 m).

8. **Phyllanthus rheedii** Wight, Ic. t. 1895, f.1.1852; Hook.f., Fl. Brit. India 5:293. 1887; Gamble, Fl Pres. Madras 1289. 1925.

Erect branched woody herbs. Leaves elliptic to obovate or oblanceate, 1-2 by 0.6-1 cm, base cuneate, apex apiculate, lower surface glaucous; stipule lanceate. Male flowers : c. 5 in fascicles, subsessile, usually associated with a long pedicelled female flower; tepals 5, oblong, disc 5-lobed. Female solitary; tepals 5, disc-glands 5, linear. Capsule globose, c. 5 mm ϕ. Seeds striate.

Fl. & Fr.: Nov.-Jan. *Distr.*: Sri Lanka, Peninsular India. Common along waysides in higher hills. *NAK 355, 828, 1367, 2695* (Kakki hills, c. 1100 m).

9. **Phyllanthus urinaria** L., Sp. Pl. 982. 1753; Roxb., Fl. Ind. 3:660. 1932; Hook.f., Fl. Brit. India 5:293. 1887; Gamble, Fl. Pres. Madras 1289. 1925; Webster, J. Arnold. Arb. 38:194. 1958; Ramamoorthy in Sald. & Nicols., Fl. Hassan Dist. 347. 1976; Rani in Matthew, Fl. Tam. Carnatic 3(2): 1469. 1983; Mohanan & Sivad., Fl. Agasthyamala 619.2002. *P. leprocarpus* Wight, Ic. t. 1895, f. 4. 1852.

Herbs. Leaves oblong to obovate, 6-15 by 3.5-8 mm, base obtuse, apex apiculate; stipules linear. Male flowers: subsessile, in upper leaf-axils; tepals 6, orbicular; stamens 3, connate; disc 6-lobed. Female in lower leaf-axils; tepals 6, obovate-oblong; disc annular. Capsule 3-valved, globose, c. 4 mm ϕ, verrucose. Seeds 3-gonous, ridged.

Fl. & Fr.: Sep.-Mar. *Distr.*: Pantropical. Common. On exposed areas in hills and lowlands. *NAK* 812 (Ranni, c.100 m), *1854* (Kodumon, c. 50 m).

10. **Phyllanthus virgatus** Forster f., Fl. Ins. Aust. 65. 1786; Ramamoorthy in Sald. & Nicols., Fl. Hassan Dist. 347. 1976; Rani in Matthew, Fl. Tam. Carnatic 3(2): 1469. 1983; Nicols., Suresh & Manilal, An Interpr. Hort. Malab.115. 1988. *P. simplex* Retz., Obs. Bot. 5:29.1788; Hook.f., Fl. Brit. India 5:295. 1887; Gamble, Fl. Pres. Madras 1289.1925. *Macraea oblongifolia* Wight, Ic.t. 1902, f.1.1852. *Niruri, Nirpulla* Rheede, Hort. Malab. 10:53, t.27. 1690.

Diffuse woody herbs. Leave oblong-lanceate, 0.7-2.5 by 0.4-0.6 cm, base round, apex acute; stipules peltate. Flowers often 2 males and 1 female, or more commonly male in lower axils and female above; tepals 6, oblong; stamens 3, free, anther dehiscence transverse; disc-glands 6, of female annular, crenulate. Capsule globose, 3-lobed, c. 3 mm ϕ, warty. Seeds 3-gonous, tubercled.

Fl. & Fr.: Most of the seasons. *Distr.*: India, Himalaya, Indo-China, Malesia, Polynesia. Most common. On waste places, along waysides, etc. *NAK 1636* (Tiruvalla, c. 10 m), *1853* (Kodumon, c. 50 m), *2418* (Kalleli, c. 100 m).

RICINUS Linnaeus
Sp. Pl. 1007. 1753.

Ricinus communis L., Sp. Pl. 1007. 1753; Roxb., Fl. Ind. 3:689. 1832; Hook.f., Fl. Brit. India 5:457. 1887; Gamble, Fl. Pres. Madras 1335. 1925; Ramamoorthy in Sald. & Nicols., Fl. Hassan Dist. 347. 1976; Matthew, Ill. Fl. Tam. Carnatic t. 652. 1982; Rani in Matthew, Fl. Tam. Carnatic 3(2):1471. 1983; Manilal, Fl. Silent Valley 254. 1988; Nicols., Suresh & Manilal, An Interpr. Hort. Malab. 118. 1988. *Avanacu, Čitavanacu, Pandi-avanacu* Rheede, Hort. Malab. 2:57-60, t.32.1679.

Monoecious shrubs. Leaves alternate, palmately 6-8-lobed, peltate, to 20 by 25 cm, lobes lanceate, 9-15 by 3-6 cm, margin coarsely serrate, apex acuminate, chartaceous. Flowers in terminal paniculate racemes; male flowers below, female ones above. Male : perianth cupular, 3-5-lobed; stamens α, filaments connate, branched. Female : tepals 5, subequal; ovary globose, echinate, 3-locular; ovule 1 per locule; styles 3, papillose. Capsule 3-lobed, c. 2 cm ϕ, prickly. Seed-coat marbled, caruncululate.

Fl. & Fr.: Most of the seaons. *Distr.*: Probably native of NE. Trop. Africa. Widely cultivated in tropics, occasionally run wild. *NAK 689* (Pandalam, c. 20 m).

Seeds powdered and eaten for curing rheumatism, liver complaints, dropsy and piles. It has laxative property. Decoction of leaves applied on breast which increases secretion of milk.

SAUROPUS Blume
Bijdr.595. 1825-1826.

Monoecious shrubs. Leaves alternate, distichous. Flowers in axillary clusters or solitary. Male : disciform or turbinate; tepals 6, connate; stamens 3, filaments combined; ovary 3-locular, ovules 2-per locule; styles 3, curved. Capsule globose, 6-valvular or rupturing irregularly.

1. Leaves to 6 cm long; tepals in female equal; pistil broader than long. ....1. **S. androgynus**
1. Leaves to 13 cm long; tepals in female unequal; pistil longer than broad. ....2. **S. saksenianus**

1. **Sauropus androgynus** (L.) Merr., Bull. Bur. Forest. Philipp. Islands 1:30. 1903; Gamble, Fl. Pres. Madras 1303. 1925; Mohanan & Sivad., Fl. Agasthyamala 621.2002. *Clutia androgyna* L., Mant. Pl. 128. 1767. *Sauropus albicans* Bl., Bijdr. 596. 1826; Hook.f., Fl. Brit. India 5:332. 187. *S. indicus* Wight, Ic. t. 1952, f.2.1853.

Erect to climbing shrubs. Leaves lanceate, 3-6 by 1.5-2.5 cm, base cuneate, apex subacute, membranous. Male flowers: greenish-yellow with reddish centre, 6-11 mm ϕ. Female : brownish, 6-8 mm ϕ; tepals equal, transversly ovate-orbicular, closely imbricating; pistil broader than long. Capsule sessile, white, with reddish tinge.

Fl. & Fr.: Aug.-Dec. *Distr.*: Sri Lanka, India, Java & Phillippines. Less common. Along forest edges. *NAK 1044* (Tiruvalla, c. 10 m), *2615* (Angamoozhy, c. 200 m).

Note : This species is most commonly an erect undershrub but one collection (*NAK 2615*) is of climbing form.

2. **Sauropus saksenianus** Manilal, Prasannakumar & Sivarajan, J. Ind. Bot. Soc. 64:294-296. 1985; Manilal, Fl. Silent Valley 255. 1988.

Erect shrubs. Leaves oblong-lanceate, 8-13(15) by 3-4 cm, base truncate-cuneate, apex gradually acute, chartaceous. Male flowers: greenish-yellow, c. 4 mm ϕ. Female : brownish, 3-4 mm ϕ; tepals unequal, inner 3 smaller, outer 3 longer, ovate-rounded, imbricating; pistil longer than broad. Capsule white, subsessile.

Fl. & Fr.: Sep.-Dec. *Distr.*: Western Ghats. Rare. In evergreen forests. *NAK 1132* (Lower Moozhiar, c. 250 m).

SEBASTIANIA K. P. J. Sprengel
Neue Entdeck. Pflanzenk. 2:118. 1820 ('1821').

Sebastiania chamaelea (L.) Muell.-Arg. in DC., Prodr. 15(2) : 1175. 1866; Hook.f., Fl. Brit. India 5:475. 1887; Gamble, Fl. Pres. Madras 1344. 1925; Airy Shaw, Kew Bull. 26:339. 1972; Ramamoorthy in Sald. & Nicols., Fl. Hassan Dsit. 348. 1976; Matthew, Ill. Fl. Tam. Carnatic t. 654. 1982; Rani in Matthew, Fl. Tam. Carnatic 3(2): 1474. 1983; Nicols., Suresh & Manilal, An Interpr. Hort. Malab. 118. 1988. *Tragia chamaelea* L., Sp. Pl. 981. 1753. *Codi-avanacu* Rheede, Hort. Malab 2:63, t.34. 1679.

Monoecious herbs. Leaves alternate, oblong-lanceate, 2-5 by 06-0.8 cm, base and apex obtuse, chartaceous. Flowers in axillary spikes. Male: minute, c. 1.5 mm ϕ, in clusters of 3; tepals 5, connate; stamens 3, exerted; anthers globose. Female: c. 1 mm ϕ, solitary, axillary or below the male flowers in spike; tepals 3, ovate; ovary 3-lobed, echinate, 3-locular; ovule 1 per locule; styles 3. Capsule 3-lobed with vertical rows of soft prickles. Seeds caruncculate.

Fl. & Fr.: Most of the seasons. *Distr.*: Sri Lanka, India to Australia. Common in exposed areas in lowlands. *NAK 1022* (Tiruvalla, c. 10 m).

SECURINEGA Commerson ex A. L. Jussieu
Gen. 388. 1789 (nom. cons.).

Securinega virosa (Roxb. ex Willd.) Baill., Adansonia 6:334. 1866; Airy Shaw, Kew Bull. 26 (2): 340. 1972; Ramamoorthy in Sald. & Nicols., Fl. Hassan Dist. 348. 1976; Matthew, Ill.Fl. Tam. Carnatic t. 655. 1982; Rani in Matthew, Fl. Tam. Carnatic 3(2): 1476. 1983; Manilal, Fl. Silent Valley 255. 1988. *Phyllanthus virosus* Roxb. ex Willd., Sp. Pl. 4:578. 1805. *Fluggea microcarpa* Bl., Bijdr. 580. 1826; Hook.f., F. Brit. India 5: 328. 1887. *F. virosa* (Willd.) Baill., Etud. Gen. Euphorb. 593. 1858; Gamble, Fl. Pres. Madras 1296. 1926.

Profusely branched dioecious shrubs. Leaves alternate, obovate, 1-3 by 1-2 cm, base cuneate, apex obtuse, thinly pubescent, chartaceous. Flowers fascicled around branchlets. Male : 5-merous; tepals 5, ovate-cancave, crenulate; stamens 5, free, exerted; pistillode 3-forked; disc-glands 5. Female : tepals 5; ovary globose, 3-locular; ovules 2 per locule; styles 3-fid, spreading; disc annular, flat, toothed. Capsule globose, white, 3-valvular. Seeds triquetrous.

Fl. & Fr.: Mar.-May. *Distr.:* Trop. Africa, Sri Lanka, India, S. Asia to Japan, Malesia, Australia, Polynesia. Not common. In evergreen forests. *NAK 1679* (Sabari hills, c.250 m).

SUREGADA Roxburgh ex Rottler
Ges. Naturf. Freunde Berlin Neue Schriften 4:206. 1803.

Suregada angustifolia (Muell.-Arg.) Airy Shaw, Kew Bull.23: 128. 1969; Matthew, Ill. Fl. Tam. Carnatic t. 656. 1982; Rani in Matthew, Fl. Tam. Carnatic 3(2): 1478. 1983; Mohanan & Sivad., Fl. Agasthyamala 622.2002. *S. angustifolia* Baill., Etude. Euphorb. 369. 1858, nom. nud. *Gelonium angustifolium* Baill., ex Muell.-Arg. in DC., Prodr. 15(2): 1128. 1866. *Gelonium lanceolatum* auct., non Willd., 1806: Roxb., Fl. Ind. 3:831. 1832; Wight, Ic. t. 1867. 1852; Hook.f., Fl. Brit. India 5:459. 1887; Gamble, Fl. Pres. Madras 1343. 1925. *Suregada lanceolata* auct., non (Willd.) Kuntze, 1891: Matthew, Ill. Fl. Tam. Carnatic t. 656. 1982.

Dioecious trees. Leaves alternate, oblanceate, 7-13 by 2-3.3 cm, base acute or cuneate, margin entire to serrate towards apex, apex acute or shortly acuminate-apiculate, lateral veins 6-8 pairs, coriaceous. Male flowers : 3-10 in axillary, subumbellate clusters, c. 5 mm ϕ; tepals 5, ovate-orbicular, concave, imbricate; stamens α, free on a receptacle mixed with glands, anthers oblong, dehiscence longitudinal. Female plants could not be collected.

Fl.: Feb. *Distr.*: Sri Lanka, Peninsular India. Rare. In semi-evergreen forests. *NAK 454* (Kodamurutti, Ranni R.F., c. 100 m).

TRAGIA Linnaeus
Sp. Pl. 980. 1753.

Monoecious stragglers with stinging hairs. Leaves alternate, palmi-or penninerved. Flowers in slender axillary spikes. Male : tepals splitting into 3-lobes; stamens 3, free, included, anthers subsessile. Female : tepals 6, ovate dentate or pinnatifid, accrescent; ovary 3-locular; ovule 1 per locule; style 3, stout. Capsule subsessile, 3-lobed. Seeds globose.

1. Fruiting tepals broadly ovate, margin dentate. ....3. **T. muelleriana**
1. Fruiting tepals linear-lanceate, margin pinnatifid.

2. Leaf-base cordate. ....1. **T. hispida**
2. Leaf-base acute or rounded. ....2. **T. involucrata**

1. **Tragia hispida** Willd., Sp. Pl. 323. 1803; Gamble, Fl. Pres. Madras 1332. 1925; Ramamoorthy in Sald. Nicols., Fl. Hassan Dist. 349. 1976.

Leaves broadly ovate, 6-12 by 3.6 cm, base cordate, margin crenate-serrate, apex acuminate, basally 3-5-nerved, hispid, chartaceous, black when dry. Spikes slender, often branched, c. 5 cm long; fruiting tepals linear, shallowly pinnatifid, teeth short, sparsely hispid. Capsule 3-lobed, c. 7 mm ϕ, hispid.

Fl. & Fr.: Jan.-Mar.(June). *Distr.*: Peninsular India (Western Ghats). Common in hills. *NAK 489* (Angamoozhy, c. 200 m).

2. **Tragia involucrata** L., Sp. Pl. 980. 1753; Hook.f., Fl. Brit. India 5:465. 1888; Gamble, Fl. Pres. Madras 1332. 1925; Ramamoorthy in Sald. & Nicols., Fl. Hassan Dist. 349. 1976; Matthew, Ill. Fl. Tam. Carnatic t. 659. 1982; Rani in Matthew, Fl. Tam Carnatic 3(2): 1479; Nicols., Suresh & Manilal, An Interpr. Hort. Malab. 118. 1988. *Schorigenam* Rheede, Hort. Malab. 2:73-74, t.39. 1679. *Valli-schoriganam* Rheede, Hort. Malab. 2:79. 1679.

Key to the varieties

1. Leaves elliptic-ovate; fruiting tepals deeply pinnatifid. ....**Tragia involucrata** var. **involucrata**
1. Leaves narrowly oblong; fruiting tepals shallowly pinnatifid. ....**Tragia involucrata** var. **angustifolia**

Tragia involucrata L. var. **involucrata**

Leaves elliptic-ovate, lower ones occasionally broadly ovate, 8-10 by 3-5 cm, base acute to rounded, margin serrate, apex acuminate, thick-chartaceous, drying dull-greenish. Spikes c. 2 cm long; fruiting tepals linear, deeply pinnatifid, teeth linear. Capsule 0.9 mm ϕ, hispid.

Fl. & Fr.: Most of the seasons. *Distr.*: Sri Lanka, India, Myanmar, China. Common in lowlands. *NAK 141* (Adoor, c. 20 m).

Juice of leaves given with honey for cough; and powdered leaves given as a remedy in fever and asthma.

Tragia involucrata L. var. **angustifolia** Hook.f., Fl. Brit. India 5:465. 1888; Gamble, Fl. Pres. Madras 1332.1925.

Slender vines. Leaves linear-oblong, 7-10 by 1.3-1.8 cm, base acute, apex gradually acute, basally 3-nerved, sparsely hispid, chartaceous. Spikes c. 2 cm long; fruiting tepals linear-oblong, teeth short, hispid. Capsule c. 7 mm, ϕ, sparesely hispid.

Fl. & Fr.: Jul.-Sept. *Distr.*: Endemic to Western Ghats. Rare. In hills. *NAK 837* (Chakkapara - Ranni R.F., c. 200 m).

3. **Tragia muelleriana** Pax and Hoffm. in Engl., Pflanzenr. IV. 147a. IX:80. 1919; Gamble, Fl. Pres. Madras 1332. 1925. *T. involucrata* L. var. *cordata* Muell.-Arg. in DC., Prodr.15(2):943. 1866; Hook.f., Fl. Brit. India 5:465. 1888.

Leaves broadly ovate to elliptic-ovate, 6-15 by 2.5-6.5 cm, base cordate, margin serrate, apex acuminate, hispid, drying black. Spikes slender, c. 5 cm long; fruiting tepals broadly ovate, margin dentate, densely villous. Capsule 3-lobed, c.12 mm ϕ.

Fl. & Fr.: Sep.-Jan. *Distr.*: Endemic to Western Ghats. Rare. In higher hills. *NAK 1445* (Arampa, c. 800 m).

TREVIA Linnaeus
Sp. Pl. 1193. 1753 ('Trewia').

Trevia polycarpa Benth. in Benth. & Hook.f., Fl. Gen. Pl. 3:318. 1880; Hook.f., Fl. Brit. India 5:424. 1887; Gamble, Fl.Pres. Madras 1319. 1925; Matthew, Ill. Fl. Tam. Carnatic t. 660. 1982; Rani in Matthew, Fl. Tam. Carnatic 3(2): 1483. 1983.

Dioecious trees. Leaves alternate, broadly ovate to deltoid, 7-15 by 4-10 cm, base truncate, apex gradually acuminate, basally 5-nerved; young leaves drying black. Flowers in axillary racemose pendulous panicles. Male : raceme to 10 cm long; tepals initially connate, splitting into 3-4 lobes; stamens α, free, exerted. Female raceme c. 8 cm long; ovary ovoid, softly pubescent, 3-locular; ovule 1 per locule. Drupe globose, c. 2 cm ϕ.

Fl. & Fr.: Sep.-Apr. *Distr.*: Peninsular India. Less common. Along streamsides. *NAK 1591, 2417 & 2598* (Along Achancovil river side, Kalleli, c. 200 m).

Note : The acceptance of spelling change from *Trewia* to *Trevia* was discussed in detail by Nicolson *et al.*, (1988).

Trevia nudiflora L., is seen planted occasionally in hills. *NAK* 338 (Upper Moozhiar-valve house compound).

Note : **Hevea brasiliensis** (Willd. ex Juss.) Muell.-Arg., the para-rubber trees have been planted in most of the forest lands of the district. *NAK 1539* (Chittar,c. 400 m). **Manihot esculenta** Crantz, the cassava is extensively cultivated even by replacing large area of forests (Maniyaar). **Manihot glaziovii** Muell.-Arg. is cultivated occasionally in the hills.

118. ULMACEAE

TREMA Loureiro
Fl. Cochinch. 539, 562. 1790.

Trema orientalis (L.) Bl., Mus. Bot. 2:61. 1856; Hook. f., Fl. Brit. India 5.483. 1888, Fischer in Gamble, Fl. Pres. Madras 1350. 1928; Ramamoorthy in Sald. & Nicols., Fl. Hassan Dist. 75. 1976; Matthew, Ill. Fl. Tam. Carnatic t.673. 1982; Matthew & Rani in Matthew, Fl. Tam. Carnatic 3(2): 1504. 1983; Manilal, Fl. Silent Valley 258. 1988; Nicols., Suresh & Manilal, An Interpr. Hort. Malab. 258. 1988; Mohanan & Sivad., Fl. Agasthyamala 623.2002. *Celtis orientalis* L., Sp. Pl. 1044. 1753; Wight, Ic.t. 1971. 1853; Bedd., Fl. Sylv. t. 311. 1873. *Mallam-toddali* Rheede, Hort. Malab. 4:83-84, t.40. 1683.

Dioecious trees. Leaves alternate, broadly lanceate, 8-13 by 3-6.5 cm, base obliquely cordate, margin serrulate, apex acute, 3-5-nerved from base, rugose above, pubescent below, thin-coriaceous. Flowers in clustered cymes, 5-merous, unisexual; tepals 5, free, scarious, hirsute; stamens 5, anthers introrse; pistillode present; ovary superior, 1-locular, ovule 1, subpendulous; styles 2.Drupe globose. Seeds ovoid to globose.

Fl. & Fr.: Aug.-Oct. *Distr.*: Trop. Africa, Sri Lanka, India, Himalayas, Indo-China, W. & S. China, Australia, Polynesia. Common. Along forest clearings. *NAK 29* (Moozhiar, c.250 m), *866* (Ranni, c. 200 m).

Leaves used for polishing wood; bark yields a fibre.

119. MORACEAE

1. Inflorescence a syconium; flowers encolsed within the receptacle; trees or scandent shrubs. .... **Ficus**
1. Inflorescence other than syconium; flowers exposed on the receptacle; unarmed trees or armed or unarmed shrubs.
    2. Leaf-margin toothed; female flowers free. .... **Streblus**
    2. Leaf-margin entire; female flowers connate. .... **Artocarpus**

ARTOCARPUS J.R. Forster *et* J.G.A. Forster Charact. Gen. 101. 1776 (nom. cons.).

Evergreen or deciduous trees. Leaves altenate; stipules amplexicaul, deciduous. Flowers unisexual, minute, very dense, aggregated in globose heads or cylindric spikes; peduncle stout, fleshy, covered with sterile bracts. Male perianth cupular, 2-lobed; stamen 1 , straight; female-perianth tubular, ovary superior, 1-locular. Fruit an anthocarp, enclosed by green, globose syncarp.

1. Branchlets glabrous; leaves obovate; fruit to 30 cm ϕ. .... 2. **A. hetrophyllus**
1. Branchlets pubescent; leaves otherwise; fruit to 15 cm ϕ.
    2. Flowers in catkin like spikes; fruits cylindric.......... 3. **A. hirsutus**
    2. Flowers in globose heads; fruits globose. .... 1. **A. gomezianus** ssp. **zeylanicus**

1. **Artocarpus gomezianus** Wall. ex Trec., Ann. Sci. Nat. Bot. Ser. 3, 8:118. 1847, ssp. **zeylanicus** Jarret, J. Arnold Arb. 41:90. 1960; Ramamoorthy & Gandhi in Sald. & Nicols., Fl. Hassan Dist. 77. 1976. *A. lakoocha* auct., non Roxb., 1832: King in Hook.f., Fl. Brit. India 5:543. 1888, p.p.; Fischer in Gamble, Fl. Pres. Madras 1369. 1925.

Leaves elliptic or elliptic-oblong, 10-16 by 5-6.5cm, base rounded or broadly acute, apex acuminate, lateral veins 6-8 pairs, reticulations very prominent, coriaceous. heads lateral, globose, 4 mm ϕ, peduncle short, pubescent; bracteoles peltate, stipitate; tepals 2, ovate; stamen 1, filament flattened. Fruit subglobose or ovoid.

Fl. & Fr.: Feb.-Apr. *Distr.*: Peninsular Inda, Andaman Islands. Rare. In semi-evergreen forests. *NAK 2639* (Mannier, c. 250 m, Konni R.F.).

2. **Artocarpus heterophyllus** Lam., Encycl. 3:210. 1789; Corner, Gard. Bull. Singapore 10:56. 1939 & in Dassan. & Fosberg, Rev. Handb. Fl. Ceylon 3:217.1981; Ramamoorthy & Gandhi in Sald. & Nicols., Fl. Hassan Dist. 77. 1976; Rani & Matthew in Matthew, Fl. Tam. Carnatic 3(2): 1509. 1983; Manilal, Fl. Silent Valley 258. 1988; Nicols., Suresh & Manilal, An Intepr. Hort. Malab. 182. 198; Mohanan & Sivad., Fl. Agasthyamala 625.2002. *A intergrifolius* auct., non L.f., 1781: Roxb., Pl. Corom. t. 250. 1815 & Fl. Ind. 3:522. 1832; Wight, Ic. t. 678. 1840; King in Hook.f., Fl. Brit. India 5.541. 1888; Fischer in in Gamble, Fl. Pres. Madras 1369. 1928. *Tsjaka-maram* Rheede, Hort. Malab. 3:17-20, t. 26-28. 1682.

Leaves obovate to elliptic-ovate, 8-15 by 3.5-8 cm, base cuneate or acute, apex obtuse or shortly acute, shiny, coriaceous. Inflorescence cauliflorous, also from main branches, enclosed by leafy deciduous scales. Male heads c. 2.5 cm ϕ; perianth 2-lobed, stamen 1, filament flattened. Female heads c. 3 cm ϕ; ovary obovoid; stigma spathulate. Syncarp oblong-globose; perianth thick-fleshy, edible. Seeds to 4 cm long.

Fl. & Fr.: Feb.-Sep. *Distr.:* Possibly native of S. India; now widely cultivated in Tropics. Less common. In evergreen forests. Failed to collect the representative specimens.

3. **Artocarpus hirsutus** Lam., Encycl. 3:210.1789, ("*hirsuta*"); Wight, Ic.t. 1957.1853; King in Hook.f., Fl. Brit. India 5:541. 1888; Fischer in Gamble, Fl. Pres. Madras 1369. 1925; Ramamoothy & Gandhi in Sald. & Nicols., Fl. Hassan Dist. 78. 1976; Nicols., Suresh & Manilal, An Interpr. Hort. Malab. 183. 1988; Mohanan & Sivad., Fl. Agasthyamala 626.2002. *Ansjeli* Rheede, Hort. Malab. 3:25-27, t.32.1682.

Leaves broadly ovate or elliptic-ovate, 14-24 by 11-14 cm, base cuneate, apex acute to shortly acuminate, lower surface hirsute, coriaceous; stipule large, foliaceous, deciduous. Spikes cylindric, slender to 15 cm long; tepals 2. Fruit cyindric, to 10 cm ϕ.

Fl. & Fr.: Jan.-May. *Distr.*: Peninsular India. Common. In evergreen forests; also widely planted for the timber. *NAK 1627* (Ranni, c.150 m). Fruit edible.

FICUS Linnaeus
Sp. Pl. 1059. 1753.

Trees or scandent shrubs with milky latex. Leaves alternate or opposite or both; stipules deciduous, leaving annular scars. Inflorescence a syconium or fig clustered, paired or solitary, axillary or cauliflorus or on branchlets, monoecious or dioecious. Flowers unisexual of different kinds, male with 1-3 stamens, female with 1-celled ovary, excentric style; gall flowers usually similar to female, with sterile pedicellate ovary, often enclosing insects and neuter flowers with tepals only. Fruit an achene, globose or ovoid, smooth or ribbed.

1. Scandent shrubs; leaf-margin undulate. .... 13. **F. travancorica**
1. Erect shrubs or trees; leaf-margin not so.
   2. Leaves decussate; figs on special leafless shoots. .... 8. **F. hispida**
   2. Leaves alternate; figs on branchlets.
      3. Figs dioecious; seeds appendaged or keeled. .... 7. **B. heterophylla**
      3. Figs monoecious; seeds smooth.
         4. Figs sessile.
            5. Branchlets and figs fulvous-tomentose. .... 5. **F. drupacea** var. **pubescens**
            5. Branchlets and figs glabrous or puberulous.
               6. Leaf- tip to 4 cm long........ 10. **F. religiosa**
               6. Leaf- tip to 2 cm long.
                  7. Petiole articulate; male flowers ostiolar. .... 14. **F. tsjakela**
                  7. Petiole not articulate; male flowers disperse.
                     8. Leaf-veins without prominent intercostals or secondary laterals; ovary white. .... 1. **F. amplissima**
                     8. Leaf-veins with prominent intercostals or secondary laterals; ovary reddish-brown or yellow.

9. Leaf-apex acuminate.
  10. Leaf with secondary lateral nerves; ovary yellowish. ....11. **F. talboti**
  10. Leaf with intercostals; ovary reddish brown. ....2. **F. arnottiana**
9. Leaf-apex rounded or acute or retuse.
  11. Figs c. 1 cm ϕ; leaves obovate. 9. **F. microcarpa**
  11. Figs c.1.5 cm ϕ.; leaves broadly elliptic. ....3. **F. benghalensis**

4. Figs peduncled.
  12. Fig(s) solitary; external bracts lateral.
    13. Leaves glabrous; figs c. 2.5 cm ϕ. ....4. **F.callosa**
    13. Leaves scabrid; figs c. 1.8 cm ϕ. ....6. **F. exasperata**
  12. Figs paired or clustered; exeternal bracts basal. ....12. **F. tinctoria** var. **parasitica**

1. **Ficus amplissima** Smith in Rees, Cyclop. 14, n. 68. 1810; Corner, Gard. Bull. Straits Settlem. 21:11. 1965; Ramamoorthy & Gandhi in Sald. & Nicols., Fl. Hassan Dist. 79. 1976; Corner in Dassan. & Fosberg, Rev. Handb. Fl. Ceylon 3:242. 1981; Matthew & Rani in Matthew, Fl. Tam. Carnatic 3(2) : 1515. 1983; Manilal, Fl. Silent Valley 259. 1988; Nicols., Suresh & Manilal, An Intepr. Hort. Malab. 183. 1988. *F. tsiela* Roxb. ex Buch.-Ham., Trans. Linn. Soc. London 15149. 1827; Roxb., Fl. Ind. 3:549. 1832; Wight, Ic.t. 668. 1840; King, Ann. Roy. Bot. Gard. (Calcutta) 1:tt. 73 & 74. 1887 & in Hook.f., Fl. Brit. India 5:515. 1888; Fischer in Gamble, Fl. Pres. Madras 1362. 1928. *Tsjela* Rheede, Hort. Malab. 3:25-27, t.32. 1682.

Leaves alternate, elliptic-oblong, 8-12 by 3.5-6.5 cm, base rounded or broadly cuneate, margin cartilaginous, apex obtusely acuminate, basally 3-nerved, lateral nerves c.10 pairs, coriaceous. Figs in pairs, sessile, monoecious, depressed gobose, c.1 cm ϕ, basal bracts 3, broadly ovate. Male flowers disperse; stamen 1, anther oblong, parallel, unequal. Female flowers sessile; ovary white-colourd. Achenes smooth.

Fl. & Fr.: Oct.-Jan. *Distr.*: Sri Lanka, Maldives, Peninsular India. Rare. Streamsides. *NAK 2231* (Thazhoor Kadavu, c. 50 m).

2. **Ficus arnottiana** (Miq.) Miq., Ann. Mus. Bot. Lugd. Batav. 3:287. 1867; King, Ann. Roy. Bot. Gard. (Calcutta) 1:tt. 68. & 84 V. 1887 & in Hook.f., Fl. Brit. India. 5:513. 1888; Fischer in Gamble, Fl. Pres. Madras 1363. 1928; Corner, Gard. Bull. Straits Settlem. 21:11. 1965; Ramamoorthy & Gandhi in Sald. & Nicols., Fl. Hassan Dist. 79. 1976; Corner in Dassan. & Fosberg, Rev. Handb. Fl. Ceylon 3:244. 1981; Matthew & Rani in Matthew, Fl. Tam. Carnatic 3(2) : 1515. 1983; Mohanan & Sivad., Fl. Agasthyamala 629.2002. *Urostigma arnottiana* Miq. in Hook.f., Lond. J. Bot. 6:564. 1847.

Leaves alternate, cordiform, 8-15 by 6-12 cm, base cordate, margin thick, apex caudate, 3-5-nerved from base, lateral nerves 6-8 pairs, subcoriaceous. *Figs* in axillary pairs or crowded near apex, sessile, globose, 0.6 mm f; bracts 3, ovate; tepals 3-4-lobed. Male flowers around the orifice; stamen 1, subessile. Female sessile, ovary depressed-globose. Achenes smooth.

Fl. & Fr.: Sep.-Dec. *Distr.*: Sri Lanka, India. Rare. In evergreen forests of upper hills. *KV 48605* (MH) (Pamba, c. 950 m).

3. **Ficus benghalensis** L., Sp. Pl. 1059. 1753; King, Ann. Roy, Bot. Gard. (Calcutta) 1: 18, t. 13. 1887 & in Hook.f., Fl. Brit. India 5:499. 1888; Fischer in Gamble, Fl. Pres. Madras 1361. 1928; Corner, Gard. Bull. Straits Settlem. 21:14. 1965 & in Dassan.& Fosberg, Rev. Handb. Fl. Ceylon 3:251. 1981; Matthew & Rani in Matthew, Fl. Tam. Carnatic 3(2): 1518. 1983; Nicols., Suresh & Manilal, An Interpr. Hort. Malab. 184. 1988; Mohanan & Sivad., Fl. Agasthyamala 630.2002. *Urostigma benghalense* (L.) Gasp., Nov. Gen. Fic. 7. 1844; Wight, Ic. t. 1989. 1853. *Peralu* Rheede, Hort. Malab. 1:49-50, t.28. 1678.

Leaves alternate, elliptic-ovate, 10-17 by 7-12 cm, base rounded to subcordate, apex obtuse to subacute, basally 3-5-nerved, lateral nerves 6-8 pairs, lower surface pubescent, coriaceous. Fig(s) 1 or 2, axillary, monoecious, depressed globose, c. 2 cm ϕ; bracts 4-5, cupular; tepals 3-5, shortly connate. Male flowers dispersed with female; stamen 1, anther oblong. Female sessile; ovary obovoid-globose. Achenes broadly ellipsoid.

Fl. & Fr.: May-Aug. *Distr.*: Sri Lanka, India, Pakistan. Planted as avenue trees. *NAK 2873* (Adoor, c. 20 m).

Tender ends of aerial roots are administered against vomiting and gonorrhoea.

4. **Ficus callosa** Willd., Mem. Acad. Roy. Sci. Hist. (Berlin) 102. 1798; King, Ann. Roy. Bot. Gard. (Calcutta) 1:64, tt. 84- V2 & 85. 1887 & in Hook.f., Fl. Brit. India 5:516. 1888; Fischer in Gamble, Fl. Pres. Madras 1364. 1928; Corner, Gard. Bull. Straits Settlem. 21:29. 1965; Ramamoorthy & Gandhi in Sald. & Nicols., Fl. Hassan Dist. 80. 1976; Nicols., Suresh & Manilal, An Interpr. Hort. Malab. 184. 1988; Mohanan & Sivad., Fl. Agasthyamala 631.2002. *Handir-alou* Rheede, Hort. Malab. 3:77-78, t.59. 1682.

Leaves alternate, elliptic or elliptic-ovate, 10-20 by 6.5-10 cm, base rounded or broadly cuneate, apex obtusely short-acuminate, basally 3-5-nerved, lateral nerves 5-12 pairs, coriaceous. Fig(s) solitary, axillary, peduncled, depressed globose, basal bracts 3, broadly ovate. Male flowers α, scattered, pedicelled; stamen(s) 1 or 2; tepals 3. Female and gall flowers : perianth divided into 3-4 lobes. Achenes smooth.

Fl. & Fr.: Sep.-Mar. *Distr.*: Sri Lanka, India, Myanmar, Java. Rare. In evergreen forests. *CNM 54312* (MH) (Moozhiar, c. 250 m).

5. **Ficus drupacea** Thunb., Diss. Fic. 11. 1786 var. **pubescens** (Roth) Corner, Gard. Bull. Straits Settlem. 17:381. 1960; Ramamoorthy & Gandhi in Sald. & Nicols., Fl. Hassan Dist. 80. 1976; Corner in Dassan. & Fosberg, Rev. Handb. Fl. Ceylon 3:247. 1981; Matthew & Rani in Matthew, Fl. Tam. Carnatic 3(2): 1519. 1983; Nicols., Suresh & Manilal, An Interpr. Hort. Malab. 184. 1988; Mohanan & Sivad., Fl. Agasthyamala 631.2002. *Ficus mysorensis* Heyne ex Roth in Roemer & Schultes, Syst. Veg. 1:508. 1817; King, Ann. Roy. Bot. Gard. (Calcutta) 1:t. 14. 1887 & in Hook.f., Fl. Brit. India 5:500 1888; Fischer in Gamble, Fl. Pres. Madras 1361. 1928, var. *pubescens* Roth in Roemer & Schultes, Syst. Veg. 1:508.1817. *Katu-alou* Rheede, Hort. Malab. 3:73-74, t. 57. 1682.

Leaves broadly elliptic, 12-18 by 8.5-10 cm, base rounded, apex obtusely acuminate, penni-nerved, lateral nerves 14-16 pairs, prominent, lower surface densely tomentose, coriaceous. Fig(s) solitary, axillary, sessile, monoecious, globose, tomentose without, c. 2 cm ϕ; basal bracts orbicular. Male flowers disperse; tepals 3-4, stamen 1, exerted. Female sessile; ovary obovoid, style filiform. Achenes smooth.

Fl. & Fr.: Dec.-Mar. *Distr.*: Sri Lanka, India, Bangladesh, Laos. Rare. In evergreen forests. *NAK 542* (Plappally, c. 350 m).

6. **Ficus exasperata** Vahl, Enum. Pl. 2:197. 1806; Corner, Gard. Bull. Straits Settlem. 21:74. 1965; Ramamoorthy & Gandhi in Sald. & Nicols., Fl. Hassan Dist. 81.1976; Corner in Dassan. & Fosberg, Rev. Handb. Fl. Ceylon 3:274. 1981; Matthew & Rani in Matthew, Fl. Tam. Carnatic 3(2): 1520. 1983; Manilal, Fl. Silent Valley 259. 1988; Nicols., Suresh & Manilal, An Interpr. Hort. Malab. 185. 1988; Mohanan & Sivad., Fl. Agasthyamala 632.2002. *F. asperrima* Roxb., (Hort. Beng. 103. 1814, nom. nud.) Fl. Ind. 3:354. 1832; Wight, Ic.t.633. 1843; King in Hook.f., Fl. Brit. India 5:522. 1888; Fischer in Gamble, Fl. Pres. Madras 1366. 1928. *Teregam* Rheede, Hort. Malab. 3:79, t-60. 1682.

Leaves alternate, oblong-elliptic to obovate, 7-14 by 4-6 cm, base rounded, margin entire to sparingly toothed, apex acuminate, 3-nerved from base, lateral nerves c. 6 pairs, scabrid, thin-coriaceous. Fig(s) solitary, axillary, dioecious, globose, c.12 mm ϕ, pubescent; bracts 2-3, lateral; tepals 3-6, free. Male flowers ostiolar, 1 or 2 seriate; stamen 1. Female sessile, ovary obovoid; gall flowers sessile or pedicellate. Achenes reticualte.

Fl. & Fr.: Feb.-May. *Distr.*: E. Africa, Arabia, S. India, Sri Lanka. Common. In hills. *NAK 298* (Moozhiar, c. 250 m), *1619* (Pampavalley, c. 100 m).

Leaves used as sand-paper for polishing furniture. Decoction of bark and root is a good medicine for leprosy.

7. **Ficus heterophylla** L.f., Suppl. Pl. 442. 1781; Roxb., Fl. Ind.3:535. 1832; Wight, Ic.t. 659. 1840; King, Ann.Roy. Bot. Gard. (Calcutta) 1:t. 94. 1888 & in Hook.f., Fl. Brit. India 5:518. 1888; Fischer in Gamble, Fl. Pres. Madras1366. 1928; Corner in Dassan. & Fosberg, Rev. Handb. Fl. Ceylon 3:272. 1981; Matthew & Rani in Matthew, Fl. Tam. Carnatic 3(2): 1521. 1983; Nicols., Suresh & Manilal, An Interpr. Hort. Malab. 185. 1988. *F. scabrella* Roxb., Fl. Ind. 1832. Wight, Ic. t. 661. 1840. *Valli-teregam* Rheede, Hort. Malab. 3:83, t. 62. 1682.

Leaves alternate, elliptic-oblong, young ones pinnately 5-7-lobed, 5-7 by 2-3 cm, base rounded, margin serrate, apex acuminate, 3-nerved from base, lateral nerves 5-7 pairs, scabrid, thin-coriaceous. Fig(s) solitary, axillary, dioecious, broadly ellipsoid, c. 1 cm ϕ; basal bracts ovate, closed by 4-6 apical bracts; tepals 3, linear-lanceate. Male flowers 2-3-seriate, pedicellate; stamen 1. Female sessile; ovary ovoid. Achenes slightly keeled or appendaged.

Fl. & Fr.: Sep.-Dec. *Distr.*: Sri Lanka, India, Myanmar, S. China to Java and Borneo. Rare. Shrubs with weak branches. Along streamsides. *NAK 2042* (Naduvathumoozhy-Konni, c. 150 m).

8. **Ficus hispida** L.f.. Suppl. Pl. 442. 1781; King, Ann. Roy. Bot. Gard. (Calcutta) 1:tt. 154 & 155. 1888. & in Hook.f., Fl. Brit. India 5:522. 1888; Fischer in Gamble, Fl. Pres. Madras 1367. 1928; Corner, Gard. Bull. Straits Settlem. 21:89. 1965; Ramamoorthy & Gandhi in Sald. & Nicols., Fl. Hassan Dist. 81. 1976; Corner in Dassan. & Fosberg, Rev. Handb. Fl. Ceylon 3:277. 1981; Matthew & Rani in Matthew, Fl. Tam. Carnatic 3(2): 1522. 1983; Manilal, Fl. Silent Valley 260. 1988; Nicols., Suresh & Manilal, An Interpr. Hort. Malab. 185. 1988; Mohanan & Sivad., Fl. Agasthyamala 633.2002. *F. oppositifolia* Roxb., Pl. Corom. t. 124. 1799. & Fl. Ind. 3: 561. 1832; Wight, Ic. t. 638. 1840. *Perin-teregam* Rheede, Hort. Malab. 3.81, t.61. 1682.

Leaves decussate, oblong-elliptic, 10-15 by 5-6 cm, base rounded to truncate, margin entire to minutely toothed, apex acute to obtusely short-acuminate, 3-nerved from base, hispid, coriaceous. Figs dioecious on leafless trailing shoots, depressed-globose, c. 2.5 cm ϕ, bracts triangular; perianth cupular, truncate; tepals 0. Male flowers ostiolar, 2-seriate, stamen 1, subsessile. Female sessile or stalked, ovary depressed-globose; gall flowers larger, stalked. Achenes lenticular, keeled.

Fl. & Fr.: Most of the seasons. *Distr.*: Sri Lanka, India, S. China to New Guinea & Queensland. The most common species of Ficus in the district. A good indicator of forest destruction. In wastelands and forest clearings. *NAK* 106 (Moozhiar).

Bark yields a fibre.

9. **Ficus microcarpa** L.f., Suppl. Pl. 442. 1781; Corner, Gard. Bull. Straits Settlem. 21:2. 1965; Ramamoorthy & Gandhi in Sald. & Nicols., Fl. Hassan Dist. 81. 1976; Corner in Dassan. & Fosberg, Rev. Handb. Fl.Ceylon 3:258. 1981; Matthew, Ill. Fl. Tam. Carnatic t. 676. 1982; Matthew & Rani in Matthew, Fl. Tam. Carnatic 3(2); 1522. 1983; Nicols.,Suresh & Manilal, An Interpr. Hort. Malab. 186. 1988; Mohanan & Sivad., Fl. Agasthyamala 634.2002. *F. retusa* auct., non L., 1767: King, Ann. Roy. Bot. Gard. (Calcutta) 1:t. 61. 1887 & in Hook.f., Fl. Brit. India 5:511. 1888; Fischer in Gamble, Fl. Pres. Madras 1362. 1928. *Itty-alu* Rheede, Hort. Malab. 1:45-46, t.26. 1678. *Itty-arealou* Rheede, Hort. Malab. 3:69-70, t.55.1682.

Leaves alternate, elliptic-ovate to obovate, 4-6 by 3-5 cm, base cuneate, apex acute to obtuse, rarely retuse,basally 3-nerved, lateral veins 12-15 pairs, closely pinnate. Figs monoecious, axillary or on leafless branchlets, globose, c. 1 cm ϕ; bracts 3, concave; tepals 3-4, free. Male flowers disperse; stamen 1, sessile; ovary ovoid-globose. Achenes smooth.

Fl. & Fr.: Mar.-May. *Distr.*: Sri Lanka, India, S. China, Ryukyu Isls. to New Britain. Rare. In evergreen forests. *NAK 1712* (Sabari hills, c.250 m).

10. **Ficus religiosa** L., Sp., Pl. 1059. 1753; Roxb., Fl. Ind. 3:547. 1832; King, Ann. Roy. Bot. Gard. (Calcutta) 1:t. 67A. 1887 & in Hook.f., Fl. Brit. Inida 5: 513. 1888; Fischer in Gamble, Fl. Pres. Madras 1363. 1928; Corner, Gard. Bull. Straits Settlem. 21:6. 1965; Ramamoorthy & Gandhi in Sald. & Nicols., Fl. Hassan Dist. 82. 1976; Corner in Dassan. & Fosberg, Rev. Handb. Fl. Ceylon 3:236. 1981; Matthew & Rani in Matthew, Fl. Tam. Carnatic 3(2) : 1527.1983; Nicols., Suresh & Manilal, An Interpr. Hort. Malab. 107. 1988. *Urostigma religiosum* (L.) Gasp., Caprif. 82, t.7, t. 1-5. 1845; Wight, Ic.t. 1967. 1853. *Arealu* Rheede, Hort. Malab. 1:47-4, t.27. 1678.

Leaves alternate, deltoid or broadly ovate, 6-15 by 5-13 cm, base truncate, apex caudate, basally 3-nerved, lateral nerves 8-10 pairs, thin-coriaceous. Figs monoecious, axillary, in pairs, globose, c. 8 mm ϕ, basal bracts cupular. Male flowers ostiolar, sessile; tepals 2, free. stamen 1. Female sessile; tepals 3-4, free, ovary ovoid-oblong. Achenes smooth.

Fl. & Fr.: Nov.-Feb. *Distr.*: Himalayan forests from Rawalpindi eastwards, Yunnan, Cochin-China, N. Thailand. Common. Planted. *NAK 2218* (Pathanamthitta town).

Sacred tree; planted near temples. Powder or decoction of bark taken with honey improves complexion, cures ulcers and purifies blood.

11. **Ficus talboti** King, Ann. Roy. Bot. Gard. (Calcutta) 1:t.63. 1887 & in Hook.f., Fl. Brit. India 5:512. 1888; Fischer in Gamble, Fl. Pres. Madras 1363. 1928; Corner, Gard. Bull. Straits Settlem. 21:19. 1965; Ramamoorthy & Gandhi in Sald. & Nicols., Fl. Hassan. Dist. 80. 1976; Corner in Dassan. & Fosberg, Rev. Handb. Fl. Ceylon 3:254. 1981; Matthew & Rani in Matthew, Fl. Tam. Carnatic 3(2) : 1528. 1983.

Leaves alternate, elliptic-oblong, 7-16 by 4-6.5 cm, base acute or cuneate, apex caudate, lateral veins c.14 pairs, thin-coriaceous. Figs monoecious, axillary, in pairs or 2-3 or rarely solitary, globose or obovoid, c. 8 mm ϕ; bracts 3, ovate; tepals 3-4, free, ovate-lanceate. Male flowers disperse; stamen 1. Female sessile; ovary ovoid-globose. Achenes smooth.

Fl. & Fr.: Jan.-Apr. *Distr.*: Sri Lanka, Peninsular India, Myanmar, Laos, Cambodia, N. Thailand. Rare. In certain sacred groves. *NAK 2327* (Aikadu grove-Kodumon, c. 50 m).

12. **Ficus tinctoria** Forster f., Fl. Ins. Austr. 76. 1786, ssp. **parasitica** (Willd.) Corner, Gard. Bull. Straits Settlem. 17:476. 1960; Ramamoorthy & Gandhi in Sald. & Nicols., Fl. Hassan Dist. 83. 1976; Corner in Dassan. & Fosberg, Rev. Handb. Fl. Ceylon 3:276. 1981; Matthew & Rani in Matthew, Fl. Tam. Carnatic 3(2): 1528. 1983; Mohanan & Sivad., Fl. Agasthyamala 636.2002. *F. parasitica* J. Koenig ex Willd., Mem. Acad. Roy. Sci. Hist. (Berlin) 102. 1798. *F. gibbosa* Bl. var. *parasitica* (Willd.) King, Ann. Roy. Bot. Gard. (Calcutta) 1: t. 2 a-b. 1887 & in Hook.f., Fl. Brit. India 5:497. 1888; Fischer in Gamble, Fl. Pres. Madras 1365. 1928. *Tsjerou-meer-alou* Rheede, Hort. Malab. 3:71, t. 56. 1682; *Athi-meer-alou* Rheede, Hort. Malab.3:75-76,t.58. 1682.

Leaves alternate, obliquely elliptic-ovate, 5-12 by 3-7 cm, base unequally acute, apex acute, lateral nerves c.5 pairs, scabrid or smooth, thin-coriaceous. Figs dioecious,usually paired clustered, axillary, globose, c. 6 mm ϕ; basal bracts ovate. Male flowers yellow, female red when mature; tepals 3-5, free, scarious. Female flowers sessile, ovary obovoid-globose. Achenes keeled or not.

Fl. & Fr.: Sep.-Feb. *Distr.*: Sri Lanka, India, Myanmar, Indo-China. Not common. In disturbed semi-evergreen forests. *CB 49220* (MH) (Near Punalur road, Ranni, c. 150 m).

13. **Ficus travancorica** King, Ann. Roy. Bot. Gard. (Calcutta) 1:28, t.26. 1887 & in Hook.f., Fl. Brit. India 5:503. 1888; Fischer in Gamble, Fl. Pres. Madras 1365. 1928.

Scandent shrubs. Leaves alternate, elliptic-lanceate to oblanceate, 6-12 by 2.5-3.5 cm, base acute, margin undulate, apex acuminate, basally 3-nerved, lateral nerves 10-12 pairs, close, coriaceous. Figs monoecious, axillary, in pairs, peduncled, globose, c. 12 mm ϕ; basal bracts 3, broady ovate to deltoid. Male flowers disperse, sessile; tepals 4 or 5, stamen 1. Female subsessile; ovary ellipsoid. Achenes smooth.

Fl. & Fr.: Mar.-May. *Dsitr.*: Hills of N. Travancore. Rare. On rocks. *NAK 1782* (Angamoozhy H.P.C. Compound, c. 250 m).

14. **Ficus tsjakela** Burm. f., Fl. Ind. 227. 1768; King in Hook.f., Fl. Brit. India 5:514. 1888; Fischer in Gamble, Fl. Pres. Madras 1362.1928; Corner, Gard. Bull. Straites Settlem. 21:7. 1965; Ramamoorthy & Gandhi in Sald. & Nicols., Fl. Hassan Dist. 83. 1976; Corner in Dassan. & Fosberg, Rev. Handb. Fl. Ceylon 3:237. 1981; Matthew & Rani in Matthew, Fl. Tam. Carnatic 3(2): 1530.1983; Manilal, Fl. Silent Valley 260. 1988; Nicols., Suresh & Manilal, An Interpr. Hort. Malab. 188. 1988; Mohanan & Sivad., Fl. Agasthyamala 637.2002. *Tsjakela* Rheede, Hort. Malab. 3:87, t.64. 1682.

Leaves alternate, elliptic-oblong, 10-18 by 3.5-6.5 cm, base subcordate or rounded, apex obtusely acuminate, 3-nerved from base, lateral veins 6-8 pairs, thin-coriaceous. Figs monoecious, axillary or lateral, in clusters, rarely 1, sessile, depressed-globose, c. 5 mm ϕ; bracts 3, ovate. Male: tepals 2-3, subsessile, 1-seriate. Female sessile; tepals 3-4; ovary obvoid, gall flowers similar to female, sessile, or stalked. Achenes smooth.

Fl. & Fr.: Mar.-May. *Distr.:* Sri Lanka, India. A common tree. In hills. *NAK 33* (Moozhiar, c. 250 m), *1615* (Chathantara, c. 100 m).

STREBLUS Loureiro
Fl. Cochinch. 599, 614. 1790.

Armed or unarmed dioecious shrubs or small trees with milky latex. Leaves alternate, penni-nerved, thin-coriaceous. Male flowers minute, sessile or stalked, aggregated in capitate axillary clusters; tepals 4, nearly free; stamens, 4, inflexed in bud; pistillode conical. Female solitary or clustered, stalked; tepals 4, free, accrescent; ovary ovoid, 1-locular, style 2-fid. Drupe enclosed by perianth.

1. Armed shrubs; female flower(s) solitary; fruiting perianth large. .... 2. **S. taxoides**

1. Unarmed shrubs; female flowers 2-5 at a node; fruiting perianth small. .... 1. **S. asper**

1. **Streblus asper** Lour., Fl. Cochinch. 615. 1790; Hook.f., Fl.Brit. India 5:489. 1888; Fischer in Gamble, Fl. Pres. Madras 1353. 1928; Corner, Gard. Bull. Straits Settlem. 19:228. 1965; Ramamoorthy & Gandhi in Sald. & Nicols., Fl. Hassan Dist. 84. 1976; Corner in Dassan. & Fosberg, Rev. Handb. Fl. Ceylon 3:281. 1981; Matthew, Ill. Fl. Tam. Carnatic t. 679. 1982; Matthew & Rani in Matthew, Fl. Tam. Carnatic 3(2): 1534. 1983; Nicols., Suresh & Manilal, An Intepr. Hort. Malab. 189. 1988. *Epicarpurus orientalis* Bl., Bijdr. 488. 1825; Wight, Ic. t. 1961. 1853. *Tinda-parua* Rheede, Hort. Malab. 1:87-88, t. 48. 1678.

Bushy shrubs or small trees with drooping branches. Leaves elliptic-obovate to subrhombic, 3-7 by 1.8-3.5 cm, base cuneate, margin sparingly toothed, apex very shortly acuminate. Male in stalked clusters. Female 2-5, in clusters; ovary ovoid; style branching from base. Drupe enclosed by c. 8 mm long perianth.

Fl. & Fr.: Jan.-Apr. *Distr.*: Sri Lanka, India, Myanmar, Indo-China, Thailand, Malesia. In lowlands and semi-evergree forests, usually along streamsides. *NAK 1643* (Tiruvalla, c. 10 m), *2817* (Kalleli, c. 150 m).

2. **Streblus taxoides** (Roth) Kurz, For. Fl. Burma 2:465. 1877; Corner in Dassan. & Fosberg, Rev. Handb. Fl. Ceylon 3:283. 1981; Matthew, Ill. Fl. Tam. Carnatic t.680. 1982. Matthew & Rani in Matthew, Fl. Tam Carnatic 3(2): 1535. 1983. *Trophis taxoides* Heyne ex Roth, Nov. Pl. Spec. 368. 1821. *T. spinosa* Roxb., Fl. Ind. 3:762. 1832. *Phyllochlamys spinosa* (Roxb.) Bureau in DC., Prodr. 17:218. 1873; Hook.f., Fl. Brit. India 5: 488. 1888; Fischer in Gamble, Fl. Pres. Madras 1352. 1928. *Epicarpus spinosus* (Roxb.) Wight, Ic.t. 1962. 1853.

Bushy armed shrubs. Leaves obovate or subrhombic, 5-8 by 3-4.5 cm, base cuneate,margin toothed, apex bluntly acuminate, lateral nerves 6-8 pairs. Male in spikes or clusters, sessile; tepals 3-4, ovate, imbricate; stamens 3-4, inflexed in bud. Female solitry, axillary, pedicelled; tepals 3-4, foliaceous, accrescent; ovary obliquely ellipsoid; style 2-partite from above the middle. Drupe obliquely ovoid, enclosed by accrescent foliaceous, c. 2.5 cm long perianth.

Fl. & Fr.: Sep.-Nov. *Distr.*: Sri Lanka, India, Myanmar, Indo-China, Thailand, Malesia. Rare. Known from only one collection. *NAK 2086* (Enathu-Adoor, c. 20 m).

120. URTICACEAE

1. Plants with stinging hairs.
   2. Leaves 3-7-lobed; female flowers usually aggregated in globose heads. .... **Girardinia**
   2. Leaves undivided; male and female flowers in cymose clusters in spikes or in panicles.

3. Herbs; cymose clusters aggregated in slender interrupted spikes. ....**Laportea**
3. Trees; cymose clusters aggregated in continuous panicles. ....**Dendrocnide**

1. Plants without stinging hairs.
4. Herbs or subshrubs.
5. Male and female flowers aggregated on a fleshy receptacle. ....**Elatostema**
5. Male and female flowers not aggregated on a fleshy receptacle.
6. Flowers in peduncled cymes.
7. Leaves opposite. ....**Pilea**
7. Leaves alternate. ....**Pellionia**
6. Flowers in sessile cymose axillary clusters. ....**Pouzolzia**
4. Trees or shrubs.
8. Cymose clusters in long spikes. ....**Boehmeria**
8. Cymose clusters in short cymes or in panicles.
9. Leaves glabrous below; cymose clusters cauliflorous. ....**Oreocnide**
9. Leaves white-pilose below; panicle clusters in leaf-axils. ....**Debregeasia**

BOEHMERIA N. J. Jacquin
Enum. Pl. Carib. 9. 1760.

Monoecious or dioecious shrubs or subshrubs. Leaves opposite or alternate, basally 3-nerved, margin coarsely toothed. Flowers small in clusters, axillary, sessile or in spikes or in panicles. Male flowers : perianth 3-5-lobed; stamens 3-5, inflexed in bud; pistillode present. Female flowers : perianth tubular, toothed; ovary superior, 1-locular; stigma filiform. Fruit an achene.

1. Flowers in sessile axillary clusters; leaves alternate; large shrubs to small trees. ....1. **B. glomerulifera**
1. Flowers in long axillary spikes; leaves opposite; subshrubs. ....2. **B. macrophylla** var. **long.ssima**

1. **Boehmeria glomerulifera** Miq. in Zoll., Syst. Verz. Ind. Archip. 101, 104. 1854; Ramamoorthy in Sald. & Nicols., Fl. Hassan Dist. 87. 1976. *B. malabarica* Wedd., Arch. Mus. Hist. Nat. 8:350. 1855-1856; Hook.f., Fl. Brit. India 5:575. 1888; Fischer in Gamble, Fl. Pres. Madras 1387. 1928.

Monoecious shrubs to small trees. Leaves alterante, ovate-lanceate to elliptic-lanceate, 7-20 by 3-9 cm, base rounded, margin serrulate, apex acuminate, lower surface tomentose, chartaceous. Flowers minute, in sessile axillary clusters. Achenes ovoid, firmly enclosed by perianth.

Fl. & Fr.: Nov.-Feb. *Distr.*: Sri Lanka, Central & S. India, Java. Rare. In evergreen forests. *NAK 272, 1260* (Kakki hills, c. 100 m).

2. **Boehmeria macrophylla** Hornem., Hort. Bot. Hafn. 2:890. 1815, var. **longissima** (Hook.f.) Ramach. & Nair, Fl. Cannanore 436. 1988. *B. platyphylla* D. Don, Prodr. Fl. Nep. 60. 1825, var. *longissima* Hook.f., Fl. Brit. India 5: 579. 1888; Fischer in Gamble, Fl. Pres. Madras 1387. 1928.

Dioecious subshrubs. Leaves opposite, unequal, elliptic-ovate, 5-13 by 2.5-6 cm, base rounded, margin coarsely dentate, apex acumiante, sparsely pubescent, membranous. Flowers minute, in clusters which are in slender, c. 15 cm long, axillary spikes. Achenes compressed.

Fl. & Fr.: Mar.-May. *Distr.*: Sri Lanka, Peninsular India. Rare. Among forest under-growth in upper hills. *NAK 2677* (Kakki hills, c. 1100 m).

DEBREGEASIA Gaudichaud-Beaupre
Voyage Bonite Bot. Atlas t. 90. 1844.

Large dioecious shrubs or small trees. Leaves alternate, 3-5-nerved from base, lower surface white-pilose; stipules intrapetiolar, connate. Flowers subsesile in stalked globose axillary paniculate clusters. Male : tepals 4, free; stamens 4, pistillode ellipsoid. Female : perianth 3- or 4- toothed, ovary obovoid, 1-locular, ovule 1, basal. Achenes initially enclosed by perianth.

1. Leaves suborbicualr or broadly ovate; peduncle of cyme c. 5 cm long..................1. **D. ceylanica**
1. Leaves oblong-lanceate; peduncle of cyme c. 7 mm long. ....2. **D. velutina**

1. **Debregeasia ceylanica** Hook.f., Fl. Brit. India 5:592. 1888; Fischer in Gamble, Fl. Pres. Madras 1389. 1928; Manilal, Fl. Silent Valley 261. 1988.

Trees. leaves suborbicular, 9-17 by 6.5-15 cm, base rounded or cordate, apex obtusely acuminate, 5-nerved from base, coriaceous. Panicle axillary; peduncle c. 5 cm long; head c. 7 mm long. Achenes ribbed.

Fl. & Fr.: Jun.-Aug. *Distr.*: Sri Lanka, Southern Western Ghats. Very rare. In evergreen forests. *NAK 1754* (Kakki hills, c. 1000 m).

2. **Debregeasia velutina** Gaud., Voy. Bonite Bot. Atlas t. 90. 1844-46; Hook.f., Fl. Brit. India 5:590. 1888; Fischer in Gamble, Fl. Pres. Madras 1389. 1928; Wilmot-Dear, Kew Bull. 43 (4): 675.1988; Mohanan & Sivad., Fl. Agasthyamala 639.2002. *Debregeasia longifolia* (Burm. f.) Wedd. in DC., Prodr. 16(1): 235. 1869; Ramamoorthy in Sald. & Nicols., Fl. Hassan Dist. 87. 1976; Matthew, Ill. Fl. Tam. Carnatic t.664. 1982; Matthew & Rani in Matthew, Fl. Tam. Carnatic 3(2): 1489. 1983; Manilal, Fl. Silent Valley 261. 1988. *Urtica longifolia* Burm.f., Fl. Ind. 197. 1768. *Conocephalus niveus* Wight, Ic.t.1959. 1853.

Large shrubs. Leaves oblong-lanceate, 9-15 by 2-4 cm, base obtuse or rounded, margin serrate, apex acuminate, basally 3-nerved, chartaceous. Panicle axillary, 5-7 mm long. Heads c. 5 mm long. Achenes mintue, ovoid.

Fl. & Fr.: Dec.-Feb. *Distr.*: Sri Lanka, India, Myanmar, Indo-China, W. China, Malesia. Common. In evergreen forest borders. *NAK 376 & 1355* (Kakki hills, c. 110 m).

DENDROCNIDE Miquel
Pl. Jungh. 29.1851.

Dendrocnide sinuata (Bl.) Chew, Gard. Bull. Straits Settlem. 21:206. 1965. *Urtica sinuata* Bl., Bijdr. 505. 1826. *Laportea crenulata* Gaud., Voy. Uranie 12:498. 1830; Hook.f., Fl. Brit. India 5:550. 1888; Fischer in Gamble, Fl. Pres. Madras 1373.1928; Mohanan & Sivad., Fl. Agasthyamala 642.2002. *Urtica crenulata* Roxb., Hort. Beng. 67.1814, Fl. Ind. 3:591. 1832; Wight, Ic. t. 686. 1840. *Ottu-plavu, Ana-iruthi.*

Small dioecious trees. Leaves alternate, oblanceate to elliptic-lanceate, 12-20 by 4-6.5 cm, base acute, margin subentire, apex acumiante, lateral veins c. 8 pairs, lower surface with small stinging hairs, thin-coriaceous. Flowers small, dense in axillary paniculate cymes with dense stinging hairs. Male : tepals 4, nearly free; stamens 4, inflexed in bud; pistillode clavate. Female: tepals 4, free; ovary superior, oblique, 1-locular. Achenes oblique, compressed, seated on persistent perianth.

Fl. & Fr.: Sep.-Dec. *Distr.*: Peninsular & Eastern India, Malaya Isls. & Sumatra. Less common. In evergreen forests. *NAK 989* (Sabari hills, c. 250 m), *1115* (Moozhiar, c. 250 m).

Stinging of the hairs of this tree is very painful and long lasting.

ELATOSTEMA J. R. Forster *et* J.G.A. Forster
Charact. Gen. 105. 1776 (nom. cons.).

Elatostema lineolatum Wight, Ic.t. 1984. 1853; Wedd., Monogr. Urtic. 312. 1856; Hook.f., Fl. Brit. India 5:565. 1888; Fischer in Gamble, Fl. Pres. Madras 1376.1928; Ramamoorthy in Sald. & Nicols., Fl. Hassan Dist. 88. 1976; Matthew, Ill. Fl. Tam. Carnatic t. 665.1982; Matthew & Rani in Matthew, Fl. Tam. Carnatic 3(2): 1490. 1983; Mohanan & Sivad., Fl. Agasthyamala 641.2002.

Monoecious herbs. Leaves alternate, distichous, oblanceate, 8-16 by 2.5-6 cm, base cuneate, margin coarsely serrate along upper half, apex long acuminate, 3-nerved from base, strongly lineolate, chartaceous. Flowers in short axillary unisexual clusters, mixed with bracts and bracteoles. Male : tepals 4, ovate, spurred at the back; stamens 4; pistillode linear. Female: tepals 3-5, linear-lanceate, spurred at the back; ovary obovoid; stigma sessile, penicellate. Achenes ellipsoid, ribbed.

Fl. & Fr.: May.-Jul. *Distr.:* Sri Lanka, India, Himalaya, Myanmar. Locally abundant. A variable herb among moist forest undergrowth. *NAK 1881* (Kadamanthara, Goodrical R.F., c. 200 m).

GIRARDINIA Gaudichaud-Beaupre
in Freycinet, Voyage Monde
Bot. 498. 1830 ('1826').

Girardinia diversifolia (Link) Friis, Kew Bull. 36:145. 1981; Matthew, Ill. Fl. Tam. Carnatic t. 1. 666. 1982; Matthew & Rani in Matthew, Fl. Tam. Carnatic 3(2):1490. 1983; Manilal, Fl. Silent Valley 262. 1988; Nicols., Suresh & Manilal, An Interpr. Hort. Malab. 258. 1988. *Urtica diversifolia* Link, Enum. Pl. Hort. Bot. Berol. Alt. 2:385. 1822, non Bl., 1825. *Girardinia heterophylla* Decne. in Jacq., Bot. Voy. Bon. 151, t.153. 1844; Hook.f., Fl. Brit. India 5:550. 1888. *G. zeylanica* Decne in Jacq., Bot. Voy. Bon. 152. 1844; Fischer in Gamble, Fl. Pres. Madras 1373. 1928; Ramamoorthy in Sald. & Nicols., Fl. Hassan Dist. 89. 1976. *Anaschorigenam* Rheede, Hort. Malab. 2:77-78, t. 41. 1679.

Subshrubs with stinging hairs. Leaves alternate, long- petioled, 3-7-lobed, 6-15 by 5-14 cm, base truncate or subcordate, margin coarsely dentate-serrate, apex acuminate, 3-nerved from base sparsely scabrous, chartaceous. Male spikes branched or unbranched. Female in globose heads, armed with dense stinging hairs. Male flowers: tepals 4, ovate, cucullate, stamens 4; pistillode globose. Female flowers : tepals connate, tubular, ventricose, 2-4-lobed; ovary flattened, ovate; style filiform. Achenes compressed.

Fl. & Fr.: Dec.-Mar. *Distr.*: Sri Lanka, India, Himalaya, Myanmar, east to China, Malesia. Common. In evergreen forests. *NAK 334* (Moozhiar, c. 250 m), (Mannarappara, c. 200 m).

Bark yields a strong fibre. Leaves made into a paste and applied as a specific medicine in very severe head-ache and in swellings on joints.

LAPORTEA Gaudichaud-Beaupre
in Freycinet, Voyage Monde-Bot. 498. 1830 (1826) (nom. cons.).

Laportea interrupta (L.) Chew, Gard. Bull. Straits Settlem. 21:200. 1965 & 25:145. 1969; Ramamoorthy in Sald. & Nicols., Fl. Hassan Dist. 89. 1976; Matthew & Rani in Matthew, Fl. Tam. Carnatic 3(2): 1491. 1983; Nicols., Suresh & Manilal, An Interpr. Hort. Malab. 258. 1988. *Urtica interrupta* L., Sp. Pl.985. 1753; Roxb., Fl. Ind. 3:585. 1832; Wight, Ic.t. 692. 1840. *Fleurya interrupta* (L.) Gaud., Voy. Bonite, Bot. 497, t.8.1826; Wight, Ic.t. 1975. 1853; Wedd., Monogr. Urtic. 115. 1856; Hook.f., Fl. Brit. Inda 5:548. 1888; Fischer in Gamble, Fl. Pres. Madras 1372. 1928. *Batti-schorigenam* Rheede, Hort. Malab. 2:75, t.40.1979.

Monoecious herbs; branchlets with stinging hairs. Leaves alternate, elliptic-ovate to broadly ovate, 3-9 by 2-7.5 cm, base rounded or truncate, margin coarsely serrate, apex acuminate, 3-nerved from base, densely bulbous-based hairy, chartaceous. Flowers in short cymose clusters, aggregated in slender, axillary, interrupted spikes. Male : tepals 4, ovate,concave stamens 4; pistillode linear, clavate. Female : tepal is 4, unequal, basally connate into a cup; ovary obliquely ovoid; style lateral, linear. Achenes ovoid, compressed. Seeds flattened.

Fl. & Fr.: Sep.-Dec. *Distr.:* Africa, Abyssinia, Sri Lanka, India, Japan, China through S.E. Asia to Queensland. Common on waysides. *NAK 850* (Ranni, c. 100 m).

OREOCNIDE Miquel
Pl. Junghuhn. 39.1851.

Oreocnide integrifolia (Gaud.) Miq., Ann.Mus. Bot. Lug.-Bat. 4:306. 1869; Matthew & Rani in Matthew, Fl. Tam. Carnatic 3(2): 264. 1983; Manilal, Fl. Silent Valley 264. 1988. *Villebrunea integrifolia* Gaud., Voy. Bonite Bot.t. 91. 1844 Wedd., Monogr. Urtic. 452. 1857; Hook.f., Fl.Brit. India 5:589. 1888; Fischer in Gamble, Fl. Pres. Madras 1288. 1928.

Dioecious evergreen trees. Leaves alternate, elliptic-lanceate, 7-18 by 2.5-5.5 cm, base acute, margin faintly crenate towards upper half, apex acuminate, penni-nerved, lateral nerves c. 8 pairs, chartaceous. Flowers subsessile in globose, cymose clusters, cauliflorous. Male : tepals 3-5; pistillode clavate. Female : tepals connate into a fleshy tube with a toothed mouth, adnate to pistil; ovary obovoid, globose; stigma discoid. Achenes covered by fleshy bracteoles.

Fl. & Fr.: Mar.-May. *Distr.*: India, Trop. Himalayas, Myanmar, Indo-China. Common In evergreen forests. *NAK 395 & 2536* (Kakki hills, c. 1100 m).

PELLIONIA Gaudichaud-Beaupre
in Freycinet, Voyage Monde Bot. 494. 1830 ('1826') (nom. cons.).

Pellionia heyneana Wedd., Monogr. Urtic. 287, t.5.1856; Hook.f., Fl. Brit. India 5:561. 1888; Fischer in Gamble, Fl. Pres. Madras 1380. 1928; Ramamoorthy in Sald. & Nicols., Fl. Hassan Dist. 90.1976; Manilal, Fl. Silent Valley 264. 1988; Mohanan & Sivad., Fl. Agasthyamala 643.2002.

Woody herbs. Leaves alternate, rarely subopposite, obliquely elliptic-ovate, 10-15 by 5-7 cm, base oblqueły cordate, apex cauminate, 3-nerved from base, closely lineolate, chartaceous. Flowers monoecious or dioceious in peduncled axillary cymes. Male: tepals 5, ovate, acute; stamens 5; pistillode conical. Female: teapls 5, nearly free; staminodes 5, scaley; ovary ellipsoid. Achenes compressed, broadly ovate.

Fl. & Fr.: Apr.-May. *Distr.:* Sri Lanka, Peninsular India. Not common. Among shady moist forest undergrowth or on dripping rocks. *NAK 609* (Thekkuthode, c. 70 m).

PILEA Lindley
Collect. Bot. *ad t.4.* 1821 (nom.cons.).

Succulent herbs. Leaves decussate, unequal, in pairs, rarely in 3s, 1- or 3- nerved from base, stipules connate, intrapetiolar. Flowers 4-merous, subsessile, in subumbellate clusters, single or in panicles. Male : tepals 4. stamens 4, inflexed in bud. Female : tepals 3 to 4, free or connate, unequal; ovary 1-locular, ovule basal; stigma penicillate. Achenes ellipsoid or subconvex.

1. Dioecious tall herbs; leaves to 14 cm long, 3-nerved. ....1. **P. melastomoides**
1. Monoecious small herbs; leaves to 7 mm long, 1-nerved. ....2. **P. microphylla**

1. **Pilea melastomoides** (Poir.) Wedd., Ann. Sci Nat. Bot. Ser.4, 1:186.1854; Matthew, Ill. Fl. Tam. Carnatic t.667.1982; Matthew & Rani in Matthew, Fl. Tam. Carnatic 3(2):1493.1983; Manilal, Fl.Silent Valley 264.1988; Mohanan & Sivad., Fl. Agasthyamala 644.2002. *Urtica melastomoides* Poir. in Lam., Encycl. (Suppl.4) 223.1816. *Pilea trinervia* Wight, Ic.t.1973.1853; Wedd., Monogr. Urtic. 224.1856; Hook.f., Fl.Brit.India 5:557.1888; Fischer in Gamble, Fl.Pres. Madras 1379.1928.

Tall herbs. Leaves opposite, rarely 3 per node, elliptic-lanceate or rhombic-lanceate, 9-18 by 4-8 cm, base acute or obtuse, margin serrate, apex acuminate, 3-nerved. raphides scattered, membranceous. Flowers dioecious in short axillary racemose panicled-cymes. Male cymes spreading; tepals 4, ovate, stamens 4. Female cymes slender; tepals 3 or 4, unequal; ovary ovoid. Achene biconvex.

Fl. & Fr.: Nov.-Dec. *Distr.*: Sri Lanka, Peninsular India, Malaya Isls. Among moist shady dorest undergrowths. *NAK 39 & 1152* (Kakki hills, c.1100 m).

2. **Pilea microphylla** (L.) Liebm., Kongel. Danske Vidensk. Selsk.Naturvidensk. Math.Afh. Ser.2, 5(2):296.1851; Hook.f., Fl.Brit. India. 5:551.1888; Fischer in Gamble, Fl. Pres. Madras 1379. 1928; Matthew & Rani in Matthew, Fl. Tam. Carnatic 3(2):1493.1983; Mohanan & Sivad., Fl. Agasthyamala 644.2002. *Parietaria microphylla* L., Syst. Nat. (ed.10) 1308. 1759. *Urtica microphylla* Sw., Fl. Ind. Occid. 1:305.1797. *Pilea muscosa* Lindley, Collect. Bot. t.4. 1821 var. *microphylla* (Sw.) Wedd., Monogr. Urtic.174.1856.

Spreading herbs. Leaves elliptic-ovate, 4-7 by 2-3 mm, base acute, apex subacute to obtuse, 1-nerved, subsucculent. Flowers in stalked umbellate clusters. Male: tepals 4, free, concave; stamens 4. Female : tepals connate, 2-4-toothed; ovary ellipsoid. Achenes ellipsoid.

Fl. & Fr.: Most of the seasons. *Distr.*: S.America, now introduced into other tropical areas. Common. Usually on moist shady walls. *NAK 870* (Pathanamthitta, c. 50 m), *1020* (Tiruvalla, c. 10 m).

POUZOLZIA Gaudichaud-Beaupre
in Freycinet, Voyage Monde Bot.502.1830 ('1826').

Monoecious herbs or shrubs. Leaves alternate, 3-9-nerved from base, cystoliths punctiform; stipules persistent. Flowers in axillary subsessile clusters, sometimes the branchlets passing into a leafy spike. Male : 4-5 merous; tepals 4 or 5 free; pistillode minute. Female subsessile; tepals connate and adnate to pistil, winged in fruit; ovary obovoid-ellipsoid; style linear, hairy. Achenes enclosed in winged perianth.

1. Shrubs; flowers in terminal leafy spikes. .... 1. **P. wightii** var. **caudata**

1. Herbs; flowers in axillary clusters. .... 2. **P. zeylanica**

1. **Pouzolzia wightii** Bennett in Bennett *et al.,* Pl. Jav. Rar. 1:66.1838; Mohanan & Sivad., Fl. Agasthyamala 646.2002. var. **caudata** (Benn.) Fischer in Gamble, Fl. Pres. Madras 1384.1928. *P.caudata* Bennett, Pl. Jav.Rar. 1:66.1838; Wight, Ic.t.2097, f.27.1853; Hook.f., Fl. Brit. India 5:585.1888.

Shrubs; branchlets tapering into leafy spikes. Leaves opposite, subsessile, oblong lanceate or elliptic-lanceate, 6-14 by 2-4 cm, base rounded, apex acuminate or gradually acute, chartaceous; upper leaves gradually reduced into floral bracts. Flowers in sessile clusters. Achenes obovoid or orbicular, prominently winged.

Fl. & Fr.: Dec.-Mar. *Distr.*: Peninsular India. Common. Moist shady areas in evergreen forests. *NAK 194* (Moozhiar, c. 250 m), *842* (Ranni, c. 100 m).

2. **Pouzolzia zeylanica** (L.) Bennett in Bennett *et al.,* Pl. Jav. Rar. 67.1838; Ramamoorthy in Sald. & Nicols., Fl.Hassan Dist. 91. 1976; Manilal., Fl. Silent Valley 265.1988; Nicols., Suresh & Manilal, An Interpr. Hort. Malab. 259.1988; Mohanan & Sivad., Fl. Agasthyamala 646.2002. *Parietaria zeylanica* L., Sp. Pl. 1052.1753. *P.indica* L., Mant. Pl. 128.1767. *Pouzolzia indica* (L.) Gaud. in Freycinet, Voy. Bonite, Bot. 503.1826; Wight, Ic.tt.1980, f.1. & t. 2100, f. 40. 1853; Hook.f., Fl. Brit. India 5:581.1888; Fischer in Gamble, Fl.Pres.Madras 1382.1928. *Bula* Rheede, Hort.Malab. 10:59, t.30.1690.

Herbs. Leaves alternate, elliptic-ovate, 1.5-6.5 by 0.5-2.5 cm, base rounded, apex acute, pubescent, chartaceous. Flowers in subsessile, axillary clusters; bracts 2; bracteoles lanceate, scarious; tepals strigose without. Achenes 2-winged.

Fl. & Fr.: Nov.-Jan. *Distr.*: Sri Lanka, India, Himalayas, Myanmar, East to China & S. Japan, Malesia. Common along shady waysides, usually after rains. *NAK 216* (Thekkuthode, c.70 m), *2460* (Arampa, c.500 m). A highly variable herb.

Whole plant is used as remedy for urinary infections.

121. HYDROCHARITACEAE

1. Stems present; leaves cauline, whorled. .... **Hydrilla**
1. Stems absent; leaves radical, other than whorled.
    2. Leaves sessile, linear. .... **Blyxa**
    2. Leaves petioled, ovate. .... **Ottelia**

BLYXA Noronha ex Thouars
Gen. Nova Madag. 4. 1806.

Stemless submerged tufted herbs. Leaves radical, linear, membranous. Spathes apically 2-fid, peduncled, articulate. Flowers bisexual. Male flowers : pedicelled, α. Female and bisexual solitary. Stamens 3 or 9, anthers linear; pistillodes 3; ovary linear, with a long filiform rostrum; ovules α, on 3 parietal placentae; styles 3. Fruits linear.

1. Seeds tubercled, obtuse or apiculate at ends. .... 1. **B. aubertii**

1. Seeds echinate, long-tailed at both ends. .... 2. **B. echinosperma**

1. **Blyxa aubertii** Rich., Mem. Cl. Sci. Math. Inst. Natl. France 12(2) : 77, t. 4.1812; Hartog in Steenis, Fl. Males. I, 5:390. 1959; Rao & Verma, Bull.Bot.Surv. India 12:141. 1970; Mohanan & Sivad., Fl. Agasthyamala 647.2002. *B. ceylanica* Hook.f., Fl. Brit. India 5:661. 1888; Fischer in Gamble, Fl. Pres. Madras 1379. 1921; Gandhi & Ramamoorthy in Sald. & Nicols., Fl. Hassan Dist. 633. 1976.

Submerged, stemless, tufted herbs. Leaves linear, 20-30 by 0.5-1.0 cm, base sheathing, apex acute, membranous; spathes tubular. Flowers bisexual; sepals 3, oblong, 1-nerved; petals linear; stamens 3. Capsule cylindric. Seeds ovate-ellipsoid, α, obscurely tubercled.

Fl. & Fr.: Nov.-Dec. *Distr.*: Madagascar, Indo-Malesia. Common in still waters and paddy fields. *NAK 2221* (Enathu, Adoor, c. 20 m).

2. **Blyxa echinosperma** (Clarke) Hook.f., Fl. Brit. India 5: 661. 1888; Fischer in Gamble, Fl. Pres. Madras 1397. 1928; Rao & Verma, Bull. Bot. Surv. India 12:140. 1970. *Hydrotrophus echinospermus* Clarke, J. Linn. Soc. Bot. 14:8, t. 1. 1875. *B. talboti* Hook.f., Fl. Brit. India 5: 661.1881; Fischer in Gamble, Fl. Pres. Madras 1397.1928; Gandhi & Ramamoorthy in Sald. & Nicols., Fl. Hassan Dist. 633. 1976.

Stemless, submerged herbs. Leaves 10-15 by 0.3-0.5 cm, 5-7-nerved, base broad, apex tapering. Scape(s) one or more. Spathes 1-flowered. Flowers bisexual, pale pinkish; sepals 3, oblong, 1-3- nerved; petals 3; stamens 3; styles compressed. Fruits torulose, ribed. Seeds α, ovate-ellipsoid, with 8 rows of blunt spines with filiform tail at both ends.

Fl. & Fr.: Oct.-Dec. *Distr.:* Tropics. Common in still waters, pools, ponds, etc. *NAK 1973* (Tiruvalla, c. 10 m).

HYDRILLA L.C. Richard
Mem. Cl. Sci. Math. Inst. Natl. France 12(2): 9, 61, 76. 1814.

Hydrilla verticillata (L.f.) Royle, Ill. Bot. Himal. Mount. 376. 1839; Hook.f., Fl. Brit. India 5:659. 1888; Fischer in Gamble, Fl. Pres. Madras 1369. 1928; Rao & Verma, Bull. Bot. Surv. India 12:140. 1970; Nicols., Suresh & Manilal, An Interpr. Hort. Malab. 293. 1988. *Serpicula verticillata* L.f., Suppl. Pl. 416. 1781; Gandhi in Ramamoorthy in Sald. & Nicols., Fl. Hassan Dist. 634. 1976. *Ene-pael* Rheede, Hort. Malab. 9:159, t. 81. 1689.

Profusely branched, submerged, tuberous, slender herbs; rooting at nodes. Leaves opposite or whorled, linear-oblong, margin entire or serrulate. Spathes solitary, axillary. Flowers unisexual, solitary, minute; sepals 3, ovate; petals 3, oblong; stamens 3. Female flowers sessile; stigma fimbriate. Fruits subcylindric. Seeds 2-6, minute.

Fl. & Fr.: Nov.-Dec. *Distr.*: Tropics & Subtropics. Common in still waters. *NAK 1603* (Chathanthara, c. 100 m).

OTTELIA Persoon
Syn. Pl. 1:400. 1805.

Ottelia alismoides (L.) Pers., Syn. Pl. 1:400. 1807; Hook.f., Fl. Brit. India 5:662. 1888; Fischer in Gamble, Fl. Pres. Madras 1398. 1928; Rao & Verma, Bull. Bot. Surv. India 12:141. 1970; Gandhi & Ramamoorthy in Sald. & Nicols., Fl. Hassan Dist. 634. 1976; Nicol., Suresh & Manilal, An Interpr. Hort. Malab. 293. 1988. *Stratiotes alismoides* L., Sp. Pl. 535. 1753. *Ottel-ambel* Rheede, Hort. Malab. 11:95, t. 46. 1692.

Flaccid, submerged, annual herbs. Submerged leaves narrow-oblong, tapering at base. Floating leaves oblong, obtuse, cordate at base, margin undulate. Spathes ovate-elliptic, with 5-10 wavy wings. Flower(s) solitary, sessile, white; sepals 3, oblong-lanceate; petals 3, obovate with yellow base; stamens 6-9, filaments hairy. Fruits fleshy, oblong. Seeds α, crowned by sepals.

Fl. & Fr.: Nov.-Dec. *Distr.*: Tropics. Not common. In pools, still waters, etc. *NAK 1960* (Tiruvalla, c. 10 m).

122. BURMANNIACEAE

BURMANNIA Linnaeus
Sp. Pl. 287. 1753.

Burmannia coelestis D. Don, Prodr. Fl. Nepal. 44. 1825; Hook.f., Fl. Brit. India 5:664. 1988; Fishcer in Gamble, Fl. Pres. Madras 1399. 1928; Rao & Verma, Bull. Bot. Surv. India 12:142. 1970; Ramamoorthy in Sald. & Nicols., Fl. Hassan Dist. 809. 1976; Mohanan & Sivad., Fl. Agasthyamala 647.2002. *Gonyanthes pusilla* Wall. ex Miers, Trans. Linn. Soc. London 18:537, t. 38, f. 3. 1841. *Burmannia pusilla* (Miers) Thw., Enum. Pl. Zeyl. 325. 1864. *B. coelestis* var. *pusilla* (Miers) Trimen, Handb. Fl. Ceylon 4:138. 1898.

Slender herbs, 10-15 cm tall. Basal leaves scale-like; cauline leaves reduced or linear-ensiform. Flower(s) solitary or 2-6, in clusters; perianth-lobes: 3 (outer) larger, winged, tubular; inner 3 inconspicuous; stamens 3; stigma shortly 3-lobed. Capsule ellipsoid. Seeds α.

Fl. & Fr.: Oct.--Dec. *Distr.*: Indo-Malesia. Common in mountain swamps. *NAK 1302* (Kakki hills, c. 1000 m).

123. ORCHIDACEAE

1. Epiphytic herbs.
   2. Leaves equitant, laterally compressed.
      3. Leaves usually (2) 4-6, to 10 cm long; flowers minute. .... **Oberonia**
      3. Leaves usually ± 40, to 1 cm long; flowers small. .... **Podochilus**
   2. Leaves not equitant or laterally compressed.
      4. Lip spurred.
         5. Column prolonged into a distinct foot. .... **Aerides**
         5. Column not prolonged into a foot.
            6. Spur septate. .... **Sarcanthus**
            6. Spur not septate.
               7. Flowers densely crowded, on short-peduncled, drooping racemes. .... **Rhynchostylis**
               7. Flowers not densely crowded on drooping racemes.
                  8. Leaves articulated on sheaths .
                     9. Spur saccate; mid-lobe of lip at rim of spur, not covering it. .... **Gastrochilus**
                     9. Spur pointed; mid-lobe of lip at halfway down spur, covering it. .... **Smithsonia**

8. Leaves not articulated on sheaths.
    10. Side-lobes of lip large. ....**Vanda**
    10. Side-lobes of lip small or absent. ....**Acampe**

4. Lip not spurred.
    11. Flowers with a mentum.
        12. Mentum formed by the fusion of lateral sepals with column-foot. ....**Dendrobium**
        12. Mentum formed by the fusion of lip with column-foot. ....**Kingidium**
    11. Flowers without a mentum.
        13. Column prolonged into a foot.
            14. Lip clawed, articulate on foot.
                15. Inflorescence a lateral raceme. ....**Bulbophyllum**
                15. Infloroescence a terminal panicle. ....**Polystachya**
            14. Lip sessile, not articulate on foot. ....**Eria**
        13. Column not prolonged into a foot.
            16. Bracts closely imbricating, ± concealing the flowers. ....**Pholidota**
            16. Bracts neither imbricating nor concealing the flowers.
                17. Lip entire.
                    18. Leaves flat. ....**Liparis**
                    18. Leaves terete. ....**Luisia**
                17. Lip ± 3-lobed.
                    19. Flowers confined to peducular tips; pollinia 2. ....**Cottonia**
                    19. Flowers dispersed almost all over the peduncle; pollinia 4 or 8. ....**Sirhookera**

1. Terrestrial herbs.
    20. Flowers with a mentum.
        21. Inflorescence scapose, lateral.
            22. Lip adnate to the top of the column, long- spurred. ....**Calanthe**
            22. Lip adnate to the base or end of column, shortly-spurred. ....**Eulophia**
        21. Inflorescence not scapose, terminal.
            23. Lip inferior.
                24. Stigma ± extended with calvate or papillate process; anther-tube prominent.
                    25. Flowers ± 3 cm ϕ; stigmas stalked. ....**Habenaria**
                    25. Flowers ± 5 cm ϕ; stigmas sessile. ....**Pecteilis**
                24. Stigma united to base of lip and auricles of column; anther-tube absent or very short. ....**Peristylus**
            23. Lip superior. ....**Satyrium**
    20. Flowers without a spur or mentum.
        26. Leaf 1, appearing after flowers (Leaves orbicular). ....**Nervilia**

26. Leaves 2 or more, apearing with flowers.
    27. Leaves purpie.
        28. Sepals connate to middle. .... **Cheirostylis**
        28. Sepals free. .... **Zeuxine**
    27. Leaves greenish.
        29. Flowers less than 10 mm long.
            30. Lip resupinate. .... **Liparis**
            30. Lip not resupinate. .... **Malaxis**
        29. Flowers more than 10 mm long. .... **Geodorum**

ACAMPE Lindley
Folia Orchid. Fasc. 4. 1853.

Epiphytic stout herbs. Leaves distichous, coriaceous, apex praemorse. Inflorescence few to α-flowered, longer or shorter than leaves. Flowers fragile, lip saccate or spurred, attached to column-base; pollinia 2, waxy; stipe long, subulate. Capsule fusiform or subcylindric.

1. Inflorescence as long as or longer than leaves; flowers lax, spur to 5 mm long. .... 1. **A. ochracea**
1. Inflorescence much shorter than leaves; flowers dense, spur to 4 mm long. .... 2. **A. praemorsa**

1. **Acampe ochracea** (Lindl.) Hochr., Bull. NY. Bot. Gard. 6:270. 1910; Sant. & Kapad., Orch. Bombay 235. 1966; Abraham & Vatsala, Intr. Orch. 452. 1981. *Saccolabium ochraceum* Lindl., Bot. Reg. Misc. 2. 1842; Hook.f., Fl. Brit. India 6:62.1890.

Leaves linear-oblong, 10-20 by 2 cm, base sheathing, articulate, apex unequally cleft. Panicle to 18 cm long, distantly branched; peduncle slender. Flowers pale yellow with reddish strips, c. 1 cm ϕ, lax; sepals subequal, obovate, lateral petals simialr to sepals, lip ovate, base 2-auricled, margin crenate; spur cylindric, to 4 mm long; column short. Capsule fusiform, c. 2 cm long.

Fl. & Fr.: Feb -Apr. *Distr.*: Sri Lanka, Peninsular & Eastern India, to Myanmar. Rare. In semi-evergreen forests. *NAK 528* (Thekkuthode, c. 70 m). Known from only one collection with both flowers and fruits.

2. **Acampe praemorsa** (Roxb.) Blatter & Mc Cann, J. Bombay Nat. Hist. Soc. 35: 495. 1932; Sant. & Kapad., Orch. Bombay 233. 1966; Jayaw. in Dassan. & Fosberg, Rev. Handb. Fl. Ceylon 2:231. 1981; Matthew, Ill. Fl. Tam. Carnatic t. 718. 1982; Seidenfaden in Matthew, Fl. Tam. Carnatic 3(3):1602. 1983; Nicols., Suresh & Manilal, An Intepr. Hort. Malab. 297. 1988; Mohanan & Sivad., Fl. Agasthyamala 652.2002. *Epidendrum praemorsum* Roxb., Pl. Corom. t. 43. 1795. *Vanda wightiana* Lindl.ex Wight, Ic. Pl. t. 1670.1851. *Acampe wightiana* (Lindl.ex Wight) Lindley, Fol. Orchid. Acampe 2.1853, Fischer in Gamble, Fl. Pres. Madras 1447. 1928. *Saccolabium praemorsum* (Roxb.) Hook.f., Fl. Brit. India 6:62. 1890. *S. wightianum* Hook.f., Fl. Brit. India 6:62. 1890. *Thalia-maravara* Rheede, Hort. Malab. 12:9, t. 4. 1693.

Leaves oblong, 12-28 by 2-3 cm, channelled, apex unequally deeply cleft, coriaceous. Inflorescence short, erect, corymbose-panicles to 10 cm long, peduncle stout. Flowers dense, yellowish; bracts broadly ovate, persistent; dorsal sepals obovate-oblong, lateral sepals oblong-ovate; lateral petals oblanceate, lip fleshy, 3-lobed, base saccate, mid-lobe ligulate, margin crenulate, spur very short, column short, stout with two terminal horns. Capsule subcylindric, ribbed.

Fl. & Fr.: May-Jun. *Distr.*: Sri Lanka, South India. The most common epiphytic species of the family. In plains. *NAK 2875* (Ranni, c. 100 m).

AERIDES Loureiro
Fl. Cochinch. 525. 1790.

Epiphytic herbs. Leaves flat or cylindric, coriaceous. Inflorescence of axillary spikes, racemes or panicles. Flowers few to a, sepals subequal; petals similar to sepals, often adnate to the foot of column, lip spurred, 3-lobed, side-lobes usually small, spur curved forwards, callus at mouth; pollinia 2, globose. Capsule ribbed.

1. Leaves cylindric; flowers 2-3, on stout unbranched peduncle. ....1. **A. cylindricum**
1. Leaves flat; flowers a, on slender branched peduncle. ....2. **A. ringens**

1. **Aerides cylindricum** Lindley, Gen. Sp. Orch. 240. 1833; Wight, Ic.t. 1744. 1852; Hook.f., Fl. Brit. India 6:44. 1890; Fischer in Gamble, Fl. Pres. Madras 1442. 1928; Abraham & Vatsala, Intr. Orch. 444. 1981.

Erect tufted herbs. Leaves cylindric, to 12 cm long. Flowers white or pink, 2-3 on a stout peduncle, 2-3 cm long; sepals elliptic-lanceate; lateral petals slightly longer, lip deep pink with yellowish throat, equally 3-lobed. Capsule fusiform.

Fl. & Fr.: May.-Sep. *Distr.*: Endemic to South India. Rare. On exposed hills slopes. *CNM 65081* (MH) (Kodumon, c. 50 m).

2. **Aerides ringens** (Lindl.) Fischer, Bull. Misc. Inform. 1928: 284. 1928 & in Gamble, Fl. Pres. Madras 1442. 1928; Sant. & Kapad., Orch. Bombay 119. 1966; Abraham & Vatsala, Intr. Orch. 446. 1981; Jayaw. in Dassan. & Fosberg, Rev. Handb. Fl. Ceylon 2:205. 1981; Seidenfaden in Matthew, Fl. Tam. Carnatic 3(3): 1593. 1983; Manilal, Fl. Silent Valley 269. 1988; Mohanan & Sivad., Fl. Agasthyamala 564.2002. *Saccolabium ringens* Lindl., Gen. Sp. Orch. 221. 1833. *S. rubrum* Wight, Ic. t. 1851. *Aerides linearis* Hook.f., Fl. Brit. India 6:47. 1890.

Robust herbs. Leaves alternate, distichous, linear-oblong, 6-20 by 1-2 cm, base sheathing, articulate, apex unequally cleft, thick. Inflorescence of panicles, in upper leaf-axils. Flowers pink or mauve, long pedicellate; bracts triangular; sepals subequal, dorsal sepals broadly oblong-elliptic, lateral ones orbicular-obovate; lateral petals spreading, obovate-orbicular, lip 3-lobed, spurred, side-lobes small, mid-lobe ovate, spur with a callus at mouth. Capsule ribbed.

Fl. & Fr.: Mar.-Nov. *Distr.*: Sri Lanka, Peninsular India. Rare. In evergreen forests. *Deb 30453* (MH) (Pamba-Vandiperiyar, c. 1100 m).

BULBOPHYLLUM Du Petit-Thouars
Hist. Pl. Orch., tabl. esp. 3, sub u. 1822 (nom. cons.).

Creeping epiphytes with uninodal pseudo-bulbs on the rhizome. Leaves 1-2. Inflorescence umbellate or racemose, on a separate scape from rhizome. Flowers without mentum; sepals unequal; petals shorter than lateral sepals, lip fleshy, sessile or clawed; column short; pollinia 4. Capsule fusiform or ellipsoid.

1. Inflorescence shorter than leaves; flowers dense, glabrous. .... 1. **B. sterile**
1. Inflorescence longer than leaves; flowers lax, hairy. .... 2 **B. tremulum**

1. **Bulbophyllum sterile** (Lam.) Suresh in Nicols., Suresh & Manilal, Interpr. Hort. Malab. 298. 1988; Mohanan & Sivad., Fl. Agasthyamala 658.2002. *Epidendrum sterile* Lam., Encycl. I: 189.1783. *Bulbophyllum neilgherrense* Wight, Ic.t. 1650. 1851; Fischer in Gamble, Fl. Pres. Madras 1418. 1928; Sant. & Kapad, Orch. Bombay 195. 1966; Sald. & Nicols., Fl. Hassan Dist. 815. 1976; Abraham & Vatsala, Intr. Orch. 334. 1981; Seidenfaden in Matthew, Fl. Tam. Carnatic 3(3) : 1591. 1988; Manilal, Fl. Silent Valley 271. 1988.

Pseudo-bulbs conical-ovoid, angled, greenish. Leaf sessile, 1, oblong. Racemes c. 10 cm long, shorter than leaves, scape sheathing at base. Flowers dull-yellow, dense; bracts small, oblong-lanceate; dorsal sepal ovate, concave, lateral sepals much larger than dorsal sepal; lateral sepals subfalcate, attached to the foot of column, form a cymbiform structure; lateral petals ovate-triangular, awned, lip tongue-shaped, side-lobes narrow, mid-lobe grooved dorsally; column produced above into 2 teeth.

Fl. & Fr.: Feb.-Apr. *Distr.*: Endemic to South India. Common. In riverine forests. *NAK 449.* (Chathanthara, c. 100 m).

2. **Bulbophyllum tremulum** Wight, Ic. t. 1749. 1841; Hook.f., Fl. Brit. India 5:763. 1890; Fischer in Gamble, Fl. Pres. Madras 1418. 1928; Sald. in Sald. & Nicols., Fl. Hassan Dist. 816. 1976; Abraham & Vatsala, Intr. Orch. 331. 1981; Manilal, Fl. Silent Valley 271. 1988.

Pseudo-bulbs ovoid-conical. Leaf 1, sessile, oblong-lanceate, 2-8 by 0.7-1.5 cm, apex obtusely acute. Racemes 12-25 cm long, much longer than leaf, scape with few sheathing bracts at base. Flowers yellowish with purple streaks; bracts minute, erect, ovate; sepals finely villous, 3-nerved; lateral petals much smaller than sepals, lanceate, margin ciliate, 1-nerved, lip ligulate, with acicular hairs on lower side, tremulus, shortly stipitate. Capsule ellipsoid.

Fl. & Fr: Sep.-Apr. *Distr.*: Endemic to Peninsular India. In evergreen forests of upper hills. *CNM 80110* (MH) (Anathode, c. 800 m).

CALANTHE R. Brown
Bot. Reg. 7: Sub t. 573. 1821 (nom. cons.).

Calanthe masuca (D.Don) Lindl., Gen. Sp. Orch. 249. 1833; Hook.f., Fl. Brit. India 5:850. 1890; Fischer in Gamble, Fl. Pres. Madras 14321. 1928; Abraham & Vatsala, Intr. Orch. 265. 1981; Manilal, Fl. Silent Valley 272. 1988; Mohanan & Sivad., Fl. Agasthyamala 659.2002. *Bletia masuca* D. Don, Prodr. Fl. Nepal. 30. 1825. *Calanthe emarginata* Wight, Ic.t. 918. 1845.

Terrestrial stout herbs. Leaves elliptic-ovate, 12-25 by 6-12 cm, base rounded, tapering into a short petiole, apex acute, plaited. Scape lateral, to 50 cm long, bracts foliaceous, ovate-lanceate. Flowers purple, c. 4 cm ϕ; sepals violet coloured, obovate; lateral petals broadly obovate, lip hardly exceeding the sepals, side-lobes short, oblong-falcate, mid-lobe much longer, broadly reniform; spur longer than sepals. Capsule broadly ellipsoid.

Fl. & Fr.: Nov.-Mar. *Distr.*: India, China, Malesia, Thailand to Australia. Rare. On moist forest shades. *NAK 1821* (Angamoozhy, c. 250 m).

CHEIROSTYLIS Blume
Bijdr. 6:t.1, t.1b; 8: 413. 1825.

Cheirostylis flabellata Wight, Ic. 5:16, t.1727. 1851 ('*Monochilus flabellatum*'); Hook.f., Fl. Brit. India 6:105. 1890; Fischer in Gamble, Fl. Pres. Madras 1454. 1928; Abraham & Vatsala, Intr. Orch. 194. 1981; Mohanan & Sivad., Fl. Agasthyamala 660.2002.

Saprophytic slender herbs, to 20 cm tall; stem covered by sheating leaf bases. Leaves 4-7, ovate, 1.5-3 by 1.5 cm, base sheathing, apex acute, membranous, pale purplish. Inflorescence racemes, to 15 cm long, glandular-pubescent. Flowers few, restricted towards the apex; bracts lanceate; sepals and petals united to form a tube; dorsal sepals lanceate, gland-dotted, lateral sepals obliquely-oblong; lateral petals falcate, lip constricted near middle, flabellate, margin deeply coarsly lobed. Capsule fusiform.

Fl. & Fr.: Feb.-May. *Distr.*: Sri Lanka, India, Bhutan to Myanmar. Rare. In moist shady forest floors. *NAK 1485* (Chittar, c. 250 m).

COTTONIA R. Wight
Icon. Pl. nd. Or 5(1): 21, t. 1755. 1851.

Cottonia peduncularis (Lindl.) Reichb.f., Cat. Orch. Schiller 52. 1857; Sant. & Kapad., Orch. Bombay 222. 1966; Sald. in Sald. & Nicols., Fl. Hassan Dist. 816. 1976; Abraham & Vatsala, Intr. Orch. 456. 1981; Jayaw. in Dassan. & Fosberg, Rev. Handb. Fl. Ceylon 2:261. 1981; Manilal, Fl. Silent Valley 274. 1988; Mohanan & Sivad., Fl. Agasthyamala 663.2002. *Vanda peduncularis* Lindl., Gen. Sp. Orch. 216. 1833. *Cottonia macrostachya* Wight, Ic.t.1755. 1851; Hook.f., Fl. Brit. India 6:26 1890; Fischer in Gamble, Fl. Pres. Madras 1439. 1928.

Epiphytic herbs. Leaves alternate, distichous, linear, 7-14 by 0.9-1.5 cm, keeled, apex unequally cleft, thick-succulent. Inflorescence of erect, long panicles. Flowers pale yellow with maroon shade, restricted to the apex of peduncular branches; bracts minute; perianth fleshy; dorsal sepal narrowly obovate, apex recurved; lateral petals spathulate, lip fleshy, with upturned upper half, faintly 3-lobed, with 2 callii at base, mid-lobe dark-purple, side-lobes yellow, pubescent. Capsule fusiform.

Fl. & Fr.: Mar.-May. *Distr.:* India, Sri Lanka. Less common. In semi-evergreen forests. *NAK 1690* (Sabar hills, c. 250 m).

Flowers are very haracteristic with the labellum having the appearance of a female wasp.

DENDROBIUM O. Swartz
Nova Acta Regiae Soc. Sci. Upsal.
Ser. 2, 6: 82.1799 (nom. cons.).

Epiphytic herbs; stem sympodial, with elongate or pseudo-bulbous shoots. Leaves alternate, or terminal, deciduous; dorsal sepals and petals free; dorsal sepals adnate to column-foot to form a mentum; lip 3-lobed or entire; column short; pollinia 4, caudicles 0. Capsule ellipsoid.

1. Leafy during flowering.
   2. Pseudo-bulbous herbs; flowers in terminal racemes. .... 3. **D. nanum**
   2. Non-pseudo-bulbous herbs; flowers in leaf-axils. .... 1. **D. aqueum**
1. Leafless during flowering.

3. Flowers a, in long racemes; lip distinctly 3-lobed. ....4. **D. ovatum**

3. Flowers 2-3, on condensed peduncle; lip entire or faintly 3-lobed. ....2. **D. macrostachyum**

1. **Dendrobium aqueum** Lindl., Bot. Reg. 6:t.59. 1843; Hook.f., Fl. Brit. India 5:739. 1890; Fischer in Gamble, Fl. Pres.Madras 1417. 1928; Sald. in Sald. & Nicols., Fl. Hassan Dist. 819. 1976; Abraham & Vatsala, Intr. Orch. 349. 1981; Seidenfaden in Matthew, Fl. Tam. Carnatic 3(3) : 1589. 1983; Manilal, Fl. Silent Valley 275. 1988. *Dendrobium album* Wight, Ic. t. 1645. 1851.

Stem clavoid, leafy during flowering. Leaves ovate-lanceate, 5-10 by 1-2 cm, base rounded, apex acuminate. Flowers white, c. 4 cm ϕ, 2-3 from leaf-axils; petals broader than sepal; lip rhombic, faintly 3-lobed, tinged with yellow, margin erose, mid-lobe triangular; mentum short, incurved.

Fl. & Fr.: Sep.-Dec. *Distr.*: Endemic to South India. Rare. In upper hills. *BDS 43942* (MH) (Pachakkanam, c. 900 m).

2. **Dendrobium macrostachyum** Lindl., Gen. Sp. Orch. 78. 1830; Wight, Ic. t. 1647. 1851; Hook.f., Fl. Brit. India 5:735. 1890; Fischer in Gamble, Fl. Pres. Madras 1416. 1928; Sald. in Sald. Nicols., Fl. Hassan Dist. 821. 1976; Abraham & Vatsala, Intr. Orch. 355. 1981; Jayaw. in Dassan. & Fosberg, Rev. Handb. Fl. Ceylon 2:90. 1981; Manilal, Fl. Silent Valley 277. 1988; Mohanan & Sivad., Fl. Agasthyamala 667.2002.

Stem slender, plurinodal, leafless when flowering. Leaves lanceate, 2-9 by 1-2.5 cm, base sheathing, apex acute. Flowers pale yellowish-green, 2-or 3 from the node, on a condensed peduncle; bracts scaly, lanceate; perianth spreading, dorsal sepal elliptic-lanceate, lateral sepals obliquely lanceate, 7-nerved; lateral petals linear-lanceate, mentum 6 mm long; lip orbicular-oblong, stipitate hairy on dorsal side, marign finely fimbriate. Capsule ellipsoid.

Fl. & Fr.: Sep.-Feb. *Distr.*: Sri Lanka, India. The common species of the genus. *NAK 552, 1584, 2375.* (Almost all forests.).

3. **Dendrobium nanum** Hook.f., Ic. Pl. 19,t. 1853. 1889 & Fl. Brit. India 5:717. 1890; Fischer in Gamble, Fl. Pres. Madras 1415. 1928.

Small herbs. Leaves 3-4, distichous, elliptic-lanceate, to 2 by 0.6 cm, base sheathing, apex obtuse. Inflorescence 4-8 cm long, terminal, filiform, zigzag. Flowers white, with green labellum, c. 1 cm ϕ; dorsal sepals narrowly lanceate-oblong, dorsally ridged, 3-nerved; mentum short; lateral petals spathulate, 3-nerved, lip included, 3-lobed, conduplicate, side-lobes narrow, erect, winged, mid-lobe suborbicular, ridged on dorsal side.

Fl. & Fr.: Sep.-Aug. *Distr.*: Endemic to Peninsular India. Very rare. On dripping rocks.*NAK 2509* (Kakki hills, c. 1100 m). Known from only a single material of a single collection.

4. **Dendrobium ovatum** (Willd.) Kraenzl. in Engler, Pflanzenr. IV. (50). II.B. (45):71. 1910; Fischer in Gamble, Fl. Pres. Madras 1416. 1928; Sant. & Kapad., Orch. Bombay 91. 1966; Abraham & Vatsala, Intr. Orch. 353. 1981; Nicols., Suresh & Mani., An Interpr. Hort. Malab. 299. 1988; Mohanan & Sivad., Fl. Agasthyamala 669.2002. *D. chlorops* Lindl., Bot. Reg. 30. (Misc.): 44. 1844; Hook.f., Fl. Brit. India 5:719. 1888. *B. barbatulum* auct., non Lindl., 1830: Wight, Ic. t. 910. 1844-1866. *Anantali-maravara* Rheede, Hort. Malab. 12:15-16, t. 7. 1693.

Leaves appears before flowering, elliptic-lanceate, 6-7 by 2-2.5 cm, base sheathing, apex acute. Racemes α-flowered, to 10 cm long, terminal or from axils fo fallen leaves. Flowers creamy-white, c. 2.5 cm ϕ; bracts ovate-lanceate; sepals elliptic-oblong; lateral petals spathulate, lip 3-lobed, yellow-hairy on dorsal side, mid-lobe large, obovate, side-lobes small, falcate, mentum c. 2 mm long. Capsule fusiform.

Fl. & Fr.: Jan.-Apr. *Distr.*: India. Common. In midlands and hills. *NAK 392, 1425 & 1508* (Chittar, c. 400 m, Moozhiar, c. 250 m).

ERIA Lindley
Bot. Reg. 11:t.904. 1825. (non. cons.).

Epiphytic small herbs; pseudobulbs densely crowded, rounded, flat, reticulate. Leaves 2-4. Inflorescence racemes, usually secund; sepals and petals free, gland-dotted, lateral sepals fastened to a column- foot, forming a mentum; pollinia 8, in two groups of 4. Capsule pyriform, ribbed.

1. Flowers usually 8-10; lip panduriform, clawless. .... 1. **E. dalzellii**

1. Flowers usually 2-6; lip lanceate, clawed. .... 2. **E. nana**

1. **Eria dalzellii** (Dalz.) Lindl., J. Linn. Soc. Bot. 3:47. 1858; Fischer in Gamble, Fl. Pres. Madras 1425. 1928; Sant. & Kapad., Orch. Bombay 151. 1966; Sald. in Sald. & Nicols., Fl. Hassan Dist. 827. 1976; Manilal, Fl. Silent Valley 279. 1988. *Dendrobium dalzellii* Hook. ex Dalz., Hooker's J. Bot. Kew Gard. Misc. 4:292. 1852. **(Fig. 54)**

Pseudo-bulbs 8-10 mm . Leaves 2-4, linear-oblong or obovate, 1-6 by 0.5-1 cm, base narrowed, sheathing, apex acute, membranous. Racemes 3-9 cm long, erect, secund. Flowers pale creamy-yellow; bracts lanceate; sepals and petals triangular-ovate, lip panduriform, margin finely crispate, incurved, with 2 thick yellowish-orange ridges dorsally.

Fl. & Fr.: Sep.-Dec. *Distr.*: Endemic to Peninsular India. Rare. In shola forests. *NAK 2503* (Kakki hills, c. 1100 m).

2. **Eria nana** A. Rich., Ann. Sci. Nat. Ser. 2, 15:19. 1841; Hook.f., Fl. Brit. India 5:789. 1890; Fischer in Gamble, Fl. Pres. Madras 1425. 1928; Abraham & Vatsala, Intr. Orch. 277. 1981; Mohanan & Sivad., Fl. Agasthyamala 672.2002.

Pseudo-bulbs c. 10 mm ϕ. Leaves 2-3, sessile, oblanceate, 0.7-5 by 0.5-1.2 cm, base sheathing, apex subacute, cuspidate. Racemes 3-8 cm long, axils flexuous-zigzag. Flowers 2-6, greenish-yellow; bracts cymbiform, triangular; dorsal sepals narrowly lanceate, lateral petals shorter and narrower than sepals, lip narrowly lanceate, calwed, conduplicate, papillose exernally, claw with a swollen callus on upper surface.

Fl. & Fr.: Jul.-Oct. *Distr.*: Endemic to Peninsular India. Rare. In upper hills. *BDS 42021* (MH) (Sabarimala slopes, c. 1000 m).

EULOPHIA R. Brown ex J. Lindley
Bot. Reg. 8: t. 686. 1823 (non. cons.).

Terrestrial tall herbs; rhizome stout, pseudo-bulbous or underground tuber-like corms. Leaves several, clustered, sheathing the lower parts of the scape, plaited. Inflorescence erect, lateral, racemose. Flowers

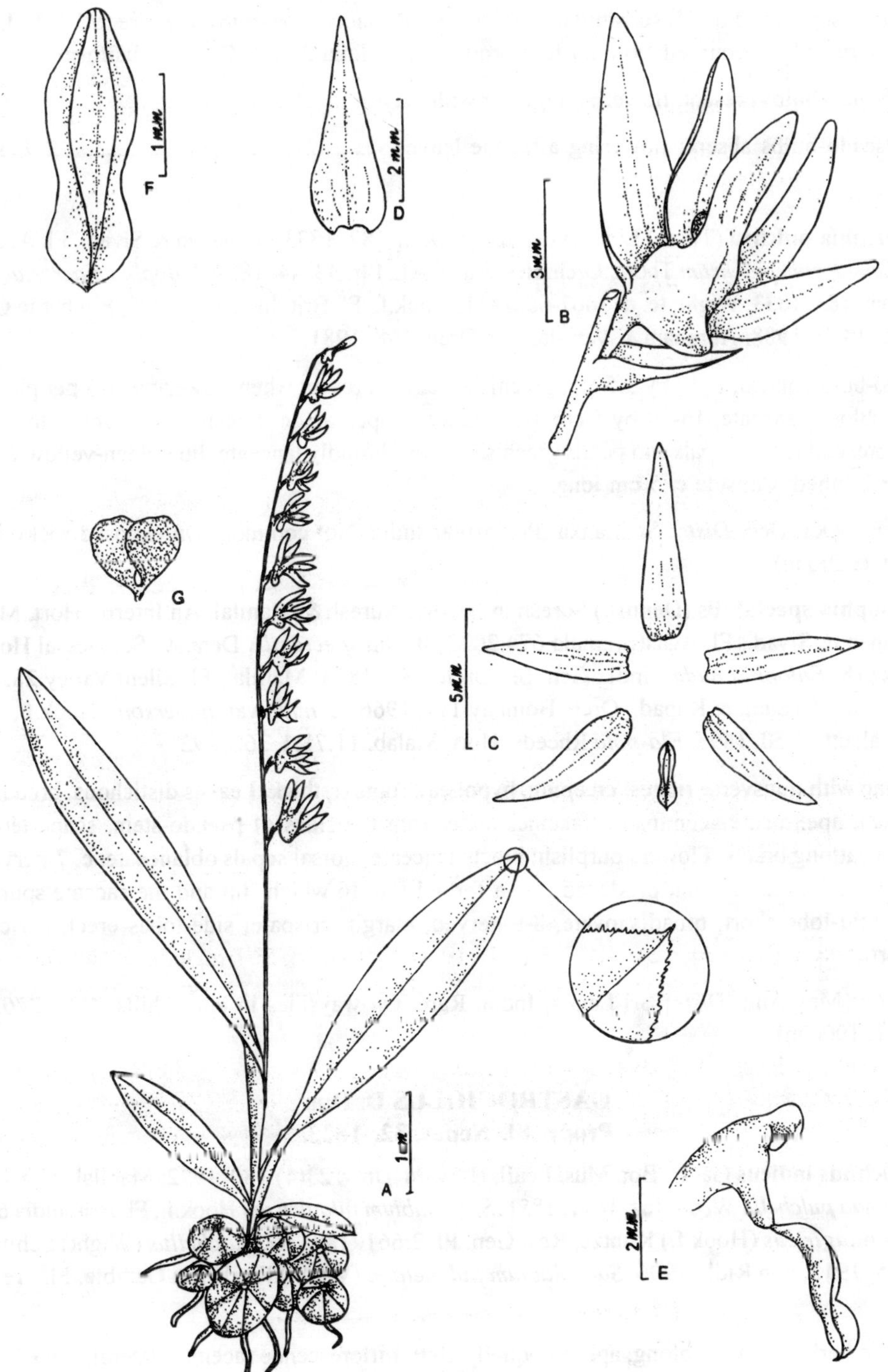

Fig. 54. ***Eria dalzellii*** **(Dalz.) Lindl.: A. Habit; B. Flower; C. Perianth parts separated; D. Bract; E. Lip with column - side view, F. Lip - front view; G. Pollinia.**

large, few to α; sepals and petals subsimilar, free, lip shortly saccate or spurred, entire to 3-lobed, mid-lobes spreading, more or less recurved; column-foot produced, pollinia 2 or 4. Capsule fusiform.

1. Pseudo-bulbs present; flowering together with leaves. .... 1. **E. pulchra**

1. Pseudo-bulbs absent; flowering after the leaves. .... 2. **E. spectabilis**

1. **Eulophia pulchra** (Thou.) Lindl., Gen. & Sp. Orch. 182. 1833; Mohanan & Sivad., Fl. Agasthyamala 674.2002. *Limodorum pulchrum* Thou., Orch. Iles. Austr. Afr. Pls. 43, 44, 1822. *Eulophia macrostachya* Lindl., Gen. Sp. Orch. 183. 1833; Wight, Ic. tt. 1667-68. 1851; Hook.f., Fl. Brit. India 6:4. 1890; Fischer in Gamble, Fl. Pres. Madras 1435. 1988; Abraham & Vatsala, Intr. Orch. 298. 1981.

Pseudo-bulb conical, c. 10 by 3.5 cm, greenish. Leaves present when flowering, 1-3 per plant, elliptic-lanceate or oblong-lanceate, 15-20 by 5 cm, base cuneate, apex acute. Racemes with scape to 70 cm long; bracts lanceate, caducous; sepals and petals greenish yellow, broadly lanceate, lip golden-yellow with reddish stripes; spur 2-lobed. Capsule c. 5 cm long.

Fl. & Fr.: Oct.-Dec. *Distr.*: Sri Lanka, Peninsular India. Not common. On exposed rocky hills. *NAK 2169* (Chiiar, c. 200 m).

2. **Eulophia spectabilis** (Dennst.) Suresh in Nicols., Suresh & Manilal, An Interpr. Hort. Malab. 300. 1988; Mohanan & Sivad., Fl. Agasthyamala 674.2002. *Wolfia spectabilis* Dennst., Schluessel Hort. Malab. 11, 25, 38. 1818. *Eulophia nuda* Lindl., Gen. Sp. Orch. 180. 1833; Manilal, Fl. Silent Valley 282. 1988. *E. andersonii* (Hook.f.) Sant. & Kapad., Orch. Bombay 117. 1966. *E. nuda* var. *andersonii* Hook.f., Ann. Roy. Bot. Gard. Calcutta 5:50. 1895. *Ela-pola* Rheede, Hort. Malab. 11:71, t. 36. 1692.

Rhizome with transverse ridges; creeping, hypogeal; roots α, thick. Leaves distichous, broadly elliptic, c. 50 by 9 cm, apex acute-acuminate. Racemes arise from the sides of pseudo-stem, scape terete, stout, covered by sheathing bracts. Flowers purplish; bracts lanceate; dorsal sepals oblanceate, c. 7-nerved; lateral petals obong, α-nerved; column produced into a forked foot to which lip and the saccate spur attached, lip 3-lobed, mid-lobe short, broadly ovate, 8-9-nerved, margin crispate; side-lobes erect, auricular, spur conical, short.

Fl. & Fr.: May-Aug. *Distr.*: Sri Lanka, India. Rare. On waysides in upper hills. *NAK 2703 & 2731* (Kakki hills, c. 1000 m).

GASTROCHILUS D. Don
Prodr. Fl. Nepal. 32. 1825.

Gastrochilus indicus Garay, Bot. Mus. Leafl. Harvard Univ. 23(4):180. 1972; Manilal, Fl. Silent Valley 284. 1988. *Vanda pulchella* Wight, Ic.t. 1671. 1851. *Saccolabium nilagiricum* Hook.f., Fl. Brit. India 6:60.1890. *Gastrochilus nilagiricus* (Hook.f.) Kuntze, Rev. Gen. Pl. 2:661. 1891. *G. pulchellus* (Wight) Schltr., Feddes Repert. 12:25. 1913, non Ridl., 1906. *Saccolabium pulchellum* (Wight) Fischer in Gamble, Fl. Pres. Madras 1446. 1928.

Epiphytic herbs. Leaves oblong, apex unequally cleft. Inflorescence racemes, lateral, often 2 at a node, sepals, petals free, sub-similar; lip saccate with transverse mid-lobe on rim of sac, sometimes hairy or fimbriate; calli or septa absent in sac; anther 2-celled; pollinia 2, globular, each with a depression; stipe narrow.

Fl. & Fr.: Oct.-Nov. *Distr.*: Western Ghats. Rare. In evergreen forests. *NAK 2866* (Sabari hills, c. 250 m).

GEODORUM G. Jackson in H.C. Andrews
Bot. Repos. 10: t.626. 1811.

Geodorrum densiflorum (Lam.) Schltr., Feddes Repert. 4:269. 1919; Fischer in Gamble, Fl. Pres.Madras 1437. 1928; Sald. in Sald. & Nicols., Fl. Hassan Dist. 831. 1976; Abraham & Vatsala, Intr. Orch. 327. 1981; Jayaw. in Dassan. & Fosberg, Rev. Handb. Fl. Ceylon 2:180. 1981; Matthew, Ill. Fl. Tam. Carnatic t. 723. 1982; Seidenfaden in Matthew, Fl. Tam. Carnatic 3(3): 1608. 1983; Manilal, Fl. Silent Valley 284. 1988; Nicols., Suresh & Manilal, An Interpr. Hort. Malab. 301. 1988; Mohanan & Sivad., Fl. Agasthyamala 675.2002. *Limodorum densiflorum* Lam., Encycl. 3:516. 1791. *Bela-pola* Rheede, Hort. Malab. 11:69-70, t. 35. 1692.

Terrestrial herbs. Leaves alternate, elliptic, 15-20 by 10-12 cm, base narrowed, sheathing, apex obtuse, cuspidate, plaited. Inflorescence racemes, lateral, to 30 cm long. Flowers dense, purplish; bracts lanceate; sepals and petals similar, lip sessile, pink, cymbiform, 3-lobed, mid-lobe 2-lobulate, side-lobes erect; column short; pollinia 2, caudicle short.

Fl. & Fr.: Mar.-Jul. *Distr.*: Sri Lanka, S. & E. India, Andaman Isls., Myanmar, Malesia, Thailand. Rare. In shade on hills. *CNM 69312* (MH) (Thekkuthode, c. 70 m).

HABENARIA Willdenow
Sp. Pl. 4(1):5, 44. 1805.

Terrestrial (rarely epiphytic as in *H. crinifera*), tuberous herbs. Leaves alternate, usually sheathing, thin-coriaceous. Inflorescence terminal, bracteate. Flowers white to greenish-white, rarely yellow, resupinate; sepals subequal, dorsal one galeate, laterals similar or dissimialr; petals divided or entire, lip spurred, 3-lobed; pollinia 2, clavate, caudicles long, enclosed in tubes; stigmatic surfaces distant. Capsule ellipsoid, mostly curved or patent, beaked.

1. Petals bipartite. .... 2. **H. digitata**
1. Petals entire.
    2. Side-lobes of lip equal or shorter than sepals.
        3. Margin of side-lobes of lip toothed. .... 5. **H. longicornu**
        3. Margin of side-lobes of lip entire.
            4. Bracts foliaceous; ovary beaked. .... 6. **H. perrottetiana**
            4. Bracts not foliaceous; ovary beakless. .... 3. **H. heyneana**
    2. Side-lobes of lip longer than sepals
        5. Mid-lobe of lip entire. .... 4. **H. longicorniculata**
        5. Mid-lobe of lip 2-furcate. .... 1. **H. crinifera**

1. **Habenaria crinifera** Lindl., Gen. Sp. Orch. 323. 1835; Hook.f., Fl. Brit. India 6:142. 1890; Fischer in Gamble, Fl. Pres. Madras 1471. 1928; Sant. & Kapad., Orch. Bombay 26.1966; Sald. in Sald. & Nicols., Fl. Hassan Dist. 832. 1976; Abraham & Vatsala, Intr. Orch.241. 1981; Jayaw. in Dassan. & Fosberg, Rev. Handb. Fl. Ceylon 2:351. 1981; Manilal, Fl. Silent Valley 285. 1988; Mohanan & Sivad., Fl. Agasthyamala 677.2002.

Leaves linear-oblong, 5-15 by 1-2 cm, base sheathing, apex acute to acuminate. Racemes to 15 cm long. Flowers white; bracts ovate, margin serrulate; dorsal sepal green, lateral sepals and petals white; lip 3-lobed, lateral lobes ending in tails, mid-lobe 2-lobulate, spur slender, to 5 cm long, flat.

Fl. & Fr.: Aug.-Oct. *Distr.*: Sri Lanka, South India. Less common. Occasionally epiphytic herbs. *NAK 838* (Ranni R.F., c. 200 m).

2. **Habenaria digitata** Lindl., Gen. Sp. Orch. 307. 1835; Hook.f., Fl. Brit. India 6:134. 1890; Fischer in Gamble, Fl. Pres. Madras 1469. 1928; Sant. & Kapad., Orch. Bombay 10. 1966; Seidenfaden in Matthew, Fl. Tam. Carnatic 3(3): 1562. 1983; Manilal, Fl. Silent Valley 286. 1988. *H. trinervia* Wight, Ic. t. 1701. 1852.

Leaves elliptic-lanceate or linear-lanceate, 3-7 by 1-2 cm, base sheathing, apex acute to shortly acuminate. Racemes to 10 cm long. Flowers greenish; bracts ovate-lanceate, foliaceous; dorsal sepal 5-nerved, lateral sepals 3-nerved, lateral petals bipartite up to the base; lip 3-lobed from base, mid-lobe longer than side-lobes, spur c. 1 cm long.

Fl. & Fr.: Aug.-Nov. *Distr.*: Deccan and Himalaya, Kashmir to Myanmar. Rare. *CNM 73473* (MH) (Chokkampatti hills, c. 800 m).

3. **Habenaria heyneana** Lindl., Gen. Sp. Orch. 320. 1835; Wight, Ic. t.923 1844; Hook.f., Fl. Brit. India 6:148. 1890; Fischer in Gamble, Fl. Pres. Madras 1471.1928; Sant. & Kapad., Orch. Bombay 32. 1966. Sald. in Sald. & Nicols., Fl. Hassan Dist. 833. 1976; Abraham & Vatsala, Intr. Orch.231.1981; Seidenfaden in Matthew, Fl. Tam. Carnatic 3(3): 1567. 1983; Manilal, Fl. Silent Valley 287. 1988. *H. glabra* A. Rich., Ann. Sci. Nat. Ser. 2, 15:75, t. 5 A. 1841; Hook.f., Fl. Brit. India 6:148. 1890.

Leaves 4 to 6, gradually transforming into flowering bracts, 2-6 by 0.5-1.2 cm, base sheathing, narrowly lanceate, apex acute. Racemes up to 10 cm long, 5-10-flowered, secund. Flowers white; bracts cymbiform, foliaceous, enclosing the ovary; sepals sub- equal, dorsal sepal 5-nerved, lateral ones 3-nerved; lateral petals falcate, united basally with the claw of lip; lip thick, 3-lobed, side-lobes shorter than mid-lobe, spur c. 12 mm long, shorter than ovary.

Fl. & Fr.: Aug.-Oct. *Distr.*: Endemic to South India. In grasslands. Not common. Locally abundant. *NAK 2510* (Ponnambalamedu, c. 1100 m).

4. **Habenaria longicorniculata** Grah., Cat. Bombay Pl.202. 1839; Sant. & Kapad., Orch. Bombay 29. 1966; Sald. in Sald. & Nicols., Fl. Hassan Dist.833. 1976; Abraham & Vatsala, Intr. Orch.238.1981; Seidenfaden in Matthew, Fl. Tam. Carnatic 3(3):1564. 1983; Manilal, Fl. Silent Valley 287. 1988. *H. longicalcarata* A. Rich., Ann. Sci. Nat. Ser. 2, 15:71, t. 38. 1841; Wight, Ic. t.925. 1844; Hook.f., Fl. Brit. India 6:141.1890; Fischer in Gamble, Fl. Pres. Madras 1470. 1928.

Leaves about 6, subradical, rarely cauline, elliptic or narrowly oblanceate, 4-14 by 1.5-2.5 cm, base sheathing, apex acute. Raceme to 6 cm long, few-flowered, usually 2. Flowers white; bracts oblanceate, most of them sterile, papillose; dorsal sepals ovate, hooded, 3-(5)- nerved, lateral sepals obliquely lanceate, spreading, 3-(5)-nerved, papillose, lateral petal spathulate, 5-nerved, lip reflexed, 3-fid up to middle, mid-lobe ligulate, ± equal to laterals; spur c.12 cm long, much longer than the ovary.

Fl. & Fr.: Sep.-Oct. *Distr.*: Peninsular India Rare. *KV 48603* (MH) (Pamba, c.1000 m).

5. **Habenaria longicornu** Lindl., Gen. Sp. Orch. 322. 1835; Hook.f., Fl. Brit. India 6:139. 1890; Fischer in Gamble, Fl. Pres. Madras 1470. 1928; Abraham & Vatsala, Intr. Orch. 229. 1981; Seidenfaden in Matthew, Fl. Tam. Carnatic 3(3): 1564. 1983; Manilal, Fl. Silent Valley 287. 1988; Mohanan & Sivad., Fl. Agasthyamala 678.2002.

Leaves 4-5, scattered, sometimes crowded towards base, lanceate, 3-8 by 0.6-2 cm, base sheathing, apex acute-mucronate. Racemes to 14 cm long. Flowers white, 3-9; bracts narrowly lanceate; dorsal sepals ovate, 5-nerved, lateral sepals longer than dorsal, ovate, 5-nerved, spreading; lateral petals shorter and much narrower than sepals, 1-nerved, lip 3-lobed, side-lobes longer and broader than mid-lobe; spur to 4 cm long, slightly exceeding the ovary.

Fl. & Fr.: Aug.-Oct. *Distr.*: Peninsular India Very rare. *CNM 58344* (MH) (Pathanamthitta hills).

6. **Habenaria perrottetiana** A. Rich., Ann. Sci. Nat. Ser. 2, 15:74, t.1841; Wight, Ic.t. 919. 1844; Hook.f., Fl. Brit. India 6:164. 1890; Fischer in Gamble, Fl. Pres. Madras 1471. 1928; Abraham & Vatsala, Intr. Orch. 236. 1981; Mohanan & Sivad., Fl. Agasthyamala 679.2002.

Leaves 5-7, broadly lanceate, 3-6 by 1-2 cm, base sheathing, amplexicaul, apex acute to acuminate. Racemes to 14 cm long. Flowers secund, yellowish-green; bracts foliaceous, cucullate, enclosing the ovary and spur, imbricating, broadly ovate; dorsal sepal ovate-oblong, 7-nerved, lateral sepals as long as or little shorter than dorsal, 4-nerved, lip shroter than mid-lobe, spur c.15 mm long, shorter than ovary.

Fl. & Fr.: Sep.-Oct. *Distr.*: Endemic to S. India. Very rare. *BDS 42002* (MH) (Uppupara).

KINGIDIUM P.F. Hunt
Kew Bull. 24: 97. 1970.

Kingidium deliciosum (Reichb. f.) Sweet, Amer. Orchid. Soc. Bull. 39: 1095. 1970; Jayaw. in Dassan. & Fosberg, Rev. Handb. Fl. cylon 2:207. 1981. *Aerides decumbens* Griff., Notul. Pl. Asiat. 3:365. 1851. *Phalaenopsis wightii* Reicho.f., Bot. Zeit. 214. 1862. *Doritis wightii* (Reich.f.) Benth., Gen. Pl.3:574. 1883. *Phalaenopsis decumbens* (Griff.) Holt., Gard. Bull. Straits Settlem. 11:286. 1947; Seidenfaden & Smitin., Orch. Thailand 544, t. 404. 1963; Abraham & Vatsala, Intr. Orch. 464. 1981.*Kingidium decumbens* (Griff.) P. F. Hunt, Kew Bull. 24:97.1970. *Kingiella decumbens* (Griff.) Rolfe, Orch. Rev.25: 197.1971; Fischer in Gamble, Fl. Pres. Madras 1439. 1928.

Epiphytic herbs. Leaves obovate to oblanceate, 4-11 by 1.5-4 cm, base acute, apex obtuse to obtusely acute, thin-chartaceous. Racemes to 18 cm long; bracts lanceate, persistent. Flowers white, marked purplish, c. 2 cm ϕ; sepals subequal, linearly elliptic-lanceate, lateral petals oblanceate, lip 3-lobed, base saccate, mid-lobe 2-fid, disc basally with a 2-fid appendage; pollinia 2, bi-partite. Capsule c. 4 cm long.

Fl. & Fr.: Aug.-Sep. *Distr.*: Sri Lanka, India, Myanmar, Malesia, Philippines, Java, Thailand. Rare. In semi-evergreen forests. *NAK 711* (Perunthenaruvi, c. 100 m). Epiphytic on Jack trees.

LIPARIS L. C. Richard
Orch. Eur. Annot. 21, 30, 38.1817;
Mem. Mus. Hist. Nat. 4:43, 52, 60.1818 (nom.cons.).

Terrestrial or epiphytic herbs. Leaves flat or plicate. Inflorscence terminal, racemose or spicate. Flowers minute, resupinate; sepals ovate-lanceate, free; petals similar to sepals but narrower, free, lip sessile on column, spurless; column usually long, curved, often winged; anther 2-celled, pollinia waxy, 4, united in pairs.

1. Terrestrial herbs; flowers purplish. .... 2. **L. nervosa**
1. Epiphytic herbs; flowers yellow or pale green.
    2. Leaves linear-lanceate, ±20 mm wide; pseudo-bulbs elongate. .... 3. **L. viridiflora**
    2. Leaves elliptic-lanceate, ± 25 mm wide; pesudo-bulbs ovate (laterally compressed). .... 1. **L. elliptica**

1. **Liparis elliptica** Wight, Ic. 5. t. 1753. 1852; Seidenfaden, Dansk Bot. Ark. 31(1):79.1976; Abraham & Vatsala, Intr. Orch. 412. 1981; Mohanan & Sivad., Fl. Agasthyamala 680.2002. *L. viridiflora* auct., non Lindl., 1830: King & Pantl., Ann. Roy. Bot. Gard.Cal. 8:35, t.47. 1898; Hook.f., Fl.Brit.India 5:704. 1890; Fischer in Gamble, Fl. Pres. Madras 1411. 1928.

Epiphytic herbs; pseudo-bulb laterally compressed. Leaves 2, elliptic-lanceate, 9-12 by 2.2-2.5 cm, apex acute. Racemes c. 12 cm long. Flowers α, greenish, ascending; bracts subulate, longer than ovary; pedicel decurved; dorsal sepal lanceate, 1-nerved, lateral sepals as long as the dorsal, lanceate, 1-nerved, lateral petal linear, much narrower than sepals; lip suborbicular, sharply infolded at the margin towards the distal end so as to appear an obscurely 3-lobed, margin crenulate.

Fl. & Fr.: Sep.-Nov. (April). *Distr.*: Sri Lanka, Peninsular & Eastern India to Java. Very rare. *NAK 2737* (Ponnambalamedu, c.1100 m).

2. **Liparis nervosa** (Thunb.) Lindl., Gen. Sp. Orch. 26. 1830; Fischer in Gamble, Fl. Pres. Madras 1410. 1928; Sant. & Kapad., Orch. Bombay 181. 1966; Abraham & Vatsala, Intr. Orch. 408.1981; Manilal, Fl. Silent Valley 291. 1988; Nicols., Suresh & Manilal, An Interpr. Hort. Malab. 301. 1988. *Ophrys nervosa* Thunb., Fl. Jap. 27.1784. *Empusa paradoxa* Lindl., Gen. Sp. Orch. 17. 1830. *Liparis paradoxa* (Lindl.) Reichb. f., Walp. Ann. 6:218. 1861; Hook.f., Fl. Brit. India 5:697. 1890. *Katou-ponnam-maravara* Rheede, Hort. Malab. 12:55, t.28.1693.

Slender terrestrial herbs. Leaves 2 or 3, narrowly elliptic-lanceate, 15-20 by 2-3.5 cm, base sheathing, apex acute to shortly acuminate. Racemes 25-35 cm long, α-flowered. Flowers pale purplish, lax, long-pedicellate; bracts ovate-lanceate; dorsal sepal linear-oblong, margins involute, lateral sepals falcate-oblong, recurved; lateral petals linear, much narrower than sepals; column short, incurved, lip deeply obcordate with 2 callosites at base.

Fl. & Fr.: Jul.-Sep. *Distr.*: Peninsular India. Rare. *Deb 30447* (MH) (Pampa, c. 1000 m).

3. **Liparis viridiflora** (Bl.) Lindl., Gen. Sp. Orch. 31.1830; Sald. in Sald. & Nicols., Fl. Hassan Dist. 835. 1976; Abraham & Vatsala, Intr. Orch. 410. 1981; Seidenfaden in Matthew, Fl. Tam. Carnatic 3(3): 1585. 1983; Manilal, Fl. Silent Valley 291. 1988; Mohanan & Sivad., Fl. Agasthyamala 680.2002. *Malaxis viridiflora* Bl., Bijdr. 392. 1825. *Liparis longipes* Lindl. in Wall., Pl. Asiat. Rar. 1: 312, t.35. 1830; Wight, Ic. t. 906. 1844; Hook.f., Fl. Brit. India 5:703. 1890; Fischer in Gamble, Fl. Pres. Madras 1411. 1928.

Epiphytic fleshy pseudo-bulbous herbs; pseudo-bulbs ovoid to cylindric, c. 8 cm long. Leaves 2, opposite, linear-oblong or oblanceate, faintly plicate, 14-24 by 1.2-2 cm, base sheathing, apex acute. Racemes to 24 cm long, α-flowered. Flowers small, pale-green; bracts lanceate; sepals sub-similar, dorsal one lanceate, revolute; petals linear, lip ovate, clawless, shallowly 3-lobed, folded lengthwise.

Fl. & Fr.: Oct.-Dec. (March) *Distr.*: Sri Lanka, Eastern India, China. Rare. *NAK 992* (Sabari hills, c. 250 m), *2358* (Kalleli, c. 330 m).

LUISIA Gaudichaud-Beaupre in Freycinet, Voyage Monde, Uranie Physicienne Bot. 426. 1829 ('1826').

Luisia zeylanica Lindl., Fl. Orchid. 3. 1853; Seidenfaden, Dansk Bot. Ark. 27(4): 62. 1971; Sald. in Sald. & Nicols., Fl. Hassan Dist. 836. 1976; Abraham & Vatsala, Intr. Orch. 486. 1981; Seidenfaden in Matthew, Fl.

Tam. Carnatic 3(3): 1595. 1983; Manilal, Fl. Silent Valley 292. 1988; Mohanan & Sivad., Fl. Agasthyamala 681.2002. *Luisia teretifolia* auct., non Gaud., 1826: Fischer in Gamble, Fl. Pres. Madras 1438. 1928; Sant. & Kapad., Orch. Bombay 213. 1966.

Epiphytic herbs. Leaves terete, 4-11 cm long, base sheathing. Racemes condensed, extra-axillary. Flowers 2-4, greenish-pink; bracts small, oblong; lateral sepals ovate, dorsal sepals ovate-lanceate, lateral petals slightly longer than sepals, sparsely gland-dotted, lip 3-lobed, with dark maroon horizontal bands, mid-lobe larger than side-lobes, 3-lobulate. Fruit winged.

Fl. & Fr.: May.-Jun. *Distr.*: Sri Lanka through Deccan to Sikkim. Common in semi-evergreen forests. *NAK 1581* (Velithode, c. 250 m).

MALAXIS Solander ex O. Swartz
Prodr. 119. 1788.

Terrestrial herbs, often pseudo-bulbous. Leaves 2-a, membranous. Inflroescence terminal, racemose. Flowers small, not resupinate; sepals and petals free, often recurved or incurved; lip sessile; not supurred, 1-3-lobed, mid-lobe incised; column without foot; anther 2-celled, pollinia 4, in pairs.

1. Plants with pseudo-bulbs; leaves 2-3; lip laciniate. .... 2. **M. densiflora**
1. Plants without pseudo-bulbs; leaves 4-5; lip 2-lobed. .... 1. **M. acuminata**

1. **Malaxis acuminata** D.Don, Prodr. Fl. Nepal. 29. 1825; Rathakr., Bull. Bot. Surv. India 13(1): 4.1971; Abraham & Vatsala, Intr. Orch. 486. 1981; Seidenfaden in Matthew, Fl. Tam. Carnatic 3(3): 1582. 1983; Manilal, Fl. Silent Valley 292. 1988; Mohanan & Sivad., Fl. Agasthyamala 682.2002. *Microstylis wallichii* Lindl., Gen. Sp. Orch. 20. 1830; Fischer in Gamble, Fl. Pres. Madras 1407. 1928.

Non-pseudo-bulbous herbs, basal portion covered with 1 or 2 sheathing bracts and bases of the leaves. Leaves 3 or 4, elliptic-lanceate, 5-12 by 1.3-3.7 cm, base sheathing, apex acute to acuminate. Raceme terminal, to 20 cm long, scape sulcate. Flowers greenish-yellow; bracts lanceate, acuminate, dorsal sepals oblong-lanceate, lateral sepals ovate, obtuse; lateral petals linear, margins revolute, lip sagittate, deeply forked at the tip.

Fl. & Fr.: Sep.-Nov. *Distr.*: India to Philippines. Rare, but locally abundant. In grasslands. *NAK 2487* (Pampa dam area, c. 1100 m).

2. **Malaxis densiflora** (A. Rich.) O.Kuntze, Rev. Gen. Pl. 2:673. 1891; Sant. & Kapad., Orch. Bombay 144. 1966. *Liparis densiflora* A. Rich., Ann. Sci. Ser. 2, 15:18, t. 18. 1841. *Microstylis versicolor* Wight, Ic. t. 901. 1844, non. Lindl., 1830; Hook.f., Fl. Brit. India 5:691. 1890. *M. luteola* Wight, Ic. t. 1632. 1852. *M. densiflora* (A. Rich.) Fischer in Gamble, Fl. Pres. Madras 1408. 1928.

Pseudo-bulbous herbs. Leaves 2, ovate-lanceate to elliptic, 4-15 by 1-3 cm, base sheathing, apex acute, plicate. Racemes 6-25 cm long; bracts lanceate. Flowers reddish, α, dense; dorsal sepal lanceate, deflexed, lateral sepals obliquely oblong, deflexed; lateral petals linear-lanceate, margins curved, held like horns, lip rhomboid, pectinate, mid-lobe shorter than laterals.

Fl. & Fr.: Jun.-Aug. *Distr.*: Sri Lanka to Peninsular India. Rare. In grasslands. *BDS 42008* (Uppupara, c. 1100 m).

NERVILIA Commerson ex Gaudichaud-Beaupre in Freycinet, Voyage Monde Bot. 421. 1829 (nom. cons.).

Nervilia aragoana Gaud. in Freycinet, Voy. Bot. 422, t.35. 1829; Fischer in Gamble, Fl. Pres. Madras 1459. 1928; Sant. & Kapad., Orch. Bombay 134. 1966; Seidenfaden, Dansk Bot. Ark. 32(2): 164. 1978; Abraham & Vatsala, Intr. Orch. 430. 1981. *Pogonia flabelliformis* Lindl., Gen. Sp. Orch.41. 1840; Hook.f., Fl. Brit. India 6:121. 1890. *P. carinata* Lindl., Gen. Sp. Orch. 144. 1840; Wight, Ic. t. 1720. 1851; Hook. f., Fl.Brit. India 6:121. 1890. *P. scottii* Reichb.f., Flora 55:276. 1872; Hookf., Fl. Brit. India 6:121. 1890.; King & Pantl., Ann. Roy. Bot. Gard. Calcutta 8: 269, t. 360. 1898.

Tuberous herbs. Leaf solitary, appearing after flowers, long-peduncled; petiole to 16 cm long. Lamina cordiform, 5-7 by 5.7-6.7 cm, margin undulate, apex acuminate, often purplish. Inflorescence a raceme, 10-30 cm long. Flowers green, pendulous, bracts deflexed, linear-lanceate, 3-nerved; sepals subequal, linear, dorsal sepal linear-oblong, faintly 3-nerved, lateral petals linear-oblanceate, 3-nerved, lip 3-lobed, mid-lobe larger, margin crenulate.

Fl. & Fr.: May-Jun. *Distr.*: India to Myanmar. Rare. Among forest undergrowth in shades. *BINU 2630* (TBGT) (Ranni, c. 150 m).

OBERONIA Lindley Gen. Sp. Orchid. Pl. 15.1830 (non. cons.).

Small, erect or pendant epiphytic stemless herbs. Leaves laterally compressed, equitant, thick. Inflorescence terminal with a minute flowers; spur 0; pollinia 4 in 2 pairs. Capsule ellipsoid.

1. Small herbs (largest leaf c. 2 cm long); flowers sunken in pits on the rachis; lip unlobed. .... 3. **O. proudlockii**

1. Comparatively large herbs (largest leaf 10-25 cm long.); flowers not sunken in the rachis; lip 3-lobed.
   2. Flowers in a continous imbricate spike; scape flat.
      3. Mid-lobe of lip larger than side-lobes, 3-lobulate. .... 2. **O. platycaulon**
      3. Mid-lobe of lip as large as side-lobes or slightly broader, 2-lobulate. .... 1. **O. brunoniana**
   2. Flowers in spaced, verticillate or subverticillate whorls; scape terete. .... 4. **O. verticillata**

1. **Oberonia brunoniana** Wight, Ic. t. 1622, 1851; Hook.f., Fl. Brit. India 5: 681. 1888; Fischer in Gamble, Fl. Pres. Madras 1406. 1928; Sant. & Kapad., Orch. Bombay 69. 1966; Seidenfaden, Dansk Bot. Ark. 25(3): 50. 1968; Sald. in Sald. & Nicols., Fl. Hassan Dist. 838. 1976; Abraham &Vatsala, Intr. Orch. 422. 1981; Seidenfaden in Matthew, Fl. Tam. Carnatic 3(3): 1580. 1983; Mohanan & Sivad., Fl. Agasthyamala 685.2002.

Leaves oblong-lanceate, falcate, 2-9 by 0.6-1.2 cm, acute. Flowers brownish, long pedicelled; bracts lanceate, sepals subequal, ovate, hyaline, sparsely gland-dotted; lateral petals linear, hyaline, obtuse, lip auricular, orbiculate, mid-lobe 2- lobulate, lobules irregularly crenulate. Capsule oblong.

Fl. & Fr.: Sep.-Oct. *Distr.*: Peninsular India. Rare. *NAK 250* (Pampa dam area, c. 1100 m, Epiphytic on *Ficus hispida* L.f.).

2. **Oberonia platycaulon** Wight, Ic. t. 1623. 1851; Hook.f., Fl. Brit. India 5:682. 1888; Fischer in Gamble, Fl. Pres. Madras 1407. 1928; Sant. & Kapad., Orch. Bombay 70, t.15 A-B. 1966; Mohanan & Sivad., Fl. Agasthyamala 686.2002.

Leaves ensiform,1.5-19 by 0.8-1.6 cm, acute; scape flattened, subtended by falcate bracts. Flowers pale orange; bracts ovate; sepals ovate-lanceate; lateral petals linear, lip 3-lobed, mid-lobe larger, 3-lobulate. Capsule oblong-ovoid.

Fl. & Fr.: Sep.-Mar. *Distr.*: S. India. Very rare. *KV 46352* (MH) (Pamba to Anathode, c. 1100 m).

3. **Oberonia produlockii** King & Pantl., J.Asiat. Soc. Bengal 66:580. 1896; Fischer in Gamble, Fl. Pres. Madras 1406. 1928; Seidenfaden, Dansk Bot. Ark. 25(3): 22. 1968; Abraham & Vatsala, Intr. Orch. 422. 1981; Seidenfaden in Matthew, Fl. Tam. Carnatic 3(3): 1578. 1989.

Leaves ovate to oblong-lanceate, 0.7-2.5 by 0.2-0.8 cm, ensiform, acute. Spike 3-7 cm long, scape fleshy, flattened. Flowers white or yellow, very small, sunk in pits in rachis; bracts lanceate; sepals subequal, dorsal sepals ovate, gland-dotted, reflexed, laterals obiquely ovate; lateral petals oblong-lanceate, gland-dotted, lip orbicular, margin crenulate-denticulate. Capsule obovoid.

Fl. & Fr.: Sep.-Feb. *Distr.*: S. India.Very rare. *PR 78009* (MH) Gavi dam area.

4. **Oberonia verticillata** Wight, Ic.t. 1626. 1851; Hook.f., Fl. Brit. India 5:677. 1888; Fischer in Gamble, Fl. Pres. Madras 1406. 1928; Sant. & Kapad., Orch. Bombay 66, t.14 A-B. 1966.

Leaves linear-lanceate, ensiform, 1-10 by 0.4-0.8 cm, acute, greenish. Inflorescence to 15 cm long. Flowers yellow in verticils; bracts lanceate; dorsal sepals ovate-oblong, concave, lateral sepals broadly ovate; lateral petals oblong, obtuse; lip larger, flabellate in outline, auriculate at base, deeply 2-lobed at apex. Capsule cylindric-ellipsoid.

Fl. & Fr.: Sep.-Dec. *Distr.*: S. India. Rare. *NAK 2767* (Kakki hills, c.1100 m).

PECTEILIS Rafinesque
Fl. Tell. 2:38. 1837.

Pecteilis gigantea (J.E.Sm.) Rafin., Fl. Tell. 2:1837; Seidenfaden, Dansk Bot. Ark. 31(3):22 1977; Abraham & Vatsala, Intr. Orch. 252. 1981; Manilal, Fl. Silent Valley 299. 1988; Mohanan & Sivad., Fl. Agasthyamala 689.2002. *Orchis gigantea* J.E.Sm., Exot. Bot.2:79, t. 100. 1805. *Platanthera susannae* auct., non (L.) Lindl., 1855: Wight, Ic. t. 920. 1844; Fischer in Gamble, Fl. Pres. Madras 1475. 1928; Sant. & Kapad., Orch. Bombay 426. 1966. *Habenaria susannae* auct., non (L.) R. Br. ex Sprengel., 1826: Hook.f., Fl. Brit. India 6:137. 1890.

Tall robust herbs. Leaves alternate, gradually transforming into bracts, ovate-oblong to oblong-lanceate, 5-15 by 2-6 cm, basally sheathing, apex acute or acuminate, coriaceous. Flowers white, 5-7 cm f, fragrant, in terminal spikes; bracts oblong-lanceate; sepals unequal, dorsal sepal very broad, obovate-rhomboid, margins revolute, 5-nerved, lateral sepals obliquely oblong, spreading; lateral petals linear-oblong, much narrower than sepals, lip flabellate, 4 by 9 cm, 3-lobed, side-lobes pectinate; spur twice as long as ovary, to 11 cm long.

Fl. & Fr.: Sep.-Nov. *Distr.:* Trop. Himalayas, Myanmar, Malesia, Thailand. Rare. In grasslands. *KV 48356* (MH) (Pamba-Anathode, c. 100 m).

PERISTYLUS Blume
Bijdr. 6: t. 1, f.30; 8: 404. 1825 (nom. cons.).

Terrestrial tuberous herbs. Leaves scattered or clustered at about middle. Inflorescence spicate or racemose, terminal. Flowers greenish-white, inconspicuous; bracts lanceate, often persistent; dorsal sepal often galeate; lip 1-3-lobed; column short, flanked by staminodes; pollinia 2. Capsule fusiform.

1. Leaves 2-4, clustered on ± the middle of stem, oblong-elliptic.
   2. Plants to 90 cm tall; leaves broadly ovate-elliptic or obovate; spur shorter than sepals. .... 3. **P. goodyeroides**
   2. Plants to 25 cm tall; leaves elliptic-lanceate; spur as long as sepals. .... 1. **P. aristatus**
1. Leaves usually 6, confined to lower half of stem. .... 2. **P. densus**

1. **Peristylus aristatus** Lindl., Gen. Sp. Orch.300. 1835; Fischer in Gamble, Fl. Pres. Madras 1474. 1928; Sald. in Sald. & Nicols., Fl. Hassan Dist. 841. 1976; Seidenfaden, Dansk Bot. Ark. 31(3): 35.1977; Abraham & Vatsala, Intr. Orch. 246. 1981; Mohanan & Sivad., Fl. Agasthyamala 690.2002. *Habenaria aristata* (Lindl.) Hook.f. in Trimen, Cat. 91. 1885, p.p.; Trimen, Handb. Fl. Ceylon 4:233. 1898; Hook.f., Fl. Brit. India 5:158. 1890, non King & Pantl., 1898.

Slender herbs. Leaves 3-4, clustered about the middle of stem, narrowly elliptic-lanceate, 8-10 by 1.5-2 cm, sparsely pubescent, membranous. Raceme long-peduncled, few-flowered. Flowers greenish-white, c. 5 mm ϕ; sepals subequal, linear-oblong; petals elliptic-lanceate, lip 3-lobed from the middle, mid-lobe longer than side-lobes; spur as long as sepals.

Fl. & Fr.: Jul.-Aug. *Distr.*: Sri Lanka, India. Not common. On forest edges. *BDS 42015*(MH) (Sabarimala, c. 250 m).

2. **Peristylus densus** (Lindl.) Sant. & Kapad., Orch. Bombay 46. 1966; Abraham & Vatsala, Intr. Orch. 243. 1981; Mohanan & Sivad., Fl. Agasthyamala 690.2002. *Coeloglossum densum* Lindl., Gen. Sp. Orch. 302. 1835. *Habenaria stenostachya* Benth. Fl. Hong. 362.1861; Hook.f., Fl. Brit. India 6:156. 1890. *Peristylus stenostachyus* (Lindl.) Kraenz., Orchid Gen. Sp. 1:502. 1898; Fischer in Gamble, Fl. Pres. Madras 1474. 1928.

Herbs to 40 cm tall. Leaves about 6, confined to lower half of the stem, linearly-elliptic-oblong, 3-6 by 1 cm, base sheathing, apex subacute, mucronate, chartaceous. Raceme slender, lax-flowered; bract lanceate, 9 by 2 mm. Flowers greenish-yellow; sepals subequal; petals larger than sepals, triangular-ovate,lip 3-lobed, lateral lobe linear, incurved, mid-lobe short, triangular, spur shorter than ovary.

Fl. & Fr.: Sep.-Nov. *Distr.:* Peninsular & Eastern India to Myanmar. Rare. In grasslands. *NAK 2530* (Kakki hills, c. 1100 m).

3. **Peristylus goodyeroides** (D. Don) Lindl., Gen. Sp. Orch. 299. 1835; Fischer in Gamble, Fl. Pres. Madras 1475. 1928; Sant. & Kapad., Orch. Bombay 52. 1966; Sald. in Sald. & Nicols., Fl. Hassan Dist. 841. 1976; Seidenfaden, Dansk Bot. Ark. 31(3):53. 1977; Abraham & Vatsala, Intr. Orch. 249. 1981; Manilal, Fl. Silent Valley 300. 1988; Mohanan & Sivad., Fl. Agasthyamala 691.2002. *Habenaria goodyeroides* D. Don, Prodr. Fl. Nepal. 25. 1825; Hook.f., Fl. Brit. India 6:161. 1980.

Herbs, to 1 m tall. Leaves 4-6, clustered more or less at the middle of stem, ovate, elliptic-obovate, 7-19 by 4-6 cm, base sheathing, apex acute. Inflorescence a long spike, α-flowered, dense; bracts lanceate;

papillose, persistent. Flowers pale yellow, fragrant; sepals subequal, ovate-oblong, 1-nerved, sparsely papillose; lateral petals broadly ovate, ventricose towards base, lip 3-lobed, mid-lobe longer than side-lobes; spur globose.

Fl. & Fr.: Jul.-Nov. *Distr.*: Sri Lanka, India, China, Malesia. Rare. On moist shady grass slopes. *CNM 69320* (MH) (Thannithode, c. 70 m).

PHOLIDOTA Lindley ex W. J. Hooker
Exot. Fl. 2: 138.1825.

Pholidota pallida Lindl., Edward's Bot. Reg. Sub. t. 1777. 1836; Sald. in Sald. & Nicols., Fl. Hassan Dist. 843. 1976; Abraham & Vatsala, Intr. Orch. 289. 1981; Matthew, Ill. Fl. Tam. Carnatic t. 704. 1982; Seidenfaden in Matthew, Fl. Tam. Carnatic 3(3): 1574. 1983; Manilal, Fl. Silent Valley 301. 1988; Nicols., Suresh & Manilal, An Interpr. Hort. Malab. 302. 1988. *P. imbricata* sensu Lindl., Bot. Reg. 14:t. 1213. 1828, non Lindl., 1825; Wight, Ic. t. 9071845; Hook.f., Fl. Brit. India 5:845. 1980; King & Pantl., Ann. Roy. Bot. Gard. (Calcutta) 8:t.201. 1898; Fischer in Gamble, Fl. Pres. Madras 1431. 1928; Sant. & Kapad., Orch. Bombay 178. 1966. *Welia-theka-maravara* Rheede, Hort. Malab. 12:47-48, t.24.1693.

Epiphytic pseudo-bulbous herbs; pseudo-bulbs in clusters, oblong-conical, 2-6 cm long. Leaves 1-2, elliptic-oblanceate, 14-30 by 3.2-7 cm, base attenuate, apex acute, strongly 3-nerved, coriaceous. Inflorescence racemes, 25-30 cm long, pendulous. Flowers white, hidden by imbricating bifarious foliaceous persistent bracts; sepals subequal, 3-nerved, dorsal sepal broadly ovate, laterals cymbiform, with a dorsal wing; lateral petals falcate-linear, 1-nerved, lip deeply saccate, sac longitudinally 3-ridged within, 3-lobed, mid-lobe 2-lobulate. Capsule ellipsoid.

Fl. & Fr.: Sep.-Jan. *Distr.*: Sri Lanka, India to Thailand, Malaya & Indo-China. Common. In all types of forests. *NAK 50, 195, 766.* (Almost all forests).

PODOCHILUS Blume
Bijdr. 6:t.5, f.12; 7:295. 1825.

Podochilus malabaricus Wight., Ic. t. 1748. 1852; Hook.f., Fl. Brit. India 6:80. 1890; Trimen, Handb. Fl. Ceylon 4: 206. 1898; Jayaw. in Dassan. & Fosberg, Rev. Handb. Fl. Ceylon 2:278. 1981. *Podochilus falcatus* Lindl., Gen. Sp. Orch. 234. 1833 var. *angustatus* Thw., nomen.; Fischer in Gamble, Fl. Pres. Madras 1450. 1928; Abraham & Vatsala, Intr. Orch. 390. 1981.

Epiphytic tufted herbs. Leaves α, (c. 40), distichous, imbricating, subfalcate, c. 1 cm long, flat, laterally compressed, apex obtuse, fleshy. Inflorescence of slender terminal racemes; bracts lanceate, persistent. Flowers white with violet tinge, minute, lateral sepals adnate to the foot or column and forming a mentum; lateral petals linear; lip small tongue-shaped, constricted near middle; column short; pollinia 4, club-form.

Fl. & Fr.: Sep.-Jan. *Distr.*: Southern Western Ghats & Sri Lanka. Rare. In riverine forests. *NAK 1833 & 2046* (Angamoozhy, c. 250 m).

POLYSTACHYA W. J. Hooker
Exot. Fl.2: t. 103. 1824 (nom. cons.).

Polystachya concreta (Jacq.) Garay & Sweet, Orchideol. 9(3): 206. 1974; Jayaw. in Dassan. & Fosberg, Rev. Handb. Fl. Ceylon 2:271. 1981; Matthew, Ill. Fl. Tam. Carantic t. 712.1982; Seidenfaden in Matthew, Fl.

Tam. Carnatic 3(3). 1592. 1983; Manilal, Fl. Sielnt Valley 301. 1988; Mohanan & Sivad., Fl. Agasthyamala 692.2002. *Epidendrum concretum* Jacq., Enum. Pl. Carib. 30. 1760. *P. flavescens* (Bl.) J.J.Sm., Fl. Buitenz. 6:284. 1905; Sant. & Kapad., Orch. Bombay 392. 1966; Sald. in Sald. & Nicols., Fl. Hassan Dist. 844. 1976; Abraham & Vatsala, Intr. Orch.392. 1981.

Epiphytic pseudo-bulbous herbs; pseudo-bulbs obovoid, serially arranged; stem short. Leaves 4 or 5, alternate, distichous, oblong-elliptic to lanceate, 6-25 by 1.2-3.2 cm, base narrowed, sheathing, apex obtuse-acute, thin-coriaceous. Inflorescence of panicles to 20 cm long, with 2 or 3 tubular sheaths. Flowers pale brown, non-resupinate; bracts lanceate, persistent; dorsal sepal ovate, lateral sepals broadly ovate, base oblique, apex recurved; lateral petals linear-oblanceate, lip 3-lobed, concealed by lateral sepals, mid-lobe crenulate; column short, prolonged into short tube.

Fl. & Fr.: Sep.-Jan. *Distr.:* Pantropical. Rare. In evergreen forests. *PR 78012* (MH) (Gavi dam area,) *KV 48395* (MH) (Pachakkanam, c. 1000 m).

RHYNCHOSTYLIS Blume
Bijdr. 6:t.49; 8:285 ('Rynchostylis') 434. 1825.

Rhynchostylis retusa (L.) Bl., Bijdr. 286, t.49. 1825; Hook.f., Fl. Brit. India 6:32. 1890; King & Pantl., Ann. Roy. Bot. Gard. (Calcutta) 8:t.284. 1825; Fischer in Gamble, Fl. Pres. Madras 1440. 1928; Sant. & Kapad., Orch. Bombay 211. 1966; Abraham & Vatsala, Intr. Orch. 454. 1981; Jayaw. in Dassan. & Fosberg, Rev. Handb. Fl. Ceylon 2:199. 1981; Seidenfadn in Matthew, Fl. Tam. Carnatic 3(3): 1597. 1983; Nicols., Suresh & Manilal, An Interpr. Hort. Malab. 12:1-4, t.1.1693. *Biti-maram-maravara* Rheede, Hort. Malab. 12:1-4, t. 1.1693.

Epiphytic stout herbs. Leaves closely alternate, distichous, linear, 15-19 by1.5-3 cm, base sheathing, articulate, apex praemorse, thick-fleshy. Inflorescence of long drooping dense, axillary racemes, to 35 cm long. Flowers pinkish, c. 2.5 cm ϕ; bracts ovate-lanceate, persistent; sepals subequal, dorsal sepal suborbicular, 5-7- nerved; lateral petals ovate-elliptic, lip spurred, bent inwards, apex deeply cleft, spur laterally compressed. Capsule cylindric-obovoid, faintly ridged.

Fl. & Fr.: Apr.-Sep. *Distr.*: Sri Lanka, India, Malesia. Less common. A beautiful orchid with thick persistent leaf-bases. *NAK 2874* (Ranni R.F., c. 100 m).

SARCANTHUS Lindley
Coll. Bot. t. 39 B. 1825.

Sarcanthus pauciflorus Wight, Ic. t. 1747. 1852; Abraham & Vatsala, Intr. Orch. 471. 1981. *Sarcanthus peninsularis* Dalz. in Hook., Kew J. Bot. 3:343. 1851; Hook.f., Fl. Brit. India 6:67. 1890; Fischer in Gamble, Fl. Pres. Madras 1447. 1928; Sant. & Kapad., Orch. Bombay 230. 32. 1966.

Epiphytic pendant herbs; branches half-terete. Leaves distichous, 3-quetrous, linear, 10-14 by 0.6-0.7 cm, base narrowed, sheathing, apex subcylindric, acute, thick-coriaceous. Inflorescence racemes, to 7 cm long, axillary. Flowers white with violet streaks; bracts subulate; sepals subequal, broadly elliptic, 3-nerved; lateral petals similar, elliptic, 3-nerved, lip thick-fleshy, 3-lobed, side-lobes small, spur channelled below, longitudinally septate inside.

Fl. & Fr.: Sep.-Mar. *Distr.*: Sri Lanka, Southern Western Ghats. Less common. In semi-evergreen forests. *NAK 1582* (Perunthenaruvi, c. 100 m).

SATYRIUM Swartz
Kongl. Vetensk. Acad. Nya Handl.
21:214. 1800 (nom. cons.).

Satyrium nepalense D. Don, Prodr. Fl. Nepal. 26.1825; Wight, Ic. t. 929. 1844-1845; Hook.f., Fl. Brit. India 6:168. 1890; Fischer in Gamble, Fl. Pres. Madras 1476. 1928; Jayaw. in Dassan. & Fosberg, Rev. Handb. Fl. Ceylon 2:383. 1981; Mohanan & Sivad., Fl. Agasthyamala 694.2002. *S. neplalense* var. *wightiana* (Lindley) Hook.f., Fl. Brit. India 6:168.1890.

Terestrial herbs, to 40 cm tall. Leaves 2 or 3, ovate-lanceate, ovate-oblong or suborbicular, 5-13 by 2-7 cm, base rounded, sheathing, apex acute, coriaceous. Inflorescence of dense spikes; bracts foliaceous, oblong-lanceate, concealing the flowers. Flowers purplish; sepals subequal, spreading; lateral petals narrowly ovate-lanceate; lip hooded, spur 2, as long as ovary; column erect; pollinia 2, caudicles recurved; stigma broad, 2-lipped.

Fl. & Fr.: Aug.-Sep. *Distr.*: Sri Lanka, Peninsular & Eastern India, Bhutan and Myanmar. Rare. In upper hills. *BDS 42005* (MH) (Uppupara, c. 1000 m).

SIRHOOKERA O. Kuntze
Rev. Gen. 2:681. 1891.

Sirhookera lanceolata (Wight) Ktze., Rev. Gen. Pl. 681. 1891; Sant. & Kapad., Orch. Bombay 163. 1966; Sald. in Sald. & Nicols., Fl. Hassan Dist. 849. 1976; Abraham & Vatsala, Intr. Orch. 258. 1981; Manilal, Fl. Silent Valley 305. 1988; Mohanan & Sivad., Fl. Agasthyamala 695.2002. *Josephia lanceolata* Wight, Ic. t. 1743. 1851; Hook.f., Fl. Brit. India 5:823. 1890; Fischer in Gamble, Fl. Pres. Madras 1428. 1928.

Epiphytic small herbs. Leaves radical, linear-oblong or lanceate, 5-12 by 2-4.5 cm, base sheathing, apex acute, lower surface purplish, coriaceous. Inflorescence a lax panicle. Flowers white, tinged with purple; sepals subequal, elliptic, concave; lateral petals narrowly elliptic, lip erect, fleshy, 3-lobed, side-lobes small, mid-lobe papillose with a basal callus; column as long as sepals, 2-keeled upwards; pollinia 4, oblong.

Fl. & Fr.: Aug.-Nov. *Distr.:* South India. Rare. In evergreen forests. *KV 46604* (MH) (Anathode, c. 100 m).

SMITHSONIA Saldanha
J. Bombay Nat. Hist. Soc. 71:73. 1974.

Smithsonia straminea (Dalz.) Saldanha, J. Bombay Nat. Hist. Soc. 71:73. 1974 & in Sald. & Nicols., Fl. Hassan Dist. 850. 1976.

Very small epiphytes. Leaves 4-5 cm long, retuse at apex. Inflorescence a lateral short raceme. Flowers few, yellow with reddish spots; sepals and petals anisomophic; lip sessile, 3-lobed, spurred, mid-lobe of lip curved, ± covering spur, spur pointed, bent upwards; column foot 0; pollina 2. Capsule narrowly ellipsoid.

Fl. & Fr.: Sep.-Dec. *Distr.:* Western Ghats. Not common. In evergreen forests. *NAK 1711* (Sabarimala, c. 250 m).

VANDA W. Jones ex R. Brown
Bot. Reg. 6: t. 506. 1820.

Vanda testacea (Lindl.) Reichb. f., Gard. Chron. 2:166. 1877; Sant. & Kapad., Orch. Bombay 219. 1966; Sald. & Nicols., Fl. Hassan Dist. 854. 1976; Abraham & Vatsala, Intr. Orch. 438. 1981; Seidenfaden in Matthew, Fl. Tam. Carnatic 3(3): 1599. 1983; Manilal, Fl. Silent Valley 307. 1988; Mohanan & Sivad., Fl. Agasthyamala 699.2002. *Aerides testacea* Lindley, Gen. Sp. Orchid. Pl. 238. 1835 ('*testaceum*'). *Vanda parviflora* Lindl., Bot. Reg. 30. Misc. 57. 1844; Hook.f., Fl. Brit. India 6:50 1890; King & Pantl., Ann. Roy. Bot. Gard. (Calcutta) 8 : t.286. 1898; Fischer in Gamble, Fl. Pres. Madras 1444. 1928; Jayaw. in Dassan. & Fosberg, Rev. Handb. Fl. Ceylon 2:215. 1981.

Epiphytic herbs, to 20 cm tall. Leaves alternate, distichous, linear, 2-20 by 0.6-1 cm, base sheathing, articulate, apex praemorse, coriaceous. Inflorescence racemes, 2 or more per-plant, axillary, erect, to 12 cm long (fruiting axis to 40 cm long); bracts small, ovate. Flowers yellow, c. 1 cm ϕ; sepals subequal, dorsal sepal oblong-oblanceate, lateral sepals obovate; lateral petals spathulate-oblanceate; lip pink, 3-lobed, spurred, side-lobes erect, mid-lobe ligulate, 2-ridged on the dorsal side, warty, apex truncate or retuse; spur short. Capsule oblong-fusiform, c. 6 cm long, winged.

Fl. & Fr.: Mar.-Apr. *Distr.*: Sri Lanka to Himalaya and eastwards to Myanmar and Thailand. Occasional. In teak forests. *NAK 1604* (Perunthenaruvi, c. 100 m), *2342* (Kalleli, c. 200 m).

ZEUXINE Lindley
Collect. Bot. Append. (n. 18). 1826 ('*Zeuxina*'); corr. J. Roeper, Linnaea 2:528. 1827 (Orth. *et* nom. cons.).

Zeuxine longilabris (Lindl.) Benth. ex Hook.f., Gen. Pl. 3:600. 1883; Hook.f., Fl. Brit. India 6:107. 1890; Fischer in Gamble, Fl. Pres. Madras 1456. 1928; Sald. in Sald. & Nicols., Fl. Hassan Dist. 855. 1976; Seidenfaden, Dansk Bot. Ark. 32(2): 91.1978 Abraham & Vatsala, Intr. Orch. 202.1981; Seidenfaden in Matthew, Fl. Tam. Carnatic 3(3): 1558. 1938; Manilal, Fl. Silent Valley 309. 1988; Mohanan & Sivad., Fl. Agasthyamala 703.2002. *Monochilus longilabris* Lindl., Gen. Sp. Orch. 487. 1840.

Saprophytic, slender, rhizomatous herbs, to 15 cm tall; rhizome swollen at node with thick roots. Leaves 3-4, ovate, 1.5-3.5 by 0.7-1.2 cm, base and apex acute, membranous, pinkish when dry. Inflorescence racemes, to 15 cm long, glandular-pubescent; bracts lanceate. Flowers white, c.1 cm ϕ; sepals subequal, oblanceate, dorsal sepal connate with the free lateral sepals; lateral petals falcate, connate at base, lip deeply lobed; lobes coarsely toothed on margin; pollinia 2.

Fl.& Fr.: Feb.-Mar. *Distr.*: Sri Lanka through Deccan, Assam & Lower Bengal, to Thailand & Cambodia. Rare. In moist forest shades. *NAK 1304* (Pampa dam area, c. 1000 m).

124. ZINGIBERACEAE

1. Inflorescence on a leafy shoot.
   2. Inflorescence racememose to paniculate.
      3. Stems slender; anthers with 2-fid wings; ovary 1-locular. .... **Globba**
      3. Stem robust; anthers without wings; ovary 3-locular. .... **Alpinia**
   2. Inflorescence spicate, congested; anthers spurred at base. .... **Curcuma**
1. Inflorescence on a leafless scape, associated with leafy stems.

4. Flowers in dense cone-like spikes; bracts usually bearing 1 flower. ....**Zingiber**

4. Flowers in elongate flexuose panicle; bracts usually bearing 6-7 flowers. ....**Elettaria**

ALPINIA Roxburgh
Asiat. Res. 11:350. 1810 (nom. cons.).

Perennial stout herbs. Flowers in racemes or panicles; bracteoles large; bracts 1-or 2-flowered; calyx-tube funnel form, 3-toothed; corolla white, tube as long as calyx-lobes; labellum with yellow and red streaks, broadly obovate, margin fimbriate, apex rounded, lateral staminodes 0 or minute; stamen 1, filament flattened, anther cells apically diverging, not crested; ovary densely pilose, inferior, 3-locular; ovules few per locule. Fruit a globose capsule. Seeds globose or angular.

1. Plants up to 1.5 m tall; leaves linear-lanceate; flowers in panicles. ....1. **A. calcarata**

1. Plants up to 3 m tall; leaves oblong-lanceate; flowers in racemes. ....2. **A. malaccensis**

1. **Alpinia calcarata** Rosc., Trans. Linn. Soc. London 8:347. 1807; Roxb., Fl. Ind. 1:69. 1832; Wight, Ic. t. 2028. 1853; Baker in Hook.f., Fl. Brit. India 6:254. 1892; Fischer in Gamble, Fl. Pres. Madras 1493. 1928; Matthew, Ill. Fl. Tam. Carnatic t. 725. 1982; Matthew & Britto in Matthew, Fl. Tam. Carnatic 3(3): 1613. 1983.

Herbs to 1.5 m tall. Leaves linear-lanceate, 10-25 by 1.5-3 cm, base tapering, apex long acuminate, glaucous below, coriaceous. Panicles terminal, to 10 cm long; bracts 1 or 2-flowered. Capsule globose, c. 12 mm ϕ. Seeds angular.

Fl. & Fr.: Jan.-Apr. *Distr.*: Sri Lanka, Peninsular India, S. Malaya Peninsula, China. Often cultivated. *NAK 2904* (Adoor, c. 20 m).

2. **Alpinia malaccensis** (Burm.f.) Rosc., Trans. Linn. Soc. London 8:345. 1807; Baker in Hook.f., Fl. Brit. India 6:255. 1892; Fischer in Gamble, Fl. Pres. Madras 1493. 1928; Rao & Verma, Bull. Bot. Surv. India 14:140. 1972 (1975); Mohanan & Sivad., Fl. Agasthyamala 705.2002. *Maranta malaccensis* Burm. f., Fl. Ind. 2. 1768. *Catimbium malaccensis* (Burm. f.) Holt., Gard. Bull. Straits Settlem. 13:152. 1950.

Herbs to 3 m high. Leaves oblong-lanceate, 30-60 by 9-12 cm, base narrowed, pubescent below, apex acuminate. Racemes terminal, to 50 cm long; bracts 1- or 2-flowered. Capsule globose, c. 18 mm ϕ, pubescent.

Fl. & Fr.: Nov.-Feb. *Distr.*: Himalayas & Peninsular India. Often cultivated. *NAK 2975* (Ranni, c. 100 m).

COSTUS Linnaeus
Sp. Pl. 2.1753.

Costus speciosus (J.Koen.) Smith, Trans. Linn. Soc. London 1:249. 1791; Roxb., Fl. Ind. 1:58. 1832; Wight, Ic. t. 2014. 1853; Hook.f., Fl. Brit. India 6:249. 1892; Fischer in Gamble, Fl. Pres. Madras 1490. 1928; Matthew, Ill. Fl. Tam. Carnatic t. 726. 1982; Matthew & Britto in Matthew, Fl. Tam. Carnatic 3(3):1613. 1983; Nicols., Suresh & Manilal, An Interpr. Hort. Malab. 316. 1988; Mohanan & Sivad., Fl. Agasthyamala 708.2002. *Banksea speciosa* J. Koen. in Retz., Obs. Bot. 3:75. 1983. *Tsjana-kua* Rheede, Hort. Malab.11:15-16, t.8. 1692.

Herbs to 2 m high with twisting stems (so leaves appears to be spiral). Leaves elliptic-obovate, 10-30 by 3-8 cm, base cuneate, apex abruptly acuminate, pubescent below, chartaceous. Flowers white in dense terminal spikes, labellum with a yellow centre; staminodes 0; stamen median, on a petaloid process, anther cells contiguous; ovary 3-locular. Fruits globose.

Fl. & Fr.: Sep.-Dec. *Distr.*: Sri Lanka, India, Himalaya, Indo-China, Malaysia to New Guinea. Very common in disturbed forests. *NAK 2973* (Ranni, c. 100 m).

Note: Some workers treat the genus *Costus* has under a separate monogeneric family, Costaceae.

CURCUMA Linnaeus
Sp. Pl. 2. 1853.

Rhizomatous herbs; roots slender or tuberous. Leaves distichous. Spikes terminal on the leafy shoot or on a separate shoot; bracts large, basally forming closed pockets, apically free and spreading, enclosing 2 or more flowers; calyx-tube split half way down; corolla funnel-form, labellum broad, entire or 2-lobed; staminodes petaloid, anthers spurred at base, or not; ovary 3-locular. Capsule globose, 3-valved.

1. Anther-lobes without a spur. ....2. **C. ecalcarata**
1. Anther-lobes with a spur at base.
    2. Rhizome conical, without sessile tubers; root tubers cyindrical. ....1.**C. coriacea**
    2. Rhizome broadly ovoid, with sessile tubers; root-tubers fusiform. ....3. **C. zedoaria**

1. **Curcuma coriacea** Mangaly & Sabu, Notes Roy. Bot. Gard. Edinburgh 45:429. 1989, Rheedea 3(2): 148.1993.

Rhizomes 3-5 by 2-5 cm. Leaves elliptic, to 15 by 6 cm, base and apex acute, upper surface uniformly green, lower surface densely pubescent, coriaceous. Spikes dense; fertile bracts oblong, white, tinged with rose colour. Flowers yellow; anther-lobes with a spur at base.

Fl. & Fr.: Mar.-Apr. *Distr.*: Endemic to Southern Western Ghats. On grasslands. *NAK 2333* (Ponnambalamedu, c. 1100 m).

2. **Curcuma ecalcarata** Sivarajan & Indu, Notes. Roy. Bot. Gard. Edinburgh 41:321. 1983; Mangaly & Sabu, Rheedea 3(2): 151.1993.

Rhizomes ovoid. Leaves elliptic-oblong or lanceate-oblong, 10-16 by 6-8 cm, base acute acuminate, glabrous or puberulous below. Spikes dense; fertile bracts green or rose- coloured. Flowers yellow; anther-lobes without a spur at base.

Fl. & Fr.: Sep.-Jan. *Distr.*: Endemic to Southern Western Ghats (Kerala). Common. along wet open areas in plains. *NAK 1923* (Tiruvalla, c. 10 m).

Note :Two types of plants, one with green fertile bracts and white coma and other with rose fertile bracts and dark rose coma and showing slight variation in flower colour also present. They are similar in all respects except for the above characters. Since these characters are not enough to treat them as variety or even forma, they are treated here as variants of a single species.

3. **Curcuma zedoaria** (Christmann) Roscoe, Trans. Linn. Soc. London 8:354. 1807 & Monandr. Pl. t. 109. 1825; Baker in Hook.f., Fl.Brit. India 6:210. 1890; Trimen, Handb. Fl. Ceylon 4:241. 1898; Schumann in Engler, Pflanzenr. 4(46): 110. 1904; Cooke, Fl. Pres. Bombay 2:732. 1907; Fischer, Rec. Bot. Surv. India 9:177. 1921 & in Gamble, Fl.Pres. Madras 1482. 1928 ("*Zeodaria*"); Ridley, Fl. Malay Penin. 4:254. 1924; Holttum, Gard. Bull. Singapore 13:71. 1950; Burtt & Smith, Notes Roy. Bot. Gard. Edinburgh 31:226. 1972 & in Dassan.

& Fosberg, Rev. Handb. Pl. Ceylon 4:501. 1983; Nicols., Suresh & Manilal, An Interpr. Hort. Malab. 317. 1988; Mangaly & Sabu, Rheedea 3(2): 168.1993. *Amomum zedoaria* Christmann in Christmann & Panzer, Linn. Pflanzensyst. 5:12. 1779. *Curcuma zerumbet* Roxb., Asiat. Res. 11:332. 1810, & Pl. Corom. 3:t. 201. 1819, Fl. Ind. 1:20,. 1820. *Kua* Rheede, Hort. Malab. 11:13. t.7. 1692.

Rhizomes broadly ovoid, 5-8 by 7-9 cm. Leaves elliptic, to 60 by 15 cm, greenish with a purple patch along the midrib on both sides on upper side. Spikes dense, fertile bracts white with rose tinge. Flowers yellow; anther-lobe with a spur at base.

Fl. & Fr.: Mar.-Apr. *Distr.:* Country of origin unknown. Cultivated and naturalized throughout India and S.E. Asia. Common on exposed midlands. *NAK 1635* (Chittar, c. 200 m).

ELETTARIA W.G. Maton
Trans. Linn. Soc. London 10:250. 1811.

Elettaria cardamomum (L.) Maton, Trans. Linn. Soc. London 10:254.1811; Baker in Hook.f., Fl. Brit. India 6:251. 1892; Fischer in Gamble, Fl. Pres. Madras 1491. 1921; Burtt & Smith, Notes Roy. Bot. Gard. Edinburgh 31:182. 1972; Matthew, Ill. Fl. Tam. Carnatic t. 728. 1982; Matthew & Britto in Matthew, Fl. Tam. Carnatic 3(3): 1616. 1983; Nicols., Suresh & Manilal, An Interpr. Hort. Malab. 317. 1988. *Amomum cardamomum* L., Sp. Pl. 1.1753. *Elettari* Rheede, Hort. Malab. 11:9-11, t. 4-6. 1692.

Perennial herbs; rootstock thick, horizontal. Leaves linear-lanceate or oblong-lanceate, 20-40 by 3-5 cm, base cuneate, apex long acuminate. Flowers in panicles which are borne on the rhizome. Inflorescence clusters 2-α-flowered; calyx tubular, split on one side; corolla white with pinkish striations, labellum emarginate at apex; anther cells contigous. Capsule subtrigonous, striate.

Fl. & Fr.: Most of the seasons. *Distr.*: Peninsular India, Sri Lanka. Often cultivated as a cash crop. *NAK 2967* (Ranni, c. 100 m).

GLOBBA Linnaeus
Mant. 2:(143): 170. 1771.

Globba ophioglossa Wight, Ic. t. 2202. 1853; Baker in Hook.f., Fl. Brit.India 6:202. 1890; Fischer in Gamble, Fl. Pres. Madras 1040. 1928; Mohanan & Sivad., Fl. Agasthyamala 710.2002.

Slender herbs; rhizome creeping. Leaves distichous, oblong or elliptic-lanceate, 9-15 by 3-5 cm, base cuneate, apex caudate. Flowers bright yellow in terminal spikes; upper bracts persistent; calyx funnel-form; corolla dark glandular, lobellum small, 2-fid; filament long, slender, anthers without wings; ovary 1-locular; ovules α. Capsule globose.

Fl. & Fr.: Jul.-Oct. *Distr.:* Peninsular India. Very common. Among forest undergrowths. *NAK 2966* (Ranni c. 100 m).

ZINGIBER Boehmer
in Ludwig, Def. Gen. ed. Boehmer 89. 1760 (nom. cons.).

Rhizomes horizontal. Leaves linear to oblong-lanceate; sheaths clasping the stem. Flowers in spikes; bracts persistent, usually with red streaks; anther-cells contigous, produced into a narrow beak; ovary 3-locular. Capsule subglobose. Seeds globose.

1. Spike-peduncle to 10 cm long. ....2. **Z. wightianum**
1. Spike-peduncle to 20-40 cm long.
    2. Leaves glabrous; spikes ovoid. ....3. **Z. zerumbet**
    2. Leaves pubescent below; spikes cylindric. ....1. **Z. neesanum**

1. **Zingiber neesanum** (Grah.) Ramamoorthy in Sald. & Nicols., Fl. Hassan Dist. 769. 1976; Manilal, Fl. Silent Valley 314.1988; Mohanan & Henry, Fl. Thiruvananthapuram 474. 1994; Mohanan & Sivad., Fl. Agasthyamala 711.2002. *Alpinia neesana* Grah., Cat. Pl. Bombay 207.1839. *Zingiber macrostachyum* Dalz. in Hook.f., Kew Journ. Bot. 4:342. 1852; Baker in Hook.f., Fl. Brit. India 6: 247. 1892; Fischer in Gamble, Fl. Pres. Madras 1490. 1928.

Pseudo-stems reddish-brown. Leaves oblong-lanceate, 30-50 by 4-10 cm, base narrowed, apex acuminate, pubescent below. Spike cylindric, to 30 cm long; peduncle elongated; bracts ovate, reddish. Flowers greenish-white. Capsule reddish, ellipsoid.

Fl. & Fr.: Nov.-Jan. *Distr.*: Peninsular India. Rare. Among forest undergrowths. *NAK 2976* (Ranni, c. 100 m).

2. **Zingiber wightianum** Thw., Enum. Pl. Zeyl. 315. 1861; Baker in Hook.f., Fl. Brit. India 6:244. 1892; Fischer in Gamble, Fl. Pres. Madras 1489. 1928; Mohanan & Sivad., Fl. Agasthyamala 711.2002. *Z. squarrosum* Wight, Ic. t. 2004. 1853, non Roxb. 1832.

Tall herbs. Leaves oblong-lanceate, 30-40 by 6-10 cm, base narrowed, apex acuminate, pubescent below. Spikes oblong; peduncle very short. Flowers pale yellow. Capsule c. 2 cm ϕ.

Fl. & Fr.: Nov.-Feb. *Distr.*: Sri Lanka, Peninsular India. Not common. Among forest undergrowth. *NAK 1828* (Ranni, c. 100 m).

3. **Zingiber zerumbet** (L.) J.E. Smith, Exot. Bot. 2:105. t. 112. 1804; Baker in Hook.f., Fl.Brit. India 6:247. 1892; Fischer in Gamble, Fl. Pres. Madras 1490. 1928; Nicols., Suresh & Manilal, An Interpr. Hort. Malab. 319. 1988; Mohanan & Sivad., Fl. Agasthyamala 712.2002. *Amomum zerumbet* L., Sp. Pl. 1.1753. *Katou-inschi-kua* Rheede, Hort. Malab. 11:27, t.13. 1692.

Herbs, to 1 m high. Leaves oblong-lanceate, 30-40 by 5-8 cm, base narrowed, apex acuminate. Spikes ovoid or oblong; bracts green or blood-red coloured. Flowers white, labellum unspotted. Capsule oblong, c. 2.5 cm long.

Fl. & Fr.: Nov.-Jan. *Distr.*: Old World tropics. Very common along forest clearings. *NAK 612* (Maniyaar, c. 70 m).

125. MARANTACEAE

1. Tall herbs; flower in panicles. ....**Schumannianthus**
1. Small herbs; flowers in capitate spikes. ....**Phrynium**

PHRYNIUM Willdenow
Sp. Pl. 1:1. 1797.

Phrynium rheedei Suresh & Nicols., Taxon 35:355. 1986; Nicols., Suresh & Manilal, An Interpr. Hort. Malab.296. 1988. *Pontederia ovata* L., Sp. Pl. 288. 1753, non *Phrynium ovatum. Phrynium capitatum* Willd.,

Sp. Pl. 1:17.1797; Baker in Hook.f., Fl. Brit. India 6:258. 1892; Fischer in Gamble, Fl. Pres. Madras 1495. 1928. *Naru-kila* Rheede, Hort. Malab. 11:67, t.34. 1692.

Erect herbs, to 1.5 m tall; aerial stem absent; root stock rhizomatous, creeping. Leaves 1-2, oblong or oblong-lanceate, 15-30 by 8-10 cm, base rounded, apex cuspidate. Spikes globose, produced from the side of the petiole above the middle; primary bracts oblong, bicuspidate at apex. Flowers purple; sepals 3, linear; corolla-tube short, lobes 3, longer, linear; staminal tube long, petaloid, anther 1-celled; ovary 3-locular; ovule 1 per locule. Capsule obconic, trigonous.

Fl. & Fr.: Apr.-Jul. *Distr.*: Sri Lanka, Peninsular India, Malesia. Very rare. Among forest undergrowths. *NAK 2932* (Thekkuthode, c. 70 m).

SCHUMANNIANTHUS Gagnepain
Bull. Soc. Bot. France 51:176.1904.

Schumannianthus virgatus (Roxb.) Rolfe, J. Bot. 14: 224. 1907; Fischer in Gamble, Fl. Pres. Madras 1494. 1928; Mohanan & Sivad., Fl. Agasthyamala 712.2002. *Phrynium virgatum* Roxb., Asiat. Res. 11: 324. 1810. *Clinogyne virgata* (Roxb.) Benth. ex Baker in Hook.f., Fl. Brit. India 6: 258. 1892.

Tall leafy herbs, to 4 m tall; stem reed-like. Leaves oblong or oblong-lanceate, 15-50 by 6-12 cm, base rounded, apex cuspidate. Flowers in pairs, in lax panicles; bracts narrow-linear; sepals ovate-lanceate; staminodes petaloid; ovary villous. Capsule subpyriform. Seeds subglobose.

Fl. & Fr.: Oct.-Mar. *Distr.:* Sri Lanka, Peninsular India. Common. Along forest borders in moist shady areas. *NAK 2931* (Thekkuthode, c. 70 m).

126. MUSACEAE

ENSETE Bruce ex Horaninow
Prodr. Monogr. Scitam. 8, 40. *t. 4 (p.p.)*. 1862.

Ensete superbum (Roxb.) Cheesman, Kew Bull. 1948: 100. 1948; Gandhi in Sald. & Nicols., Fl. Hassan Dist. 761. 1976; Mohanan & Sivad., Fl. Agasthyamala 713.2002. *Musa superba* Roxb., Pl. Corom. t. 223. 1811; Wight, Ic. tt. 2017 & 2018. 1853; Baker in Hook.f., Fl. Brit. India 6:261. 1892; Fischer in Gamble, Fl. Pres. Madras 1497. 1928.

Stout tall rhizomatous herbs. Leaves spirally arranged, oblong 1-2 m by 60-90 cm, base rounded or narrowed, apex obtuse; petiole c. 30 cm long. Flowers monoecious, on stout elongated drooping spikes; bracts foliaceous, brownish, orbicular. Flowers ϕ, in 2 dense rows. Fruits cylindric, to 8 cm long, in clusters.

Fl. & Fr.: Aug.-Oct. *Distr.*: Endemic to Western Ghats. Rare. On rocky hills slopes.

Note : Few plants were observed near Pampa dam in Kakki hills but failed to collect.

127. AMARYLLIDACEAE

PANCRATIUM Linnaeus
Sp. Pl. 290.1753.

Pancratium triflorum Roxb., Fl. Ind. 2: 126. 1832; Hook.f., Fl. Brit. India 6: 285. 1892; Fischer in Gamble, Fl. Pres. Madras 1505. 1928; Matthew & Britto in Matthew, Fl Tam. Carnatic 3(3):1626.1983; Nicols., Suresh

& Manilal, An Interpr. Hort. Malab. 273. 1988. *P. verecundum* auct., non Aiton, 1789 : Wight, Ic. t. 2023. 1853; Hook.f., Fl. Brit. India 6: 286. 1892. *Cattulli-pola* Rheede, Hort. Malab. 11:77, t. 40. 1692.

Bulbous herbs. Leaves linear-lanceate, 13-30 by 1.5-2.5 cm, flat, base narrow, apex subacute. Flowers 2-8, in scapose umbels; 2-bracteate; perianth-lobes 6, white; stamens 6, connate at base into a petaloid, 2-dentate, membranous cup; ovary inferior, 3-locular; ovules a per locule, 2-seriate. Capsule subglobose.

Fl. & Fr.: Jul.-Aug. *Distr.*: Sri Lanka, India. Ephemerals on exposed fields. *NAK 2906* (Adoor, c. 20 m).

128. AGAVACEAE

1. Leaf-margin armed; peduncles more than 4 m long. .... **Furcraea**
1. Leaf-margin unarmed; peduncles less than 1 m long. .... **Dracaena**

DRACAENA Vandelli ex Linnaeus
Syst. Nat. ed. 12.2:246, Mant.9, 63. 1767.

Dracaena terniflora Roxb., Fl. Ind.2:159. 1832; Hook.f., Fl. Brit. India 6:328. 1892; Fischer in Gamble, Fl. Pres. Madras 1521. 1928; Gandhi in Sald. & Nicols., Fl. Hassan Dist. 803. 1976; Mohanan & Sivad., Fl. Agasthyamala 725. 2002. *D. terminalia* Wight, Ic. t. 2054. 1853.

Erect subshrubs, rarely straggling. Leaves crowded at apex, elliptic-lanceate, 10-25 by 3-7 cm, base attenuate, amplexicaule, apex acuminate. Flowers white, in terminal or rarely axillary racemes; perianth funnel-form, lobes 6; stamens 6; ovary superior, 3-locular; ovule 1-per locule. Berry globose. Seeds globose or angular.

Fl. & Fr.: Jan.-Apr. *Distr.:* Indian & S.E. Asian regions. Not common. Among forest undergrowths in upper hills. *NAK 2968* (Ranni, c. 100 m).

FURCRAEA Ventenat
Bull. Sci. Soc. Philom. Paris 1:65. 1793.

Furcraea foetida (L.) Haworth, Syn. Pl. Succ. 73. 1819; Gandhi in Sald. & Nicols., Fl. Hassan Dist. 803. 1976. *Agave foetida* L., Sp. Pl. 323.1753. *Furcraea gigantea* Vent., Bull. Soc. Philom. Paris 1:65. 1793; Fischer in Gamble, Fl. Pres. Madras 1056. 1928.

Perennial shrubs with rhizomatous stems. Leaves c. 1.5 m by 15 cm, in rosette. Flowers dull white; perianth-lobes broad, free; staminal filament flattened; ovary inferior, 3-locular; style swollen below; ovule a per locule; floral buds transformed into bulbils.

Fl. & Fr.: Feb.-Apr. *Distr.:* Native to Central America, now introduced in other parts of the tropics. Planted, as well as running wild. *NAK 2836* (Adoor, c. 20 m).

129. HYPOXIDACEAE

1. Perianth-tube produced above ovary; stamens perigynous; seeds beaked. .... **Curculigo**
1. Perianth-tube not produced above ovary; stamens epigynous; seeds not beaked. .... **Molineria**

CURCULIGO J. Gaertner
Fruct. 1:63, t.16, f.11. 1788.

Curculigo orchioides Gaertner, Fruct. 1:63, t.16, f. 11. 1788; Roxb., Pl. Corom. t. 13. 1795; Hook.f., Fl. Brit. India 6:279. 1892; Fischer in Gamble, Fl. Pres. Madras 1502. 1928; Matthew, Ill. Fl. Tam. Carnatic t.733. 1982; Matthew & Britto in Matthew, Fl. Tam. Carnatic 3(3) : 1627. 1983; Nicols., Suresh & Manilal, An Interpr. Hort. Malab. 294. 1988; Mohanan & Sivad., Fl. Agasthyamala 716.2002. *Nella-pana-kelengu* Rheede, Hort. Malab. 12:111, t.59. 1693.

Small tuberous rooted herbs. Leaves radical, plicate, lanceate, 5-20 by 0.8-1.5 cm, base narrowed, apex acute, villous below or glabrescent, sheath persistent. Raceme(s) often solitary, scape subterranean; bracts spathaceous, densely imbricating. Flowers bisexual; perianth-lobes 6; stamens 6; ovary 3-locular. Capsule oblong. Seeds globose, beaked.

Fl. & Fr.: Most of the seasons. *Distr.*: Sri Lanka, India, Himalayas, Japan, Malesia, Australia. Abundant on forest floors and in plains. *NAK 2907* (Adoor, c. 20 m).

MOLINERIA Colla
Mem. Reale Accad. Sci. Torino 31 (Hortus Ripul. App. 2) : 331. 1826.

Molineria trichocarpa (Wight) Balakr., J. Bombay Nat. Hist. Soc. 53:330.1966; Matthew, Ill. Fl. Tam. Carnatic t.734. 1982; Matthew & Britto in Matthew, Fl. Tam. Carnatic 3(3) : 1628. 1983; Mohanan & Sivad., Fl. Agasthyamala 716.2002. *Hypoxis trichocarpa* Wight, Ic.t.2045. 1853, p.p.; *H. latifolia* Wight, Ic. t. 2044. 1853. *H. leptostachya* Wight, Ic.t. 2045. 1853. *H. pauciflora* Wight, Ic.t. 2046. 1853. *Molineria finlaysoniana* Wallich ex Baker, J. Linn. Soc. Bot. 17:121. 1878; Fischer in Gamble, Fl. Pres. Madras 1502. 1928. *Curculigo finlaysoniana* Wallich ex Hook.f., Fl. Brit. India 6:279. 1892.

Tuberous rooted herbs. Leaves lanceate, 9-25 by 1.5-3.5 cm, flat, plicate, prominently 7-nerved; petiole to 20 cm long; scape hidden among leaf bases. Flowers regular, bisexual; perianth-lobes 6, yellow; stamens 6; ovary 3-locular; ovules a per locule. Berry oblong-ellipsoid. Seeds 10, globose, striate, not beaked.

Fl. & Fr.: Most of the seasons. *Distr.*: Sri Lanka, Peninsular India. Common in florest floors of upper hills. *NAK 2969* (Ranni, c. 100 m).

130. DIOSCOREACEAE

DIOSCOREA Linnaeus
Sp. Pl. 1032. 1753.

Dioecious twiners; stem twining clockwise or anticlockwise. Flowers unisexual, usually in spikes. Male flowers : perianth-lobes 6; stamens 6 or 3; pistillode present or 0. Female flowers : perianth-lobes 6, free; staminodes 3-6 or 0; ovary inferior, triquetrous, 3-locular; styles 3. Fruit a 3-winged capsule.

1. Leaves simple.
    2. Leaves cordiform; stem twining to left. .... 1. **D. bulbifera**
    2. Leaves elliptic-oblong; stem twining to right. .... 3. **D. oppositifolia**
1. Leaves compound.
    3. Leaves velvetty-tomentose below. .... 5. **D. tomentosa**

3. Leaves glabrous below.
   4. Fertile stamens 6; leaf-venation externally very prominent. ....2. **D. hispida**
   4. Fertile stamens 3; leaf-venation externally not prominent. ....4. **D. pentaphylla**

1. **Dioscorea bulbifera** L., Sp. Pl. 1033. 1753; Wight, Ic. t. 878. 1844; Fischer in Gamble, Fl. Pres. Madras 1511. 1828; Matthew & Britto in Matthew, Fl. Tam. Carnatic 3(3): 1634. 1983; Nicols., Suresh & Manilal, An Interpr. Hort. Malab. 290. 1988. *D. sativa* Thunb., Fl. Jap. 15. 1784, non L., 1753; Hook.f., Fl. Brit. India 6: 295. 1892. *Katu-katsjil* Rheede, Hort. Malab. 7:69, t. 36. 1688. *Kattu-kelengu* Rheede, Hort. Malab. 7:71, t. 37. 1688.

Vines with tuberous root-stocks; stem twining to left. Leaves simple, ovate-suborbicular, c. 15 by 12 cm, base deeply cordate, apex acuminate, to shortly caudate. Inflorescence a raceme or panicle. Flowers unisexual; perianth-lobes 3+3; stamens 3 or 6; staminodes 3; ovary inferior, triquetrous, 3-locular; ovules 2 per locule. Capsule oblong, winged.

Fl. & Fr.: Sep.-Nov. *Distr.*: Old World tropics. Along forest borders. *NAK 2908* (Adoor, c. 20 m).

2. **Dioscorea hispida** Dennst., Schluess. Hort. Malab. 15. 1818; Fischer in Gamble, Fl. Pres. Madras 1511. 1928; Prain & Burkill, Ann. Roy. Bot. Gard. Calcutta 14:188, t. 77. 1936. *D. daemona* auct., non Roxb., 1832: Wight, Ic.t. 811. 1844; Hook.f., Fl. Brit. India 6:289. 1892.

Stem twining to left. Leaves 3-foliolate, broadly ovate or obovate, 8-20 by 5-15 cm, base cuneate, apex cuspidate, glabrous or finely hispid, venation very prominent externally. Male flowers in dense cylindric clustered spikes. Female spike(s) solitary. Capsule cylindric, quadrately winged.

Fl. & Fr.: Nov.-Dec. *Distr.*: In evergreen forests. Not common. *NAK 2970* (Ranni, c. 100 m).

3. **Dioscorea oppositifolia** L., Sp. Pl. 1033. 1753; Roxb., Fl. Ind. 3: 804. 1832; Wight, Ic. t. 813. 1844; Hook.f., Fl. Brit. India 6: 292. 1892; Fischer in Gamble, Fl. Pres. Madras 1512. 1928; Matthew & Britto in Matthew, Fl. Tam. Carnatic 3(3): 1634. 1983;Mohanan & Sivad., Fl. Agasthyamala 718.2002.

Stem twining to right. Leaves simple, opposite, ovate or oblong-elliptic, 3-10 by 1.5-6 cm, base obtuse, apex subacute, mucronate, coriaceous. Male spikes panicled, clustered. Female spike(s) solitary or paired. Capsule obovoid. Seeds orbicular, winged.

Fl. & Fr.: Jul.-Dec. *Distr.*: Sri Lanka, India. Common. On thickets along forest borders. *NAK 133* (Moozhiar, c. 250 m).

4. **Dioscorea pentaphylla** L., Sp. Pl. 1032. 1753; Roxb., Fl. Ind. 3: 806. 1832; Wight, Ic. t. 814. 1844; Hook.f., Fl. Brit. India 6: 289. 1892; Fischer in Gamble, Fl. Pres. Madras 1511. 1928; Matthew & Britto in Matthew, Fl. Tam. Carnatic 3(3): 1635. 1983, var. *rheedei* Prain & Burkill, J. Asiat. Soc. Bengal 10: 23. 1914, ("*rheedii*"); Nicols., Suresh & Manilal, An Interpr. Hort. Malab. 292. 1988. *Tjsageri-nuren* Rheede, Hort. Malab. 7:63, t. 33. 1688. *Katu-nuren* Rheede, Hort. Malab.7:65 ("63") t. 34. 1688. *Nurven-kelengu* Rheede, Hort. Malab. 7:67, t. 35. 1688.

Stem twining to left. Leaves 3-5-foliolate; leaflets elliptic-ovate, 3-7 by 1.5-3 cm, base acute or attenuate, apex acute, glabrous or pubescent below, drying black. Male raceme(s) axillary, solitary or clustered, woolly. Female spike(s) solitary or paired. Capsule oblong, winged. Seed 4, apically winged.

Fl. & Fr.: Aug.-Feb. *Distr.*: India, Himalayas, Myanmar, Thailand, Indo-China, China, Malesia. Common along forest borders. *NAK 784* (Ranni, c. 100 m).

5. **Dioscorea tomentosa** J. Koeing ex Sprengel, Pl.Minus Cognit. Pug. 2: 92. 1815; Roxb., Fl. Ind. 3:805. 1832; Hook.f., Fl.Brit. India 6: 289. 1892; Fischer in Gamble, Fl. Pres. Madras 1511. 1928; Matthew, Ill. Fl.Tam.Carnatic t. 738.1982; Matthew & Britto in Matthew, Fl. Tam. Carnatic 3(3): 1636. 1983;Mohanan & Sivad., Fl. Agasthyamala 718.2002.

Stem twining to left. Leaves 3-(5)-foliolate, ovate, broadly elliptic or lanceate, 4-13 by 1.5-7.5 cm, coriaceous, base obtuse-subacute, apex rotund-acute. Male spike(s) solitary or paired; female spikes axillary, paired. Capsule oblong, tomentose. Seeds 3, obovoid, apically winged.

Fl. & Fr.: Jul.-Dec. *Distr.:* Peninsular India. Not common. In interior evergreen forest. *NAK 1218* (Moozhiar, c.250 m).

131. LILIACEAE

1. Climbers.
    2. Cladodes present; leaves scaley. .... **Asparagus**
    2. Cladodes absent; leaves well developed. .... **Gloriossa**
1. Non-climbers.
    3. Stem aerial; leaves ensiform. .... **Dianella**
    3. Stem underground; leaves linear. .... **Chlorophytum**

ASPARAGUS Linnaeus
Sp. Pl. 313. 1753.

Scandent shrubs. Leaves scaley; cladodes leaf-like. Flowers white, fragrant, minute, regular, in fascicled-racemes. Flowers white; perianth-lobes 6; stamens 6; ovary 3-gonous, 3-locular; ovules usually 2-per locule. Fruit a globose berry.

1. Cladodes ribbed. .... 2. **A. racemosus**
1. Cladodes smooth. .... 1. **A. gonoclados**

1. **Asparagus gonoclados** Baker, J. Linn. Soc. Bot. 14: 627. 1875; Hook.f., Fl. Brit. India 6: 318.1982; Fischer in Gamble, Fl. Pres. Madras 1517.1928;Mohanan & Sivad., Fl. Agasthyamala 721.2002.

Branchlets angled. Cladodes 3-5 mm long, flat, falcate, ensiform 2-6 nate, spinous tipped. Berry c. 1 cm ϕ.

Fl. & Fr.: Aug.-Oct. *Distr.:* Endemic to Western Ghats. Less common. In forest borders. *NAK 84* (Moozhiar, 250 m).

2. **Asparagus racemosus** Willd., Sp. Pl. 2: 152. 1799; Roxb., Fl. Ind. 2: 151.1832; Wight, Ic. t. 2056.1853; Fischer in Gamble, Fl. Pres. Madras 1517.1928; Matthew, Ill. Fl. Tam. Carnatic t. 741. 1982; Matthew & Britto in Matthew, Fl. Tam. Carnatic 3(3): 1639. 1983; Nicols., Suresh & Manilal, An Interpr. Hort. Malab. 295. 1988;Mohanan & Sivad., Fl. Agasthyamala 721.2002. *Schada-veli-kelangu* Rheede, Hort. Malab. 10:19-20, t.10. 1690.

Branchlets glabrous. Scale-leaves triangular; cladodes 2-6, linear-oblong, ribbed. Raceme(s) solitary or 3 in a cluster. Berry globose or angled.

Fl. & Fr.: Jul.-Oct. *Distr.*: Africa through S.Asia to China, S. Malesia and N. Australia. Common. Forest borders and in sacred groves. *NAK 2909* (Adoor, c. 20 m).

CHLOROPHYTUM Ker-Gawler
Bot. Mag. t. 1071. 1807.

Erect tuberous rooted herbs. Leaves radical, linear or elliptic-lanceate. Scape usually one or 2-3. Flowers close or distant, bisexual; perianth shallowly 6-lobed; stamens 6; anthers versatile; ovary subglobose, 3-lobed, 3-locular; ovules 4 or more per locule. Capsule globose-oblong, 3-winged or 3-angled. Seeds suborbicular, warty.

1. Leaves elliptic-lanceate or oblanceate. .... 2. **C. heynei**
1. Leaves narrow-linear.
   2. Flowers in dense racemes. .... 1. **C. attenuatum**
   2. Flowers in lax racemes. .... 3. **C. laxum**

1. **Chlorophytum attenuatum** (Wight) Baker, J. Linn. Soc. Bot. 15: 332. 1876; Hook.f., Fl. Brit. India 6: 335. 1892; Fischer in Gamble, Fl. Pres. Madras 1526. 1928. *Phalangium attenuatum* Wight, Ic. t. 2037. 1853.

Leaves linear, 15-30 by 1-2.5 cm. Scape to 30 cm long. Flowers white, dense; pedicels jointed above the middle; perianth-lobes 3-5-nerved. Capsule obcordate.

Fl. & Fr.: Sep.-Dec. *Distr.*: Peninsular India. Common in rock-crevices. *NAK 2432* (Arampa, c. 300 m).

2. **Chlorophytum heynei** Rottler ex Baker, J. Linn. Soc.Bot. 15: 322. 1876; Fischer in Gamble, Fl.Pres. Madras 1525. 1928; Matthew & Britto in Matthew, Fl. Tam. Carnatic 3(3): 1641.1983; Mohanan & Sivad., Fl. Agasthyamala 722.2002. *C. heyneanum* Wallich ex Hook.f., Fl. Brit. India 6: 333. 1892.

Leaves elliptic-lanceate, 8-30 by 0.7-5 cm, flat, α-nerved, base attenuate, apex acuminate, shiny. Scape solitary; racemes dense-flowered. Capsule triquetrous.

Fl. & Fr.: Jul.-Aug. *Distr.*: India, Sri Lanka. Rare. Among forest undergrowths. *NAK 2526* (Arampa, c. 300 m).

3. **Chlorophytum laxum** R. Br., Prodr. 277. 1810; Hook.f., Fl. Brit. India 6: 336. 1892; Fischer in Gamble, Fl. Pres. Madras 1526.1928; Gandhi in Sald. & Nicols., Fl. Hassan Dist. 795. 1976;Mohanan & Sivad., Fl. Agasthyamala 723.2002.

Leaves linear-oblong, subdistichous, 15-50 by 0.6-1 cm. Flowers white, distant, on filiform scape; pedicels slender, articulate at middle. Capsule subrotund. Seeds angled.

Fl. & Fr.: Sep.-Dec. *Distr.*: Tropics. Rare. Among wet grassy hill slopes in upper hills *NAK 2971* (Ranni, c. 100 m).

DIANELLA Lamarck
Encycl. Meth. Bot. 2:276.1786.

Dianella ensifolia (L.) DC. in Reddute, Liliac. I: t. 1. 1802; Hook.f., Fl. Brit. India 6: 337.1892; Fischer in Gamble, Fl. Pres. Madras 1521. 1928; Mohanan & Sivad., Fl. Agasthyamala 724.2002. *Dracaena ensifolia*

L., Mant. Pl.63.1767. *Dracaena ensata* Thunb., Diss. Bot. Drac. 4. 1808. *Dianella ensata* (Thunb.) Henderson, Taxon 26: 136: 136. 1977.

Stout, stoloniferous herbs. Leaves distichous, equitant, linear-ensiform, 30-90 by 1-2.5 cm, apex acute, coriaceous. Flowers white, greenish or bluish in cymose panicles; scape slender, angled, sulcate; perianth-lobes 6, spreading; stamens 6, filaments thickened, anthers poricidal; ovary 3-locular; ovules 4-8 per locule. Fruit a globose berry. Seeds ovoid to compressed.

Fl. & Fr.: Sep.-Nov. *Distr.*: Western Ghats. Under shade in evergreen forests of upper hills. *NAK 56* (Moozhiar, c. 250 m).

GLORIOSSA Linnaeus
Sp. Pl. 305. 1753.

Gloriosa superba L., Sp. Pl. 305. 1753; Roxb., Fl. Ind. 2: 143. 1832; Wight, Ic. t. 2047. 1853; Hook.f., Fl. Brit. India 6: 358. 1842; Fischer in Gamble, Fl. Pres. Madras 1519. 1928; Matthew, Ill. Fl. Tam. Carnatic t. 743. 1982; Matthew & Britto in Matthew, Fl. Tam. Carnatic 3(3): 1643. 1983; Nicols., Suresh & Manilal, An Interpr. Hort.Malab. 295. 1988. *Mendoni* Rheede, Hort. Malab. 7:107-108. t. 57. 1688.

Climbers. Leaves alternate, opposite, or whorled, oblong-lanceate, sessile, 5-13 by 1-4 cm, base cordate, amplexicaul, apex acuminute, tendrilled, α-nerved, chartaceous. Flowers large, axillary, solitary; perianth-lobes 6, reflexed or spreading, yellowish below and reddish above, margin wavy; ovary 3-locular; ovules α per locule. Capsule ellipsoid-oblong. Seeds α, globose, warty.

Fl. & Fr.: Oct.-Mar. *Distr.*: Old World tropics. Not common. On thickets and in sacred groves in plains. *NAK 2910* (Adoor, c. 20 m).

132. SMILACACEAE

SMILAX Linnaeus
Sp. Pl. 1028. 1753.

Smilax zeylanica L., Sp. Pl. 1029. 1753; Hook.f., Fl. Brit. India 6: 309. 1892; Fischer in Gamble, Fl. Pres. Madras 1518. 1928; Matthew, Ill. Fl. Tam. Carnatic t. 746. 1982; Matthew & Britto in Matthew, Fl. Tam. Carnatic 3(3): 1648. 1983; Nicols., Suresh & Manilal, An Interpr. Hort. Malab. 296. 1988; Mohanan & Sivad., Fl. Agasthyamala 727 2002. *S. macrophylla* Roxb., Fl. Ind. 3: 793. 1832; Hook.f., Fl. Brit. India 6:310. 1892. *Cari-villandi* Rheede, Hort. Malab. 7:59-60, t. 31. 1688.

Climbers; branchlets usually prickled. Leaves with petiole bearing 2 tendrils; lamina oblong, elliptic or ovate-lanceate, 5-10 by 2-5 cm, base obtuse, apex obtuse-cuspidate, 3-5-nerved, coriaceous. Flowers unisexual. Male in axillary umbels; perianth-lobes greenish; 3-nerved; stamens 6, filaments flat; ovary 3-lobed, 3-locular; ovules(s) 1 or 2 per locule; staminodes 3-6, linear. Berry globose.

Fl. & Fr.: Jul.-Aug. *Distr.*: India, Indo-China, Myanmar, Malesia. Common. On thickets along forest borders. *NAK 136* (Moozhiar, c. 250 m).

133. PONTEDERIACEAE

1. Petiole swollen to form floats; perianth-lobes connate below. .... **Eichornia**
1. Petiole not swollen to form floats; perianth-lobes free. .... **Monochoria**

EICHHORNIA A. Richard
in Sagra, Hist. Fis. Cuba 11: 273. 1850.

Eichhornia crassipes (C. Martius) Solms-Laub. in A. DC., Monogr. Phan. 4:527.1883; Fischer in Gamble, Fl. Pres. Madras 1530. 1928; Backer in Steenis, Fl. Males. I, 4: 259. 1951; Gopal & Sharma, Water Hyacinth 6. 1981; Matthew, Ill. Fl. Tam. Carnatic t. 747. 1982; Matthew & Rani in Matthew, Fl. Tam. Carnatic 3(2): 1649. 1983. *Pontederia crassipes* C. Martius, Nov. Gen. Sp. Pl. 9, t.4. 1823.

Aquatic free-floating or rooted herbs. Leaves radical, rosulate, broadly ovate-rhomboid, 5-10 by 4-9 cm, base cuneate, apex obtuse, nerves curved, faint, chartaceous; petiole 6-30 cm long, spongy, swollen about the middle. Inflorescence of terminal spikes. Flowers dense, irregular, lilac or blue with yellow median blotch; lobes 3+3; stamens 3+3, glandular pubescent; ovary 3-locular; ovules a per locule; stigma globose. Capsule ovoid, enveloped by perianth. Seeds α, ribbed.

Fl. & Fr.: Jan.-May. *Distr.*: S. America, now naturalized in tropics. An aggressive weed which is steadily replacing the native aquatic flora. *NAK 2982* (Tiruvalla, c. 10 m).

MONOCHORIA K. B. Persl
Rel. Haenk. 1: 127. 1827.

Monochoria vaginalis (Burm.f.) Presl, Rel. Haenk. 1: 128. 1827; Hook.f., Fl. Brit. India 4: 363. 1882; Fischer in Gamble, Fl. Pres. Madras 1529. 1928; Backer in Steenis, Fl. Males. I, 4: 256. 1951; Ramamoorthy in Sald. & Nicols., Fl. Hassan Dist. 791. 1976; Matthew, Ill. Fl. Tam. Carnatic t. 748. 1982; Matthew & Rani in Matthew, Fl. Tam. Carnatic 3(2): 1650. 1983; Nicols., Suresh & Manilal, An Interpr. Hort. Malab. 314.1988; Mohanan & Sivad., Fl. Agasthyamala 727.2002. *Pontederia vaginalis* Burm. f., Fl. Ind. 80. 1768; Roxb., Pl. Corom. t. 110. 1799. *Carim-gola* Rheede, Hort.Malab.11:91-92, t. 44. 1692.

Aquatic rooted herbs. Leaves ovate, 3-8 by 1.5-5 cm, base cordate, apex abruptly acuminate, chartaceous; petiole 10-30 cm long, sheathing at base. Inflorescence of solitary racemes (s); bracts membranous. Flowers regular, blue, perianth deeply lobed, lobes 3+3; stamens 6; ovary 3-locular; ovules α-per locule. Seeds transversely striate.

Fl. & Fr.: Nov.-Mar. *Distr.*: India, China, Japan, Malesia. Common in marshy fields in plains. *NAK 1307* (Tiruvalla, c. 10 m).

134. XYRIDACEAE

XYRIS Linnaeus
Sp. Pl. 42. 1753.

Xyris pauciflora Willd., Phytographia 1:2, t.1, f.1. 1794; Hook.f., Fl. Brit. India 6:365. 1892; Fischer in Gamble, Fl. Pres. Madras 1532. 1928; Royen in Steenis, Fl. Males. I,4:371. 1953; Gandhi in Sald. & Nicols., Fl. Hassan Dist. 640. 1976; Matthew & Rani in Matthew, Fl. Tam. Carnatic 3(2): 1653. 1983.

Scapigerous herbs. Leaves linear, to 10 cm long, paplillose. Flowers in terminal heads on scapes; bracts orbicular, chest-nut brown, mucronate; sepals 3, deciduous; petals 3, yellow, obovate, clawed; stamens 3; staminodes 3; ovary superior, 1-locular; ovules α, on 3 parietal placentae. Capsule loculicidal. Seeds ellipsoid, ribbed.

Fl. & Fr.: Jan.-Mar. *Distr.*: Sri Lanka, India, Myanmar, China, Malesia, Australia. Occasional in marshy fields. *NAK 2972* (Ranni, c. 100 m).

135. COMMELINACEAE

1. Fruit a berry. .... **Aclisia**
1. Fruit a capsule.
    2. Inflorescence enclosed in leaf-sheaths. .... **Amischophacelus**
    2. Inflorescence exceeding leaf-sheaths.
        3. Fertile stamens 3.
            4. Cymes from spathaceous bracts. .... **Commelina**
            4. Cymes not from spathaceous bracts.
                5. Flowers regular. .... **Murdannia**
                5. Flowers irregular. .... **Aneilema**
        3. Fertile stamens 6.
            6. Inflorescence a cyme; bracteoles foliaceous; capsule 3-locular. .... **Cyanotis**
            6. Inflorescence a panicle; bracteoles not foliaceous; capsule 2-locular. .... **Floscopa**

ACLISIA E. H. F. Meyer ex K. B. Presl
Rel. Haenk. 1:137.1827.

Aclisia secundiflora (Bl.) Bakh. f. in Backer & Bakh. f., Fl. Java 3: 658. 1968; Gandhi in Sald. & Nicols., Fl. Hassan Dist. 641. 1976. *Commelina secundiflora* Bl., Enum. Pl. Jav. 1:5.1827. *Aclisia indica* Wight, Ic. t. 2068. 1853. *Pollia sorzogonensis* (E.Meyer) Endlicher ex Steud. var. *indica* (Wight) Clarke in A. & C.DC, Monogr. Phan. 3:127. 1881; Hook.f., Fl. Brit. India 6:368. 1892; Fischer in Gamble, Fl. Pres. Madras 1534.1931.

Stout subscandent herbs. Leaves elliptic-lanceate, to 16 by 5 cm, base rounded, apex caudate; lower leaves with long sheaths; upper ones sessile, without sheaths. Flowers blue, in terminal panicles; stamens 3 or 6; staminodes 0 or 3; ovary sessile, 3-locular; ovules 2-10 per locule. Berry globose, bluish. Seeds compressed.

Fl. & Fr.: Nov.-Dec. *Distr.:* Peninsular India. Very rare. Among forest undergrowths in upper hills. *NAK 1447 & 2087* (Arampa, c. 400 m).

AMISCHOPHACELUS R. S. Rao *et* R. V. Kammathy
J. Linn. Soc., Bot. 59:305.1966.

Amischophacelus axillaris (L.) Rao & Kamm., J.Linn. Soc. Bot. 59: 306.1966; Gandhi in Sald. & Nicols., Fl. Hassan Dist. 642. 1976; Matthew, Ill. Fl. Tam. Carnatic t. 750. 1982; Matthew & Britto in Matthew, Fl. Tam. Carnatic 3(3):1655.1983; Nicols., Suresh & Manilal, An Interpr. Hort. Malab. 282.1988; Mohanan & Sivad., Fl. Agasthyamala 729.2002. *Commelina axillaris* L., Sp.Pl.42.1753. *Cyanotis axillaris* (L.) D.Don, Prodr. Fl. Nepal.46.1825; Hook.f., Fl. Brit. India 6:388.1892; Fischer in Gamble, Fl. Pres. Madras 1550.1931.

Tradescantia axillaris (L.) L., Mant. Pl.2:371.1771; Roxb., Pl.Corom. t.107.1799. *Nirpulli* Rheede, Hort.Malab. 10:25, t.13.1690.

Ascending herbs. Leaves alternate, narrow-lanceate, 5-10 by 0.5-0.8 cm, succulent, usually flat, base obtuse, apex acuminate, sheaths pouched. Cymes in axillary clusters, enclosed in leaf-sheaths. Flowers irregular, bisexual; sepals 3, greenish; petals 3, pinkish purple; stamens 6, filaments pink with pilose hairs; ovary superior, 3-locular; ovules 2 per locule. Capsule ellipsoid, loculicidal. Seeds 6, cylilndric, pitted.

Fl. & Fr.: Nov.-Jan. *Distr.*: Himalayas, India to China, Malesia, Australia. Common in marshy lowlands. *NAK 670, 903* (Pandalam, Tiruvalla).

ANEILEMA R. Brown
Prodr. 270. 1810.

Decumbent or erect herbs. Flowers irregular, usually in terminal panicels or axillary cymes; sepals 3, free, membranous; petals 3, free, subequal; stamens 2 or 3; staminodes 2-4, filaments bearded or not. Capsule globose. Seeds rugose or reticulate.

1. Leaves crowded below the panicle. .... 1. **A. montanum**
1. Leaves scattered.
   2. Leaves glabrous; bracts caducous. .... 2. **A. ovalifolium**
   2. Leaves hispid above; bracts persistent. .... 3. **A. scaberrimum**

1. **Aneilema montanum** (Wight) Clarke in A. & C.DC., Monogr. Phan. 3:217.1881; Hook.f., Fl. Brit. India 6:382.1892; Fischer in Gamble, Fl.Pres.Madras 1546.1931; Rao, Notes Roy. Bot. Gard. Edinburgh 25:182. 1964; Gandhi in Sald. & Nicols., Fl. Hassan Dist. 643. 1976; Mohanan & Sivad., Fl. Agasthyamala 730.2002. *Dictyospermum montanum* Wight, Ic. t. 2069. 1853.

Erect or subscandent herbs. Leaves ovate to elliptic, 8-13 by 3-4 cm, base cuneate or rounded, apex acuminate. Flowers white in long-peduncled panicles; bracts leafy. Capsule globose. Seeds rugose.

Fl. & Fr.: Most of the seasons. *Distr.*: Peninsular India. Most common species of the genus. Along moist or water logged areas of hills. *NAK 201, 616, 714 & 742* (Almost all forests).

2. **Aneilema ovalifolium** (Wight) Hook.f. ex Clarke in A. & C.DC., Monogr. Phan. 3:218.1881; Hook. f., Fl. Brit. India 6:382. 1892; Fischer in Gamble, Fl. Pres. Madras 1546.1931; Rao, Notes Roy. Bot. Gard. Edinburgh 25:188.1964; Gandhi in Sald. & Nicols., Fl. Hassan Dist. 643. 1976; Mohanan & Sivad., Fl. Agasthyamala 731.2002. *Dictyospermum ovalifolium* Wight, Ic. t.2070.1853.

Erect stout herbs. Leaves elliptic-lanceate, 6-10 by 3-4 cm, base cuneate, apex caudate. Flowers white on short peduncled panicles. Capsule globose with recurved pedicels

Fl. & Fr.: Oct.-Dec. *Distr.*: Peninsular India. Rare. Among forest undergrowths. *NAK 89* (Moozhiar, c. 250 m).

3. **Aneilema scaberrimum** (Bl.) Kunth, Enum. 4:69. 1843; Hook. f., Fl.Brit. India 6:382.1892; Fischer in Gamble, Fl. Pres. Madras 1546.1931; Mohanan & Sivad., Fl. Agasthyamala 731.2002. *Commelina scaberrima* Bl., Enum. 1:4.1827; Gandhi in Sald. & Nicols., Fl. Hassan Dist. 643. 1976. *Dictyospermum protensum* Wight, Ic. t.2071.1853.

Trailing herbs. Leaves oblong-lanceate or elliptic-lanceate, to 10 by 4 cm, base and apex acute, hispid above. Flowers pale pink in panicles of umbelled-racemes; bracts persistent. Capsule globose. Seeds 3, rugose.

Fl. & Fr.: Sep.-Dec. *Distr.*: Peninsular India. Less common. Among forest undergrowths. *NAK 320, 1167* (Moozhiar, c. 250 m).

COMMELINA Linnaeus
Sp. Pl.40.1753.

Herbs. Leaves subsessile, sheathing at base. Inflorescence of dichotomous cymes, axillary and/or terminal; spathes hooded, funnel-form or ovate-lanceate, plicate. Flowers of upper cymes small and deciduous; sepals 3, outer one cucullate; petals pink, clawed, unequal; stamens 3, filaments glabrous; ovary 3-locular; ovule(s) 1 or 2 in anticous cells, solitary or 0 in posticuous cell. Capsule loculicidal. Seeds smooth, rugose, reticulate or pitted.

1. Bract ovate-lanceate.
    2. Seeds appendaged; posticous cell of capsule obtuse on margins. ....1. **C. attenuata**
    2. Seeds not appendaged; posticous cell of capsule keeled on margins. ....3. **C. diffusa**
1. Bract funnel-form or cucullate.
    3. Capsule 2-locular; leaves linear-lanceate. ....4. **C. ensifolia**
    3. Capsule 3-locular; leaves other than linear lanceate.
        4. Two anticous cells of ovary 2-ovuled, the posticous cell 1-ovuled or obsolute. ....2. **C. benghalensis**
        4. All cells of ovary 1-ovuled.
            5. Capsule 3-valved; leaves elliptic-lanceate. ....6. **C. paludosa**
            5. Capsule 2-valved; leaves narrow lanceate. ....5. **C. erecta**

1. **Commelina attenuata** J. Koenig ex Vahl, Enum. Pl. 2:168. 1806; Hook. f., Fl. Brit. India 6:372.1892; Fischer in Gamble, Fl. Pres. Madras 1539. 1931; Matthew & Britto in Matthew, Fl. Tam. Carnatic 3(3): 1656. 1983.

Erect herbs. Leaves linear-lanceate, to 10 by 1 cm, base obtuse, apex acute to acuminate; sheath margin ciliate; bract narrowly lanceate, base sagittate, softly pilose within, apex caudate. Flowers blue. Capsule cylindric, 2-locular. Seeds appendaged.

Fl. & Fr.: Nov.-Jan. *Distr.*: India, Sri Lanka. Less common. Moist fields. *NAK 394* (Adoor, c. 20 m).

2. **Commelina benghalensis** L., Sp.Pl. 41.1753; Hook.f., Fl. Brit. India 6:370 1892; Fischer in Gamble, Fl. Pres. Madras 1539.1931; Gandhi in Sald. & Nicols., Fl. Hassan Dist. 644. 1976; Matthew, Ill. Fl. Tam. Carnatic t.751.1982; Matthew & Britto in Matthew, Fl. Tam. Carnatic 3(3): 1657. 1983.

Ascending or suberect herbs. Leaves ovate or elliptic-ovate, 3-5 by 2-3 cm, base rounded or subtruncate, apex acute-obtuse, sheath-apex with rufous hairs; bract clustered, funnel-shaped. Flowers blue, sepals (outer ones) linear, inner orbicular. Capsule ellipsoid, 3-locular. Seeds pitted.

Fl. & Fr.: Nov.-Jan. *Distr.:* Africa, Himalayas, India to China, Japan, Malesia. Common in exposed arable lands. *NAK 88* (Moozhiar, c. 250 m).

3. **Commelina diffusa** Burm. f., Fl. Ind. 18, t.7, f.2. 1768; Rao, Notes Roy. Bot.Gard. Edinburgh 25:179.1964; Gandhi in Sald. & Nicols., Fl. Hassan Dist. 644. 1976. Matthew & Britto in Matthew, Fl. Tam. Carnatic 3(3): 1658. 1983; Mohanan & Sivad., Fl. Agasthyamala 733.2002. *C. nudiflora* auct., non L.,1753: Hook.f., Fl.Brit. India 6:369. 1892; Fischer in Gamble, Fl. Pres. Madras 1538.1931.

Trailing or diffuse herbs. Leaves ovate-lanceate, 4-6 by 0.5-0.8 cm, base subcordate, apex acuminate; sheath ciliate; bract complicate, oblong-lanceate, base cordate; peduncle to 3.5 cm long. Capsule 3-locular. Seeds tuberculate, appendaged.

Fr. & Fr.: Nov.-Dec. *Distr.*: Pantropics. Common among hills in moist wayside thickets. *NAK 1486, 2016, 2624* (Moozhiar, Arampa, Konni, etc.).

4. **Commelina ensifolia** R.Br., Prodr. 269.1810; Hook.f., Fl. Brit. India 6:374.1892; Fischer in Gamble, Fl. Pres. Madras 1540. 1931; Rao, Blumea 14:352. 1966; Gandhi in Sald. & Nocols., Fl. Hassan Dist. 645. 1976; Matthew & Britto in Matthew, Fl. Tam. Carnatic 3(3): 1658. 1983. *C. undulata* var. *setosa* Clarke in A. & C.DC., Monogr. Phan. 3:179. 1881; Hook.f., Fl. Brit. India 6:373.1892; Fischer in Gamble, Fl.Pres. Madras 1539.1931; Rao, Notes Roy. Bot. Gard. Edinburgh 26: 352. 1965.

Trailing herbs. Leaves linear or linear-lanceate, 4-11 by 0.6-0.8 cm, base narrow-obtuse, apex acute, glabrescent; bract solitary, cucullate, pubescent; peduncle to 1 cm long. Capsule 2-locular. Seeds ovoid, smooth.

Fl. & Fr.: Nov.-Jan. *Distr.*: Sri Lanka, India, Australia. Rare. Among thickets in hills. *NAK 845, 2015* (Ranni, Arampa etc.).

5. **Commelina erecta** L., Sp. Pl.41.1753; Morton, J.Linn. Soc. Bot. 60:183.1967; Rao, Maharashtra Vidhan Mandir Patrika 6:53. 1971; Matthew & Britto in Matthew, Fl. Tam. Carnatic 3(3): 1659.1983. *C. undulata* R. Br., Prodr. 270. 1810; Hook. f., Fl. Brit. India 6:373.1892; Rao, Blumea 14:351.1966. *C. kurzii* Clarke, J.Linn.Soc.Bot. 11:444.1870; Hook. f., Fl.Brit.India 6:373. 1892; Fischer in Gamble, Fl.Pres. Madras 1540.1931.

Erect herbs. Leaves lanceate or oblong-lanceate, to 13 by 3.5 cm, base oblique, apex gradually acute, chartaceous; sheath-margin ciliate; bract 3-5, in a terminal cluster, cucullate, broadly cordate, subsessile. Capsule 3-locular, 2-valved. Seeds 3, adnate to capsule wall.

Fl. & Fr.: Nov.-Jan. *Distr.*: India, Africa, Australia. Not common. Among forest undergrowth. *NAK 823, 851* (Ranni, Moozhiar, etc.).

6. **Commelina paludosa** Bl., Enum. Pl. Jav. 1:2.1827. *C. obliqua* Buch.-Ham. ex D.Don, Prodr. Fl. Nepal. 45.1825, non Vahl, 1805; Hook. f., Fl. Brit. India 6:372. 1892, p.p.; Fischer in Gamble, Fl. Pres. Madras 1539. 1931; Gandhi in Sald. & Nicols., Fl. Hassan Dist. 645. 1976.

Stout herbs. Leaves lanceate to elliptic-lanceate, 6-20 by 1.5-5 cm, base and apex acute to caudate, scabrous, sessile or shortly petioled; bract(s) solitary or crowded, funnel-form, glabrous or scabrid. Capsule obovoid. Seeds cylindric, smooth.

Fl. & Fr.: Nov.-Dec. *Distr.*: Himalayas, India. Not common. Among forest undergrowth in upper hills. *NAK 38, 1154 & 1500* (Moozhiar & Arampa, 200-500 m).

CYANOTIS D. Don
Prodr. Fl. Nepal. 45.1825.

Ascending herbs. Leaves sheathing at base. Inflorescence of scorpioid cymes, axillary and/or terminal; bracts prominent; bracteoles 2-seriate, secund, foliaceous; sepals subequal; petals purple, unequal; stamens 6, fertile, filaments hairy; ovary 3-locular; ovules 2 per locule. Capsule loculicidal. Seed(s) 1 or 2 per locule.

1. Branchlets cottony or cobwebby. .... 1. **C. arachnoidea**
1. Branchlets not so.
   2. Bracteoles prominently cross-nerved; plants purplish. .... 2. **C. arcotensis**
   2. Bracteoles not cross-nerved; plants greenish.
      3. Roots tuberous, fascicled. .... 5. **C. tuberosa**
      3. Roots not so.
         4. Glabrous herbs; cymes soliltary, deeply recurved. .... 3. **C. cristata**
         4. Pubescent herbs; cymes clustered, straight or obscurely recurved.
            5. Leaves silky-villous below; cymes terminal; seeds rugose. .... 6. **C. villosa**
            5. Leaves appressed-pilose below; cymes terminal and/or axillary; seeds pitted. .... 4. **C. pilosa**

1. **Cyanotis arachnoidea** Clarke in A. & C. DC., Monogr. Phan. 3:250. 1881; Hook.f., Fl. Brit. India 6:383. 1892; Fischer in Gamble, Fl. Pres. Madras 1550. 1931; Matthew & Britto in Matthew, Fl. Tam. Carnatic 3(3): 1661. 1983; Mohanan & Sivad., Fl. Agasthyamala 735.2002.

Stem cottony or cobwebby. Leaves elliptic-lanceate to linear-lancate, to 10 by 0.8 cm, base obtuse, apex acute, cottony. Cymes terminal and/or axillary; bracts ovate-lanceate, exceeding the cyme. Flowers pink or purple; ovary pilose. Capsule ellipsoid. Seeds pitted.

Fl. & Fr.: Nov.-Jan. *Distr.*: Peninsular India, Sri Lanka. Not common. In upper hill slopes. *NAK 2215* (Chittar, c.300 m).

2. **Cyanotis arcotensis** Rao, Blumea 14:345.1966; Matthew & Britto in Matthew, Fl. Tam. Carnatic 3(3): 1661. 1983; Mohanan & Sivad., Fl. Agasthyamala 735.2002. *C. papilionacea* auct., non (L.) Schult.f., 1853, p.p.: Hook.f., Fl. Brit. India 6:384.1892; Fischer in Gamble, Fl. Pres. Madras 1549. 1931; Rao, Notes Roy. Bot. Gard. Edinburgh 25:185. 1965.

Plants purplish. Leaves lanceate, 2-5.5 by 0.2-0.3 cm, base cordate, apex acute; sheath pubescent. Cymes terminal, solitary or 2-4 in a cluster; peduncle 3-6 cm long; bracts as long as cyme, cross-nerved, ciliate. Capsule ellipsoid. Seeds 6, pitted.

Fl. & Fr.: Nov.-Jan. *Distr.*: Peninsular India. Not common. On exposed rocky hills. *NAK 897 & 1090* (Plappally, c. 400 m).

Much ressembling **C. cristata**, but differs in having purplish colour of plant and prominently corss-nerved bracteoles.

3. **Cyanotis cristata** (L.) D.Don, Prodr. Fl. Nepal. 46.1825; Hook.f., Fl.Brit. India 6:385.1892; Fischer in Gamble, Fl. Pres. Madras 1549.1931; Gandhi in Sald. & Nicols., Fl. Hassan Dist. 646. 1976; Matthew & Britto

in Matthew, Fl. Tam. Carnatic 3(3): 1662.1983; Nicols., Suresh & Manilal, An Interpr. Hort. Malab. 282.1988; Mohanan & Sivad., Fl. Agasthyamala 736.2002. *Commelina cristata* L., Sp. Pl. 42.1753. *Veetla-caitu* Rheede, Hort. Malab. 7:109, t.58. 1688.

Plants greenish. Leaves ovate or oblong-lanceate, 2-3 by 0.8-2 cm, base cordate, margin ciliate, apex obtuse. Cymes deeply falcate, recurved; bracts ovate-lanceate, exceeding the cyme. Flowers rose to purple; ovary globose. Capsule to 2 mm long. Seeds 6, trigonous with 2 large pits on 2-sides.

Fl. & Fr.: Most of the seasons. *Distr.*: Tropics. Common on exposed lands. *NAK 831 & 1020* (Ranni in Tiruvalla).

4. **Cyanotis pilosa** Schult. f. in L., Syst. Veg. 7:1155. 1830; Hook.f., Fl. Brit. India 6:387. 1892; Fischer in Gamble, Fl. Pres. Madras 1549. 1931, p.p.; Matthew & Britto in Matthew, Fl. Tam. Carnatic 3(3): 1663. 1983; Mohanan & Sivad., Fl. Agasthyamala 737.2002.

Pilose herbs. Leaves narrowly lanceate or elliptic-lanceate, 5-10 by 1-2.5 cm, base subcordate, apex acute, lower surface appressed pilose; sheath pilose. Cymes terminal and/or axillary; bracts lanceate, exceeding the cymes. Flowers purplish. Capsule to 3 mm long. Seeds pyramidal, pitted.

Fl. & Fr.: Nov.-Dec. *Distr.*: Peninsular India, Sri Lanka. Less common. Among forest undergrowths in upper hills. *NAK 1153, 1451* (Moozhiar, Arampa etc.).

5. **Cyanotis tuberosa** (Roxb.) Schult. & Schult. f.in L., Syst. Veg. 7(2): 1153. 1830; Hook.f., Fl. Brit. India 6:386. 1892; Fischer in Gamble, Fl. Pres. Madras 1549. 1931; Gandhi in Sald. & Nicols., Fl. Hassan Dist. 646. 1976; Matthew, & Britto in Matthew, Fl. Tam. Carnatic 3(3): 1663. 1983. *Tradescantia tuberosa* Roxb., Pl. Corom. t. 108. 1799 & Fl. Ind. 2:119. 1832.

Tuberous rooted herbs; basal leaves oblong, 10-30 by 1-1.5 cm, upper ones linear ensiform, base obtuse, apex acute, lower surface sericeous. Cymes terminal and/or axillary; peduncle 2-5 cm long; bracts ovate, equal or slightly exceeding the cyme. Capsule to 2.5 mm long. Seeds 6, pyramidal.

Fl. & Fr.: Oct.-Dec. Distr.: Peninsular India. Rare. On exposed hills slopes. *NAK 2755* (Kakki hills, c. 1000 m).

6. **Cyanotis villosa** (Spreng.) Schult. f. in L., Syst. Veg. 7:1155. 1830; Hook.f., Fl. Brit. India 6:387. 1892; Fischer in Gamble, Fl. Pres. Madras 1550. 1931; Matthew & Britto in Matthew, Fl. Tam. Carnatic 3(3): 1664. 1983. *Tradescantia villosa* Spreng. in L., Syst. Veg. (ed. 16) 2:116. 1925. *Cyanotis lanceata* Wight, Ic. t. 2085. 1953.

Villous herbs. Leaves ovate-lanceate to narrowly lanceate, 3-6 by 0.5-1 cm, base cordate, apex acuminate, lower surface villous. Cymes terminal, 2 or 3 in a cluster; bracts exceeding the cyme; corolla purple. Capsule to 3 mm long. Seeds 5, rugose.

Fl. & Fr.: Nov.-Jan. *Distr.*: Peninsular India, Sri Lanka. Not common; Among forests undergrowth in upper hills. *NAK* 1184 (Arampa, c. 400 m).

FLOSCOPA Loureiro
Fl. Cochinch. 193. 1790.

Floscopa scandens Lour., Fl. Cochinch. 193. 1790; Hook.f., Fl. Brit. India 6:390. 1892; Fischer in Gamble, Fl. Pres. Madras 1552. 1931; Mohanan & Sivad., Fl. Agasthyamala 737.2002. *Dithyrocarpus rothii* Wight, Ic. t.208. 1839.

Subscandent or ascending herbs. Leaves elliptic-lanceate, to 7 by 2 cm, base cuneate, apex acute, upper surface scaberulous; sheath-mouth fringed with villous hairs. Flowers white or pink in terminal or axillary thyrsoid secund-panicles; sepals 3; petals 3; stamens 6; staminodes 0; ovary 2-locular; ovule 1 per locule. Capsule subglobose. Seeds wrinkled.

Fl. & Fr.: Nov.-Jan. *Distr.:* Peninsular India. Common. Along streamsides. Locally abundant. *NAK 274, 1370, 1499* (Almost all forests).

MURDANNIA Royle
Ill. Bot. Himalayan Mts.
403. 1840 (nom. cons.).

Annual or perennial herbs. Leaves often alternate. Inflorescence of terminal or axillary panicles. Flowers regular, pink to violet; sepals membranous; petals subequal; stamens 3; staminodes 1-3; ovary 3-locular. Capsule usually 3-angled. Seeds 1 or 2-seriate, rugose or pitted.

1. Seeds 1-seriate in each locule.
    2. Leaves radical; scape present, very long.
        3. Scape c. 50 cm or more long; leaves ensiform. ....2. **M. edulis**
        3. Scape c. 15 cm long; leaves oblong. ....3. **M. glauca**
    2. Leaves cauline; scape very short or absent.
        4. Leaves broadly elliptic or ovate-elliptic, margin hyaline. ....4. **M. japonica**
        4. Leaves linear-lanceate, margin not so.
            5. Panicle sessile. ....8. **M. zeylanica**
            5. Panicle pedunculate.
                6. Stems decumbent; mouth of sheath straight.
                    7. Flowers rose coloured; leaves ovate. ....6. **M. pauciflora**
                    7. Flowers purple coloured; leaves oblong-lanceate. ....5. **M. nudiflora**
                6. Stem erect; mouth of sheath oblique. ....1. **M. dimorpha**
1. Seeds 2-seriate in each locule. ....7. **M. semiteres**

1. **Murdannia dimorpha** (Dalz.) Brueck. in Engler & Prantl, Pflanzenf. (ed 2) 15a:173. 1930. *Aneilema dimorphum* Dalz. in Hooker's J. Bot. Kew Gard. Misc. 3:138. 1851; Hook.f., Fl. Brit. India 6:377. 1892; Fischer in Gamble, Fl. Pres. Madras 1545. 1931.

Erect branched herbs. Leaves lanceate to linear-oblong, 3-8 by 0.6-1.2 cm, base obtuse, apex acute; sheath-mouth oblique. Panicle very lax, few-flowered; bracts ovate to orbicular. Seeds subcubical, dark brown.

Fl. & Fr.: Nov.-Jan. *Distr.:* Sri Lanka, Peninsular India. Common. in shady hills. *NAK 811, 844, 902* (Almost all evergreen forests).

2. **Murdannia edulis** (Stokes) Faden, Taxon 29:77. 1980. *Commelina edulis* Stokes, Bot. Materia Med. 1:184. 1812. *Commelina scapiflora* Roxb., Fl. Ind. I:178.1820. *Aneilema scapiflorum* (Roxb.) Kostel., Alleg.

Med.-Pharm. Fl. 1:127. 1831; Hook.f., Fl. Brit. India 6:375. 1892; Fischer in Gamble, Fl. Pres. Madras 1544. 1931.

Scapigerous herbs. Leaves radical, linear-ensiform, 30 by 1 cm, base rounded, apex acuminate; scape to 50 cm long; bracts foliaceous, sheathing. Flowers c. 2.5 cm ϕ. Seeds angled, minutely reticulate.

Fl. & Fr.: Nov.-Jan. *Distr.*: Himalayas, Peninsular India, Sri Lanka. Not common. On exposed grassy hills. *NAK 2171* (Chittar, c. 400 m).

3. **Murdannia glauca** (Thw. ex Clarke) Brueck. in Engler & Prantl, Pflanzenf. (ed.2) 15 a: 173. 1930; Mohanan & Sivad., Fl. Agasthyamala 739.2002. *Aneilema glaucum* Thw. ex Clarke in A. & C. DC., Monogr. Phan. 3:200. 1881; Hook.f., Fl. Brit. India 6:375. 1892; Fischer in Gamble, Fl. Pres. Madras 1544. 1931.

Scapigerous herbs. Leaves radical, oblong, flat, c. 14 by 3 cm, base rounded, margin crispate, apex acute, glaucous beneath; scape c. 20 cm long. Panicle spreading. Flowers small, purple. Capsule oblanceoid. Seeds minutely reticulate.

Fl. & Fr.: Oct.-Dec. *Distr.*: Southern Western Ghats, Sri Lanka. Rare. On dripping rocks. Collected only once. *NAK 597* (Thekkuthode, c. 70 m).

4. **Murdannia japonica** (Thunb.) Faden, Taxon 26:142. 1977. *Commelina japonica* Thunb., Trans. Linn. Soc. London 2:332. 1794; Mohanan & Sivad., Fl. Agasthyamala 739.2002. *Aneilema lineolatum* Kunth, Enum. Pl. 4:69. 1843; Hook.f., Fl. Brit. India 6:376. 1892; Fischer in Gamble, Fl. Pres. Madras 1544. 1931.

Erect robust herbs; roots tuberous. Leaves ovate-lanceate or elliptic, 10-16 by 3.5-6 cm, base rounded or cordate, margin undulate, hyaline, apex acute. Flowers blue, in terminal dichotomously branched panicles. Capsule subglobose. Seeds reticulate, glandular-pubescent.

Fl. & Fr.: Sept.-Oct. *Distr.*: Peninsular India. Common in forest undergrowths of upper hills. *NAK 726, 795 & 986* (Ranni, Sabarimala, etc.)

5. **Murdannia nudiflora** (L.) Brenan, Kew Bull. 7:189. 1952; Gandhi in Sald. & Nicols., Fl. Hassan Dist. 648. 1976; Matthew & Britto in Matthew, Fl. Tam. Carantic 3(3): 1666. 1983; Nicols., Suresh & Manilal, An Interpr. Hort. Malab. 283. 1988. *Commelina nudiflora* L., Sp. Pl. 41. 1753, p.p.; Roxb., Fl. Ind. 1·177. 1820. *Aneilema nudiflorum* (L.) R. Br., Prodr. 271. 1810; Hook.f., Fl. Brit.India 6:378. 1892; Fischer in Gamble, Fl.Pres. Madras 1545. 1931. *Tali-pullu* Rheede, Hort. Malab. 9:123, t. 58. 1688.

Procumbent herbs; rooting at lower nodes. Leaves oblong-lanceate, 8-15 by 1-2 cm, base rounded or cordate, apex acute; sheaths ciliate. Flowers violet, in lax terminal cymose-panicles. Capsule globose. Seeds 2 per locule, pitted.

Fl. & Fr.: Aug.-Dec. *Distr.*: Tropics. Less common. Among forest undergrowths. *NAK 243 & 730* (Moozhiar, Ranni, etc.).

6. **Murdannia pauciflora** (Wight) Brueck. in Engler & Prantl, Pflanzenf. (ed. 2) 15a: 173. 1930; Gandhi in Sald. & Nicols., Fl. Hassan Dsit. 649. 1976; Mohanan & Sivad., Fl. Agasthyamala 740.2002. *Aneilema pauciflorum* Wight, Ic. t. 2077. 1853, non Dalz., 1851; Hook.f., Fl. Brit. India 6:378. 1892; Fischer in Gamble, Fl. Pres. Madras 1545. 1931. *Murdannia wightii* Rao & Kammathy, Notes Roy. Bot. Gard. Edinburgh 25:184. 1964, nom. superfl.

Decumbent, spreading herbs. Leaves ovate, 2-3 by 1-1.5 cm, base cordate, apex acute. Flowers rose-coloured or brownish-yellow, 1-5 in axillary and terminal cymes; filaments glabrous. Capsule ellipsoid. Seeds pitted.

Fl. & Fr.: Nov.-Jan. *Distr.:* Peninsular India. On exposed moist-shady areas of upper hills. *NAK 821* (Plappally, c. 400 m).

7. **Murdannia semiteres** (Dalz.) Santapau, Poona Agric. Coll. Mag. 41:284. 1951; Brenan, Kew Bull. 7:184. 1952; Gandhi in Sald. & Nicols., Fl. Hassan Dist. 649. 1976; Matthew & Britto in Matthew, Fl. Tam. Carnatic 3(3): 1666. 1983; Nicols., Suresh & Mani, An Interpr. Hort. Malab. 283. 1988; Mohanan & Sivad., Fl. Agasthyamala 740.2002. *Aneilema semiteres* Dalz. in Hook. J. Bot. Kew Gard. Misc. 3:138. 1851. *Dichaespermum juncoides* Wight, Ic. t. 2078, f.2.1853. *A. paniculatum* Wall. ex Clarke in A. & C.DC., Monogr. Phan. 3:21. 1881, non Wight, 1853; Gamble, Fl. Pres. Madras 1546. 1931. *Dichaespermum paniculatm* Hook.f. & Thoms. ex Hook.f., Fl. Brit. India 6:381. 1892, pro syn. *Nelam-pullu* Rheede, Hort. Malab. 10:37, t.19. 1690.

Erect tufted slender herbs. Leaves linear, filiform, apex acute-apiculate. Flowers blue in branched lax panicles; bracts minute, ocreate. Capsule subglobose, angular. Seeds compressed, smooth.

Fl. & Fr.: Oct.-Dec. *Distr.*: Peninsular India. Not common. On moist shady upper hills. *NAK 1093* (Sabarimala, c. 200 m).

8. **Murdannia zeylanica** (Clarke) Brueck. var. **longicapsa** (Clarke) Rao & Kammathy, Bull. Bot. Surv. India 3:394. 1961; Mohanan & Sivad., Fl. Agasthyamala 741.2002. *Aneilema zeylanica* var. *longicapsa* Clarke in A. & C.DC., Monogr. Phan. 3:204.1881; Hook.f., Fl. Brit. India 6:376. 1892; Fischer in Gamble, Fl. Pres. Madras 1544. 1931.

Slender decumbent herbs. Leaves ovate-lanceate, 4-8 by 2-3 cm, base rounded or cuneate, apex acute, hairy. Flowers white in terminal sessile panicles. Capsule narrowly ellipsoid. Seeds smooth.

Fl. & Fr.: Aug.-Nov. *Distr.:* Peninsular India. Common. Among forests undergrowth. *NAK 320* (Moozhiar, c. 250 m).

136. FLAGELLARIACEAE

FLAGELLARIA Linnaeus
Sp. Pl. 333. 1753.

Flagellaria indica L., Sp. Pl. 333. 1753; Hook.f., Fl. Brit. India 6: 391. 1892; Henry & Swamin., Ind. J. For. 3: 140 . 1980; Nicols., Suresh & Manilal, An Interpr. Hort. Malab. 293. 1988. *Panambu-valli* Rheede, Hort. Malab. 7:99, t. 53. 1688.

Climbing shrubs. Leaves sessile, lanceate, 10-25 by 3-5 cm, base sheathing, 2-auricled, apex cirrhose, α-nerved, coriaceous. Flowers bisexual, sessile, small in terminal panicles; perianth-lobes 6, cream-coloured, persistent; stamens 6, exerted; ovary superior; 3-locular; ovule 1per locule; styles 3. Fruit a Drupe.

Fl. & Fr.: Aug.-Oct. *Distr.*: Trop. Asia, Africa, Rare. Occasional in certain sacred groves. *NAK 1320* (Kodumon, c. 40 m).

137. ARECACEAE
(PALMAE)

1. Stragglers; spadices as long as leaves. .... **Calamus**
1. Trees or shrubs; spadices shorter than leaves.
   2. Leaves 2-pinnate; leaflets obliquely cuneate. .... **Caryota**

2. Leaves 1-pinnate; leaflets linear-oblong.
   3. Leaflets glaucous beneath, base auricled. ....**Arenga**
   3. Leaflets not so, base other than auricled.
      4. Lowest leaflets modified into spines; ovary of 3 free carpels. ....**Phoenix**
      4. Lowest leaflets non-spinous; ovary of 1 carpel (3-celled). ....**Bentinckia**

ARENGA Labillardiere
in A.P. de Candolle, Bull. Sci. Soc. Philom.
Paris 2:162. Nov. 1800 (nom. cons.).

Arenga wightii Griff., Calcutta J. Nat. Hist. 5: 475. 1845; Becc. & Hook.f. in Hook.f., Fl.Brit. India 6: 422. 1892; Fischer in Gamble, Fl. Pres. Madras 1558. 1931; Mohanan & Sivad., Fl. Agasthyamala 742.2002.

Erect, stout monoecious palms, up to 10 m tall. Leaves pinnatisect, to 7 m long; leaflets alternate, linear-ensiform, base unequally auricled, white-glaucous beneath. Spadices much branched, to 1 m long. Spathes a or absent; ovary 3-locular; ovule 1 per locule. Fruit subglobose. Seeds compressed.

Fl. & Fr.: Apr. *Distr.*: Peninsular India. Rare. In evergreen forests. *NAK 1653* (Kalleli, c. 250 m).

BENTINCKIA A. Berry
in Roxburgh, Fl. Indica ed.2, 3:621. 1832.

Bentinckia condapanna Berry ex Roxb., Fl. Ind. 3: 621. 1832; Becc. & Hook.f. in Hook. f., Fl. Brit. India 6: 418. 1892; Fischer in Gamble, Fl. Pres. Madras 155. 1931; Mohanan & Sivad., Fl. Agasthyamala 743.2002.

Slender palms, to 20 m tall. Leaves c. 1.5 m long, leaflets oblong, ribbed. Spadices from axils of fallen leaves; spathes α. Flowers minute, sunken on axis. Male flowers gluacous; stamens 6. Female flowers, large, ovoid; staminodes 6, minute; ovary 3-locular; ovule 1. Fruit subglobose.

*Fl.:& Fr.:*Throughout the year. *Distr.:* Endemic to Western Ghats. Very rare. Failed to collect, but observed a good population of this species on the steep cuttings of Kakki hills near Ponnambalamedu.

CALAMUS Linnaeus
Sp. Pl. 325. 1753.

Dioecious scramblers. Leaves sheathy, cirrate or ecirrate; sheath spinescent, often flagellate and/or cirrate; leaflets variously arranged. Inflorescence, both male and long-flagellate morphologically similar; bracts tubular at base. Male flowers : calyx cupular, distinctly 3-lobed; petals 3; stamens 6, pistillode minute or 0. Female flowers : larger than male; calyx faintly 3-lobed; petals 3; staminodes 6, basally connate to form a cup like ring; style 0 or very short; stigmas 3; ovary scaley without, 3-locular; ovule 1 per locule. Capsule spherical or ellipsoid, densely scaly without. Seed 1, pitted or groved.

Note :Most of the species of this genus are gradually vanishing from their habitats because of over exploitation for cane industry.

1. Leaflets 6-veined, oblong. ....6. **C. vattayila**
1. Leaflets 3-veined, linear-lanceate.
   2. Leaflets clustered on rachis.
      3. Stem with sheath 3-6 cm f; capsule ellipsoid. ....4. **C. thwaitesii**

3. Stem with sheath ± 1 cm f; capsule globose. ....5. **C. travancoricus**

2. Leaflets not clustered on rachis.

4. Leaf-sheath, rachis and inflorescence-axis glabrous. ....3. **C. rotang**

4. Leaf-sheath, rachis and inflorescence-axis with brown tomentum.

5. Leaf-sheath green; spines to 1.5 cm long; mouth with small stiff spines. ....1. **C. gamblei**

5. Leaf-sheath brownish green, spines to 2.5 cm long; mouth with long papery spines. ....2. **C. hookerianus**

1. **Calamus gamblei** Becc. ex Becc. & Hook.f. in Hook.f., Fl. Brit. India 6: 493. 1893; Becc., Ann. Roy. Bot. Gard. Calcutta 11: 96, 316, t. 123. 1908; Fischer in Gamble, Fl. Pres. Madras 1568. 1928; Renuka, Rattans of Western Ghats 26. 1992. *Pacha chooral.*

Stem with sheaths c. 2.5 cm f; sheath green; spines 0.5-1.5 cm long, horizontal, bulbous-based; flagellum to 4 m long. Leaves ecirrate, c. 1.2 m long; petiole-spines c. 2 cm long; rachis with hooked spines; leaflets c. 40 by 2.5 cm, apex long acuminate with short bristles; veins ciliate. Inflorescence flagellate, c. 3 m long; partial inflorescence 5-6, smaller, male inflorescence with branches of 3 orders; female with branches of 2 orders; involucre disci-form. Capsule globose-obpyriform, c. 2 cm ϕ; scales in ± 23 rows.

Fl. & Fr.: Dec.-Jun. *Distr.:* Endemic to Western Ghats. Less common. In evergreen forests of upper hills. Large scramblers. *Renuka 3173* (KFRI) (Pampa, Kakki hills, c. 1100 m).

Cane is of good quality and widely used in furniture industry.

2. **Calamus hookerianus** Becc., Ann. Roy. Bot. Gard. Calcutta 11: 83, 226, t.70. 1908; Fischer in Gamble, Fl. Pres. Madras 1568. 1928; Renuka, Rattans of Western Ghats 30. 1992.

Stem with sheaths ± 4 cm ϕ; sheaths brownish-green, densely triangular spinescent, to 2.5 cm long, interspersed with small spines and tomentum; mouth with long papery yellow spines; flagellum to 5 cm long. Leaves ecirrate, c. 2 m long; petiole-spines 1-10 cm long, brown or black, triangular, rachis with single row of spines; leaflets c. 50 by 2.5 cm, mid-vein ciliate beneath from middle upwards. Inflorescence to 5 m long; male branching in 3 orders. Female branching in 2 orders; involucre cup-shaped. Capsule subglobose, c. 1 cm ϕ, scales in ± 18 rows.

Fl. & Fr.: Jan.-Jul. Distr.: Western Ghats. Less common. In evergreen forests of upper hills. Moderate sized canes climbing high into the canopy. *NAK 1668* (Velithode, Moozhiar, c. 250 m).

Cane extensively used in furniture industry and basket making.

3. **Calamus rotang** L., Sp. Pl. 325. 1753; Mart., Hist. Nat. Palm. 3:334, t.116: f.8, t. 122:f. 12. 1823-1853; Becc. & Hook.f.,in Hook. f., Fl. Brit. India 6: 447. 1892; Becc., Ann. Roy. Bot.Gard. Calcutta 11:98, 269. 1908; Fischer in Gamble, Fl. Pres. Madras 1568. 1928; Renuka, Rattans of Western Ghats 49. 1992.

Stem with sheath c. 3 cm ϕ; sheath green, spinescent, spines c. 1 cm long, needle-like, yellow with black tip; leaflets c. 35 by 2 cm, apex long acuminate, margin spinulose; mid-vein ciliate from middle upwards. Inflorescence to 3 m long; partial inflorescence 4-5, to 70 cm long. Male-rachillae c. 3 cm long; female-rachilla to 8 cm long; involucre cup-shaped. Capsule ovoid; scales ± 21 rows.

Fl. & Fr.: Oct.-May. *Distr.*: Western Ghats. Occasional in sacred groves. *NAK 110* (Adoor, Kootungal Kavu, c. 20 M). This is usually a coastal species.

4. **Calamus thwaitesii** Becc. ex Becc. & Hook.f. in Hook.f., Fl. Brit. India 6: 441. 1892; Becc., Ann. Roy. Bot. Gard. Calcutta 11:137, 1908, t. 10.11 (appendix). 1913.*C.thwaitesii* Becc. var. *canaranus* Becc., Ann. Roy. Bot. Gard. Calcutta 11:138, 1908; Fischer in Gamble, Fl. Pres. Madras 1567. 1928; Renuka, Rattans of Western Ghats 51. 1992. *Vandichooral.*

Stem with sheath c. 6 cm ϕ; sheath yellow, densely spinescent, spines arising from a raised rim-like surface, flat, to 3 cm long; smaller spines black with yellow base; flagellum c. 9 m long. Leaf c. 3 m long, ecirrate; petiole and rachis yellow, spines black, flat, arranged in oblique whorls; leaflets α, usually grouped, to 80 cm long, mid-vein long-bristly. Inflorescence to 6 m long; male-rachillae to 8 cm long. Flowers distichous; male-rachilla c. 15 cm long; involucre cup-shaped. Capsule ovoid; scales in ± 12 vertical rows.

Fl. & Fr.: Nov.-May. *Distr.*: Endemic to Western Ghats. Less common. Robust rattan in evergreen and semi-evergreen forests. *NAK 1669* (Sabari hills, c. 250 m).

This is the thickest cane available in Western Ghats. So, it is being highly exploited and becoming rare.

5. **Calamus travancoricus** Bedd. ex Becc. & Hook.f., in Hook.f., Fl. Brit. India 6: 452. 1893; Becc., Ann. Roy. Bot. Gard. Calcutta 11: 95, 310. 1908; Fischer in Gamble, Fl. Pres. Madras 1567. 1928; Nicols., Suresh & Manilal, An Interpr. Hort. Malab. 279. 1988; Mohanan & Sivad., Fl. Agasthyamala 745.2002. Renuka, Rattans of Western Ghats 53. 1992. *Katu-tsjurel* Rheede, Hort. Malab. 12:121, t.64. 1693. *Arichooral.*

Stem with sheath c. 0.8 cm ϕ; sheath green, spines 0.5 cm long, mouth of sheath with long slender spines; ocrea papery; flagellum to 1.3 m long. Leaves to 45 cm long, ecirrate; rachis with claw- like spines; leaflets grouped, to 25 cm long. Male inflorescence to 1 m long; partial inflorescence 8-10, to 10 cm long. Female inflorescence to 40 cm long; partial inflorescence 3-4, c. 12 cm long; involucre shallow, concave. Capsule globose, c. 1 cm ϕ; scales in ± 27 rows.

Fl. & Fr.: Oct.-Jun. *Distr.*: Western Ghats. Not common. In evergreen forests. Very slender rattan. *NAK 504* (Adoor-Kootungal kavu, c. 20 m), *1103* (Moozhiar, c. 250 m).

Cane used in handicrafts and furniture industry.

6. **Calamus vattayila** Renuka, Curr. Sci. 56: 1012. 1987 & Rattans of Western Ghats 53. 1992. *Vattayilayan chooral.*

Stem with sheath c. 2.5 cm f; sheath dark green, sparingly spiny; flagellum c. 4 cm long. Leaf ecirrate, to 1 m long; rachis-spines in 3-rows; leaflets alternate, c. 40 by 10 cm, 6-veined, tip short-bristly. Female inflorescence c. 1 m long, in bunches; partial inflorescence to 40 cm long; involucre cup-shaped. Capsule narrowly ellipsoid, scales in ± 27 rows.

Fl. & Fr.: May-Aug. *Distr.*: Endemic to Western Ghats. Not common. In evergreen forests. Rattan characterised by 6-veined broad leaves. *Renuka & Nambiar 3029* (KFRI) (Moozhiar, c. 250 m).

CARYOTA Linnaeus
Sp. Pl. 1189. 1753.

Caryota urens L., Sp. Pl. 1189. 1753; Roxb., Fl. Ind. 3: 625. 1832; Hook.f., Fl. Brit. India 6: 422. 1892; Blatter, Palms Brit. India 339. 1926; Fischer in Gamble, Fl. Pres. Madras 1560. 1931; Matthew, Ill. Fl. Tam.

Carnatic t. 757. 1982; Matthew & Britto in Matthew, Fl. Tam. Carnatic 3(3): 1672. 1983; Nicols., Suresh & Manilal, An Interpr. Hort. Malab. 280. 1988; Mohanan & Sivad., Fl. Agasthyamala 746.2002 *Schunda-pana* Rheede, Hort. Malab. 1: 15-16, t. 11. 1678.

Stout tall palms. Leaves 2-pinnate; pinnae c. 10 pairs, subopposite, curved; leaflets clustered or alternate, 15-20 by 7-10 cm, base cuneate, oblique, margin praemorse, irregulary serrate, upper serration caudate. Spadices interfoliar. Flowers α, monoecious; sepals 3; petals linear-oblong, rounded in female; stamens c. 40; ovary 3-gonous, 3-locular; ovule 1 per locule. Fruit globose. Seeds small.

Fl. & Fr.: Jan.-Apr. *Distr.*: India, Sri Lanka, Myanmar, Malaya. Common in sacred groves. *NAK 2911* (Adoor, c. 20 m).

PHOENIX Linnaeus
Sp. Pl.1188. 1753.

Phoenix loureirii Kunth, Enum. Pl. 3: 257. 1841; Matthew, Ill. Fl. Tam. Carnatic tt. 760, 761. 1982; Matthew & Britto in Matthew, Fl. Tam. Carnatic 3(3): 1674. 1983; Mohanan & Sivad., Fl. Agasthyamala 747.2002. *P. humilis* Royle, Ill. Himal. 394. 397 & 399. 1840, nomen, var. *pedunculata* Becc., Malesia 3: 379 & 387, t.44, ff. 13-15, 18-21 & 25-27. 1890; Becc. & Hook.f.in Hook.f., Fl. Brit. India 6: 427. 1892; Blatter, Palms Brit. Inida 20. 1926; Fischer in Gamble, Fl. Pres. Madras 1560. 1931.

Stout small palms; trunk often absent. Leaves to 2.5 m long;leaflets 15-30 by 0.8-1 cm, base narrow, apex spinous. Spadices interfoliar. Male spikes in clusters; calyx-lobes triangular; stamens 6. Female spadix to 1 m long. Drupe oblong.

Fl. & Fr.:Jan.-May. *Distr.*: India, Myanmar, Indo-China, China. On exposed base hills slopes. Locally abundant in Kakki hills. *NAK 1321* (Kakki hills, c. 1100 m).

Cocos nucifera L. and **Areca catechu** L. are widely cultivated.

138. PANDANACEAE

PANDANUS S. Parkinson
J. Voyage South Seas 46. 1773.

Dioecious shrubs, prop-rooted. Leaves linear-ensiform, trifariously spiralled, prickly on margins and midrib beneath. Inflorescence terminal. Female inflorescence pendulous in age, solitary or a spike of several heads (Cephalia). Cephalium composed of simple or poly drupes; carples 1-locular. Male inflorescence spicate, the spike composed of many densely packed floral units; stamens few to α, in phalanges. Ripe fruits 1-seeded or several seeded.

1. Leaves glaucous beneath; drupe cylindric; montane species. .... 1. **P. thwaitesii**

1. Leaves not glaucous beneath; drupe club-shaped; lowland species. .... 2. **P. unipapillatus**

1. **Pandanus thwaitesii** Martelli, Webbia 1: 369. 1905; Fischer in Gamble, Fl. Pres. Madras 1570. 1931; Mohanan & Sivad., Fl. Agasthyamala 748.2002.

Undershrubs. Leaves narrow-linear, c. 1 m by 4 cm, margins and keels ascendingly spinulose, glaucous beneath; veins tessellated, chartaceous. Male spathes boat-shaped, margins spinulose. Drupes connate.

Fl. & Fr.: Jul.-Aug. *Distr.*: Endemic to Peninsular India. Less common. An evergreen species among forest undergrowth. Locally abundant. *NAK 1624* (Arampa, c. 500 m).

2. **Pandanus unipapillatus** Dennst., Schluessel Hort. Malab. 15, 23, 27. 1818; Nicols., Suresh & Manilal, An Interpr. Hort. Malab. 238. 1988. *P. canaranus* Warb. in Engl., Pflanzenr. IV, 9: 75. 1900; Sald. & Nicols., Fl. Hassan Dist. 778. 1976; Manilal & Suresh, New Botanist 11: 123. 1984. *Perin-kaida-taddi* Rheede, Hort. Malab. 2:5, t.71679. *Thazha kaida.*

Shrubs or small trees. Leaves linear-oblong, c. 2 m by 5 cm; prickles greenish to brownish. Female head oblong-ovoid, composed of many simple drupes. Male inflorescence spicate, lower bracts with spinulose flagelliform tips; stamens α, together.

Fl. & Fr.: Jul.-Aug. *Distr.:* Peninsular India. Common. Along streamsides. *NAK 2983* (Tiruvalla, c. 10 m).

139. ARACEAE

1. Aquatic, free-floating herbs. .... **Pistia**
1. Terrestrial, epiphytic or marshy riverine rooted herbs.
    2. Riverine or marshy herbs.
        3. Leaves linear; pistillate flowers united. .... **Cryptocoryne**
        3. Leaves oblong-elliptic; pistillate flowers free. .... **Lagenandra**
    2. Terrestrial or epiphytic herbs.
        4. Epiphytic herbs; bulbiliferous shoots present. .... **Remusatia**
        4. Terrestrial, erect or climbing herbs; bulbiferous shoots absent.
            5. Climbing herbs. .... **Pothos**
            5. Crenum or tuberous; erect herbs.
                6. Leaves simple.
                    7. Stem tuberous; leaves peltate. .... **Colocasia**
                    7. Stem cormous; leaves not peltate.
                        8. Spadix included in the spathe; ovules basal and apical .... **Theriophonum**
                        8. Spadix exerted from the spathe; ovules basal. .... **Typhonium**
                6. Leaves compound.
                    9. Leave trichotomously compound; inflorescence appearing before leaves........................ **Amorphophallus**
                    9. Leaves radiately, pedately or pinnately compound; inflorescence appearing with leaves.
                        10. Cormous tall herbs; leaves radiately or pedately lobed.................................. **Arisaema**
                        10. Rhizomatous herbs; leaves pinnately compound.................................. **Anaphyllum**

AMORPHOPHALLUS Blume ex Decaisne
Nouv.Ann. Mus. Hist. Nat. 3: 366. 1834. (nom. cons.).

Cormous herbs. Leaves solitary, trichotomously forked, usually appearing after flowering. Inflorescence short or long peduncled; spathe tubular below, expanded above into a limb portion. Spadix with female flowers at base, male flowers above, contiguous or separated by neuter flowers; spadix-appendix sterile; ovary 1-4-locular; ovule 1, basal in each locule.

1. Peduncle more than 30 cm long; spathe ovate acuminate; neuter flowers present between female and male zones. ....1. **A. hohenackeri**

1. Peduncle less than 10 cm long; spathe campanulate; neuter flowers absent between female and male zones. ....2. **A paeoniifolius**

1. **Amorphophallus hohenackeri** (Schott) Engl. *et* Gehrm. in Engler, Pflanzenr. 48 IV.23C: 103. 1911; Mohanan & Sivad., Fl. Agasthyamala 750.2002.

Leaflets oblong or elliptic-oblong, 3-20 by 1-4 cm, base decurrent, apex acute or acuminate; petiole 20-25 cm long, spathe ± 3 cm wide; limb lanceate, acute; spadix-appendix cylindric, tapering to tip. Berries ellipsoid.

Fl. & Fr.: May.- Aug. *Distr.*: Western Ghats. Rare. Along hills floors in shade. *NAK 1764* (Ranni, c. 100 m).

2. **Amorphophallus paeoniifolius** (Dennst.) Nicols., Taxon 26:338.1977; Suresh, Sivad. & Manilal, Taxon 32(1): 128. 1983. *Dracontnium peoniifolius* Dennst., Schluess. Hort. Malab.13, 38.1818. *Arum campanulatum sensu* Dillwyn, Rev. Hort. Malab 59.1839, non Roxb., 1819; *Arum rumphii* sensu Gaud.in Freycinet, ,Voy.Uranic 2: 4.1826, as to name; Dillwyn, Rev. Hort. Malab. 59.1839.

Leaflets oblong or elliptic-oblong, 5-10 cm long, base oblique, margin wavy, apex acuminate; petiole 60-80 cm long, stout, dark green with pale green blotches. Spathes campanulate, 15-25 cm wide, margin recurved, undulate and crisped; spadix-appendix elongate conoid.Berries globose.

Fl. & Fr.: May.-Sep. *Distr.*: India, Sri Lanka & Pacific Islands. Among forests shades. *NAK 1230* (Moozhiar, c. 200 m).

ANAPHYLLUM H. W. Schott
Bonplandia 5:126. 1857.

Anaphyllum wightti Schott, Gen. Aroid. t. 83. 1858; Hook.f., Fl. Brit. India 6: 551. 1893; Fischer in Gamble, Fl. Pres. Madras 1589. 1931; Mohanan & Sivad., Fl. Agasthyamala 753.2002. *Keeri-kizhangu.*

Slender herbs with creeping root stock. Leaves: juvenile ones simple, ovtae, hastate at base, mature ones pinnati-sect, to 60 cm long; leaflets elliptic-lanceate or oblong, base decurrent, apex acuminate; petioles 60 cm to 1 m long. Spathes ovate-lanceate, 8-10 cm wide, spirally twisted. Flowers bisexual, covering the whole spadix, appendage 0; stamens 4-6; ovary ovoid, 1-locular; ovule 1. Berries obovoid.

Fl. & Fr.: Aug.-Nov. *Distr.*: Endemic to Peninsular India. Not common. Under shade in upper hills. *NAK 1475* (Arampa, c. 400 m).

ARISAEMA C. F. P. von Martius
Flora 14:459. 1831.

Cormous, paradioecious herbs. Leaves 1-2, compound, radiately or pedately divided; leaflets sessile. Spathe tubular below, expanded into a limb above; spadix monoecious or dioecious. Male flowers α, lax, sessile or stalked; anthers 2-5, on simple or branched filaments. Female flowers with 1-locular ovary; ovules ± 9, basal. Berry ovoid.

1. Leaves radiately divided; spadix with appendix erect and shorter than the spathe. .... 1. **A. leschenaultii**

1. Leaves pedately divided; spadix with appendix sigmoid, longer than the spathe. .... 2. **A. tortuosum**

1. **Arisaema leschenaultii** Bl., Rumphia 1:93.1836; Hook.f., Fl. Brit. India 6:504.1893; Fischer in Gamble, Fl. Pres. Madras 1585. 1931; Nicols. in Sald. & Nicols., Fl. Hassan Dist.784. 1976; Matthew, Ill. Fl. Tam. Carnatic t.765. 1982; Sivadas. & Nicols. in Matthew, Fl. Tam. Carnatic 3(3) :1688. 1983; Mohanan & Sivad., Fl. Agasthyamala 756.2002. *A. pulchrum* N.E. Br., J. Linn. Soc. Bot. 18:252, t. 6. 1880; Hook.f., Fl.Brit. India 6:605. 1893; Fischer in Gamble, Fl. Pres. Madras 1585. 1931.

Corms c. 4 cm ϕ; leaf single; petiole to 80 cm long; leaflets 5-12, elliptic-oblong to obovate-lanceate, 8-30 by 2-11 cm, base cuneate, apex acuminate, peduncle ± equal to petiole. Spathe vertically stripped with white and green; limb ovate-lanceate, c. 8 cm long, apex long-tailed; spadix usually dioecious, rarely monoecious. Female spadix: female flowers at base; rarely neuter flowers present; appendix subcylindric, c. 6 cm long; in male spadix basal portion with group of male flowers, neuter flowers 0; monoecious spadix with basal female flowers and apical male flowers. Seeds 2-3.

Fl. & Fr.: Jul.-Sep. *Distr.*: South India. Common in forest shades of upper hills. *NAK 85, 2704* (Moozhiar & Kakki hills).

2. **Arisaema tortuosum** (Wall.) Schott in Schott & Endl., Melet. Bot. 17. 1832; Hook.f., Fl. Brit. India 6: 502. 1893; Fischer in Gamble, Fl. Pres. Madras 1584. 1931; Nicols. in Sald. & Nicols., Fl. Hassan Dist. 785. 1976; Sivadasan & Nicols. in Matthew, Fl. Tam. Carnatic 3(3): 1689. 1983; Mohanan & Sivad., Fl. Agasthyamala 757.2002. *Arum tortuosum* Wall., Pl. Asiat. Rar. 2: 10, t.114. 1830.

Corms c. 4 cm ϕ. Leaves two; petiole to 40 cm long; leaflets 5-7, pedatisect, elliptic-oblong, 8-10 by 3-4 cm, base oblique, apex acuminate. Spathe c. 12 cm long; limb horizontally bent, ovate, c. 4.5 cm wide, greenish; spadix usually monoecious, ± 17 cm long, basally female flowers, apically male flowers rarely, few neuters present between female and male; appendix subulate, long-tailed. Berry -orange-red. Seed(s) 1-2.

Fl. & Fr.:-Jun.-Aug. *Distr.:* Himalayas, South India & Sri Lanka. Rare. Among shades in upper hills. *NAK 2708* (Kakki hills, c. 110 m).

COLOCASIA Schott
in Schott et Endlicher, Melet. Bot. 18.1832.(nom.cons.).

Colocasia esculenta (L.) Schott in Schott & Endl., Melet. Bot. 18. 1832; Purseglove, Trop. Crops, Monocots 61. 1975; Nicols. in Sald. & Nicols., Fl. Hassan Dist. 786. 1976; Matthew, Ill. Fl. Tam. Carnatic tt. 766 & 979 c.1982; Sivad. & Nicols. in Matthew, Fl. Tam. Carnatic 3(3): 1691. 1983; Nicols., Suresh & Manilal, An Interpr. Hort. Malab. 275. 1988. *Arum esculentum* L., Sp. Pl. 965. 1753. *Arum colocasia* L., Sp. Pl. 965.

1753; Wight, Ic. t. 786, f.1. 1844. *Arum esculentum* (L.) Vent., Mag. Encycl. 4: 471. 1801. *Colocasia antiquorum* Schott in Schott & Endl., Melet. Bot. 18. 1832; Wight, Ic. t. 785, f.1. 1844; Hook.f., Fl. Brit. India 6: 523. 1893; Fischer in Gamble, Fl. Pres. Madras 1580.1931. *Weli-ila* Rheede, Hort. Malab. 11:43, t.22. 1692.

Perennial tuberous herbs. Leaves peltate; petiole sheathing at base; lamina ovate, 40 by 30 cm, base cordate-sagittate, apex acute. Spathe axillary, to 35 cm long, basally convolute, upper portion lanceate, yellow; spadix sessile, shorter than spathe, female flowers basal folowed by neuter flowers and male flowers; ovary 1-locular; ovules α; male flowers yellow, reduced to 3-6-angled synandria. Berry oblong. Seeds α.

Fl. & Fr.: Nov.-Jan. *Distr.*: Native to Asia, now pantropical. Forming gregarious populations along marshy areas and streamsides. *NAK 210* (Plappally, c. 300 m).

CRYPTOCORYNE Fischer ex Wydler
Linnaea 5: 428. 1830.

Cryptocoryne retrospiralis (Roxb.) Kunth, Enum. Pl. 3:12. 1841; Wight, Ic. t. 772. 1844; Hook.f., Fl. Brit. India 6: 493. 1893; Fischer in Gamble, Fl. Pres. Madras 1575. 1931; Nicols. in Sald. & Nicols., Fl. Hassan Dist. 786. 1976; Matthew, Ill. Fl. Tam. Carnatic tt.767 & 979. 1982; Sivad. & Nicols. in Matthew, Fl. Tam. Carnatic 3(3): 1693. 1983. *Ambrosinia retrospiralis* Roxb., (Hort. Beng. 65. 1814, nom. nud.) Fl. Ind. 3: 492. 1832. *Ambrosinia unilocularis* Roxb., Fl.Ind. 493.1832. *Cryptocoryne unilocularis* (Roxb.) Kunth, Enum. Pl. 3:13.1841; Wight, Ic.t. 774. 1844; Fischer in Gamble, Fl. Pres. Madras 1575. 1931.

Rheophytic creeping herbs. Leaves simple, tufted, linear-lanceate, 10-20 by 1-4 mm, base cuneate, margin wavy, apex acute. Inflorescene with a tubular spathe, c. 18 cm long; limb flat, ovate-lanceate, long-acuminate, spirally twisted; spadix with female portion c. 6 mm long, male portion c. 3 mm long. Female flowers 4-7, in a single whorl, united to form a syncarp. Male flowers closely arranged, forming a subglobose mass; neuter flowers clavate. Fruit a α-seeded syncarp.

Fl. & Fr.: Jan.-Mar. *Distr.*: South and NE India. Locally abundant along streamsides. *NAK 477, 2233* (Perunthenaruvi, c. 100 m).

LAGENANDRA Dalzell
in Hooker's J. Bot.Kew. Gard. Misc. 4: 289. 1852.

Marshy or rheophytic herbs. Leaves simple, long-petioled. Spathe tubular-ellipsoid, basally forming a chamber, enclosing the spadix; limb expanded above, terminating in a subulate appendage; spadix slender, female flowers below, male above, both widely sperated by slender interstice. Male flowers α, forming a cylindric mass; stamens 1-2; female flowers a, spirally arranged, ovary 1-locular; ovules 1-2, or many basal. Berries free.

1. Leaf-base usually acute.
    2. Spathe dark-purplish, limb warty; leaves to 50 cm long. ....2. **L. ovata**
    2. Spathe pale-purple or violet, limb smooth; leaves to 20 cm long ....3. **L. toxicaria**
1. Leaf-base usually rounded or subcordate. ....1. **L. meeboldii**

1. **Lagenandra meeboldii** (Engler) Fischer in Gamble, Fl. Pres. Madras 1576. 1931; Nicols. in Sald. & Nicols., Fl. Hassan Dist. 787. 1976. *Cryptocoryne meeboldii* Engler, Pflanzenr. IV. 23 f (73): 234. 1920.

Rhizome slender, c. 1 cm ϕ. Leaves ovate or elliptic 5-15 by 3-6 cm, base rounded or subcordate, apex acute or subacute, lateral veins α, archcd. Spathe pale purple, smooth; limb triangular or broadly ovate, tip slender caudatc, 3.5 cm long.

Fl. & Fr.: Sep.-Dec. *Distr.*: South India. Rare. Streamside marshes. *NAK 2691* (Arampa, c. 400 m).

2. **Lagenandra ovata** (L.) Thw., Enum. Pl. Zeyl. 334. 1864; Fischer in Gamble, Fl. Pres. Madras 1576. 1931; Nicols. in Sald. & Nicols., Fl. Hassan Dist. 787. 1976, App.2: 7. 1976. Nicols., Suresh & Manilal, An Interpr. Hort. Malab. 275. 1988. *Arum ovatum* L., Sp. Pl. 967. 1753. *Karin-pola* Rheede, Hort. Malab. 11:45, t.23. 1692.

Rhizome stout, c. 5 cm ϕ. Leaves elliptic-oblong, 12-50 by 5-12 cm, base acute, margin slightly wavy, apex acute, greenish; petiole ± equal to lamina. Spathe dark-purple, warty out side the limb; tip thick, to 6 cm long.

Fl. & Fr.: Sep.-Dec. *Distr.*: SW India & Sri Lanka. Rare. Locally abundant. Along streamside marshes or mountain marshes. *NAK 2027* (Manniera, c. 200 m).

3. **Lagenandra toxicaria** Dalz. in Hooker, J. Bot. Kew. Gard. Misc. 4:289. 1852; Hook.f., Fl. Brit. India 6: 495. 1892.

Rhizome slender, 1-2 cm ϕ. Leaves elliptic or elliptic-ovate, 10-16 by 3-6 cm, base and apex acute, margin crispate, lateral veins α, faint. Spathe pale brownish or purplish coloured, c. 5 cm long, , limb triangular-ovate, smooth, abruptly tapering into a slender limb.

Fl. & Fr.: Sep.-Oct. *Distr.*: South India. Common. Along streamsides. *NAK 473*. (Ranni, c.100 m).

PISTIA Linnaeus
Sp. Pl. 963. 1753.

Pistia stratiotes L., Sp. Pl. 963. 1753; Roxb., Pl. Corom. t. 268. 1820 & Fl. Ind. 3: 131.1832; Hook.f., Fl. Brit. India 6: 497. 1893; Fischer in Gamble, Fl. Pres. Madras 1573. 1931; Nicols. in Sald. & Nicols., Fl. Hassan Dist. 788. 1976; Sivad. & Nicols. in Matthew, Fl.Tam. Carnatic 3(3): 1696. 1983; Nicols., Suresh & Manilal, An Interpr. Hort. Malab. 276. 1988. *Kodda-pail* Rheede, Hort. Malab. 11:63-64, t.32. 1692.

Aquatic floating herbs. Leaves sessile, tufted, obovate, 2-10 by 1-4 cm, veins prominent, white soft pubescent. Spathe small, to 10 mm long, basal portion convoluted, upper portion expanded into an ovate-acute limb, pale yellow; spadix with pistillate portion adnate to the spathe; female and male flowers separated by a disc-like structure. Male flowers with 4-6 anthers forming a synandrium; ovary 1-locular; ovules α, on parietal placenta. Berry small, ovoid.

Fl. & Fr.: Dec.-Feb. *Distr.*: Tropics and subtropics. Locally abundant in shallow pools. *NAK 312* (Adoor, c. 20 m).

POTHOS Linnaeus
Sp. Pl. 968. 1753.

Pothos scandens L., Sp. Pl. 968. 1753; Hook.f., Fl. Brit. India 6: 551. 1893; Fischer in Gamble, Fl. Pres. Madras 1592. 1931; Nicols. in Sald. & Nicols., Fl. Hassan Dist. 788. 1976; Nicols., Suresh & Manilal, An Interpr. Hort. Malab. 276.1988; Mohanan & Sivad., Fl. Agasthyamala 761.2002. *Ana-paruva* Rheede, Hort. Malab. 7:75, t.40. 1688.

Climbers. Leaves simple, distichous; petiole broadly winged, apex rounded or emarginate; lamina ovate or ovate-lanceate, rarely linear-lanceate (highly variable), 4-8 by 2-4 cm, base rounded, apex acute, veins prominent, close, coriaceous. Inflorescence small, axillary, solitary, c.1 cm long; spathe cymbiform, 5-8 mm long; spadix globose or ovoid. Flowers bisexual; perianth-lobes 6; stamens 6; ovary 3-locular; ovule 1-per locule. Berries ellipsoid. Seed(s) 1-3, ellipsoid.

Fl. & Fr.: Aug.-Oct. *Distr.*: Madagascar to New Guinea. Very common in plains and hills. *NAK 439* (Adoor, c. 20 m), *2260* (Moozhiar, c.250 m).

Note : One collection of Pothos (*NAK 2386*), with only fruits resembled with recently described species viz., **Pothos crassipedunculatus** Sivadasan & Mohanan (1992). But the flowering materials are needed for the confirmation of identity.

REMUSATIA Schott
in Schott et Endlicher Melet. Bot. 18. 1832.

Remusatia vivipara (Roxb.) Schott in Schott & Endl., Melet. Bot. 18. 1832. Hook.f., Fl. Brit. India 6:521. 1893; Fischer in Gamble, Fl. Pres. Madras 1583. 1931; Nicols. in Sald. & Nicols., Fl. Hassan Dist. 788. 1976; Matthew, Ill. Fl. Tam. Carnatic t. 770.1982; Sivad. & Nicols. in Matthew, Fl. Tam. Carnatic 3(3); 1697.1983; Nicols., Suresh & Manilal, An Interpr. Hort. Malab. 276. 1988; Mohanan & Sivad., Fl. Agasthyamala 761.2002. *Arum viviparum* Roxb., (Hort. Beng. 65. 1814, "*viviparum*", nom. nud.) Fl. Ind. 3: 496. 1832; Wight, Ic. t. 798. 1844. *Maravara-tsjembu* Rheede, Hort. Malab. 12:19-20, t.9. 1693.

Cormous epiphytic herbs. Leaf usually solitary, ovate, peltate, 20-35 by 10-15 cm, base subcordate, apex acute, chartaceous. Leafless bulbiferous shoots c. 40 cm long. Inflorescence appears after leaves, to 15 cm long, basal portion convolute, oblong, upper limb portion broadly obovate, c. 9 cm wide; spadix sessile, 7 cm long. Female flowers below, male flowers above, neuters in middle; ovary 1-locular; ovules 3-4. Fruit a berry, ovoid.

Fl. & Fr.: Feb.-Apr. *Distr.*: Indo-Malesia, Africa, Madagascar, Australia. Common as epiphytes, rarely in rock crevices. *NAK 55* (Moozhiar, c. 250 m).

THERIOPHONUM Blume
Rumphia 1: 127. 1837 ('1835').

Theriophonum infaustum N.E. Br., J. Linn. Soc. Bot. 18:260. 1881; Sivad. & Nicols., Kew Bull. 37: 282. 1982; Suresh, Sivad.& Manilal, Taxon 32 : 129 1983; Nicols., Suresh & Manilal, An Interpr. Hort. Malab. 277. 1988; Hook. f., Fl. Brit. India 6:513. 1893; Gamble, Fl. Pres. Madras 1579. 1931. *Nir-tsjembu* Rheede, Hort. Malab. 11:33-34, t.17. 1692.

Cormous herbs. Leaves long-petioled, ovate, 6-12 by 3-6 cm, base hastate, apex acute. Spathe tubular below, slightly contricted at mouth, limb lanceate. Spadix with female flowers at base, male flowers above separated from the female by a barren zone, few neuters just above the female flowers; ovary 1-locular; placentation basal and apical. Berries ovoid.

Fl. & Fr.: Sep.-Dec. *Distr.*: Southern Western Ghats. Not common. Along wet areas in plains. *NAK 2035* (Vallicode-Kottayam, c. 100 m).

TYPHONIUM Schott
Wiener Z. Kunst. 1829 (3): 732. 1829.

Typhonium flagelliforme (Lodd.) Bl., Rumphia 1:134. 1837; Nicols. & Sivad., Blumea 27:489. 1981; Suresh, Sivad.& Manilal, Taxon 32:129. 1983; Nicols., Suresh & Manilal, An Interpr. Hort. Malab. 278. 1988. *Arum flagelliforme* Lodd., Bot. Cab. 4:t. 396. 1819. *Nelenchena major* Rheede, Hort. Malab. 11:39, t. 20. 1692.

Cormous herbs. Leaves simple, rarely divided, ovate, 3-15 by 1-6 cm, base cordate, apex acute, chartaceous. Spathe 6-15 cm long,tubular below, mouth constricted, expanded above into an apically twisted limb; spadix exerted; male and female flowers widely separated, neuter flowers just above the female, heteromorphic; ovary 1-locular; ovules 1-2, basal. Berry ovoid. Seeds 1 or 2, globose.

Fl. & Fr.: Jul.-Sep. *Distr.*: Southern Western Ghats. Not common. In moist areas in plains. *NAK 2981* (Tiruvalla, c. 10 m).

Caladium bicolor (Aiton) Vent., a native of South America is occasionally seen running wild. (*NAK 2445*).

140. LEMNACEAE

LEMNA Linnaeus
Sp. Pl. 970. 1753.

Lemna perpusilla J. Torrey, Fl. N.Y. 2: 245. 1843; van der Plas in Steenis, Fl. Males. I, 7: 231. 1971; Matthew & Britto in Matthew, Fl. Tam.Carnatic 3(3): 1705. 1983. *L. paucicostata* Hegelm., Lemnac. 139, t.8. 1868; Hook.f., Fl. Brit. India 6: 556. 1893; Fischer in Gamble, Fl. Pres. Madras 1593. 1931.

Minute floating herbs. Fronds flat, in groups of 2 or 3, broadly ovate to suborbicular, to 2 mm long, ± asymmetric, base obtuse to subacute, apex obtuse-rounded, 3-nerved; roots from lower surface or margins of leaf, sheath elongate. Inflorescence spatheate with 1 female and 2 male flower(s); perianth 0; stamen 1, anthers transversely dehiscent; ovary sessile, globose, 1-locular. Fruit an utricle. Seeds ribbed.

Fl. & Fr.: Jul.-Sep. *Distr.:* Tropics. Common in still water. *NAK 2721* (Tiruvalla, c. 10 m).

141. ALISMATACEAE

SAGITTARIA Linnaeus
Sp. Pl. 993. 1753.

Sagittaria guaynensis H.B.K., Nov. Gen.Sp.1: 250. 1816; Hook. f., Fl. Brit. India 6: 561. 1893, p.p.; Fischer in Gamble, Fl. Pres. Madras 1596. 1931. ssp. **lappula** (D.Don). Bogin, Mem. New York Bot. Gard. 9: 192, f. 5. 1955. *S. lappula* D. Don, Prodr. Fl. Nepal. 22. 1825.

Aquatic herbs. Leaves floating, broadly ovate, 2-10 by 1.5-6 cm, base deeply cordate, apex obtuse, chartaceous. Flowers 1-sexual or polygamous; lower bisexual flowers, upper usually male flowers; sepals 3, herbaceous, persistent; petals 3, white, caducous; stamens 9-12; staminodes often present in female flowers; carpels 3-6 or more, superior, laterally flattened, 1-locular; ovule 1, basal. Fruit a globose cluster of achenes.

Fl. & Fr: Nov.-Jan. *Distr.:* Southern Western Ghats. Not common. In shallow pools, abandoned paddy fields, etc. *NAK 1974 & 2088* (Tiruvalla, c. 10 m).

142. NAJADACEAE

NAJAS Linnaeus
Sp. Pl. 1015. 1753.

Najas graminea Del., Descr. Egypte Hist. Nat. 282, t.50, f.3. 1813; Hook.f., Fl. Brit. India 6: 569. 1893; Fischer in Gamble, Fl. Pres. Madras 1604. 1931; Wilde in Steenis, Fl. Males. I, 6: 169. 1962; Matthew, Ill. Fl. Tam. Carnatic t.776. 1922; Matthew & Britto in Matthew, Fl. Tam. Carnatic 3(3): 1710. 1983.

Submerged herbs. Leaves whorled, acicular, 1-2 by 0.2-0.5 cm, flat, translucent, margin softly spinescent; sheath spinescent on margins. Flowers unisexual, monoecious or dioecious; male flower(s) 1-3; perianth-lobes globose-oblong; anthers 4-celled. Female flower(s) solitary, axillary; perianth-lobes closely appressed to ovary; ovary ellipsoid. Fruit subterete or subtrigonous.

Fl. & Fr.: Sep.-Dec.*Distr.*: Pantropical. In lowlands. Common. *NAK 2905* (Adoor, c. 20 m).

143. ERIOCAULACEAE

ERIOCAULON Linnaeus
Sp. Pl. 87. 1753.

Annual scapose herbs. Leaves radical, linear-lanceate, base sheathing. Flowers minute; unisexual, closely aggregated in solitary capitulum. Capitula bisexual, rarely unisexual; receptacle flat, convex or hemispherical; involucral bracts α, peduncle often twisted, usually ribbed. Male flowers : stipitate; outer tepals 2 or 3, connate, split down on one side; inner tepals often connate to form a tube, rarely free; lobes often bearing a subapical gland; stamens 4 or 6, anthers often black, white or yellow. Female flowers : sessile or stipitate; outer tepals 2 or 3, rarely 1 or 0; tepals 3, free, bearing an apical black gland, pilose papillate hairs; ovary superior, 2-3-locular; ovule 1 per locule. Achenes minute, smooth or reticulate.

1. Anthers white or yellow. .... 1. **E. cinereum**
1. Anthers black or dark green.
    2. Floral bracts glabrous or glabrescent. .... 5. **E. truncatum**
    2. Floral bracts pilose.
        3. Involucral bracts much larger than floral bracts.
            4. Floral bracts truncate or rounded at apex. .... 6. **E. xeranthemum**
            4. Floral bracts cuspidate at apex. .... 2. **E. dianae** var. **dianae**
        3. Involucral bracts shorter than floral bracts.
            5. Tepals of male connate, hyanline or whitish. .... 3. **E. quinquangulare**
            5. Tepals of male free, dark. .... 4. **E. thwaitesii**

1. **Eriocaulon cinereum** R.Br., Prodr. 254. 1810; Gandhi in Sald. & Nicols., Fl. Hassan Dist. 650. 1976; Matthew & Britto in Matthew, Fl. Tam. Carnatic 1719. 1983. *E. sieboldianum* Sleb. & Zucc. ex Steud., Syn. Pl. Glumac. 2:272. 1855; Hook.f., Fl. Brit. India 6:577. 1893; Fischer in Gamble, Fl. Pres. Madras 1619. 1931.

Leaves narrow-linear, to 5 cm long, apex acuminate; peduncles α, slender, ribbed, to 12 cm long. Capitula c. 5 mm ϕ, whitish; involucral bracts scarious, outer broadly ovate, inner lanceate; floral bracts linear-oblong, hyaline; receptacle elongated. Male flowers: outer tepals spathaceous, apex 3-toothed, split down one side, inner tepals minute with an apical black gland; stamens with white or yellow anthers. Female flowers: tepals 2, rarely 0, 1, or 3, filiform.

Fl. & Fr.: Sep.-Jan. *Distr.*: Sri Lanka, India. Himalayas, Myanmar, China, Japan, Australia. Common in lowlands and hills. *NAK 465, 441, 1611 & 2129* (Ranni, Chittar, etc.).

2. **Eriocaulon dianae** Fyson, J. Ind. Bot. Soc. 2:259, tt.1, 12, 12. 1921; Fischer in Gamble, Fl. Pres. Madras 1620. 1931.

Leaves linear, to 10 cm long, apex acute or subacute; peduncles α, slender, ribbed, to 15 cm long; bracts obovate. Capitula hemispheric, c. 5 mm ϕ; involucral bracts lanceate, acute or obtuse; receptacle conical, pilose. Male flowers: lobes 3, small, papillose, inner tepals glandular at apex. Female flowers : outer petals free, heteromerous, tip white hairy, inner tepals eglandular.

Fl. & Fr.: Nov.-Dec. *Distr.*: South India. Less common. In mountain swamps. *NAK 420, 2130* (Ranni & Chittar).

3. **Eriocaulon quinquangulare** L., Sp. Pl. 87. 1753; Hook.f., Fl. Brit. India 6: 582. 1893; Fischer in Gamble, Fl. Pres. Madras 1620. 1931; Gandhi in Sald. & Nicols., Fl. Hassan Dist. 654. 1976; Matthew & Britto in Matthew, Fl. Tam. Carnatic 3(3): 1720. 1983.

Leaves linear-ensiform, to 10 cm long; purplish beneath, 5-11-nerved; peduncles α, 5-ribbed, 3-25 cm long, sheath often purplish. Capitula subglobose, greyish or whitish; involucral bracts obovate, obtuse, scarious; floral bracts obovate or oblanceate, acuminate, hairy at apex. Male flowers : tepals oblong or linear, apex white-papillose, with or without a black gland. Female flowers: outer tepals 3, free, inner oblanceate, pilose, apex black-glandular. Seeds smooth or transversely striate.

Fl. & Fr.: Nov.-Jan. *Distr.*: Sri Lanka, India, Nepal & Maynmar. Common in plains and hills. *NAK* 2578, 2602 (Moozhiar, c. 250 m).

4. **Ericaulon thwaitesii** Koern., Linnaea 27 : 627. 1854; Hook.f., Fl. Brit. India 6:583. 1893; Fischer in Gamble, Fl. Pres. Madras 1620. 1931; Matthew, Ill. Fl. Tam. Carnatic t. 777. 1982; Matthew & Britto in Matthew, Fl. Tam. Carnatic 3(3): 1721. 1983; Mohanan & Sivad., Fl. Agasthyamala 766. 2002.

Leaves linear or ensiform, to 7 cm long, faintly 5-veined, apex acute; peduncles α, ribbed; involucral bracts obovate, apex rounded or subtruncate, yellowish-brown; floral bracts obovate-spathulate, concave, apex rounded or shortly cuspidate, dark; receptacle flat or shortly conical, pilose. Male flowers : tepals free, concave, with or without glands. Female flowers: outer tepals 2 or 3, keel-shaped, inner tepals 3, minute, filiform, apically tufted hairy, eglandular.

Fl. & Fr.: Sep.-Dec. *Distr.*: Peninsular India, Sri Lanka. Less common. In hills. *NAK 1916* (Arampa, c. 400 m).

5. **Eriocaulon truncatum** Buch.-Ham ex Mart. in Wall., Pl. Asiat. Rar. 3:29. 1832; Hook.f., Fl. Brit. India 6:578. 1893; Fischer in Gamble, Fl. Pres. Madras 1619. 1931; Gandhi in Sald. & Nicols., Fl. Hassan Dist. 654. 1976; Matthew & Britto in Matthew, Fl. Tam. Carnatic 3(3) : 1722. 1983; Mohanan & Sivad., Fl. Agasthyamala 767.2002.

Leaves ensiform, to 5 cm long, apex subacute or obtusely acuminate; peduncles α, 5-8-ribbed; to 5-20 cm tall. Capitula hemispheric, c. 4 mm ϕ; involucral bracts spathulate-obovate, glabrous or sparsely puberulous; receptacle subconic. Male flowers: outer tepals 2, shortly connate on one side, concave, inner tepals with black apical gland. Female flowers: outer tepals 2, free, linear or spathulate, sparsely puberulous or glabrous; inner tepals 3, glabrescent with or without apical gland.

Fl. & Fr.: Sep.-Dec. *Distr.*: Sri Lanka, India, Myanmar, China, Malaya. Common. In mountain swamps. *NAK 18 & 2601* (Moozhiar, c. 250 m).

6. **Eriocaulon xeranthemum** Mart. in Wall., Pl. Asiat. Rar. 3:29. 1832; Hook.f., Fl. Brit. India 6:584. 1893; Fischer in Gamble, Fl. Pres. Madras 1620. 1931.

Leaves linear-lanceate, 1-3 cm long, apex acute, 7-11-nerved, peduncles α, striate. Capitula hemispheric, small; involucral bracts lanceate, much longer than flroal bracts; floral bracts oblong-obovate, truncate or rounded, hyaline, apex white-hairy. Flowers few; male: outer tepals free or ± connate, apex white-pilose; inner tepals apically black-glandular. Female: outer tepals 2 or 3, linear, apex dark coloured; inner tepals 3, oblanceate, tip hairy, black-glandular.

Fl. & Fr.: Sep.-Dec. *Distr.*: South India. In plains and hills. The common species of the genus in the district. *NAK 809, 1910* (Ranni, Tiruvalla, etc.)

144. CYPERACEAE

1. Flowers unisexual.
    2. Ovary enclosed in an utricle; hypogynous disc 0. .... **Carex**
    2. Ovary not enclosed in an utricle; hypogynous disc present. .... **Scleria**
1. Flowers bisexual.
    3. Leaves reduced to sheaths.
        4. Involucral bracts well developed; inflorescence pseudolateral. .... **Schoenoplectus**
        4. Involucral bracts much reduced or absent; inflorescence terminal.
            5. Hypogynous bristles present; style base persistent. .... **Eleocharis**
            5. Hypogynous bristles absent; style base deciduous. .... **Fimbristylis** *p.p.* **(F. tetragona)**
    3. Leaves with well developed blades.
        6. Hypogynous bristles or scales present.
            7. Nut with elongated beak; style-base dilated .... **Rhynchospora**
            7. Nut without elongated beak; style-base not dilated.
                8. Leaves distant; scales intermixed with hypogynous bristles; glumes aristate. .... **Fuirena**
                8. Leaves basal, scales not intermixed with hypogynous bristles; glumes not aristate. .... **Lipocarpha**
        6. Hypogynous scales or bristles 0.
            9. Glumes spiral.
                10. Style deciduous, leaving a persistent bulbiform tumour on nut. .... **Bulbostylis**
                10. Style-base persistent or if deciduous not leaving a tumour on nut. .... **Fimbristylis** *p.p*

9. Glumes distichous.

11. Rachilla of spikelet articulate at base.

12. Stigmas 2. .... **Kyllinga**

12. Stigmas 3. .... **Mariscus**

11. Rachilla of spikelet not articulate.

13. Stigmas 2. .... **Pycreus**

13. Stigmas 3. .... **Cyperus**

BULBOSTYLIS Kunth
Enum. Pl. 2:205. 1837 (nom. cons.).

Bulbostylis barbata (Rottb.) Clarke in Hook. f., Fl. Brit. India 6: 651. 1893; Fischer in Gamble, Fl. Pres. Madras 1662. 1931; Kern in Steenis, Fl. Males. I, 7:539. 1974; Hooper in Sald. & Nicols., Fl. Hassan Dist. 659. 1976; Matthew, Ill. Fl.Tam. Carnatic t.778. 1982; Britto & Matthew in Matthew, Fl. Tam. Carnatic 3(3): 1725. 1983; Mohanan & Sivad., Fl. Agasthyamala 769.2002. *Scirpus barbatus* Rottb., Descr. Pl. Rar. 27. 1772 & Descr. Ic. Rar.52, t. 17. & 4.1773.

Tufted annuals. Leaves basal, filiform. Inflorescence terminal, capitate, hemispherical, involucral bracts overtopping inflorescence; spikelets sessile; glumes ovate, mucronate; rachilla flexuous, persistent, narrowly winged; hypogynous bristles 0; stamens 1-3, apiculate; style articulate, basally bulbiform, persistent. Nut 3-gonous, shortly stipitate, smooth.

Fl. & Fr.: Nov.-Feb. *Distr.*: Subtropics and Old World tropics. Less common. On sandy soil from plains to hills. *NAK 2002* (Tiruvalla, c.10 m).

CAREX Linnaeus
Sp. Pl. 972. 1753.

Perennials; rhizomes woody; stem triquetrous. Leaves flat or canaliculate, coriaceous. Flowers unisexual, in spikes or panicles; involucral bracts foliar; spikelets distant, bisexual, apically female, male at base; rachilla persistent; glumes membranous, mucro erect, stiff; perianth bristly or 0; stamens 2 or 3 to α, .ully enclosed in flask-shaped utricle. Nut biconvex or trigonous.

1. Spikelets 20-60 mm long; utricle (mature) bright red, obovoid to subglobose. .... 1. **C. baccans**

1. Spikelets 4-15 mm long; utricle (mature) greenish-stramineous, ellipsoid. .... 2. **C. filicina**

1. **Carex baccans** Nees in Wight, Contrib. Bot. Ind.122. 1834; Clarke in Hook.f., Fl. Brit. India 6 : 722. 1894; Fischer in Gamble, Fl. Pres. Madras 1687. 1931; Nelms, Reinwardtia 1(3): 322. 1951; Mohanan & Sivad., Fl. Agasthyamala 770.2002.

Tall herbs. Leaves scattered, exceeding the stem; sheath membranous, blackish-red. Panicle up to 50 cm long; spikelets linear; glume reddish, ovate-obolong, mucronate. Utricle subsessile, inflated, broadly obovoid to subglobose, mature one reddish to ultimately black. Nuts triquetrous, ellipsoid, rostrate.

Fl. & Fr.: Sep.-Feb. *Distr.:* Western Ghats. Not common. Along forest edges in upper hills. *NAK 361* (Kakki hills, 1100 m).

2. **Carex filicina** Nees in Wight, Contrib. Bot. Ind. 123. 1834; Clarke in Hook.f., Fl. Brit. India 6:717. 1894; Fischer in Gamble, Fl. Pres. Madras 1686. 1931; Nelms, Reinwardtia 1(3) : 304. 1951; Matthew, Ill. Fl. Tam. Carnatic t. 779. 1982; Britto & Matthew in Matthew, Fl. Tam. Carnatic 3(3): 1727. 1983; Mohanan & Sivad., Fl. Agasthyamala 771.2002.

Herbs. Leaves (basal ones) as long as stem, flat or canaliculate; sheath membranous. Panicle pyramidal, distant, up to 12 cm long; spikelets linear, glumes greenish-brown with pale reddish-brown streaks, ovate-lanceate, long mucronate. Utricle ellipsoid, mature greenish-stramineous, long-beaked. Nuts trigonous, ellipsoid, apiculate.

Fl. & Fr.: Sep.-Mar. *Distr.*: Tropics. Common along forest floors and forest outskirts. *NAK 2569* (Kakki hills, c. 1100 m).

CYPERUS Linnaeus
Sp. Pl. 44. 1753.

Annuals or perennials; rhizome 0 to stoloniferous; stem trigonous, triquetrous or terete, tufted or solitary. Leaves basal, flat or canaliculate. Inflorescence terminal, anthelate or capitate; involucral bracts foliar; spikes oblong to ovoid; spikelets usually compressed or subcylindric, 1 to α-flowered; rachilla straight or flexuous, winged or not. Flowers usualy all bisexual; glumes distichous; perianth 0; stamens 1-3; style 2- or 3-fid or 3-notched. Nut trigonous or biconvex.

1. Aquatic herbs; style entire or 3-notched. .... 3. **C. cephalotes**
1. Terrestrial (Marshy) herbs; style 3-partite.
    2. Spikelets digitately or stellately arranged or capitate.
        3. Leaves and bracts filiform; glumes aristate.
            4. Glumes purplish-brown; nut linear-cylindric. .... 1. **C. castaneus**
            4. Glumes pale-brown; nut narrowly obovoid. .... 5. **C. cuspidatus**
        3. Leaves and bracts not filiform; glumes mucronulate or shortly cuspidate.
            5. Glumes cuspidate, rather distant. .... 7. **C. diffusus**
            5. Glumes mucronulate, closely imbricating.
                6. Secondary rays normally present.
                    7. Anther-connective with bristly appendages. .... 10. **C. haspan**
                    7. Anther-connective without appendages. .... 15. **C. tenuispica**
                6. Secondary rays absent. .... 6. **C. difformis**
    2. Spikelets spicate or raceme-like.
        8. Rachilla of spikelet not or hardly winged.
            9. Glumes aristate; nut half as long as glume. .... 2. **C. compressus**
            9. Glumes muticous or mucronate; nut not so.
                10. Glumes orbicular-ovate. .... 11. **C. iria**
                10. Glumes oblong or elliptic-oblong.
                    11. Margins of glumes hyaline. .... 13. **C. procerus**
                    11. Margins of glumes not or litile hyaline.

12. Rachilla (very) slender; leaves well developed. ....9. **C. distans**
12. Rachilla not (very) slender;
leaves 0 or very few, not well developed. ....12. **C. malaccensis**

8. Rachilla of spikelets distinctly winged.

13. Leaves reduced to sheaths. ....4. **C. corymbosus**
13. Leaves well developed.
14. Spikes narrowly ovoid; rhizome stoloniferous. ....14. **C. rotundus**
14. Spikes oblong; rhizome short. ....10. **C. exaltatus**

1. **Cyperus castaneous** Willd., Sp. Pl. 1:278. 1979; Clarke in Hook.f., Fl. Brit. India 6:598. 1893; Fischer in Gamble, Fl. Pres. Madras 1639. 1931; Mohanan & Sivad., Fl. Agasthyamala 773.2002.

Tufted annuals. Leaves filiform, to 10 cm long. Inflorescence head-like or anthelate; involucral bracts filiform, shorter or longer than inflorescence; spikelets c. 2 mm long; glumes aristate, purplish-brown; rachilla flexuose, wingless. Nut narrowly obovoid, trigonous.

Fl. & Fr.: Nov.-Feb. *Distr.*: Peninsular India. Less common. On marshes in plains and hills. *NAK 9, 20, 681* & *1099* (Moozhiar, Tiruvalla, etc.)

2. **Cyperus cephalotes** Vahl, Enum. Pl. 2:311. 1805; Clarke in Hook.f., Fl.Brit. India 6:597. 1893; Fischer in Gamble, Fl. Pres. Madras 1639. 1931.

Aquatic floating perennials, rhizome stoloniferous; stem subtrigonous. Leaves 2-3, canaliculate, gradually narrowed into a triquetrous scabrid tip. Inflorescence capitate; bracts 3-5, to 13 cm long; spikelets congested, ovate-lanceate, compressed; rachilla ridged, hardly winged. Nut triquetrous, stipitate; stipe corky.

Fl. & Fr.: Nov.-Mar. *Distr.*: Peninsular India. On pools, lakes shallow fields, etc. Locally abundant. *NAK 1958* (Tiruvalla, c. 10 m).

3. **Cyperus compressus** L., Sp. Pl. 46. 1753, emend. Dandy in Exell, Cat. S. Tome 357. 1944; Fischer in Gamble, Fl. Pres. Madras 1640. 1931; Kern in Steenis, Fl. Males. I, 7:617. 1974; Hooper in Sald. & Nicols , Fl. Hassan Dist. 663. 1976; Koyama, Gard. Bull. Straits Settlem. 30:138. 1977; Britto & Matthew in Matthew, Fl. Tam. Carnatic 3(3):1734. 1983.

Tufted annuals; stem triquetrous. Leaves flat or channelled, to 15 by 0-..3 cm; sheath purplish. Inflorescence anthelate; involucral bracts 3-5, up to 15 cm long; spikes racemose or digitately clustered; spikelets compressed, pale green; rachilla flexuous, winged when young. Nut obovoid, trigonous, stipitate.

Fl. & Fr.: Most of the seasons. *Distr.*: Pantropical. Common in plains. *NAK 1960* (Tiruvalla, c. 10 m).

4. **Cyperus corymbosus** Rottb., Descr. Ic. Rar. 42, t.7, f.4. 1773; Clarke in Hook.f., Fl. Brit. India 6:612. 1893; Fischer in Gamble, Fl. Pres. Madras 1641. 1931; Koyama, Gard. Bull. Straits Settlem. 30:131. 1977.

Herbs to 1 m high; stem triquetrous. Leaves reduced to sheaths. Inflorescence anthelate, anthela compound; bracts 2-5, shorter than anthela; spikelets clustered, suberect or horizontal to rachis; rachilla persistent, straight, winged. Nut obovoid, triquetrous.

Fl. & Fr.: Sep.-Jan. *Distr.*: Pantropical. Not common. In mountain marshes. *NAK 10* (Moozhiar, c.250 m).

5. **Cyperus cuspidatus** Kunth, Nov. Gen. Pl. 1:204. 1815; Clarke in Hook.f., Fl. Brit. India 6:598.1893; Kern in Steenis, Fl. Males. I, 7: 629. 1974; Hooper in Sald. & Nicols., Fl. Hassan Dist. 664. 1976; Britto & Matthew in Matthew, Fl. Tam. Carnatic 3(3) : 1935. 1983. *C. uncinatus* auct., non Poiret, 1806: Fischer in Gamble, Fl. Pres. Madras 1639. 1931.

Tufted small herbs; stem capillary. Leaves setaceous, as long as or half as long as stem. Inflorescence simple or subcompound; spikes ovoid, digitately spikeletted; spikelets in clusters, purple; rachilla persistent, flexuous, hardly winged; glumes ovate, aristate. Nut obovoid, triquetrous, winged, stipitate.

Fl. & Fr.: Nov.-Feb. *Distr.;* Pantropical. Not common. In plains. *NAK 679, 876 & 2131* (Adoor & Tiruvalla).

6. **Cyperus difformis** L., Cent. Pl. 2:6. 1756; Roxb., Fl. Ind. 1:195. 1832; Clarke in Hook. f., Fl. Brit. India 6:599. 1893; Fischer in Gamble, Fl. Pres. Madras 1640. 1931; Kern in Steenis, Fl. Males. I, 7:629. 1974; Hooper in Sald. & Nicols., Fl. Hassan Dist. 664. 1976; Matthew, Ill. Fl. Tam. Carnatic t. 781. 1982; Britto & Matthew in Matthew, Fl. Tam. Carnatic 3(3) : 1736. 1983.

Tufted annuals; stem slender, triquetrous. Leaves usually flat. Inflorescence simple or compound, involucral bracts 3, overtopping; primary rays 5-7; spikelets narrowly linear, clustered in globose or capitate heads; rachilla wingless; glumes membranous. Nut obovoid, triquetrous, stipitate.

Fl. & Fr.: Nov.-Feb. *Distr.*: Tropics & Subtropics. Common as a weed in paddy fields. *NAK 604, 1015 & 2091* (Adoor, Tiruvalla).

7. **Cyperus diffusus** Vahl, Enum. Pl.2:321. 1805; Clarke in Hook.f., Fl. Brit. India 6:599. 1893; Fischer in Gamble, Fl. Pres. Madras 1639. 1931. **(Fig. 55)**

Slender perennials. Leaves 12-35 cm long, rigid. Inflorescence anthelate; bracts 4-12, foliaceous; spikelets usually in clusters, rarely few, solitary, compressed; rachilla hardly winged; glumes purplish, broadly ovate to suborbicular. Nut ellipsoid, triquetrous, dark purple.

Fl. & Fr.: Sep-Feb. *Distr.*: Tropics. Common in mountain marshes. *NAK 603, 996, 2472 & 2581* (Moozhiar, c. 250 m).

8. **Cyperus distans** L.f., Suppl. Pl. 103. 1781; Clarke in Hook.f., Fl. Brit. India 6:607. 1883; Fischer in Gamble, Fl. Pres. Madras 1640. 1931; Hooper in Sald. & Nicols., Fl. Hassan Dist. 665. 1976; Britto & Matthew in Matthew, Fl. Tam. Carnatic 3(3): 1737. 1983.

Key to the varieties

1. Glumes half-imbricate, 3-5-nerved. .... **C. distatum** var. **distans**

1. Glumes very distant, nerveless. .... **C. distatum** var. **pseudonutans**

Cyperus distans L.f., var. **distans**

Herbs to 1 m tall; stem not tufted, trigonous. Leaves flat, as long as the stem. Inflorescence decompound, anthelate; involucral bracts 5-7, primary rays 7-9, secondary rays 1-3, slender; spikes ovate with horizontal linear spikelets at right angles to the rachis; rachilla persistent, wing caducous; glumes half-imbricate, purplish, 3-5-nerved. Nut ellipsoid, trigonous, puncticulate.

Fl. & Fr.: Most of the seasons. *Distr.:* Tropics. & Subtropics. Common in hills. *NAK 590.* (Thekkuthode, c. 70 m).

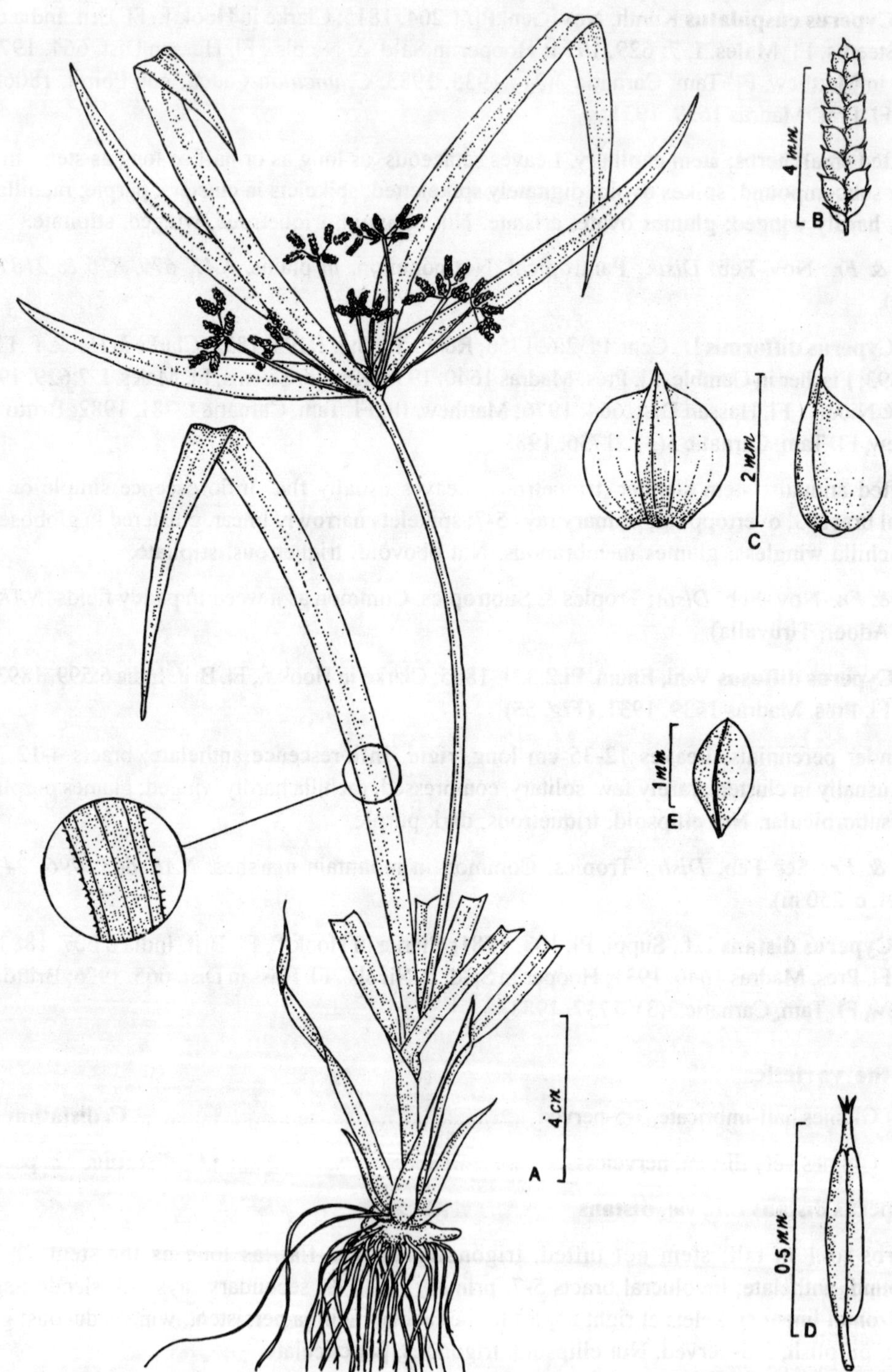

Fig. 55. ***Cyperus diffusus*** **Vahl: A. Habit; B. Spike; C. Glume – different views; D. Stamen; E. Nut.**

Cyperus distans L.f., var. **pseudonutans** Kuek. in Engler, Pflanzenr. 101:140. 1935; Hooper in Sald. & Nicols., Fl. Hassan Dist. 665. 1976; *C. nutans* sensu auct., non Vahl, 1805: Clarke in Hook. f., Fl. Brit. India 6:607. 1893, p.p.

Perennial herbs; spikelets suberect; glumes very distant with nerveless sides.

Fl. & Fr.: Most of the seasons. *Distr.*: Tropics. Common. In plains. *NAK 953* (Adoor, c. 20 m).

9. **Cyperus exaltatus** Retz., Obs. Bot. 5:11.1789; Clarke in Hook.f., Fl. Brit. India 6:616. 1893; Fischer in Gamble, Fl. Pres. Madras 1642. 1931; Kern in Steenis, Fl. Males. I, 7:602. 1974; Hooper in Sald. & Nicols., Fl. Hassan Dist. 665. 1976; Britto & Mathew in Matthew, Fl. Tam. Carnatic 3(3) : 1737. 1983; Nicols., Suresh & Manilal, An Intepr. Hort. Malab. 284. 1988. *Wara-pullu* Rheede, Hort. Malab. 12:77, t.42. 1693.

Tufted perennials. Leaves flat or folded, to 90 cm long. Inflorescence decompound; involucral bracts 5-7, overtopping; spikes digitate; spikelets horizontal; rachilla persistent, winged; glumes ovate, mucronate. Nut obovoid to ellipsoid.

Fl. & Fr.: Dec.-Mar. *Distr.*: Pantropical. Not common. In plains. *NAK 921* (Tiruvalla, c. 10 m).

10. **Cyperus haspan** L., Sp. Pl.45. 1753; Roxb., Fl.Ind. 1:213. 1820; Clarke in Hook.f., Fl. Brit. India 6:606. 1893; Fischer in Gamble, Fl. Pres. Madras 1640. 1931; Kern in Steenis, Fl. Males. I, 7:616. 1974; Hooper in Sald. & Nicols., Fl. Hassan Dist. 665. 1976; Britto & Mathew in Matthew, Fl. Tam. Carnatic 3(3): 1738. 1983; Mohanan & Sivad., Fl. Agasthyamala 774.2002. ssp. **haspan. (Fig.56, H-J).**

Tufted herbs; stems slender, triquetrous. Leaves flat. Inflorescence compound or decompound; involucral bracts 5, primary rays 10-15, secondary rays 5-7; spikelets spicate, linear; rachilla persistent, wingless; glumes broadly obovate, shortly mucronate; stamens 3, anthers appendaged at apex. Nut obovoid, triquetrous.

Fl. & Fr.: Most of the seasons. *Distr.*: Pantropical. Very common in plains and hills. *NAK 596, 938 & 1279* (Tiruvalla, Adoor, Moozhiar, etc.).

11. **Cyperus iria** L., Sp. Pl. 45. 1753; Roxb., Fl. Ind. 1:204. 1820; Clarke in Hook.f., Fl. Brit. India 6:606. 1893; Fischer in Gamble, Fl. Pres. Madras 1640. 1931; Kern in Steenis, Fl. Males I,7:616. 1974; Hooper in Sald. & Nicols., Fl. Hassan Dist. 666. 1976; Britto & Matthew in Matthew, Fl. Tam. Carnatic 3(3) : 1783. 1983.

Tufted or solitary herbs; stem triquetrous. Leaves flat or channelled, to 15 cm long. Inflorescence compound or decompound; involucral bracts 5; primary rays 5-7, secondary rays 3-5; spikelets spicate; rachilla peristent, wingless; glumes broadly obovate, shortly mucronulate. Nut obovoid or ellipsoid, trigonous, stipitate.

Fl. & Fr.: Nov.-Feb. *Distr.*: Tropics & Subtropics. Common as a weed in paddy fields. *NAK 990* (Tiruvalla, c. 10 m)

12. **Cyperus malaccensis** Lam., Tab. Encycl. 1:146. 1791; Clarke in Hook.f., Fl. Brit. India 6:608. 1893; Fischer in Gamble, Fl. Pres. Madras 1641. 1931; Kern in Steenis, Fl. Males. I, 7:613. 1974; Nicols., Suresh & Manilal, An Intepr. Hort. Malab. 284. 1988. *Pota pullu* Rheede, Hort. Malab. 12:93, t.50. 1693.

Tall perennials; rhizome thick, stoloniferous; stem trigonous. Leaves absent or very few, to 15 cm long. Inflorescence compound or congested; involucral bracts 3-5, coriaceous; primary rays 3-12; secondary rays 5-7; spikelets linear, subterete. Nut narrolwy oblong, faintly trigonous.

Fl. & Fr.: Nov.-Feb. *Distr.*: Southern Western Ghats. Occasional. Along hill tracts. *NAK 2424* (Moozhiar, c.250 m).

13. **Cyperus procerus** Rottb., Descr. Ic. Rar. 29, t.5, f.3. 1773; Roxb., Fl. Ind. 1:206. 1820; Baker in Hook.f., Fl. Brit. India 6:610. 1893; Fischer in Gamble, Fl. Pres. Madras 1641. 1931; Kern in Steenis, Fl. Males. I, 7.:61. 1974; Britto & Matthew in Matthew, Fl. Tam. Carnatic 3(3): 1741. 1983.

Tall herbs; stem triquetrous, stout. Leaves basally channelled, apically flat, as long as stem, c. 1 cm ϕ, spongy, thick. Inflorescence compound; involucral bracts 3 or 4, much exceeding; spikes pyramidal; spikelets spicate, horizontal to rachis, purplish; rachilla persistent, hardly winged; glumes ovate, margin hyaline. Nut obovoid, triquetrous.

Fl. & Fr.: Nov.-Feb. *Distr.*: Sri Lanka, India, Malesia & China. Not common. On exposed hill tops. *NAK 11, 576* (Moozhiar, c. 250 m).

14. **Cyperus rotundus** L., Sp. Pl. 45. 1753; Roxb., Fl. Ind. 1:201. 1820; Clarke in Hook.f., Fl. Brit. India 6:614. 1893; Fischer in Gamble, Fl. Pres. Madras 1641. 1931; Kern in Steenis, Fl. Males. 1, 7:604. 1974; Hooper in Sald. & Nicols., Fl. Hassan Dist. 669. 1976; Britto & Matthew in Matthew, Fl. Tam. Carnatic 3(3):1742. 1983; Mohanan & Sivad., Fl. Agasthyamala 775.2002.

Small herbs; stems not tufted or sparsely tufted. Leaves flat, to 15 by 0.2 cm. Inflorescence simple, or compound; involucral bracts 3, unequal, exceeding or not; spikelets spicate, purplish or stramineous; rachilla persistent, winged; glumes closely imbricating, ovate, membranous. Nut oblong, trigonous.

Fl. & Fr.: Most of the seasons. *Distr.*: Pantropical. Common in plains. *NAK 1025* (Tiruvalla, c. 10 m).

15. **Cyperus tenuispica** Steudel, Syn. Pl. Glum. 2:11.1855; Fischer in Gamble, Fl. Pres. Madras 1650. 1931; Kern in Steenis, Fl. Males. I, 7:625. 1974; Hooper in Sald. & Nicols., Fl. Hassan Dist. 670. 1976; Britto & Matthew in Matthew, Fl. Tam. Carnatic 3(3): 1744. 1983; Mohanan & Sivad., Fl. Agasthyamala 776.2002. *C. flavidus* auct., non. Retz.,1789 : Baker in Hook.f., Fl. Brit. India 6:600. 1893. **(Fig. 56, A-G)**

Tufted herbs; stem triquetrous. Leaves flat, to 20 cm long. Inflorescence compound or decompound; involucral bracts 3-5, exceeding; primary rays 7-10, secondary rays 5-7; spilkelets narrowly linear, pale brown or stramineous; rachilla flexuous, wingless; glumes ovate, often purplish, anthers without a crest at apex. Nut ovoid, trabeculate.

Fl. & Fr.: Most of the seasons. *Distr.*: Tropical and Subtropical Africa and Asia. Common weed in paddy fields and waste places. *NAK 1074* (Tiruvalla, c. 10 m).

ELEOCHARIS R. Brown
Prodr. 224. 1810.

Annuals or perennials; stem triquetrous or ribbed. Leaves reduced. Inflorescence a terminal, solitary spike, spikelets terete, globose or ovoid; rachilla persistent. Flowers bisexual; hypogynous bristles 4-10, retrorsely scabrid; style 2 or 3-fid, base persistent. Nut obovoid or orbicular, triquetrous or biconvex, smooth or cancellate.

1. Annuals; stem 5-ribbed; spikelets ovoid. ....1. **E. retroflexa**
1. Perennials; stem triquetrous; spikelets terete. ....2. **E. spiralis** ssp. **chaetaria**

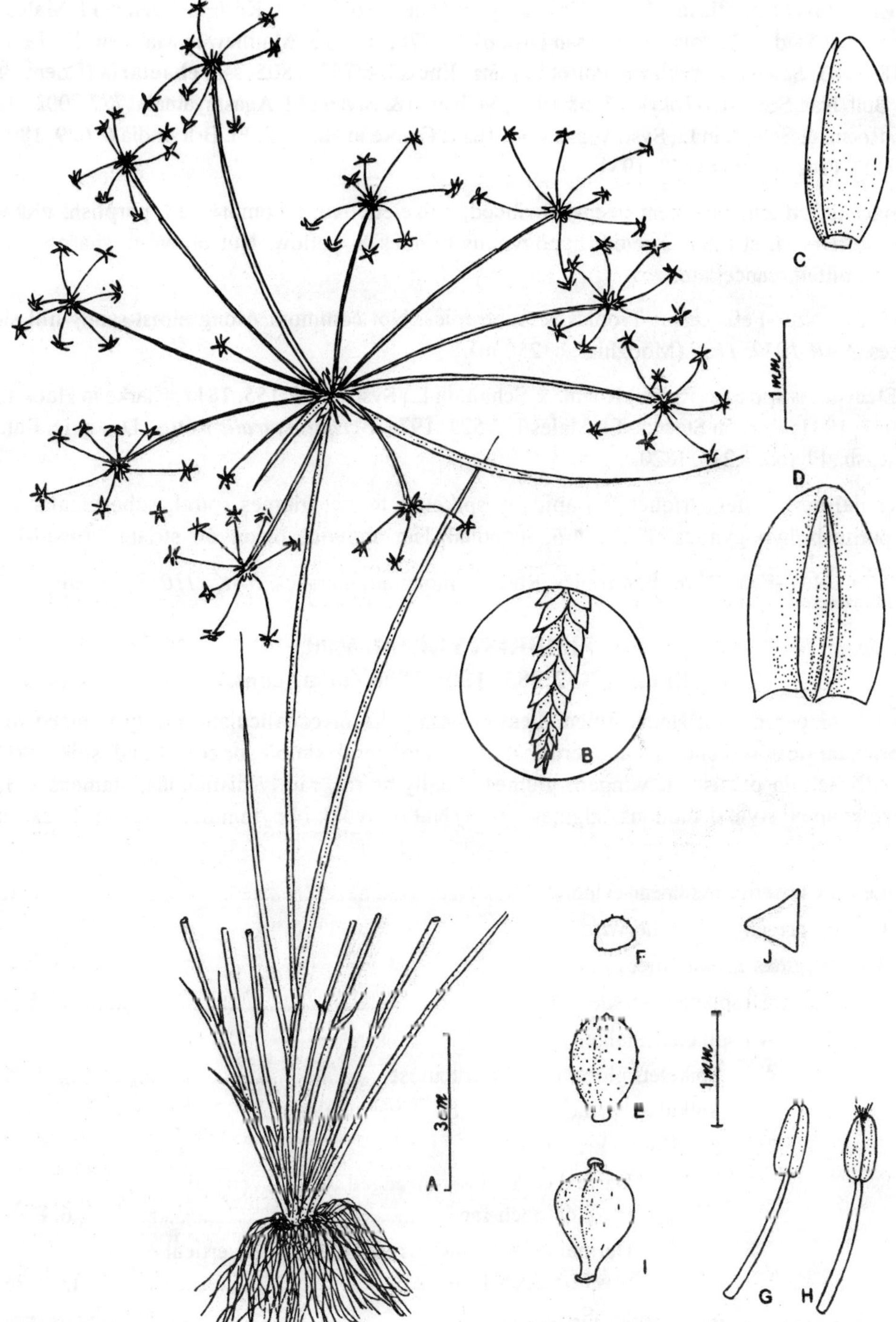

Fig. 56. ***Cyperus tenuispica*** **Steudel (A-G) : A. Habit; B. Spike; C. Glume – side view; D. Glume – abaxial view; E. Nut; F. Nut - C.S.; G. Stamen.** ***Cyperus haspan*** **L. (H-J) : H. Stamen; I. Nut; J. Nut – C.S.**

1. **Eleocharis retroflexa** (Poiret) Urban, Symb. Ant. 2:165. 1900; Kern in Steenis, Fl. Males. I, 7:534. 1974; Hooper in Sald. & Nicols., Fl. Hassan Dist. 673. 1976; Britto & Matthew in Matthew, Fl. Tam. Carnatic 3(3); 1748. 1983. *Scirpus retroflexus* Poiret in Lam., Encycl. 6:753. 1805. ssp. **chaetaria** (Roem. & Schult.) Koyama, Bull. Nat. Sci. Mus. Tokyo 17: 68.1974; Mohanan & Sivad., Fl. Agasthyamala 777.2002. *Eleocharis chaetaria* Roem. & Schult. in L., Syst. Veg. 2:154. 1817; Clarke in Hook. f., Fl. Brit. India 6: 629. 1893; Fischer in Gamble, Fl. Pres. Madras 1648. 1931.

Slender tufted annuals; stem deeply 5-ribbed; spikelets ovoid, compressed, purplish; glume basally distichous, α-nerved, strongly keeled; hypogynous bristles 8, yellow. Nut obovoid, sharply triquetrous, stramineous, pitted, cancellate.

Fl. & Fr.: Nov.-Feb. *Distr.*: Tropics & Subtropics. Not common. Along moist sandy hill slopes and streamsides. *NAK 1079, 1129* (Moozhiar, c. 250 m).

2. **Eleocharis spiralis** (Rottb.) Roem. & Schult. in L., Syst. Veg.2:155. 1817; Clarke in Hook.f., Fl. Brit. India 3:1647. 1931; Kern in Steenis, Fl. Males I, 7:527. 1974. *Scripus spiralis* Rottb., Descr. Ic. Rar. 45, t.15, f.1.1773; Roxb., Fl. Ind. 1:215. 1820.

Perennial herbs; stem triquetrous apically spikelets terete; glumes spiral, suborbicular to broadly obovate, purplish, hypogynous bristles 4-6, subequal. Nut obovoid, biconvex, striate, brownish.

Fl. & Fr.: Nov.-Feb. *Distr.*: Pantropics. Rare. In mountain marshes. *NAK 2110* (Kakki hills, c. 1000 m).

FIMBRISTYLIS M. Vahl
Enum. Pl. 2:285. 1805-1806 (nom. cons.).

Annuals or perennials; stems tufted. Leaves basal, flat or canaliculate, rarely reduced to sheaths. Flowers bisexual. Inflorescence usually terminal or pseudolateral, simple, or compound; spikelet(s) solitary or clustered; rachilla persistent, winged; glumes usually spiral, rarely distichous; stamens 1-3, anthers appendaged at apex; style deciduous, stigmas 2 or 3. Nut obovoid, obpyramidal, terete or linear, trigonous or biconvex, smooth or ornamented.

1. Leaves absent; nut linear-cyindric. .... 12. **F. tetragona**
1. Leaves present; nut otherwise.
    2. Stigmas 2; nuts biconvex.
        3. All spikelets sessile. .... 3. **F. argentea**
        3. All spikelets stalked.
            4 Spikelets globose; nut verruculose. .... 5. **F. cymosa**
            4. Spikelets ovoid-terete; nut not verruculose.
                5. Ligule present.
                    6. Dermal cells of nut arranged in 5-10 vertical rows on each face. .... 6. **F. dichotoma**
                    6. Dermal cells of nuts arranged in 12-24 vertical rows on each face. .... 13. **F. tomentosa**
                5. Ligule absent. .... 1. **F. aestivalis**
    2. Stigmas 3; nuts trigonous.
        7. Flowering stems leafy.

8. Stem bearing 1-3 spikelets; nuts up to 3 mm long. ....10. **F. ovata**

8. Stem bearing 1-α spikelet(s); nuts up to 1.5 mm long.

9. Leaves equitant, laterally flattened. ....7. **F. littoralis**

9. Leaves not equitant, dorso-ventrally flattened.

10. Spikelets narrow-linear; glumes distichous at base only. ....4. **F. cinnamometorum**

10. Spikelets lanceate; glumes distichous throughout.

11. Glumes loosely imbricating, apex subacute; nut obpyramidal. ....9. **F. narayanii**

11. Glumes closely imbricating, apex mucronate; nut obvoid. ....11. **F. pseudonarayanii**

7. Flowering stem leafless; glumes all spiral.

12. Stem 4-5-angled; spikelets ovoid or globose; stamens 1-2; lowland species. ....8. **F. miliacea**

12. Stem 3-angled; spikelets lanceoid; stamens 3; montane species. ....2. **F. angamoozhiensis**

1. **Fimbristylis aestivalis** (Retz.) Vahl, Enum. Pl. 2: 288. 1806; Clarke in Hook.f., Fl. Brit. India 6:637. 1893, p.p.; Kern in Steenis, Fl. Males. I, 7:584. 1974; Hooper in Sald. & Nicols., Fl. Hassan Dist. 675. 1976; Britto & Matthew in Matthew, Fl. Tam. Carnatic 3(3): 1751. 1983; Mohanan & Sivad., Fl. Agasthyamala 778.2002. *Scirpus aestivalis* Retz., Obs. Bot. 4:12. 1768.

Tufted pubescent herbs; stems slender, setaceous. Leaves flat or folded, filiform, ligule 0. Inflorescence compound; involucral bracts overtopping; primary rays filiform. Spikelets terete or angular; rachilla narrolwy winged; glumes strongly 3-nerved; stamens 1; style 2-fid. Nut obovoid, biconvex, faintly reticulate, umbonulate.

Fl. & Fr.: Nov.-Feb. *Distr.:* Tropics & Subtropics. Not common. In paddy fields from plains to hills. *NAK 1624, 2425* (Tiruvalla, c. 10 m & Moozhiar).

2. **Fimbristylis angamoozhiensis** Ravi & Anil Kumar, Rheedea 3 (2):108.1993; Mohanan & Sivad., Fl. Agasthyamala 779.2002 **(Fig. 57)**

Robust perennial herbs. Rhizome short, thick and woody. Roots fibrous and black. Culms triquetrous towards the base and more or less flattened towards the apex, glabrous, up to 50 cm tall. Leaves many, basal and spiral; lower most 3 to 4 cataphylloid, coriaceous, tapering to an acute apex, pale brownish, rest foliaceos; sheath coriaceous, membranous and hyaline, opposite the blade, glabrous, mouth oblique; blade eligulate, flat, coriaceous, tapering to an acute to subacute apex, recurved on the margins, minutely scaberulous towards the apex, glabrous otherwise, up to 70 cm long, 8 mm wide. Inflorescence a compound umbel, up to 25 cm wide; bracts several, outer 3 or 4 leaf-like, much exceeding the umbel, gradually tapering to an acuminate apex, minutely scaberulous on the margins, glabrous otherwise, the inner narrower to filiform; bracteoles filiform; broader towards the base, minutely puberulous, shorter or longer than the umbellets, up to 5 cm long; primary and secondary rays a, unequal, slender, scaberulous on the margins, up to 10 cm long. Spikelets a, solitary on slender, 5-8 mm long peduncles, sessile at the forkings, linear-lanceolate, acute at both ends, 5 mm long, 1 mm wide, up to 10-flowered; rachilla prominently winged, wings brownish-black,

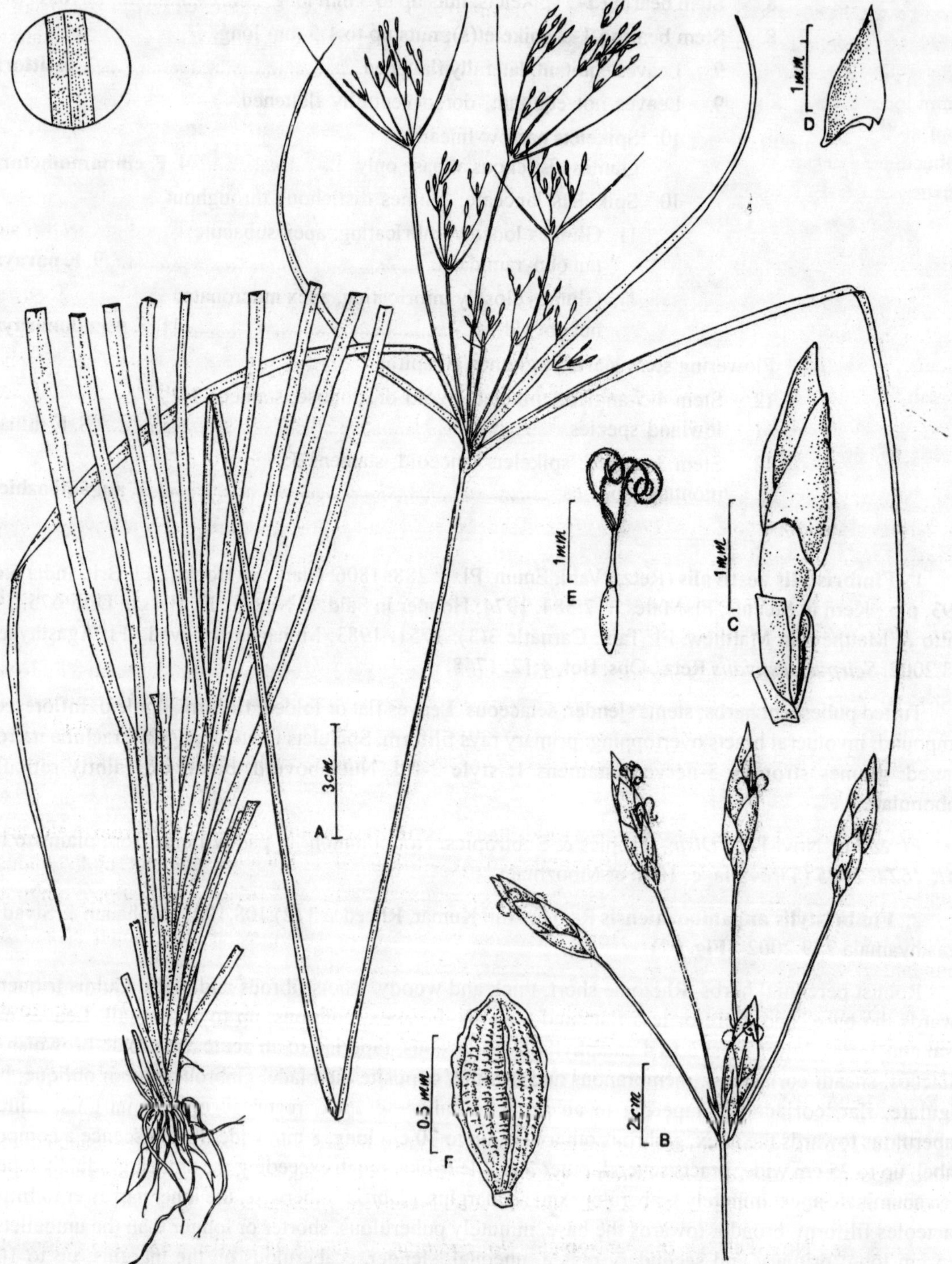

Fig. 57. ***Fimbristylis angamoozhiensis*** **Ravi & Anil Kumar: A. Habit; B. An anthela; C. Spikelet; D. Glume; E. Pistil; F. Nut.**

persistent. Glumes about 10, lower most 1 or 2 empty, upper most 1 to 3 tabescent; fertile ones triangular-ovate, boat-shaped, keeled, keel 3-nerved, sparsely scaberulous on the back towards the mucronate apex, mucro shortly spinulose, margins hyaline, brownish-yellow, glabrous, veinless. Stamens 3, anthers linear, c. 1 mm long, crest acuminate. Style less than 1 mm long, sparsely ciliolate towards the apex. Stigmas 3, longer than style. Nuts oblanceoid to ellipsoid, trigonus, cuneate at base, shortly stipitate, subacute at apex, trabeculate, surface cells thin-walled, transverely hexagonal in 3 to 6 vertical rows on each face, sparsely verrucose, mostly towards the upper half, pale white, c. 1 mm long, 0.5 mm wide.

Fl. & Fr.: Sep.- Apr. *Distr.*: Hitherto known only from the type locality. On exposed grassy hill slopes. *NAK 1799* (Angamoozhy c 250 m)

3. **Fimbristylis argentea** (Rottb.) Vahl, Enum. Pl. 2:294. 1806; Clarke in Hook.f., Fl. Brit. India 6:640. 1893; Fischer in Gamble, Fl. Pres. Madras 1659. 1931; Kern in Steenis, Fl. Males. I, 7:586. 1974; Hooper in Sald. & Nicols., Fl. Hasssan Dist.676. 1976; Britto & Matthew in Matthew, Fl. Tam. Carnatic 3(3): 1752.1983; Nicols., Suresh & Manilal, An Interpr. Hort. Malab. 286. 1988; Mohanan & Sivad., Fl. Agasthyamala 779.2002. *Scirpus argenteus* Rottb., Descr. Pl. Rar. 27.1777 & Descr. Ic. Rar. 51, t.17, f.6.1773. *Mulen-pullu* Rheede, Hort. Malab. 12:101, t.54.1693.

Tufted herbs; stem setaceos, trigonous. Leaves flat, filiform; ligule 0. Inflorescence capitate, 5-12 in a cluster, involucral bracts 3, overtopping; spikelets sessile; rachilla narrowly winged; glumes ovate, 3-nerved; stamen 1. Nut obovoid, biconvex, obscurely vertically striate, umbonulate.

Fl. & Fr.: Most of the seasons. *Distr.*: S. & S.E. Asia, Sri Lanka, India to Malesia. Common. In plains. *NAK, 1607, 1962* (Tiruvalla, c. 10 m).

4. **Fimbristylis cinnamometorum** (Vahl) Kunth, Enum. Pl. 2:229. 1837; Kern in Steenis, Fl. Males. I, 7:565. 1974; Hooper in Sald. & Nicols., Fl. Hassan Dist. 676. 1976; Britto & Matthew in Matthew, Fl. Tam. Carnatic 3(3): 1753. 1983. *Scirpus cinnamometorum* Vahl, Enum. Pl. 2:278. 1806. *Fimbristylis cyperoides* R. Br., Prodr. 228. 1810, var. *cinnamometorum* (Vahl) Clarke in Hook.f., Fl. Brit. India 6:650. 1893; Fischer in Gamble, Fl. Pres. Madras 1659. 1931.

Tufted herbs. Leaves flat or folded, rigid; ligule 0. Inflorescence compound or decompound; primary rays 3-7, secondary rays 3-5; spikelet(s) solitary, narrow-linear; rachilla winged; glumes subdistichous, red gland-dotted; stamens 3; style 3-fid. Nut obovoid or ellipsoid, trigonous, ridged, verruculose, umbonulate.

Fl. & Fr.: Nov.-Mar. *Distr.*: Tropics. Among grasses in mountain swamps of upper hills. Locally abundant. *NAK 810, 910, 2038 & 2580* (Moozhiar, Kakki hills, etc.).

5. **Fimbristylis cymosa** R. Br., Prodr. 228.1810; Kern in Steenis, Fl. Males. I,7; 557. 1974; Hooper in Sald. & Nicols., Fl. Hassan Dist. 677. 1976; Britto & Matthew in Matthew, Fl. Tam. Carnatic 3(3): 1754. 1983. *F. spathacea* Roth, Nov. Pl. Spec. 24. 1821; Clarke in Hook.f., Fl. Brit. India 6:640. 1893; Fischer in Gamble, Fl. Pres. Madras 1659. 1931.

Tufted herbs; stem trigonous or subterate. Leaves flat or folded, rigid, basal. Inflorescence compound or decompound; involucral bracts filiform; spikelet(s) solitary or clustered, subglobose; rachilla winged; glumes orbicular to ovate; stamens 2; style 2-fid. Nut obovoid, biconvex, umbonulate.

Fl. & Fr.: Nov.-Feb. *Distr.*: Pantropical. Not common. In plains and hills. *NAK 2323* (Moozhiar, c. 250 m).

6. **Fimbristylis dichotoma** (L.) Vahl, Enum. Pl. 2:287. 1805; Clarke in Hook.f., Fl. Brit. India 6:635. 1893, p.p.; Fischer in Gamble, Fl. Pres. Madras 1658. 1931; Kern in Steenis, Fl. Males. I, 7:574. 1974; Hooper in Sald. & Nicols., Fl. Hassan Dist. 677. 1976; Britto & Matthew in Matthew, Fl. Tam. Carnatic 3(3): 1754. 1983; Mohanan & Sivad., Fl. Agasthyamala 780.2002. *Scirpus dichotomus* L., Sp. Pl. 50. 1753. *Scripus diphyllus* Retz., Obs. 5:15.1789. *Fimbristylis diphylla* (Retz.) Vahl, Enum. Pl. 2:289. 1805; Clarke in Hook.f., Fl. Brit. India 6:636. 1893.

Tufted herbs. Leaves folded or rarely flat, ligulate. Inflorescence simple or decompound; primary rays 3-5, secondary rays 3; rachilla winged; glumes broadly ovate to orbicular; stamen 1; style 2-fid. Nut obovoid, biconvex, faintly cross-barred.

Fl. & Fr.: Most of the seasons. *Distr.*: Pantropics. The most common species of the genus. From plains to hills. *NAK 1006, 1071* (Tiruvalla, c. 10 m). A morphologically highly variable herb.

7. **Fimbristylis littoralis** Gaud. in Frey., Voy. Bot. 413. 1826; Kern in Steenis, Fl. Males. I, 7:551. 1974. *F. miliacea* Vahl, Enum. Pl. 2:287. 1805, quoad descr., excl. basionym; Clarke in Hook. f., Fl. Brit. India 6:644. 1893, p.p.; Fischer in Gamble, Fl. Pres. Madras 1660. 1981.

Tufted herbs, stem 4-5-angled. Leaves several, equitant, usually ± distichous, flat. Inflorescence decompound or rarely compound; spikelet(s) solitary; glumes ovate-oblong, muticous, 3-nerved; stamens 1-2; style 3-fid. Nut narrowly obovoid, trigonous, verruculose.

Fl. & Fr.: Most of the seasons. *Distr.*: Tropics & Subtropics. Common in moist swampy places in plains. *NAK 1050* (Tiruvalla, c.10 m).

8. **Fimbristylis miliacea** (L.) Vahl, Enum. Pl. 2:287. 1806; Kern in Steenis, Fl. Males. I, 7:552. 1974. Hooper in Sald. & Nicols., Fl. Hassan Dist. 679. 1976; Mohanan & Sivad., Fl. Agasthyamala 781.2002. *Scirpus miliaceus* L., Syst. Veg. ed. 10, 868. 1759. *F. quinquangularis* Kunth, Enum. Pl. 2:229. 1837; Clarke in Hook. f., Fl. Brit. India 6:644. 1893.

Tufted annuals; stem 4-5-angled; flowering stem leafless or with a solitary basal leaves. Leaves flat, margins cartilaginous. Inflorescence decompound; spikelet(s) solitary, ovoid, subterete or angular; rachilla winged; glumes ovate, mucronulate; stamens 1-2; style 3-fid. Nut obovoid or suborbicular, trigonous, densely verruculose.

Fl. & Fr.: May-Feb. *Distr.*: Tropics & Subtropics. Common as a weed in paddy fields. *NAK 1014* (Tiruvalla, c. 10 m).

9. **Fimbristylis narayanii** Fischer, Kew Bull. 1931:46. 1931 & in Gamble, Fl. Pres. Madras 1660. 1931; Mohanan & Sivad., Fl. Agasthyamala 782.2002.

Annual tufted herbs; stem slender, faintly 4-angular. Leaves narrowly ensiform, flat. Inflorescence decompound, primary rays 1-4, capillary; spikelet(s) solitary, lanceate, flat; glumes distichous, broadly triangular, subacute, weakly keeled; stamens 1-2; style 3-fid. Nut obpyramidal, trigonous, faintly verruculose.

Fl. & Fr.: Nov.-Feb. *Distr.*: Southern Western Ghats. Rare. On mountain swamps. *NAK 1093, 808, 2037* (Moozhiar, c. 250 m).

10. **Fimbristylis ovata** (Burm. f.) Kern, Blumea 15:126. 1967 & in Steenis, Fl. Males. I, 7:565. 1974; Hooper in Sald. & Nicols., Fl. Hassan Dist. 679. 1976; Britto & Matthew in Matthew, Fl. Tam. Carnatic 3(3): 1758. 1983. *Carex ovata* Burm. f., Fl. Ind. 194. 1768. *Cyperus monostachyos* L., Mant.Pl. 180. 1771. *Fimbristylis monostachyos* (L.) Hassk., Pl. Jav. Rar. 61. 1848; Clarke in Hook.f., Fl. Brit. India 6:649. 1893.

Tufted herbs; stem trigonous. Leaves flat, margins inrolled. Inflorescence 1(-2) spikeletted; spikelets terete; rachilla winged; glumes basally distichous, apically spiral, mucronate, keel strongly 3-5-nerved; stamens 3; style 3-fid. Nut obovoid, trigonous, stramineous, glossy, tubercled.

Fl. & Fr.: Sep.-Jan. *Distr.*: Pantropical. Less common. In marshes on hills. *NAK 2583* (Moozhiar, c. 250 m).

11. **Fimbristylis pseudonarayanii** Ravi & Anil Kumar, Rheedea 3 (2): 111.1992: Mohanan & Sivad., Fl. Agasthyamala 782.2002. **(Fig. 58)**

Tufted annual herbs. Rhizome nil. Roots fibrous. Culms slender, smooth, more or less quadrangular, up to 20 cm tall, c. 0.5 mm wide. Leaves a, basal, lowest 1 or 2 sheathing, acuminate or with a reduced blade, others foliaceous; sheath membranous, glabrous, mouth truncate, up to 2 cm long; blade linear-flattened, ensiform, acute to subacute at apex, 0.5 to 2 mm wide, usually reaching the middle of the culm. Inflorescence a simple umbel of 3-5-spikelets, rarely 2 or 1 spikelet per culm; bracts 2-5, glumaceous; rays slender, triquetrous, up to 2 cm long, base sheathing. Spikelets compressed, ellipsoid to lanceoid, up to 2 cm long, 2 mm wide; rachilla zig-zag, winged, wing pale yellow to brownish, hyaline, elliptic, deciduous, c. 1 mm long. Glumes distichous, up to 30, closely imbricating at maturity, broadly triangular (when unfloded), strongly keeled on the back, keel sharply angled, broader towards the apex, exceeding the glume apex and abruptly acuminate into a slightly excurved mucro, glabrous to scaberulous towards the middle, sides coriaceous, pale yellow to brownish, veinless, glabrous or minutely scaberulous in the middle towards the keel, margins hyaline, 2.75-3 mm long, 1 mm wide (when folded). Stamens 3, anthers linear, sagittate at the base, crest subacute, 0.75 to 1 mm long. Style glabrous, base conical, brownish gland-dotted, 1.75-2 mm long. Stigmas 3, c.1 mm long. Nut obconical to obovoid, trigonous, subtruncate to obtuse and subumbonulate at apex, broadly cuneate at base, shortly stipitate, prominently ridged at angles, white to pale yellowish-brown, sparsely verrucose in the upper half, surface-cells hexagonal to rectangular in 5-7 well defined vertical rows, 0.8-0.9 mm long and 0.5-0.7 mm wide.

Fl. & Fr.: Nov.-Feb. *Distr.*: Endemic to southern Western Ghats. Widely distributed in the hilly tracts of Kerala. *Ravi 2254* (Thenmala, c. 100 m), *NAK 2037* (Ranni R.F., c. 100 m).

12. **Fimbristylis tetragona** R. Br., Prodr. 226. 1810; Clarke in Hook.f., Fl. Brit. India 6:631. 1893; Fischer in Gamble, Fl. Pres. Madras 1658. 1931; Kern in Steenis, Fl. Males I, 7: 590. 1974; Hooper in Sald. & Nicols., Fl. Hassan Dist. 681. 1976; Britto & Matthew in Matthew, Fl. Tam. Carnatic 3(3): 1761. 1983.

Tufted herbs; stem 4-angular. Leaves reduced to membranous sheath; spikelet(s) solitary, subglobose; rachilla narrowly winged; glumes ovate to cuneate; keel 3-nerved. stamens 2; style 3-fid. Nut terete, vertically ridged, tuberculate.

Fl. & Fr.: Nov.-Feb. *Distr.*: Tropics. Rare. In mountain marshes. *NAK 1883* (Chittar, c. 400 m).

13. **Fimbristylis tomentosa** Vahl, Enum. Pl. 2:290. 1806; Kern in Steenis, Fl.Males. I, 7:576. 1974. Hooper in Sald. & Nicols., Fl. Hassan Dist. 680. 1976. *F. podocarpa* Need & Meyen ex Nees in Wight, Contrib. Bot. India 98. 1834, p.p.; Clarke in Hook.f., Fl. Brit. India 6:638. 1893.

Tufted herbs; stems slender, trigonous. Leaves falcate, ligulate. Inflorescence compound, primary rays 4-6, spreading; spikelet(s) solitary, ovoid; glumes broadly ovate; stamens 2; style 2-fid. Nut suborbicular, long-stipitate, smooth or rarely verruculose.

Fl. & Fr.: Nov.-Feb. *Distr.*: Tropics. Not common. Along marshes in plains. *NAK 619, 1072* (Pandalam, Tiruvalla, etc.)

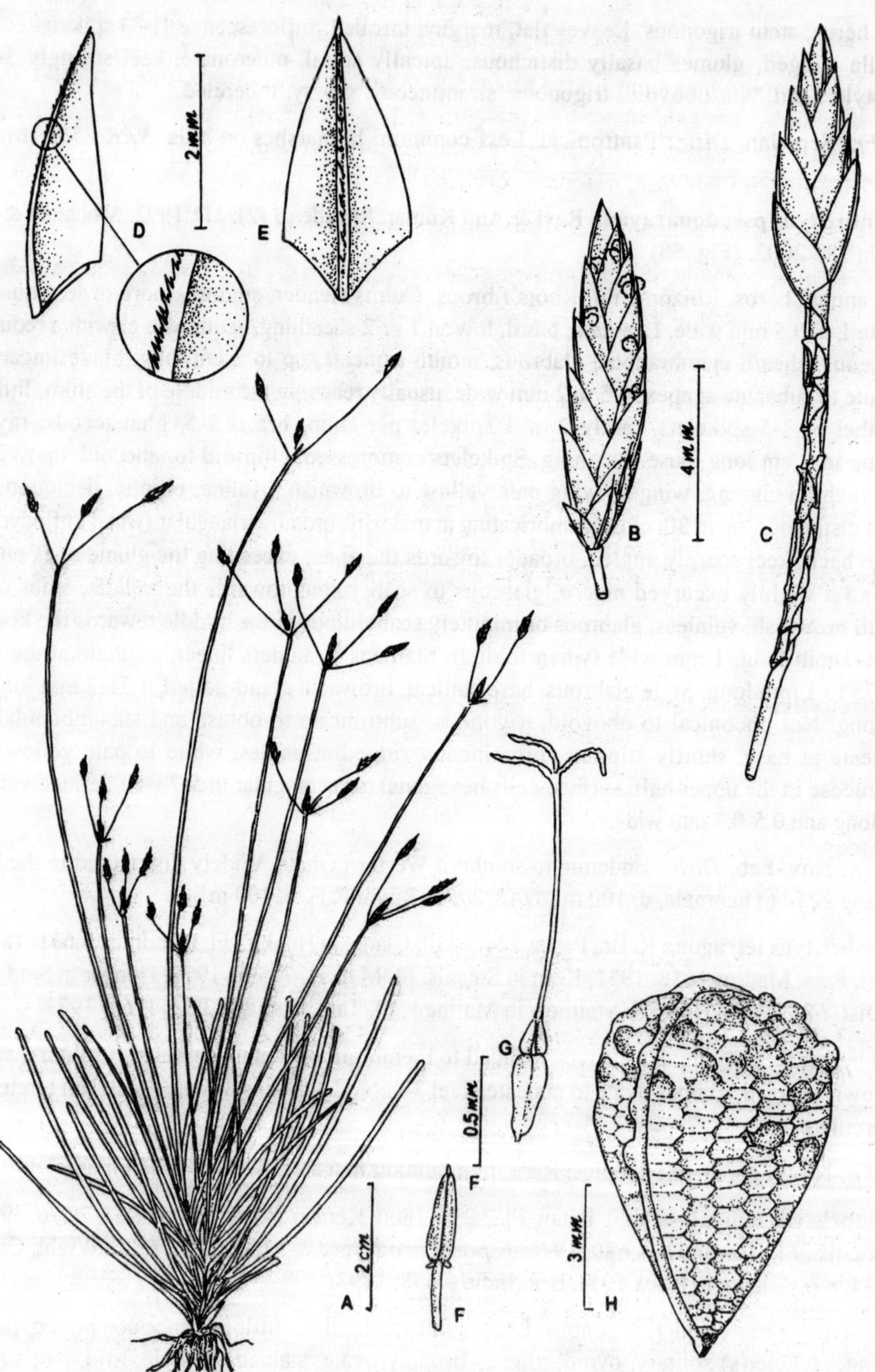

Fig. 58. ***Fimbristylis pseudonarayanii*** **Ravi & Anil Kumar: A. Habit; B. An young spikelet; C. A mature spikelet; D. Glume – side view; E. Glume – abaxial view; F. Stamen; G. Pistil; H. Nut.**

FUIRENA Rottboell
Descript. Icon. Nov. Pl. 70. 1773.

Annuals or perennials; stem noded, strongly ribbed. Leaves flat, distant; sheaths loose; ligule membranous. Inflorescence paniculate; spikelets clustered; rachilla persistent, wingless; glumes spiral, hairy, 3-nerved, aristate. Flowers bisexual; perianth 6-partite, the outer 3 bristle-like, the inner 3 scale-like; stamens 2-3; styles 3-partite. Nut obovoid, trigonous, beaked.

1. Densely pubescent annuals; spikelet-clusters congested at the tips of stem. ....1. **F. ciliaris**
1. Glabrous perennials; spikelet-clusters scattered all along the stem. ....2. **F. umbellata**

1. **Fuirena ciliaris** (L.) Roxb., Fl. Ind. 1:184. 1820; Kern in Steenis, Fl. Males. I, 7:519. 1974; Hooper in Sald. & Nicols., Fl. Hassan Dist. 683. 1976; Britto & Matthew in Matthew, Fl. Tam. Carnatic 3(3): 1763. 1983. *Scripus ciliaris* L., Mant. Pl. 182. 1771. *Fuirena glomerata* Lam., Tabl. Encycl. 1:150. 1791; Clarke in Hook.f., Fl. Brit. India 6:666. 1893; Fischer in Gamble, Fl. Pres. Madras 1669. 1931. **(Fig. 59)**

Annuals. Leaves flaccid, c. 7-nerved. Inflorescence umbellate; spikelets in clusters, ellipsoid, squarrulose, α-flowered; glumes oblong-obovate, pilose, arista to 1 mm long; bristles 4-6, filiform; scales 3, quadrate, 3-ridged, 3-dentate at apex. Nut fuscous, smooth.

Fl. & Fr.: Sep.-Feb. *Distr.*: Trop. Africa, SE. Asia, Malesia. Common in wastelands in plains. *NAK 2902* (Adoor, c. 20 m).

2. **Fuirena umbellata** Rottb., Descr. Ic.Rar. Pl. 70, t.19, f.3. 1773; Clarke in Hook.f., Fl. Brit. India 6:666. 1893; Fischer in Gamble, Fl. Pres. Madras 1669. 1931; Hooper in Sald. & Nicols., Fl. Hassan Dist. 683. 1976.

Perennials; stem 4-6-angled. Leaves to 25 cm long, 3(5-) nerved. Panicles congested, α, in lower axils; spikelets in clusters, ovoid-oblong; glumes obovate-oblong, hairy, awn c. 1 mm long, recurved; bristles 0; scales subquadrate, mucronate, 3-nerved. Nut obovoid, triquetrous, stramineous-fuscous, smooth.

Fl. & Fr.: Jul.-Dec. *Distr.*: Tropics. Not common. In mountain marshes. *NAK 2374* (Ranni R.F., c. 100 m).

HYPOLYTRUM Persoon
Syn. Pl. 1:70. 1805.

Hypolytrum nemorum (Vahl) Spreng., Syst. 1:233. 1825; Nicols., Suresh & Manilal, An Interpr. Hort. Malab. 286. 1988; Mohanan & Sivad., Fl. Agasthyamala 784.2002. *Schoenus nemorum* Vahl, Symb. Bot. 3:8. 1794. *Hypolyptrum latifolium* Pers., Syn. Pl. 1:70. 1805; *Hypolytrum latifolium* Clarke in Hook.f., Fl. Brit. India 6:678. 1894; Fischer in Gamble, Fl. Pres. Madras 1673. 1931. *Beera-kaida* Rheede, Hort. Malab. 12:109, t.58. 1693.

Stout perennials; rhizomes woody, purplish; stem to 60 cm tall, triquetrous. Leaves α, usually in basal clusters or little above the middle, to 1 m long, margin scabrous, I-nerved, coriaceous. Inflorescence a terminal corymbose-panicle; bracts 2-4, foliaceous; spikelets obovoid (fruiting one subglobose); glumes obovate-suborbicular, closely imbricating, obtuse. Flowers bisexual, α; scales 2; stamens 3; style 3-partite, persistent. Nut ellipsoid, trigonous, smooth, glossy.

Fl. & Fr.: Sep.-Mar. *Distr.*: Sri Lanka, India, Malesia, China, Australia, Polynesia. Common. Among forest undergrowth. *NAK 482,1106,2147* (Most of the evergreen forests).

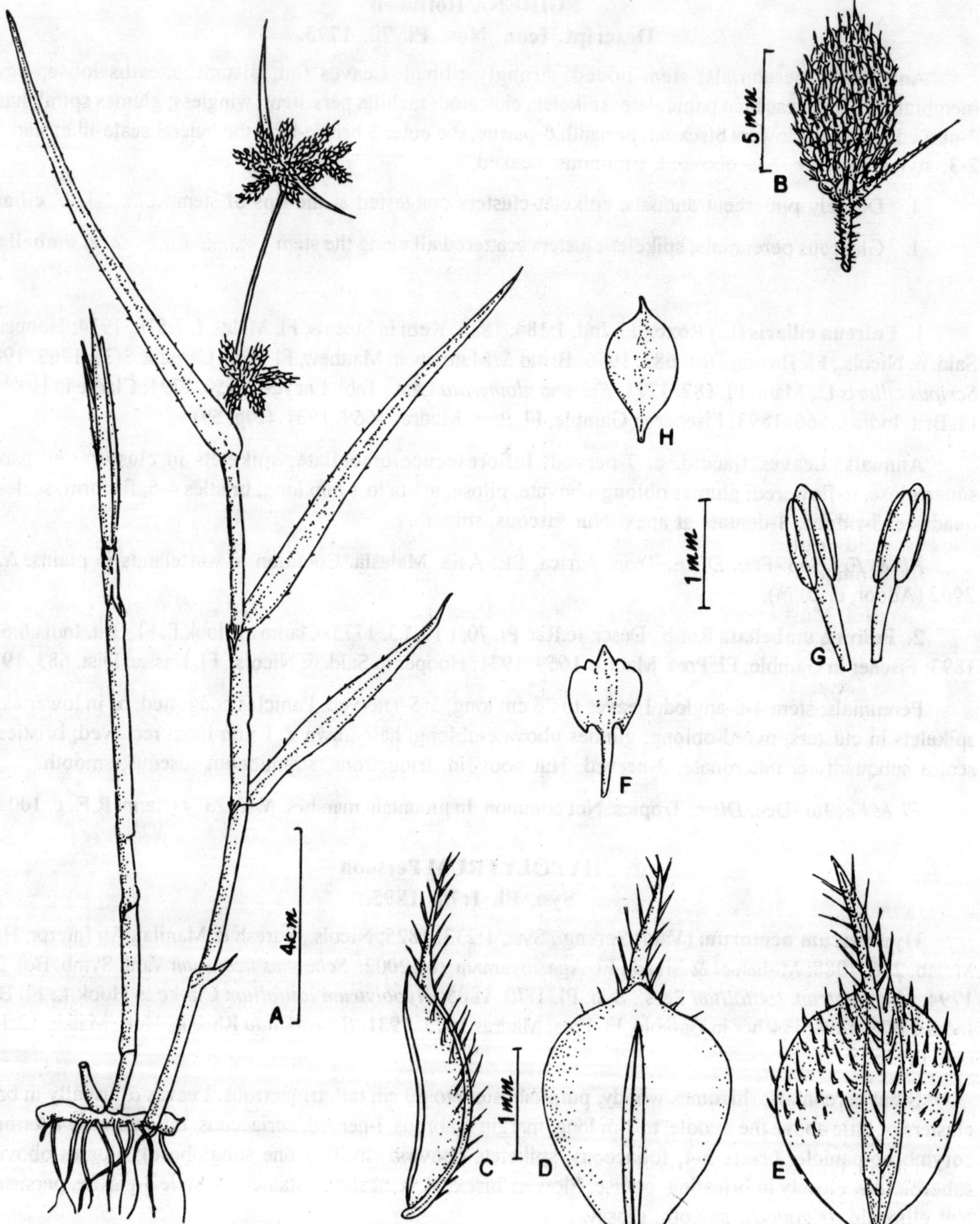

Fig. 59. ***Fuirena ciliaris*** **(L.) Roxb.: A. Habit; B. Spike; C. Glume - side view; D. Glume - adaxial view; E. Glume – abaxial view; F. Scale; G. Stamens; H. Nut.**

KYLLINGA Rottboell
Descript. Icon. Nov. Pl. 12. 1773 (nom. cons.).

Perennial herbs. Leaves basal. Inflorescence of capitate spikes ('Heads'); involucral bracts 3; spikelets a; glumes distichous; stamens 3; style 2-partite. Nut flattened, ellipsoid.

1. Glumes winged on keel. .... 3. **K. nemoralis**
1. Glumes not winged on keel.
 2. Rhizome densely tufted; head 3 or 4-lobed. .... 4. **K. triceps**
 2. Rhizome not tufted; heads 3 or 4-lobed.
  3. Rhizome stoloniferous; slender herbs. .... 1. **K. brevifolia**
  3. Rhizome not stoloniferous; robust herbs. .... 2. **K. melanosperma**

1. **Kyllinga brevifolia** Rottb., Descr. Ic. Nov. Pl. 13, t.4, f.3. 1773; Clarke in Hook.f., Fl. Brit. India 6:588. 1893; Fischer in Gamble, Fl. Pres. Madras 1624. 1931; Hooper in Sald. & Nicols., Fl. Hassan Dist. 686. 1976; Koyama, Gard. Bull. Straits Settlem. 30:163. 1977; Britto & Matthew in Matthew, Fl. Tam. Carnatic 3(3): 1765. 1983; Mohanan & Sivad., Fl. Agasthyamala 785.2002. *Cyperus brevifolius* (Rottb.) Hassk., Cat. Hort. Bot. Bogor. 24. 1844; Kern in Steenis, Fl. Males. I, 7:656. 1974.

Slender herbs; rhizome stoloniferous; stem spaced, to 20 cm tall. Leaves flat or canaliculate; spike solitary, ovoid or globose; spikelets α, in a cluster; glumes 4, keel 1-3-nerved. Nut biconvex, obovoid or ellipsoid, stipitate.

Fl. & Fr.: Most of the seasons. *Distr.:* Tropics & Subtropics. Common in lowlands. *NAK 408 & 1012* (Tiruvalla, c. 10 m).

2. **Kyllinga melanosperma** Nees in Wight, Contrib. Bot. India 91. 1834; Clarke in Hook.f., Fl. Brit. India 6:588. 1893; Fischer in Gamble, Fl. Pres. Madras 1624. 1931; Hooper in Sald. & Nicols., Fl. Hassan Dist. 687. 1976; Koyama, Gard. Bull. Straits Settlem. 30:162. 1977; Britto & Matthew in Mattew, Fl. Tam. Carnatic 3(3): 1767. 1983; Mohanan & Sivad., Fl. Agasthyamala 786.2002. *Cyperus melanospermus* (Nees) Valcken., Gesl. Cyp. Mal. Arch. 50, t.2, f.8. 1898; Kern in Steenis, Fl. Males. I, 7:655. 1974.

Robust herbs; rhizome short, thick; stem close, to 60 cm tall. Leaves flat or folded; spike ovoid, solitary; spikelets ellipsoid or oblong, α in a cluster; glumes ovate-elliptic, keel smooth or spinulose. Nut obovoid, biconvex, fulvous, stipitate.

Fl. & Fr.: Sep.-Dec. *Distr.*: Tropical & Subtropical Africa, S. Asia, Fiji & Malesia. Common in montane marshes. *NAK 100, 336*. (Moozhiar, c. 250 m).

3. **Kyllinga nemoralis** (Forster & Forster f.) Dandy ex Hutch. & Dalz., Fl. W. Trop. Africa 2:486, 487. 1936; Hooper in Sald. & Nicols., Fl. Hassan Dist. 687. 1976; Koyama, Gard. Bull. Straits Settlem. 30:163. 1977; Britto & Matthew in Matthew, Fl. Tam. Carnatic 3(3): 1767. 1983; Mohanan & Sivad., Fl. Agasthyamala 786.2002. *Thyrocephalon nemoralis* Forster & Forster f., Char. Gen Pl. 130. 1776. *Kyllinga monocephala* Rottb., Descr. Ic. Nov. Pl.13, t.4, f.4. 1773, *nom. illegit.;* Clarke in Hook.f., Fl. Brit. India 6:588. 1893; Fischer in Gamble, Fl. Pres. Madras 1624. 1931. *Cyperus kyllinga* Endl., Cat. Horti Vindob. 1:94. 1842; Kern in Steenis, Fl. Males. I, 7:659. 1974.

Slender herbs; stem close or distant, to 40 cm tall. Leaves α, canaliculate; spikes globose, white; spikelets ovoid or ellipsoid; rachilla wingless; glumes linear, strongly keeled. Nut obovoid, biconvex, fuscous or yellow, stipitate.

Fl. & Fr.: Most of the seasons. *Distr.*: Pantropical. Very common. In plains. *NAK* 1026 (Tiruvalla, c. 10 m).

4. **Kyllinga triceps** Rottb., Descr. Ic. Nov. Pl. 14, t.4, f.6. 1773; Clarke in Hoook f., Fl. Brit. India 6:587. 1893; Fischer in Gamble, Fl. Pres. Madras 1623. 1931; Koyama, Gard. Bull. Straits Settlem. 30:162. 1977; Hooper in Sald. & Nicols., Fl. Hassan Dist. 688. 1976; Britto & Matthew in Matthew, Fl. Tam. Carnatic 3(3): 1768. 1983. *Cyperus triceps* Endl., Cat. Horti Vindob. 1:94. 1842; Kern in Steenis, Fl. Males. I, 7:659. 1974; Matthew, Ill. Fl. Tam. Carnatic t. 785. 1982.

Tufted herbs; stem to 10 cm tall. Leaves flat or folded. Inflorescence of usually 3-(5)-lobed spikes; spikelets oblong, 1-flowered; glumes hyaline, ovate to narrowly linear; keel 5-nerved. Nut biconvex, apiculate, stipitate.

Fl. & Fr.: Nov.-Jan. *Distr.*: Old World tropics. Not common. On open lowlands. *NAK 2926* (Adoor, c. 20 m).

LIPOCARPHA R. Brown in Tuckey Narr. Exped. Congo 5:459. 1818 (nom. cons.).

Lipocarpha chinensis (Osbeck) Kern, Blumea (suppl.) 4:167.1958 & in Steenis, Fl. Males.1, 7:521. 1974; Hooper in Sald. & Nicols., Fl. Hassan Dist. 688. 1976; Britto & Matthew in Matthew, Fl. Tam. Carnatic 3(3): 1769. 1983; Mohanan & Sivad., Fl. Agasthyamala 788.2002. *Scirpus chinensis* Osbeck, Dagb. Ostind. Resa 220. 1757. *Lipocarpha argentea* R. Br. in Tuckey, Narr.Exp. Congo, App. 477. 1818, *nom. illegit.;* Clarke in Hook.f., Fl. Brit. India 6:667. 1893; Fischer in Gamble, Fl. Pres. Madras 1670. 1931.

Annuals; stamens tufted, triquetrous. Leaves basal, flat. Inflorescence terminal, capitate; bracts 3, foliaceous; spikelets in a cluster, ovoid. Flowers bisexual; glumes spiral, fuscous, elliptic, hypogynous, scales 2, 5-7-nerved; stamen(s) 1 or 2; style 2 or 3-partite. Nut ellipsoid, trigonous, smooth or punctulate.

Fl. & Fr: Sep.-Feb. *Distr.:* Tropics & Subtropics. Less common. In mountain marshes. *NAK 909* (Plappally, c. 400 m).

MARISCUS Vahl Enum. Pl 2:372. 1805-1806 (nom. cons.).

Perennials; stems tufted or not. Leaves α. Inflorescence a simple or compound umbel; spikes oblong. Flowers bisexual; glume distichous; rachilla deciduous above the prophyll and falling before glumes; stamens 3; style 3-fid. Nut 3-gonous.

1. Stem-base swollen by sheaths.
   2. Spike 1; spikelets 2-6-flowered. .... **4. M. dubius**
   2. Spikes 5-7; spikelets 1-flowered. .... **1. M. clarkei**
1. Stem-base not swollen by sheaths.
   3. Spikelets 1-flowered; spikes c. 7 mm broad. .... **5. M. paniceus**
   3. Spikelets 2-4-flowered; spikes c. 1 cm broad.
      4. Nut-bearing glume cuspidate. .... **2. M. concinnus**
      4. Nut-bearing glume not cuspidate. .... **3. M. cyperinus**

1. **Mariscus clarkei** (T. Cooke) Koyama, J. Jap. Bot. 51:313. 1976 & Gard. Bull. Straits Settlem. 30:158.1977; Britto & Matthew in Matthew, Fl. Tam. Carnatic 3(3): 1770.1983. *Cyperus clarkei* T. Cooke, Fl. Bombay 2:873. *Mariscus bulbosus* Clarke in Hook.f., Fl. Brit. India 6:620. 1893, non Steudel, 1855; Fischer in Gamble, Fl. Pres. Madras 1644. 1931; Matthew, Ill. Fl. Tam. Carnatic t.792. 1982.

Rhizome short, stoloniferous; stems not tufted. Leaves as long as stem, flat or canaliculate, flaccid; sheaths purplish-brown. Inflorescence simple; spikes 5-7 in a cluster, spicate, spikelets terete; glumes ovate to lanceate. Nut ellipsoid, trigonous.

Fl. & Fr.: Sep.-Dec. *Distr.:* Sri Lanka, Peninsular India. Common along hills slopes. *NAK 568, 659* (Ranni, Konni, etc.).

2. **Mariscus concinnus** Schrader ex Nees in Mart., Fl. Bras. 2(1):47. 1843; Nicols., Suresh & Manilal, An Interpr. Hort. Malab. 287. 1988. *Scirpus cyperoides* L., Mant. 2:181. 1771. *Kyllinga sumatrensis* Retz., Obs.Bot. 4:13.1786. *Mariscus sumatrensis* (Retz.) Koyama, Gard. Bull. Straits Settlem. 30:154. 1977; Britto & Matthew in Matthew, Fl. Tam. Carnatic 3(3): 1775. 1983. *Mariscus sieberianus* Nees ex Clarke in Hook. f., Fl. Brit. India 6:622. 1893; *Cyperus cyperoides* (L.) Kuntze, Revis. Gen.Pl. 3(2): 333. 1898; Kern in Steenis, Fl. Males. I, 7:642. 1974. Fischer in Gamble, Fl. Pres. Madras 164. 1931. *Kolpullu* Rheede, Hort. Malab. 12:119, t.68.1693.

Rhizome short; stem sparsely tufted, triquetrous. Leaves α, flat to canaliculate, flaccid or rigid. Inflorescence simple; involucral bracts 5-7; spikes cylindric; spikelets 50-70, narrowly linear, 2-flowered; rachilla straight. Nut ellipsoid.

Fl. & Fr.: Sep.-Dec. *Distr.*: Tropics and Subtropics. Not common. In hills. *NAK 2483 & 2523* (Kakki hills, c. 1000 m).

3. **Mariscus cyperinus** (Retz.) Vahl, Emum. Pl. 2:377. 1806; Clarke in Hook.f., Fl. Brit. India 6:621. 1893; Fischer in Gamble, Fl. Pres. Madras 1644. 1931; Hooper in Sald. & Nicols., Fl. Hassan Dist. 689. 1976; Koyama, Gard. Bull. Straits Settlem. 30:157. 1977; Britto & Matthew in Matthew, Fl. Tam. Carnatic 3(3): 1771. 1983. *Kyllinga cyperina* Retz., Obs. 6: 21. 1791. *Cyperus cyperinus* (Retz.) J. V. Suringar, Gesl. Cyp. Mal. Arch. 154, t.6, f.10. 1898; Kern in Steenis, Fl. Males. I, 7: 1974.

Rhizome absent or very short; stem triquetrous. Leaves as long as stem or shorter. Inflorescence of simple umbels; spikes oblong; spikelets linear, transversely spreading, 2-4-fid; rachilla prominent. Nut cylindric, angular, slightly curved.

Fl. & Fr.: Sep.-Dec. *Distr.*: Tropics & Subtropics. Not common. Along hillways. *NAK 2059 & 2477* (Kakki hills, c. 1000 m).

4. **Mariscus dubius** (Rottb.) Kuek. ex Fischer in Gamble, Fl. Pres. Madras 1644. 1931; Hooper in Sald. & Nicols., Fl Hassan Dist. 689. 1789. 1976; Koyama, Gard. Bull. Straits Settlem. 30:158. 1977; Britto & Matthew, in Matthew Fl. Tam. Carantic 3(3): 1772. 1983; Mohanan & Sivad., Fl. Agasthyamala 789.2002. *Cyperus dubius* Rottb., Descr. Ic. Nov.Pl. 20, t.4, f.5. 1773; Roxb., Fl. Ind. 1:192. 1820; Kern in Stenis, Fl. Males. I, 7:643. 1974. *Mariscus dregeanus* Kunth, Enum. Pl. 2:120. 1837; Clarke in Hook.f., Fl. Brit. India 6:620. 1893, p.p.

Rhizome short; stem tufted, slender, triquetrous, base bulbous. Leaves flat or canaliculate; sheath pale or pinkish. Inflorescence simple, capitate; involucral bracts 3; spikelets α in a cluster, lanceoid; glumes broadly ovate, α-nerved, mucronulate, keel 1-nerved. Nut ellipsoid, trigonous.

Fl. & Fr.: Sep.-Dec. *Distr.*: Tropics. Common along hills slopes. *NAK 658* (Ranni, c.200 m).

5. **Mariscus paniceus** (Rottb.) Vahl, Enum. Pl. 2:373. 1806; Clarke in Hook.f., Fl. Brit. India 6:620. 1893; Fischer in Gamble, Fl. Pres. Madras 1644. 1931; Hooper in Sald. & Nicols., Fl. Hassan Dist. 690. 1976; Koyama, Gard. Bull. Straits Settlem. 30:157. 1977; Britto & Matthew in Matthew, Fl. Tam. Carnatic 3(3): 1774. 1983; Mohanan & Sivad., Fl. Agasthyamala 789.2002. *Kyllinga panicea* Rottb., Descr. Ic. Rar. 15, t.4, f.1. 1773. *Cyperus paniceus* (Rottb.) Boeckeler, Linnaea 36:381. 1870, p.p.; Kern in Steenis, Fl. Males. I, 7:643.1974.

Rhizome creeping, stoloniferous. Leaves α, flat; sheaths pale. Inflorescence simple spikes, 3-9, terete; spikelets α, linear, 1-flowered; glumes 4, oblong, pale green, muticous. Nut narrowly ellipsoid, trigonous, stramineous.

Fl. & Fr.: Sep.-Dec. *Distr.*: Sri Lanka, India, Nepal & Malesia. Common in plains. *NAK 1027* (Tiruvalla, c. 10 m).

PYCREUS Palisot de Beauvois
Fl. Oware 2:48. 1816.

Annuals or Perennials; stems usually tufted. Leaves flat or canaliculate. Inflorescence simple or compound, anthelate or capitate; spikelets α, flattened, α-flowered; glumes α, distichous, keeled. Flowers bisexual; stamens 2; style 2-partite. Nut dark brown to black.

1. Nuts wrinkled with wavy or broken transverse ridges; epidermal cells linear-oblong. .... 5. **P. stramineus**
1. Nuts not wrinkled; epidermal cells isodiametric, roundish or hexagonal.
  2. Robust herbs; leaves c. 8 mm wide. .... 4. **P. puncticulatus**
  2. Slender herbs; leaves 2-4 mm wide.
    3. Rachilla winged.
      4. Glumes c. 1.5 mm long; spikelets, c. 2 mm wide. .... 3. **P. pumilus**
      4. Glumes c. 2 mm long; spikelets 1. 5 mm wide. .... 2. **P. polystachyos**
    3. Rachilla not wingd. .... 1. **P. flavidus**

1. **Pycreus flavidus** (Retz.) Koyama, J.Jap. Bot. 51: 313. 1976 & Gard. Bull. Straits Settlem. 30:150. 1977; Britto & Matthew in Matthew, Fl. Tam. Carnatic 3(3): 1777. 1983. *Cyperus flavidus* Retz., Obs. Bot. 5:13. 1788; Kern in Steenis, Fl. Males. I, 7:648. 1974. *Cyperus globosus* Allioni, Fl. Pedem. Auct., 49. 1789. *Pycreus globosus* (Allioni) Reichb., Fl. Germ. Excurs. 1:140(10). 1830; Fischer in Gamble, Fl. Pres. Madras 1627. 1931. Hooper in Sald. & Nicols., Fl. Hassan Dist. 692. 1976; *Pycreus capillaris* Nees, Linnaea 9: 283. 1834; Clarke in Hook.f., Fl. Brti. India 6:591. 1893.

Rhizome short, thick. Leaves basal, flat or folded. Inflorescence compound; bracts 3; spikes fuscous, spicate; spikelet 10-15, oblong; rachilla straight, persistent, wingless; glumes ovate, dark-purple, margin hyaline, keel 3-nerved. Nut biconvex, ellipsoid, laterally compressed, stipitate.

Fl. & Fr.: Oct.-Dec. *Distr.*: S. Europe, Africa, W. & C. Asia, India, China, Japan, Malesia. Common. On exposed hills. *NAK 2480* (Moozhiar, c. 250 m).

2. **Pycreus polystachyos** (Rottb.) P. Beauv., Pl. Oware 2:48, t.86, f.2. 1807; Hook.f., Fl. Brit. India 6:592. 1893; Hooper in Sald. & Nicols., Fl. Hassan Dist. 693. 1976; Koyama, Gard. Bull. Straits Settlem. 30:149. 1977; Mohanan & Sivad., Fl. Agasthyamala 790.2002. *Cyperus polystachyos* Rottb., Descr. Pl. Rar. 21. 1772;

Roxb., Fl. Ind. 1:197. 1820; Kern in Steenis, Fl. Males. I, 7:649. 1794. *Pycreus odoratus* Urban, Symb. Antill. 2:164. 1900; Fischer in Gamble, Fl. Pres. Madras 1627. 1931.

Key to the varieties

1. Inflorescence anthelate. ....**Pycreus polystachyos** var. **laxiflorus**

1. Infloresceце densely clustered or capit.............................**Pycreus polystachyos** var. **polystachyos**

Pycreus polystachyos (Rottb.) P. Beauv. var. **polystachyos**

Stem tufted. Leaves flat or canaliculate. Inflorescence simple or subcompound, densely clustered; spikelets oblong, stramineous; rachilla strongly flexuous, persistent, narrowly winged; glumes ovate or oblong; stamen(s) 1 or 2. Nut ovoid or obovoid, biconvex, stipitate.

Fl. & Fr.: Most of the seasons. *Distr.*: Cosmopolitan. Very common in plains. *NAK 1007* (Tiruvalla, c. 10 m).

Pycreus polystachyos (Rottb.) P. Beauv. var. **laxiflorus** (Benth.) Clarke in Hook.f., Fl. Brit.India 6:592. 1893. *Cyperus polystachyos* Rottb. var. *laxiflorus* Benth., Fl. Austral. 7:261. 1878; Kern in Steenis, Fl. Males. I, 7:649. 1949. 1974; Hooper in Sald. & Nicols., Fl. Hassan. Dist. 693. 1976;

Inflorescence anthelate; spikelets horizontal on spikes, purplish-brown; rachilla straight, winged. Nut symmetrical.

Fl. & Fr.: Most of the seasons. *Distr.*: Peninsular India, Sri Lanka. Less common. *NAK 1013* (Tiruvalla, c. 10 m).

3. **Pycreus pumilus** (L.) Nees, Hook.f., Fl. Brit. India 6:591. 1893; Hooper in Sald. & Nicols., Fl. Hassan Dist. 693. 1976; Britto & Matthew in Matthew, Fl. Tam. Carnatic Nicols., Fl. Hassan Dist. 693. 1976; Britto & Matthew, in Matthew, Fl. Tam. Carnatic 3(3): 1778. 1983; Mohanan & Sivad., Fl. Agasthyamala 791.2002. *Cyperus pumilus* L., Cent. Pl. 2:6. 1756; Roxb., Fl. Ind. 1:200. 1820; Kern in Steenis, Fl. Males.I, 7:650. 1974. *C. nitens* Retz., Obs. Bot. 5:13. 1789. *Pycreus nitens* (Retz.) Nees, Nov. Acta Acad. Caes. Leop.-Carol. Nat. Cur. 19 (Suppl.1): 53. 1843; Clarke in Hook. f., Fl. Brit. India 6:591. 1893.

Stems slender, filiform, erect or spreading, tufted. Leaves α, flat or canaliculate. Inflorescence compound, horizontal or close; spikes horizontal; spikelets 5-15 per spike, oblong; rachilla flexuous, winged; glumes ovate, mucro c.1 mm long, erect or recurved, keel 3-5-nerved. Nut biconvex, laterally compressed, stipitate.

Fl. & Fr.: Sep.-Dec. *Distr.*: Sri Lanka, India, Nepal, Indo-China, S.E. China, Malesia. Common. In mountain marshes and lowlands. *NAK 860, 1008* (Tiruvalla, c. 10 m), *2572* (Moozhiar, c. 250 m).

4. **Pycreus puncticulatus** (Vahl) Nees in Martius, Fl. Bras. 2(1):10, in nota. 1842; Clarke in Hook.f., Fl. Brit. India 6:593. 1893, Fischer in Gamble, Fl. Pres. Madras 1628. 1931; Koyama, Gard. Bull. Straits Settlem. 30:149. 1977; Britto & Matthew in Matthew, Fl. Tam. Carnatic 3(3): 1779. 1983. *Cyperus puncticulatus* Vahl, Enum. Pl. 2:348. 1806.

Stems robust, to 1 m tall, erect. Leaves α, flat or canaliculate. Inflorescence decompound; involucral bracts 5-7, primary rays 5-7, secondary rays 3-5 in cluster. Spikelets oblong, purple-brown; rachilla clearly flexuous, winged; glumes ovate or suborbicular, keel strong, 3-nerved. Nut obovoid, laterally compressd, truncate, stipitate.

Fl. & Fr.: Nov.-Feb. *Distr.*: Sri Lanka, Peninsular India, Malaya Peninsula. Common. Usually as a weed in paddy fields. *NAK 2876* (Adoor, c. 20 m).

5. **Pycreus stramineus** Clarke in Hook.f., Fl. Brit. India 6:589. 1893; Fischer in Gamble, Fl. Pres. Madras 1627. 1931; Hooper in Sald. & Nicols., Fl. Hassan Dist. 694. 1976. *Cyperus stramineus* Nees in Wight, Contrib. Bot. India 74. 1834.

Stems tufted, 10-30 cm tall. Leaves filiform; involucral bracts 1-2(4). Inflorescence simple, contracted; spikelets 2-6 together, linear, compressed; rachilla broad, wingless, ovate, mucronulate, stramineous, keel 3-nerved. Nuts obovid, compressed, blackish-brown with undulating lines.

Fl. & Fr.; Aug.-Dec. Not common. On hills. *NAK 1961* (Arampa, c. 400 m).

RHYNCHOSPORA Vahl
Enum. Pl. 2:229. 1806 ('Rynchospora') corr. Willdenow, Enum. Pl. Hort. Berol. 71.1809 (Orth. *et* nom. cons.).

Rhynchospora corymbosa (L.) Britton, Trans. New York Acad. Sci. 11: 84. 1892; Fischer in Gamble, Fl. Pres. Madras 1672. 1931; Hooper in Sald. & Nicols., Fl. Hassan Dist. 695. 1976; Kern in Steenis, Fl. Males. I, 7:713. 1974; Matthew, Ill. Fl. Tam. Carnatic tt. 794. & 795. 1982; Brito & Matthew, in Matthew, Fl. Tam. Carnatic 3(3): 1781. 1983; Mohanan & Sivad., Fl. Agasthyamala 791.2002. *Scirpus corymbosus* L., Cent. Pl. 2:7. 1756. *Rhynchospora aurea* Vahl, Enum. Pl. 2:229. 1806; Clarke in Hook.f., Fl. Brit. India 6:670. 1893.

Perennial robust herbs; stem trigonous. Leaves basal and cauline, apically flat; basally complicate, to 1 m long. Panicles terminal and axillary, corymbiform; bracts foliar; spikelets 2-7, in a cluster, subterete; glumes: lower-ones empty, upper most glumes female, mid-ones bisexual; stamens 3; style shortly 2-fid. Nut obovoid, compressed, punctulate, plicate-rugose.

Fl. & Fr.: Sep.-Dec. *Distr.*: Pantropical. Less common. Forming dense populations both in lowlands and hills. *NAK 1271* (Moozhiar, c. 20 m).

SCHOENOPLECTUS (H.G.L. Reichenbach) Palla
Bot. Jahrb. Syst. 10:298. 1888 (nom. cons.).

Annuals or perennials; stems caespitose or sparsely tufted, terete or triquetrous. Leaves reduced to sheaths or very short blades. Inflorescence capitate or compact, pseudolateral; involucral bract solitary, foliar, continous with the stem; spikelet(s) solitary or clustered; rachilla wingless. Flowers bisexual; glumes spiral, keeled, mucronate; hypogynous bristles 6 or 7; stamens (2-) 3; style 2- or 3- partite. Nut obovoid, biconvex, smooth or transversely undulate.

1. Slender herbs; stem half-terete. .... **1. S. juncoides**
1. Robust herbs; stem triquetrous. .... **2. S. mucronatus**

1. **Schoenoplectus juncoides** (Roxb.) Palla, Bot. Jahrb. Syst. 10:299. 1888; Hooper in Sald. & Nicols., Fl. Hassan Dist. 697. 1976; Britto & Matthew in Matthew, Fl. Tam. Carnatic 3(3): 1784. 1983. *Scirpus juncoides* Roxb., Fl. Ind. 1:218. 1820; Kern in Steenis, Fl. Males. I, 7:512. 1974. *Scirpus erectus* auct., non Poret, 1829: Clarke in Hook.f., Fl. Brit. India 6:656. 1893; Fischer in Gamble, Fl. Pres. Madras 1666. 1931.

Stems densely caespitose; sheath grey-fuscous, obliquely truncate; spikelets 1-3 in a cluster, ovoid, fulvous; glumes broadly ovate, concave, mucronate, α-nerved, keeled; hypogynous bristles 6 or 7. Nut broadly obovoid, unequally biconvex.

Fl. & Fr.: Nov.-Feb. *Distr.*: India, Indo-China, China, Japan, Malesia. In certain abandoned paddy fields, in midlands. Locally abundant. *NAK 2724* (Konni, c. 40 m).

2. **Schoenoplectus mucronatus** (L.) Palla, Verh. K.K. Zool.-Bot. Ges. Wien 38. Sitzb.: 49. 1888 & Bot. Jahrb. Syst. 10:299. 1889; Hooper in Sald. & Nicols., Fl. Hassan Dist. 699. 1976; Britto & Matthew in Matthew, Fl. Tam. Carnatic 3(3): 1785. 1983. *Scirpus mucronatus* L., Sp. Pl. 50. 1753; Clarke in Hook.f., Fl. Brit. India 6:657. 1893; Fischer in Gamble, Fl. Pres. Madras 1666. 1931; Kern in Steenis, Fl. Males. I, 7:510. 1974.

Stem sparsely tufted, triquetrous, spongy; sheath stramineous or brown; spikelets 10-25, in a cluster, terete, ovoid; glumes broadly ovate, concave, α-nerved, mucro short, keeled; hypogynous bristles 6 or 7. Nut broadly obovate, biconvex or subtrigonous.

Fl. & Fr.: Oct.-Dec. *Dsitr.*: Eurasia. Not common. In mountain marshes. *NAK 2516* (Kakki hills, c. 1000 m).

SCLERIA Bergius
Kongl. Ventensk. Acad. Handl. 26:142. 1765.

Perennials; stem triquetrous. Leaves flat; sheath apically broad, often winged. Inflorescence racemes or paniculate; bracts foliar, prophylls filiform; spikelets 1-3, all bisexual, female or unisexual; glumes distichous, keeled; stamen(s) 1-3; styles 3-partite; disc usually developed. Nut trigonous, ovoid or broadly obovoid or globose.

1. Hypogynous disc 0 or obsolete; sheaths narrow. .... 2. **S. lithosperma**
1. Hypogynous disc well develped, 3-lobed.
    2. Nuts smooth; stem to 1 m tall. .... 1. **S. levis**
    2. Nuts reticulate; stems to 2 m high. .... 3. **S. terrestris**

1. **Scleria levis** Retz., Obs. Bot. 4:13. 1786; Blake, J. Arnold Arb. 35:226. 1954, non Clarke, 1894; Mohanan & Sivad., Fl. Agasthyamala 792.2002. *S. zeylanica* Poir., Encycl. Wight, Contr. Bot. India 117. 1834; Clarke in Hook.f., Fl. Brit. India 6:689. 1894; Fischer in Gamble, Fl. Pres. Madras 1677. 1931.

Rhizome creeping; stem(s) solitary or sparsely tufted. Leaves flat; sheath inflated upwards, winged. Panicle terminal and lateral; involucral bracts foliaceous; spikelets unisexual; glumes ovate, stramineous to pale brown; stamens 3. Nut subglobose, glossy-white, hypogynous disc deeply 3-lobed.

Fl. & Fr.: Most of the seasons. *Distr.*: Tropics. Common along forest clearings. *NAK* 238, 581 & 1891. (Moozhiar, Konni, etc.).

2. **Scleria lithosperma** (l.) Sw., Nov. Gen. Pl. 18.1788; Nees in Wight, Contr. Bot. India 117. 1834; Clarke in Hook.f., Fl. Brit. India 6:685. 1894; Fischer in Gamble, Fl. Pres. Madras 1677. 1931; Hooper in Sald. & Nicols., Fl. Hassan Dist. 700. 1976; Kern in Steenis, Fl. Males. I, 7:740. 1974; Britto & Matthew in Matthew, Fl. Tam. Carnatic 3(3): 1887. 1983; Nicols., Suresh & Manilal, An Interpr. Hort. Malab. 289. 1988; Mohanan & Sivad., Fl. Agasthyamala 793.2002. *Scirpus lithospermus* L., Sp. Pl. 51. 1753. *Kaden pullu* Rheede, Hort. Malab. 12:89, t. 48. 1693.

Stem tufted. Leaves clustered on ± the middle of stem, canaliculate; sheaths pubescent, not winged. Panicle narrow, spicate; spikelets bisexual; glumes lanceate, purplish. Nut trigonous, ovoid, minutely umbonulate with 3 basal depressions and 3 glandular patches.

Fl. & Fr.: Most of the seasons. *Distr.*: Pantropical. The most common forest species. In forest clearings & hills slopes. *NAK 205, 582 & 604* (In almost all forests).

3. **Scleria terrestris** (L.) Fassett, Rhodora 26:159. 1924; Kern in Steenis, Fl. Males. I, 7:733. 1974; Hooper in Sald. & Nicols., Fl. Hassan Dist. 701. 1976; Matthew, Ill. Fl. Tam. Carnatic tt. 798 & 799. 1982; Britto & Matthew in Matthew, Fl. Tam. Carnatic 3(3): 1788. 1983; Nicols., Suresh & Manilal, An Intepr. Hort. Malab. 289. 1988; Mohanan & Sivad., Fl. Agasthyamala 793.2002. *Zizania terrestris* L., Sp. Pl. 991. 1753. *Scleria elata* Thw., Enum. Pl. Zeyl. 353. 1864; Clarke in Hook. f., Fl. Brit. India 6:690. 1894. *S. cochinchinensis* (Lour.) Druce, Bot. Exch. Club. Soc. Brit. Isles 4:644.1917; Fischer in Gamble, Fl. Pres. Madras 1678. 1931. *S. melanostoma* Nees ex Boeck., Linnaea 38:514. 1874; Clarke in Hook.f., Fl. Brit. India 6:692. 1894. *Katou-tsjolam* Rheede, Hort. Malab. 12:113, t.60.1693.

Rhizome thick; stem solitary, stout, to 2 m tall, triquetrous, scaberulous. Leaves scattered, flat; sheaths apically broad, 3-winged. Panicle spreading, terminal and lateral; spikelet(s) solitary or clustered, unisexual; glumes broadly ovate, keel mucronate. Nut ovoid or globose, glossy, sparsely scaberulous, cancellate; disc 3-lobed.

Fl. & Fr.: Oct.-Dec. *Distr.*: Tropics & Subtropics. Not common. In upper hills. *NAK 99, 206, 2060* (Moozhiar, Kakki hills, Arampa, etc.).

145. POACEAE (GRAMINEAE)

1. Woody shrubs or arborescent.
   2. Thorny trees; palea 2-keeled, 5-7-nerved. .... **Bambusa**
   2. Thornless shrubs; palea not keeled, faintly nerved. .... **Ochlandra**
1. Herbs.
   3. Female flower(s) enclosed in a stony bract. .... **Coix**
   3. Female flower(s) not enclosed in a stony bract.
      4. Mature spikelets usually falling entire from or with their pedicel; often the spikelet with 2 heteromorphous florets.
         5. Spikelets usually in pairs (one sessile, the other or both pedicelled); (upper) lemma membranous.
            6. Joints of the rachis and pedicels often swollen; raceme terete. .... **Rottboellia**
            6. Joints of the rachis or pedicels not swollen; raceme not terete.
               7. All spikelets homogamous; if heterogamous, the paleas are short or 0.
                  8. Spikes a in compound-panicle or in racemes.

9. Spikelets (all) pedicelled, rachis rigid. .... **Imperata**
9. Spikelets 2-nate, one sessile, the other pedicelled. .... **Spodiopogon**

8. Spikes solitary or digitate.
10. Rachis articulate.
11. Spike(s) solitary; spikelets 2-awned. .... **Pogonatherum**
11. Spikes digitate; spikelets 1-2-awned.
12. Spikelets in pairs, pedicelled. .... **Microstegium**
12. Spikelet in pairs, one pedicelled and the other sessile. .... **Eulalia**
10. Rachis not articulate. .... **Dimeria**

7. All spikelets heterogamous, sessile & pedicelled, (sessile-bisexual, pedicelled-male or empty).
13. Fertile spikelet 2-flowered. .... **Ischaemum**
13. Fertile spikelet 1-flowered.
14. Panicle or raceme subtended by spathes.
15. Spikes reduced, involucred.
16. Involucre present in modified spikelet. .... **Themeda**
16. Involucre absent in modified spikelet.
17. Callus of spikelet inflated. .... **Apluda**
17. Callus of spikelet not inflated. .... **Pseudanthistiria**
15. Spikes not reduced, not involucred.
18. Spike(s) 1 in each spatheole. .... **Schizachyrium**
18. Spikes 2 in each spatheole; aromatic herbs; inflorescence a panicle. .... **Cymbopogon**
14. Panicle not subtended by spathes.
19. All spikelet-pairs alike.
20. Awn of upper lemma arising below the middle. .... **Arthraxon**
20. Awn of upper lemma arising from tip or the sinus.
21. Upper lemma reduced; joints and pedicels filiform. .... **Capillipedium**
21. Upper lemma not reduced; joints and pedicels not filiform (always spikelets 1 sessile and other pedicelled). .... **Chrysopogon**

19. All spikelet-pairs not alike (lowest 1 or more pairs different)................**Heteropogon**

5. Spikelets usually in continous spikes, racemes or panicles; (upper) lemma chartaceous or crustaceous.

22. Inflorescence usually a spiciform or digitate or very rarely solitary raceme(s); margin of fruiting lemma flat.

23. Spikelets muticous; lower lemma 5-7-nerved. ....**Digitaria**

23. Spikelets awned; lower lemma nerveless. ....**Alloteropsis**

22. Inflorescence otherwise; margin of fruiting lemma convolute.

24. Spikelets falling entire and singly.

25. Spikelets muticous; fruiting lemma crustaceous.

26. Inflroescence an open spiciform panicle or raceme.

27. Spikelets, usually the upper one supported by bristle like branchlets. ....**Setaria**

27. Spikelets not supported by bristle like appendages.

28. Spikelets distinctly gibbous; inflorescence a spiciform-raceme. ....**Cyrtococcum**

28. Spikelets not gibbous; inflorescence a spiciform-panicle. ....**Panicum**

26. Inflorescence not an open spiciform- raceme or panicle.

29. Dorsal side of fruit facing towards the axis.

30. Lower glume 0 or reduced to a scale.

31. Lower glume 0; spikelets-lanceoid; culms stoloniferous. ....**Axonopus**

31. Lower glume reduced to a minute scale (very rarely 0); spikelets planoconvex, ovate-elliptic or orbicular; culms not stoloniferous. ....**Paspalum**

30. Lower glume present.

32. Glumes muticous. .... **Urochloa**

32. Glumes awned

33. Glume-apex acute; lower glumes awned. ....**Oplismenus**

33. Glume-apex retuse; lower glumes awnless. .... **Echinochloa**

29. Dorsal side of fruit not facing the axis.

34. Spikelets laterally compressed; lower glume herbaceous. .... **Pseudechinolaena**

34. Spikelets dorsally compressed; lower glume not herbaceous. .... **Brachiaria**

25. Spikelets finely awned; fruiting lemma membranous or chartaceous. .... **Rhynchelytrum**

24. Spikelets falling in groups (involucre of feathery bristles). .... **Pennisetum**

4. Mature spikelets not usually falling entire, if falling entire then without 2 heteromorphous florets.

35. Leaf-blade usually transversely veined between main veins. .... **Centotheca**

35. Leaf-blade not so.

36. Awn of fertile floret kneed and twisted below the knee.

37. Spikelets 2-flowered.

38. Panicles contracted, spiciform, densely spiculate. .... **Jansenella**

38. Panicles lax, not spiciform, loosely spiculate. .... **Arundinella**

37. Spikelets 1-flowered.

39. Inflorescence a panicle.

40. Lemma coriaceous, firmly enclosing the grain. .... **Aristida**

40. Lemma membranous, not firmly enclosing the grain. .... **Garnotia**

39. Inflorescence a spiciform-raceme. .... **Perotis**

36. Awn of fertile floret never kneed.

41. Lemma 3-nerved.

42. Inflorescence usually panicles.

43. Spikelet 1-flowered; glumes and lemmas similar. .... **Sporobolus**

43. Spikelet 2-α-flowered; glume and lemmas dissimilar. .... **Eragrostis**

42. Inflorescence usually racemose spikes.

44. Spike(s) solitary (lower glume usually in a furrow in the rachis). .... **Tripogon**

44. Spike(s) digitate, umbelled or racemose.

45. Spikes digitate; spikelet 1-flowered. .... **Cynodon**

45. Spikes not digitate; spikelet 2-α-flowered.

46. Imperfect florets absent; spikes never umbelled.

47. Glumes and lemmas muticous or mucronulate. .... **Eleusine**

47. Glumes and lemmas distinctly mucronate. ....**Dactyloctenium**

46. Imperfect florets present; spikes umbelled. ....**Chloris**

41. Lemma 5-α-nerved.

48. Glumes scale-like or subulate; lower florets reduced to scales or bristles. ....**Oryza**

48. Glumes an entire or 2-lobed rim; lower florets suppressed. ....**Leersia**

ALLOTEROPSIS J.S. Presl ex C.B. Presl Rel. Haenk. 343. 1830.

Alloteropsis cimicina (L.) Stapf in Prain, Fl. Trop. Africa 9: 487. 1919; Fischer in Gamble, Fl. Pres. Madras 1766. 1934; Bor, Grass. Burma, Ceylon, India & Pakistan 276. 1960; Ramamoorthy in Sald. & Nicols., Fl. Hassan Dist. 706. 1976; Matthew, Ill. Fl. Tam. Carantic t. 801. 1982; Britto & Matthew in Matthew, Fl. Tam. Carnatic 3(3): 1799. 1983; Mohanan & Sivad., Fl. Agasthyamala 799.2002. *Milium cimicinus* L., Mant. Pl. 184. 1771. *Axonopus cimicinus* (L.) P. Beauv., Ess. Agrost. 12. 1812; Hook.f., Fl. Brit. India 7: 64. 1896; Rang. & Tad., Handb. S. Ind. Grass. 108, ff. 107 & 108. 1921.

Tufted slender annuals; culms canaliculate, tuberculate-pilose. Leaves narrowly lanceate, 2-5 by 0.7-1 5 cm, base amplexicaul, margin tuberculate-ciliate, apex abruptly acute; sheath 7-15 cm long, ligule fimbriate. Racemes whorled, rachis scaberulous. Spikelet(s) solitary, ellipsoid, 3-6 mm long, 2-flowered; glumes unequal, lower glume 3-nerved, upper one 5-nerved, cartilaginous, ciliate; lower lemma to 3 mm long, upper lemma 5-nerved, awn erect, scaberulous, c. 3 mm long; palea papillose; anthers 3 to 1 mm long; ovary lenticular.

Fl. & Fr.: Apr.-May. *Distr.*: Sri Lanka, India, Myanmar, Tropical Africa to Australia. Common in waste places, abandoned fields, etc. of lowlands. *NAK 2808* (Adoor, c. 20 m).

APLUDA Linnaeus Sp. Pl. 82. 1753.

Apluda mutica L., Sp. Pl. 82. 1753; Fischer in Gamble, Fl. Pres. Madras 1750. 1934; Bor, Grass. Burma, Ceylon, India & Pakistan 93. 1960; Ramamoorthy in Sald. & Nicols., Fl. Hassan Dist. 707. 1976; Matthew, Ill. Fl. Tam. Carnatic tt. 804 & 805. 1982; Britto & Matthew in Matthew, Fl. Tam. Carnatic 3(3): 1800. 1983; Manilal, Fl. Silent Valley 36. 1988; Mohanan & Sivad., Fl. Agasthyamala 800.2002. *A. aristata* L., Amoen. Acad. 303. 1756; Fischer in Gamble, Fl. Pres. Madras 750. 1750. 1934. *A. varia* Hackel ssp. *mutica* (L.) Hackel & ssp. *aristata* Hackel in A.DC., Monogr. Phan. 6 : 196. 1889; Hook.f., Fl. Brit. India 7: 150. 1896; Rang. & Tad., Handb. S. Ind. Grass. 171, ff. 12 & 13. 1921.

Tufted tall weak herbs. Leaves oblong, 6-20 by 0.5-1 cm, base attenuate, apex setaceous, softly pilose, sheath to 7 cm long, ligule membranous. Panicles terminal and/or axillary, interrupted, to 20 cm long, racemes α; spatheoles pinkish. Spikelets 3-per raceme, 1 sessile, 2-pedicelled, sessile spikelet 2-flowered, lower male or bisexual, upper bisexual, lower glume elliptic-ovate, 11-13-nerved, scaberulous without, apex 2-dentate, upper glumes broadly ovate, c. 13-nerved, keeled, apex 2-dentate; lower lemma membranous, 3-nerved, 2-keeled; palea subscarious, upper lemma hyaline, deeply 2-fid. Caryopsis ellipsoid.

Fl. & Fr.: Sep.-Nov. *Distr.*: Tropical Asia, Pakistan, India, Asutralia & New Caledonia. Common among wayside bushes in lowlands and hills. *NAK 2207* (Ranni, c. 100 m).

ARISTIDA Linnaeus
Sp. Pl. 82. 1753.

Aristida setacea Retz., Obs. Bot. 4: 22. 1786; Roxb., Fl. Ind. 1: 350. 1820; Hook.f., Fl. Brit. India 7: 225. 1896; Rang. & Tad., Handb. S. Ind. Grass. 225, f. 173. 1921; Fischer in Gamble, Fl. Pres. Madras 1809. 1934; Bor, Grass. Burma, Ceylon, India & Pakistan 412. 1960; Ramamoorthy in Sald. & Nicols., Fl. Hassan Dist. 708. 1976; Britto & Matthew in Matthew, Fl. Tam. Carnatic 3(3): 1803. 1983; Mohanan & Sivad., Fl. Agasthyamala 801. 2002.

Tufted stout herbs. Leaves cauline, oblong, to 30 cm long, convolute, sheath to 7.5 cm long. Panicle contracted, 15-30 cm, erect. Spikelet 1-flowered, rachilla short, disarticulating above the upper glumes, linear convolute, arista to 2.5 mm long; setae 3, subequal, to 3 cm long; ovary linear.

Fl. & Fr.: Most of the seasons. *Distr.*: Sri Lanka, India, Myanmar. Common along waysides in lowlands. *NAK 2840* (Adoor, c. 20 m).

ARTHRAXON Palisot de Beauvois
Essai Agrost. 111. 1812.

Annuals or perennials; culms often bearded at nodes. Leaves ovate-oblong to lanceate, flat, base amplexicaul, sheaths lax, ligule fimbriate. Racemes terminal, subdigitate; callus stiff-pilose. Spikelets often in pairs, sessile ones bisexual; pedicelled one sterile or male laterally compressed; glumes oblong, nerves with spicules, upper glumes keeled, spiculed without; lower lemma lanceate; palea 0, upper lemma deeply lobed, with an awn arising at the base of its back, geniculate; anthers 2 or 3. Caryopsis ovoid-ellipsoid.

1. Pedicelled spikelets (at least in the upper part of raceme) developed; stamens 2; spikelets to 3 mm long. .... 2. **A. lancifolius**
1. Pedicelled spikelets not developed; stamens 3; spikelets 3-8 mm long.
    2. Anthers 0.5 mm or less long; spikelets 3-4 mm long. .... 3. **A. quartinianus**
    2. Anthers 3.5 mm or more long; spikelets 6-8 mm long. .... 1. **A. castratus**

1. **Arthraxon castratus** (Griff.) Naray. ex Bor in Kanj. *et al.*, Fl. Assam 5: 376. 1940 & Grass. Burma, Ceylon, India & Pakistan 99. 1960; Welzen, Blumea 27: 263. 1981; Matthew, Ill. Fl. Tam. Carnatic 3(3): 1804. 1983. *Andropogon castratus* Griff., Not. Pl. Asiat. 3: 89. 1851. *Arthraxon rudis* (Steudel) Hochst., Flora 39: 188. 1856; Hook.f., Fl.Brit. India 7: 144. 1896; Fischer in Gamble, Fl. Pres. Madras 1728. 1934.

Culms to 60 cm long, nodes glabrous. Leaves ovate-lanceate, 3-6 by 1-1.5 cm, margin tuberculate-ciliate, sheaths lax, compressed; ligule c. 2.5 mm long. Racemes 4-7 cm, joints pilose; sessile spikelet c. 6.5 mm long, glumes unequal, chitinous, lower glume 7-nerved, 2-keeled, upper glume elliptic, 3-nerved, 1-keeled, spiculed; lemmas long ciliate, obscurely 1-nerved, paleate; awn to 1.5 cm long, palea notched at apex, stipe long-pilose.

Fl. & Fr.: Dec.-Mar. *Distr.*: Sri Lanka, India, S. Vietnam, Malesia. Common in moist hilly tracts. *NAK 350 & 1315* (Moozhiar, c. 250 m).

2. **Arthraxon lancifolius** (Trin.) Hochst., Flora 39: 188. 1856; Fischer in Gamble, Fl. Pres. Madras 1729. 1934; Bor, Grass. Burma, Ceylon, India & Pakistan 100. 1960; Ramamoorthy in Sald. & Nicols., Fl. Hassan Dist. 709. 1976; Manilal, Fl. Silent Valley 347. 1988. *Andropogon lancifolius* Trin., Mem. Acad. Sci. St. Petersburg, Ser. 6, 2:271. 1832. *Arthraxon microphyllus* (Trin.) Hochst., Flora 39: 188. 1856; Hook.f., Fl. Brit. India 7: 147. 1896. *Andropogon microphyllus* Trin., Mem. Acad. Imp. Sci. St. Petersburg, Ser. 6, 2: 275. 1832.

Culms very slender, straggling, nodes bearded; leaves ovate to linear-lanceate, 1.5-3 by 0.5-1 cm, margin tuberculate-ciliate. Racemes c. 4 cm long, joints densly long-ciliate; sessile spikelets lanceoid, c. 3 m long; pedicel flat, lower glume 2-dentate, upper glume c. 2 mm long, mucronate, awn to 1.3 cm long; anthers 2.

Fl. & Fr.: Dec.-Mar. *Distr.*: Tropical Africa, Sri Lanka, India, Myanmar, Malesia. Less common. On rocky hills. *NAK 1253* (Moozhiar, c. 250 m).

3. **Arthraxon quartinianus** (A. Rich.) Nash, N. Amer. Flora 17: 99. 1912; Fischer in Gamble, Fl. Pres. Madras 1729. 1934; Bor, Grass. Burma, Ceylon, India & Pakistan 102. 1960; Mohanan & Sivad., Fl. Agasthyamala 802.2002. *Alectoridia quartiniana* A. Rich., Tent. Fl. Abyss. 2: 488, t. 99. 1851. *Arthraxon ciliaris* auct., non Beauv., 1812: Hook.f., Fl. Brit. India 7: 146. 1896.

Culms slender, straggling, nodes bearded. Leaves elliptic-lanceate to linear, 1.5-4 by 0.8-1.5 cm, margin sparsely ciliate; sheaths lax. Racemes slender, c. 6 cm long, pedicelled spikelets not developed; lower glume of sessile spikelet 3-4 mm long, compressed, not keeled; minutely scaberulous; awn to 1.2 cm long.

Fl. & Fr.: Feb.-Apr. *Distr.*: Endemic to Peninsular India. Not common. Along hill slopes. *NAK 405* (Ranni, c. 100 m).

ARUNDINELLA Raddi
Agrost. Brasil. 36, t.1, f.3. 1823.

Annuals or perennials; clums tufted, erect. Leaves flat or rolled, ligule scarious. Panicles open or contracted. Spikelets paired, 2-flowered, disarticulating between florets, lower male or sterile, upper bisexual; glumes unequal, membranous, 3-5-nerved, lower lemma 3-7-nerved, paleate, upper lemma shortly emarginate with a persistent geniculate awn, callus short; anthers 3. *Caryopsis* ellipsoid.

1. Upper lemma with two seta, one on each side of awn.
 2. Perennials; culms glabrous or sparsely pubescent on nodes. ....2. **A. mesophylla**
 2. Annuals; ..lms usually with tubercled hairs. ....3. **A. nervosa**
1. Upper lemma with a single seta (rarely absent).
 3. Spikelets glabrous or sparsely hairy on the keels of the glumes; plants drying yellowish green. ....4. **A. purpurea** var. **purpurea**
 3. Spikelets densely tuberculate-ciliate; plants drying not so. ....1. **A. ciliata**

1. **Arundinella ciliata** (Roxb.) Nees ex Miq., Nieuwe Verh. Eerste Kl. Kon. Ned. Inst. Wetensch. 3(4): 30. 1851; Bor, Grass. Burma, Ceylon, India & Pakistan 421. 1960; Ramamoorthy in Sald. & Nicols., Fl. Hassan Dist. 710. 1976; Matthew, Ill. Fl. Tam. Carnatic tt. 810 & 811. 1982; Britto & Matthew in Matthew, Fl. Tam. Carnatic 3(3): 1806. 1983; Manilal, Fl. Silent Valley 348. 1988; Mohanan & Sivad., Fl. Agasthyamala 803.2002. *Holcus ciliatus* Roxb., Fl. Ind. 1: 321. 1820. *Arundinella agrostoides* auct., non Trin., 1836: Hook.f., Fl. Brit. India 7: 71. 1896. *A. holcoides* auct., non (Kunth) Trin., 1836: Fischer in Gamble, Fl. Pres. Madras 1801. 1934.

Culms to 50 cm tall, tuberculate-pilose, node bearded. Leaves oblong, 5-12 by 0.5-0.7 cm, tuberculate-pilose. Panicles cylindric, contracted, 7-15 cm long; spikelets oblong; glumes 5-nerved, pilose to 1/3 at base, lower glume awned, upper glume caudate, lower lemma acute, paleate; anthers c. 1.5 mm long, awn to 4 mm log. Caryopsis ellipsoid.

Fl. & Fr.: Mar.-Oct. *Distr.*: Endemic to Peninsular India. Common in hills. *NAK 413* (Moozhiar, c. 250 m).

2. **Arundinella mesophylla** Nees ex Steud., Syn. Pl. Glum. 1: 115. 1854; Hook. f., Fl. Brit. India 7: 69. 1896; Fischer in Gamble, Fl. Pres. Madras 1801. 1934; Bor, Grass. Burma, Ceylon, India & Pakistan 423. 1960; Manilal, Fl. Silent Valley 348. 1988.

Perennials; culms slender to 50 cm tall, glabrous or sparsely pubescent. Leaves oblong-lanceate, 3-12 by 0.5-1 cm. Panicle with crowded spicate-racemes, 4-12 cm long; spikelets 4-5 mm long; sessile or shortly stipitate; glumes tuberculate-ciliate; upper lemma scaberulous at apex with two setae; lower lemma acute; awns c. 4 mm long, setae 2, subequal.

Fl. & Fr.: Dec.-Jan. *Distr.*: Endemic to Peninsular India. Common in moist hills slopes. *NAK 1258* (Moozhiar, c. 250 m).

3. **Arundinella nervosa** (Roxb.) Nees ex Hook. & Arn., Bot. Beech. Voy. 237. 1836; Hook.f., Brit. India 7: 70 1896; Bor, Kew Bull. 1955: 395. 1955 & Grass. Burma, Ceylon, India & Pakistan 424. 1960. *Holcus nervosus* Roxb., Fl. Ind. 1: 320. 1820.

Slender annuals. Leaves oblong-lanceate, 10 by 5 mm. Panicle of dense racemes, to 15 cm long; spikelets 5-6 mm long; pedicel 3-6 mm long, glumes oblong-lanceate; long-acuminate; lower glumes c. 3.5 mm long; upper glume c. 6 mm long; lower lemma 3-nerved; upper lemma densely scaberulous; awn to 1 cm long; setae 2, subequal; palea narrowly lanceate.

Fl. & Fr.: Dec.-Jan. *Distr.*: Endemic to Western Peninsular India. Not common. *NAK 412* (Moozhiar, c. 250 m).

4. **Arundinella purpurea** Hochst. ex Steud., Syn. Pl. Glum. 1: 115. 1854; Bor, Grass. Burma, Ceylon, India & Pakistan 424. 1960; Ramamoorthy in Sald. & Nicols., Fl. Hassan Dist. 711. 1976; Manilal, Fl. Silent Valley 349. 1988; Mohanan & Sivad., Fl. Agasthyamala 804.2002. *A. fuscata* auct., non Nees ex Buese, 1854: Hook. f., Fl. Brit. India 7: 74. 1896; Fischer in Gamble, Fl. Pres. Madras 1802. 1934, var. **purpurea.**

Perennials; culms 10-22 cm tall. Leaves oblong, 5-20 by 0.6-1.3 cm, base rounded or subcordate, margin tuberculate-ciliate, drying yellowish. Panicles short, lower glumes to 3.5 mm long, upper lemma c. 2 mm long, awns c. 3 mm long.

Fl. & Fr.: Oct.-Nov. *Distr.*: Endemic to Peninsular India. Not common. On hills. *NAK 411 & 417* (Moozhiar, c. 250 m).

AXONOPUS Palisot de Beauvois
Essai Agrost. 12:154. 1812.

Axonopus compressus (Sw.) P. Beauv., Ess. Agrost. 12: 154. 1812; Bor, Grass. Burma, Ceylon, India & Pakistan 278. 1960; Ramamoorthy in Sald. & Nicols., Fl. Hassan Dist. 712. 1976; Mohanan & Sivad., Fl. Agasthyamala 805.2002. *Milium compressum* Sw., Prodr. Veg. Ind. Occi. 24. 1788.

Perennials, creeping, stoloniferous. Leaves rosette-like, oblong to linear-lanceate, 2-5 by 0.3-1.5 cm, base rounded or subcordate, apex acute; ligule fimbriate. Racemes 2-6, binate, digitate, 2-10 cm long; rachis 3-gonous, winged. Spikelet ellipsoid, sessile or subsessile, green or purplish; lower glume 0; upper glume 5-7- nerved, softly hairy on nerves; lower lemma 5-nerved, villous on dorsal side; upper lemma faintly 3-5-nerved; bearded at apex; florets bisexual; stamens 3; stigmas white.

Fl. & Fr.: Most of the seasons. *Distr.*: Tropics & Subtropics. Common along waste places, wet places, etc. *NAK 2866* (Moozhiar, c. 250 m).

The common 'carpet grass'. Morphology of this grass is highly variable.

BAMBUSA Schreber
Gen. 236. 1789; 828. 1791 (nom. cons.).

Bambusa bambos (L.) Voss in Vilmorin, Blumengartnerei 1:1189.1896; McClure, Blumea, Suppl. 3:95. 1946; Nicols., Suresh & Manilal, An Interpr. Hort. Malab. 306 1988; Mohanan & Sivad., Fl. Agasthyamala 805.2002. *Arundo bambos* L., Sp. Pl. 81. 1753. *Bambos arundinacea* Retz., Obs. Bot. 5:24. 1788. . *Bambusa arundinacea* (Retz.) Roxb., Pl. Corom. t. 79. 1796; Fischer in Gamble, Fl. Pres. Madras 1859. 1934; Ramamoorthy in Sald. Nicols., Fl. Hassan Dist. 712. 1976. *Ily* Rheede, Hort. Malab. 1:25-26, t. 16. 1678. *Mullanmula*

Thorny tree-like shrubs; culms 15-25 cm f; spines 2-3, recurved; culm-sheaths broadly triangular, densely brownish-hairy within. Leaves to 20 cm long. Inflorescence a compound panicle with the spikelets in heads. Spikelets 1-a-flowered; glumes 1-3, lemmas ovate-lanceate, mucronate; petals 2-keeled; lodicules 3, membranous; stamens 6. Caryopsis linearly cylindric.

Fl. & Fr.: Flowering or fruiting could not be observed during the present study. *Distr.*: Peninsular India. Common in dry exposed hills. *NAK 2832* (Vallicode-Kottayam, c. 150 m).

Culms are used for making basket, mats, etc. and as rafts in thatched houses; grains are occasionally eaten by the poor.

BRACHIARIA (Trinius) Grisebach
in Ledebour, Fl. Rossica 4: 469. 1853.

Brachiaria eruciformis (Smith) Griseb. in Ledeb., Fl. Ross. 4: 469. 1853; Fischer in Gamble, Fl. Pres. Madras 1769. 1934; Bor, Grass. Burma, Ceylon, India & Pakistan 283. 1960; Ramamoorthy in Sald. & Nicols., Fl. Hassan Dist. 713. 1976; Britto & Matthew in Matthew, Fl. Tam. Carnatic 3(3): 1812. 1983. *Panicum eruciforme* Smith in Sibth. & Smith, Fl. Graeca 1: 44, t. 59. 1806. *P. isachne* Roth ex Roem. & Schult. in L., Syst. Veg. 2: 458. 1817; Hook.f., Fl. Brit. India 7: 28. 1896; Rang. & Tad., Handb. S. Ind. Grass. 66, ff. 80 & 81. 1921.

Erect herbs. Leaves lanceate, 1.5-5 by 0.3-0.5 cm, pilose, margin scabrid, apex acuminate, sheath 2-4 cm long. Racemes terminal, cylindric, 7-9 cm long, branches triquetrous, winged, scabrid. Spikelet ellipsoid, secund-imbricate, densely pubescent, shortly stipitate; glumes unequal, membranous, lower glume a minute scale, upper glume as long as spikelet, 5-nerved, lower lemma 3-nerved, upper lemma crustaceous, smooth, 3-nerved; stamens 3. Caryopis ovoid.

Fl. & Fr.: Dec.-Mar. *Distr.*: India, westwards to Spain & N. Africa. Common. Along moist hill slopes. *NAK 2821 & 2826* (Kakki hills, c. 1100 m).

CAPILLIPEDIUM Stapf
in Prain, Fl. Trop. Africa 9:169. 1917.

Annuals or perennials. Leaves flat. Inflorescence a profusely branched panicle, branches delicate and capillary. Racemes jointed, joints and pedicels slender, longitudinally grooved. Spikelets in pairs, one sessile, other pedicelled. Sessile spikelets: glumes subequal, membranous, lower 2-keeled, upper boat-shaped, ± grooved on each side of keel; lemmas unequal, lower hyaline, empty, upper reduced to the base of a geniculate awn, enclosing a bisexual floret; stamens 2. Caryopsis cylindric; pedicelled spikelet : glume usually one, the second if present, shorter or much reduced; lemmas present, hyaline, awnless.

1. Culms decumbent (much geniculately branched); nodes sparingly bearded. .... 1. **C. filiculme**
1. Culms erect, usually simple; nodes densely bearded. .... 2. **C. huegelii**

1. **Capillipedium filiculme** (Hook.f.) Stapf in Hooker's Ic. Pl. Sub. t. 3085. 1922; Fischer in Gamble, Fl. Pres. Madras 1730. 1934, ("*filiculmis*"); Bor, Grass. Burma, Ceylon, India & Pakistan 111. 1960. *Andropogon filiculmis* Hook.f., Fl. Brit. India 7: 181. 1896. *Dicanthium filiculme* (Hook.f.) Jain & Deshpande, Bull. Bot. Surv. India 20: 134. 1978.

Culms weak, decumbent and trailing, much geniculately branched; rooting from lower nodes; nodes sparingly hairy. Panicles 2-5 cm long; pedicels terete; callus of sessile spikelet densely villous, lower glume glabrous or sparsely hairy below the middle; awns to 1.5 cm long.

Fl. & Fr.: Dec.-Mar. *Distr.*: Peninsular India. Not common. On hills slopes. *NAK 2044* (Chittar, c. 200 m).

2. **Capillipedium huegelii** (Hack.) A. Camus, Rev. Int. Bot. Appl. Agric. Trop. 1: 308. 1921; Fischer in Gamble, Fl. Pres. Madras 1730. 1934; Bor, Grass. Burma, Ceylon, India & Pakistan 111. 1960; Ramamoorthy in Sald. & Nicols., Fl. Hassan Dist. 714. 1976. *Andropogon huegelii* Hack. in A.DC., Monogr. Phan. 6: 492. 1889; Hook.f., Fl. Brit. India 7: 180. 1896. *Dicanthium huegelii* (Hack.) Jain & Deshpande, Bull. Bot. Surv. India 20: 135. 1978.

Culms robust, erect, sparingly branched, nodes densely bearded. Panicles contracted or lax, pedicels terete; callus of sessile spikelet densely villous, lower glume usually villous all over or only below the middle, rarely glabrous, awns 1.2-2.3 cm long.

Fl. & Fr.: Dec.-Mar. *Distr.*: Endemic to Peninsular India. Rare. Among waysides bushes in hills. *NAK 2195* (Moozhiar, c. 250 m).

CENTOTHECA Palisot de Beauvois
Essai Agrost. 69. 1812

Centotheca lappacea (L.) Desv., Nouv. Bull. Sci. Soc. Philom. Paris 2: 189. 1810; Hook.f., Fl. Brit. India 7:332. 1896; Fischer in Gamble, Fl. Pres. Madras 1848. 1934, Bor, Grass. Burma, Ceylon, India & Pakistan 457. 1960; Britto & Matthew in Matthew, Fl. Tam. Carnatic 3(3): 1817.1983; Manilal, Fl. Silent Valley 349. 1988; Mohanan & Sivad., Fl. Agasthyamala 807.2002. *Cenchrus lappaceus* L., Sp. Pl. ed. 2,1488.1763.

Perennials; culms decumbent, to 60 cm tall. Leaves broadly elliptic-oblong, 6-10 by 2.5 cm, base and apex acute; sheaths imbricating, to 6 cm long, apex tuberculate ciliate; ligule membraneous. Panicle lax, to 10 cm long. Spikelets secund, laterally compressed, to 6 mm long. with 2 bisexual florets and one sterile floret, rachilla slender, articulate; glumes unequal, persistent, lower glume ovate, 3-or 5-nerved, upper glume lanceate, 5-nerved, lemmas oblong, 9-11-nerved; upper lemmas with deflexed tuberculate bristles; stamens 3. Caryopsis obovoid-oblong.

Fl. & Fr.: Dec.-Mar. *Distr.*: S.E. Asia, China, Polynesia & Trop. Africa. Not common. Along waysides on upper hills. *NAK 398, 2200* (Moozhiar, c. 250 m).

Note : For nomenclatural discussion, refer Saldanha and Nicolson (1976). (Fl. Hassan Dist. 715).

CHLORIS O. Swartz
Prodr. 1:25. 1788.

Chloris barbata Sw., Fl.Ind. Occid. 1: 200. 1797; Hook. f., Fl.Brit.India 7: 292.1896; Rang. & Tad., Handb. S. Ind. Grass. 264, ff. 199, 200 & 201. 1921; Fischer in Gamble, Fl. Pres. Madras 1838. 1934; Bor, Grass. Burma, Ceylon, India & Pakistan 465. 1960; Ramamoorthy in Sald. & Nicols., Fl. Hassan Dist. 715. 1976; Britto & Matthew in Matthew, Fl. Tam. Carnatic 3(3): 1820. 1983; Nicols., Suresh & Manilal, An Interpr. Hort. Malab.307. 1988. *Konda-pullu* Rheede, Hort. Malab. 12:95, t.51. 1693.

Rhizomatous perennials. Leaves oblong, 4-20 by 0.4 cm, flat, rarely plicate, densely scabrid, sheaths to 8 cm long, densely ciliate at apex. Spikes digitate; spikelets subsessile, unilateral, 2-seriate, each with 1 fertile floret and 1 or more sterile florets, glumes unequal, membraneous, lanceate, 1-nerved, mucornate, fertile lemma obovate, densely eiliate on margins and on keel; palea 2-nerved, 2-keeled; callus rounded, ciliate; sterile lemmas awned; stamens 3. Caryopsis ellipsoid.

Fl. & Fr.: Sep.-Feb. *Distr.*: Tropics. Not common. Along waysides. *NAK 16* (Moozhiar, c. 250 m).

CHRYSOPOGON Trinius
Fund. Agrost. 187. 1822. ('1820') (nom. cons.).

Perennial erect tufted herbs. Leaves flat or folded, sheaths glabrous, ligule a fimbriate rim. Panicles terminal, lax, branches whorled, bearing terminal racemes with 1 sessile and 2 pedicelled-spikelets. Sessile spikelet: usually laterally compressed; glumes subequal, lower glume cartilaginous, rounded; upper glume awned, faintly 1-keeled; lemmas hyaline, lower empty, upper one 2-dentate or entire, awned; palea small or 0, enclosing a bisexual floret; stamens 3. Caryopsis linear, laterally compressed. Pedicelled spikelets : pedicel shorter or more than ½ as long as sessile spikelets; spikelets dorsally compressed, not awned.

1. Culms creeping and ascending; pedicels glabrescent. .... 1. **C. aciculatus**
1. Culms always erect; pedicels villous.
    2. Pedicels less than ½ as long as the sessile spikelets; upper glume of pedicelled spikelet distinctly aristate. .... 2. **C. hackelii**
    2. Pedicels ½ as long as the sessile spikelets; upper glume of pedicelled spikelet not or very shortly aristate. .... 3. **C. zeylanicus**

1. **Chrysopogon aciculatus** (Retz.) Trin., Fund. Agrost. 181. 1820; Fischer in Gamble, Fl. Pres. Madras 1738. 1934; Bor, Grass. Burma, Ceylon, India & Pakistan 115. 1960; Ramamoorthy in Sald. & Nicols., Fl.

Hassan Dist. 716. 1976; Matthew, Ill.Fl. Tam. Carnatic tt.829 & 830. 1982; Britto & Matthew in Matthew, Fl. Tam. Carnatic 3(3): 1822. 1983; Nicols., Suresh & Manilal, An Interpr. Hort.Malab. 307. 1988; Mohanan & Sivad., Fl. Agasthyamala 808.2002 *Andropogon aciculatus* Retz., Obs. Bot. 5:22. 1789; Hook.f., Fl. Brit. India 7:188. 1896. *Kudira-pullu* Rheede, Hort. Malab.12 : 79, t. 43.1683. *Sneha-pullu.*

Creeping and decumbent herbs. Leaves lanceate or ovate-lanceate, c. 10 by 0.6 cm, margin serrulate, apex acute to obtuse; ligule annular. Panicle pyramidal, 5-10 cm long; sessile spikelets: lanceoid, callus bearded, pungent; upper glume 3-nerved; stamens 3; stigmas golden-yellow; pedicelled spikelet: lanceoid or linear-lanceoid, awnless; lower glume 5-nerved; upper glume 3-nerved; stamens 3.

Fl. & Fr.: Most of the seasons. *Distr.*: Tropics. 'The love grass'. Common along streamsides and degraded forest floors. *NAK 2864* (Ranni, c. 150 m). Culms form thick carpets and act as a soil binder.

2. **Chrysopogon hackelii** (Hook.f.) Fischer in Gamble, Fl.Pres.Madras 1739. 1834; Bor, Grass. Burma, Ceylong, India & Pakistan 117. 1960. Ramamoorthy in Sald. & Nicols., Fl. Hassan Dist. 716. 1976; Mohanan & Sivad., Fl. Agasthyamala 809.2002. *Andropogon hackelii* Hook. f., Fl. Brit. India 7: 194. 1896.

Robust, densely tufted herbs; culms to 1.8 m tall. Leaves linear-oblong, 15-90 by 0.3-0.8 cm, margin often silky ciliate. Panicle 10-20 cm long, pyramidal, hairs chocolate brown; branches spreading; upper glume of sessile spikelets glabrous; glumes of pedicelled spikelets muticous, glabrous, arista of upper glume c. 5 mm long.

Fl. & Fr.: Dec.-Mar. *Distr.*: Endemic to South India. Not common. Locally abundant along waysides of upper hills. *NAK 371* (Moozhiar, c. 250 m).

3. **Chrysopogon zeylanicus** (Nees ex Steud.) Thw., Enum. Pl. Zeyl. 366. 1864; Fischer in Gamble, Fl. Pres. Madras 1739. 1934; Bor, Grass. Burma, Ceylon, India & Pakistan 119. 1960. *Andropogon zeylanicus* Nees ex Stued., Syn. Pl. Glumac. 1: 397. 1854; Hook. f., Fl. Brit. India 7: 192. 1896.

Tufted herbs to 1 m tall. Leaves linear-oblong, 20-35 by 0.9-1.3, folded, coriaceous, densely imbricate towards base. Panicles to 15 cm long, branches erect or spreading; glumes of pedicelled spikelets acuminate, of sessile spikelets aristate; awn to 3 cm long.

Fl. & Fr.: Dec.-Feb. *Distr.*: Sri Lanka, South India. Rare. On rocky upper hills. *NAK 2192* (Chittar, c. 450 m).

COIX Linnaeus
Sp. Pl. 972. 1753.

Coix lacryma-jobi L., Sp. Pl. 972. 1753; Roxb., Fl. Ind. 3: 568. 1832; Hook. f., Fl. Brit. India 7: 100. 1896; Rang. & Tad., Handb. S. Ind. Grass. 141, t. 126. 1921; Fischer in Gamble, Fl. Pres. Madras 1705. 1934; Bor, Grass. Burma, Ceylon, India & Pakistan 264. 1960; Ramamoorthy in Sald. & Nicols., Fl. Hassan Dist. 718. 1976; Britto & Matthew in Matthew, Fl. Tam. Carnatic 3(3): 1826. 1983; Manilal, Fl. Silent Valley 350. 1988; Nicols., Suresh & Manilal, An Interpr. Hort. Malab. 308. 1988. *Cutri-conda* Rheede, Hort. Malab. 12:133, t. 70. 1693.

Tufted robust herbs. Leaves flat, linear-oblong, 15-50 by 1 cm, margin scaberulous, sheaths loose, to 6 cm long, glabrous. Inflorescence terminal and axillary of 2 racemes, to 4 cm long, consisting of one female spikelet completely enclosed in a globose or ovoid basal cupule like bract. Male racemes exerted from the mouth of cupule. Female spikelet: glumes orbicular, beaked, chartaceous, lemmas lanceate, hyaline. Caryopsis subglobose, furrowed in the middle. Male spikelet: lanceate, lower glume 2-keeled, winged on keels, lower lemma and palea nearly as long as spikelets.

Fl. & Fr.: Oct.-Nov. *Distr.:* Native of trop. Asia; now throughout the tropics. Common on waste places in lowlands. *NAK 1052* (Tiruvalla, c. 10 m).

Fruits are used for making chains and necklaces by local people.

CYMBOPOGON K. P. J. Sprengel
Pl. Min. Cogn. Pugil. 2: 14. 1815.

Perennial tall aromatic grasses. Leaves linear-lanceate; ligules membranous. Panicle lax, 10-80 cm long; joints of racemes villous; spikelets 2-nate; sessile spikelets elliptic or linear-lanceoid, awned or not; glumes 3-5-nerved; lower floret empty; upper floret bisexual; pedicelled spikelet linear-lanceoid or elliptic-lanceoid; stamens 3.

1. Sessile spikelets linear-lanceoid, awnless. .... 1. **C. citratus**

1. Sessile spikelets ellipsoid, awned. .... 2. **C. flexuosus**

1. **Cymbopogon citratus** (DC.) Stapf, Kew Bull. 1906: 357. 1906; Fischer in Gamble, Fl. Pres. Madras 1756. 1934; Bor, Grass. Burma, Ceylon, India & Pakistan 126. 1960. *Andropoggon citratus* DC., Cat. Hort. Monsp. 78. 1813; Hook.f., Fl. Brit. India 7:210. 1896.

Leaves linear-lanceate, 15-60 by 0.8-1.5 cm, glaucous. Panicles lax, 10-15 cm. Racemes 1-1.5 cm long; glumes ovate-lanceate, lower 5-nerved; upper 3-nerved; lemmas 2-nerved; pedicel 2-4 mm long, densely villous.

Fl. & Fr.: Jul.-Sep. *Distr.:* Tropics. Locally abundant. *NAK 2865* (Ponnambalamedu, c. 1100 m).

The valuable aromatic 'lemon grass oil' is extracted from the leaves of this grass.

2. **Cymbopogon flexuosus** (Nees ex Steud.) Wats. in Atk. Gaz. N.W. Prov. Ind. 392. 1882; Fischer in Gamble, Fl. Pres. Madras 1756. 1934; Bor, Grass.Burma, Ceylon, India & Pakistan 127. 1960; Ramamoorthy in Sald. & Nicols., Fl. Hassan Dist. 719. 1976; Mohanan & Sivad., Fl. Agasthyamala 809.2002. *Andropogon flexuosus* Nees ex Steud., Syn. Pl. Glumac. 1: 388. 1854. *A. nardus* L., var. *flexuosus* (Nees ex Steud.) Hack. in A.DC., Monogr. Phan. 6: 603. 1889; Hook. f., Fl. Brit. India 7: 207. 1896. *Mechil-pullu, Pura-pullu.*

Leaves linear, 20 cm-1.25 m long, flat; sheaths glabrous; ligule membranous. Inflorescence a large complex greyish panicle; branches slender flexuous, often drooping. Racemes paired, with a spatheole beneath. Spikelets binate, one sessile, other pedicelled; lowest pair of one or both racemes homogamous. Sessile spikelet dorsally compressed, callus obtuse, lower floret empty, upper one bisexual; lower glume as long as spikelet, 2-keeled, keel winged; upper glume boat- shaped, 1-keeled; lower lemma glabrous, hyaline, epaleate; upper lemma hyaline, deeply 2-fid; stamens 3. Caryopsis oblong. Pedicelled spikelet: lower glume as long as spikelets, 7-or more-nerved; upper glume boat-shaped, 1-keeled, lower lemma hyaline, epaleate; stamens 3.

Fl. & Fr.: Nov.-Feb. *Distr.*: South India. Common in exposed forest areas. *NAK 2193* (Chittar, c. 200 m).

The plants yield an inferior kind of lemon-grass oil.

CYNODON L. C. Richard
in Persoon, Syn. Pl. 1: 85. 1805 (nom. cons.).

Cynodon dactylon (L.) Pers., Syn. Pl. 1: 85. 1805; Hook. f., Fl. Brit. India 7: 288. 1896; Rang. & Tad., Handb. S. Ind. Grass. 250, ff. 190 & 191. 1921; Fischer in Gamble, Fl. Pres. Madras 1835. 1934; Bor, Grass. Burma, Ceylon, India & Pakistan 469. 1960; Matthew, Ill.Fl. Tam. Carnatic t. 838. 1982; Britto & Matthew in Mattew, Fl. Tam. Carnatic 3(3): 1829. 1983; Nicols., Suresh & Manilal, An Interpr. Hort. Malab. 308. 1988; Mohanan & Sivad., Fl. Agasthyamala 810.2002. *Panicum dactylon* L., Sp. Pl. 58. 1753. *Beli-caraga* Rheede, Hort. Malab. 12:87, t.47. 1693. *Karuka-pullu.*

Slender perennials with stolons. Leaves linear-oblong, to 10 cm long, folded, softly scaberulous; sheaths to 3.5 cm long, ligule membranous, shortly ciliate. Inflorescence of terminal, digitate 3-4 spikes. Spikes 1-sided, oblong to 5 cm long. Spikelets sessile, laterally compressed, alternately 2-seriate, imbricate, 1-flowered; glumes lanceate, 1-nerved, persistent; lemma pubescent on keel; palea as long as lemma; stamens 3. Caryopsis linear.

Fl. & Fr.: Dec.-Jan. *Distr.*: Tropical and warm temperate regions throughout the world. Very common in moist and warm areas of lowlands and hills. *NAK 2796* (Moozhiar, c. 250 m).

CYRTOCOCCUM Stapf
in D. Prain, Fl. Trop. Africa 9:15. 1917; 9: 745. 1920.

Perennials or annuals. Leaves oblong-lanceate or ovate-lanceate. Panicle contracted or lax. Spikelets long or short-pedicelled, laterally compressed; glumes membranous; lemmas disimilar, 5-(7)-nerved, margin involute; lower floret empty; upper floret bisexual; lodicules 2; stamens 3; styles 2. Caryopsis obovoid.

1. Panicle lax; pedicels much longer than spikelets. ....1. **C. muricatum**

1. Panicle contracted; pedicels ± equal to spikelets. ....2. **C. oxyphyllum**

1. **Cyrtococcum muricatum** (Retz.) Bor, Grass. Burma, Ceylon, India & Pakistan 291. 1960. *Panicum muricatum* Retz., Obs. Bot. 4:18. 1786.

Annuals; culms creeping, rooting at lower nodes. Leaves ovate-lanceate, elliptic-lanceate, 1-10 by 0.4-1.8 cm, base ciliate on margin. Panicles 5-20 cm long, lax. Spikelets obovate, brownish or purplish; lower lemma ovate-oblong, upper lemma deltoid, 5-7-nerved; lower palea 2-nerved.

Fl. & Fr.: Jul.-Nov. *Distr.*: Tropics. Common along waysides in hills. *NAK 2883* (Moozhiar, c. 250 m).

2. **Cyrtococcum oxyphyllum** (Steud.) Stapf in Hook., Ic. Pl. Sub. t. 3096. 1922; Fischer in Gamble, Fl. Pres. Madras 1786. 1934; Bor, Grass. Burma, Ceylon, India & Pakistan 291. 1960; Ramamoorthy in Sald. & Nicols., Fl. Hassan Dist. 720. 1976; Britto & Matthew in Matthew, Fl. Tam. Carnatic 3(3): 1830. 1983; Manilal, Fl. Silent Valley 357. 1988. *Panicum oxyphyllum* Steud., Syn. Pl. Glum. 1: 65. 1854. *P. pilipes* Nees & Arn. ex Buse in Miq., Pl. Jungh. 3; 376. 1854; Hook. f., Fl. Brit. India 7: 57. 1896.

Perennials; culms slender, creeping, nodes glabrous or bearded. Leaves oblong, 5-15 by 0.6-1 cm, softly pilose or not; sheaths to 4 cm long, densely ciliate along margins; ligule membranous. Panicle loose and open, spreading, to 8 cm long; branch-axils densely tufted pilose; pedicels with long white hairs. Spikelets laterally compressed, c. 2 mm long; glumes subequal, membranous, 3-nerved; lower and upper glumes ovate, lower lemma suborbicular, 5-nerved, palea narrow-linear, as long as lemma; stamens 3. Caryopsis obovoid.

Fl. & Fr.: Dec.-Feb. *Distr.*: Sri Lanka, India, Myanmar, China, Malesia. Less common. On moist hills slopes. *NAK 307, 1230* (Moozhiar, c. 250 m).

DACTYLOCTENIUM Willdenow
Enum. Pl. Horti Berol. 1029. 1809.

Dactyloctenium aegyptium (L.) P. Beauv., Essai Agrost. expl. t. 15. 1812; Fischer in Gamble, Fl. Pres. Madras 1840. 1934; Bor., Grass. Burma, Ceylon, India & Pakistan 489. 1960; Ramamoorthy in Sald. & Nicols., Fl. Hassan Dist. 721. 1976; Matthew, Ill. Fl. Tam. Carnatic tt. 841 & 842. 1982; Britto & Matthew in Matthew, Fl. Tam. Carnatic 3(3): 1831. 1983; Nicols., Suresh & Manilal, An Interpr. Hort. Malab. 308. 1988. *Cynosurus aegyptius* L., Sp. Pl. 72. 1753. *Eleusine aegyptia* (L.) Desf., Fl. Atlant. 1: 85. 1798; Hook. f., Fl. Brit. India 7: 295. 1896; Rang. & Tad., Handb. S. Ind. Grass. 276, ff. 5, 208 & 209. 1921. *Kavara-pullu* Rheede, Hort. Malab.12:131, t. 69. 1693.

Annuals, culms to 60 cm tall, erect, geniculate. Leaves linear, 5-15 by 0.4-0.6 cm, flat, tuberculate-hispid without; sheaths to 5 cm long, ligule membranous, ciliolate. Inflorescence of digitate, 2-7 oblong spikes. Spikelets 2- or 3-flowered, sessile, broadly ovate, to 3.5 mm long, laterally compressed, densely imbricating in 2 rows, rachis produced as a sharp point obove; rachilla disarticulating obove lower glime; glumes subequal, folded, lower glume scabrid along keel; keel of upper glume smooth or hispidulous, produced into an awn, keel of lemmas gibbous, paleate stamens 3.

Fl. & Fr.: Most of the seasons. *Distr.*: In tropical and warm temperate regions of Old World. Common in lowlands. *NAK 2941* (Adoor, c. 20 m).

DICHAETARIA Nees ex Steudel
Syn. Pl. Glum. 1. 145. 1854.

Dichaetaria wightii Nees ex Steud., Syn. Pl. Glumac. 1:145. 1854; Hook. f., Fl. Brit. India 7: 300. 1896; Fischer in Gamble, Fl. Pres. Madras 1829. 1934; Bor, Grass. Burma, Ceylon, India & Pakistan 471. 1960.

Perennial slender herbs, to 60 cm tall. Leaves narrow, linear, flat, 30-60 cm long. Panicles of few racemes on a long simple axis. Spikelets few, distant, 1-flowered, narrow, terete; rachilla jointed at the base, produced into an empty arista; glumes 2, unequal, lanceate, 3-nerved; callus elongate, bearded, palea minutely 2-dentate, 2-nerved, containing a bisexual floret; lodicules 2, fleshy; stamens 3. Caryopsis linear.

Fl. & Fr.: Jan.-Mar. *Distr.:* Sri Lanka, South India. Rare. Among moist shady areas of interior forests. *NAK 2854* (Moozhiar, c. 250 m).

DIGITARIA Heister ex Fabricius
Enum. 207. 1759.

Annuals or perennials. Leaves oblong, flat, flaccid, tuberculate pubescent; sheaths usually lax and open. Inflorescence terminal, spike-like racemes, digitate, corymbose; rachis winged, serrate, rarely smooth. Spikelets oblong, 2-nate or 3-nate, rarely solitary, convex on back; lower glume a rim or a triangular scale, sometimes 0; upper glume variable, 1-3-or 5-nerved, rarely nerveless; lower lemma membranous, 5-7-nerved, as long as spikelet; palea reduced to scale; upper floret bisexual, upper lemma cartilaginous, concealing palea by margins, 3-nerved; palea 2-nerved; stamens 3; styles 2, apical. Caryopsis narrow-ellipsoid.

1. Spikelets ternate.
   2. Racemes on an obscure common axis; spicules absent in distal portion of pedicel. ....2. **D. longiflora**
   2. Racemes on a well developed common axis; spicules present throughout the pedicel. ....5. **D. ternata**
1. Spikelets binate.
   3. Rachis margin smooth. ....3. **D. radicosa**
   3. Rachis margin serrate.
      4. Lower glume 0 or very obscure. ....4. **D. setigera**
      4. Lower glime rim-like or a triangular scale. ....1. **D. ciliaris**

1. **Digitaria ciliaris** (Retz.) Koeler, Descr. Garm. 27.1802; Veldk., Blumea 21: 32. 1973; Ramamoorthy in Sald. & Nicols., Fl. Hassan Dist. 723. 1976; Britto & Matthew in Matthew, Fl. Tam. Carnatic 3(3): 1837. 1983; Mohanan & Sivad., Fl. Agasthyamala 813.2002. *Panicum ciliare* Retz., Obs. Bot. 4: 16. 1786. *Panicum adscendens* Kunth, Nov. Gen. Pl. 1: 97. 1816. *Digitaria adscendens* (Kunth) Henrard, Blumea 1: 92. 1934; Bor, Grass. Burma, Ceylon, India & Pakistan 298. 1960. *Digitaria marginata* Link, Enum. Hort. Berol. 1: 102. 1821; Fischer in Gamble, Fl. Pres. Madras 1764. 1934. *D. sanguinalis* (Link) Scop., var *extensum* Rang. & Tad., Handb. S. Ind. Grass. 56, tt. 74 & 75. 1921.

Annuals; culms to 60 m tall. Leaves oblong-linear, 3-8 by 0.3-1 cm, flat, pubescent, sheaths lax. Racemes 3-6 to 15 cm long, digitate or in 1-3 whorl(s) on a common axis, rachis serrate. Spikelets oblong, binate; lower glume a triangular scale; upper glume linear-lanceate, 3-5-nerved; lower lemma 5-nerved, marginal ones densely pubescent; upper lemma obscurely 3-nerved.

Fl. & Fr.: Sep.-Nov. *Distr.*: Old World tropics. Common along waysides and open places. *NAK 97, 2203* (Moozhiar, Ranni etc.).

2. **Digitaria longiflora** (Retz.) Pres., Syn. Pl. 1: 85. 1805; Fischer in Gamble, Fl.Pres. Madras 1765. 1934; Bor, Grass. Burma, Ceylon, India & Pakistan 302. 1960; Veldk., Blumea 21: 66. 1973; Ramamoorthy in Sald. & Nicols., Fl. Hassan Dist. 724. 1976; Britto & Matthew in Matthew, Fl. Tam. Carnatic 3(3): 1838. 1983; Manilal, Fl. Silent Valley 351. 1988; Mohanan & Sivad., Fl. Agasthyamala 813.2002. *Paspalum longiflorum* Retz., Obs. Bot. 4: 15. 1786; Hook. f., Fl. Brit. India 7: 17. 1896, p.p.

Annuals; culms 10-40 cm tall. Leaves ovate-lanceate, 2-6 by 0.3-0.7 cm, flat, margin spiculed, apex acute; sheaths lax at basal nodes. Racemes 2 or 3, digitate, common axis 0 or shortly developed; rachis serrate. Spikelets ternate, pubescent; lower glume 0; upper glume 3-5-nerved; lower lemma as long as spikelet with 5-7 anastomosing nerves; upper lemma coriaceous, 3-nerved.

Fl. & Fr.: Sep.-Nov. *Distr.*: Sri Lanka, Peninsular India. Common in waste places and open fields. *NAK 13* (Moozhiar, c. 250 m).

3. **Digitaria radicosa** (C. Presl) Miq., Fl. Ned. Ind. 3: 437. 1857; Veldkamp, Blumea 21: 35. 1973; Britto & Matthew in Matthew, Fl. Tam. Carnatic 3(3): 1838. 1983. *Panicum radicosum* C. Presl, Rel. Haenk. 1: 297.1830. *Panicum timorense* Kunth, Enum. Pl. 1: 83. 1833. *Digitaria timorensis* (Kunth) Bal., J. Bot. (Morot) 4 : 138. 1890; Bor, Grass. Burma, Ceylon, India & Pakistan 306. 1960. *Paspalum sanguinale* Lam., var. *debile* Hook. f., Fl. Brit. India 7: 16. 1896.

Annuals or perennials; culms to 30 cm tall. Leaves linear-oblong, 2-7 by 0.3-0.5 cm, flat, appressed-pilose; sheaths lax. Racemes 2 or 3, 5-10 cm long; rachis smooth; spikelets oblong, binate, pubescent; lower glume a minute rim or scale, upper glume 3-nerved with overtopping pubescence; lower lemma 3-5-nerved, marginal ones densely pubescent; upper lemma 3-nerved.

Fl. & Fr.: Dec.-Mar. *Distr.*: South-East Asia, Polynesia. Common in hills. *NAK 2820* (Moozhiar, c. 250 m).

4. **Digitaria setigera** Roth ex Roem. & Schult., Syst. Veg. 2:474. 1817; Veldkamp, Blumea 21:37. 1973; Bor, Grass. Burma, Ceylon, India & Pakistan 305. 1960. *Paspalum sanguinale* (L.) Lam. var. *extensum* Hook.f., Fl.Brit. India 7:15. 1896. *Digitaria sanguinalis* (L.) Scop. var *extensum* (Hook.f.) Rang. & Tad., Handb. S. Ind. Grass. 56, ff. 74-75. 1921.

Annuals; culms to 1 m long, creeping; nodes glabrous. Leaves elliptic-lanceate, linear-lanceate or linear, base rounded, apex acute; ligule to 5 mm long. Racemes 3-15, digitate or subdigitate; rachis flat, winged; pedicels triquetrous; spikelets ellipsoid or lanceoid, upper glume 0-3-nerved, lower glume 0 or minute, lower lemma 7-nerved; palea 2-keeled, 2-nerved.

Fl. & Fr.: Most of the seasons. *Distr.*: Pantropics. Common along waysides, waste places, forest clearing, etc. *NAK 2934* (Adoor, c. 20 m).

5. **Digitaria ternata** (A. Rich.) Stapf ex Dyer in Harv. & Sonder, Fl. Cap. 376. 1898; Fischer in Gamble, Fl. Pres. Madras 1765. 1934; Bor, Grass. Burma, Ceylon, India & Pakistan 306. 1960; Ramamoorthy in Sald. & Nicols., Fl. Hassan Dist. 724. 1976; Britto & Matthew in Matthew, Fl. Tam. Carnatic 3(3): 1840. 1983. *Cynodon ternatus* A. Rich., Tent. Fl. Abyss. 2: 405. 1851. *Paspalum ternatum* Hook.f., Fl. Brit. India 7: 17. 1896.

Perennials; culms to 1 m long, rooted at lower nodes. Leaves oblong-linear, 6-15 cm long, flat; sheaths to 7 cm long. Racemes 4 or 5, to 12 cm long; common axis to 3.5 cm long; rachis serrate. Spikelets ovoid, ternate; lower glume 0; upper glume 3-nerved, lower lemma 5-7-nerved; upper lemma fuscose.

Fl. & Fr.: Sep.-Feb. *Distr.*: Tropical E. and S. Africa, Peninsular & Eastern India, Myanmar, China. Not Common. Along waste places in hills. *NAK 2823* (Moozhiar, c. 250 m).

DIMERIA R. Brown
Prodr. 204. 1810.

Annuals or perennials; culms extremely slender to rigid, tufted, erect or ascending, nodes often bearded. Leaves narrow-lanceate or linear; sheaths tuberculate-pilose. Inflorescence of 1-3 terminal spike-like racemes; rachis straight, rarely flexuous, flat or trigonous, winged or wingless. Spikelet solitary, secund, laterally compressed, 2-flowereds; reddish; glumes subequal, keel variously winged, apex acute to aristate; lower lemma linear; palea 0; upper floret bisexual, lemma notched at apex with a geniculate awn; palea linear or 0; stamens 2. Caryopsis linear-ellipsoid.

1. Rachis clearly angular or trigonous.
   2. Racem(s) 1-3; spikelets 1-3 mm long. .... 6. **D. ornithopoda**
   2. Racemes usually 2; spikelets 4-5 mm long. .... 1. **D. idukkiensis**
1. Rachis faintly angular or flat.
   3. Upper and lower glumes winged.

4. Annuals; often found in rocky hills. ....4. **D. mooneyi**
4. Perennials; often found in mountain marshes. ....5. **D. namboodiriana**
3. Upper glumes only winged or both not winged, or narrowly winged.
5. Upper glume winged towards apex only. ....2. **D. kurumthotticalana**
5. Upper glume winged all along.
6. Raceme solitary. ....3. **D. lawsonii**
6. Racemes 1-2.
7. Rachis broadly winged. ....7. **D. sivarajanii**
7. Rachis wingless. ....8. **D. thwaitesii**

1. Dimeria idukkiensis Ravi *et* Anil Kumar, Rheedea 2: 104. 1992.

Annuals. Leaves scattered, linear-lanceate to 6.5 cm long, tuberculate-pilose, membranous or chartaceous. Racemes usually 2, rarely 1. Spikelets elliptic-oblong, 4-5 mm long; rachis 3-gonous, c. 0.5 mm wide; lower glume linear-elliptic, narrowly keeled; upper glume winged in the upper 1/2 to 2/5 of the keel; upper lemma elliptic-lanceate, lower lemma oblanceate; apex acute.

Fl. & Fr.: Sep.-Dec. *Distr.*: Hitherto known only from type localities (Peerumedu) and Kakki hills. Rare. *NAK 2893* (Kakki hills, c. 1100 m).

2. Dimeria kurumthotticalana Jacob, J. Bombay Nat. Hist. Soc. 47(1): 49. 1947; Sreekumar & Nair, Fl. Kerala, Grass. 96. 1991, emend descr. Ravi *et* Anil Kumar, Rheedea 2: 106. 1992.

Culms tufted, to 40 cm tall, nodes bearded. Leaves linear-lanceate, 2-10 x 0.1-0.4 cm, tuberculate-pilose; ligules ovate, membranous. Racemes 1-2, 2-5 cm long, flat, 1-1.5 mm wide, margin ciliate. Spikelets elliptic, 5-6 mm long, callus hairs to 1 mm long; lower glume lanceate, upper glume elliptic, acuminate, keel broadly winged towards apex; upper lemma oblanceate, 1-nerved, paleate; palea hyaline, sparsely short ciliate; lower lemma notched at apex, awned; awn c. 15 mm long.

Fl. & Fr.: Dec.-Feb. *Distr.*: Endemic to Western Ghats. Rare. On exposed rocks. *NAK 2122* (Vallicode-Kottayam, c. 150 m). This is the first report after type collection.

3. Dimeria lawsonii (Hook.f.) Fischer in Gamble, Fl. Pres. Madras 1713. 1934; Bor, Grass. Burma, Ceylon, India & Pakistan 142. 1960; Mohanan & Sivad., Fl. Agasthyamala 815.2002. *D. pusilla* Thw., Enum. Pl. Zeyl. 369. 1864 var. *lawsonii* Hook. f., Fl. Brit. India 7: 103. 1869.

Culms tufted, slender, usually geniculate or erect. Leaves lanceate or linear-lanceate, 1-10 x 0.1-0.3 cm, tuberculate-hairy. Receme solitary, 1-6 cm long, rachis flat, 0.5-0.75 wide, margin ciliate. Spikelets oblong-lanceate, 5-7 mm long; callus hairs 0.5-1 mm long; lower glumes linear-lanceate, acuminate or awned; upper glumes elliptic-lanceate, awned, narrowly winged; lower lemma oblanceate, 1-nerved, upper lemma notched, awned, hyaline, 1-nerved; awn 8-10 mm long, geniculate.

Fl. & Fr.: Oct.-Mar. *Distr.*: Endemic to Western Ghats. Rare. On rocky hills. *NAK 2199* (Vallicode-Kottayam, c. 150 m).

4. Dimeria mooneyi Raizada ex Mooney, Suppl. Bot. Bihar & Orissa 263. 1950; Bor, Grass. Burma, Ceylon, India & Pakistan 142. 1960; Mohanan & Sivad., Fl. Agasthyamala 816.2002.

Culms tufted, stiff, erect, to 20 cm tall, nodes bearded. Leaves lancate or linear, 2-6 x 0.2-0.3 mm, margins thickened, scabrid, sparsely tuberculate hairy. Racemes 2, to 5.5 cm long, rachis flat, linear, c. 1.5 mm wide, margin ciliate; spikelets much compressed, 4-6 mm long; lower glume with a thick wing on back in lower two-thirds; upper glume with a thick wing all along the backside, lower floret empty, lemma hyaline; palea 0; upper floret bisexual; lemma hyaline, notched at apex; palea 0; awn 13-15 mm long, geniculate.

Fl. & Fr.: Sep.-Dec. *Distr.*: Endemic to South India. Rare. On rocky hills. *NAK 2835* (Kalleli R.F.). This is the first report from Kerala State.

5. **Dimeria namboodiriana** Ravi *et* Mohanan, Rheedea 7(1): 1-4. 1997.

Marshy perennials with runners; culms solitary or densely tufted, simple or branched up to 50 cm tall. Leaves elliptic lanceate, 20 cm x 7 mm, base rounded or acute, apex acuminate, densely hairy throughout. Racemes 2, rachis fuscous, straight, flattened on the back, shortly raised and convex on the face, broadly winged, sparsely ciliate on the margins. Spikelets fuscous, ellipsoid, 4-6 mm long; callus thickly bearded; upper glume narrowly elliptic with winged keel; lower lemma hyaline, oblanceate, 2-nerved, sparsely ciliate; upper lemma 1-nerved, apex bifid; awn up to 10 mm long. Grains cylindric to ellipsoid, biconvexly flattened.

Fl. & Fr.: Oct.-Dec. *Distr.*: Hitherto known from only the type locality - Ponnambalamedu. *Ravi 24050,* Holotype - TBGT (Ponnambalamedu, c. 1200 m).

6. **Dimeria ornithopoda** Trin., Fund. Agrost. 167, t. 14. 1820; Hook. f., Fl. Brit. India 7: 104. 1869; Fischer in Gamble, Fl. Pres. Madras 1713. 1934; Bor, Grass. Burma, Ceylon, India & Pakistan 142. 1960; Ramamoorthy in Sald. & Nicols., Fl. Hassan Dist. 725. 1976; Manilal, Fl. Silent Valley 352. 1988; Mohanan & Sivad., Fl. Agasthyamala 816.2002.

Culms very slender, geniculate. Leaves lanceate, 1-6 x 0.2-0.4 cm, sparsely pubescent. Racemes 2, rarely 1 or 3, 1-5 cm long; rachis trigonous, 0.25 mm wide. Spikelets lanceate, 1-2 mm long; lower glume linear-lanceate, upper glume lanceate; lower floret empty; upper floret bisexual; lower lemma oblanceate; hyaline; upper lemma notched, awned; awn 5-12 mm long.

Fl. & Fr.: Sep.-Dec. *Distr.*: Sri Lanka, India, Myanmar, China, Malesia. Common. On hills slopes, rocky places, etc. *NAK* 2791 (Kakki hills, c. 1000 m).

7. **Dimeria sivarajanii** Mohanan *et* Ravi, Rheedea 6(2): 47-50. 1996.

Annual erect herbs; culms densely tufted, branched, up to 40 cm tall; nodes plumose. Leaves linear-elliptic or elliptic-lanceate, 15 cm x 5 mm, base rounded or acute, apex acuminate, densely hairy throughout. Racemes 1 or 2, rachis flattened on the back, longitudinally ridged on the face, broadly winged, sparsely ciliate on the margin. Spikelets ellipsoid, 5-8 mm long; callus thickly bearded, sparsely ciliate on the margin; lower glume subcoriaceous; upper glume aristate, narrowly elliptic with winged keel; lower lemma hyaline, oblanceate, 2-nerved, epaleate, sparsely ciliate; upper lemma 1-nerved, apex bifid; awn up to 14 mm long. Grains cylindric.

Fl.: & Fr.: Oct.-Dec. *Distr.*: Hitherto known only from the type locality - Kochu Pampa hills. *Ravi 24041,* Holotype - TBGT (Kochu Pampa hills, c. 1200 m).

8. **Dimeria thwaitesii** Hack. in DC., Monogr. Phan. 6: 78. 1889; Fischer in Gamble, Fl. Pres. Madras 1713. 1934; Bor, Grass. Burma, Ceylon, India & Pakistan 144. 1960. *D. pusilla* var. *pallidia* Thw. ex Hook.f., Fl. Brit. India 7: 103. 1896.

Culms slender, 15-40 cm tall, erect or geniculate. Leaves linear-lanceate, 2-12 x 0.1-0.3 cm, tuberculate-hairy. Raceme(s) solitary or very rarely two, 3-6 cm long; rachis flat, c. 0.5 mm wide. Spikelets oblanceate, 4-5 mm long, acuminate; lower glume linear lanceate, upper glume lanceate, acuminate, densely hairy; lower floret empty, upper floret bisexual; lower lemma 1-nerved, hyaline; upper lemma notched; awn 8-12 mm long.

Fl. & Fr.: Sep.-Dec. *Distr.*: Sri Lanka, India. Less common. On hill slopes. *NAK 2208* (Chittar, c. 300 m).

ECHINOCHLOA Palisot de Beauvois
Essai Agrost. 53. 1812.

Echinochloa stagnina (Retz.) P. Beauv., Essai Agrost. 53: 161 & 171. 1812; Fischer in Gamble, Fl. Pres. Madras 1731. 1934; Bor, Grass. Burma, Ceylon, India & Pakistan 31. 1960; Ramamoorthy in Sald. & Nicols., Fl. Hassan Dist. 726. 1976. *Panicum stagninum* Retz., Obs. Bot. 5:17. 1789; Rang. & Tad., Handb. S. Ind. Grass. f. 83. 1921. *P. crussgalli* auct., non L., 1753: Hook. f., Fl. Brit India 7: 30. 1896.

Annuals; culms ascending, rooting from lower nodes. Leaves flat, linear, 7-30 x 0.6-1 cm. Inflorescence of racemes, rachis setose, scabrid. Spikelets ovoid-ellipsoid, c. 5 mm long; glumes membranous, unequal; lower glume shorter; upper one as long as spikelet; lower lemma empty; awn to 5 cm long; upper lemma crustaceous; palea as long, with rounded sides; stamens 3. Caryopsis broadly ellipsoid.

Fl. & Fr.: Sep.-Feb. *Distr.*: Trop. Asia and Africa, Common in low-lying wet areas. *NAK 1957* (Tiruvalla, c. 10 m).

ELEUSINE J. Gaertner
Fruct. 1: 7. 1788.

Eleusine indica (L.) Gaertn., Fruct. 1:8. 1788; Hook. f., Fl. Brit. India 7: 293. 1896; Rang. & Tad., Handb. S. Ind. Grass. f. 206. 1921; Fischer in Gamble, Fl. Pres. Madras 1839. 1934; Bor, Grass. Burma, Ceylon, India & Pakistan 493. 1960; Ramamoorthy in Sald. & Nicols., Fl. Hassan Dist. 727. 1976; Matthew, Ill. Fl. Tam. Carnatic tt. 855 & 856. 1982; Britto & Matthew in Matthew, Fl. Tam. Carnatic 3(3): 1848. 1983; Manilal, Fl. Silent Valley 353. 1988; Mohanan & Sivad., Fl. Agasthyamala 817.2002. *Cynosurus indicus* L., Sp. Pl. 72. 1753.

Perennials; culms tufted. Leaves flat, linear, 10-25 by 0.4-0.6 cm, ligule membranous, ciliate; spikes digitate, secund, 2-4 in whorls. Spikelets sessile, ovoid, laterally compressed, 5-7-flowered; glumes persistent, lanceate, 1-nerved, lemmas shorter, 3-nerved, paleate; stamens 3. Caryopsis subglobose, deeply grooved on one side, enclosed by loose hyaline pericarp.

Fl. & Fr.: Dec.-Feb. *Distr.*: Tropical and subtropical regions. Common along waysides in lowlands. *NAK 918, 2822* (Tiruvalla, c.10 m).

ERAGROSTIS N.M. Wolf
Gen. Pl. 23. 1776; Gen. Sp. 63, 65. 1781.

Annuals; culms slender. Leaves oblong, flat or inrolled, often glandular; ligule a fimbriate rim. Inflorescence a terminal, open or contracted or spike-like panicle; spikelets laterally compressed; glumes subequal, shorter than the lowest lemma, 1-nerved; lemmas 3-nerved; palea slightly shorter than lemma, 2-keeled; stamens 1-3. Caryopsis minute, globose to ellipsoid.

1. Rachilla articulate, breaking up from the apex downwards. ....2. **E. tenella**
1. Rachilla non-articulate, persistent; lemmas falling off from the base upwards.
    2. Spikelets ovate or ovate-oblong, much-compressed; leaf-margin non-glandular. ....3. **E. uniloides**
    2. Spikelets linear or linear-oblong, slightly compressed; leaf-margin glandular. ....1. **E. cilianensis**

1. **Eragrostis cilianensis** (All.) Vign., Malpighia 18: 386. 1904; Fischer in Gamle, Fl. Pres. Madras 1827. 1934; Bor, Grass. Burma, Ceylon, India & Pakistan 503. 1960. *Poa cilianensis* All., Fl. Edem. 2: 246, t. 91, f. 2. 1785. *Eragrostis major* Host, Ic. Descr. Gram. Austriac. 4:14, t.24. 1809.

Culms to 1 m tall. Leaves flat, 15-30 by 0.5-1 cm, margin glandular. Panicle open, to 30 cm long. Spikelets linear-oblong, flat, c. 1.5 cm long; lemmas laterally strong-nerved.

Fl. & Fr.: Dec.-Mar. *Distr.:* Widely distributed in warmer regions of the world. Not common. In mountain marshes. *NAK 2571* (Kakki hills, c. 1000 m).

2. **Eragrostis tenella** (L.) P. Beauv. ex Roem. & Schult., Syst. 2: 576. 1817; Stapf in Hook.f., Fl.Brit. India 7:315. 1896, incl. var. *plumosa;* Bor, Grass. Burma, Ceylon, India & Pakistan 513. 1960; Mohanan & Sivad., Fl. Agasthyamala 819.2002. *Poa tenella* L., Sp. Pl. 69. 1753. *Poa plumosa* Retz., Obs.Bot. 4: 20. 1786. *Eragrostis plumosa* (Retz.) Link, Hort. Berol. 1: 192. 1827; Fischer in Gamble, Fl. Pres. Madras 1826. 1934.

Culms very slender, non-glandular. Leaves lanceate, 3-18 by 0.2-0.5 cm, glabrescent. Panicle loose, pyramidal, 5-10 cm long. Spikelets ellipsoid, c. 1 mm long; glumes obtuse, lemmas broadly ovate, palea 0.8 mm long.

Fl. & Fr.: Most of the seasons. *Distr.*: Old World tropics, now introduced in America. Common. From plains to hills. *NAK 570* (Ranni, c. 100 m).

3. **Eragrostis uniloides** (Retz.) Nees ex Steud., Syn. Pl. Glum. 1: 264. 1854; Fischer in Gamble, Fl. Pres. Madras 1826. 1934; Bor, Grass. Burma, Ceylon, India & Pakistan 515. 1960; Ramamoorthy in Sald. & Nicols., Fl. Hassan Dist. 730. 1976; Britto & Matthew in Matthew, Fl. Tam. Carnatic 3(3): 1858. 1983; Manilal, Fl. Silent Valley 354. 1988; Mohanan & Sivad., Fl. Agasthyamala 820.2002. *Poa uniloides* Retz., Obs. Bot. 5: 19. 1788. *Eragrostis amabilis* auct., non (L) Wight & Arn. ex Hook & Arn., 1838: Stapf in Hook.f., Fl. Brit. India 7: 317. 1896; Rang. & Tad., Handb. S. Ind. Grass. f. 218. 1921.

Culms slender, 15-50 cm tall. Leaves oblong, 2-8 cm long, flat or inrolled. Panicles narrowly oblong or broadly ovate, 6-30 cm long. Spikelets ovoid to oblong, 1 by 3.5 mm, straw-coloured, tinged with purple; lemma punctulate; palea falling with lemma.

Fl. & Fr.: Most of the seasons. *Distr.*: India, S.E. Asia. Very common. *NAK 12, 415, 908.* (Almost all places).

EULALIA Kunth
Rev. Gram. 1: 160. 1829.

Eulalia trispicata (Schult. & Schult. f.) Henr., Blumea 3: 453. 1940; Bor, Grass. Burma, Ceylon, India & Pakistan 157. 1960; Ramamoorthy in Sald. & Nicols., Fl. Hassan Dist. 732. 1976; Matthew, Ill. Fl. Tam. Carnatic tt. 870 & 871. 1982; Britto & Matthew in Matthew, Fl. Tam. Carnatic 3(3): 1860. 1983; Manilal, Fl. Silent

Valley 354. 1988; Mohanan & Sivad., Fl. Agasthyamala 822.2002. *Andropogon trispicatus* Schult. & Schult. f., Syst. Veg. 2: 452. 1824. *A. tristachyos* Roxb., Fl. Ind.1: 261. 1820, non Kunth, 1816. *Eulalia tristachya* (Roxb.) Kuntze, Rev. Gen. Pl. 2: 775. 1891; Fischer in Gamble, Fl. Pres. Madras 1715. 1934. *Pollinia argentia* Trin., Bull. Sci. Acad. Imp. Sci. 1: 71. 1836; Hook.f., Fl. Brit. India 7: 111. 1896.

Perennials; clums to 1 m tall, nodes glabrous. Leaves linear-lanceate, 20 by 0.5 cm, corrolate. Racemes 3-7, subdigitate, densely silky. Spikelets binate, similar; lower glume of sessile spikelet narrowly elliptic, subcoriaceous, 2-nerved, 2-keeled, long-setose without; upper glume 1-nerved; lower lemma epaleate; upper lemma deeply 2-fid, awn to 1.5 cm long; stamens 3. Caryopsis ellipsoid.

Fl. & Fr.: Feb.-Apr. *Distr.*: Sri Lanka, India, Pakistan, Myanmar, Nepal, China, Malesia to Australia. Not common. On rocky hills of upper hills. *NAK 419, 2040* (Chittar, Moozhiar etc.).

GARNOTIA A.T. Brongniart
in Duperrey, Voyage Monde Coquille, Bot. (Phan.) 2:132. 1832 ('1829').

Erect perennials. Leaves flat or convolute. Inflorescence of terminal panicles. Spikelets solitary or twin, articulated on the pedicel, 1-flowered; glumes 2, subequal, lanceate, cuspidate or awned; lemma entire or retuse at apex, awned, rarely muticous; palea often auricled at base, containing a bisexual floret; stamens 3. Caryopsis narrowly cylindric.

1. Panicles lax; racemes in each fascicle few, spreading. ....1. **G. courtallensis**

1. Panicles contracted; racemes in each fascicle a, ascending. ....2. **G. tenella**

1. Garnotia courtallensis (Arn. & Nees) Thw., Enum. Pl. Zeyl. 363. 1864; Hook.f., Fl. Brit. India 7: 244. 1869; Fischer in Gamble, Fl. Pres. Madras 1812. 1934; Bor, Grass. Burma, Ceylon, India & Pakistan 567. 1960; Manilal, Fl. Silent Valley 355. 1988. *Miquelia courtallensis* Arn. & Nees, Nov. Act. Nat. Cur. 19, Supp. 1: 179. 1843.

Culms tufted; nodes glabrous or sparsely bearded. Leaves flat or convolute, to 8 cm long, base narrowed, pubescent. Panicles lax, racemes in each fascicle few, spreading. Spikelets narrow-lanceoid, base bearded; glumes muticous or apiculate. Caryopsis ellipsoid.

Fl. & Fr.: Dec.-Feb. *Distr.*: Sri Lanka, S. India. Not common. On moist-shady hills. *NAK 2148* (Upper Moozhiar, c. 500 m).

2. **Garnotia tenella** (Arn. ex Miq.) Jan., Repert. Sp. Nov. Regni Veg.17:86. 1921; Matthew, Ill. Fl. Tam. Carnatic tt. 872 & 873. 1982; Britto & Matthew in Matthew, Fl. Tam. Carnatic 3(3): 1861. 1983; Manilal, Fl. Silent Valley 355.1988; Mohanan & Sivad., Fl. Agasthyamala 823.2002. *Berghustia tenella* Arn. ex Miq., Nieuwe Verh. Eerste Kl. Ned. Inst. Wetensch. Amsterdam 3(4): 34 1851. *Garnotia stricta* Hook.f., Fl. Brit. India. 7: 243. 1896, p.p., non Brongn., 1831; Fischer in Gamble, Fl. Pres. Madras 1812. 1934; Bor, Grass. Burma, Ceylon, India & Pakistan 569. 1960.

Culms tufted; nodes densely bearded. Leaves oblong-lanceate, 4-15 by 0.5-1 m. Panicles contracted; racemes in each fascicle α, ascending, spikelets oblong-linear, lower glume muticous, upper cuspidate; lemma α-dentate; palea oblong. Caryopsis ellipsoid.

Fl. & Fr.: Oct.-Nov. *Distr.*: Widely distributed in S.E. Asia. Less common. On exposed hills. *NAK* 2121 2198 (Chittar, c. 300 m).

HETEROPOGON Persoon
Syn. Pl. 2:533. 1807.

Heteropogon contortus (L.) P. Beauv. ex Roem. & Schult. in L., Syst. Veg. 2: 836. 1817; Fischer in Gamble, Fl. Pres. Madras 1743. 1934; Bor, Grass. Burma, Ceylon, India & Pakistan 163. 1960; Ramamoorthy in Sald. & Nicols., Fl. Hassan Dist. 734. 1976; Matthew, Ill. Fl. Tam. Carnatic tt. 876 & 877. 1982; Britto & Matthew in Matthew, Fl. Tam. Carnatic 3(3): 1862. 1983. *Andropogon contortus* L., Sp. Pl. 1045. 1753; Hook.f., Fl.Brit. India 7: 199. 1896; Rang. & Tad., Handb. S. Ind.Grass 207, ff. 162 & 163. 1921.

Perennials, to 60 cm tall. Leaves lanceate, 4-15 by 0.2-0.5 cm, scaberulous; ligule submembranous, fimbriate. Inflorescence a solitary raceme, to 10 cm long, awns forming a twisted spire; lower spikelets homogamous, awnless and male; upper ones awned, heterogamous, sessile ones female, pedicelled ones male or sterile. Sessile spikelet: callus oblique, bearded, lower glume involute, 7-nerved, hispid, upper glume 3-nerved, pilose; lower lemma nerveless, epaleate; upper lemma linear, awn to 7 cm long, epaleate; stamens 3. Caryopsis linear-cylindric.

Fl. & Fr.: Oct.-Nov. *Distr.*: Widely distributed in the tropics. Common in exposed midlands. *NAK 2042* (Ranni, c. 100 m).

IMPERATA Cyrillo
Pl. Rar. Neap. 2: (26). 1792.

Imperata cylindrica (L.) Raeusch., Nom. Bot. (ed. 3), 10.1797 var. **major** (Nees) Hubb. ex Hubb. & Vaughan, Grass. Maur. 96. 1940; Bor, Grass. Burma, Ceylon, India & Pakistan 170. 1960; Ramamoorthy in Sald. & Nicols., Fl. Hassan Dist. 734. 1976; Matthew, Ill. Fl. Tam. Carnatic tt. 880 & 881. 1982; Britto & Matthew in Matthew, Fl. Tam. Carnatic 3(3): 1864. 1983; Manilal, Fl. Silent Valley 355. 1988; Nicols., Suresh & Manilal, An Interpr. Hort. Malab. 310. 1988; Mohanan & Sivad., Fl. Agasthyamala 825.2002. *Lagurus cylindricus* L., Syst. Nat. ed. 10, 878. 1759. *Imperata koenigii* (Retz.) P. Beauv. var. *major* Nees, Fl. Afr. austral. ill. 89. 1841. *I. cylindrica* (L.) P. Beauv. var. *koenigii* (Retz.) Durand. & Schinz, Consp. Fl. Afric. 5: 694. 1894; Fischer in Gamble, Fl. Pres. Madras 1707. 1934. *I. arundinacea* Cyrillo, Pl. Rar. Neapol. 2: 27. 1792; Hook.f., Fl. Brit. India 7: 106. 1896, p.p.; Rang. & Tad., Handb. S. Ind. Grass. 147, f. 128. 1921. *Kodi-pullu* Rheede, Hort. Malab. 12:107, t.57. 1693.

Perennials; root stock stoloniferous, creeping, nodes bearded. Leaves mostly basal, lanceate, 3-10 by 0.2-0.5 cm, convolute, glabrescent; ligule membranous. Panicles to 15 cm long; rachis flexuous, jointed; callus bearded. Spikelets elliptic-lanceate, α-flowered, enveloped in long silky hairs; lower glume membranous, long pilose without; upper glume fringed at apex; lower lemma denticulate, epaleate; palea inrolled; stamens 3. Caryopsis linear.

Fl. & Fr.: Dec.-Mar. *Distr.*: Old World tropics. Not common. On waysides in hills. *NAK 2276* (Moozhiar, c. 250 m).

ISCHAEMUM Linnaeus
Sp. Pl. 1049. 1753.

Annuals or perennials with tufted culms. Racemes 1-10; digitate or paniculate; spikelets sessile or pedicelled and winged, with or without awns; racemes exserted or included; glumes winged or not; lodicules 2; stamens 3. Grains small, ellipsoid.

1. Raceme(s) 1-2, hairy, up to 4 cm long. ....2. **I. timorense**

1. Racemes always 2, rarely 3, glabrous, up to 10 cm long. ....1. **I. quilonensis**

1. **Ischaemum quilonensis** Ravi *et* Shaju, Rheedea 8(2): 152. 1998.

Creeping perennials with tufted culms. Leaves linear lanceate to linear oblanceate, to 35 mm long. Racemes 2, very rarely 3, up to 10 cm long; rachis glabrous to sparsely ciliate on the margins; sessile spikelets ovoid, subglabrous; lower glume keeled, 11-13-nerved; upper glume with scaberulous keel, 3-5-nerved. Pedicelled spikelet oblong ellipsoid; lower glume thickly coriaceous, 11-13-nerved; upper glume 5-7-nerved; lower lemma 3-nerved.

Fl. & Fr. Aug.-Mar. *Distr.:* Hitherto known only from Kollam district of Kerala. Collections from Adoor area are available at TBGT.

Common in marshy lands. This speceis is closely allied to *I. barbatum* Retz. and *I. mangaluricum* (Hack.) Stapf ex Fischer.

2. **Ischaemum timorense** Kunth, Revis. Gramin. 1: 369, t. 98. 1830; Hook.f., Fl. Brit. India 7: 136. 1896; Fischer in Gamble, Fl. Pres. Madras 1934, incl. var. *villosa* Bor, Grass. Burma, Ceylon, India & Pakistan 185. 1960.

Perennials; culms to 90 cm tall. Leaves convolute when young, eventually flat, 3-8 (15) x 0.6-1 cm; glabrous; leaf-sheaths auricled, petioled. Inflorescence of paired or digitate racemes. Spikelets binate; sessile spikelet: lower glume ribbed towards apex, hirsute, upper glume obscurely 2-lobed, keeled upwards, mucronate or cuspidate; lower lemma and palea as long as the glumes; upper lemma 2-fid; pedicelled spikelet equal to sessile; stamens 3. Caryopsis cylindric.

Fl. & Fr.: Feb.-Mar. *Distr.*: Sri Lanka, India to Malayan Isls. Common along waysides on hills. *NAK 2793* (Kakki hills, c. 1000 m).

JANSENELLA Bor
Kew Bull. 1955: 96. 1955.

Jansenella griffithiana (Muell.) Bor, Kew Bull. 1955: 98. 1955 & in Grass. Burma, Ceylon, India & Pakistan 426. 1960; Ramamoorthy in Sald. & Nicols., Fl. Hassan Dist. 738. 1976; Manilal, Fl. Silent Valley 357. 1988; Mohanan & Sivad., Fl. Agasthyamala 829.2002. *Danthonia griffithiana* Muell., Bot. Zeitung (Berlin) 14: 347. 1856. *Arundinella avenacea* Munro ex Thw., Eunm. Pl. Zeyl. 362. 1864; Hook.f., F. Brit. India 7: 69. 1896; Fischer in Gamble, Fl. Pres. Madras 1801. 1934.

Annuals; culms tufted, slender. Leaves lanceate, 2.5- x 0.6-1 cm, base cordate, apex acute, margin ciliate. Panicle congested into a crowded ovoid head. Spikelets 1-2-flowered; glumes membranous; lower shorter 3-5(7)-nerved, setose; lemmas unequal, lower 3-7-nerved; palea linear, 2-keeled, empty or enclosing a male or female floret; upper lemma pilose, involute with 2 apical setae, geniculate; palea enclosed in the lemma, 2-keeled; stamens 3. Caryopsis obovoid.

Fl. & Fr.: Sep.-Dec. *Distr.*: Sri Lanka, Peninsular and Eastern India. Not common. In upper hills. *NAK 2410* (Kakki hills, c. 1000 m).

LEERSIA O. Swartz
Prodr. 1: 21, 1788 (nom. cons.)

Leersia hexandra Sw., Nov. Gen. Pl. 21. 1788; Hook. f., Fl. Brit. India 7: 94. 1896; Rang. & Tad., Handb. S. Ind. Grass. 125, f. 119. 1921; Fischer in Gamble, Fl. Pres. Madras 1845. 1934; Bor, Grass. Burma, Ceylon, India & Pakistan 599. 1960; Matthew, Ill. Fl. Tam. Carnatic tt. 888 & 889. 1982; Britto & Matthew in Matthew, Fl. Tam. Carnatic 3(3): 1871. 1983.

Perennials; culms slender, ascending, to 1 m tall, nodes bearded. Leaves lanceate, 5-20 by 0.5-1 cm, scaberulous, setaceous; ligule to 4 mm long. Inflorescence of panicles. Spikelet(s) solitary, laterally compressed, 1-flowered; glumes absent or represented by an obscure frill; lemma ovate, coriaceous, 3-nerved, ciliate, 1-keeled; palea linear, 3-keeled, mid keel scabrid; stamens usually 6, rarely 3. Caryopsis compressed, oblong.

Fl. & Fr.: Oct-Feb. *Distr.*: Old World tropics & New World. Less common. In marshy lowlands and occasionally in hills. *NAK 2000* (Chittar, c. 300 m).

MICROSTEGIUM Nees
in Lindley, Nat. Syst. ed. 2. 447. 1836.

Microstegium ciliatum (Trin.) A. Camus, Ann. Soc. Linn. Lyon 68: 201. 1921; Fischer in Gamble, Fl. Pres. Madras 1716; 1934; Bor, Grass. Burma, Ceylon, India & Pakistan 193. 1960; Mohanan & Sivad., Fl. Agasthyamala 830. 2002. *Pollinia ciliata* Trim., Mem. Acad. Imp. Sci. St. Petersburg, Ser. 6, Sci. Math. 2: 307. 1832; Hook. f., Fl. Brit. India 7: 116. 1896.

Perennials; culms erect or decumbent, to 1 m tall. Leaves flat, 2.5-10 by 0.5-1.2 cm, base narrowed, petioled; ligule membranous. Inflorescence of fascicled spiciform racemes. Racemes 2-10, each to 7 cm long, rachis readily disarticulating, joints and pedicels long-ciliate. Spikelets binate, one sessile, other pedicelled; glumes of sessile spikelet c. 5 mm long, lower longitudinally chanelled, margin keeled, upper boat-shaped, keeled; lemmas hyaline, upper shorter, 2-fid, awn from sinus to 1 cm long; stamens 1, 2 or 3. Caryopsis cylindric.

Fl. & Fl.: Dec.-Feb. *Distr.*: India to Myanmar. Rare. On exposed hill tops. *NAK 2201* (Moozhiar, c. 250 m).

OCHLANDRA Thwaites
Enum. Pl. Zeyl. 376. 1864.

Ochlandra travancorica (Bedd.) Benth. ex Gamble, Ann. Roy. Bot. Gard. (Calcutta) 7: 125, t. 111. 1896; Mohanan & Sivad., Fl. Agasthyamala 831. 2002. *Beesha travancorica* Bedd., Fl. Sylv. 3: ccxxxiv, t. 324. 1873; Hook. f., Fl. Brit. India 7: 418. 1896; Fischer in Gamble, Fl. Pres. Madras 1863. 1934; Nicols., Suresh & Manilal, An Interpr. Hort. Malab. 311. 1988, *Nola-ily* Rheede, Hort. Malab. 5: 119-12 (no plate). 1685.

Shrubs; culms woody, reed-like, erect; sheath persistent. Leaves oblong, flat, 15-40 by 5-10 cm, a-veined, margin cartilaginous; sheaths striate, fringed. Inflorescence subverticillate spicate-panicle. Spikelets 5-6 cm long, 1-flowered; glumes 2-5, mucronate at apex. Lemmas similar to uppermost glume; paleas membranous; lodicules a; stamens 6-a, filaments united. Caryopsis ovoid, long-beaked, supported by persistent glumes.

Fl.: At long intervals and not observed during the present study. *Distr.*: Endemic to South India (Kerala), *NAK 2606* (Moozhiar, c. 250 m).

Form very dense populations along the forest floors, destructed forests, etc. A good indicator species of forest destruction. A heavily exploited species, particularly in Angamoozhy, Moozhiar and Kakki hills. The reeds are extensively used in paper industry and for making country baskets and mats.

Note: **Ochlandra scriptoria** (Dennst.) Fischer was reported from the district, but could not collect during the present study.

OPLISMENUS Palisot de Beauvois
Fl. Oware 2: 14. 1810 ('1807').

Annuals or perennials; culms diffuse or procumbent, rooting at lower nodes. Leaves flat, ovate to lanceate, base obliaque-veined. Inflorescence of simple or panicled spiciform-racemes. Spikelets solitary or fascicled, secund; glumes subequal, both or only lower awned; lemmas unequal, muticous, mucronate or aristatc; palca present or absent containing male floret or empty; upper lemma as long as lower, muticous palea containing a female floret; stamens 3. Caryopsis oblong-cylindric.

1. Annuals; leaves to 5 cm long; awns of glumes capillary. .... 1. **O. burmannii**
1. Perennials; leaves to 12 cm long; awns of glumes needle-like. .... 2. **O. compositus**

1. **Oplismenus burmannii** (Retz.) P. Beauv., Essai. Agrost. 54: 168. & 169. 1812; Hook.f., Fl. Brit. India 7: 68. 1896; Fischer in Gamble, Fl. Pres. Madras 1778. 1934; Bor, Grass. Burma, Ceylon, India & Pakistan 317. 1960; Ramamoorthy in Sald. & Nicols., Fl. Hassan Dist. 742. 1976. *Panicum burmannii* Retz., Obs. Bot. 3: 10. 1783.

Annuals; culms slender. Leaves ovate-lanceate, 2-5 by 1-1.5 cm, base rounded, apex acute. Racemes few, distant or close, forming short panicles. Spikelets c. 2 mm long, awns capillary, minutely scaberuolus.

Fl. & Fr.: Sep.-Dec. *Distr.*: Tropics. On moist exposed sandy areas in plains. Occasional. *NAK 2964* (c. 20 m).

2. **Oplismenus compositus** (L.) P. Beauv., Essai Agrost. 54: 168. 1812; Hook.f., Fl. Brit. India 7: 66. 1896; Fischer in Gamble, Fl. Pres. Madras 1778. 1934; Bor, Grass. Burma, Ceylon, India & Pakistan 317. 1960; Ramamoorthy in Sald. & Nicols., Fl. Hassan Dist. 742. 1976; Matthew, Ill. Fl. Tam. Carnatic tt. 901. & 902. 1982; Britto & Matthew in Matthew, Fl. Tam. Carnatic 3(3): 1877. 1983; Manilal, Fl. Silent Valley 358. 1988; Mohanan & Sivad., Fl. Agasthyamala 832. 2002. *Panicum compositum* L., Sp. Pl. 57. 1753.

Annuals; culms creeping. Leaves oblong or ovate-lanceate, 2.5-7 by 0.8-2 cm, sparsely pilose. Racemes c. 8, distant; rachis hollow. Spikelets secund, paired, distant, subsessile, dorsally compressed, c. 3.5 mm long; awns needle like. Caryopsis ellipsoid.

Fl. & Fr.: Sep.-Dec. *Distr.:* Old World & New World Tropics. Common along hill slopes. *NAK 2206* (Moozhiar, c. 250 m).

ORYZA Linnaeus
Sp. Pl. 333. 1753.

Oryza meyeriana (Zoll. & Mor.) Baill., Hist. des pl. xii: 166.1894. *Padia meyeriana* Zoll. & Mor., Syst. Verz. Zoll. 103. 1845-46. ssp. **granulata** (Nees & Arn. ex Watt) Tateoka, Bot. Mag. Tokyo 75: 460. 1962; Mohanan & Sivad., Fl. Agasthyamala 833.2002. *O. granulata* Nees et Arn. ex Watt, Dict. Econ. Prod. India 5:500. 1891; Hook.f., Fl. Brit. India 7: 93. 1896; Bor, Grass. Burma, Ceylon, India & Pakistan 604. 1960. *O. meyeriana* Baill., Hist. Pl. 12: 166. 1894; Fischer in Gamble, Fl. Pres. Madras 1845. 1934.

Annual herbs, to 90 cm tall. Leaves flat or convolute, 7-20 by 1-2 cm. Inflorescence of panicles, 4-10 cm long. Spikelets few, strongly laterally compressed, 1-flowered; glumes minute; lemma chartaceous, densely irregularly granulate, strongly α- nerved; palea as long, containing a bisexual floret; stamens 6. Caryopsis slender, cylindric.

Fl. & Fr.: Dec.-Feb. *Distr.*: S. India. Not common. In water-logged areas. *NAK* 65 (Moozhiar, c. 250 m).

Note: Different high-yielding hybrid varieties of **Oryza sativa L.,** are seen widely cultivated in the plains and hills replacing the traditional low-yielding but disease and pest resistant cultivars.

PANICUM Linnaeus
Sp. Pl. 55. 1753.

Annuals or perennials; culms often tufted. Leaves oblong-lanceate, flat; sheaths hispid; ligule membranous, fimbriate or with a tuft of pilose hairs. Inflorescence of panicles. Spikelet lanceate-ellipsoid, 2-flowered; glumes unequal, dissimilar, membranous; lower glume shorter than the upper glume; lower floret male or sterile; lemma similar to upper glume; palea often reduced or 0; upper floret bisexual, lemma crustaceous; stamens 3. Caryopsis ellipsoid.

1. Spikelets 4-5 mm long; leaves ovate-lanceate. ....1. **P. gardneri**

1. Spikelets to 3.5 mm long; leaves linear-lanceate. ....2. **P. repens**

1. **Panicum gardneri** Thw., Enum. Pl. Zeyl. 359. 1864; Fischer in Gamble, Fl. Pres. Madras 1783. 1934; Bor, Grass. Burma, Ceylon, India & Pakistan 326. 1960; Manilal, Fl. Silent Valley 358. 1988; Mohanan & Sivad., Fl. Agasthyamala 834.2002.

Slender herbs, to 1.5 m tall. Leaves ovate-lanceate to broadly lanceate, 6-16 by 1.5-5.25 cm, base rounded or cordate, glabrous. Spikelets to 5 mm long, dense; glumes silvery hyaline on the margins; upper glume and lower lemma subequal, larger than the lower glume and upper lemma.

Fl. & Fr.: Sep.-Oct. *Distr.:* Sri Lanka, S. India. Common. Along hills slopes. *NAK 342* (Ranni, c. 100 m).

2. **Panicum repens** L., Sp. Pl. (ed. 2), 8. 1762; Hook.f., Fl Brit. India 7: 49. 1896; Rang. & Tad., Handb. S. Ind. Grass. 99, ff. 102 & 103. 1921; Fischer in Gamble, Fl. Pres. Madras 1783. 1934; Bor, Grass. Burma, Ceylon, India & Pakistan 330. 1960; Ramamoorthy in Sald. & Nicols., Fl. Hassan Dist. 745. 1976; Britto & Matthew in Matthew, Fl. Tam. Carnatic 3(3): 1882. 1983; Mohanan & Sivad., Fl. Agasthyamala 835.2002.

Herbs to 1 m tall. Leaves linear-lanceate, strongly distichous, 7-15 by 0.5-0.7 cm, sparsely pubescent. Panicles 5-10 cm long, alternate. Spikelets ellipsoid, 2.5-3 mm long, lower glume suborbicular, 3-nerved, apex truncate, upper glume lanceate, 9-nerved, lower lemma 10-nerved, upper lemma indurated.

Fl. & Fr.: Jul.-Sep. *Distr.*: Tropics and subtropics of both hemispheres. Common along waysides in lowlands. *NAK 2812* (Adoor, c. 20 m).

PASPALUM Linnaeus
Syst. Nat. ed. 10, 855. 1359. 1759.

Annuals or perennials. Leaves linear-lanceate, flat or convolute. Inflorescence of 2 or 3 racemes, digitate or borne on a central axis; rachis flat with slender ridge in the middle. Spkelets 2-4- ranked, secund; lower glume usually absent, upper glume as long as spikelet, lower florets sterile; upper floret bisexual; lemma and palea crustaceous; stamens 3. Caryopsis plano-convex, enclosed by lemma and palea.

1. Spikelets to 2 mm long, fringed with fine white hairs from the margins. ....1. **P. conjugatum**
1. Spikelets to 2.5 long, glabrous or sparsely ciliate on the margins.
    2. Upper glume puberulous; spikelets elliptic; subaquatics. ....2. **P. distichum**
    2. Upper glume glabrous; spikelets ovoid-orbicular; terrestrials. ....3. **P. scrobiculatum**

1. **Paspalum conjugatum** Berg., Acta Helv. Phys.-Math. 7: 129, t.8. 1772; Hook.f., Fl. Brit. India 7: 11. 1896; Fischer in Gamble, Fl. Pres. Madras 1772. 1934; Bor, Grass. Burma, Ceylon, India & Pakistan 306. 1960; Ramamoorthy in Sald. & Nicols., Fl. Hassan Dist. 746. 1976; Manilal, Fl. Silent Valley 359. 1988; Mohanan & Sivad., Fl. Agasthyamala 837.2002.

Perennials. Leaves lanceate, glabrous. Racemes paired, rarely 3; rachis flat, margins cartilaginous, smooth or scaberulous. Spikelets apiculate or shortly acute, glumes ciliate with long pilose hairs.

Fl. & Fr.: Mar.-Oct. *Distr.:* Old World tropics. Common along waysides in hills. *NaK 95* (Moozhiar, c. 250 m), 856 (Plappally, c. 300 m).

2. **Paspalum distichum** L., Syst. Nat., ed. 10, 855. 1759; Hook.f., Fl. Brit. India 7: 12. 1896; Bor, Grass. Burma, Ceylon, India & Pakistan 338. 1960; Brummitt, Taxon 32: 281. 1983; Nicols., Suresh & Manilal, An Interpr. Hort. Malab. 312. 1988. *Digitaria paspaloides* Michaux, Fl. Bor.-Amer. 1: 46. 1803. *Paspalum paspaloides* (Michaux) Scribner, Mem. Torrey Bot. Club 5: 29. 1894. *Tereta-pullu* Rheede, Hort. Malab. 12:81, t.44. 1693.

Perennials. Leaves oblong-linear, 2-15 by 0.4-0. cm, flat or convolute. Racemes 2 or 3, to 9 cm long. Spikelets broadly ellipsoid to ovate-ellipsoid, 2.5-4.5 mm long; upper glume shortly pubescent without, 5-7-nerved; upper lemma 3-nerved.

Fl & Fr.: Mar.-Oct. *Distr.*: Tropics and subtropics of the world. Less common. In water-logged areas. *NAK 404* (Ranni, c. 100 m).

3. **Paspalum scrobiculatum** L., Mant. Pl. 1:29. 1767; Hook.f., Fl. Brit. India 7: 10. 1896; Fischer in Gamble, Fl. Pres. Madras 1772. 1934; Bor, Grass. Burma, Ceylon, India & Pakistan 340. 1960; Ramamoorthy in Sald. & Nicols., Fl. Hassan Dist. 747. 1976; Britto & Matthew in Matthew, Fl. Tam. Carnatic 3(3): 1887. 1983; Manilal, Fl. Silent Valley 360. 1988; Mohanan & Sivad., Fl. Agasthyamala 838.2002.

Annuals. Leaves lanceate or linear, convolute, glabrous. Racemes 2-5, rarely solitary, rachis broad, ridge produced at apex into 2 short rounded or acute lobes. Spikelets suborbicular, lower lemma 7-nerved, palea with wide membranous basal auricles.

Fl. & Fr.: Sep.-Dec. *Distr.*: Old World Tropics. Common along waysides in shades. *NAK 96* (Moozhiar), *661* (Pampa Valley, c. 100 m).

PENNISETUM L. C. Richard ex Persoon
Syn. Pl. 1: 72. 1805.

Pennisetum polystachyon (L.) Schult., Syst. Veg. Mant. 2: 146. 1824; Fischer in Gamble, Fl. Pres. Madras 1792. 1934; Bor, Grass. Burma, Ceylon, India & Pakistan 346. 1960; Manilal, Fl. Silent Valley 360. 1988; Mohanan & Sivad., Fl. Agasthyamala 838.2002. *Panicum polystachyon* L., Syst. Nat., ed. 10. 2: 870. 1759.

Stout annuals; culms erect, to 1.5 m tall. Leaves linear, flat. Inflroscence a spike-like panicle, to 20 cm long, rachis slender, angled. Spikelets 2-5 in a cluster, subtended by scabrid, involucral bracts; lower glumes often reduced or 0; upper glumes as long as spikelet; lower lemma with a male or sterile floret; upper lemma with a bisexual floret. Caryopsis cylindric, dorsally compressed.

Fl. & Fr.: Nov.-Dec. *Distr.*: Old World tropics. Common in lowlands and hills. *NAK 92* (Moozhiar, c. 250 m).

PEROTIS W. Aiton
Hortus Kew. 1: 85. 1789.

Perotis indica (L.) Kuntze, Rev. Gen. Pl. 2: 787. 1891; Fischer in Gamble, Fl. Pres. Madras 1814. 1934; Bor, Grass. Burma, Ceylon, India & Pakistan 611. 1960; Ramamoorthy in Sald. & Nicols., Fl. Hassan Dist. 748. 1976; Matthew, Ill. Fl. Tam. Carnatic tt. 912. & 913. 1982; Britto & Matthew in Matthew, Fl. Tam. Carnatic 3(3): 1889. 1983; Nicols., Suresh & Manilal, An Interpr. Hort. Malab. 312. 1988; Mohanan & Sivad., Fl. Agasthyamala 838.2002. *Anthoxanthum indicum* L., Sp. Pl. 28. 1753. *Perotis latifolia* Aiton, Hort. Kew. 1: 85. 1789; Hook.f., Fl. Brit. India 7: 98. 1896; Rang. & Tad., Handb. S. Ind. Grass. 137, ff. 124 & 125. 1921. *Tsjeria-kuren-pullu* Rheede, Hort. Malab. 12:117,t.62. 1693.

Annuals; culms slender, tufted, to 40 cm tall. Leaves lanceate, 1.5-3 by 0.2-0.5 cm, flat; ligule a membranous rim. Inflorescence a slender spike to 12 cm long. Spikelets linear, awned, spiral on rachis, 1-flowered; glumes subequal, narrow, 1-nerved; lower glume scabrid without, awn to 1.5 cm long; upper glume glabrous, awn to 1 cm long; lemma 1-nerved; stamens 3. Caryopsis linear.

Fl. & Fr.: Most of the seasons. *Distr.*: Sri Lanka, India, Nepal, Myanmar, S.E. Asia. Common in lowlands. *NAK 2167* (Tiruvalla, c. 10 m).

POGONATHERUM Palisot de Beauvois
Eassai Agrost. 56, 176. 1812.

Pogonatherum crinitum (Thunb.) Kunth, Enum. Pl. 1: 478. 1833; Hook.f., Fl. Brit. India 7: 14. 1896; Bor, Grass. Burma, Ceylon, India & Pakistan 200. 1960; Matthew, Ill. Fl. Tam. Carnatic tt. 918 & 919. 1982; Britto & Matthew in Matthew, Fl. Tam. Carnatic 3(3): 1891. 1983; Mohanan & Sivad., Fl. Agasthyamala 840.2002. *Andropogon crinitus* Thunb., Fl. Jap. 40, t. 7. 1784.

Slender annuals, nodes bearded. Leaves lanceate, flat, 1.5-5 by 0.2-0.4 cm, minutely scaberulous. Inflorescence of slender spiciform-racemes. Spikelets 2-nate, one sessile, the other pedicelled, 1-2-flowered. Sessile spikelet : callus bearded; lower glume 2-nerved, setose; upper glume ovate, 1-nerved, 1-keeled, setose, scaberulous without, awn to 1.5 cm long, upper lemma ovate, deeply 2-fid, awn to 1.5 cm long; stamen(s) 1(2). Pedicelled spikelet similar to sessile.

Fl. & Fr.: Feb.-Mar. *Distr.*: Afghanistan, India, Pakistan, Malesia & China. Not common. On rocky hills. *NAK 2721* (Moozhiar, c. 250 m).

PSEUDANTHISTIRIA (Hack.) Hooker f.
Fl. Brit. India 7: 219. 1896.

Pseudanthistiria umbellata (Hack.) Hook.f., Fl. Brit. India 7: 220. 1896; Fischer in Gamble, Fl. Pres. Madras 1749. 1934; Bor, Grass. Burma, Ceylon, India & Pakistan 204. 1960; Manilal, Fl. Silent Valley 361. 1988; Mohanan & Sivad., Fl. Agasthyamala 840.2002. *Andropogon umbellatus* Hack. in DC., Monogr. Phan. 6: 401. 1889.

Annuals; culms decumbent to erect, rooting at lower nodes. Leaves narrowly elliptic, 2-6 by 0.6-0.9 cm, glabrous. Inflorescence of subumbellate fascicle or racemes; subtended by a spathe. Spikelets 2-jointed, the basal with one sessile and one pedicelled spikelet; the upper with one sessile and 2-pedicelled spikelets. Sessile spikelet: callus bearded; glumes subequal, lower 7-nerved, back slightly sulcate on each side of the midrib; lemmas dissimilar, lower quadrate, upper stipitiform, enclosing a bisexual floret; stamens 3. Caryopsis linear. Pedicelled spikelet: glumes subequal, membranous; lemmas 0.

Fl. & Fr.: Nov.-Dec. *Distr.*: Sri Lanka, South India. Common in upper hills. *NAK 239, 370, 403* (Most of the hills).

PSEUDECHINOLAENA Stapf
in Prain, Fl. Trop. Africa 9: 494. 1919.

Pseudechinolaena polystachya (Kunth) Stapf in Oliver, Fl. Trop. Africa 9: 495. 1919; Fischer in Gamble, Fl. Pres. Madras 1766. 1934; Bor, Grass. Burma, Ceylon, India & Pakistan 352. 1960; Ramamoorthy in Sald. & Nicols., Fl. Hassan Dist. 749. 1976; Matthew, Ill. Fl.Tam. Carnatic tt. 920 & 921. 1982; Britto & Matthew in Matthew, Fl. Tam. Carnatic 3(3): 1892. 1983; Manilal, Fl. Silent Valley 362. 1988. *Echinolaena polystachya* Kunth, Nov. Gen. Pl. 1: 119. 1816. *Panicum uncinatum* Raddi, Agrost. Bras. 41. 1823; Hook.f., Fl. Brit. India 7: 58. 1896.

Slender annuals; nodes appressed pubescent. Leaves ovate-lanceate, 1.5-4.5 by 0.5-1 cm, base oblique, apex acute, pilose. Inflorescences of distant racemes. Spikelet ovoid, solitary, dorsally compressed; lower glume broadly ovate, 3-nerved, aristate; upper glume boat-shaped, 7-nerved, with rows of hooked bristles in between; lower lemma ovate, 3-nerved; palea convolute; uppr lemma crustaceous, 5-nerved, paleate; stamens 3. Caryopsis oblong, gibbous on the back.

Fl. & Fr.: Sep.-Dec. *Distr.*: Old World & New World tropics. Not common. In mountain marshes. *NAK 2197* (Moozhiar, c. 250 m).

RHYNCHELYTRUM C.G.D. Nees
in Lindley, Nat. Syst. ed. 2, 378, 446. 1836 ('*Rhynchelythrum*') corr. C.G.D. Nees, Fl. Afr. Austr. (xxi) (Errata). 64, 488. 1841.

Rhynchelytrum repens (Willd.) Hubbard, Kew Bull. 1934: 110. 1934; Bor, Grass. Burma, Ceylon, India & Pakistan 355. 1960; Cope in Nasir & Ali, Fl. Pakistan 143: 176. 1982; Matthew, Ill. Fl. Tam. Carnatic tt. 924 & 925. 1982; Britto & Matthew in Matthew, Fl. Tam. Carnatic 3(2): 1893. 1983; Mohanan & Sivad., Fl. Agasthyamala

841.2002. *Saccharum repens* Willd., Sp. Pl. ed. 4, 1(1): 322. 1798. *Monachyron villosum* Parl. in Hook., Niger Fl. 191. 1849. *Tricholaena wightii* Nees & Arn. ex Steud., Syn. Pl. Glumac. 1: 93. 1854, pro syn; Hook. f., Fl. Brit. India 7: 65. 1896. *Rhynchelytrum villosum* (Parl.) Chiov. in Annuario, Reale Inst. Bot. Roman. 8:310. 1908; Fischer in Gamble, Fl. Pres. Madras 1791. 1934; Bor, Grass. Burma, Ceylon, India & Pakistan 355. 1960.

Perennials; culms to 1 m tall. Leaves oblong, narrow-lanceate, 5-20 by 0.3-0.5 cm, flat. Inflorescence of compound or decompound panicles. Spikelets ovoid, laterally compressd, purple, long-villous, callus bearded; lower glume linear, bristly, without; upper glume gibbous, 5-nerved, keeled, densely bristly long ciliate along margins, beaked, emarginate, awn to 1.5 mm long. Lower lemma densely bristly without, ciliate, arista c. 2 mm long, paleate, upper lemma retuse, arista 0, paleate; stamens 3. Caryopsis narrow-ellipsoid.

Fl. & Fr.: Nov.-Dec. *Distr.*: Africa, India, Pakistan. Introduced to most other tropical countries. Along waysides in upper hills. Locally abundant. *NAK 1394* (Kakki hills, c. 100 m).

ROTTBOELLIA Linnaeus f.
Dissert. Nov. Gram. Gen. 22. 1779 (nom. cons.).

Rottboellia cochinchinensis (Lour.) Clayton, Kew Bull. 35: 817. 1981; Mohanan & Sivad., Fl. Agasthyamala 842.2002. *Stegosia cochinchinensis* Lour., Fl. Cochinch. 1: 51. 1790. *Rottboellia exaltata* L. f., Nov. Gram. Gen. 40, t.I. 1781, non (L.) L. f., 1779; Hook.f., Fl. Brit. India 7: 156. 1896; Rang. & Tad., Handb. S. Ind. Grass. f. 145. 1921; Fischer in Gamble, Fl. Pres. Madras 159. 1934; Bor, Grass. Burma, Ceylon, India & Pakistan 206. 1960. *Stegosia exaltata* (L.f.) Nash, North Amer. Fl.17, 84. 1909.

Annual or perennial tall herbs, rooting at lower nodes. Leaves oblong-linear, to 1 m long, flat, sheath densely pubescent. Inflorescence of solitary slender cylindric spiciform-racemes, rachis obliquely fragile, jointed. Spikelets 2-nate, sessile and pedicelled. Sessile spikelet: glumes paleate, subeqaul, coriaceous, lower convex and closing the cavity; upper boat-shaped; lemmas hyaline, paleate, lower usually enclosing a male floret, upper a bisexual one; stamens 3. Caryopsis ellipsoid. Pedicelled spikelet : glumes green, lemmas hyaline, both or the upper only enclosing a male floret

Fl. & Fr.: Sep.-Feb. *Distr.*: Tropical Asia, Africa, Malesia & Australia. Less common. On moist waysides in lowlands. *NAK 2041* (Tiruvalla, c. 10 m).

SCHIZACHYRIUM C.G.D. Nees
Agrost. Brasil. 331. 1829.

Schizachyrium brevifolium (Sw.) Nees ex Buese in Miq., Pl. Jungh. 359. 1854; Fischer in Gamble, Fl. Pres. Madras 1752. 1934; Bor, Grass. Burma, Ceylon, India & Pakistan 215. 1960; Matthew, Ill. Fl. Tam. Carnatic tt. 932. 1982., Britto & Matthew in Matthew, Fl. Tam. Carnatic 3(3): 1898. 1983; Mohanan & Sivad., Fl. Agasthyamala 844.2002. *Andropogon brevifolius* Sw., Nov. Gen. Pl. 26. 1788; Hook.f., Fl. Brit. India 7: 165. 1896.

Annuals; culms tufted, geniculate, nodes glabrous. Leaves linear, 1.5-3.5 by 0.3-0.5 cm, convolute. Inflorescence of single, spatheate slender racemes; internodes and pedicels linear. Spikelets binate, pedicelled and sessile, both falling together. Sessile spikelet: dorsally compressed, lower glume oblong-lanceate, ± 2-keeled, upper glume linear, lower lemma hyaline, upper lemma deeply 2-lobed, awn to 8 mm long, palea c. 2 mm long. Caryopsis linear. Pedicelled spikelet: male or sterile, smaller than the sessile.

Fl. & Fr.: Sep.-Nov. *Distr.*: Old World tropics. Less common. In moist rocky upper hills. *NAK 2196* (Chittar, c. 300 m).

SETARIA Palisot de Beauvois
Essai Agrost. 51, 178. 1812 (nom. cons.).

Annuals or perennials. Leaves flat or convolute or plicate. Inflorescence of open panicles or spike-like. Spikelets oblong to ovoid, ± planoconvex, awnless, usually subtended by 1 or more persisting bristles; glumes unequal, membranous, lower floret sterile or male; lower lemma thin; upper floret bisexual; upper lemma crustaceous; stamens 3. Caryopsis ellipsoid.

1. Leaves plicate; panicles to 20 cm long, effuse wih elongated lax branches. .... 1. **S. palmifolia**

1. Leaves flat; panicles to 12 cm long, spiciform with short contracted branches. .... 2. **S. pumila**

1. **Setaria palmifolia** (J. Koenig) Stapf, J. Linn. Soc. 42: 186. 1914; Fischer in Gamble, Fl. Pres. Madras 1789. 1934; Bor, Grass. Burma, Ceylon, India & Pakistan 363. 1960; Britto & Matthew in Matthew, Fl. Tam. Carnatic 3(3): 1900. 1983; Manilal, Fl. Silent Valley 363. 1988; Mohanan & Sivad., Fl. Agasthyamala 844.2002. *Panicum palmaefolium* J. Koenig, Naturforscher 23:208. 1788. *P. plicatum* Willd., Enum. Hort. Berol. 1033. 1809, non Lam., 1791; Hook.f., Fl. Brit. India 7: 55. 1896.

Perennials; culms stout to 1.5 m tall. Leaves oblong-lanceate, 20-25 by 2-4 cm, plicately folded, prominently nerved, pubescent. Panicles pyramidal; involucral bristle 1or 0. Spikelets ellipsoid or ovoid; lower glume ovate, 5-7- nerved, upper glume 5-9-nerved, lower lemma 5-nerved; palea hyaline, upper lemma indurated.

Fl. & Fr.: Nov.-Dec. *Distr.*: Old World tropics. Common in forest clearing. *NAK 407* (Adoor, c. 20 m).

2. **Setaria pumila** (Poiret) Roem. & Schult in L., Syst. Veg. 2: 891. 1817; Britto & Matthew in Matthew, Fl. Tam. Carnatic 3(3): 1901. 1983; Manilal, Fl. Silent Valley 363. 1988. *Panicum pumilum* Poir., Encycl. (Suppl.) 4: 273. 1816. *Panicum pallide-fuscum* Schum., Beskr. Guin. Pl. 58. 1827. *Setaria glauca* Hack., Bot. Soc. Brot. 3: 135. 1884, non P. Beauv., 1812; Hook.f., Fl. Brit. India 7: 78. 1896. *Setaria pallide-fusca* (Schum.) Stapf & Hubb., Bull. Misc. Inform.1930: 259. 1930; Fischer in Gamble, Fl. Pres. Madras 1789. 1934; Bor, Grass Burma, Ceylon, India & Pakistan 363. 1960.

Annuals; culms 10-50 cm, geniculate. Leaves lanceate, 4.5-12 by 0.4-0.7 cm, flaccid, flat. Inflorescence of spikes or spiciform panicles, compact; involucral birstles 9-11 per whorl. Spikelets 3 in a cluster or solitary; lower glume ovate, 1-3- nerved; upper glume 5-7-nerved, lower lemma 5-nerved, upper lemma indurated.

Fl. & Fr.: Apr.-Aug. *Distr.*: Old World tropics. Common on hills slopes. *NAK 2570* (Moozhiar, c. 250 m).

SPODIOPOGON Trinius
Fund. Agrost. 192, t.1. 1822 ('1820').

Spodiopogon rhizophorus (Steud.) Pilger in Engler & Prantl, Pfanzenf. ed. 2. 14 e: 119. 1940; Bor, Grass. Burma, Ceyon, India & Pakistan 246. 1960; Ramamoorthy in Sald. & Nicols., Fl. Hassan Dist. 753. 1976;

Mohanan & Sivad., Fl. Agasthyamala 845.2002. *Andropogon rhizophorus* Steud., Syn. Pl. Glumac. 1: 381. 1854. *Spodiopogon albidus* Benth., J. Linn. Soc. Bot. 19: 66. 1881; Hook.f., Fl. Brit India 7: 108. 1896. Fischer in Gamble, Fl. Pres. Madras 1934.

Perennials; culms tufted, often straggling, to 1 m tall. Leaves flat, 5-25 x 1-3 cm, flaccid, apex acuminate, to aristate, base narrowed into a petiole. Inflorescence a panicle. Spikelets 2-3-nate, one sessile, 1-2-pedicelled, 1-2-flowered; lower male or empty rachis and pedicels densely silky hairy; glumes subequal, membranous; lemmas hyaline; lower plicate, paleate or not; upper deeply 2-fid, with a long slender awn from the sinus; palea short; stamens 3. Caryopsis narrow-fusiform.

Fl. & Fr.: Nov.-Dec. *Distr.*: South India. On moist hill slopes of upper ghats. *NAK 2200* (Upper Moozhiar, c. 250 m).

SPOROBOLUS R. Brown
Prodr. 169. 1810.

Sporobolus indicus (L.) R. Br., Prodr. 170.1810. *Agrostris indica* L., Sp. Pl. 63.1753, var. **diander** (Retz.) ; Mohanan & Sivad., Fl. Agasthyamala 846.2002. *Agrostis diandra* Retz., Obs. Bot. 5: 18. 1788. *Sporobolus diander* (Retz.) P. Beauv., Ess. Agrost. 26: 147, 178. 1812; Jovet & Guedes, Taxon 22: 163. 1973. Hook.f., Fl. Brit. India 7: 247. 1896; Fischer in Gamble, Fl. Pres. Madras 1817. 1934; Bor, Grass. Burma, Ceylon, India & Pakistan 629. 1960; Ramamoorthy in Sald. & Nicols., Fl. Hassan Dist. 754. 1976; Britto in Matthew, Fl. Tam. Carnatic 3(3): 1905. 1983; Manilal, Fl. Silent Valley 364. 1988.

Perennial slender herbs; culms tufted, 20-60 cm tall. Leaves oblong, flat or convolute, 7-20 cm long. Inflorescence a loose pyramidal panicle. Spikelets ellipsoid, small, densely clustered, 1-flowered, rachilla disarticulating below the floret; glumes membranous, unequal, 1-nerved; lemmas ovate or oblong, muticous; palea as long, 2-nerved; stamens 2 or 3. Caryopsis oblong.

Fl. & Fr.: Sep-Oct. *Distr.*: Sri Lanka, India, Pakistan, Myanmar and Australia. Very common along waysides in lowlands and hills. *NAK 877* (Plappally, c. 300 m).

THEMEDA Forsskal
Fl. Aegypt.-Arab. 178. 1775.

Perennials; culms tufted, erect, stout. Leaves narrow-lanceate. Racemes clustered in large panicles with 2 homogamous pairs of sessile, persistent, involucral spikelets and 1-4 sessile spikelets with pedicelled ones. Sessile spikelet 2-flowered; glumes equal, coriaceous; lemmas dissimmilar, unequal, lower hyaline, epaleate, upper awned. Caryopsis linear-obovoid.

1. Bisexual spikelets often 2 in the raceme; involucral spikelets in superposed pairs. .... 1. **T. tremula**
1. Bisexual spikelet only 1 in the raceme; involucral spikelets all on the same level.
   2. Panicle decompound; awn of the spikelet up to 3 cm long. .... 3. **T. sabarimalayana**
   2. Panicle simple; awn of the spikelet up to 6 cm long. .... 2. **T. triandra**

1. **Themeda tremula** (Nees ex Steud.) Hack. in A. DC., Monogr. Phan. 6: 667. 1889; Fischer in Gamble, Fl. Pres. Madras 1746. 1934; Bor, Grass. Burma, Ceylon, India & Pakistan 254. 1960; Ramamoorthy in Sald.&

Nicols., Fl. Hassan Dist. 756. 1976. *Anthistiria tremula* Nees ex Steud., Syn. Pl. Glumac. 1:401. 1855; Hook.f., Fl. Brit. India 7: 214. 1896; Rang. & Tad., Handb. S. Ind. Grass. f. 165. 1921.

Culms slender, to 1 m tall. Leaves 7-50 x 0.3-0.5 cm. Panicle racemiform; involucral spikelets in superposed pairs, often 2 bisexual spikelets in the raceme; spathules finely tubercled-setose; lower glumes oblong, coriaceous, tubercled-setose; awn 2-3 cm long.

Fl. & Fr.: Sep.-Dec. *Distr.*: Sri Lanka, South India. Not common. Hills of upper ghats. *NAK 418* (Moozhiar, c. 250 m).

2. **Themeda triandra** Forssk., Fl. Aegypt.-Arab. 123, 178 & Florula 123. 1775; Fischer in Gamble, Fl. Pres. Madras 1746. 1934; Bor, Grass. Burma, Ceylon, India & Pakistan 254. 1960; Ramamoorthy in Sald. & Nicols., Fl. Hassan Dist. 756. 1976; Matthew, Ill. Fl. Tam. Carnatic tt. 943 & 944. 1982; Britto & Matthew in Matthew, Fl. Tam. Carnatic 3(3): 1908. 1983; Manilal, Fl. Silent Valley 366. 1988; Mohanan & Sivad., Fl. Agasthyamala 848.2002. *Anthistiria imberbis* Retz., Obs. Bot. 3: 11. 1783; Hook.f., Fl. Brit. India 7: 211. 1896.

Culms solid, to 2 m long. Leaves 15-20 x 0.4-0.7 cm, subcoriaceous. Panicle dense, elongate, to 60 cm long; spatheoles tuberculate-pilose; involucral spikelets all on the same level; glumes coriaceous; upper glume pubescent and scabrid without; awn 3-7 cm long.

Fl. & Fr. Nov.-Dec. *Distr.*: Old World tropics. Locally very abundant. On grassy hill slopes. *NAK 364* & *1346* (Kakki hills, c. 1000 cm).

3. **Themeda sabarimalayana** Sreekumar & Nair, Bull. Bot. Surv. India 29 (1-4): 127-128. 1987(1989); Sreekumar & Nair, Fl. Kerala Grasses 198. 1991.

Perennials, up to 2 m tall, erect; culms tufted; nodes glabrous. Leaves linear-lanceate, 15-40 x 0.5-1 cm. Panicles 30-60 cm long, sparsely hairy or glabrous, drooping. Involucral spikelets 2-pairs, lanceoid, 4-6 mm long; lower glume 9-13-nerved; upper glume 3-5-nerved. Sessile spikelets solitary, awned; pedicelled spikelets 3 or 4 in a raceme, narrowly ellipsoid, 4-6 mm long; grain slender.

Fl. & Fr.: Oct.-Dec. *Distr.*: So far known only from the type locality. In grasslands along the way to Sabarimala. The species could not be located during the present study.

TRIPOGON J. J. Roemer *et* J.A. Schultes
Syst. Veg. 2: 34, 600. 1817.

Tripogon bromoides Roem. & Schult. in L., Syst. Veg. 2: 600. 1817; Hook.f., Fl. Brit. India 7: 287. 1896; Fischer in Gamble, Fl. Pres. Madras 1834. 1934; Bor, Grass. Burma, Ceylon, India & Pakistan 521. 1960; Ramamoorthy in Sald. & Nicols., Fl. Hassan Dist. 757. 1976; Matthew, Ill. Fl. Tam. Carnatic t. 951. 1982; Britto & Matthew in Matthew, Fl. Tam. Carnatic 3(3): 1911. 1983; Manilal, Fl. Silent Valley 366. 1988.

Perennials; culms 20-50 cm tall. Leaves linear, 2-6 cm long, convolute. Inflorescence of solitary spikes, subsecund. Spikelets 2-seriate, sessile, lanceoid, c. 7-flowered, rachilla, jointed; glumes unequal, persistent, lower membranous, 1-nerved, oblique; upper shortly 2-toothed, aristulate, lemmas ovate, 3-nerved, apically 4-lobed, arista 1.5 mm long, palea with winged keels; stamens 3. Caryopsis narrow-ellipsoid.

Fl. & Fr.: Sep.-Oct. *Distr.*: Nepal, S. India. Less common. On rocky hills of upper ghats. *NAK 2123* (Chittar, c. 300 m).

UROCHLOA Palisot de Beauvois
Essai Agrost. 52. 1812.

Urochloa panicoides P. Beauv., Essai Agrost. 53, t.11, f.1. 1812; Fischer in Gamble, Fl.Pres. Madras 1775. 1934; Bor, Grass. Burma, Ceylon,India & Pakistan 372. 1960; Ramamoorthy in Sald. & Nicols., Fl. Hassan Dist. 757. 1976; Matthew, Ill. Fl. Tam. Carnatic tt. 953 & 954. 1982; Britto & Matthew in Matthew, Fl. Tam. Carnatic 3(3): 1912. 1983.

Annuals; culms to 30 cm long. Leaves linear to narrolwy lanceate, 4-12 by 1 cm. Racemes lax, pedicels softly bristly. Spikelet solitary, ovoid flat; lower glume broadly ovate, 3-5-nerved; upper glume ovate, margins clasping lemmas, membranous, 9-11-nerved; palea lanceate, keels winged; upper lemma ovate-suborbiculate, crustaceous, warty; palea crustaceous; stamens 3. Caryopsis linear.

Fl. & Fr.: Sep.-Dec. *Distr.:* Africa, India, Pakistan. Introduced in Australia. Less common. Moist open areas of lowlands. *NAK* 579 (Ranni, c. 100 m).

Saccharum officinarum L. is extensively cultivated in Pandalam, Tiruvalla, Vallicode-Kottayam, etc. Occasionally **Vetiveria zizanioides** (L.) Nash is cultivated by local people in hills.

GYMNOSPERMAE

146. CYCADACEAE

CYCAS Linnaues
Sp. Pl. 1188. 1753; Mant. Pl. 2: 166. 1771.

Cycas circinalis L., Sp. Pl. 1188. 1753; Hook.f., Fl. Brit. India 5: 656. 1888; Fischer in Gamble, Fl. Pres. Madras 1394. 1928; Gandhi in Sald. & Nicols., Fl. Hassan Dist. 31. 1976; Matthew & Rani in Matthew, Fl.Tam. Carnatic 3(3): 1917. 1983; Nicols., Suresh & Mani., An Intepr. Hort. Malab. 34. 1988. *Todda-panna* Hort. Malab. 3:9-14, t.13-21. 1682.

Trees with terminal crown of large pinnately compound leaves; dioecious. Microsporophylls densely aggregated to form large terminal cones. Megasporophylls loosely arranged, crowded round the apex of stem; ovules 1-5 on either side of sporophyll. Seeds globose.

Fl. & Fr.: Dec.-Feb. *Distr.:* Tropics. Locally abundant in certain moist deciduous and semi-evergreen forests. *NAK 723* (Maniyaar, c. 70 m).

147. PODOCARPACEAE

NAGEIA J.Gaertner
Fruct. 1:191. 1788.

Nageia wallichiana (Presl) Kuntze, Rev. Gen. Pl. 2:800. 1891; Laubenf., Blumea 32: 210. 1987. *Podocarpus wallichianus* Presl, Bot. Bemerk. 110. 1844; Fischer in Gamble, Fl. Pres. Madras 1393. 1931. *P. latifolia* Wall., Pl. Asiat. Rar. 1: 26, t. 30. 1830, non (Thunb.) R. Br., 1825; Hook.f., Fl. Brit. India 5: 649. 1888. *Decussocarpus wallichianus* (Presl) Laubenf. J. Arnold.Arb. 50: 349. 1969.

Large trees. Leaves opposite or subopposite, lanceate, to 15 cm by 3 cm, coriaceous.

'Flowering' or 'Fruiting' stage couldn't be observed during the present study. Rare. 3 to 4 trees located in Kakki hills. *NAK 2318* (Kakki hills c. 1000 m).

148. GNETACEAE

GNETUM Linnaeus

Syst. Nat. ed. 12, 2: 637. 1767; Mant. Pl. 1: 18. 125. 1767.

Gnetum edule (Willd.) Bl., Nov. Pl. Expos. 31. 1833; Robinson, Philipp. J. Sci. Bot. 7: 419. 1912; Nicols., Suresh & Mani., An Interpr. Hort. Malab. 34, 1988. *Thoa edulis* Willd., Sp. Pl. 4:477. 1805. *Gnetum scandens* Roxb., Hort. Beng. 66. 1814, nom. illegit (incl. type of *Thoa edulis* Willd., 1805). *Gnetum ula* Brongn. in Duperrey, Voy. Monde 7:12. 1829; Fischer in Gamble, Fl. Pres. Madras 1885. 1936; Maheswari & Vasil, Gnetum 10. 1961; Gandhi in Sald. & Nicols., Fl. Hassan Dist. 32. 1976; Matthew, Ill. Fl. Tam. Carnatic t. 961. 1982; Matthew, & Rani in Matthew, Fl. Tam. Carnatic 3(3): 1916. 1983. *Ula* Rheede, Hort. Malab. 7:41, t.22. 1688.

Dioecious liane. Leaves simple, decussate, elliptic-ovate or broadly elliptic, 6-12 by 3-8 cm, base rounded, apex acute, coriaceous. Strobili axillary from the axils of a pairs of basal, opposite bracts; collars cupular. Male strobilus: 4-5 cm long, stalked, 20-30 in a ring, exerted at maturity; stamen 1, microsporangia 2. Female strobilus with. ± 8 ovules around a node. 'Fruit' drupaceous, ellipsoid.

Fl. & Fr.: Mar.-Apr. *Distr.*: Peninsular India. Common in evergren forests and sacred groves. *NAK 2976* (Ranni, c. 100 m).

REFERENCES

Ahmedullah, M. & M. P. Nayar. 1986. *Endemic plants of the Indian region.* Botanical Survey of India, Calcutta.

Anilkumar, N., M. Sivadasan & M. K. Ratheesh Narayanan. 2001. A new species of *Dysoxylum* Blume (Meliaceae) from India. *Rheedea* 11(2): 115-118.

Arora, R. K. & E. Roshni Nayar. 1986. Wild relatives and related species of crop plants in India - their diversity and distribution. *Bull. Bot. Surv. India* 25:35-45.

Balgooy, M. M. J. van. 1987. *Collecting. In*: E.F. De Vogel (Ed.), *Manual of Herbarium Taxonomy: Theory and Practice*. Rijksherbarium, Leiden, The Netherlands.

Beddome, R. H. 1868-1874. *Icones Plantarum Indiae Orientalis,* Parts 1-15. Gantz Bros., Madras. (Repr. ed , 1972).

Bentham, G. & J. D. Hooker. 1862-1883. *Genera Plantarum*, Vols. 1-3. L. Reeve & Co., London.

Bor, N. L. 1960. *The Grasses of Bruma, Ceylon, India and Pakistan*. Pergamon Press, London.

Britto, S. J. & K. M. Matthew. 1983. *Crateva. In*: K.M. Matthew, *Flora of Tamilnadu Carnatic.* Rapinat Herbarium, Thiruchirapally. 3(1): 50-51.

Borssum Waalkes, J. van. 1966. Malesian Malvaceae revised. *Blumea* 14:1-213.

Bourdillon, T. F. 1908. *The Forest Trees of Travancore*. Government Press, Trivandrum.

Champion, H. G. & S. K. Seth. 1968. *A revised Survey of the Forest Types of India*. Delhi.

Clarke, C. B. 1879. *Ammunia ritchei. In*: J.D. Hooker, *Flora of British India*. L. Reeve & Co , London Vol. 2: 566.

Dassanayake, M. D. & F. R. Fosberg (Eds.). 1980-1991. *A revised Handbook to the Flora of Ceylon*. Vols. 1-7. Oxford & IBH Publishing Co. Pvt. Ltd., New Delhi.

Farr, E. R., J.A. Leussink, & F. A. Stafleu (Eds.). 1979. *Index Nominum Genericorum.* 3 Vols. Regnum Vegetabile, Utrecht, Netherlands.

Fosberg, F.R. & M.H. Sachet. 1965. *Manual for Tropical Herbaria.* *Regnum Vegetabile* 39. International Bureau for Plant Taxonomy and Nomenclature, The Netherlands.

Gamble, J. S. & C. E. C. Fischer, 1915-1936. *Flora of the Presidency of Madras.* 11 parts. Adlard & Son Ltd., London.

Gamble, J. S. 1916. *Salacia malabarica* – A new species of Hippocrateaceae. *Kew Bull.* 1916: 133.

Garcia de Orta. 1565. *Coloquios dos simples e drogas he cousas medicinais da India*. Goa.

Henry, A. N. & P. Daniel. 1988. In defence of the much-maligned (Alpha) taxonomy in India. *J. Bombay Nat. Hist. Soc.* 86:206-212.

Henry, A. N., M. Chandrabose, M. S. Swaminathan & N. C. Nair. 1984. Agastyamalai and its environs: A potential area for a biosphere reserve. *J. Bombay Nat. Hist. Soc.* 81: 282-290.

Hooker, J. D. 1872-1897. *The Flora of British India*, 7 Vols. L. Reeve & Co., London.

Hutchinson, J. 1926. *The families of Flowering Plants* - I. Macmillan & Co. Ltd., London.

Hutchinson, J. 1934. *The families of Flowering Plants* - II. Macmillan & Co., Ltd., London.

Hutchinson, J. 1959. *The Families of Flowering Plants* - Vol. I & II (Ed. 2). Clarendon Press, Oxford.

Hutchinson, J. 1973. *The Families of Flowering Plants* - II. *Monocotyledons*, (Ed. 3). Clarendon Press, Oxford.

Jacob, K. C. 1947. Some new species of South Indian plants. *J. Bombay Nat. Hist. Soc.* 47:50.

Jain, S. K. & R. R. Rao. 1977. *A handbook of Field and Herbarium methods*. Today & Tomorrow, New Delhi.

Janardhanan, K. P. 1979. Rediscovery of *Rotala ritchei* (C.B.Cl.) Koehne (Lythraceae) after one hundred years. *Bull. Bot. Surv. India* 21(1-4): 230-231.

Lawrence, G. H. M., A. F. G. Buchheium, G. S. Daniels & H. Dotezal. 1968. *Botanico Periodicum Huntianum* (B-P-H). Hunt Institute of Botanical Documentation, Pittsburgh, USA.

Manilal, K. S. 1988. *Flora of Silent Valley*. Mathubhoomi Press, Calicut.

Masters, M. T. 1874. *Grewia. In:* J.D. Hooker, *Flora of British India.* L. Reeve & Co., London. 1:383-393.

Matthew, K. M. 1983. *Flora of the Tamil Nadu Carnatic*, Parts 1-3. Rapinat Herbarium, Tiruchirapalli.

Mehr-Homji, 1982. A new classification of the phytogeographic zones of India. *Ind. J. Bot.* 7:224-233.

Mohanan, C. N. & N. C. Nair. 1982. *Kunstleria* Prain - A new genus record for India and a new species in the genus. *Proc. Ind. Acad. Sci. (Plant Sci.)* 90:207-209.

Mohanan, C. N. 1984. *Studies on the Flora of Quilon District, Kerala*. Ph.D. thesis, Madras University, Madras.

Mohanan, M.& N. Ravi. 1996. *Dimeria sivarajanii* (Poaceae) a new species from Kerala, India. *Rheedea* 6(2): 47-50.

Mohanan, M. & M. Sivadasan. 2002. *Flora of Agasthayamala*. Bishan Singh Mahendra Pal Singh, Dehra Dun, India.

Myers, Norman. 1990. The biodiversity challenge expanded: Hot-spots analysis. *The Environmentalist* 10(4): 243-255.

Nair, N. C. & P. Daniel. 1986. The floristic diversity of the Western Ghats and its conservation: A review. *Proc. Indian Acad. Sci. (Anim. Sci./Plant Sci.) Suppl. Nov.* 127-163.

Nayar, M. P. & A. R. K. Sastry, (Eds.). 1987. *Red Data Book of indian Plants,* Vol.1. Botanical Survey of India, Calcutta.

Nayar, M. P. 1977. Changing patterns of the Indian flora. *Bull. Bot. Surv. India* 19(1-4):145-155.

Nayar, M. P. 1980 (1982). Endemic flora of Peninsular India and its significance. *Bull. Bot. Surv. India* 22:12-23.

Nayar, M. P. 1996. *Hotspots of endemic plants of India, Nepal and Bhutan.* Tropical Botanic Garden and Research institute, Palode, Thiruvananthapuram, India.

Nicolson, D. H., C. R. Suresh, & K. S. Manilal, 1988. *An Interpretation of van Rheede's Hortus Indicus Malabaricus.* Regunum Vegetabile 119. Koeltz Scientific Books,Germany.

Paul, T. K. & M. P. Nayar, 1988. Malvaceae. *In*: M. P. Nayar *et al.* (Eds.), *Fascicles of Flora of India.* Botanical Survey of India, Calcutta.19: 64-233.

Radford, A. E., W. C. Dickinson, J. R. Massey & C. R. Bell, 1974. *Vascular Plant Systematics.* Harper & Row, New York.

Rama Rao, M. 1914. *Flowering Plants of Travancore.* Government Press, Trivandrum.

Raven, Peter H. 1988. Our diminishing tropical forests. *In*: E.O. Wilson & Francis M. Peter (Eds.), *Biodiversity.* National Academy Press, Washington, DC. pp.119-122.

Ravi, N. & N. Anil Kumar. 1990. *Julostylis polyandra* (Malvaceae) - A new species from India. *J. Bombay Nat. Hist. Soc.* 87: 260-262.

Ravi, N. & N. Anil Kumar. 1992. New and interesting species of *Dimeria* R.Br. (Poaceae) from Kerala, India. *Rheedea* 2(2): 101-107.

Ravi, N. & N. Mohanan. 1997. *Dimeria namboodiriana*, another new species of Poacaeae from Kerala, India. *Rheedea* 7(1): 1-4.

Ravi, N., N. Mohanan, T. Shaju, M. S. Kiran Raj & R. Rajesh. 1998. Three new species of *Ischaemum* L. (Poaceae) from Kerala, India. *Rheedea* 8(2): 149-158.

Reid, Walter. V, R. Kenton Miller, (Eds). 1989.Keeping the options alive: The scientific basis for conserving Biodiversity. World Resouce Institute. pp. 31-32.

Rheede tot Draakenstein, H.A. van, 1678-1703. *Hortus Indicus Malabaricus*, Vols. 1-12. Amsterdam (Repr. ed., 1983- Dehra Dun).

Saldanha, C. J. & D. H. Nicolson. 1976. *Flora of Hassan District, Karnataka, India.* Amerind Publishing Co. Pvt. Ltd., New Delhi.

Sasidharan, N. 2004. *Biodiversity documentation of Kerala.* Part 6. *Flowering plants.* KFRI Handbook. No. 17. Kerala Forest Research Institute, Peechi.

Sasidharan, N. & W. Vink. 1991. A new species of *Palaquium* Blanco (Sapotaceae) from India. *Blumea* 35: 385-387.

Sivadasan, M. & N. Mohanan, & C. Sathish Kumar. 1989. *Pothos crassipedunculatus* - a new species of *Pothos* Sect. *Allopothos* (Araceae) from India. *Pl. Syst. Evol.* 168:221-225.

Sivadasan, M. & N. Mohanan. 1991. *Ixora agasthyamalayana,* a new species of Rubiaceae from India. *Bot. Bull. Academia Sinica* 32: 313-316.

Sivadasan, M. 1989. *Amorphophallus smithsonianus* (Araceae), a new species from India, and a note on *A.* sect. *Synantherias. Willdenowia* 18: 435-440.

Sivadasan, M., N. Mohanan, & G. Rajkumar. 1993. *Amorphophallus bonaccordensis,* a new species of Araceae from India. *Blumea* 39: 295-299.

Sreekumar, P. V. & V. J. Nair. (1987) 1989. *Themeda sabarimalayana* (Poaceae), A new species from Kerala. *Bull. Bot. Surv. India* 29(1-4): 127-128.

Sreekumar, P. V. & V. J. Nair. 1991. *Flora of Kerala- Grasses*. Botanical Survey of India, Calcutta.

Stafleu, F. A. 1999. *International Code of Botanical Nomenclature*. Regnum Vegetabile, 97. Utrecht, The Netherlands.

Steenis, C. G. G. J. van, (Ed.). 1948-58. *Flora Malesiana* (Vols. 1,4,5,6). Sijthoff & Noordhoff International Publishers, The Netherlands.

Steenis, C. G. G. J. van, (Ed.). 1972-76. *Flora Malesiana* (Vols. 7,8). Sijthoff & Noordhoff International Publishers, The Netherlands.

Subramanyam, K. & M. P. Nayar. 1971. Plant taxonomy - Its past role and future lines of action in India. *Bull. Bot. Surv. India* 13:147.

Swaminathan, M. S. 1991. Biological diversity and biological productivity. *Biology Education*. (Jan.-Mar). 5-19.

Urban, I. 1883. *Monographie der Turneraceen.* Gebrueder Borntraeger, Berlin.

Wight, R. 1839-1853. *Icones Plantarum Indiae Orientalis*, Vols.1-6. Messr. Franck & Co. Madras. (Repr. ed.1963, Vols. 1-3, New York).

Wilson, E. O. 1988 . The current state of Biological diversity. *In* : E.O. Wilson & Francis M. Peter (Eds.), *Biodiversity.* National Academy Press, Washington, DC. pp.3-18.

INDEX

Ageratum conyzoides 21
A. integrifolia 8
A. lawii 157
A. nervosa 17
A. purpurea 17
Aana kotti-maram 99
Aattu-ilippa 295
Aattu-vanchi 447
Abelmoschus manihot 9, 78
Abelmoschus moschatus 9, 78
Abrus precatorius 10
Abutilon indicum 8, 78
Abutilon persicum 80
Abutilon polyandrum 80
Acacia 23
Acacia auriculiformis 23, 206
Acacia caesia 202
Acacia intsia 202
Acacia pennata 202
Acacia speciosa 203
Acacia torta 202
Acalypha paniculata 434
Acalypha racemosa 434
Acalypha spiciflora 440
Acampe ochracea 477
Acampe pracemorsa 477
Acampe wightiana 477
Acanthospermum hispidum 273
Acara-patsjotti Rheede 36
Achyranthes aspera 405
Achyranthes lanata 405
Achyranthes porphyristachya 405
Achyranthes prostrata 407
Aclisia indica 509
Aclisia secundiflora 509
Acorus calamus 303
Acroblastom ambavanense 431
Acrocephalus capitatus 397
Acrocephalus hispidus 397
Acrocephalus indicus 397
Acronychia laurifolia 110
Acronychia pedunculata 110
Acronychia pedunculata 9
Actephila excelsa 435
Actephila neilgherrensis 435
Actinodaphne campnulata 422
Actinodaphne hirsuta 422
Actinodaphne hookeri 422
Actinodaphne malabarica 422
Adaca-manjen 287
Adambea glabra 227
Adambea glabra 227
Adambea hirsuta 227
Adekku Panal 177
Adelia resinosa 450
Adenanthera pavonina 206
Adenia hondala 232
Adenosacme lawii 258
Adenosma capitatum 354
Adenosma indiana 354
Adenosma subrepens 354
Adenostemma lavenia 273

Adenostemma viscosum 273
Adhatoda wyanaadensis 383
Adina cordifolia 250
Adumbuvalil 343
Aegiceras minus 151
Aeginetia indica 365
Aerides cylindricum 478
Aerides decumbens 487
Aerides linearis 478
Aerides ringens 478
Aerides testacea 496
Aerva lanata 405
Aeschynanthus ceylanica 369
Aeschynanthus perrottettii 369
Aeschynanthus planiculmis 369
Aeschynomene americana glandulosa 22
Agalaia pervirdis 9
Aganosma roxburghii 311
Agave foetida 502
Ageratina adenophora 274
Ageratum conyzoides 182, 274
Ageratum haustonianum 21
Ageratum houstonianum 275
Aglaia elaeagnoidea 116
Aglaia polystachya 116
Aglaia roxburghiana 116
Agrostis diandra 588
Agrostris indica 588
Ahnathondi 96
Aidia gardneri 247
Akara-puda 208
Alakku-Cheru 148
Alangium salviifolium 244
Albizia chinensis 10, 203
Albizia lebbeck 203
Albizia marginata 203
Albizia stipulata 203
Alectoridia quartiniana 562
Aleurites moluccana 435
Allamanda cathartica 315
Allmania nodiflora 406
Alloteropsis cimicina 18, 560
Alpinia calcarata 497
Alpinia malaccensis 497
Alpinia neesana 500
Alseodaphne semecarpifolia 423
Alsina prostrata 67
Alstonia scholaris 8, 307
Alternanthera polygonoides 406
Alternanthera sessilis 406
Alternanthera tenella 406
Alternanthera triandra 406
Alysicarpus buplerifolius 11
Alysicarpus monilifer 17
Alysicarpus parviflorus 173
Amaranthus spinosus 407
Amaranthus viridis 407
Ambrosinia retrospiralis 525
Ambrosinia unilocularis 525
Ambulia aromatica 357
Amel-podi 265
Amischophacelus axillaris 8, 509
Ammania baccifera 8, 227
Ammania leptopetala 229
Ammania nana 228
Ammania pentandra 229
Ammania peploides 228
Ammania rosea 229
Ammania rotundifolia 229
Ammannia ritchiei 229
Amolago Rheede 418
Amomum cardamomum 499
Amomum zerumbet 500
Amorphophallus hohenackeri 523
Amorphophallus paeoniifolius 523
Amvetti 437
Ana-iruthi 469
Anakurunthoti 87

Anantali-maravara 481
Ana-paruva 526
Anaphalis aristata 275
Anaphalis beddomei 275
Anaphalis lawii 276
Anaphalis oblonga 276
Anaphyllum wightti 523
Anaschorigenam 470
Ana-schovadi 281
Ana-schunda 352
Ana-vinga 57
Anavinga ovata 57
Ancistrocladus heyneunus 75
Androgopon aciculatus 567
Andrographis atropurpurea 374
Andrographis echioides 380
Andrographis lineata 374
Andrographis macrobotrys 375
Andrographis neesiana 375
Andrographis producta 375
Andrographis wighitana 374
Andropoggon citratus 568
Andropogon brevifolius 586
Andropogon castratus 561
Andropogon contortus 578
Andropogon crinitus 584
Andropogon filiculmis 565
Andropogon flexuosus 568
Andropogon hackelii 567
Andropogon huegelii 565
Andropogon lancifolius 562
Andropogon microphyllus 562
Andropogon nardus 568
Andropogon rhizophorus 588
Andropogon trispiculus 577
Andropogon tristachyos 577
Andropogon umbellatus 585
Andropogon zeylanicus 567
Aneilema glaucum 516
Aneilema lineolatum 516
Aneilema montanum 510
Aneilema nudiflorum 516
Aneilema ovalifolium 510
Aneilema paniculatum 517
Aneilema pauciflorum 516
Aneilema scaberrimum 510
Aneilema scapiflorum 515
Aneilema semiteres 517
Aneilema zyelanica 517
Angelonia biflora 354
Angolam 244
Aniseia martinicensis 333
Anisochilus carnosus 397
Anisochilus verticullatus 398
Anisomeles indica 8, 10, 398
Anisomeles ovata 398
Annonaceae 25
Anogeissus latifolia 209
Anomospermum excelsum 435
Anotis decipiens 259
Anotis leschenaultiana 259
Ansjeli 461
Anthistiria imberbis 589
Anthistiria tremula 589
Anthocephalus cadamba 259
Anthocephalus chinensis 259
Anthoxanthum indicum 584
Antidesma acidum 436
Antidesma alexiteria 136
Antidesma ghaesembilla 436
Antidesma menasu 136
Antidesma menasu 9
Antidesma paniculatum 436
Antidesma pubescens 436
Antidesma zeylanicum 18, 436
Antidsma diandrum 436
Apama siliquosa 416
Aphanamixis polystachya 116

Apluda aristata 560
Apluda mutica 560
Apluda varia 560
Apocynum frutescens 311
Aporosa acuminata 437
Aporosa lindleyana 437
Arachis fruticosa 190
Aralia malabarica 243
Archidendron monadelphum 204
Archimonis sesamoides 355
Ardisia pauciflora 10, 294
Arealu 465
Areca catichu 521
Arenga wightii 518
Argostemma calycinum 260
Argyreia elliptica 335
Argyreia hirsuta 335
Argyreia malabarica 338
Aria-vela 56
Arisaema leschenaultii 524
Arisaema pulchrum 524
Arisaema tortuosum 524
Aristida setacea 561
Aristolochia indica 414
Aristolochia krisagathra 22, 415
Aristolochia lanceolata 414
Aristolochic roxburghiana 415
Aristolochia tagala 415
Arsitata 560
Artabotrys zeylanicus 9
Artanema longifolium 355
Artanema sesamoides 355
Artemisia indica 276
Artemisia nilagirica 276
Artemisia vulgaris 276
Arthraxon castratus 561
Arthraxon ciliaris 562
Arthraxon lancifolius 562
Arthraxon quartinianus 562
Arthraxon rudis 561
Artocarpus gomezianus 460
Artocarpus heterophyllus 460
Artocarpus hirsuta 8, 9
Artocarpus hirsutus 10, 461
Artocarpus intergrifolius 460
Artocarpus lakoocha 460
Artocarpus ponga 75
Arum colocasia 524
Arum companulatum sensu 523
Arum esculentum 524, 525
Arum flagelliforme 528
Arum ovatum 526
Arum tortuosum 524
Arum viviparum 527
Arun rumphii 523
Arundinella agrostoides 562
Arundinella avenacea 579
Arundinella ciliata 562
Arundinella fuscata 563
Arundinella holcoides 562
Arundinella mesophylla 17, 563
Arundinella nervosa 563
Arundinella purpurea 10, 563
Arundo bambos 564
Asclepias curassavica 322
Asclepias gigantea 315
Asclepias volubilis 322
Asjogam 200
Asparagus gonoclados 505
Asparagus racemosus 505
Asteracantha longifolia 380
Asystasia chelonoides 375
Asystasia dalzelliana 10, 376
Asystasia gangetica 376
Asystasia violacea 376
Asytasia coromandeliana 376
Atalantia racemosa 9, 111
Athampu-valli 159

Athi-meer-alou 466
Atragene zeylanica L. 35
Atylosia lineata 17, 21
Atylosia lineata 9
Avanacu 455
Avukkaram 148
Axonopus cimicinus 560
Axonopus compressus 563
B. barbatulum 481
B. latifolia 145
B. mollis 10
B. purpurea 17
Baccaurea courtallensis 9, 18, 437
Bacopa monnieri 355
Baeobotrys indica 294
Bahel suchulli 380
Bahel-schulli 380
Bahel-tsjulli 355
Balam-pulli 201
Balanophora 13
Balanophora abbreviata 22, 431
Balanophora fungosa 431
Balanophora indica 431
Balanophora polyandra 431
Baldingera glandulosa 265
Baleria prionitis 377
Baliospermum axilara 438
Baliospermum montanum 438
Baliospermum polyandrum 438
Baliospermum solanifolium 18
Baliospermum solanifolium 21, 438
Ballel Rheede 340
Ballota suaveolens 400
Balsamina minor 108
Bambos arundinacea 564
Bambusa urundinacea 564
Bambusa arundinacea giguntea 17
Bambusa bambos 564
Banalia thyrsiflora 408
Banisteria benghalensis 104
Banksea speciosa 497
Barleria courtallica 9, 377
Barleria cristata 377
Barleria longifolia 380
Barringtonia acutangula 219
Bassia elliptica 296
Bassia longifolia 295
Bassia malabarica 296
Bassia neriifolia 296
Batatas paniculata 342
Batta-valli 51
Batti-schorigenam 471
Bauhinia phoenicea 195
Bauhinia pupurea 201
Beenel 110
Beera-kaida 547
Beesha travancorica 580
Begonia malabarica 239
Begonia subpeltata 239
Beilschmiedia bourdillonii 423
Beilschmiedia fagifolia 423
Bela-pola 485
Bel-ericu 315
Beli-caraga 569
Belilla Rheede 258
Belutta tsjori-valli Rheede 134
Belutta-kaka-kodi 309
Belutta-modela-muccu 409
Bem-curini 381
Bem-schelli 254
Ben-patsja Rheede 330
Ben-Pavel 236
Bentinckia condapanna 518
Ben-tiru-tali 335
Berghustia tenella 577
Bhasma-valli 100
Bhootham-kolli 279
Bidens biternata 276

Bidens pilosa 276
Bignonia colais 371
Bingonia xylocarpa 371
Biophytum candolleanun: 105
Biophytum reinwardtii 104
Biophytum sensitivum 105
Bischofia javanica 438
Biti-maram-maravara 494
Bletia masuca 479
Blumea alata 278
Blumea barbata 278
Blumea lanceolaria 278
Blumea membranacea 10, 279
Blumea mollis 8, 279
Blumea neilgherrensis 279
Blumea spectabilis 279
Blumea virens 279
Blumea wightiana 279
Blyxa aubertii 474
Blyxa ceylanica 474
Blyxa echinosperma 8, 11, 474
Boehmeria glomerulifera 468
Boehmeria macrophylla 468
Boehmeria malabarica 468
Boehmeria platyphylla 468
Boerhavia diffusa 404
Boerhavia erecta 404
Boerhavia punarnava 404
Boerhavia repens 404
Bombax ceiba 9
Bombax ceiba 93
Bombax malabaricum 93
Bombax scopulorum 93
Bonamia semidigyna 336
Bonnaya brachiata 360
Bonnaya reptans 361
Bonnaya veronicifolia 359
Borago zeylanica 331
Borreria articularis 267
Borreria eradii 267
Borreria hispida 267
Borreria latifolia 267
Borreria ocymoides 268
Borreria pusilla 268
Borreria repens 268
Borreria rosea 268
Borreria stricta 268
Brachiaria eruciformis 564
Brachiaria ramosa 18
Bragantia wallichii 416
Brami Rheede 355
Brassica juncea 54
Breweria cordata 336
Breweria roxburghii 336
Breynia patens 439
Breynia retusa 439
Breynia rhamnoides 439
Breynia vitis-idaea 439
Bridelia retusa 439
Bridelia scandens 440
Bridelia zeylanica 447
Brucea sumatrana 114
Brunella 397
Bryonia amplexicaulis 236
Bryonia grandis 234
Bryonia maysorensis 238
Bryonia palmata 232, 235
Bryonia scabrella 236
Bryonopsis laciniosa 235
Bryophyllum calycinum 207
Bryophyllum pinnatum 207
Buchnera asiatica 364
Buchnera hispida 355
Bula 472
Bulbophyllum neilgherrense 9, 479
Bulbophyllum sterile 479
Bulbophyllum tremulum 479
Bulbostylis barbata 532

Burmannia coelestis 475
Burmannia coelestis pusilla 475
Bursinophyllum arboreum 244
Butea parviflora 17
Butea purpurea 17
C. aciculatus 10
C. apetalum 69
C. clarkei 9, 11
C. decipiens 69
C. esculenta 57
C. evolvuloides 11, 17
C. filicina 10
C. gummi-gutta 70
C. hirsuta 50
C. iria 8, 11
C. macrocarpus 51
C. magna 56
C. multiflora 17
C. nana 17
C. religiosa 56
C. salicifolia 17
C. semperflorens 167
C. subperfoliata 17
C. umbellata 17
C. vattayila 9
C. walkeri 17
C. wightianum 69
C. zeylanica 57
Cacalia sonchifolia 282
Cactus indicus 240
Cactus monocanthos 240
Cadel-avanacu 112
Caesalpinia cucullata 195
Caicotten-pala 445
Cajanus heynei 178
Cajanus lineatus 157
Cajanus scarabaeoides 157
Cajennaem 281
Calamus gamblei 519
Calamus hookerianus 9, 519
Calamus rotang 519
Calamus thwaitesii 9, 520
Calamus travancoricus 520
Calamus vattayila 9, 520
Calanthe emarginata 479
Calanthe masuca 479
Calcalia aurantiaca 284
Callicarpa lanata 390
Callicarpa tomentosa 390
Calonyction bona-nox 340
Calophyllum akara Burm. 36
Calophyllum calaba 11, 69
Calophyllum nagassarium 72
Calopogonium mucunoides 17
Calotropis gigantea 315
Calycopteris floribunda 10, 211
Cambogia G. gutta 70
Cambogia gummi-gutta 70
Camellia sinensis 73
Canarium strictum 115
Canavalia virosa 17, 160
Candolianum 105
Caniram Rheede 324
Canscora diffusa 325
Canscora grandiflora 326
Canscora lawii 325
Canscora perfoliata 326
Canscora roxburghii 326
Canscora wallichii 326
Cansjan-cora 326
Canthium angustifolium 248
Canthium coromandelicum 248
Canthium dicoccum 248
Canthium didymum 248
Canthium leschenaultii 248
Canthium parviflorum 248
Canthium parviflorum 248
Canthium rheedei 248

Canthium thyrsoideum 270
Canthium umbellatum 248
Capillipedium filiculme 565
Capitulum 30
Caplillipedium huegelii 565
Capraria crustacea 360
Carallia brachiata 208
Carallia integerrima 209
Carambu 231
Cardiospermum halicacabum 8
Careloe-vegon 414
Carex baccans 10, 532
Carex filicina 533
Carex ovata 544
Careya arborea 220
Carim curini 378
Carim-gola 508
Carim-tumba 354
Carissa spinarum 307
Carivillandi 507
Carpopogon monospermum 186
Caryea arborea 10
Caryota urens 520
Cascabela thevetia 315
Casearia anavinga 57
Casearia ovata 57
Cassia absus 10, 197
Cassia alata 10, 197
Cassia fistula 197
Cassia florrida 199
Cassia hirsuta 198
Cassia intermedia 17, 198
Cassia kleinii 198
Cassia mimosoides 198
Cassia occidentalis 199
Cassia rhombifolia 197
Cassia siamea 199
Cassia sophera 199
Cassia tora 199
Catharanthus pusillus 308
Catharanthus roseus 308
Catimbium malaccensis 497
Catri-conda 567
Cattu-carambu 231
Cattu-gasturi 78
Cattulli-pola 502
Cattu-molago 419
Cattu-picinna 235
Cattu-tirpali 418
Cattu-valli 51
Catunaregam spinosa 266
Cavalam 96
Cedrela toona 121
Celastrus opposita 126
Celastrus paniculatus 125
Celosia nodiflora 406
Celtis orientalis 459
Cenchrus lappaceus 565
Cenella asiatica 241
Centotheca lappacea 565
Centranthera hispida 356
Centranthera indica 356
Centranthera procumbens 356
Centratherum rangacharii 286
Centrosema pubescens 17
Cephalandru indica 234
Ceropegia biflora 316
Ceropegia candelabrum 316
Ceropegia fimbriifera 317
Ceropegia metziana 317
Ceropegia stocksii 317
Ceropegia tuberosa 316
Chadachi 99
Chailletia gelonioides 122
Chalcas paniculata 113
Chamatha 159
Champacam 39
Chasalia curviflora 249

Chassalia ophioxyloides 249
Chathura-mulla 304
Chavica roxburghii 418
Chavica sphaerostachya 418
Cheirostylis flabellat 480
Chemmarom 116
Chempoovam 141
Chen-mulli 377
Chennelii 420
Chenopodium ambrosioides 409
Cherkulathi 223
Cheru chokla 116
Cheru Rheede 146
Cherumaram 122
Cheru-mulli 307
Cheru-Pachotti 176
Cheru-pulladi 176
Cheru-thekku 390
Chilmoria pentandra 59
Chilocarpus atro-viridis 308
Chilocarpus malabaricus 9, 308
Chiococa malabarica 254
Chionanthus linocieroides 302
Chionanthus malabaricus 302
Chionanthus mala-elengi 9, 302
Chionanthus ramiflora 302
Chionanthus wightii 302
Chittila-madakku 142
Chloranthus brachystachya 419
Chloranthus grandifolius 420
Chloris barbata 566
Chlorophytum attenuatum 506
Chlorophytum heyneanum 506
Chlorophytum heynei 506
Chlorophytum laxum 506
Chomelia asiatica 268
Chonemorpha fragrans 309
Chonemorpha macrophylla 309
Chora-payin 420
Christisonia aurantiaca 366
Christisonia bicolor 366
Christisonia tubulosa 366
Chromolaena odorata 21, 182, 280
Chrysopogon aciculatus 8, 18, 566
Chrysopogon hackelii 567
Chrysopogon hacklei 10, 17
Chrysopogonzeylanicus 567
Chunnambu-valli 135
Ciadessa baccifera 118
Ciadessa fruticosa 118
Cicendia fastigiata 328
Cinnamomum iners 423
Cinnamomum malabatrum 9
Cinnamomum sulphuratum 424
Cinnmomum malabatrum 423
Cionnamometorum 543
Cipadessa baccifera 10
Cissus discolor Bl. 135
Cissus javana 9
Cissus leucostaphyla Dennst. 136
Cit-amerdu 52
Citavanacu 455
Citharexylum fruticosum 390
Citharexylum subserratum 390
Cladopus hookerianus 411
Claoxylon muricatum 450, 451
Clausena austroindica 111
Clausena heptaphylla 111
Cleidion javanicum 440
Cleidion spiciflorum 440
Cleome burmanni 55
Cleome monophylla 55
Cleome viscosa 8, 11, 55
Clerodendrum fragrans 391
Clerodendrum infortunatum 392
Clerodendrum paniculatum 391
Clerodendrum philippinum 391
Clerodendrum serratum 391

Clerodendrum viscosum 391
Clidemia hirta 21
Clidemia hirta 221
Clinogyne virgata 501
Clitoria virginiana 160
Tandale-cotti 164
Clutia androgyna 456
Clutia retusa 439
Clutia scandens 440
Clypea hermandiifolia 52
Coccinia grandis 234
Coccinia indica 234
Cocos nucifera 521
Codaga-pala 309
Codagen Rheede 242
Codariocalyx motorius 175
Coddam-pulli 70
Codi-avanacu 456
Coeloglossum densum 492
Coffea conephora 270
Coix lacryma-jobi 567
Coldenia procumbens 11, 329
Coletta-veeta 377
Coleus malabaricus 398
Colingil 191
Colocasia antiquorum 525
Colocasia esculenta 524
Columnea chinensis 357
Columnea longifolia 355
Coluppa Rheede 406
Combretum extensum 211
Combretum latifolium 11, 211
Comemlina obliqua 512
Commelina attenuata 511
Commelina axillaris 509
Commelina benghalensis 511
Commelina cristata 514
Commelina diffusa 512
Commelina edulis 515
Commelina ensifolia 512
Commelina erecta 512
Commelina japonica 516
Commelina kurzii 512
Commelina nudiflora 512, 516
Commelina paludosa 512
Commelina scaberrima 510
Commelina scapiflora 515
Commelina secundiflora 509
Commelina undulata 512
Communist-pascha 280
Conarus santaloides 151
Conna 197
Conocarpus latifolia 209
Conocephalus niveus 469
Convolvulus alsinoides 337
Convolvulus bicolor 338
Convolvulus cairicus 340
Convolvulus cymosus 348
Convolvulus dentatus 344
Convolvulus dissectus 345
Convolvulus emarginatus 333
Convolvulus hastatus 347
Convolvulus indicus 342
Convolvulus malabaricus 338
Convolvulus marginatus 342
Convolvulus martinicensis 333
Convolvulus nil 343
Convolvulus nummularius 338
Convolvulus obscurus 343
Convolvulus paniculatus 342
Convolvulus semidigynus 336
Convolvulus sublobatus 338
Convolvulus tridentatus 347
Convolvulus turpethum 347
Convolvulus umbellatus 348
Convolvulus uniflorus 333
Convolvulus vitifolius 348
Conyza cinerea 290

Conyza japonica 280
Corchorus acutangulus 98
Corchorus aestuans 98
Corchorus olitorius 98
Cordia cylindristachya 332
Cordia octandra 332
Coreopsis biternata 276
Corinti-panel 48
Corosinam Rheede 356
Cosmostigma racemosum 317
Costus speciosus 9, 497
Cottam Rheede 403
Cottonia macrostachya 480
Cottonia peduncularis 480
Cotylendon pinnata 207
Couradi 99
Couro-moelli 58
Cracea tinctoria 191
Crassocephalum crepidioides 284
Crateva nurvala 56
Crepis acaulis 280
Crotalaria 17
Crotalaria barbata 17
Crotalaria calycina 11
Crota'aria clarkei 17
Crotalaria macrophylla 179
Crotalaria pallida 10
Crotalaria sericea 17
Crotalaria striata 164
Crotalaria walkeri 9
Croton aromaticus 441
Croton bonpulndianus 441
Croton bonplandianus 21
Croton caudatus 441
Croton klotzschianus 441
Croton laccifer 441
Croton macrophyllus 444
Croton malabaricus 18
Croton malabaricus 442
Croton reticulatus 442
Croton solanifolius 438
Croton sparsiflorus 441
Croton tiglium 442
Croton zeylanicus 442
Cryptocoryne 525
Cryptocoryne retrospiralis 11, 525
Cryptolepis buchanani 319
Cryptophragmium canescens 379
Crytocoryne unilocularis 525
Cucculus glaucescens 51
Cucumella silentvalleyii 22, 234
Cucumis maderaspaanus 236
Cullenia exarillata 9, 94
Cullenia excelsa 94
Cumbulu Rheede 392
Cupa-vela 308
Cupi Rheede *268*
Curculigo finlaysoniana 503
Curcuma coriacea 498
Curcuma ecalcarata 498
Curcuma zedoaria 498
Curucligo orchioides 503
Curutu-pala 313
Cuscuta chinensis 336
Cuscuta reflexa 336
Cyanotis arachnoidea 513
Cyanotis arcotensis 513
Cyanotis axillaris 509
Cyanotis cristata 513
Cyanotis lanceata 514
Cyanotis papilionacea 513
Cyanotis pilosa 514
Cyanotis tuberosa 514
Cyanotis villosa 514
Cyathula prostrata 407
Cycas circinalis 590
Cyclea peltata 8
Cyclostemon macrophyllus 444

Cylicodaphne floribunda 425
Cymbopogon citratus 21, 568
Cymbopogon flexuosus 568
Cymburus urticaefolius 394
Cynanchum callialatum 319
Cynanchum indicum 320
Cynodon dactylon 569
Cynodon ternatus 572
Cynosurus aegyptius 570
Cynosurus indicus 575
Cyoperus cyperinus 551
Cyper difformis 535
Cyper distans 537
Cyperus brevifolius 549
Cyperus castaneous 534
Cyperus cephalotes 534
Cyperus clarkei 551
Cyperus compressus 534
Cyperus corymbosus 534
Cyperus cuspidatus 535
Cyperus cyperoides 551
Cyperus difformis 11
Cyperus diffusus 535
Cyperus distans 535
Cyperus dubius 551
Cyperus exaltatus 537
Cyperus flavidus 538, 552
cyperus globosus 552
Cyperus haspan 8, 10
Cyperus iria 8, 537
Cyperus kyllinga 549
Cyperus malaccensis 537
Cyperus melanospermus 549
Cyperus monostachyos 544
Cyperus nitens 553
Cyperus nutans 537
Cyperus paniceus 552
Cyperus polystachyos 552, 553
Cyperus procerus 538
Cyperus pumilus 553
Cyperus puncticulatus 553
Cyperus rotundus 538
Cyperus stramineus 554
Cyperus tenuispica 538
Cyperus triceps 550
Cyperus uncinatus 535
Cyperushaspan 537
Cyrtococcum muricatum 569
Cyrtococcum oxyphyllum 569
Cystis pinnatus 187
D. brevipes 17
D. championi 168
D. heterocarpon 9
D. heterophyllum 9
D. heyneana 169
D. kurumthotticalana 18
D. lawsoni 18
D. microphyllum 17
D. mooneyi 18
D. namboodiriana 18
D. ornithopoda 18
D. paniculata 9
D. rostrata 168
D. rufescens 173
D. scandens 17
D. sivarajanii 18
D. thothathrii 17
Desmodium triflorum 8, 10
D. velutinum 9
D. wynaadense 174
Dactyloctenium aegyptium 8, 18, 570
Daedalacanthus montanus 378
Dalbergia latifolia 8, 17
Dalbergia pseudo-sisso 11,17, 22
Dalzellia ramosissima 411
Dalzellia zeylanica 412
Dandha-pathri 126
Danthonia griffithiana 579

Datura arborea 349
Datura stramonium 349
Debregeasia ceylanica 10, 469
Debregeasia longifolia 469
Debregeasia velutina 469
Decaneurum dendigulense 291
Decussocarpus wallichianus 590
Deloni regia 201
Dendrobium album 481
Dendrobium aqueum 481
Dendrobium chlorops 481
Dendrobium dalzellii 482
Dendrobium macrostachyum 481
Dendrobium nanum 481
Dendrobium ovatum 481
Dendrocnide sinuata 469
Dendrophthoe elasticus 428
Derris brevipes 17
Derris canarensis 11,17, 22
Derris thothathrii 17
Desmodium 17
Desmodium cephalotes 176
Desmodium congestum 176
Desmodium ferrugienum waynadense 17
Desmodium ferrugineaum ferrugineum 17
Desmodium ferrugineum 9
Desmodium gyrans 175
Desmodium heterophyllum 8, 10
Desmodium latifolium 177
Desmodium microphyllum 9
Desmodium motorium 17
Desmodium parviflorum 173, 175
Desmodium polycarpum 174
Desmos ramarowii 45
Dialium coromandelicum 147
Dianella ensifolia 506
Diatoma brachiata 209
Dicanthium filiculme 565
Dicanthium huegelii 565
Dicerma pulchellum 176
Diceros caespitosus 360
Dichaespermum juncoides 517
Dichaespermum paniculatm 517
Dichaetaria wightii 570
Dichapetalum gelonioides 9, 121
Dichopsis elliptica 297
Dichrocephala integrifolia 281
Dichrocephala latifolia 281
Dicliptera cuneata 377
Dicraea algaeformis 412
Dicraea stylosa 412
Dictyospermum montanum 510
Dictyospermum ovalifolium 510
Dictyospermum protensum 510
Didymocarpus humboldtianus 370
Digitaria adscendens 8, 571
Digitaria ciliaris 571
Digitaria longiflora 571
Digitaria marginata 571
Digitaria paspaloides 583
Digitaria radicosa 571
Digitaria sanguinalis 571, 572
Digitaria setigera 572
Digitaria ternata 572
Digitaria timorensis 571
Dillenia pentagyna 10
Dimeria idukkiensis 17, 18
Dimeria idukkiensis 573
Dimeria kurumthotticalana 18, 573
Dimeria lawsonii 573
Dimeria mooneyi 18, 22, 573
Dimeria namboodiriana 574
Dimeria ornithopoda 10, 574
Dimeria pusilla 573, 574
Dimeria sivarajanii 574
Dimeria thwaitesii 574
Dimocarpus longan 9, 10
Dimorphocalyx lawianus 442

Dioscorea bulbifera 504
Dioscorea daemona 504
Dioscorea hispida 504
Dioscorea oppositifolia 504
Dioscorea pentaphylla 504
Dioscorea sativa 504
Dioscorea tomentosa 505
Diospyros bourdilloni 9
Diospyros bourdillonii 297
Diospyros canarica 298
Diospyros candolleana 9, 298
Diospyros embryopteris 298
Diospyros hirsuta 298
Diospyros malabarica 298
Diospyros nilgirica 298
Diospyros paniculata 300
Diospyros peregrina 298
Diplocyclos palmatus 235
Dipteracanthus prostatus 378
Dipterocarpus bourdilloni 74
Dipterocarpus bourdilloni 9
Dipterospermum personatum 371
Dithyrocarpus rothii 514
Dolichos falcatus 177
Dolichos gigantea 186
Dolichos phaseoloides 188
Dolichos pruriens 186
Dolichos scarabaeoides 157
Dolichos trilobus 17
Dolichos umbellatus 193
Dolichos virosa 160
Dondisia leschenaultii 248
Dopatrium junceum 356
Doritis wightii 487
Dracaena ensifolia 506
Dracaena terminalia 502
Dracaena terniflora 502
Dracontnium poeniifolius 523
Dregea volubilis 322
Drosera burmanni 11
Drosera burmannii 21, 208
Drosera indica 208
Drosera peltata 208
Drupatris cochinchinensis 300
Drymaria cordata 67
Drymaria diandra 67
Drypetes oblongifolia 18, 444
Dumasia congesta 178
Dysoxylum malabaricum 118
Dysoxylum swaminathanianum 22, 119
Dyzoxylum malabaricum 9
E. quinquangulare 11
E. tetragonum 11
E. thwaitesii 11
E. uniloides 10
Easwara-mulla 414
Ebermaiera glauca 388
Ebermaiera zeylanica 388
Ecbolium linneanum 378
Ecbolium viride 378
Echinochloa stagnina 575
Echinolaena polystachya 585
Echites caryophyllata 311
Echites fragrans 309
Echites macrophylla 309
Echites pubescens 309
Echites scholaris 307
Eclipta alba 11, 281
Eclipta prostrata 281
Edangkorana 371
Ehretia canarensis 332
Ehretia laevis 332
Ehretica cuneata 331
Eichhornia crassipes 8, 11, 21, 508
Elaeagnus conferta 427
Elaeagnus indica 427
Elaeagnus latifolia 427
Elaeocarpus munronii 102

Elaeocarpus serratus 102
Elaeocarpus tuberculatus 9, 102
Ela-pola 484
Elatostema lineolatum 470
Elengi Rheede 296
Eleocharis chaetaria 540
Eleocharis retroflexa 11, 540
Eleocharis spiralis 540
Elephantopus scaber 281
Elettaria cardamomum 499
Eleusine aegyptia 570
Eleusine indica 8, 575
Ellertonia rheedei 311
Elsholtzia paniculata 403
Elytranthe loniceroides 429
Embelia glandulifera 294
Embelia ribes 294
Emblica officinalis 453
Emilia sonchifolia 282
Empusa paradoxa 488
Ene-pael 474
Ensete superbum 501
Entada rheedei 11, 204
Entada scandens 204
Epaltes divaricata 282
Epicarpurus orientalis 467
Epicarpus spinosus 467
Epidendrum concretum 494
Epidendrum praemorsum 477
Epidendrum sterile 479
Eragrostis amabilis 576
Eragrostis cilianensis 10, 576
Eragrostis major 576
Eragrostis plumosa 576
Eragrostis tenella 576
Eragrostis uniloides 18, 576
Eranthemum capense 378
Eranthemum malabaricum 386
Eranthemum montanum 378
Eria dalazellii 10, 482
Eria nana 482
Ericaulon thwaitesii 530
Ericu 315
Erigeron alatum 278
Erigeron canadensis 283
Erigeron japonicum 280
Erigeron molle 279
Erima-pavel 236
Eriocaulon cinereum 529
Eriocaulon dianae 11, 530
Eriocaulon quinquangulare 530
Eriocaulon sieboldianum 529
Eriocaulon truncatum 530
Eriocaulon xeranthemum 531
Ervatamia caudata 314
Ervatamia heyneana 313
Erycibe paniculata 337
Erycibe wightiana 337
Erythrina stricta 17
Erythropalum populifolium 9, 122
Erythroxylum monogynum 103
Ethulia divaricata 282
Ethulia sparganophora 288
Eucalyptus 23
Eucalyptus grandis 213
Eucalyptus tereticornis 213
Eugenia arnottiana 216
Eugenia caryophyllaea 215
Eugenia hemispherica 216
Eugenia jambolana 216
Eugenia laeta 217
Eugenia lissophylla 219
Eugenia malabarica 217
Eugenia mooniana 214
Eugenia mundagam 217
Eugenia occidentalis 217
Eugenia pauciflora 217
Eugenia rottleriana 214

Eugenia rubicunda 219
Eugenia spicata 219
Eugenia thwaitesii 214
Eugeniodes wynadense 301
Eulalia trispicata 10, 576
Eulalia tristachya 577
Eulophia andersonii 484
Eulophia macrostachya 484
Eulophia nuda 484
Eulophia nuda andersonii 484
Eulophia pulchra 484
Eulophia spectabilis 484
Euodia lunu-ankenda 111
Euonymus crenulatus 10, 126
Eupatorium adenophorum 274
Eupatorium cordatum 286
Eupatorium divergens 290
Eupatorium glandulosum 274
Eupatorium odoratunm 280
Euphorbia geniculata 444
Euphorbia heterophylla 444
Euphorbia heyneana 445
Euphorbia hirta 445
Euphorbia microphylla 445
Euphorbia nivulea 446
Euphorbia pilulifera 445
Euphorbia prunifolia 444
Euphorbia pulcherrima 446
Euphorbia thymifolia 445
Euphorbia tirucalli 446
Euphorbia tithymaloides 451
Euphorbia vajravelui 445
Euphoria longana 141
Eurya japonica 73
Evodia roxburghiana 111
Evolvulus alsinoides 337, 338
Evolvulus alsinoides var. decumbens 31
Evolvulus decumbens 338
Evolvulus hederaceus 347
Evolvulus hirsutus 338
Evolvulus nummularius 338
Exacum bicolor 328
Exacum courtallense 326, 327
Exacum lawii 327
Exacum sessile 11, 327
Exacum tetragonum 328
F. congesta 9
F. littoralis 11
F. miliacea 11
F. virosa 457
Faboideae 17
Fagara lunu-akenda 111
Fagraea ceilanica 323
Ficus amplissima 462
Ficus arnottiana 462
Ficus asperrima 464
Ficus benghalensis 463
Ficus calosa 463
Ficus drupacea 463
Ficus exasperata 464
Ficus gibbosa 466
Ficus heterophylla 11, 464
Ficus hispida 8, 10, 464
Ficus microcarpa 465
Ficus mysorensis 463
Ficus oppositifolia 464
Ficus parasitica 466
Ficus religiosa 465
Ficus retusa 465
Ficus scabrella 464
Ficus talboti 465
Ficus tinctoria 466
Ficus travancorica 466
Ficus tsjakela 8, 466
Fimbristylis aestivalis 541
Fimbristylis angamoozhiensis 541
Fimbristylis argentea 543
Fimbristy'is cinnamometorum 11

Fimbristylis connamometorum 543
Fimbristylis cymosa 543
Fimbristylis cyperoides 543
Fimbristylis dichotoma 8, 10, 11, 544
Fimbristylis difforinis 8
Fimbristylis diphylla 544
Fimbristylis hospan 8
Fimbristylis littoralis 544
Fimbristylis miliacea 544
Fimbristylis monostachyos 544
Fimbristylis narayanii 544
Fimbristylis ovata 544
Fimbristylis podocarpa 545
Fimbristylis pseudonarayanii 545
Fimbristylis quinquangularis 544
Fimbristylis spathacea 543
Fimbristylis tenuispica 8
Fimbristylis tetragona 545
Fimbristylis tomentosa 545
Fioria vitifolia 8, 80
Flacourtia indica 58
Flacourtia montana 58
Flacourtia ramontchi 58
Flacourtia sepiaria 58
Flagellaria indica 517
Flemingia congesta 179, 180
Flemingia macrophylla 9
Flemingia semialata 17, 180
Flemingia strobilifera 17, 179
Fleurya interrupta 471
Floscopa scandens 11, 514
Fluggea microcarpa 457
Forsythia mala-elengi 302
Fuirena ciliaris 547
Fuirena glomerata 547
Fuirena umbellata 547
Furcraea foetida 502
Furcraea gigantea 502
G. ellipticum 18
G. pentaphylla 10
G. tomentosum 18
G. wightii 20
Galinsoga parviflora 283
Garcinia cambogia 70
Garcinia gummi-gutta 8, 70
Garcinia malabarica 298
Garcinia ovalidolia 72
Garcinia spicata 9, 72
Garcinia wightii 72
Gardenia dumetorum 266
Gardenia spinosa 266
Garnotia courtallensis 18, 577
Garnotia stricta 577
Garnotia tenella 577
Gastrochilus indicus 484
Gastrochilus nilagiricus 484
Gastrochilus pulchellus 484
Gautteria malabarica 49
Geissaspis cristata 17
Gelonium angustifolium 457
Gelonium lanceolatum 457
Gendarussa wynaadensis 383
Gentiana diffusa 325
Geodorrum densiflorum 485
Geophila herbacea 249
Geophila reniformis 249
Geophila repens 249
Gerardia delphinifolia 363
Getonia floribunda 211
Girardinia diversifolia 470
Girardinia heterophylla 470
Glinus oppositifolius 240
Gliricidia sepium 194
Globba ophioglossa 499
Glochidion ellipticum 446
Glochidion malabaricum 18, 446
Glochidion tomentosum 447
Glochidion zeylanicum 447

Gloriosa superba 21, 507
Glycine lineata 157
Glycosmis cauritiana 112
Glycosmis cochinchinensis 112
Glycosmis mauritiana 10
Glycosmis pentaphylla 112
Gmelina arborea 392
Gmelina coromandelica 248
Gmelina indica 58
Gnaphalium 275
Gnaphalium indicum 283
Gnaphalium polycaulon 283
Gnetum edule 591
Gnetum scandens 591
Gnetum ula 591
Gomphandra coriacea 9, 123
Gomphandra polymorpha 123
Gomphia angustifolia 114
Gomphia serrata 114
Gomphostemma eriocarpon 399
Gomphostemma heyneanum 399
Gomphosiemma oblongum 399
Gomphostemma strobilinum 399
Gomphrena celosioides 408
Gomphrena decumbens 408
Gomphrena globosa 408
Gomphrena hispida 397
Gomphrena sessilis 406
Goniothalamus rhyncantherus 20
Gonyanthes pusilla 475
Gopuram-thangi 380
Gordonia obtusa 10
Granadilla hondala 232
Grangea latifolia 281
Gratiola ciliata 360
Gratiola grandiflora 359
Gratiola hyssopioides 361
Gratiola juncea 356
Gratiola parviflora 361
Gratiola pusilla 360
Gratiola reptans 361
Gratiola rotundifolia 361
Gratiola ruellioides 361
Gratiola serrata 360
Gratiola veronicaefolia 359
Gratiola viscosa 362
Grewia abutilifolia 99
Grewia aspera 99
Grewia microcos 100
Grewia obtusa 99
Grewia salviifolia 244
Grewia tiliifolia 10, 99
Grewia umbellifera 11, 100
Griffithella hookeriana 411
Griffithia regulosa 262
Griffithia speciosa 262
Griffthia gardneri 247
Grumilea congesta 265
Grumilea elongata 265
Grumilea nigra 265
Grweia glabra 99
Grwia disperma 99
Guatteria fragrans 48
Guatteria korinti 48
Gymnacranthera canarica 420
Gymnacranthera eugeniifolia 420
Gymnacranthera farquhariana 420
Gymnostachyum canescens 379
Gymnostachyum polyanthum 379
Gynura aurantiaca 284
Gynura nitida 284
Gynura pseudo-china 284
H. longicorniculata 11
H. longifolia 146
H. nitida 10
H. wallichiana 9
Habenaria aristata 492
Habenaria crinifera 485

Habenaria digitata 486
habenaria goodyeroides 492
Habenaria heyneana 11, 486
Habenaria longicalcarata 486
Habenaria longicorniculata 9, 486
Habenaria longicornu 486
Habenaria perrottetiana 487
Habenaria stenostachya 492
Habenaria susannae 491
Habenaria trinervia 486
Haldenia cordifolia 8, 250
Hammatu Rheede 349
Handir-alou 463
Hapalosia loeiflingiae 67
Harpullia arborea 9
Harpullia cupanoides 142
Harpullia imbricata 142
Heckeri subpeltata 416
Hedyotis auricularia 8, 10, 251
Hedyotis brachypoda 251
Hedyotis corymbosa 251
Hedyotis diffusa 8, 251
Hedyotis herbacea 252
Hedyotis membranacea 252
Hedyotis nitida 252
Hedyotis pruinosa 254
Hedysarum bracteatum 179
Hedysarum bupleurifolium 156
Hedysarum gangeticum 174
Hedysarum gyrans 175
Hedysarum heterocarpon 174
Hedysarum heterophyllum 174
Hedysarum latifolium 177
Hedysarum leutescens 188
Hedysarum microphyllum 175
Hedysarum moniliferum 156
Hedysarum motorium 175
Hedysarum polycarpum 174
Hedysarum pulchellum 175
Hedysarum strobiliferum 180
Hedysarum triangulare 176
Hedysarum triflorum 176
Hedysarum triquetrum 177
Hedysarum velutinum 177
Hedysarum viscidum 187
Hedysaum vaginale 156
Helecteres isora 10
Helicanthes elastica 428
Helicteres isora 94
Heliotropium indicum 11, 330
Heliotropium marifolium 330
Heliotropium scabrum 330
Helixanthera obtusata 9, 428
Helixanthera wallichiana 428
Hemichoriste montana 382
Hemidesmus indicus 10, 319
Hemigyrosa canescens 142
Hemigyrosa deficiens 142
Heptapleurum micranthum 243
Heptapleurum rostratum 243
Heptapleurum venulosum 244
Heracleum sprengelianum 242
Herpestis monnieria 355
Heteropogon contortus 8, 10, 578
Hevea braziliensis 459
Hewittia bicolor 338
Hewittia malabarica 338
Hewittia sublobata 338
Hexacentris mysorensis 389
Heynea trijuga 121
Heyneana 399
Hibiscus abelmoschus 78
Hibiscus aculeatus 81
Hibiscus caesius 82
Hibiscus hispidissimus 81
Hibiscus ketmia 82
Hibiscus lunarifolius 9, 81, 82
Hibiscus manihot 78

Hibiscus obtusifolius 82
Hibiscus sreenarayanianus 22, 82
Hibiscus surattensis 8, 84
Hibiscus tetraphyllus 78
HIbiscus tiliaceus 8, 84
Hibiscus vitifolius 80
Hippia integrifolia 281
Hiptage benghalensis 11, 104
Hiptage madablota 104
Holarrehan pubescens 10
Holarrhena antidysenterica 309
Holarrhena codaga 309
Holarrhena pubescens 309
Holcus ciliatus 562
Holcus nervosus 563
Holcusnervosus 563
Holigarna arnottiana 8, 9
Holigarna wightii 146
Holosteum cordatum 67
Homonoia riparia 11, 447
Hopea glabra 11, 74
Hopea parviflora 9, 74
Hopea ponga 8, 75
Hoppea fastigiata 328
Hortus Indicus Malabaricus 12
Hottonia indica 358
Hoya ovalifolia 320
Humboldtia vahilana 11, 17, 200
Hybanthus enneaspermus 56
Hybanthus enneaspermus 8
Hydnocarpus laurifolia 59
Hydnocarpus pendulus 22, 59
Hydnocarpus pentandra 8, 11, 59
Hydnocarpus wightiana 59
Hydrilla verticillata 8, 11, 474
Hydrobryum olivaceum 414
Hydrocera triflora 105
Hydrocotyle asiatica 241
Hydrocotyle conferta 242
Hydrocotyle javanica 242
Hydrotrophus echinospermus 474
Hygrophila angustifolia 11, 380
Hygrophila longifolia 380
Hygrophila ringens 380
Hygrophila salicifolia 380
Hygrophila schulli 380
Hygrophila spinosa 380
Hypolyptrum latifolium 547
Hypolytrum nemorum 547
Hypoxis latifolia 503
Hypoxis pauciflora 503
Hypoxis trichocarpa 503
Hyptis rhomboidea 400
Hyptis suaveolens 10, 21, 400
I. galegoides 17
I. longiracemosa 17
Ichnocarpus frutescens 8, 311
Idampiri-valampiri 95
Idou-moulli Rheede 430
Ilapongu 74
Ilippa 296
Ilysanthes hyssopioides 361
Ilysa::thes parviflora 361
Ilysanthes rotundifolia 361
Ilysanthes serrata 360
Ilysanthes veronicifolia 359
Ilysnthes reptans 361
Impatiens anamallayensis 107
Impatiens balsamina 106
Impatiens chinensis 107
Impatiens flaccida 107
Impatiens goughii 107
Impatiens herbicola 107
Impatiens inconspicua 107
Impatiens kleinii 108
Impatiens minor 108
Impatiens pusilla 11, 107
Impatiens scapiflora 108

Impatiens tomentosa 108
Impatiens triflora 105
Impatiens verticillata 108
Impatiens viscosa 110
Imperata arundinacea 578
Imperata cylindrica 578
Imperata koenigii 578
Indigofera benthamiana 183
Indigofera constricta 17, 182
Indigofera endecaphylla 183
Indigofera flaccida 182
Indigofera galegoides 17
Indigofera longiracemosa 183
Indigofera spicata 9, 183
Indigofera teysmannii 183
Indigofera zollingeriana 183
Indobanalia thyrsiflora 408
Indoneesiella echioides 380
Indotristicha ramosissima 411
Inota-inodien 350
Instsia Rheede 202
Inula indica 291
Ionidium suffruticosum 56
Ipomoea alba 340
Ipomoea angulata 341
Ipomoea aquatica 340
Ipomoea blancoi 344
Ipomoea bona-nox 340
Ipomoea bracteata 341
Ipomoea cairica 340
Ipomoea carnea 341
Ipomoea chryseides 347
Ipomoea coccinea 342
Ipomoea congesta 342
Ipomoea cymosa 348
Ipomoea deccana 341
Ipomoea digitata 342
Ipomoea elliptica 335
Ipomoea eriocarpa 341
Ipomoea fistulosa 341
Ipomoea hederacea 343
Ipomoea hederifolia 341
Ipomoea hispida 341
Ipomoea indica 342
Ipomoea malabarica 338
Ipomoea marginata 342
Ipomoea mauritiana 342
Ipomoea maxima 342
Ipomoea nil 343
Ipomoea obscura 343
Ipomoea paniculata 342
Ipomoea pes-caprae 343
Ipomoea pes-tigridis 343
Ipomoea phoenicea 341
Ipomoea pileata 344
Ipomoea pulchella 340
Ipomoea reptans 340
Ipomoea sepiaria 342
Ipomoea sessiliflora 341
Ipomoea sinuata 345
Ipomoea tridentata 347
Ipomoea triloba 344
Ipomoea turpethum 348
Ipomoea uniflora 333
Ipomoea vitifolia 348
Isanthera permollis 370
Ischaemum quilonensis 18, 579
Ischaemum timorense 579
Isonandra lanceolata 295
Isora-murri 95
Itti-canni 429
Itty-alu 465
Itty-arealou 465
Ixora alba 254
Ixora barbata 254
Ixora brachiata 255
Ixora coccinea 255
Ixora cuneifolia 255

Ixora finalaysoniana 256
Ixora lanceolaria 254, 255
Ixora lanceolatea 254
Ixora leucantha 9, 255
Ixora malabarica 254
Ixora nigricans 9, 256
Ixora polyantha 256
Ixora tomentosa 262
J. procumbens 10
J. prostrata 10
Jambolifera pedunculata 110
Jambosa hemispherica 216
Jambosa laeta 217
Jambosa mundagam 217
Jambosa occidentalis 217
Jambosa pauciflora 217
Janam-kolli 239
Jansenella griffithiana 579
Jasminum angustifolium 303
Jasminum azoricum 9, 303
Jasminum flexile 303
Jasminum rottlerianum 10, 303
Jatropha curcas 448
Jatropha gossypifolia 448
Jatropha moluccana 435
Jatropha montana 438
Jeeraka-pullu 325
Jonesia asoca 200
Josephia lanceolata 495
Julostylis 22
Julostylis angustifolia 85
Julostylis polyandra 18, 22, 84
Jussiaea adscendens 230
Jussiaea hyssopifolia 230
Jussiaea linifolia 230
Jussiaea repens 230
Jussiaea suffruticosa 231
Justicia adhatoda 383
Justicia andersonii 383
Justicia atropurpurea 374
Justicia betonica 381
Justicia canscens 383
Justicia diffusa 382
Justicia echioides 380
Justicia gangetica 376
Justicia japonica 8, 9, 10, 382
Justicia montana 378, 382
Justicia nasuta 387
Justicia parviflora 387
Justicia pectinata 388
Justicia procumbens 382
Justicia prostrata 382
Justicia santapaui 382
Justicia simplex 382
Justicia viridis 378
Justicia wynaadensis 383
Jymphaea nouchali 54
Kaatu-champa 217
Kadaladi 394
Kadalavanakku 448
Kaden pullu 555
Kaikka-thetti 226
Kaipa-isjira 240
Kaka-kalingi-valli 195
Kaka-pu 365
Kakka-Thodali 113
Kaku-valli 186
Kalam-potti 221
Kalanchoe pinnata 207
Kalesjam Rheede 147
Kal-ilavu 93
Kallurukki 363
Kal-rudraksham 102
Kal-toddavaddi 195
Kaltsjerou - panel 49
Kalu-polapen 364
Kambili 111
Kamettia caryophyllata 311

Kamettia malabarica 311
Kametti-valli 311
Kampakam 74
Kampili-chedi 123
Kana-kappalam 139
Kanala 111
Kanden-kara 248
Kani-Conna 197
Kanneli-itti-canni 428
Kara-angolam 244
Kar-Agil 113
Karayani 94
Kare-kandel 209
Karenda-valli 414
Kareta-tsjori-valli Rheede 135
Karin Chora 298
Kari-njara 217
Karin-kara 64
Karin-pola 526
Karinta-kali 249
Kari-vella 300
Kari-vetti 304
Karivi-valli 237
Karjavo-maram 225
Karuka-pullu 569
Karuppa kodi 35
Katou karua 423
Katou tsjaka Rheede 259
Katou-belluta-ampelpodi 249
Katouinschi-kua 500
Katou-kadali 222
Katou-karua 423
Katoukonna Rheede 204
Katou-nirouri 454
Katou-ponnam-maravara 488
Katou-tsjeroe 146
Katoutsjolam 556
Katsjan-panel 48
Kattukaranaordonia obtusa 733
Kattu-kelengu 338, 504
Kattu-Moovila 187
Kattu-munthiri 207
Kattu-muthira 177
Kattu-paruthi 92
Kattu-pokayila 293
Kattu-veppu 119
Katu-alou 463
Katu-baramareca 160
Katu-beloeren 80
Katu-karivi 374
Katu-katsjil 504
Katu-kurka 397
Katu-nuren 504
Katu-pal-valli 319
Katu-pee-tsjanga-puspam 360
Katu-pitsjegam-Mulla 303
Katu-thulasi 402
Katu-tsjetti-Pu 276
Katu-tsjurel 520
Katu-vistna-clandi 61
Kavara-pullu 570
Kayyala-puliyan 239
Keeri-kizhangu 523
Keezhar-nelli 452
Kili-thinipanji 125
Kilukki-chedi 164
Kingidium decumbens 487
Kingidium deliciosum 487
Kirganelia reticulata 454
Klugia notoniana 370
Knema attenuata 420
Knoxia corymbosa 256
Knoxia sumatrensis 256
Kodangal 242
Kodassari 415
Kodatsjeri 61
Kodda-pail 526
Kodi-pullu 578

Konda-pullu 566
Kopullu 551
Kora kkadi 121
Kudici-valli 347
Kudira-pullu 567
Kuduka-mooli 414
Kula-punna 142
Kunstleria keralensis 17, 183
Kuppa-keera 407
Kuriel 149
Kurundoti 90
Kyllinga brevifolia 549
Kyllinga cyperina 551
Kyllinga melanosperma 549
Kyllinga monocephala 549
Kyllinga nemoralis 549
Kyllinga panicea 552
Kyllinga sumatrensis 551
Kyllinga triceps 550
L. arborea 112
L. hirta 10
L. pentaphylla 112
L. wightii Clarke 137
Lagenandra meeboldii 525
Lagenandra ovata 526
Lagenandra toxicaria 11, 526
Lagerstroemia lanceolata 228
Lagerstroemia microcarpa 228
Lagerstroemia reginae 227
Lagerstroemia thomsonii 228
Lagerstromia flos-reginae 227
Lagerstromia glabra 227
Lagerstromia hirsuta 227
Laggera alata 278
Lagurus cylindricus 578
Laneasagum oblngifolium 444
Lannea coromandelica 8
Lantana aculeata 392
Lantana camara 392
Laportea crenulata 469
Laportea interrupta 471
Lasianthus strigillosus 256
Laurus malabtrum 423
Lavendula carnosa 397
Lawia acuminata 258
Lawia zeylanica 412
Lawsonia inermis 229
Leea indica 9, 10
Leea sambucina 137
Leersia hexandra 580
Lemna paucicostata 528
Lemna perpusilla 11
Lemna perpusilla 528
Leonurus indicus 401
Lepianthes umbellata 416
Lepidagathis hyalina 383
Lepidagathis incurva 383
Lepidagathis mucronata 383
Lepisanthes deficiens 142
Lepistemon leiocalyx 344
Lepistemon verdcortii 345
Lettsomia elliptica 335
Leucas biflora 10, 400
Leucas hirta 401
Leucas indica 401
Leucas lavandulifolia 401
Leucas linifolia 401
Leucas vestita 401
Leucas zeylanica 401
Ligustrum robustum 304
Ligustrum walkeri 304
Limnanthemum indicum 329
Limnophila aromatica 8, 357
Limnophila chinensis 357
Limnophila conferta 358
Limnophila gratioloides 358
Limnophila gratissima 357
Limnophila hirsuta 357

Limnophila indica 8, 357
Limnophila repens 358
Limodorum densiflorum 485
Limodorum pulchrum 484
Limonia mauritiana 112
Limonia pentaphylla 112
Lindernia anagallis 359
Lindernia angustifolia 359
Lindernia antipoda 359
Lindernia caespitosa 360
Lindernia ciliata 360
Lindernia crustacea 360
Lindernia hirsuta 362
Lindernia hyssopioides 361
Lindernia japonica 362
Lindernia parviflora 361
Lindernia pusilla 360
Lindernia rotundifolia 361
Lindernia ruellioides 361
Lindernia viscosa 362
Linociera malabarica 302
Linociera ramiflora 302
Linociera wightii 302
Linum mysorense 10, 103
Liparis densiflora 489
Liparis elliptica 488
Liparis nervosa 488
Liparis paradoxa 488
Liparis viridiflora 488
Lipocarpha argentea 550
Lipocarpha chinensis 550
Litsea bourdillonii 425
Litsea coriacea 9, 425
Litsea floribunda 9, 425
Litsea insignis 425
Litsea oleoides 425
Litsea scrbiculata 426
Litsea travancorica 425
Litsea venulosa 426
Litsea wightiana 425
Litsea zeylanica 426
Lobelia alsinoides 292
Lobelia heyneana 292
Lobelia nicot ianifolia 293
Lobelia pumila 362
Lobelia trialata 293
Lobelia trigona 292
Lochnera pusilla 308
Lochnera rosea 308
Lonecera parasitica 429
Longsdorffia indica 431
Lophopetalum wightianum 9, 126
Loranthus elasticus 428
Loranthus loniceroides 429
Loranthus obtusatus 428
Loranthus wallichianus 428
Lranthus buddlejoids 429
Ludwigia adscendens 230
Ludwigia hyssopifolia 230
Ludwigia octovalvis 231
Ludwigia parviflora 231
Ludwigia perennis 11, 231
Luffa aegyptiaca 21, 235
Luffa cylindrica 235
Luisia teretifolia 489
Luisia zeylanica 488
Lycianthes laevis 349
Lysimachia monnieri 355
M. hirsuta 17
M. japanica 10
M. monosperma 17
M. pinnata 144
M. radiatum 52
M. resinosus muricatus, 18
M. rohifolia 144
M. zeylanica 9
Macaranga indica 449
Macaranga peltata 449

Macaranga roxburghii 449
Machilus macrantha 426
Mackaya populifolia 122
Macraea gardneriana 454
Macraea oblongifolia 455
Macrophylla 507
Macrsolen parasiticus 429
Madhuca longifolia 11, 295
Madhuca neriifolia 21, 296
Madiram-pulli 201
Madukka 64
Maesa dubia 294
Maesa indica 294, 428
Maesa perottetiana 294
Magnoliaceae 25
Mail-elou 395
Makum-ko-ka 204
Malacka-pela 214
Mala-elengi 302
Malamavu 145
Malam-kuratha 126
Malampavetta 57
Malampunna 36
Malamtodda-vadi 198
Mala-muringa 371
Malankoova-paruthi 81
Malankuruvi 301
Malathangi 50
Malaxis acuminata 489
Malaxis densiflora 489
Malaxis luteola 489
Malaxis viridiflora 488
Mallam-toddali 459
Mallothus rhamnifolius 9
Mallotus albus 451
Mallotus intermedius 450
Mallotus muricatus 450
Mallotus philippensis 449, 450
Mallotus resinosus 450, 451
Mallotus resinosus resinosus 18
Mallotus rhamnifolius 451
Mallotus tetracoccus 451
Mallotus walkerae 450
Malva coromandaliana 85
Malvastrum coromandelianum 85
Malvastrum tricuspidata 85
Malvastrum tricuspidatum 85
Manam-padam 403
Manga-nari 357
Mangostana cambogia 70
Manihot esculenta 459
Manihot glaziovii 459
Mani porandi 137
Manilkara zapota 297
Manirouri Rheede 439
Manjakadambu 250
Manjanathi 257
Man-onapu 108
Manoranjini valli 40
Manulea indiana 354
Mappia foetida 123
Mara vetti-thali 425
Marali 139
Maranta malaccensis 497
Maravara-tsjembu 527
Mareta-inali 320
Mariscus bulbosus 551
Mariscus clarkei 551
Mariscus concinnus 551
Mariscus cyperinus 551
Mariscus dregeanus 551
Mariscus dubius 551
Mariscus sieberianus 551
Mariscus sumatrensis 551
Marotti 59
Marscus paniceus 552
Marsdenia volubilis 322
Mastixia arborea 11, 244

Mastixia meziana 244
Mastixia pentandra 244
Mathagiri-vembu 121
Mavu Rheede 147
Mazus japonicus 362
Mazus pumilus 362
Mazus rugosus 362
Mechil-pullu 568
Medinilla beddomei 221
Medinilla radicans 221
Meesia serrata 114
Meiogyne pannosa 9
Melanthesa obliqua 439
Melanthesa rhamnoides 439
Melanthesa turbinata 439
Melastoma asperum 222
Melastoma hirta 221
Melastoma malabathricum 10, 221
Melia azedarah 119
Melia baccifera 118
Meliosma arnottiana 144
Melochia corchorifolia 95
Melochia cordata 87
Melothria amplexicaulis 237
Melothria heterophylla 237
Melothria maderaspatana 236
Melothria perpusilla 238
Melothria subtruncata 238
Melothria thwaitesii 238
Melothria zeylanica 238
Memecylon amplexicaule 226
Memecylon edule 225
Memecylon heyneanum 225
Memecylon jambosoides 225
Memecylon lawsoni 226
Memecylon malabarica 226
Memecylon malabaricum 226
Memecylon umbellatum 226
Mendoni 507
Menispermum acuminatum 52
Menispermum cocculus 50
Menispermum cordifolium 52
Menispermum japonicum 52
Menispermum peltatum 51
Menyanthes indica 329
Merremia chryseides 347
Merremia convolvulacea 347
Merremia dissecta 345
Merremia hastata 347
Merremia hederacea 345
Merremia tridentata 347
Merremia turpethum 347
Merremia umbellata 348
Merremia vitifolia 348
Mesua ferrea 72
Mesua nagassarium 9, 72
Mezoneuron cucullatum 195
Micranthus oppositifolius 386
Microcarpaea minima 362
Microcos paniculata 8, 100
Microelus roeperianus 438
Microstegium ciliatum 580
Microstylis versicolor 489
Microstylis wallichii 489
Mikania cordata 21, 286
Mikania scandens 286
Milium cimicinus 560
Milium compressm 563
Millingtonia arnottiana 144
Millingtonia simplicifolia 144
Mimosa caesia 202
Mimosa chinensis 203
Mimosa inermis 205
Mimosa invisa 205
Mimosa lebbeck 203
Mimosa marginata 203
Mimosa monadelpha 204
Mimosa pennata 202

Mimosa pudica 8, 205
Mimosa torta 202
Mimosa xylocarpa 206
Mimusoideae 17
Mimusops elengi 296
Minangani 397
Minari 187
Miquelia courtallensis 577
Miquelia dentata 123
Mirabilis jalapa 404
Mitracarpus hirtus 257
Mitracarpus scaber 257
Mitracarpus verticillatus 257
Mitracarpus verticllatus 8
Mitragyna tubulosa 257
Mitrasacme alsinoides 323
Mitrasacme indica 323
Mitrasacme malaccensis 323
Mitrasacme polymorpha 323
Mitrasacme pygmaea 323
Mniopsis hookeriana 411
Mniopsis selaginoides 412
Moacurra gelonioides 121
Modecca 232
Modecca palmata 232
Modira-vali 77
Modogam Rheede 323
Molago-codi 419
Molinaea canescens 142
Molineria finlaysoniana 503
Molineria trichocarpa 503
Mollugo oppositofolia 240
Mollugo paniculata 241
Mollugo pentaphylla 11, 241
Mollugo stricta 241
Momordica charantia 236
Momordica dioica 21, 236
Monachyron villosum 586
Moniera cuneifolia 355
Monocera munronii 102
Monochilus flabellatum 480
Monochilus longilabris 496
Monochoria vaginalis 8, 11, 21
Monochoria vaginalis 8, 508
Monosis wightiana 290
Moonga-pezhu 145
Mootti-pazham 437
Mootti-thoori 437
Morinda pubescens 257
Morinda tinctoria 257
Morinda tomentosa 257
Motta-modecca 232
Moul-elavou 93
Mucca-piri 236
Mucuna gigantea 17, 186
Mucuna hirsuta 17, 21, 187
Mucuna monosperma 17, 186
Mucuna pruriens 186
Muel-schevi 282
Mukia manderaspatana 236
Mukia scabrella 236
Mukkutti 104
Muldera trichostachya 419
Mulen-pullu 543
Mullanmula 564
Mullen-Keera 407
Mullu venga 439
Mundagam 217
Munda-valli 340
Munnicksia laurifolia 59
Murdannia dimorpha 9, 10, 515
Murdannia edulis 515
Murdannia glauca 516
Murdannia japonica 516
Murdannia nudiflora 516
Murdannia pauciflora 516
Murdannia semiteres 517
Murdannia wightii 516

Murdannia zeylanica 517
Murraya exotica 113
Murraya paniculata 9, 113
Mussaenda belilla 258
Mussaenda frondosa 258
Mussaenda laxa 258
Mycetia acuminata 258
Myristica attenuata 420
Myristica beddomei 421
Myristica contorta 421
Myristica corticosa 420
Myristica dactyloides 421
Myristica eugeniaefolia 420
Myristica farquhariana 420
Myristica fraagrans 421
Myristica laurifolia 421
Myristica malabarica 21
Myristica malabarica 421
Myrobalanus bellirica 212
Myrtus caryophyllata 215
Myrtus cumini 216
Myrtus laurinus 300
Myrtus zeylanica 219
Myxoprum serratulum 304
Myxoprum smilacifolium 9, 21, 304
N. colebrookeana 147
N. heyneanus 9
N. malabarica 54
N. stellata 54
Naga-danti 438
Nageia wallichiana 590
Naikadambu 257
Nai-kumbil 390
Najas graminea 11, 529
Nansjera-patsja 322
Naravelia zeylanica 9
Naregamia alata 10, 119
Narinam-poulli 84
Nari-patsja 279
Nauclea cadamba 259
Nauclea cordifolia 250
Nauclea missionis 260
Nauclea tubulosa 257
Neanotis decipiens 259
Neanotis indica 259
Nedel-ambel 329
Nedunar 48
Neergramp 230
Neer-vetti 447
Neesiella echioids 380
Nelam-mari 194
Nelamparenda 57
Nelam-pullu 517
Nela-naregam 119
Nela-vagu 87
Nelem-pala 314
Nelenchena major 528
Nelipu Rheede 368
Nella-pana-kelengu 503
Nelli 453
Nellikka-puli 447
Nelsonia campestris 383
Nelsonia canescens 383
Nelumbium speciosum 54
Nelumbo nucifera 11
Nelumbo nucifera 54
Nemedra elaegnoidea 116
Nennal valli 36
Neolamarckia cadamba 259
Neolitsea scrobiculata 426
Nepeta indica 398
Nephelium longana 141
Nephelium stipulaceum 143
Nerium oleandr 315
Nerium tinctorium 314
Nervilia aragoana 490
Neurocalyx calycinus 260
Neurocalyx wightii 260

Nialel 116
Niirvala 56
Nikida kodi 35
Nilgirianthus barbatus 384
Nilgirianthus beddomei 384
Nilgirianthus ciliatus 9, 385
Nilgirianthus heyneanus 385
Nilgirianthus punctatus 385
Nilicamaram 453
Nir-carambu 230
Nir-kurunthu 244
Nirpulla 455
Nir-pullari 188
Nirpulli 511
Nir-tsjembu 527
Niruren 95
Niruri 455
Niruri Rheede 439
Njeringil 273
Njota-njodien-valli 316
Noeli-tali 436
Noel-valli 169
Nola-ily 580
Nominum Genericorum 14
Nothapodytes foetida 123
Nothapodytes nimmoniana 123
Nstsjatam 51
Nuli-tali 436
Nurven-kelengu 504
Nyalel 116
Nyctanthes angustifolia 303
Nymphaea nelumbo 54
Nymphaea nouchali 11
Nymphaeaceae 25
Nymphoides indica 329
O. platycaulon 9
O. proudlockii 9
Oberonia brunoniana 9, 490
Oberonia platycaulon 491
Oberonia produlockii 491
Oberonia verticillata 491
Ochlandra scriptoria 581
Ochlandra travancorica 11, 18, 580
Ochreinauclea missionis 260
Ochrenauclea missionis 11
Ocimum americanum 402
Ocimum canum 402
Ocimum capitatum 397
Odal 125
Odina wodier 147
Odiya-Madantha 378
Oenothera octovalvis 231
Oldenlandia auricularia 251
Oldenlandia brachypoda 251
Oldenlandia corymbosa 251, 252
Oldenlandia herbacea 252
Oldenlandia nitida 252
Oldlandia membranacea 252
Oldnlandia diffusa 251
Oldnlandia pruinosa 254
Olea dioica 304
Olea linocieroides 302
Oligopholis tubulosa 366
Onapu 107
Operculina turpethum 348
Ophelia corymbosa 328
Ophiorrhiza eriantha 261
Ophiorrhiza harrisiana 261
Ophiorrhiza mungos 261
Ophiorrhiza pectinata 261
Ophiorrhiza prostrata 261
Ophiorrhiza rugosa 261
Ophioxylon serpentinum 313
Ophrys nervosa 488
Oplismenus burmannii 581
Oplismenus compositus 18, 581
Opuntia monocantha 240
Opuntia vulgaris 240

Orchis gigantea 491
Oreocnide integrifolia 471
Origanum benghalense 403
Oroxylum indicum 372
Orygia portulacifolia 68
Oryza granulata 582
Oryza meyeriana 582
Osbeckia aspera 222, 223
Osbeckia courtallensis 222
Osbeckia kleinii 222
Osbeckia muralis 223
Osbeckia travancorica 222
Osbeckia truncata 223
Osbeckia virgata 223
Osbeckia wightiana 223
Otonychium imbricatum 142
Ottel-ambel 474
Ottelia alismoides 11, 474
Ottu-plavu 469
Ouratia angustifolia 114
Ouratia serrata 114
Oxalis corniculata 105
Oxalis reinwardtii 104
Oxyceros rugulosus 262
P. chinensis 10, 61
P. elongata 10
P. gageana 18
P. glabrum 11
P. glomerata 61
P. ramaswamiuru 61
P. urinaria 9
P. virgatus 9
Pacha chooral 519
Pada-kelengu 51
Pada-valli 51
Padia meyeriana 582
Padiri Rheede 371
Paederota minima 362
Pai-paroea 99
Pala Rheede 307
Palaquium ellipticum 9, 296
Palmodecca 232
Pal-Modecca 342
Pal-valli 311
Panambu-valli 517
Panam-palca 421
Pancratium triflorum 501
Pancratium verecundum 501
Pandanus canaranus 522
Pandanus thwaitesii 521
Pandanus unipapillatus 522
Pandiavanacu 455
Panel 112
Panicum adscendens 571
Panicum burmannii 581
Panicum ciliare 571
Panicum compositum 581
Panicum crussgalli 575
Panicum dactylon 569
Panicum eruciforme 564
Panicum gardneri 582
Panicum isachne 564
Panicum oxyphyllum 569
Panicum pallide-fuscum 587
Panicum palmaefolium 587
Panicum plicatum 587
Panicum polystachyon 584
Panicum pumilum 587
Panicum radicosum 571
Panicum repens 11
Panicum repens 582
Panicum stagninum 575
Panicum timorense 571
Panicum uncinatum 585
Panicummuricatum 569
Panitsjiku-maram 298
Panni-valli 169
Panthal maram 43

Parand-ka-valli 204
Paratropica venulosa 244
Parietaria microphylla 472
Parietaria zeylanica 472
Pariti 84
Parpadagam 252
Pasa-kotta 143
Paspalum conjugatum 583
Paspalum distichum 583
Paspalum longiflorum 571
Paspalum paspaloides 583
Paspalum sanguinale 571, 572
Paspalum scrobiculatum 583
Paspalum ternatum 572
Passiflora foetida 10, 22, 232, 233
Passiflora hispida 233
Patsjotti Rheede 300
Paullinia asiatica 113
Pavetta 262
Pavetta canarica 270
Pavetta hispidula 264
Pavetta indica 9, 262
Pavetta tomentosa 262
Pavetta zeylanica 9, 264
Pecteilis gigantea 491
Pedicularis zeylanica 363
Pedilanthus tithymaloides 451
Pee-coipa 406
Pee-cupameni 434
Pee-jnota-jnodien 350
Pee-naari 96
Pee-ponnagam 450
Pee-tandale-cotti 167
Pee-tardavel 406
Pee-tsjerou-ponnagam 451
Pee-tumba 380
Pee-vetti 447
Pellionia heyneana 471
Peltophorum pterocarpum 201
Pennisetum polystachyon 584
Penone 75
Peperomia pellucida 417
Peperomia portulacoides 417
Peplis indica 228
Peragu Rheede 392
Peralu 463
Perim-kaku-valli 204
Perim-kurigil 149
Perinjara 216
Perin-kaida-toddi 522
Perin-kara 102
Perin-nirouri 439
Perin-panel 110
Perin-teregam 464
Perin-toddali Rheede 131
Periploca arborea 314
Periploca indica 320
Peristylus aristatus 492
Peristylus densus 492
Peristylus goodyeroides 492
Peristylus stenostachyus 492
Perotis indica 584
Perotis latifolia 584
Persea macrantha 426
Persea macrantha 9
Persicaria barbata 409
Persicaria chinensis 410
Persicaria glabra 410
Pevetti 447
Phaeanthus malabaricus 20
Phalaenopsis decumbens 487
Phalaenopsis wightii 487
Phalangium attenuatum 506
Phaselous calcaratus 193
Phaseolus dalzelli 193
Phaseolus dalzellianus 193
Phaseolus pauciflorus 193
Phaulopsis dorsiflora 386

Phaulopsis imbricata 386
Phaulopsis parviflora 386
Phcnospora hedysaroides 188
Phelbophyllum lawsonii 386
Phlomis biflora 400
Phlomis hirta 401
Phlomis linifolia 401
Phlomis zeylanica 401
Phoberos crenatus 59
Phoenix humilis 521
Phoenix loureirii 521
Pholidota imbricata 9, 493
Pholidota pallida 493
Phrynium capitatum 500
Phrynium ovatum 500
Phrynium rheedei 500
Phrynium virgatum 501
Phyllanthus airy-shawi 453
Phyllanthus amarus 21
Phyllanthus amarus 452
Phyllanthus bailloniana 18, 453
Phyllanthus debilis 453
Phyllanthus emblica 453
Phyllanthus fraternus 452
Phyllanthus gageana 454
Phyllanthus gardnerianus 454
Phyllanthus leprocarpus 455
Phyllanthus mukerjeeanus 453
Phyllanthus niruri 452, 453
Phyllanthus niruri debilis 453
Phyllanthus patens 439
Phyllanthus reticulatus 454
Phyllanthus retusus 439
Phyllanthus rhamnoides 439
Phyllanthus rheedii 9, 454
Phyllanthus simplex 454, 455
Phyllanthus urinaria 455
Phyllanthus virgatus 455
Phyllanthus virosus 457
Phyllocephalum rangacharii 286
Phyllochlamys spinosa 467
Phyllodium pulchellum 176
Physalis angulata 350
Physalis minima 350
Physalis peruviana 350
Pierardia courtallensis 437
Pierardia macrostachys 437
Pilea melastomoides 472
Pilea microphylla 472
Pilea muscosa 472
Pilipes 569
Piper argyrophyllum 21, 418
Piper attenuatu 419
Piper betle 419
Piper brachystachyum 418
Piper hymenophyllum 418
Piper longum 418
Piper mullesua 418
Piper nigrum 418
Piper pellucidum 417
Piper portulacoides 417
Piper subpeltatum 416
Piper sylvestre 419
Piper trichostachyon 419
Piper trioicum 419
Piper umbellatum 416
Piripu Rheede 410
Pistia stratiotes 526
Pithecellobium bigeminum 204
Pithecellobium monadelphum 204
Pittipunna 36
Platanthera susannae 491
Plectranthus wightii 402
Plectronia angustifolia 248
Plectronia didyma 248
Plectronia rheedei 248
Plectronia rheedii 248
Plectronia umbellata 248

Pleurostylia opposita 126
Pleurostylia wightii 126
Plumeria rubra 315
Poa cilianensis 576
Poa plumosa 576
Poa tenella 576
Poa uniloides 576
Podocarpus latifolia 590
Podocarpus wallichianus 590
Podochilus falcatus 493
Podochilus malabaricus 493
Podostemon algaeformis 412
Podostemon hokerianus 411
Podostemon olivaceum 414
Podostemon selaginoides 412
Podostemon stylosus 412
Poea-tsjetti 114
Pogonatherum crinitum 584
Pogonia carinata 490
Pogonia flabelliformis 490
Pogonia scottii 490
Pogostemon benghalensis 403
Pogostemon heyneanus 403
Pogostemon paniculatus 403
Pogostemon petchouly 403
Pogostemon travancoricus 403
Pogostmon plectranthoides 403
Polla 50
Pollia sorzogonensis 509
Pollinia argentia 577
Pollinia ciliata 580
Polycarpon loeflingiae 67
Polycarpon prostratum 67
Polygala arvensis 61
Polygala bolbothrix 10, 61
Polygala chinensis 61
Polygala ciliata 64
Polygala elongata 62
Polygala javana 62
Polygala rosmarinifolia 62
Polygala telephioides 62
Polygonum barbatum 11, 409
Polygonum chinense 410
Polygonum glabrum 410
Polygonum indicum 410
Polygonum plebeium 410
Polyplerum stylosum 11, 412
Polystachya concreta 493
Polystachya flavescens 494
Ponga 75
Pongam 187
Pongamia canarensis 169
Pongamia glabra 187
Pongamia pinnata 11, 17, 187
Pongu 75
Ponnagam Rheede 450
Ponnam-tagera 199
Pontederia crassipes 508
Pontederia ovata 500
Pontederia vaginalis 508
Poovam 143
Poovanam 143
Pori-poovam 143
Pori-punna 141
Pori-vatta 451
Portulaca cuneifolia 68
Portulaca oleracea 68
Pota-pullu 537
Pothomorphe subpeltata 416
Pothos crassipedunculatus 22
Pothos pentandrus 430
Pothos scandens 8, 9, 526
Potta-cavalam 96
Potta-vaga 203
Poutaletsje 254
Pouzolzia caudata 472
Pouzolzia indica 472
Pouzolzia wightii 473

Pouzolzia zeylanica 472
Premna coriacea 10, 393
Premna glaberrima 393
Premna nimnoiana 123
Pre-tsjnga-puspam 360
Pristrophe montana 385
Prunella indica 397
Pseudanthistiria umbellata 585
Pseudarthria viscida 187
Pseudechinolaena polystachya 585
Pseuderanthemum malabaricum 386
Psidium guajava 214
Psuedarthria viscida 17
Psychoria longifolia 249
Psychotria anamallayana 264
Psychotria congesta 265
Psychotria elongata 265
Psychotria glandulosa 265
Psychotria herbacea 249
Psychotria macrocarpa 265
Psychotria nigra 265
Psychotria nilgiriensis 265
Psychotria nudiflora 265
Psychotria ophioxyloides 249
Psychotria thwaitesii 265
Ptelea arborea 142
Pterocarpus marsupium 10, 17, 188
Pterospermum diversifolium 95
Pterospermum glabrescens 95
Pterygota alata 9, 96
Puam-curundala 290
Pubescens 463
Pueraria phaseoloides 17, 188
Pul-colli 387
Pulippan-cheli 118
Puliyarila 105
Pullanji 211
Pulli-schovadi 343
Pura-pullu 568
Puria trilobata (Lam.) 135
Putoria indica 259
Puzhukkadi-conna 197
Pycreus capillaris 552
Pycreus flavidus 552
Pycreus globosus 552
Pycreus nitens 553
Pycreus polystachyos 8, 552, 553
Pycreus pumilus 553
Pycreus puncticulatus 553
Pycreus stramineus 554
Quamoclit phoenicea 341
Quisqualis malabarica 211
Radermachera xylocarpa 371
Radia gardneri 247
Ramanama-pacha 175
Randia brandisii 266
Randia dumetorum 266
Randia rugulosa 262
Randia spinosa 266
Randia tomentosa 266
Rauvolfia micrantha 313
Rauvolfia serpentina 21, 313
Reidia bailloniana 453
Reidia gageana 454
Remusatia vivipara 9, 527
Rhabdia lycioides 331
Rhamnus vitis idaea 439
Rheedei 504
Rhinacanthus communis 387
Rhinacanthus nasuta 387
Rhinanthus indica 356
Rhombifolia scabrida 90
Rhychospora aurea 554
Rhynchelytrum repens 10, 585
Rhynchelytrum villosum 586
Rhynchoglossum notonianum 370
Rhynchosia acutissima 17, 189
Rhynchospora corymbosa 554

Rhynchostylis retusa 9, 494
Rhynchotechum permolle 370
Ricinus communis 455
Rivea hirsuta 335
Rondeletia asiatica 268
Rosaceae 29
Rostellularia procumbens 382
Rostellularja japonica 382
Rotala indica 228
Rotala leptopetala 229
Rotala ritchei 20, 22, 229
Rotala rosea 229
Rotala rotundifolia 229
Rottboellia cochinchinensis 586
Rottboellia exaltata 586
Rottlera tetracocca 451
Rotula aquatica 11
Rotula aquatica 331
Rourea santaloides 151
Rourea sclerocarpa 149
Rubus fairholmianus 206
Rubus fulvus 207
Rubus glomeratus 207
Rubus molluccanus 206, 207
Ruellia anagallis 359
Ruellia antipoda 359
Ruellia dorsiflora 386
Ruellia imbricata 386
Ruellia prostrataa 378
Ruellia punctata 385
Ruellia ringens 380
Ruellia salicifolia 380
Rungia longifolia 387
Rungia parviflora 387, 388
S. blanda 17
S. ciliata 10
S. guttata 10
S. hispida 10
S. humilis 87
S. lithosperma 10
S. paniculata 10
S. paynii 96
S. polyandra 80
S. rhombifolia 8, 10
S. rhombifolia retusa 86
S. wightianum 477
Saccharum officinarum 590
Saccharum repens 586
Saccolabium nilagiricum 484
Saccolabium ochraceum 477
Saccolabium praemorsum 477
Saccolabium pulchellum 484
Saccolabium ringens 478
Sagina procumbens 67
Sagina saginoides 67
Sagittaria guaynensis 528
Sagittaria lappula 528
Salacia fruticosa 127
Salacia malabarica 20, 127
Salamonia oblongifolia 64
Salmalia malabarica 93
Salomonia ciliata 64
Salvia officinalis 403
Samanea saman 206
Sambiri Then-Kotta 148
Sapindus deficiens 142
Sapindus tetraphylla 142
Sapindus trifoliata 143
Saraca asoca 17, 21,
Saraca indica 200
Sarcandra chloranthoides 419
Sarcandra grandifolia 420
Sarcandra irvingbaileyi 419
Sarcanthu pauciflorus 494
Sarcanthus peninsularis 494
Sarcostigma kleinii 125
Satyrium nepalense 495
Satyrium neplalense wightiana 495

Sauropus albicans 456
Sauropus androgynus 456
Sauropus indicus 456
Sauropus saksenianus 18, 22, 456
Scabrida 90
Scepa lindleyana 437
Schada-veli-kelangu 505
Schageri-cottom 100
Schanganam-pullu 252
Schefflera rostrata 243
Schefflera venulosa 243
Scher-padavalam 237
Scheru-bula 405
Scheru-cadelari 407
Scherukatu-valli-kaniram 324
Scherunam-cottam 440
Scheru-schunda 352
Schfflera micrantha 243
Schizachyrium brevifolium 586
Schleichera oleosa 10
Schleichera trijuga 143
Schoenoplectus juncoides 554
Schoenoplectus mucronatus 555
Schoenus nemorum 547
Schorigenam 458
Schovanna-modela-muccu 410
Schumannianthus virgatus 501
Scirpus aestivalis 541
Scirpus argenteus 543
Scirpus barbatus 532
Scirpus chinensis 550
Scirpus cinnamometorum 543
Scirpus corymbosus 554
Scirpus cyperoides 551
Scirpus dichotomus 544
Scirpus erectus 554
Scirpus juncoides 554
Scirpus lithospermus 555
Scirpus miliaceus 544
Scirpus mucronatus 555
Scirpus retroflexus 540
Scleria cochinchinensis 556
Scleria elata 556
Scleria levis 10, 555
Scleria lithosperma 10, 555, 556
Scleria terrestirs 556
Scleria zeylanica 555
Scleropyrum pentandrum 430
Scleropyrum wallichianum 430
Scolopia crenata 10, 59
Scoparia dulcis 8, 21, 363
Scripus ciliaris 547
Scripus diphyllus 544
Scripus spiralis 540
Scurrula parasitica 9, 429
Sebastiania chamaelea 456
Securinega virosa 457
Semecarpus anacardium 9
Semecarpus grahamii 146
Semecarpus travancorica 9
Sendera-clanti 347
Senecio corymbosus 286
Senecio pseudo-china 284
Senna occidentalis 199
Serpicula verticillata 474
Sesamum mulayanum 21, 372
Sesamum orientale 372
Setaria glauca 587
Setaria pallide-fusca 587
Setaria palmifolia 10, 587
Sgaeraea grandiflora 19
Shuteria vestita 17, 189
Sida acuta 8, 86
Sida beddomei 87
Sida cordata 87
Sida cordifolia 87, 89
Sida fryxellii 10, 22, 89
Sida glutinosa 89

Sida indica 80
Sida mysorensis 89
Sida radicans 87
Sida ravii 22, 89
Sida rhombifolia 90
Sida rhombifolia ssp. rhombifolia 90
Sida rhombifolia ssp. retusa 21, 90
Sida scabrida 90
Sida veronicifolia 87
Sidae 90
Sinapis juncea 55
Sirhookera lanceolata 495
Smilax zeylanica 507
Smithia bigemina 17, 190
Smithia blanda 17
Smithsonia straminea 22, 495
Sneha-pullu 567
Solanum aculeatissimum 351
Solanum anguivi 352
Solanum capsicoides 351
Solanum ferox 352
Solanum giganteum 351
Solanum indicum 352
Solanum laeve 349
Solanum lasiocarpum 352
Solanum stramonifolium 352
Solanum torvum 10, 352
Solanum violaceum 352
Solda 95
Solena amplexicaulis 236
Solena heterophylla 236
Soneri-ila 224
Sonerila brunonis 224
Sonerila rheedii 224
Sonerila sahyadrica 224
Sonerila versicolor 224
Sonerila wallichii 224
Sopubia delphinifolia 363
Sopubia trifida 364
Spatholobus purpuresus 159
Spatholobus roxburghii 159
Spergula saginoides 67
Spermacoce articularis 10, 267
Spermacoce decandollei 268
Spermacoce hispida 267
Spermacoce latifolia 267
Spermacoce mauritiana 268
Spermacoce pusilla 268
Spermacoce repens 268
Spermacoce sumatrensis 256
Sphaerathus indicus 287
Sphaerocarya wallichiana 430
Sphenodesme involucrata 393
Sphenodesme paniculata 393
Spilanthes acmella 288
Spilanthes calva 10, 287
Spilanthes ciliata 288
Spilanthes paniculata 288
Spilanthes radicans 288
Spodiopogon albidus 588
Spodiopogon rhizophorus 587
Sporobolus diander 588
Sporobolus indicus 588
Srobilanthes adenophorus 384
Stachytarpheta cayennensis 394
Stachytarpheta indica 394
Stachytarpheta jamaicensis 394
Stachytarpheta uricaefolia 394
Stachytarphetta indica 394
Stalagmitis cambogioides 72
Staphylea indica Burm.f. 137
Staurogyne glauca 388
Staurogyne zeylanica 388
Staurospermum verticillatus 257
Stegosia cochinchinensis 586
Stegosia exaltata 586
Stemodia hirsuta 357
Stemodia repens 358

Stemonurus foetidus 123
Stephania hermandiifolia 52
Stephania japonica 9, 52
Stephegyne tubulosa 257
Sterculia alata 96
Sterculia balanghas 10, 96
Sterculia guttata 96
Stereopermum colais 10, 371
Stereospermum personatum 371
Stereospermum tetragonum 371
Stereospermum xylocarpum 371
Stilago diandra 436
Stratiotes alismoides 474
Streblus asper 467
Streblus taxoides 467
Striga asiatica 11, 364
Striga lutea 364
Strobilanthes anceps 385
Strobilanthes barbatus 384
Strobilanthes ciliatus 385
Strobilanthes heyneanus 385
Strobilanthes lawsonii 386
Strobilanthes rugosus 385
Strombosia ceylanica 122
Struchium sparganophorum 288
Strychnos bicirrhosa 324
Strychnos bourdilloni 325
Strychnos cinnamomifolia 325
Strychnos colubrina sensu 324
Strychnos colurbrina 325
Strychnos malaccensis 325
Strychnos minor 324
Strychnos nux-vomica 324
Strychnos rheedei 325
Strychnos wallichiana 325
Strychnos wightii 325
Stylocoryne canarica 270
Stylosanthes fruticosa 190
Stylosanthes mucronata 190
Suregada angustifolia 457
Suregada lanceolata 457
Swertia corymbosa 328
Symplocos acuminata 301
Symplocos barberi 301
Symplocos beddomei 301
Symplocos cochinchinensis 300
Symplocos laurina 300
Symplocos macrophylla 10, 301
Symplocos racemosa 301
Symplocos rosea 301
Symplocos spicata 300
Symplocos wynadense 301
Synedrella nodiflora 21
Synedrella nodiflora 289
Syzygium arnottianum 216, 219
Syzygium caryophyllatum 215
Syzygium cumini 216
Syzygium densiflorum 216
Syzygium hemisphericum 216
Syzygium jambolanum 216
Syzygium laetum 9, 217
Syzygium malabaricum 9, 10, 217
Syzygium mundagam 9
Syzygium occidentale 217
Syzygium rubicundum 219
Syzygium travancoricum 219
Syzygium zeylanicum 8, 219
T. involucrata angustifolia 18
T. paniculata 10
T. rheedei DC. 36
T. spinosa 467
Tabernaemontana alternifolia 10, 313
Tabernaemontana caudata 314
Tabernaemontana crispa 313
Tabernaemontana divaricata 315
Tabernaemontana gamblei 314
Tabernaemontana heyneana 313
Taddalia aculeata 113

Tala-neli 347
Talinum cuneifolium 68
Talinum protulacifolium 68
Tali-pariti 84
Tali-pullu 516
Talir kkara 58
Talu-dama 404
Tamara 54
Tamarindus indica 201
Tandale-cotti 165
Tani Rheede 212
Taravel 267
Tarenna asiatica 268
Tarenna canarica 270
Taxillus tomentosus 429
Tectona grandis 8, 395
Tephroia tinctoria 191
Tephrosia pulcherrima 11,17, 191
Tephrosia purpurea 10, 191
Tephrosia tinctoria 17
Teramnus mollis 17, 192
Teregam 464
Tereta-pullu 583
Terminalia bellirica 212
Terminalia catappa 213
Terminalia crenulata 10, 212
Terniola ramosissima 411
Terniola zeylanica 412
Tetracera laevis auct. 36
Tetrameles nudiflora 239
Tetranthera coriacea 425
Tetranthera oleoides 425
Tetranthera venulosa 426
Tetranthera wightiana 425
Tetrastigma muricatum Wall. 136
Thalia-maravara 477
Tharavu 267
Thattan-chavana 314
Thazha kaida 522
Theka Rheede 395
Themeda sabarimalayana 18, 589
Themeda tremula 588
Themeda triandra 10, 589
Thempavu 212
Theriophonum infaustum 527
Thespesia lampas 92
Thoa edulis 591
Thodali 132
Thol-najaval 216
Thora-paeru 159
Thotta-vaadi 205
Thottea dinghoui 415
Thottea duchartrei 415
Thottea siliquosa 416
Thottea siliquosa 9
Thrikolpa konna 348
Thunbergia fragrans 389
Thunbergia mysorensis 389
Thutthi 80
Thyrocephalon nemoralis 549
Tiliacora acuminata 52
Tiliacora racemosa 52
Tinda-parua 467
Tinospora 50
Tinospora cordifolia 52
Tiri-itti-canni 430
Tiru-tali 342
Tjsageri-nuren 504
Toddalia asiatica 113
Toddalia bilocularis 113
Todda-panna 590
Todda-Vaddi 105
Tomex tomentosa 390
Tona ciliata 121
Tondi-teregam 390
Torenia asiatica 365
Torenia bicolor 365
Torenia hirta 360

Torenia travancorica 365
Torrenia bicolor 9
Tota-piri 237
Tournefortia heyneana 9, 331
Tradescantia axillaris 510
Tradescantia tuberosa 514
Tradescantia villosa 514
Tragia chamaelea 456
Tragia hispida 10, 458
Tragia involucrata 458
Tragia involucrata angustifolia 18, 458
Tragia involucrata cordata 458
Tragia muelleriana 18, 458
Trema orientalis 459
Trevia nudiflora 459
Trevia polycarpa 11, 18, 459
Trichilia connaroides 121
Trichodesma zeylanicum 331
Tricholaena wightii 586
Trichosanthes cucumerina 237
Trichosanthes cuspidata 237
Trichosanthes nervifolia 237
Tridax procumbens 8, 21, 289
Trigonostemon lawianus 442
Tripogon bromoides 589
Tristicha ramosissima 411
Tristicha zeylanica 412
Triumfetta annua 10, 101
Triumfetta pilosa 9, 101
Triumfetta rhomboidea 101
Trophis taxoides 467
Tsjadaen 398
Tsjaka-maram 460
Tsjakela 466
Tsjana-kua 497
Tsjanga-puspam 361
Tsjemcumulu 365
Tsjeria-kuren-pullu 584
Tsjeria-samastravadi 219
Tsjeri-manga-nari 358
Tsjerou-kara 248
Tsjerou-meer-alou 466
Tsjerou-panel 48
Tsjerouponna 69
Tsjerou-ponnagam 450
Tsjerou-theka 391
Tsjerujonganam 241
Tsjeru-parua 86
Tsjerutardevel Rheede 382
Tsjeru-uren 95
Tsjieria-narinampuli 239
Tsjocatti 114
Tsjudan-tsjera 358
Tsjvanna-ampelodi 313
Turnera elagans 231
Turnera subulata 231
Turnera trioniflora 231
Turnera ulmifolia 231
Turpinia malabarica 9
Tylophora asthmatica 320
Tylophora indica 320
Tylophora pauciflora 322
Typhonium flagelliforme 528
Ula Rheede 591
Ulinga Rheede 141
Unda-payin 420
Unnam 99
Unona narum 49
Unona ponnosa 43
Unona ramarowii 45
Uraria hamosa 192
Uraria rufescens 192
Urena lobata 10
Urena lobata 92
Urena sinuata 92
Uritica diversifolia 470
Urochloa panicoides 18, 590
Urostigma arnottiana 462

Urostigma benghalense 463
Urostigma religiosum 465
Urtica crenulata 469
Urtica interrupta 471
Urtica longifolia 469
Urtica melastomoides 472
Urtica microphylla 472
Urtica sinuata 469
Utricularia affinis 369
Utricularia aurea 367
Utricularia bifida 367
Utricularia caerulea 367
Utricularia caerulea 368
Utricularia fasciculata 367
Utricularia flexuosa 11, 367
Utricularia nivea 367
Utricularia orbiculata 368
Utricularia racemosa 367, 368
Utricularia reticulata 368
Utricularia roseo-purpurea 368
Utricularia striatula 368
Utricularia subramanii 368
Utricularia uliginosa 369
Uvaria hookeri 49
Uvaria suberosa 48
V. himalayana Brandis 136
V. tenuifolia Wight & Arn. 132
Vaccinium neilgherrense 10, 293
Valla-kurinji 381
Vallia-tsjori-valli Rheede 136
Valli-caniram 52
Valli-kanjiram 325
Valli-kara 248
Valli-schoriganam 458
Valli-teregam 464
Valli-upu-dali 376
Vanda parviflora 496
Vanda peduncularis 480
Vanda pulchella 484
Vanda testacea 496
Vanda wightiana 477
Vandelia caespitosa 360
Vandelia scabra 360
Vandellia angustifolia 359
Vandellia angustifolia 359
Vandellia crustacea 360
Vandellia hirsuta 362
Vandellia pedunculata 359
Vandichooral 520
Vareca zeylanica 57
Vateria indica 75
Vatham kolli 35
Vatta-kumbil 451
Vazhappunna 36
Vazhukkal 84
Vediplavu 94
Veetla-caitu 514
Vella devadaram 103
Vella-akil 118
Velladambu 244
Velli-oeren 80
Velutta-itti-canni 428
Velutta-modela-muccu 409
Vempu 119
Venkotta 126
Ven-kurinji 377
Venthiruthali 335
Ventilago bombaiensis Dalz. 130
Vepris bilocularis 113
Verbena jamaicensis 394
Verbesina alba 281
Verbesina lavenia 273
Verbesina nodiflora 289
Verbesina prostrata 281
Vernonia arborea 290
Vernonia cinerea 290
Vernonia divergens 290
Vernonia indica 291

Vernonia monosis 290
Vernonia nilgheryensis 290
Vetiveria zizanioides 590
Vetti-tali 437
Viburnum acuminatum **245**
Viburnum punctatum **245**
Vicoa auriculata **291**
Vicoa indica 291
Vigna dalzelliana 193
Vigna pilosa 17, 21, 193
Vigna umbellata 193
Villebrunea integrifolia 471
Vinca pusilla 308
Vinca rosea 308
Viola enneasperma 56
Viola suffruticosa 56
Viscum orientale 9, 430
Vistnu-clandi 337
Vitex altissima 9, 395
Vitex arborea 396
Vitex leucoxylon 11, 395
Vitex peduncularis 396
Vitex pinnata 396
Vitex pubescens 396
Vitis discolor (Bl.) Dalz. 135
Vitis glauca (Roxb.) 135
Vitis glyptocarpa Thw. 135
Vitis japonica Thumb. 132
Vitis lanceolaria 136
Vitis muricata Wall. 136
Vitis neilgherriensis Wight 136
Vitis pedata (Lam.) 134
Vitis rheedei Wight & Arn. 135
Vitis sulcata 136
Vitis tenuifolia Wight & Arn. 134
Vizhalari 294
Vokameria fragrans 391
Volkameria serrata 391
Volvulopsis nummularius 338
Wadapu Rheede 408
Waltheria americana 97
Waltheria indica 97
Wara-pullu 537
Wattakaka volubilis 322
Watta-kaka-kodi 322
Watta-kurinji 322
Wattou-valli 317
Webera canarica 270
Webera tetrandra 248
Wedelia urticifolia 291
Welia-cupameni 434
Welia-theka-maravara 493
Weli-ila 525
Wendlandia notoniana 270
Wendlandia thyrsoidea 270
Willisia selaginoides 412
Wolfia spectabilis 484
Wrightia arborea 314
Wrightia tinctoria 314
Wrightia tomentosa 314
Wulfenia notoniana 370
Xanthium indicum 292
Xanthium strumarium 292
Xanthochymus ovalifolius 72
Xanthochymus spicatus 72
Xanthophyllum angustifolium 64
Xanthophyllum arnottianum 64
Xylia dolabriformis 206
Xylia xylocarpa 10
Xylia xylocarpa 206
Xyris pauciflora 508
Z. *jujuba* (L.) Gaertn. 131
Zanthoxylon 121
Zanthoxylum connaroides 121
Zehneria maysorensis 238
Zehneria thwaitesii 238
Zehneria umbelleta 237
Zeodaria 498

Zeuxine longilabris 496
Zeylanidium olivaceum 11, 414
Zingiber macrostachyum 500
Zingiber macrostachyum 9
Zingiber neesanum 50
Zingiber squarrosum 500
Zingiber wightianum 500
Zingiber zerumbet 500
Zizania terrestris 556
Ziziphus oenoplia 8, 10
Ziziphus rugosa 10
Zornia diphylla 194
Zornia gibbosa 194
Zornia quilonensis 17, 194